Engineering Mathematics-I

ENGINEERING MATHEMATICS-II

In Accordance With Prescribed Syllabus of AMIETE–ET/CS/IT (New Scheme)

Code: AE–56 / AC–56 / AT–56

Also for engineering, technology and applied sciences examinations of various universities and engineering colleges of India and abroad

(Including Over 500 Problems With Solutions completely solved in detail)

CONTENTS

Engineering Mathematics-I

In Accordance With Prescribed Syllabus of AMIETE–ET/CS/IT (New Scheme)
Code: AE–51 / AC–51 / AT–51

Also for engineering, technology and applied sciences examinations of various universities and engineering colleges of India and abroad

(Including Over 600 Problems With Solutions completely solved in detail)

I.J.S. SARNA
B.E. Tech. (Mech. Engg.)
Delhi College of Engineering
Delhi University, Delhi

CBS PUBLISHERS & DISTRIBUTORS PVT. LTD.
New Delhi • Bengaluru • Pune • Kochi • Chennai

ISBN : 978-81-239-2053-5

First Edition : 2012

Published by Satish Kumar Jain and produced by V.K. Jain for
CBS Publishers & Distributors Pvt. Ltd.,
CBS Plaza, 4819/XI Prahlad Street, 24 Ansari Road, Daryaganj,
New Delhi - 110002, India. • Website: www.cbspd.com
e-mail: delhi@cbspd.com, cbspubs@vsnl.com, cbspubs@airtelmail.in
Ph.: 23289259, 23266861, 23266867 • Fax: 011-23243014

Branches:

- ***Bengaluru:*** Seema House, 2975, 17th Cross, K.R. Road, Bansankari 2nd Stage, Bangalore - 560070 Ph.: 26771678/79 Fax: 080-26771680 • e-mail: bangalore@cbspd.com
- ***Pune:*** Bhuruk Prestige, Sr. No. 52/12/2+1+3/2, Narhe, Haveli (Near Katraj-Dehu Road by Pass), Pune-411051 Ph.: +91-20-32404169 • Fax: 020-24464059 e-mail: pune@cbspd.com
- ***Kochi:*** 36/14, Kalluvilakam, Lissie Hospital Road, Cochin - 682018, Kerala • e-mail: cochin@cbspd.com Ph.: 0484-4059061-65 • Fax: 0484-4059065
- ***Chennai:*** 20, West Park Road, Shenoy Nagar, Chennai - 600030 e-mail: chennai@cbspd.com Ph.: 044-26260666-26202620 Fax: 044-45530020

Printed at:
Somya Printers, Delhi-110053

Dedicated to the memory of

My Respected Parents
Late Saran Kaur and Late Rawel Singh Sarna
Who are responsible for what I am today

and

My Esteemed Professor Late G.S. Sharma
Who taught Mathematics to Engineering Students at Delhi College of Engineering (Delhi University), Delhi for over four decades

Preface

I have the pleasure in presenting to my readers the first of a two-volume set written to meet the requirements of degree students of engineering, technology, applied sciences and equivalent professional examinations of various universities and engineering colleges of India and abroad. The text of this volume has been divided into thirteen chapters. The text presented deals with advanced topics in engineering mathematics usually included in the syllabi of degree courses of engineering, technology, applied sciences and equivalent professional examinations of most of the universities.

While writing this book I have made constant endeavour to include large number of typical solved problems along with illustrations. The book also contains a number of supplementary problems with answers (hints have been provided wherever necessary).

A remarkable thing about this book is that it includes solved and unsolved problems collected from the past years' examination papers of various universities and engineering colleges in order to familiarize the students with the types of problems asked in these examinations. So far as unsolved problems are concerned, vital hints have been provided wherever necessary so as to enable the readers to arrive at correct answer without any difficulty. Some major features of this book are:

- The text has been presented in such a form that it can be followed by the students with minimum guidance from others.
- Over 600 typical solved problems, from easy to difficult, supplemented with neat sketches wherever necessary have been provided.
- In each chapter maximum possible emphasis has been laid on theory.
- The notations used are as per latest standards.

It is hoped that the book will adequately meet the requirements and expectations of the readers.

Most sincere gratitude to various esteemed authors whose books I have consulted while accomplishing the task of writing this book. I am grateful to my publishers for their hard work and enthusiasm in publishing this book within a short period of time.

For further improvement of this book, I welcome suggestions from my readers.

New Delhi

I.J.S. SARNA

Syllabus for AMIETE–ET/CS/IT (New Scheme)
Code: AE51/AC51/AT51

ENGINEERING MATHEMATICS-I

UNIT I

PARTIAL DIFFERENTIATION AND ITS APPLICATIONS

Introduction to function of two or more variables; Partial derivatives; Homogeneous functions–Euler's theorem; Total derivatives; Differentiation of implicit functions; change of variables; Jacobians; properties of Jacobians; Taylor's theorem for functions of two variables (only statement); Maxima and Minima of functions of two variables; Lagrange's Method of undetermined Multipliers; Rule of differentiation under integral sign

UNIT II

MULTIPLE INTEGRALS

Introduction to Double Integrals; Evaluation of Double Integrals; Evaluation of Double Integrals in polar coordinates; Change of order of integration; Triple Integrals; Evaluation of Triple Integrals; Area by Double Integration; volume as Double Integral; volume as Triple Integral.

UNIT III

LINEAR ALGEBRA

Introduction to determinants and matrices; Elementary row operations on a matrix; Rank of a matrix; Consistency of system of linear equation; Gauss elimination method to solve system of Linear equations; Eigen Values and Eigen Vectors of Matrix; Properties of Eigen values; Solution of a system of linear equations.

UNIT IV

NUMERICAL METHODS

Introduction; Solution of algebraic and transcendental equations; Regula-falsi method; Newton-Raphson's method; Numerical solution of ordinary differential equation; Taylor's Series method; Euler's Method; Modified Euler's Method; IV order Runge Kutta method; Gauss-Siedel method to solve system of linear equations; Power method to obtain the dominant Eigen value of a Matrix and its corresponding Eigen Vector.

UNIT V

SOLUTIONS OF FIRST ORDER AND FIRST DEGREE ORDINARY DIFFERENTIAL EQUATION

Definition and Practical approach to differential equation; Solution of differential equation and geometrical meaning of differential equation; Solution by the method of variable separable; Homogeneous differential equation; Equation reducible to homogeneous differential equation; Linear equations; Bernoulli's equation; Exact equation; To find orthogonal trajectories of the given family of curves; Physical applications.

UNIT VI

LINEAR DIFFERENTIAL EQUATIONS OF HIGHER ORDER

Definition and General form of Linear differential equation of higher order, the operator D; complete solution of Linear differential equation as a sum of complementary function (C.F.) and particular integral (P.I.); Rules for finding the complementary function; the inverse operator 1/ f (D), Rules for finding Particular integral; method of variation of parameter to find the Particular integral; Cauchy and Legendre Homogenous Linear equations; Simultaneous Linear equations with constant coefficients.

UNIT VII

SPECIAL FUNCTIONS AND SERIES SOLUTIONS OF DIFFERENTIAL EQUATIONS

Gamma functions; Recurrence formula for Γ (n); value of Γ (½); Beta Function; Symmetry of Beta Function; Relation between Gamma and Beta Function; Illustrative examples; Series solution of Differential equations (Method of Frobenius); Validity of series solution; series solution when X = 0 is an ordinary point of the equation; series solution when X = 0 is a regular singularity of the equation.

UNIT VIII

BESSEL AND LEGENDRE DIFFERENTIAL EQUATION

Bessel equation-Bessel functions, Recurrence formula for $J_n(x)$; Generating function for $J_n(x)$; Equations Reducible to Bessel's equation; Orthogonality of Bessel functions; Legendre's differential equation; Legendre Polynomials; Rodrigue's formula; Orthogonality of Legendre polynomials.

Note: Students have to answer **FIVE** full questions out of **EIGHT** questions to be set from each unit carrying 16 marks.

Contents

CHAPTER

1

Partial Derivatives

1.1 INTRODUCTION

In many applied problems we come across functions which depend on two or more independent variables. For example the area of a rectangle depend on its sides of length x and y; volume of a rectangular parallelopiped depends on its edges of length x, y, z; and the ideal gas law in physics states that under appropriate conditions the pressure exerted by a gas is a function of the volume of the gas and its temperature.

If the value of a quantity z depends upon the values of two variables x and y then z is called a function of x and y and, as in the case of function of single variable defined by $y = f(x)$, we write function of two variables as

$$z = f(x, y). \qquad ...(1.1)$$

The variables x and y are called independent variables while z is called the dependent variable. One can inquire how the value of z changes if y is held fixed and x is allowed to vary, or if x is held fixed and y is allowed to vary.

A similar definition can be given for functioning of more than two variables. In general, we define a real valued function of n variables as

$$z = f(x_1, x_2, ..., x_n). \qquad ...(1.2)$$

The function as defined by equation (1.2) is called an *explicit* function, whereas a function defined by $\phi(z, x_1, x_2, ..., x_n) = 0$ is called an *implicit* function.

We shall discuss the calculus of the functions of two variables in detail and then generalize to the case of several variables.

1.2 FUNCTIONS OF TWO VARIABLES

Consider the function of two variables $z = f(x, y)$. The set of points (x, y) in the xy-plane for which function $f(x, y)$ is defined is called the *domain* of definition or simply *domain* of the function and is denoted by D. The domain may be the entire xy plane or a part of the xy plane. The collection of the corresponding values of z is called the *range* of the function.

For example if $z = \sqrt{1-(x^2+y^2)}$, the domain for which z is real consists of the set of points (x, y) such that $x^2 + y^2 \le 1$. The range is the set of all real, positive numbers.

Neighborhood of a Point

For a function of two independent variables, two kinds of neighbourhoods are in common use – the *circular* and the *square* neighbourhoods. A circular neighbourhood of the point $P_0(x_0, y_0)$ is the set of all points (x, y) which lie inside a circle of radius δ with centre at the point (x_0, y_0). The number δ is arbitrary and can be chosen as small as we like but not zero. We usually denote this neighborhood by $N_\delta(P)$. Thus the δ-neighbourhood of $P_0(x_0, y_o)$ is the circular region defined by

$$N_\delta(P) = \left\{(x, y): \sqrt{(x-x_0)^2+(y-y_0)^2} < \delta\right\}. \qquad ...(1.3)$$

Since $\quad |x - x_0| \le \sqrt{(x-x_0)^2+(y-y_0)^2}$ and $|y - y_0| \le \sqrt{(x-x_0)^2+(y-y_0)^2}$,

the neighbourhood of the point $P(x_0, y_0)$ can also be defined as

$$N_\delta(P) = \left\{(x, y): |x - x_0| < \delta \text{ and } |y - y_0| < \delta\right\} \qquad ...(1.4)$$

that is, the set of all points which lie inside a square of side 2δ with centre at (x_0, y_0) and sides parallel to the coordinate axes (Fig. 1.1).

If the point $P(x_0, y_0)$ is not included in the set, then it is called the *deleted δ-neighborhood* of the point, that is, the set of points which satisfy

$$0 < \sqrt{(x-x_0)^2 + (y-y_0)^2} < \delta \qquad ...(1.5)$$

is called the deleted neighborhood of $P(x_0, y_0)$.

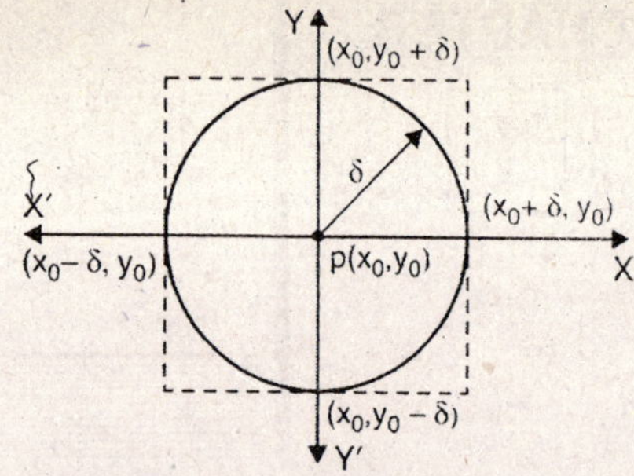

Fig. 1.1: Neighborhood of a point $P(x_0, y_0)$

1.3 LIMITS

Let $z = f(x, y)$ be a function of two variables defined in a deleted δ neighborhood of (x_0, y_0) [that is, $f(x, y)$ may be undefined at (x_0, y_0)]. We say that L is the limit of $f(x, y)$ as (x, y) approaches (x_0, y_0) and write

$\lim_{(x, y)\to(x_0, y_0)} f(x, y) = L$ if for every number $\varepsilon > 0$ there exists a corresponding number $\delta > 0$ such that for all (x, y) in the domain of f satisfies

$|f(x, y) - L| < \varepsilon$, whenever the distance between (x, y) and (x_0, y_0) satisfies

$$0 < \sqrt{(x-x_0)^2 + (y-y_0)^2} < \delta . \qquad ...(1.6)$$

Remarks

1. As in the case of function of one variable, the existence of the limit of $f(x, y)$ as $(x, y) \to (x_0, y_0)$ is in no way dependent on the existence of a value of $f(x, y)$ at (x_0, y_0).
2. *It must be noted that in order* for $\lim_{(x, y)\to(x_0, y_0)} f(x, y)$ or $\lim_{\substack{x\to x_0 \\ y\to y_0}} f(x, y)$ *to exist, it must have the same value regardless of the approach of* (x, y) *to* (x_0, y_0). The limit is same along all the paths, that is, limit should be independent of the path of (x, y) approaching (x_0, y_0). *Thus if two different approaches give different values, the limit can not exist. In other words if the limit is dependent on a path, then the limit does not exist.*
3. Sometimes it is possible to determine the limit by changing the given function to polars form.

1.4 PROPERTIES OF LIMITS OF FUNCTIONS OF TWO VARIABLES

Let $u = f(x, y)$ and $v = g(x, y)$ be two real valued functions defined in a domain D. Let

$$\lim_{(x, y)\to(x_0, y_0)} f(x, y) = L_1 \text{ and } \lim_{(x, y)\to(x_0, y_0)} g(x, y) = L_2 .$$

Then, the following rules hold

1. ***Sum Rule:*** $\lim_{(x, y)\to(x_0, y_0)} [f(x, y) + g(x, y)] = L_1 + L_2$.
2. ***Difference Rule:*** $\lim_{(x, y)\to(x_0, y_0)} [f(x, y) - g(x, y)] = L_1 - L_2$.
3. ***Constant multiple Rule:*** $\lim_{(x, y)\to(x_0, y_0)} [k f(x, y)] = kL_1$ for any real constant k.
4. ***Product Rule:*** $\lim_{(x, y)\to(x_0, y_0)} [f(x, y) g(x, y)] = L_1 L_2$.
5. ***Quotient Rule:*** $\lim_{(x, y)\to(x_0, y_0)} [f(x, y)/g(x, y)] = L_1/L_2$, $L_2 \neq 0$.
6. ***Power Rule:*** If p and q are integers, then $\lim [f(x, y)]^{p/q} = L_1^{p/q}$, provided $L_1^{p/q}$ is a real number.

ILLUSTRATIVE EXAMPLES

EXAMPLE 1.1 Evaluate the following limits:

(*i*) $\lim\limits_{(x,y)\to(3,1)}\left(\dfrac{2xy^2}{6+y^2}+\dfrac{1}{2}xy\right)$, (*ii*) $\lim\limits_{(x,y)\to(0,0)}\dfrac{x\sin(x^2+y^2)}{x^2+y^2}$,

(*iii*) $\lim\limits_{(x,y)\to(2,1)}\dfrac{\sin^{-1}(xy-2)}{\tan^{-1}(3xy-6)}$, (*iv*) $\lim\limits_{(x,y)\to(0,0)}\dfrac{y-x}{y+x}$.

Solution (*i*) Using standard laws for limits, we see that

(*i*) $$\lim_{(x,y)\to(3,1)}\left(\frac{2xy^2}{6+y^2}+\frac{1}{2}xy\right)=\lim_{(x,y)\to(3,1)}\frac{2xy^2}{6+y^2}+\lim_{(x,y)\to(3,1)}\frac{1}{2}xy=\frac{2(3)(1)}{6+1}+\frac{1}{2}(3)(1)=\frac{6}{7}+\frac{3}{2}=\frac{33}{14}.$$

(*ii*) $$\lim_{(x,y)\to(0,0)}\frac{x\sin(x^2+y^2)}{x^2+y^2}=\lim_{(x,y)\to(0,0)}x\cdot\lim_{(x,y)\to(0,0)}\frac{\sin(x^2+y^2)}{x^2+y^2}=0.1=0.$$

(*iii*) Let $xy-2=t$. Therefore $t\to 0$ as $(x,y)\to(2,1)$.We can now write

$$\lim_{(x,y)\to(2,1)}f(x,y)=\lim_{t\to 0}\frac{\sin^{-1}t}{\tan^{-1}3t}=\lim_{t\to 0}\left[\frac{(\sin^{-1}t)/t}{\tan^{-1}(3t)/(3t)}\right]\left[\frac{t}{3t}\right]=1\cdot\frac{1}{3}=\frac{1}{3}.$$

(*iv*) In some cases the standard laws donot suffice. The limit does not exist, if it is not finite or if it depends on a particular path.
Consider the path $y=mx$. As $(x,y)\to(0,0)$, we get $x\to 0$.

Therefore $\lim\limits_{(x,y)\to(0,0)}\dfrac{y-x}{y+x}=\lim\limits_{x\to 0}\dfrac{mx-x}{mx+x}=\dfrac{m-1}{m+1}$, which depends on m. For different values of m, we obtain different limits. Hence, the limit does not exist.

EXAMPLE 1.2 Show that the following limits

(*i*) $\lim\limits_{(x,y)\to(0,0)}\dfrac{x}{\sqrt{x^2+y^2}}$, (*ii*) $\lim\limits_{(x,y)\to(0,0)}\dfrac{x+\sqrt{y}}{x^2+y}$,

(*iii*) $\lim\limits_{(x,y)\to(0,1)}\tan^{-1}(y/x)$, (*AMIETE Dec. 2006*) (*iv*) $\lim\limits_{(x,y)\to(0,0)}\dfrac{xy^3}{x^2+y^6}$

do not exist.

Solution (*i*) Consider the path $y=mx$. As $(x,y)\to(0,0)$ we get $x\to 0$.

Therefore, $\lim\limits_{(x,y)\to(0,0)}\dfrac{x}{\sqrt{x^2+y^2}}=\lim\limits_{x\to 0}\dfrac{x}{x\sqrt{1+m^2}}=\dfrac{1}{\sqrt{1+m^2}}$ which depends on m. For different values of m, we obtain different limits. Hence, the limit does not exist.

Alternative: Setting $x=r\cos\theta$, $y=r\sin\theta$, we obtain $\lim\limits_{(x,y)\to(0,0)}\dfrac{x}{\sqrt{x^2+y^2}}=\lim\limits_{r\to 0}\dfrac{r\cos\theta}{r}=\cos\theta$,

which depends on θ. The limit is dependent on different radial paths θ = constant. Hence, the limit does not exist.

(*ii*) Choose the path $y=mx^2$. As $(x,y)\to(0,0)$, we get $x\to 0$.

Therefore, $$\lim_{(x,y)\to(0,0)}\frac{x+\sqrt{y}}{x^2+y}=\lim_{x\to 0}\frac{x(1+\sqrt{m})}{x^2(1+m)}=\lim_{x\to 0}\frac{1+\sqrt{m}}{x(1+m)}=\infty.$$

Since the limit is not finite, so limit does not exist.

(*iii*) We have $\lim\limits_{(x,\,y)\to(0,\,1)} \tan^{-1}(y/x) = \tan^{-1}(\pm\infty) = \pm\pi/2$ depending on whether the point (0, 1) is approached from left or from right along the line $y = 1$. If we approach from left, we obtain the limit as $-\pi/2$ and if we approach from right, we obtain the limit as $\pi/2$. Since the limit is not unique, the limit does not exist as $(x, y) \to (0,1)$.

(*iv*) Choose the path $x = my^3$. As $(x, y) \to (0, 0)$, we get $y \to 0$.

Therefore $$\lim_{(x,y)\to(0,0)} \frac{xy^3}{x^2+y^6} = \lim_{y\to 0}\frac{my^6}{(1+m^2)y^6} = \frac{m}{1+m^2}$$

which depends on m. For different values of m, we obtain different limits. Hence, the limit does not exist.

EXAMPLE 1.3 Examine the limits of $f(x, y) = x^2y/(x^4 + y^2)$, $(x, y) \neq (0, 0)$ as $(x, y) \to (0, 0)$ along the line $y = mx$ and along the parabola $y = x^2$. Does $f(x, y)$ have a limit as $(x, y) \to (0, 0)$? *(AMIETE; W-2004)*

Solution *First consider the path* $y = mx$. *As* $(x, y) \to (0, 0)$, we get $x \to 0$.

Therefore, $$\lim_{(x,y)\to(0,0)} \frac{x^2y}{x^4+y^2} = \lim_{x\to 0}\frac{x^2(mx)}{x^4+m^2x^2} = \lim_{x\to 0}\frac{mx}{x^2+m^2} = 0.$$

Now consider the path $y = x^2$, when $(x, y) \to (0, 0)$, we get $x \to 0$.

Therefore, $$\lim_{(x,y)\to(0,0)} \frac{x^2y}{x^4+y^2} = \lim_{x\to 0}\frac{x^2(x^2)}{x^4+x^4} = \lim_{x\to 0}\frac{1}{2} = \frac{1}{2}.$$

Since the limits along the two paths are different, so limit does not exist.

1.5 CONTINUITY

Def. A function $z = f(x, y)$ of the two real independent variable x and y is said to be *continuous* at a point (x_0, y_0), if

(*i*) $f(x, y)$ is defined at the point (x_0, y_0),

(*ii*) $\lim\limits_{(x,\,y)\to(x_0,\,y_0)} f(x, y) = L$, that is, the limit exists as $(x, y) \to (x_0, y_0)$, and

(*iii*) $\lim\limits_{(x,\,y)\to(x_0,\,y_0)} f(x, y) = L = f(x_0, y_0)$.

A function is *continuous* if it is continuous at every point of its domain.

Therefore, a function $f(x, y)$ is continuous at (x_0, y_0) if

$$|f(x,y) - f(x_0,y_0)| < \varepsilon, \quad \text{whenever } \sqrt{(x-x_0)^2 + (y-y_0)^2} < \delta. \qquad ...(1.7)$$

If any one of the above conditions is not satisfied, then the function is said to be *discontinuous* at the point (x_0, y_0) which is then called a *point of discontinuity*.

EXAMPLE 1.4 (*i*) Investigate for continuity at the point (1, 2).

$$f(x, y) = \begin{cases} x^2 + 2y, & (x, y) \neq (1, 2) \\ 0, & (x, y) = (1, 2). \end{cases}$$

(*ii*) Show that the given function is continuous at the point (0, 0).

$$f(x,y) = \begin{cases} \dfrac{\sin^{-1}(x+2y)}{\tan^{-1}(2x+4y)}, & (x, y) \neq (0,0) \\ 1/2, & (x, y) = (0,0). \end{cases}$$

Solution (*i*) $\lim\limits_{(x,y)\to(1,2)} f(x,y) = \lim\limits_{(x,\,y)\to(1,\,2)} (x^2+2y) = \lim\limits_{(x,\,y)\to(1,\,2)} x^2 + \lim\limits_{(x,y)\to(1,\,2)} 2y = 1 + 4 = 5.$

Since $\lim\limits_{(x,\,y)\to(1,\,2)} f(x,y) \neq f(1,2)$, the function is not continuous at (1, 2).

The point (1, 2) is a *point of removable discontinuity*.

(*ii*) Let $x + 2x = t$. Therefore, $t \to 0$ as $(x, y) \to (0,0)$. We can now write

$$\lim_{(x,y)\to(0,0)} f(x,y) = \lim_{t\to 0} \frac{\sin^{-1} t}{\tan^{-1} 2t} = \lim_{t\to 0}\left[\frac{(\sin^{-1} t)/t}{(\tan^{-1} 2t)/2t}\right]\left[\frac{t}{2t}\right] = \frac{1}{2}.$$

Since $\lim_{(x,y)\to(0,0)} f(x,y) = f(0,0) = 1/2$, the given function is continuous at $(x, y) = (0, 0)$.

EXAMPLE 1.5 Show that the following functions are discontinuous at the given points

(*i*) $f(x, y) = \begin{cases} \dfrac{x^2 - y^2}{x^2 + y^2}, & (x, y) \neq (0, 0) \\ 0 \quad , & (x, y) = (0, 0) \end{cases}$ at the point (0, 0).

(*ii*) $f(x, y) = \begin{cases} \dfrac{x^2 - x\sqrt{y}}{x^2 + y}, & (x, y) \neq (0, 0) \\ 0 \quad , & (x, y) = (0, 0) \end{cases}$ at the point (0, 0).

Solution (*i*) Choose the path $y = mx$. As $(x, y) \to (0, 0)$, we get $x \to 0$.

Therefore, $$\lim_{(x,y)\to(0,0)} \frac{x^2 - y^2}{x^2 + y^2} = \lim_{x\to 0} \frac{x^2(1 - m^2)}{x^2(1 + m^2)} = \frac{1 - m^2}{1 + m^2}$$

which depends on m. Since, the limit does not exist, the function is not continuous at (0. 0).

(*ii*) Choose the path $y = m^2x^2$. As $(x, y) \to (0, 0)$, we get $x \to 0$.

Therefore, $$\lim_{(x,y)\to(0,0)} \frac{x^2 - x\sqrt{y}}{x^2 + y} = \lim_{x\to 0} \frac{x^2(1 - m)}{x^2(1 + m^2)} = \frac{1 - m}{1 + m^2}$$

which depends on m. Since the limit does not exist, the function is not continuous at (0, 0).

EXERCISE 1.1

Evaluate each of the following limits where they exists.

1. $\lim_{(x,y)\to(\pi,0)} x\cos\left(\dfrac{x - y}{4}\right)$.

2. $\lim_{(x,y)\to(4,\pi)} x^2 \sin\dfrac{y}{x}$.

3. $\lim_{(x,y)\to(1,1)} \dfrac{x^3 - y^3}{x^2 - y^2}$. (*AMIETE, S-2005*) [**Hint:** Factorize and cancel $(x - y)$]

4. $\lim_{\substack{x\to 0\\ y\to 0}} \dfrac{xy}{x^2 + y^2}$. (*AMIETE Dec. 2010*)

5. $\lim_{(x,y)\to(0,0)} \dfrac{(1 + x^2)\sin y}{y}$.

6. $\lim_{(x,y)\to(0,0)} \dfrac{x^2}{x^2 + y^2}$.

7. $\lim_{(x,y)\to(0,0)} \dfrac{xy}{\sqrt{(x^2 + y^2)}}$. (*AMIETE June 2007*)

8. $\lim_{(x,y)\to(0,0)} \dfrac{2x^2 y}{x^4 + y^2}$. (*AMIETE, June 2006 Dec., 2004*)

9. $\lim_{(x,y)\to(0,0)} \dfrac{x^3 y}{x^6 + y^2}$. [**Hint:** Choose the path $y = mx^3$]

10. $\lim_{(x,y)\to(0,0)} \dfrac{x^2 - y^2}{x^2 + y^2}$.

11. $\lim_{(x,y)\to(0,0)} \dfrac{x^2 - xy}{\sqrt{x} - \sqrt{y}}$.

12. $\lim_{(x,y)\to(0,0} (x^2 + y^2)\log(x^2 + y^2)$.

[**Hint:** Use polar coordinates (r, θ) of the point (x, y) with $r \geq 0$.

$$\lim_{(x,y)\to(0,0)} (x^2+y^2)\log(x^2+y^2) = \lim_{r\to 0^+} r^2 \log r^2$$

$$= \lim_{r\to 0^+} \frac{2\log r}{1/r^2}\left(\frac{\infty}{\infty}\text{ form}\right) = \lim_{r\to 0^+}\frac{2/r}{-2/r^3} = \lim_{r\to 0^+}(-r^2) = 0.]$$

13. $\displaystyle\lim_{(x,y)\to(2,-2)} \frac{x^2+xy+x+y}{x+y}$. *(AMIETE S-2008)*

14. $\displaystyle\lim_{(x,y)\to(0,0)} \frac{x+\sqrt{y}}{\sqrt{x^2+y}}$. *(AMIETE, June 2010, Dec. 2007)*

Ans. **1.** $\pi\sqrt{2}/2$ **2.** $8\sqrt{2}$, **3.** 3/2. **4.** limit does not exist, **5.** 1, **6.** no limit, **7.** 0, **8.** no limit, **9.** limit does not exist, **10.** limit does not exist, **11.** 0, **12.** 0, **13.** 3, **14.** limit does not exist.

15. Discuss the continuity of the following functions at the given points.

(*i*) $f(x, y) = \begin{cases} \dfrac{xy}{x^2+y^2}, & (x, y) \neq (0, 0) \\ 0, & (x, y) = (0, 0) \end{cases}$ at (0, 0)

(*ii*) $f(x, y) = \begin{cases} \dfrac{2x-y}{x+y}, & (x, y) \neq (0, 0) \\ 0, & (x, y) = (0, 0) \end{cases}$ at (0, 0)

(*iii*) $f(x, y) = \begin{cases} \dfrac{2xy}{\sqrt{x^2+y^2}}, & (x, y) \neq (0, 0) \\ 0, & (x, y) = (0, 0) \end{cases}$ at (0, 0)

(*iv*) $f(x, y) = \begin{cases} \dfrac{x^2+y^2}{\tan xy}, & (x, y) \neq (0, 0) \\ 0, & (x, y) = (0, 0) \end{cases}$ at (0, 0)

[Hint: (*iv*) Choose the path $y = mx$]

Ans. (*i*) Discontinuous, (*ii*) Discontinuous, (*iii*) Continuous, (*iv*) Discontinuous.

16. Show that the given function is discontinuous at the given point.

$$f(x, y) = \begin{cases} \dfrac{x^2+xy+x+y}{x+y}, & (x, y) \neq (2, 2) \\ 4 ; & (x, y) = (2, 2) \end{cases}$$ *(AMIETE, June 2009)*

at the point (2, 2).

[Hint: $\displaystyle\lim_{(x,y)\to(2,2)} f(x, y) = \lim_{(x,y)\to(2,2)} \frac{(x+y)(x+1)}{x+y} = \lim_{(x,y)\to(2,2)} (x+1) = 3.$

Since $\displaystyle\lim_{(x,y)\to(2,2)} f(x, y) \neq f(2, 2)$, the function is not contnuous at (2, 2).]

1.6 PARTIAL DERIVATIVES

Let $z = f(x, y)$ be a function of two independent variables x and y. If x varies while y is held fixed then z becomes a function of x. The derivative of f with respect to x, treating y as constant is called *first partial derivative* of f with respect to x and is denoted by symbols $\dfrac{\partial z}{\partial x}\left(\text{or } \dfrac{\partial f}{\partial x} \text{ or } f_x(x, y)\right)$.

Thus, at any point (x, y)

$$\frac{\partial f}{\partial x} = f_x = \lim_{\Delta x\to 0} \frac{f(x+\Delta x, y) - f(x, y)}{\Delta x}, \qquad ...(1.8)$$

provided the limit exists.

Similarly, if y varies while x is held fixed, the *first partial derivative* of f with respect to y is

$$\frac{\partial f}{\partial y} = f_y = \lim_{\Delta y\to 0} \frac{f(x, y+\Delta y) - f(x, y)}{\Delta y} \qquad ...(1.9)$$

provided the limit exists.

The derivative of $f(x, y)$ at (x_0, y_0) with respect to x is defined as $\lim_{\Delta x \to 0} \frac{f(x_0 + \Delta x, y_0) - f(x_0, y_0)}{\Delta x}$ and denoted by $\frac{\partial f}{\partial x}(x_0, y_0)$ or simply $f_x(x_0, y_0)$. Similarly $f_y(x_0, y_0)$ or $\left[\frac{\partial f}{\partial y}(x_0, y_0)\right]$ is defined.

In general, if z is a function of a number of independent variables, then the partial derivative of z with respect to any one of those variables is obtained by differentiating z with respect to that variable while all other variables are held constant.

ILLUSTRATIVE EXAMPLES

EXAMPLE 1.6 Find the first order partial derivatives of the following functions at the specified point from the first principles.

(*i*) $f(x, y) = 2x^2 - xy + y^2$ at (x_0, y_0).

(*ii*) $f(x, y) = ye^{-x}$ at (4, 2).

(*iii*) $f(x, y) = \sin(3x + 2y)$ at $(0, \pi/3)$.

Solution We have

(*i*)

$$\left.\frac{\partial f}{\partial x}\right|_{(x_0, y_0)} = f_x(x_0, y_0) = \lim_{\Delta x \to 0} \frac{f(x_0 + \Delta x, y_0) - f(x_0, y_0)}{\Delta x}$$

$$= \lim_{\Delta x \to 0} \frac{\left[2(x_0 + \Delta x)^2 - (x_0 + \Delta x)y_0 + y_0^2\right] - \left[2x_0^2 - x_0 y_0 + y_0^2\right]}{\Delta x}$$

$$= \lim_{\Delta x \to 0} \frac{(4x_0 - y_0)\Delta x + 2(\Delta x)^2}{\Delta x} = \lim_{\Delta x \to 0} 4x_0 - y_0 + 2\Delta x = 4x_0 - y_0.$$

$$\left.\frac{\partial f}{\partial y}\right|_{(x_0, y_0)} = f_y(x_0, y_0) = \lim_{\Delta y \to 0} \frac{f(x_0, y_0 + \Delta y) - f(x_0, y_0)}{\Delta y}$$

$$= \lim_{\Delta y \to 0} \frac{\left[2x_0^2 - x_0(y_0 + \Delta y) + (y_0 + \Delta y)^2\right] - \left[2x_0^2 - x_0 y_0 + y_0^2\right]}{\Delta y}$$

$$= \lim_{\Delta y \to 0} \frac{(-x_0 + 2y_0)\Delta y + (\Delta y)^2}{\Delta y} = \lim_{\Delta y \to 0} (-x_0 + 2y_0 + \Delta y) = -x_0 + 2y_0.$$

(*ii*) At the point (x, y)

$$\frac{\partial f}{\partial x} = \lim_{\Delta x \to 0} \frac{ye^{-(x+\Delta x)} - ye^{-x}}{\Delta x} = \lim_{\Delta x \to 0} -\frac{ye^{-x}(1 - e^{-\Delta x})}{\Delta x} = -ye^{-x} \lim_{\Delta x \to 0} \frac{1 - e^{-\Delta x}}{\Delta x} = -ye^{-x}.$$

$$\frac{\partial f}{\partial y} = \lim_{\Delta y \to 0} \frac{(y + \Delta y)e^{-x} - ye^{-x}}{\Delta y} = e^{-x}$$

$$\therefore \quad \left.\frac{\partial f}{\partial x}\right|_{(4, 2)} = -2e^{-4}; \quad \left.\frac{\partial f}{\partial y}\right|_{(4, 2)} = e^{-4}.$$

(*iii*)

$$\frac{\partial f}{\partial x} = \lim_{\Delta x \to 0} \frac{\sin\{3(x + \Delta x) + 2y\} - \sin(3x + 2y)}{\Delta x}$$

$$= \lim_{\Delta x \to 0} \frac{2\cos[3x + 2y + (3\Delta x/2)]\sin(3\Delta x/2)}{\Delta x}$$

$$= \lim_{\Delta x \to 0} \frac{3\cos[3x + 2y + (3\Delta x/2)]\sin(3\Delta x/2)}{3\Delta x/2}$$

$$= 3 \lim_{\Delta x \to 0} \cos[3x + 2y + (3\Delta x/2)] \cdot \lim_{\Delta x \to 0} \frac{\sin(3\Delta x/2)}{(3\Delta x/2)} = 3\cos(3x + 2y).$$

$$\frac{\partial f}{\partial y} = \lim_{\Delta y \to 0} \frac{\sin\{3x + 2(y + \Delta y)\} - \sin(3x + 2y)}{\Delta y} = \lim_{\Delta y \to 0} \frac{2\cos(3x + 2y + \Delta y)\sin \Delta y}{\Delta y}$$

$$= 2 \lim_{\Delta y \to 0} \cos(3x + 2y + \Delta y) \cdot \lim_{\Delta y \to 0} \frac{\sin \Delta y}{\Delta y} = 2\cos(3x + 2y).$$

$$\therefore \quad \left.\frac{\partial f}{\partial x}\right|_{(0,\, \pi/3)} = 3\cos(2\pi/3) = -\frac{3}{2}, \text{ and } \left.\frac{\partial f}{\partial y}\right|_{(0,\, \pi/3)} = 2\cos(2\pi/3) = -1.$$

EXAMPLE 1.7 Show that the function

$$f(x, y) = \begin{cases} \dfrac{x^2 - y^2}{x - y}, & (x, y) \neq (1, -1) \\ 0, & (x, y) = (1, -1) \end{cases}$$

(*i*) is continuous at (0, 0),
(*ii*) possesses partial derivatives $f_x(1, -1)$ and $f_y(1, -1)$.

Solution (*i*) We have

$$\lim_{(x, y) \to (1, -1)} \frac{x^2 - y^2}{x - y} = \lim_{(x, y) \to (1, -1)} (x + y) = 0 = f(1, -1).$$

Therefore, the function is continuous at (1, –1).

(*ii*) The partial derivatives are given by

$$f_x(1, -1) = \lim_{\Delta x \to 0} \frac{f(1 + \Delta x, -1) - f(1, -1)}{\Delta x} = \lim_{\Delta x \to 0} \frac{1}{\Delta x}\left[\frac{(1 + \Delta x)^2 - 1}{(1 + \Delta x) + 1} - 0\right] = \lim_{\Delta x \to 0} \frac{2 + \Delta x}{2 + \Delta x} = 1.$$

$$f_y(1, -1) = \lim_{\Delta y \to 0} \frac{f(1, -1 + \Delta y) - f(1, -1)}{\Delta y} = \lim_{\Delta y \to 0} \frac{1}{\Delta y}\left[\frac{1 - (-1 + \Delta y)^2}{1 - (-1 + \Delta y)} - 0\right] = \lim_{\Delta y \to 0} \frac{2 - \Delta y}{2 - \Delta y} = 1.$$

Therefore, the first order partial derivatives exist at (1, –1).

EXAMPLE 1.8 Show that the function

$$f(x, y) = \begin{cases} \dfrac{xy}{x^2 + 2y^2}, & (x, y) \neq (0, 0) \\ 0, & (x, y) = (0, 0) \end{cases}$$

is discontinuous at (0, 0) but its partial derivatives f_x and f_y exist at (0, 0).

Solution Choose the path $y = mx$. We see that

$$\lim_{(x, y) \to (0, 0)} f(x, y) = \lim_{x \to 0} \frac{mx^2}{(1 + 2m^2)x^2} = \frac{m}{1 + 2m^2}.$$

Since the limit depends on the value of m, that is, on the path of approach and is different for the different paths followed and therefore the limit does not exist. Hence, the function $f(x, y)$ is not continuous at (0, 0). We now have

$$f_x(0, 0) = \lim_{\Delta x \to 0} \frac{f(\Delta x, 0) - f(0, 0)}{\Delta x} = \lim_{\Delta x \to 0} \frac{0 - 0}{\Delta x} = 0.$$

$$f_y(0, 0) = \lim_{\Delta y \to 0} \frac{f(0, \Delta y) - f(0, 0)}{\Delta y} = \lim_{\Delta y \to 0} \frac{0 - 0}{\Delta y} = 0.$$

Therefore, the partial derivatives f_x and f_y exist at (0, 0).

EXAMPLE 1.9 Show that for the function $f(x,y)=\sqrt{|xy|}$, partial derivatives f_x and f_y both exist at the origin and have the value 0. Also show that these two partial derivatives are continuous except at the origin.

(AMIETE, Dec. 2005)

Solution For $(x, y) = (0, 0)$

$$f_x(0,0) = \lim_{\Delta x\to 0}\frac{f(\Delta x,0)-f(0,0)}{\Delta x} = \lim_{\Delta x\to 0}\frac{0-0}{\Delta x}=0$$

$$f_y(0,0) = \lim_{\Delta y\to 0}\frac{f(0,\Delta y)-f(0,0)}{\Delta y} = \lim_{\Delta y\to 0}\frac{0-0}{\Delta y}=0 .$$

For $(x, y) \neq (0, 0)$

$$f_x(x,y) = \lim_{\Delta x\to 0}\frac{f(x+\Delta x,y)-f(x,y)}{\Delta x} = \lim_{\Delta x\to 0}\frac{\sqrt{|x+\Delta x|\,|y|}-\sqrt{|x|\,|y|}}{\Delta x}$$

$$= \lim_{\Delta x\to 0}\sqrt{|y|}\,\frac{|x+\Delta x|-|x|}{\Delta x\left[\sqrt{|x+\Delta x|}+\sqrt{|x|}\right]}.$$

Now as $\Delta x \to 0$, we can take $x + \Delta x > 0$, i.e., $|x + \Delta x| = x + \Delta x$, when $x > 0$.

And $x + \Delta x < 0$ or $|x + \Delta x| = -(x + \Delta x)$, when $x < 0$.

$$\therefore \quad f_x(x,y) = \begin{cases}\sqrt{|y|}/2\sqrt{|x|}, & \text{when } x>0,\\ -\sqrt{|y|}/2\sqrt{|x|}, & \text{when } x<0.\end{cases}$$

Similarly,
$$f_y(x,y) = \begin{cases}\sqrt{|x|}/2\sqrt{|y|}, & \text{when } y>0,\\ -\sqrt{|x|}/2\sqrt{|y|}, & \text{when } y<0\end{cases}$$

which are not continuous at the origin.

EXAMPLE 1.10 Find the first order partial derivatives of (*i*) $z = \log(x^2 + y^2)$, (*ii*) $u = 3x^2y - x\sin xy$.

Solution (*i*) we have $z = \log(x^2 + y^2)$. Treating y as a constant and differentiating w.r.t. x, then

$$\frac{\partial z}{\partial x}=\frac{1}{x^2+y^2}\cdot 2x=\frac{2x}{x^2+y^2}.$$

Now treating x as constant and differentiating w.r.t y, we have

$$\frac{\partial z}{\partial y}=\frac{1}{x^2+y^2}\cdot 2y=\frac{2y}{x^2+y^2}.$$

(*ii*) We have $u = 3x^2y - x\sin xy.$

$\therefore$ $\partial u/\partial x = 6xy - [xy\cos(xy) + \sin xy] = 6xy - xy\cos(xy) + \sin(xy),$

and $\partial u/\partial y = 3x^2 - x^2\cos(xy).$

EXAMPLE 1.11 (*i*) If $z=e^{2x^2+xy+3y^2}$, find $\partial z/\partial x$, $\partial z/\partial y$.

(*ii*) If $f(x, y) = x^3y - xy^3$, find $\left[\frac{1}{\partial f/\partial x}+\frac{1}{\partial f/\partial y}\right]_{\substack{x=1\\y=2}}$.

(*iii*) The relation $\left(P+\frac{a}{v^2}\right)(v-b)=RT$, in which a, b and R are constants is given. Find an expression for $\left(\frac{\partial P}{\partial T}\right)_v$ and show that $T\left(\frac{\partial P}{\partial T}\right)_v - P=\frac{a}{v^2}$.

(AMIE, W-2003)

Solution (*i*) We have $\partial z/\partial x=(4x+y)e^{2x^2+xy+3y^2}$. (Here y is treated as a constant).

And $\partial z/\partial y=(x+6y)e^{2x^2+xy+3y^2}$. (Here x has been treated as a constant).

(*ii*) We have $f(x, y) = x^3y - xy^3$.

$\therefore \quad \partial f/\partial x = 3x^2y - y^3$ and $\partial f/\partial y = x^3 - 3xy^2$.

$$\left[\frac{1}{\partial f/\partial x} + \frac{1}{\partial f/\partial y}\right]_{\substack{x=1\\y=2}} = \left[\frac{1}{3x^2y - y^3} + \frac{1}{x^3 - 3xy^2}\right]_{\substack{x=1\\y=2}} = \left[\frac{1}{6-8} + \frac{1}{1-12}\right] = -\frac{13}{22}.$$

(*iii*) We have $\left(P + \frac{a}{v^2}\right)(v - b) = RT.$ ...(*i*)

Differentiating partially w.r.t T, treating v as constant.

$$\left\{\left(\frac{\partial P}{\partial T}\right)_v + 0\right\}(v-b) + \left(P + \frac{a}{v^2}\right)0 = R \text{ or } \left(\frac{\partial P}{\partial T}\right)_v (v-b) = R \Rightarrow \left(\frac{\partial P}{\partial T}\right)_v = \frac{R}{v-b}. \quad ...(ii)$$

Now from (*ii*), $T\left(\frac{\partial P}{\partial T}\right)_v = \frac{R \cdot T}{v-b} = \left(P + \frac{a}{v^2}\right)$, from (*i*) or $T\left(\frac{\partial P}{\partial T}\right)_v - P = \frac{a}{v^2}$.

EXAMPLE 1.12 If $u = (1 - 2xy + y^2)^{-1/2}$, prove that $x\frac{\partial u}{\partial x} - y\frac{\partial u}{\partial y} = y^2u^3$. (*MDU-2001*)

Solution We have

$$\frac{\partial u}{\partial x} = -\frac{1}{2}(1 - 2xy + y^2)^{-3/2} \cdot (-2y) = y/(1 - 2xy + y^2)^{3/2},$$

$$\frac{\partial u}{\partial y} = -\frac{1}{2}(1 - 2xy + y^2)^{-3/2} \cdot (-2x + 2y) = (x - y)/(1 - 2xy + y^2)^{3/2}$$

$$\therefore \quad x\frac{\partial u}{\partial x} - y\frac{\partial u}{\partial y} = (xy - xy + y^2) / (1 - 2xy + y^2)^{3/2}$$

$= y^2(1 - 2xy + y^2)^{-3/2} = y^2u^3$, which is same as required.

EXAMPLE 1.13 If $x = r\cos\theta$, $y = r\sin\theta$, find

(*a*) $(\partial x/\partial r)_\theta$, $(\partial x/\partial r)_y$, $(\partial r/\partial x)_y$, $(\partial\theta/\partial y)_x$ (*b*) $\partial^2\theta/\partial x^2 + \partial^2\theta/\partial y^2$ (*UPTU 2001; MDU-2001*)

Note: *Before we solve the problem let us clearly understand the meaning of* $(\partial x/\partial r)_\theta$, $(\partial x/\partial r)_y$ *etc.*

In the present problem we have four variables x, y, r, θ connected by two relations x = r cos θ, y = r sin θ. Any one of the four variables can be expressed in terms of any two of the remaining three. For example x can be expressed in terms of r, θ or r, y or θ, y.

Now $(\partial x/\partial r)_\theta$ *means the partial derivative of x w.r.t. r treating θ as constant in a relation expressing x as function of r and θ.*

Solution (*a*) To get $(\partial x/\partial r)_\theta$, we have to express x in terms of r and θ which in fact is already given to us because $x = r\cos\theta$. Therefore, $(\partial x/\partial r)_\theta = \cos\theta$. ...(*i*)

To find $(\partial x/\partial r)_y$, we have to express x as a function of r and y.

Since $x = r\cos\theta$, $y = r\sin\theta$. Therefore, $x^2 + y^2 = r^2$. Thus $x = \sqrt{r^2 - y^2}$.

$$\therefore \quad \left(\frac{\partial x}{\partial r}\right)_y = \frac{r}{\sqrt{r^2 - y^2}} = \frac{r}{x} = \sec\theta \quad ...(ii)$$

[From (*i*) and (*ii*) we note that $\left(\frac{\partial x}{\partial r}\right)_\theta \neq \left(\frac{\partial x}{\partial r}\right)_y$]

To determine $(\partial r/\partial x)_y$ we know that $r = \sqrt{x^2 + y^2}$.

$$\therefore \quad \left(\frac{\partial r}{\partial x}\right)_y = \frac{x}{\sqrt{x^2 + y^2}} = \frac{x}{r} = \cos\theta.$$

As $\theta = \tan^{-1}(y/x)$, ...(*iii*)

$$\therefore \quad \left(\frac{\partial \theta}{\partial y}\right)_x = \frac{x}{x^2 + y^2}. \quad ....(iv)$$

(*b*) From (*iii*), $\left(\frac{\partial\theta}{\partial x}\right)_y = -\frac{y}{x^2+y^2}$.

$$\therefore \quad \frac{\partial^2\theta}{\partial x^2} = \frac{\partial}{\partial x}\left(\frac{\partial\theta}{\partial x}\right) = \frac{\partial}{\partial x}\left(-\frac{y}{x^2+y^2}\right) = \frac{2xy}{(x^2+y^2)^2}.$$

From (*iv*), $\frac{\partial^2\theta}{\partial y^2} = \frac{\partial}{\partial y}\left(\frac{x}{x^2+y^2}\right) = -\frac{2xy}{(x^2+y^2)^2}$.

Hence, $\frac{\partial^2\theta}{\partial x^2} + \frac{\partial^2\theta}{\partial y^2} = 0$.

EXERCISE 1.2

1. Find f_x and f_y when
   (*i*) $f(x, y) = x^4 + x^2y^2 + y^4$, (*ii*) $f(x, y) = e^{ax} \sin by$,
   (*iii*) $f(x, y) = \tan^{-1}\left(\frac{x^2+y^2}{x+y}\right)$, (*iv*) $f(x, y) = x^y$. (*AMIETE Dec 2007*)

Ans. (*i*) $4x^3 + 2xy^2,\ 2x^2y + 4y^3$. (*ii*) $ae^{ax} \sin by,\ be^{ax} \cos by$.
(*iii*) $(x^2 + 2xy - y^2)/\{(x+y)^2 + (x^2+y^2)^2\}$;
$(y^2 + 2xy - x^2)/\{(x+y)^2 + (x^2+y^2)^2\}$. (*iv*) $yx^{y-1};\ x^y \log x$.

2. If $f(x, y) = \begin{cases} \frac{xy}{x^2+y^2}, & (x, y) \neq (0, 0) \\ 0, & (x, y) = (0, 0). \end{cases}$

   Show that both the partial derivatives exist at (0, 0) but the function is not continuous thereat.

3. Show that the function $f(x, y) = \begin{cases} \frac{x^2+y^2}{x-y}, & (x, y) \neq (0, 0) \\ 0, & (x, y) = (0, 0) \end{cases}$

   possesses partial derivatives at (0, 0), though it is not continuous at (0, 0).

4. If $f(x, y) = 2x^2 - 3xy + 4y^2$, show that $f_x(2, 3) = -1$, $f_y(2, 3) = 18$.

5. (*i*) If $u = \sin^{-1}(x/y)$, show that $x\frac{\partial u}{\partial x} + y\frac{\partial u}{\partial y} = 0$.

   (*ii*) If $z = \log\sqrt{x^2+y^2}$, show that $x\frac{\partial z}{\partial x} + y\frac{\partial z}{\partial y} = 1$. (*GGSIPU 2006*)

6. Given that $x = r\cos\theta$, $y = r\sin\theta$, find $r_x, r_y, \theta_x, \theta_y$.

   [**Hint:** $r^2 = x^2 + y^2$. Note $r_x \equiv \left(\frac{\partial r}{\partial x}\right)$, $r_y \equiv \left(\frac{\partial r}{\partial y}\right)$, $\theta_x \equiv \left(\frac{\partial\theta}{\partial x}\right)$, $\theta y \equiv \left(\frac{\partial\theta}{\partial y}\right)$]

Ans. $r_x = \cos\theta$, $r_y = \sin\theta$, $\theta_x = (\sin\theta)/r$, $\theta_y = (\cos\theta)/r$.

7. If $z = f(ax + by)$, show that $b\frac{\partial z}{\partial x} - a\frac{\partial z}{\partial y} = 0$.

8. If $z(x + y) = x^2 + y^2$, show that $\left(\frac{\partial z}{\partial x} - \frac{\partial z}{\partial y}\right)^2 = 4\left(1 - \frac{\partial z}{\partial x} - \frac{\partial z}{\partial y}\right)$. (*VTU 2003*)

9. Find the first order partial derivatives with respect to each variable, if $f(r, \theta, z) = \dfrac{r(r - \cos 2\theta)}{r^2 + z^2}$.

Ans. $\dfrac{\partial f}{\partial r} = \dfrac{2rz^2 + (r^2 - z^2)\cos 2\theta}{(r^2 + z^2)^2}$, $\dfrac{\partial f}{\partial \theta} = \dfrac{2r \sin 2\theta}{(r^2 + z^2)}$; $\dfrac{\partial f}{\partial z} = -\dfrac{2z(r^2 - r\cos 2\theta)}{(r^2 + z^2)^2}$.

10. If $z = e^{ax + by} . f(ax - by)$, prove that $b\dfrac{\partial z}{\partial x} + a\dfrac{\partial z}{\partial y} = 2abz$.

(*GGSIPU 2006; PTU 2006; AMIE, S-2004; AMIETE, S-2004; VTU-2004*)

11. If $f(x, y) = (1 - 2xy + y^2)^{-1/2}$, show that $\dfrac{\partial}{\partial x}\left\{(1 - x^2)\dfrac{\partial f}{\partial x}\right\} + \dfrac{\partial}{\partial y}\left(y^2 \dfrac{\partial f}{\partial y}\right) = 0.$

(*AMIE, S-2009; MDU, 2006*)

[**Hint**: $\dfrac{\partial f}{\partial x} = y(1 - 2xy + y^2)^{-3/2}$, $\dfrac{\partial f}{\partial y} = (x - y)(1 - 2xy + y^2)^{-3/2}$.

$$\therefore \quad \frac{\partial}{\partial x}\left[(1 - x^2)y(1 - 2xy + y^2)^{-3/2}\right] = 3y^2(1 - x^2)(1 - 2xy + y^2)^{-5/2} - 2xy(1 - 2xy + x^2)^{-3/2},$$

and $\dfrac{\partial}{\partial y}\left[(y^2x - y^3)(1 - 2xy + y^2)^{-3/2}\right] = 3y^2(x - y)^2(1 - 2xy + y^2)^{-5/2} + (2xy - 3y^2)(1 - 2xy + y^2)^{-3/2}$

Adding these, we get

$$(1 - 2xy + y^2)^{-5/2}\left[3y^2(1 - x^2) - 2xy(1 - 2xy + y^2) + 3y^2(x - y)^2 + (2xy - 3y^2)(1 - 2xy + y^2) = 0\right]$$

12. If $\phi = x\{1 + (a^2/r^2)\}$, $\psi = y\{1 - (a^2/r^2)\}$, where $r^2 = x^2 + y^2$. Show that $\partial\phi/\partial x = -\partial\psi/\partial y$ and $\partial\phi/\partial y = -\partial\psi/\partial x$.

(*AMIW, S-2003*)

13. If $\dfrac{x^2}{a^2 + u} + \dfrac{y^2}{b^2 + u} + \dfrac{z^2}{c^2 + u} = 1$, show that

$$\left(\frac{\partial u}{\partial x}\right)^2 + \left(\frac{\partial u}{\partial y}\right)^2 + \left(\frac{\partial u}{\partial z}\right)^2 = 2\left(x\frac{\partial u}{\partial x} + y\frac{\partial u}{\partial y} + z\frac{\partial u}{\partial z}\right). \qquad \text{(UPTU 2003)}$$

[**Hint**: We have $x^2 \cdot (a^2 + u)^{-1} + y^2(b^2 + u)^{-1} + z^2(c^2 + u)^{-1} = 1$,

where u is a function of x, y, z. Differentiating this partially w.r.t. x, we obtain

$$2x \cdot (a^2 + u)^{-1} - \left\{x^2(a^2 + u)^{-2} + y^2(b^2 + u)^{-2} + z^2(c^2 + u)^{-2}\right\}\partial u/\partial x = 0$$

$$\Rightarrow \quad \frac{\partial u}{\partial x} = \frac{2x}{(a^2 + u)} \div \left[\frac{x^2}{(a^2 + u)^2} + \frac{y^2}{(b^2 + u)^2} + \frac{z^2}{(c^2 + u)^2}\right] = \frac{2x}{a^2 + u} \cdot \frac{1}{K} \qquad ...(i)$$

where
$$K = \left(\frac{x^2}{(a^2 + u)^2} + \frac{y^2}{(b^2 + u)^2} + \frac{z^2}{(c^2 + u)^2}\right).$$

Similarly
$$\frac{\partial u}{\partial y} = \frac{2y}{b^2 + u} \cdot \frac{1}{K} \quad ...(ii) \quad \text{and} \quad \frac{\partial u}{\partial z} = \frac{2z}{c^2 + u} \cdot \frac{1}{K}. \qquad ...(iii)$$

On squaring (*i*), (*ii*), (*iii*) and adding, we obtain

$$\left(\frac{\partial u}{\partial x}\right)^2 + \left(\frac{\partial u}{\partial y}\right)^2 + \left(\frac{\partial u}{\partial z}\right)^2 = \frac{4}{K^2}\left[\frac{x^2}{(a^2 + u)^2} + \frac{y^2}{(b^2 + u)^2} + \frac{z^2}{(c^2 + u)^2}\right] = \frac{4}{K^2} \cdot K = \frac{4}{K}. \qquad ...(iv)$$

On mulitplying both sides of (*i*) by $2x$, (*ii*) by $2y$ and (*iii*) by $2z$ and adding, we obtain

$$2\left(x\frac{\partial u}{\partial x}+y\frac{\partial u}{\partial y}+z\frac{\partial u}{\partial z}\right)=\frac{4}{k}\left[\frac{x^2}{(a^2+u)}+\frac{y^2}{(b^2+u)}+\frac{z^2}{(c^2+u)}\right]=\frac{4}{K}.1=\frac{4}{k} \quad ...(v)$$

From (*iv*) and (*v*), we obtain the desired result.]

14. If $u=\log(x^3+y^3-x^2y-xy^2)$. Show that $\frac{\partial u}{\partial x}+\frac{\partial u}{\partial y}=\frac{2}{x+y}$. (*AMIE, S-2007*)

[Hint: $u=\log(x^3+y^3-x^2y-xy^2)=\log\left[x^2(x-y)-y^2(x-y)\right]$

$=\log\left[(x^2-y^2)(x-y)\right]=\log\left[(x-y)^2(x+y)\right]=2\log(x-y)+\log(x+y)]$.

15. If $u=f(y/x)$. Show that $x\frac{\partial u}{\partial x}+y\frac{\partial u}{\partial y}=0$. (*AMIETE., S-2008*)

16. If $f(x, y)=\tan^{-1}(y/x)$, find $\partial f/\partial x$ and $\partial f/\partial y$. (*AMIE., W-2009*) **Ans.** $-y/(x^2+y^2)$, $x/(x^2+y^2)$.

1.7 HIGHER ORDER PARTIAL DERIVATIVES

Let the first-order partial derivatives of a function of two variables $z=f(x, y)$ that is, $\partial f/\partial x$ and $\partial f/\partial y$ exist at all the points in the domain D. Then, the first order partial derivatives are also functions of x and y and thus can be further differentiated partially with respect to these independent variables that is. w.r.t. x or w.r.t. y. This gives rise to four possible **second order** partial derivatives of f or z, which are defined as

$$\frac{\partial^2 f}{\partial x^2}=\frac{\partial}{\partial x}\left(\frac{\partial f}{\partial x}\right)=f_{xx}(x, y)=\lim_{\Delta x\to 0}\left[\frac{f_x(x+\Delta x, y)-f_x(x, y)}{\Delta x}\right]$$

$$\frac{\partial^2 f}{\partial y\partial x}=\frac{\partial}{\partial y}\left(\frac{\partial f}{\partial x}\right)=f_{yx}(x, y)=\lim_{\Delta y\to 0}\left[\frac{f_x(x, y+\Delta y)-f_x(x, y)}{\Delta y}\right]$$

(differentiate partialiy first with respect to x and then with respect to y)

$$\frac{\partial^2 f}{\partial x\partial y}=\frac{\partial}{\partial x}\left(\frac{\partial f}{\partial y}\right)=f_{xy}(x, y)=\lim_{\Delta x\to 0}\left[\frac{f_y(x+\Delta x, y)-f_y(x, y)}{\Delta x}\right]$$

(differentiate partially first with respect to y and then with respect to x)

$$\frac{\partial^2 f}{\partial y^2}=\frac{\partial}{\partial y}\left(\frac{\partial f}{\partial y}\right)=f_{yy}(x, y)=\lim_{\Delta y\to 0}\left[\frac{f_y(x, y+\Delta y)-f_y(x, y)}{\Delta y}\right],$$

provided the limits exists. The derivatives f_{xy} and f_{yx} are called *mixed derivatives*. If f_{xy} and f_{yx} are continuous at a point $P(x, y)$, then at this point $f_{xy}=f_{yx}$. If all the second order partial derivatives exists at all points in D, then these derivatives are also functions of x and y can be further differentiated.

ILLUSTRATIVE EXAMPLES

EXAMPLE 1.14 Find all the second order partial derivatives of the function $f(x, y)=x^2y^3+x^4y$.

Solution We have $f_x(x, y)=\partial f/\partial x=2xy^3+4x^3y$, $f_y(x, y)=\partial f/\partial y=3x^2y^2+x^4$.

$$f_{yx}(x, y)=\frac{\partial}{\partial y}\left(\frac{\partial f}{\partial x}\right)=\frac{\partial}{\partial y}(f_x)=\frac{\partial}{\partial y}(2xy^3+4x^3y)=6xy^2+4x^3,$$

$$f_{xy}(x, y)=\frac{\partial}{\partial x}\left(\frac{\partial f}{\partial y}\right)=\frac{\partial}{\partial x}(f_y)=\frac{\partial}{\partial x}(3x^2y^2+x^4)=6xy^2+4x^3,$$

$$f_{xx}(x, y)=\frac{\partial}{\partial x}\left(\frac{\partial f}{\partial x}\right)=\frac{\partial}{\partial x}(f_x)=\frac{\partial}{\partial x}(2xy^3+4x^3y)=2y^3+12x^2y,$$

$$f_{yy}(x, y) = \frac{\partial}{\partial y}\left(\frac{\partial f}{\partial y}\right) = \frac{\partial}{\partial y}(f_y) = \frac{\partial}{\partial y}(3x^2y^2 + x^4) = 6x^2y.$$

We note that $f_{xy} = f_{yx}$.

EXAMPLE 1.15 For the function $f(x, y) = \begin{cases} \dfrac{xy(x^2 - y^2)}{x^2 + y^2}, & (x, y) \neq (0, 0) \\ 0 \quad , & (x, y) = (0, 0) \end{cases}$ show that $f_{xy}(0, 0) \neq f_{yx}(0, 0)$.

Solution We obtain the required derivatives as

$$f_x(0, 0) = \lim_{\Delta x \to 0} \frac{f(\Delta x, 0) - f(0, 0)}{\Delta x} = 0, \; f_y(0, 0) = \lim_{\Delta y \to 0} \frac{f(0, \Delta y) - f(0, 0)}{\Delta y} = 0$$

$$f_x(0, y) = \lim_{\Delta x \to 0} \frac{f(\Delta x, y) - f(0, y)}{\Delta x} = \lim_{\Delta x \to 0} \frac{y\left[(\Delta x)^2 - y^2\right]\Delta x}{\left[(\Delta x)^2 + y^2\right]\Delta x} = -y$$

$$f_y(x, 0) = \lim_{\Delta y \to 0} \frac{f(x, \Delta y) - f(x, 0)}{\Delta y} = \lim_{\Delta y \to 0} \frac{x\left[x^2 - (\Delta y)^2\right]\Delta y}{\left[x^2 + (\Delta y)^2\right]\Delta y} = x$$

Now,

$$f_{xy}(0, 0) = \frac{\partial}{\partial x}\left(\frac{\partial f}{\partial y}\right)_{(0,0)} = \lim_{\Delta x \to 0} \frac{f_y(\Delta x, 0) - f_y(0, 0)}{\Delta x} = \lim_{\Delta x \to 0} \frac{\Delta x - 0}{\Delta x} = 1.$$

$$f_{yx}(0,0) = \frac{\partial}{\partial y}\left(\frac{\partial f}{\partial x}\right)_{(0,0)} = \lim_{\Delta y \to 0} \frac{f_x(0, \Delta y) - f_x(0, 0)}{\Delta y} = \lim_{\Delta y \to 0} \frac{-\Delta y - 0}{\Delta y} = -1.$$

Hence, $\quad f_{xy}(0, 0) \neq f_{yx}(0, 0)$.

EXAMPLE 1.16 Compute $f_{xy}(0, 0)$ and $f_{yx}(0, 0)$ for the function

$$f(x, y) = \begin{cases} \dfrac{xy^3}{x + y^2}, & (x, y) \neq (0, 0) \\ 0 \quad , & (x, y) = (0, 0). \end{cases}$$

(AMIETE, June 2008; Dec. 2004)

Also discuss the continuity of f_{xy} and f_{yx} at (0, 0).

Solution We have

$$f_x(0, 0) = \lim_{\Delta x \to 0} \frac{f(\Delta x, 0) - f(0, 0)}{\Delta x} = 0, \; f_y(0, 0) = \lim_{\Delta y \to 0} \frac{f(0, \Delta y) - f(0, 0)}{\Delta y} = 0$$

$$f_x(0, y) = \lim_{\Delta x \to 0} \frac{f(\Delta x, y) - f(0, y)}{\Delta x} = \lim_{\Delta x \to 0} \frac{y^3 \Delta x}{\left[\Delta x + y^2\right]\Delta x} = y$$

$$f_y(x, 0) = \lim_{\Delta y \to 0} \frac{f(x, \Delta y) - f(x, 0)}{\Delta y} = \lim_{\Delta y \to 0} \frac{x(\Delta y)^3}{\left[x + (\Delta y)^2\right]\Delta y} = 0$$

$$f_{xy}(0, 0) = \lim_{\Delta x \to 0} \frac{f_y(\Delta x, 0) - f_y(0, 0)}{\Delta x} = 0.$$

$$f_{yx}(0, 0) = \lim_{\Delta y \to 0} \frac{f_x(0, \Delta y) - f_x(0, 0)}{\Delta y} = \lim_{\Delta y \to 0} \frac{\Delta y}{\Delta y} = 1.$$

Since $f_{xy}(0, 0) \neq f_{yx}(0, 0)$, f_{xy} and f_{yx} are not continuous at (0, 0).

Alternative We have $f_x(x, y) = \dfrac{y^3(x + y^2) - xy^3}{(x + y^2)^2} = \dfrac{y^5}{(x + y^2)^2}$.

We find that for $(x, y) \neq (0, 0)$

$$f_{yx}(x, y) = \frac{y^6 + 5xy^4}{(x + y^2)^3} = f_{xy}(x, y).$$

Along the path $x = my^2$, we obtain

$$\lim_{(x,y) \to (0,0)} f_{yx}(x, y) = \lim_{y \to 0} \frac{y^6(1+5m)}{y^6(1+m)^3} = \frac{1+5m}{(1+m)^3}.$$

Since the limit does not exist, f_{yx} is not continuous at (0, 0).

EXAMPLE 1.17 Verify that $\frac{\partial^2 u}{\partial y \partial x} = \frac{\partial^2 u}{\partial x \partial y}$ for the following cases:

(*i*) $u = ax^3 + 3\,bx^2y + 3\,cxy^2 + dy^3$, (*ii*) $u = \tan^{-1}(x/y)$.

Solution (*i*) We have $u = ax^3 + 3\,bx^2y + 3\,cxy^2 + dy^3$.

$$\partial u / \partial x = 3ax^2 + 6bxy + 3cy^2, \quad \partial u / \partial y = 3bx^2 + 6cxy + 3dy^2.$$

$$\therefore \quad \frac{\partial^2 u}{\partial y \partial x} = \frac{\partial}{\partial y}\left(\frac{\partial u}{\partial x}\right) = \frac{\partial}{\partial y}(3ax^2 + 6bxy + 3cy^2) = 6bx + 6cy, \qquad \text{...}(a)$$

and

$$\frac{\partial^2 u}{\partial x \partial y} = \frac{\partial}{\partial x}\left(\frac{\partial u}{\partial y}\right) = \frac{\partial}{\partial x}(3bx^2 + 6cxy + 3dy^2) = 6bx + 6cy. \qquad \text{...}(b)$$

From (*a*) and (*b*), follows the required result.

(*ii*) We have $u = \tan^{-1}\left(\frac{x}{y}\right)$.

$$\therefore \quad \frac{\partial u}{\partial x} = \frac{1}{1+(x/y)^2} \cdot \frac{1}{y} = \frac{y}{x^2+y^2}, \quad \frac{\partial u}{\partial y} = \frac{1}{1+(x/y)^2} \frac{-x}{y^2} = \frac{-x}{x^2+y^2}.$$

$$\frac{\partial^2 u}{\partial y \partial x} = \frac{\partial}{\partial y}\left(\frac{\partial u}{\partial x}\right) = \frac{\partial}{\partial y}\left(\frac{y}{x^2+y^2}\right) = \frac{(x^2+y^2) - y(2y)}{(x^2+y^2)^2} = \frac{x^2-y^2}{(x^2+y^2)^2}, \qquad \text{...}(c)$$

and

$$\frac{\partial^2 u}{\partial x \partial y} = \frac{\partial}{\partial x}\left(\frac{\partial u}{\partial y}\right) = \frac{\partial}{\partial x}\left(\frac{-x}{x^2+y^2}\right) = \frac{(x^2+y^2)(-1) - (-x)(2x)}{(x^2+y^2)^2} = \frac{x^2-y^2}{(x^2+y^2)^2}. \qquad \text{...}(d)$$

From (*c*) and (*d*) follows the required result.

EXAMPLE 1.18 If $u = x^y$, show that $\frac{\partial^3 u}{\partial x^2 \partial y} = \frac{\partial^3 u}{\partial x \partial y \partial x}$. (*UPTU-2001*)

Solution We have $u = x^y$. Therefore, $\partial u/\partial y = x^y \log_e x$.

$$\frac{\partial^2 u}{\partial x \partial y} = \frac{\partial}{\partial x}\left(\frac{\partial u}{\partial y}\right) = yx^{y-1} \log_e x + x^y \cdot \frac{1}{x} = x^{y-1}(1 + y \log_e x),$$

and

$$\frac{\partial^3 u}{\partial x^2 \partial y} = \frac{\partial}{\partial x}\left(\frac{\partial^2 u}{\partial x \partial y}\right) = \frac{\partial}{\partial x}\left[x^{y-1}(1 + y \log_e x)\right]. \qquad \text{...}(i)$$

Also $\quad \partial u/\partial x = yx^{y-1}$

Therefore,

$$\frac{\partial^2 u}{\partial y \partial x} = \frac{\partial}{\partial y}\left(\frac{\partial u}{\partial x}\right) = x^{y-1} + y(x^{y-1} \log_e x) = x^{y-1}(1 + y \log_e x),$$

and

$$\frac{\partial^3 u}{\partial x \partial y \partial x} = \frac{\partial}{\partial x}\left(\frac{\partial^2 u}{\partial y \partial x}\right) = \frac{\partial}{\partial x}\left\{x^{y-1}(1 + y \log_e x)\right\}. \qquad \text{...}(ii)$$

From (*i*) and (*ii*) follows the result.

EXAMPLE 1.19 (*a*) Show that at the point on the surface $x^x y^y z^z = c$, where $x = y = z$, we have

$$\frac{\partial^2 z}{\partial x \partial y} = -[x \log(ex)]^{-1}.$$ (*AMIE, W-2005; DCE 2004, AMIETE, S-2004; UPTU 2001*)

(*b*) If $z = x^2 \tan^{-1}(y/x) - y^2 \tan^{-1}(x/y)$, prove that $\partial^2 z/\partial y \partial x = (x^2 - y^2) / (x^2 + y^2)$. (*AMIE, S-2003*)

Solution (*a*) We have $x^x y^y z^z = c$.

Taking logarithms, we get $\log x^x + \log y^y + \log z^z = \log c$

or $x \log x + y \log y + z \log z = \log c$.

Differentiating partially w.r.t. y, we obtain

$$y \cdot \frac{1}{y} + \log y + \left(z \cdot \frac{1}{z} + \log z \right) \frac{\partial z}{\partial y} = 0 \quad \text{or} \quad \frac{\partial z}{\partial y} = -\frac{1 + \log y}{1 + \log z}$$

or $$\frac{\partial z}{\partial y} = -\frac{\log e + \log y}{\log e + \log z} = -\frac{\log ey}{\log ez}$$

Similarly, we obtain $$\frac{\partial z}{\partial x} = -\frac{\log ex}{\log ez} \qquad ...(i)$$

Now, we have $$\frac{\partial^2 z}{\partial x \partial y} = \frac{\partial}{\partial x}\left(\frac{\partial z}{\partial y}\right) = \frac{\partial}{\partial x}\left(-\frac{\log ey}{\log ez}\right) = -\log ey \cdot \frac{\partial}{\partial x}(\log ez)^{-1}$$

$$= \log ey (\log ez)^{-2} \cdot \frac{e}{ez} \cdot \frac{\partial z}{\partial x} = \frac{-\log ey}{(\log ez)^2} \cdot \frac{1}{z} \cdot \frac{\log ex}{\log ez}. \qquad \text{[Using } (i)]$$

Thus at $y = z = x$ we obtain

$$\frac{\partial^2 z}{\partial x \partial y} = -\frac{\log ex}{(\log ex)^2} \cdot \frac{1}{x} \cdot \frac{\log(ex)}{\log(ex)} = -\frac{1}{x \log(ex)}.$$

Hence $$\frac{\partial^2 z}{\partial x \partial y} = -[x \log(ex)]^{-1}.$$

(*b*) We have $z = x^2 \tan^{-1}(y/x) - y^2 \tan^{-1}(x/y)$.

$$\frac{\partial z}{\partial x} = 2x \tan^{-1}\left(\frac{y}{x}\right) + x^2 \cdot \frac{1}{1 + (y/x)^2} \cdot \left(\frac{-y}{x^2}\right) - y^2 \cdot \frac{1}{1 + (x/y)^2} \cdot \frac{1}{y}$$

$$= 2x \tan^{-1}\left(\frac{y}{x}\right) - \frac{x^2 y}{x^2 + y^2} - \frac{y^3}{x^2 + y^2} = 2x \tan^{-1}\left(\frac{y}{x}\right) - y$$

Therefore, $$\frac{\partial^2 z}{\partial y \partial x} = \frac{\partial}{\partial y}\left(\frac{\partial z}{\partial x}\right) = \frac{\partial}{\partial y}\left(2x \tan^{-1}\left(\frac{y}{x}\right) - y\right)$$

$$= 2x \cdot \frac{1}{1 + (y/x)^2} \cdot \frac{1}{x} - 1 = \frac{2x^2}{x^2 + y^2} - 1 = \frac{x^2 - y^2}{x^2 + y^2}.$$

EXAMPLE 1.20 If $v = (x^2 + y^2 + z^2)^{-1/2}$, show that

$$\frac{\partial^2 v}{\partial x^2} + \frac{\partial^2 v}{\partial y^2} + \frac{\partial^2 v}{\partial z^2} = 0.$$ (*VTU, 2006; Osmania, 2003S; JNTU, 2000*)

Solution We have $$\frac{\partial v}{\partial x} = -\frac{1}{2}(x^2 + y^2 + z^2)^{-3/2} \cdot 2x = -x(x^2 + y^2 + z^2)^{-3/2}.$$

$$\frac{\partial^2 v}{\partial x^2} = \frac{\partial}{\partial x}\left(\frac{\partial v}{\partial x}\right) = \frac{\partial}{\partial x}\left[-x(x^2+y^2+z^2)^{-3/2}\right]$$
$$= -[x \cdot (-3/2)(x^2+y^2+z^2)^{-5/2} \cdot 2x + (x^2+y^2+z^2)^{-3/2}]$$
$$= -[3x^2(x^2+y^2+z^2)^{-5/2} + (x^2+y^2+z^2)^{-3/2}]$$
$$= -(x^2+y^2+z^2)^{-5/2}[-3x^2 + (x^2+y^2+z^2)]$$
$$= -(x^2+y^2+z^2)^{-5/2}(-2x^2+y^2+z^2).$$

Similarly, we obtain $\partial^2 v/\partial y^2 = -(x^2+y^2+z^2)^{-5/2}(x^2-2y^2+z^2)$,

and $\partial^2 v/\partial z^2 = -(x^2+y^2+z^2)^{-5/2}(x^2+y^2-2z^2)$.

$$\therefore \quad \frac{\partial^2 v}{\partial x^2}+\frac{\partial^2 v}{\partial y^2}+\frac{\partial^2 v}{\partial z^2} = -(x^2+y^2+z^2)^{-5/2}(0) = 0.$$

EXAMPLE 1.21 If $z = f(x+ct) + \phi(x-ct)$ where f and ϕ are twice differentiable functions and c is a real constant. Show that $\dfrac{\partial^2 z}{\partial t^2} = c^2 \cdot \dfrac{\partial^2 z}{\partial x^2}$. *(JNTU, 2006; VTU, 2003S; AMIE, W-2000)*

Solution We have $\partial z/\partial t = cf'(x+ct) - c\phi'(x-ct)$.

$$\frac{\partial^2 z}{\partial t^2} = \frac{\partial}{\partial t}\left(\frac{\partial z}{\partial t}\right) = c^2 f''(x+ct) + c^2\phi''(x-ct)$$
$$= c^2[f''(x+ct)+\phi''(x-ct)]. \qquad \ldots(i)$$

Similarly, we obtain $\partial^2 z/\partial x^2 = f''(x+ct)+\phi''(x-ct)$ ...(*ii*)

From (*i*) and (*ii*) it follows that $\dfrac{\partial^2 z}{\partial t^2} = c^2 \cdot \dfrac{\partial^2 z}{\partial x^2}$.

EXAMPLE 1.22 If $\theta = t^n e^{-r^2/4t}$, what value of n will make $\dfrac{1}{r^2}\dfrac{\partial}{\partial r}\left(r^2\dfrac{\partial\theta}{\partial r}\right) = \dfrac{\partial\theta}{\partial t}$?

(AMIE, S-2011, Kurukshetra, 2006; UPTU, 2006; AMIETE, S-2003)

Solution We have $\partial\theta/\partial r = t^n \cdot e^{-r^2/4t}(-r/2t)$.

$$\therefore \quad r^2\frac{\partial\theta}{\partial r} = \frac{-r^3}{2t}t^n \cdot e^{-r^2/4t} = \frac{-r^3}{2}t^{n-1}\cdot e^{-r^2/4t},$$

and
$$\frac{\partial}{\partial r}\left(r^2\frac{\partial\theta}{\partial r}\right) = \frac{\partial}{\partial r}\left(\frac{-r^3}{2}t^{n-1}\cdot e^{-r^2/4t}\right) = \frac{-t^{n-1}}{2}\left[3r^2 e^{-r^2/4t} + r^3\cdot e^{-r^2/4t}\cdot\left(\frac{-r}{2t}\right)\right]$$
$$= \frac{-t^{n-1}}{2}r^2 e^{-r^2/4t}\left(3-\frac{r^2}{2t}\right).$$

$$\frac{1}{r^2}\frac{\partial}{\partial r}\left(r^2\frac{\partial\theta}{\partial r}\right) = \frac{-t^{n-1}}{2}e^{-r^2/4t}\left(3-\frac{r^2}{2t}\right) = t^{n-2}\cdot e^{-r^2/4t}\left(\frac{-3t}{2}+\frac{r^2}{4}\right). \qquad \ldots(i)$$

Also
$$\frac{\partial\theta}{\partial t} = nt^{n-1}e^{-r^2/4t} + t^n\cdot e^{-r^2/4t}\left(\frac{r^2}{4t^2}\right) = t^{n-2}e^{-r^2/4t}\left(nt+\frac{r^2}{4}\right). \qquad \ldots(ii)$$

Comparing (*i*) and (*ii*), we see that $n = -3/2$ will make them equal.

EXAMPLE 1.23 If $u = f(r)$ and $x = r\cos\theta$, $y = r\sin\theta$, prove that

$$\frac{\partial^2 u}{\partial x^2}+\frac{\partial^2 u}{\partial y^2} = f''(r)+\frac{1}{r}f'(r).$$ *(Rajasthan, 2006; UPTU 2005; MDU 2001)*

Solution We have $u = f(r)$ and $r^2 = x^2 + y^2$.

Therefore, $$\frac{\partial u}{\partial x} = f'(r)\cdot\frac{\partial r}{\partial x} = f'(r)\cdot\frac{x}{r},\quad \frac{\partial^2 u}{\partial x^2} = \frac{\partial}{\partial x}\left(\frac{\partial u}{\partial x}\right) = \frac{\partial}{\partial x}\left[f'(r)\cdot\frac{x}{r}\right]$$

$$= f''(r)\cdot\frac{\partial r}{\partial x}\cdot\frac{x}{r} + f'(r)\left[\frac{r - x(\partial r/\partial x)}{r^2}\right] = f''(r)\cdot\frac{x^2}{r^2} + f'(r)\cdot\frac{r - x(x/r)}{r^2}$$

$$= f''(r)\cdot\frac{x^2}{r^2} + f'(r)\cdot\frac{r^2 - x^2}{r^3} \qquad ...(i)$$

Similarly, we obtain $$\frac{\partial^2 u}{\partial y^2} = f''(r)\cdot\frac{y^2}{r^2} + f'(r)\cdot\frac{r^2 - y^2}{r^3}. \qquad ...(ii)$$

Adding (i) and (ii), we get

$$\frac{\partial^2 u}{\partial x^2} + \frac{\partial^2 u}{\partial y^2} = f''(r)\left(\frac{x^2 + y^2}{r^2}\right) + f'(r)\cdot\frac{2r^2 - (x^2 + y^2)}{r^3}$$

$$= f''(r) + \frac{1}{r}f'(r), \text{ the required result.}$$

EXAMPLE 1.24 If $u = e^{r\cos\theta}\cos(r\sin\theta)$ and $v = e^{r\cos\theta}\sin(r\sin\theta)$ prove that $\frac{\partial u}{\partial r} = \frac{1}{r}\cdot\frac{\partial v}{\partial \theta}$ and $\frac{\partial v}{\partial r} = -\frac{1}{r}\frac{\partial u}{\partial \theta}$.

Hence deduce that $\frac{\partial^2 u}{\partial r^2} + \frac{1}{r}\frac{\partial u}{\partial r} + \frac{1}{r^2}\frac{\partial^2 u}{\partial \theta^2} = 0.$ *(NIT Kurukshetra, 2009; DCE 2004)*

Solution We have

$$\partial u/\partial r = \cos\theta\, e^{r\cos\theta}\cos(r\sin\theta) - \sin\theta\, e^{r\cos\theta}\sin(r\sin\theta)$$
$$= e^{r\cos\theta}\cos(\theta + r\sin\theta). \qquad ...(i)$$
$$\partial u/\partial\theta = -r\sin\theta\, e^{r\cos\theta}\cos(r\sin\theta) - r\cos\theta\, e^{r\cos\theta}\sin(r\sin\theta)$$
$$= -re^{r\cos\theta}\sin(\theta + r\sin\theta). \qquad ...(ii)$$
$$\partial v/\partial r = \cos\theta\, e^{r\cos\theta}\sin(r\sin\theta) + \sin\theta\, e^{r\cos\theta}\cos(r\sin\theta)$$
$$= e^{r\cos\theta}\sin(\theta + r\sin\theta). \qquad ...(iii)$$

and $$\partial v/\partial\phi = -r\sin\theta\, e^{r\cos\theta}\sin(r\sin\theta) + r\cos\theta\, e^{r\cos\theta}\cos(r\sin\theta)$$
$$= re^{r\cos\theta}\cos(\theta + r\sin\theta). \qquad ...(iv)$$

From (i) and (iv) $\frac{\partial u}{\partial r} = \frac{1}{r}\frac{\partial v}{\partial \theta}$, and from (ii) and (iii) $\frac{\partial v}{\partial r} = -\frac{1}{r}\frac{\partial u}{\partial \theta}$.

Now $$\frac{\partial^2 u}{\partial r^2} = \frac{\partial}{\partial r}\left(\frac{\partial u}{\partial r}\right) = \frac{\partial}{\partial r}\left(\frac{1}{r}\frac{\partial v}{\partial \theta}\right) = -\frac{1}{r^2}\frac{\partial v}{\partial \theta} + \frac{1}{r}\frac{\partial^2 v}{\partial r\partial\theta},$$

and $$\frac{\partial^2 u}{\partial \theta^2} = \frac{\partial}{\partial \theta}\left(\frac{\partial u}{\partial \theta}\right) = \frac{\partial}{\partial \theta}\left(-r\frac{\partial v}{\partial r}\right) = -r\frac{\partial^2 v}{\partial\theta\partial r}.$$

Therefore, $$\frac{\partial^2 u}{\partial r^2} + \frac{1}{r}\frac{\partial u}{\partial r} + \frac{1}{r^2}\frac{\partial^2 u}{\partial \theta^2} = -\frac{1}{r^2}\frac{\partial v}{\partial \theta} + \frac{1}{r}\frac{\partial^2 v}{\partial r\partial\theta} + \frac{1}{r^2}\frac{\partial v}{\partial \theta} - \frac{1}{r}\frac{\partial^2 v}{\partial\theta\partial r} = 0.$$

1.8 HOMOGENEOUS FUNCTIONS

A function $f(x, y)$ is said to be *homogeneous* of degree n in x and y, if it can be written in any one of the following forms

(i) $f(\lambda x, \lambda y) = \lambda^n f(x, y)$, $\lambda > 0$. ...(1.10)

(ii) $f(x, y) = x^n g(y/x)$. ...(1.11)

(iii) $f(x, y) = y^n h(x/y)$. ...(1.12)

Similarly, a function $f(x, y, z)$ of three variables is said to be homogeneous, of degree n, if it can be written as

$$f(\lambda x, \lambda y, \lambda z) = \lambda^n f(x, y, z) \qquad \text{or} \qquad f(x, y, z) = x^n g(y/x, z/x) \text{ etc.}$$

Some examples of homogeneous functions are the following:

Function, f	Degree of homogeneity
1. $x^3 + 3x^2y$	3
2. $x^2 \cos(y/x)$	2
3. $\sin^{-1}(y/x)$	0
4. $\dfrac{x+y}{\sqrt{x}+\sqrt{y}}$	1/2
5. $\sqrt{ax^2+2hxy+by^2}$	1
6. $\dfrac{1}{xy+x^2}$	–2
7. $1/(x^3 + y^3 + z^3)$	–3

The function $f(x, y) = (x + y^2)/(y + x^2)$ is not homogeneous.

EXAMPLE 1.25 Determine the degree of homogeneity of each of the following functions

(i) $f(x,y)=\dfrac{x^3+y^3+3x^2y}{2x+3y}$. (ii) $f(l, k) = (al^4 + bk^4)^{1/2}$.

(iii) $f(x, y) = (x^2, y)^{1/3}$. (iv) $f(x, y) = Ax^{\alpha} y^{\beta}$.

Solution (i) We have $f(x,y)=\dfrac{x^3+y^3+3x^2y}{2x+3y}$. Replacing x by λx and y by λy, $\lambda > 0$, we obtain

$$f(\lambda x, \lambda y) = \frac{(\lambda x)^3+(\lambda y)^3+3(\lambda x)^2(\lambda y)}{2\lambda x+3\lambda y} = \lambda^2\left(\frac{x^3+y^3+3x^2y}{2x+3y}\right) = \lambda^2 f(x, y).$$

Hence the degree of homogeneity is 2.

(ii) Setting $l = \lambda l$, $k = \lambda k$, $\lambda > 0$, we obtain

$$f(\lambda l, \lambda k) = [a(\lambda l)^4 + b(\lambda k)^4]^{1/2} = \lambda^2 (al^4 + bk^4)^{1/2} = \lambda^2 f(l, k).$$

Hence the degree of homogeneity is 2.

(iii) We have
$$f(x, y) = (x^2y)^{1/3}.$$
$$\therefore \quad f(\lambda x, \lambda y) = [(\lambda x)^2(\lambda y)]^{1/3} = \lambda(x^2y)^{1/3} = \lambda f(x, y).$$

Hence the degree of homogeneity is 1.

(iv)
$$f(x, y) = A\, x^{\alpha}y^{\beta}.$$
$$\therefore \quad f(\lambda x, \lambda y) = A\,(\lambda x)^{\alpha}(\lambda y)^{\beta} = \lambda^{\alpha+\beta}(Ax^{\alpha}y^{\beta}) = \lambda^{\alpha+\beta} f(x, y).$$

So that the degree of homogeneity is $\alpha + \beta$.

1.9 EULER'S THEOREM ON HOMOGENEOUS FUNCTIONS

If $f(x, y)$ is a homogeneous function of degree n in x and y and has continuous first and second order partial derivatives, then

(a) $$x\frac{\partial f}{\partial x} + y\frac{\partial f}{\partial y} = nf. \qquad \text{...(1.13) (AMIETE, June 2009; AMIE, W-2003, W-2002)}$$

(b) $$x^2\frac{\partial^2 f}{\partial x^2} + 2xy\frac{\partial^2 f}{\partial x \partial y} + y^2\frac{\partial^2 f}{\partial y^2} = n(n-1)f. \qquad \text{...(1.14) (AMIETE, June 2009; VTU 2007; UPTU 2006)}$$

Proof (a) Since $f(x, y)$ is homogeneous function of degree n in x and y, we can write $f(x, y) = x^n g(y/x)$. Differentiating partially with respect to x and y, we obtain

$$\frac{\partial f}{\partial x} = nx^{n-1}g\left(\frac{y}{x}\right)+x^n\cdot g'\left(\frac{y}{x}\right)\left(-\frac{y}{x^2}\right)=nx^{n-1}g\left(\frac{y}{x}\right)-yx^{n-2}g'\left(\frac{y}{x}\right).$$

$$\frac{\partial f}{\partial y} = x^n g'\left(\frac{y}{x}\right)\cdot\frac{1}{x}=x^{n-1}g'\left(\frac{y}{x}\right).$$

Hence, we obtain

$$x\frac{\partial f}{\partial x}+y\frac{\partial f}{\partial y} = nx^n g\left(\frac{y}{x}\right)-yx^{n-1}g'\left(\frac{y}{x}\right)+yx^{n-1}g'\left(\frac{y}{x}\right)=nx^n g\left(\frac{y}{x}\right)=nf.$$

Note [In general if $f(x_1, x_2, ..., x_k)$ be a homogeneous function of order n, then $x_1\frac{\partial f}{\partial x_1}+x_2\frac{\partial f}{\partial x_2}+......+x_k\frac{\partial f}{\partial x_k}=nf$.]

(*b*) Differentiating (1.13) partially w.r.t x and y, we get

$$x\frac{\partial^2 f}{\partial x^2}+\frac{\partial f}{\partial x}+y\frac{\partial^2 f}{\partial x\partial y}=n\frac{\partial f}{\partial x}, \quad ...(1.15)$$

and
$$x\frac{\partial^2 f}{\partial y\partial x}+\frac{\partial f}{\partial y}+y\frac{\partial^2 f}{\partial y^2}=n\frac{\partial f}{\partial y}. \quad ...(1.16)$$

Multiplying (1.15) by x and (1.16) by y and adding, we obtain.

$$x^2\frac{\partial^2 f}{\partial x^2}+\left(x\frac{\partial f}{\partial x}+y\frac{\partial f}{\partial y}\right)+xy\left(\frac{\partial^2 f}{\partial x\partial y}+\frac{\partial^2 f}{\partial y\partial x}\right)+y^2\frac{\partial^2 f}{\partial y^2}=n\left(x\frac{\partial f}{\partial x}+y\frac{\partial f}{\partial y}\right)$$

or
$$x^2\frac{\partial^2 f}{\partial x^2}+2xy\frac{\partial^2 f}{\partial x\partial y}+y^2\frac{\partial^2 f}{\partial y^2}=n(n-1)f.$$

ILLUSTRATIVE EXAMPLES

EXAMPLE 1.26 Verify Euler's theorem for the following functions.

(*i*) $f(x, y) = ax^2 + 2hxy + by^2$, (*ii*) $f(x, y)=\dfrac{(x^{1/4}+y^{1/4})}{(x^{1/5}+y^{1/5})}$. (*UPTU 2005*)

(*iii*) $f(x, y, z) = 3x^2yz + 5xy^2z + 4z^4$. (*PTU 2006*)

Solution (*i*) We have $\partial f/\partial x = 2ax + 2hy$, $\partial f/\partial y = 2hx + 2by$.

Therefore, $$x\frac{\partial f}{\partial x}+y\frac{\partial f}{\partial y} = x(2ax + 2hy) + y(2hx + 2by) = 2(ax^2 + 2\,hxy + by^2) = 2f(x, y).$$

Since $f(x, y)$ is homogeneous of degree 2, Euler's theorem is verified.

(*ii*) We have $$f(x, y) = (x^{1/4}+y^{1/4})/(x^{1/5}+y^{1/5}).$$

Both the numerator and denominator are homogeneous. Their degrees are 1/4 and 1/5 respectively. So the degree of $f(x, y)$ is 1/4 – 1/5 = 1/20.

Now
$$\frac{\partial f}{\partial x} = \frac{\frac{1}{4}x^{-3/4}(x^{1/5}+y^{1/5})-\frac{1}{5}x^{-4/5}(x^{1/4}+y^{1/4})}{(x^{1/5}+y^{1/5})^2}$$

and
$$\frac{\partial f}{\partial y} = \frac{\frac{1}{4}y^{-3/4}(x^{1/5}+y^{1/5})-\frac{1}{5}y^{-4/5}(x^{1/4}+y^{1/4})}{(x^{1/5}+y^{1/5})^2}.$$

Therefore, $$x\frac{\partial f}{\partial x}+y\frac{\partial f}{\partial y} = \frac{\left(\frac{1}{4}-\frac{1}{5}\right)(x^{1/5}+y^{1/5})(x^{1/4}+y^{1/4})}{(x^{1/5}+y^{1/5})^2}$$

$$= \frac{1}{20}\cdot\frac{x^{1/4}+y^{1/4}}{x^{1/5}+y^{1/5}} = \frac{1}{20}f(x,y),$$ which verifies Euler's theorem.

(*iii*) $\partial f/\partial x = 6xyz + 5y^2z$, $\partial f/\partial y = 3x^2z + 10xyz$ and $\partial f/\partial z = 3x^2y + 5xy^2 + 16z^3$.

Therefore, $$x\frac{\partial f}{\partial x}+y\frac{\partial f}{\partial y}+z\frac{\partial f}{\partial z} = 6x^2yz + 5y^2zx + 3x^2zy + 10xy^2z + 3x^2yz + 5xy^2z + 16z^4$$

$$= 12\,x^2yz + 20\,xy^2z + 16\,z^4$$

$$= 4[3x^2yz + 5xy^2z + 4z^4] = 4f(x, y, z).$$

Since the degree of homogeneity is 4 so Euler's theorem is verified.

EXAMPLE 1.27 Find the value of $x\frac{\partial u}{\partial x}+y\frac{\partial u}{\partial y}$ when

(*i*) $u = \sin^{-1}(x/y) + \tan^{-1}(y/x)$, (*AMIETE-Dec. 2010, June 2010; AMIE, S-2003*)

(*ii*) $u = x^3\log\frac{\sqrt[3]{y}-\sqrt[3]{x}}{\sqrt[3]{y}+\sqrt[3]{x}}$, (*iii*) $u = \sin\{(x-y)/(x+y)\}^{1/2}$,

(*iv*) $u = 3\log x - \log y^3 + \tan^{-1}(y/x)$.

Solution (*i*) Since u is a homogeneous function of degree 0, thus by Euler's theorem $x(\partial u/\partial x) + y(\partial u/\partial y) = 0.u = 0$.

(*ii*) $u = x^3\log\left[\frac{(y/x)^{1/3}-1}{(y/x)^{1/3}+1}\right]$ is homogeneous function of degree 3. Therefore, by Euler's theorem

$$x\frac{\partial u}{\partial x}+y\frac{\partial u}{\partial y} = nu = 3u = 3x^3\log\frac{\sqrt[3]{y}-\sqrt[3]{x}}{\sqrt[3]{y}+\sqrt[3]{x}}.$$

(*iii*) $u = \sin\left(\frac{x-y}{x+y}\right)^{1/2} = \sin\left(\frac{1-(y/x)}{1+(y/x)}\right)^{1/2}$ is a homogeneous function of degree zero.

Therefore, $x(\partial u/\partial x) + y(\partial u/\partial y) = 0$.

(*iv*) $u = 3\log x - \log y^3 + \tan^{-1}(y/x) = \log(x^3/y^3) + \tan^{-1}(y/x)$ is a homogeneous function of degree 0.

Therefore, $x(\partial u/\partial x) + y(\partial u/\partial y) = 0$.

EXAMPLE 1.28 If $F(x, y) = x^4y^2\sin^{-1}(y/x)$, then find the value of $x(\partial F/\partial x) + y(\partial F/\partial y)$ by actual differentiation. (*AMIETE, Dec. 2001*)

Solution We have $F(x, y) = x^4y^2\sin^{-1}(y/x)$.

Differentiating partially w.r.t. x, we obtain

$$\frac{\partial F}{\partial x} = y^2\left[4x^3\sin^{-1}\left(\frac{y}{x}\right)+x^4\cdot\frac{1}{\sqrt{1-(y/x)^2}}\cdot\left(-\frac{y}{x^2}\right)\right].$$

Therefore, $$x\frac{\partial F}{\partial x} = 4x^4y^2\sin^{-1}\left(\frac{y}{x}\right)-\frac{x^4y^3}{\sqrt{x^2-y^2}}. \quad ...(i)$$

Now, differentiating partially w.r.t. y, we obtain

$$\frac{\partial F}{\partial y} = x^4\left[2y\sin^{-1}\left(\frac{y}{x}\right)+y^2\cdot\frac{1}{\sqrt{1-(y/x)^2}}\cdot\frac{1}{x}\right].$$

Therefore, $$y\frac{\partial F}{\partial y} = 2x^4y^2\sin^{-1}\left(\frac{y}{x}\right)+\frac{x^4y^3}{\sqrt{x^2-y^2}}. \quad ...(ii)$$

Adding (*i*) and (*ii*), we obtain

$$x\frac{\partial F}{\partial x}+y\frac{\partial F}{\partial y} = 6[x^4y^2 \sin^{-1}(y/x)] = 6F(x, y).$$

EXAMPLE 1.29 (*a*) If $u=\sin^{-1}\left(\frac{x^2+y^2}{x+y}\right)$, show that $x\frac{\partial u}{\partial x}+y\frac{\partial u}{\partial y}=\tan u$.

(*AMIETE, June 2010; GGSIPU 2007; AMIE, W-2005; VTU 2003*)

(*b*) If $f(x, y)=\frac{1}{x^2}+\frac{1}{xy}+\frac{\log x-\log y}{x^2+y^2}$, show that $x\frac{\partial f}{\partial x}+y\frac{\partial f}{\partial y}+2f(x,y)=0$. (*AMIE, S-2007; PTU 2004*)

(*c*) If $u=\cos\left(\frac{xy+yz+zx}{x^2+y^2+z^2}\right)$, prove that $x\frac{\partial u}{\partial x}+y\frac{\partial u}{\partial y}+z\frac{\partial u}{\partial z}=0$.

(*d*) If $u=\log\left(\frac{x^4+y^4}{x+y}\right)$, show that $x\frac{\partial u}{\partial x}+y\frac{\partial u}{\partial y}=3$. (*MDU 2004; UPTU 2001*)

Solution (*a*) The given function can be written as

$$\sin u=\frac{x^2+y^2}{x+y}=x\left\{\frac{1+(y/x)^2}{1+(y/x)}\right\}.$$

Therefore, sin u is a homogeneous function of degree 1.

Using the Euler's theorem for $f = \sin u$ and $n = 1$, we obtain

$$x\frac{\partial}{\partial x}(\sin u)+y\frac{\partial}{\partial y}(\sin u)=\sin u \qquad ...(i)$$

or

$$x\cos u\frac{\partial u}{\partial x}+y\cos u\frac{\partial u}{\partial y}=\sin u \;\text{ or }\; x\frac{\partial u}{\partial x}+y\frac{\partial u}{\partial y}=\tan u.$$

(*b*) $f(x, y)=\frac{1}{x^2}\left\{1+\frac{x}{y}+\frac{\log(x/y)}{1+(y^2/x^2)}\right\}=x^{-2}\left\{1+(y/x)^{-1}-\frac{\log(y/x)}{1+(y^2/x^2)}\right\}.$

The degree of homogeneity is –2. Thus by Euler's theorem, we have

$$x\frac{\partial f}{\partial x}+y\frac{\partial f}{\partial y}=-2f(x, y) \;\text{ or }\; x\frac{\partial f}{\partial x}+y\frac{\partial f}{\partial y}+2f(x, y)=0.$$

(*c*) $u=f(x, y,z)=\cos\left(\frac{xy+yz+zx}{x^2+y^2+z^2}\right).$

The degree of homogeneity is 0. Thus by Euler's theorem, we have

$$x\frac{\partial u}{\partial x}+y\frac{\partial u}{\partial y}+z\frac{\partial u}{\partial z}=0(u)=0.$$

(*d*) Let $z=e^u=\frac{x^4+y^4}{x+y}=x^3\frac{\left[1+(y/x)^4\right]}{\left[1+(y/x)\right]}$. The degree of homogeneity is 3.

∴ By Euler's theorem, $x(\partial z/\partial x) + y(\partial z/\partial y) = 3z$

or

$$x\cdot\frac{\partial z}{\partial u}\cdot\frac{\partial u}{\partial x}+y\frac{\partial z}{\partial u}\cdot\frac{\partial u}{\partial y}=3z \;\text{ or }\; x\cdot e^u\cdot\frac{\partial u}{\partial x}+y\cdot e^u\cdot\frac{\partial u}{\partial y}=3e^u$$

or

$$x(\partial u/\partial x) + y(\partial u/\partial y) = 3.$$

EXAMPLE 1.30 Use Euler's theorem to prove

$$x\frac{\partial v}{\partial x}+y\frac{dv}{\partial y}+\frac{1}{2}\cot v=0, \text{ where } v=\cos^{-1}\left(\frac{x+y}{\sqrt{x}+\sqrt{y}}\right), 0<x, y<1.$$

(*AMIETE, Dec. 2009; GGSIPU 2006; VTU 2004; AMIE; W-2004*)

Solution The given function can be written as

$$\cos v = \frac{x+y}{\sqrt{x}+\sqrt{y}} = \frac{x\{1+(y/x)\}}{\sqrt{x}\{1+(\sqrt{y/x})\}} = x^{1/2}\left\{\frac{1+(y/x)}{1+(\sqrt{y/x})}\right\}.$$

Therefore, $\cos v$ is a homogeneous function of degree 1/2.

Using the Euler Theorem for $f = \cos v$ and $n = 1/2$, we obtain

$$x\frac{\partial}{\partial x}(\cos v) + y\frac{\partial}{\partial y}(\cos v) = \frac{1}{2}\cos v \quad \text{or} \quad -x(\sin v)\frac{\partial v}{\partial x} - y(\sin v)\frac{\partial v}{\partial y} = \frac{1}{2}\cos v$$

or $$x\frac{\partial v}{\partial x} + y\frac{\partial v}{\partial y} + \frac{1}{2}\cot v = 0, \text{ which is same as required.}$$

EXAMPLE 1.31 If z is a homogeneous function of x, y of degree n and $z = f(u)$ then

(*a*) $x\dfrac{\partial u}{\partial x} + y\dfrac{\partial u}{\partial y} = n\cdot\dfrac{f(u)}{f'(u)}$, (*b*) $x^2\dfrac{\partial^2 u}{\partial x^2} + 2xy\dfrac{\partial^2 u}{\partial x\partial y} + y^2\dfrac{\partial^2 u}{\partial y^2} = g(u)\{g'(u)-1\}$ where $g(u) = n\dfrac{f(u)}{f'(u)}$.

Solution (*a*) Since z is a homogeneous function of x, y of degree n, we have, by Euler's theorem

$$x\frac{\partial z}{\partial x} + y\frac{\partial z}{\partial y} = nz. \qquad ...(i)$$

As $z = f(u)$, we have $\dfrac{\partial z}{\partial x} = \dfrac{\partial z}{\partial u}\dfrac{\partial u}{\partial x} = f'(u)\dfrac{\partial u}{\partial x}$.

Similarly, we obtain $\dfrac{\partial z}{\partial y} = f'(u)\dfrac{\partial u}{\partial y}$. Substituting in (*i*), we get

$$x\frac{\partial u}{\partial x}f'(u) + y\frac{\partial u}{\partial y}f'(u) = n\cdot f(u) \quad \text{or} \quad x\frac{\partial u}{\partial x} + y\frac{\partial u}{\partial y} = n\frac{f(u)}{f'(u)}.$$

(*b*) Set $$x\frac{\partial u}{\partial x} + y\frac{\partial u}{\partial y} = n\frac{f(u)}{f'(u)} = g(u). \qquad ...(ii)$$

Differentiating (*ii*) partially w.r.t., x we obtain

$$x\frac{\partial^2 u}{\partial x^2} + \frac{\partial u}{\partial x} + y\frac{\partial^2 u}{\partial x\partial y} = g'(u)\frac{\partial u}{\partial x} \quad \text{or} \quad x\frac{\partial^2 u}{\partial x^2} + y\frac{\partial^2 u}{\partial x\partial y} = \{g'(u)-1\}\frac{\partial u}{\partial x}. \qquad ...(iii)$$

Similarly, we obtain $y\dfrac{\partial^2 u}{\partial y^2} + x\dfrac{\partial^2 u}{\partial y\partial x} = \{g'(u)-1\}\dfrac{\partial u}{\partial y}$. ...(*iv*)

Multiplying (*iii*) by x and (*iv*) by y and then adding, we obtain

$$x^2\frac{\partial^2 u}{\partial x^2} + 2xy\frac{\partial^2 u}{\partial x\partial y} + y^2\frac{\partial^2 u}{\partial y^2} = \{g'(u)-1\}\left(x\frac{\partial u}{\partial x} + y\frac{\partial u}{\partial y}\right) = g(u)\{g'(u)-1\} \ \{\text{by }(ii)\}$$

EXAMPLE 1.32 If $U = \sec^{-1}\left(\dfrac{x^3-y^3}{x-y}\right)$, then evaluate $x^2U_{xx} + 2xyU_{xy} + y^2U_{yy}$. (*AMIETE Dec. 2001*)

Solution The given function can be written as

$$\sec U = \frac{x^3-y^3}{x-y} = \frac{x^3[1-(y/x)^3]}{x[1-(y/x)]} = x^2\left\{\frac{1-(y/x)^3}{1-(y/x)}\right\}.$$

Therefore sec U is a homogeneous function of degree 2.

Using Euler's theorem for $f = \sec U$ and $n = 2$, we obtain

$$x\frac{\partial}{\partial x}(\sec U) + y\frac{\partial}{\partial y}(\sec U) = 2\sec U$$

or $$x\sec U\tan U\frac{\partial U}{\partial x}+y\sec U\tan U\frac{\partial U}{\partial y}=2\sec U$$

or $$x\frac{\partial U}{\partial x}+y\frac{\partial U}{\partial y}=\frac{2\sec U}{\sec U\tan U}=2\cot U.$$

In the above result, if we set 2 cot $U = G(U)$, then

$$x^2\frac{\partial^2 U}{\partial x^2}+2xy\frac{\partial^2 U}{\partial x\partial y}+y^2\frac{\partial^2 U}{\partial y^2} = G(U)\,[G'(U)-1] = 2\cot U\,[-2\operatorname{cosec}^2 U-1]$$

$$= 2\cot U\,[-2\,(1+\cot^2 U)-1] = -2\cot U\,[2\cot^2 U+3].$$

Thus the value of $x^2U_{xx} + 2\,xy\,U_{xy} + y^2U_{yy}$ is equal to $-2\cot U\,(2\cot^2 U + 3)$.

EXAMPLE 1.33 If $u = \sin^{-1}\left\{\frac{x^{1/3}+y^{1/3}}{x^{1/2}+y^{1/2}}\right\}^{1/2}$ or $\operatorname{cosec}^{-1}\left\{\frac{x^{1/2}+y^{1/2}}{x^{1/3}+y^{1/3}}\right\}^{1/2}$, prove that

$x^2u_{xx} + 2xyu_{xy} + y^2u_{yy} = \frac{\tan u}{144}(13+\tan^2 u)$. *(Rohtak, 2006S; MDU 2006; DCE, 2004; AMIE, S-2000)*

Solution The given function can be written as

$$\sin u=\left\{\frac{x^{1/3}+y^{1/3}}{x^{1/2}+y^{1/2}}\right\}^{1/2}=\frac{x^{1/6}}{x^{1/4}}\left\{\frac{1+(y/x)^{1/3}}{1+(y/x)^{1/2}}\right\}^{1/2}=x^{-1/12}\left\{\frac{1+(y/x)^{1/3}}{1+(y/x)^{1/2}}\right\}^{1/2}.$$

Therefore, sin u is a homogeneous function of degree $-1/12$.

Using Euler's theorem for $f = \sin u$ and $n = -1/12$, we obtain

$$x\frac{\partial}{\partial x}(\sin u)+y\frac{\partial}{\partial y}(\sin u)=-\frac{1}{12}\sin u \text{ or } x\cos u\frac{\partial u}{\partial x}+y\cos u\frac{\partial u}{\partial y}=-\frac{1}{12}\sin u$$

or $$x\frac{\partial u}{\partial x}+y\frac{\partial u}{\partial y}=-\frac{1}{12}\tan u. \quad ...(i)$$

We know that if $x\frac{\partial u}{\partial x}+y\frac{\partial u}{\partial y}=G(u)$ then

$$x^2\frac{\partial^2 u}{\partial x^2}+2xy\frac{\partial^2 u}{\partial x\partial y}+y^2\frac{\partial^2 u}{\partial y^2} = G(u)\{G'(u)-1\}. \quad ...(ii)$$

Set $G(u)=-\frac{1}{12}\tan u$ and $G'(u)=-\frac{1}{12}\sec^2 u=-\frac{1}{12}(1+\tan^2 u)$, we obtain

$$x^2\frac{\partial^2 u}{\partial x^2}+2xy\frac{\partial^2 u}{\partial x\partial y}+y^2\frac{\partial^2 u}{\partial y^2} = -\frac{1}{12}\tan u\left\{-\frac{1}{12}(1+\tan^2 u)-1\right\}$$

or $$x^2u_{xx} + 2\,xy\,u_{xy} + y^2u_{yy} = \frac{\tan u}{144}(13+\tan^2 u), \text{ which is same as required.}$$

EXAMPLE 1.34 (*a*) If $z = x\phi\left(\frac{y}{x}\right)+\psi\left(\frac{y}{x}\right)$, where ϕ and ψ are differentiable functions, then prove that

(*a*) $x^2\frac{\partial^2 z}{\partial x^2}+2xy\frac{\partial^2 z}{\partial x\partial y}+y^2\frac{\partial^2 z}{\partial y^2}=0.$ *(AMIE. S-2008; UPTU 2006; AMIETE S-2002)*

(*b*) If $z = x^m f(y/x) + x^n g(x/y)$, show that

$$x^2\frac{\partial^2 z}{\partial x^2}+2xy\frac{\partial^2 z}{\partial x\partial y}+y^2\frac{\partial^2 z}{\partial y^2}+mnz=(m+n-1)\left(x\frac{\partial z}{\partial x}+y\frac{\partial z}{\partial y}\right).$$ *(AMIE S-2000)*

Solution (*a*) We have $z = x\phi(y/x) + \psi(y/x)$, then

$$\partial z/\partial x = \phi(y/x) + \{x\phi'(y/x) + \psi'(y/x)\}(-y/x^2),$$

and
$$\partial z/\partial y = \{x\phi'(y/x) + \psi'(y/x)\}(1/x).$$

Therefore, $x\dfrac{\partial z}{\partial x} + y\dfrac{\partial z}{\partial y} = x\phi\left(\dfrac{y}{x}\right)$.

Differentiating this partially with respect to x and y, we obtain

$$x\frac{\partial^2 z}{\partial x^2} + \frac{\partial z}{\partial x} + y\frac{\partial^2 z}{\partial x\,\partial y} = x\phi'\left(\frac{y}{x}\right)\cdot\left(\frac{-y}{x^2}\right) + \phi\left(\frac{y}{x}\right), \qquad ...(i)$$

and
$$x\frac{\partial^2 z}{\partial y\,\partial x} + \frac{\partial z}{\partial y} + y\frac{\partial^2 z}{\partial y^2} = x\phi'\left(\frac{y}{x}\right)\left(\frac{1}{x}\right). \qquad ...(ii)$$

Multiplying (*i*) by x and (*ii*) by y and then adding, we get

$$x^2\frac{\partial^2 z}{\partial x^2} + 2xy\frac{\partial^2 z}{\partial x\partial y} + y^2\frac{\partial^2 z}{\partial y^2} + \left\{x\frac{\partial z}{\partial x} + y\frac{\partial z}{\partial y}\right\} = x\phi\left(\frac{y}{x}\right)$$

or
$$x^2\frac{\partial^2 z}{\partial x^2} + 2xy\frac{\partial^2 z}{\partial x\,\partial y} + y^2\frac{\partial^2 z}{\partial y^2} = 0, \left[\because x\frac{\partial z}{\partial x} + y\frac{\partial z}{\partial y} = x\phi\left(\frac{y}{x}\right)\right].$$

(*b*) Let $x^m f(y/x) = u$ and $x^n g(x/y) = v$

Then
$$z = u + v. \qquad ...(i)$$

Now $u = x^m f(y/x)$ is a homogeneous function of degree m. We know that

$$x^2\frac{\partial^2 u}{\partial x^2} + 2xy\frac{\partial^2 u}{\partial x\,\partial y} + y^2\frac{\partial^2 u}{\partial y^2} = m(m-1)u. \qquad ...(ii)$$

Also $v = x^n g(x/y)$ is a homogeneous function of degree n, so we have

$$x^2\frac{\partial^2 v}{\partial x^2} + 2xy\frac{\partial^2 v}{\partial x\,\partial y} + y^2\frac{\partial^2 v}{\partial y^2} = n(n-1)v. \qquad ...(iii)$$

Adding (*ii*) and (*iii*), we have

$$x^2\frac{\partial^2}{\partial x^2}(u+v) + 2xy\frac{\partial^2}{\partial x\,\partial y}(u+v) + y^2\frac{\partial^2}{\partial y^2}(u+v) = m(m-1)u + n(n-1)v,$$

or
$$x^2\frac{\partial^2 z}{\partial x^2} + 2xy\frac{\partial^2 z}{\partial x\,\partial y} + y^2\frac{\partial^2 z}{\partial y^2} = m(m-1)u + n(n-1)v \qquad ...(iv)$$

Since u and v, are homogeneous functions in x and y of degree m and n respectively, therefore by Euler's theorem

$$x\frac{\partial u}{\partial x} + y\frac{\partial u}{\partial y} = mu \text{ and } x\frac{\partial v}{\partial x} + y\frac{\partial v}{\partial y} = nv.$$

Adding these and with the help of (*i*), we obtain

$$x\frac{\partial z}{\partial x} + y\frac{\partial z}{\partial y} = mu + nv. \qquad ...(v)$$

Now
$$\begin{aligned} m(m-1)u + n(n-1)v &= (m^2u + n^2v) - (mu + nv) \\ &= m(m+n)u + n(m+n)v - mn(u+v) - (mu+nv) \\ &= (m+n)(mu+nv) - mn(u+v) - (mu+nv) \\ &= (m+n-1)(mu+nv) - mn\,z, \quad \text{from } (i). \end{aligned}$$

Substituting the value in (*iv*), we obtain

$$x^2\frac{\partial^2 z}{\partial x^2} + 2xy\frac{\partial^2 z}{\partial x\partial y} + y^2\frac{\partial^2 z}{\partial y^2} = (m+n-1)(mu+nv) - mnz$$

$$\text{or } x^2 \frac{\partial^2 z}{\partial x^2} + 2xy \frac{\partial^2 z}{\partial x \partial y} + y^2 \frac{\partial^2 z}{\partial y^2} + mnz = (m + n - 1)(mu + nv)$$

$$= (m+n-1)\left(x\frac{\partial z}{\partial x} + y\frac{\partial z}{\partial y}\right), \text{ from } (v). \text{ Hence proved.}$$

EXAMPLE 1.35 If $u(x, y) = \dfrac{(x^2 + y^2)^n}{2n(2n-1)} + xf\left(\dfrac{y}{x}\right) + g\left(\dfrac{y}{x}\right)$, where f, g are arbitrary functions, prove by using Euler's theorem that $\left(x\dfrac{\partial}{\partial x} + y\dfrac{\partial}{\partial y}\right)^2 u(x, y) = (x^2 + y^2)^n$. (*AMIE S-2003*)

Solution Set $\dfrac{(x^2 + y^2)^n}{2n(2n-1)} = v$, $xf\left(\dfrac{y}{x}\right) = w$ and $g\left(\dfrac{y}{x}\right) = s$.

Then $\qquad u = v + w + s \qquad$ ...(*i*)

$$v = \frac{(x^2 + y^2)^n}{2n(2n-1)} = \frac{x^{2n}\{1 + (y/x)^2\}^n}{2n(2n-1)} \text{ is a homogeneous function of degree } 2n.$$

We know that If v be a homogeneous function of degree $2n$, then

$$x^2 \frac{\partial^2 v}{\partial x^2} + 2xy \frac{\partial^2 v}{\partial x \partial y} + y^2 \frac{\partial^2 v}{\partial y^2} = 2n(2n-1)v. \qquad \text{...(ii)}$$

Also $w = xf(y/x)$ is a homogeneous function of degree one

$$\therefore \quad x^2 \frac{\partial^2 w}{\partial x^2} + 2xy \frac{\partial^2 w}{\partial x \partial y} + y^2 \frac{\partial^2 w}{\partial y^2} = 1(1-1)w = 0. \qquad \text{...(iii)}$$

$s = g(y/x)$ is a homogeneous function of degree zero, so we have

$$x^2 \frac{\partial^2 s}{\partial x^2} + 2xy \frac{\partial^2 s}{\partial x \partial y} + y^2 \frac{\partial^2 s}{\partial y^2} = 0. \qquad \text{...(iv)}$$

Adding (*ii*), (*iii*) and (*iv*), we obtain

$$x^2 \frac{\partial^2}{\partial x^2}(v + w + s) + 2xy \frac{\partial^2}{\partial x \partial y}(v + w + s) + y^2 \frac{\partial^2}{\partial y^2}(v + w + s) = 2n(2n-1)v$$

$$\text{or} \quad \left(x^2 \frac{\partial^2}{\partial x^2} + 2xy \frac{\partial^2}{\partial x \partial y} + y^2 \frac{\partial^2}{\partial y^2}\right) u(x, y) = \frac{2n(2n-1)(x^2 + y^2)^n}{2n(2n-1)}$$

$$\text{or} \quad \left(x\frac{\partial}{\partial x} + y\frac{\partial}{\partial y}\right)^2 u(x, y) = (x^2 + y^2)^n.$$

EXERCISE 1.3

1. If $f(x, y) = x^4 - 4x^3y + 8xy^3 - y^4$, show that $f_{xx}(2, -1) = 96$, $f_{xy}(2, -1) = -24$, $f_{yy}(2, -1) = -108$.
2. If $z = \tan^{-1}(y/x)$, show that $\dfrac{\partial^2 z}{\partial x^2} + \dfrac{\partial^2 z}{\partial y^2} = 0$.
3. If $u = \dfrac{1}{2}\log(x^2 + y^2)$, then show that $\dfrac{\partial^2 u}{\partial x^2} + \dfrac{\partial^2 u}{\partial y^2} = 0$.
4. If $f(x, y) = \log(x^2 + y^2) + \tan^{-1}(y/x)$, then show that $\dfrac{\partial^2 f}{\partial x^2} + \dfrac{\partial^2 f}{\partial y^2} = 0$.
5. If $u = x^3 - 3xy^2$, show that $\dfrac{\partial^2 u}{\partial x^2} + \dfrac{\partial^2 u}{\partial y^2} = 0$.

6. Show that the function $u = \log 1/r$, where $r = \sqrt{(x-a)^2 + (y-b)^2}$ satisfies the partial differential equation $\dfrac{\partial^2 u}{\partial x^2} + \dfrac{\partial^2 u}{\partial y^2} = 0.$ (DCE, 2004)

7. If $v = (ax + by)^2 - (x^2 + y^2)$ where $a^2 + b^2 = 2$, show that $\dfrac{\partial^2 v}{\partial x^2} + \dfrac{\partial^2 v}{\partial y^2} = 0.$

8. If $V = \sqrt{x^2 + y^2 + z^2}$, show that $V_{xx} + V_{yy} + V_{zz} = 2/v$. (AMIE, W-2000)

9. Find the value of a, if $V = x^3 + axy^2$ satisfies the equation $\dfrac{\partial^2 V}{\partial x^2} + \dfrac{\partial^2 V}{\partial y^2} = 0$. Taking this value of V, show that If $u = r^n V$, and $r^2 = x^2 + y^2$, then $\dfrac{\partial^2 u}{\partial x^2} + \dfrac{\partial^2 u}{\partial y^2} = n(n+6)r^{n-2}V.$ **Ans.** $a = -3$

10. If $u = x^2 \tan^{-1}(y/x) - y^2 \tan^{-1}(x/y)$, show that $\dfrac{\partial^2 u}{\partial x \partial y} = \dfrac{x^2 - y^2}{x^2 + y^2} = \dfrac{\partial^2 u}{\partial y \partial x}$. (AMIE, 2003; 2000, Madras 2000)

11. Verify that $\dfrac{\partial^2 u}{\partial x \partial y} = \dfrac{\partial^2 u}{\partial y \partial x}$, when u is equal to

(i) $ax^2 + 2hxy + by^2$, (ii) $\log \dfrac{x^2 + y^2}{xy}$, (iii) $\sin^{-1}(x/y)$, (iv) $\log(y \sin x + x \sin y)$.

12. If $u = \log_e r^2$, where $r^2 = x^2 + y^2 + z^2$ show that $r^2(u_{xx} + u_{yy} + u_{zz}) = 1$.

13. (i) For the function $f(x, y) = \begin{cases} x^2 y\left(\dfrac{x-y}{x^2+y^2}\right), & (x, y) \neq (0, 0) \\ 0, & (x, y) = (0, 0). \end{cases}$

Show that $f_{xy}(0, 0) \neq f_{yx}(0, 0)$.

(ii) For the function $f(x, y) = \begin{cases} \dfrac{xy(2x^2 - 3y^2)}{x^2 + y^2}, & (x, y) \neq (0, 0) \\ 0, & (x, y) = (0, 0). \end{cases}$

Show that $f_{xy}(0, 0) \neq f_{yx}(0, 0)$. (AMIETE, June 2007)

14. For the function $f(x, y) = \begin{cases} x^2 \tan^{-1}\left(\dfrac{y}{x}\right) - y^2 \tan^{-1}\left(\dfrac{x}{y}\right), & x \neq 0,\ y \neq 0 \\ 0, & \text{elsewhere.} \end{cases}$

Show that $f_{xy} \neq f_{yx}$ at (0, 0).

15. If $V = x \log_e (x + r) - r$, where $r^2 = x^2 + y^2$, show that $\dfrac{\partial^2 V}{\partial x^2} + \dfrac{\partial^2 V}{\partial y^2} = \dfrac{1}{x+r}$.

16. Let $r = \sqrt{x^2 + y^2}$ be the radius vector from the origin to the point (x, y) in the xy-plane and θ be the angle which r makes with x-axis. Show that

(i) $\dfrac{\partial^2 r}{\partial y^2} = \dfrac{\cos^2 \theta}{r}$, (ii) $\dfrac{\partial^2 r}{\partial x^2} + \dfrac{\partial^2 r}{\partial y^2} = \dfrac{1}{r}\left\{\left(\dfrac{\partial r}{\partial x}\right)^2 + \left(\dfrac{\partial r}{\partial y}\right)^2\right\}$. (AMIE, S-2008)

17. If $u = x/r^3$ and $r^2 = x^2 + y^2 + z^2$, show that $\dfrac{\partial^2 u}{\partial x^2} + \dfrac{\partial^2 u}{\partial y^2} + \dfrac{\partial^2 u}{\partial z^2} = 0.$

18. If $f(x,y)=\frac{1}{\sqrt{y}}e^{-(x-a)^2/4y}$, show that $f_{xy}(x, y) = f_{yx}(x, y)$.

19. If $V = r^m$ where $r^2 = x^2 + y^2 + z^2$, show that $V_{xx} + V_{yy} + V_{zz} = m(m + 1)\, r^{m-2}$.

(AMIE, S-2007; PTU, 2006; Raipur, 2005)

20. If $V = f(r)$ and $r^2 = x^2 + y^2 + z^2$, show that $V_{xx} + V_{yy} + V_{zz} = f''(r) + \frac{2}{r}f'(r)$, where the notations have their usual meaning.

21. If $u = Ae^{-gx}\sin(nt - gx)$, where A, g, n are positive constants, satisfies the heat conduction equation $\frac{\partial u}{\partial t}=\mu\frac{\partial^2 u}{\partial x^2}$, then show that $g=\sqrt{n/2\mu}$.

22. If $u=\tan^{-1}\frac{xy}{\sqrt{(1+x^2+y^2)}}$, show that $\frac{\partial^2 u}{\partial x\partial y}=\frac{1}{(1+x^2+y^2)^{3/2}}$.

23. If $z(z^2 + 3x) + 3y = 0$, show that $\frac{\partial^2 z}{\partial x^2}+\frac{\partial^2 z}{\partial y^2}=\frac{2z(x-1)}{(z^2+x)^3}$.

24. If z is a function of x and y determined by the equation $z^3 - 3yz - 3x = 0$, show that $z\frac{dz}{dx}=\frac{dz}{dy}$ and $z\left\{\frac{\partial^2 z}{\partial x\partial y}+\left(\frac{\partial z}{\partial x}\right)^2\right\}=\frac{\partial^2 z}{\partial y^2}$.

25. If $u = e^{xyz}$, show that $\frac{\partial^3 u}{\partial x\partial y\partial z}=(1+3xyz+x^2y^2z^2)e^{xyz}$. *(AMIE, W-2000)*

[Hint: $\partial u/\partial z = xye^{xyz}$, $\partial^2 u/\partial y\partial z = x^2yze^{xyz}+xe^{xyz}=(x^2yz+x)e^{xyz}$.

$\therefore\quad \frac{\partial^3 u}{\partial x\partial y\partial z}=yz(x^2yz+x)e^{xyz}+(2xyz+1)e^{xyz}=(x^2y^2z^2+3xyz+1)e^{xyz}$.]

26. If $u = \log(e^x + e^y + e^z)$, show that $\frac{\partial^3 u}{\partial x\partial y\partial z}=2e^{x+y+z-3u}$.

27. If $u = e^{xy}\sin z$, show that $\frac{\partial^3 u}{\partial x\partial y\partial z}=\frac{\partial^3 u}{\partial y\partial z\partial x}=e^{xy}(1+xy)\cos z$.

28. If $u = \log(x^3 + y^3 + z^3 - 3xyz)$, show that

(a) $\frac{\partial u}{\partial x}+\frac{\partial u}{\partial y}+\frac{\partial u}{\partial z}=\frac{3}{x+y+z}$,

(b) $\left(\frac{\partial}{\partial x}+\frac{\partial}{\partial y}+\frac{\partial}{\partial z}\right)^2 u=\frac{-9}{(x+y+z)^2}$, *(AMIETE, Dec. 2010, Dec. 2009; PTU, 2008; Kurukshetra, 2006; UPTU 2006; MDU 2004, JNTU 2003)*

(c) $\frac{\partial^2 u}{\partial x^2}+\frac{\partial^2 u}{\partial y^2}+\frac{\partial^2 u}{\partial z^2}+2\frac{\partial^2 u}{\partial x\partial y}+2\frac{\partial^2 u}{\partial y\partial z}+2\frac{\partial^2 u}{\partial z\partial x}=\frac{-9}{(x+y+z)^2}$.

[Hint: (b) $\frac{\partial u}{\partial x}=\frac{3x^2-3yz}{x^3+y^3+z^3-3xyz}$; $\frac{\partial u}{\partial y}=\frac{3y^2-3xz}{x^3+y^3+z^3-3xyz}$; $\frac{\partial u}{\partial z}=\frac{3z^2-3xy}{x^3+y^3+z^3-3xy}$.

Adding, $\sum\frac{\partial u}{\partial x}=\frac{3}{x+y+z}$. $(\because x^3+y^3+z^3-3xyz=(x+y+z)(x^2+y^2+z^2-xy-yz-zx))$

Now $\left(\frac{\partial}{\partial x}+\frac{\partial}{\partial y}+\frac{\partial}{\partial z}\right)^2 u=\left(\frac{\partial}{\partial x}+\frac{\partial}{\partial y}+\frac{\partial}{\partial z}\right)\left(\frac{\partial}{\partial x}+\frac{\partial}{\partial y}+\frac{\partial}{\partial z}\right)u=\left(\frac{\partial}{\partial x}+\frac{\partial}{\partial y}+\frac{\partial}{\partial z}\right)\left(\frac{\partial u}{\partial x}+\frac{\partial u}{\partial y}+\frac{\partial u}{\partial z}\right)$

$$=\left(\frac{\partial}{\partial x}+\frac{\partial}{\partial y}+\frac{\partial}{\partial z}\right)\left(\frac{3}{x+y+z}\right)=\frac{-3}{(x+y+z)^2}-\frac{3}{(x+y+z)^2}-\frac{3}{(x+y+z)^2}=\frac{-9}{(x+y+z)^2}\Bigg].$$

29. If $z = \log(e^x + e^y)$, show that $rt - s^2 = 0$, where $r \equiv \partial^2 z/\partial x^2$, $t \equiv \partial^2 z/\partial y^2$, $s \equiv \partial^2 z/\partial y\partial x$.

30. (i) If $u = e^{xyz} f(yx/z)$, show that $x\frac{\partial u}{\partial x}+z\frac{\partial u}{\partial z}=2xyzu$; $y\frac{\partial u}{\partial y}+z\frac{\partial u}{\partial z}=2xyzu$.

Also deduce that $x\frac{\partial^2 u}{\partial z\,\partial x}=y\frac{\partial^2 u}{\partial z\,\partial y}$. *(AMIE, S-2001)*

(ii) Find the value of n so that the equation $v = r^n(3\cos^2\theta - 1)$ satisfies the relation

$$\frac{\partial}{\partial r}\left(r^2\frac{\partial v}{\partial r}\right)+\frac{1}{\sin\theta}\frac{\partial}{\partial\theta}\left(\sin\theta\frac{\partial v}{\partial\theta}\right)=0.$$ **Ans.** $n = 2, -3$

31. If $u(x, y) = \frac{x^3+y^3}{x+y}$, $(x, y) \neq 0$, then evaluate $x\frac{\partial^2 u}{\partial x^2}+y\frac{\partial^2 u}{\partial x\partial y}-\frac{\partial u}{\partial x}$. **Ans.** 0

32. Show that $\frac{\partial^2 z}{\partial x^2}-\frac{2\partial^2 z}{\partial x\partial y}+\frac{\partial^2 z}{\partial y^2}=0$, where $z = xf(x+y) + yg(x+y)$.

33. Verify Euler's theorem in the following cases

(i) $f(x, y, z) = axy + byz + czx$, *(ii)* $u=\frac{(x^3+y^3)}{(x+y)}$,

(iii) $z = e^{x/y}$, *(iv)* $f(x,y)=\frac{(x^2+y^2)}{\sqrt{x+y}}$. *(AMIE, S-2002)*

34. If $u=f\left(\frac{x}{y},\frac{y}{z},\frac{z}{x}\right)$, show that $x\frac{\partial u}{\partial x}+y\frac{\partial u}{\partial y}+z\frac{\partial u}{\partial z}=0$. *(AMIE, S-2001)*

35. *(a)* If $z = \log(x^2 + xy + y^2)$, show that $x\frac{\partial z}{\partial x}+y\frac{\partial z}{\partial y}=2$. *(AMIE, S-2005)*

(b) Show that $x\frac{\partial u}{\partial x}+y\frac{\partial u}{\partial y}+z\frac{\partial u}{\partial z}=2\tan u$, where $u=\sin^{-1}\left(\frac{x^3+y^3+z^3}{ax+by+cz}\right)$. *(UPTU 2003)*

36. If $u=\tan^{-1}\left(\frac{x^3+y^3}{x-y}\right)$, show that

(i) $x\frac{\partial u}{\partial x}+y\frac{\partial u}{\partial y}=\sin 2u$. *(GGSIPU 2006; MDU 2002, Bhopal 2002, UPTU 2001, AMIE, W-2001)*

(ii) $x^2u_{xx} + 2xyu_{xy} + y^2u_{yy} = \sin 4u - \sin 2u = 2\sin u\cos 3u$.
(SVTU 2007; VTU 2005; DCE, 2004; PTU-2002; Andhra, 2000)

37. If $z = x^4y^2\sin^{-1}(x/y) + \log x - \log y$, then show that

$$x\frac{\partial z}{\partial x}+y\frac{\partial z}{\partial y}=6x^4y^2\sin^{-1}(x/y).$$ *(AMIETE, June 2003; UPTU 2003)*

38. Show that $x\frac{\partial u}{\partial x}+y\frac{\partial u}{\partial y}=2u\log u$ where $\log u=\frac{x^3+y^3}{3x+4y}$.

39. If $u(x, y) = x^2\tan^{-1}(y/x) - y^2\tan^{-1}(x/y)$, $x > 0$, $y > 0$, then show that $x^2\frac{\partial^2 u}{\partial x^2}+2xy\frac{\partial^2 u}{\partial x\partial y}+y^2\frac{\partial^2 u}{\partial y^2}=2u$.
(AMIE, S-2009, AMIETE, W-2008)

40. If $u = \tan^{-1}\left(\dfrac{x^2 - y^2}{x - y}\right)$, show that $x\dfrac{\partial u}{\partial x} + y\dfrac{\partial u}{\partial y} = \dfrac{1}{2}\sin 2u$ and

$$x^2\frac{\partial^2 u}{\partial x^2} + 2xy\frac{\partial^2 u}{\partial x\partial y} + y^2\frac{\partial^2 u}{\partial y^2} = \frac{1}{4}(\sin 4u - 2\sin 2u).$$

41. If $u = \sin^{-1}\left(\dfrac{x + y}{\sqrt{x} + \sqrt{y}}\right)$. Prove that $x^2\dfrac{\partial^2 u}{\partial x^2} + 2xy\dfrac{\partial^2 u}{\partial x\partial y} + y^2\dfrac{\partial^2 u}{\partial y^2} = -\dfrac{\sin u \cos 2u}{4\cos^3 u}$.

(AMIETE, June 2011; PTU 2006; DCE 2005; MDU 2004; UPTU 2001)

42. If $u = \dfrac{x^2y^2}{x^2 + y^2}\log\dfrac{y}{x}$ and $v = \cos^{-1}\left(\dfrac{xy}{x^2 + y^2}\right)$ and if $z = u + v$, prove that $x\dfrac{\partial z}{\partial x} + y\dfrac{\partial z}{\partial y} = 2u$.

(AMIETE, Dec.-2003)

43. If $x = e^u \tan v$, $y = e^u \sec v$, find the value of $\left(x\dfrac{\partial u}{\partial x} + y\dfrac{\partial u}{\partial y}\right)\left(x\dfrac{\partial v}{\partial x} + y\dfrac{\partial v}{\partial y}\right)$. **Ans.** 0.

(AMIETE, June 2001)

[**Hint:** Eliminate u, we obtain $x/y = \sin v$ or $v = \sin^{-1}(x/y)$. Since v is a homogeneous function of degree zero, thus by Euler Theorem $x\dfrac{\partial v}{\partial x} + y\dfrac{\partial v}{\partial y} = 0$.]

44. Let $u(x, y) = \dfrac{x^3 + y^3}{x + y} + x\tan^{-1}\left(\dfrac{y}{x}\right)$, $(x, y) \neq (0, 0)$, then $x\dfrac{\partial u}{\partial x} + y\dfrac{\partial u}{\partial y} = \dfrac{2(x^3 + y^3)}{x + y} + x\tan^{-1}\dfrac{y}{x}$.

(AMIETE, June-2005)

45. If $u = \tan^{-1}(y^2/x)$, show that $x^2\dfrac{\partial^2 u}{\partial x^2} + 2xy\dfrac{\partial^2 u}{\partial x\partial y} + y^2\dfrac{\partial^2 u}{\partial y^2} = -\sin^2 u \sin 2u$. *(PTU 2006)*

46. Given that $u(x, y) = 5 + x\sin(x/y) + y^2\sin(y/x)$. Compute $x^2\dfrac{\partial^2 u}{\partial x^2} + 2xy\dfrac{\partial^2 u}{\partial x\partial y} + y^2\dfrac{\partial^2 u}{\partial y^2}$.

Ans. $2y^2\sin(y/x)$.

47. Find $x^2u_{xx} + 2xy\,u_{xy} + y^2u_{yy}$ when

(*a*) $u = \sin^{-1}(x^3 + y^3)^{2/5}$,

(*b*) $u = \log(x^3 + y^3 - x^2y - xy^2)$,

(*c*) $u = \log(x^4 + x^2y^2 + y^4)$, *(AMIE, S-2000)*

(*d*) $u = \tan^{-1}\left\{\dfrac{(x^3 + y^3)^{1/2}}{\sqrt{x} + \sqrt{y}}\right\}$,

(*e*) $u = \sin^{-1}((x^2 + y^2)/(x + y))$,

(*f*) $u = (x^2 + y^2)^{1/3}$

Ans. (*a*) $\dfrac{6}{5}\tan u\left(\dfrac{6}{5}\sec^2 u - 1\right)$, (*b*) –3, (*c*) –4, (*d*) $-2\sin^3 u\cos u$, (*e*) $\tan^3 u$, (*f*) $-2u/9$.

48. If $u = \dfrac{x^2y^2}{x + y}$, show that $x\dfrac{\partial^2 u}{\partial x^2} + y\dfrac{\partial^2 u}{\partial x\partial y} = 2\dfrac{\partial u}{\partial x}$. *(GGSIPU 2007)*

49. If $e^u = x^3 + y^3 + z^3 - 3xyz$, then prove that $\left(\dfrac{\partial}{\partial x} + \dfrac{\partial}{\partial y} + \dfrac{\partial}{\partial z}\right)^3 u = \dfrac{54}{(x + y + z)^3}$.

50. If $z = (x + y) + (x + y)\,\phi\left(\dfrac{y}{x}\right)$, then prove that $x\left(\dfrac{\partial^2 z}{\partial x^2} - \dfrac{\partial^2 z}{\partial y\partial x}\right) = y\left(\dfrac{\partial^2 z}{\partial y^2} - \dfrac{\partial^2 z}{\partial x\partial y}\right)$. *(AMIETE, W-2010)*

[**Hint:** We have $\dfrac{\partial z}{\partial x} = 1 - \dfrac{y^2}{x^2}$, $\dfrac{\partial z}{\partial y} = 2 + \dfrac{2y}{x}$, $\dfrac{\partial^2 z}{\partial x^2} = \dfrac{2y^2}{x^3}$, $\dfrac{\partial^2 z}{\partial y^2} = \dfrac{2}{x}$, $\dfrac{\partial^2 z}{\partial y\partial x} = \dfrac{-2y}{x^2}$, $\dfrac{\partial^2 z}{\partial x\partial y} = \dfrac{-2y}{x^2}$.

L.H.S. $= x\left(\dfrac{\partial^2 z}{\partial x^2} - \dfrac{\partial^2 z}{\partial y\partial x}\right) = x\left(\dfrac{2y^2}{x^3} + \dfrac{2y}{x^2}\right) = y\left(\dfrac{2}{x} + \dfrac{2y}{x^2}\right) = y\left(\dfrac{\partial^2 z}{\partial y^2} - \dfrac{\partial^2 z}{\partial x\partial y}\right)$ = R.H.S. Hence the result.]

1.10 TOTAL DIFFERENTIAL

Let a function of two variables $z = f(x, y)$ be defined in some domain D in the xy-plane. Let $P(x, y)$ be any point in D and $(x + \Delta x, y + \Delta y)$ be a point in the neighborhood of (x, y) in D. Then $\Delta z = f(x + \Delta x, y + \Delta y) - f(x, y)$ is called the *total increment* in z corresponding to the increments Δx in x and Δy in y. Adding and subtracting $f(x, y + \Delta y)$ in the second member, we have

$$\Delta z = \{f(x + \Delta x, y + \Delta y) - f(x, y + \Delta y)\} + \{f(x, y + \Delta y) - f(x, y)\}$$

Applying the mean value theorem for functions of one variable, we obtain

$$\Delta z = \Delta x f_x(x + \theta_1 \Delta x, y + \Delta y) + \Delta y f_y(x, y + \theta_2 \Delta y) \qquad 0 < \theta_1 < 1, 0 < \theta_2 < 1.$$

Since f_x and f_y are continuous at (x, y), it follows that

$$f_x(x + \theta_1 \Delta x, y + \Delta y) = f_x(x, y) + \epsilon_1, \qquad f_y(x, y + \theta_2 \Delta y) = f_y(x, y) + \varepsilon_2$$

where $\varepsilon_1 \to 0$, $\varepsilon_2 \to 0$ as $\Delta x \to 0$ and $\Delta y \to 0$.

Thus, $\quad \Delta z = f_x \Delta x + f_y \Delta y + \varepsilon_1 \Delta x + \varepsilon_2 \Delta y.$

Defining $\quad \Delta x = dx$, $\Delta y = dy$, we have $\Delta z = f_x dx + f_y dy + \varepsilon_1 dx + \varepsilon_2 dy.$

We call $\quad dz = f_x dx + f_y dy$ ...(1.17), $\quad$ the *differential* of z (or f) or the *principal part* of Δz (or Δf).

Remark: For a function of n variables $z = f(x_1, x_2, ..., x_n)$, we write the total differential as

$$dz = f_{x_1} dx_1 + f_{x_2} dx_2 + ... + f_{x_n} dx_n. \qquad ...(1.18)$$

1.11 SECOND ORDER TOTAL DIFFERENTIALS

For the function $z = f(x, y)$, the first order total differential of z is $dz = f_x dx + f_y dy.$

The second order total differential of z, denoted by d^2z, is given by

$$\begin{aligned} d^2z &= d(dz) = d(f_x dx + f_y dy) \\ &= \frac{\partial}{\partial x}(f_x dx + f_y dy)dx + \frac{\partial}{\partial y}(f_x dx + f_y dy)dy \\ &= \left[f_{xx}dx + f_x \frac{\partial}{\partial x}(dx) + f_{yx}dy + f_y \frac{\partial}{\partial x}(dy)\right]dx + \left[f_{yx}dx + f_x \frac{\partial}{\partial y}(dx) + f_{yy}dy + f_y \frac{\partial}{\partial y}(dy)\right]dy \end{aligned}$$

Since dx and dy are considered as constants so

$$\frac{\partial}{\partial x}(dx) = 0, \frac{\partial}{\partial x}(dy) = 0, \frac{\partial}{\partial y}(dx) = 0, \frac{\partial}{\partial y}(dy) = 0.$$

Therefore, $\quad d^2z = (f_{xx}dx + f_{yx}dy)dx + (f_{yx}dx + f_{yy}dy)dy$

or $\quad d^2z = f_{xx}(dx)^2 + 2f_{xy}dx\,dy + f_{yy}(dy)^2. \qquad (\because f_{xy} = f_{yx})$ ...(1.19)

In abbreviated notation, it may be written as $d^2z = \left(\frac{\partial}{\partial x}dx + \frac{\partial}{\partial y}dy\right)^2 z.$

Similarly, $\quad d^3z = \left(\frac{\partial}{\partial x}dx + \frac{\partial}{\partial y}dy\right)^3 z.$

EXAMPLE 1.36 Find the total differential of the following functions

(*i*) $z = xy^2 \log(y/x)$, $\qquad$ (*ii*) $u = x^2 e^{y/x}$, $\qquad$ (*iii*) $u = \left(xz + \frac{x}{z}\right)^y$, $z \neq 0$.

Solution (*i*) We have, $f(x, y) = xy^2 \log(y/x)$.

$$f_x = y^2 \log\left(\frac{y}{x}\right) + xy^2 \cdot \frac{x}{y}\left(\frac{-y}{x^2}\right) = y^2 \log\left(\frac{y}{x}\right) - y^2,$$

and

$$f_y = 2xy \log\left(\frac{y}{x}\right) + xy^2 \cdot \frac{x}{y} \cdot \frac{1}{x} = 2xy \log\left(\frac{y}{x}\right) + xy.$$

Therefore, we obtain the total differential as

$$dz = f_x dx + f_y dy = \{y^2 \log(y/x) - y^2\}dx + \{2xy \log(y/x) + xy\}dy.$$

(*ii*) $f(x, y) = x^2 e^{y/x}, \quad f_x = x^2 e^{y/x} \cdot (-yx^{-2}) + 2xe^{y/x}, \quad f_y = x^2 e^{y/x} \cdot \left(\frac{1}{x}\right) = xe^{y/x}.$

Therefore, we obtain the total differential as $du = f_x\,dx + f_y\,dy = (2xe^{y/x} - ye^{y/x})dx + xe^{y/x}dy.$

(*iii*) $$f(x, y, z) = \left(xz + \frac{x}{z}\right)^y, \; f_x = y\left(xz + \frac{x}{z}\right)^{y-1}\left(z + \frac{1}{z}\right),$$

$$f_y = \left(xz + \frac{x}{z}\right)^y \log\left(xz + \frac{x}{z}\right), \; f_z = y\left(xz + \frac{x}{z}\right)^{y-1} \cdot \left(x - \frac{x}{z^2}\right).$$

Therefore, we obtain the total differential as

$$du = \left(xz + \frac{x}{z}\right)^{y-1}\left[y\left(z + \frac{1}{z}\right)dx + xy\left(1 - \frac{1}{z^2}\right)dz\right] + \left[\left(xz + \frac{x}{z}\right)^y \log\left(xz + \frac{x}{z}\right)\right]dy.$$

EXAMPLE 1.37 If $z = f(x, y) = x^2y - 3y$, find (*a*) Δz, (*b*) dz, (*c*) $f(5.12, 6.85)$ without direct computation.

Solution (*a*) $\Delta z = f(x + \Delta x, y + \Delta y) - f(x, y) = \{(x+\Delta x)^2(y+\Delta y) - 3(y + \Delta y)\} - (x^2y - 3y)$

$$= \underbrace{2xy\,\Delta x + (x^2 - 3)\,\Delta y}_{(A)} + \underbrace{(\Delta x)^2 y + 2x\Delta x \Delta y + (\Delta x)^2\,\Delta y}_{(B)}.$$

(*b*) The sum (A) is the principal part of Δz and is the differential of z that is, dz.
Thus $dz = 2xy\,\Delta x + (x^2 - 3)\,\Delta y = 2\,xydx + (x^2 - 3)\,dy.$

Alternative: $$dz = \frac{\partial f}{\partial x}dx + \frac{\partial f}{\partial y}dy = f_x dx + f_y dy = 2\,xy\,dx + (x^2 - 3)\,dy.$$

(*c*) We have to find $f(x + \Delta x, y + \Delta y)$, when $x + \Delta x = 5.12$ and $y + \Delta y = 6.85$. Choosing $x = 5$, $\Delta x = 0.12$, $y = 7$, $\Delta y = -0.15$. Since Δx and Δy are small, we use the fact that

$$f(x + \Delta x, y + \Delta y) = f(x, y) + \Delta z \approx f(x, y) + dz \text{ that is, } z + dz.$$

Now $$z = f(x, y) = f(5, 7) = 5^2(7) - 3(7) = 154.$$

$$dz = f_x dx + f_y\,dy = 2xy\,dx + (x^2 - 3)\,dy$$

$$= 2(5)(7)(0.12) + (5^2 - 3)(-0.15) = 5.1.$$

Then the required value is $154 + 5.1 = 159.1$ (approx.)

EXAMPLE 1.38 Find the second order total differentials of the following function

(*i*) $z = 7y \log(1 + x)$. (*ii*) $z = xe^{xy}$.

Solution (*i*) We have $z = f(x, y) = 7y \log(1 + x)$.

$$f_x = 7y/(1 + x), \quad f_y = 7 \log(1 + x).$$

Also $$f_{xx} = -7y/(1 + x)^2, \; f_{yy} = 0, f_{xy} = 7/(1 + x).$$

Therefore, the second order total differential of z is

$$d^2z = f_{xx}(dx)^2 + 2f_{xy}\,dx\,dy + f_{yy}(dy)^2 = -\frac{7y}{(1+x)^2}(dx)^2 + \frac{14}{1+x}dxdy.$$

(*ii*) We have $$f(x,y) = xe^{xy}, \quad f_x = \partial f/\partial x = e^{xy} + xye^{xy},$$

$$f_y = \partial f/\partial y = x^2e^{xy}; \quad f_{xx} = \partial^2 f/\partial x^2 = ye^{xy} + y\{e^{xy} + xye^{xy}\},$$

$$f_{yy} = \frac{\partial^2 f}{\partial y^2} = \frac{\partial}{\partial y}\left(\frac{\partial f}{\partial y}\right) = \frac{\partial}{\partial y}(x^2e^{xy}) = x^3e^{xy},$$

$$f_{xy} = \frac{\partial}{\partial x}\left(\frac{\partial f}{\partial y}\right) = \frac{\partial}{\partial x}(x^2e^{xy}) = 2xe^{xy} + x^2ye^{xy} = xe^{xy}(xy + 2).$$

Therefore, the second order total differential of z is

$$d^2z = f_{xx}\,(dx)^2 + 2f_{xy}\,dx\,dy + f_{yy}(dy)^2$$
$$= \{ye^{xy} + ye^{xy} + xy^2\,e^{xy}\}\,(dx)^2 + (4xe^{xy} + 2x^2ye^{xy})dxdy + x^3\,e^{xy}(dy)^2$$

or
$$d^2z = (xy + 2)\,ye^{xy}(dx)^2 + 2(xy + 2)\,xe^{xy}dxdy + x^3e^{xy}\,(dy)^2.$$

EXAMPLE 1.39 If $x^2 + y^2 + z^2 - 2xyz = 1$, show that $\dfrac{dx}{\sqrt{1-x^2}} + \dfrac{dy}{\sqrt{1-y^2}} + \dfrac{dz}{\sqrt{1-z^2}} = 0.$

(NIT, Kurukshetra, 2009)

Solution We have

$$x^2 + y^2 + z^2 - 2xyz = 1.$$

$\therefore$
$$2xdx + 2\,ydy + 2\,zdz - 2(xydz + yzdx + zxdy) = 0$$

or
$$(x - yz)\,dx + (y - zx)\,dy + (z - xy)\,dz = 0 \qquad ...(i)$$

Now
$$(x - yz)^2 = x^2 + y^2z^2 - 2\,xyz = (x^2 - 2\,xyz) + y^2z^2$$
$$= (1 - y^2 - z^2) + y^2z^2 = 1 - y^2 - z^2 + y^2z^2 = (1 - y^2)(1 - z^2)$$

$\therefore$
$$x - yz = \sqrt{(1-y^2)(1-z^2)}.$$

Similarly,
$$y - zx = \sqrt{(1-x^2)(1-z^2)}, \text{ and } z - xy = \sqrt{(1-x^2)(1-y^2)}.$$

Putting these values of $x - yz$, $y - zx$ and $z - xy$ in (i), we get

$$\sqrt{(1-y^2)(1-z^2)}\,dx + \sqrt{(1-x^2)(1-z^2)}\,dy + \sqrt{(1-x^2)(1-y^2)}\,dz = 0.$$

Dividing throughout by $\sqrt{(1-x^2)(1-y^2)(1-z^2)}$, we obtain

$$\frac{dx}{\sqrt{1-x^2}} + \frac{dy}{\sqrt{1-y^2}} + \frac{dz}{\sqrt{1-z^2}} = 0, \text{ the required result.}$$

1.12 EXACT DIFFERENTIAL

If $z = f(x, y)$ is a function of two variables with continuous first order partial derivatives in a region R of the xy-plane, then its *total or exact differential, df,* of $f(x, y)$ is expressed as

$$df = \frac{\partial f}{\partial x}dx + \frac{\partial f}{\partial y}dy. \qquad ...(1.20)$$

Let $M(x, y)$ and $N(x, y)$ be continuous and have continuous first order partial derivatives in a region in the xy-plane.

A differential expression $M(x, y)dx + N(x, y)dy$ is an *exact differential* in a region R of the xy-plane if it corresponds to the differential of some function $f(x, y)$.

Now,
$$\frac{\partial f}{\partial x}dx + \frac{\partial f}{\partial y}dy = M(x,y)dx + N(x,y)dy$$

Therefore,
$$\frac{\partial f}{\partial x} = M(x,y) \text{ and } \frac{\partial f}{\partial y} = N(x,y) \qquad ...(1.21)$$

From eqs (1.21), we obtain
$$\frac{\partial}{\partial y}\left(\frac{\partial f}{\partial x}\right) = \frac{\partial^2 f}{\partial y \partial x} = \frac{\partial M}{\partial y} \text{ and } \frac{\partial}{\partial x}\left(\frac{\partial f}{\partial y}\right) = \frac{\partial^2 f}{\partial x \partial y} = \frac{\partial N}{\partial x}.$$

By the assumption of continuity the two partial derivatives of order 2 are equal. Thus $\dfrac{\partial M}{\partial y} = \dfrac{\partial N}{\partial x}$.

Thus a necessary and sufficient condition that the expression $M(x, y)dx + N(x, y)dy$ be an exact differential is $\dfrac{\partial M}{\partial y} = \dfrac{\partial N}{\partial x}$. ...(1.22)

EXAMPLE 1.40 State whether following expressions are exact differential or not

(*i*) $2x \sin y\, dx + x^2 \cos y\, dy$, (*ii*) $(x^2y + 2y^3)dx - (2x^3 + 3xy^2)dy$,

(*iii*) $\left(\dfrac{x^2}{y} - \dfrac{y^2}{x}\right)dx - \left(\dfrac{x^3}{3y^2} + 2y \log x\right)dy.$

Solution (*i*) Here $M = 2x \sin y$ and $N = x^2 \cos y$.

$$\frac{\partial M}{\partial y} = 2x\cos y, \qquad \frac{\partial N}{\partial x} = 2x\cos y. \quad \text{Since } \frac{\partial M}{\partial y} = \frac{\partial N}{\partial x}.$$

Therefore, the given expression is an exact differential.

(*ii*) $M = x^2y + 2y^3$, $N = -(2x^3 + 3xy^2)$,

$$\frac{\partial M}{\partial y} = x^2 + 6y^2, \qquad \frac{\partial N}{\partial x} = -\left(6x^2 + 3y^2\right). \quad \text{Since } \frac{\partial M}{\partial y} \neq \frac{\partial N}{\partial x}.$$

Thus, the given expression is not an exact differential.

(*iii*) $M = \dfrac{x^2}{y} - \dfrac{y^2}{x}$, $N = -\left(\dfrac{x^3}{3y^2} + 2y \log x\right)$.

$$\frac{\partial M}{\partial y} = -\frac{x^2}{y^2} - \frac{2y}{x}; \qquad \frac{\partial N}{\partial x} = -\frac{x^2}{y^2} - \frac{2y}{x}. \quad \text{Since } \frac{\partial M}{\partial y} = \frac{\partial N}{\partial x}.$$

Thus, the given expression is exact.

EXAMPLE 1.41 Show that expression $(3x^2y - 2y^2)\, dx + (x^3 - 4xy + 6y^2)\, dy$ can be written as an exact differential of a function $\phi(x, y)$ and find this function.

Solution Suppose that $(3x^2y - 2y^2)\, dx + (x^3 - 4xy + 6y^2)dy = d\phi = \dfrac{\partial \phi}{\partial x}dx + \dfrac{\partial \phi}{\partial y}dy$.

Comparing, we get $\dfrac{\partial \phi}{\partial x} = 3x^2y - 2y^2$ ...(*i*) and $\dfrac{\partial \phi}{\partial y} = x^3 - 4xy + 6y^2$ ...(*ii*).

Integrating (*i*) w.r.t. x keeping y constant, we have $\phi(x, y) = x^3y - 2xy^2 + g(y)$, where $g(y)$ is the constant of integration.

Substituting this into (*ii*), we obtain $\partial\phi/\partial y = x^3 - 4xy + (dg/dy) = x^3 - 4xy + 6y^2$ from which $dg/dy = 6y^2$ giving $g(y) = 2y^3 + c$.

Hence, the required function is $\phi(x, y) = x^3y - 2xy^2 + 2y^3 + c$, where c is an arbitrary constant.

1.13 TOTAL DERIVATIVES

1.13.1 Derivatives of Composite Function (Chain Rule)

Let $w = f(x, y)$ be a function of two independent variables x and y. Suppose that x and y are in turn functions of a single variable t, say $x = \phi(t)$, $y = \psi(t)$. Then $w = f[\phi(t), \psi(t)]$ is a composite function of the independent variable t.

Now, assume that the partial derivatives f_x, f_y are continuous functions of x, y and $\phi(t)$, $\psi(t)$ are differentiable functions of t.

Let Δx, Δy and Δw be the increments respectively in x, y and w corresponding to the increment Δt in t.

Then we have $\quad \Delta w = \dfrac{\partial f}{\partial x}\Delta x + \dfrac{\partial f}{\partial y}\Delta y + \varepsilon_1 \Delta x + \varepsilon_2 \Delta y.$

Dividing both sides by Δt, we obtain

$$\frac{\Delta w}{\Delta t} = \frac{\partial f}{\partial x}\frac{\Delta x}{\Delta t} + \frac{\partial f}{\partial y}\cdot\frac{\Delta y}{\Delta t} + \varepsilon_1 \frac{\Delta x}{\Delta t} + \varepsilon_2 \cdot \frac{\Delta y}{\Delta t}. \qquad ...(1.23)$$

CHAIN RULE

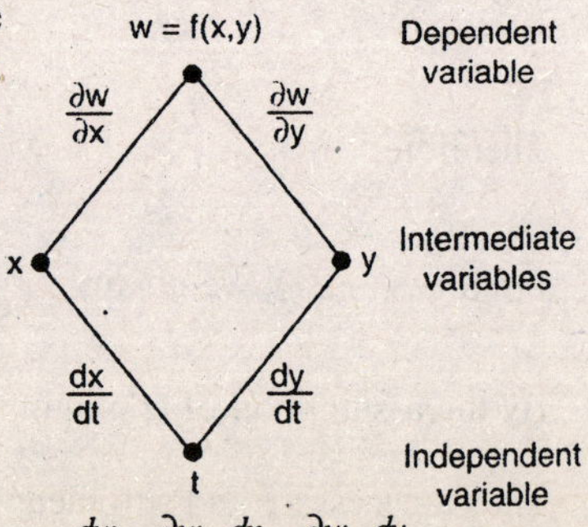

w = f(x,y) — Dependent variable

$\dfrac{\partial w}{\partial x}$, $\dfrac{\partial w}{\partial y}$

x, y — Intermediate variables

$\dfrac{dx}{dt}$, $\dfrac{dy}{dt}$

t — Independent variable

$$\frac{dw}{dt} = \frac{\partial w}{\partial x}\frac{dx}{dt} + \frac{\partial w}{\partial y}\frac{dy}{dt}.$$

Fig. 1.2

Now as $\Delta t \to 0$; $\Delta x \to 0$; $\Delta y \to 0$ and $\varepsilon_1\left(\dfrac{\Delta x}{\Delta t}\right) \to 0$, $\varepsilon_2\left(\dfrac{\Delta y}{\Delta t}\right) \to 0$.

Therefore, taking limits on both sides in Eq. (1.23) as $\Delta t \to 0$, we obtain

$$\frac{dw}{dt} = \frac{\partial f}{\partial x}\frac{dx}{dt} + \frac{\partial f}{\partial y}\frac{dy}{dt} \quad \text{or} \quad \frac{dw}{dt} = \frac{\partial w}{\partial x}\frac{dx}{dt} + \frac{\partial w}{\partial y}\frac{dy}{dt}.$$

1.13.2 Chain Rule for Functions of *n* Independent Variables

Consider a function of n variables $w = f(x_1, x_2, \ldots\ldots, x_n)$ and $x_1, x_2, \ldots\ldots, x_n$ are differentiable functions of some independent variable t, then

$$\frac{dw}{dt} = \frac{\partial f}{\partial x_1}\frac{dx_1}{dt} + \frac{\partial f}{\partial x_2}\frac{dx_2}{dt} + \ldots\ldots + \frac{\partial f}{\partial x_n}\frac{dx}{dt}. \qquad \ldots(1.24)$$

ILLUSTRATIVE EXAMPLES

EXAMPLE 1.42 If $z = e^{xy^2}$, $x = t \cos t$, $y = t \sin t$, compute dz/dt at $t = \pi/2$. (*AMIE, S-2006*)

Solution When $t = \pi/2$, we get $x = 0$, $y = \pi/2$. Using the chain rule, we obtain

$$\frac{dz}{dt} = \frac{\partial z}{\partial x}\frac{dx}{dt} + \frac{\partial z}{\partial y}\frac{dy}{dt} = y^2 e^{xy^2}(\cos t - t\sin t) + 2xy\,e^{xy^2}(\sin t + t\cos t).$$

Subtituting $t = \pi/2$, $x = 0$ and $y = \pi/2$, we obtain $(dz/dt)_{t=\pi/2} = -\pi^3/8$.

EXAMPLE 1.43 (*a*) If $u = \sin^{-1}(x - y)$, $x = 3t$, $y = 4t^3$, show that $du/dt = 3(1 - t^2)^{-1/2}$. (*PTU 2005*)

(*b*) If $u = x^3 + y^3$, where $x = a \cos t$, $y = b \sin t$. Find du/dt and verify the result by direct substitution.

(*c*) If $u = xe^y z$, where $y = \sqrt{a^2 - x^2}$, $z = \sin^2 x$. Find du/dx.

Solution (*a*) We have

$$\partial u/\partial x = 1/\sqrt{1-(x-y)^2}, \quad dx/dt = 3,$$

$$\partial u/\partial y = -1/\sqrt{1-(x-y)^2}, \quad dy/dt = 12t^2.$$

Using the chain rule, we obtain $\dfrac{du}{dt} = \dfrac{\partial u}{\partial x}\dfrac{dx}{dt} + \dfrac{\partial u}{\partial y}\dfrac{dy}{dt}$

$$= \frac{1}{\sqrt{1-(x-y)^2}}\cdot 3 - \frac{1}{\sqrt{1-(x-y)^2}}\cdot 12t^2$$

$$= \frac{3}{\sqrt{1-(3t-4t^3)^2}} - \frac{12t^2}{\sqrt{1-(3t-4t^3)^2}}, \text{ (Substitute for the intermediate variables)}$$

$$= \frac{3(1-4t^2)}{\sqrt{1-(3t-4t^3)^2}} = \frac{3(1-4t^2)}{\sqrt{1-9t^2-16t^6+24t^4}} = \frac{3(1-4t^2)}{\sqrt{(1-t^2)(1-8t^2+16t^4)}} = \frac{3}{\sqrt{1-t^2}}.$$

Alternative

$u = \sin^{-1}(x - y) = \sin^{-1}(3t - 4t^3)$. Setting $t = \sin\theta$, we obtain
$u = \sin^{-1}(3\sin\theta - 4\sin^3\theta) = \sin^{-1}(\sin 3\theta) = 3\theta = 3\sin^{-1}t$.

Therefore, $du/dt = 3/\sqrt{1-t^2}$.

(*b*) We have $\partial u/\partial x = 3x^2$, $\partial u/\partial y = 3y^2$, and $dx/dt = -a \sin t$, $dy/dt = b \cos t$.

Using the chain rule, we obtain $\dfrac{du}{dt} = \dfrac{\partial u}{\partial x}\dfrac{dx}{dt} + \dfrac{\partial u}{\partial y}\dfrac{dy}{dt}$

$= 3x^2(-a \sin t) + 3y^2(b \cos t) = 3a^2\cos^2 t(-a \sin t) + 3b^2 \sin^2 t(b \cos t)$

(Substitute for the intermediate variables.)

$= -3a^3 \cos^2 t \sin t + 3b^3 \sin^2 t \cos t$.

Verification

$$u = x^3 + y^3 = a^3 \cos^3 t + b^3 \sin^3 t.$$

Therefore,
$$du/dt = 3\,a^3 \cos^2 t\,(-\sin t) + 3b^3 \sin^2 t\,(\cos t)$$
$$= -3\,a^3 \cos^2 t \sin t + 3b^3 \sin^2 t \cos t.$$

(c) $\dfrac{\partial u}{\partial x} = e^y z,\ \dfrac{\partial u}{\partial y} = xe^y z,\ \dfrac{\partial u}{\partial z} = xe^y$ and $\dfrac{dy}{dx} = \dfrac{1}{2}(a^2 - x^2)^{-1/2} \cdot (-2x) = -\dfrac{x}{\sqrt{a^2 - x^2}}$,

$$\frac{dz}{dx} = 2 \sin x \cos x \;=\; -x/\sqrt{a^2 - x^2},\ dz/dx = 2\sin x \cos x.$$

Using the chain rule, we obtain $\dfrac{du}{dx} = \dfrac{\partial u}{\partial x} + \dfrac{\partial u}{\partial y}\cdot\dfrac{dy}{dx} + \dfrac{\partial u}{\partial z}\cdot\dfrac{dz}{dx}$

$$= e^y z + xe^y z\left(\frac{-x}{\sqrt{a^2 - x^2}}\right) + xe^y(\sin 2x) \;=\; e^y\left[z - \frac{x^2 z}{\sqrt{a^2 - x^2}} + x \sin 2x\right].$$

EXAMPLE 1.44 If $u = \sin(x/y)$, $x = e^t$ and $y = t^2$ then find du/dt as function of t. *(PTU, 2006)*

Solution $\dfrac{\partial u}{\partial x} = \dfrac{1}{y}\cos\dfrac{x}{y},\quad \dfrac{\partial u}{\partial y} = -\dfrac{x}{y^2}\cos\dfrac{x}{y};\quad \dfrac{dx}{dt} = e^t,\quad \dfrac{dy}{dt} = 2t.$

Using the chain rule, we obtain $\dfrac{du}{dt} = \dfrac{\partial u}{\partial x}\dfrac{dx}{dt} + \dfrac{\partial u}{\partial y}\dfrac{dy}{dt}$

$$= \cos\left(\frac{x}{y}\right)\left(\frac{1}{y}\right)(e^t) + \cos\left(\frac{x}{y}\right)\left(-\frac{x}{y^2}\right)2t$$

$$= \frac{e^t}{t^2}\cos\left(\frac{e^t}{t^2}\right) - \frac{2e^t}{t^3}\cos\left(\frac{e^t}{t^2}\right) \quad \text{(Substitute for the intermediate variables).}$$

$$= (t-2)\frac{e^t}{t^3}\cos\left(\frac{e^t}{t^2}\right).$$

EXAMPLE 1.45 Given $u = e^{ax}(y - z)$, $y = a \sin x$, $z = \cos x$ find du/dx.

Solution $\partial u/\partial x = ae^{ax}(y - z)$, $\partial u/\partial y = e^{ax}$, $\partial u/\partial z = -e^{ax}$, $dy/dx = a \cos x$, $dz/dx = -\sin x$.

Using the chain rule, we obtain $\dfrac{du}{dx} = \dfrac{\partial u}{\partial x} + \dfrac{\partial u}{\partial y}\cdot\dfrac{dy}{dx} + \dfrac{\partial u}{\partial z}\cdot\dfrac{dz}{dx}$

$$= ae^{ax}(y - z) + ae^{ax}\cos x + e^{ax}\sin x$$
$$= ae^{ax}(a \sin x - \cos x) + ae^{ax}\cos x + e^{ax}\sin x \quad \text{(Substitute for the intermediate variables)}$$
$$= e^{ax}(a^2 + 1)\sin x.$$

RELATED RATES PROBLEMS

EXAMPLE 1.46 At a certain instant the three dimensions of a rectangular parallelopiped are 15 cm, 20 cm, 25 cm, and these are increasing at the respective rates of 5 mm per second, 8 mm per second, 3 mm per second. How fast is the volume increasing?

Solution Let x, y, and z be the edges of a rectangular parallelopiped. Then $V = xyz$. Considering x, y and z as functions of time t, we have

$$\frac{dV}{dt} = \frac{\partial V}{\partial x}\frac{dx}{dt} + \frac{\partial V}{\partial y}\cdot\frac{dy}{dt} + \frac{\partial V}{\partial z}\cdot\frac{dz}{dt} = yz\frac{dx}{dt} + zx\frac{dy}{dt} + xy\frac{dz}{dt}.$$

It is given that $x = 15$, $dx/dt = 0.5$, $y = 20$, $dy/dt = 0.8$, $z = 25$ and $dz/dt = 0.3$,

we have
$$dV/dt = (20)(25)(0.5) + (25)(15)(0.8) + (15)(20)(0.3)$$
$$= 250 + 300 + 90 = 640 \text{ cm}^3 \text{ per sec.}$$

Thus the volume is increasing at the rate of 640 cu cm per sec.

EXAMPLE 1.47 The altitude of a right circular cone is 15 units and is increasing at the rate of 0.2 unit per min. The radius of the base is 10 units and is decreasing at the rate of 0.3 unit per min. At what rate is the volume changing?

Solution Let r be the radius and h the altitude of the cone. Then $V = \pi r^2 h/3$, considering r and h as functions of time t. We have

$$\frac{\partial V}{\partial r} = \frac{2}{3}\pi r h,\ \frac{\partial V}{\partial h} = \frac{\pi r^2}{3},\ h = 15, \frac{dh}{dt} = +0.2,\ r = 10,\ \frac{dr}{dt} = -0.3.$$

Therefore, $$\frac{dV}{dt} = \frac{\partial V}{\partial r}\cdot\frac{dr}{dt} + \frac{\partial V}{\partial h}\cdot\frac{dh}{dt} = \frac{\pi}{3}\left(2rh\frac{dr}{dt} + r^2\frac{dh}{dt}\right)$$

$$= \frac{\pi}{3}[2(10)(15)(-0.3) + 10^2(0.2)] = -\frac{70\pi}{3}\text{units}^3/\text{min}.$$

The negative sign indicate that volume is decreasing.

EXAMPLE 1.48 A point is moving along the curve of intersection of $x^2 + 3xy + 3y^2 = z^2$ and the plane $x - 2y + 4 = 0$. When $x = 2$ and is increasing at 3 units/sec, find (*a*) how y is changing, (*b*) how z is changing, and (*c*) the speed of the point.

Solution From $z = \pm(x^2 + 3xy + 3y^2)^{1/2}$, we obtain $\dfrac{dz}{dt} = \dfrac{\partial z}{\partial x}\dfrac{dx}{dt} + \dfrac{\partial z}{\partial y}\dfrac{dy}{dt}$

$$= \frac{1}{2}\frac{(2x+3y)}{\sqrt{x^2+3xy+3y^2}}\frac{dx}{dt} + \frac{1}{2}\frac{(3x+6y)}{\sqrt{x^2+3xy+3y^2}}\frac{dy}{dt}.$$

Since $x - 2y + 4 = 0$, $y = 3$, when $x = 2$; also differentiation yields $\dfrac{dx}{dt} = 2\cdot\dfrac{dy}{dt}$.

(*a*) when $x = 2$, $dx/dt = +3$ then $dy/dt = 3/2$ units/sec, increasing.

(*b*) $z = \pm(4 + 18 + 27)^{1/2} = \pm 7$. At (2, 3, 7), $\dfrac{dz}{dt} = \dfrac{13}{14}\cdot 3 + \dfrac{24}{14}\cdot\dfrac{3}{2} = \dfrac{75}{14}$ units/sec, increasing.

At (2, 3, –7), $\dfrac{dz}{dt} = -\dfrac{39}{14} - \dfrac{36}{14} = -\dfrac{75}{14}$ units/sec that is, decreasing.

(*c*) Speed of the point $= \sqrt{\left(\dfrac{dx}{dt}\right)^2 + \left(\dfrac{dy}{dt}\right)^2 + \left(\dfrac{dz}{dt}\right)^2} = \sqrt{9 + \dfrac{9}{4} + \left(\dfrac{75}{14}\right)^2} = \sqrt{39.9489} = 6.32$ units/sec.

EXAMPLE 1.49 The radius of the base of a certain cone is increasing at the rate of 8 cm/minute and the altitude is decreasing at the rate of 10 cm/minute. Find the rate of change of the total surface of the cone when the radius is 18 cm and the altitude 60 cm.

Solution Let r be the radius, and h the altitude of the cone.

Then the total surface area of the cone is

$$S = \pi r^2 + \pi r l = \pi r^2 + \pi r\sqrt{h^2 + r^2} = f(r, h).$$

Considering r and h as functions of time t,

$$\frac{dS}{dt} = \frac{\partial S}{\partial r}\cdot\frac{dr}{dt} + \frac{\partial S}{\partial h}\cdot\frac{dh}{dt}.$$

We also have $$\frac{\partial S}{\partial r} = 2\pi r + \pi\left[\sqrt{h^2+r^2} + \frac{r^2}{\sqrt{h^2+r^2}}\right]$$

$$= 2\pi r + \frac{\pi(h^2 + 2r^2)}{\sqrt{h^2+r^2}} \text{ and } \frac{\partial S}{\partial h} = \frac{\pi r h}{\sqrt{h^2+r^2}}.$$

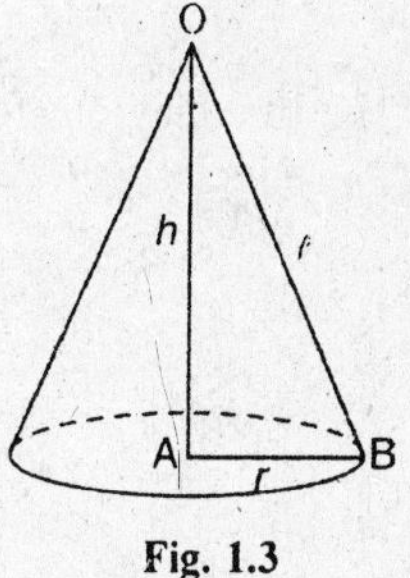

Fig. 1.3

Therefore, $$\frac{dS}{dt} = 2\pi r\frac{dr}{dt} + \frac{\pi(h^2+2r^2)}{\sqrt{h^2+r^2}}\frac{dr}{dt} + \frac{\pi r h}{\sqrt{h^2+r^2}}\frac{dh}{dt}. \quad ...(i)$$

When $r = 18$, $h = 60$, $dr/dt = +8$ and $dh/dt = -10$, then

$$\frac{dS}{dt} = 2\pi(18)(8) + \frac{\pi(4248)}{62.64}(8) + \frac{\pi(18)(60)}{62.64}(-10) = 288\,\pi + 370\,\pi = 658\,\pi \text{ cm}^2/\text{min}.$$

Thus the total surface of the cone is increasing at the rate of 658 π cm^2/minute.

1.14 CHANGE OF VARIABLES: FUNCTIONS DEFINED ON SURFACES

Suppose that $$w = f(x, y) \quad \text{... (1.25)}$$
is a function of two independent variables x, y and x, y are functions of two new independent variables r, s given by $x = \phi(r, s)$, $y = \psi(r, s)$. By chain rule, we have

$$\frac{\partial w}{\partial r} = \frac{\partial w}{\partial x}\frac{\partial x}{\partial r} + \frac{\partial w}{\partial y}\frac{\partial y}{\partial r} \quad \text{...(1.26)}$$

and
$$\frac{\partial w}{\partial s} = \frac{\partial w}{\partial x}\frac{\partial x}{\partial s} + \frac{\partial w}{\partial y}\frac{\partial y}{\partial s}. \quad \text{...(1.27)}$$

Tree diagrams for these relations are shown in Figure 1.4.

CHAIN RULE

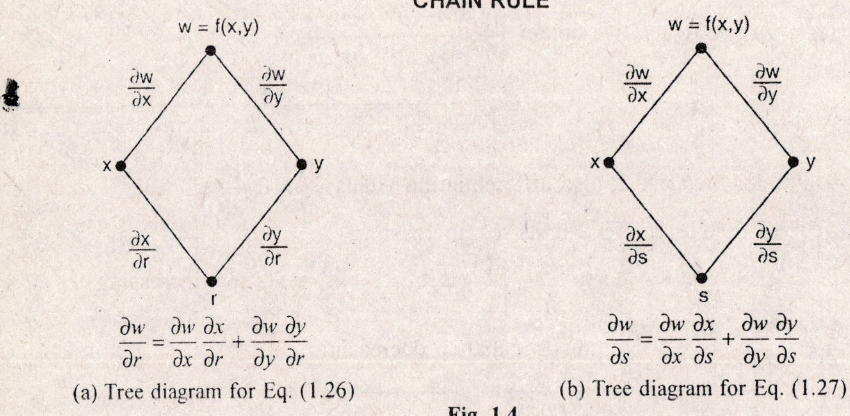

(a) Tree diagram for Eq. (1.26) (b) Tree diagram for Eq. (1.27)

Fig. 1.4

Similarly, if $w = f(x, y, z)$ is a function of three independent variable x, y, z and x, y, z are functions of two new independent variables r, s, given by $x = \phi(r, s)$, $y = \psi(r, s)$, $z = h(r, s)$, then by chain rule, we have

$$\frac{\partial w}{\partial r} = \frac{\partial w}{\partial x}\frac{\partial x}{\partial r} + \frac{\partial w}{\partial y}\frac{\partial y}{\partial r} + \frac{\partial w}{\partial z}\frac{\partial z}{\partial r} \quad \text{...(A)}$$

$$\frac{\partial w}{\partial s} = \frac{\partial w}{\partial x}\frac{\partial x}{\partial s} + \frac{\partial w}{\partial y}\frac{\partial y}{\partial s} + \frac{\partial w}{\partial z}\frac{\partial z}{\partial s} \quad \text{...(B)}$$

The tree diagram for Eqs (A) and (B) are shown in Fig. 1.5.

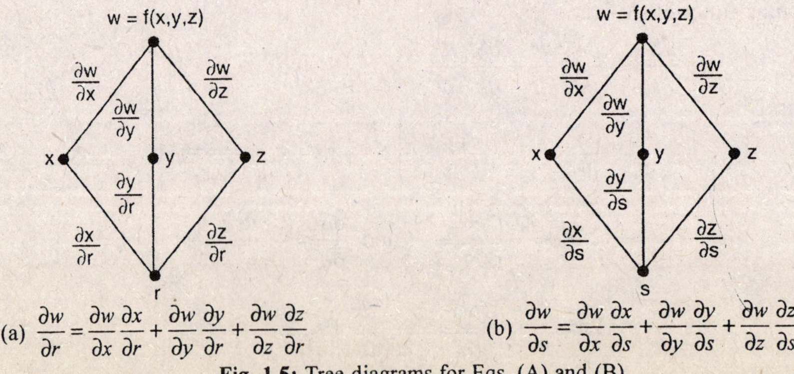

Fig. 1.5: Tree diagrams for Eqs. (A) and (B).

LLUSTRATIVE EXAMPLES

EXAMPLE 1.50 (*a*) If $F(x, y) = (2x + y)/(y - 2x)$, $x = 2u - 3v$, $y = u + 2v$, find (*i*) $\partial F/\partial u$, (*ii*) $\partial F/\partial v$, where $u = 2$, $v = 1$.

(*b*) If $z = f(x, y)$, $x = e^u + e^{-v}$ and $y = e^{-u} - e^v$, then show that

$$\frac{\partial f}{\partial u} - \frac{\partial f}{\partial v} = x\frac{\partial f}{\partial x} - y\frac{\partial f}{\partial y}.$$ (*VTU, 2006; AMIE, W-2001; MDU 2000*)

(*c*) If $u = f(r, s)$, $r = x + at$, $s = y + bt$ and x, y, t are independent variables, show that $\dfrac{\partial u}{\partial t} = a\dfrac{\partial u}{\partial x} + b\dfrac{\partial u}{\partial y}$.

Solution (*a*) Using the chain rule, we obtain $\dfrac{\partial F}{\partial u} = \dfrac{\partial F}{\partial x}\cdot\dfrac{\partial x}{\partial u} + \dfrac{\partial F}{\partial y}\cdot\dfrac{\partial y}{\partial u}$

$$= \left\{\frac{2(y-2x)-(2x+y)(-2)}{(y-2x)^2}\right\}\cdot 2 + \left\{\frac{(y-2x)-(2x+y)}{(y-2x)^2}\right\}\cdot 1 = \frac{4(2y-x)}{(y-2x)^2},$$

and
$$\frac{\partial F}{\partial v} = \frac{\partial F}{\partial x}\cdot\frac{\partial x}{\partial v} + \frac{\partial F}{\partial y}\cdot\frac{\partial y}{\partial v} = \frac{4y}{(y-2x)^2}(-3) + \frac{(-4x)}{(y-2x)^2}\cdot 2 = -\frac{4(2x+3y)}{(y-2x)^2}.$$

When $u = 2$, $v = 1$; then $x = 1$, $y = 4$.

Therefore,
$$\left(\frac{\partial F}{\partial u}\right)_{\substack{x=1\\y=4}} = 4\left[\frac{(2y-x)}{(y-2x)^2}\right]_{\substack{x=1\\y=4}} = 7 \text{ and } \left(\frac{\partial F}{\partial V}\right)_{\substack{x=1\\y=4}} = -14.$$

(*b*) We have
$$\frac{\partial x}{\partial u} = e^u,\ \frac{\partial x}{\partial v} = -e^{-v},\ \frac{\partial y}{\partial u} = -e^{-u} \text{ and } \frac{\partial y}{\partial v} = -e^v.$$

Using the chain rule, we obtain
$$\frac{\partial f}{\partial u} = \frac{\partial f}{\partial x}\cdot\frac{\partial x}{\partial u} + \frac{\partial f}{\partial y}\cdot\frac{\partial y}{\partial u} = e^u\frac{\partial f}{\partial x} - e^{-u}\frac{\partial f}{\partial y}, \quad \text{...(i)}$$

and
$$\frac{\partial f}{\partial v} = \frac{\partial f}{\partial x}\cdot\frac{\partial x}{\partial v} + \frac{\partial f}{\partial y}\cdot\frac{\partial y}{\partial v} = -e^{-v}\frac{\partial f}{\partial x} - e^v\frac{\partial f}{\partial y}. \quad \text{...(ii)}$$

Therefore,
$$\frac{\partial f}{\partial u} - \frac{\partial f}{\partial v} = (e^u + e^{-v})\frac{\partial f}{\partial x} - (e^{-u} - e^v)\frac{\partial f}{\partial y} = x\frac{\partial f}{\partial x} - y\frac{\partial f}{\partial y}.$$

(*c*) We have by the chain rule

$$\frac{\partial u}{\partial x} = \frac{\partial u}{\partial r}\cdot\frac{\partial r}{\partial x} + \frac{\partial u}{\partial s}\cdot\frac{\partial s}{\partial x} = 1\cdot\frac{\partial u}{\partial r} + 0\cdot\frac{\partial u}{\partial s} = \frac{\partial u}{\partial r} \quad \text{...(i)}$$

$$\frac{\partial u}{\partial y} = \frac{\partial u}{\partial r}\cdot\frac{\partial r}{\partial y} + \frac{\partial u}{\partial s}\cdot\frac{\partial s}{\partial y} = 0\cdot\frac{\partial u}{\partial r} + 1\cdot\frac{\partial u}{\partial s} = \frac{\partial u}{\partial s} \quad \text{...(ii)}$$

$$\frac{\partial u}{\partial t} = \frac{\partial u}{\partial r}\cdot\frac{\partial r}{\partial t} + \frac{\partial u}{\partial s}\cdot\frac{\partial s}{\partial t} = a\frac{\partial u}{\partial r} + b\frac{\partial u}{\partial s}. \quad \text{...(iii)}$$

From (*i*), (*ii*) and (*iii*), we find

$$\frac{\partial u}{\partial t} = a\frac{\partial u}{\partial x} + b\frac{\partial u}{\partial y}.$$

EXAMPLE 1.51 (*a*) If $w = f(u, v)$, where $u = \sqrt{x^2 + y^2}$ and $v = \tan^{-1}(y/x)$, find $\partial f/\partial x$ and $\partial f/\partial y$.

(*b*) If $f(x, y, z) = z\sin(y/x)$, where $x = 3r^2 + 2s$, $y = 4r - 2s^3$ and $z = 2r^2 - 3s^2$, find $\partial f/\partial r$ and $\partial f/\partial s$.

Solution (*a*) We have $u = \sqrt{x^2 + y^2}$ and $v = \tan^{-1}(y/x)$. We obtain

$$\frac{\partial u}{\partial x} = \frac{x}{\sqrt{x^2+y^2}},\ \frac{\partial u}{\partial y} = \frac{y}{\sqrt{x^2+y^2}},\ \frac{\partial v}{\partial x} = \frac{-y}{x^2+y^2} \text{ and } \frac{\partial v}{\partial y} = \frac{x}{x^2+y^2}.$$

Using the chain rule, we obtain

$$\frac{\partial f}{\partial x} = \frac{\partial f}{\partial u}\cdot\frac{\partial u}{\partial x}+\frac{\partial f}{\partial v}\cdot\frac{\partial v}{\partial x} = \frac{x}{\sqrt{x^2+y^2}}\frac{\partial f}{\partial u}-\frac{y}{x^2+y^2}\frac{\partial f}{\partial v}=\frac{1}{x^2+y^2}\left(x\sqrt{x^2+y^2}\frac{\partial f}{\partial u}-y\frac{\partial f}{\partial v}\right)$$

and
$$\frac{\partial f}{\partial y} = \frac{\partial f}{\partial u}\cdot\frac{\partial u}{\partial y}+\frac{\partial f}{\partial v}\cdot\frac{\partial v}{\partial y} = \frac{y}{\sqrt{x^2+y^2}}\frac{\partial f}{\partial u}+\frac{x}{x^2+y^2}\frac{\partial f}{\partial v}=\frac{1}{x^2+y^2}\left(y\sqrt{x^2+y^2}\frac{\partial f}{\partial u}+x\frac{\partial f}{\partial v}\right).$$

(*b*) Using the chain rule, we obtain $\frac{\partial f}{\partial r}=\frac{\partial f}{\partial x}\cdot\frac{\partial x}{\partial r}+\frac{\partial f}{\partial y}\cdot\frac{\partial y}{\partial r}+\frac{\partial f}{\partial z}\cdot\frac{\partial z}{\partial r}$

$$= \left\{-\frac{y}{x^2}z\cos\left(\frac{y}{x}\right)\right\}6r+\left\{\frac{z}{x}\cos\left(\frac{y}{x}\right)\right\}4+\left\{\sin\left(\frac{y}{x}\right)\right\}(4r)$$

$$= \frac{-6ryz}{x^2}\cos\left(\frac{y}{x}\right)+\frac{4z}{x}\cos\left(\frac{y}{x}\right)+4r\sin\left(\frac{y}{x}\right),$$

and
$$\frac{\partial f}{\partial s}=\frac{\partial f}{\partial x}\cdot\frac{\partial x}{\partial s}+\frac{\partial f}{\partial y}\cdot\frac{\partial y}{\partial s}+\frac{\partial f}{\partial z}\cdot\frac{\partial z}{\partial s}$$

$$= \left\{-\frac{yz}{x^2}\cos\left(\frac{y}{x}\right)\right\}2+\left\{\frac{z}{x}\cos\left(\frac{y}{x}\right)\right\}(-6s^2)+\left\{\sin\left(\frac{y}{x}\right)\right\}\cdot(-6s)$$

$$= \frac{-2yz}{x^2}\cos\left(\frac{y}{x}\right)-6\frac{s^2z}{x}\cos\left(\frac{y}{x}\right)-6s\sin\left(\frac{y}{x}\right).$$

EXAMPLE 1.52 If $u=f(y-z, z-x, x-y)$, show that $\frac{\partial u}{\partial x}+\frac{\partial u}{\partial y}+\frac{\partial u}{\partial z}=0.$

(AMIE; W-2009, W-2006; UPTU 2005; VTU, 2003S; AMIETE June 2002)

Solution Let $r=y-z$, $s=z-x$, $t=x-y$, so that $u=f(r, s, t)$. Then u is a function of the three variables r, s, t, which in turn are functions of the three variables x, y, z.

Using the chain rule, we obtain $\frac{\partial u}{\partial x} = \frac{\partial u}{\partial r}\cdot\frac{\partial r}{\partial x}+\frac{\partial u}{\partial s}\cdot\frac{\partial s}{\partial x}+\frac{\partial u}{\partial t}\cdot\frac{\partial t}{\partial x}$

$$= 0\cdot\frac{\partial u}{\partial r}+(-1)\frac{\partial u}{\partial s}+(1)\frac{\partial u}{\partial t}=-\frac{\partial u}{\partial s}+\frac{\partial u}{\partial t} \quad \text{...(i)}$$

$$\frac{\partial u}{\partial y} = \frac{\partial u}{\partial r}\cdot\frac{\partial r}{\partial y}+\frac{\partial u}{\partial s}\cdot\frac{\partial s}{\partial y}+\frac{\partial u}{\partial t}\cdot\frac{\partial t}{\partial y}=(1)\frac{\partial u}{\partial r}+0\left(\frac{\partial u}{\partial s}\right)+(-1)\frac{\partial u}{\partial t}=\frac{\partial u}{\partial r}-\frac{\partial u}{\partial t}, \quad \text{...(ii)}$$

and
$$\frac{\partial u}{\partial z} = \frac{\partial u}{\partial r}\cdot\frac{\partial r}{\partial z}+\frac{\partial u}{\partial s}\cdot\frac{\partial s}{\partial z}+\frac{\partial u}{\partial t}\cdot\frac{\partial t}{\partial z}=(-1)\frac{\partial u}{\partial r}+(1)\frac{\partial u}{\partial s}+(0)\frac{\partial u}{\partial t}=-\frac{\partial u}{\partial r}+\frac{\partial u}{\partial s} \quad \text{...(iii)}$$

Adding (*i*), (*ii*) and (*iii*), we obtain

$$\frac{\partial u}{\partial x}+\frac{\partial u}{\partial y}+\frac{\partial u}{\partial z} = 0.$$

EXAMPLE 1.53 If $\phi(cx-az, cy-bz)=0$, show that $ap+bq=c$, where $p=\partial z/\partial x$ and $q=\partial z/\partial y$.

Solution Let $u=cx-az$ and $v=cy-bz$. ...(*i*)

Then the given function is $\phi(u, v)=0$.

$$\therefore \quad \frac{\partial\phi}{\partial u}\frac{\partial u}{\partial x}+\frac{\partial\phi}{\partial v}\frac{\partial v}{\partial x}=0, \qquad \frac{\partial\phi}{\partial u}\frac{\partial u}{\partial y}+\frac{\partial\phi}{\partial v}\frac{\partial v}{\partial y}=0.$$

Eliminating $\frac{\partial \phi}{\partial u}$ and $\frac{\partial \phi}{\partial v}$, we obtain $\begin{vmatrix} \frac{\partial u}{\partial x} & \frac{\partial v}{\partial x} \\ \frac{\partial u}{\partial y} & \frac{\partial v}{\partial y} \end{vmatrix} = 0.$ ...(*ii*)

From (*i*) $\partial u/\partial x = c - ap$, $\partial u/\partial y = -aq$; $\partial v/\partial x = -bp$, $\partial v/\partial y = c - bq$.

Substituting these values in (*ii*), we obtain

$$\begin{vmatrix} c-ap & -bp \\ -aq & c-bq \end{vmatrix} = 0 \text{ or } (c-ap)(c-bq) - abpq = 0 \text{ or } ap + bq = c.$$

EXAMPLE 1.54 Let u be a function of (x, y) and x, y are functions of (θ, ϕ) defined by $x + y = 2e^{\theta} \cos \phi$, and $x - y = 2ie^{\theta} \sin \phi$, where $i = \sqrt{-1}$. Show that

(*i*) $x\frac{\partial u}{\partial x} + y\frac{\partial u}{\partial y} = \frac{\partial u}{\partial \theta}$, (*AMIETE Dec. 2004*)

(*ii*) $\frac{\partial^2 u}{\partial \theta^2} + \frac{\partial^2 u}{\partial \phi^2} = 4xy\frac{\partial^2 u}{\partial x \partial y}$. (*AMIETE June 2010, Dec. 2009, 2007; MDU 2002; UPTU 2002*)

Solution We have $x = e^{\theta}(\cos \phi + i \sin \phi) = e^{\theta} \cdot e^{i\phi}$,

and $y = e^{\theta}(\cos \phi - i \sin \phi) = e^{\theta} \cdot e^{-i\phi}$.

Here u is a composite function of θ and ϕ.

$$\therefore \quad \frac{\partial u}{\partial \theta} = \frac{\partial u}{\partial x}\cdot\frac{\partial x}{\partial \theta} + \frac{\partial u}{\partial y}\cdot\frac{\partial y}{\partial \theta} = \frac{\partial u}{\partial x}(e^{\theta} \cdot e^{i\phi}) + \frac{\partial u}{\partial y}(e^{\theta} \cdot e^{-i\phi}) \quad ...(i)$$

$$= x\frac{\partial u}{\partial x} + y\frac{\partial u}{\partial y}, \text{ the required result.}$$

(*i*) may be written in operator form

$$\frac{\partial}{\partial \theta} = e^{\theta}\left(\frac{\partial}{\partial x}e^{i\phi} + \frac{\partial}{\partial y}e^{-i\phi}\right). \quad ...(ii)$$

Also $$\frac{\partial u}{\partial \phi} = \frac{\partial u}{\partial x}\cdot\frac{\partial x}{\partial \phi} + \frac{\partial u}{\partial y}\cdot\frac{\partial y}{\partial \phi} = \frac{\partial u}{\partial x}(e^{\theta} \cdot ie^{i\phi}) + \frac{\partial u}{\partial y}(e^{\theta} \cdot -ie^{-i\phi})$$

or $$\frac{\partial}{\partial \phi} = ie^{\theta}\left(\frac{\partial}{\partial x}\cdot e^{i\phi} - \frac{\partial}{\partial y}e^{-i\phi}\right). \quad ...(iii)$$

Using the operator (*ii*), we have

$$\frac{\partial^2 u}{\partial \theta^2} = \frac{\partial}{\partial \theta}\left(\frac{\partial u}{\partial \theta}\right) = e^{2\theta}\left(\frac{\partial}{\partial x}e^{i\phi} + \frac{\partial}{\partial y}e^{-i\phi}\right)\left(\frac{\partial u}{\partial x}e^{i\phi} + \frac{\partial u}{\partial y}e^{-i\phi}\right)$$

$$= e^{2\theta}\left(e^{2i\phi}\frac{\partial^2 u}{\partial x^2} + 2\frac{\partial^2 u}{\partial x \partial y} + e^{-2i\phi}\frac{\partial^2 u}{\partial y^2}\right), \quad ...(iv)$$

and $$\frac{\partial^2 u}{\partial \phi^2} = \frac{\partial}{\partial \phi}\left(\frac{\partial u}{\partial \phi}\right) = (ie^{\theta})^2\left(\frac{\partial}{\partial x}e^{i\phi} - \frac{\partial}{\partial y}e^{-i\phi}\right)\left(\frac{\partial u}{\partial x}e^{i\phi} - \frac{\partial u}{\partial y}e^{-i\phi}\right)$$

$$= -e^{2\theta}\left(e^{2i\phi}\frac{\partial^2 u}{\partial x^2} - 2\frac{\partial^2 u}{\partial x \partial y} + e^{-2i\phi}\frac{\partial^2 u}{\partial y^2}\right). \quad ...(v)$$

Adding (*iv*) and (*v*), we get

$$\frac{\partial^2 u}{\partial \theta^2} + \frac{\partial^2 u}{\partial \phi^2} = e^{2\theta}\left[4\frac{\partial^2 u}{\partial x \partial y}\right] = 4xy\frac{\partial^2 u}{\partial x \partial y}. \quad [\because xy = e^{2\theta}]$$

EXAMPLE 1.55 If $x = e^v \sec u$, $y = e^v \tan u$ and $\phi = \phi(x, y)$, show that

$$\cos u\left(\frac{\partial^2\phi}{\partial u\,\partial v} - \frac{\partial\phi}{\partial u}\right) = xy\left(\frac{\partial^2\phi}{\partial x^2} + \frac{\partial^2\phi}{\partial y^2}\right) + (x^2 + y^2)\frac{\partial^2\phi}{\partial x\,\partial y}. \qquad \textit{(AMIE, W-2000)}$$

Solution We have $x = e^v \sec u$, $y = e^v \tan u$ and $\phi = \phi(x, y)$.

Using the chain rule, we obtain

$$\frac{\partial\phi}{\partial u} = \frac{\partial\phi}{\partial x}\cdot\frac{\partial x}{\partial u} + \frac{\partial\phi}{\partial y}\cdot\frac{\partial y}{\partial u} = \frac{\partial\phi}{\partial x}(e^v \sec u \tan u) + \frac{\partial\phi}{\partial y}(e^v \sec^2 u)$$

$$= (\sec u)\, y\frac{\partial\phi}{\partial x} + (\sec u)\, x\frac{\partial\phi}{\partial y} \text{ or } \cos u\cdot\frac{\partial\phi}{\partial u} = y\frac{\partial\phi}{\partial x} + x\frac{\partial\phi}{\partial y}$$

$$\Rightarrow \qquad \cos u\frac{\partial}{\partial u} = y\frac{\partial}{\partial x} + x\frac{\partial}{\partial y} \qquad \text{...}(i)$$

$$\frac{\partial\phi}{\partial v} = \frac{\partial\phi}{\partial x}\cdot\frac{\partial x}{\partial v} + \frac{\partial\phi}{\partial y}\cdot\frac{\partial y}{\partial v} = \frac{\partial\phi}{\partial x}(e^v \sec u) + \frac{\partial\phi}{\partial y}(e^v \tan u) = x\frac{\partial\phi}{\partial x} + y\frac{\partial\phi}{\partial y}. \qquad \text{...}(ii)$$

Now

$$\cos u\left(\frac{\partial^2\phi}{\partial u\,\partial v}\right) = \cos u\frac{\partial}{\partial u}\left(\frac{\partial\phi}{\partial v}\right) = \left(y\frac{\partial}{\partial x} + x\frac{\partial}{\partial y}\right)\left(x\frac{\partial\phi}{\partial x} + y\frac{\partial\phi}{\partial y}\right) \qquad \text{from } (i) \text{ and } (ii)$$

$$= y\frac{\partial}{\partial x}\left(x\frac{\partial\phi}{\partial x} + y\frac{\partial\phi}{\partial y}\right) + x\frac{\partial}{\partial y}\left(x\frac{\partial\phi}{\partial x} + y\frac{\partial\phi}{\partial y}\right)$$

$$= y\left[1\cdot\frac{\partial\phi}{\partial x} + x\frac{\partial^2\phi}{\partial x^2} + y\frac{\partial^2\phi}{\partial x\partial y}\right] + x\left(x\frac{\partial^2\phi}{\partial y\,\partial x} + 1\cdot\frac{\partial\phi}{\partial y} + y\frac{\partial^2\phi}{\partial y^2}\right)$$

$$= xy\left(\frac{\partial^2\phi}{\partial x^2} + \frac{\partial^2\phi}{\partial y^2}\right) + (x^2 + y^2)\frac{\partial^2\phi}{\partial x\,\partial y} + \left(x\frac{\partial\phi}{\partial y} + y\frac{\partial\phi}{\partial x}\right),$$

$$= xy\left(\frac{\partial^2\phi}{\partial x^2} + \frac{\partial^2\phi}{\partial y^2}\right) + (x^2 + y^2)\frac{\partial^2\phi}{\partial x\,\partial y} + \cos u\frac{\partial\phi}{\partial u}, \qquad \text{from } (i)$$

or $\cos u\left(\dfrac{\partial^2\phi}{\partial u\,\partial v} - \dfrac{\partial\phi}{\partial u}\right) = xy\left(\dfrac{\partial^2\phi}{\partial x^2} + \dfrac{\partial^2\phi}{\partial y^2}\right) + (x^2 + y^2)\dfrac{\partial^2\phi}{\partial x\,\partial y}$, which is same as required.

EXAMPLE 1.56 If $x = \dfrac{\cos\theta}{u}$, $y = \dfrac{\sin\theta}{u}$ and $z = f(x, y)$, then show that

$$\frac{\partial^2 z}{\partial x^2} + \frac{\partial^2 z}{\partial y^2} = u^4\frac{\partial^2 z}{\partial u^2} + u^3\frac{\partial z}{\partial u} + u^2\frac{\partial^2 z}{\partial\theta^2}. \qquad \textit{(AMIETE, Dec. 2002)}$$

Solution We have $\dfrac{\partial x}{\partial u} = \dfrac{-\cos\theta}{u^2}$, $\dfrac{\partial x}{\partial\theta} = \dfrac{-\sin\theta}{u}$ and $\dfrac{\partial y}{\partial u} = \dfrac{-\sin\theta}{u^2}$ and $\dfrac{\partial y}{\partial\theta} = \dfrac{\cos\theta}{u}$.

Using the chain rule, we obtain $\dfrac{\partial z}{\partial u} = \dfrac{\partial z}{\partial x}\cdot\dfrac{\partial x}{\partial u} + \dfrac{\partial z}{\partial y}\cdot\dfrac{\partial y}{\partial u} = -\dfrac{\cos\theta}{u^2}\dfrac{\partial z}{\partial x} - \dfrac{\sin\theta}{u^2}\dfrac{\partial z}{\partial y}$,

and $\dfrac{\partial z}{\partial\theta} = \dfrac{\partial z}{\partial x}\cdot\dfrac{\partial x}{\partial\theta} + \dfrac{\partial z}{\partial y}\cdot\dfrac{\partial y}{\partial\theta} = \dfrac{-\sin\theta}{u}\dfrac{\partial z}{\partial x} + \dfrac{\cos\theta}{u}\dfrac{\partial z}{\partial y}$.

Solving these equations, we get

$$\frac{\partial z}{\partial x} = -\left(u^2\cos\theta\frac{\partial z}{\partial u} + u\sin\theta\frac{\partial z}{\partial\theta}\right) \Rightarrow \frac{\partial}{\partial x} = -\left(u^2\cos\theta\frac{\partial}{\partial u} + u\sin\theta\frac{\partial}{\partial\theta}\right), \qquad \text{(in operator form)}$$

and

$$\frac{\partial z}{\partial y} = \left(u\cos\theta\frac{\partial z}{\partial\theta} - u^2\sin\theta\frac{\partial z}{\partial u}\right) \Rightarrow \frac{\partial}{\partial y} = \left(u\cos\theta\frac{\partial}{\partial\theta} - u^2\sin\theta\frac{\partial}{\partial u}\right), \qquad \text{(in operator form)}$$

Hence $$\frac{\partial^2 z}{\partial x^2} = \frac{\partial}{\partial x}\left(\frac{\partial z}{\partial x}\right) = \left(u^2\cos\theta\frac{\partial}{\partial u} + u\sin\theta\frac{\partial}{\partial\theta}\right)\left(u^2\cos\theta\frac{\partial z}{\partial u} + u\sin\theta\frac{\partial z}{\partial\theta}\right)$$

$$= u^4\cos^2\theta\frac{\partial^2 z}{\partial u^2} + 2u^3\sin\theta\cos\theta\frac{\partial^2 z}{\partial u\partial\theta} + u^2\sin^2\theta\frac{\partial^2 z}{\partial\theta^2}$$

$$+(2u^3\cos^2\theta - u^3\sin^2\theta)\frac{\partial z}{\partial u} + 2u^2\cos\theta\sin\theta\frac{\partial z}{\partial\theta} \qquad ...(i)$$

Similarly, we obtain $$\frac{\partial^2 z}{\partial y^2} = -2u^3\sin\theta\cos\theta\frac{\partial^2 z}{\partial u\partial\theta} + u^2\cos^2\theta\frac{\partial^2 z}{\partial\theta^2} + u^4\sin^2\theta\frac{\partial^2 z}{\partial u^2}$$

$$+(2u^3\sin^2\theta - u^3\cos^2\theta)\frac{\partial z}{\partial u} - 2u^2\sin\theta\cos\theta\frac{\partial z}{\partial\theta}. \qquad ...(ii)$$

Adding (*i*) and (*ii*), we obtain

$$\frac{\partial^2 z}{\partial x^2} + \frac{\partial^2 z}{\partial y^2} = u^4\frac{\partial^2 z}{\partial u^2} + u^3\frac{\partial z}{\partial u} + u^2\frac{\partial^2 z}{\partial\theta^2}$$, the required result.

EXAMPLE 1.57 If $u = f(x, y)$ and $x = r\cos\theta$, $y = r\sin\theta$, show that $\frac{\partial^2 u}{\partial x^2} + \frac{\partial^2 u}{\partial y^2} = 0$ is transformed to the form

$$\frac{\partial^2 u}{\partial r^2} + \frac{1}{r}\frac{\partial u}{\partial r} + \frac{1}{r^2}\frac{\partial^2 u}{\partial\theta^2} = 0.$$ (*GGSIPU 2007; DCE 2004; UPTU 2004*)

Solution We have $x = r\cos\theta$, $y = r\sin\theta$. Therefore,

$r = \sqrt{(x^2 + y^2)}$, and $\theta = \tan^{-1}(y/x)$.

$$\frac{\partial r}{\partial x} = \frac{x}{\sqrt{(x^2 + y^2)}} = \cos\theta \text{ and } \frac{\partial\theta}{\partial x} = -\frac{y}{x^2 + y^2} = -\frac{\sin\theta}{r}.$$

Using the chain rule, we obtain $$\frac{\partial u}{\partial x} = \frac{\partial u}{\partial r}\cdot\frac{\partial r}{\partial x} + \frac{\partial u}{\partial\theta}\cdot\frac{\partial\theta}{\partial x} = \cos\theta\frac{\partial u}{\partial r} - \frac{\sin\theta}{r}\frac{\partial u}{\partial\theta}$$

Therefore, $$\frac{\partial}{\partial x} = \cos\theta\frac{\partial}{\partial r} - \frac{\sin\theta}{r}\frac{\partial}{\partial\theta}$$ (in operator form)

Similarly, we obtain $$\frac{\partial}{\partial y} = \sin\theta\frac{\partial}{\partial r} + \frac{\cos\theta}{r}\frac{\partial}{\partial\theta}.$$ (in operator form)

$$\therefore \quad \frac{\partial^2 u}{\partial x^2} = \frac{\partial}{\partial x}\left(\frac{\partial u}{\partial x}\right) = \left(\cos\theta\frac{\partial}{\partial r} - \frac{\sin\theta}{r}\frac{\partial}{\partial\theta}\right)\left(\cos\theta\frac{\partial u}{\partial r} - \left(\frac{\sin\theta}{r}\right)\frac{\partial u}{\partial\theta}\right)$$

$$= \cos^2\theta\frac{\partial^2 u}{\partial r^2} - \frac{2\sin\theta\cos\theta}{r}\frac{\partial^2 u}{\partial r\partial\theta} + \frac{\sin^2\theta}{r^2}\frac{\partial^2 u}{\partial\theta^2} + 2\frac{\sin\theta\cos\theta}{r^2}\frac{\partial u}{\partial\theta} + \frac{\sin^2\theta}{r}\frac{\partial u}{\partial r} \qquad ...(i)$$

We also have $$\frac{\partial^2 u}{\partial y^2} = \frac{\partial}{\partial y}\left(\frac{\partial u}{\partial y}\right) = \left(\sin\theta\frac{\partial}{\partial r} + \frac{\cos\theta}{r}\frac{\partial}{\partial\theta}\right)\left(\sin\theta\frac{\partial u}{\partial r} + \frac{\cos\theta}{r}\frac{\partial u}{\partial\theta}\right)$$

$$= \sin^2\theta\frac{\partial^2 u}{\partial r^2} + \frac{2\sin\theta\cos\theta}{r}\frac{\partial^2 u}{\partial r\partial\theta} + \frac{\cos^2\theta}{r^2}\frac{\partial^2 u}{\partial\theta^2} + \frac{\cos^2\theta}{r}\frac{\partial u}{\partial r} - \frac{2\sin\theta\cos\theta}{r^2}\frac{\partial u}{\partial\theta}. \qquad ...(ii)$$

Adding (*i*) and (*ii*), we obtain

$$\frac{\partial^2 u}{\partial x^2} + \frac{\partial^2 u}{\partial y^2} = \frac{\partial^2 u}{\partial r^2} + \frac{1}{r^2}\frac{\partial^2 u}{\partial\theta^2} + \frac{1}{r}\frac{\partial u}{\partial r}. \qquad ...(iii)$$

$\therefore$ $\frac{\partial^2 u}{\partial x^2} + \frac{\partial^2 u}{\partial y^2} = 0$ transforms (*iii*) into $\frac{\partial^2 u}{\partial r^2} + \frac{1}{r}\frac{\partial u}{\partial r} + \frac{1}{r^2}\frac{\partial^2 u}{\partial\theta^2} = 0.$

1.15 DIFFERENTIATION OF IMPLICIT FUNCTIONS

An equation of the form $$f(x, y) = 0 \qquad ...(1.28)$$
in which y is not expressible directly in terms of x, is known as an implicit function of x and y. The above equation defines either x or y as an implicit function of the other.

To find dy/dx for the implicit function $f(x, y) = 0$, we proceed as follows:

Let $z = f(x, y)$, where $z = 0$ for all (x, y).

Then $$dz = \frac{\partial f}{\partial x}dx + \frac{\partial f}{\partial y}dy \quad \text{or} \quad dz = f_x\,dx + f_y\,dy.$$

On the other hand, since $z = 0$, for all (x, y) implies $dz = 0$ and hence

$$f_x\,dx + f_y\,dy = 0 \quad \text{or} \quad \frac{dy}{dx} = -\frac{\partial f/\partial x}{\partial f/\partial y} = -\frac{f_x}{f_y},\ (f_y \neq 0). \qquad ...(1.29)$$

We can find d^2y/dx^2 by differentiating Eq. (1.29) with respect to x, and assuming that $f_{xy} = f_{yx}$.

$$\frac{d^2y}{dx^2} = \frac{d}{dx}\left(\frac{dy}{dx}\right) = -\frac{d}{dx}\left(\frac{f_x}{f_y}\right) = -\frac{f_y\frac{d}{dx}(f_x) - f_x\frac{d}{dx}(f_y)}{(f_y)^2}$$

$$= -\frac{f_y\left(f_{xx} + (f_{yx})y'\right) - f_x\left(f_{xy} + (f_{yy})y'\right)}{(f_y)^2}$$

$$= -\frac{(f_y f_{xx} - f_x f_{xy}) + (f_y f_{yx} - f_x f_{yy})y'}{(f_y)^2}$$

Substituting $y' = -f_x/f_y$, we obtain

$$\frac{d^2y}{dx^2} = -\frac{(f_x)^2 f_{yy} - 2f_x f_y f_{xy} + (f_y)^2 f_{xx}}{(f_y)^3}, \text{ since } f_{yx} = f_{xy}. \qquad ...(1.30)$$

The equation $$F(x, y, z) = 0 \qquad ...(1.31)$$
defines z as an implicit function of the two independent variables x and y. To find the partial derivatives of z with respect to x and to y, proceed as follows:

Let $$u = F(x, y, z).$$

Then $du = \frac{\partial F}{\partial x}dx + \frac{\partial F}{\partial y}dy + \frac{\partial F}{\partial z}dz$, and this holds no matter what the independent variables are. Now let z be chosen as that function of the independent variables x and y which satisfied (1.31). Then $u = 0$, $du = 0$, and we have

$$\frac{\partial F}{\partial x}dx + \frac{\partial F}{\partial y}dy + \frac{\partial F}{\partial z}dz = 0 \qquad ...(1.32)$$

But now $dz = \frac{\partial z}{\partial x}dx + \frac{\partial z}{\partial y}dy.$

Substituting this value in (1.32) and simplifying gives

$$\left(\frac{\partial F}{\partial x} + \frac{\partial F}{\partial z}\frac{\partial z}{\partial x}\right)dx + \left(\frac{\partial F}{\partial y} + \frac{\partial F}{\partial z}\frac{\partial z}{\partial y}\right)dy = 0.$$

Since x and y are independent, we have

$$\frac{\partial F}{\partial x} + \frac{\partial F}{\partial z}\frac{\partial z}{\partial x} = 0 \text{ and } \frac{\partial F}{\partial y} + \frac{\partial F}{\partial z}\frac{\partial z}{\partial y} = 0,$$

from which $$\frac{\partial z}{\partial x} = -\frac{\partial F/\partial x}{\partial F/\partial z} \text{ and } \frac{\partial z}{\partial y} = -\frac{\partial F/\partial y}{\partial F/\partial z}.$$

This also can be written as $\frac{\partial z}{\partial x} = -\frac{F_x}{F_z}$ ...(1.33) and $\frac{\partial z}{\partial y} = -\frac{F_y}{F_z}$, where $F_z \neq 0$. ...(1.34)

ILLUSTRATIVE EXAMPLES

EXAMPLE 1.58 Find dy/dx, when

(*i*) $f(x, y) = e^{xy} + y^2 - \cos\sqrt{x} = 0$.

(*ii*) $f(x, y) = \log(x^2 + y^2) + \tan^{-1}(y/x) = 0$. (*AMIE, S-2002*)

Solution (*i*) $$\frac{\partial f}{\partial x} = ye^{xy} + \frac{1}{2}\frac{\sin\sqrt{x}}{\sqrt{x}} = \frac{(2\sqrt{x}ye^{xy} + \sin\sqrt{x})}{2\sqrt{x}}$$

and $\frac{\partial f}{\partial y} = xe^{xy} + 2y$. Therefore, $$\frac{dy}{dx} = -\frac{\partial f/\partial x}{\partial f/\partial y} = -\left[\frac{2\sqrt{x}\,ye^{xy} + \sin\sqrt{x}}{2\sqrt{x}\,(xe^{xy} + 2y)}\right].$$

(*ii*) $$\frac{\partial f}{\partial x} = \frac{2x}{x^2 + y^2} + \frac{1}{1 + (y/x)^2}\left(\frac{-y}{x^2}\right) = \frac{2x}{x^2 + y^2} - \frac{y}{x^2 + y^2} = \frac{2x - y}{x^2 + y^2},$$

and $$\frac{\partial f}{\partial y} = \frac{2y}{x^2 + y^2} + \frac{1}{1 + (y/x)^2}\left(\frac{1}{x}\right) = \frac{2y}{x^2 + y^2} + \frac{x}{x^2 + y^2} = \frac{x + 2y}{x^2 + y^2}.$$

Therefore, $$\frac{dy}{dx} = -\frac{\partial f/\partial x}{\partial f/\partial y} = -\left(\frac{2x - y}{x + 2y}\right) = \frac{y - 2x}{2y + x}.$$

EXAMPLE 1.59 If $f(x, y) = x^3 + y^3 - 3\,axy = 0$, find dy/dx and d^2y/dx^2. (*AMIE, W-2005*)

Solution We have $f(x, y) = x^3 + y^3 - 3\,axy = 0$.

$f_x = \partial f/\partial x = 3x^2 - 3ay$ and $f_y = \partial f/\partial y = 3y^2 - 3ax$.

$$\therefore \quad \frac{dy}{dx} = -\frac{\partial f/\partial x}{\partial f/\partial y} = -\left(\frac{f_x}{f_y}\right) = -\left(\frac{3x^2 - 3ay}{3y^2 - 3ax}\right) = \frac{ay - x^2}{y^2 - ax}.$$

Now $f_{xx} = 6x$, $f_{yy} = 6y$, $f_{xy} = f_{yx} = -3a$.

$$\therefore \quad \frac{d^2y}{dx^2} = -\left\{\frac{(f_y)^2 f_{xx} - 2f_x f_y f_{xy} + (f_x)^2 f_{yy}}{(f_y)^3}\right\}$$

$$= -\left\{\frac{(3y^2 - 3ax)^2\,6x - 2(3x^2 - 3ay)(3y^2 - 3ax)(-3a) + (3x^2 - 3ay)^2\,6y}{(3y^2 - 3ax)^3}\right\}$$

$$= -\left\{\frac{2x(y^2 - ax)^2 + 2a(x^2 - ay)(y^2 - ax) + 2y(x^2 - ay)^2}{(y^2 - ax)^3}\right\} = -\frac{2a^3xy}{(y^2 - ax)^3}.$$

EXAMPLE 1.60 If x increases at the rate of 2 cm per second at the instant when $x = 3$ cm and y = 1 cm, at what rate must y be changing in order that the function $2xy^2 - 3x^2y$ shall be neither increasing nor decreasing?

Solution Let $u = 2xy^2 - 3x^2y$. Using the chain rule, we obtain

$$\frac{du}{dt} = \frac{\partial u}{\partial x}\frac{dx}{dt} + \frac{\partial u}{\partial y}\cdot\frac{dy}{dt} = (2y^2 - 6xy)\frac{dx}{dt} + (4xy - 3x^2)\frac{dy}{dt}. \quad ...(i)$$

As u is neither increasing nor decreasing so $du/dt = 0$.

Thus, $$\frac{dy}{dt} = -\left(\frac{2y^2 - 6xy}{4xy - 3x^2}\right)\frac{dx}{dt}.$$

When $x = 3$, $y = 1$, $\frac{dx}{dt} = +2$, then $\frac{dy}{dt} = -\left(\frac{2-18}{12-27}\right)2 = -2\frac{2}{15}$ cm/sec.

Thus y is decreasing at the rate of $2\frac{2}{15}$ cm/sec.

EXAMPLE 1.61 If $u = x^3 y$, find du/dt if $x^5 + y = t$ ...(*i*)

and $x^2 + y^3 = t^2$. ...(*ii*)

Solution Equations (*i*) and (*ii*) define x and y as implicit functions of t. Now differentiating these equations w.r.t. t, we obtain

$$5x^4\left(\frac{dx}{dt}\right) + \left(\frac{dy}{dt}\right) = 1 \quad ...(iii) \qquad \text{and} \qquad 2x\left(\frac{dx}{dt}\right) + 3y^2\left(\frac{dy}{dt}\right) = 2t. \quad ...(iv)$$

Solving (*iii*) and (*iv*) simultaneously for dx/dt and dy/dt, we obtain

$$\frac{dx}{dt} = \frac{\begin{vmatrix} 1 & 1 \\ 2t & 3y^2 \end{vmatrix}}{\begin{vmatrix} 5x^4 & 1 \\ 2x & 3y^2 \end{vmatrix}} = \frac{3y^2 - 2t}{15x^4y^2 - 2x}, \qquad \frac{dy}{dt} = \frac{\begin{vmatrix} 5x^4 & 1 \\ 2x & 2t \end{vmatrix}}{\begin{vmatrix} 5x^4 & 1 \\ 2x & 3y^2 \end{vmatrix}} = \frac{10x^4t - 2x}{15x^4y^2 - 2x}.$$

Using the chain rule, we obtain $$\frac{du}{dt} = \frac{\partial u}{\partial x}\frac{dx}{dt} + \frac{\partial u}{\partial y}\frac{dy}{dt} = 3x^2y\left(\frac{3y^2 - 2t}{15x^4y^2 - 2x}\right) + x^3\left(\frac{10x^4t - 2x}{15x^4y^2 - 2x}\right).$$

EXAMPLE 1.62 If $u = x\log(xy)$, where $x^3 + y^3 + 3xy = 1$, find du/dx. (*UPTU 2001*)

Solution We have $f(x, y) = x^3 + y^3 + 3xy - 1 = 0$.

$$\partial f/\partial x = 3x^2 + 3y \quad \text{and} \quad \partial f/\partial y = 3y^2 + 3x.$$

Therefore, $$\frac{dy}{dx} = -\frac{\partial f/\partial x}{\partial f/\partial y} = -\frac{x^2 + y}{y^2 + x}$$ Given $u = x\log(xy)$.

Therefore, $$\frac{\partial u}{\partial x} = \log(xy) + \frac{xy}{xy} = 1 + \log(xy), \text{ and } \frac{\partial u}{\partial y} = \frac{x^2}{xy} = \frac{x}{y}.$$

Using the chain rule, we obtain $$\frac{du}{dx} = \frac{\partial u}{\partial x} + \frac{\partial u}{\partial y}\cdot\frac{dy}{dx} = 1 + \log(xy) - \frac{x(x^2 + y)}{y(x + y^2)}.$$

EXAMPLE 1.63 Use implicit differentiation to find $\partial z/\partial x$ and $\partial z/\partial y$, given

(*a*) $F(x, y, z) = x^2 + 3xy - 2y^2 + 3xz + z^2 = 0$. (*b*) $z = e^x\cos(y + z)$.

Solution (*a*) $F_x = \partial F/\partial x = 2x + 3y + 3z$, $F_y = \partial F/\partial y = 3x - 4y$, and $F_z = \partial F/\partial z = 3x + 2z$.

$$\therefore \quad \frac{\partial z}{\partial x} = -\frac{F_x}{F_z} = -\frac{2x + 3y + 3z}{3x + 2z} \quad \text{and} \quad \frac{\partial z}{\partial y} = -\frac{F_y}{F_z} = -\frac{3x - 4y}{3x + 2z}.$$

(*b*) Set $F(x, y, z) = z - e^x\cos(y + z)$; then

$\partial F/\partial x = -e^x\cos(y + z)$, $\partial F/\partial y = e^x\sin(y + z)$, and $\partial F/\partial z = 1 + e^x\sin(y + z)$.

Therefore, $$\frac{\partial z}{\partial x} = -\frac{\partial F/\partial x}{\partial F/\partial z} = \frac{e^x\cos(y + z)}{1 + e^x\sin(y + z)} = \frac{z}{1 + e^x\sin(y + z)},$$

$$\frac{\partial z}{\partial y} = -\frac{F_y}{F_z} = -\frac{e^x\sin(y + z)}{1 + e^x\sin(y + z)}.$$

EXAMPLE 1.64 By the equation $x + 2y + z - 2\sqrt{xyz} = 10$, z is defined as an implicit function of x and y. Find the partial derivatives of this function.

Solution Set $F(x, y, z) = x + 2y + z - 2\sqrt{xyz} - 10 = 0$.

$$\frac{\partial F}{\partial x} = 1 - \frac{yz}{\sqrt{xyz}} = \frac{\sqrt{xyz} - yz}{\sqrt{xyz}}; \quad \frac{\partial F}{\partial y} = 2 - \frac{zx}{\sqrt{xyz}} = \frac{2\sqrt{xyz} - zx}{\sqrt{xyz}},$$

$$\frac{\partial F}{\partial z} = 1 - \frac{xy}{\sqrt{xyz}} = \frac{\sqrt{xyz} - xy}{\sqrt{xyz}}.$$

Therefore,
$$\frac{\partial z}{\partial x} = -\frac{\partial F/\partial x}{\partial F/\partial z} = \frac{yz - \sqrt{xyz}}{\sqrt{xyz} - xy}, \quad \frac{\partial z}{\partial y} = -\frac{\partial F/\partial y}{\partial F/\partial z} = \frac{zx - 2\sqrt{xyz}}{\sqrt{xyz} - xy}.$$

EXAMPLE 1.65 If u and v are defined as functions of x and y by the equations $2u - v = -x^2 - xy$ and $u + 2v = y^2 - xy$, find

(*a*) $\partial u/\partial x$, (*b*) $\partial v/\partial x$, (*c*) $\partial u/\partial y$, (*d*) $\partial v/\partial y$.

Solution Differentiate the given equations with respect to x, considering u and v as functions of x and y.

We have
$$2\frac{\partial u}{\partial x} - \frac{\partial v}{\partial x} = -2x - y, \quad ...(i) \qquad \frac{\partial u}{\partial x} + 2\frac{\partial v}{\partial x} = -y. \quad ...(ii)$$

Solving (*i*) and (*ii*) simultaneously for $\partial u/\partial x$ and $\partial v/\partial x$, we obtain

$$\frac{\partial u}{\partial x} = \frac{\begin{vmatrix} -2x - y & -1 \\ -y & 2 \end{vmatrix}}{\begin{vmatrix} 2 & -1 \\ 1 & 2 \end{vmatrix}} = \frac{1}{5}(4x + 3y), \quad \frac{\partial v}{\partial x} = \frac{\begin{vmatrix} 2 & -2x - y \\ 1 & -y \end{vmatrix}}{\begin{vmatrix} 2 & -1 \\ 1 & 2 \end{vmatrix}} = \frac{1}{5}(2x - y).$$

Now differentiating, the given equations with respect to y, we have

$$2\frac{\partial u}{\partial y} - \frac{\partial v}{\partial y} = -x; \quad ...(iii) \qquad \frac{\partial u}{\partial y} + 2\frac{\partial v}{\partial y} = 2y - x.$$

Solving,
$$\frac{\partial u}{\partial y} = \frac{\begin{vmatrix} -x & -1 \\ 2y - x & 2 \end{vmatrix}}{\begin{vmatrix} 2 & -1 \\ 1 & 2 \end{vmatrix}} = \frac{1}{5}(2y - 3x), \quad \frac{\partial v}{\partial y} = \frac{\begin{vmatrix} 2 & -x \\ 1 & 2y - x \end{vmatrix}}{\begin{vmatrix} 2 & -1 \\ 1 & 2 \end{vmatrix}} = \frac{1}{5}(4y - x).$$

EXERCISES 1.4

1. Find the total differential of the following functions

(*i*) $f(x, y, z) = x^2z - yz^3 + x^4$. (*ii*) $u = (x^2 + y^2 + z^2)^{-1/2}$.

(*iii*) $w = xyz + (xyz)^{-1}$. (*iv*) $z = \log_e(\cos y/x)$. (*v*) $z = x^{\log y}$.

Ans. (*i*) $df = (2xz + 4x^3)dx - z^3dy + (x^2 - 3yz^2)\,dz$. (*ii*) $du = -(x^2 + y^2 + z^2)^{-3/2}(x\,dx + y\,dy + z\,dz)$.

(*iii*) $dw = [1 - (xyz)^{-2}]\,[yz\,dx + zx\,dy + xy\,dz]$.

(*iv*) $dz = \frac{1}{x^2}\tan\frac{y}{x}(ydx - xdy)$. (*v*) $dz = \frac{x^{(\log y - 1)}}{y}(y\log y\,dx + x\log x\,dy)$.

2. (*a*) If $f(x, y) = e^{xy^2}$, find the total differential of the function at the point (1, 2).

(*AMIETE W-2005*) **Ans.** $e^4(dx + dy)$.

(*b*) If $f(x, y) = \tan^{-1}(x/y)$, find the total differential of the function at the point (1, 1).

(*AMIETE S-2005*) **Ans.** $(dx - dy)/2$.

3. Find the second order total differentials of the following functions.

(*i*) $z = xy(x + y)$. (*ii*) $u = xyz$. (*iii*) $z = x + y + \log(xy)$.

Ans. (*i*) $d^2z = 2y(dx)^2 + 2x(dy)^2 + 4(x + y)\, dxdy$.

(*ii*) $d^2u = 2(z\, dx\, dy + xdydz + ydzdx)$. (*iii*) $d^2z = -\dfrac{1}{x^2}(dx)^2 - \dfrac{1}{y^2}(dy)^2$.

4. Compute Δu and du for the function $u = 2x^2 + 3y^2$, when $x = 10$, $y = 8$, $\Delta x = 0.2$, $\Delta y = 0.3$, and compare the results.

Ans. $\Delta u = 22.75$, $du = 22.4$, $\Delta u - du = 1.6\%$ of Δu.

5. Determine whether each of the following are exact differentials of a function and if so, find the function

(*i*) $(2xy + y^2)\, dx + (x^2 + 2\, xy)\, dy$.

(*ii*) $(2xy^2 + 3y \cos 3x)\, dx + (2x^2y + \sin 3x)\, dy$. (*iii*) $(5xy - y^2)\, dx + (3\, xe^y - x^2)\, dy$.

Ans. (*i*) $x^2y + xy^2 + c$. (*ii*) $x^2y^2 + y \sin 3x + c$. (*iii*) not exact.

6. (*i*) If $z = u^2 + v^2$ and $u = at^2$, $v = 2at$, find dz/dt. **Ans.** $4a^2t(t^2 + 2)$.

(*ii*) Find dz/dt, when $z = xy^2 + yx^2$ and $x = at^2$, $y = 2\, at$. Verify by direct substitution.

Ans. $a^3(16\, t^3 + 10\, t^4)$.

(*iii*) If $u = x^4y^5$, where $x = t^2$ and $y = t^3$, find du/dt. **Ans.** $23\, t^{22}$.

(*iv*) If $u = x^2y^3$, $x = \log t$, $y = e^t$, find du/dt. **Ans.** $2\log t \cdot \dfrac{e^{3t}}{t} + 3(\log t)^2\, e^{3t}$.

7. Find du/dx of the following functions.

(*i*) $u = \sin(x^2 + y^2)$, where $a^2x^2 + b^2y^2 = c^2$. **Ans.** $\dfrac{2x}{b^2}(b^2 - a^2)\cos(x^2 + y^2)$.

(*ii*) $u = y^2 + yz + z^2$, where $z = \sin x$, $y = e^x$. **Ans.** $e^x(2e^x + \cos x) + \sin x(e^x + 2\cos x)$.

8. If $u = x^2 - y^2 + \sin yz$, where $y = e^x$ and $z = \log_e x$, find du/dx. Verify by direct substitution.

(*UPTU 2005*) **Ans.** $2(x - e^{2x}) + e^x \cos(e^x \log x)\, [\log x + x^{-1}]$.

9. Find dw/dt when

(*i*) $w = x^2 + y^2 + z^2$, $x = e^t \cos t$, $y = e^t \sin t$, $z = e^t$.

(*ii*) $w = e^x \sin yz$, $x = t^2$, $y = t - 1$, $z = 1/t$.

(*iii*) $w = e^{x+y} \cos 2z$, $x = \log t$, $y = \log(t^2 + 1)$, $z = t$.

(*iv*) $w = \dfrac{x}{z} + \dfrac{y}{z}$, $x = \cos^2 t$, $y = \sin^2 t$, $z = 1/t$; $t = 3$.

Ans. (*i*) $4\, e^{2t}$ (*ii*) $e^{t^2}\left(2t \sin \dfrac{t-1}{t} + \dfrac{1}{t^2} \cos \dfrac{t-1}{t}\right)$.

(*iii*) $(3t^2 + 1)\cos 2t - 2(t^3 + t)\sin 2t$. (*iv*) $dw/dt = 1$; dw/dt at $t = 3$ is 1.

10. The length, width and height of a rectangular box are increasing at rates of 1 unit/sec, 2 unit/sec and 3 units/sec respectively.

(*i*) At what rate is the volume increasing when the length is 2 units, the width is 3 units and the height is 6 units?

(*ii*) At what rate is the length of the diagonal increasing at that instant?

(*iii*) At what rate is the surface area of the box increasing at the given instant?

Ans. (*i*) 60 cubic units/sec (*ii*) 26/7 unit per sec.

11. A point is moving on the curve of intersection of the surface $x^2 + xy + y^2 - z^2 = 0$ and the plane $x - y + 2 = 0$. When x is 3 and is increasing 2 units per second, find

(*i*) the rate at which y is changing,

(*ii*) the rate at which z is changing,

(*iii*) the speed with which the point is moving.

Ans. (*i*) 2 units per second; (*ii*) 24/7 units per sec; (*iii*) 4.44 units per second.

12. At a certain instant the radius of a right circular cylinder is 6 units and is increasing at the rate 0.2 unit/sec, while the altitude is 8 units and is decreasing at the rate of 0.4 unit/sec. Find the time rate of change (*i*) of the volume and (*ii*) of the surface at that instant.

Ans. (*i*) 4.8 π unit^3/sec, (*ii*) 3.2 π units^2/sec.

13. Find the total differential coefficient of $u = x^2y$ with respect to x, when x and y are connected by the relation $x^2 + xy + y^2 = 1$. (*DCE*, 2004) **Ans.** $\dfrac{du}{dx} = 2xy - \dfrac{x^2(2x+y)}{x+2y}$.

14. (*i*) If $u = f(r, s)$, $r = x + y$, $s = x - y$, show that $\dfrac{\partial u}{\partial x} + \dfrac{\partial u}{\partial y} = 2\dfrac{\partial u}{\partial r}$.

(*ii*) If $f(u) = \sin u$ and $u = \sqrt{x^2 + y^2}$, show that $\left(\dfrac{\partial f}{\partial x}\right)^2 + \left(\dfrac{\partial f}{\partial y}\right)^2 = \cos^2 u$. (*AMIE, S-2001*)

15. (*i*) Given $V = f(x, y, z)$ where $x = r\cos\theta$, $y = r\sin\theta$, $z = t$; obtain expressions for

$$\frac{\partial V}{\partial r}, \frac{\partial V}{\partial \theta}, \frac{\partial V}{\partial t} \text{ in terms of } \frac{\partial V}{\partial x}, \frac{\partial V}{\partial y}, \frac{\partial V}{\partial z}.$$

(*ii*) If $u = f(r, s, t)$, and $r = x/y$, $s = y/z$, $t = z/x$, then show that

$$x\frac{\partial u}{\partial x} + y\frac{\partial u}{\partial y} + z\frac{\partial u}{\partial z} = 0.$$

(*iii*) If $z = x^3 - xy + y^3$, and $x = r\cos\theta$, $y = r\sin\theta$, find $\partial z/\partial r$ and $\partial z/\partial \theta$.

Ans. $(3x^2 - y)\cos\theta + (3y^2 - x)\sin\theta$; $(3x^2 - y)(-r\sin\theta) + (3y^2 - x)r\cos\theta$.

[**Hint:** (*i*) $\dfrac{\partial x}{\partial r} = \cos\theta$, $\dfrac{\partial x}{\partial \theta} = -r\sin\theta$, $\dfrac{\partial x}{\partial t} = 0$, $\dfrac{\partial y}{\partial r} = \sin\theta$, $\dfrac{\partial y}{\partial \theta} = r\cos\theta$, $\dfrac{\partial y}{\partial t} = 0$

$$\frac{\partial z}{\partial r} = \frac{\partial z}{\partial \theta} = 0, \quad \frac{\partial z}{\partial t} = 1.$$

$$\frac{\partial V}{\partial r} = \frac{\partial V}{\partial x}\frac{\partial x}{\partial r} + \frac{\partial V}{\partial y}\frac{\partial y}{\partial r} + \frac{\partial V}{\partial z}\frac{\partial z}{\partial r} = \cos\theta\frac{\partial V}{\partial x} + \sin\theta\frac{\partial V}{\partial y}$$

$$\frac{\partial V}{\partial \theta} = \frac{\partial V}{\partial x}\frac{\partial x}{\partial \theta} + \frac{\partial V}{\partial y}\frac{\partial y}{\partial \theta} + \frac{\partial V}{\partial z}\frac{\partial z}{\partial \theta} = -r\sin\theta\frac{\partial V}{\partial x} + r\cos\theta\frac{\partial V}{\partial y}$$

$$\frac{\partial V}{\partial t} = \frac{\partial V}{\partial x}\frac{\partial x}{\partial t} + \frac{\partial V}{\partial y}\frac{\partial y}{\partial t} + \frac{\partial V}{\partial z}\cdot\frac{\partial z}{\partial t} = \frac{\partial V}{\partial z}].$$

16. If $u = f\left[\dfrac{y-x}{xy}, \dfrac{z-x}{xz}\right]$, show that $x^2\dfrac{\partial u}{\partial x} + y^2\dfrac{\partial u}{\partial y} + z^2\dfrac{\partial u}{\partial z} = 0$. (*UPTU 2005; DCE 2004*)

17. If $x^2 = au + bv$, $y^2 = au - bv$ and V is a function of x, y, then show that $x\dfrac{\partial V}{\partial x} + y\dfrac{\partial V}{\partial y} = 2\left(u\dfrac{\partial V}{\partial u} + v\dfrac{\partial V}{\partial v}\right)$.

18. If $\phi = f(x, y, z)$ and $x = \sqrt{vw}$, $y = \sqrt{wu}$, $z = \sqrt{uv}$, then show that

$$u\frac{\partial \phi}{\partial u} + v\frac{\partial \phi}{\partial v} + w\frac{\partial \phi}{\partial w} = x\frac{\partial \phi}{\partial x} + y\frac{\partial \phi}{\partial y} + z\frac{\partial \phi}{\partial z}.$$

19. If $u = \dfrac{x^2 - y^2}{x^2 + y^2}$ and $z = \sin^{-1}\sqrt{u}$, find dz, $\dfrac{\partial z}{\partial x}$ and $\dfrac{\partial z}{\partial y}$.

Ans. $\sqrt{2}\cdot x(ydx - xdy)/(x^2 + y^2)(\sqrt{(x^2 - y^2)})$; $\sqrt{2}\,xy/(x^2 + y^2)\sqrt{(x^2 - y^2)}$; $-\sqrt{2}\,x^2/(x^2 + y^2)\sqrt{(x^2 - y^2)}$.

20. If $x = u + v + w$, $y = vw + wu + uv$, $z = uvw$ and F is a function for x, y, z, then show that

$$u\frac{\partial F}{\partial u}+v\frac{\partial F}{\partial v}+w\frac{\partial F}{\partial w}=x\frac{\partial F}{\partial x}+2y\frac{\partial F}{\partial y}+3z\frac{\partial F}{\partial z}.$$

21. Given $z = xy^2 + x^3y$, where $x = r^2 - 2s$, $y = 2r - s^2$, find $\partial z/\partial r$ and $\partial z/\partial s$.

Ans. $2\{(y^2 + 3x^2y)r + (2xy + x^3)\}$; $-2\{(y^2 + 3x^2y) + s(2xy + x^3)\}$.

22. If $u = x^2 + y^2$, $x = s + 3t$ and $y = 2s - t$, obtain the values of $\partial^2u/\partial s^2$ and $\partial^2u/\partial t^2$ *(AMIE, W-2003)*

Ans. 10, 20.

23. Given $w = (x + y + z)^2$, $x = r - s$, $y = \cos(r + s)$, $z = \sin(r + s)$. Find $\partial w/\partial r$ at $r = 1$, $s = -1$. **Ans.** 12.

24. If $f(x, y) = f(u, v)$ and $u = x^2 - y^2$ and $v = 2xy$ then show that $\dfrac{\partial^2 f}{\partial x^2}+\dfrac{\partial^2 f}{\partial y^2}=4(x^2+y^2)\left(\dfrac{\partial^2\phi}{\partial u^2}+\dfrac{\partial^2\phi}{\partial v^2}\right)$.

25. If $z = f(x, y)$, $x = e^u \cos v$, $y = e^u \sin v$, show that $\dfrac{\partial^2 z}{\partial x^2}+\dfrac{\partial^2 z}{\partial y^2}=e^{-2u}\left(\dfrac{\partial^2 z}{\partial u^2}+\dfrac{\partial^2 z}{\partial v^2}\right)$.

26. If $u = f(x^2 + 2yz, y^2 + 2zx)$, prove that

$$(y^2-zx)\frac{\partial u}{\partial x}+(x^2-yz)\frac{\partial u}{\partial y}+(z^2-xy)\frac{\partial u}{\partial z}=0.$$

27. If $f(x + y + z, x^2 + y^2 + z^2) = 0$, show that $(y - z)p + (z - x)q = x - y$, where $p = \partial z/\partial x$ and $q = \partial z/\partial y$.

28. If $\phi\left(\dfrac{z}{x^3},\dfrac{y}{x}\right)=0$, show that $px + qy = 3z$.

29. Find p and q if $x=\sqrt{a}(\sin u+\cos v)$, $y=\sqrt{a}(\cos u-\sin v)$, $z = -1 + \sin(u - v)$, where p and q mean $\partial z/\partial x$ and $\partial z/\partial y$ respectively.

[**Hint:** $x^2/a = \sin^2 u + \cos^2 v + 2\sin u \cos v$, ...(*i*)

$y^2/a = \cos^2 u + \sin^2 v - 2\cos u \sin v$. ...(*ii*)

Adding (*i*) and (*ii*), we obtain

$(x^2/a) + (y^2/a) = 2 + 2\sin(u - v) = 2 + 2(1 + z)$ [$\because z = -1 + \sin(u - v)$]

or $z=\dfrac{1}{2a}(x^2+y^2)-2$. Therefore, $\dfrac{\partial z}{\partial x}=\dfrac{x}{a}$ and $\dfrac{\partial z}{\partial y}=\dfrac{y}{a}$].

30. Using implicit differentiation, obtain the following

(*i*) dy/dx, when $x^y + y^x = a^b$, a and b are constant,

(*ii*) dy/dx, when $x^3 + 3x^2y + 6xy^2 + y^3 = 1$,

(*iii*) dy/dx, when $(\cos x)^y = (\sin y)^x$,

(*iv*) dy/dx, when $\tan^{-1}(x/y) + y^3 + 1 = 0$; $x > 0$, $y > 0$. *(AMIE, W-2002)*

Ans. (*i*) $-\dfrac{(yx^{y-1}+y^x\log y)}{(xy^{x-1}+x^y\log x)}$, (*ii*) $-\dfrac{(x^2+2xy+2y^2)}{x^2+4xy+y^2}$,

(*iii*) $\dfrac{y\tan x+\log\sin y}{\log\cos x-x\cot y}$, (*iv*) $\dfrac{y}{(x-3x^2y^2-3y^4)}$.

31. If $f(x, y) = 0$ and $\phi(y, z) = 0$ then show that

$$\frac{\partial f}{\partial y}\cdot\frac{\partial\phi}{\partial z}\cdot\frac{dz}{dx}=\frac{\partial f}{\partial x}\cdot\frac{\partial\phi}{\partial y}$$ *(AMIE, W-2005; UPTU 2005)*

32. Find $\partial w/\partial v$, when $u = 0$, $v = 0$ if $w = (x^2 + y - 2)^4 + (x - y + 2)^3$, $x = u - 2v + 1$ and $y = 2u + v - 2$.

Ans. 99.

[**Hint:** Using the chain rule: $\dfrac{\partial w}{\partial v}=\dfrac{\partial w}{\partial x}\cdot\dfrac{\partial x}{\partial v}+\dfrac{\partial w}{\partial y}\cdot\dfrac{\partial y}{\partial v}$. When $u = v = 0$, then $x = 1, y = -2$. Find $\left(\dfrac{\partial w}{\partial v}\right)_{\substack{x=1\\y=-2}}$].

33. If $u = f(x, y)$, $x = r\cos\theta$, $y = r\sin\theta$, then show that

$$\left(\frac{\partial u}{\partial x}\right)^2 + \left(\frac{\partial u}{\partial y}\right)^2 = \left(\frac{\partial u}{\partial r}\right)^2 + \frac{1}{r^2}\left(\frac{\partial u}{\partial \theta}\right)^2.$$ *(AMIETE, June 2008; VTU, 2000)*

34. If z be a function of x and y, and u and v be two other variables such that $u = lx + my$, $v = ly - mx$,

show that $\dfrac{\partial^2 z}{\partial x^2} + \dfrac{\partial^2 z}{\partial y^2} = (l^2 + m^2)\left(\dfrac{\partial^2 z}{\partial u^2} + \dfrac{\partial^2 z}{\partial v^2}\right)$. *(PTU, 2004)*

[**Hint:** Using the chain rule, we obtain $\dfrac{\partial z}{\partial x} = \dfrac{\partial z}{\partial u}\cdot\dfrac{\partial u}{\partial x} + \dfrac{\partial z}{\partial v}\cdot\dfrac{\partial v}{\partial x} = l\dfrac{\partial z}{\partial u} - m\dfrac{\partial z}{\partial v}$

$\Rightarrow \quad \dfrac{\partial}{\partial x} = l\dfrac{\partial}{\partial u} - m\dfrac{\partial}{\partial v}$. Similarly, we obtain $\dfrac{\partial}{\partial y} = m\dfrac{\partial}{\partial u} + l\dfrac{\partial}{\partial v}$.

$$\therefore \quad \frac{\partial^2 z}{\partial x^2} = \frac{\partial}{\partial x}\left(\frac{\partial z}{\partial x}\right) = \left(l\frac{\partial}{\partial u} - m\frac{\partial}{\partial v}\right)\left(l\frac{\partial z}{\partial u} - m\frac{\partial z}{\partial v}\right)$$

$$= l^2\frac{\partial^2 z}{\partial u^2} - 2lm\frac{\partial^2 z}{\partial u \partial v} + m^2\frac{\partial^2 z}{\partial v^2}. \qquad ...(i)$$

Similarly, we obtain

$$\frac{\partial^2 z}{\partial y^2} = m^2\frac{\partial^2 z}{\partial u^2} + 2lm\frac{\partial^2 z}{\partial u \partial v} + l^2\frac{\partial^2 z}{\partial v^2}. \qquad ...(ii)$$

Adding (*i*) and (*ii*), we get the required result.]

35. (*i*) If $x = u + v$, $y = uv$ and z be any function of x and y, show that $u\dfrac{\partial z}{\partial u} + v\dfrac{\partial z}{\partial v} = x\dfrac{\partial z}{\partial x} + 2y\dfrac{\partial z}{\partial y}$.

(*ii*) If $z = uv$ and $u^2 + v^2 - x - y = 0$, $u^2 - v^2 + 3x + y = 0$. Find $\partial z/\partial x$.

[**Hint:** (*ii*) Solving the last two equations for u^2 and v^2, we get $u^2 = -x$, $v^2 = 2x + y$.

Now $\quad \dfrac{\partial z}{\partial x} = u\dfrac{\partial v}{\partial x} + v\dfrac{\partial u}{\partial x} = u\left(\dfrac{1}{v}\right) + v\left(-\dfrac{1}{2u}\right) = \dfrac{2u^2 - v^2}{2uv} \quad (\because \partial u/\partial x = -1/2u \text{ and } \partial v/\partial x = 1/v)$].

36. If u and v are defined as functions of x and y by the equations $u^2 - v = 3x + y$ and $u - 2v^2 = x - 2y$, find

$$\frac{\partial u}{\partial x}, \frac{\partial v}{\partial x}, \frac{\partial u}{\partial y} \text{ and } \frac{\partial v}{\partial y}.$$

Ans. $\dfrac{1-12v}{1-8uv}, \dfrac{2u-3}{1-8uv}, -\dfrac{(2+4v)}{1-8uv}, -\dfrac{1+4u}{1-8uv}$ assumed that $1 - 8uv \neq 0$.

37. Suppose that u and v are two functions of x and y and satisfy the relations $u^2 - v^2 = x$, $2uv = y$.

Then compute $\dfrac{\partial u}{\partial x}, \dfrac{\partial u}{\partial y}, \dfrac{\partial v}{\partial x}$ and $\dfrac{\partial v}{\partial y}$, assuming that they exist. *(AMIE.; S-2008)*

[**Hint:** $\dfrac{2u\partial u}{\partial x} - 2v\dfrac{\partial v}{\partial x} = 1$, $2v\dfrac{\partial u}{\partial x} + 2u\dfrac{\partial v}{\partial x} = 0$. Solving simultaneously for $\partial u/\partial x$ and $\partial v/\partial x$. We have

$$\frac{\partial u}{\partial x} = \frac{\begin{vmatrix} 1 & -2v \\ 0 & 2u \end{vmatrix}}{\begin{vmatrix} 2u & -2v \\ 2v & 2u \end{vmatrix}} = \frac{u}{2(u^2+v^2)}, \quad \frac{\partial v}{\partial x} = \frac{\begin{vmatrix} 2u & 1 \\ 2u & 0 \end{vmatrix}}{\begin{vmatrix} 2u & -2v \\ 2v & 2u \end{vmatrix}} = -\frac{v}{2(u^2+v^2)}.$$]

Similarly, the other partial derivatives are obtained.

Ans. $\dfrac{u}{2(u^2+v^2)}, \dfrac{v}{2(u^2+v^2)}, -\dfrac{v}{2(u^2+v^2)}, \dfrac{u}{2(u^2+v^2)}$.

38. If $z^3 - xz - y = 0$, prove that $\dfrac{\partial^2 z}{\partial x \partial y} = -\dfrac{x+3z^2}{(3z^2-x)^3}$.

39. If $u = f(x, y)$ and $x = \xi \cos\alpha - \eta \sin\alpha$, $y = \xi \sin a + \eta \cos\alpha$, where α is a constant, show that

$$\frac{\partial^2 u}{\partial x^2} + \frac{\partial^2 u}{\partial y^2} = \frac{\partial^2 u}{\partial \xi^2} + \frac{\partial^2 u}{\partial \eta^2}.$$ (*AMIE, S-2000*)

[**Hint:** Using the chain rule, we obtain

$$\frac{\partial u}{\partial \xi} = \frac{\partial u}{\partial x}\cdot\frac{\partial x}{\partial \xi} + \frac{\partial u}{\partial y}\cdot\frac{\partial y}{\partial \xi} = \frac{\partial u}{\partial x}\cos\alpha + \frac{\partial u}{\partial y}\sin\alpha = \left(\cos\alpha\frac{\partial}{\partial x} + \sin\alpha\frac{\partial}{\partial y}\right)u$$

$$\Rightarrow \quad \frac{\partial}{\partial \xi} = \left(\cos\alpha\frac{\partial}{\partial x} + \sin\alpha\frac{\partial}{\partial y}\right).$$

$$\therefore \quad \frac{\partial^2 u}{\partial \xi^2} = \left(\cos\alpha\frac{\partial}{\partial x} + \sin\alpha\frac{\partial}{\partial y}\right)^2 u = \cos^2\alpha\frac{\partial^2 u}{\partial x^2} + 2\sin\alpha\cos\alpha\frac{\partial^2 u}{\partial x \partial y} + \sin^2\alpha\frac{\partial^2 u}{\partial y^2} \quad \text{...}(i)$$

Similarly, we obtain $\dfrac{\partial^2 u}{\partial \eta^2} = \sin^2\alpha\dfrac{\partial^2 u}{\partial x^2} - 2\sin\alpha\cos\alpha\dfrac{\partial^2 u}{\partial x \partial y} + \cos^2\alpha\dfrac{\partial^2 u}{\partial y^2}$. ...(*ii*)

Adding (*i*) and (*ii*), we obtain $\dfrac{\partial^2 u}{\partial \xi^2} + \dfrac{\partial^2 u}{\partial \eta^2} = \dfrac{\partial^2 u}{\partial x^2} + \dfrac{\partial^2 u}{\partial y^2}$.]

40. If $x + y = (u + v)^2$, $x - y = (u - v)^2$, show that $4(x^2 - y^2)\left(\dfrac{\partial^2 f}{\partial x^2} - \dfrac{\partial^2 f}{\partial y^2}\right) = (u^2 - v^2)\left(\dfrac{\partial^2 f}{\partial u^2} - \dfrac{\partial^2 f}{\partial^2 v}\right)$.

41. If $x = r\cos\theta$ and $y = r\sin\theta$, then for any function $u = f(x, y)$, prove that

$$\frac{\partial u}{\partial x} = \cos\theta\frac{\partial u}{\partial r} - \frac{\sin\theta}{r}\frac{\partial u}{\partial \theta}.$$ (*AMIE, S-2002*)

[**Hint:** We have $r = \sqrt{x^2 + y^2}$ and $\theta = \tan^{-1}(y/x)$. Using the chain rule,

$$\frac{\partial u}{\partial x} = \frac{\partial u}{\partial r}\cdot\frac{\partial r}{\partial x} + \frac{\partial u}{\partial \theta}\cdot\frac{\partial \theta}{\partial x} = \frac{x}{\sqrt{x^2+y^2}}\frac{\partial u}{\partial r} + \left(\frac{-y}{x^2+y^2}\right)\frac{\partial u}{\partial \theta} = \cos\theta\frac{\partial u}{\partial r} - \frac{\sin\theta}{r}\frac{\partial u}{\partial \theta}].$$

42. If $u = x + ay$ and $v = x + by$, transform the equation $2\dfrac{\partial^2 z}{\partial x^2} - 5\dfrac{\partial^2 z}{\partial x \partial y} + 3\dfrac{\partial^2 z}{\partial y^2} = 0$ into $\dfrac{\partial^2 z}{\partial u \partial v} = 0$, and find the possible values of a and b. (*AMIE, W-2004*)

[**Hint:** We have $\partial u/\partial x = 1$, $\partial u/\partial y = a$, $\partial v/\partial x = 1$ and $\partial v/\partial y = b$. Using the chain rule, we obtain

$$\frac{\partial z}{\partial x} = \frac{\partial z}{\partial u}\cdot\frac{\partial u}{\partial x} + \frac{\partial z}{\partial v}\cdot\frac{\partial v}{\partial x} = \frac{\partial z}{\partial u} + \frac{\partial z}{\partial v} \Rightarrow \frac{\partial}{\partial x} = \frac{\partial}{\partial u} + \frac{\partial}{\partial v}.$$

Similarly, we obtain $\dfrac{\partial}{\partial y} = a\dfrac{\partial}{\partial u} + b\dfrac{\partial}{\partial v}$,

$$\therefore \quad \frac{\partial^2 z}{\partial x^2} = \frac{\partial}{\partial x}\left(\frac{\partial z}{\partial x}\right) = \left(\frac{\partial}{\partial u} + \frac{\partial}{\partial v}\right)\left(\frac{\partial z}{\partial u} + \frac{\partial z}{\partial v}\right) = \frac{\partial^2 z}{\partial u^2} + 2\frac{\partial^2 z}{\partial u \partial v} + \frac{\partial^2 z}{\partial v^2},$$

$$\frac{\partial^2 z}{\partial y^2} = a^2\frac{\partial^2 z}{\partial u^2} + 2ab\frac{\partial^2 z}{\partial u \partial v} + b^2\frac{\partial^2 z}{\partial v^2},$$

and $$\frac{\partial^2 z}{\partial x \partial y} = \frac{\partial}{\partial x}\left(\frac{\partial z}{\partial y}\right) = a\frac{\partial^2 z}{\partial u^2} + (a+b)\frac{\partial^2 z}{\partial u \partial v} + b\frac{\partial^2 z}{\partial v^2}.$$

Thus $$2\frac{\partial^2 z}{\partial x^2} - 5\frac{\partial^2 z}{\partial x \partial y} + 3\frac{\partial^2 z}{\partial y^2} = \frac{\partial^2 z}{\partial u^2}(2 - 5a + 3a^2) + \frac{\partial^2 z}{\partial u \partial v}(4 - 5a - 5b + 6ab) + \frac{\partial^2 z}{\partial v^2}(2 - 5b + 3b^2).$$

It is given that left hand expression be zero, and only term $\partial^2 z/\partial u \partial v$ on right hand expression should exist so coefficients of $\partial^2 z/\partial u^2$ and $\partial^2 z/\partial v^2$ must be equal to 0.

Therefore $2 - 5a + 3a^2 = 0$ and $2 - 5b + 3b^2 = 0$.

On solving, we get the solution sets (a, b) either $(1, 2/3)$ or $(2/3, 1)$.]

43. (*a*) If $z = f(x, y)$, where $x = e^u \cos v$ and $y = e^u \sin v$, show that $y\frac{\partial z}{\partial u} + x\frac{\partial z}{\partial v} = e^{2u}\frac{\partial z}{\partial y}$.

(*b*) If $u = u(x, y)$ and $x = e^r \cos\theta$, $y = e^r \sin\theta$, show that $\left(\frac{\partial u}{\partial x}\right)^2 + \left(\frac{\partial u}{\partial y}\right)^2 = e^{-2r}\left[\left(\frac{\partial u}{\partial r}\right)^2 + \left(\frac{\partial u}{\partial \theta}\right)^2\right]$.

44. If $z = f(u, v)$ where $u = x^2 - y^2$ and $v = 2xy$, with $(x, y) \neq (0, 0)$. Show that the differential equation $x\frac{\partial z}{\partial x} - y\frac{\partial z}{\partial y} = 0$ is equivalent to $\partial z/\partial u = 0$. (*AMIE, S-2007*)

1.16 JACOBIANS

Def. If $u = f(x, y)$ and $v = \phi(x, y)$ be two continuous functions of the independent variables x and y such that $\partial u/\partial x, \partial u/\partial y, \partial v/\partial x, \partial v/\partial y$ are also continuous in x and y, then the *Jacobian* of u and v with respect to x and y is defined by the determinant $\begin{vmatrix} \partial u/\partial x & \partial u/\partial y \\ \partial v/\partial x & \partial v/\partial y \end{vmatrix}$.

The above determinant is often denoted by the symbol $\frac{\partial(u,v)}{\partial(x, y)}$ or $J\left(\frac{u, v}{x, y}\right)$.

Similarly if u, v, w are functions of three independent variables x, y, z then the determinant

$$\begin{vmatrix} \partial u/\partial x & \partial u/\partial y & \partial u/\partial z \\ \partial v/\partial x & \partial v/\partial y & \partial v/\partial z \\ \partial w/\partial x & \partial w/\partial y & \partial w/\partial z \end{vmatrix}$$

is called the Jacobian of u, v, w with respect to x, y, z. This determinant is denoted by the symbol

$$\frac{\partial(u, v, w)}{\partial(x, y, z)} \text{ or } J\left(\frac{u, v, w}{x, y, z}\right).$$

In general if $x_1, x_2, ..., x_n$ be n functions of the n variables $y_1, y_2 ..., y_n$ then

$$\frac{\partial(x_1, x_2, ..., x_n)}{\partial(y_1, y_2, ..., y_n)} = \begin{vmatrix} \partial x_1/\partial y_1 & \partial x_1/\partial y_2 & \cdots & \partial x_1/\partial y_n \\ \partial x_2/\partial y_1 & \partial x_2/\partial y_2 & \cdots & \partial x_2/\partial y_n \\ \vdots & \vdots & & \vdots \\ \partial x_n/\partial y_1 & \partial x_n/\partial y_2 & \cdots & \cdots\partial x_n/\partial y_n \end{vmatrix}$$

is the Jacobian of $x_1, x_2, ..., x_n$ with respect to $y_1, y_2, ..., y_n$.

An important application of Jacobians is in connection with the change of variables in multiple integrals.

1.16.1 Properties of Jacobians

1. *If J_1 is the Jacobian of u, v with respect to x, y and J_2 the Jacobian of x, y with respect to u, v then* $J_1 J_2 = 1$. (*AMIETE, June-2005; AMIE, W-2003*)

Proof.

$$J_1J_2 = \begin{vmatrix} \partial u/\partial x & \partial u/\partial y \\ \partial v/\partial x & \partial v/\partial y \end{vmatrix} \times \begin{vmatrix} \partial x/\partial u & \partial x/\partial v \\ \partial y/\partial u & \partial y/\partial v \end{vmatrix} = \begin{vmatrix} \partial u/\partial x & \partial u/\partial y \\ \partial v/\partial x & \partial v/\partial y \end{vmatrix} \times \begin{vmatrix} \partial x/\partial u & \partial y/\partial u \\ \partial x/\partial v & \partial y/\partial v \end{vmatrix}$$

(interchanging rows and columns of the second determinant)

$$= \begin{vmatrix} \dfrac{\partial u}{\partial x}\cdot\dfrac{\partial x}{\partial u}+\dfrac{\partial u}{\partial y}\cdot\dfrac{\partial y}{\partial u} & \dfrac{\partial u}{\partial x}\cdot\dfrac{\partial x}{\partial v}+\dfrac{\partial u}{\partial y}\cdot\dfrac{\partial y}{\partial v} \\ \dfrac{\partial v}{\partial x}\cdot\dfrac{\partial x}{\partial u}+\dfrac{\partial v}{\partial y}\cdot\dfrac{\partial y}{\partial u} & \dfrac{\partial v}{\partial x}\cdot\dfrac{\partial x}{\partial v}+\dfrac{\partial v}{\partial y}\cdot\dfrac{\partial y}{\partial v} \end{vmatrix}. \qquad ...(1.35)$$

Let $u = f(x, y)$ and $v = g(x, y)$. Suppose, on solving for x and y, we get $x = \phi(u, v)$ and $y = \psi(u, v)$.

Then

$$\left.\begin{aligned} \frac{\partial u}{\partial u} &= 1 = \frac{\partial u}{\partial x}\cdot\frac{\partial x}{\partial u}+\frac{\partial u}{\partial y}\cdot\frac{\partial y}{\partial u} \qquad & \frac{\partial u}{\partial v} &= 0 = \frac{\partial u}{\partial x}\cdot\frac{\partial x}{\partial v}+\frac{\partial u}{\partial y}\cdot\frac{\partial y}{\partial v} \\ \frac{\partial v}{\partial u} &= 0 = \frac{\partial v}{\partial x}\cdot\frac{\partial x}{\partial u}+\frac{\partial v}{\partial y}\cdot\frac{\partial y}{\partial u} \qquad & \frac{\partial v}{\partial v} &= 1 = \frac{\partial v}{\partial x}\cdot\frac{\partial x}{\partial v}+\frac{\partial v}{\partial y}\cdot\frac{\partial y}{\partial v} \end{aligned}\right\}. \qquad ...(1.36)$$

With the help of Eq. (1.36) we can write Eq. (1.35) in the form

$$\begin{vmatrix} 1 & 0 \\ 0 & 1 \end{vmatrix} = 1. \qquad \therefore \qquad J_1J_2 = 1, \text{ that is, } \frac{\partial(u,v)}{\partial(x,y)}\cdot\frac{\partial(x,y)}{\partial(u,v)} = 1.$$

Note. The result can be easily extended.

Thus $$\frac{\partial(u, v, w, ...)}{\partial(x, y, z, ...)}\cdot\frac{\partial(x, y, z, ...)}{\partial(u, v, w, ...)} = 1.$$

2. *If u, v are functions of r, s where r, s are functions of x, y, then*

$$\frac{\partial(u, v)}{\partial(x, y)} = \frac{\partial(u, v)}{\partial(r, s)} \times \frac{\partial(r, s)}{\partial(x, y)}.$$

Proof.

$$\textbf{R.H.S.} = \frac{\partial(u,v)}{\partial(r,s)} \times \frac{\partial(r,s)}{\partial(x,y)} = \begin{vmatrix} \partial u/\partial r & \partial u/\partial s \\ \partial v/\partial r & \partial v/\partial s \end{vmatrix} \times \begin{vmatrix} \partial r/\partial x & \partial r/\partial y \\ \partial s/\partial x & \partial s/\partial y \end{vmatrix} = \begin{vmatrix} \partial u/\partial r & \partial u/\partial s \\ \partial v/\partial r & \partial v/\partial s \end{vmatrix} \times \begin{vmatrix} \partial r/\partial x & \partial s/\partial x \\ \partial r/\partial y & \partial s/\partial y \end{vmatrix}$$

(interchanging rows and columns in second determinant)

$$= \begin{vmatrix} \dfrac{\partial u}{\partial r}\cdot\dfrac{\partial r}{\partial x}+\dfrac{\partial u}{\partial s}\cdot\dfrac{\partial s}{\partial x} & \dfrac{\partial u}{\partial r}\cdot\dfrac{\partial r}{\partial y}+\dfrac{\partial u}{\partial s}\cdot\dfrac{\partial s}{\partial y} \\ \dfrac{\partial v}{\partial r}\cdot\dfrac{\partial r}{\partial x}+\dfrac{\partial v}{\partial s}\cdot\dfrac{\partial s}{\partial x} & \dfrac{\partial v}{\partial r}\cdot\dfrac{\partial r}{\partial y}+\dfrac{\partial v}{\partial s}\cdot\dfrac{\partial s}{\partial y} \end{vmatrix} = \begin{vmatrix} \partial u/\partial x & \partial u/\partial y \\ \partial v/\partial x & \partial v/\partial y \end{vmatrix} = \frac{\partial(u,v)}{\partial(x,y)} = \text{L.H.S.}$$

3. *If functions u, v, w of three independent variables x, y, z are not independent, then the jacobian of u, v, w with respect to x, y, z vanishes, i.e.*

$$\frac{\partial(u,v,w)}{\partial(x,y,z)} = 0.$$

Proof.

In case u, v, w are not independent, then there must be a relation $f(u, v, w) = 0$ connecting them. Differentiating $f(u, v, w) = 0$ with respect to x, y, z respectively, we get

$$\frac{\partial f}{\partial u}\frac{\partial u}{\partial x}+\frac{\partial f}{\partial v}\frac{\partial v}{\partial x}+\frac{\partial f}{\partial w}\frac{\partial w}{\partial x} = 0,$$

$$\frac{\partial f}{\partial u}\frac{\partial u}{\partial y}+\frac{\partial f}{\partial v}\frac{\partial v}{\partial y}+\frac{\partial f}{\partial w}\frac{\partial w}{\partial y} = 0,$$

$$\frac{\partial f}{\partial u}\frac{\partial u}{\partial z}+\frac{\partial f}{\partial v}\frac{\partial v}{\partial z}+\frac{\partial f}{\partial w}\frac{\partial w}{\partial z} = 0.$$

Eliminating $\partial f/\partial u$, $\partial f/\partial v$, $\partial f/\partial w$ from the above equations, we have

$$\begin{vmatrix} \partial u/\partial x & \partial v/\partial x & \partial w/\partial x \\ \partial u/\partial y & \partial v/\partial y & \partial w/\partial y \\ \partial u/\partial z & \partial v/\partial z & \partial w/\partial z \end{vmatrix} = 0 \text{ or } \frac{\partial(u,v,w)}{\partial(x,y,z)} = 0.$$

ILLUSTRATIVE EXAMPLES

EXAMPLE 1.66 If $x = r\cos\theta$, $y = r\sin\theta$, find the Jacobians $J = \dfrac{\partial(x,y)}{\partial(r,\theta)}$ and $J' = \dfrac{\partial(r,\theta)}{\partial(x,y)}$. Show that $JJ' = 1$.

(AMIETE, S-2008; AMIE, W-2002; Andhra 2000)

Solution We have $x = r\cos\theta$, $y = r\sin\theta$.

$\partial x/\partial r = \cos\theta$, $\partial x/\partial\theta = -r\sin\theta$, $\partial y/\partial r = \sin\theta$, $\partial y/\partial\theta = r\cos\theta$.

$$\therefore \quad J = \frac{\partial(x,y)}{\partial(r,\theta)} = \begin{vmatrix} \partial x/\partial r & \partial x/\partial\theta \\ \partial y/\partial r & \partial y/\partial\theta \end{vmatrix} = \begin{vmatrix} \cos\theta & -r\sin\theta \\ \sin\theta & r\cos\theta \end{vmatrix} = r(\cos^2\theta + \sin^2\theta) = r.$$

Now $r^2 = x^2 + y^2$, $\theta = \tan{-1}\,(y/x)$.

$$\frac{\partial r}{\partial x} = \frac{x}{r}, \quad \frac{\partial r}{\partial y} = \frac{y}{r}, \quad \frac{\partial\theta}{\partial x} = \frac{-y}{x^2+y^2} = \frac{-y}{r^2}, \quad \frac{\partial\theta}{\partial y} = \frac{x}{x^2+y^2} = \frac{x}{r^2}.$$

$$\therefore \quad J' = \frac{\partial(r,\theta)}{\partial(x,y)} = \begin{vmatrix} \partial r/\partial x & \partial r/\partial y \\ \partial\theta/\partial x & \partial\theta/\partial y \end{vmatrix} = \begin{vmatrix} x/r & y/r \\ -y/r^2 & x/r^2 \end{vmatrix} = \frac{x^2}{r^3} + \frac{y^2}{r^3} = \frac{r^2}{r^3} = \frac{1}{r}.$$

Hence $JJ' = r(1/r) = 1$.

EXAMPLE 1.67 If $u = x^2 - 2y$, $v = x + y$, prove that $\dfrac{\partial(u,v)}{\partial(x,y)} = 2x + 2$.

Solution We have $\partial u/\partial x = 2x$, $\partial u/\partial y = -2$, $\partial v/\partial x = 1$, $\partial v/\partial y = 1$,

$$\therefore \quad \frac{\partial(u,v)}{\partial(x,y)} = \begin{vmatrix} \dfrac{\partial u}{\partial x} & \dfrac{\partial u}{\partial y} \\ \dfrac{\partial v}{\partial x} & \dfrac{\partial v}{\partial y} \end{vmatrix} = \begin{vmatrix} 2x & -2 \\ 1 & 1 \end{vmatrix} = 2x + 2.$$

EXAMPLE 1.68 If in cylindrical co-ordinates $x = r\cos\theta$, $y = r\sin\theta$, $z = z$, find $\partial(x, y, z)/\partial(r, \theta, z)$.

Solution We have $\partial x/\partial r = \cos\theta$, $\quad \partial x/\partial\theta = -r\sin\theta$, $\quad \partial x/\partial z = 0$;

$$\frac{\partial y}{\partial r} = \sin\theta, \quad \frac{\partial y}{\partial\theta} = r\cos\theta, \quad \frac{\partial y}{\partial z} = 0; \quad \frac{\partial z}{\partial r} = 0, \quad \frac{\partial z}{\partial\theta} = 0, \quad \frac{\partial z}{\partial z} = 1$$

$$\therefore \quad \frac{\partial(x,y,z)}{\partial(r,\theta,z)} = \begin{vmatrix} \cos\theta & -r\sin\theta & 0 \\ \sin\theta & r\cos\theta & 0 \\ 0 & 0 & 1 \end{vmatrix} = 1(r\cos^2\theta + r\sin^2\theta) = r.$$

EXAMPLE 1.69 If $F = xu + v - y$, $G = u^2 + vy + w$, $H = zu - v + vw$. Compute $\dfrac{\partial(F,G,H)}{\partial(u,w,v)}$.

Solution

$$\frac{\partial(F,G,H)}{\partial(u,w,v)} = \begin{vmatrix} \partial F/\partial u & \partial F/\partial w & \partial F/\partial v \\ \partial G/\partial u & \partial G/\partial w & \partial G/\partial v \\ \partial H/\partial u & \partial H/\partial w & \partial H/\partial v \end{vmatrix} = \begin{vmatrix} x & 0 & 1 \\ 2u & 1 & y \\ z & v & -1+w \end{vmatrix}$$

$$= x\,(w - 1 - vy) + 1(2uv - z) = xw - x - xyv + 2uv - z.$$

EXAMPLE 1.70 In spherical polar coordinates, $x = \rho \sin\phi \cos\theta$, $y = \rho \sin\phi \sin\theta$, and $z = \rho \cos\phi$, show that

$$J\left(\frac{x, y, z}{\rho, \phi, \theta}\right) = \rho^2 \sin\phi. \qquad \textit{(AMIE, S-2004)}$$

Solution We have $\dfrac{\partial x}{\partial \rho} = \sin\phi\cos\theta,\ \dfrac{\partial x}{\partial \phi} = \rho\cos\phi\cos\theta,\ \dfrac{\partial x}{\partial \theta} = -\rho\sin\phi\sin\theta,$

$$\frac{\partial y}{\partial \rho} = \sin\phi\sin\theta,\ \frac{\partial y}{\partial \phi} = \rho\cos\phi\sin\theta,\ \frac{\partial y}{\partial \theta} = \rho\sin\phi\cos\theta,$$

$$\frac{\partial z}{\partial \rho} = \cos\phi,\ \frac{\partial z}{\partial \phi} = -\rho\sin\phi,\ \frac{\partial z}{\partial \theta} = 0.$$

$$\therefore \quad J\left(\frac{x, y, z}{\rho, \phi, \theta}\right) = \frac{\partial(x, y, z)}{\partial(\rho, \phi, \theta)} = \begin{vmatrix} \partial x/\partial\rho & \partial x/\partial\phi & \partial x/\partial\theta \\ \partial y/\partial\rho & \partial y/\partial\phi & \partial y/\partial\theta \\ \partial z/\partial\rho & \partial z/\partial\phi & \partial z/\partial\theta \end{vmatrix} = \begin{vmatrix} \sin\phi\cos\theta & \rho\cos\phi\cos\theta & -\rho\sin\phi\sin\theta \\ \sin\phi\sin\theta & \rho\cos\phi\sin\theta & \rho\sin\phi\cos\theta \\ \cos\phi & -\rho\sin\phi & 0 \end{vmatrix}$$

$$= \frac{\rho^2}{\cos\theta}\begin{vmatrix} \sin\phi\cos^2\theta & \cos\phi\cos^2\theta & -\sin\phi\sin\theta\cos\theta \\ \sin\phi\sin\theta & \cos\phi\sin\theta & \sin\phi\cos\theta \\ \cos\phi & -\sin\phi & 0 \end{vmatrix}_{(R_1 \to R_1 + R_2 \sin\theta)}$$

$$= \frac{\rho^2}{\cos\theta}\begin{vmatrix} \sin\phi & \cos\phi & 0 \\ \sin\phi\sin\theta & \cos\phi\sin\theta & \sin\phi\cos\theta \\ \cos\phi & -\sin\phi & 0 \end{vmatrix}$$

$$= \frac{\rho^2}{\cos\theta}(-\sin\phi\cos\theta)(-\sin^2\phi - \cos^2\phi) = \rho^2\sin\phi.$$

EXAMPLE 1.71 If $x = a \cosh\xi \cos\eta$, $y = a \sinh\xi \sin\eta$, show that

$$\frac{\partial(x, y)}{\partial(\xi, \eta)} = \frac{a^2}{2}(\cosh 2\xi - \cos 2\eta). \qquad \textit{(SVTU, 2007)}$$

Solution $\dfrac{\partial(x, y)}{\partial(\xi, \eta)} = \begin{vmatrix} \partial x/\partial\xi & \partial x/\partial\eta \\ \partial y/\partial\xi & \partial y/\partial\eta \end{vmatrix} = \begin{vmatrix} a\sinh\xi\cos\eta & -a\cosh\xi\sin\eta \\ a\cosh\xi\sin\eta & a\sinh\xi\cos\eta \end{vmatrix}$

$$= a^2 [\sinh^2\xi \cos^2\eta + \cosh^2\xi \sin^2\eta]$$

$$= a^2[(\cosh^2\xi - 1)\cos^2\eta + \cosh^2\xi(1 - \cos^2\eta)] = a^2[\cosh^2\xi - \cos^2\eta]$$

$$= a^2\left[\frac{1 + \cosh 2\xi}{2} - \frac{1 + \cos 2\eta}{2}\right] = \frac{a^2}{2}(\cosh 2\xi - \cos 2\eta).$$

EXAMPLE 1.72 If $y_1 = x_2 x_3/x_1$, $y_2 = x_3 x_1/x_2$, $y_3 = x_1 x_2/x_3$. Show that the Jacobian of y_1, y_2, y_3 with respect to x_1, x_2, x_3 is 4. *(UPTU 2006; AMIE, W-2001)*

Solution $$\frac{\partial(y_1, y_2, y_3)}{\partial(x_1, x_2, x_3)} = \begin{vmatrix} \partial y_1/\partial x_1 & \partial y_1/\partial x_2 & \partial y_1/\partial x_3 \\ \partial y_2/\partial x_1 & \partial y_2/\partial x_2 & \partial y_2/\partial x_3 \\ \partial y_3/\partial x_1 & \partial y_3/\partial x_2 & \partial y_3/\partial x_3 \end{vmatrix} = \begin{vmatrix} \dfrac{-x_2x_3}{x_1^2} & \dfrac{x_3}{x_1} & \dfrac{x_2}{x_1} \\ \dfrac{x_3}{x_2} & \dfrac{-x_3x_1}{x_2^2} & \dfrac{x_1}{x_2} \\ \dfrac{x_2}{x_3} & \dfrac{x_1}{x_3} & \dfrac{-x_1x_2}{x_3^2} \end{vmatrix}$$

$$= \frac{1}{x_1^2 x_2^2 x_3^2}\begin{vmatrix} -x_2x_3 & x_3x_1 & x_2x_1 \\ x_3x_2 & -x_3x_1 & x_1x_2 \\ x_2x_3 & x_3x_1 & -x_1x_2 \end{vmatrix} = \frac{x_1^2x_2^2x_3^2}{x_1^2 x_2^2 x_3^2}\begin{vmatrix} -1 & 1 & 1 \\ 1 & -1 & 1 \\ 1 & 1 & -1 \end{vmatrix} \begin{matrix} \\ (R_2 \to R_2 + R_1) \\ (R_3 \to R_3 + R_1) \end{matrix}$$

$$= \begin{vmatrix} -1 & 1 & 1 \\ 0 & 0 & 2 \\ 0 & 2 & 0 \end{vmatrix} = -1(0-4) = 4$$

EXAMPLE 1.73 If $u = x^2 - y^2$, $v = 2xy$, $x = r\cos\theta$ and $y = r\sin\theta$, evaluate $\dfrac{\partial(u,v)}{\partial(r,\theta)}$.

(VTU 2007; Madras 2006)

Solution We know that

$$\frac{\partial(u,v)}{\partial(r,\theta)} = \frac{\partial(u,v)}{\partial(x,y)}\cdot\frac{\partial(x,y)}{\partial(r,\theta)} \qquad \text{[Art. 1.16.1, Property 2]}$$

$$= \begin{vmatrix} \dfrac{\partial u}{\partial x} & \dfrac{\partial u}{\partial y} \\ \dfrac{\partial v}{\partial x} & \dfrac{\partial v}{\partial y} \end{vmatrix}\begin{vmatrix} \dfrac{\partial x}{\partial r} & \dfrac{\partial x}{\partial \theta} \\ \dfrac{\partial y}{\partial r} & \dfrac{\partial y}{\partial \theta} \end{vmatrix} = \begin{vmatrix} 2x & -2y \\ 2y & 2x \end{vmatrix}\begin{vmatrix} \cos\theta & -r\sin\theta \\ \sin\theta & r\cos\theta \end{vmatrix}$$

$$= 4(x^2 + y^2).\, r = 4r\,(r^2\cos^2\theta + r^2\sin^2\theta) = 4r^3.$$

EXAMPLE 1.74 If $u = xyz$, $v = xy + yz + zx$ and, $w = x + y + z$. Compute $\dfrac{\partial(u,v,w)}{\partial(x,y,z)}$.

Solution

$$J\left(\frac{u,v,w}{x,y,z}\right) = \begin{vmatrix} \partial u/\partial x & \partial u/\partial y & \partial u/\partial z \\ \partial v/\partial x & \partial v/\partial y & \partial v/\partial z \\ \partial w/\partial x & \partial w/\partial y & \partial w/\partial z \end{vmatrix} = \begin{vmatrix} yz & xz & xy \\ y+z & x+z & y+x \\ 1 & 1 & 1 \end{vmatrix} \begin{matrix} C_1 \to C_1 - C_3 \\ C_2 \to C_2 - C_3 \end{matrix}$$

$$= \begin{vmatrix} yz-xy & xz-xy & xy \\ z-x & z-y & x+y \\ 0 & 0 & 1 \end{vmatrix} = \begin{vmatrix} y(z-x) & x(z-y) & xy \\ z-x & z-y & x+y \\ 0 & 0 & 1 \end{vmatrix} = (z-x)(z-y)\begin{vmatrix} y & x & xy \\ 1 & 1 & x+y \\ 0 & 0 & 1 \end{vmatrix}$$

$$= (z-x)(z-y)(y-x) = (x-y)(y-z)(z-x).$$

EXAMPLE 1.75 (*a*) For the transformation $x = e^u\cos v$, $y = e^u\sin v$, prove that $\dfrac{\partial(x,y)}{\partial(u,v)}\cdot\dfrac{\partial(u,v)}{\partial(x,y)} = 1$.

(VTU 2005)

(*b*) For the transformations $x = a(u+v)$, $y = b(u-v)$ and $u = r^2\cos 2\theta$, $v = r^2\sin 2\theta$, find $\dfrac{\partial(x,y)}{\partial(r,\theta)}$.

Solution (*a*) The inverse transformation for $x = e^u\cos v$, $y = e^u\sin v$,

are $$u = \frac{1}{2}\log(x^2+y^2),\ v = \tan^{-1}\frac{y}{x}.$$

$$\therefore \quad \frac{\partial(x,y)}{\partial(u,v)} = \begin{vmatrix} \partial x/\partial u & \partial x/\partial v \\ \partial y/\partial u & \partial y/\partial v \end{vmatrix} = \begin{vmatrix} e^u\cos v & -e^u\sin v \\ e^u\sin v & e^u\cos v \end{vmatrix} = e^{2u}, \qquad \text{...(}i\text{)}$$

and $$\frac{\partial(u,v)}{\partial(x,y)} = \begin{vmatrix} \partial u/\partial x & \partial u/\partial y \\ \partial v/\partial x & \partial v/\partial y \end{vmatrix} = \begin{vmatrix} \dfrac{x}{x^2+y^2} & \dfrac{y}{x^2+y^2} \\ \dfrac{-y}{x^2+y^2} & \dfrac{x}{x^2+y^2} \end{vmatrix} = \frac{1}{x^2+y^2} = e^{-2u}. \qquad \text{...(}ii\text{)}$$

Thus, from (*i*) and (*ii*), we have

$$\frac{\partial(x,y)}{\partial(u,v)}\cdot\frac{\partial(u,v)}{\partial(x,y)} = e^{2u}\cdot e^{-2u} = e^{2u-2u} = e^{0} = 1.$$

(*b*) $$\frac{\partial(x,y)}{\partial(u,v)} = \begin{vmatrix} a & a \\ b & -b \end{vmatrix} = -2ab, \qquad \text{...}(iii)$$

and $$\frac{\partial(u,v)}{\partial(r,\theta)} = \begin{vmatrix} 2r\cos 2\theta & -2r^2\sin 2\theta \\ 2r\sin 2\theta & 2r^2\cos 2\theta \end{vmatrix} = 4r^3. \qquad \text{...}(iv)$$

∴ Using the Property 2, we get

$$\frac{\partial(x,y)}{\partial(r,\theta)} = \frac{\partial(x,y)}{\partial(u,v)}\cdot\frac{\partial(u,v)}{\partial(r,\theta)} = (-2ab)\cdot(4r^3) = -8abr^3.$$

EXERCISE 1.5

1. Given $x = u^2 - v^2$, $y = 2uv$, calculate $J\left(\frac{x,y}{u,v}\right)$. **Ans.** $4(u^2 + v^2)$.

2. If $x = r\sin^2\theta$, $y = r\cos^2\theta$, find $J\left(\frac{x,y}{r,\theta}\right)$. **Ans.** $2r\sin\theta\cos\theta$.

3. Find $\frac{\partial(u,y)}{\partial(x,y)}$, if $u = x - y^{2/3}$ and $v = x + y^2$ **Ans.** $2y + (2/(3y^{1/3}))$.

4. If $x = \cos u$, $y = \cos u \sin u$, $z = \cos w \sin v \sin u$, show that $\frac{\partial(x,y,z)}{\partial(u,v,w)} = -\sin^3 u \sin^2 v \sin w$.

5. If $x = u(1 - v)$, $y = uv$, evaluate $J = \frac{\partial(x,y)}{\partial(u,v)}$ and $J' = \frac{\partial(u,v)}{\partial(x,y)}$ and hence verify the result $JJ' = 1$.

(*AWIE, W-2004; VTU 2000S*)

[**Hint:** $x = u - uv = u - y \Rightarrow u = x + y$ and $v = y/u = y/(x + y)$, $J = u$, $J' = 1/(x + y) = 1/u$. \ $JJ' = 1$].

6. (*i*) If $u = x + y + z$, $uv = y + z$, $uvw = z$ then show that $\frac{\partial(x,y,z)}{\partial(u,v,w)} = u^2 v$.

(*PTU, 2009; KUK 2009; AMIE, W-2003; VTU, 2003*)

(*ii*) If $u = yz/x$, $v = zx/y$ and $w = xy/z$, show that $\frac{\partial(u,v,w)}{\partial(x,y,z)} = 4$. (*AMIE, W-2001*)

[**Hint:** (*i*) $x = u - uv$, $y = uv - uvw$, $z = uvw$.

$$\therefore \frac{\partial(x,y,z)}{\partial(u,v,w)} = \begin{vmatrix} 1-v & -u & 0 \\ v-vw & u-uw & -uv \\ vw & uw & uv \end{vmatrix} \text{ (operate } R_2 \to R_2 + R_3) = \begin{vmatrix} 1-v & -u & 0 \\ v & u & 0 \\ vw & uw & uv \end{vmatrix} = u^2 v].$$

7. If $u = x(1 - r^2)^{-1/2}$, $v = y(1 - r^2)^{-1/2}$, $w = z(1 - r^2)^{-1/2}$, where $r^2 = x^2 + y^2 + z^2$, then show that

$$\frac{\partial(u,v,w)}{\partial(x,y,z)} = (1 - r^2)^{-5/2}.$$

8. If $x = u^2 - v^2$, $y = 2uv$, find $\frac{\partial(u,v)}{\partial(x,y)}$. **Ans.** $1/4\,(u^2 + v^2)$.

Remark While solving the following problems, we should remember that if $\frac{\partial(u,v)}{\partial(x,y)} = 0$ then there exists a functional relationship between u and v.

9. If $u = x + 3y^2 - z^3$, $v = 4x^2yz$, $w = 2z^2 - xy$, evaluate $\dfrac{\partial(u, v, w)}{\partial(x, y, z)}$ at (1, –1, 0). (*VTU, 2006*) **Ans.** 20

10. Show that the functions

$f(x, y) = x^2 + xy + y^2 = 0$, $\phi(x, y) = x^4 + 2x^3y + 3x^2y^2 + 2xy^3 + y^4 = 0$ are functionally dependent.

[**Hint:** Show that $\begin{vmatrix} f_x & f_y \\ \phi_x & \phi_y \end{vmatrix} = 0$].

11. If $u = \dfrac{x+y}{x-y}$ and $v = \dfrac{xy}{(x-y)^2}$, find $\dfrac{\partial(u,v)}{\partial(x,y)}$. Are u and v functionally related? If so, find the relationship.

Ans. $u^2 = 1 + 4v$.

12. If $u = y + z$, $v = x + 2z^2$, $w = x - 4yz - 2y^2$, find $\dfrac{\partial(u, v, w)}{\partial(x, y, z)}$. Are u, v, w functionally related? If so find relationship. **Ans.** $w = v - u^2$.

13. If $u = xy + yz + zx$, $v = x^2 + y^2 + z^2$ and $w = x + y + z$, determine whether there is a functional relationship between u, v, w and if so, find it. **Ans.** $w^2 = v + 2u$.

14. If $u = \dfrac{x+y}{1-xy}$ and $v = \tan^{-1} x + \tan^{-1} y$. Evaluate $\dfrac{\partial(u,v)}{\partial(x,y)}$ when $xy \neq 1$. State whether u and v are functionally related. If so find the relationship between them. (*AMIE, S-2005, S-2003*)

[**Hint:** $\dfrac{\partial(u,v)}{\partial(x,y)} = \begin{vmatrix} \dfrac{1+y^2}{(1-xy)^2} & \dfrac{1+x^2}{(1-xy)^2} \\ \dfrac{1}{1+x^2} & \dfrac{1}{1+y^2} \end{vmatrix} = 0.$ $(xy \neq 1)$.

Therefore, u, v are functionally related. $\tan v = \tan(\tan^{-1} x + \tan^{-1} y) = \dfrac{x+y}{1-xy} = u.$]

15. If $u = \sin^{-1} x + \sin^{-1} y$ and $v = x\sqrt{1-y^2} + y\sqrt{1-x^2}$, determine whether there is a functional relationship between u and v, if so find it. **Ans.** $\sin u = v$. (*AMIE, W-2001*)

[**Hint:** We find $\dfrac{\partial(u,v)}{\partial(x,y)} = 0$. Since Jacobian u, v with respect to x, y vanishes, therefore u and v are functionally related.

We have $u = \sin^{-1} x + \sin^{-1} y$.

$\therefore \quad \sin u = \sin(\sin^{-1}x + \sin^{-1}y) = \sin(\sin^{-1}x)\cos(\sin^{-1}y) + \cos(\sin^{-1}x)\sin(\sin^{-1}y)$

$= x\sqrt{1-\sin^2(\sin^{-1} y)} + \sqrt{1-\sin^2(\sin^{-1} x)} \cdot y$

$= x\sqrt{1-\{\sin(\sin^{-1} y)\}^2} + \sqrt{1-\{\sin(\sin^{-1} x)\}^2} \cdot y = x\sqrt{1-y^2} + y\sqrt{1-x^2} = v.$

$\therefore$ $v = \sin u$ is the required relationship between u and v].

16. Let $u = x + y - z$, $v = x - y + z$, $w^2 = x^2 + y^2 + z^2 + ayz$. Find the value of a if u, v, w are functionally dependent. Hence find the relation. **Ans.** $a = 2$; $u^2 + v^2 = 2w$.

CHAPTER

2

Applications of Partial Derivatives

2.1 GEOMETRICAL INTERPRETATION OF PARTIAL DERIVATIVES OF A FUNCTION OF TWO VARIABLES

The partial derivatives have simple geometric interpretations. Consider the surface S represented by $z = f(x, y)$ in Fig. 2.1. Through the point $P(x, y, z)$, there is a curve APB that is, the intersection with the surface of the plane through P parallel to the xz plane. Similarly CPD is the curve through P *i.e.*, the intersection with the surface of the plane through P parallel to the yz plane. As x varies while y is held fixed, P moves along the curve APB and the value of $\partial z/\partial x$ at (x, y) is the slope of the tangent line to the curve APB at P. Similarly, as y varies while x is held fixed, P moves along the curve CPD, and the value of $\partial z/\partial y$ at (x, y) is the slope of the tangent line to the curve CPD at P.

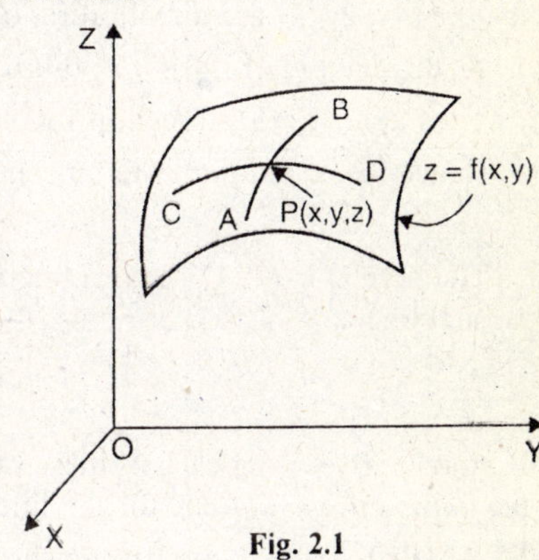

Fig. 2.1

2.2 APPLICATIONS TO GEOMETRY

2.2.1 Tangent Plane and Normal Line to a Surface

Let the equation of the given surface be $F(x, y, z) = 0$, and let $P_0\ (x_0, y_0, z_0)$ be the point on the given surface then direction ratios of a normal to the tangent plane at any point $P_0\ (x_0, y_0, z_0)$ on the surface are given by $(\partial F/\partial x)_{P_0}$, $(\partial F/\partial y)_{P_0}$, $(\partial F/\partial z)_{P_0}$.

The equation of the tangent plane to the surface $F(x, y, z) = 0$ at one of its points $P_0(x_0, y_0, z_0)$ is

$$\left.\frac{\partial F}{\partial x}\right|_{P_0}(x - x_0) + \left.\frac{\partial F}{\partial y}\right|_{P_0}(y - y_0) + \left.\frac{\partial F}{\partial z}\right|_{P_0}(z - z_0) = 0, \qquad ...(2.1)$$

and the equations of the normal line that is, perpendicular to the tangent plane at the point $P(x_0, y_0, z_0)$ are

$$\frac{x - x_0}{\left.\partial F/\partial x\right|_{P_0}} = \frac{y - y_0}{\left.\partial F/\partial y\right|_{P_0}} = \frac{z - z_0}{\left.\partial F/\partial z\right|_{P_0}} \qquad ...(2.2)$$

with the understanding that the partial derivatives have been evaluated at the point P_0.

Setting each of these ratios equal to a parameter (such as t or u) and solving for x, y, and z yields the *parametric equations* of the normal line.

2.2.2 Tangent Line and Normal Plane to a Space Curve

A space curve may also be defined by the pair of equations

$$F(x, y, z) = 0, \quad G(x, y, z) = 0.$$

At the point $P_0(x_0, y_0, z_0)$ of the curve, the equations of the tangent line are

$$\frac{x - x_0}{\begin{vmatrix} \partial F/\partial y & \partial F/\partial z \\ \partial G/\partial y & \partial G/\partial z \end{vmatrix}_{P_0}} = \frac{y - y_0}{\begin{vmatrix} \partial F/\partial z & \partial F/\partial x \\ \partial G/\partial z & \partial G/\partial x \end{vmatrix}_{P_0}} = \frac{z - z_0}{\begin{vmatrix} \partial F/\partial x & \partial F/\partial y \\ \partial G/\partial x & \partial G/\partial y \end{vmatrix}_{P_0}} \qquad ...(2.3)$$

and the equation of the normal plane is

$$\begin{vmatrix} \partial F/\partial y & \partial F/\partial z \\ \partial G/\partial y & \partial G/\partial z \end{vmatrix}_{P_0} (x-x_0) + \begin{vmatrix} \partial F/\partial z & \partial F/\partial x \\ \partial G/\partial z & \partial G/\partial x \end{vmatrix}_{P_0} (y-y_0)$$

$$+\begin{vmatrix} \partial F/\partial x & \partial F/\partial y \\ \partial G/\partial x & \partial G/\partial y \end{vmatrix}_{P_0} (z-z_0)=0 . \qquad ...(2.4)$$

Fig. 2.2

Note: All partial derivatives are to be evaluated at the point P_0.

EXAMPLE 2.1 Find the equations of the tangent plane and normal line to the surface $2x^2 + y^2 = 3 - 2z$ at the point (2, 1, –3).

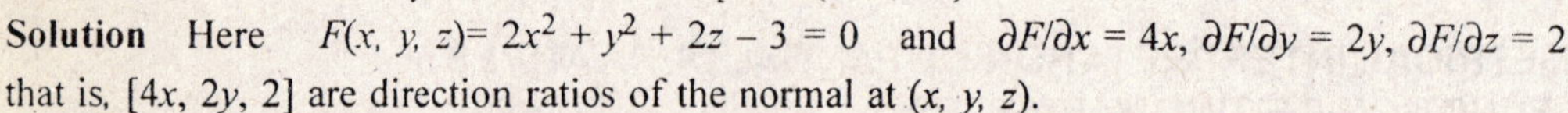

Solution Here $F(x, y, z) = 2x^2 + y^2 + 2z - 3 = 0$ and $\partial F/\partial x = 4x$, $\partial F/\partial y = 2y$, $\partial F/\partial z = 2$ that is, $[4x, 2y, 2]$ are direction ratios of the normal at (x, y, z).

At the point (2, 1, –3), $\partial F/\partial x = 8$, $\partial F/\partial y = 2$, $\partial F/\partial z = 2$.

Thus at (2, 1, –3), direction ratios of the normal are (8, 2, 2) or (4, 1, 1).

Hence the equation of the tangent plane at (2, 1, –3) is

$$4(x-2) + 1\,(y-1) + 1\,(z+3) = 0 \text{ or } 4x + y + z = 6.$$

The normal at (2, 1, –3) is a line through (2, 1, –3) perpendicular to the above plane. The equations of the normal line at (2, 1, –3) are thus

$$\frac{x-2}{4} = \frac{y-1}{1} = \frac{z+3}{1}.$$

Remark By setting each of these ratios equal to the parameter t, we have $x = 2 + 4t$, $y = 1 + t$, $z = t - 3$ called the *parametric equations* for the line.

EXAMPLE 2.2 Show that the surfaces $F(x, y, z) = x^2 + 4y^2 - 4z^2 - 4 = 0$ and $G(x, y, z) = x^2 + y^2 + z^2 - 6x - 6y + 2z + 10 = 0$ are tangent at the point (2, 1, 1).

Solution We have to prove that the two surfaces have the same tangent plane at the given point.

At (2, 1, 1), $\partial F/\partial x = 2x = 4$, $\partial F/\partial y = 8y = 8$, $\partial F/\partial z = -8z = -8$

and $\partial G/\partial x = 2x - 6 = -2$, $\partial G/\partial y = 2y - 6 = -4$, $\partial G/\partial z = 2z + 2 = 4$.

Since the sets of direction numbers [4, 8, –8] and [–2, –4, 4] of the normal lines of the two surfaces are proportional, the surfaces have the common tangent plane

$$1(x-2) + 2(y-1) - 2(z-1) = 0 \text{ or } x + 2y - 2z = 2.$$

EXAMPLE 2.3 Show that the surface $f(x, y, z) = x^2 - 2yz + y^3 - 4 = 0$ is perpendicular to any member of the family of surfaces $F(x, y, z) = x^2 + 1 - (2 - 4a)\,y^2 - az^2 = 0$ at the point of intersection (1, –1, 2).

Solution We have $f(x, y, z) = x^2 - 2yz + y^3 - 4 = 0$.

Therefore, $\partial f/\partial x = 2x$, $\partial f/\partial y = -2z + 3y^2$, $\partial f/\partial z = -2y$.

At the point (1, –1, 2) $\partial f/\partial x = 2$, $\partial f/\partial y = -1$, $\partial f/\partial z = 2$.

Also, we have $F(x, y, z) = x^2 + 1 - (2 - 4a)y^2 - az^2 = 0$.

∴ $\partial F/\partial x = 2x$, $\partial F/\partial y = -2(2 - 4a)y$, $\partial F/\partial z = -2az$.

At the point (1, –1, 2), the values of $\partial F/\partial x$, $\partial F/\partial y$ and $\partial F/\partial z$ are respectively 2, $4 - 8a$, $-4a$.

Now applying $$\left(\frac{\partial f}{\partial x}\right)\left(\frac{\partial F}{\partial x}\right) + \left(\frac{\partial f}{\partial y}\right)\left(\frac{\partial F}{\partial y}\right) + \left(\frac{\partial f}{\partial z}\right)\left(\frac{\partial F}{\partial z}\right) = 0,$$

we obtain $(2)(2) + (-1)(4 - 8a) + (2)(-4a) = 0$,

it follows that given surfaces are perpendicular for all 'a', and so the required result follows.

EXAMPLE 2.4 Find the equations of the tangent line and the normal plane to the curve $x^2 + 2y^2 + 2z^2 = 5$, $3x - 2y - z = 0$ at the point (1, 1, 1).

Solution The equations of the surfaces intersecting in the curve are

$$F(x, y, z) = x^2 + 2y^2 + 2z^2 - 1 - 5 = 0, \qquad G(x, y, z) = 3x - 2y - z = 0.$$

At (1, 1, 1), $\begin{vmatrix} \partial F/\partial y & \partial F/\partial z \\ \partial G/\partial y & \partial G/\partial z \end{vmatrix} = \begin{vmatrix} 4y & 4z \\ -2 & -1 \end{vmatrix} = \begin{vmatrix} 4 & 4 \\ -2 & -1 \end{vmatrix} = 4,$

$$\begin{vmatrix} \partial F/\partial z & \partial F/\partial x \\ \partial G/\partial z & \partial G/\partial x \end{vmatrix} = \begin{vmatrix} 4z & 2x \\ -1 & 3 \end{vmatrix} = \begin{vmatrix} 4 & 2 \\ -1 & 3 \end{vmatrix} = 14,$$

$$\begin{vmatrix} \partial F/\partial x & \partial F/\partial y \\ \partial G/\partial x & \partial G/\partial y \end{vmatrix} = \begin{vmatrix} 2x & 4y \\ 3 & -2 \end{vmatrix} = \begin{vmatrix} 2 & 4 \\ 3 & -2 \end{vmatrix} = -16.$$

With [2, 7, –8] as a set of direction numbers of the tangent, its equations are $\dfrac{x-1}{2} = \dfrac{y-1}{7} = \dfrac{z-1}{-8}$.

The equation of the normal plane is $2(x - 1) + 7(y - 1) - 8(z - 1) = 2x + 7y - 8z - 1 = 0$.

EXERCISE 2.1

1. Find the equations of the tangent plane and normal line to the given surface at the given point:
   (*i*) $x^2 + 2y^2 + 3z^2 = 12$, (1, 2, –1), (*AMIETE, Dec 2010*) (*ii*) $z = xy$, (3, –4, –12),
   (*iii*) $2x^2 + 2xy + y^2 + z + 1 = 0$, (1, –2, –3),
   (*iv*) $x^2 + 3y^2 - 4z^2 + 3xy - 10yz + 4x - 5z - 22 = 0$, (1. –2, 1).

Ans. (*i*) $x + 4y - 3z = 12$; $x - 1 = (y-2)/4 = (z+1)/3$,
(*ii*) $4x - 3y + z = 12$; $(x-3)/4 = (y+4)/-3 = z+12$,
(*iii*) $z - 2y = 1$; $(x-1)/0 = (y+2)/2 = (z+3)/-1$,
(*iv*) $19y - 7z + 45 = 0$; $x = 1$, $7y + 19z - 5 = 0$.

2. Find the equations of the tangent planes and the normal line to the following surfaces at the specified points:
   (*i*) $x^2 + y^2 = 4z$ at (2, –4, 5), (*ii*) $z = 3x^2 + 2y^2 - 11$ at (2, 1, 3),
   (*iii*) $x^2yz + 3y^2 = 2xz^2 - 8z$ at (1, 2, –1), (*iv*) $x^2y + xz^2 = z - 1$ at (1, –3, 2) (*AMIE; W-2008*)

Ans. (*i*) $x - 2y - z = 5$; $\dfrac{(x-2)}{-1} = \dfrac{(y+4)}{2} = z - 5$, (*ii*) $12x + 4y - z = 25$; $\dfrac{(x-2)}{12} = \dfrac{(y-1)}{4} = \dfrac{(z-3)}{-1}$,
(*iii*) $6x - 11y - 14z + 2 = 0$; $\dfrac{(x-1)}{-6} = \dfrac{(y-2)}{11} = \dfrac{(z+1)}{14}$,
(*iv*) $2x - y - 3z + 1 = 0$; $\dfrac{x-1}{-2} = y + 3 = \dfrac{z-2}{3}$.

3. Show that the surfaces $F(x, y, z) = xy + yz - 4zx = 0$ and $G(x, y, z) = 3z^2 - 5x + y = 0$ intersect at right angles at the point (1, 2, 1).
4. Show that the surfaces $F(x, y, z) = 3x^2 + 4y^2 + 8z^2 - 36 = 0$ and $G(x, y, z) = x^2 + 2y^2 - 4z^2 - 6 = 0$ intersect at right angles.
5. The surfaces $x^2y^2 + 2x + z^3 = 16$ and $3x^2 + y^2 - 2z = 9$ intersect in a curve which passes through the point (2, 1, 2). What are the equations of the respective tangent planes to the two surfaces at that point?
   Ans. $3x + 4y + 6z = 22$; $6x + y - z = 11$.
6. Find the condition that the plane $ax + by + cz + d = 0$ should touch the surface $px^2 + qy^2 + 2z = 0$.
   Ans. $a^2/p + b^2/q + 2cd = 0$.
7. Find equations for the (*i*) tangent line and (*ii*) normal plane to the curve $3x^2y + y^2z = -2$, $2xz - x^2y = 3$ at the point (1, –1, 1).

Ans. (*i*) $\dfrac{(x-1)}{3} = \dfrac{(y+1)}{16} = \dfrac{(z-1)}{2}$ or $x = 1 + 3t$, $y = 16t - 1$, $z = 1 + 2t$. (*ii*) $3x + 16y + 2z + 11 = 0$.

8. Find the equations of the tangent line and normal plane to the curve, $x^2 + 2y^2 + 3z^2 = 3$, $2x + 3y + 4z = 5$ at (1, 1, 0). (*AMIE, W-2007*) **Ans.** $(x - 1)/-8 = (y - 1)/4 = z$, $8x - 4y - z - 4 = 0$.

2.3 ENVELOPES, EVOLUTES

2.3.1 Envelope Definition

Let $f(x, y, \alpha) = 0$ be an equation of a family of curves whose parameter is α. If the neighbouring members of the family *viz.* $f(x, y, \alpha) = 0$ and $f(x, y, \alpha + \delta\alpha) = 0$ cut in a point P which tends to another point Q, as $\delta\alpha \to 0$, the locus of Q is called the envelope of the family.

Method of finding the Envelope

Let $f(x, y, \alpha) = 0$ be an equation of a family of curves, whose parameter is α. If α be supposed to be a particular value of the parameter then $f(x, y, \alpha) = 0$ ...(2.5) represents one member of the family.

Let another equation of member of the family be $f(x, y, \alpha + \delta\alpha) = 0$ ...(2.6)

Let these two equations (2.5) and (2.6) intersect at P.

The coordinates of P will satisfy the equation $f(x, y, \alpha + \delta\alpha) - f(x, y, \alpha) = 0$,

and therefore also the equation

$$\frac{f(x, y, \alpha + \delta\alpha) - f(x, y, \alpha)}{\delta\alpha} = 0.$$

Taking limit as $\delta\alpha \to 0$, we see that the coordinates of the point Q to which P tends as $\delta\alpha \to 0$, satisfy the equation

$$\lim_{\delta\alpha \to 0} \frac{f(x, y, \alpha + \delta\alpha) - f(x, y, \alpha)}{\delta\alpha} = 0,$$

or

$$\frac{\partial}{\partial x} f(x, y, \alpha) = 0. \qquad \text{...(2.7)}$$

Since Q is a point on (2.5) hence the coordinates of Q must satisfy (2.5) also.

Therefore by eliminating α between (2.5) and (2.7), we get the equation of the locus of Q which is the equation of the envelope.

Rule for finding the envelope

(*i*) *Let the equation of a family of curves be $f(x, y, \alpha) = 0$, where α is the parameter.*

(*ii*) *Find $\frac{\partial}{\partial \alpha} f(x, y, \alpha) = 0$, that is, differentiate $f(x, y, \alpha) = 0$ partially with respect to parameter α.*

(*iii*) *Eliminate α between $f(x, y, \alpha) = 0$ and equate $\partial f/\partial\alpha$ equal to 0. The result is the required equation of the envelope of the family of curves $f(x, y, \alpha) = 0$.*

Remark In the following examples $f_\alpha(x, y, \alpha) = 0$ denotes the partial derivative of $f(x, y, \alpha)$ with respect to α.

ILLUSTRATIVE EXAMPLES

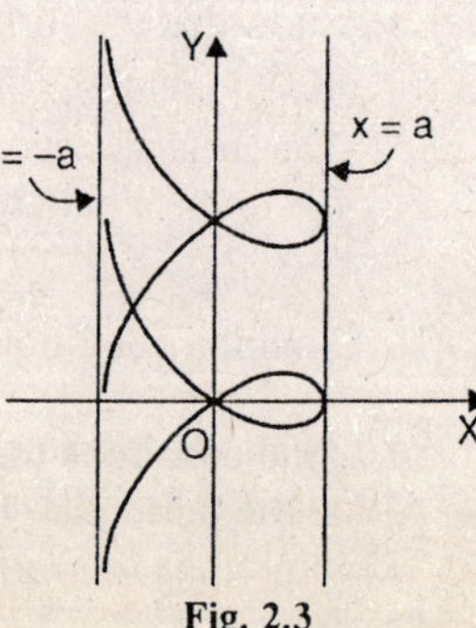

Fig. 2.3

EXAMPLE 2.5 Find the envelope of the family $x^2 (x - a) + (x + a)(y - m)^2 = 0$, where '$a$' is a constant and m is a parameter.

Solution Differentiating with respect to m, we get $-2(x + a)(y - m) = 0$.

Eliminating m, we get $x^2 (x - a) = 0$, which is the envelope.

Thus the envelope consists of two lines $x = 0$ and $x = a$.

EXAMPLE 2.6 Find the envelope of the following families of lines:

(*i*) $y = mx + (1/m)$; the parameter being m, (*ii*) $y = mx + am^3$,

(*iii*) $y = mx - 2am - am^3$; (*iv*) $y = mx + \sqrt{a^2m^2 + b^2}$

Solution (*i*) We have $f(x, y, m) = y - mx - (1/m) = 0$ or $-m^2x + my - 1 = 0$. ...(1)

Differentiating (1) partially with respect to m, we have $-2mx + y = 0$ or $m = y/2x$.

Substituting the value of m in (1), and simplifying we obtain the equation of the envelope as $y^2 = 4x$, which is the equation of a parabola.

(*ii*) We have $f(x, y, m) = y - mx - am^3 = 0$ ...(2)

Therefore $f_m(x, y, m) = -x - 3am^2 = 0$ or $m^2 = -x/3a$.

Substituting $m^2 = -x/3a$ in $y^2 = m^2(x + am^2)^2$, that is, the square of the equation (2) and simplifying, we get

$27ay^2 + 4x^3 = 0$, the required envelope.

(*iii*) Here $f_m(x, y, m) = x - 2a - 3am^2 = 0$ or $m = \pm\sqrt{(x-2a)/3a}$.

Substituting the value of m in $y = m(x - 2a - am^2)$, we get

$$y = \pm\sqrt{\frac{x-2a}{3a}}\left[x - 2a - a\left(\frac{x-2a}{3a}\right)\right].$$

Squaring both sides and simplifying, we get

$27ay^2 = 4(x - 2a)^3$, the equation of the envelope.

(*iv*) We have $f(x, y, m) = y - mx - \sqrt{a^2m^2 + b^2} = 0.$ ...(3)

Therefore, $f_m(x, y, m) = -x - \dfrac{-a^2m}{\sqrt{a^2m^2+b^2}} = 0$ or $x = \dfrac{-a^2m}{\sqrt{a^2m^2+b^2}}$. ...(4)

Substituting the values of x in the given equation, we get

$$y = -\frac{a^2m^2}{\sqrt{a^2m^2+b^2}} + \sqrt{a^2m^2+b^2} = \frac{b^2}{\sqrt{a^2m^2+b^2}} \quad \text{or} \quad \sqrt{a^2m^2+b^2} = \frac{b^2}{y}.$$

Now replacing $\sqrt{a^2m^2+b^2}$ by b^2/y in Eqns. (3) and (4), we get

$$y = mx + \frac{b^2}{y}, \quad \text{or} \quad y - \frac{b^2}{y} = mx$$

and $x = -\dfrac{a^2m}{b^2/y}$, that is $-b^2x = ma^2y$.

Dividing these and simplifying, we obtain $x^2/a^2 + y^2/b^2 = 1$, which is the required envelope.

EXAMPLE 2.7 Find the envelope of the following families.

(*i*) $x \cos\alpha + y \sin\alpha = p$, α being the variable parameter, (*ii*) $\dfrac{ax}{\cos\alpha} - \dfrac{by}{\sin\alpha} = a^2 - b^2$,

(*iii*) $x \sec\alpha + y \operatorname{cosec}\alpha = a$, (*iv*) $x \cos\alpha + y \sin\alpha = l \sin\alpha \cos\alpha$.

Solution (*i*) We have

$$f(x, y, \alpha) = x \cos\alpha + y \sin\alpha - p = 0. \quad ...(1)$$

Differentiating with respect to α,

$$f_\alpha(x, y, \alpha) = -x \sin\alpha + y \cos\alpha = 0. \quad ...(2)$$

Multiply (1) by $\cos\alpha$, and (2) by $\sin\alpha$ and subtracting, we get $x = p \cos\alpha$.

Similarly eliminating x between these equations, we obtain $y = p \sin\alpha$.

The parametric equations of the envelope are therefore,

$$\begin{cases} x = p\cos\alpha \\ y = p\sin\alpha \end{cases} \quad ...(3)$$

α being the parameter.

Squaring Eq. (3) and adding, we obtain $x^2 + y^2 = p^2$, the rectangular equation of the envelope, a circle.

(*ii*) We have

$$f(x, y, \alpha) = \frac{ax}{\cos\alpha} - \frac{by}{\sin\alpha} - (a^2 - b^2) = 0 \quad \text{...(4)}$$

Therefore, $$f_\alpha(x, y, \alpha) = \frac{ax\sin\alpha}{\cos^2\alpha} + \frac{by\cos\alpha}{\sin^2\alpha} = 0 \quad \text{or} \quad \tan\alpha = -(by/ax)^{1/3}.$$

$$\sin\alpha = \mp\frac{(by)^{1/3}}{\sqrt{(ax)^{2/3} + (by)^{2/3}}}, \qquad \cos\alpha = \pm\frac{(ax)^{1/3}}{\sqrt{(ax)^{2/3} + (by)^{2/3}}}.$$

Substituting the values in Eq. (4), we obtain

$$\pm[(ax)^{2/3} + (by)^{2/3}][(ax)^{2/3} + (by)^{2/3}]^{1/2} = a^2 - b^2$$

or $$\pm[(ax)^{2/3} + (by)^{2/3}]^{3/2} = (a^2 - b^2)$$

or $$(ax)^{2/3} + (by)^{2/3} = (a^2 - b^2)^{2/3},$$ which is the required envelope.

(*iii*) We have

$$f(x, y, \alpha) = x\sec\alpha + y\,\mathrm{cosec}\,\alpha - a = 0$$
$$= x\sin\alpha + y\cos\alpha - a\sin\alpha\cos\alpha = 0. \quad \text{...(5)}$$

Therefore, $$f_\alpha(x, y, \alpha) = x\cos\alpha - y\sin\alpha + a\sin^2\alpha - a\cos^2\alpha = 0. \quad \text{...(6)}$$

Multiplying (5) by $\sin\alpha$ and (6) by $\cos\alpha$ and adding, we get

$$x(\sin^2\alpha + \cos^2\alpha) - a\cos^3\alpha = 0 \Rightarrow x = a\cos^3\alpha.$$

Similarly, eliminating x between (5) and (6), we obtain $y = a\sin^3\alpha$.

The parametric equations of the envelope are $$\left.\begin{aligned} x &= a\cos^3\alpha, \\ y &= a\sin^3\alpha \end{aligned}\right\} \quad \text{...(7)}$$

α being the parameter. the corresponding rectangular equation is found from equations (7) by eliminating α as follows.

$$x^{2/3} = a^{2/3}\cos^2\alpha; \quad y^{2/3} = a^{2/3}\sin^2\alpha.$$

Adding these, we get

$$x^{2/3} + y^{2/3} = a^{2/3},$$ the rectangular equation of the hypo cycloid.

(*iv*) Writing $x\cos\alpha + y\sin\alpha = l\sin\alpha\cos\alpha$ as

$$\frac{x}{\sin\alpha} + \frac{y}{\cos\alpha} = l, \quad \text{...(8)}$$ and differentiating w.r.t., α we get

$$-\frac{x\cos\alpha}{\sin^2\alpha} + \frac{y\sin\alpha}{\cos^2\alpha} = 0 \quad \text{or} \quad \frac{\cos^3\alpha}{y} = \frac{\sin^3\alpha}{x}, \quad \text{or} \quad \frac{\sin^2\alpha}{x^{2/3}} = \frac{\cos^2\alpha}{y^{2/3}} = \frac{1}{x^{2/3} + y^{2/3}}.$$

Substituting for $\sin\alpha$ and $\cos\alpha$ from this in (8), we get the equation of the envelope as

$$x^{2/3}\sqrt{(x^{2/3} + y^{2/3})} + y^{2/3}\sqrt{(x^{2/3} + y^{2/3})} = l \quad \text{or} \quad (x^{2/3} + y^{2/3})\sqrt{(x^{2/3} + y^{2/3})} = l$$

or $x^{2/3} + y^{2/3} = l^{2/3}$, which is the equation of an astroid.

EXAMPLE 2.8 Find the envelope of the family of lines $x\cos^3\alpha + y\sin^3\alpha = a$, the parameter being α.

Solution We have $x\cos^3\alpha + y\sin^3\alpha = a$. ...(*i*)

Differentiating (*i*) w.r.t. α we get $-3x\cos^2\alpha\sin\alpha + 3y\sin^2\alpha\cos\alpha = 0$

or $$x\cos\alpha - y\sin\alpha = 0 \quad \text{or} \quad \tan\alpha = x/y.$$

$$\therefore \quad \sin\alpha = x\Big/\sqrt{x^2 + y^2} \quad \text{and} \quad \cos\alpha = y\Big/\sqrt{x^2 + y^2}.$$

Substituting the values of $\cos\alpha$ and $\sin\alpha$ in (*i*), we get

$$x\cdot\frac{y^3}{(x^2 + y^2)^{3/2}} + y\cdot\frac{x^3}{(x^2 + y^2)^{3/2}} = a \quad \text{or} \quad xy^3 + yx^3 = a(x^2 + y^2)^{3/2},$$

or $\quad xy(x^2 + y^2) = a(x^2 + y^2)^{3/2}$ that is, $xy = a(x^2 + y^2)^{1/2}$.

Squaring both sides, we have $x^2y^2 = a^2(x^2 + y^2)$, which is the required envelope.

EXAMPLE 2.9 Find the envelope of the family of straight lines $x/a + y/b = 1$ where the parameters a, b are connected by the relation (*i*) $a + b = c$, (*ii*) $ab = c^2$, (*iii*) $a^n + b^n = c^n$, c being a constant.

Solution (*i*) We have $x/a + y/b = 1$ ...(1) and $a + b = c$ ...(2)

Differentiating (1) and (2), with respect to the parameter b regarding a as a function of b, we obtain

$$\frac{x}{a^2}\frac{da}{db} + \frac{y}{b^2} = 0 \quad ...(3) \quad \text{and} \quad \frac{da}{db} + 1 = 0 \quad ...(4)$$

Equating the values of da/db in (3) and (4), we get $x/a^2 = y/b^2 = k$ (say).

Therefore, $x/a + y/b = k(a + b) = ck = 1$, or $k = 1/c$. That is, $a^2 = cx$, $b^2 = cy$.

Substituting these values of a and b in (2), we obtain

$$\sqrt{cx} + \sqrt{cy} = c \quad \text{or} \quad \sqrt{x} + \sqrt{y} = \sqrt{c}.$$

(*ii*) Since $ab = c^2$, we have $b(da/db) + a = 0$. ...(5)

Equating the values of da/db in (3) and (5), we get $x/a = y/b = k$ (say).

Therefore, $x/a + y/b = 2k = 1$, or $k = 1/2$. That is $a = 2x$, $b = 2y$.

Substituting these values in $ab = c^2$, we get $4xy = c^2$,the required equation of the envelope.

(*iii*) Here $na^{n-1}(da/db) + nb^{n-1} = 0$. ...(6)

Equating the values of da/db in (3) and (6), we get,

$$x/a^{n+1} = y/b^{n+1} = k(\text{say}).$$

Therefore, $x/a + y/b = k(a^n + b^n) = kc^n = 1$ or $k = 1/c^n$.

That is, $a = (xc^n)^{1/n+1}$ and $b = (yc^n)^{1/n+1}$.

Putting these values of a and b in $a^n + b^n = c^n$, we get

$$(xc^n)^{n/n+1} + (yc^n)^{n/n+1} = c^n \quad \text{or} \quad x^{n/n+1} + y^{n/n+1} = c^{n/n+1},$$ the required envelope.

EXAMPLE 2.10 Find the envelope of a line of constant length 'a' whose extremities move along the fixed rectangular axes.

Solution Let AB = a in length, and let $x \cos\alpha + y \sin\alpha - p = 0$ ...(*i*) be its equation. Now as AB moves, both α and p will vary but p may be found in terms of α.

For $AO = AB \cos\alpha = a\cos\alpha$, and therefore $p = AO \sin\alpha = a\sin\alpha\cos\alpha$. Substituting in (*i*), we get

$$x\cos\alpha + y\sin\alpha - a\sin\alpha\cos\alpha = 0 \quad ...(ii)$$

where α is the variable parameter. This equation is of the form $f(x, y, \alpha) = 0$.

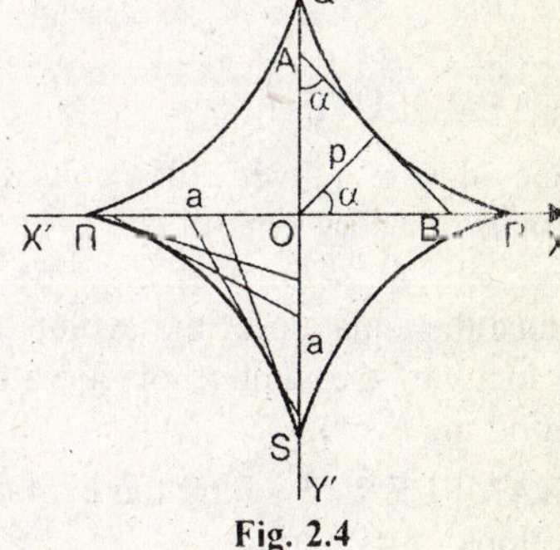

Fig. 2.4

Differentiating with respect to α, the equation $f_\alpha(x, y, \alpha) = 0$, we obtain

$$-x\sin\alpha + y\cos\alpha + a\sin^2\alpha - a\cos^2\alpha = 0. \quad ...(iii)$$

Solving (*ii*) and (*iii*) for x and y in terms of α, the result is

$$\left.\begin{array}{l} x = a\sin^3\alpha, \\ y = a\cos^3\alpha \end{array}\right\}$$ ...(*iv*), the parametric equations of the envelope.

From (*iv*), we get $\left.\begin{array}{l} x^{2/3} = a^{2/3}\sin^2\alpha, \\ y^{2/3} = a^{2/3}\cos^2\alpha. \end{array}\right\}$. Adding these, we obtain $x^{2/3} + y^{2/3} = a^{2/3}$, the rectangular equation of envelope.

EXAMPLE 2.11 Find the envelope of the family of ellipses whose axes coincide and whose area is constant.

Solution Let the equation of the ellipse be

$$x^2/a^2 + y^2/b^2 = 1, \quad \text{...}(i)$$

where a and b are such that $\pi\, ab = k$ or $ab = k/\pi = c^2$. ...(*ii*)

Differentiating (*i*) and (*ii*) w.r.t. b regarding a as a function of b, we get

$$-\frac{2x^2}{a^3}\cdot\frac{da}{db} - \frac{2y^2}{b^3} = 0 \quad \text{...}(iii) \quad \text{and} \quad a + b\frac{da}{db} = 0. \quad \text{...}(iv)$$

Equating values of da/db from (*iii*) and (*iv*)

$$y^2a^3/x^2b^3 = a/b \quad \text{or} \quad x^2/a^2 = y^2/b^2. \quad \text{...}(v)$$

From (*i*) and (*v*) $x^2/a^2 = 1/2$ and $y^2/b^2 = 1/2$ or $a = \sqrt{2}\,x$ and $b = \sqrt{2}\,y$.

Substituting these values in (ii), the envelope is $2xy = c^2$.

EXAMPLE 2.12 Show that the envelope of the circles whose centres lie on the parabola $y^2 = 4ax$ and which pass through its vertex is

$$2ay^2 + x(x^2 + y^2) = 0.$$

Solution Let $P(\alpha, \beta)$ be any point on the parabola, then we have

$$\beta^2 = 4a\alpha. \quad \text{...}(i)$$

The equation of the circle with centre at P and passing through origin O is

$$(x - \alpha)^2 + (y - \beta)^2 = \alpha^2 + \beta^2 \quad \text{or} \quad x^2 - 2\alpha x + y^2 - 2\beta y = 0. \quad \text{...}(ii)$$

From (*i*) and (*ii*), eliminating α, we get

$$x^2 - \frac{x\beta^2}{2a} + y^2 - 2y\beta = 0. \quad \text{...}(iii)$$

Differentiating partially (*iii*) with respect to β, we have

$$-\frac{2x\beta}{2a} - 2y = 0 \quad \text{or} \quad \beta = -\frac{2y}{x}.a. \quad \text{...}(iv)$$

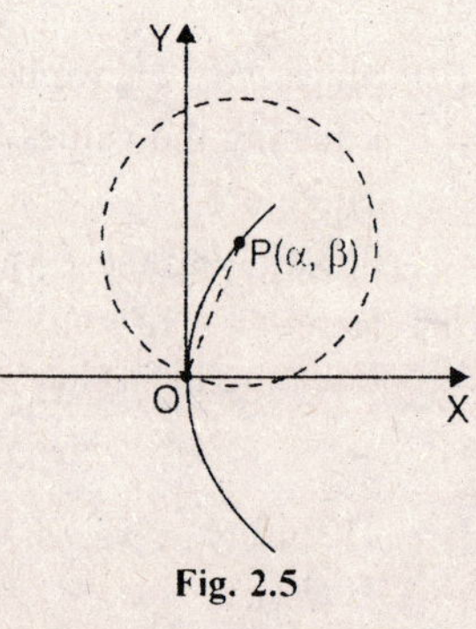

Fig. 2.5

Substituting the value of β in (iii), we obtain the equation of the envelope as

$$x^2 - \frac{4y^2}{x^2}\cdot\frac{a^2x}{2a} + y^2 - 2y\left(-\frac{2y}{x}\cdot a\right) = 0 \quad \text{or} \quad x^2 + y^2 + 2\frac{ay^2}{x} = 0$$

or $\quad 2ay^2 + x(x^2 + y^2) = 0$ or $y^2(x + 2a) + x^3 = 0$. (The cissoid)

2.4 EVOLUTE

The evolute of a given curve considered as the envelope of its normals. Since the normals to a curve are all tangent to the evolute, it is evident that the *evolute of a curve may be defined as the envelope of its normals*. It is generally easier to deduce the equation of the evolute as the envelope of the normals rather than as the locus of the centers of curvature. We have thus an alternative method of finding the evolute.

Evolute envelope of normals

Given curve

Fig. 2.6

EXAMPLE 2.13 Find the evolute of the parabola $y^2 = 4ax$ considered as the envelope of its normals. (*Madras, 2003*)

Solution The equation of normal to the parabola $y^2 = 4ax$ in the parametric form is $y = mx - 2am - am^3$, ...(*i*) where m is the parameter.

Differentiating w.r.t. to m, we have

$$0 = x - 2a - 3\,am^2, \quad \text{or} \quad m = \pm\left(\frac{x - 2a}{3a}\right)^{1/2}.$$

Substituting this in (*i*), we get

$$y = \sqrt{\frac{x - 2a}{3a}}\left[x - 2a - a\left(\frac{x - 2a}{3a}\right)\right] = \sqrt{\frac{x - 2a}{3a}}\left[\frac{2}{3}(x - 2a)\right] = \frac{2}{\sqrt{a}}\left(\frac{x - 2a}{3}\right)^{3/2}.$$

Squaring both sides, we get the evolute $27ay^2 = 4(x - 2a)^3$.

EXAMPLE 2.14 Prove that the evolute of the tractrix $x = a\left(\cos t + \log \tan \frac{t}{2}\right)$, $y = a \sin t$, is the catenary $y = a \cosh (x/a)$. *(Anna, 2005S)*

Solution Here $\dfrac{dx}{dt} = a\left(-\sin t + \left(\dfrac{1}{2}\right)\dfrac{1}{\tan\frac{t}{2}}\cdot\sec^2\dfrac{t}{2}\right) = a\left(-\sin t + \dfrac{1}{\sin t}\right) = \dfrac{a\cos^2 t}{\sin t}$, and $\dfrac{dy}{dt} = a\cos t$.

Therefore, $\dfrac{dy}{dx} = \dfrac{dy/dt}{dx/dt} = \dfrac{a\cos t}{a\cos^2 t/\sin t} = \tan t$.

The equation of the normal to the tractrix at 't' is given by

$$y - a\sin t = -\frac{1}{\tan t}\left[x - a\cos t - a\log\tan\frac{t}{2}\right]$$

or
$$x + y\tan t - a\sec t - a\log\tan\frac{t}{2} = 0. \qquad ...(i)$$

Differentiating (i) w.r.t. parameter t, we get

$$y\sec^2 t - a\sec t\tan t - \frac{1}{2}a\cdot\frac{1}{\tan\frac{t}{2}}\cdot\left(\sec^2\frac{t}{2}\right) = 0$$

or $\quad y\sec^2 t - a\sec t\tan t - a\,\mathrm{cosec}\, t = 0 \quad$ or $\quad y\sec^2 t = a\left[\dfrac{\sin t}{\cos^2 t} + \dfrac{1}{\sin t}\right] = \dfrac{1}{\sin t\cos^2 t}$

or $\quad y = a/\sin t$. Putting this in (i), we get $x + a\sec t - a\sec t - a\log\tan\dfrac{t}{2} = 0$

or $\quad x = a\log\tan(t/2)\quad$ or $\quad e^{x/a} = \tan(t/2)\quad$ and $\quad e^{-x/a} = \cot(t/2)$.

Therefore, $e^{x/a} + e^{-x/a} = \dfrac{\sin(t/2)}{\cos(t/2)} + \dfrac{\cos(t/2)}{\sin(t/2)} = \dfrac{1}{\cos(t/2)\sin(t/2)}$,

or $\quad \cosh(x/a) = 1/\sin t = y/a$. Hence, the evolute is $y = a\cosh(x/a)$.

EXERCISE 2.2

1. Find the envelopes of the following straight lines:

   (i) $y = cx + \dfrac{a}{c}$, the parameter being c. (ii) $y = m^2x - 2m^3$ the parameter being m.

   (iii) $y = \dfrac{x}{m} + m^3$. (iv) $\dfrac{x}{c} - \dfrac{y}{c^3} = 2$.

Ans. (i) $y^2 = 6x$, (ii) $27y - x^3 = 0$, (iii) $27x^2 = 4y^3$, (iv) $x^3 = 27y$.

2. Find the envelopes of the following systems of circles:

   (i) $(x - \alpha)^2 + y^2 = 4\alpha$, where α is the parameter. (ii) $x^2 + (y - \alpha)^2 = 2\alpha$.

Ans. (i) $y^2 = 4x + 4$, a parabola. (ii) $x^2 = 2y + 1$.

3. Find the envelope of the following family of straight lines:

   (i) $\dfrac{x}{\sqrt{\sin\theta}} + \dfrac{y}{\sqrt{\cos\theta}} = a$, where θ is the parameter.

   [**Hint:** Equating $f_\theta(x, y, \theta) = 0$, we get $\tan\theta = (x/y)^{2/3}$

   Therefore, $\sin\theta = \pm\dfrac{x^{2/5}}{\sqrt{x^{4/5} + y^{4/5}}}$, $\cos\theta = \pm\dfrac{y^{2/5}}{\sqrt{x^{4/5} + y^{4/5}}}$].

(*ii*) $x \sin\theta - y\cos\theta = a\theta$.

(*iii*) $x\cos\alpha + y\sin\alpha = d$, where $d = a(1 + \cos\alpha)$, the parameter being α.

Ans. (*i*) $x^{4/5} + y^{4/5} = a^{4/5}$, (*ii*) $x = a(\cos\theta + \theta\sin\theta)$, $y = a(\sin\theta - \theta\cos\theta)$, the parametric equation of the envelope, θ being the parameter, (*iii*) $(x - a)^2 + y^2 = a^2$.

4. Show that the family of circles $(x - a)^2 + y^2 = a^2$ has no envelope.

5. Show that the envelope of the straight line which makes on the coordinate axes intercepts whose sum is c, is the parabola.

$$\sqrt{x} + \sqrt{y} = \sqrt{c}.$$

6. Show that the envelope of the family of circles whose diameters are double ordinates of the ellipse $x^2/a^2 + y^2/b^2 = 1$, is the ellipse $x^2/(a^2 + b^2) + y^2/b^2 = 1$.

7. Prove that the envelope of the system of lines $(x/\alpha) + (y/\beta) = 1$, where α, β are connected by the equation $\alpha/a + \beta/b = 1$, is the parabola $(x/a)^{1/2} + (y/b)^{1/2} = 1$.

8. Find the envelope of the family of ellipse $b^2a^2 + a^2y^2 = a^2b^2$ when the sum of its semi-axis equals c.
Ans. $x^{2/3} + y^{2/3} = c^{2/3}$.

9. Prove that the evolute of the rectangular hyperbola $xy = c^2$ is $(x + y)^{2/3} - (x - y)^{2/3} = (4c)^{2/3}$.

10. Show that the evolute of the cycloid $x = a(\theta - \sin\theta)$, $y = a(1 - \cos\theta)$ is another equal cycloid.
(*Madras 2006*)

11. Find the evolute of the ellipse $x^2/a^2 + y^2/b^2 = 1$. **Ans.** $(ax)^{2/3} + (by)^{2/3} = (a^2 - b^2)^{2/3}$.

2.5 GRADIENT OF A SCALAR FIELD AND DIRECTIONAL DERIVATIVE

Before defining gradient of a scalar field and directional derivative, we shall introduce the concepts of a scalar functions and vector functions.

Point Functions: A variable quantity the value of which at any point in a region of space depends on the position of the point, is called a *point function* or *function of position.*

There are two types of point functions.

(*a*) *Scalar point function* (*b*) *Vector point function.*

2.5.1 Scalar Point Function

If to each point P in a region of space R, there corresponds a scalar $f(P)$, then f is called a *scalar function of position* or *scalar point function* and we say that scalar field f has been defined in R. Some of the examples of scalar fields are:

(*i*) The distance $f(P)$ of any point P from a fixed point P_0 in space is a scalar field. This is certainly independent of the choice of the coordinate system.

(*ii*) The temperature T at any point on the earth's surface at a certain time.

(*iii*) The density at any point of a certain body occupying given region.

(*iv*) $\phi(x, y, z) = 3xy^2z + x^2y$ defines a scalar field.

2.5.2 Vector Point Function

If to each point P in a region R of space there corresponds a vector $\mathbf{v}(P)$, then $\mathbf{v}$ is called a *vector function of position* or *vector point function* and we say that a vector field $\mathbf{v}$ has been defined in R. Some examples of vector fields are:

(*i*) velocity at each point of a fluid in motion.

(*ii*) tangent vector to a curve in space.

(*iii*) normal vector to a surface.

(*iv*) gravitational and electromagnetic force fields.

(*v*) $\mathbf{v} = \mathbf{v}(P) = v_1\mathbf{i} + v_2\mathbf{j} + v_3\mathbf{k}$ defined at each point P.

If we introduce Cartesian co-ordinates x, y, z then the value of the scalar point function f at a point $P(x, y, z)$ may be written as $f(x, y, z)$ but we should bear in mind that this value is independent of particular choice of axes of co-ordinates and depends only on the position of P. In order to indicate this fact we sometimes write $f(P)$ instead $f(x, y, z)$.

Similarly at a point $P(x, y, z)$ the vector point function $\mathbf{v}$ may be written in the form:

$\mathbf{v}(x, y, z) = v_1(x, y, z)\mathbf{i} + v_2(x, y, z)\mathbf{j} + v_3(x, y, z)\mathbf{k}$ bearing in mind that $\mathbf{v}$ depends only on the position of P and defines the same vector for every choice of co-ordinate system. This fact is indicated symbolically by writing $\mathbf{v}(P)$ in place of $\mathbf{v}(x, y, z)$.

If the scalar and vector fields depend on time also, then we denote them as $f(P, t)$ and $\mathbf{v}(P, t)$ respectively. Both the fields are independent of the choice of the coodinate systems.

Differentiation of these scalar and vector point functions follow the same rules as those of ordinary calculus. Thus if $f(x, y, z)$ and $\mathbf{v}(x, y, z)$ are respectively scalar and vector point functions, then

$$df = \frac{\partial f}{\partial x}dx + \frac{\partial f}{\partial y}dy + \frac{\partial f}{\partial z}dz, \qquad \frac{df}{dt} = \frac{\partial f}{\partial x}\frac{dx}{dt} + \frac{\partial f}{\partial y}\frac{dy}{dt} + \frac{\partial f}{\partial z}\frac{dz}{dt},$$

and

$$d\mathbf{v} = \frac{\partial \mathbf{v}}{\partial x}dx + \frac{\partial \mathbf{v}}{\partial y}dy + \frac{\partial \mathbf{v}}{\partial z}dz, \qquad \frac{d\mathbf{v}}{dt} = \frac{\partial \mathbf{v}}{\partial x}\frac{dx}{dt} + \frac{\partial \mathbf{v}}{\partial y}\frac{dy}{dt} + \frac{\partial \mathbf{v}}{\partial z}\frac{dz}{dt}.$$

Some of the vector fields can be obtained from scalar fields. The relation between the two types of fields is accomplished by the "*gradient*". To define the gradient of a scalar field, we first introduce a vector differential operator called *del* operator denoted by ∇.

2.5.3 The vector differential operator Del

We define the vector differential operator ∇ (based on partial differentiations) in two and three dimensions as

$$\nabla = \mathbf{i}\frac{\partial}{\partial x} + \mathbf{j}\frac{\partial}{\partial y} \text{ and } \nabla = \mathbf{i}\frac{\partial}{\partial x} + \mathbf{j}\frac{\partial}{\partial y} + \mathbf{k}\frac{\partial}{\partial z}.$$

The del operator is analogous to the derivative operator d/dx, which when applied to $f(x)$ produces the derivative $f'(x)$. The symbol ∇, an inverted capital Greek delta is also known as *nabla*.

Gradient of a scalar field. *When the vector differential operator ∇ is applied to a differentiable function $z = f(x, y)$ or $w = f(x, y, z)$, defining a scalar field we say that the vectors*

$$\nabla f(x, y) = \left(\frac{\partial}{\partial x}\mathbf{i} + \frac{\partial}{\partial y}\mathbf{j}\right)f = \frac{\partial f}{\partial x}\mathbf{i} + \frac{\partial f}{\partial y}\mathbf{j} \qquad \text{...(2.8)}$$

and

$$\nabla f(x, y, z) = \left(\frac{\partial}{\partial x}\mathbf{i} + \frac{\partial}{\partial y}\mathbf{j} + \frac{\partial}{\partial z}\mathbf{k}\right)f = \frac{\partial f}{\partial x}\mathbf{i} + \frac{\partial f}{\partial y}\mathbf{j} + \frac{\partial f}{\partial z}\mathbf{k} \qquad \text{...(2.9)}$$

are gradient of the respective functions.

Note that the del operator ∇ operates on a "scalar field and produce a vector field. The vector ∇f is usually read "grad f" as well as "gradient of f" and "del f".

2.5.4 Geometrical Interpretation of Gradient

Let $f(P) = f(x, y, z)$ be a differential scalar function. Let $f(x, y, z) = c$ = constant be a level surface* through P. Vector normal to tangent plane, to the surface at the point P, is called the *normal vector* to the surface at this point.

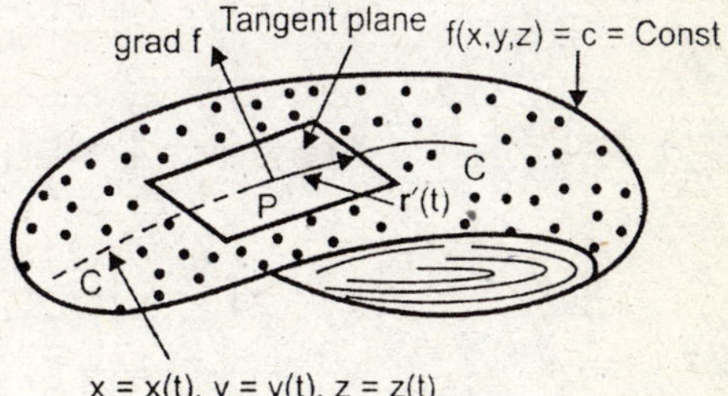

Fig. 2.7: Gradient f as a normal to level surface

* If a surface $f(x, y, z) = c$ = constant, is drawn through any point P of the region such that at each point on the surface, the function has the same value as at P then such a surface is known as *level surface* through P. For different values of c, we obtain different surfaces. For example, if $f(x, y, z)$ represents temperature in a medium, then $f(x, y, z) = c$ represents a surface on which the temperature is a constant c. Such surfaces are called *isothermal* surfaces.

Let $x = x(t)$, $y = y(t)$, $z = z(t)$ be the parametric representation of the smooth curve C on the surface passing through a point P on the surface. The curve C in the space can be represented by a vector function $\mathbf{r}(t) = x(t)\mathbf{i} + y(t)\mathbf{j} + z(t)\mathbf{k}$. Since the curve lie on the surface, we have

$f(x(t), y(t), z(t)) = c$. Then, $\dfrac{d}{dt} f(x(t), y(t), z(t)) = 0$.

By chain rule, we have
$$\frac{\partial f}{\partial x}\frac{dx}{dt} + \frac{\partial f}{\partial y}\frac{dy}{dt} + \frac{\partial f}{\partial z}\frac{dz}{dt} = 0 \qquad \text{...(2.10)}$$

or
$$\left(\mathbf{i}\frac{\partial}{\partial x} + \mathbf{j}\frac{\partial}{\partial y} + \mathbf{k}\frac{\partial}{\partial z}\right) f \bullet \left(\mathbf{i}\frac{dx}{dt} + \mathbf{j}\frac{dy}{dt} + \mathbf{k}\frac{dz}{dt}\right) = 0 \quad \text{or} \quad \nabla f \bullet \mathbf{r}'(t) = 0 \qquad \text{...(2.11)}$$

$\mathbf{r}'(t)$ is a tangent vector to curve C at the point P and lies in the tangent plane to the surface at P. Hence grad f is orthogonal to all the tangent vector at P. Thus gradient as surface normal vector.

LLUSTRATIVE EXAMPLES

EXAMPLE 2.15 Find the gradient of the following scalar fields

(*i*) $f(x, y) = x^3 - 3x^2y^2 + y^3$ at (1, 2).

(*ii*) $\phi(x, y, z) = xy^2 + 3x^2 - z^3$ at (2, –1, 4).

Solution (*i*) We have
$$\nabla f(x, y) = \left(\mathbf{i}\frac{\partial}{\partial x} + \mathbf{j}\frac{\partial}{\partial y}\right)(x^3 - 3x^2y^2 + y^3)$$
$$= \mathbf{i}\frac{\partial}{\partial x}(x^3 - 3x^2y^2 + y^3) + \mathbf{j}\frac{\partial}{\partial y}(x^3 - 3x^2y^2 + y^3)$$
$$= (3x^2 - 6xy^2)\mathbf{i} + (-6x^2y + 3y^2)\,\mathbf{j}.$$

At (1, 2), we obtain $\nabla f(x, y) = -21\,\mathbf{i}$.

(*ii*) We have
$$\nabla\phi(x, y, z) = \left(\mathbf{i}\frac{\partial}{\partial x} + \mathbf{j}\frac{\partial}{\partial y} + \mathbf{k}\frac{\partial}{\partial z}\right)(xy^2 + 3x^2 - z^3) = (y^2 + 6x)\,\mathbf{i} + 2\,xy\,\mathbf{j} - 3z^2\mathbf{k}.$$

At (2, –1, 4), we obtain $\nabla\phi(2, -1, 4) = 13\,\mathbf{i} - 4\mathbf{j} - 48\mathbf{k}$.

EXAMPLE 2.16 (*a*) *If* $\mathbf{R} = x\mathbf{i} + y\mathbf{j} + z\mathbf{k}$, $r = |\mathbf{R}|$ and $\hat{\mathbf{R}} = \mathbf{R}/r$, then show that

(*i*) grad $r = \hat{\mathbf{R}}$, (*ii*) grad $(1/r) = -\hat{\mathbf{R}}/r^2 = -\mathbf{R}/r^3$. *(PTU 2003; UPTU 2002)*

(*b*) Find $\nabla\phi$ if $\phi = \log|\mathbf{R}|$. *(AMIE, W-2008)*

Solution. (*a*) (*i*) We have $r^2 = x^2 + y^2 + z^2$. Therefore grad r, that is,
$$\nabla r = \mathbf{i}\frac{\partial r}{\partial x} + \mathbf{j}\frac{\partial r}{\partial y} + \mathbf{k}\frac{\partial r}{\partial z} = \mathbf{i}\frac{x}{r} + \mathbf{j}\frac{y}{r} + \mathbf{k}\frac{z}{r} = \frac{x\mathbf{i} + y\mathbf{j} + z\mathbf{k}}{r} = \frac{\mathbf{R}}{r} = \hat{\mathbf{R}}.$$

(*ii*)
$$\text{grad}\,\frac{1}{r} = \left(\mathbf{i}\frac{\partial}{\partial x} + \mathbf{j}\frac{\partial}{\partial y} + \mathbf{k}\frac{\partial}{\partial z}\right)\left(\frac{1}{r}\right) = \mathbf{i}\left(-\frac{1}{r^2}\frac{\partial r}{\partial x}\right) + \mathbf{j}\left(-\frac{1}{r^2}\frac{\partial r}{\partial y}\right) + \mathbf{k}\left(-\frac{1}{r^2}\frac{\partial r}{\partial z}\right)$$
$$= -\frac{1}{r^2}\left(\frac{x}{r}\mathbf{i} + \frac{y}{r}\mathbf{j} + \frac{z}{r}\mathbf{k}\right) = -\frac{1}{r^2}\left(\frac{\mathbf{R}}{r}\right) = -\frac{\mathbf{R}}{r^3} = -\frac{\hat{\mathbf{R}}}{r^2}.$$

(*b*) We have $|\mathbf{r}| = |\mathbf{R}| = \sqrt{x^2 + y^2 + z^2}$ and $\phi = \log|\mathbf{R}| = \dfrac{1}{2}\log(x^2 + y^2 + z^2)$.
$$\therefore\ \nabla\phi = \frac{1}{2}\nabla\log(x^2 + y^2 + z^2) = \frac{1}{2}\left\{\mathbf{i}\frac{\partial}{\partial x}\log(x^2 + y^2 + z^2) + \mathbf{j}\frac{\partial}{\partial y}\log(x^2 + y^2 + z^2) + \mathbf{k}\frac{\partial}{\partial z}\log(x^2 + y^2 + z^2)\right\}$$
$$= \frac{1}{2}\left\{\mathbf{i}\frac{2x}{x^2 + y^2 + z^2} + \mathbf{j}\frac{2y}{x^2 + y^2 + z^2} + \mathbf{k}\frac{2z}{x^2 + y^2 + z^2}\right\} = \frac{x\mathbf{i} + y\mathbf{i} + z\mathbf{k}}{x^2 + y^2 + z^2} = \frac{\mathbf{R}}{r^2}.$$

EXAMPLE 2.17 Prove that

(*i*) $\nabla f(r) = \dfrac{f'(r)}{|\mathbf{R}|}\mathbf{R}$, where $\mathbf{R} = x\mathbf{i} + y\mathbf{j} + z\mathbf{k}$ and $r = |\mathbf{R}|$, (*AMIETE, June 2009*) (*ii*) $\nabla r^n = nr^{n-2}\,\mathbf{R}$.

Solution (*i*) $\nabla f(r) = \left(\mathbf{i}\dfrac{\partial}{\partial x} + \mathbf{j}\dfrac{\partial}{\partial y} + \mathbf{k}\dfrac{\partial}{\partial z}\right) f(r) = \mathbf{i}\dfrac{\partial}{\partial x} f(r) + \mathbf{j}\dfrac{\partial}{\partial y} f(r) + \mathbf{k}\dfrac{\partial}{\partial z} f(r)$

$$= \mathbf{i}\, f'(r)\frac{\partial r}{\partial x} + \mathbf{j}\, f'(r)\frac{\partial r}{\partial y} + \mathbf{k}\, f'(r)\frac{\partial r}{\partial z}, \text{ where } f'(r) = \frac{d}{dr} f(r).$$

Since $r = |\mathbf{R}|\sqrt{x^2 + y^2 + z^2}$. $\therefore \dfrac{\partial r}{\partial x} = \dfrac{\partial}{\partial x}(x^2 + y^2 + z^2)^{1/2} = \dfrac{x}{\sqrt{x^2 + y^2 + z^2}} = \dfrac{x}{|\mathbf{R}|}$.

Similarly $\dfrac{\partial r}{\partial y} = \dfrac{y}{|\mathbf{R}|}$ and $\dfrac{\partial r}{\partial z} = \dfrac{z}{|\mathbf{R}|}$.

$$\therefore \quad \nabla f(r) = \mathbf{i}\, f'(r)\frac{x}{|\mathbf{R}|} + \mathbf{j}\, f'(r)\frac{y}{|\mathbf{R}|} + \mathbf{k}\, f'(r)\frac{z}{|\mathbf{R}|} = \frac{f'(r)}{|\mathbf{R}|}(x\mathbf{i} + y\mathbf{j} + z\mathbf{k}) = \frac{f'(r)}{|\mathbf{R}|}\mathbf{R}.$$

(*ii*) Using the result of part (*i*), we have $f(r) = r^n$. Therefore $f'(r) = nr^{n-1}$.

Hence $$\nabla r^n = \frac{nr^{n-1}}{|\mathbf{R}|}\mathbf{R} = nr^{n-2}\mathbf{R}.$$

Alternative $$\nabla r^n = \mathbf{i}\frac{\partial}{\partial x} r^n + \mathbf{j}\frac{\partial}{\partial y} r^n + \mathbf{k}\frac{\partial}{\partial z} r^n.$$

Now $$\frac{\partial}{\partial x} r^n = \frac{\partial}{\partial x}(x^2 + y^2 + z^2)^{n/2} = \frac{n}{2}(x^2 + y^2 + z^2)^{(n-2)/2} \cdot 2x$$

$$= nx(x^2 + y^2 + z^2)^{(n-2)/2} = nxr^{n-2}.$$

Similarly $\dfrac{\partial}{\partial y} r^n = nyr^{n-2}$ and $\dfrac{\partial}{\partial z} r^n = nzr^{n-2}$.

$$\therefore \quad \nabla r^n = nxr^{n-2}\mathbf{i} + nyr^{n-2}\mathbf{j} + nzr^{n-2}\mathbf{k} = nr^{n-2}(x\mathbf{i} + y\mathbf{j} + z\mathbf{k}) = nr^{n-2}\mathbf{R}.$$

EXAMPLE 2.18 (*i*) If $\mathbf{R} = x\mathbf{i} + y\mathbf{j} + z\mathbf{k}$ and $r = |\mathbf{R}|$, then find $\nabla|\mathbf{R}|^3$,

(*ii*) Find $\nabla\phi$ if $\phi = (x^2 + y^2 + z^2)e^{-\sqrt{(x^2+y^2+z^2)}}$.

Solution (*i*) We have $|\mathbf{R}| = \sqrt{(x^2 + y^2 + z^2)}$, therefore $|\mathbf{R}|^3 = (x^2 + y^2 + z^2)^{3/2}$ and

$$\nabla|\mathbf{R}|^3 = \left(\mathbf{i}\frac{\partial}{\partial x} + \mathbf{j}\frac{\partial}{\partial y} + \mathbf{k}\frac{\partial}{\partial z}\right)(x^2 + y^2 + z^2)^{3/2}$$

$$= \mathbf{i}\frac{\partial}{\partial x}(x^2 + y^2 + z^2)^{3/2} + \mathbf{j}\frac{\partial}{\partial y}(x^2 + y^2 + z^2)^{3/2} + \mathbf{k}\frac{\partial}{\partial z}(x^2 + y^2 + z^2)^{3/2}$$

$$= \mathbf{i}\left[\frac{3}{2}(x^2 + y^2 + z^2)^{1/2} \cdot 2x\right] + \mathbf{j}\left[\frac{3}{2}(x^2 + y^2 + z^2)^{1/2} \cdot 2y\right] + \mathbf{k}\left[\frac{3}{2}(x^2 + y^2 + z^2)^{1/2} \cdot 2z\right]$$

$$= 3(x^2 + y^2 + z^2)^{1/2}[x\mathbf{i} + y\mathbf{j} + z\mathbf{k}] = 3r\,\mathbf{R}.$$

(*ii*) $\mathbf{R} = x\mathbf{i} + y\mathbf{j} + z\mathbf{k}$, $r = |\mathbf{R}| = \sqrt{(x^2 + y^2 + z^2)}$.

$$\therefore \ \nabla\phi = \left(\mathbf{i}\frac{\partial}{\partial x} + \mathbf{j}\frac{\partial}{\partial y} + \mathbf{k}\frac{\partial}{\partial z}\right)(x^2 + y^2 + z^2)e^{-\sqrt{(x^2+y^2+z^2)}} = \sum \mathbf{i}\frac{\partial}{\partial x}\left[(x^2 + y^2 + z^2)e^{-\sqrt{(x^2+y^2+z^2)}}\right]$$

$$= \sum \mathbf{i}\left[2xe^{-\sqrt{(x^2+y^2+z^2)}} - (x^2+y^2+z^2)e^{-\sqrt{(x^2+y^2+z^2)}} \cdot \frac{1}{2}(x^2+y^2+z^2)^{-1/2} \cdot 2x\right]$$

$$= \sum \mathbf{i}\left[e^{-\sqrt{(x^2+y^2+z^2)}} \cdot x\left\{2-\sqrt{(x^2+y^2+z^2)}\right\}\right]$$

$$= e^{-\sqrt{(x^2+y^2+z^2)}}\left\{2-\sqrt{(x^2+y^2+z^2)}\right\}(x\mathbf{i}+y\mathbf{j}+z\mathbf{k}) = e^{-r}(2-r)\mathbf{R}.$$

EXAMPLE 2.19 Find a unit normal vector to the surface $x^2y + 2xz = 4$ at the point (2, –2, 3).

(*AMIETE, Dec 2007; AMIE, W-2005*)

Solution Let us regard the given surface as a particular level surface of the function $\phi = x^2y + 2xz$. Then the gradient of this function at the point (2, –2, 3) will be perpendicular to the level surface through (2, –2, 3), which is the given surface.

Now $\phi(x, y, z) = x^2y + 2xz - 4$. $\partial\phi/\partial x = 2xy + 2z$, $\partial\phi/\partial y = x^2$ and $\partial\phi/\partial z = 2x$.

Then, the normal vector is given by

$$\nabla\phi = \mathbf{i}\frac{\partial\phi}{\partial x} + \mathbf{j}\frac{\partial\phi}{\partial y} + \mathbf{k}\frac{\partial\phi}{\partial z} = (2xy+2z)\mathbf{i} + x^2\mathbf{j} + 2x\mathbf{k}.$$

At (2, –2, 3), we obtain the normal vector as $\nabla\phi(2, -2, 3) = -2\mathbf{i} + 4\mathbf{j} + 4\mathbf{k}$.

Hence a unit vector normal to the given surface is given by

$$\mathbf{n} = \frac{\nabla\phi}{|\nabla\phi|} = \frac{(-2\mathbf{i}+4\mathbf{j}+4\mathbf{k})}{\sqrt{4+16+16}} = -\frac{1}{3}\mathbf{i} + \frac{2}{3}\mathbf{j} + \frac{2}{3}\mathbf{k}.$$

EXAMPLE 2.20 Find the normal vector and the equation of the tangent plane to the surface

$xz^2 + x^2y - z + 1 = 0$ at the point (1, –3, 2).

Solution Let $f(x, y, z) = xz^2 + x^2y - z + 1 = 0$ be the surface. Then, the normal vector is given by

$$\mathbf{N} = \nabla f = \left(\mathbf{i}\frac{\partial}{\partial x} + \mathbf{j}\frac{\partial}{\partial y} + \mathbf{k}\frac{\partial}{\partial z}\right)(xz^2 + x^2y - z + 1) = (z^2 + 2xy)\mathbf{i} + x^2\mathbf{j} + (2xz - 1)\mathbf{k}.$$

At (1, –3, 2), the normal vector is given by $\mathbf{N} = \nabla f(1, -3, 2) = -2\mathbf{i} + \mathbf{j} + 3\mathbf{k}$.

The tangent plane at the point (1, –3, 2) is given by

$$-2(x-1) + 1(y+3) + 3(z-2) = 0 \quad \text{or} \quad 2x - y - 3z + 1 = 0.$$

EXAMPLE 2.21 Find the angle between the surfaces $x^2 + y^2 + z^2 = 9$ and $z + 3 = x^2 + y^2$ at the point (2, –1, 2).

(*Kottayam, 2005; UPTU 2003; MDU, 2001*)

Solution The angle between two surfaces at a common point is the angle between their normals at that point. We have

$$f_1(x, y, z) = x^2 + y^2 + z^2 - 9 = 0, \ \nabla f_1(x, y, z) = 2x\mathbf{i} + 2y\mathbf{j} + 2z\mathbf{k}.$$

$\nabla f_1(2, -1, 2) = 4\mathbf{i} - 2\mathbf{j} + 4\mathbf{k}$, which is normal $\mathbf{N}_1$ to $x^2 + y^2 + z^2 = 9$ at (2, –1, 2).

$$f_2(x, y, z) = x^2 + y^2 - z - 3 = 0, \ \nabla f_2(x, y, z) = 2x\mathbf{i} + 2y\mathbf{j} - \mathbf{k}.$$

$\mathbf{N}_2 = \nabla f_2(2, -1, 2) = 4\mathbf{i} - 2\mathbf{j} - \mathbf{k}$. If the required angle is θ then

$$\cos\theta = \frac{\mathbf{N}_1 \cdot \mathbf{N}_2}{|\mathbf{N}_1||\mathbf{N}_2|} = \frac{(4\mathbf{i}-2\mathbf{j}+4\mathbf{k})\cdot(4\mathbf{i}-2\mathbf{j}-\mathbf{k})}{\sqrt{(4)^2+(-2)^2+(4)^2}\sqrt{(4)^2+(-2)^2+(-1)^2}}$$

$$= \frac{16+4-4}{6\sqrt{21}} = \frac{8\sqrt{21}}{63}, \quad \text{or} \quad \theta = \cos^{-1}(8\sqrt{21}/63) = 54°25'.$$

EXAMPLE 2.22 If $\nabla f = (x + 2y + 4z)\,\mathbf{i} + (2x - 3y - z)\mathbf{j} + (4x - y + 2z)\,\mathbf{k}$, find the scalar function f.

Solution Since $\nabla f = \frac{\partial f}{\partial x}\mathbf{i} + \frac{\partial f}{\partial y}\mathbf{j} + \frac{\partial f}{\partial z}\mathbf{k}$. Comparing, we obtain

$$\frac{\partial f}{\partial x} = x + 2y + 4z \quad \cdots (i), \quad \frac{\partial f}{\partial y} = 2x - 3y - z \quad \cdots (ii), \quad \frac{\partial f}{\partial z} = 4x - y + 2z \quad \cdots (iii).$$

Integrating the first equation, we obtain $f(x, y, z) = \frac{x^2}{2} + 2xy + 4zx + g(y, z)$. ...(A)

Substituting in the second equation, we get $\frac{\partial f}{\partial y} = 2x - 3y - z = 2x + \frac{\partial g}{\partial y}$, or $\frac{\partial g}{\partial y} = -3y - z$.

Therefore, $g(y, z) = -3\frac{y^2}{2} - yz + g(z)$.

Now (A) become $f(x, y, z) = (x^2/2) + 2xy + 4zx + (-3/2)y^2 - yz + g(z)$. ...(B)

Substituting in the third equation, we get

$$\frac{\partial f}{\partial z} = 4x - y + 2z = 4x - y + \frac{dg}{dz} \text{ or } \frac{dg}{dz} = 2z.$$ Integrating, we get $g = z^2 + c_1$, constant.

Thus from (B), we obtain

$$f(x, y, z) = (x^2/2) + 2xy + 4xz - yz + (-3/2)y^2 + z^2 + c_1.$$

Remark A vector field **F** derived from a scalar field f such that $\mathbf{F} = \nabla f$ is called *a conservative vector field* and f is called the *scalar potential.*

EXAMPLE 2.23 Show that the vector field defined by the vector function $\mathbf{v} = xyz(yz\mathbf{i} + zx\mathbf{j} + xy\mathbf{k})$ is conservative. *(AMIETE, June 2011, 2008)*

Solution If the given vector field is conservative, then it can be expressed as the gradient of a scalar function $f(x, y, z)$. Therefore

$$\nabla f = \left(\mathbf{i}\frac{\partial f}{\partial x} + \mathbf{j}\frac{\partial f}{\partial y} + \mathbf{k}\frac{\partial f}{\partial z}\right) = \mathbf{v} = xyz(yz\mathbf{i} + zx\mathbf{j} + xy\mathbf{k}).$$

Comparing, we obtain

$$\frac{\partial f}{\partial x} = xy^2z^2 \quad \ldots (i), \quad \frac{\partial f}{\partial y} = x^2yz^2 \quad \ldots (ii), \quad \frac{\partial f}{\partial z} = x^2y^2z \quad \ldots (iii).$$

Integrating the first equation, we obtain

$f(x, y, z) = (1/2)\, x^2y^2z^2 + g(y, z)$. ...(A) Substituting in the second equation, we get

$$\frac{\partial f}{\partial y} = x^2yz^2 = x^2yz^2 + \frac{\partial g}{\partial y} \text{ or } \frac{\partial g}{\partial y} = 0 \Rightarrow g = g(z).$$

Now Eq. (A) become $f(x, y, z) = (1/2)\, x^2y^2z^2 + g(z)$. ...(B)

Substituting in the third equation, we get

$$\frac{\partial f}{\partial z} = x^2y^2z = x^2y^2z + \frac{dg}{dz} \text{ or } \frac{dg}{dz} = 0 \Rightarrow g = c_1, \text{ constant.}$$

Hence, $f(x, y, z) = x^2y^2z^2 + c$. Therefore, there exists a scalar function $f(x, y, z)$ such that $\nabla f = \mathbf{v}$ and the vector field **v** is conservative.

2.5.5 Directional Derivative

Let $f(P) = f(x, y, z)$ be a differential scalar field. Then $\partial f/\partial x$, $\partial f/\partial y$, $\partial f/\partial z$ denote the rates of change of f in the directions of x, y and z axis, respectively. Let $\hat{\mathbf{b}} = b_1\mathbf{i} + b_2\mathbf{j} + b_3\mathbf{k}$ be any unit vector. Let P_o *be any point, its position vector is* $\mathbf{a} = a_1\,\mathbf{i} + a_2\,\mathbf{j} + a_3\,\mathbf{k}$. Refer Fig. 2.8. Let P_oC is parallel to $\hat{\mathbf{b}}$ and Q is a variable point on the directed line segment P_oC. Then the position vector of point Q is given by

$$\mathbf{r}(s) = \mathbf{a} + s\hat{\mathbf{b}}, \quad s \ge 0 = (a_1 + b_1s)\mathbf{i} + (a_2 + b_2s)\mathbf{j} + (a_3 + b_3s)\mathbf{k} = x(s)\mathbf{i} + y(s)\mathbf{j} + z(s)\mathbf{k}.$$

The vector $\overrightarrow{P_0Q}$ is given by $s\hat{\mathbf{b}}$. Since $|\hat{\mathbf{b}}| = 1$, so the distance from P_0 to Q is s.

Then $\dfrac{\partial f}{\partial s} = \lim\limits_{s \to 0} \dfrac{f(Q) - f(P_0)}{s}$

if it exists, is called the *directional derivative* of f at the point P_0 in the direction of $\hat{\mathbf{b}}$. It is the rate of change of f in the direction of $\mathbf{b}$.

Therefore, $\dfrac{\partial}{\partial s} f(x(s), y(s), z(s))$ is the rate of change of f with respect to the distance s.

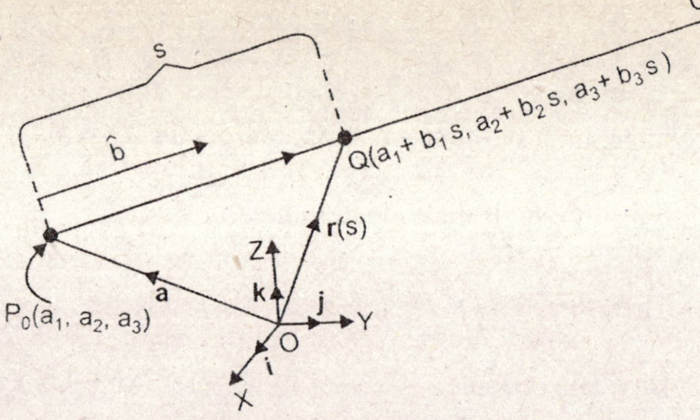

Fig. 2.8: Directional derivative

We have by the chain rule $\dfrac{\partial f}{\partial s} = \dfrac{\partial f}{\partial x}\dfrac{dx}{ds} + \dfrac{\partial f}{\partial y}\dfrac{dy}{ds} + \dfrac{\partial f}{\partial z}\dfrac{dz}{ds}$ ...(*i*)

where dx/ds, dy/ds, dz/ds are calculated at the point P_0 (where $s = 0$). We write (*i*) as

$$\frac{\partial f}{\partial s} = \left(\mathbf{i}\frac{\partial}{\partial x} + \mathbf{j}\frac{\partial}{\partial y} + \mathbf{k}\frac{\partial}{\partial z}\right) f \cdot \left(\mathbf{i}\frac{dx}{ds} + \mathbf{j}\frac{dy}{ds} + \mathbf{k}\frac{dz}{ds}\right) = \nabla f \cdot \frac{d\mathbf{r}}{ds}$$

But $d\mathbf{r}/ds = \hat{\mathbf{b}}$ (a unit vector). Therefore, the directional derivative of f in the direction of $\hat{\mathbf{b}}$, denoted by $D_{\mathbf{b}}(f)$, is given by

$$D_{\mathbf{b}}(f) = \nabla f \cdot \hat{\mathbf{b}} = grad(f) \cdot \hat{\mathbf{b}} \qquad ...(2.12)$$

Remark If the direction is specified by a vector $\mathbf{u}$, then we have $\hat{\mathbf{b}} = \mathbf{u}/|\mathbf{u}|$ and $D_{\mathbf{b}}(f) = \nabla f \cdot \mathbf{u}/|\mathbf{u}|$.

Maximum/Minimum value of the Directional derivative

Let f represent a function of either two or three variables. Since the directional derivative is a dot product, we see from the definition of dot or scalar product that

$$D_{\mathbf{b}}(f) = \nabla f \cdot \hat{\mathbf{b}} = |\nabla f|\,|\hat{\mathbf{b}}| \cos\alpha = |\nabla f| \cos\alpha, \qquad \text{as } (|\hat{\mathbf{b}}| = 1),$$

where α is the angle between the vector ∇f and $\mathbf{b}$. Because $0 \le \alpha \le \pi$, we have $-1 \le \cos\alpha \le 1$ and consequently $-|\nabla f| \le D_{\mathbf{b}}(f) \le |\nabla f|$.

Thus the maximum value of the directional derivative is $|\nabla f|$ and it occurs when $\alpha = 0$ *i.e.*, $\hat{\mathbf{b}}$ has the same direction as $\nabla\phi$. This direction is the direction of the normal vector. The minimum value of the directional derivative is $-|\nabla\phi|$ and it occurs when $\theta = \pi$ that is $\hat{\mathbf{b}}$ and ∇f have opposite directions. We may also say that the gradient vector ∇f points in the direction in which f increases most rapidly, whereas $-\nabla f$ points in the direction of the most rapid decrease of f.

ILLUSTRATIVE EXAMPLES

EXAMPLE 2.24 Find the directional derivative of the function $f(x, y, z) = xy^2 + yz^3$ at the point $(2, -1, 1)$ in the direction of the vector $\mathbf{i} + 2\mathbf{j} + 3\mathbf{k}$? *(AMIETE, Dec. 2009; Kurukshetra, 2006; AMIE, S-2005)*

Solution We have $\nabla f = \nabla(xy^2 + yz^3) = y^2\mathbf{i} + (2xy + z^3)\mathbf{j} + 3yz^2\mathbf{k}$.

$\nabla f(2, -1, 1) = \mathbf{i} - 3\mathbf{j} - 3\mathbf{k}$. The projection of this in the direction of the given vector will be the required directional derivative.

The unit vector in the given direction is $\hat{\mathbf{b}} = \dfrac{(\mathbf{i} + 2\mathbf{j} + 3\mathbf{k})}{\sqrt{1 + 4 + 9}} = \dfrac{\mathbf{i} + 2\mathbf{j} + 3\mathbf{k}}{\sqrt{14}}$.

Therefore, $D_{\mathbf{b}} f = \nabla f \cdot \hat{\mathbf{b}} = (\mathbf{i} - 3\mathbf{j} - 3\mathbf{k}) \cdot \dfrac{(\mathbf{i} + 2\mathbf{j} + 3\mathbf{k})}{\sqrt{14}} = \dfrac{(1)(1) + (-3)(2) + (-3)(3)}{\sqrt{14}} = \dfrac{-14}{\sqrt{14}} = -\sqrt{14}$.

The negative sign indicates that f decreases at P in the direction of $\mathbf{b}$.

EXAMPLE 2.25 Find the directional derivative of $f(x, y) = 2x^2y^3 + 6xy$ at (1, 1) in the direction of a unit vector whose angle with the positive x-axis is $\pi/6$.

Solution We have $\nabla f(x, y) = (4xy^3 + 6y)\,\mathbf{i} + (6x^2y^2 + 6x)\mathbf{j}$. At the point (1, 1), $\nabla f = 10\mathbf{i} + 12\mathbf{j}$.

$$\hat{\mathbf{b}} = (\cos\theta)\,\mathbf{i} + (\sin\theta)\mathbf{j} = \frac{\sqrt{3}}{2}\mathbf{i} + \frac{1}{2}\mathbf{j}.$$

Therefore $\quad D_{\mathbf{b}} f = \nabla f \cdot \hat{\mathbf{b}} = (10\mathbf{i} + 12\mathbf{j}) \cdot \left(\frac{\sqrt{3}}{2}\mathbf{i} + \frac{1}{2}\mathbf{j}\right) = 5\sqrt{3} + 6.$

EXAMPLE 2.26 (*i*) In what direction from the point (3, 1, –2) is the directional derivative of $f(x, y, z) = x^2y^2z^4$ maximum and (*ii*) what is its magnitude? (*AMIE, W-2005; MDU 2004*)

Solution We have $\nabla f = \nabla(x^2y^2z^4) = 2xy^2z^4\mathbf{i} + 2x^2yz^4\mathbf{j} + 4x^2y^2z^3\mathbf{k}$.

At the point (3, 1, –2), we have $\nabla f = 96\mathbf{i} + 288\mathbf{j} - 288\mathbf{k}$.

(*i*) The maximum value of the directional derivative is $|\nabla f|$ and it occurs when $\theta = 0$, that is $\hat{\mathbf{b}}$ has the direction of ∇f.

(*ii*) The magnitude of the maximum directional derivative

$$= \sqrt{(96)^2 + (288)^2 + (-288)^2} = \sqrt{175104} = 96\sqrt{19}.$$

EXAMPLE 2.27 (*i*) Find the directional derivative of the function $f(x, y, z) = 2x^3y - 3y^2z$ at $P(1, 2, -1)$ in a direction toward $Q(3, -1, 5)$.

(*ii*) In what direction from P is the directional derivative a maximum?

(*iii*) What is the magnitude of the maximum directional derivative?

Solution (*i*) We have $\nabla f = 6x^2y\mathbf{i} + (2x^3 - 6yz)\mathbf{j} - 3y^2\mathbf{k}$.

At the point (1, 2, –1), $\nabla f = 12\mathbf{i} + 14\mathbf{j} - 12\mathbf{k}$.

The direction of vector = $(3 - 1)\mathbf{i} + (-1-2)\mathbf{j} + [5 -(-1)]\mathbf{k} = 2\mathbf{i} - 3\mathbf{j} + 6\mathbf{k}$.

The unit vector in the given direction is $\hat{\mathbf{b}} = \dfrac{2\mathbf{i} - 3\mathbf{j} + 6\mathbf{k}}{\sqrt{(2)^2 + (-3)^2 + (6)^2}} = \dfrac{2\mathbf{i} - 3\mathbf{j} + 6\mathbf{k}}{7}$.

Therefore, directional derivative at P, $D_{\mathbf{b}} f = \nabla f \cdot \hat{\mathbf{b}}$

$$= (12\mathbf{i} + 14\mathbf{j} - 12\mathbf{k}) \cdot \left(\frac{2\mathbf{i} - 3\mathbf{j} + 6\mathbf{k}}{7}\right) = \frac{24 - 42 - 72}{7} = -\frac{90}{7},$$

that is, f is decreasing in the given direction.

(*ii*) The directional derivative is a maximum in the direction $\nabla f = 12\mathbf{i} + 14\mathbf{j} - 12\mathbf{k}$.

(*iii*) The magnitude of the maximum directional derivative is $|\nabla f| = |12\mathbf{i} + 14\mathbf{j} - 12\mathbf{k}| = \sqrt{144 + 196 + 144} = 22.$

EXAMPLE 2.28 If the directional derivative of $f(x, y, z) = axy^2 + byz + cz^2x^3$ at (1, 2, –1) has a maximum magnitude of 64 in the direction parallel to z-axis, find the values of constants a, b, c. (*AMIETE, June 2001*)

Solution $\nabla f = \dfrac{\partial f}{\partial x}\mathbf{i} + \dfrac{\partial f}{\partial y}\mathbf{j} + \dfrac{\partial f}{\partial z}\mathbf{k} = (ay^2 + 3cz^2x^2)\mathbf{i} + (2axy + bz)\mathbf{j} + (by + 2czx^3)\mathbf{k}$.

$\therefore \qquad \nabla f(1, 2, -1) = (4a + 3c)\mathbf{i} + (4a - b)\mathbf{j} + (2b - 2c)\mathbf{k}$.

Unit vector $\hat{\mathbf{n}}$ is a direction parallel to the z-axis is $o\mathbf{i} + o\mathbf{j} + \mathbf{k}$. Therefore, $\hat{\mathbf{n}} \cdot \nabla f = 2b - 2c = 64$ (given). Directional derivative is maximum along $\bar{\mathbf{n}}$; this means that $\bar{\mathbf{n}}$ is parallel to ∇f at (1, 2, –1).

Thus $4a + 3c = 0$ and $4a - b = 0$. Solving hese equations, we get $a = 6$, $b = 24$, $c = -8$.

EXAMPLE 2.29 Find the directional derivative of $f(x, y, z) = x^2yz^3$ along the curve $x = e^{-u}$, $y = 2\sin u + 1$, $z = u - \cos u$ at the point P where $u = 0$.

Solution The point P corresponding to $u = 0$ is $(1, 1, -1)$.

$$\nabla f = 2xyz^3\mathbf{i} + x^2z^3\mathbf{j} + 3x^2yz^2\mathbf{k}. \qquad \therefore \nabla f(1, 1, -1) = -2\mathbf{i} - \mathbf{j} + 3\mathbf{k}.$$

The position vector $\mathbf{r}$ of any point on the given curve is given by

$$\mathbf{r} = e^{-u}\,\mathbf{i} + (2\sin u + 1)\mathbf{j} + (u - \cos u)\mathbf{k}.$$

A tangent vector to the curve is

$$\mathbf{r}'(u) = \frac{d\mathbf{r}}{du} = \frac{d}{du}\left\{e^{-u}\mathbf{i} + (2\sin u + 1)\mathbf{j} + (u - \cos u)\mathbf{k}\right\}$$

$$= -e^{-u}\,\mathbf{i} + 2\cos u\,\mathbf{j} + (1 + \sin u)\mathbf{k}. \qquad \therefore (d\mathbf{r}/du) \text{ at } P, \text{ that is, } \mathbf{r}'(0) = -\mathbf{i} + 2\mathbf{j} + \mathbf{k}.$$

The unit tangent vector in this direction $= \dfrac{-\mathbf{i}+2\mathbf{j}+\mathbf{k}}{\sqrt{1+4+1}} = \dfrac{-\mathbf{i}+2\mathbf{j}+\mathbf{k}}{\sqrt{6}}$.

Then, directional derivative = $\nabla f \cdot$ (unit tangent vector in this direction)

$$= (-2\mathbf{i} - \mathbf{j} + 3\mathbf{k}) \cdot \left(\frac{-\mathbf{i}+2\mathbf{j}+\mathbf{k}}{\sqrt{6}}\right) = \frac{1}{\sqrt{6}}(2 - 2 + 3) = \frac{3}{\sqrt{6}} = \frac{\sqrt{6}}{2}.$$

Since this is positive, f is increasing in this direction.

2.6 PROPERTIES OF GRADIENT

If f and g are differentiable scalar functions of position (x, y, z), then

(*i*) $\nabla(k f) = k \nabla f$ (any number k), (*ii*) $\nabla(f \pm g) = \nabla f \pm \nabla g$,

(*iii*) $\nabla(c_1 f + c_2 g) = c_1 \nabla f + c_2 \nabla g$, c_1, c_2 arbitrary constants, (*iv*) $\nabla(fg) = f\nabla g + g\nabla f$,

(*v*) $\nabla\left(\dfrac{f}{g}\right) = \dfrac{g\nabla f - f\nabla g}{g^2}$, $g \neq 0$.

The proof of properties (*ii*), (*iv*) and (*v*) is given below.

(*ii*)
$$\nabla(f \pm g) = \left(\mathbf{i}\frac{\partial}{\partial x} + \mathbf{j}\frac{\partial}{\partial y} + \mathbf{k}\frac{\partial}{\partial z}\right)(f \pm g) = \mathbf{i}\frac{\partial}{\partial x}(f \pm g) + \mathbf{j}\frac{\partial}{\partial y}(f \pm g) + \mathbf{k}\frac{\partial}{\partial z}(f \pm g)$$

$$= \left(\mathbf{i}\frac{\partial f}{\partial x} + \mathbf{j}\frac{\partial f}{\partial y} + \mathbf{k}\frac{\partial f}{\partial z}\right) \pm \left(\mathbf{i}\frac{\partial g}{\partial x} + \mathbf{j}\frac{\partial g}{\partial y} + \mathbf{k}\frac{\partial g}{\partial z}\right) = \nabla f \pm \nabla g.$$

(*iv*)
$$\nabla(fg) = \left(\mathbf{i}\frac{\partial}{\partial x} + \mathbf{j}\frac{\partial}{\partial y} + \mathbf{k}\frac{\partial}{\partial z}\right)fg = \mathbf{i}\left(f\frac{\partial g}{\partial x} + g\frac{\partial f}{\partial x}\right) + \mathbf{j}\left(f\frac{\partial g}{\partial y} + g\frac{\partial f}{\partial y}\right) + \mathbf{k}\left(f\frac{\partial g}{\partial z} + g\frac{\partial f}{\partial z}\right)$$

$$= f\left(\mathbf{i}\frac{\partial g}{\partial x} + \mathbf{j}\frac{\partial g}{\partial y} + \mathbf{k}\frac{\partial g}{\partial z}\right) + g\left(\mathbf{i}\frac{\partial f}{\partial x} + \mathbf{j}\frac{\partial f}{\partial y} + \mathbf{k}\frac{\partial f}{\partial z}\right) = f\nabla g + g\nabla f.$$

(*v*)
$$\nabla\left(\frac{f}{g}\right) = \left(\mathbf{i}\frac{\partial}{\partial x} + \mathbf{j}\frac{\partial}{\partial y} + \mathbf{k}\frac{\partial}{\partial z}\right)\left(\frac{f}{g}\right)$$

$$= \mathbf{i}\frac{\left(g\dfrac{\partial f}{\partial x} - f\dfrac{\partial g}{\partial x}\right)}{g^2} + \mathbf{j}\frac{\left(g\dfrac{\partial f}{\partial y} - f\dfrac{\partial g}{\partial y}\right)}{g^2} + \mathbf{k}\frac{\left(g\dfrac{\partial f}{\partial z} - f\dfrac{\partial g}{\partial z}\right)}{g^2}, \text{ if } g \neq 0$$

$$= \frac{1}{g^2}\left[g\left(\mathbf{i}\frac{\partial f}{\partial x} + \mathbf{j}\frac{\partial f}{\partial y} + \mathbf{k}\frac{\partial f}{\partial z}\right) - f\left(\mathbf{i}\frac{\partial g}{\partial x} + \mathbf{j}\frac{\partial g}{\partial y} + \mathbf{k}\frac{\partial g}{\partial z}\right)\right] = \frac{g\nabla f - f\nabla g}{g^2}.$$

EXERCISE 2.3

1. Compute the gradient of the following scalar fields

(*i*) $f(x, y) = y^2 - 4xy$, (*ii*) $f(x, y) = 5y - x^3y^2$,

(*iii*) $f(x, y, z) = xy^2 + 3x^2 - z^3$, (*iv*) $f(x, y, z) = xy^2/z^3$, (*v*) $f(x, y, z) = x^2 + yz$,

(*vi*) $\phi(x, y, z) = x^3 + y^3 + z^3 - 3xyz$, (*vii*) $\phi(x, y, z) = \log(x^2 + y^2 + z^2)$.

Ans. (*i*) $-4y\mathbf{i} + (2y - 4x)\mathbf{j}$, (*ii*) $-3x^2y^2\mathbf{i} + (5 - 2x^3y)\mathbf{j}$, (*iii*) $(y^2 + 6x)\mathbf{i} + 2xy\mathbf{j} - 3z^2\mathbf{k}$,

(*iv*) $(y^2/z^3)\mathbf{i} + (2xy/z^3)\mathbf{j} - (3xy^2/z^4)\mathbf{k}$, (*v*) $2x\mathbf{i} + z\mathbf{j} + y\mathbf{k}$, (*vi*) $3(x^2 - yz)\mathbf{i} + 3(y^2 - zx)\mathbf{j} + 3(z^2 - xy)\mathbf{k}$.

[**Hint:** (*vii*) $\nabla \log(x^2 + y^2 + z^2) = \nabla \log r^2 = \frac{2}{r} \cdot \frac{\mathbf{r}}{r} = \frac{2\mathbf{r}}{r^2} = \frac{2(x\mathbf{i} + y\mathbf{j} + 3\mathbf{k})}{x^2 + y^2 + z^2}$.]

2. Find the gradient of the given function at the indicated point

(*i*) $f(x, y) = x^2 - 4y^2$, (2, 4). (*ii*) $f(x, y, z) = 3x^2y - y^3z^2$, (1, –2, –1).

(*iii*) $f(x, y, z) = 2xz^4 - x^2y$, (2, –2, –1). (*iv*) $f(x, y, z) = x^2y + y^2x + z^2$, (1, 1, 1).

(*v*) $f(x, y, z) = x^3 - y^3 + xz^2$, (1, 1, 2). (*AMIETE, Dec. 2010*)

Ans. (*i*) $4\mathbf{i} - 32\mathbf{j}$, (*ii*) $-12\mathbf{i} - 9\mathbf{j} - 16\mathbf{k}$, (*iii*) $10\mathbf{i} - 4\mathbf{j} - 16\mathbf{k}$, (*iv*) $3\mathbf{i} + 3\mathbf{j} + 2\mathbf{k}$, (*v*) $7\mathbf{i} - 3\mathbf{j} + 4\mathbf{k}$.

3. (*a*) Evaluate $\nabla\left(3r^2 - 4\sqrt{r} + \frac{6}{\sqrt[3]{r}}\right)$. (*b*) Show that grad $\left(\frac{1}{r^2}\right) = \frac{-2\mathbf{r}}{r^4}$. (*MDU 2002*)

Ans. (*a*) $(6 - 2r^{-3/2} - 2r^{-7/3})\mathbf{r}$.

4. (*a*) Find the normal vector and the unit normal vector to the given curve/surface at the indicated point

(*i*) $x^2 - y^2 = 12$, (4, 2). (*ii*) $z = x^2 + y^2$, (1, 2, 5).

(*iii*) $xy^3z^2 = 4$, (–1, –1, 2). (*iv*) $x^4 - 3xyz + z^2 + 1 = 0$, (1, 1, 1).

(*v*) $x^3 + y^3 + 3xyz = 3$, (1, 2, –1). (*AMIETE, June 2002*) (*vi*) $x^3 - xyz = 1$, (1, 1, 1). (*AMIETE, Dec. 2004*)

(*vii*) $f(x, y, z) = xy^2 + 2yz - 8 = 0$ at (3, –2. 1) (*AMIE, W-2007*)

(*b*) Find the magnitude of the vector drawn perpendicular to the surface $x^2 + 2y^2 + z^2 = 7$ at the point (1, –1, 2). (*AMIE, S 2009*)

Ans. (*i*) $4(2\mathbf{i} - \mathbf{j})$, $(2\mathbf{i} - \mathbf{j})/\sqrt{5}$, (*ii*) $(2\mathbf{i} + 4\mathbf{j} - \mathbf{k})$; $(2\mathbf{i} + 4\mathbf{j} - \mathbf{k})/\sqrt{21}$,

(*iii*) $4(\mathbf{i} + 3\mathbf{j} - \mathbf{k})$; $-(\mathbf{i} + 3\mathbf{j} - \mathbf{k})/\sqrt{11}$, (*iv*) $\mathbf{i} - 3\mathbf{j} - \mathbf{k}$; $(\mathbf{i} - 3\mathbf{j} - \mathbf{k})/\sqrt{11}$,

(*v*) $(-\mathbf{i} + 3\mathbf{j} + 2\mathbf{k})$; $(-\mathbf{i} + 3\mathbf{j} + 2\mathbf{k})/\sqrt{14}$, (*vi*) $2\mathbf{i} - \mathbf{j} - \mathbf{k}$; $(2\mathbf{i} - \mathbf{j} - \mathbf{k})/\sqrt{6}$,

(*vii*) $4\mathbf{i} - 10\mathbf{j} - 4\mathbf{k}$, $(2\mathbf{i} - 5\mathbf{j} - 2\mathbf{k})/\sqrt{33}$. (*b*) 6.

5. If $u = x + y + z$, $v = x^2 + y^2 + z^2$, $w = xy + yz + zx$, show that $\nabla u \bullet [\nabla v \times \nabla w] = 0$. (*UPTU, 2002*)
[**Alternative:** If $u = x + y + z$, $v = x^2 + y^2 + z^2$, $w = yz + zx + xy$, prove that grad u grad v and grad w are coplanar.]

6. What is the angle between the normals to the surface $xy = z^2$ at the points (1, 4, 2) and (–3, –3, 3)?

Ans. $\cos^{-1}(1/\sqrt{22})$.

7. Find the angle between the two surfaces at the indicated point of intersection.

(*i*) $xy^2z = 3x + z^2$, $3x^2 - y^2 + 2z = 1$; (1, –2, 1).

(*ii*) $x \log z = y^2 - 1$, $x^2y = 2 - z$, (1, 1, 1). (*AMIETE, Dec. 2006, JNTU 2003*)

Ans. (*i*) $\cos^{-1}(\sqrt{6}/14)$, (*ii*) $\cos^{-1}(1/\sqrt{30})$.

8. (*a*) Find the normal vector and the equation to tangent plane to surface $z = \sqrt{x^2 + y^2}$ at point (3, 4, 5). (*AMIE, W-2010; AMIETE, Dec. 2005*)

(*b*) Find the equation of the tangent plane to the ellipsoid $x^2 + 4y^2 + z^2 = 18$ at the point (1, 2, 1), and determine the acute angle that this plane makes with the xy-plane.

Ans. (*a*) $-\frac{3}{5}\mathbf{i} - \frac{4}{5}\mathbf{j} + \mathbf{k}$; $3x + 4y - 5z = 0$. (*b*) $x + 8y + z = 18$; $\approx 83°$.

9. Find the value of λ and μ so that the surfaces $\lambda x^2 - \mu yz = (\lambda + 2)x$ and $4x^2y + z^3 = 4$ intersect orthogonally at (1, –1, 2). *(AMIETE, S-2009, W-2008, W-2005; MDU 2005)* **Ans.** $\lambda = 5/2$, $\mu = 1$.

10. Find the constants a and b such that the surface $ax^2 - byz = (a + 2)x$ will be orthogonal to the surface $4x^2y + z^3 = 4$ at the point (1, –1, 2). *(AMIE, S-2007; Madras 2004)*

Ans. $a = 5/2$, $b = 1$ [**Hint:** Since (1, –1, 2) lies on $ax^2 - 2byz = (a + 4)x$, we get $b = 1$. Use the condition of orthogonality to get a.]

11. If $\nabla f = 2xy\,z^3\mathbf{i} + x^2z^3\,\mathbf{j} + 3x^2yz^2\,\mathbf{k}$, find $f(x, y, z)$ if $f(1, -2, 2) = 4$. **Ans.** $f(x, y, z) = x^2yz^3 + 20$.

12. Find the directional derivative of

(*i*) $f(x, y) = xe^y + \cos(xy)$ at (2, 0) in the direction of $3\mathbf{i} - 4\mathbf{j}$.

(*ii*) $f(x, y, z) = 4xz^3 - 3x^2y^2z^2$ at (2, –1, 2) along z-axis.

(*iii*) $f(x, y, z) = x^2yz + 4xz^2$ at (1, 2, –1) in the direction $2\mathbf{i} - \mathbf{j} - 2\mathbf{k}$.

(AMIE, W-2008; VTU 2007; JNTU 2006; MDU 2006)

(*iv*) $f(x, y, z) = xy^2 - 4x^2y + z^2$ at (1, –1, 2) in the direction of $6\mathbf{i} + 2\mathbf{j} + 3\mathbf{k}$.

(*v*) $f(x, y, z) = 4xz^3 - 3x^2y^2z$ at (2, –1, 2) in the direction $2\mathbf{i} - 3\mathbf{j} + 6\mathbf{k}$.

(*vi*) $\phi(x, y, z) = 4\,e^{2x - y + z}$ at (1, 1, –1) in the direction of vector $\mathbf{j}$.

(*vii*) $\phi = 3e^{2x-y+z}$ at $A(1, 1, -1)$ in the direction $\overline{AB}$ where B is the point (–3, 5, 6). *(AMIETE, Dec.2001)*

(*viii*) $\phi(x, y) = x/(x^2 + y^2)$ in the direction of a line making an angle of 30° with the positive x-axis at the point (0, 1). *(AMIETE, Dec. 2002)*

Ans. (*i*) –1, (*ii*) 48, (*iii*) 37/3, (*iv*) 54/7, (*v*) 376/7, (*vi*) –4 (*vii*) –5/3, (*viii*) $\sqrt{3}/2$.

13. What is the directional derivative of the scalar function $\phi = xy^2 + yz^2$ at the point (2, –1, 1) in the direction of normal to the surface $x \log z - y^2 + 4 = 0$ at (–1, 2, 1)? *(MDU 2007; AMIETE, W-2004; Andhra, 2000)*

Ans. $15/\sqrt{17}$.

14. Find the directional derivative of $\vec{V}^2$ where $\vec{V} = xy^2\mathbf{i} + zy^2\mathbf{j} + xz^2\mathbf{k}$ at the point (2, 0, 3) in the direction of the outward normal to the sphere $x^2 + y^2 + z^2 = 14$ at the point (3, 2, 1). *(AMIETE Dec. 2001)*

Ans. $702\sqrt{14}/7$.

15. (*i*) Find the directional derivative of the function $\phi = 4e^{2x - y + z}$ at the point $A(1, 1, -1)$ in the direction towards the point $B(-3, 5, 6)$.

(*ii*) In what direction from the point (1, 1, –1) is the directional derivative of $\phi = x^2 - 2y^2 + 4z^2$ a maximum? Also find the value of this maximum directional derivative.

Ans. (*i*) –20/9, (*ii*) $2\mathbf{i} - 4\mathbf{j} - 8\mathbf{k}$; $2\sqrt{21}$.

16. What is the greatest rate of increase of

(*i*) $\phi(x, y, z) = x^2 + yz^2$ at (1, –1, 3)? *(AMIE, S-2004)* (*ii*) $f(x, y, z) = xyz^2$ at (1, 0, 3)?

Ans. (*i*) 11; (*ii*) 9.

17. Find the directions in which $\phi(x, y) = (x^2 + y^2)/2$ increases most rapidly and decreases most rapidly at the point (1, 1). **Ans.** $\frac{1}{\sqrt{2}}\mathbf{i} + \frac{1}{\sqrt{2}}\mathbf{j}$; $-\frac{1}{\sqrt{2}}\mathbf{i} - \frac{1}{\sqrt{2}}\mathbf{j}$.

18. (*i*) Find the directional derivative of $f(x, y, z) = 2xy - z^2$ at (2, –1, 1) in a direction toward (3, 1, –1).

(*ii*) In what direction is the directional derivative a maximum?

(*iii*) What is the value of this maximum?

Ans. (*i*) 10/3, (*ii*) $-2\mathbf{i} + 4\mathbf{j} - 2\mathbf{k}$ (*iii*) $2\sqrt{6}$.

19. (*i*) Given that the directional derivative of $f(x, y, z)$ at the point (3, –2, 1) in the direction of $\mathbf{a} = 2\mathbf{i} - \mathbf{j} - 2\mathbf{k}$ is –5 and that $|\nabla f(3, -2, 1)| = 5$, find $\nabla f(3, -2, 1)$.

(*ii*) Find the directional derivative of $\phi = 5x^2y - 5y^2z + 2.5z^2x$ at the point $P(1, 1, 1)$ in the direction of the line $(x-1)/2 = (y-3)/-2 = z$. *(UPTU 2004)*

Ans. (*i*) $-5(2\mathbf{i} - \mathbf{j} - 2\mathbf{k})/3$, (*ii*) 35/3.

20. The temperature at a point (x, y, z) in a space is given by $T(x, y, z) = x^2 + y^2 - z$. A mosquito located at (1, 1, 2) desires to fly in such a direction that it will get warm as soon as possible. In what direction should it fly? *(AMIE, S-2005)* **Ans.** $(2\mathbf{i} + 2\mathbf{j} - \mathbf{k})/3$.

21. The temperature at a point (x, y, z) in a space is given by $T(x, y, z) = x^2 + y^2 - z$. A fly located at (4, 4, 2) desires to fly in a direction that gets cooler faster. Find the direction in which it should fly. Also find the rate decrease of temperature in the direction of flight. *(AMIETE, June 2005)*

Ans. In the direction of maximum rate of decrease, $-(8\mathbf{i} + 8\mathbf{j} - \mathbf{k})$.

22. Find the values of a, b, c so that the directional derivative of $\phi = ax^2 + by^2 + cz^2$ at the point (1, 1, 2) has a maximum magnitude 4 in the direction of y-axis. [**Hint:** $\nabla\phi$ at $(1, 1, 2) = 4\mathbf{j}$]

Ans. $a = 0, b = 2, c = 0$.

23. If the directional derivative of $\phi = ax^2y + by^2z + cz^2x$ at (1, 1, 1) has maximum magnitude 15 in the direction parallel to line $\dfrac{x-1}{2} = \dfrac{y-3}{-2} = z$, find the value of a, b, c. *(AMIETE June 2007, 2003)*

[**Hint:** The maximum directional derivative of ϕ at any point is the value of $\nabla\phi$ at the point.

$$\nabla\phi = (2axy + cz^2)\mathbf{i} + (2byz + ax^2)\mathbf{j} + (by^2 + 2czx)\mathbf{k}.$$

$$\nabla\phi\,(1, 1, 1) = (2a + c)\mathbf{i} + (2b + a)\mathbf{j} + (b + 2c)\mathbf{k} = 15\frac{(2\mathbf{i} - 2\mathbf{j} + \mathbf{k})}{\sqrt{4+4+1}} = 10\mathbf{i} - 10\mathbf{j} + 5\mathbf{k}.$$

$\Rightarrow$ $2a + c = 10$, $2b + a = -10$, $b + 2c = 5$. Solving, we get $a = 20/9$; $b = -55/9$, $c = 50/9$].

24. Find $\phi(r)$ such that $\nabla\phi = \mathbf{r}/r^5$ and $\phi(1) = 0$. **Ans.** $\dfrac{1}{3}\left(1 - \dfrac{1}{r^3}\right)$.

25. If $f(x, y, z) = x^2z + e^{y/x}$ and $g(x, y, z) = 2z^2y - xy^2$, find (*a*) $\nabla(f + g)$ and (*b*) $\nabla(fg)$ at the point (1, 0, - 2).

Ans. (*a*) $-4\mathbf{i} + 9\mathbf{j} + \mathbf{k}$, (*b*) $-8\mathbf{j}$.

26. If $f(x, y) = \log_e \sqrt{x + y^2}$, show that grad $f = \dfrac{\mathbf{r} - (\mathbf{k}\cdot\mathbf{r})\mathbf{k}}{\{\mathbf{r} - (\mathbf{k}\cdot\mathbf{r})\mathbf{k}\}\cdot\{\mathbf{r} - (\mathbf{k}\cdot\mathbf{r})\mathbf{k}\}}$, where $\mathbf{r} = x\mathbf{i} + y\mathbf{j} + z\mathbf{k}$.

[**Hint:** We have $\mathbf{k} \bullet \mathbf{r} = \mathbf{z}$. Therefore, $\mathbf{r} - (\mathbf{k}\cdot\mathbf{r})\,\mathbf{k} = (x\mathbf{i} + y\mathbf{j} + z\mathbf{k}) - z\mathbf{k} = x\mathbf{i} + y\mathbf{j} = \mathbf{r}_1$ say)

We have to show that $\nabla f = \dfrac{\mathbf{r}_1}{|\mathbf{r}_1^2|} = \dfrac{x\mathbf{i} + y\mathbf{j}}{x^2 + y^2}$. Now $f(x, y) = \dfrac{1}{2}\log(x^2 + y^2)$.

Therefore, $\nabla f = \left(\mathbf{i}\dfrac{\partial}{\partial x} + \mathbf{j}\dfrac{\partial}{\partial y}\right)\dfrac{1}{2}\log(x^2 + y^2) = \dfrac{1}{2}\left[\dfrac{2x\mathbf{i}}{x^2 + y^2} + \dfrac{2y\mathbf{j}}{x^2 + y^2}\right] = \dfrac{x\mathbf{i} + y\mathbf{j}}{x^2 + y^2}$.]

2.7 TAYLOR'S THEOREM FOR A FUNCTION OF TWO VARIABLES

We know by Taylor's theorem that for a function $f(x)$ of single variable x,

$$f(x + h) = f(x) + hf'(x) + \frac{h^2}{2!}f''(x) + \frac{h^3}{3!}f'''(x) + \ldots\ldots\ldots\ldots$$

Now let $f(x, y)$ be a function of two independent variables x and y. *If y is kept constant,* then by Taylor's theorem for a function of single variable x, we have

$$f(x+h, y+k) = f(x, y+k) + h\frac{\partial}{\partial x}f(x, y+k) + \frac{h^2}{2!}\frac{\partial^2}{\partial x^2}f(x, y+k) + \frac{h^3}{3!}\frac{\partial^3}{\partial x^3}f(x, y+k) + \ldots \quad \text{...(2.13)}$$

Now *keeping x constant* and applying Taylor's theorem for a function of single variable y, we have

$$f(x, y+k) = f(x, y) + k\frac{\partial}{\partial y}f(x, y) + \frac{k^2}{2!}\frac{\partial^2}{\partial y^2}f(x, y) + \frac{k^3}{3!}\frac{\partial^3}{\partial y^3}f(x, y) + \ldots \quad \text{...(2.14)}$$

Using (2.14), we can write (2.13) in the form

$$f(x+h, y+k) = \left\{f(x, y)+k\frac{\partial}{\partial y}f(x, y)+\frac{k^2}{2!}\frac{\partial^2}{\partial y^2}f(x, y)+\frac{k^3}{3!}\frac{\partial^3}{\partial y^3}f(x, y)+\ldots\right\}$$

$$+h\frac{\partial}{\partial x}\left\{f(x, y)+k\frac{\partial}{\partial y}f(x, y)+\frac{k^2}{2!}\cdot\frac{\partial^2}{\partial y^2}f(x, y)+\ldots\right\}$$

$$+\frac{h^2}{2!}\frac{\partial^2}{\partial x^2}\left\{f(x, y)+k\frac{\partial}{\partial y}f(x, y)+\ldots\right\}+\frac{h^3}{3!}\frac{\partial^3}{\partial x^3}[f(x, y)+\ldots]+\ldots$$

$$= \left\{f(x, y)+k\frac{\partial}{\partial y}f(x, y)+\frac{k^2}{2!}\frac{\partial^2}{\partial y^2}f(x, y)+\frac{k^3}{3!}\frac{\partial^3}{\partial y^3}f(x, y)+\ldots\right\}$$

$$+\left\{h\frac{\partial}{\partial x}f(x, y)+hk\frac{\partial^2}{\partial x\partial y}f(x, y)+\frac{hk^2}{2!}\frac{\partial^3}{\partial x\partial y^2}f(x, y)+\ldots\right\}$$

$$+\left\{\frac{h^2}{2!}\frac{\partial^2}{\partial x^2}f(x, y)+\frac{h^2k}{2!}\frac{\partial^3}{\partial x^2\partial y}f(x, y)+\ldots\right\}+\frac{h^3}{3!}\frac{\partial^3}{\partial x^3}f(x, y)+\ldots$$

$$= f(x,y)+\left\{h\frac{\partial}{\partial x}f(x,y)+k\frac{\partial}{\partial y}f(x,y)\right\}+\left\{\frac{h^2}{2!}\frac{\partial^2}{\partial x^2}f(x, y)+hk\frac{\partial^2}{\partial x\partial y}f(x, y)+\frac{k^2}{2!}\frac{\partial^2}{\partial y^2}f(x, y)\right\}$$

$$+\left\{\frac{h^3}{3!}\frac{\partial^3}{\partial x^3}f(x,y)+\frac{h^2k}{2!}\frac{\partial^3}{\partial x^2\partial y}f(x,y)+\frac{hk^2}{2!}\frac{\partial^3}{\partial x\partial y^2}f(x,y)+\frac{k^3}{3!}\frac{\partial^3}{\partial y^3}f(x, y)\right\}+\ldots$$

$$= f(x, y)+\left\{h\frac{\partial}{\partial x}f(x, y)+k\frac{\partial f}{\partial y}(x, y)\right\}+\frac{1}{2!}\left\{h^2\frac{\partial^2}{\partial x^2}f(x, y)+2hk\frac{\partial^2}{\partial x\partial y}f(x, y)+k^2\frac{\partial^2}{\partial y^2}f(x, y)\right\}$$

$$+\frac{1}{3!}\left\{h^3\frac{\partial^3}{\partial x^3}f(x, y)+3h^2k\frac{\partial^3}{\partial x^2\partial y}f(x, y)+3hk^2\frac{\partial^3}{\partial x\partial y^2}f(x, y)+k^3\frac{\partial^3}{\partial y^3}f(x, y)\right\}+\ldots \quad \ldots(2.15)$$

The above result can be written symbolically as

$$f(x+h, y+k) = f(x, y)+\left(h\frac{\partial}{\partial x}+k\frac{\partial}{\partial y}\right)f(x, y)+\frac{1}{2!}\left(h\frac{\partial}{\partial x}+k\frac{\partial}{\partial y}\right)^2 f(x, y)+\frac{1}{3!}\left(h\frac{\partial}{\partial x}+k\frac{\partial}{\partial y}\right)^3 f(x, y)+\ldots$$

where $\left(h\frac{\partial}{\partial x}+k\frac{\partial}{\partial y}\right)^2$ stands for the operator $h^2\frac{\partial^2}{\partial x^2}+2hk\frac{\partial^2}{\partial x\partial y}+k^2\frac{\partial^2}{\partial y^2}$. Similarly for $\left(h\frac{\partial}{\partial x}+k\frac{\partial}{\partial y}\right)^3$ and etc.

Remarks

1. Putting $x = a$ and $y = b$ and denoting the value of $\partial f/\partial x$ for $x = a$, $y = b$ by $f_x(a, b)$, and etc. We get from (2.15)

$$f(a+h, b+k) = f(a, b) + [hf_x(a, b) + kf_y(a, b)] + \frac{1}{2!}[h^2 f_{xx}(a, b) + 2hkf_{xy}(a, b) + k^2 f_{yy}(a, b)]$$

$$+\frac{1}{3!}\left[h^3 f_{xxx}(a, b)+3h^2kf_{xxy}(a, b)+3hk^2 f_{xyy}(a, b)+k^3 f_{yyy}(a, b)\right]+\ldots \quad (2.16)$$

2. If we put $a + h = x$ and $b + k = y$ so that $h = (x - a)$ and $k = (y - b)$, the result is *Taylor's Theorem for a function of two variables,*

$$f(x, y) = f(a, b) + [(x-a) f_x (a, b) + (y-b) f_y (a, b)] + $$
$$+\frac{1}{2!}\left[(x-a)^2 f_{xx}(a, b)+2(x-a)(y-b)f_{xy}(a, b)+(y-b)^2 f_{yy}(a, b)\right]+\ldots \quad \ldots(2.17)$$

3. Setting $a = b = 0$ in eqn. (2.17), we obtain

$$f(x, y) = f(0, 0)+\{x f_x(0, 0)+y f_y(0, 0)\}+\frac{1}{2!}\{x^2 f_{xx}(0, 0)+2xy f_{xy}(0, 0)+y^2 f_{yy}(0, 0)\}$$
$$+\frac{1}{3!}\{x^3 f_{xxx}(0, 0)+3x^2 y f_{xxy}(0, 0)+3xy^2 f_{xyy}(0, 0)+y^3 f_{yyy}(0, 0)\}+\ldots \quad \ldots(2.18)$$

This expansion is known as Maclaurin's expansion for functions of two variables $f(x, y)$.

4. The expansion (2.17) is called Taylor's expansion of $f(x, y)$ at (a, b) or in the neighbourhood of (a, b) or in powers of $(x-a)$ and $(y-b)$. Obviously (2.18) is expansion of $f(x, y)$ at (0, 0) (or in the neighbourhood of (0, 0) or in powers of x and y).

ILLUSTRATIVE EXAMPLES

EXAMPLE 2.30 Expand $e^x \sin y$ by Taylor's series in powers of x and y as the terms of third degree.

(*Kottaym 2005; AMIE, S-2002, AMIETE, June 2002*)

Solution We have $f(x, y) = e^x \sin y$; $f(0, 0) = 0$

$f_x(x, y) = e^x \sin y$; $f_x(0, 0) = 0$

$f_y(x, y) = e^x \cos y$; $f_y(0, 0) = 1$

$f_{xx}(x, y) = e^x \sin y$; $f_{xx}(0, 0) = 0$

$f_{xy}(x, y) = e^x \cos y$; $f_{xy}(0, 0) = 1$

$f_{yy}(x, y) = -e^x \sin y$; $f_{yy}(0, 0) = 0$

$f_{xxx}(x, y)$ {i.e., $f_{x^3}(x, y)$} $= e^x \sin y$; $f_{x^3}(0, 0)\} = 0$

$f_{xxy}(x, y)$ {i.e., $f_{x^2y}(x, y)$} $= e^x \cos y$; $f_{x^2y}(0, 0)\} = 1$

$f_{xy^2}(x, y) = -e^x \sin y$; $f_{xy^2}(0, 0)\} = 0$

$f_{y^3}(x, y) = -e^x \cos y$; $f_{y^3}(0, 0)\} = -1$

....

....

Substituting these values in Taylor's series, we obtain

$$e^x \sin y = 0+x(0)+y(1)+\frac{1}{2!}\left[x^2(0)+2xy(1)+y^2(0)\right]+\frac{1}{3!}\left[x^3(0)+3x^2y(1)+3xy^2(0)+y^3(-1)\right]+\ldots$$
$$= y+xy+\frac{1}{2}x^2y-\frac{1}{6}y^3+\ldots$$

EXAMPLE 2.31 Expand $\dfrac{(x+h)(y+k)}{x+h+y+k}$ in powers of h and k upto and inclusive of the second degree terms.

(*AMIE, S-2001*)

Solution We have $f(x+h, y+k)=\dfrac{(x+h)(y+k)}{x+h+y+k}$. Therefore $f(x, y)=\dfrac{xy}{x+y}$.

$$f_x=\frac{(x+y)y-xy}{(x+y)^2}=\frac{y^2}{(x+y)^2}; \quad f_y=\frac{x^2}{(x+y)^2}; \quad f_{xx}=-\frac{2y^2}{(x+y)^3};$$
$$f_{xy}=\frac{(x+y)^2\cdot 2x-x^2\cdot 2(x+y)}{(x+y)^4}=\frac{2xy}{(x+y)^3}; \quad f_{yy}=-\frac{2x^2}{(x+y)^3}.$$
$$\therefore \quad \frac{(x+h)(y+k)}{x+h+y+k}=f(x+h, y+k)=f(x, y)+(hf_x+kf_y)+\frac{1}{2!}(h^2 f_{xx}+2hkf_{xy}+k^2 f_{yy})+\ldots\ldots$$

$$= \frac{xy}{x+y} + \left(\frac{hy^2}{(x+y)^2} + \frac{kx^2}{(x+y)^2} \right) + \frac{1}{2!}\left(\frac{-2h^2y^2}{(x+y)^3} + \frac{4hkxy}{(x+y)^3} - \frac{2k^2x^2}{(x+y)^3} \right) + \ldots\ldots$$

$$= \frac{xy}{x+y} + \frac{1}{(x+y)^2}(hy^2 + kx^2) - \frac{1}{(x+y)^3}(h^2y^2 - 2hkxy + k^2x^2) + \ldots$$

EXAMPLE 2.32 Expand $f(x, y) = x^2y + 3y - 2$ in powers of $(x - 1)$ and $(y + 2)$ upto 3rd degree terms.
(UPTU, 2006; Anna 2005; AMIE, S-2004; AMIETE, W-2003)

Solution The expansion of $f(x, y)$ at the point (a, b) or in powers of $(x - a)$ and $(y - b)$ is given by

$$f(x, y) = f(a, b) + \{(x - a) f_x(a, b) + (y - b) f_y(a, b)\}$$

$$+ \frac{1}{2!}\left[(x-a)^2 f_{xx}(a, b) + 2(x-a)(y-b) f_{xy}(a, b) + (y-b)^2 f_{yy}(a, b) \right]$$

$$+ \frac{1}{3!}\left[(x-a)^3 f_{xxx}(a, b) + 3(x-a)^2(y-b) f_{xxy} + 3(x-a)(y-b)^2 f_{xyy}(a, b) + (y-b)^3 f_{yyy} \right] + \ldots \quad \ldots(1)$$

We have

$f(x, y) = x^2y + 3y - 2;\ a = 1, b = -2$; $f(1, -2) = -10$
$f_x(x, y) = 2xy$; $f_x(1, -2) = -4$
$f_y(x, y) = x^2 + 3$; $f_y(1, -2) = 4$
$f_{xx}(x, y) = 2y$; $f_{xx}(1, -2) = -4$
$f_{xy}(x, y) = 2x$; $f_{xy}(1, -2) = 2$
$f_{yy}(x, y) = 0$; $f_{yy}(1, -2) = 0$
$f_{xxx}(x, y) = 0$; $f_{xxx}(1, -2) = 0$
$f_{xxy}(x, y) = 2$; $f_{xxy}(1, -2) = 2$
$f_{xyy}(x, y) = 0$; $f_{xyy}(1, -2) = 0$
$f_{yyy}(x, y) = 0$; $f_{yyy}(1, -2) = 0.$

Substituting these values in (1), we obtain

$$f(x, y) = x^2y + 3y - 2 = -10 + [(x - 1)(-4) + (y + 2)(4)]$$

$$+ \frac{1}{2!}\left[(x-1)^2(-4) + 2(x-1)(y+2)(2) + (y+2)^2(0) \right] + \frac{1}{3!}\left[0 + 3(x-1)^2(y+2)(2) + 0 + 0 \right]$$

$$= -10 - 4(x - 1) + 4(y + 2) - 2(x - 1)^2 + 2(x - 1)(y + 2) + (x - 1)^2(y + 2).$$

EXAMPLE 2.33 Expand $f(x, y) = \tan^{-1}(y/x)$ in powers of $(x - 1)$ and $(y - 1)$ up to the third degree terms. Hence compute $f(1.1, 0.9)$ approximately. *(AMIETE, Dec 2009; JNTU 2006; UPTU, 2006; AMIE, S-2002)*

Solution We have $f(x, y) = \tan^{-1}(y/x)$; $f(1, 1) = \pi/4$
$f_x(x, y) = -y/(x^2 + y^2)$; $f_x(1, 1) = -1/2$
$f_y(x, y) = x/(x^2 + y^2)$; $f_y(1, 1) = 1/2$
$f_{xx}(x, y) = 2xy/(x^2 + y^2)^2$; $f_{xx}(1, 1) = 1/2$
$f_{xy}(x, y) = (y^2 - x^2)/(x^2 + y^2)^2$; $f_{xy}(1, 1) = 0$
$f_{yy}(x, y) = -2xy/(x^2 + y^2)^2$; $f_{yy}(1, 1) = -1/2$
$f_{xxx}(x, y) = (2y^3 - 6x^2y)/(x^2 + y^2)^3$; $f_{xxx}(1, 1) = -1/2$
$f_{xxy}(x, y) = (2x^3 - 6xy^2)/(x^2 + y^2)^3$; $f_{xxy}(1, 1) = -1/2$
$f_{xyy}(x, y) = (6x^2y - 2y^3)/(x^2 + y^2)^3$; $f_{xyy}(1, 1) = 1/2$
$f_{yyy}(x, y) = (6xy^2 - 2x^3)/(x^2 + y^2)^3$; $f_{yyy}(1, 1) = 1/2$

Taylor's expansion of $f(x, y)$ in powers of $(x - 1)$ and $(y - 1)$ is given by

$$f(x, y) = f(1, 1) + [(x - 1)f_x(1, 1) + (y - 1)f_y(1, 1)]$$

$$+\frac{1}{2!}\left[(x-1)^2 f_{xx}(1,1)+2(x-1)(y-1)f_{xy}(1,1)+(y-1)^2 f_{yy}(1,1)\right]$$

$$+\frac{1}{3!}\left[(x-1)^3 f_{xxx}(1,1)+3(x-1)^2(y-1)f_{xxy}(1,1)+3(x-1)(y-1)^2 f_{xyy}(1,1)+(y-1)^3 f_{yyy}(1,1)\right]+\ldots$$

$$\therefore\ \tan^{-1}\left(\frac{y}{x}\right)=\frac{\pi}{4}+\left[(x-1)\left(-\frac{1}{2}\right)+(y-1)\left(\frac{1}{2}\right)\right]+\frac{1}{2!}\left[(x-1)^2\left(\frac{1}{2}\right)+2(x-1)(y-1)(0)+(y-1)^2\left(-\frac{1}{2}\right)\right]$$

$$+\frac{1}{3!}\left[(x-1)^3\left(-\frac{1}{2}\right)+3(x-1)^2(y-1)\left(-\frac{1}{2}\right)+3(x-1)(y-1)^2\left(\frac{1}{2}\right)+(y-1)^3\left(\frac{1}{2}\right)\right]+\ldots$$

$$=\frac{\pi}{4}-\frac{1}{2}[(x-1)-(y-1)]+\frac{1}{4}\left[(x-1)^2-(y-1)^2\right]-$$

$$\frac{1}{12}\left[(x-1)^3+3(x-1)^2(y-1)-3(x-1)(y-1)^2-(y-1)^3\right]+\ldots$$

Putting $x = 1.1$ and $y = 0.9$, we obtain

$$f(1.1, 0.9)\approx\frac{\pi}{4}-\frac{1}{2}(0.2)+\frac{1}{4}(0)-\frac{1}{12}\left[(0.1)^3-3(0.1)^3-3(0.1)^3-(-0.1)^3\right]$$

$$= 0.7854 - 0.1 + 0.0003 = 0.6857.$$

EXAMPLE 2.34 (*i*) Expand $f(x, y) = \sin xy$ in powers of $(x - 1)$ and $(y - \pi/2)$ upto second degree terms. *(AMIE, W 2004; VTU, 2004)*

(*ii*) Find the linear and the quadratic Taylor's series polynomial approximations to the function $f(x, y) = 2x^3 - 4x^2y + 3y^3$, about the point (1, 2) *(AMIETE, Dec. 2006; AMIE, W-2002)*

Solution (*i*) We have

$f(x, y) = \sin xy$	;	$f(1, \pi/2) = 1$
$f_x(x, y) = y\cos(xy)$	;	$f_x(1, \pi/2) = 0$
$f_y(x, y) = x\cos(xy)$	;	$f_y(1, \pi/2) = 0$
$f_{xx}(x, y) = -y^2\sin(xy)$	;	$f_{xx}(1, \pi/2) = -\pi^2/4$
$f_{xy} = \cos(xy) + xy(-\sin xy)$	;	$f_{xy}(1, \pi/2) = -\pi/2$
$f_{yy} = -x^2\sin(xy)$	;	$f_{yy}(1, \pi/2) = -1.$
....		....
....		....

Substituting the values of partial derivatives obtained above in Taylor's series expansion, we obtain

$$\sin(xy)=1+(x-1)(0)+\left(y-\frac{\pi}{2}\right)(0)+\frac{1}{2!}\left[(x-1)^2\left(-\frac{\pi^2}{4}\right)+2(x-1)\left(y-\frac{\pi}{2}\right)\left(-\frac{\pi}{2}\right)+\left(y-\frac{\pi}{2}\right)^2(-1)\right]+\ldots$$

or $$\sin(xy)=1-\frac{\pi^2}{8}(x-1)^2-\frac{\pi}{2}(x-1)\left(y-\frac{\pi}{2}\right)-\frac{1}{2}\left(y-\frac{\pi}{2}\right)^2+\ldots$$

(*ii*) We have

$f(x, y) = 2x^3 - 4x^2y + 3y^3;$	$f(1, 2) = 2 - 8 + 24 = 18$
$f_x(x, y) = 6x^2 - 8xy;$	$f_x(1, 2) = 6 - 16 = -10$
$f_y(x, y) = -4x^2 + 9y^2;$	$f_y(1, 2) = -4 + 36 = 32$
$f_{xx}(x, y) = 12x - 8y;$	$f_{xx}(1, 2) = 12 - 16 = -4$
$f_{xy}(x, y) = -8x;$	$f_{xy}(1, 2) = -8$
$f_{yy}(x, y) = 18y;$	$f_{yy}(1, 2) = 36$

$f_{xxx}(x, y) = 12,\ f_{xxy}(x, y) = -8,\ f_{xyy}(x, y) = 0,\ f_{yyy}(x, y) = 18.$

The linear approximation in given by

$$f(x, y) \approx f(1, 2) + [(x-1) f_x(1, 2) + (y-2) f_y(1, 2)] = 18 + (x-1)(-10) + (y-2)(32)]$$
$$= 18 - 10(x-1) + 32(y-2).$$

The quadratic approximation is given by

$$f(x, y) \approx f(1, 2) + [(x-1) f_x(1, 2) + (y-2) f_y(1, 2)]$$
$$+\frac{1}{2}\Big[(x-1)^2 f_{xx}(1,2) + 2(x-1)(y-2) f_{xy}(1,2) + (y-2)^2 f_{yy}(1,2)\Big]$$
$$= 18 - 10(x-1) + 32(y-2) + \frac{1}{2}[-4(x-1)^2 - 16(x-1)(y-2) + 36(y-2)^2]$$
$$= 18 - 10(x-1) + 32(y-2) - 2[(x-1)^2 + 4(x-1)(y-2) - 9(y-2)^2].$$

EXAMPLE 2.35 If $f(x, y) = \tan^{-1}(xy)$, find an approximate value of $f(1.1, 0.8)$ using the Taylor's series quadratic approximation. *(AMIETE, June 2007)*

Solution Let $(a, b) = (1.0, 1.0)$, $h = 0.1$, $k = -0.2$. Then $f(1.1, 0.8) = f(1 + 0.1, 1 - 0.2)$.

Using the Taylor's series quadratic approximation, we have

$$f(1.1, 0.8) \approx f(1, 1) + [hf_x(1, 1) + kf_y(1, 1)] + \frac{1}{2!}[h^2 f_{xx}(1, 1) + 2hkf_{xy}(1, 1) + k^2 f_{yy}(1, 1)] \quad ...(1)$$

We have

$$f(x, y) = \tan^{-1}(xy); \qquad f(1, 1) = \tan^{-1}(1) = \pi/4 \approx 0.7854$$

$$f_x(x, y) = \frac{y}{1+x^2y^2}; \qquad f_x(1, 1) = 0.5$$

$$f_y(x, y) = \frac{x}{1+x^2y^2}; \qquad f_y(1, 1) = 0.5$$

$$f_{xx}(x, y) = -\frac{2xy^3}{(1+x^2y^2)^2}; \qquad f_{xx}(1, 1) = -0.5$$

$$f_{xy}(x, y) = \frac{(1+x^2y^2) - 2x^2y^2}{(1+x^2y^2)^2} = \frac{1-x^2y^2}{(1+x^2y^2)^2}; \qquad f_{xy}(1, 1) = 0$$

$$f_{yy}(x, y) = -\frac{2x^3y}{(1+x^2y^2)^2}, \qquad f_{yy}(1, 1) = -0.5$$

Therefore by (1), we obtain,

$$f(1.1, 0.8) \approx 0.7854 + \{(0.5)(0.1) + (0.5)(-0.2)\} +$$
$$\frac{1}{2!}\{(0.01)(-0.5) + 2(0.1)(-0.2)(0) + (0.04)(-0.5)\} + ... = 0.7854 - 0.05 - 0.0125 = 0.7229.$$

EXERCISE 2.4

1. (*i*) Expand $e^x \cos y$ by Taylor's theorem in powers of x and y as far as terms of the third degree.
   Alternative: Expand $e^x \cos y$ at the point (0, 0) up to terms of the third degree.

   (*ii*) Expand the function $f(x, y) = 1/(xy)$ in powers of $(x - 2)$ and $(y + 1)$ up to terms of the second degree using Taylor's theorem.

   (*iii*) Find a quadratic $f(x, y) = \sin x \sin y$ near the origin.

Ans. (*i*) $1 + x + \frac{1}{2!}(x^2 - y^2) + \frac{1}{3!}(x^3 - 3xy^2) + ...$

(*ii*) $\left(-\frac{1}{2}\right) + \frac{1}{4}(x-2) - \frac{1}{2}(y+1) - \frac{1}{8}(x-2)^2 + \frac{1}{4}(x-2)(y+1) - \frac{1}{2}(y+1)^2 + ...$ (*iii*) $\sin x \sin y = xy$.

2. Show that $e^y \log(1+x) = x + xy - \frac{x^2}{2} + \frac{1}{2}(xy^2 - x^2y) + \frac{1}{3}x^3 + \ldots$

3. (*i*) Expand $e^x \cos y$ in the neighbourhood of $(1, \pi/4)$.

(*ii*) Expand $e^x \cos y$ in powers of x and $\left(y - \frac{\pi}{2}\right)$ as far as 3rd degree terms, using Taylor's series expansion. (*AMIETE, S-2005*)

Ans. (*i*) $\frac{e}{\sqrt{2}}\left[1 + (x-1) - \left(y - \frac{\pi}{4}\right) + \frac{1}{2}(x-1)^2 - (x-1)\left(y - \frac{\pi}{4}\right) - \frac{1}{2}\left(y - \frac{\pi}{4}\right)^2 + \ldots\right]$

(*ii*) $-\left(y - \frac{\pi}{2}\right) - x\left(y - \frac{\pi}{2}\right) - \frac{1}{2}x^2\left(y - \frac{\pi}{2}\right) + \frac{1}{6}\left(y - \frac{\pi}{2}\right)^3$.

4. (*i*) Expand y^2/x^3 at the point $(1, -1)$.

(*ii*) Expand $f(x, y) = e^{xy}$ at the point $(1, 1)$ up to terms of the third degree.

Ans. (*i*) $1 - 3(x-1) - 2(y+1) + 6(x-1)^2 + 6(x-1)(y+1) + (y+1)^2 + \ldots$

(*ii*) $e + (x-1)\,e + (y-1)\,e + \frac{1}{2}(x-1)^2 e + 2\,(x-1)\,(y-1)\,e + \frac{1}{2}(y-1)^2 e +$

$\frac{1}{6}(x-1)^3 e + \frac{3}{2}(x-1)^2(y-1)\,e + \frac{3}{2}(x-1)(y-1)^2 e + \frac{1}{6}(y-1)^3 e + \ldots$

5. Find the first six terms of the expansion of the function $e^x \log(1+y)$ in a Taylor series in the neighbourhood of point $(0, 0)$. (*JNTU 2006; PTU 2006*)

Ans. $y + xy - \frac{y^2}{2} + \frac{1}{2}x^2y - \frac{1}{2}xy^2 + \frac{1}{3}y^3 + \ldots$

6. (*i*) Expand $f(x, y) = y^x$ in neighbourhood of $(1, 1)$ upto the terms of second degree.

(*ii*) Expand the function $f(x, y) = x^y$ in powers of $(x-1)$ and $(y-1)$, finding the terms upto the third order included. Use the result to compute (without tables) $(1.02, 1.1)$ and $1.1^{1.02}$. (*AMIETE, W-2004*)

Ans. (*i*) $1 + (y-1) + (x-1)(y-1) + \ldots$

(*ii*) $1 + (x-1) + (x-1)(y-1) + (1/2)(x-1)^2(y-1) + \ldots$; $f(1.02, 1.1) \approx 1.022$; $1.1^{1.02} = 1.1021$.

7. Verify the following expansion.

(*i*) $\sin(x+y) = x + y - \frac{(x+y)^3}{3!} + \ldots\ldots$

(*ii*) $a^x \log(1+y) = y + \frac{1}{2}(2xy\log a - y^2 + x^2y\log^2 a - xy^2\log a) + \frac{1}{3}y^3 + \ldots$

8. If $f(x, y) = \tan^{-1}xy$, compute $f(0.9. -1.2)$ approximately. **Ans.** 2.306

9. Expand $xy^2 + \sin xy$ at the point $(1, \pi/2)$ upto terms of the third degree.

Ans. $1 + \frac{\pi^2}{4} + \frac{\pi^2}{4}(x-1) + \pi\left(y - \frac{\pi}{2}\right) + \frac{1}{2!}\left[-\frac{\pi^2}{4}(x-1)^2 + \pi(x-1)\left(y - \frac{\pi}{2}\right) + \left(y - \frac{\pi}{2}\right)^2\right] + \ldots\ldots$

10. Expand $x^2y + \sin y + e^x$ in powers of $(x-1)$ and $(y-\pi)$ upto quadratic terms by using Taylor's series.

Ans. $(\pi + e) + (2\pi + e)(x-1) + \frac{1}{2!}(x-1)^2(2\pi + e) + 2(x-1)(y-\pi)$.

11. (*i*) Expand $\sin\left(\frac{x+\pi}{y+2}\right)$ in powers of x and y up to the terms of second degree and hence find the approximate value of $f(x, y)$ at $x = 0.1$ and $y = 0.1$ by taking $\pi = 3.1416$.

(*ii*) Expand $f(x,y) = e^{2x}\cos 3y$, in Taylor's power series of x and y upto quadratic terms.

Ans. (*i*) $1-\frac{1}{8}x^2+\pi\frac{xy}{8}-\frac{\pi^2}{32}y^2+......$; 0.9996 (*ii*) $1+2x+2x^2-9y^2/2+......$

12. Expand $f(x, y) = 21 + x - 20y + 4x^2 + xy + 6y^2$ in Taylor series of maximum order about the point (–1, 2).

Ans. $f(x, y) = 6 - 5(x + 1) + 3(y - 2) + 4(x + 1)^2 + (x + 1)(y - 2) + 6(y - 2)^2$.

13. Obtain the second order Taylor's series approximation to the function $f(x,y) = xy^2 + y\cos(x-y)$ about the point (1, 1).

Ans. $f(x,y) = 2+(x-1)+3(y-1)+\frac{1}{2}\left[-(x-1)^2+6(x-1)(y-1)+(y-1)^2\right]$.

2.8 MAXIMUM AND MINIMUM VALUES OF A FUNCTION

Let $z = f(x, y)$ be a function of two independent variables x and y which is defined and continuous in some closed and bounded region R. Let (a, b) be an interior point of R and $(a + h, b + k)$ be a point in its neighbourhood and lies inside R. We define the following:

Maximum Value: *A function $f(x, y)$ is said to have a relative (or local) maximum value for $x = a$, $y = b$, if there exists a small neighbourhood of (a, b) such that for every point $(a + h, b + k)$ of this neighbourhood,* $f(a + h, b + k) \le f(a, b)$ for all h and k. The point (a, b) is called a point of *relative* (or local) *maximum*.

Minimum value: *A function $f(x, y)$ is said to have a relative (or local) minimum value for $x = a$, $y = b$ if there exists a small neighbourhood of (a, b) such that for every point $(a + h, b + k)$ of this neighbourhood* $f(a + h, b + k) \ge f(a, b)$ for all h and k. The point (a, b) is called a point of *relative* (or local) *minimum*.

The points at which maximum/minimum values of the function occur are also called *point of extrema* or the *stationary points* and the maximum and the minimum values taken together are called the *extreme or extremum values* of the function.

We can represent $z = f(x, y)$ graphically by surface. From the figure 2.9 (*a*) it is clear that $z = f(x, y)$ has a maximum value for $x = a$, $y = b$ because $f(a, b)$ *i.e.*, PQ is greater than $f(x, y)$ for all points (x, y) close to (a, b). Thus the maximum value of $z = f(x, y)$ occurs at the top of an elevation.

Similarly, we can easily see from figure 2.9 (*b*) that minimum value occurs at the bottom of a depression.

Thus at a local maximum point the surface *descends* in every direction while at a local minimum point it *ascends* in every direction.

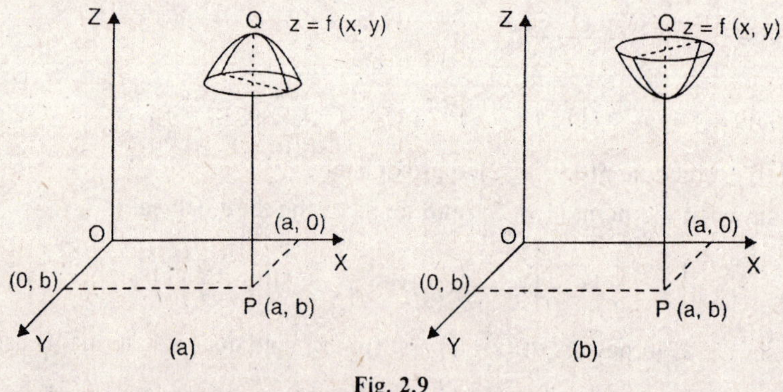

Fig. 2.9

2.8.1 Absolute Maximum and Absolute Minimum

A function $f(x, y)$ may also attain its maximum or minimum values on the boundary of the region. The greatest and the least values attained by a function over the entire region including the boundary are called the *absolute maximum and absolute minimum* values respectively.

2.9 CONDITIONS FOR A FUNCTION TO HAVE AN EXTREMUM

By definition, a function $f(x, y)$ has an extreme value at (x, y) if $f(x+h, y+k) - f(x, y)$ does not change sign for small values of h, k. If this sign is negative $f(x, y)$ is maximum and in case it is positive $f(x, y)$ is minimum.

By Taylor's Theorem

$$f(x+h, y+k) = f(x, y) + \left(h\frac{\partial f}{\partial x} + k\frac{\partial f}{\partial y}\right) + \frac{1}{2!}\left(h^2\frac{\partial^2 f}{\partial x^2} + 2hk\frac{\partial^2 f}{\partial x \partial y} + k^2\frac{\partial^2 f}{\partial y^2}\right) + \ldots$$

$$\therefore \quad f(x+h, y+k) - f(x, y) = \left(h\frac{\partial f}{\partial x} + k\frac{\partial f}{\partial y}\right) + \frac{1}{2!}\left(h^2\frac{\partial^2 f}{\partial x^2} + 2hk\frac{\partial^2 f}{\partial x \partial y} + k^2\frac{\partial^2 f}{\partial y^2}\right) + \ldots \quad \ldots(2.19)$$

When h and k are small, the sign on the R.H.S. and, therefore of the L.H.S. is governed by $h\frac{\partial f}{\partial x} + k\frac{\partial f}{\partial y}$ which changes sign with h and k.

Thus, (x, y) is a point of extreme value if $h\frac{\partial f}{\partial x} + k\frac{\partial f}{\partial y} = 0$.

Since h and k are independent $\partial f/\partial x = 0$, $\partial f/\partial y = 0$, $\quad \ldots(2.20)$
which is the *necessary condition* for extreme value.

By solving equations (2.20) we get points which may be points of maximum or minimum value. Let (a, b) be one such point.

Now with $x = a$, $y = b$ and using (2.20), we have from (2.19)

$$f(a+h, b+k) - f(a, b) = \frac{1}{2!}[h^2 f_x^2(a, b) + 2hk\, f_{xy}(a, b) + k^2 f_y^2(a, b)] + \ldots$$

$$= \frac{1}{2!}(rh^2 + 2shk + tk^2) + \ldots \quad \ldots(2.21)$$

where $\quad r = f_x^2(a, b),\ s = f_{xy}(a, b),\ t = f_y^2(a, b)$.

For sufficiently small values of h and k, the sign on the R.H.S. of (2.21) is the same as that of $(rh^2 + 2shk + tk^2)$.

Case I: $rt - s^2 > 0$.

In this case $r \neq 0$, $t \neq 0$ and, we have $rh^2 + 2shk + tk^2 = \frac{1}{r}[(rh + sk)^2 + (rt - s^2)k^2]$.

Since $rt - s^2 > 0$, the sign of $rh^2 + 2shk + tk^2$ is always the same as that of r.

In this case $f(a, b)$ has an extreme value which is maximum if r is negative and minimum if r is positive.

Case II: $rt - s^2 < 0$.

In case $r \neq 0$, then $\quad rh^2 + 2shk + tk^2 = \frac{1}{r}[(rh + sk)^2 + (rt - s^2)k^2]$.

Since $rt - s^2 < 0$, the expression has different signs when $k = 0$, and when $rh + sk = 0$. Hence $f(a, b)$ is not an extreme value.

The proof when $t \neq 0$ is similar.

When $r = 0$, $t = 0$, then $rh^2 + 2shk + tk^2 = 2shk$ which assume values with different signs. Hence $f(a, b)$ is not an extreme value in this case.

Case III: $rt - s^2 = 0$.

In this case $\quad rh^2 + 2shk + tk^2 = \frac{1}{r}[(rh + sk)^2 + (rt - s^2)k^2] = (rh + sk)^2/r$

which is zero for values of h, k such that $h/k = -s/r$.

This is, therefore, a doubtful case. To determine the nature of sign, further investigation is required.

Working rule to find extremum values of a function z = f (x, y).

To determine the extremum values for $z = f(x, y)$, we proceed as follows:

(*i*) *Find* $\partial z/\partial x$ *and* $\partial z/\partial y$ *and equate each to zero. Solve the equations thus obtained, i.e.,* $\partial z/\partial x = 0$ *and* $\partial z/\partial y = 0$ *simultaneously. Let* $(x_1, y_1), (x_2, y_2)$ ..., *be the pairs of roots of these equations.*

(*ii*) *Find* $r = \partial^2 f/\partial x^2$, $s = \partial^2 f/\partial x\partial y$ *and* $t = \partial^2 f/\partial y^2$ *for each of pair of roots.*

(*iii*) (*a*) *If* $rt - s^2 > 0$, *and* $r < 0$, *then* $z = f(x, y)$ *has a relative maximum value for that pair.*

(*b*) *If* $rt - s^2 > 0$ *and* $r > 0$ *then the function f has a relative minimum value at that point.*

(*c*) *If* $rt - s^2 < 0$, *the function f has no local minimum or local maximum value at that point. In this case, the point* $P(a, b)$ *is called a saddle point. At a saddle point, the tangent plane to the surface* $z = f(x, y)$ *is horizontal, but near the point the surface is partly above and partly below this tangent plane.*

(*d*) *If* $rt - s^2 = 0$ *no conclusion about an extremum can be drawn and further investigation is needed.*

Remark The points $(x_1, y_1), (x_2, y_2)$... which are the roots of $\partial f/\partial x = 0$ and $\partial f/\partial y = 0$, are called **stationary points.** The function $f(x, y)$ is said to be stationary at $(x_1, y_1), (x_2, y_2)$ etc. and its values there are called **stationary values.** Thus every extreme value is a stationary value but the converse may not be true.

ILLUSTRATIVE EXAMPLES

EXAMPLE 2.36 Determine the points where $f(x, y) = x^3 + y^3 - 3axy$ has a maximum or minimum values. (*UPTU, 2004*)

Solution We have $\partial f/\partial x = 3x^2 - 3ay$, $\partial f/\partial y = 3y^2 - 3ax$,

$r = \partial^2 f/\partial x^2 = 6x$, $s = \partial^2 f/\partial x\partial y = -3a$, $t = \partial^2 f/\partial y^2 = 6y$. Now $\partial f/\partial x = 0$ and $\partial f/\partial y = 0$ give.

$$3x^2 - 3ay = 0 \text{ or } x^2 - ay = 0, \quad ...(1) \quad \text{and} \quad 3y^2 - 3ax = 0 \text{ or } y^2 - ax = 0. \quad ...(2)$$

We solve (1) and (2) to get the stationary points.

Putting $y = x^2/a$ from (1) in (2), we get $x^4/a^2 - ax = 0$ or $x(x - a)(x^2 + ax + a^2) = 0$.

The real roots of this equation are $x = 0$, $x = a$. Putting these values of x in (1) and (2), we get $y = 0$ and $y = a$.

$\therefore$ (0, 0) and (a, a) are the stationary points.

At (0, 0) $r = 0$, $s = -3a$ and $t = 0$.

Since $rt - s^2 = -9a^2 < 0$, (for all values of 'a'), while $r = 0$. Hence it is a saddle point.

At (a, a) $rt - s^2 = 36\,a^2 - 9a^2 = 27a^2 > 0$ (for all values of a) and $r = 6a$.

Now if 'a' is –ve, r is –ve, therefore, the function has maximum value, when $a < 0$.

If 'a' is +ve, r is +ve, therefore, the function has minimum value, when $a > 0$.

Hence, point (a, a) gives a maximum if a is negative and minimum if 'a' is positive, for $r > 0$.

EXAMPLE 2.37 Find the extreme values of $xy(a - x - y)$ (*AMIE, S-2003*)

Alternative Decompose a positive number 'a' into three parts so that their sum is 'a' and product is maximum. (*AMIE, W-2009*)

Solution Let x = first part, y = second part then $(a - x - y)$ = third part, and function to be examined is $f(x, y) = xy(a - x - y) = axy - x^2y - xy^2$.

$\therefore$ $\partial f/\partial x = ay - 2xy - y^2$, $\partial f/\partial y = ax - x^2 - 2xy$,

$r = \partial^2 f/\partial x^2 = -2y$, $s = \partial^2 f/\partial x\partial y = a - 2x - 2y$, $t = \partial^2 f/\partial y^2 = -2x$.

We now solve the simultaneous equations $\partial f/\partial x = 0$ and $\partial f/\partial y = 0$.

We get as one pair of values $x = a/3$, $y = a/3$.

Now $rt - s^2 = 4xy - (a - 2x - 2y)^2$. When $x = a/3$ and $y = a/3$, $rt - s^2 = a^2/3 > 0$ and $\partial^2 f/\partial x^2 = -2a/3 < 0$.

It is seen that product is a maximum when $x = a/3$ and $y = a/3$. Therefore, the third part is also $a/3$, and the maximum value of the product $= a^3/27$.

EXAMPLE 2.38 Find the extreme values of $u = x^2y^2 - 5x^2 - 8xy - 5y^2$.

Solution We have $\partial u/\partial x = 2xy^2 - 10x - 8y$, $\partial u/\partial y = 2x^2y - 8x - 10y$.

Solving these two simultaneous equations $\partial u/\partial x = 0$ and $\partial u/\partial y = 0$ that is,

$$xy^2 - 5x - 4y = 0 \quad ...(1) \qquad \text{and} \qquad x^2y - 4x - 5y = 0 \quad ...(2)$$

Subtracting (1) from (2) we get $xy(x - y) + x - y = 0$ or $(x - y)(xy + 1) = 0$.

Substituting $x = y$ in (1) we get $y^3 - 9y = 0$ $y = 0, \pm 3$.

Again putting $x = -1/y$ in (1) we have $-y^2 + 1 = 0$ or $y = \pm 1$.

Hence (0, 0), (3, 3), (–3, –3), (–1, 1), (1, –1) are five stationary points.

Now, $r = \partial^2 u/\partial x^2 = 2y^2 - 10$, $\quad s = \partial^2 u/\partial y \partial x = 4xy - 8$, $\quad t = \partial^2 u/\partial y^2 = 2x^2 - 10$.

At (0, 0), $r = -10$, $s = -8$, $t = -10$. $\quad \therefore rt - s^2 = 100 - 64 > 0$ and $r < 0$.

Hence (0, 0) is a point of maximum value. The maximum value, obviously, is zero.

At (3, 3), $r = 8$, $s = 28$, $t = 8$. $\quad \therefore rt - s^2 < 0$.

Thus there is no extreme value at (3, 3).

Similarly, there is no extreme value at (–3, –3), (–1, 1), (1, –1).

EXAMPLE 2.39 Determine the points where the function $x^4 + y^4 - 2x^2 + 4xy - 2y^2$ has a maximum or minimum. *(AMIETE, June 2009; JNTU 2003)*

Solution Let $u = x^4 + y^4 - 2x^2 + 4xy - 2y^2$.

$\therefore \quad \partial u/\partial x = 4x^3 - 4x + 4y$, $\quad \partial u/\partial y = 4y^3 + 4x - 4y$,

$r = \partial^2 u/\partial x^2 = 12x^2 - 4$, $\quad s = \partial^2 u/\partial x \partial y = 4$, $\quad t = \partial^2 u/\partial y^2 = 12y^2 - 4$.

Now the points of maximum or minimum values are given by

$$\partial u/\partial x = 4x^3 - 4x + 4y = 0 \quad ...(1) \qquad \partial u/\partial y = 4y^3 + 4x - 4y = 0. \quad ...(2)$$

Adding (1) and (2) we get $x^3 + y^3 = 0$ or $x = -y$. Substituting $x = -y$ in (2) we have $y^3 - 2y = 0$.

$\therefore \quad y = 0, \pm\sqrt{2}$ and the coresponding values of x are $0, \mp\sqrt{2}$.

Hence the stationary points are (0, 0), $(-\sqrt{2}, \sqrt{2})$, and $(\sqrt{2}, -\sqrt{2})$.

At $(-\sqrt{2}, \sqrt{2})$; $r = 20$, $s = 4$, $t = 20$. As $rt - s^2 = 384 > 0$ and $r > 0$.

$\therefore$ At $(-\sqrt{2}, \sqrt{2})$, u has a minimum value.

Similarly at $(\sqrt{2}, -\sqrt{2})$, the function has a minimum value.

Now at **(0, 0),** $r = -4$, $s = 4$, $t = -4$, therefore $rt - s^2 = 0$.

Hence at (0, 0) further investigation is needed.

Now $\quad u = x^4 + y^4 - 2x^2 + 4xy - 2y^2 = x^4 + y^4 - 2(x - y)^2$.

At (0, 0), $u = 0$. When x, y are small, $u = -2(x - y)^2$ approximately which is negative when $x \neq y$. When $x = y$, $u = x^4 + y^4$ which is + ve.

Thus in a small neighbourhood of (0, 0) there are points where the function is –ve that is, $< u(0, 0)$ and also there are points where the function is +ve that is, $> u(0, 0)$.

Hence at (0, 0) there is no exterme value. (0, 0) is a saddle point.

EXAMPLE 2.40 Find the absolute maxima and minima of the function $f(x, y) = x^2 + xy + y^2 - 6x$ on the rectangular plate $0 \le x \le 5$, $-3 \le y \le 3$.

Solution The function f can attain maximum/minimum values at the critical points or on the boundary of the rectangle ABCD (Refer Fig. 2.10).

We have $f_x = 2x + y - 6 = 0$, $f_y = x + 2y = 0$. The critical point is (4, –2). Now $r = f_{xx} = 2$, $s = f_{xy} = 1$, $t = f_{yy} = 2$, $rt - s^2 = 3$.

Since $rt - s^2 > 0$ and $r > 0$, the point (4, –2) is a point of relative minimum. The minimum value is $f(4, -2) = -12$.

On the boundary line AB, we have $y = -3$ and $f(x, y) = f(x, -3) = g(x) = x^2 - 9x + 9$, which is a function of one variable. Setting $dg/dx = 0$, we get $2x - 9 = 0$ or $x = 9/2$. Now $d^2g/dx^2 = 2 > 0$. Therefore, at $x = 9/2$, the function has a minimum. The minimum value is $g(9/2) = (81/4) - (81/2) + 9 = -45/4$.

Also at the corners (0, –3), (5, –3), we have $f(0, -3) = 9$, $f(5, -3) = -11$.

Similarly, along the other boundary lines, we have the following results.

At $x = 5$: $h(y) = y^2 + 5y - 5$; $dh/dy = 2y + 5 = 0$, gives $y = -5/2$; $d^2h/dy^2 = 2 > 0$. Therefore $y = -5/2$ is a point of minimum. The minimum value is $f(5, -5/2) = -45/4$. At the corner (5, 3) we have $f(5, 3) = 19$.

At $y = 3$: $g(x) = x^2 - 3x + 9$; $dg/dx = 2x - 3 = 0$ gives $x = 3/2$; $d^2g/dx^2 = 2 > 0$. Therefore $x = 3/2$ is a point of minimum. The minimum value is $f(3/2, 3) = 27/4$. At the corner point (0, 3), we have $f(0, 3) = 9$.

At $x = 0$: $h(y) = y^2$, $dh/dy = 2y = 0$, gives $y = 0$; $d^2h/dy^2 = 2 > 0$. Therefore $y = 0$ is a point of minimum. The minimum value is $f(0, 0) = 0$. Therefore, the absolute minimum value is –12 which occur at (4, 2) and the absolute maximum value is 19 which occurs at the point (5, 3).

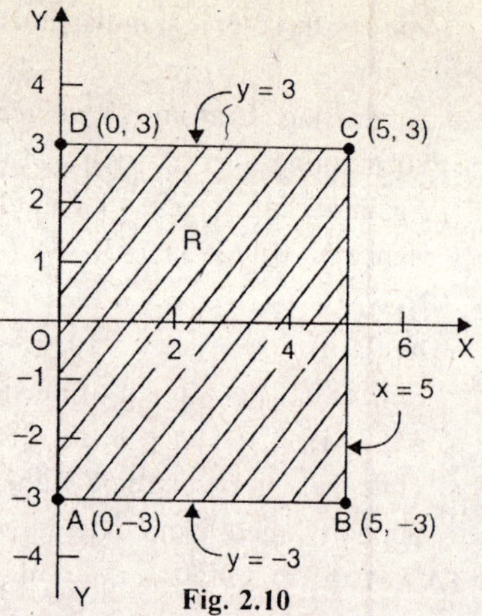

Fig. 2.10

EXAMPLE 2.41 Find the absolute maximum and minimum values of the function $f(x, y) = 3x^2 + y^2 - x$ over the region $2x^2 + y^2 \le 1$. *(AMIETE, W-2008; AMIE-W-2007)*

Solution We have $f_x = 6x - 1 = 0$ and $f_y = 2y = 0$. Therefore, the critical point is $(1/6, 0)$. We find that $r = f_{xx} = 6$, $s = f_{xy} = 0$, $t = f_{yy} = 2$ and $rt - s^2 = 12 > 0$. Therefore, (1/6, 0) is a point of minimum. The minimum value at this point is $f(1/6, 0) = 3(1/6)^2 - (1/6) = -1/12$.

On the **boundary,** we have $y^2 = 1 - 2x^2$, $-1/\sqrt{2} \le x \le 1/\sqrt{2}$. Substituting in $f(x, y)$, we obtain

$$f(x, y) = 3x^2 + (1 - 2x^2) - x = 1 - x + x^2 = h(x).$$

which is a function of one variable. Setting $dh/dx = 0$. $dh/dx = -1 + 2x = 0$ or $x = 1/2$. Also $d^2h/dx^2 = 2 > 0$.

For $x = 1/2$, we get $y^2 = 1 - 2x^2 = 1/2$ or $y = \pm 1/\sqrt{2}$ Hence, the points $(1/2, \pm 1/\sqrt{2})$ are points of minimum. The minimum value is

$$f(1/2, \pm 1/\sqrt{2}) = 3\,(1/4) + (1/2) - (1/2) = 3/4.$$

At the vertices, we have
$$f(1/\sqrt{2}, 0) = (3/2) - (\sqrt{2}/2) = (3 - \sqrt{2})/2,$$
$$f(-1/\sqrt{2}, 0) = (3/2) + (\sqrt{2}/2) = (3 + \sqrt{2})/2, \quad f(0, \pm 1) = 1.$$

Therefore, the given function has absolute minimum value $-1/12$ at (1/6, 0) and absolute maximum value $(3 + \sqrt{2})/2$ at $(-1/\sqrt{2}, 0)$.

EXAMPLE 2.42 Given $f(x, y, z) = \dfrac{5xyz}{x + 2y + 4z}$, find the values of x, y, z for which $f(x, y, z)$ is maximum subject to $xyz = 8$. *(SVTU, 2007; Kurukshetra, 2005; AMIE, W-2002)*

Solution Since $xyz = 8$, we obtain

$$f(x, y, z) = \frac{5xyz}{x + 2y + 4z} = \frac{(5)(8)}{x + 2y + (32/xy)} \quad \text{or} \quad \frac{40}{f} = x + 2y + \frac{32}{xy}.$$

Now f is a maximum when $40/f$ is a minimum.

$$\frac{\partial(40/f)}{\partial x} = 1 - \frac{32}{x^2 y} = 0 \text{ and } \frac{\partial(40/f)}{\partial y} = 2 - \frac{32}{xy^2} = 0.$$

Solving these two simultaneous equations, we obtain $x = 4$, $y = 2$.

At (4, 2) We obtain $r = \dfrac{\partial^2}{\partial x^2}\left(\dfrac{40}{f}\right) = \dfrac{64}{x^3 y} = \dfrac{64}{(4^3)(2)} = \dfrac{1}{2}$,

$$t = \frac{\partial^2}{\partial y^2}\left(\frac{40}{f}\right) = \frac{64}{xy^3} = \frac{64}{(4)(2^3)} = 2, \text{ and } s = \frac{\partial^2}{\partial x \partial y}\left(\frac{40}{f}\right) = \frac{32}{x^2 y^2} = \frac{32}{(4^2)(2^2)} = \frac{1}{2}.$$

Therefore $rt - s^2 = \frac{1}{2} \cdot 2 - \left(\frac{1}{2}\right)^2 = \frac{3}{4} > 0,\ r > 0.$

Hence, $x = 4$, $y = 2$ and $z = \frac{8}{xy} = \frac{8}{(4)(2)} = 1$ that is, (4, 2, 1) gives a minimum $40/f$ and a maximum $f(x, y, z)$.

EXAMPLE 2.43 A long piece of tin 12 unit wide is to be made into a trough by bending up two sides to form equal angles with the base. Find the amount to be bent up and the angle of inclination of each side that will make the carrying capacity a maximum.

Solution Let x unit be the amount bent up and θ the inclination of the sides.

The carrying capacity will be maximum when the area K of the cross section shown in the figure must be a maximum. The cross-section is a trapezoid of upper base EF = $12 - 2x + 2x\cos\theta$, lower base $CD = 12 - 2x$ and altitude $= x\sin\theta$. The area, K is given by

$$K = 12x\sin\theta - 2x^2\sin\theta + x^2\sin\theta\cos\theta.$$

Fig. 2.11

By differentiation, we have

$$\partial K/\partial\theta = 12x\cos\theta - 2x^2\cos\theta + x^2(\cos^2\theta - \sin^2\theta)$$
$$= x[12\cos\theta - 2x\cos\theta + x\cos^2\theta - x\sin^2\theta].$$
$$\partial K/\partial x = 12\sin\theta - 4x\sin\theta + 2x\sin\theta\cos\theta = 2\sin\theta(6 - 2x + x\cos\theta).$$

Setting the partial derivatives equal to zero, we have the two equations

$$x[12\cos\theta - 2x\cos\theta + x(\cos^2\theta - \sin^2\theta)] = 0 \quad ...(1) \quad \text{and} \quad 2\sin\theta(6 - 2x + x\cos\theta) = 0. \quad ...(2)$$

One solution of this system is $\theta = 0$, $x = 0$, which has no meaning in the physical problem. Assuming $\theta \neq 0$, $x \neq 0$, and solving the equations, we get $\cos\theta = 1/2 = \cos 60°$ that is, $\theta = 60°$ and $x = 4$.

A consideration of the physical problem shows that there must exist a maximum value of the area. Hence the maximum value occurs when $\theta = 60°$ and $x = 4$ units.

2.10 LAGRANGE'S METHOD OF MULTIPLIERS

This method is useful when we have to find the extremum of the function of a number of variables which are not independent but are connected by some given relation or relations.

Supposing we have to find the extremum of the function $f(x, y, z)$ under the condition

$$\phi(x, y, z) = 0. \quad ...(2.22)$$

In such cases we construct an auxiliary function of the form

$$F(x, y, z, \lambda) = f(x, y, z) + \lambda\phi(x, y, z) \quad ...(2.23)$$

where λ is an undetermined parameter and is known as *Lagrange multipliers*.

Then, to determine the stationary points of F, we have the necessary conditions

$\partial F/\partial x = 0$, $\partial F/\partial y = 0$, $\partial F/\partial z = 0$, which give the equations

$$\frac{\partial f}{\partial x} + \lambda\frac{\partial\phi}{\partial x} = 0,\quad \frac{\partial f}{\partial y} + \lambda\frac{\partial\phi}{\partial y} = 0,\quad \frac{\partial f}{\partial z} + \lambda\frac{\partial\phi}{\partial z} = 0. \quad ...(2.24)$$

From these equations and $\phi(x, y, z) = 0$ determine x, y, z and λ, the latter only played an auxiliary role and will not be needed any more.

While this method is very convenient in finding the stationary points, it does not help in finding the nature of those stationary points. In applied problems this can be done from physical considerations.

Remark If we wish to find the relative maximum or minimum values of a function $f(x_1, x_2, x_3)$ subject to the constraint conditions $\phi_1(x_1, x_2, x_3) = 0$, $\phi_2(x_1, x_2, x_3) = 0$, we form the auxiliary function.

$F(x_1, x_2, x_3, \lambda_1, \lambda_2) = f(x_1, x_2, x_3) + \lambda_1\theta_1(x_1, x_2, x_3) + \lambda_2\theta_2(x_1, x_2, x_3)$ that is,

$F(x_1, x_2, x_3, \lambda_1, \lambda_2) = f + \lambda_1\theta_1 + + \lambda_2\theta_2$ subject to the necessary conditions $\partial F/\partial x_1 = 0$, $\partial F/\partial x_2 = 0$, $\partial F/\partial x_3 = 0$, where λ_1, λ_2 are independent of x_1, x_2, x_3 are the *Lagrange multipliers*.

ILLUSTRATIVE EXAMPLES

EXAMPLE 2.44 A rectangular box, open at the top is to have a given capacity. Find the dimensions of the box requiring least material for its construction. (*AMIETE, June 2009; AMIE, W-2004*)

Solution Let x, y, z be the dimensions of the box, S its surface area and V its capacity.

Now $V = xyz$ and $S = xy + 2yz + 2xz$.

Consider the auxiliary function $F(x, y, z, \lambda) = xy + 2yz + 2zx + \lambda(xyz - V)$.

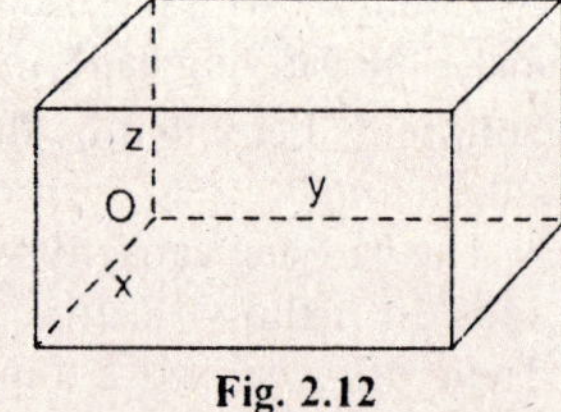

Fig. 2.12

The necessary conditions for extremum as

$$\partial F/\partial x = y + 2z + \lambda(yz) = 0, \quad ...(1)$$

$$\partial F/\partial y = x + 2z + \lambda(xz) = 0, \quad ...(2)$$

$$\partial F/\partial z = 2y + 2x + \lambda(xy) = 0. \quad ...(3)$$

and $$xyz = V. \quad ...(4)$$

Multiplying (1) by x and (2) by y and substracting we get

$2zx - 2zy = 0$ or $2z(x - y) = 0$. Hence $x = y$. $[z \neq 0]$.

Similarly, from (2) and (3), we get $y = 2z$. Substituting $x = y = 2z$ in (4) we get $2z = (2V)^{1/3}$.

$\therefore\ x = y = 2z = (2V)^{1/3}$ are the required dimensions which, from physical considerations, give the minimum surface.

Note. When $V = 32$ cubic units, $x = y = 4$ unit, $z = 2$ unit.

EXAMPLE 2.45 A tent of a given volume has a square base of side $2a$, has its four sides vertical of height b and is surmounted by a regular pyramid of height h. Find the values of a and b in terms of h such that the canvas required for its construction is minimum.

Solution Let V be the volume and S the surface area of the tent.

Now $V = 4a^2b + \{(4a^2)h/3\}$, and $S = 8ab + 4a\sqrt{a^2+h^2}$.

Consider the auxiliary function

$$F(a,b,h,\lambda) = 8ab + 4a\sqrt{a^2+h^2} + \lambda[4a^2b + (4a^2h/3) - V].$$

For the extremum, we have the necessary conditions

$$\frac{\partial F}{\partial a} = 8b + 4\sqrt{a^2+h^2} + \frac{4a^2}{\sqrt{a^2+h^2}} + \lambda\left[8ab + \frac{8ah}{3}\right] = 0, \quad ...(1)$$

$$\frac{\partial F}{\partial b} = 8a + 4\lambda a^2 = 0, \quad ...(2) \qquad \frac{\partial F}{\partial h} = \frac{4ah}{\sqrt{a^2+h^2}} + \frac{4\lambda}{3}a^2 = 0. \quad ...(3)$$

From (2) $\lambda a + 2 = 0$. ...(4). From (3), $3h + \lambda a\sqrt{a^2+h^2} = 0$ $(\because a \neq 0)$. ...(5)

Substituting the value of λa from (4) in (5) and simplifying, we get $a = \sqrt{5}h/2$.

Substituting $\lambda a = -2$ and $a = \sqrt{5}h/2$. in (1) and simplifying, we get $b = h/2$.

Thus when $a = \sqrt{5}h/2$ and $b = h/2$, we get the stationary value of S.

From physical considerations this stationary value is a minimun.

EXAMPLE 2.46 A torpedo has the shape of a cylinder with conical ends. For given surface area S, show that the dimensions which give maximum volume are $l = h = 2r/\sqrt{5}$, where l is the length of the cylinder, r its radius and h the altitude of the cone.

Solution Here $S = 2\pi rl + 2\pi r\sqrt{r^2+h^2}$, ...(*a*) and $V = \pi r^2 l + 2\pi r^2 h/3$. ...(*b*)

We have to find the maximum value of V subject to the condition (*a*).

Considering the auxiliary function

$$F(l,h,r,\lambda) = \pi r^2 l + (2\pi r^2 h/3) + \lambda\left[2\pi rl + 2\pi r\sqrt{r^2+h^2} - S\right].$$

For the extremum, we have the necessary conditions

$$\frac{\partial F}{\partial l}=\pi r^2+\lambda(2\pi r)=0, \qquad ...(1)$$

$$\frac{\partial F}{\partial h}=\frac{2}{3}\pi r^2+\frac{2\pi rh\lambda}{\sqrt{r^2+h^2}}=0, \qquad ...(2)$$

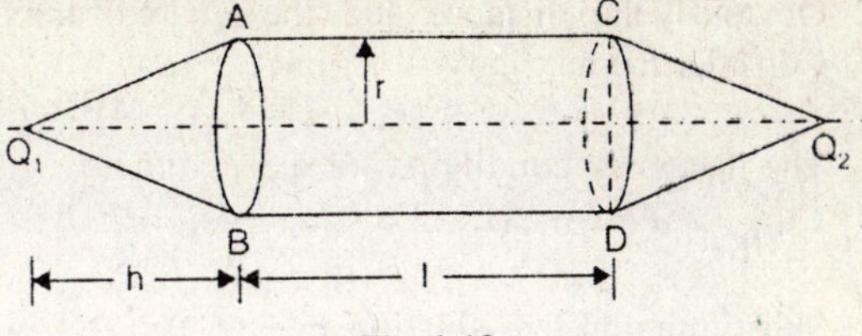

Fig. 2.13

$$\frac{\partial F}{\partial r}=2\pi rl+\frac{4}{3}\pi rh+2\pi\lambda\left[l+\sqrt{r^2+h^2}+\frac{r^2}{\sqrt{r^2+h^2}}\right]=0. \qquad ...(3)$$

From (1) $r+2\lambda=0$ ($\because r\neq 0$), or $2\lambda=-r$.

Putting $2\lambda=-r$ in (2) and simplifying, we get $h=2r/\sqrt{5}$.

Substituting $h=2r/\sqrt{5}$. and $2\lambda=-r$ in (3) and simplifying, we get $l=2r/\sqrt{5}$.

The maximum value of V is when $l=h=2r/\sqrt{5}$.

EXAMPLE 2.47 Find the volume of the greatest rectangular parallelopiped with edges parallel to the coordinate axes that can be inscribed inside the ellipsoid

$$\frac{x^2}{a^2}+\frac{y^2}{b^2}+\frac{z^2}{c^2}=1.$$ *(AMIE, W-2008; Madras 2006; MDU, 2005; AMIETE S-2004; UPTU, 2003)*

Solution Let the edges of the parallelopiped which are parallel to the coordinate axes be $2x$, $2y$ and $2z$. Then its volume is $V=8xyz$. Now we have to find the maximum value of $V=8xyz$ subject to the condition

$$x^2/a^2+y^2/b^2+z^2/c^2=1. \qquad ...(i)$$

Consider the auxiliary function

$$F(x,y,z,\lambda)=8xyz+\lambda\left(\frac{x^2}{a^2}+\frac{y^2}{b^2}+\frac{z^2}{c^2}-1\right).$$

For the extremum, we have the necessary conditions

$$\frac{\partial F}{\partial x}=8yz+\lambda\left(\frac{2x}{a^2}\right)=0, \qquad ...(ii)$$

$$\frac{\partial F}{\partial y}=8zx+\lambda\left(\frac{2y}{b^2}\right)=0, \qquad ...(iii)$$

$$\frac{\partial F}{\partial z}=8xy+\lambda\left(\frac{2z}{c^2}\right)=0, \qquad ...(iv)$$

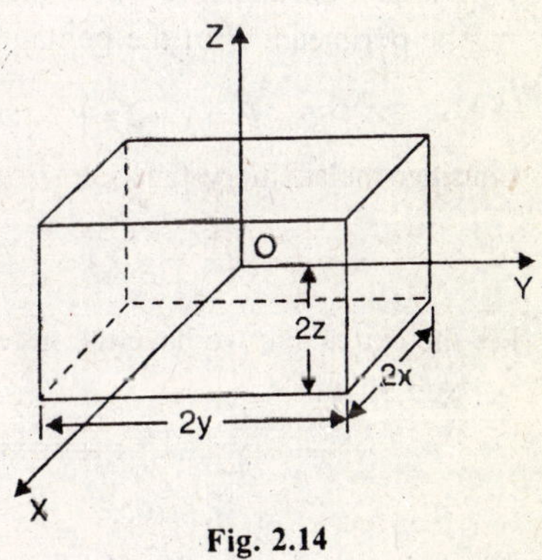

Fig. 2.14

Equating the values of λ from (*ii*) and (*iii*), we get $x^2/a^2=y^2/b^2$.

Similarly from (*iii*) and (*iv*) we obtain $y^2/b^2=z^2/c^2$. $\therefore x^2/a^2=y^2/b^2=z^2/c^2$.

Substituting $x^2/a^2=y^2/b^2=z^2/c^2$ in (*i*) we get $z^2/c^2=1/3$.

$$\therefore \quad \frac{x^2}{a^2}=\frac{y^2}{b^2}=\frac{z^2}{c^2}=\frac{1}{3} \text{ which gives } x=\frac{a}{\sqrt{3}},\ y=\frac{b}{\sqrt{3}},\ z=\frac{c}{\sqrt{3}}.$$

Hence the greatest volume $=(2a/\sqrt{3})\,(2b/\sqrt{3})\,(2c/\sqrt{3})=8abc/3\sqrt{3}$.

EXAMPLE 2.48 Find the maximum and minimum distances of the point (3, 4, 12) from the sphere $x^2+y^2+z^2=1$. *(AMIETE, June 2010; PTU, 2005; UPTU, 2002)*

Solution Let (x, y, z) be any point on the sphere, then its distance from (3, 4, 12) is given by

$$\sqrt{(x-3)^2+(y-4)^2+(z-12)^2}\,.$$

Obviously the 'distance' and 'the square of the distance have extremum values for the same values of x, y, z. Consider the auxiliary function

$$F(x, y, z, \lambda) = \{(x-3)^2 + (y-4)^2 + (z-12)^2\} + \lambda(x^2 + y^2 + z^2 - 1).$$

The necessary conditions for extremum as

$$\partial F/\partial x = 2(x-3) + 2\lambda x = 0, \quad \partial F/\partial y = 2(y-4) + 2\lambda y = 0, \quad \partial F/\partial z = 2(z-12) + 2\lambda y = 0, \text{ give}$$

$$x = 3/(1+\lambda), \qquad y = 4/(1+\lambda), \qquad z = 12/(1+\lambda).$$

Substituting in the equation $x^2 + y^2 + z^2 = 1$, we obtain

$$\frac{9}{(1+\lambda)^2} + \frac{16}{(1+\lambda)^2} + \frac{144}{(1+\lambda)^2} = 1, \quad \text{or} \quad 1 + \lambda = \pm 13 \quad \Rightarrow \quad \lambda = 12, -14.$$

Hence the points are $\left(\frac{3}{13}, \frac{4}{13}, \frac{12}{13}\right)$ and $\left(-\frac{3}{13}, -\frac{4}{13}, -\frac{12}{13}\right)$.

$\therefore$ The minimum distance $= \sqrt{\left(3-\frac{3}{13}\right)^2 + \left(4-\frac{4}{13}\right)^2 + \left(12-\frac{12}{13}\right)^2} = 12$ units,

and the maximum distance $= \sqrt{\left(3+\frac{3}{13}\right)^2 + \left(4+\frac{4}{13}\right)^2 + \left(12+\frac{12}{13}\right)^2} = 14$ units.

EXAMPLE 2.49 A pentagon is made by mounting an issoceles triangle on top of a rectangle. What dimensions minimise the perimeter P for a given area K?

Solution Let x and y be the length and breadth of rectangle and z, the altitude of an issoceles triangle, as shown in the Figure 2.15.

Then area of pentagon $K = xy + (xz)/2$.

The perimeter, P of the pentagon is given by

$$P = x + 2y + 2\sqrt{(x^2/4) + z^2} = x + 2y + \sqrt{x^2 + 4z^2}.$$

Consider the auxiliary function

$$F(x, y, z, \lambda) = x + 2y + \sqrt{x^2 + 4z^2} + \lambda\left(xy + \frac{1}{2}xz - K\right).$$

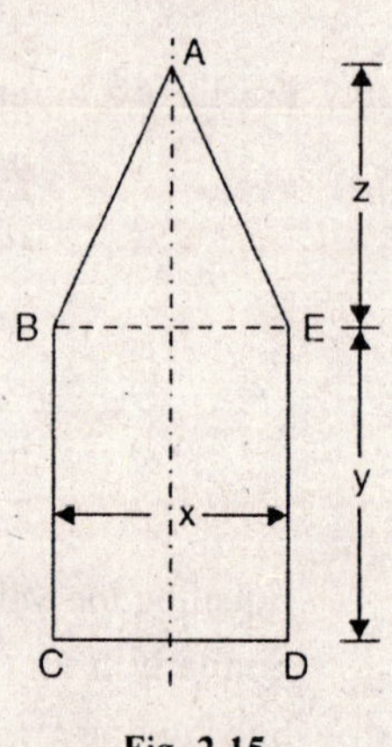

Fig. 2.15

For the extremum, we have the necessary conditions

$$\frac{\partial F}{\partial x} = 1 + \frac{x}{\sqrt{x^2 + 4z^2}} + \lambda\left(y + \frac{z}{2}\right) = 0, \qquad \text{...(1)}$$

$$\frac{\partial F}{\partial y} = 2 + \lambda x = 0, \qquad \text{...(2)}$$

$$\frac{\partial F}{\partial z} = \frac{4z}{\sqrt{x^2 + 4z^2}} + \frac{\lambda x}{2} = 0. \qquad \text{...(3)}$$

From (2), $\lambda = -2/x$. Putting this value in (3), we obtain

$$\frac{4z}{\sqrt{x^2 + 4z^2}} - 1 = 0 \Rightarrow 2\sqrt{3}z = x. \qquad \text{...(4)}$$

From (1), we obtain $(\sqrt{3} + 1)z = y$ ...(5). From (4) and (5), we have $\dfrac{x}{2\sqrt{3}} = \dfrac{y}{\sqrt{3}+1} = \dfrac{z}{1}$

or $\qquad x : y : z = 2\sqrt{3} : 1 + \sqrt{3} : 1$, for the perimeter to be minimum.

EXAMPLE 2.50 If $u = a^3x^2 + b^3y^2 + c^3z^2$ where $x^{-1} + y^{-1} + z^{-1} = 1$, show that the stationary value of u is given by $x = (\Sigma a)/a,\ y = (\Sigma b)/b,\ z = (\Sigma c)/c$.

(AMIETE, Dec 2009; PTU, 2008;, KUK, 2007; Kerala, 2005; UPTech, 2004)

Solution We have $u = a^3x^2 + b^3y^2 + c^3z^2$. Let $\phi = x^{-1} + y^{-1} + z^{-1} - 1$.

Now $\partial u/\partial x = 2a^3x,\quad \partial u/\partial y = 2b^3y,\quad \partial u/\partial z = 2c^3z.$

$\partial\phi/\partial x = -1/x^2,\quad \partial\phi/\partial y = -1/y^2,\quad \partial\phi/\partial z = -1/z^2.$

By Lagrange's multiplier

$$\frac{\partial u}{\partial x} + \lambda\frac{\partial\phi}{\partial x} = 0 \qquad \frac{\partial u}{\partial y} + \lambda\frac{\partial\phi}{\partial y} = 0 \qquad \frac{\partial u}{\partial z} + \lambda\frac{\partial\phi}{\partial z} = 0.$$

That is, $2a^3x + \lambda(-1/x^2) = 0$ or $2a^3x^3 - \lambda = 0 \Rightarrow a^3x^3 = \lambda/2.$

Similarly, $b^3y^3 = \lambda/2$ and $c^3z^3 = \lambda/2.$

$a^3x^3 = b^3y^3 = c^3z^3 = \lambda/2 \Rightarrow ax = by = cz = (\lambda/2)^{1/3} = k(\text{say}).$

$\therefore$ $x = k/a,\quad y = k/b,\quad z = k/c.$ Given that $x^{-1} + y^{-1} + z^{-1} = 1.$ ...(*i*)

Substituting the values of x, y and z in (*i*) we obtain $a + b + c = k$.

Therefore $x = k/a = \dfrac{(a+b+c)}{a} = \Sigma a/a,\quad y = k/b = \dfrac{(a+b+c)}{b} = \Sigma b/b,\quad z = k/c = \dfrac{(a+b+c)}{c} = \Sigma c/c.$

EXAMPLE 2.51 Find the shortest distance from (0, 0) to hyperbola $x^2 + 8\,xy + 7\,y^2 = 225$ in XY plane.

(AMIE, S-2006; AMIETE, W-2005)

Solution We have to find the minimum value of $x^2 + y^2$ (the square of the distance from the origin to any point in the XY plane) subject to the constraint $x^2 + 8xy + 7y^2 = 225$.

Consider the auxiliary function, $F(x, y, \lambda) = (x^2 + y^2) + \lambda\,(x^2 + 8xy + 7y^2 - 225)$.

For the extremum, we have the necessary conditions:

$\partial F/\partial x = 2x + 2\lambda x + 8\lambda y = 0$ or $(1 + \lambda)x + 4\lambda y = 0$ ...(1)

$\partial F/\partial y = 2y + 8\lambda x + 14\lambda y = 0$ or $4\lambda x + (1 + 7\lambda)\,y = 0.$ ...(2)

From (1) and (2), Since $(x, y) \neq (0, 0)$, we must have

$$\begin{vmatrix} 1+\lambda & 4\lambda \\ 4\lambda & 1+7\lambda \end{vmatrix} = 0 \text{ that is, } 9\lambda^2 - 8\lambda - 1 = 0 \text{ or } \lambda = 1, -1/9.$$

Case 1: $\lambda = 1$.

From (1) or (2), $x = -2y$ and substitution in $x^2 + 8xy + 7y^2 = 225$ gives $y^2 = -45$, for which no real solution exists.

Case 2: $\lambda = -1/9$.

From (1) or (2), $y = 2x$ and substitution in $x^2 + 8\,xy + 7y^2 = 225$ gives $x^2 = 5$. Then $y^2 = 4x^2 = (4)\,(5) = 20$ and so $x^2 + y^2 = 25$. Thus the required shortest distance is $\sqrt{25} = 5$ units.

EXAMPLE 2.52 If r is the distance of a point on the conic $ax^2 + by^2 + cz^2 = 1$, $lx + my + nz = 0$ from the origin, show that the stationary values of r are given by $\dfrac{l^2}{1-ar^2} + \dfrac{m^2}{1-br^2} + \dfrac{n^2}{1-cr^2} = 0.$ *(AMIETE, Dec. 2002)*

Solution We have to find the stationary values of

$$r^2 = x^2 + y^2 + z^2 \qquad ...(i)$$

subject to the conditions that $ax^2 + by^2 + cz^2 - 1 = 0$ ...(*ii*) and $lx + my + nz = 0.$ ...(*iii*)

In this case, we use two Lagrange multipliers λ_1, λ_2. Consider the auxiliary function

$$F(x, y, z, \lambda_1, \lambda_2) = (x^2 + y^2 + z^2) + \lambda_1(ax^2 + by^2 + cz^2 - 1) + \lambda_2(lx + my + nz).$$

For the extremum, we have the necessary conditions, set the partial derivatives of F w.r.t. x, y, z equal to zero. Thus, we obtain

$\partial F/\partial x = 2x + 2\lambda_1 ax + l\lambda_2 = 0,$ ...(*iv*) $\quad \partial F/\partial y = 2y + 2\lambda_1 by + m\lambda_2 = 0,$ ...(*v*)

$\partial F/\partial z = 2z + 2\lambda_1 cz + n\lambda_2 = 0.$ ...(*vi*)

On multiplying (*iv*), (*v*), (*vi*) by *x*, *y*, and *z* respectively and then on adding, we get

$$2(x^2 + y^2 + z^2) + 2\lambda_1 (ax^2 + by^2 + cz^2) + \lambda_2 (lx + my + nz) = 0$$

or $$2(r^2) + 2\lambda_1(1) + \lambda_2(0) = 0 \text{ or } \lambda_1 = -r^2.$$

Substituting for λ_1 in (*iv*), we get

$$2x - r^2(2\,ax) + \lambda_2(l) = 0 \text{ or } 2x(1 - ar^2) = -\lambda_2 l \quad \text{or} \quad \frac{l}{1-ar^2} = -\frac{2x}{\lambda^2} \text{ or } \frac{l^2}{1-ar^2} = -\frac{2lx}{\lambda_2}.$$

Similarly, $$\frac{m^2}{1-br^2} = -\frac{2my}{\lambda_2} \text{ and } \frac{n^2}{1-cr^2} = -\frac{2nz}{\lambda_2}.$$ Adding these, we get

$$\frac{l^2}{1-ar^2} + \frac{m^2}{1-br^2} + \frac{n^2}{1-cr^2} = -\frac{2}{\lambda_2}(lx + my + nz) = \frac{-2}{\lambda_2}(0) = 0, \text{ the required result.}$$

EXAMPLE 2.53 Find the maximum and minimum values of $x^2 + y^2 + z^2$ subject to the constraint conditions $x^2/4 + y^2/5 + z^2/25 = 1$, and $z = x + y$.

Solution We must find the extrema of $f(x, y, z) = x^2 + y^2 + z^2$ subject to the constraint conditions

$$\phi_1 = \frac{x^2}{4} + \frac{y^2}{5} + \frac{z^2}{25} - 1 = 0 \text{ and } \phi_2 = x + y - z = 0.$$

Consider the auxiliary function $F(x, y, z, \lambda_1, \lambda_2) = x^2 + y^2 + z^2 + \lambda_1\left(\frac{x^2}{4} + \frac{y^2}{5} + \frac{z^2}{25} - 1\right) + \lambda_2(x + y - z)$.

For the extremum, we have the necessary conditions

$$\left.\begin{aligned} \frac{\partial F}{\partial x} &= 2x + \frac{\lambda_1}{2}x + \lambda_2 = 0; \\ \frac{\partial F}{\partial y} &= 2y + \frac{2\lambda_1}{5}y + \lambda_2 = 0; \\ \frac{\partial F}{\partial z} &= 2z + \frac{2\lambda_1}{25}z - \lambda_2 = 0. \end{aligned}\right\} \qquad ...(1)$$

Solving these equations for *x*, *y*, *z*, we find $x = -2\lambda_2/(\lambda_1 + 4)$, $y = -5\lambda_2/(2\lambda_1 + 10)$, $z = 25\lambda_2/(2\lambda_1 + 50)$. ...(2)

Substituting in second constraint condition, $x + y - z = 0$, we get

$$\frac{2}{\lambda_1 + 4} + \frac{5}{2\lambda_1 + 10} + \frac{25}{2\lambda_1 + 50} = 0,\ \lambda_2 \neq 0. \qquad ...(3)$$

For if, $\lambda_2 = 0$, $x = y = z = 0$, but (0, 0, 0) does not satisfy the other condition of constraint. Hence, from (3), $17\lambda_1^2 + 245\lambda_1 + 750 = 0$ or $(\lambda_1 + 10)(17\lambda_1 + 75) = 0$ from which $\lambda_1 = -10$ or $-75/17$.

Case 1: $\lambda_1 = -10$.

From (2), $x = \lambda_2/3$, $y = \lambda_2/2$, $z = 5\lambda_2/6$.

Substituting in the first constraint condition, $x^2/4 + y^2/5 + z^2/25 = 1$, yields

$\lambda_2^2 = 180/19$ or $\lambda_2 = \pm 6\sqrt{5/19}$. This gives the two critical points.

$(2\sqrt{5/19}, 3\sqrt{5/19}, 5\sqrt{5/19}), (-2\sqrt{5/19}, -3\sqrt{5/19}, -5\sqrt{5/19})$.

The value of $x^2 + y^2 + z^2$ corresponding to these critical points is $(20 + 45 + 125)/19 = 10$.

Case 2: $\lambda_1 = -75/17$.

From (2), $x = \frac{34}{7}\lambda_2$, $y = -\frac{17}{4}\lambda_2$, $z = \frac{17}{28}\lambda_2$. Substituting in the first constraint condition, $\frac{x^2}{4} + \frac{y^2}{5} + \frac{z^2}{25} = 1$

yield $\lambda_2 = \pm\frac{140}{17\sqrt{646}}$ which gives the critical points $\left(\frac{40}{\sqrt{646}}, -\frac{35}{\sqrt{646}}, \frac{5}{\sqrt{646}}\right), \left(-\frac{40}{\sqrt{646}}, \frac{35}{\sqrt{646}}, \frac{-5}{\sqrt{646}}\right)$.

The value of $x^2 + y^2 + z^2$ corresponding to these is $(1600 + 1225 + 25)/646 = 75/17$. Thus, the required maximum value is 10 and the minimum value is 75/17.

EXAMPLE 2.54 Find the shortest distance between the line $y = 10 - 2x$ and the ellipse $x^2/4 + y^2/9 = 1$.
(AMIETE, June, 2011; AMIE, W-2010)

Solution Let (x, y) be a point on the ellipse and (a, b) be a point on the line Then the shortest distance between the line and the ellipse is the square root of the minimum value of

$f(x, y, a, b) = (x - a)^2 + (y - b)^2$ subject to the constraints

$\phi_1(x, y) = 9x^2 + 4y^2 - 36 = 0$ and $\phi_2(a, b) = 2a + b - 10 = 0$.

Consider the auxiliary fnction

$$F(x, y, a, b, \lambda_1, \lambda_2) = (x - a)^2 + (y - b)^2 + \lambda_1(9x^2 + 4y^2 - 36) + \lambda_2(2a + b - 10).$$

For extremum, we have the necessary conditions

$$\partial F/\partial x = 2(x - a) + 18x\lambda_1 = 0, \quad \text{or} \quad \lambda_1 x = (a - x)/9$$
$$\partial F/\partial y = 2(y - b) + 8y\lambda_1 = 0, \quad \text{or} \quad \lambda_1 y = (b - y)/4$$
$$\partial F/\partial a = -2(x - a) + 2\lambda_2 = 0, \quad \text{or} \quad \lambda_2 = x - a$$
$$\partial F/\partial b = -2(y - b) + \lambda_2 = 0, \quad \text{or} \quad \lambda_2 = 2(y - b).$$

Eliminating λ_1 and λ_2 from the above equations, we obtain $4(a - x)y = 9(b - y)x$ and $x - a = 2(y - b)$. Dividing the two equations, we get $8y = 9x$. Substituting in the equation of the ellipse, we obtain

$$9x^2 + 4\left(\frac{9x}{8}\right)^2 = 36, \text{ or } x^2 = 64/25 \Rightarrow x = \pm 8/5 \text{ and } y = \pm 9/5.$$

Corresponding to $x = 8/5$, $y = 9/5$, we obtain $\frac{8}{5} - a = 2\left(\frac{9}{5} - b\right)$ or $a = 2b - 2$.

Substituting in the equation of the line $2a + b - 10 = 0$, we get $a = 18/5$ and $b = 14/5$.

Hence, an extremum is obtained when $(x, y) = (8/5, 9/5)$ and $(a, b) = (18/5, 14/5)$. The distance between the two points is

$$\sqrt{\left(\frac{18}{5} - \frac{8}{5}\right)^2 + \left(\frac{14}{5} - \frac{9}{5}\right)^2} = \sqrt{5} \text{ units.}$$

Corresponding to $x = -8/5$, $y = -9/5$, we get $a - 2b = 2$. Substituting in the equation $2a + b - 10 = 0$, we get $a = 22/5$, $b = 6/5$. Hence another extremum is obtained when $(x, y) = (-8/5, -9/5)$ and $(a, b) = (22/5, 6/5)$.

The distance be4tween these two points $= \sqrt{\left(\frac{22}{5} + \frac{8}{5}\right)^2 + \left(\frac{6}{5} + \frac{9}{5}\right)^2} = 3\sqrt{5}$ units.

Hence, the shortest distance between the line and the ellipse is $\sqrt{5}$ units.

EXERCISE 2.5

1. Discuss for maxima and minima the following functions:

(i) $x^3 + y^3 + 3xy$.
(ii) $x^2 + y^2 + 6x + 12$.
(iii) $x^3 + 3xy^2 - 15x^2 - 15y^2 + 72x$. *(AMIE, S-2001)*
(iv) $3x^2 + xy + y^2 + x + y$. *(AMIE, S-2002)*
(v) $x^3y^2(1 - x - y)$, $x > 0$, $y > 0$. *(AMIE, S-2004)*
(vi) $\sin x \sin y \sin(x + y)$. $(0 \le x \le \pi, 0 \le y \le \pi)$.
(vii) $x^2y + xy^2 - axy$. *(AMIE, S-2003)*
(viii) $x^3 + y^3 - 3x - 12y + 20$.

Ans. (i) Maximum value 1 at $(-1, -1)$.
(ii) Minimum value 3 at $(-3, 0)$.
(iii) Maximum value 112 at $(4, 0)$, Minimum value 108 at $(6, 0)$.
(iv) Minimum value $-33/121$ at $(-1/11, -5/11)$.
(v) Maximum value $1/432$ at $(1/2, 1/3)$; no extreme value at $(0, 0)$.
(vi) Maximum value $3\sqrt{3}/8$ at $(\pi/3, \pi/3)$; Minimum value $-3\sqrt{3}/8$ at $(2\pi/3, 2\pi/3)$.

(*vii*) Maximum or Minimum exist at $(a/3, a/3)$; Minimum value $= -a^3/27$; Maximum value $= a^3/27$.

(*viii*) Minimum value 2 at (1, 2); Maximum value 38 at (–1, –2); (–1, 2) and (1, –2) are saddle points.

2. (*i*) Test for local maxima or minima, given $f(x, y) = x^2 + 2xy + 2y^2 + 2x + y$.

(*ii*) Discuss the maxima or minima of the function $z = xy + a^3/x + a^3/y$. (*AMIE, W-2004,*)

Ans. (*i*) Minimum at (–3/2, 1/2); (*ii*) Minimum at (a, a).

3. (*i*) Find the stationary points of $f(x, y) = x^3 + y^2 + 3x^2 + 4xy$. Examine them for the extreme value of the function.

(*ii*) Test for maximum and minimum the function: $z = x^2 + xy + y^2 + x^{-1} + y^{-1}$.

Ans. (*i*) (0, 0), (2/3, –4/3); Minimum value = –4/27. (*ii*) Minimum z at $x = y = 1/\sqrt[3]{3}$.

4. Show that the maximum value of $\sin x + \sin y + \sin(x + y)$ $[0 \le x \le (\pi/2),\ 0 \le y \le (\pi/2)]$ is $3\sqrt{3}/\sqrt{2}$.

5. Find the absolute maximum and minimum values of the function $f(x, y) = 4x^2 + 9y^2 - 8x - 12y + 4$ over the rectangle in first quadrant bounded by the lines $x = 2$, $y = 3$ and the coordinate axes. (*AMIETE, June-2007*)

Ans. Absolute minimum value is –4 which occurs at (1, 2/3) and the absolute maximum value is 49 which occurs at the point (2, 3) and (0, 3).

6. (*i*) Find the point in the plane $2x + y + 2z = 16$ nearest to the origin.

(*ii*) Find the point $P(x, y, z)$ on the plane $2x - y - z - 5 = 0$ that lies closest to the origin.

(*iii*) Find the point on the plane $x + 2y + 3z = 13$ closest to the point (1, 1, 1).

Ans. (*i*) [(32/9), –(16/9), (32/9)]; (*ii*) [(5/3), (5/6), –(5/6)]; (*iii*) [(3/2), 2, (5/2)].

7. (*i*) Find points on the surface $z^2 = xy + 1$ that is nearest to the origin. (*Andhra 2000;*)

(*ii*) Find the points on the surface $z^2 = xy + 4$ closest to the origin.

Ans. (*i*) (0, 0, ±1); (*ii*) (2, –2, 0) and (–2, 2, 0).

8. (*i*) Find the minimum distance from the surface $x^2 + y^2 - z^2 = 1$ to the origin.

(*ii*) Find the minimum distance from the origin to the surface $x^2 - z^2 - 1 = 0$.

(*iii*) Find the shortest and largest distance from the point (1, 2, –1) to the sphere $x^2 + y^2 + z^2 = 24$.

Ans. (*i*) Minimum distance = 1. (*ii*) points on the surface closest to the origin (± 1, 0, 0), Minimum distance = 1. (*iii*) Minimum distance $= \sqrt{6}$; Maximum distance $= \sqrt{54}$.

9. Find the stationary values of $x^2 + y^2 + z^2$ given that $ax + by + cz = p$. (*UPTU, 2006*)

Ans. $p^2/(a^2 + b^2 + c^2)$.

10. Prove that the rectangular solid of maximum value which can be inscribed in a sphere is cube.

11. Divide 120 into 3 parts such that the sum of the products taken two at a time shall be minimum.

Ans. 40, 40, 40.

12. Given $x + y + z = a$, find the maximum value of $x^m y^n z^p$. (*AMIETE, Dec 2010; K.U.Dec 2006*)

[**Hint:** Consider the auxiliary functions $F(x, y, z, \lambda) = x^m y^n z^p + \lambda\,(x + y + z - a)$.

For the extremum, we have the necessary conditions

$$\partial F/\partial x = m x^{m-1} y^n z^p + \lambda = 0,\quad \partial F/\partial y = n x^m y^{n-1} z^p + \lambda = 0,\quad \partial F/\partial z = p x^m y^n z^{p-1} + \lambda = 0.$$

or $$-\lambda = m x^{m-1} y^n z^p = n x^m y^{n-1} z^p = p x^m y^n z^{p-1},$$

that is, $$\frac{m}{x} = \frac{n}{y} = \frac{p}{z} = \frac{m+n+p}{x+y+z} = \frac{m+n+p}{a}.$$

Therefore, the maximum value of $x^m y^n z^p$ occurs when

$$x = am/(m + n + p),\quad y = an/(m + n + p),\quad z = ap/(m + n + p).$$

Hence, the maximum value of $x^m y^n z^p = \dfrac{a^{m+n+p} \cdot m^m n^n p^p}{(m+n+p)^{m+n+p}}$].

13. Find the maximum value of $u = x^p y^q z^r$ when the variable x, y, z are subject to the condition $ax + by + cz = p + q + r$. **Ans.** $(p^p q^q r^r)/a^p b^q c^r$.

14. In a plane triangle ABC, find the maximum value of cos A cos B cos C. **Ans.** 1/8.

[**Hint:** Let $F = \cos A \cos B \cos C + \lambda (A + B + C - \pi)$.

$\partial F/\partial A = -\sin A \cos B \cos C + \lambda$. Also find $\partial F/\partial B$ and $\partial F/\partial C$ and equate them to zero, we get

$\sin A \cos B \cos C = \sin B \cos A \cos C = \sin C \cos A \cos B$

$\Rightarrow$ $\tan A = \tan B = \tan C \Rightarrow A = B = C = \pi/3$.

Maximum value of cos A cos B cos C = (1/2) (1/2) (1/2) = 1/8].

15. Find the dimensions of the rectangular parallelopiped of maximum volume which has three faces in the co-ordinates planes and one vertex in the plane $x/a + y/b + z/c = 1$. **Ans.** $(a/3, b/3, c/3)$; $V = abc/27$.

16. The temperature T at any point (x, y, z) in space is $T = 400\,xyz^2$. Find the highest temperature on the surface of the unit sphere $x^2 + y^2 + z^2 = 1$. *(AMIE, W-2005)* **Ans.** 50.

17. Find the dimensions of the rectangular box open at the top of maximum capacity whose surface is 432 sq. cm. *(AMIETE, Dec 2010; K.U. Dec 2006; Madras, 2005)* **Ans.** (12, 12, 6 cm)

18. A tent having the form of a cylinder surmounted by a cone is to contain a given volume. Find its dimensions if the convas required is a minimum. **Ans.** radius: cylinder height: cone height $= \sqrt{5} : 1 : 2$.

19. Find the largest product the positive numbers x, y and z can have if $x + y + z^2 = 16$.

20. If $3x^{-1} + 4y^{-1} + 5z^{-1} = 6$. Find the values of x, y, z which make $x + y + z$ a minimum.

Ans. $x = \lambda\sqrt{3},\ y = 2\lambda,\ z = \lambda\sqrt{5}$, where $\lambda = (\sqrt{3} + 2 + \sqrt{5})/6$.

21. A solid consists of a cylinder surmounted by a cone. Let r = radius, h = altitude of the cylinder, H = altitude of the cone; S = surface area, and V = volume of the solid. Show that the minimum value of S for a given value of V is given by

$$r^3 = \frac{3V}{\pi(3+\sqrt{5})},\ H = \frac{2r}{\sqrt{5}},\ S = \frac{3V}{r}.$$

22. Find the stationary value of $u = xyz$ subject to $x + y + z = 5$, $xy + yz + zx = 8$.

Ans. $x = 4/3$, $y = 4/3$, $z = 7/3$.

[**Hint:** Find the stationary value of $u = xy(5 - x - y)$ subject to condition $yx + (x + y)(5 - x - y) = 8$].

23. Using Lagrange's method of multipliers, find the critical points (stationary values) of the function $f(x, y, z) = x^2 + y^2 + z^2$ given that $z^2 = xy + 1$. **Ans.** $(0, 0, \pm 1)$.

24. Find the extreme values of $f(x, y, z) = 2x + 3y + z$ such that $x^2 + y^2 = 5$ and $x + z = 1$.

(AMIETE, June 2010; Dec 2007) **Ans.** Maximum value is $1 + 5\sqrt{2}$ and minimum value $1 - 5\sqrt{2}$.

25. If $u = ax^2 + by^2 + cz^2$ where $x^2 + y^2 + z^2 = 1$ and $lx + my + nz = 0$, prove that the stationary values of u satisfy the equation:

$$\frac{l^2}{a-u} + \frac{m^2}{b-u} + \frac{n^2}{c-u} = 0.$$

26. Find the relative maximum and minimum values of the function $f(x, y) = x^4 - y^4 - 2x^2 + 2y^2$. *(AMIETE, June 2005)*

Ans. (0, 1) and (0, –1) are points of relative minimum and minimum value at each point is –1; (–1, 0) and (1, 0) are points of relative maximum and maximum value at each point is 1; Points (0, 0), (±1, ±1) are neither the points of maximum nor minimum.

27. Divide 24 into three parts such that the continued product of the first, the square of the second and the cube of the third may be maximum. *(AMIE, S-2011, 2007; Bhillai 2005)* **Ans.** $x = 12$, $y = 8$, $z = 4$.

[**Hint:** $x + y + z = 24$ and xy^2z^3 is a maximum.]

28. A point P moves on a plane containing a triangle ABC. Prove that $(AP)^2 + (BP)^2 + (CP)^2$ will be minimum when P is at the centroid of the triangle ABC. *(AMIE, W-2009, S-2005)*

29. Find the extreme value of $x^2 + y^2 + z^2 + xy + yz + zx$ subject to the conditions $x + y + z = 1$ and $x + 2y + 3z = 3$ *(AMIETE, Dec. 2004)* **Ans.** Extreme value is 11/12 at (–1/6, 1/3, 5/6).

30. Show that the stationary values of $u = \frac{x^2}{a^4} + \frac{y^2}{b^4} + \frac{z^2}{c^4}$, where $lx + my + nz = 0$ and $\frac{x^2}{a^2} + \frac{y^2}{b^2} + \frac{z^2}{c^2} = 1$ are the roots of the equation $\frac{l^2a^4}{1-a^2u} + \frac{m^2b^4}{1-b^2u} + \frac{n^2c^4}{1-c^2u} = 0$. (*AMIETE, Dec 2003*)

31. Show that the maximum and minimum distances from the origin to section of the surface $(x^2 + y^2 + z^2)^2 = x^2 + 2y^2 + 3z^2$ made by the plane $x + y + z = 0$ are the roots of the equation $3r^4 - 12r^2 + 11 = 0$, where r is the distance from the origin to any point on the section.

32. Using Lagrange's multipliers, find a point (x, y, z) on the unit sphere $x^2 + y^2 + z^2 = 1$ which minimizes the function $x + y^2 + yz + 2z^2$. **Ans.** $(-1, 0, 0)$

33. Find the minimum value of $x^2 + y^2 + z^2$ subject to the condition $xyz = a^3$.
(*AMIETE, June 2009, June 2008; AMIE., S-2008*) **Ans.** $3a^2$.

34. Find the extreme values of $f(x, y, z) = x^2 + 2xy + z^2$ subject to the constraints $g(x, y, z) = 2x + y = 0$ and $h(x, y, z) = x + y + z = 1$. (*AMIETE., June 2006*). **Ans.** Extreme value is 3/2 at $(1/2, -1, 3/2)$.

35. Show that $f(x, y, z) = (x + y + z)^3 - 3(x + y + z) - 24xyz + a^3$ has a minima at $(1, 1, 1)$ and a maxima at $(-1, -1, -1)$. (*AMIE, W-2007*)

36. Find the greatest and the least value of the function $f(x, y) = x^2 + 2xy - 4x + 8y$ in the rectangle bounded by the straight lines $x = 0, y = 0, x = 1, y = 2$.

Ans. The greatest value $f(x, y) = 17$ is at the point $(1, 2)$; the least value $f(x, y) = -3$ at the point $(1, 0)$; the stationary point $(-4, 6)$ lies outside the given domain.

37. Find the absolute maxima and minima of the $f(x, y) = 2x^2 - 4x + y^2 - 4y + 1$ on the closed triangular plate bounded by the lines $x = 0, y = 0, y = 2x$ in the first quadrant.

Ans. The absolute maxima value $f(x, y) = 1$ is at the point $(0, 0)$; the absolute minima value $f(x, y) = -5$ at the point $(1, -2)$.

38. Find the absolute maximum and minimum values of $f(x, y) = 3xy - 6x - 3y + 7$ on the closed triangular region R with vertices $(0, 0)$, $(3, 0)$, and $(0, 5)$.

Ans. The absolute maximum value of $f(x, y) = 7$ is at the point $(0, 0)$; the absolute minimum value $f(x, y) = -11$ at the point $(3, 0)$.

2.11 APPROXIMATION BY TOTAL DIFFERENTIALS

Let the function $z = f(x, y)$ be differentiable at the point (x, y). The total increment Δz of the function z corresponding to the increments Δx in x and Δy in y is defined by

$$\Delta z = f(x + \Delta x, y + \Delta y) - f(x, y),$$

or

$$f(x + \Delta x, y + \Delta y) = f(x, y) + \Delta z. \quad \text{...(2.25)}$$

We know the approximate formula $\Delta z \approx dz$, ...(2.26) where $dz = \frac{\partial f}{\partial x}\Delta x + \frac{\partial f}{\partial y}\Delta y$.

Substituting into (2.25) the expanded expression for dz in place of Δz, we get the approximate formula

$$f(x+\Delta x, y+\Delta y) \approx f(x, y) + \frac{\partial}{\partial x} f(x, y)\cdot \Delta x + \frac{\partial}{\partial y} f(x, y)\cdot \Delta y. \quad \text{...(2.27)}$$

This result has applications in estimating errors in calculations.

EXAMPLE 2.55 A certain function $z = f(x, y)$ has values $f(2, 3) = 5$, $f_x(2, 3) = 3$ and $f_y(2, 3) = 7$. Find an approximate value of $f(1.98, 3.01)$.

Solution We have $x = 2, y = 3, \Delta x = -0.02, \Delta y = 0.01$.

Therefore,

$$f(1.98, 3.01) \approx f(2, 3) + f_x(2, 3)\cdot \Delta x + f_y(2, 3)\cdot \Delta y$$
$$= 5 + (3)(-0.02) + (7)(0.01) = 5.01.$$

EXAMPLE 2.56 Using differentials, find an approximate value of

(*i*) $f(4.1, 4.9)$, where $f(x, y) = \sqrt{x^3 + x^2 y}$. (*AMIE, W-2005*)

(*ii*) $f(1.94, 0.98, 2.01)$, where $f(x, y, z) = \sqrt{x^2 + y^2 + z^2}$.

(*iii*) $f(4.05, 7.97)$, where $f(x, y) = (x)^{1/2}(y)^{1/3}$.

Solution (*i*) Let $(x, y) = (4, 5)$, $\Delta x = 0.1$, $\Delta y = -0.1$. We have

$$f(x, y) = \sqrt{x^3 + x^2 y}, \qquad f(4, 5) = \sqrt{64 + (16)(5)} = 12,$$

$$f_x(x, y) = \frac{3x^2 + 2xy}{2\sqrt{x^3 + x^2 y}}, \qquad f_x(4, 5) = \frac{48 + 40}{2\sqrt{64 + 80}} = \frac{11}{3},$$

$$f_y(x, y) = \frac{x^2}{2\sqrt{x^3 + x^2 y}}, \qquad f_y(4, 5) = \frac{16}{(2)(12)} = \frac{2}{3}.$$

Therefore,

$$f(4.1, 4.9) \approx f(4, 5) + f_x(4, 5)\,\Delta x + f_y(4, 5)\Delta y$$
$$= 12 + (11/3)(0.1) + (2/3)(-0.1) = 12.3.$$

(*ii*) Let $(x, y, z) = (2, 1, 2)$, $\Delta x = -0.06$, $\Delta y = -0.02$, $\Delta z = 0.01$.

Take

$$f(x, y, z) = \sqrt{x^2 + y^2 + z^2}, \qquad f(2, 1, 2) = 3,$$

$$f_x(x, y, z) = \frac{x}{\sqrt{x^2 + y^2 + z^2}}, \qquad f_x(2, 1, 2) = \frac{2}{3},$$

$$f_y(x, y, z) = \frac{y}{\sqrt{x^2 + y^2 + z^2}}, \qquad f_y(2, 1, 2) = \frac{1}{3},$$

$$f_z(x, y, z) = \frac{z}{\sqrt{x^2 + y^2 + z^2}}, \qquad f_z = \frac{2}{3}.$$

Therefore,

$$f(1.94, 0.98, 2.01) \approx f(2, 1, 2) + f_x(2, 1, 2)\,\Delta x + f_y(2, 1, 2)\,\Delta y + f_z(2, 1, 2)\Delta z$$
$$= 3 + (2/3)(-0.06) + (1/3)(-0.02) + (2/3)(0.01) = 2.96.$$

(*iii*) Let $(x, y) = (4, 8)$, $\Delta x = 0.05$, $\Delta y = -0.03$, we have

$$f(x, y) = x^{1/2}, y^{1/3}, \qquad f(4, 8) = 4^{1/2}.\ 8^{1/3} = 4,$$

$$f_x(x, y) = \frac{y^{1/3}}{2x^{1/2}}, \qquad f_x(4, 8) = \frac{1}{2},$$

$$f_y(x, y) = \frac{x^{1/2}}{3y^{2/3}}, \qquad f_y(4, 8) = \frac{1}{6}.$$

Therefore

$$f(4.05, 7.97) \approx f(4, 8) + f_x(4, 8) \cdot \Delta x + f_y(4, 8) \cdot \Delta y$$
$$= 4 + (1/2)(0.05) + (1/6)(-0.03) = 4.02.$$

2.12 USE OF A DIFFERENTIALS TO ESTIMATE ERRORS IN CALCULATIONS

Consider a function of n variables $x_1, x_2, \ldots, x_n$. Let the function $z = f(x_1, x_2, \ldots, x_n)$ be differentiable at the point $P(x_1, x_2, \ldots, x_n)$.

Let there be errors $\Delta x_1, \Delta x_2, \ldots, \Delta x_n$ in measuring the values of $x_1, x_2, \ldots, x_n$ respectively. Then, the computed value of z using the inexact values of the arguments will be obtained with an error

$$\Delta z = f(x_1 + \Delta x_1, x_2 + \Delta x_2, \ldots, x_n + \Delta x_n) - f(x_1, x_2 \ldots, x_n).$$

When the errors $\Delta x_1, \Delta x_2, \ldots, \Delta x_n$ are small in magnitude, we can replace, approximately, the total increment by the total differential, we obtain

$$f(x_1+\Delta x_1, x_2+\Delta x_2, \ldots x_n+\Delta x_n) \approx f(x_1, x_2, \ldots, x_n) + f_{x_1}\Delta x_1 + f_{x_2}\Delta x_2 + \ldots + f_{x_n}\Delta x_n,$$

where the partial derivatives are evaluated at the point $(x_1, x_2, \ldots, x_n)$. This is the generalization of the result for functions of two variables.

Since the values of the partial derivatives and the errors in arguments may be either positive or negative, we define the *absolute error* as

$$|\Delta z| \approx |dz| = |df| = \left|f_{x_1}\Delta x_1 + f_{x_2}\Delta x_2 + \ldots\ldots f_{x_n}\Delta x_n\right|.$$

Then $|df| \le |f_{x_1}||\Delta x_1| + |f_{x_2}||\Delta x_2| + \ldots\ldots + |f_{x_n}||\Delta x_n|$ gives the *maximum absolute error* in z.

Relative and Percentage error. The expression $\dfrac{|df|}{|f|}$ is called the *maximum relative error* and $\dfrac{|df|}{|f|} \times 100$ is called the *percentage error*.

The maximum relative error can also be written as

$$\frac{|df|}{|f|} \le \left|\frac{\partial f/\partial x_1}{f}\right||\Delta x_1| + \left|\frac{\partial f/\partial x_2}{f}\right||\Delta x_2| + \ldots + \left|\frac{\partial f/\partial x_n}{f}\right||\Delta x_n|.$$

But $$\frac{\partial f/\partial x_1}{f} = \frac{\partial}{\partial x_1}\log|f|,\ \frac{\partial f/\partial x_2}{f} = \frac{\partial}{\partial x_2}\log|f|, \ldots \text{ etc.}$$

$$\therefore \quad \frac{|df|}{|f|} \le \left|\frac{\partial}{\partial x_1}[\log|f|]\right||\Delta x_1| + \left|\frac{\partial}{\partial x_2}[\log|f|]\right||\Delta x_2| + \ldots\ldots + \left|\frac{\partial}{\partial x_n}[\log|f|]\right||\Delta x_n|.$$

Remark

(1) We may write $\Delta x_1, \Delta x_2, \ldots$as dx_1, dx_2 respectively.

(2) While solving problems, relative error may be found directly by logarithmic differentiation.

(3) In practical problems, error in calculation are due to small errors in the data upon which calculation is based. The latter may arise from lack of precision in the measurements or from other causes.

ILLUSTRATIVE EXAMPLES

EXAMPLE 2.57 The period T of a simple pendulum is $T = 2\pi\sqrt{l/g}$. Find the maximum error in T due to possible error upto 1% in l and 2.5% in g. *(AMIETE, Dec 2009; UPTU, 2004)*

Solution We have $T = 2\pi\sqrt{l/g}$. Taking logarithms, we get

$$\log T = \log 2\pi + \frac{1}{2}\log l - \frac{1}{2}\log g. \text{ Therefore, } \frac{1}{T}dT = 0 + \frac{1}{2}\cdot\frac{dl}{l} - \frac{1}{2}\cdot\frac{dg}{g}$$

or $$100\frac{dT}{T} = \frac{1}{2}\left[100\left(\frac{dl}{l}\right) - 100\left(\frac{dg}{g}\right)\right].$$

The percentage error will be maximum when 100 (dl/l) = 1 and 100 (dg/g) = –2.5 that is, the error in l is positive and that in g is negative.

$$\therefore \quad 100\frac{dT}{T} = \frac{1}{2}[1-(-2.5)] = 1.75.$$ Hence maximum error in T = 1.75%.

EXAMPLE 2.58 The work that must be done to propel a ship of displacement D for a distance s in time t is proportional to $s^2D^{2/3}t^2$. Find approximately the percentage increase of work necessary when the displacement is increased by 1%, the time is diminished by 1% and the distance is increased by 3%.

Solution Let the work done be W. Therefore, $W = ks^2D^{2/3}t^2$, where k is a constant of proprotionality.

Taking logarithms, we obtain $\log W = \log k + 2\log s + (2/3)\log D + 2\log t$.

Therefore, $$\frac{dW}{W} = 2\frac{ds}{s} + \frac{2}{3}\frac{dD}{D} + 2\frac{dt}{t}.$$ Multiplying both sides by 100, we obtain

$$100\frac{dW}{W}=2\left(100\frac{ds}{s}\right)+\frac{2}{3}\left(100\frac{dD}{D}\right)+2\left(100\frac{dt}{t}\right).\quad \text{Since } 100\left(\frac{ds}{s}\right)=3,\ 100\left(\frac{dD}{D}\right)=1 \text{ and } 100\left(\frac{dt}{t}\right)=-1.$$

Therefore, $$100\frac{dW}{W}=\text{Percentage increase of work}=2(3)+\frac{2}{3}(1)+2(-1)=4\frac{2}{3}\%.$$

EXAMPLE 2.59 The indicated horse power I of an engine is calculated from the formula $I = PLAN/33000$, where $A = \pi D^2/4$. Assuming that error of λ percent may have been made in measuring P, L, N and D, find the greatest possible error in I.

Solution We have $I = PLAN/33000$. Putting $A = \pi D^2/4$, we get $I=\dfrac{\pi PLD^2N}{(4)(33000)}=\pi PL\dfrac{D^2N}{132000}$.

$\therefore$ $$\log I = \log \pi + \log P + \log L + 2\log D + \log N - \log 132000.$$

Differentiating and multiplying by 100, we obtain

$$100\frac{dI}{I}=100\frac{dP}{P}+100\frac{dL}{L}+(2)(100)\frac{dD}{D}+100\frac{dN}{N}.$$

Putting $$100\left(\frac{dP}{P}\right)=100\left(\frac{dL}{L}\right)=100\left(\frac{dD}{D}\right)=100\left(\frac{dN}{N}\right)=\lambda.\ \text{Therefore, } 100\left(\frac{dI}{I}\right)=5\lambda.$$

Thus the greatest possible error in $I = 5\lambda\%$.

EXAMPLE 2.60 With the usual meaning for a, b, c and s, if Δ be the area of a triangle prove that error in Δ resulting from a small error in the measurement of c, is given by

$$d\Delta=\frac{\Delta}{4}\left[\frac{1}{s}+\frac{1}{s-a}+\frac{1}{s-b}-\frac{1}{s-c}\right]dc.$$

Solution We know that

$$\Delta=\sqrt{s(s-a)(s-b)(s-c)},\ \text{where } s=(a+b+c)/2.$$

$\therefore$ $$\log\Delta=\frac{1}{2}[\log s+\log(s-a)+\log(s-b)+\log(s-c)].$$

Hence, $$\frac{d\Delta}{\Delta}=\frac{1}{2}\left[\frac{ds}{s}+\frac{ds}{s-a}+\frac{ds}{s-b}+\frac{ds-dc}{s-c}\right].\quad\ldots(i)$$ Now $s = (a + b + c)/2$, therefore $ds = dc/2$.

Hence puting $ds = dc/2$ in (i) we get

$$\frac{d\Delta}{\Delta}=\frac{1}{2}\left[\frac{dc}{2s}+\frac{dc}{2(s-a)}+\frac{dc}{2(s-b)}-\frac{dc}{2(s-c)}\right]\quad\text{or}\quad d\Delta=\frac{\Delta}{4}\left[\frac{1}{s}+\frac{1}{s-a}+\frac{1}{s-b}-\frac{1}{s-c}\right]dc.$$

EXAMPLE 2.61 In estimating the cost of a pile of bricks measured as 2 m × 15 m × 1.2 m, the tape is stretched 1 per cent beyond the standard length. If the count is 450 bricks to 1 cu, m, and bricks cost ₹ 530.00 per thousand, find the aproximate error in the cost. *(AMIE, S-2004, VTU-2001)*

Solution Let x, y and z be the length, breadth and height of the pile so that its volume, $V = xyz$.

Therefore, by logarithmic differentiation, we have $\dfrac{dV}{V}=\dfrac{dx}{x}+\dfrac{dy}{y}+\dfrac{dz}{z}$.

Given $dx/x = dy/y = dz/z = 1/100$, and $V = 2 \times 15 \times 1.2 = 36\text{m}^3$.

Therefore, $dV = 36(3/100) = 1.08\ \text{m}^3$. Number of bricks in $dV = 1.08 \times 450 = 486$.

$\therefore$ Error in the cost = ₹ 486 × (530/1000) = ₹ 257.58, which is a loss to the seller of bricks.

EXAMPLE 2.62 In a plane triangle ABC, if the sides a, b, be kept constant, show that the variations of its angles are given by the relation

$$\frac{dA}{\sqrt{a^2-b^2\sin^2 A}}=\frac{dB}{\sqrt{b^2-a^2\sin^2 B}}=-\frac{dC}{c},$$

the letters having their usual significance. *(AMIETE, W-2005)*

Solution In any triangle ABC, $\frac{\sin A}{a} = \frac{\sin B}{b}$ or $b \sin A = a \sin B$.

Differentiating, $b \cos A \frac{dA}{dB} = a \cos B$, ($a$, b being constants.)

or $b \cos A \, dA - a \cos B \, dB + 0. \, dC = 0$. ...(*i*) But $A + B + C = \pi$ Therefore, $dA + dB + dC = 0$. ...(*ii*)

From (*i*) and (*ii*) by cross-multiplication, we obtain

$$\frac{dA}{-a\cos B} = \frac{dB}{-b\cos A} = \frac{dC}{b\cos A + a\cos B} \quad \text{or} \quad \frac{dA}{a\cos B} = \frac{dB}{b\cos A} = -\frac{dC}{b\cos A + a\cos B}.$$

Using $\cos\theta = \sqrt{1-\sin^2\theta}$ and projection formula $c = a \cos B + b \cos A$,

we obtain $\frac{dA}{\sqrt{a^2 - a^2\sin^2 B}} = \frac{dB}{\sqrt{b^2 - b^2\sin^2 A}} = -\frac{dC}{c}$. Since $a \sin B = b \sin A$.

Therefore, $\frac{dA}{\sqrt{a^2 - b^2\sin^2 A}} = \frac{dB}{\sqrt{b^2 - a^2\sin^2 B}} = -\frac{dC}{c}$.

EXAMPLE 2.63 The area of a triangle is calculated from angles A and C and the side b. If dA is the error in measuring $\angle A$, show that the relative error in measuring area is $\frac{\sin C}{\sin A \sin(C+A)} dA$.

Solution Let K be the area of the triangle = (1/2) $ab \sin C$ (Refer Fig. 2.16)

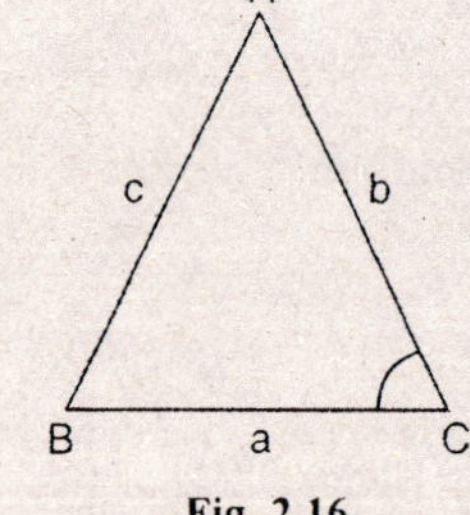

Fig. 2.16

Using the Sine Laws: $\frac{a}{\sin A} = \frac{b}{\sin B} = \frac{c}{\sin C}$. We have

$$a = \frac{b\sin A}{\sin B} \text{ and } \angle B = \pi - (\angle C + \angle A).$$

$$\therefore \quad K = \frac{1}{2}\cdot\frac{b^2 \sin A \sin C}{\sin B} = \frac{t \sin A}{\sin(C+A)}, \text{ where } t = \frac{1}{2}b^2 \sin C.$$

Taking logarithms on both sides, we have

$\log K = \log t + \log \sin A - \log \sin (C + A)$. Differentiating, we get

$$\frac{dK}{K} = \left[\frac{\cos A}{\sin A} - \frac{\cos(C+A)}{\sin(C+A)}\right] dA = \left[\frac{\cos A \sin(C+A) - \cos(C+A)\sin A}{\sin A \sin(C+A)}\right] dA$$

$$= \left\{\frac{\sin(\overline{C+A}-A)}{\sin A\sin(C+A)}\right\} dA = \frac{\sin C}{\sin A \sin(C+A)} dA.$$

Therefore, relative error $= \frac{\sin C}{\sin A \sin(C+A)} dA$.

EXAMPLE 2.64 Find the percentage error in the computed area of an ellipse when an error of +1 percent is made in measuring the major and minor axes. (*AMIETE, June 2009; V.T.U., 2005*)

Solution Let the major and minor axes of the ellipse be $2a$ and $2b$ respectively. The errors Δa and Δb in computing the lengths of the semi-major and semi-minor axes are, $\Delta a = a(0.01) = 0.01a$ and $\Delta b = b(0.01) = 0.01\,b$. The area of the ellipse is given by $A = \pi ab$. The error in computing the area of the ellipse is

$$dA = \left|\frac{\partial A}{\partial a}\right| \cdot \Delta a + \left|\frac{\partial A}{\partial b}\right| \cdot \Delta b = \pi b(0.01a) + \pi a(0.01b) = 0.02\pi ab.$$

Relative error is $\left|\frac{dA}{A}\right| = \frac{0.02\pi ab}{\pi ab} = 0.02$. Thus, percentage error in $A = \left|\frac{dA}{A}\right| \times 100 = 0.02 \times 100 = 2\%$.

EXAMPLE 2.65 In measuring a rectangular block of wood, the dimensions were found to be 10, 12 and 20 units with a possible error of 0.05 unit in each of the measurements. Find (approximately) the greatest error in the surface area of the block and the percentage error in the area caused by the errors in the individual measurements.

Solution The surface area $S = 2(xy + yz + zx)$, where x, y, z are the lengths of the edges. Using the relation $dS = \frac{\partial S}{\partial x}\Delta x + \frac{\partial S}{\partial y}\Delta y + \frac{\partial S}{\partial z}\Delta z$ to get $dS = 2(y + z)\Delta x + 2(x + z)\Delta y + 2(y + x)\Delta z$.

The greatest error in S occurs when the errors in the lengths are of the same sign, say positive. Then

$$dS = 2(12 + 20)(0.05) + 2(10 + 20)(0.05) + 2(12 + 10)(0.05) = 8.4 \text{ unit}^2.$$

The surface area, $S = 2(120 + 240 + 200) = 1120 \text{ unit}^2$.

$\therefore$ The percentage error in the surface area $= \frac{dS}{S} \times 100 = \frac{8.4}{1120} \times 100 = 0.75\%$

EXAMPLE 2.66 The power consumed in an electrical resistor is given by $P = E^2/R$ (in watts). If $E = 200$ volts and $R = 8$ ohms, by how much does the power change if E is decreased by 5 volts and R is decreased by 0.2 ohm?

Solution We have $\frac{\partial P}{\partial E} = \frac{2E}{R}, \frac{\partial P}{\partial R} = -\frac{E^2}{R^2}$. Therefore $dP = \frac{2E}{R} \cdot \Delta E - \frac{E^2}{R^2}\Delta R$.

When $E = 200$, $R = 8$, $\Delta E = dE = -5$ and $\Delta R = dR = -0.2$, then

$$dP = \frac{2(200)}{8}(-5) - \left(\frac{200}{8}\right)^2 (-0.2) = -250 + 125 = -125.$$

Thus the power is reduced by approximately 125 watts.

EXAMPLE 2.67 In a triangle ABC, right angled at C, leg $a = 32$ units and leg $c = 75$ units, determined with maximum absolute errors of 0.2 unit and 0.1 unit in measuring c and a respectively. Determine the angle A from the formula $\sin A = a/c$; and compute the maximum absolute error in the calculation of angle A.

Solution We have $\sin A = a/c$ or $\angle A = \sin^{-1}(a/c)$.

Differentiating partially, we obtain $\frac{\partial A}{\partial a} = \frac{1}{\sqrt{c^2 - a^2}}$ and $\frac{\partial A}{\partial c} = -\frac{a}{c\sqrt{c^2 - a^2}}$.

Employing the relation $|dA| = \left|\frac{\partial A}{\partial a}\right||\Delta a| + \left|\frac{\partial A}{\partial c}\right||\Delta c|$ and substituting the proper values, we get

$$|dA| = \left|\frac{1}{\sqrt{75^2 - 32^2}}\right||0.1| + \left|\frac{-32}{75\sqrt{75^2 - 32^2}}\right||0.2| = 0.00273 \text{ radian} = 9'24''$$

Thus $\angle A = \sin^{-1}(32/75) \pm 9'24''$.

EXAMPLE 2.68 Suppose $u = xe^y + y \sin z$ and that x, y, z can be measured with maximum possible errors of $\pm$ 0.2, $\pm$ 0.6 and $\pm\ \pi/180$ respectively. Estimate the maximum possible error in calculating u from the measured values $x = 2$, $y = \log 3$, $z = \pi/2$.

Solution We have $u = xe^y + y \sin z$. Therefore, $\partial u/\partial x = e^y$, $\partial u/\partial y = xe^y + \sin z$ and $\partial u/\partial z = y \cos z$.

Employing $du = \frac{\partial u}{\partial x} \cdot \Delta x + \frac{\partial u}{\partial y} \cdot \Delta y + \frac{\partial u}{\partial z} \cdot \Delta z$, we obtain

$$du = e^y \cdot \Delta x + (xe^y + \sin z)\,\Delta y + (y \cos z)\,\Delta z. \quad ...(i)$$

Substituting the values of $x, y, z, \Delta x, \Delta y$ and Δz in (i), we get

$$du = e^{\log 3}(\pm 0.2) + \left[2 \cdot (e^{\log 3}) + \sin\frac{\pi}{2}\right](\pm 0.6) + (\log 3)\left(\cos\frac{\pi}{2}\right)\left(\pm\frac{\pi}{180}\right)$$

$= 3(\pm 0.2) + (6 + 1)(\pm 0.6) + 0 = \pm 4.8$, which are the maximum possible positive and negative errors in u.

EXAMPLE 2.69 The sides of a triangle are measured as 12 cm and 15 cm and the angle included between them as 60°. If the lengths can be measured within 1% accuracy while the angle can be measured within 2% accuracy, find the percentage error in determining (*i*) area of the triangle (*ii*) length of the opposite side of the triangle. (*AMIETE; Dec. 2002*)

Solution (*i*) Refer Fig. 2.17. Area of a triangle, $K = (1/2)\ bc \sin A$. Differentiating partially, we get

$$\frac{\partial K}{\partial b} = \frac{1}{2} c \sin A,\ \frac{\partial K}{\partial c} = \frac{1}{2} b \sin A \text{ and } \frac{\partial K}{\partial A} = \frac{1}{2} bc \cos A.$$

Using the formula: $dK = \frac{\partial k}{\partial b}\Delta b + \frac{\partial k}{\partial c}\Delta c + \frac{\partial k}{\partial A}\Delta A$, we get

$$dK = \frac{1}{2} c(\sin A)\cdot \Delta b + \frac{1}{2} b(\sin A)\cdot \Delta c + \frac{1}{2} bc(\cos A)\cdot \Delta A. \qquad ...(i)$$

Substituting $b = 12$, $c = 15$, $\angle A = 60°$, $\Delta b = 0.12$, $\Delta c = 0.15$ and $\Delta A = (\pi/150)^c$ in (*i*), we obtain

$$dK = \frac{1}{2}(15)(\sin 60°)(0.12) + \frac{1}{2}(12)(\sin 60°)(0.15) + \frac{1}{2}(12)(15)\cos 60°\left(\frac{\pi}{150}\right)$$

$$= \frac{15\sqrt{3}\,(0.12)}{4} + 3\sqrt{3}\,(0.15) + \frac{3\pi}{10}$$

$$= 0.7794 + 0.7794 + 0.9429 = 2.5017 \text{ cm}^2.$$

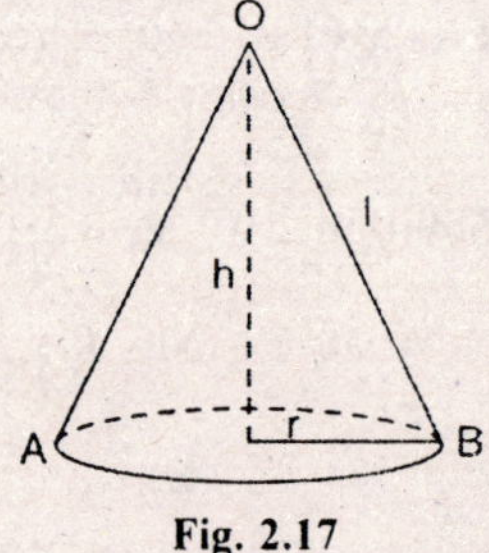

Fig. 2.17

Area of triangle, $K = \frac{1}{2}(15)(12)\sin 60° = 77.94 \text{ cm}^2$.

$\therefore$ Percentage error in computing area $= \frac{dK}{K}\times 100 = \frac{2.5017}{77.94}\times 100 = 3.21\%$.

(*ii*) Using the law of cosines,

$$\cos A = \frac{b^2 + c^2 - a^2}{2bc} \Rightarrow a^2 = b^2 + c^2 - 2bc\cos A \qquad ...(ii)$$

Differentiating (*ii*) partially, we get

$$\frac{\partial a}{\partial b} = \frac{b - c\cos A}{a},\quad \frac{\partial a}{\partial c} = \frac{c - b\cos A}{a},\quad \frac{\partial a}{\partial A} = \frac{bc\sin A}{a}$$

$$\therefore \quad da = \frac{\partial a}{\partial b}\cdot\Delta b + \frac{\partial a}{\partial c}\cdot\Delta c + \frac{\partial a}{\partial A}\cdot\Delta A = \frac{(b - c\cos A)\Delta b + (c - b\cos A)\Delta c + (bc\sin A)\Delta A}{a}. \qquad ...(iii)$$

Substituting $b = 12$, $c = 15$, $\angle A = 60°$, $\Delta b = 0.12$, $\Delta c = 0.15$ and $\Delta A = \pi/150$ in (*iii*), we obtain.

$$da = \left\{\left(12 - 15\cdot\frac{1}{2}\right)0.12 + (15 - 6)\,0.15 + (12)\,(15)\left(\frac{\sqrt{3}}{2}\right)\cdot\frac{\pi}{150}\right\} \div 13.748$$

$$= (0.54 + 1.35 + 3.266)/13.748 = 5.156/13.748 = 0.375.$$

Note: Side '*a*' is determined by employing the formula $a^2 = b^2 + c^2 - 2\,bc\cos A$ to get $a^2 = 189$ or $a = 13.748$.

Therefore % error in third side '*a*' $= \frac{da}{a}\times 100 = \frac{0.375\times 100}{13.748} = 2.73\%$.

EXAMPLE 2.70 The height h and semi-vertical angle α of a cone are measured and the total area A of surface of cone including that of base is calculated in terms of h, α. If h and α are in error by small quantities δh and $\delta\alpha$ respectively, find the corresponding error in the area. Show further that if $\alpha = \pi/6$, an error of +1% in h will be approximately compensated by an error of – 0.33 degree in α. (*AMIETE, June 2003*)

Solution Let r be the base radius and l the slant height of the cone, then total surface area A = area of base + area of curved surface

$$= \pi r^2 + \pi rl = \pi r(r + l) = \pi h\tan\alpha(h\tan\alpha + h\sec\alpha)$$

$$= \pi h^2(\tan^2\alpha + \tan\alpha\sec\alpha).$$

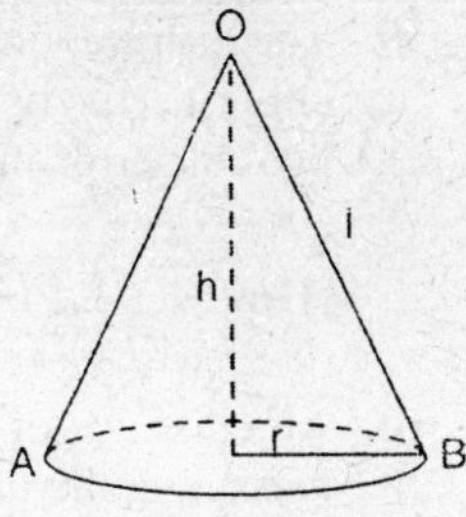

Fig. 2.18

$\therefore \quad dA = \dfrac{\partial A}{\partial h}\delta h + \dfrac{\partial A}{\partial \alpha}\delta\alpha = 2\pi h(\tan^2\alpha + \tan\alpha\sec\alpha)\delta h +$

$\pi h^2\,(2\tan\alpha\sec^2\alpha + \sec^3\alpha + \tan^2\alpha\sec\alpha)\,\delta\alpha \quad ...(i)$

which is the error dA in A corresponding to errors δh and $\delta\alpha$ in h and α respectively. Putting $\alpha = \pi/6$ and $\delta h = h/100$ in (i), we have

$$dA = \frac{2\pi h^2}{100}\left(\frac{1}{3}+\frac{1}{\sqrt{3}}\cdot\frac{2}{\sqrt{3}}\right)+\pi h^2\left(\frac{2}{\sqrt{3}}\cdot\frac{4}{3}+\frac{8}{3\sqrt{3}}+\frac{1}{3}\cdot\frac{2}{\sqrt{3}}\right)\delta\alpha$$

$$= \frac{2\pi h^2}{100}(1) + \pi h^2\left(\frac{8}{3\sqrt{3}}+\frac{8}{3\sqrt{3}}+\frac{2}{3\sqrt{3}}\right)\delta\alpha = 2\pi h^2(0.01) + \pi h^2\,(3.464)\,\delta\alpha.$$

The error in h will be compensated by the error in α, when $dA = 0$ that is, $2\pi h^2(0.01) + \pi h^2\,(3.464)\,\delta\alpha = 0$

or $$\delta\alpha = -\frac{2\,(0.01)}{3.464}\text{ radian} = -\frac{0.02}{3.464}\times 57.3° = -0.33°.$$

EXERCISE 2.6

1. Evaluate $\log(\sqrt[3]{1.03}+\sqrt[4]{0.98}-1)$.
[**Hint:** Let z = $\log(x^{1/3}+y^{1/4}-1)$. Take $x = 1, y = 1, dx = 0.03, dy = -0.02$. Then $z = \log 1 = 0$.
$\therefore\ dz = \left(\frac{1}{3}x^{-2/3}dx + \frac{1}{4}y^{-3/4}dy\right)\Big/(x^{1/3}+y^{1/4}-1)$. Substituting the proper values, we get
$dz = 3^{-1}\,(0.03) + 4^{-1}\,(-0.02) = 0.005$, which is the approximate value required].

2. (i) If $f = x^2 y^3 z^{1/10}$, find the approximate value of f, when $x = 1.99$, $y = 3.01$ and $z = 0.98$.
(ii) Find an approximate value of $f(2.1, 3.2)$, where $f(x, y) = x^y$. Given $\log 2 = 0.3010$.
Ans. (i) 107.784, (ii) 9.6816.

3. Using differential, compute $\sqrt[5]{(3.8)^2+2(2.1)^3}$ approximately. **Ans.** 0.01

4. If the power, P required to propel a steamer varies as the cube of the velocity and the square of length, prove that a 3% increase in velocity and 4% increase in length will require an increase of about 17% H.P. *(AMIE, W-2000)*
[**Hint:** $P = cv^3 l^2$. Therefore, $\log P = \log c + 3\log v + 2\log l$.
Thus $100\frac{dP}{P} = 3\left(100\frac{dv}{v}\right)+2\left(100\frac{dl}{l}\right) = 3\times 3 + 2\times 4 = 17\%.$]

5. The period T of a simple pendulum of length l with small oscillations is given by $T = 2\pi\sqrt{l/g}$. If T is computed using $l = 2\ m$ and $g = 9.8\ m/\text{sec}^2$, find the approximate error in T if the true length are $l = 2.015m$ and $g = 9.82\ m/\text{sec}^2$. Find also the percentage error.
[**Hint:** $dT = T\left(\frac{dl}{2l}-\frac{dg}{2g}\right)$. Substituting $l = 2$, $g = 9.8$, $dl = 0.015$, $dg = 0.02$ and $T = 2\pi\sqrt{2/9.8} = 2.84$,
we get $dT = 0.00775$ sec. to be added to computed value and $\left(\frac{dT}{T}\right)(100) = 0.273\%$]

6. If $\theta = \lambda/v^2$ and $\phi = \mu/v$, where λ, μ are constants, show that the percentage error of θ is nearly double of that of ϕ. *(AMIE, W-2003)*

7. Show that the relative error of a quotient does not exceed the sum of the relative errors of the dividend and the divisor. *(AMIE, W-2001)*

8. If the kinetic energy k is given by the formula $k = wv^2/2g$. Use total differential to find approximately the change in k as w changes from 49 to 49.5 and v changes from 1600 to 1590. *(VTU, 2006)*
[**Hint:** $dk = (v^2\,dw + 2wv dv)/2g = wv dv/g, + (v^2 dw)/2g$. Here $w = 49$, $v = 1600$, $dw = 0.5$, $dv = -10$.
$\therefore\ dk = -4500$ units].

9. The diameter and altitude of a can in the shape of a right circular cylinder are measured as 4 cm and 6 cm respectively. The possible error in each measurements is 0.1 cm. Find approximately the maximum possible error in the values computed for the volume and the lateral surface.

(AMIETE, Dec. 2008; AMIE, W-2004)

[**Hint:** $V = \pi x^2 y/4$, $S = \pi xy$, where x is diameter and y is altitude of the cylinder. $dV = \pi(2xydx + x^2dy)/4$, $dS = \pi(xdy + ydx)$. Since $x = 4$, $y = 6$, $dx = dy = 0.1$, we get $dV = 1.6\ \pi\ \text{cm}^3$ and $dS = \pi\ \text{cm}^2$].

10. At a distance of 50 metres from foot of a tower, the elevation of its top is 30°. If the possible errors in measuring the distance and the elevation are 2 cm and 0.05 degree respectively, find the approximate error in the calculated height. *(UPTech, 2004)*

[**Hint:** Let $h = x \tan \alpha$. $\therefore$ $dh = (\tan \alpha)\, dx + (x \sec^2 \alpha)\, d\alpha$.

Given $\alpha = \pi/6$, $x = 50$, $dx = 1/50$, $d\alpha = (0.05)\ (\pi/180)$ radian.

$$\therefore\ dh = \frac{1}{\sqrt{3}} \cdot \frac{1}{50} + 50\left(\frac{2}{\sqrt{3}}\right)^2 (0.05)\left(\frac{\pi}{180}\right) = \frac{\sqrt{3}}{150} + \frac{11}{189} = 0.07 \text{ m} = 7 \text{ cm.}]$$

11. A triangle ABC is inscribed in a circle of radius R. Prove that if the triangle be slightly varied so as to remain inscribed in the same circle, then

$$\frac{da}{\cos A} + \frac{db}{\cos B} + \frac{dc}{\cos C} = 0.$$

[**Hint:** $R = \dfrac{a}{2\sin A} = \dfrac{b}{2\sin B} = \dfrac{c}{2\sin C}$. $\therefore$ $a = 2R \sin A$, $b = 2R \sin B$, $c = 2R \sin C$.

Hence $da = 2R \cos A\ dA$, $db = 2R \cos B\ dB$, $dc = 2R \cos C\ dC$

That is, $dA = da/2R \cos A$, $dB = db/2R \cos B$, $dC = dc/2R \cos C$.

In any triangle, ABC, $\angle A + \angle B + \angle C = \pi$ or $dA + dB + dC = 0$. Hence etc.]

12. A metal box is 4 unit long, 3 unit wide and 2 unit deep. The thickness of the metal is 0.05 unit. Find approximately the volume of the metal in the box. **Ans.** 1.3 cubic unit.

13. The voltage V across a resistor is measured with an error h, and resistance R is measured with an error k. Show that the error in calculating the $W(V, R) = V^2/R$ generated in the resistor is $V(2Rh - Vk)/R^2$.

14. The time of swing t of a pendulum of length l under certain condition is given by $t = 2\pi\sqrt{l/g'}$, where $g' = g\left(\dfrac{r}{r+h}\right)^2$. Find the percentage error in t due to the errors of p percent in h and q per cent in l. **Ans.** $\dfrac{1}{2}\left(q + \dfrac{2ph}{r+h}\right)$.

15. Approximate the area of rectangle of dimensions 35.02 by 24.97 units.

[**Hint:** Let area of rectangle be denoted by K. Then $K = xy$, $dK = \dfrac{\partial K}{\partial x}dx + \dfrac{\partial K}{\partial y}dy = ydx + xdy$.

Take $x = 35$, $y = 25$, $dx = 0.02$, $dy = -0.03$. Therefore,

$K = (35)(25) = 875$. $dK = 25(0.02) + 35(-0.03) = -0.55$.

The area is approx. $K + dK = 875 - 0.55 = 874.45$ unit^2]

16. For the formula $R = E/C$, find the maximum error and the percentage error in R if $C = 20$ Amp. with a possible error of 0.1 Amp and E = 120 volts with a possible error of 0.05 volt. **Ans.** 0.0325 ohm, 0.54%.

[**Hint:** The maximum error will occur when $dE = 0.05$ and $dC = -0.1$.

Find dR, using $dR = \dfrac{E}{C}\left(\dfrac{dE}{E} - \dfrac{dC}{C}\right)$ and hence find $\dfrac{dR}{R}(100)$.]

17. Approximate the change in the hypotenuse of a right triangle of legs 6 and 8 units when the shorter leg is lengthened by 1/4 unit and the longer leg is shortened by 1/8 unit.

[**Hint:** Let x, y, and z be the shorter leg, the longer leg and the hypotenuse of the triangle. Then $z=\sqrt{x^2+y^2}$, $\partial z/\partial x = x/\sqrt{x^2+y^2}$, $\partial z/\partial y = y/\sqrt{x^2+y^2}$. $\therefore\ dz = \frac{\partial z}{\partial x}dx + \frac{\partial z}{\partial y}dy = \frac{x\,dx + y\,dy}{\sqrt{x^2+y^2}}$.

When $x = 6$, $y = 8$, $dx = 1/4$, $dy = -1/8$, then $dz = \frac{6(1/4)+8(-1/8)}{\sqrt{6^2+8^2}} = \frac{1}{20}$ unit.

Thus the hypotenuse is lengthened by approximetely 1/20 unit].

18. A quantity u can be computed by the formula $u = 4\alpha^{1/2}\beta^{3/4}\gamma^{-1/5}$. If an error of ±0.5%, ±1% and ±0.2% is made in measuring α, β and γ respectively, compute maximum percentage error in u, when the above formula is used. **Ans.** 1.04%

19. If specific gravity is determined by the formula $s = \frac{A}{A-W}$, where A is the weight in the air and W the weight in water. What is (a) approximately the largest error in s if A can be read within 0.01 gm and W within 0.02 gm, the actual readings being A = 1.1 gm; W = 0.6 gm? (b) the largest relative error?

[**Hint:** $ds = \frac{\partial s}{\partial A}dA + \frac{\partial s}{\partial W}dW = -\frac{W dA}{(A-W)^2} + \frac{A dW}{(A-W)^2}$. The greatest error in s will occur when the error in A is of negative sign. Put $dA = -0.01$, $dW = 0.02$, $A = 1.1$, $W = 0.6$, we obtain

$$ds = +\frac{0.6(0.01)}{0.25} + \frac{1.1(0.02)}{0.25} = 0.112,\ \text{hence}\ \frac{ds}{s} = \frac{(0.112)(5)}{11} = \frac{56}{1100}\Bigg].$$

20. Two sides of a triangle are found by measurement to be 18.9 m and 23.4 m and the included angle to be 60°. These measurements were subject to errors whose maximum values are 3 cm in each length and 1° in the angle.

(*i*) Find the approximate maximum error and the percentage error made in calculating the third side from these measurements.

(*ii*) What is approximately the greatest possible error in the computed value of the area?

Ans. (*i*) 0.34, 1.58%, (*ii*) 8.22 m^2.

[**Hint:** (*i*) Differentiating $x^2 = a^2 + b^2 - 2ab\cos\alpha$ partially with respect to a, b and α respectively, where a, b are the given sides and α the included angle, we get

$$\frac{\partial x}{\partial \alpha} = \frac{a - b\cos\alpha}{x},\quad \frac{\partial x}{\partial b} = \frac{b - a\cos\alpha}{x},\quad \frac{\partial x}{\partial \alpha} = \frac{ab\sin\alpha}{x}.$$

$$\therefore\quad dx = \frac{\partial x}{\partial a}\cdot da + \frac{\partial x}{\partial b}\cdot db + \frac{\partial x}{\partial \alpha}\cdot d\alpha \quad ...(i) \quad = \frac{(a-b\cos\alpha)\,da + (b-a\cos\alpha)\,db + (ab\sin\alpha)\,d\alpha}{x}$$

Put $a = 18.9$, $b = 23.4$, $\alpha = 60°$, $da = db = 0.03$, $d\alpha = 1° = \pi/180$. We get from (*i*)

$dx = 0.34$ and hence $\frac{dx}{x}(100) = 1.58\%$.

$$(ii)\quad \Delta = \frac{1}{2}ab\sin\alpha,\quad \frac{\partial\Delta}{\partial a} = \frac{1}{2}b\sin\alpha,\quad \frac{\partial\Delta}{\partial b} = \frac{1}{2}a\sin\alpha,\quad \frac{\partial\Delta}{\partial\alpha} = \frac{1}{2}ab\cos\alpha.$$

$$\therefore\quad d\Delta = \frac{\partial\Delta}{\partial a}da + \frac{\partial\Delta}{\partial b}\cdot db + \frac{\partial\Delta}{\partial\alpha}\cdot d\alpha\,]$$

21. Two sides of a triangle were measured as 150 and 200 units and the included angle is 60°. If the possible errors are 0.2 unit in measuring the sides and 1° in the angle, what is the greatest possible error in the computed area? **Ans.** 161.21 units^2.

22. Show that the approximate change in the angle A of a triangle ABC due to small changes δa, δb, δc in the sides a, b, c respectively is given by $\delta A = \frac{a}{2\Delta}(\delta a - \delta b\cos C - \delta c\cos B)$, where Δ is the area of the triangle ABC. Verify that $\delta A + \delta B + \delta C = 0$. (*AMIETE, June 2010, Dec 2007, June 2001*).

[**Hint:** Using cosine and projection formula. We know that

$\cos A = (b^2 + c^2 - a^2)/2bc$ or $2\, bc \cos A = b^2 + c^2 - a^2$.

$\therefore \quad (\delta b)c \cos A + (\delta c) \cdot b \cos A - bc \sin A(\delta A) = b(\delta b) + c(\delta c) - a(\delta a)$

$= (c \cos A + a \cos C)\delta b + [a \cos B + b \cos A]\delta c - a\, \delta\, a,$

or $\quad bc (\sin A)\, \delta A = a[(\delta a) - (\delta b) \cos C - (\delta c) \cos B)]$

or $\quad 2\Delta\, \delta A = a[\delta a - \delta b \cos C - \delta\, c \cos B] \qquad [\because \Delta = (bc \sin A)/2]$

or $\quad \delta A = \dfrac{a}{2\Delta}(\delta a - \delta b \cos C - \delta c \cos B)$. Similarly, $\delta B = \dfrac{b}{2\Delta}(\delta b - \delta c \cos A - \delta a \cos C)$ and δC etc.

Thus $\delta A + \delta B + \delta C = 0$].

23. The deflection at the centre of a rod of length l and diameter d supported at its ends and loaded at the centre with a weight w varies as wl^3d^{-4}. What is the increase in the deflection corresponding to $p\%$ increase in w, $q\%$ decrease in l and $r\%$ increase in d? **Ans.** $(p - 3q - 4r)\%$.

2.13 DIFFERENTIATION UNDER THE INTEGRAL SIGN

Very often we have to integrate functions which contain not only the variables of integration but other quantities called parameters. The integral in such cases, is a function of the parameters.

For example, $\int_a^b f(x, \alpha)\, dx = F(\alpha)$, where a, b are constants and α the parameter.

Now F is function of α having, in general, a derivative w.r.t, α given by

$$\frac{dF}{d\alpha} = \frac{d}{d\alpha}\int_a^b f(x, \alpha)\, dx. \tag{2.28}$$

Thus to get $dF/d\alpha$ we have to first integrate $f(x, \alpha)$ w.r.t. x and then find the derivative w.r.t. α. Evaluation of the integral in many cases may be difficult. Under certain circumstances it is more convenient to find $dF/d\alpha$ by interchanging the processes of differentiation and integration in (2.28). In this connection we have the following rule:

2.13.1 Leibnitz Rule

When limits of integration are constants

If $f(x, \alpha)$ and $\dfrac{\partial}{\partial \alpha} f(x, \alpha)$ be continuous functions of x and α, then

$$\frac{d}{d\alpha}\left[\int_a^b f(x,\alpha)dx\right] = \int_a^b \frac{\partial}{\partial \alpha} f(x,\alpha)\, dx \ldots\ldots\ldots\ldots\ldots \tag{2.29}$$

where a, b are constants independent of α.

Proof. Let $\quad F(\alpha) = \int_a^b f(x,\alpha)dx$.

Then $\quad F(\alpha + \delta\alpha) - F(\alpha) = \int_a^b f(x, \alpha + \delta\alpha)dx - \int_a^b f(x, \alpha)\, dx$

$$= \int_a^b [f(x, \alpha + \delta\alpha) - f(x, \alpha)]\, dx$$

$$= \int_a^b \left[f(x, \alpha) + \delta\alpha \frac{\partial f}{\partial \alpha}(x, \alpha + \theta\, \delta\alpha) - f(x, \alpha)\right] dx, \quad \text{where } 0 < \theta < 1.$$

[$\because$ By mean value theorem $f(x + h) = f(x) + hf'(x + \theta h)$, $0 < \theta < 1$]

$$= \delta\alpha \int_a^b \frac{\partial f}{\partial \alpha}(x, \alpha + \theta\delta\alpha)\, dx.$$

Proceeding to the limit as $\delta\alpha \to 0$, $\lim\limits_{\delta\alpha \to 0} \dfrac{F(\alpha + \delta\alpha) - F(\alpha)}{\delta\alpha} = \displaystyle\int_a^b \frac{\partial f}{\partial \alpha}(x, \alpha)\, dx,$

or $$\frac{\mathbf{dF}}{\mathbf{d}\alpha} = \int_a^b \frac{\partial \mathbf{f}}{\partial \alpha}(\mathbf{x}, \alpha)\, dx\,.$$

Remarks Leibnitz rule can be used

(i) to evaluate certain definite integrals,

(ii) to deduce the values of many difficult integrals from the values of simpler ones.

EXAMPLE 2.71 Using differentiation under integral sign, evaluate

$$\int_0^1 \frac{x^\alpha - 1}{\log x}\, dx,\ \alpha \geq 0.$$ *(AMIETE June 2010; MDU, 2004)*

Solution Let $$F(\alpha) = \int_0^1 \frac{x^\alpha - 1}{\log x}\, dx\,. \qquad ...(i)$$

By Leibnitz rule $$F'(\alpha) = \frac{dF}{d\alpha} = \int_0^1 \frac{\partial}{\partial \alpha}\left(\frac{x^\alpha - 1}{\log x}\right) dx = \int_0^1 \frac{x^\alpha \log x}{\log x}\, dx$$

$$= \int_0^1 x^\alpha\, dx = \left[\frac{x^{\alpha+1}}{\alpha+1}\right]_0^1 = \frac{1}{\alpha+1}.$$

Integrating, we get $F(\alpha) = \log(\alpha + 1) + c$. ...(ii)

Since from (i) we have $F(0) = 0$, therefore from (ii) $F(0) = 0 = \log 1 + c$ or $c = 0$.

Hence $\displaystyle\int_0^1 \frac{x^\alpha - 1}{\log x}\, dx = \log(1 + \alpha).$

EXAMPLE 2.72 Given $\displaystyle\int_0^\pi \frac{dx}{a + b\cos x} = \frac{\pi}{\sqrt{a^2 - b^2}}$ $(a > b)$, evaluate the integrals $\displaystyle\int_0^\pi \frac{dx}{(a + b\cos x)^2}$ and $\displaystyle\int_0^\pi \frac{\cos x}{(a + b\cos x)^2}\, dx.$ (Madras, 2006)

Solution We have $$\int_0^\pi \frac{dx}{a + b\cos x} = \pi(a^2 - b^2)^{-1/2}. \qquad ...(i)$$

Differentiating both sides of (i) w.r.t. a,

$$\int_0^\pi \frac{\partial}{\partial a}\left(\frac{1}{a + b\cos x}\right) dx = \frac{\partial}{\partial a}\{\pi(a^2 - b^2)^{-1/2}\}$$

or $$-\int_0^\pi \frac{dx}{(a + b\cos x)^2} = -\frac{1}{2}\pi(a^2 - b^2)^{-3/2} \cdot 2a$$

or $$\int_0^\pi \frac{dx}{(a + b\cos x)^2} = \frac{\pi a}{(a^2 - b^2)^{3/2}}.$$

Now differentiating both sides of (i) w.r.t. b, we obtain

$$\int_0^\pi -(a + b\cos x)^{-2} \cdot \cos x\, dx = \pi\left(-\frac{1}{2}\right)(a^2 - b^2)^{-3/2} \cdot (-2b).$$

$\therefore$ $$\int_0^\pi \frac{\cos x}{(a + b\cos x)^2}\, dx = -\frac{\pi b}{(a^2 - b^2)^{3/2}}.$$

EXAMPLE 2.73 Using the result $\int_0^{\infty} e^{-x^2}\,dx = \frac{\sqrt{\pi}}{2}$, evaluate the integral $\int_0^{\infty} e^{-bx^2} \cos 2ax\,dx, (b > 0)$.

Solution Let
$$\Phi(a) = \int_0^{\infty} e^{-bx^2} \cos 2axdx. \quad ...(i)$$

By Leibnitz rule, we have

$$\frac{d}{da}\phi(a) = \int_0^{\infty} \frac{\partial}{\partial a}(e^{-bx^2} \cos 2ax)\,dx = \int_0^{\infty} e^{-bx^2}(-2x \sin 2ax)\,dx$$

$$= \frac{1}{b}\int_0^{\infty} \underset{\text{II}}{(-2bx)\, e^{-bx^2}} \cdot \underset{\text{I}}{\sin 2ax}\,dx \qquad \text{(Integrating by parts)}$$

$$= \frac{1}{b}\left[\left(\sin 2\,ax \cdot e^{-bx^2}\right)_0^{\infty} - \int_0^{\infty} 2a \cos 2ax \cdot e^{-bx^2}\,dx\right]$$

$$= \frac{1}{b}\left\{0 - 2a\int_0^{\infty} e^{-bx^2} \cos 2axdx\right\} = \frac{-2a}{b}\phi(a). \qquad \{\text{by } (i)\}$$

Therefore, $\frac{\phi'(a)}{\phi(a)} = -\frac{2a}{b}$. Integrating w.r.t. '$a$' we have $\log \phi(a) = -\frac{a^2}{b} + c_1$

or
$$\phi(a) = e^{-\frac{a^2}{b} + c_1} = e^{c_1} \cdot e^{\frac{-a^2}{b}} = ce^{\frac{-a^2}{b}}. \quad ...(ii)$$

Putting $a = 0$, we have $\phi(0) = c$.

Now, from (i) we have

$$\phi(0) = \int_0^{\infty} e^{-bx^2}\,dx. \text{ To get } \phi(0) \text{ we put } \sqrt{b}\,x = t.$$

Then $\phi(0) = \frac{1}{\sqrt{b}}\int_0^{\infty} e^{-t^2}\,dt = \frac{1}{\sqrt{b}} \cdot \frac{\sqrt{\pi}}{2}$. ...$(iii)$ Therefore, (iii) gives $\frac{1}{2}\sqrt{\frac{\pi}{b}} = c$.

Hence, from (ii), we have

$$\phi(a) = \frac{1}{2} \cdot \sqrt{\frac{\pi}{b}} \cdot e^{-a^2/b}, \text{ which is the required result.}$$

EXAMPLE 2.74 Prove that $\int_0^{\infty} \frac{\tan^{-1}(ax)}{x(1+x^2)}\,dx = \frac{\pi}{2}\log(1+a)$, where $a \geq 0$.

(Rohtak, 2006 S; Anna, 2005 S; Bhillai, 2005; Kuk., 2005)

Solution Let $F(a) = \int_0^{\infty} \frac{\tan^{-1}(ax)}{x(1+x^2)}\,dx$. ...$(i)$

Differentiating both sides with respect to a, we have

$$F'(a) = \frac{dF}{da} = \int_0^{\infty} \frac{\partial}{\partial a}\left[\frac{\tan^{-1}(ax)}{\{x(1+x^2)\}}\right]dx = \int_0^{\infty} \frac{1}{x(1+x^2)} \cdot \frac{1}{(1+a^2x^2)} \cdot xdx$$

$$= \frac{1}{a^2}\int_0^\infty \frac{dx}{(x^2+1)\{x^2+(1/a^2)\}} = \frac{1}{a^2}\cdot\frac{a^2}{1-a^2}\int_0^\infty\left[\frac{1}{(x^2+1)} - \frac{1}{\{x^2+(1/a^2)\}}\right]dx$$

(By resolving into partial fractions)

$$= \frac{1}{1-a^2}\left[\tan^{-1} x - a\tan^{-1} ax\right]_0^\infty = \frac{1}{1-a^2}\left(\frac{\pi}{2} - \frac{\pi}{2}a\right) = \frac{\pi}{2}\cdot\left(\frac{1}{1+a}\right).$$

Integrating w.r.t. 'a', we have

$F(a) = \frac{\pi}{2}\log(1+a) + c.$ Putting $a = 0$, we get $F(0) = c$.

But, from (*i*), $F(0) = 0$, therefore, $c = 0$.

Hence $F(a) = \int_0^\infty \frac{\tan^{-1}(ax)}{x(1+x^2)}dx = \frac{\pi}{2}\log(1+a)$, the required value.

2.13.2 Leibnitz Rule

When limits of integration are functions of the parameter.

If $f(x, \alpha)$, $\frac{\partial f}{\partial \alpha}(x, \alpha)$ be continuous functions of x and α then

$$\frac{d}{d\alpha}\left[\int_{\phi(\alpha)}^{\psi(\alpha)} f(x,\alpha)dx\right] = \int_{\phi(\alpha)}^{\psi(\alpha)} \frac{\partial}{\partial \alpha} f(x,\alpha)dx + f\{\psi(\alpha), \alpha\}\frac{d\psi}{d\alpha} - f\{\phi(\alpha), \alpha\}\frac{d\phi}{d\alpha} \quad ...(2.30)$$

provided $\phi(\alpha)$ and $\psi(\alpha)$ possess continuous first order derivatives w.r.t. α

Remark If the integrand f is independent of α, we obtain from Eq. 2.30.

$$\frac{d}{d\alpha}\int_{\phi(\alpha)}^{\psi(\alpha)} f(x)\,dx = f\{\psi(\alpha)\}\frac{d}{d\alpha}\psi(\alpha) - f\{\phi(\alpha)\}\frac{d}{d\alpha}\phi(\alpha).$$

EXAMPLE 2.75 Evaluate $\int_0^{\pi/2} \frac{\log(1+\cos a\cos x)}{\cos x}dx$.

Solution Let $\qquad F(a) = \int_0^{\pi/2} \frac{\log(1+\cos a\cos x)}{\cos x}dx. \qquad ...(i)$

Differentiating w.r.t. a. $\qquad F'(a) = \int_0^{\pi/2}\frac{\partial}{\partial a}\left[\frac{\log(1+\cos a\cos x)}{\cos x}\right]dx$

$$= \int_0^{\pi/2}\frac{-\sin a\cos x}{\cos x(1+\cos a\cos x)}dx = -\sin a\int_0^{\pi/2}\frac{1}{1+\cos a\cos x}dx$$

$$= -\sin a\int_0^{\pi/2}\frac{1}{(\cos^2 x/2+\sin^2 x/2)+\cos a(\cos^2 x/2-\sin^2 x/2)}dx$$

$$= -\sin a\int_0^{\pi/2}\frac{\sec^2 x/2}{(1+\cos a)+(1-\cos a)\tan^2 x/2}dx$$

$$= -\frac{\sin a}{1-\cos a}\int_0^{\pi/2} \frac{\sec^2 x/2}{\{(1+\cos a)/(1-\cos a)\}+\tan^2 (x/2)}\,dx$$

$$= -\frac{\sin a}{1-\cos a}\cdot 2\sqrt{\frac{1-\cos a}{1+\cos a}}\left[\tan^{-1}\left(\sqrt{\frac{1-\cos a}{1+\cos a}}\cdot\tan\frac{x}{2}\right)\right]_0^{\pi/2}$$

$$= -\frac{2\sin a}{\sqrt{1-\cos^2 a}}\tan^{-1}\sqrt{\frac{2\sin^2 a/2}{2\cos^2 a/2}} = -2\cdot\frac{a}{2} = -a.$$

Integrating w.r.t a, we obtain $F(a) = -\frac{a^2}{2}+c$ ...(*ii*) When $a = \pi/2$, $F(\pi/2) = 0$, from (*i*).

Now from (*ii*), $$F(\pi/2) = 0 = -\frac{\pi^2}{8}+c \Rightarrow c = \frac{\pi^2}{8}.$$

Hence, $$F(a) = \int_0^{\pi/2}\frac{\log(1+\cos a\cos x)}{\cos x}\,dx = -\frac{a^2}{2}+\frac{\pi^2}{8} = \frac{1}{2}\left(\frac{\pi^2}{4}-a^2\right).$$

EXAMPLE 2.76 Evaluate $\int_0^a \frac{\log_e(1+ax)}{1+x^2}\,dx$ and hence, show that

$$\int_0^1 \frac{\log_e(1+x)}{1+x^2}\,dx = \frac{\pi}{8}\log_e 2.$$ *(K.U., Dec, 2006; Hissar, 2005 S)*

Solution Let $$F(a) = \int_0^a \frac{\log_e(1+ax)}{1+x^2}\,dx. \qquad ...(i)$$

Differentiating with respect to a, we obtain

$$F'(a) = \int_0^a \frac{\partial}{\partial a}\left[\frac{\log_e(1+ax)}{1+x^2}\right]dx + \frac{\log_e(1+a^2)}{1+a^2}\frac{d}{da}(a) - \frac{\log_e(1+a0)}{1+0}\cdot\frac{d}{da}(0).$$

$$= \int_0^a \frac{x}{(1+ax)(1+x^2)}\,dx + \frac{\log_e(1+a^2)}{1+a^2}. \qquad ...(ii)$$

To resolve the integrand into partial fractions, proceed as under:

Let $$\frac{x}{(1+ax)(1+x^2)} = \frac{A}{1+ax}+\frac{Bx+c}{1+x^2} \Rightarrow x = A(1+x^2)+(Bx+C)(1+ax).$$

Putting $x = -1/a$, we get $A = -a/(1+a^2)$.

Comparing coefficient of x^2, we obtain $0 = A + aB \Rightarrow B = 1/(1+a^2)$.

Comparing constant terms, we get $0 = A + C \Rightarrow C = -A = a/(1+a^2)$.

$$\therefore \quad \frac{x}{(1+ax)(1+x^2)} = \frac{1}{1+a^2}\left[\frac{-a}{1+ax}+\frac{x+a}{1+x^2}\right].$$

$$\int_0^a \frac{x}{(1+ax)(1+x^2)}\,dx = \frac{a}{1+a^2}\int_0^a\frac{dx}{1+ax}+\frac{1}{2(1+a^2)}\int_0^a\frac{2x}{1+x^2}dx+\frac{a}{1+a^2}\int_0^a\frac{dx}{1+x^2}$$

$$= -\frac{1}{1+a^2}\left[\log_e(1+ax)\right]_0^a+\frac{1}{2(1+a^2)}\left[\log_e(1+x^2)\right]_0^a+\frac{a}{1+a^2}\left[\tan^{-1}x\right]_0^a$$

$$= -\frac{\log_e(1+a^2)}{1+a^2} + \frac{\log_e(1+a^2)}{2(1+a^2)} + \frac{a\tan^{-1}a}{1+a^2}.$$

Substituting this value in (*ii*), we obtain

$$F'(a) = \frac{\log_e(1+a^2)}{2(1+a^2)} + \frac{a\tan^{-1}a}{1+a^2}.$$

Now integrating both sides with respect to a,

$$F(a) = \frac{1}{2}\int \underset{\text{I}}{\log_e(1+a^2)} \cdot \underset{\text{II}}{\frac{1}{1+a^2}}\, da + \int \frac{a\tan^{-1}a}{1+a^2}\, da \qquad \text{[Integrating by parts]}$$

$$= \frac{1}{2}\left[\log_e(1+a^2)\cdot\tan^{-1}a - \int\frac{2a}{1+a^2}\cdot\tan^{-1}a\, da\right] + \int\frac{a\tan^{-1}a}{1+a^2}\, da + c$$

$$= \frac{1}{2}\log_e(1+a^2)\cdot\tan^{-1}a + c. \qquad \text{...(iii)}$$

Putting $a = 0$ in (*i*), we obtain $F(0) = 0$. Therefore, from (*iii*) $F(0) = 0 = c$ or $c = 0$.

Hence (*iii*) gives, $\displaystyle\int_0^\infty \frac{\log_e(1+ax)}{1+x^2}\, dx = \frac{1}{2}\log_e(1+a^2)\tan^{-1}a.$

Putting $a = 1$, we obtain $\displaystyle\int_0^1 \frac{\log_e(1+x)}{1+x^2}\, dx = \frac{\pi}{8}\log_e 2.$

EXAMPLE 2.77 Evaluate the following integrals:

(*i*) $\displaystyle\int_0^{\pi/2}\frac{\log(1+a\sin^2 x)}{\sin^2 x}\, dx,$ (*NIT, Kurukshetra, 2002*) (*ii*) $\displaystyle\int_0^\infty \frac{\log(1+a^2x^2)}{1+b^2x^2}\, dx.$

Solution (*i*) Let $F(a) = \displaystyle\int_0^{\pi/2}\frac{\log(1+a\sin^2 x)}{\sin^2 x}\, dx.$ ...(1)

Differentiating w.r.t a, we obtain

$$F'(a) = \int_0^{\pi/2}\frac{\partial}{\partial a}\left[\frac{\log(1+a\sin^2 x)}{\sin^2 x}\right] dx$$

$$= \int_0^{\pi/2}\frac{1}{\sin^2 x}\left(\frac{1}{1+a\sin^2 x}\right)\sin^2 x\, dx = \int_0^{\pi/2}\frac{1}{1+a\sin^2 x}\, dx$$

$$= \int_0^{\pi/2}\frac{1}{\cos^2 x+\sin^2 x+a\sin^2 x}\, dx = \int_0^{\pi/2}\frac{1}{\cos^2 x+(1+a)\sin^2 x}\, dx$$

$$= \int_0^{\pi/2}\frac{\sec^2 x}{1+(1+a)\tan^2 x}\, dx. \qquad \left(\begin{array}{l}\textit{Put } \tan = z. \textit{ Then } (\sec^2 x)\, dx = dz\\ \textit{when } x=0,\ z=0 \textit{ and when } x=\pi/2,\ z=\infty.\end{array}\right)$$

$$\therefore\ F'(a) = \int_0^\infty\frac{dz}{1+(1+a)z^2} = \frac{1}{1+a}\int_0^\infty\frac{dz}{\frac{1}{1+a}+z^2} = \frac{\sqrt{1+a}}{1+a}\left[\tan^{-1}\frac{z}{1/\sqrt{1+a}}\right]_0^\infty = \frac{\pi}{2}(1+a)^{-1/2}.$$

Integrating w.r.t a, we obtain $F(a) = \pi\sqrt{1+a} + c.$ ...(2)

Put $a = 0$, in (1), we get $F(a) = 0$, therefore (2), imply $c = -\pi$.

Hence $\displaystyle\int_0^{\pi/2} \frac{\log(1 + a\sin^2 x)}{\sin^2 x}\,dx = \pi(\sqrt{1+a} - 1).$

(*ii*) Let $$F(a, b) = \int_0^\infty \frac{\log(1 + a^2x^2)}{1 + b^2x^2}\,dx. \qquad ...(3)$$

Differentiating w.r.t. a, we get

$$\frac{\partial F}{\partial a} = \int_0^\infty \frac{1}{1+b^2x^2}\left(\frac{2ax^2}{1+a^2x^2}\right)dx = 2a\int_0^\infty \frac{x^2}{(1+b^2x^2)(1+a^2x^2)}\,dx$$

$$= 2a\int_0^\infty \frac{1}{b^2-a^2}\left(\frac{1}{1+a^2x^2} - \frac{1}{1+b^2x^2}\right)dx$$

$$= \frac{2a}{b^2-a^2}\left[\frac{1}{a}\tan^{-1}ax - \frac{1}{b}\tan^{-1}bx\right]_0^\infty = \frac{2a}{b^2-a^2}\left(\frac{1}{a} - \frac{1}{b}\right)\frac{\pi}{2} = \frac{\pi}{b(a+b)}.$$

Integrating w.r.t a, we get

$$F(a, b) = \frac{\pi}{b}\int \frac{1}{a+b}\,da = \frac{\pi}{b}\log(a+b) + c. \qquad ...(4)$$

Put $a = 0$ in (3). Then $F(0, b) = 0$. Therefore, from (4), $0 = \dfrac{\pi}{b}\log b + c \Rightarrow c = -\dfrac{\pi}{b}\log b$.

$$\therefore \quad F(a, b) = \frac{\pi}{b}\log(a+b) - \frac{\pi}{b}\log b = \frac{\pi}{b}\log\left(\frac{a+b}{b}\right) = \frac{\pi}{b}\log\left(1 + \frac{a}{b}\right).$$

Hence, $\displaystyle\int_0^\infty \frac{\log(1+a^2x^2)}{1+b^2x^2}\,dx = \frac{\pi}{b}\log\left(1 + \frac{a}{b}\right).$

ILLUSTRATIVE EXAMPLES

EXAMPLE 2.78 If $y = \displaystyle\int_0^x f(t)\sin[k(x-t)]\,dt$, prove that y satisfies the differential equation

$$\frac{d^2y}{dx^2} + k^2y = kf(x).$$

Solution We have

$$\frac{dy}{dx} = \int_0^x \frac{\partial}{\partial x}[f(t)\sin\{k(x-t)\}\,dt + f(t)\sin\{k(x-x)\}\frac{d}{dx}(x) - 0]$$

$$= \int_0^x f(t)\cos\{k(x-t)\}\,kdt + 0 - 0 = k\int_0^x f(t)\cos\{k(x-t)\,dt\}.$$

Therefore, $$\frac{d^2y}{dx^2} = k\left[\int_0^x \frac{\partial}{\partial x}f(t)\cos\{k(x-t)\}\,dt + f(x)\cos k(x-x)\cdot\frac{d}{dx}(x) - 0\right]$$

$$= k\left[\int_0^x -f(t)\sin\{k(x-t)\cdot k\,dt\} + f(x)\right] = -k^2\int_0^x f(t)\sin\{k(x-t)\}\,dt + k\,f(x)$$

$$= -k^2 y + kf(x) \Rightarrow \frac{d^2y}{dx^2} + k^2 y = kf(x).$$

EXAMPLE 2.79 Differentiating $\int_0^x \frac{dx}{x^2+a^2} = \frac{1}{a}\tan^{-1}\frac{x}{a}$ under the integral sign, find the value of $\int_0^x \frac{dx}{(x^2+a^2)^2}$.

Solution Differentiating both sides the given relation w.r.t. 'a' (parameter), we get

$$\frac{d}{da}\int_0^x \frac{dx}{x^2+a^2} = \frac{d}{da}\left(\frac{1}{a}\tan^{-1}\frac{x}{a}\right)$$

$$\Rightarrow \int_0^x \frac{\partial}{\partial a}\frac{dx}{(x^2+a^2)} = -\frac{1}{a^2}\tan^{-1}\frac{x}{a} + \frac{1}{a}\cdot\frac{1}{1+\frac{x^2}{a^2}}\left(\frac{-x}{a^2}\right)$$

$$\Rightarrow \int_0^x \frac{-2a}{(x^2+a^2)^2}dx = -\frac{1}{a^2}\tan^{-1}\frac{x}{a} - \frac{1}{a}\frac{x}{x^2+a^2}$$

$$\Rightarrow \int_0^x \frac{dx}{(x^2+a^2)^2} = \frac{1}{2a^3}\tan^{-1}\frac{x}{a} + \frac{x}{2a^2(x^2+a^2)},$$ which is the required value.

EXAMPLE 2.80 Evaluate $\lim_{x\to 0}\frac{\int_0^x \cos t^2 dt}{x}$.

Solution Let $f(x) = \int_0^x \cos t^2 dt$ and $\phi(x) = x$. Obviously $f(0) = \phi(0) = 0$.

$$\therefore \lim_{x\to 0}\frac{f(x)}{\phi(x)}\left(\frac{0}{0}\text{ form}\right) = \lim_{x\to 0}\frac{f'(x)}{\phi'(x)}$$ (By L' Hospital's Rule)

$$= \lim_{x\to 0}\left[\frac{d}{dx}\int_0^x \cos t^2 dt\right]\Big/1 = \lim_{x\to 0}\frac{(\cos x^2)\cdot 1 - 0}{1} = \lim_{x\to 0}\cos x^2 = 1.$$

EXAMPLE 2.81 If $f(a) = \int_a^{a^2}\frac{\sin ax}{x}dx$, find $f'(a)$ where $a \neq 0$.

Solution Using Leibnitz's rule, we obtain

$$f'(a) = \int_a^{a^2}\frac{x\cos ax}{x}dx + 2a\cdot\frac{\sin a^3}{a^2} - 1\cdot\frac{\sin a^2}{a}$$

$$= \int_a^{a^2}\cos ax\,dx + \frac{2}{a}\sin a^3 - \frac{1}{a}\sin a^2 = \left[\frac{\sin ax}{a}\right]_a^{a^2} + \frac{2}{a}\sin a^3 - \frac{1}{a}\sin a^2$$

$$= \frac{1}{a}\sin a^3 - \frac{1}{a}\sin a^2 + \frac{2}{a}\sin a^3 - \frac{1}{a}\sin a^2 = \frac{3}{a}\sin a^3 - \frac{2}{a}\sin a^2.$$

EXERCISE 2.7

1. Find $df/d\alpha$, if

(*i*) $f(\alpha) = \int_0^{\pi} (1 - \alpha \cos x)^2 \, dx$, (*ii*) $f(\alpha) = \int_0^{\alpha^2} \tan^{-1}\left(\frac{x}{\alpha}\right) dx$,

(*iii*) $f(\alpha) = \int_0^{\alpha} \tan (x - \alpha) \, dx$, (*iv*) $f(\alpha) = \int_1^{2} \log (\alpha x) \, dx$.

Ans. (*i*) $\alpha\pi$, (*ii*) $2\alpha \tan^{-1} \alpha - \frac{1}{2} \log (\alpha^2 + 1)$, (*iii*) $-\tan \alpha$, (*iv*) $1/\alpha$.

2. If $\int_0^{\pi} \frac{dx}{\alpha - \cos x} = \frac{\pi}{\sqrt{\alpha^2 - 1}}$, $\alpha > 1$, find the integral $\int_0^{\pi} \frac{dx}{(\alpha - \cos x)^2}$. Hence find the value of $\int_0^{\pi} \frac{dx}{(2 - \cos x)^2}$. **Ans.** $\frac{\pi\alpha}{(\alpha^2 - 1)^{3/2}}$; $\frac{2\pi}{3\sqrt{3}}$.

3. Evaluate the integral $\int_0^{\infty} \frac{e^{-ax} \sin x}{x} dx$, $a > 0$ and deduce that $\int_0^{\infty} \frac{\sin x}{x} = \frac{\pi}{2}$. **Ans.** $\frac{\pi}{2} - \tan^{-1} a$.

4. Prove that

(*i*) $\int_0^{\infty} \frac{e^{-x}}{x} (1 - e^{-ax}) \, dx = \log (1 + a)$, where $a > -1$. (*Rohtak, 2003*)

(*ii*) $\int_0^{\infty} \frac{x^{\alpha - 1}}{1 - x} dx = \pi \cot \alpha \pi$, where $0 < \alpha < 1$.

Evaluate the following integrals, using the method of differentiation under the integral sign.

5. $\int_0^{\pi} \frac{\log (1 + a \cos x)}{\cos x} dx, |a| < 1$. (*VTU, 2000*) **Ans.** $\pi \sin^{-1} a$

6. $\int_0^{\pi} \log (1 + a \cos x) \, dx$, $|a| \le 1$. **Ans.** $\pi \log \left(\frac{1}{2} + \frac{1}{2}\sqrt{1 - a^2}\right)$, $|a| < 1$.

7. $\int_0^{\pi/2} \frac{dx}{(a^2 \sin^2 x + b^2 \cos^2 x)^2}$. **Ans.** $\frac{\pi(a^2 + b^2)}{4a^3 b^3}$.

8. $\int_0^{\pi/2} \log (a^2 \cos^2 \theta + b^2 \sin^2 \theta) \, d\theta$, $(a > 0, b > 0)$. **Ans.** $\pi \log \left[\frac{a + b}{2}\right]$.

9. $\int_0^{\pi} \log \left(\frac{a + b \sin \theta}{a - b \sin \theta}\right) \operatorname{cosec} \theta \, d\theta$, where $0 < |b| < a$. **Ans.** $2\pi \sin^{-1} (b/a)$

10. Use the rule of differentiation under the integral sign to show that

$$\int_0^{\infty} \frac{e^{-x}}{x} \left[a - \frac{1}{x} + \frac{1}{x} e^{-ax}\right] dx = (a + 1) \log (a + 1) - a.$$

[**Hint:** Let $\phi(a) = \int_0^{\infty} \frac{e^{-x}}{x} \left[a - \frac{1}{x} + \frac{1}{x} e^{-ax}\right] dx$. Find $\phi'(a)$ and then $\phi''(a)$. Integrate twice and determine arbitrary constants using $\phi(0) = 0$.]

11. Evaluate the following integrals:

(*i*) $\int_0^{\infty} \frac{e^{-ax} - e^{-bx}}{x} dx$, $a > 0, b > 0$. **Ans.** $\log (b/a)$ (*ii*) $\int_0^{1} \frac{x^a - x^b}{\log x} dx$, $a > b > -1$. **Ans.** $\log \left(\frac{a + 1}{b + 1}\right)$.

12. Obtain the following differential coefficients:

(*i*) $\frac{d}{da} \int_a^{a^2} \cos x^2 dx$, (*ii*) $\frac{d}{dx} \int_{x^2}^{2} \log (1 + t^2) \, dt$, (*iii*) $\frac{d}{da} \int_0^{a/2} (a^2 - u^2)^{1/2} du$.

Ans. (*i*) $2a \cdot \cos a^4 - \cos a^2$, (*ii*) $-\{\log (1 + x^4)\} \cdot 2x$, (*iii*) $\sqrt{3}\, a/4$.

CHAPTER

3

Elements of Curve Tracing and Standard Curves

3.1 INTRODUCTION

In the course of next few chapters, we shall be dealing with applications of integral calculus to areas of plane curves, lengths of curves, volumes and surfaces of solids of revolution etc. In these applications, which ultimately reduce to evaluation of definite integrals, we shall be required to write proper limits of integration. For this purpose it is essential that we have an idea of the general form of curves represented by given equations which in turn requires knowledge of curve tracing.

In this chapter we shall first discuss the general problem of curve tracing and then trace some important curves which frequently occur in engineering applications.

3.2 CURVES IN CARTESIAN COORDINATES: $f(x, y) = 0$

To trace the graph of a curve whose equation is given in cartesian coordinates, consideration of some or all the following points is very helpful.

A. *Symmetry.*
B. *Tangents to the curve at the origin, if it passes through it.*
C. *Asymptotes (particularly those that are parallel to the axes).*
D. *Transformation of axes.*

We discuss these one by one.

3.2.1 Symmetry

(i) Symmetry about x-axis

If the equation of a curve remains unchanged when y is changed into –y, the curve is symmetrical about x axis. In this case, if the point $P(x, y)$ lies on a curve, then $Q(x, -y)$ also lies on it. For example, the curve $y^2 = 4ax$ is symmetrical about x-axis. (Fig. 3.1)

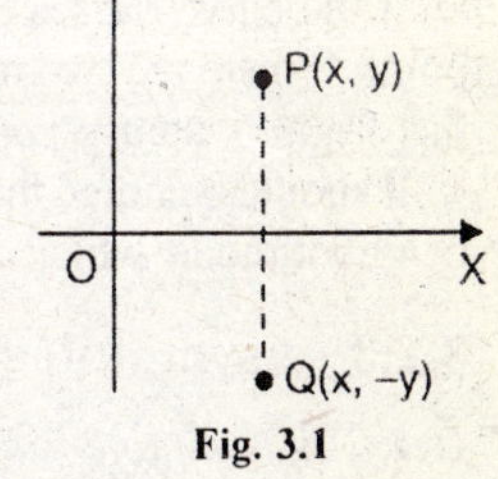

Fig. 3.1

(ii) Symmetry about y-axis

If *the equation of a curve remains unchanged when x is changed into –x, the curve is symmetrical about y-axis.* In this case if (x, y) lies on the curve, then $(-x, y)$ also lies on its.

Obviously $x^2 = 4ay$ is symmetrical about y-axis. (Fig. 3.2)

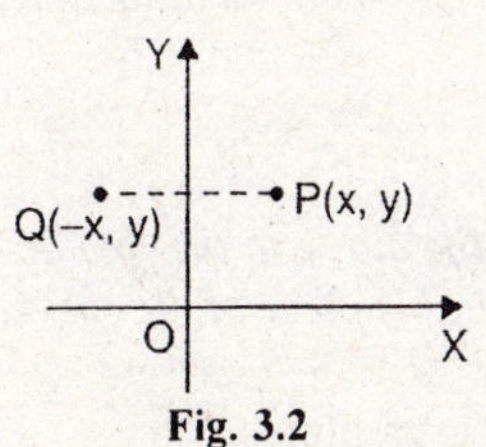

Fig. 3.2

Note. It follows from (*i*) and (*ii*) that in the case of algebraic curves, the curve is symmetrical about x-axis, if the equation of the curve involves only even powers of y. Similarly if only even powers of x, occur, the curve is symmetrical about *y-axis*.

(iii) Symmetry in opposite quadrants

If the equation of the curve remains unchanged when x is changed into –x and y is changed into –y, then the curve is symmetrical in opposite quadrants.

For example the curve $xy = c^2$ is symmetrical in opposite quadrants.

(iv) Symmetry about the line y = x

If the equation remains unchanged when x and y are interchanged, then the curve is symmetrical about the line y = x. (Fig. 3.3)

Clearly the curve $x^3 + y^3 = 3axy$ is symmetrical about the line $y = x$.

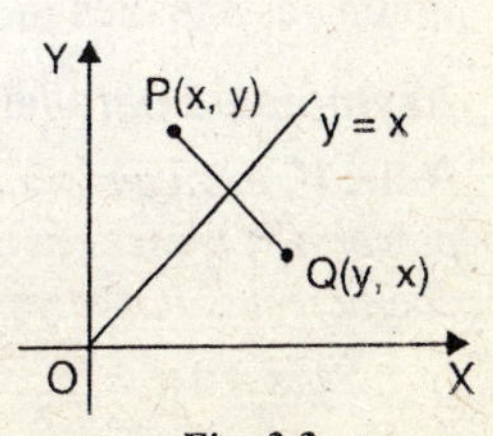

Fig. 3.3

Imp. note: If we know that a curve is symmetrical about x-axis, then its form need be known above x-axis only. The part below x-axis, (on basis of symmetry) can be easily drawn.

Similarly in case of symmetry about y-axis, we have to know the shape on the one side of y-axis. The shape on the other side is known. In case the curve is symmetrical about both axes, then it is sufficient to find the shape in the first quadrant. The remaining part can be, then, easily completed.

3.2.2 Tangents to the Curve at the Origin

Rule. *If a curve passes through the origin, then the tangents at the origin are obtained by equating to zero, the terms of the lowest degree in the equation.*

For example the parabola $y^2 = 4ax$ passes through the origin. The lowest degree term being $4ax$, the equation of the tangent to the curve at the origin is $4ax = 0$ or $x = 0$, (*i.e.*, the y-axis).

Similarly in the curve $x^2 = 4ay$, $y = 0$, (*i.e.*, the x-axis) is the tangent at the origin.

The curve $x^3 + y^3 - 3\ axy = 0$ also passes through the origin. Therefore the tangents there are given by $3\ axy = 0$ or $x = 0$ and $y = 0$. Thus both axes touch the curve at the origin.

When there are two tangents to a curve at a point, as in the last case above, it means two branches of the curve passes through that point. Such a point is called *double point.*

Types of double points

(*i*) *If the two tangents at a double point on a curve are real and different, the double point is called a NODE [Fig. 3.4 (a)].*

(*ii*) *If the two tangents coincide, the double point is called a CUSP [Fig. 3.4 (b) (i) & (ii).*

(*iii*) *In case the two tangents are imaginary, the double point is called a Conjugate Point or Isolated Point [Fig. 3.4 (c)].*

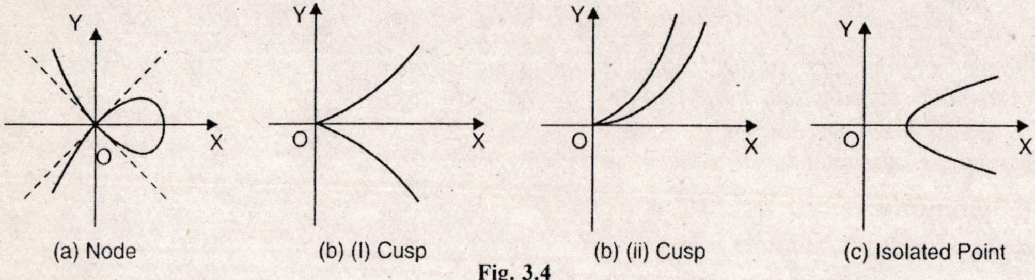

(a) Node (b) (I) Cusp (b) (ii) Cusp (c) Isolated Point

Fig. 3.4

If the curve passes through the origin, we can find the shape of the curve there by finding tangents at that point. If there are two real and different tangents, the origin is a node. In case there are two co-incident tangents, the origin is a cusp and if the two tangents at the origin are imaginary, the origin is on isolated point.

It should be noted that a point is called isolated point on the curve if there are no other points in its immediate neighbourhood which lie on the curve.

3.2.3 Asymptotes

Def. *An asymptote is a line which has the property that the distance from a point P on the curve to the line approaches zero as the distance from P to the origin increases without bound and P is on suitable piece of the curve.*

Asymptotes are usually classified as the following:

(*i*) Horizontal asymptotes, (*ii*) Vertical asymptotes, and (*iii*) Inclined or oblique asymptotes.

An asymptote which is not parallel to y-axis is called an **oblique asymptote.** We state below the rules for finding asymptotes to given curves.

Asymptotes parallel to x-axis

Rule: If, *in an equation of nth degree, the term containing x^n is absent, then the co-efficient of the next highest power of x present in the equation when equated to zero, gives the asymptotes parallel to x-axis (provided the coefficient is not merely a constant).*

Asymptotes parallel to y-axis.

Rule *If, in an equation of nth degree, the term containing y^n is absent, then the co-efficient of the next highest power of y present in the equation when equated to zero, gives the asymptotes parallel to y-axis (provided the co-efficient is not merely a constant).*

EXAMPLE 3.1 Find the asymptotes parallel to co-ordinate axes, of the following curves:

(*i*) $x^3 + 3xy^2 + y^2 + 2x + y = 0$; (*ii*) $x^2y^3 + x^3y^2 = x^3 + y^3$; (*iii*) $y^2x - a^2(x - a) = 0$.

Solution (*i*) The degree of the equation is 3.

(*a*) Since x^3 is present, there is no asymptote parallel to x-axis.

(*b*) y^3 is absent. The coefficient of next highest power of y that is, y^2 is $3x + 1$. Therefore, asymptote parallel to y-axis is $3x + 1 = 0$.

(*ii*) The equation is of 5th degree. Here x^5 and y^5 are both absent.

(*a*) The next highest power of x is x^3. The coefficient of x^3 being $y^2 - 1$, asymptotes parallel to x-axis are given by $y^2 - 1 = 0$. These are $y = \pm 1$.

(*b*) Similarly asymptotes parallel to y-axis are given by $x^2 - 1 = 0$. These are $x = \pm 1$.

(*iii*) The degree of the equation is 3. Here both x^3 and y^3 are absent.

(*a*) The next highest power of x is x. The coefficient of x is $y^2 - a^2$ and, therefore, asymptotes parallel to x-axis are $y^2 - a^2 = 0$ *i.e.*, $y = \pm a$.

(*b*) Also the next highest power of y is y^2 and its coefficient is x. Therefore, $x = 0$ (*i.e.*, y-axis) is an asymptote.

Oblique asymptotes

Notation. Suppose the equation of the curve is of nth degree. Let $\phi_n(m)$ denote the nth degree terms when we put $x = 1$ and $y = m$ in them. Similarly $\phi_{n-1}(m)$ is obtained by putting $x = 1$ and $y = m$ in $(n - 1)$th degree terms and so on.

Rule to get oblique asymptotes

In the highest degree terms put $x = 1$, $y = m$ and get $\phi_n(m)$. Solve $\phi_n(m) = 0$. Let $m_1, m_2, \ldots$ be its roots. These roots are the slopes of the asymptotes.

Next get $\phi_{n-1}(m)$ by putting $x = 1$, $y = m$ in $(n - 1)$th degree terms.

Put different values of m, that is, $m_1, m_2 \ldots$ in $c = -\dfrac{\phi_{n-1}(m)}{\phi'_n(m)}$ $\left[\text{where } \phi'_n(m) = \dfrac{d}{dm}\phi_n(m)\right]$ and get $c_1, c_2, \ldots$

Then $y = m_1x + c_1$, $y = m_2x + c_2$, ...are the oblique asymptotes.

EXAMPLE 3.2 Find the asymptotes of $y^3 - x^2y + 2y^2 + 4y + 1 = 0$.

Solution Putting $x = 1$ and $y = m$ in 3rd degree and second degree terms we get

$$\phi_3(m) = m^3 - m, \quad \phi_2(m) = 2m^2.$$

The slopes of the asymptotes are roots of $\phi_3(m) = m^3 - m = 0$. $\therefore$ $m = 0, 1, -1$ are the slopes of the asymptotes.

Now
$$c = -\frac{\phi_2(m)}{\phi'_3(m)} = -\frac{2m^2}{3m^2 - 1}. \qquad \ldots(i)$$

For $m = 0$, $c = -\dfrac{0}{-1} = 0$. [Putting $m = 0$ in (*i*)]

For $m = 1$, $c = -\dfrac{2}{3-1} = -1$. For $m = -1$, $c = -\dfrac{2}{3-1} = -1$.

Hence the three asymptotes are: $y = 0$, $y = x - 1$, $y = -x - 1$.

3.2.4 Shifting the Origin to a Point (*h*, *k*) without Changing Direction of Axes

When the origin is shifted to a point (h, k) without changing the direction of axes, the equation $f(x, y) = 0$ to a curve becomes $f(x + h, y + k) = 0$.

Such a transformation is useful in finding the tangents to a curve at a given point.

EXAMPLE 3.3 What will the equation $x^2 + y^2 - 6x + 4y - 12 = 0$ become when the origin of co-ordinates is shifted to $(3, -2)$?

Solution Here $h = 3$, $k = -2$. Therefore, the transformed equation is

$$(x + 3)^2 + (y - 2)^2 - 6(x + 3) + 4(y - 2) - 12 = 0$$

or simplifying, $$x^2 + y^2 - 25 = 0.$$

3.3 RULES FOR CURVE TRACING–CARTESIAN EQUATIONS

We now give below some of the main rules which usually suffice to obtain the general form of curves from their equations.

1. **Symmetry:**
   (*i*) If the equation remains unchanged when y is changed into $-y$, the curve is symmetrical about x-axis.
   (*ii*) If the equation does not change when x is changed into $-x$, the cuve is symmetrical about y-axis.
   (*iii*) If the equation does not change when signs of both x and y are changed, there is symmetry in opposite quadrants.
2. **Origin:** See whether the curve passes through the origin. If it does, write down the equations of the tangents there, by equating to zero the lowest degree terms.
3. **Intersection with axes:** Find the points of intersection with co-ordinate axes and, if necessary, the tangents there by shifting the origin to those points.
4. **Asymptotes:** Find the asymptotes to the curve, if any, particularly those which are parallel to the axes.
5. **Region:** If possible, express the equation of the curve in the form $y = f(x)$. See how y varies when x varies continuously. Find when x or y are imaginary and thus determine the regions where no part of the curve lies.

For example, consider the equation $y^2 = \dfrac{(x-1)^3}{(4-x)}$.

When $x < 1$, y^2 is negative so that y is imaginary.

Similarly when $x > 4$, y is imaginary. Thus the curve lie entirely in the region between the two parallel lines $x = 1$ and $x = 4$.

ILLUSTRATIVE EXAMPLES

EXAMPLE 3.4 Trace the curve $y^2 = 4ax$. (Parabola)

Solution We have the following information about the curve:

(*i*) When y is changed into $-y$, the equation remains unchanged. Hence the curve is symmetrical about x-axis.

(*ii*) The curve passes through (0, 0) and the tangent there is $4ax = 0$ or $x = 0$ *i.e.*, the y-axis.

(*iii*) When x is negative, y is imaginary. Thus no part of the curve lies to the left of y-axis.

(*iv*) Taking $y = \sqrt{4ax}$, we note that when x increases continuously y also increases continuously and when $x \to \infty$, y also tends to infinity.

When the shape above x-axis is known, then by symmetry, the shape below x-axis is also known. The graph of the curve is given in fig. 3.5.

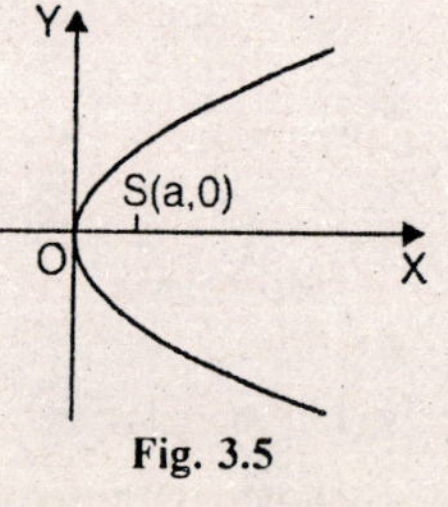

Fig. 3.5

EXAMPLE 3.5 Trace the curve $a^2y^2 = x^2(a^2 - x^2)$.

Solution We have the following information about the curve:

(*i*) Since there are only even powers of x and y in the equation, the curve is symmetrical about both the axes.

(*ii*) The curve passes through the origin and the tangents there are given by $a^2y^2 - a^2x^2 = 0$ or $y^2 = x^2$. So the tangents are $y = \pm x$. Hence the origin is a node.

(*iii*) Putting $x = 0$ we get $y^2 = 0$ or $y = 0$. Therefore, the curve meets the y-axis at (0, 0) only.

Putting $y = 0$, we have $x^2(a^2 - x^2) = 0$.

$\therefore\ x = 0, \pm a$. Hence the curve cuts the x-axis at (0, 0), (a, 0) ($-a$, 0).

The curve being symmetrical about both the axes, an idea about the shape of the curve in the first quadrant is sufficient to trace the curve completely.

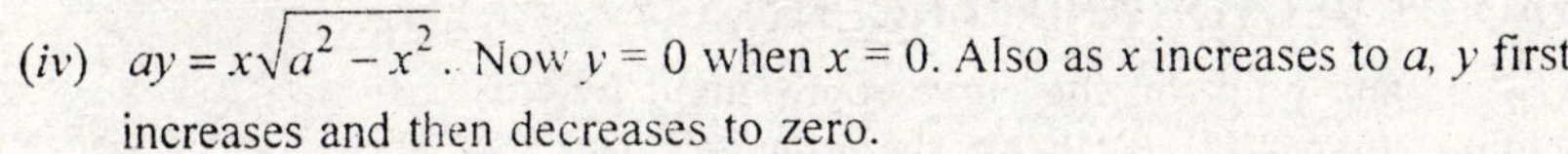

(*iv*) $ay = x\sqrt{a^2 - x^2}$. Now $y = 0$ when $x = 0$. Also as x increases to a, y first increases and then decreases to zero.

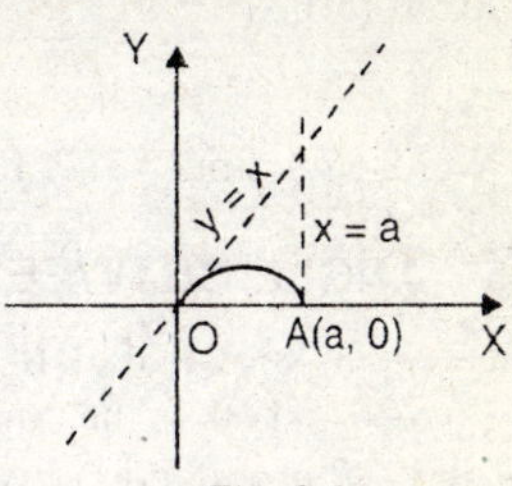

Fig. 3.6

Shifting the origin to $A(a, 0)$ the equation of the curve becomes

$$a^2y^2 = (x + a)^2\,[a^2 - (x + a)^2]$$

or

$$a^2y^2 = (x + a)^2\,(-x^2 - 2xa).$$

The tangent to the curve at new origin that is, A is $x = 0$ which is new y-axis.

The shape of the curve in the first quadrant is given in Fig. 3.6.

Since the curve is symmetrical about both the axes, the graph of the curve is given in Fig. 3.7.

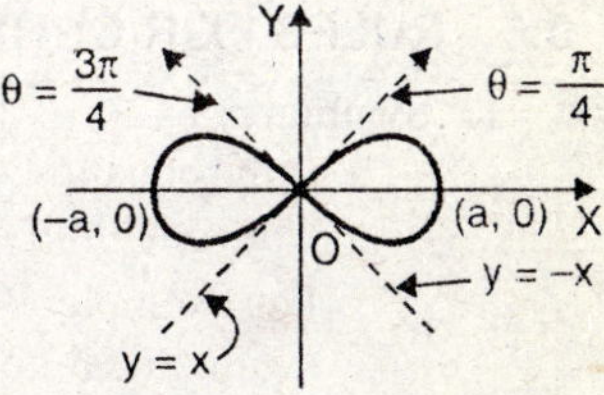

Fig. 3.7

EXAMPLE 3.6 Trace the curve $xy^2 = a^2\,(a - x)$.

Solution We have the following information about the curve:

(*i*) The curve is symmetrical about x-axis.

(*ii*) It does not pass through the origin.

(*iii*) It meets the x-axis in (a, 0) only. It does not intersect the y-axis.

(*iv*) y is imaginary when x is negative and also when $x > a$. Thus the curve lies entirely in the region between the parallel lines $x = 0$ and $x = a$.

(*v*) $x = 0$ that is, y-axis is an asymptote to the curve.

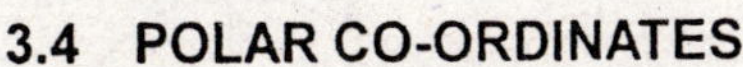

(*vi*) From $y^2 = \dfrac{a^2\,(a - x)}{x}$, we note that $y = \infty$ when $x = 0$ and y decreases as x increases. It is zero when $x = a$.

(*vii*) Shifting the origin to the point (a, 0), the equation reduces to

$$(x + a)\,y^2 = a^2\,[a - (x + a)]$$

or

$$(x + a)y^2 = -a^2x.$$

The tangent to the curve at the new origin (that is, (a, 0)] is $x = 0$ or the new y-axis. Keeping in view all the above points, the graph of the curve is given in Fig. 3.8.

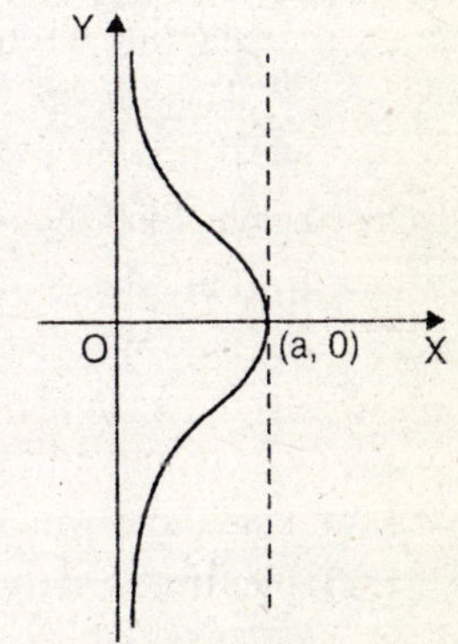

Fig. 3.8

3.4 POLAR CO-ORDINATES

We know how to represent points and curves in cartesian co-ordinates. Another system for such representation is the **polar system.** In this system, the position of a point P on a plane is known, if we are given

(*i*) its distance r from a fixed point O.

(*ii*) the inclination θ of OP to OA, a fixed line through O.

Here r is called the **radius vector** of P and θ is called the **vectorial angle.** The two r and θ, are called the polar co-ordinates of P and written as (r, θ).

O is called the **pole** and OA is called the **initial line.**

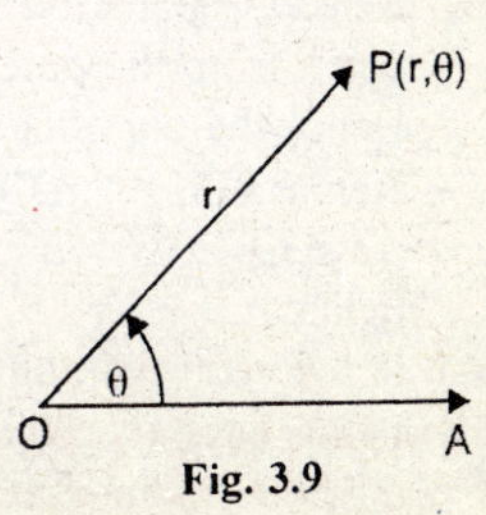

Fig. 3.9

For changing from cartesian to polar co-ordinates, or vice versa, we take origin as pole, the x-axis as initial line.

Then $\quad x = r\cos\theta, \quad y = r\sin\theta$ ...(3.1)

are the formulae for conversion from cartesian to polar co-ordinates.

From (3.1) $x^2 + y^2 = r^2$ and $\tan\theta = y/x$

or $$r = \sqrt{x^2 + y^2} \text{ and } \theta = \tan^{-1}(y/x) \qquad ...(3.2)$$

We use relations (3.2) for conversion from polar to cartesian co-ordinates.

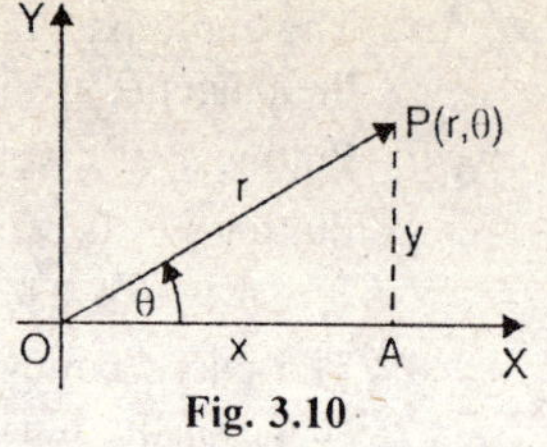

Fig. 3.10

3.5 ANGLE BETWEEN RADIUS VECTOR AND TANGENT

The reader knows that if $P(r, \theta)$ is any point on the curve represented by $f(r, \theta) = 0$ and PT is the tangent to the curve at $P(r, \theta)$, then ϕ, the angle between PT and OP produced is given by

$$\tan\phi = r \cdot (d\theta/dr) \qquad ...(3.3)$$

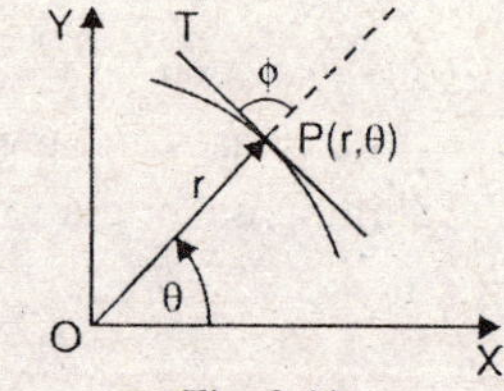

Fig. 3.11

3.6 RULES FOR CURVE TRACING–POLAR CO-ORDINATES

1. **Symmetry**
   (*i*) If the equation of the curve remains unchanged when θ is changed into $-\theta$, the curve is symmetrical about the initial line.
   For example $r = a(1 - \cos\theta)$ remains unchanged when θ is changed to $-\theta$. Hence the curve is symmetrical about the initial line.
   (*ii*) If the powers of r in the equation of a curve are all even, then the curve is symmetrical about the pole.
   The curve $r^2 = a^2 \cos 2\theta$, for example, is symmetrical about the pole.
   (*iii*) If the equation remains unchanged when θ is changed into $\pi - \theta$, the curve is symmetrical about the line through the pole and perpendicular to the initial line.
   The curve $r = a\sin 3\theta$ is an example of this type.
   (*iv*) If there is a value of θ for which $r = 0$, the curve passes through the pole.
2. **Region:** Find the regions in which a curve does not lie. This can be determined as follow:
   (*i*) Find the value of θ which makes r^2 negative and therefore, r imaginary.
   (*ii*) Find whether the values of r and θ are confined between certain limits.
   For example, in the curve $r = a\sin 2\theta$, the values of r lie between the limits 0 and a. In this case the curve lies within the circle with centre 0 and radius a.
3. ϕ. Find $\tan\phi$ which will give the angle that the tangent makes with the radius vector.
4. **Table:** If necessary, form a table of values of r for both positive and negative values of θ and note how r varies with θ.

EXAMPLE 3.7 Trace the curve

(*i*) $r = a$ (*ii*) $\theta = \alpha$ (*iii*) $r = 2a\cos\theta$.

Solution (*i*) If (r, θ) be the co-ordinates of a point, then r is the distance of the point from the pole.

Since the equation of the curve is $r = a$, the distance of any point on the curve from the pole is a.

Hence the curve is a circle with radius a and centre at the pole.(Fig. 3.12)

Remark: If $r = a$ then $r^2 = a^2$. Changing this equation in the cartesian form we get $x^2 + y^2 = a^2$ which, obviously, is a circle with centre at the pole and radius a.

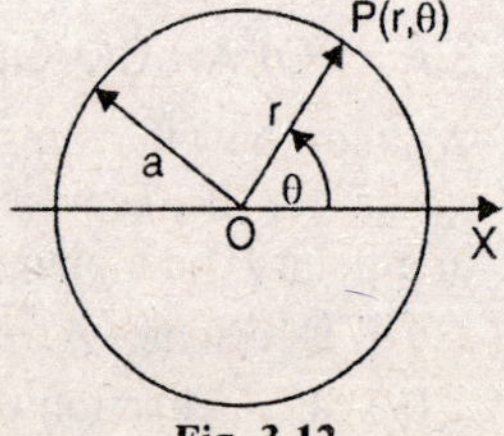

Fig. 3.12

(*ii*) θ, the vectorial angle of a point $P(r, \theta)$ is the angle which OP makes with the initial line.

In this case every point on the curve has the same vectorial angle α. Hence the curve is the line through the pole inclined at an angle α to the initial line. (Refer Fig. 3.13)

Note Changing this equation in the cartesian form we get $\tan^{-1}(y/x) = \alpha$ or $y = x \tan \alpha$, which is a line through the origin with slope $\tan \alpha$.

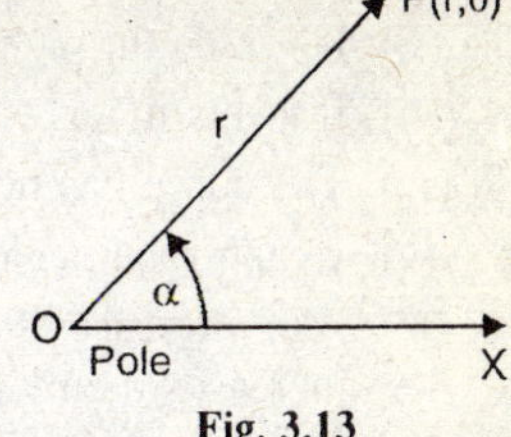

Fig. 3.13

(*iii*) (*a*) The equation $r = 2a \cos \theta$ remains unchanged when θ is changed into $-\theta$. Therefore, the curve is symmetrical about the initial line.

(*b*) $r = 0$ when $\theta = \pi/2$. Hence the curve passes through the pole. Also when $\theta = 0$, $r = 2a$. This shows that the curve crosses the initial line at $A(2a, 0)$.

(*c*) The greatest value of r is $2a$. Hence the curve lies entirely within a circle with centre at the pole and radius $2a$.

(*d*) When $\pi/2 < \theta < 3\pi/2$, r is negative. Hence, no part of the curve lies to the left of the line perpendicular to the initial line through the pole.

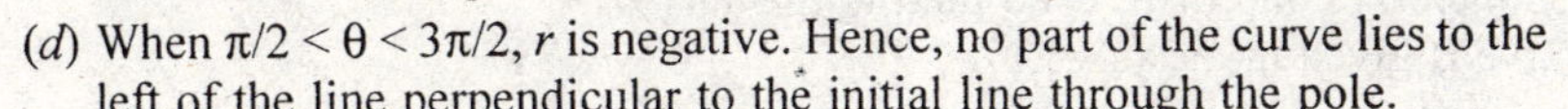

(*e*) $\tan \phi = r \dfrac{d\theta}{dr} = (2a \cos\theta)\left(\dfrac{-1}{2a \sin\theta}\right) = -\cot\theta$, where ϕ is the angle between the tangent at (r, θ) and the radius vector.

When $\theta = 0$, $\phi = \pi/2$. Hence tangent at A is perpendicular to the initial line.

When $\theta = \pi/2$, $\phi = \pi$. Hence tangent at the pole coincides with the line perpendicular to the initial line.

Table 3.1

θ:	0	$\pi/6$	$\pi/4$	$\pi/3$	$\pi/2$
r:	$2a$	$\sqrt{3}a$	$\sqrt{2}a$	a	0

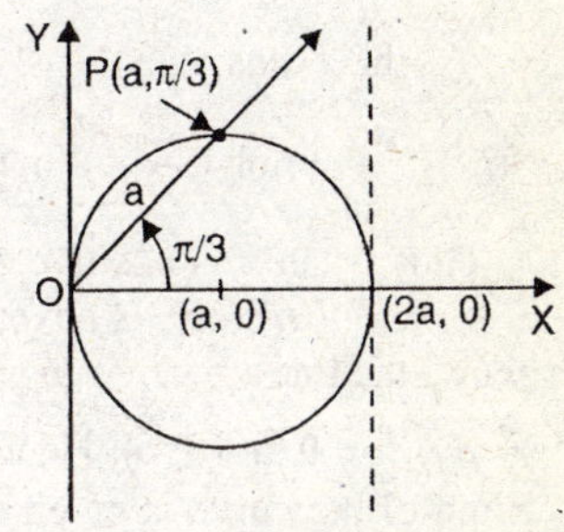

Fig. 3.14

Considering all the points in Table 3.1, the graph of the curve is given in Fig. 3.14. From the graph of curve, it is obvious that the curve is a circle with centre at $(a, 0)$ and radius a.

3.7 SOME IMPORTANT CURVES

3.7.1 Cartesian Equations

EXAMPLE 3.8 Trace the curve $y^2 = ax^3$ (Semi-cubical parabola).

Solution We have the following information about the curve:

(*i*) The equation remains unchanged when y is changed to $-y$. Hence the curve is symmetrical about x-axis.

(*ii*) The curve passes through $(0, 0)$, the origin and tangents there are $y^2 = 0$, that is, two co-incident tangents. Thus the origin is a cusp.

(*iii*) The curve meets the axes only in $(0, 0)$, and no other point.

(*iv*) x cannot be negative. Therefore, no part of the curve lies to the left of y-axes.

(*v*) There are no asymptotes.

(*vi*) As x increases, y increases and when x is very large, y is also very large. The graph of the curve is given in Fig. 3.15.

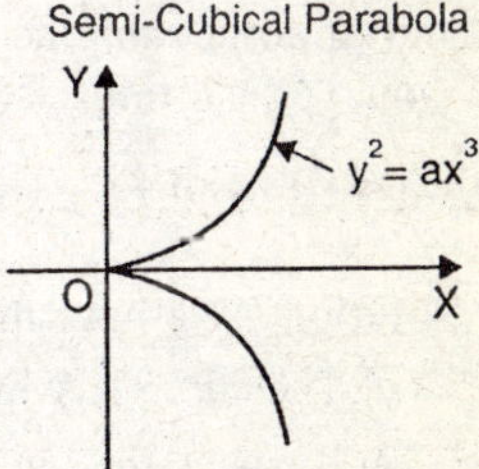

Fig. 3.15

EXAMPLE 3.9 Trace the curve $y = ax^3$ (Cubical Parabola).

Solution We have the following information about the curve:

(*i*) The equation remains unchanged when signs of both x and y are changed. Therefore, the curve is symmetrical in opposite quadrants.

(*ii*) The curve passes through the origin. The tangent there (by equating to zero the lowest degree term) is $y = 0$, that is, x-axis.

(*iii*) The curve meets the axes only at $(0, 0)$.

(*iv*) When x is negative, y is negative and when x is positive, y is positive. Hence the curve lies in first and third quadrants only.

(*v*) When x increases y increases. There is no limit to the increase of either. The graph of the curve is given in Fig. 3.16.

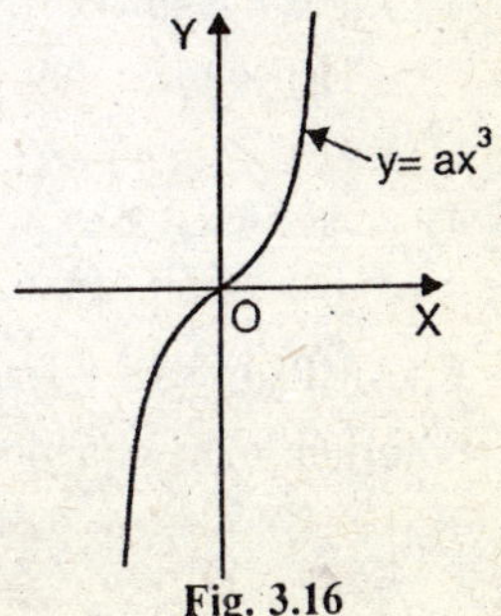

Fig. 3.16

EXAMPLE 3.10 Trace the curve $9ay^2 = x(x-a)^2$. (*MDU 2000*)

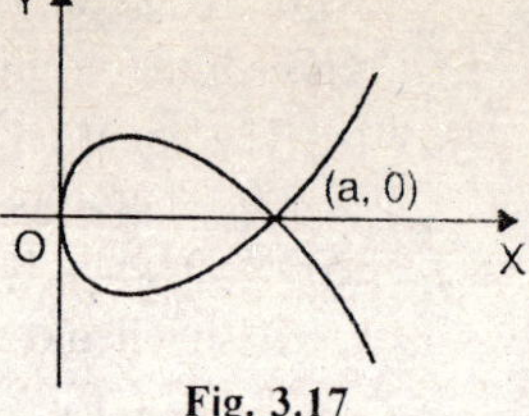

Fig. 3.17

Solution We have the following information about the curve:

(*i*) The curve is symmetrical about x-axis only.

(*ii*) It passes through the origin and tangent there is $x = 0$, that is, y-axis.

(*iii*) It meets the axes at (0, 0) and $(a, 0)$.

(*iv*) The curve has no asymptotes.

(*v*) x cannot be negative. Hence no part of the curve lies to the left of y-axis.

(*vi*) As x increases from 0 to a, y first increase and then decreases to zero. When x is greater than a and increases then y also increases and there is no limit to this increase.
The graph of the curve is given in the fig. 3.17.

EXAMPLE 3.11 Trace the curve $y = c \cosh(x/c)$. (*Catenary*)

Solution We have the following information about the curve:

The equation to the curve can be written in the form $y = c\,(e^{x/c} + e^{-x/c})/2$.

(*i*) The equation remains unchanged when x is changed into $-x$. Hence the curve is symmetrical about y-axis.

(*ii*) When $x = 0,\ y = c\cdot\dfrac{e^0 + e^0}{2} = c\cdot\dfrac{1+1}{2} = c$. Hence the curve intersects the y-axis at $(0, c)$. It does not cut the x-axis.

(*iii*) $\dfrac{dy}{dh} = \sinh\left(\dfrac{x}{c}\right)$. When x is positive, $\dfrac{dy}{dx}$ is positive.

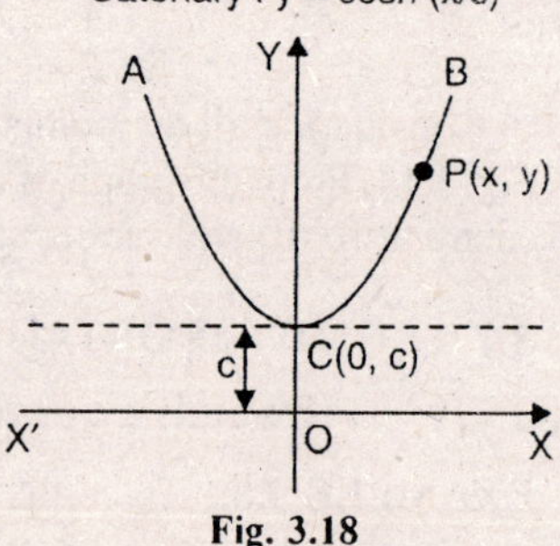

Fig. 3.18

Thus when x is positive and increases, y also increases. Also as $x \to \infty$, $y \to \infty$. y, therefore, is never less than c and hence no part of the curve lies below the line $y = c$.

$dy/dx = 0$ at $(0, c)$. Hence the tangent at $(0, c)$ is parallel to x-axis. The graph of the curve is given in Fig. 3.18.

Note: The curve represented by $y = c \cosh(x/c)$ is called **Catenary.** Point C is called the **vertex** of the catenary. In fact the curve in which a perfectly flexible and uniform cable hangs freely under gravity between two points (not in the same vertical line), is called a catenary.

EXAMPLE 3.12 Trace the curve $y^2(2a - x) = x^3$ $(a > 0)$.
(*UPTU 2005; JNTU 2003; VTU 2003S*)

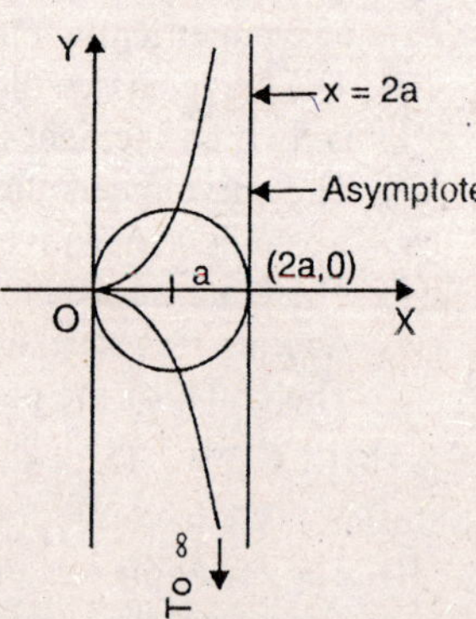

Fig. 3.19

Solution We have the following information about the curve:

(*i*) The curve is symmetrical about x-axis.

(*ii*) It passes through the origin and tangents there are given by $y^2 = 0$.
There are two coincident tangents. Hence the origin is a cusp.

(*iii*) The curve cuts the axis only at the origin.

(*iv*) y is imaginary when x is negative or $x > 2a$.
Hence the entire curve lies between the line $x = 0$ and $x = 2a$.

(*v*) $x = 2a$ is an asymptote to the curve.

(*vi*) As x increases, y increases and as x approaches $2a$, y becomes very large.
The graph of the curve is given in Fig. 3.19.

EXAMPLE 3.13 Trace the curve $x\,(x^2 + y^2) = a\,(x^2 - y^2)$. (*VTU, 2000*)

Solution The equation of the curve can be written in the form

$$y^2\,(a+x) = x^2\,(a-x) \text{ or } y^2 = \frac{a-x}{a+x}x^2.$$

We have the following information about the curve:

(*i*) The curve is symmetrical about x-axis.

(*ii*) It passes through the origin. Tangents there are given by $y^2 = x^2$. Hence the tangents are $y = \pm x$. Since there are two real and different tangents, the origin is a node.

(*iii*) The curve meets the axes only at (0, 0) and $(a, 0)$.

(*iv*) When $x > a$, y is imaginary. Also when $x < -a$, y is imaginary ($\because x + a$ is negative). Hence the entire curve lies in the region between the parallel lines $x = -a$ and $x = a$.

(*v*) By shifting the origin to the point $(a, 0)$, the equation becomes

$$y^2(x + 2a) = (x + a)^2\,[a - (x + a)]$$

or $$y^2(x + 2a) = -x(x + a)^2$$

or $$y^2(x + 2a) = -x(x^2 + 2ax + a^2).$$

The tangent at the new origin is $x = 0$, that is, the new y-axis.

(*vi*) $x + a = 0$ is an asymptote.

(*vii*) When $x \to -a$, $y \to \infty$.

Considering the above particulars, the graph of the curve is given in Fig. 3.20.

Fig. 3.20

EXAMPLE 3.14 Trace the curve $(x/a)^{2/3} + (y/b)^{2/3} = 1$. (Hypocycloid)

(*GGSIPU 2006; VTU 2003; MDU 2003*)

Solution The equation of the curve can be rewritten as

$$(x^2/a^2)^{1/3} + (y^2/b^2)^{1/3} = 1 \quad \ldots(i)$$

We have the following information about the curve:

(*i*) The above form shows that there is symmetry about both the axes.

(*ii*) The curve does not passes through the origin.

(*iii*) Putting $y = 0$, we get $x = \pm a$ that is, the curve crosses the x-axis at $(a, 0)$ and $(-a, 0)$. Similarly the curve crosses the y-axis at $(0, b)$ and $(0, -b)$.

(*iv*) From the equation (*i*) of the curve we find that when $x > a$, y^2 is negative that is, y is imaginary so the curve does not exist for values of $x > a$. Similarly, the curve does not exist for values of $y > b$.

(*v*) There are no asymptotes.

(*vi*) For the given curve, $dy/dx = -b^{2/3}y^{1/3} / a^{2/3}x^{1/3}$.

Tangents at $(a, 0)$ is along x-axis and at $(0, b)$ is at right angles to x-axis. The graph of the curve is given in Fig. 3.21.

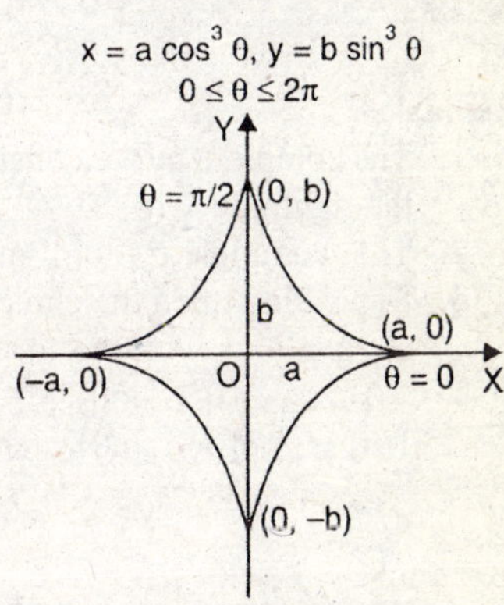

Fig. 3.21

3.7.2 Curves Represented in Parametric Form

Very often the co-ordinates of any point on a curve can be expressed in terms of one variable, called the **parameter.** For example $x = at^2$, $y = 2at$ are the parametric equations of the parabola $y^2 = 4ax$. Here 't' is the parameter. We also know that $x = a\cos\theta$, $y = b\sin\theta$ are parametric equations of the ellipse $x^2/a^2 + y^2/b^2 = 1$, θ being the parameter in this case. We shall now trace some other curves when their parametric equations are given.

EXAMPLE 3.15 Trace the Cycloid $x = a(\theta - \sin\theta)$, $y = a(1 - \cos\theta)$.

Solution We shall trace the curve when θ lies in the interval $(0, 2\pi)$.

We have the following information about the curve:

(*i*) **Symmetry:** When θ is changed into $-\theta$, x changes to $-x$ but y remains unchanged. Hence the curve is symmetrical about y-axis.

(*ii*) **Origin:** When $\theta = 0$, $x = 0$, $y = 0$. Therefore, the curve passes through the origin.

(*iii*) **Intersection with axes:** It intersects the x-axis where, putting $y = 0$ we get $1 - \cos\theta = 0$, or $\theta = 0, 2\pi$. For these values of θ, $x = 0, 2\pi a$. Therefore, the point of intersection with x-axis are (0, 0) $(2\pi a, 0)$.

To get points of intersection with y-axis, we put $x = 0$.

$$\therefore \qquad a(\theta - \sin\theta) = 0 \text{ or } \theta = 0.$$

For this value of θ, $y = 0$. Hence the point of intersection with y-axis is $(0, 0)$.

(*iv*) $dx/d\theta = a(1 - \cos\theta)$, $dy/d\theta = a\sin\theta$.

$$\therefore \quad \frac{dy}{dx} = \frac{dy}{d\theta}\cdot\frac{d\theta}{dx} = \frac{a\sin\theta}{a(1-\cos\theta)} = \frac{2\sin(\theta/2)\cos(\theta/2)}{2\sin^2(\theta/2)} = \cot\frac{\theta}{2}.$$

Thus $\dfrac{dy}{dx} = 0$ when $\cot\dfrac{\theta}{2} = 0$ or $\theta = \pi$. For this value of θ

$x = \pi a$, $y = 2a$. Because tangent at $(\pi a, 2a)$ is parallel to x-axis.

$dy/dx = \infty$ when $\cot(\theta/2) = \infty$ or $\theta = 0, 2\pi$. Hence tangent is perpendicular to x-axis at $(0, 0)$ and $(2\pi a, 0)$.

(*v*) **Region** Now y cannot be negative ($\because \cos\theta \le 1$) and also y cannot be greater than $2a$. Therefore, the curve lies between the lines $y = 0$ and $y = 2a$. The graph of the curve is given in Fig. 3.22.

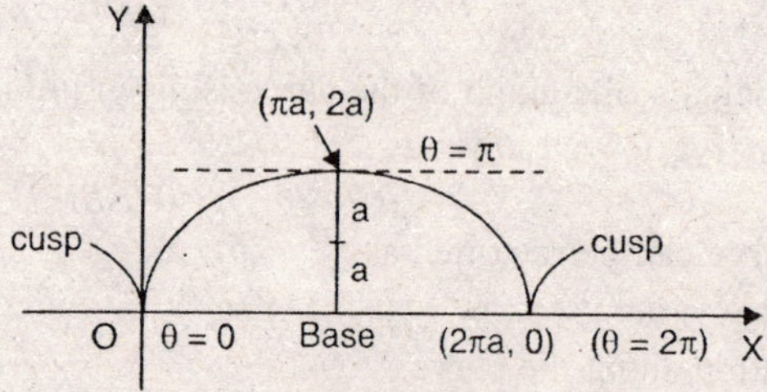

Fig. 3.22

Remarks

1. The complete curve consists of the portion from $\theta = 0$ to $\theta = 2\pi$ and endless repetitions of that part to the right and left.
2. The 'complete cycloid' means one arch of the cycloid between two consecutive cusps.
3. Depending upon the choice of axes, we have different forms of equations to the cycloid. The three types of equations with the forms of the curves they represent are given here:

   (*i*) $x = a(\theta + \sin\theta)$, $y = a(1 + \cos\theta)$ (*ii*) $x = a(\theta - \sin\theta)$, $y = a(1 + \cos\theta)$

   (*iii*) $x = a(\theta + \sin\theta)$, $y = a(1 - \cos\theta)$ (cycloid, vertex at origin (that is, Inverted cycloid)

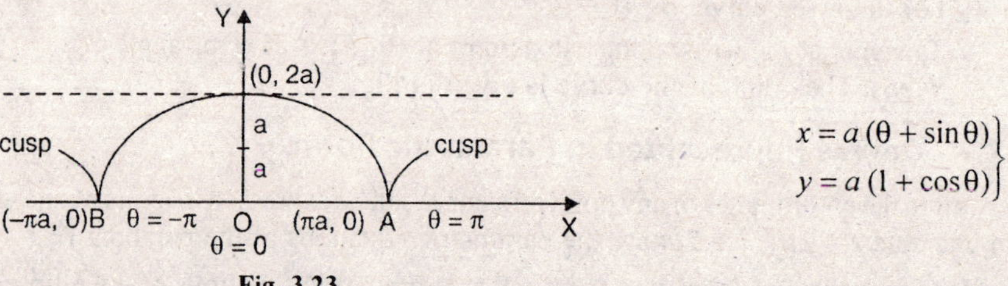

Fig. 3.23

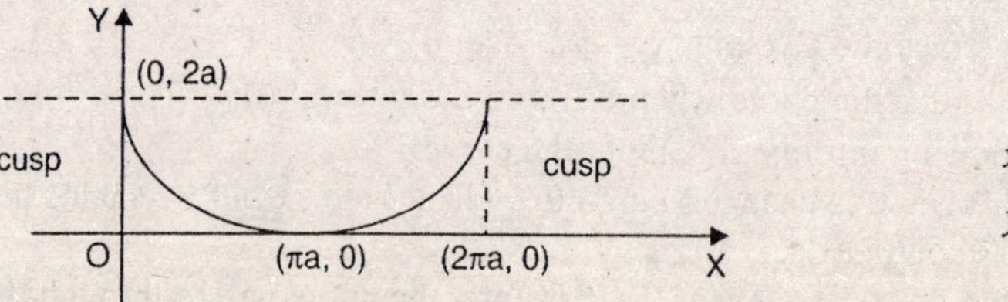

Fig. 3.24

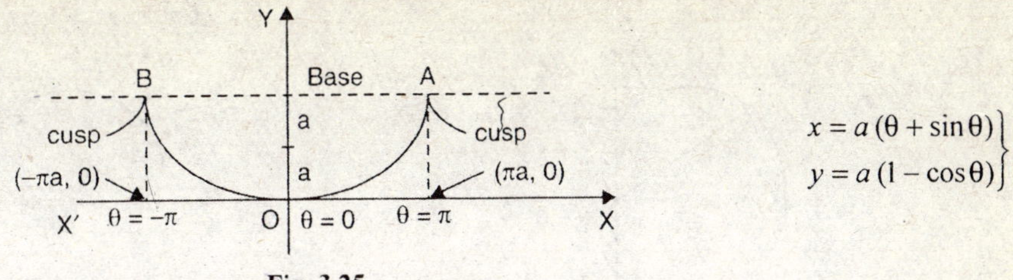

$$\left.\begin{aligned} x &= a(\theta + \sin\theta) \\ y &= a(1 - \cos\theta) \end{aligned}\right\}$$

Fig. 3.25

EXAMPLE 3.16 Trace the curve $x^{2/3} + y^{2/3} = a^{2/3}$ (Hypocycloid of four cusps - Astroid).

(MDU 2003; VTU 2003;Osmania 2002)

Solution The parametric equations of this curve are $x = a\cos^3\theta$, $y = a\sin^3\theta$. We have the following information about the curve:

(*i*) When θ is changed into $-\theta$, y changes to $-y$ but x remains unchanged. Hence the curve is symmetrical about x-axis.

(*ii*) When θ is changed to $\pi - \theta$ $(0 \le \theta \le \pi/2)$, x changes into $-x$ but y remains unchanged. The curve, therefore, is symmetrical about y-axis also.

(*iii*) $y = 0$ when $\theta = 0$ and then $x = a$. Hence the curve intersects the x-axis at $A(a, 0)$.
$x = 0$ when $\theta = \pi/2$ and then $y = a$. Therefore, the curve cuts the y-axis at $(0, a)$.

(*iv*) $\dfrac{dy}{dx} = \dfrac{dy}{d\theta}\Big/\dfrac{dx}{d\theta} = (3a\sin^2\theta\cos\theta)/(-3a\cos^2\theta\sin\theta) = -\tan\theta$.

Now $dy/dx = 0$, when $\theta = 0$. Thus the tangent at $A(a, 0)$ coincides with x-axis.

$dy/dx = -\infty$, when $\theta = \pi/2$. Therefore the tangent is perpendicular to x-axis at $(0, a)$. In other words y-axis is the tangent.

(*v*) As θ increases from 0 to $\pi/2$, $\cos\theta$ and hence x decreases from a to 0 while y increases from 0 to a. The part of the curve in the first quadrant is now known.

Since there is symmetry about both axes, we can complete the curve in other quadrants.

The graph of the curve is given in Fig. 3.26.

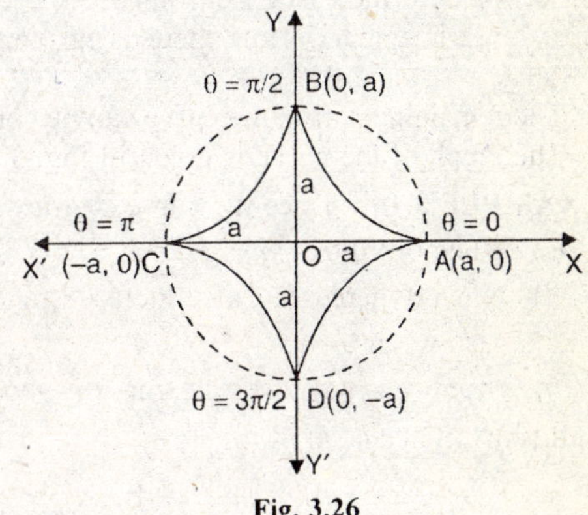

Fig. 3.26

3.7.3 Curve Represented in Polar Form

EXAMPLE 3.17 Trace the curve, $r = a(1 - \cos\theta)$. (Cardioid)

Solution We have the following information about the curve:

(*i*) When θ is changed to $-\theta$, the equation remains unchanged. The curve, therefore, is symmetrical about the initial line.

(*ii*) $r = 0$ when $\theta = 0$. Hence the curve passes through the pole.

(*iii*) When θ increases from 0 to $\pi/2$, $\cos\theta$ decreases form 1 to 0 and, therefore, r increases from 0 to a.

When θ increases from $\pi/2$ to π, $\cos\theta$ decreases from 0 to -1 and thus r increases from a to $2a$.

Now the part of the curve when θ increases from π to 2π is known because of symmetry about the initial line. The curve is given in Fig. 3.27.

Note The cardioid with equation in the form $r = a(1 + \cos\theta)$ can similarly be traced. The graph of the curve is given in the Fig. 3.28.

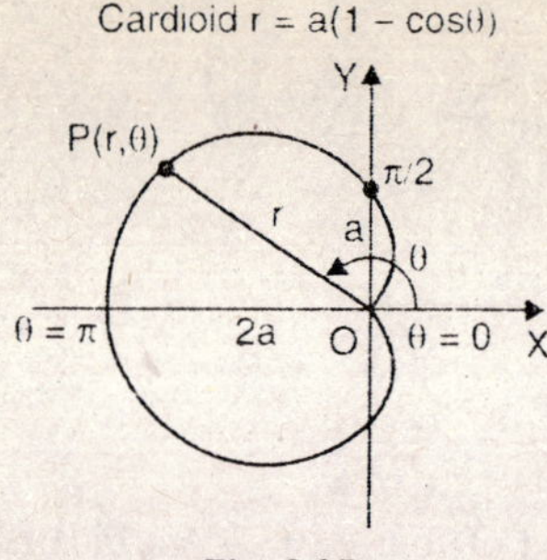

Fig. 3.27

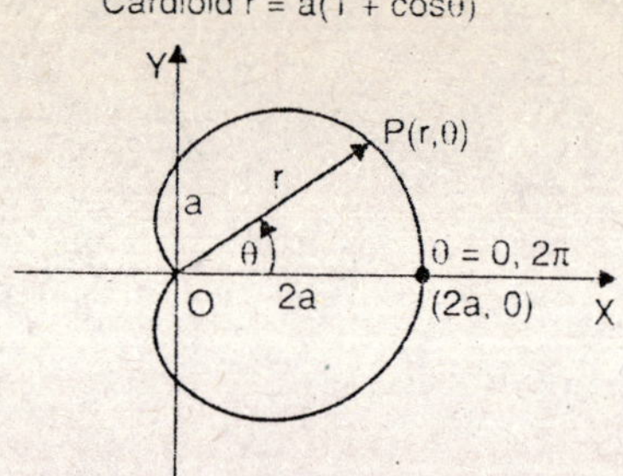

Fig. 3.28

EXAMPLE 3.18 Trace the curve $r^2 = a^2 \cos(2\theta)$ *(Lemniscate of Bernoulli)*. *(GGSIPU 2006; JNTU 2003;)*

Solution We have the following information about the curve:

(*i*) It is symmetrical about the initial line.

(*ii*) It is symmetrical about the line through the pole and perpendicular to the initial line.

Because of the aforesaid symmetry we need consider the change in r as θ varies from 0 to $\pi/2$.

(*iii*) When θ increases from 0 to $\pi/4$, 2θ changes from 0 to $\pi/2$ so that $\cos 2\theta$ decreases from 1 to 0 and thus r decreases from a to 0. For $\theta = \pi/4$, $r = 0$. When θ increases from $\pi/4$ to $\pi/2$, $\cos 2\theta$ is negative and, therefore, r is imaginary. Thus there is no part of the curve between the lines $\theta = \pi/4$ and $\theta = \pi/2$.

The remaining part of the curve can be completed on the basis of symmetry. The graph of the curve is given in Fig. 3.29.

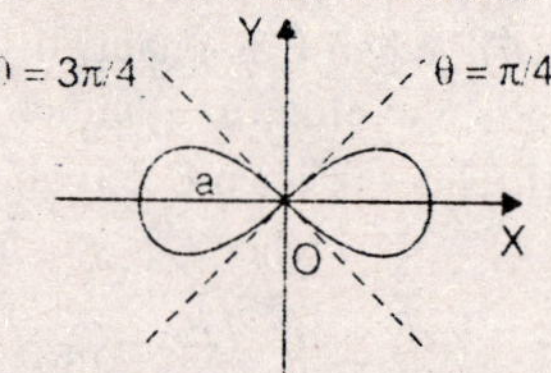

Fig. 3.29

EXAMPLE 3.19 Trace the curve $r = ae^{b\theta}$ $(a, b > 0)$ *(Equiangular spiral)*.

(*i*) When $\theta = 0$, $r = a$.

(*ii*) When θ increases r also increases and as $\theta \to \infty$, $r \to \infty$. Also when $\theta \to -\infty$, $r \to 0$.

(*iii*) r is always positive. The curve is shown in Fig. 3.30.

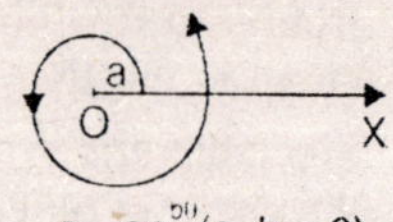

Fig. 3.30

Solution Since $dr/d\theta = abe^{b\theta}$.

$$\therefore \quad \tan\phi = r\frac{d\theta}{dr} = \frac{ae^{b\theta}}{abe^{b\theta}} = \frac{1}{b}, \quad \Rightarrow \quad \phi = \tan^{-1}\left(\frac{1}{b}\right), \text{ a constant.}$$

Because the angle between the radius vector of any point and tangent there is always constant, the curve $r = ae^{b\theta}$ is called an equiangular spiral.

EXAMPLE 3.20 Trace the curve $r = a\sin(3\theta)$ $(a > 0)$. *(Three leaved rose)*

Solution We have the following information about the curve:

(*i*) $r = 0$ when $\theta = 0$. Hence the curve passes through the pole.

(*ii*) The value of r cannot exceed a. Therefore the curves lies inside the circle with center at the pole and radius a.

(*iii*) When θ increases from 0 to $\pi/6$, 3θ increases from 0 to $\pi/2$. Consequently r increases from 0 to a.

When θ increases from $\pi/6$ to $\pi/3$, 3θ increases from $\pi/2$ to π and r decreases from a to 0.

When θ increases from $\pi/3$ to $\pi/2$, r is negative but its magnitude increases from 0 to a.

When θ increases from $\pi/2$ to $2\pi/3$, r is still negative and numerically decreases from a to 0.

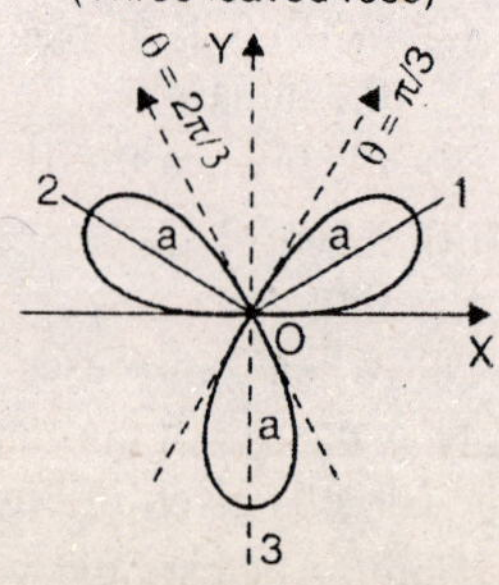

Fig. 3.31

Considering, similarly the variations of r. As θ increases from $2\pi/3$ to $5\pi/6$ and then from $5\pi/6$ to π, we get the second loop above the initial line.

As θ increases from π onward the same three loops of the curve are repeated. The graph of the curve is given in Fig. 3.31.

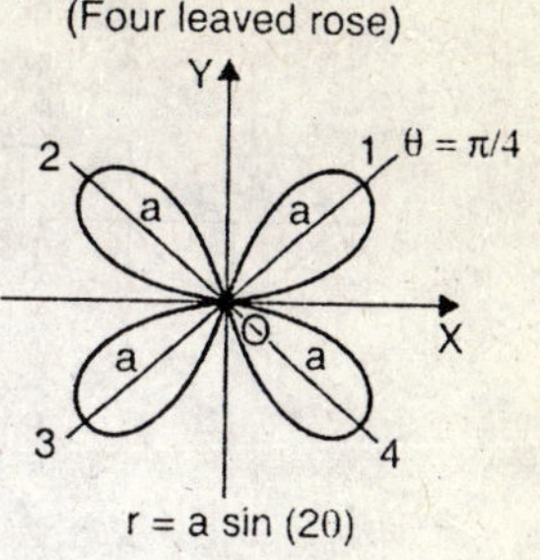

Fig. 3.32

EXAMPLE 3.21 Trace the curve $r = a \sin(2\theta)$ *(Four leaved rose)*

Proceeding as in the above example, we get the graph of curve given in Fig. 3.32.

Example 3.22 Trace the curve $x^3 + y^3 = 3\,axy$. *(Folium of Descartes).*

(UPTU 2003; MDU 2001)

Solution To trace this curve, we require both Cartesian and Polar forms of its equation. We have the following information about the curve:

(*i*) The equation remains unchanged when x and y are interchanged. Hence the curve is symmetrical about the line $y = x$.

(*ii*) It passes through (0, 0) that is, the origin. The tangents there are $xy = 0$ that is, $x = 0$ and $y = 0$. Thus the origin is a node on the curve.

(*iii*) It does not meet the axes at any other point.

(*iv*) Putting $y = x$ in $x^3 + y^3 = 3\,axy$, we get

$$2x^3 - 3ax^2 = 0 \text{ or } x^2(2x - 3a) = 0.$$

Therefore, $x = 0, 3a/2$.

Now when $x = 0, y = 0$;

and when $x = 3a/2, y = 3a/2$.

$\therefore$ The curve meets the line $y = x$ at (0, 0) and $(3a/2, 3a/2)$.

And x and y cannot both be negative.

($\because$ in that case R.H.S. is +ve and L.H.S. is negative).

Thus no part of the curve lies in the third quadrant.

(*v*) $x + y + a = 0$ is an asymptote to the curve.

(*vi*) On transforming to polar co-ordinates, the polar form of the equation is

$$r = (3a \sin\theta \cos\theta)/(\cos^3\theta + \sin^3\theta).$$

Obviously $r - 0$ when $\theta = 0$ and $\theta = \pi/2$.

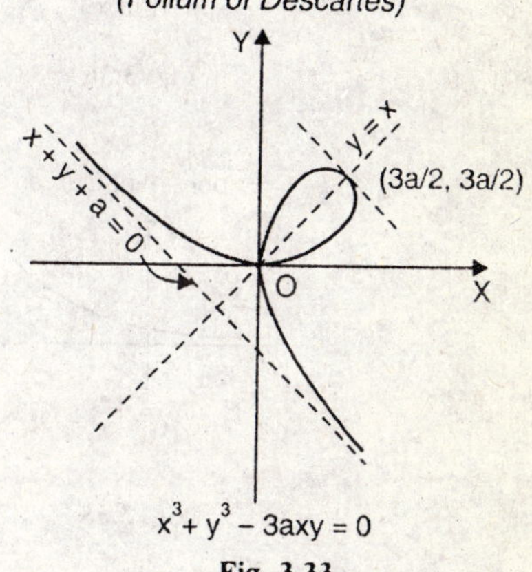

Fig. 3.33

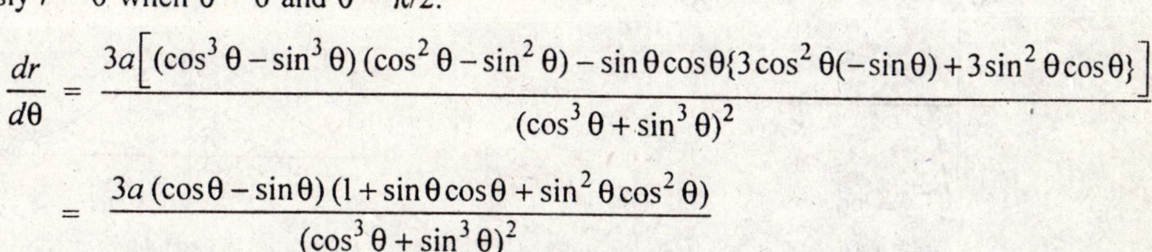

$$\frac{dr}{d\theta} = \frac{3a\left[(\cos^3\theta - \sin^3\theta)(\cos^2\theta - \sin^2\theta) - \sin\theta\cos\theta\{3\cos^2\theta(-\sin\theta) + 3\sin^2\theta\cos\theta\}\right]}{(\cos^3\theta + \sin^3\theta)^2}$$

$$= \frac{3a(\cos\theta - \sin\theta)(1 + \sin\theta\cos\theta + \sin^2\theta\cos^2\theta)}{(\cos^3\theta + \sin^3\theta)^2}$$

Now $dr/d\theta = 0$ when $\cos\theta - \sin\theta = 0$ that is, $\tan\theta = 1 \Rightarrow \theta = \pi/4$ or $5\pi/4$.

Since $dr/d\theta$ is + ve when θ increases from 0 to $\pi/4$, r increases from 0 to $3a/\sqrt{2}$. As θ increases from $\pi/4$ to $\pi/2$, r decreases from $3a\sqrt{2}$ to 0. As θ increases from $\pi/2$ to $3\pi/4$, r is negative and numerical increases from 0 to ∞ so that the point (r, θ) describes the part of the curve shown in the fourth quadrant. Because of symmetry about $y = x$, the part in sc ɔnd quadrant can be drawn. The graph of the curve is given in Fig. 3.33.

3.8 CURVES FOR REFERENCE

For the convenience of the readers a number of the more common curves are given here.

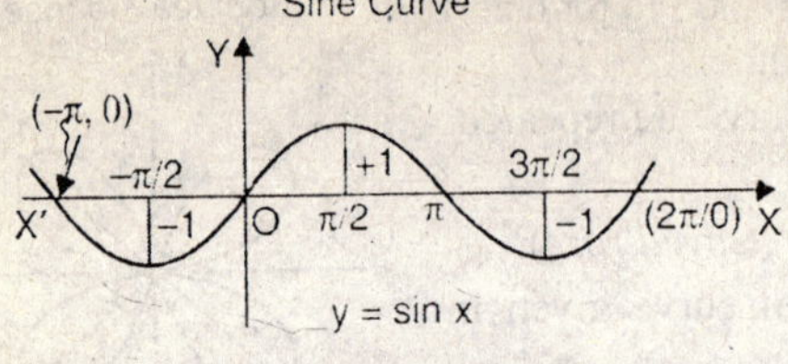

Fig. 3.34

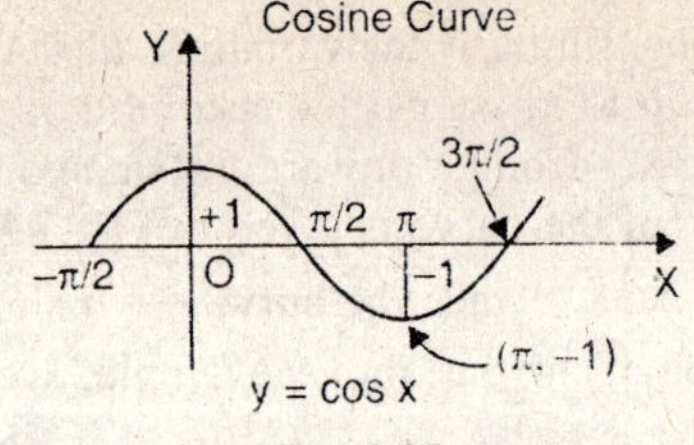

Fig. 3.35

Tangent Curve

y = tan x

Fig. 3.36

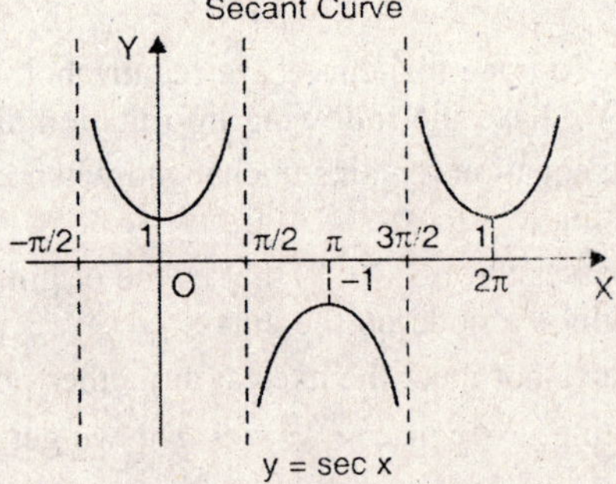

Fig. 3.37

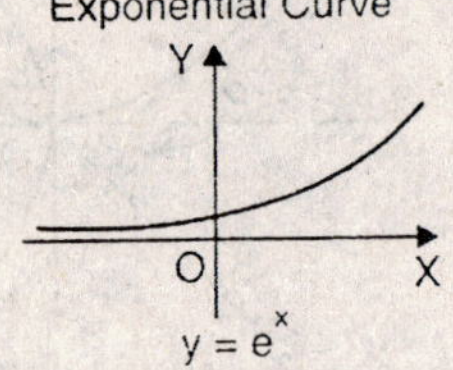

Fig. 3.38

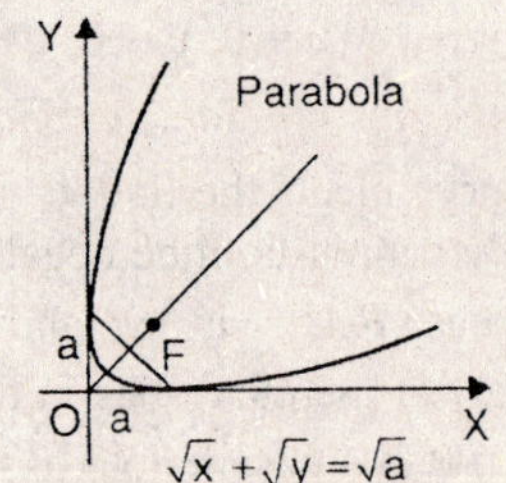

Fig. 3.39

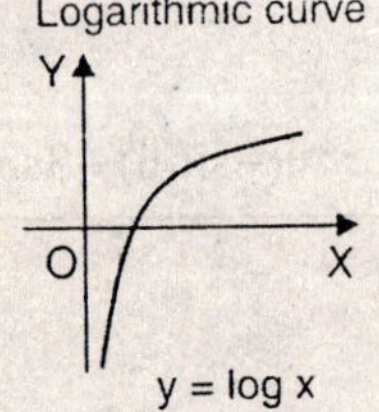

Fig. 3.40

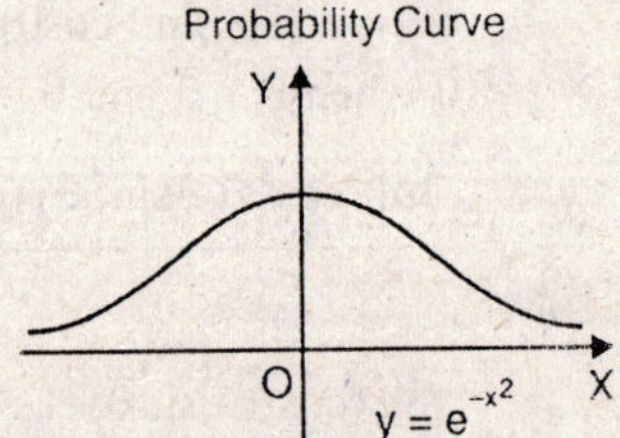

Fig. 3.41

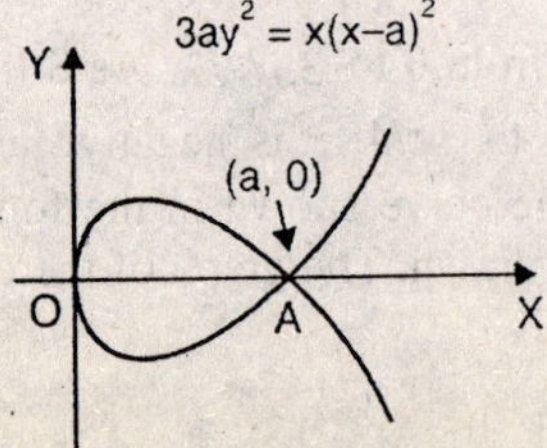

Fig. 3.42

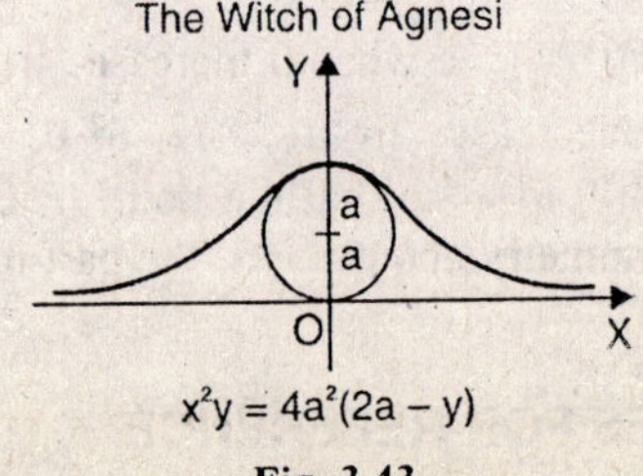

Fig. 3.43

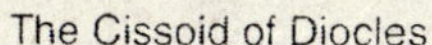

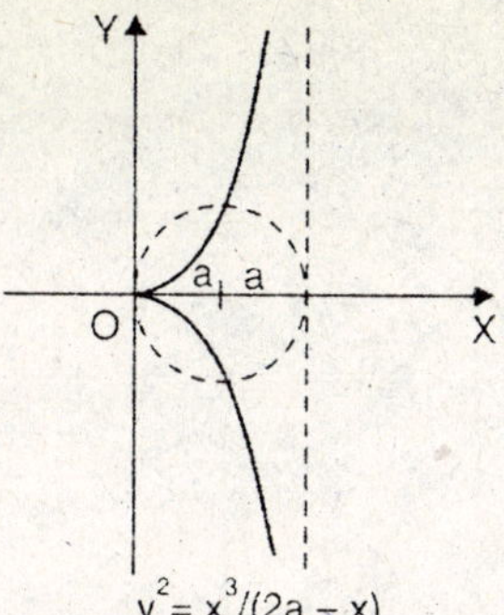

$y^2 = x^3/(2a - x)$

Fig. 3.44

Strophoid

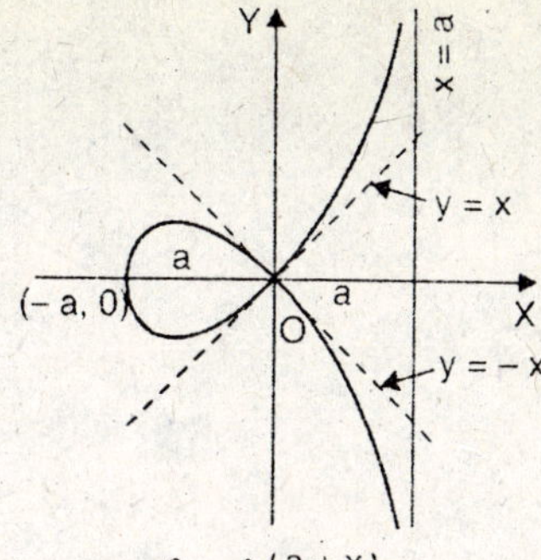

$y^2 = x^2\left(\frac{a + x}{a - x}\right)$

Fig. 3.45

Equilateral Hyperbola

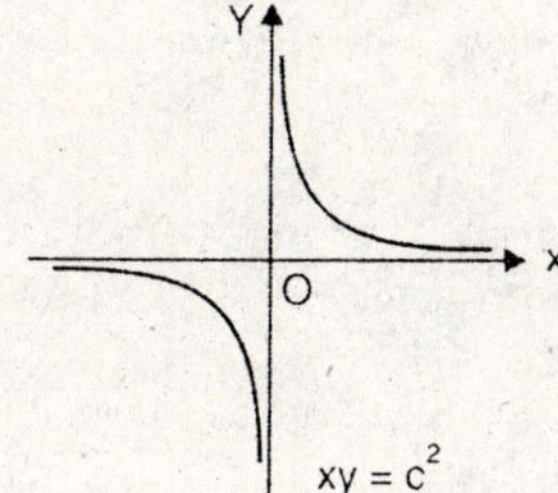

$xy = c^2$

Fig. 3.46

Limacon

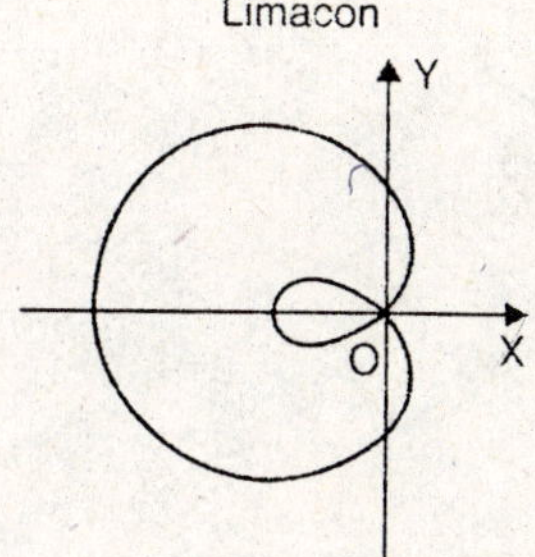

$r = b - a \cos\theta$, $(b < a)$

Fig. 3.47

Logarithmic or
Equiangular Spiral

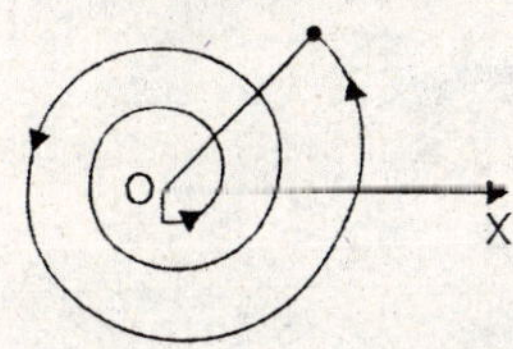

$r = e^{a\theta}$, or $\log r = a\theta$

Fig. 3.48

Spiral of Archimedes

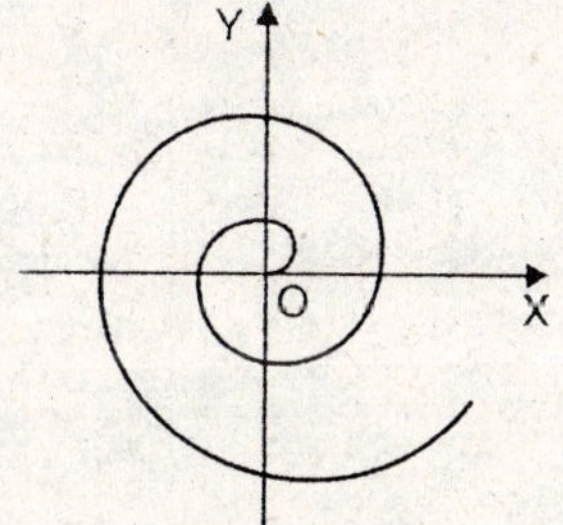

$r = a\theta$, $(a > 0)$

Fig. 3.49

Cardioid: $r = 1 + \sin\theta$

$(2, \pi/2)$

1

1

1

O

X

Fig. 3.50

Four leaved rose

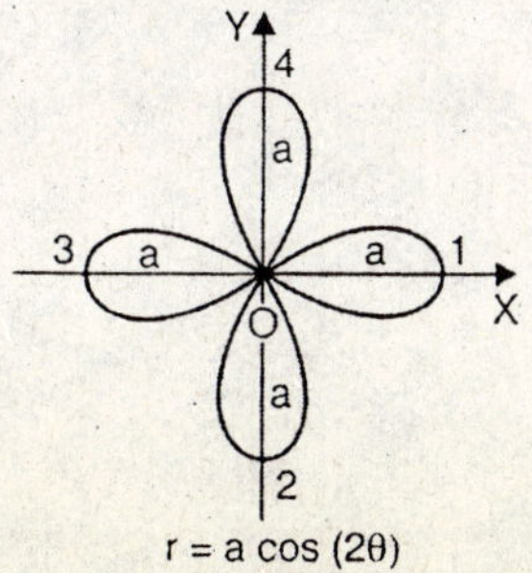

$r = a \cos(2\theta)$

Fig. 3.51

Three leaved rose

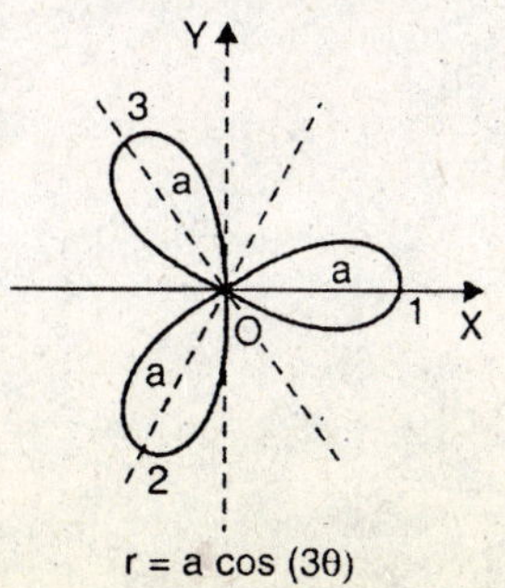

$r = a \cos(3\theta)$

Fig. 3.52

CHAPTER

4

Multiple Integrals

4.1 DOUBLE INTEGRALS

Notation: The symbol $\int_a^b \int_c^d f(x, y)\, dy\, dx$ is called a double integral. It is evaluated as follows:

We first integrate f(x, y) with respect to y (considering x as a constant) between the limits c and d. The result is then integrated with respect to x between the limits a and b.

Thus the double integral

$$\int_a^b \int_c^d f(x, y)\, dy\, dx = \int_a^b \left(\int_c^d f(x, y)\, dy \right) dx = \int_a^b dx \int_c^d f(x, y)\, dy.$$

The limits c and d may be constants or functions of x.

EXAMPLE 4.1 Evaluate (*i*) $\int_0^3 \int_1^2 (x^2 + 3y^2)\, dy\, dx$, (*ii*) $\int_{x=0}^{1} \int_{y=0}^{2} (x^2 + 3xy^2)\, dx dy$. (*AMIETE June 2009*)

Solution (*i*) The given integral $= \int_0^3 \left(\int_1^2 (x^2 + 3y^2)\, dy \right) dx$

$$= \int_0^3 \left[x^2 y + 3 \cdot \frac{y^3}{3} \right]_1^2 dx \qquad \text{(integrating w.r. to } y\text{, treating } x \text{ as a constant)}$$

$$= \int_0^3 \left[(2x^2 + 2^3) - (x^2 + 1) \right] dx = \int_0^3 (x^2 + 7)\, dx = \left[\frac{x^3}{3} + 7x \right]_0^3 = \left(\frac{3^3}{3} + 7(3) \right) = 30.$$

(*ii*) $\int_{x=0}^{1} \int_{y=0}^{2} (x^2 + 3xy^2)\, dx dy = \int_{x=0}^{1} dx \int_{y=0}^{2} (x^2 + 3xy^2)\, dy = \int_0^1 \left[x^2 y + xy^3 \right]_0^2 dx$

$$= 2\int_0^1 (x^2 + 4x)\, dx = 2\left[\frac{x^3}{3} + 2x^2 \right]_0^1 = \frac{14}{3}.$$

EXAMPLE 4.2 Evaluate $\int_0^1 \int_0^{\sqrt{1+x^2}} \frac{dydx}{1 + x^2 + y^2}$. (*PTU 2006; Rajasthan, 2005; UPTU 2001; Madras 2000*)

Solution Since the limits of y are functions of x, the integration will first be performed with respect to y (treating x as a constant). Thus

$$\mathrm{I} = \int_0^1 \left[\int_0^{\sqrt{1+x^2}} \frac{1}{(1 + x^2) + y^2} dy \right] dx = \int_0^1 \frac{1}{\sqrt{1 + x^2}} \left[\tan^{-1} \frac{y}{\sqrt{1 + x^2}} \right]_0^{\sqrt{1+x^2}} dx$$

$$= \int_0^1 \frac{1}{\sqrt{1+x^2}}\left[\tan^{-1}1 - \tan^{-1}0\right]dx = \frac{\pi}{4}\int_0^1 \frac{dx}{\sqrt{1+x^2}}$$

$$= \frac{\pi}{4}\left[\log_e(x+\sqrt{1+x^2})\right]_0^1 = \frac{\pi}{4}\left[\log_e(1+\sqrt{2}) - \log 1\right] = \frac{\pi}{4}\log_e(\sqrt{2}+1).$$

EXAMPLE 4.3 Evaluate $\int_0^2 \int_0^{x^2} x(x^2+y^2)\,dy\,dx.$

Solution The given integral $= \int_0^2 \left[\int_0^{x^2} (x^3+xy^2)\,dy\right]dx = \int_0^2 \left[x^3 y + x\frac{y^3}{3}\right]_0^{x^2} dx = \int_0^2 \left(x^3\cdot x^2 + x\cdot\frac{(x^2)^3}{3}\right)dx$

$$= \int_0^2 \left(x^5 + \frac{x^7}{3}\right)dx = \left[\frac{x^6}{6} + \frac{x^8}{24}\right]_0^2 = \frac{2^6}{6} + \frac{2^8}{24} = \frac{64}{3}.$$

EXAMPLE 4.4 Evaluate $\int_0^{\pi} \int_0^{x} x\sin(x+y)\,dy\,dx.$

Solution The given integral $= \int_0^{\pi} x\left[-\cos(x+y)\right]_0^x dx$

$$= -\int_0^{\pi} x(\cos 2x - \cos x)\,dx = -\left[\left[x\left(\frac{\sin 2x}{2} - \sin x\right)\right]_0^{\pi} - \int_0^{\pi}\left(\frac{\sin 2x}{2} - \sin x\right)dx\right]$$

$$= \int_0^{\pi}\left(\frac{\sin 2x}{2} - \sin x\right)dx = \left[-\frac{\cos 2x}{4} + \cos x\right]_0^{\pi} = \left[\left(-\frac{1}{4} - 1\right) - \left(-\frac{1}{4} + 1\right)\right] = -2.$$

4.1.1 Definition

We know the definition of the definite integral $\int_a^b f(x)\,dx$ as the limit of a sum.

We shall now define the double integral of a function $f(x, y)$ over a closed region R, assuming that $f(x, y)$ is given for all points of the region.

Sub-divide the region R into rectangles by drawing lines parallel to co-ordinate axes. Number the rectangles which lie entirely inside R, from 1 to n.

Let $P(x_i, y_i)$ be any point inside ith rectangle whose area is ΔA_i and form the sum

$$S_n = \sum_{i=1}^{n} f(x_i, y_i)\Delta A_i.$$

Fig. 4.1: Sub-division of R

Let $n \to \infty$, such that the length of the largest diagonal of the rectangles approaches zero. If $\lim_{n\to\infty} S_n$ exists, independent of the choice of the subdivision and the point (x_i, y_i), then we say that $f(x, y)$ is integrable over R. This limit is called the *double integral* of $f(x, y)$ over the region R and is denoted by $\iint_R f(x, y)\,dA$. Other symbols used for this limit are

$$\iint_R f(x,y)\,dy\,dx \text{ or } \iint_R f(x,y)\,dx\,dy.$$

4.1.2 A Useful Example

We shall now find the mass of plane lamina with variable density in the form of a double integral. This will indicate how double integrals are evaluated in practice.

Let the plane lamina be bounded by the curves $y = f_1(x)$, $y = f_2(x)$ and the ordinates $x = a$ and $x = b$. Let the density σ at any point $P(x, y)$ be given by $\sigma = f(x, y)$.

We divide the area of the lamina into strips parallel to y-axis. One such strip is $ABCD$. We divide the strip into elementary rectangular meshes like $PQRS$ with dimensions Δx and Δy (as shown).

The mass of this mesh $= f(x, y)\,\Delta y\,\Delta x$.

The mass of the strip = Sum of masses of such meshes

$$= \lim_{\Delta y \to 0} \sum f(x, y)\,\Delta y\,\Delta x$$

(x and Δx remain constant for this summation)

$$= \left[\int_{f_1(x)}^{f_2(x)} f(x, y)\,dy\right]\Delta x = \phi(x)\Delta x,$$

where
$$\phi(x) = \int_{f_1}^{f_2} f(x, y)\,dy. \qquad \text{...(4.1)}$$

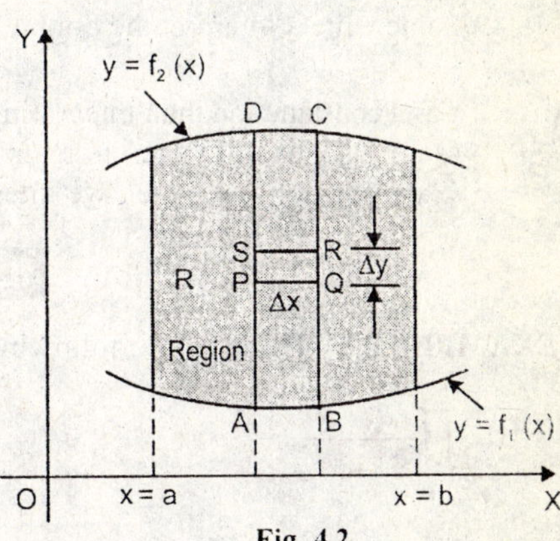

Fig. 4.2

The mass of the lamina = the sum of masses of such strips

$$= \lim_{\Delta x \to 0} \sum \phi(x)\,\Delta x = \int_a^b \phi(x)\,dx$$

$$= \int_a^b \int_{f_1}^{f_2} f(x, y)\,dy\,dx \qquad \text{by (4.1)...(4.2).}$$

In this we first integrate $f(x, y)$ w.r. to y, treating x as a constant and the resulting expression is integrated with respect to x.

Similarly, if the lamina is bounded by the curves $x = \Psi_1(y)$, $x = \psi_2(y)$ and the lines $y = c$, $y = d$, we can divide the area, thus bounded, into strips parallel to x-axis. Proceeding as above, we can show that the mass of the lamina

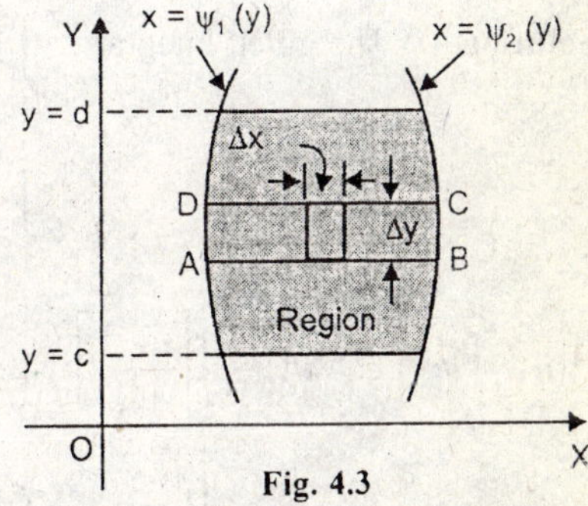

Fig. 4.3

$$= \int_c^d \int_{\psi_1}^{\psi_2} f(x, y)\,dx\,dy. \qquad \text{...(4.3)}$$

In this we first integrate $f(x, y)$ w.r. to x, treating y as a constant and then the resulting expression is integrated w.r.t. y.

Remarks

1. The double integral $\int_a^b \int_{f_1(x)}^{f_2(x)} f(x, y)\,dy\,dx$ is also called the value of $\iint f(x, y)\,dy\,dx$ over the region bounded by the curves $y = f_1(x)$, $y = f_2(x)$, and the lines $x = a$ and $x = b$.
2. The area of the adjoining figure may be considered as bounded by the curves, $y = f_1(x)$, $y = f_2(x)$ and the lines $x = a$, $x = b$.

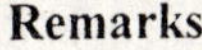

Hence $\iint f(x, y)\,dy\,dx$ over this region $= \int_a^b \int_{f_1}^{f_2} f(x, y)\,dy\,dx.$

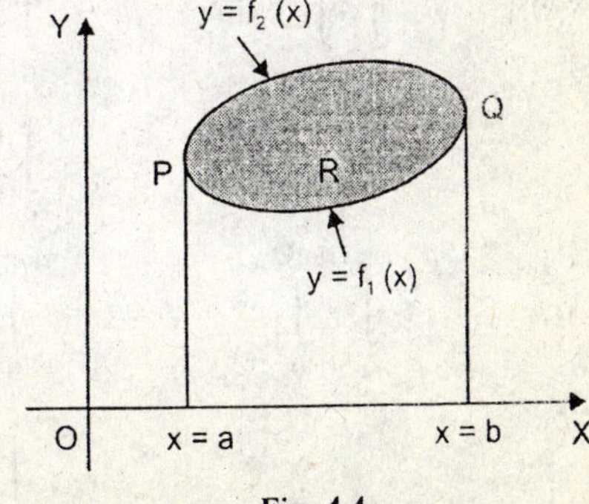

Fig. 4.4

3. *We should carefully note that when the area is divided into strips parallel to y-axis, then the first integral is w.r. to y, treating x as a constant.*
The limits of integration for y are the values of y (as functions of x) at the lower end of the strip and the upper end of the strip. These values are given by the equations of the curves on which the lower extremity and the upper extremity lie.
Then the integration w.r. to x is between the limits which are values of x at the extreme left and extreme right of the region.
Similar remarks apply when the region is divided into strips parallel to x-axis.

4. Some writers interpret the double integral $\int_a^b\int_{f_1}^{f_2} f(x,y)\,dx\,dy$ to mean integral of $f(x, y)$ w.r. to y, treating x as a constant and then integrating the resulting expression w.r.t. x. We should bear in mind that if f_1 and f_2 are functions of x then it is understood that we integrate w.r. to y first and then w.r. to x and if f_1 and f_2 are functions of y then we integrate w.r. to x first and then w.r. to y.

ILLUSTRATIVE EXAMPLES

EXAMPLE 4.5 Evaluate the following double integrals:

(*i*) $\int_1^2\int_0^x \frac{dy\,dx}{y^2+x^2}$, (*ii*) $\int_0^4 dx\int_0^{x^2} e^{y/x}dy$, (*iii*) $\int_0^{\pi}\int_0^{\cos y} x\sin y\,dx\,dy$,

(*iv*) $\int_0^{\log 3}\int_0^{\log 2} e^{x+y}dy\,dx$, (*v*) $\int_0^1 dy\int_0^1 \frac{x-y}{(x+y)^3}dx$.

Solution (*i*) The given integral $= \int_1^2\left(\int_0^x \frac{dy}{y^2+x^2}\right)dx$

$$= \int_1^2 \frac{1}{x}\left[\tan^{-1}\frac{y}{x}\right]_0^x dx = \int_1^2 \frac{1}{x}(\tan^{-1}1-\tan 0)\,dx = \frac{\pi}{4}\int_1^2\frac{1}{x}dx = \frac{\pi}{4}[\log x]_1^2 = \frac{\pi}{4}\log 2.$$

(*ii*) $$\int_0^4 dx\int_0^{x^2} e^{y/x}dy = \int_0^4\left[\frac{e^{y/x}}{1/x}\right]_0^{x^2} dx = \int_0^4\left[xe^{y/x}\right]_0^{x^2} dx$$

$$= \int_0^4\left(\underset{\text{I}}{x}\,\underset{\text{II}}{e^x}-x\right)dx = \left[xe^x\right]_0^4 - \int_0^4 e^x\,dx - \left[\frac{x^2}{2}\right]_0^4 = \left[xe^x-e^x\right]_0^4 - \frac{1}{2}[x^2]_0^4 = 3e^4-7.$$

(*iii*) $$\int_0^{\pi}\int_0^{\cos y} x\sin y\,dx\,dy = \int_0^{\pi}\left(\int_0^{\cos y} x\sin y\,dx\right)dy = \int_0^{\pi}\left[\frac{x^2}{2}\sin y\right]_{x=0}^{\cos y} dy$$

$$= \int_0^{\pi}\frac{1}{2}\cos^2 y\sin y\,dy = \left[-\frac{1}{6}\cos^3 y\right]_0^{\pi} = \frac{1}{3}.$$

(*iv*) $$\int_0^{\log 3}\int_0^{\log 2} e^{x+y}dy\,dx = \int_0^{\log 3}\left[\int_0^{\log 2} e^{x+y}\,dy\right]dx = \int_0^{\log 3}\left[e^{x+y}\right]_0^{\log 2} dx$$

$$= \int_0^{\log 3}(e^{x+\log 2}-e^x)\,dx = \int_0^{\log 3}(2e^x-e^x)\,dx = \int_0^{\log 3} e^x\,dx = \left[e^x\right]_0^{\log 3} = 3-1 = 2.$$

(*v*) $$\int_0^1 dy \int_0^1 \frac{x-y}{(x+y)^3}\,dx = \int_0^1 dy \int_0^1 \left\{\frac{1}{(x+y)^2} - \frac{2y}{(x+y)^3}\right\} dx = \int_0^1 \left[-\frac{1}{x+y} + \frac{y}{(x+y)^2}\right]_0^1 dy$$

$$= \int_0^1 \left[-\frac{1}{1+y} + \frac{y}{(1+y)^2}\right] dy = -\int_0^1 \frac{dy}{(1+y)^2} = \left[\frac{1}{1+y}\right]_0^1 = -\frac{1}{2}.$$

EXAMPLE 4.6 (*i*) Evaluate $\iint_R (x^2+y^2)\,dy\,dx$, where R is the region in the positive quadrant for which $x + y \le 1$.

(*ii*) Evaluate $\iint_R xy\,dx\,dy$, where R is the quadrant of the circle $x^2 + y^2 = a^2$ where $x \ge 0$, $y \ge 0$.

(*Rajasthan, 2006; VTU, 2001; Madras 2000*)

(*iii*) Evaluate $\iint_S \sqrt{xy - y^2}\,dx\,dy$ over the region S, triangle with vertices (0, 0), (10, 1) and (1, 1).

Solution (*i*) Refer Fig. 4.5 Here the region of integration is the area bounded by the two axes and straight line $x + y = 1$. We can consider it as the area bounded by the lines, $y = 0$, $y = 1 - x$, $x = 0$, and $x = 1$.

Hence, $$\iint_R (x^2+y^2)\,dy\,dx = \int_{x=0}^{1}\int_{y=0}^{1-x} (x^2+y^2)\,dy\,dx = \int_0^1 \left[x^2 y + \frac{y^3}{3}\right]_0^{1-x} dx$$

$$= \int_0^1 \left[x^2(1-x) + \frac{1}{3}(1-x)^3\right] dx = \left[\frac{x^3}{3} - \frac{x^4}{4} + \frac{1}{3}\frac{(1-x)^4}{4(-1)}\right]_0^1$$

$$= \left[\frac{1}{3}(1-0) - \frac{1}{4}(1-0) - \frac{1}{12}(0-1)\right]$$

$$= \frac{1}{3} - \frac{1}{4} + \frac{1}{12} = \frac{1}{6}.$$

Fig. 4.5

(*ii*) Refer Fig. 4.6 The region *R* is bounded by the curves $x = 0$, $x = \sqrt{a^2 - y^2}$, $y = 0$, and $y = a$.

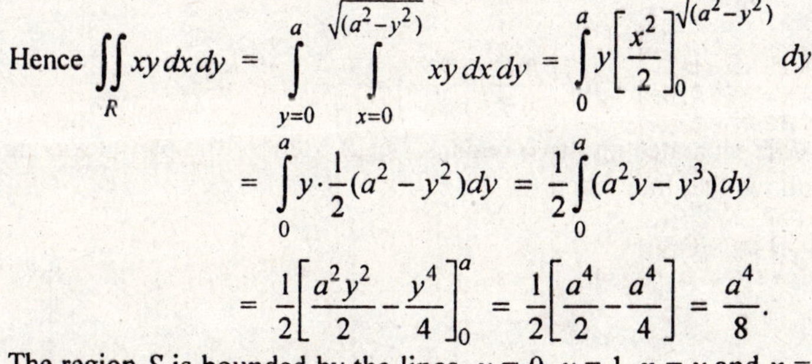

Hence $$\iint_R xy\,dx\,dy = \int_{y=0}^{a}\int_{x=0}^{\sqrt{(a^2-y^2)}} xy\,dx\,dy = \int_0^a y\left[\frac{x^2}{2}\right]_0^{\sqrt{(a^2-y^2)}} dy$$

$$= \int_0^a y\cdot\frac{1}{2}(a^2-y^2)\,dy = \frac{1}{2}\int_0^a (a^2 y - y^3)\,dy$$

$$= \frac{1}{2}\left[\frac{a^2y^2}{2} - \frac{y^4}{4}\right]_0^a = \frac{1}{2}\left[\frac{a^4}{2} - \frac{a^4}{4}\right] = \frac{a^4}{8}.$$

Fig. 4.6

(*iii*) The region *S* is bounded by the lines, $y = 0$, $y = 1$, $x = y$ and $x = 10y$.

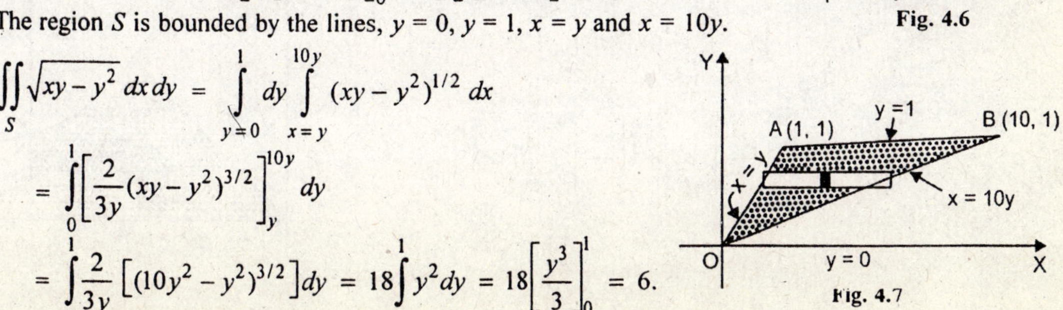

$$\iint_S \sqrt{xy - y^2}\,dx\,dy = \int_{y=0}^{1} dy \int_{x=y}^{10y} (xy - y^2)^{1/2}\,dx$$

$$= \int_0^1 \left[\frac{2}{3y}(xy - y^2)^{3/2}\right]_y^{10y} dy$$

$$= \int_0^1 \frac{2}{3y}\left[(10y^2 - y^2)^{3/2}\right] dy = 18\int_0^1 y^2\,dy = 18\left[\frac{y^3}{3}\right]_0^1 = 6.$$

Fig. 4.7

EXAMPLE 4.7 (*i*) Evaluate $\iint (x+y)^2\,dy\,dx$ over the area bounded by the ellipse $x^2/a^2 + y^2/b^2 = 1$.
(*KUK, 2009; UPTU, 2005*)

(*ii*) Evaluate $\iint_R x^2 dx\,dy$, where R is the region in the first quadrant bounded by the hyperbola $xy = 16$ and the lines $y = x$, $y = 0$ and $x = 8$. (*UPTU-2002; AMIE., S-2001*)

(*iii*) Evaluate $\iint_R dx\,dy$, where R is the region between $y = 2x$ and $y = x^2$ lying to the left of $x = 1$.

Solution (*i*) For the ellipse

$y/b = \pm\sqrt{\left(1-\dfrac{x^2}{a^2}\right)}$, the region of integration can be considered as bounded by the curves

$$y = -b\sqrt{[1-(x^2/a^2)]},\ y = b\sqrt{[1-(x^2/a^2)]},\ x = -a \text{ and } x = a.$$

Therefore, the given double integral

$$= \int_{-a}^{a}\int_{-b\sqrt{1-x^2/a^2}}^{b\sqrt{(1-x^2/a^2)}} (x^2+2xy+y^2)\,dy\,dx = 2\int_{-a}^{a}\int_{0}^{b\sqrt{(1-x^2/a^2)}} (x^2+y^2)\,dy\,dx$$

($\because$ xy is an odd function of y and $x^2 + y^2$ is an even function of y)

$$= 2\int_{-a}^{a}\left[x^2y+(y^3/3)\right]_0^{b\sqrt{(1-x^2/a^2)}} dx = 4\int_0^a\left\{x^2b\sqrt{1-\frac{x^2}{a^2}}+\frac{1}{3}b^3\left(1-\frac{x^2}{a^2}\right)^{3/2}\right\}dx$$

$$= 4b\int_0^{\pi/2}\left\{a^2\sin^2\theta\cos\theta+\frac{1}{3}b^2\cos^3\theta\right\}a\cos\theta\,d\theta \quad [\text{On putting } x = a\sin\theta \text{ and } dx = a\cos\theta\,d\theta.]$$

$$= 4ab\int_0^{\pi/2}\left(a^2\sin^2\theta\cos^2\theta+\frac{1}{3}b^2\cos^4\theta\right)d\theta$$

$$= 4ab\left[a^2\cdot\frac{1}{4}\cdot\frac{1}{2}\cdot\frac{\pi}{2}+\frac{1}{3}b^2\cdot\frac{3}{4}\cdot\frac{1}{2}\cdot\frac{\pi}{2}\right] = 4ab\left[\frac{\pi a^2}{16}+\frac{\pi b^2}{16}\right] = \frac{\pi ab}{4}(a^2+b^2).$$

(*ii*) Refer Fig. 4.8 Here the region R is separated into two regions. Let R_1 denote the part of R lying below the line $y = 2$ and R_2 the part above that line. Then

$$\iint_R x^2 dx\,dy = \iint_{R_1} x^2 dx\,dy + \iint_{R_2} x^2 dx\,dy$$

$$= \int_{y=0}^{2}\int_{x=y}^{8} x^2 dx\,dy + \int_{2}^{4}\int_{x=y}^{16/y} x^2 dx\,dy$$

$$= \int_0^2\left[\frac{x^3}{3}\right]_y^8 dy + \int_2^4\left[\frac{x^3}{3}\right]_y^{16/y} dy$$

$$= \frac{1}{3}\int_0^2 (8^3-y^3)\,dy + \frac{1}{3}\int_2^4\left(\frac{16^3}{y^3}-y^3\right)dy$$

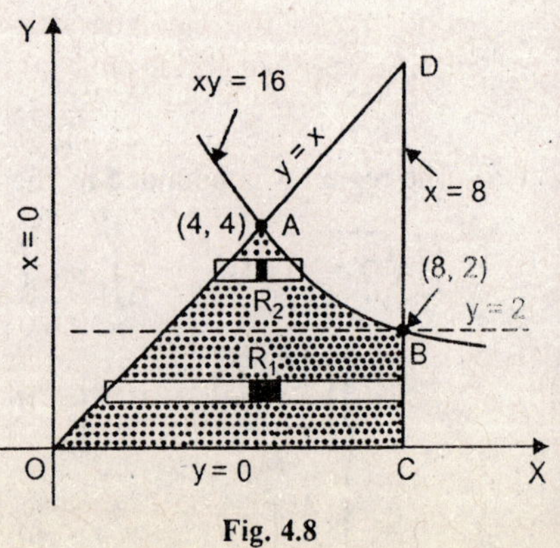

Fig. 4.8

$$= \frac{1}{3}\left[8^3 y - \frac{y^4}{4}\right]_0^2 + \frac{1}{3}\left[-\frac{(16)^3}{2y^2} - \frac{y^4}{4}\right]_2^4$$

$$= \frac{1}{3}\left(8^3 \cdot 2 - \frac{2^4}{4}\right) + \frac{1}{3}\left[-\frac{(16)^3}{2}\cdot\left(\frac{1}{16} - \frac{1}{4}\right) - \frac{1}{4}(4^4 - 2^4)\right]$$

$$= \frac{1020}{3} + \frac{1}{3}\left[\frac{3}{2}(16)^2 - 60\right] = 340 + 108 = 448.$$

[**Alternative.** Using the vertical strips, separate R with the line $x = 4$ and we obtain

$$\iint_R x^2 dy\, dx = \int_0^4 \int_{y=0}^{x} x^2 dy\, dx + \int_4^8 \int_{y=0}^{16/x} x^2 dy\, dx \,]$$

(*iii*) Refer Fig. 4.9, it is obvious that the region R is separated into two regions and an iterated integral is evaluated for each. Let R_1 denote the part R lying below CD and R_2 the part above CD. Then

$$\iint_R dx\, dy = \iint_{R_1} dx\, dy + \iint_{R_2} dx\, dy$$

$$= \int_{y=0}^{1} \int_{x=y/2}^{\sqrt{y}} dx dy + \int_{y=1}^{2} \int_{x=y/2}^{1} dx dy$$

$$= \int_0^1 \left(\sqrt{y} - \frac{y}{2}\right) dy + \int_1^2 \left(1 - \frac{y}{2}\right) dy$$

$$= \left[\frac{2}{3} y^{3/2} - \frac{y^2}{4}\right]_0^1 + \left[y - \frac{y^2}{4}\right]_1^2$$

$$= \left(\frac{2}{3} - \frac{1}{4}\right) + \left\{(2-1) - \frac{1}{4}(4-1)\right\} = \frac{5}{12} + \frac{1}{4} = \frac{2}{3}.$$

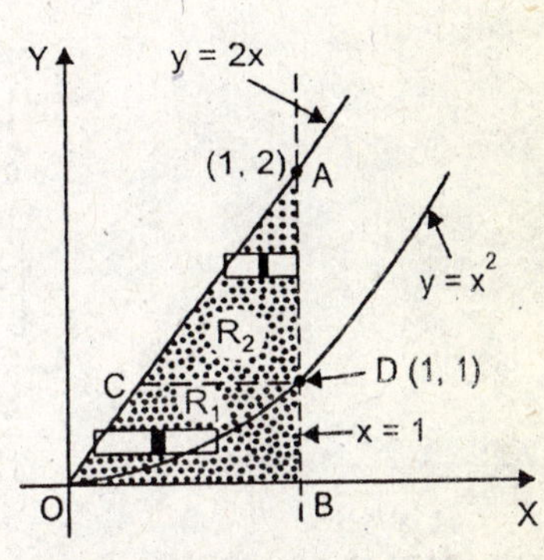

Fig. 4.9

4.2 CHANGE OF ORDER OF INTEGRATION

In Art 4.1 we have seen that a double integral may be evaluated with respect to y first and then with respect to x or with respect to x first and then with respect to y. Sometimes one order of evaluation may be more convenient as compared to the other. For this we may have to change the order of integration to evaluate a given integral.

1. In case the region of integration is a rectangle with sides parallel to axes, then all the limits of integration are constants. In this case the change in the order of integration does not, ordinarily, require change in the limits of integration. From Fig. 4.10 it is clear that

$$\int_{x=a}^{b} \int_{y=c}^{d} f(x,y)\, dy\, dx = \int_{y=c}^{d} \int_{x=a}^{b} f(x,y)\, dx\, dy.$$

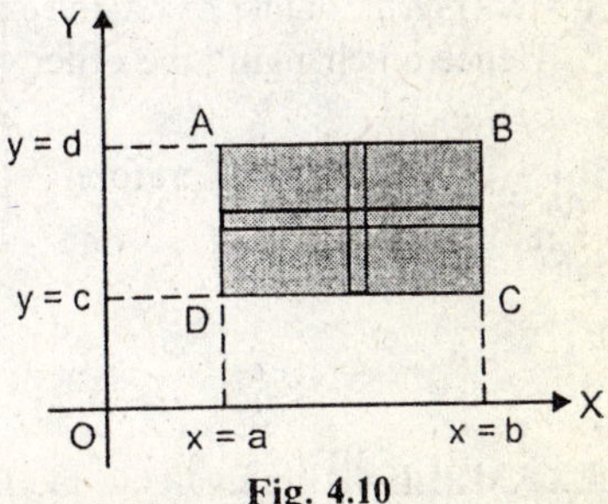

Fig. 4.10

2. But if the limits of integration are variable, a change in the order of integration requires change in the limits of integration. In such cases a rough sketch of the region of integration is essential. This helps in fixing the new limits of integration. Sometimes in changing the order of integration we may have to split up the region of integration in two or more parts and the given integral will be expressed as the sum of two or more integrals. The following solved examples will make the ideas clear.

LLUSTRATIVE EXAMPLES

EXAMPLE 4.8 Change the order of integration and then evaluate it

$$\int_0^a \int_{x/a}^{\sqrt{x/a}} (x^2 + y^2)\, dy\, dx. \qquad (\textit{AMIETE, June 2002})$$

Solution Refer Fig. 4.11(a) the region is bounded by the curves

$y = x/a$ and $y = \sqrt{x/a}$ between $x = 0$ and $x = a$.

To find limits for integrating in the reverse order, we imagine a horizontal line passing from left to right through the region. It enters at $x = ay^2$ and leaves at $x = ay$. To include all such lines, we let y run from $y = 0$ to $y = 1$ [Fig. 4.11 (b)]. The integral is

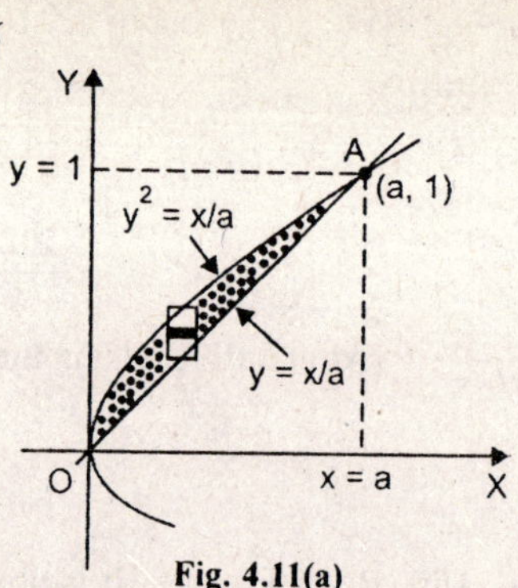

Fig. 4.11(a)

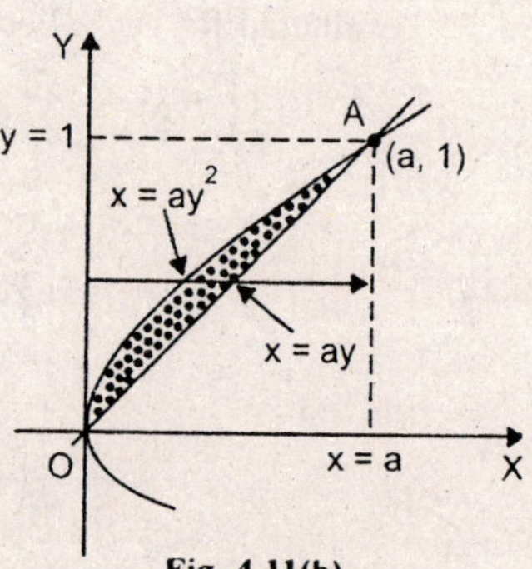

Fig. 4.11(b)

$$\int_{y=0}^{1} \int_{x=ay^2}^{ay} (x^2 + y^2)\, dx\, dy = \int_0^1 \left[\frac{x^3}{3} + xy^2\right]_{ay^2}^{ay} dy$$

$$= \int_0^1 \left(\frac{a^3y^3}{3} + ay^3 - \frac{a^3y^6}{3} - ay^4\right) dy$$

$$= \left[\frac{a^3y^4}{12} + \frac{ay^4}{4} - \frac{a^3y^7}{21} - \frac{ay^5}{5}\right]_0^1$$

$$= \frac{a^3}{12} + \frac{a}{4} - \frac{a^3}{21} - \frac{a}{5} = \frac{a}{4}\left(\frac{a^2}{7} + \frac{1}{5}\right).$$

EXAMPLE 4.9 Change the order of integration in $\displaystyle\int_{y=0}^{a} \int_{x=y}^{a} \frac{x\, dx\, dy}{x^2 + y^2}$, and hence evaluate the same.

(*AMIETE, June 2010*)

Solution From the limits of integration, it is clear that the region of integration is bounded by $x = y$, $x = a$, $y = 0$ and $y = a$ that is, the triangle $0AB$.

The given order of integration indicates that the region is supposed to be divided into horizontal strips.

For changing the order of integration, the area is supposed to be divided into vertical strips which are bounded from below by $y = 0$ and from above $y = x$. Then the values of x at the extreme left and right are 0 and a respectively.

Hence on changing the order of integration the given integral becomes

$$\int_{x=0}^{a} \int_{y=0}^{x} \frac{x\, dy\, dx}{x^2 + y^2}. \quad \text{Therefore,} \quad \int_{y=0}^{a} \int_{x=y}^{a} \frac{x}{x^2 + y^2}\, dx dy = \int_{x=0}^{a} \int_{y=0}^{x} \frac{x}{x^2 + y^2}\, dy dx$$

$$= \int_0^a x \cdot \left|\frac{1}{x}\tan^{-1}\frac{y}{x}\right|_0^x dx = \int_0^a \frac{\pi}{4}\, dx = \frac{\pi a}{4}.$$

Fig. 4.12

EXAMPLE 4.10 Evaluate the following integrals by changing the order of integration.

(*i*) $\displaystyle\int_0^{4a} \int_{x^2/4a}^{2\sqrt{(ax)}} dy\, dx$, (*PTU 2009; MDU-2001*) (*ii*) $\displaystyle\int_0^{a} \int_{x^2/a}^{2a-x} xy\, dy\, dx$, (*PTU 2006; AMIETE June 2005*) (*iii*) $\displaystyle\int_0^1 \int_x^{\sqrt{(2-x^2)}} \frac{x\,dy\,dx}{\sqrt{(x^2+y^2)}}$. (*UPTU, 2006; JNTU, 2005; MDU 2004*)

Solution

(*i*) The given limits show that the region of integration is the area bounded by the curves $y = x^2/4a$, $y = 2\sqrt{(ax)}$, $x = 0$, and $x = 0$, and $x = 4a$. (See Fig. 4.13). The first two curves are parabolas $x^2 = 4ay$ and $y^2 = 4ax$, which intersect at (0, 0) and (4*a*, 4*a*). We can also consider the same area enclosed by $x = y^2/4a$, $x = 2\sqrt{(ay)}$, $y = 0$, and $y = 4a$.

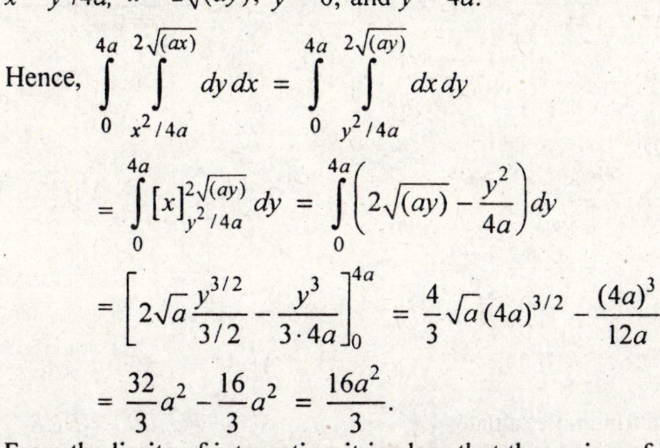

Hence, $\displaystyle\int_0^{4a}\int_{x^2/4a}^{2\sqrt{(ax)}} dy\,dx = \int_0^{4a}\int_{y^2/4a}^{2\sqrt{(ay)}} dx\,dy$

$$= \int_0^{4a} [x]_{y^2/4a}^{2\sqrt{(ay)}} dy = \int_0^{4a}\left(2\sqrt{(ay)} - \frac{y^2}{4a}\right)dy$$

$$= \left[2\sqrt{a}\frac{y^{3/2}}{3/2} - \frac{y^3}{3\cdot 4a}\right]_0^{4a} = \frac{4}{3}\sqrt{a}(4a)^{3/2} - \frac{(4a)^3}{12a}$$

$$= \frac{32}{3}a^2 - \frac{16}{3}a^2 = \frac{16a^2}{3}.$$

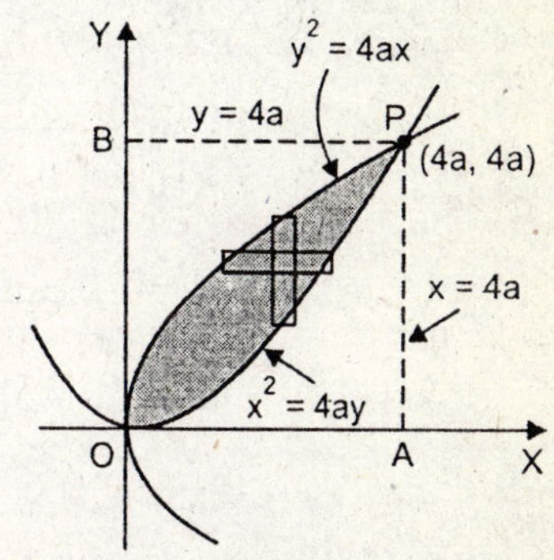

Fig. 4.13

(*ii*) From the limits of integration it is clear that the region of integration is bounded by the curves $x^2 = ay$, $x + y = 2a$, $x = 0$ and $x = a$ (shown shaded).

To cover this area by horizontal strips we have to split it into two parts, one lying below the line *BA* and the other above it.

(This is necessitated by the fact that the right extremities of strips below *BA* lie on the parabola $x^2 = ay$ while in the case of strips above *BA*, the right extremities lie on the line $x + y = 2a$).

Therefore, $\displaystyle\int_{x=0}^{a}\int_{y=x^2/a}^{2a-x} xy\,dy\,dx = \int_{y=0}^{a}\int_{x=0}^{\sqrt{ay}} (xy\,dx\,dy) + \int_{y=a}^{2a}\int_{x=0}^{2a-y} xy\,dx\,dy$

$$= \int_0^a y\left[\frac{x^2}{2}\right]_0^{\sqrt{ay}} dy + \int_a^{2a} y\left[\frac{x^2}{2}\right]_0^{2a-y} dy$$

$$= \frac{a}{2}\int_0^a y^2\,dy + \frac{1}{2}\int_a^{2a} y(2a-y)^2\,dy$$

$$= \frac{a}{2}\left[\frac{y^3}{3}\right]_0^a + \frac{1}{2}\int_a^{2a}(4a^2y + y^3 - 4ay^2)\,dy$$

$$= \frac{a^4}{6} + \frac{1}{2}\left[2a^2y^2 + \frac{y^4}{4} - \frac{4ay^3}{3}\right]_a^{2a}$$

$$= \frac{a^4}{6} + \frac{1}{2}\left[2a^2(3a^2) + \frac{1}{4}(16a^4 - a^4) - \frac{4a}{3}(8a^3 - a^3)\right]$$

$$= \frac{a^4}{6} + \frac{1}{2}\left(6a^4 + \frac{15}{4}a^4 - \frac{28a^4}{3}\right) = \frac{a^4}{6} + \frac{5a^4}{24} = \frac{3}{8}a^4.$$

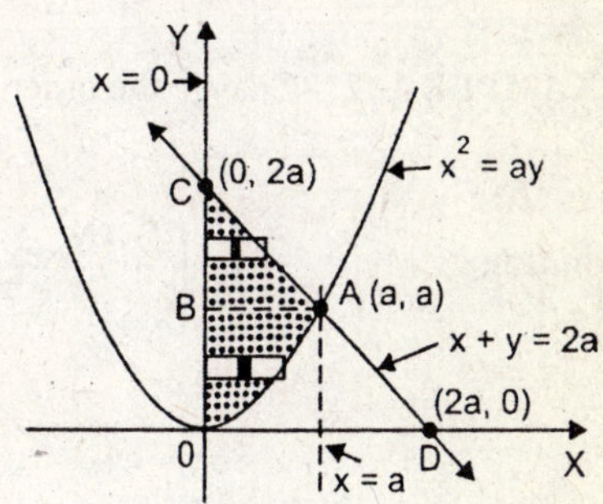

Fig. 4.14

(*iii*) The given limits show that the region of integration is bounded by the curves $y = x,\ y = \sqrt{(2-x^2)},\ x = 0$, and $x = 1$ (See Fig. 4.15).

Thus the area $OABCO$ lies between the lines $y = x,\ x = 0$, and the circle $x^2 + y^2 = 2$.

To cover this area by horizontal strips, we shall split it into two parts; one lying below the line CA and the other above CA; then

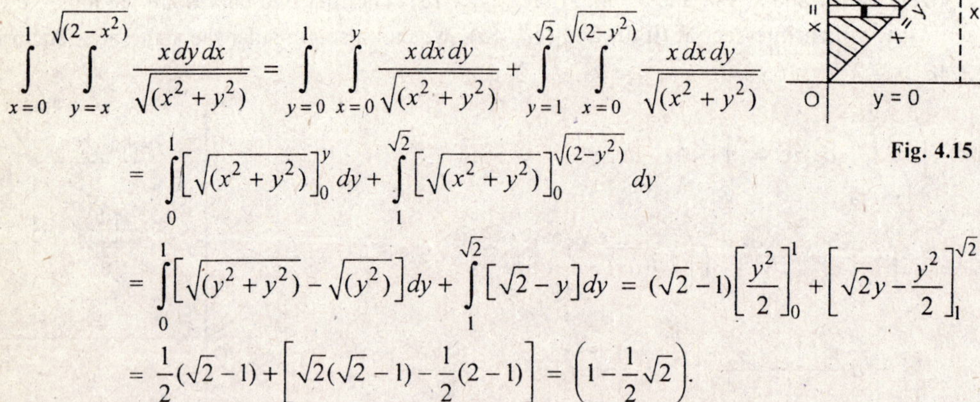

$$\int_{x=0}^{1}\int_{y=x}^{\sqrt{(2-x^2)}} \frac{x\,dy\,dx}{\sqrt{(x^2+y^2)}} = \int_{y=0}^{1}\int_{x=0}^{y} \frac{x\,dx\,dy}{\sqrt{(x^2+y^2)}} + \int_{y=1}^{\sqrt{2}}\int_{x=0}^{\sqrt{(2-y^2)}} \frac{x\,dx\,dy}{\sqrt{(x^2+y^2)}}$$

$$= \int_0^1 \left[\sqrt{(x^2+y^2)}\right]_0^y dy + \int_1^{\sqrt{2}} \left[\sqrt{(x^2+y^2)}\right]_0^{\sqrt{(2-y^2)}} dy$$

$$= \int_0^1 \left[\sqrt{(y^2+y^2)} - \sqrt{(y^2)}\right] dy + \int_1^{\sqrt{2}} \left[\sqrt{2} - y\right] dy = (\sqrt{2}-1)\left[\frac{y^2}{2}\right]_0^1 + \left[\sqrt{2}y - \frac{y^2}{2}\right]_1^{\sqrt{2}}$$

$$= \frac{1}{2}(\sqrt{2}-1) + \left[\sqrt{2}(\sqrt{2}-1) - \frac{1}{2}(2-1)\right] = \left(1 - \frac{1}{2}\sqrt{2}\right).$$

Fig. 4.15

EXAMPLE 4.11 Change the order of integration and evaluate $\int_0^{\infty}\int_x^{\infty} \frac{e^{-y}}{y}\,dy\,dx.$ (*VTU, 2007; DCE, 2004*)

Solution The first integration is by vertical strips extending from $y = x$ to $y = \infty$; the strips starting from $x = 0$ and going to $x = \infty$. So the region of integration is the upper half of the first quadrant. If the same region is covered by horizontal strips, their ends will be on $x = 0$ and $x = y$, while the strips will start from $y = 0$ and go to $y = \infty$. Hence

$$\int_0^{\infty}\int_x^{\infty} (e^{-y}/y)\,dy\,dx = \int_0^{\infty}\int_0^{y} (e^{-y}/y)\,dx\,dy = \int_0^{\infty} (e^{-y}/y)[x]_0^y\,dy$$

$$= \int_0^{\infty} e^{-y}\,dy = \left[-e^{-y}\right]_0^{\infty} = 1.$$

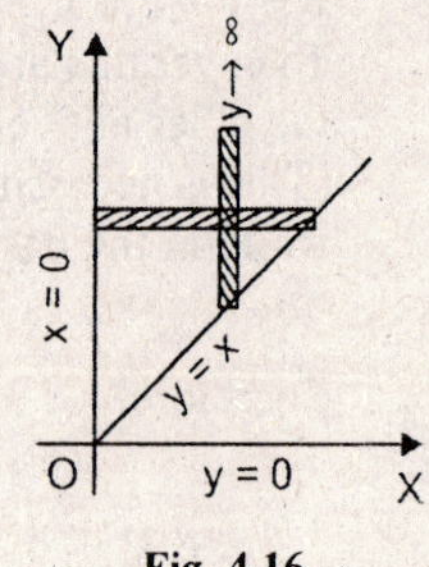

Fig. 4.16

EXAMPLE 4.12 Change the order of integration of $\int_0^{\infty}\int_0^{\infty} e^{-xy}\sin nx\,dx\,dy$. Show that $\int_0^{\infty} \frac{\sin nx}{x}\,dx = \frac{\pi}{2}$.

(*AMIETE-Dec, 2010*)

Solution

$$\int_0^{\infty}\int_0^{\infty} e^{-xy}\sin nx\,dx\,dy = \int_0^{\infty} dy\left(\int_0^{\infty} e^{-xy}\sin nx\,dx\right)$$

$$= \int_0^{\infty} dy\left[-\frac{e^{-xy}}{n^2+y^2}(y\sin nx + n\cos nx)\right]_0^{\infty}$$

$$= \int_0^{\infty} \frac{n}{y^2+n^2}\,dy = n\cdot\frac{1}{n}\left[\tan^{-1}\frac{y}{n}\right]_0^{\infty} = \frac{\pi}{2} \qquad \ldots(i)$$

On changing the order of integration, we have

$$\int_0^{\infty}\int_0^{\infty} e^{-xy}\sin nx\,dx\,dy = \int_0^{\infty} \sin nx\,dx\left(\int_0^{\infty} e^{-xy}\,dy\right)$$

$$= \int_0^\infty \sin nx\, dx \left[\frac{e^{-xy}}{-x}\right]_0^\infty = \int_0^\infty \frac{\sin nx\, dx}{x}\left[-\frac{1}{e^{xy}}\right]_0^\infty$$

$$= \int_0^\infty \frac{\sin nx}{x}\, dx\,[-0+1] = \int_0^\infty \frac{\sin nx}{x}\, dx \qquad ...(ii)$$

From (*i*) and (*ii*), we obtain

$$\int_0^\infty \frac{\sin nx}{x}\, dx = \frac{\pi}{2}.$$

4.3 DOUBLE INTEGRALS IN POLAR FORM

The evaluation of the integral $\int_{\theta_1}^{\theta_1}\int_{r_1}^{r_2} f(r,\theta)\, dr\, d\theta$ is done exactly as in the case of double integrals in Cartesian co-ordinates that is, we first integrate $f(r,\theta)$ w.r.t. r (treating θ as a constant), between the limits r_1 and r_2. The resulting expression is integrated w.r.t. θ between the limits θ_1 and θ_2. Here r_1 and r_2 may be constants or functions of θ.

Similarly to evaluate $\int_{r_1}^{r_2}\int_{\theta_1}^{\theta_2} f(r,\theta)\, d\theta\, dr$, we first integrate w.r.t. θ between the limits θ_1 and θ_2. The resulting expression is integrated w.r.t. r between the limits r_1 and r_2. The limits θ_1, θ_2 may be constants or functions of r.

4.3.1 Geometrical Interpretation of $\int_{\theta_1}^{\theta_2}\int_{r_1}^{r_2} f(r,\theta)\, dr\, d\theta$

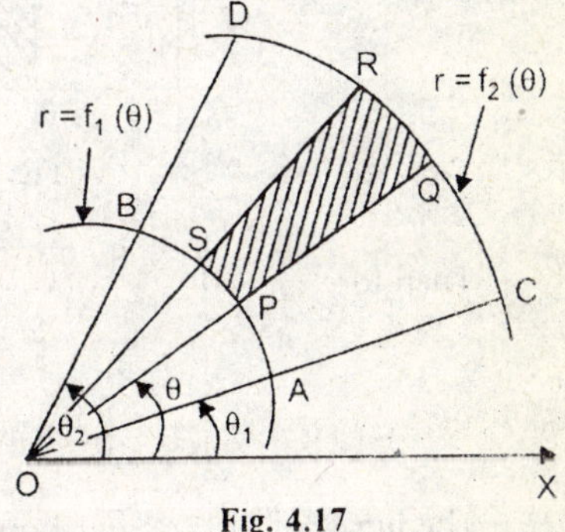

Fig. 4.17

Refer Fig. 4.17, the region *ACDB* is bounded by the curves $r = f_1(\theta)$ and $r = f_2(\theta)$ and the radii vectors $\theta = \theta_1$ and $\theta = \theta_2$.

Now $\int_{r_1}^{r_2} f(r,\theta)\, dr$ shows that the integration is along the shaded portion from *PS* to *QR* and then integration w.r.t. θ shows that the shaded portion slides from *AC* to *BD*.

EXAMPLE 4.13 Evaluate

(*i*) $\int_0^{\pi/2}\int_0^a r^2 \sin\theta\, dr\, d\theta$, (*ii*) $\int_{0=\theta}^{\pi}\int_{r=0}^{a\sin\theta} r\, dr\, d\theta$.

Solution (*i*) The given integral

$$\int_0^{\pi/2}\int_0^a r^2 \sin\theta\, dr\, d\theta = \int_0^{\pi/2}\left[\frac{r^3}{3}\right]_0^a \sin\theta\, d\theta = \frac{a^3}{3}\int_0^{\pi/2} \sin\theta\, d\theta = \frac{a^3}{3}[-\cos\theta]_0^{\pi/2} = \frac{a^3}{3}.$$

(*ii*)
$$\int_0^{\pi}\int_0^{a\sin\theta} r\, dr\, d\theta = \int_0^{\pi}\left[\frac{r^2}{2}\right]_0^{a\sin\theta} d\theta = \frac{1}{2}\int_0^{\pi} a^2\sin^2\theta\, d\theta = \frac{a^2}{4}\int_0^{\pi}(1-\cos 2\theta)\, d\theta$$

$$= \frac{a^2}{4}\left[\theta - \frac{\sin 2\theta}{2}\right]_0^{\pi} = \frac{a^2}{4}(\pi) = \frac{\pi a^2}{4}.$$

4.4 CHANGING CARTESIAN INTEGRALS INTO POLAR INTEGRALS

From the limits of the given integral sketch the region of integration. The equations of the bounding curves can now be expressed in polar co-ordinates. Then change $dy\,dx$ into $r\,dr\,d\theta$ and put $x = r\cos\theta$ and $y = r\sin\theta$ in the integrand. The limits of integration can be determined from the equations of the bounding curves.

The following solved examples will clarify the procedure.

EXAMPLE 4.14 Evaluate the following integrals by changing to polar co-ordinates

(*i*) $\displaystyle\int_{x=0}^{a}\int_{y=0}^{\sqrt{a^2-x^2}} y^2\sqrt{x^2+y^2}\,dy\,dx$, (*AMIE, S-2001*) (*ii*) $\displaystyle\int_{0}^{2}\int_{0}^{\sqrt{2x-x^2}} \frac{x\,dy\,dx}{\sqrt{x^2+y^2}}$.

Solution (*i*) The given limits of integration show that the region of integration lies between the curves

$$y = 0,\; y = \sqrt{(a^2 - x^2)},\; x = 0 \text{ and } x = a. \qquad \text{[See Fig. 4.18(}a\text{)].}$$

Thus the region of integration is the part of the circle $x^2 + y^2 = a^2$ in the first quadrant. In polar co-ordinates, the region of integration is bounded by the curves:

$$r = 0,\; r = a,\; \theta = 0,\; \theta = \pi/2. \qquad \text{[See Fig. 4.18(}b\text{)]}$$

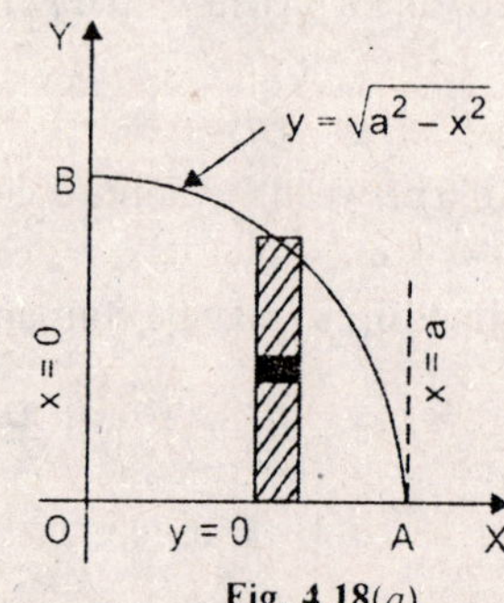

Fig. 4.18(*a*)

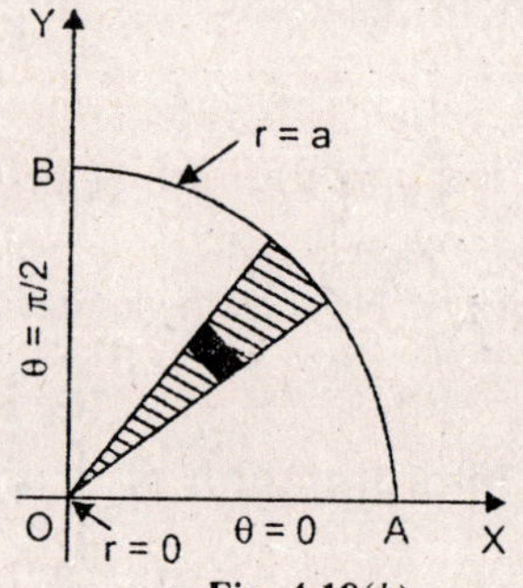

Fig. 4.18(*b*)

Therefore, $\displaystyle\int_{0}^{a}\int_{0}^{\sqrt{a^2-x^2}} y^2\sqrt{(x^2+y^2)}\,dy\,dx = \int_{0}^{\pi/2}\int_{0}^{a} r^2\sin^2\theta\cdot r\cdot r\,dr\,d\theta$

$$= \int_{0}^{\pi/2}\sin^2\theta\left[\frac{r^5}{5}\right]_0^a d\theta = \frac{1}{5}a^5\int_{0}^{\pi/2}\sin^2\theta\,d\theta = \frac{1}{5}a^5\cdot\frac{1}{2}\cdot\frac{\pi}{2} = \frac{\pi a^5}{20}.$$

(*ii*) The limits of integration show that the region of integration lies between the curves

$$y = 0,\; y = \sqrt{(2x - x^2)}, x = 0 \text{ and } x = 2.$$

Thus the region of integration is the semi-circular area shown in the Fig. 4.19(*a*), The circle $y = \sqrt{2x - x^2}$, that is, $x^2 + y^2 = 2x$, transforms into $r^2 = 2r\cos\theta$, or $r = 2\cos\theta$.

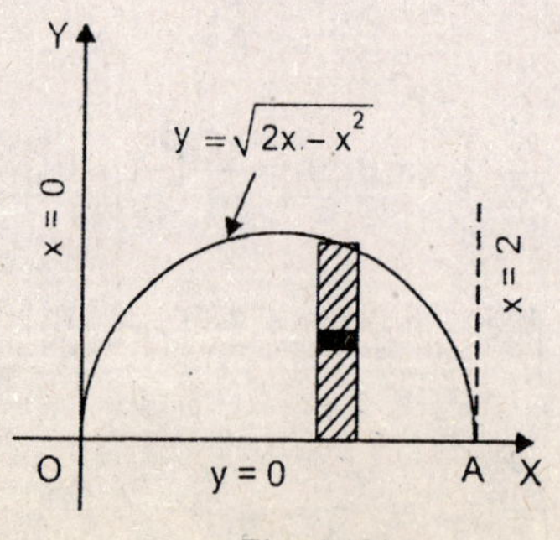

Fig. 4.19(*a*)

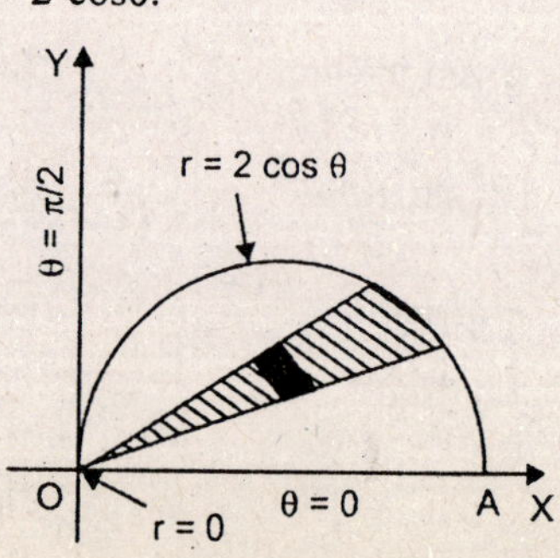

Fig. 4.19(*b*)

Hence, in polar co-ordinates, the region of integration is bounded by the curves

$$r = 0,\ r = 2\cos\theta,\ \ \theta = 0,\ \theta = \pi/2.$$

Therefore,

$$\int_0^2 \int_0^{\sqrt{(2x-x^2)}} \frac{x\,dy\,dx}{\sqrt{(x^2+y^2)}} = \int_0^{\pi/2}\int_0^{2\cos\theta} \frac{r\cos\theta}{r}\cdot r\,dr\,d\theta = \int_0^{\pi/2}\int_0^{2\cos\theta} r\cos\theta\,dr\,d\theta = \int_0^{\pi/2} \cos\theta\left[\frac{r^2}{2}\right]_0^{2\cos\theta} d\theta$$

$$= \frac{1}{2}\int_0^{\pi/2}(4\cos^2\theta)\cos\theta\,d\theta = 2\int_0^{\pi/2}\cos^3\theta\,d\theta = 2\cdot\frac{2}{3} = \frac{4}{3}.$$

EXAMPLE 4.15 Change into polar co-ordinates and evaluate.

(*i*) $\int_0^a\int_y^a \frac{x\,dx\,dy}{x^2+y^2}$, (*ii*) $\int_0^\infty\int_0^\infty e^{-(x^2+y^2)}\,dy\,dx$. Hence show that $\int_0^\infty e^{-x^2}\,dx = \frac{\sqrt{\pi}}{2}$. *(AMIETE, June, 2010; MDU 2005)*

Solution (*i*) The given limits of integration show that the region of integration lies between the curves

$$x = y,\ x = a,\ y = 0, \text{ and } y = a.$$

Thus the region of integration is the triangular area shown in the diagram 4.20(*a*).

The given area can be covered by radial strips as in the Fig. 4.20(*b*). One end of these strips is at $r = 0$ and the other end at $x = a$ that is, $r\cos\theta = a$ or $r = a\sec\theta$.

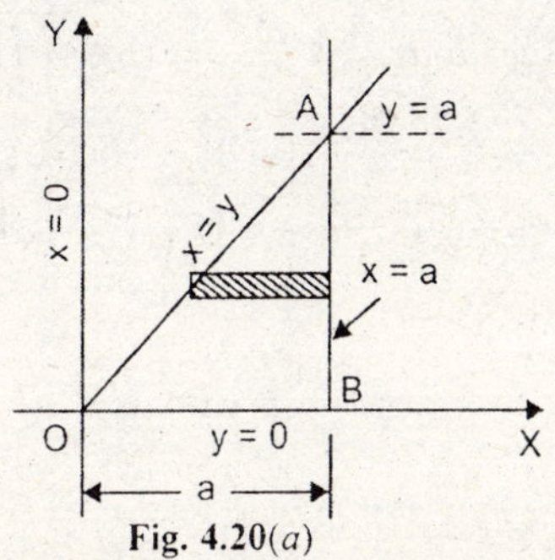

Fig. 4.20(*a*)

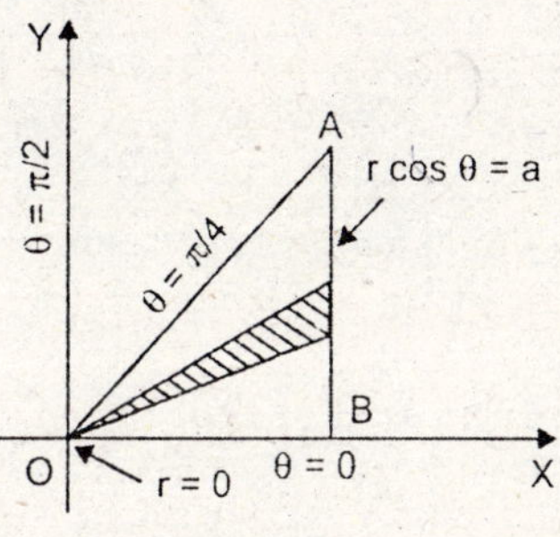

Fig. 4.20(*b*)

The strips start from $\theta = 0$ and end on $\theta = \pi/4$. Hence

$$\int_0^a\int_y^a \frac{x\,dx\,dy}{x^2+y^2} = \int_0^{\pi/4}\int_0^{a/\cos\theta}\frac{r\cos\theta}{r^2}r\,dr\,d\theta = \int_0^{\pi/4}\int_0^{a\sec\theta}\cos\theta\,dr\,d\theta$$

$$= \int_0^{\pi/4}\cos\theta[r]_0^{a\sec\theta}d\theta = \int_0^{\pi/4}\cos\theta\, a\sec\theta\,d\theta = \int_0^{\pi/4} a\,d\theta = a[\theta]_0^{\pi/4} = \frac{\pi a}{4}.$$

(*ii*) From the limits of integration we find that the first integration is along a vertical strip extending from $y = 0$ to $y = \infty$. The strip slides from $x = 0$ and goes to ∞. Thus the region of integration is the whole of first quadrant.

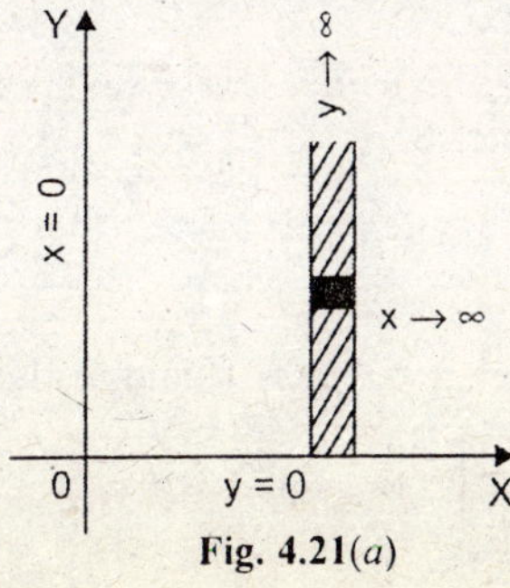

Fig. 4.21(*a*)

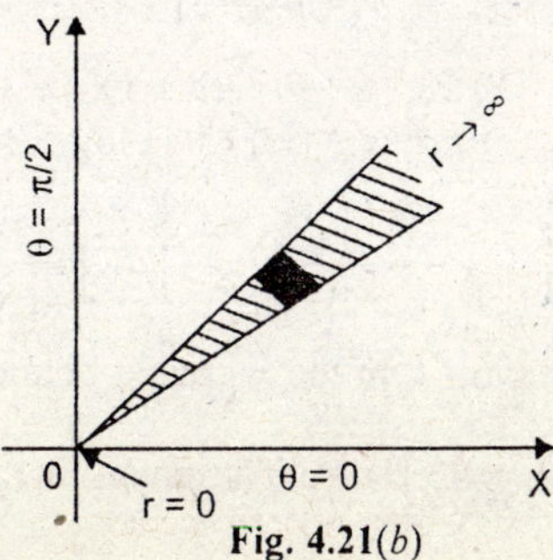

Fig. 4.21(*b*)

This region can be covered by radial strips extending from $r = 0$ to $r = \infty$. The strips start from $\theta = 0$ and go upto $\theta = \pi/2$.

Hence $$\int_0^\infty\int_0^\infty e^{-(x^2+y^2)}dy\,dx = \int_0^{\pi/2}\int_0^\infty e^{-r^2}\cdot r\,dr\,d\theta = -\frac{1}{2}\int_0^{\pi/2}\int_0^\infty (-2r)e^{-r^2}\,dr\,d\theta$$

$$= -\frac{1}{2}\int_0^{\pi/2}\left[e^{-r^2}\right]_0^\infty d\theta = -\frac{1}{2}\int_0^{\pi/2}(0-1)d\theta = \frac{1}{2}\int_0^{\pi/2}1d\theta = \frac{\pi}{4}.$$

The above result may be written as $\left(\int_0^\infty e^{-x^2}dx\right)\cdot\left(\int_0^\infty e^{-y^2}dy\right) = \frac{\pi}{4}$.

or $$\left(\int_0^\infty e^{-x^2}dx\right)\left(\int_0^\infty e^{-x^2}dx\right) = \frac{\pi}{4} \text{ or } \left(\int_0^\infty e^{-x^2}dx\right)^2 = \frac{\pi}{4} \Rightarrow \int_0^\infty e^{-x^2}dx = \frac{\sqrt{\pi}}{2}.$$

EXERCISE 4.1

1. Evaluate the following double integrals.

(*i*) $\int_0^2\int_1^2 (x^2+y^2)\,dx\,dy$, (*ii*) $\int_{-1}^2\int_{-3}^3 (y^2-3xy)\,dx\,dy$, (*iii*) $\int_1^2\int_0^2 e^{x+y}\,dx\,dy$,

(*iv*) $\iint_R f(x, y)\,dx\,dy$ for $f(x, y) = 1-6x^2y$ and $R: 0 \le x \le 2, -1 \le y \le 1$. (*v*) $\int_0^1\int_x^{2x}(x^2+y^2)\,dy\,dx$,

(*vi*) $\int_0^{\pi/2}\int_0^\pi \cos(x+y)\,dx\,dy$, (*AMIE, W-2009*) (*vii*) $\int_0^1\int_x^{\sqrt{x}}(x^2+y^2)\,dy\,dx$ (*VTU, 2000*),

(*viii*) $\int_0^1\int_{x^2}^{2-x} xy\,dx\,dy$, (*AMIETE, June 2010, W-2009*) (*ix*) $\int_0^1\int_0^1 \frac{dx\,dy}{\sqrt{(1-x^2)(1-y^2)}}$,

(*x*) $\int_0^a\int_0^{\sqrt{a^2-x^2}}\sqrt{a^2-x^2-y^2}\,dy\,dx$,

[Hint: (*x*) The given integral may be re-written as $\int_0^a dx\int_0^{\sqrt{a^2-x^2}}\sqrt{(a^2-x^2)-y^2}\,dy$]

(*xi*) $\int_0^2 dx\int_{\sqrt{2x}}^2 \frac{y}{\sqrt{x^2+y^2+1}}dy$, (*xii*) $\int_0^{2a}\int_0^{\sqrt{2ax-x^2}} x^2\,dy\,dx$.

Ans. (*i*) 22/3, (*ii*) 18, (*iii*) $e(e+1)(e-1)^2$, (*iv*) 4, (*v*) 5/6, (*vi*) –2, (*vii*) 3/35, (*viii*) 3/8, (*ix*) $\pi^2/4$, (*x*) $\pi a^3/6$, (*xi*) (5/4) log 5 – 1, (*xii*) $5\pi a^4/8$.

2. Show that $\int_0^1 dx\int_0^1 \frac{x-y}{(x+y)^3}dy \ne \int_0^1 dy\int_0^1 \frac{x-y}{(x+y)^3}dx$. (*AMIE, W-2009*)

3. Sketch the region R in the xy plane bounded by $y = x^2$, $x = 2$, $y = 1$, and evaluate the double integral $\iint_R (x^2+y^2)\,dx\,dy$. **[Hint.** The given integral $= \int_{y=1}^4\int_{x=\sqrt{y}}^2 (x^2+y^2)\,dx\,dy$.] **Ans.** 1006/105.

4. Evaluate $\iint_R xy\,dy\,dx$ over the region R enclosed between $y = x/2$, $y = \sqrt{x}$, $x = 2$ and $x = 4$. **Ans.** 11/6.

5. Evaluate the double integral $\iint_D (4 - x^2 - y^2)\,dx\,dy$, if the domain D is bounded by the straight lines $x = 0$, $x = 1$, $y = 0$ and $y = 3/2$. **Ans.** 35/8.

6. Calculate $\iint_R \frac{\sin x}{x}\,dy\,dx$, where R is the triangle in the xy-plane bounded by the x-axis, the line $y = x$ and the line $x = 1$. **Ans.** 0.46.

7. Evaluate the double integral of the function $f(x, y) = 1 + x + y$ over the region bounded by the lines $y = -x$, $x = \sqrt{y}$, $y = 2$ and $y = 0$. **Ans.** $(44\sqrt{2} + 65)/15$.

8. Evaluate $\iint e^{2x+3y}\,dx\,dy$ over the triangle bounded by $x = 0$, $y = 0$ and $x + y = 1$. **Ans.** $(e - 1)^2(2e + 1)/6$.

9. Find the value of $\iint_R dy\,dx$ and $\iint_R dx\,dy$ where R is the region in the first quadrant bounded by the semi-cubical parabola $y^2 = x^3$ and the line $y = x$. **Ans.** 1/10, 1/10.

10. Evaluate $\iint_A xy\,dx\,dy$, where A is the region bounded by x-axis, ordinate $x = 2a$ and the curve $x^2 = 4ay$.

(*AMIE, S-2006, MDU 2000*) **Ans.** $a^4/3$.

11. Evaluate $\iint xy\,dy\,dx$, over the region in the positive quadrant for which $x + y \le 1$. **Ans.** 1/24

12. Find the value of $\iint (x^2 + y^2)\,dx\,dy$ over the region bounded by the curves $y = 4x$, $x + y = 3$, $y = 0$ and $y = 2$. **Ans.** 9.646.

13. (*a*) Evaluate $\iint \frac{xy}{\sqrt{1 - y^2}}\,dy\,dx$, over the positive quadrant of the circle $x^2 + y^2 = 1$.

(*b*) Evaluate $\iint xy(x + y)\,dx\,dy$ over the area between $y = x^2$ and $y = x$. (*AMIETE, Dec 2010, June 2003*)

Ans. (*a*) 1/6, (*b*) 3/56.

14. Evaluate $\iint_R (2x - y^2)\,dx\,dy$ over the triangular region R enclosed between the lines $y = -x + 1$, $y = x + 1$ and $y = 3$. **Ans.** 68/3.

15. Evaluate the double integral $\iint_R xy\,dy\,dx$ where R is the region bounded by the x-axis, the line $y = 2x$ and the parabola $y = x^2/4a$. (*AMIETE, S-2009, W-2006*)

[Hint: $\iint_R xy\,dy\,dx = \int_{x=0}^{8a} \int_{y=x^2/4a}^{2x} xy\,dy\,dx = \frac{2048\,a^4}{3}$].

[Alternative: $\iint_R xy\,dx\,dy = \int_0^{16a} \int_{x=y/2}^{\sqrt{4ay}} xy\,dx\,dy = \frac{2048}{3}a^4$].

16. Evaluate $\int\int_R (x^2 + y^2)\,dy\,dx$, where R is the region bounded by the x-axis, the line $y = 2x$ and the parabola $x^2 = 4y$. (*AMIE, S-2004*) **Ans.** 165888/105.

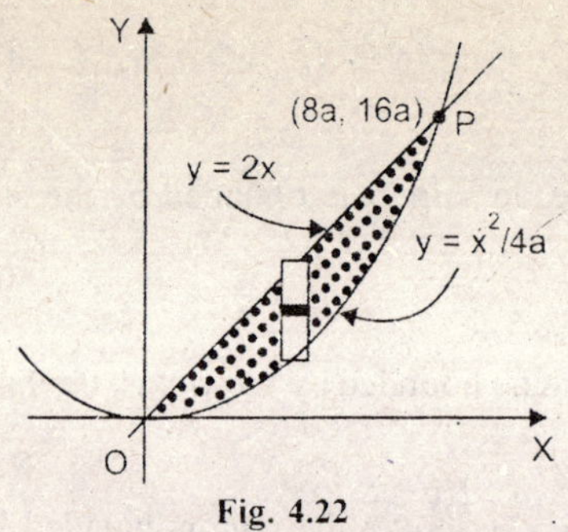

Fig. 4.22

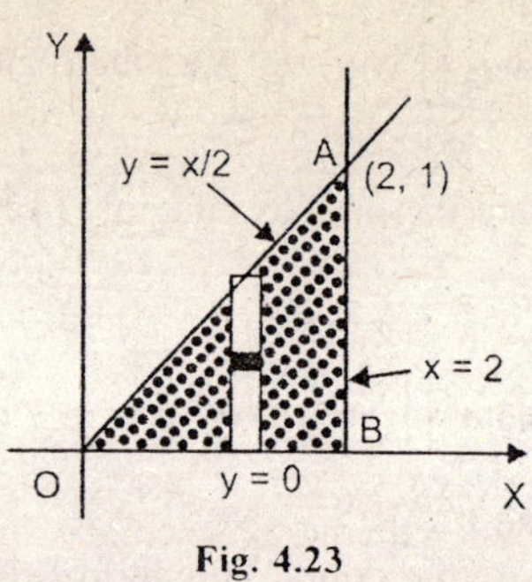

Fig. 4.23

17. Evaluate the double integral $\iint_R e^{x^2}\,dy\,dx$, where the region R is given by $R : 2y \le x \le 2$ and $0 \le y \le 1$.

[Hint: $\iint_R e^{x^2}\,dy\,dx = \int_{x=0}^{2}\int_{y=0}^{x/2} e^{x^2}\,dy\,dx = \frac{1}{4}(e^4 - 1)$.] (*AMIE, W 2010; AMIETE, Dec., 2009; PTU, 2005*)

18. Sketch the region of integration for the following integrals and write an equivalent double integral with the order of integration reversed.

(*i*) $\int_0^4\int_x^{2\sqrt{x}} f(x, y)\,dy\,dx$, (*ii*) $\int_0^{2a}\int_{y^2/4a}^{3a-y} f(x, y)\,dx\,dy$, (*iii*) $\int_0^2\int_{x^2/4}^{3-x} \phi(x, y)\,dy\,dx$,

(*iv*) $\int_0^2\int_{x^2}^{2x} (4x+2)\,dy\,dx$, (*v*) $\int_{-1}^{1} dx \int_0^{\sqrt{1-x^2}} f(x, y)\,dy$, (*vi*) $\int_0^5\int_{2-x}^{2+x} f(x, y)\,dy\,dx$

(*vii*) $\int_0^a\int_{\sqrt{a^2-y^2}}^{y+a} f(x, y)\,dx\,dy$, (*viii*) $\int_{x=-a}^{a}\int_{y=0}^{\sqrt{a^2-x^2}} (a-x)^2\,dy\,dx$.

Ans. (*i*) $\int_0^4\int_{y^2/4}^{y} f(x, y)\,dx\,dy$, (*ii*) $\int_0^a\int_0^{2\sqrt{ax}} f(x, y)\,dy\,dx + \int_0^{3a}\int_0^{3a-x} f(x, y)\,dy\,dx$,

(*iii*) $\int_0^1\int_0^{2\sqrt{y}} \phi(x, y)\,dx\,dy + \int_1^3\int_0^{3-y} \phi(x, y)\,dx\,dy$, (*iv*) $\int_0^4\int_{y/2}^{\sqrt{y}} (4x+2)\,dx\,dy$,

(*v*) $\int_0^1 dy \int_{-\sqrt{1-y^2}}^{\sqrt{1-y^2}} f(x, y)\,dx$, (*vi*) $\int_{y=-3}^{2}\int_{x=2-y}^{5} dx\,dy + \int_{y=2}^{7}\int_{x=y-2}^{5} dx\,dy$

(*vii*) $\int_0^a\int_{\sqrt{a^2-x^2}}^{a} f(x, y)\,dy\,dx + \int_a^{2a}\int_{x-a}^{a} f(x, y)\,dy\,dx$ (*viii*) $\int_{y=0}^{a}\int_{x=-\sqrt{a^2-y^2}}^{\sqrt{a^2-y^2}} (a-x)^2\,dx\,dy$

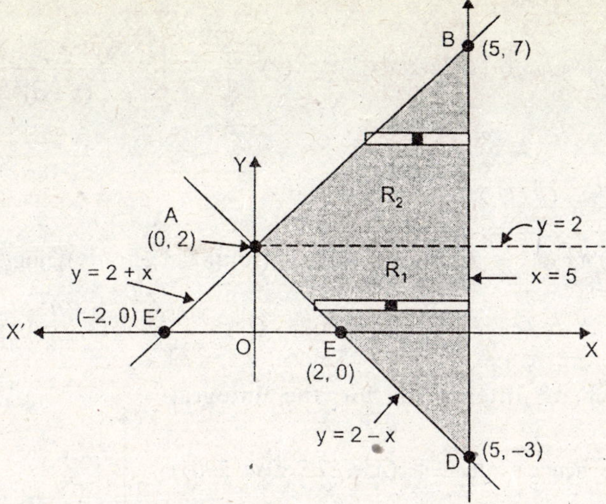

Fig. 4.24: (Problem 18(*vi*))

19. Evaluate the following integrals by changing the order of integration.

(*i*) $\int_0^1 \int_{x^2}^{2-x} xy\,dy\,dx$, (*SVTU, 2007; UPTU, 2005; Cochin, 2005*), (*ii*) $\int_0^a \int_{\sqrt{ax}}^a \frac{y^2\,dy\,dx}{\sqrt{y^4 - a^2x^2}}$, (*SVTU, 2006S*)

(*iii*) $\int_0^1 \int_{3y}^3 e^{x^2}\,dx\,dy$, (*iv*) $\int_0^1 \int_{e^x}^e \frac{dy\,dx}{\log y}$ (*v*) $\int_0^1 \int_x^{2-x} \frac{x}{y}\,dy\,dx$, (*AMIETE, June 2004*)

[**Hint.** (*v*) The given integral $= \int_0^1 \int_0^y \frac{x}{y}\,dx\,dy + \int_1^2 \int_0^{2-y} \frac{x}{y}\,dx\,dy = \frac{1}{4} + 2\log_e 2 - \frac{5}{4} = 2\log_e 2 - 1$]

(*vi*) $\int_0^\infty \int_0^x x e^{-x^2/y}\,dy\,dx$ (*NIT Kurukshetra, 2007; SVTU, 2006; UPTU, 2005; MDU 2005; VTU-2004*)

(*vii*) $\int_0^1 \int_y^1 x^2 e^{xy}\,dx\,dy$, (*AMIETE, June 2006*) (*viii*) $\int_0^1 \int_{y^2}^y \frac{y\,dx\,dy}{(1-x)(x-y^2)^{1/2}}$ (*UPTU 2001*)

[**Hint:** (*viii*) The given integral $= \int_0^1 \int_x^{\sqrt{x}} \frac{y\,dy\,dx}{(1-x)(x-y^2)^{1/2}}$

$$= \int_0^1 \left[\frac{-(x-y^2)^{1/2}}{1-x}\right]_x^{\sqrt{x}} dx = \int_0^1 \frac{(x-x^2)^{1/2}}{1-x}\,dx = \int_0^1 \left(\frac{x}{1-x}\right)^{1/2} dx$$

$$= \int_0^{\pi/2} \left(\frac{\sin^2\theta}{1-\sin^2\theta}\right)^{1/2} \cdot 2\sin\theta\cos\theta\,d\theta \text{ (on putting } x = \sin^2\theta) = \int_0^{\pi/2} 2\sin^2\theta\,d\theta = \frac{\pi}{2}].$$

(*ix*) $\int_0^{2a}\int_{x^2/4a}^{3a-x}(x^2+y^2)dy\,dx$ (*DCE-2004*), (*x*) $\int_0^{a}\int_0^{a-\sqrt{a^2-y^2}}\frac{xy\log(x+a)}{(x-a)^2}dxdy$ (*DCE-2004*)

(*xi*) $\int_0^{1}\int_0^{\sqrt{1-x^2}}y^2dydx$ (*PTU 2004*)

Ans. (*i*) 3/8, (*ii*) $\pi a^2/6$, (*iii*) $(e^9-1)/6$, (*iv*) $(e-1)$, (*v*) $2\log_e 2-1$, (*vi*) 1/2,

(*vii*) $\int_{x=0}^{1}\int_{y=0}^{x}x^2e^{xy}\,dydx;(e-2)/2$, (*viii*) $\pi/2$, (*ix*) $\frac{314a^4}{35}$, (*x*) $\frac{a^2}{8}(1+2\log a)$, (*xi*) $\frac{\pi}{16}$.

20. Change the order of integration for the integral $\int_0^{2a}\int_{\sqrt{2ax-x^2}}^{\sqrt{2ax}}dydx$, and hence evaluate it. (*AMIETE, Dec. 2003*)

Ans. $(16-3\pi)a^2/6$.

[**Hint:** The region of integration is bounded by the circle $x^2+y^2=2ax$, the parabola $y^2=2ax$, $x=0$ and $x=2a$. The region of integration is divided into three parts OAB, BCD and AEC. The equation of the circle is $x^2-2ax+y^2=0$. Therefore, $x=a\pm\sqrt{(a^2-y^2)}$. Out of the two values of x the lessor one is $a-\sqrt{a^2-y^2}$ and the greater one is $a+\sqrt{a^2-y^2}$.

Hence, the given integral

$$=\iint_{R_1}dx\,dy+\iint_{R_2}dx\,dy+\iint_{R_3}dx\,dy$$

$$=\int_{y=0}^{a}\int_{x=y^2/2a}^{a-\sqrt{a^2-y^2}}dx\,dy+\int_{y=0}^{a}\int_{x=a+\sqrt{a^2-y^2}}^{2a}dx\,dy+\int_{y=a}^{2a}\int_{x=y^2/2a}^{2a}dxdy].$$

Fig. 4.25

21. Change the order of integration in the double integral $\int_0^a\int_0^x\frac{\phi'(y)dydx}{\sqrt{(a-x)(x-y)}}$ and hence find its value.

(*AMIE, S-2007*)

[**Hint:** The region of integration is bounded by $y=0$, $y=x$ and $x=a$. Thus OAB is the region of integration. Consider an elementary strip parallel to x-axis. The double integral after changing the order of integration is

$$I=\int_{y=0}^{a}\int_{x=y}^{a}\frac{\phi'(y)dxdy}{\sqrt{(a-x)(x-y)}}.$$ To find its value, proceed as under:

Put $x=a\sin^2\theta+y\cos^2\theta$, $dx=2(a-y)\sin\theta\cos\theta\,d\theta$.

$a-x=(a-y)\cos^2\theta$, $x-y=(a-y)\sin^2\theta$.

When $x=y$, $\sin^2\theta=0$, that is $\theta=0$ and when $x=a$, $\theta=\pi/2$.

Thus the limits of x are from 0 to $\pi/2$.

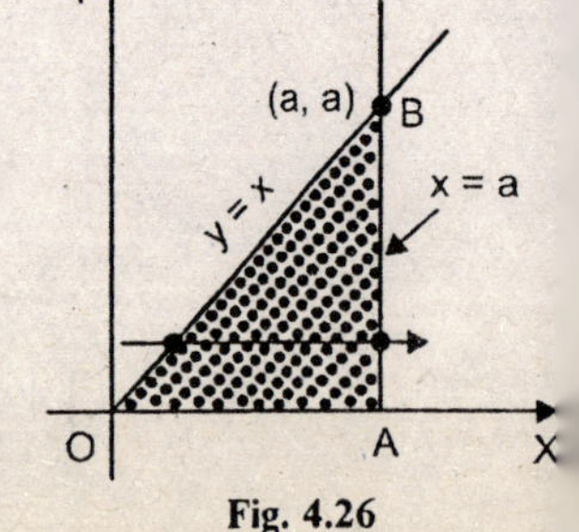

Fig. 4.26

$$\therefore \int_y^a \frac{dx}{\sqrt{(a-x)(x-y)}} = \int_0^{\pi/2} \frac{2(a-y)\sin\theta\cos\theta d\theta}{(a-y)\sin\theta\cos\theta} = 2\int_0^{\pi/2} d\theta = \pi.$$

Hence $I = \int_0^a \pi\phi'(y)dy = \pi[\phi(y)]_0^a = \pi[\phi(a)-\phi(0)].$

22. Evaluate the following integrals

(*i*) $\int_0^{2\pi}\int_0^{1-\cos\theta} r^3\cos^2\theta\, dr\, d\theta$, (*ii*) $\int_0^{\pi/4}\int_0^{\tan\theta\sec\theta} r^3\cos^2\theta\, dr\, d\theta$ (*iii*) $\int_0^{\pi}\int_0^{a(1+\cos\theta)} r^2\sin\theta\, dr\, d\theta$,

(*iv*) $\int_0^{\pi/2}\int_0^{2a\cos\theta} \frac{r^3}{a} dr\, d\theta$, (*v*) $\int_0^{\pi/2}\int_{a\cos\theta}^{a} r^4 dr\, d\theta$.

Ans. (*i*) $49\pi/32$, (*ii*) $1/20$, (*iii*) $4a^3/3$, (*iv*) $3\pi a^3/4$, (*v*) $\left(\pi - \frac{16}{15}\right)\frac{a^5}{10}$.

23. Compute the following integrals by changing to polar co-ordinates:

(*i*) $\int_0^1\int_0^x \sqrt{x^2+y^2}\, dy\, dx$, (*ii*) $\int_0^a\int_y^a \frac{x^2 dx\, dy}{\sqrt{x^2+y^2}}$, (*iii*) $\int_0^a\int_0^{\sqrt{a^2-x^2}} (x^2+y^2)dx\, dy$,

(*iv*) $\int_0^{2a}\int_0^{\sqrt{2ax-x^2}} dy\, dx$, (*v*) $\int_0^1\int_x^{\sqrt{2x-x^2}} (x^2+y^2)dy\, dx$,

[**Hint:** (*v*) In polar co-ordinates, the region of integration is bounded by $r = 0$, $r = 2\cos\theta$, $\theta = \pi/4$, $\theta = \pi/2$. Refer Figure 4.27 (*a*) and (*b*).]

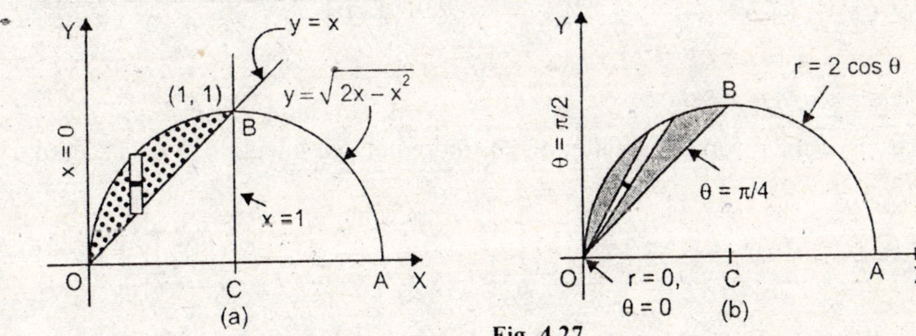

Fig. 4.27

(*vi*) $\int_0^{2a}\int_0^{\sqrt{2ax-x^2}} (x^2+y^2)dy\, dx$, [**Hint:** the given integral $= \int_0^{\pi/2}\int_0^{2a\cos\theta} r^2 \cdot r\, dr\, d\theta$] (*AMIE, S-2003*)

Ans. (*i*) $\frac{1}{6}[\sqrt{2}+\log(1+\sqrt{2})]$, (*ii*) $\frac{1}{3}a^3\log(\sqrt{2}+1)$, (*iii*) $\frac{\pi a^4}{8}$, (*iv*) $\frac{\pi a^2}{2}$, (*v*) $\left(\frac{3}{8}\pi - 1\right)$, (*vi*) $\frac{3\pi a^4}{4}$.

24. Let $\int_0^1\int_1^2 \frac{1}{x^2+y^2} dx\, dy + \int_1^2\int_y^2 \frac{1}{x^2+y^2} dx\, dy = \iint_R \frac{1}{x^2+y^2} dy\, dx$. Recognise the region R of integration on the r.h.s. and then evaluate the integral on the right in the order indicated. (*AMIETE, Dec. 2004*)

Ans. $\iint_R \frac{1}{x^2+y^2} dy\, dx = \int_{x=1}^{2}\int_{y=0}^{x} \frac{1}{x^2+y^2} dy\, dx = \frac{\pi}{4}\log_e 2$

25. Express $\int_0^{a/\sqrt{2}} \int_0^{x} x\,dy\,dx + \int_{a/\sqrt{2}}^{a} \int_0^{\sqrt{a^2-x^2}} x\,dy\,dx$, as a single integral and then evaluate it.

Ans. $\int_0^{\pi/4} \int_0^{a} r^2 \cos\theta\, dr\, d\theta$. (*AMIETE, Dec. 2005*)

26. Express the sum of the following integrals as a single integral and hence evaluate by changing the order of integration:

$$\int_0^{a/\sqrt{2}} \int_0^{x} \cos\left[k(x^2+y^2)\right] dy\,dx + \int_{a/\sqrt{2}}^{a} \int_0^{\sqrt{a^2-x^2}} \cos\left[k(x^2+y^2)\right] dy\,dx.$$ (*AMIETE, Dec. 2002*)

[**Hint:** The given integral $= \int_{\theta=0}^{\pi/4} d\theta \int_{r=0}^{a} r\cos kr^2\, dr = \frac{\pi}{8k}\sin(ka^2)$].

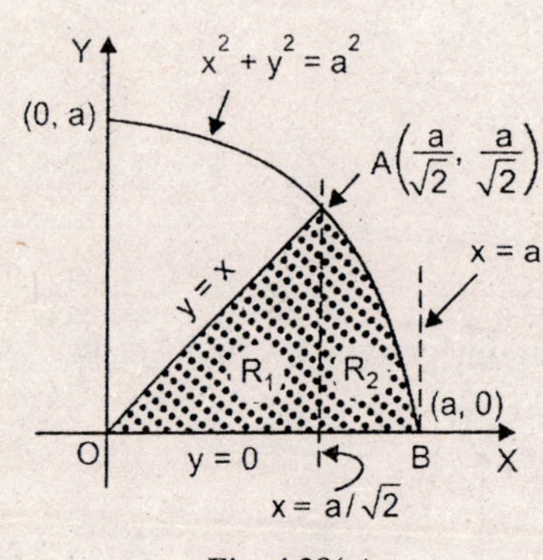

Fig. 4.28(*a*)

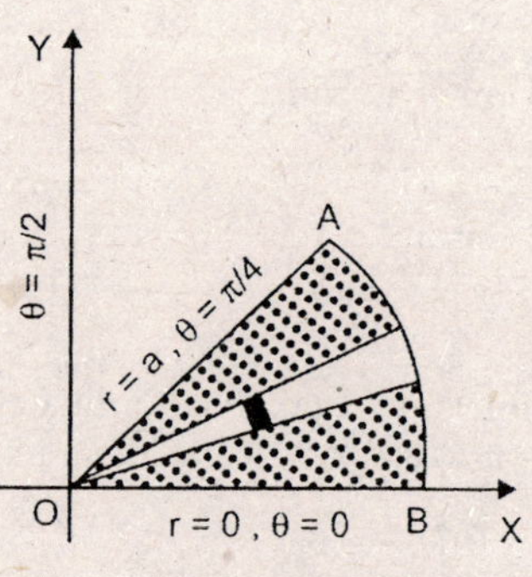

Fig. 4.28(*b*)

27. Transform each of the given integrals to one or more iterated integrals in polar coordinates

(*i*) $\int_0^1 \int_0^x f(x, y)\,dy\,dx$, (*ii*) $\int_0^1 dx \int_0^1 f(x, y)\,dy$, (*iii*) $\int_0^1 \left[\int_0^{x^2} f(x, y)\,dy\right] dx$.

Ans. (*i*) $\int_0^{\pi/4} \left[\int_0^{\sec\theta} f(r\cos\theta, r\sin\theta)\, r\,dr\right] d\theta$,

(*ii*) $\int_0^{\pi/4} \left[\int_0^{\sec\theta} f(r\cos\theta, r\sin\theta)\, r\,dr\right] d\theta + \int_{\pi/4}^{\pi/2} \left[\int_0^{\operatorname{cosec}\theta} f(r\cos\theta, r\sin\theta)\, r\,dr\right] d\theta$,

(*iii*) $\int_0^{\pi/4} \left[\int_{\tan\theta\sec\theta}^{\sec\theta} f(r\cos\theta, r\sin\theta)\, r\,dr\right] d\theta$.

28. Compute the integral $\int_0^a \int_0^{\sqrt{a^2-x^2}} \sqrt{a^2-x^2-y^2}\, dy\, dx$ by changing to polar coordinates.

[**Hint.** The region of integration lies between the curves
$y=0$, $y=\sqrt{a^2-x^2}$, $x=0$ and $x=a$ (Fig. 4.29).
In polar coordinates, the region is bounded by the curves
$r=0$, $r=a$, $\theta=0$ and $\theta=\pi/2$.

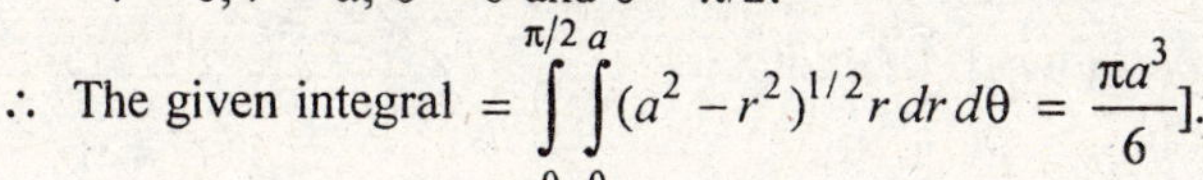

$\therefore$ The given integral $= \int_0^{\pi/2}\int_0^a (a^2-r^2)^{1/2} r\, dr\, d\theta = \frac{\pi a^3}{6}$].

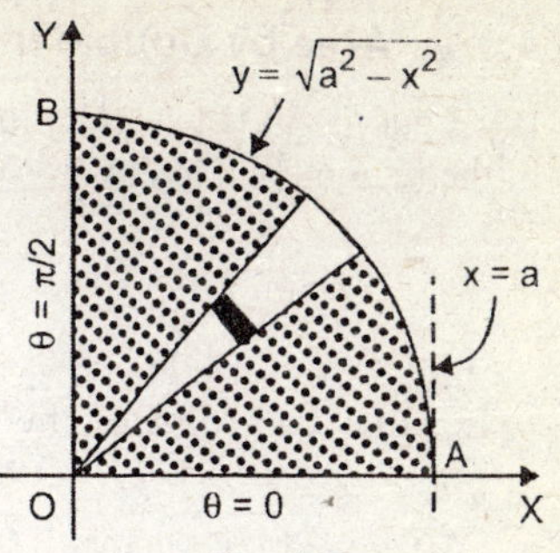

Fig. 4.29

29. Change the cartesian integral into an equivalent polar integral and then evaluate $\int_0^2 \int_0^{\sqrt{1-(x-1)^2}} \frac{x+y}{x^2+y^2} dy dx.$

Ans. $\left(\frac{\pi}{2}+1\right)$

30. Evaluate the integral $\iint \sqrt{4a^2-x^2-y^2}\, dy\, dx$ taken over the upper half of the circle $x^2+y^2-2ax=0$, by changing to polar co-ordinates.

[**Hint.** We have to evaluate $\int_0^{2a}\int_0^{\sqrt{2ax-x^2}} \sqrt{4a^2-x^2-y^2}\, dy dx$

Changing to polar co-ordinates, we have the given integral

$$= \int_{\theta=0}^{\pi/2}\int_{r=0}^{2a\cos\theta} \sqrt{4a^2-r^2}\cdot r\, dr\, d\theta$$

$$= \frac{4a^3}{9}(3\pi-4).$$

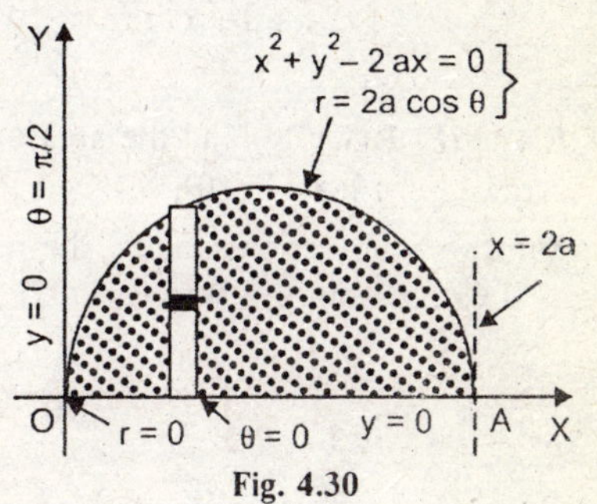

Fig. 4.30

31. Evaluate the integral $\iint_R (a^2-x^2-y^2)\, dy\, dx$, where R is the region $x^2+y^2 \le a^2$. **Ans.** $\pi a^4/2$

32. Evaluate the integral $\iint_R \sqrt{x^2+y^2}\, dx\, dy$ by changing to polar coordinates, where R is the region in the xy-plane bounded by the circles $x^2+y^2=a^2$ and $x^2+y^2=b^2$, $b>a$.

[**Hint:** The given integral $= \int_0^{2\pi}\int_a^b r(r\, dr\, d\theta) = \frac{2\pi}{3}(b^3-a^3)$.]

33. Evaluate the integral $\iint_R e^{-x^2-y^2}\, dx\, dy$, where R is the annulus bounded by $x^2+y^2=1$ and $x^2+y^2=4$ by changing to polar coordinates.

Ans. $\pi(e^3-1)/e^4$.

4.5 APPLICATIONS OF DOUBLE INTEGRALS

Double integrals have large number of applications. We state some of them.

4.5.1 Area by Double Integration

If we put $f(x, y) = 1$ in equation (4.2) of Art 4.1, we find that the area A of the region bounded by $y = f_1(x)$, $y = f_2(x)$, $x = a$ and $x = b$ is given by

$$A = \int_a^b \int_{f_1}^{f_2} dy\,dx.$$

In polar co-ordinates the area bounded by the curves $r = f_1(\theta)$, $r = f_2(\theta)$, the radii vectors $\theta = \theta_1$, $\theta = \theta_2$ is given by $\int_{\theta_1}^{\theta_2} \int_{f_1}^{f_2} r\,dr\,d\theta$.

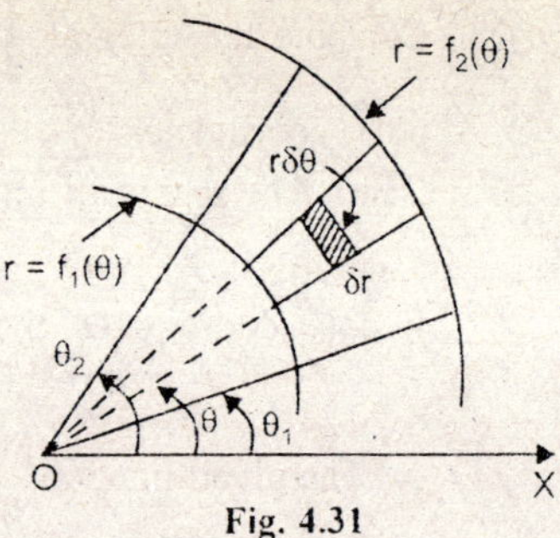

Fig. 4.31

EXAMPLE 4.16 Find the area bounded by the parabola $y = x^2$ and the line $y = 2x + 3$.

Solution Ref. Fig. 4.32, using vertical strips.

$$\text{Required area} = \int_{-1}^{3} \int_{x^2}^{2x+3} dy\,dx$$

$$= \int_{-1}^{3} (2x + 3 - x^2)\,dx = \left[x^2 + 3x - \frac{x^3}{3}\right]_{-1}^{3} = 32/3 \text{ sq. units.}$$

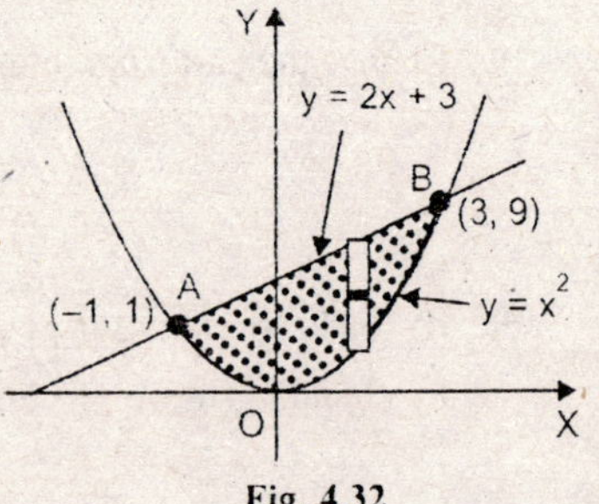

Fig. 4.32

EXAMPLE 4.17 Find the area outside the circle $r = 2$ and inside the cardioid $r = 2(1 + \cos\theta)$.

Solution Due to symmetry, the required area is twice that swept over as θ varies from $\theta = 0$ to $\theta = \pi/2$. Thus, area,

$$= 2\int_0^{\pi/2} \int_2^{2(1+\cos\theta)} r\,dr\,d\theta = 2\int_0^{\pi/2} \left[\frac{1}{2}r^2\right]_2^{2(1+\cos\theta)} d\theta$$

$$= 4\int_0^{\pi/2} (2\cos\theta + \cos^2\theta)\,d\theta = 4\left[2\sin\theta + \frac{\theta}{2} + \frac{\sin 2\theta}{4}\right]_0^{\pi/2}$$

$= (\pi + 8)$ square units.

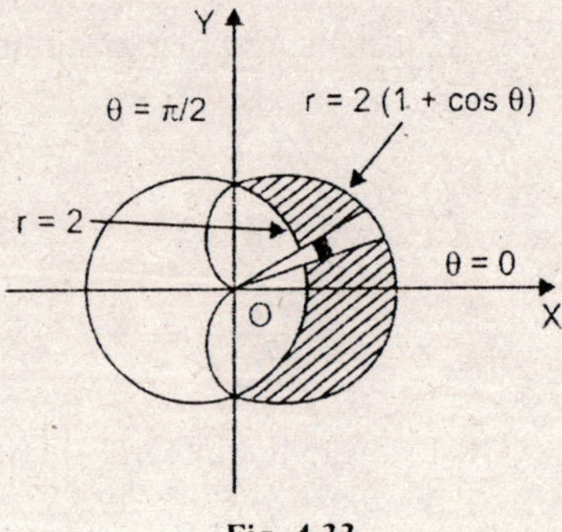

Fig. 4.33

4.5.2 Mass of a Plate

As shown in Art. 4.1, the mass of plate occupying an area A and having a density $\rho = f(x, y)$ where A is bounded by the curves $y = f_1(x)$, $y = f_2(x)$ and the lines $x = a$, $x = b$, is given by

$$M = \int_a^b \int_{f_1(x)}^{f_2(x)} f(x, y)\,dy\,dx.$$

In polar coordinates, the mass is given by $M = \int_{\theta_1}^{\theta_2} \int_{f_1(\theta)}^{f_2(\theta)} F(r, \theta)\,r\,dr\,d\theta$, where the area A is bounded by $r = f_1(\theta)$, $r = f_2(\theta)$, $\theta = \theta_1$ and $\theta = \theta_2$ and density $\rho = F(r, \theta)$.

EXAMPLE 4.18: Find the mass of a square plate of side a if the density varies as the square of the distance from a vertex.

Solution (Ref. Fig. 4.34). Let the vertex from which distances are measured be at the origin.

Then $\rho(x, y) = k(x^2 + y^2)$

and
$$M = \iint_A \rho(x, y)\, dy\, dx = \int_0^a \int_0^a k(x^2 + y^2)\, dy\, dx$$

$$= k\int_0^a \left[x^2 y + \frac{y^3}{3}\right]_0^a dx = k\int_0^a \left(x^2 a + \frac{a^3}{3}\right) dx$$

$$= k\left[\frac{a x^3}{3} + \frac{a^3}{3} x\right]_0^a = \frac{2}{3} k\, a^4 \text{ units.}$$

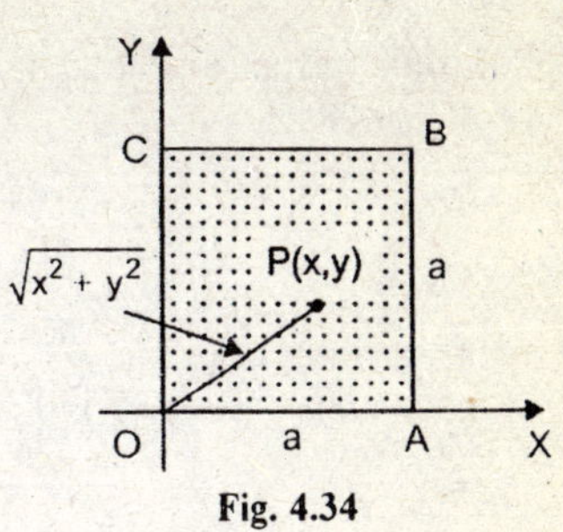

Fig. 4.34

EXAMPLE 4.19 Find the mass of a plate in the form of a cardioid $r = a(1 + \cos\theta)$, if the density of the mass varies as the distance from the pole.

Solution At the point $P(r, \theta)$ the density $\rho = kr$. Since total mass M is twice the mass above the initial line,

$$M = 2\int_0^{\pi} \int_0^{a(1+\cos\theta)} (kr)\cdot r\, dr\, d\theta = 2k\frac{a^3}{3}\int_0^{\pi} (1 + \cos\theta)^3\, d\theta$$

$$= \frac{2}{3} k a^3 \int_0^{\pi} 2^3 \cdot \cos^6 \frac{\theta}{2}\, d\theta$$

$$= \frac{16}{3} k a^3 \int_0^{\pi/2} 2\cos^6 t\, dt \quad \left(\text{where } \frac{\theta}{2} = t\right)$$

$$= \frac{32}{3} k a^3 \cdot \frac{5\cdot 3\cdot 1}{6\cdot 4\cdot 2}\cdot\frac{\pi}{2} = \frac{5}{3}\pi k a^3.$$

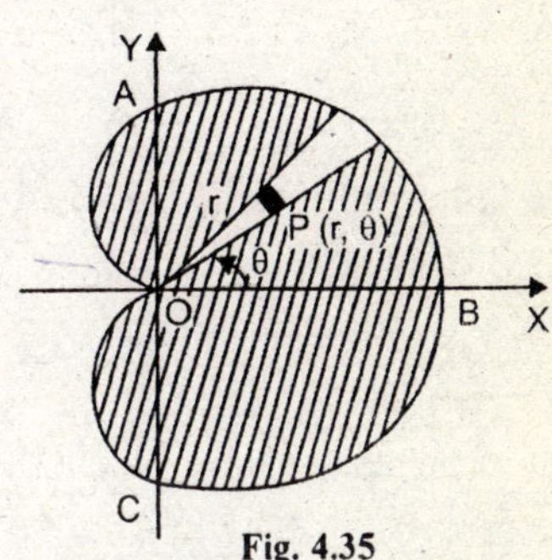

Fig. 4.35

4.5.3 Volume under a Surface (Double Integration)

The volume of a vertical column whose upper end is in the surface $z = f(x, y)$ and whose lower base is in the XOY plane is defined by the double integral $V = \iint_R z\, dy\, dx$ or $\iint_R f(x, y)\, dy\, dx$ the region R being the lower base of the column.

Fig. 4.36

EXAMPLE 4.20 Find by double integration the volume of the sphere

$$x^2 + y^2 + z^2 = a^2.$$

Solution The required volume is eight times the volume in the first octant.

Since the volume in the first octant is bounded from above by the surface $z = \sqrt{a^2 - x^2 - y^2}$ and has its base R as the quadrant of the circle $x^2 + y^2 = a^2$ in the xy-plane. (Refer Fig. 4.36)

$$V = 8\iint_R z\, dy\, dx = 8\int_0^a \int_0^{\sqrt{a^2 - x^2}} \sqrt{(a^2 - x^2) - y^2}\, dy\, dx$$

$$= 8\int_0^a \left[\frac{y}{2}\sqrt{(a^2 - x^2) - y^2} + \frac{(a^2 - x^2)}{2}\sin^{-1}\frac{y}{\sqrt{a^2 - x^2}}\right]_0^{\sqrt{a^2 - x^2}} dx = 8\int_0^a \left[0 + \frac{a^2 - x^2}{2}\cdot\frac{\pi}{2}\right] dx$$

$$= 2\pi\int_0^a (a^2 - x^2)\, dx = 2\pi\left[a^2 x - \frac{x^3}{3}\right]_0^a = 2\pi\left[a^3 - \frac{a^3}{3}\right] = \frac{4\pi a^3}{3} \text{ cubic units.}$$

EXAMPLE 4.21 Find the volume bounded by the cylinder $x^2 + y^2 = 4$, the planes $y + z = 3$ and $z = 0$.

(*Madras 2000S; AMIE, W-2000*)

Solution From Fig. 4.37 it is evident that $z = 3 - y$ is to be integrated over the circle $x^2 + y^2 = 4$ in the xy-plane. Hence

$$V = \int_{-2}^{2} \int_{-\sqrt{4-y^2}}^{\sqrt{4-y^2}} (3 - y)\,dx\,dy$$

$$= 2\int_{-2}^{2} \int_{0}^{\sqrt{4-y^2}} (3 - y)\,dx\,dy = 2\int_{-2}^{2} (3 - y)\sqrt{4 - y^2}\,dy$$

$$= 6\int_{-2}^{2} \sqrt{4 - y^2}\,dy - 2\int_{-2}^{2} y\sqrt{4 - y^2}\,dy = 12\int_{0}^{2} \sqrt{4 - y^2}\,dy$$

(The second term vanishes as the integrand is an odd function)

$$= 12\left[\frac{y\sqrt{4 - y^2}}{2} + \frac{4}{2}\sin^{-1}\frac{y}{2}\right]_0^2 = 12 \cdot 2 \cdot \frac{\pi}{2} = 12\pi \text{ cubic units.}$$

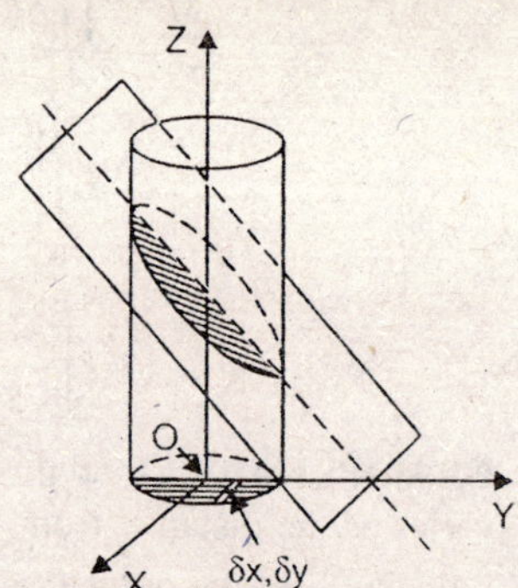

Fig. 4.37

4.5.4 Volume of Solid of Revolution

The volume of the solid formed by the revolution of the area A about the x-axis

$$= \iint_R 2\pi(y)(dx\,dy). \quad ...(4.4)$$

If the area revolves about the y-axis, we have

$$V = \iint_R 2\pi\, x\, dx\, dy. \quad ...(4.5)$$

In case of *polar coordinates* if the area bounded by the curve $r = f(\theta)$ and radii vectors $\theta = \alpha$, $\theta = \beta$ revolves about the initial line $\theta = 0$, the volume of the solid generated

$$= \iint_R 2\pi\,(r \sin\theta)\;(r\,dr\,d\theta) = 2\pi \iint_R r^2 \sin\theta\, dr\, d\theta$$

$$= 2\pi \int_{\theta=\alpha}^{\beta} \sin\theta\, d\theta \int_{r=0}^{f(\theta)} r^2\, dr = \frac{2\pi}{3}\int_{\alpha}^{\beta} r^3 \sin\theta\, d\theta.$$

Similarly, volume of solid generated (for polar curves) about line y-axis, that is $\theta = \pi/2$.

$$= \frac{2\pi}{3}\int_{\alpha}^{\beta} r^3 \cos\theta\, d\theta.$$

EXAMPLE 4.22 Calculate by double integration, the volume generated by the revolution of the cordioids $r = a(1 - \cos\theta)$ about its axis. (*AMIETE, June 2010; KUK, 2009*)

Solution Consider the upper half of the cordioid.

$$\text{The required volume} = \int_{\theta=0}^{\pi} \int_{r=0}^{a(1-\cos\theta)} 2\,\pi r^2 \sin\theta\, dr\, d\theta$$

$$= 2\pi\int_0^{\pi}\left[\frac{r^3}{3}\right]_0^{a(1-\cos\theta)} \sin\theta\, d\theta = \frac{2\pi}{3}a^3\int_0^{\pi}(1 - \cos\theta)^3 \sin\theta\, d\theta$$

$$= \frac{2\pi a^3}{3}\left[\frac{(1 - \cos\theta)^4}{4}\right]_0^{\pi} = \frac{2\pi a^3}{3}\left[\frac{2^4}{4}\right] = \frac{8\pi a^3}{3} \text{ cubic units.}$$

4.5.5 Mass and Moment Formulas for thin Plates (Lamina) Covering Regions in the xy-plane

Center of mass: We give below the formula to find the mass of a lamina from its density function.

Let $f(x, y) = \rho(x, y)$ be a density function (mass per unit area) occupies a region R in the xy-plane, then

Mass: $M = \iint_R f(x, y)\, dy\, dx = \iint_R \rho(x, y)\, dy\, dx.$

First moments of the lamina about the x-axis $= M_x = \iint_R y\rho(x, y)\, dy\, dx$

First moments of the lamina about the y-axis $= M_y = \iint_R x\rho(x, y)\, dy\, dx.$

Center of Mass:

$$\bar{x} = M_y / M = \frac{1}{\text{mass of } R} \iint_R x\rho(x, y)\, dy\, dx$$

$$= \iint_R x\rho(x, y)\, dy\, dx \Big/ \iint_R \rho(x, y)\, dy\, dx \qquad \text{...(4.6)}$$

$$\bar{y} = M_x / M = \frac{1}{\text{mass of } R} \iint_R y\rho(x, y)\, dy\, dx$$

$$= \iint_R y\rho(x, y)\, dy\, dx \Big/ \iint_R \rho(x, y)\, dy\, dx \qquad \text{...(4.7)}$$

Equations (4.6) and (4.7) gives the coordinates of the *centre of gravity* $(\bar{x}, \bar{y})$ of the mass M in R.

Centroids of Geometric Figures

For a homogeneous lamina, the density function $\rho(x, y)$ is constant and it cancels out the numerator and denominator of the formulas for $\bar{x}$ and $\bar{y}$. Thus when ρ is constant, the location of the center of mass becomes a feature of the object's shape and not of the material of which it in made. In such cases, the center of mass is called the **centroid** of the shape. To determine a centroid, we set $\rho(x, y) = 1$ and proceed to find $\bar{x}$ and $\bar{y}$ as before.

4.5.6 Moment of Inertia (Second Moments)

About the x-axis : $I_x = \iint_R y^2 \rho(x, y)\, dy\, dx$

About the y-axis : $I_y = \iint_R x^2 \rho(x, y)\, dy\, dx$

About the origin (polar moment) : $I_o = \iint_R (x^2 + y^2)\, \rho(x, y)\, dy\, dx = I_x + I_y.$

The moment I_o in sometimes called I_z, the moment of inertia about the z-axis.
About the lines $x = a$ and $y = b$:

$$I_x = \iint_R (y - b)^2\, \rho(x, y)\, dy\, dx, \quad I_y = \iint_R (x - a)^2\, \rho(x, y)\, dy\, dx.$$

The **radius of gyration** k_x is defined by the equation $I_x = Mk_x^2$. It tells how for from the x-axis the entire mass of the plate might be concentrated to give the same I_x. The radii k_y and k_o are defined in a similar way.

Radii of gyration

About the x-axis : $k_x = \sqrt{I_x / M}$

About the y-axis : $k_y = \sqrt{I_y / M}$

About the Origin : $k_o = \sqrt{I_o / M}$.

EXAMPLE 4.23 Find the center of mass of a thin triangular plate bounded by the y-axis and the lines $y = x$ and $y = 2 - x$. The plate's density at the point (x, y) is $\rho(x, y) = 6x + 3y + 3$.

Solution Refer Fig. 4.38, the triangular region OAB covered by the plate in Example 4.23.

The plate's mass is

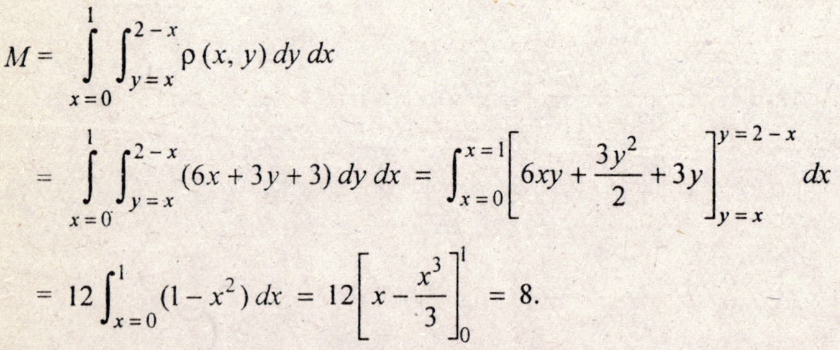

$$M = \int_{x=0}^{1} \int_{y=x}^{2-x} \rho(x, y)\, dy\, dx$$

$$= \int_{x=0}^{1} \int_{y=x}^{2-x} (6x + 3y + 3)\, dy\, dx = \int_{x=0}^{x=1} \left[6xy + \frac{3y^2}{2} + 3y \right]_{y=x}^{y=2-x} dx$$

$$= 12 \int_{x=0}^{1} (1 - x^2)\, dx = 12 \left[x - \frac{x^3}{3} \right]_0^1 = 8.$$

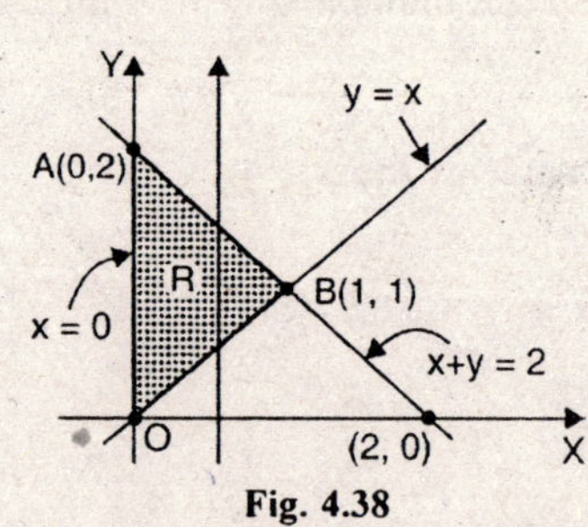

Fig. 4.38

The first moment about the x-axis is

$$M_x = \int_0^1 \int_x^{2-x} y\,\rho(x, y)\, dy\, dx = \int_0^1 \int_x^{2-x} (6xy + 3y^2 + 3y)\, dy\, dx = \int_0^1 \left[3xy^2 + y^3 + \frac{3y^2}{2} \right]_{y=x}^{y=2-x} dx$$

$$= \int_0^1 (14 - 6x - 6x^2 - 2x^3)\, dx = \left[14x - 3x^2 - 2x^3 - \frac{x^4}{2} \right]_0^1 = \frac{17}{2}.$$

The first moment about the y-axis is

$$M_y = \int_0^1 \int_x^{2-x} x\,\rho(x, y)\, dy\, dx = \int_0^1 \int_x^{2-x} (6x^2 + 3xy + 3x)\, dy\, dx$$

$$= \int_0^1 \left[6x^2 y + \frac{3xy^2}{2} + 3xy \right]_{y=x}^{y=2-x} dx = \int_0^1 (12x - 12x^3)\, dx = \left[6x^2 - 3x^4 \right]_0^1 = 3.$$

The coordinates of the center of mass are therefore

$$\bar{x} = \frac{M_y}{M} = \frac{3}{8}, \qquad \bar{y} = \frac{M_x}{M} = \frac{17}{2(8)} = \frac{17}{16}.$$

EXAMPLE 4.24 A thin plate covers the triangular region bounded by the x-axis and the lines $x = 1$ and $y = 2x$. The plate's density at the point (x, y) is $\rho(x, y) = 6x + 6y + 6$. Find the moments of inertia and radii of gyration about the coordinate axes.

Solution The boundary of the plate is shown in Fig. 4.39.

The plate's mass is $M = \int_{x=0}^{x=1} \int_{y=0}^{y=2x} \rho(x, y)\, dy\, dx$

$$= \int_0^1 \int_0^{2x} (6x + 6y + 6)\, dy\, dx = \int_0^1 [6xy + 3y^2 + 6y]_{y=0}^{2x}\, dx = \int_0^1 (24x^2 + 12x)\, dx = \left[8x^3 + 6x^2 \right]_0^1 = 14.$$

The moment of inertia about the x-axis is

$$I_x = \int_{x=0}^{x=1}\int_{y=0}^{y=2x} y^2\rho(x,y)\,dy\,dx = \int_0^1\int_0^{2x}(6xy^2+6y^3+6y^2)\,dy\,dx$$

$$= \int_0^1\left[2xy^3+\frac{3}{2}y^4+2y^3\right]_{y=0}^{y=2x}dx$$

$$= \int_0^1(40x^4+16x^3)dx = \left[8x^5+4x^4\right]_0^1 = 12.$$

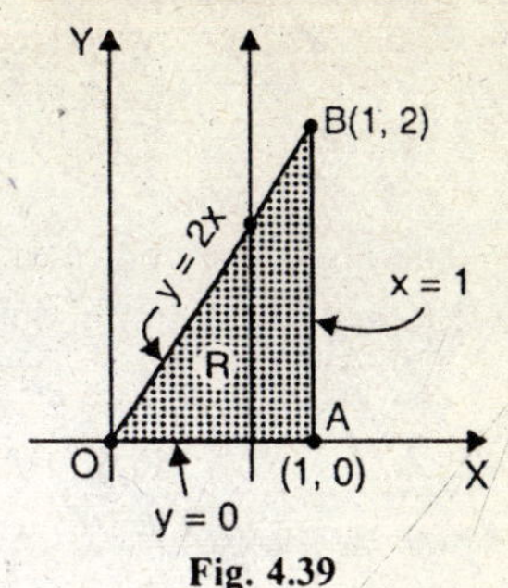

Fig. 4.39

The moment of inertia about the y-axis is

$$I_y = \int_{x=0}^{x=1}\int_{y=0}^{y=2x} x^2\rho(x,y)\,dy\,dx = \int_0^1\int_0^{2x}(6x^3+6x^2y+6x^2)\,dy\,dx$$

$$= \int_0^1\left[6x^3y+3x^2y^2+6x^2y\right]_0^{2x}dx = \int_0^1(24x^4+12x^3)dx = \left[\frac{24}{5}x^5+3x^4\right]_0^1 = \frac{39}{5}.$$

Employing the equation $I_o = I_x + I_y$ we get $I_o = 12+\frac{39}{5} = \frac{99}{5}$. There is no need to evaluate an integral to find I_o.

The three radii of gyration are

$$k_x = \sqrt{I_x/M} = \sqrt{12/14} = \sqrt{6/7},\ k_y = \sqrt{I_y/M} = \sqrt{39/70},\ k_o = \sqrt{I_o/M} = \sqrt{99/70}.$$

EXAMPLE 4.25 Find the centre of gravity of a plate whose density $\rho(x, y)$ is constant and is bounded by the curves $y = x^2$ and $y = x + 2$. Also, find the moments of inertia about the axes.

Solution The mass of the plate is given by

$$M = \iint_R \rho(x,y)\,dy\,dx = k\iint_R dy\,dx, \qquad [\rho(x,y) = k \text{ constant}]$$

The boundary of the plate is shown in Fig 4.40. The line $y = x + 2$ intersects the parabola $x^2 = y$ at the points $(-1, 1)$ and $(2, 4)$. Therefore,

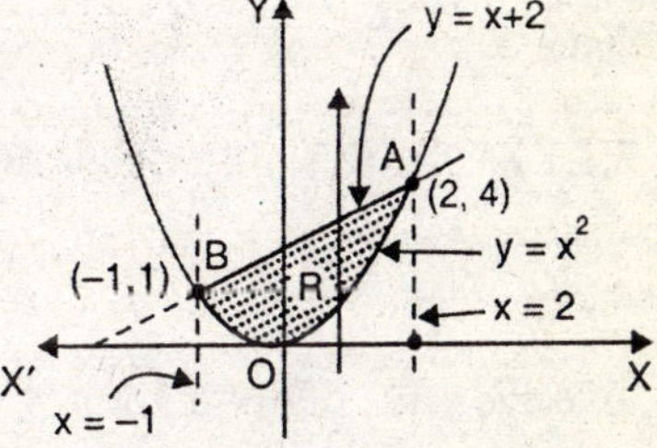

Fig. 4.40

$$M = k\int_{x=-1}^{x=2}\int_{y=x^2}^{y=x+2} dy\,dx = k\int_{-1}^{2}(x+2-x^2)\,dx$$

$$= k\left[\frac{x^2}{2}+2x-\frac{x^3}{3}\right]_{-1}^{2} = k\left[\frac{1}{2}(3)+2(3)-\frac{1}{3}(8+1)\right] = \frac{9}{2}k.$$

The center of gravity $(\bar{x}, \bar{y})$ is given by

$$\bar{x} = \frac{1}{M}\iint_R x\,\rho(x,y)\,dy\,dx = \frac{2}{9}\int_{-1}^{2}\int_{x^2}^{x+2} x\,dy\,dx = \frac{2}{9}\int_{-1}^{2}[xy]_{x^2}^{x+2}dx$$

$$= \frac{2}{9}\int_{-1}^{2}(x^2+2x-x^3)\,dx = \frac{2}{9}\left[\frac{x^3}{3}+x^2-\frac{x^4}{4}\right]_{-1}^{2} = \frac{2}{9}\cdot\frac{9}{4} = \frac{1}{2}.$$

$$\bar{y} = \frac{1}{M}\iint_R y\,\rho(x,y)\,dy\,dx = \frac{2}{9}\int_{-1}^{2}\int_{x^2}^{x+2} y\,dy\,dx = \frac{2}{9}\int_{-1}^{2}\left[\frac{y^2}{2}\right]_{x^2}^{x+2}dx$$

$$= \frac{2}{9}\int_{-1}^{2}\frac{1}{2}(x^2+4x+4-x^4)\,dx = \frac{1}{9}\left[\frac{x^3}{3}+2x^2+4x-\frac{x^5}{5}\right]_{-1}^{2} = \frac{1}{9}\cdot\frac{72}{5} = \frac{8}{5}.$$

Therefore, the centre of gravity is located at $(1/2, 8/5)$.

Moment of inertia about the x-axis is given by

$$I_x = \iint_R y^2\,\rho(x,y)\,dy\,dx = k\int_{-1}^{2}\int_{x^2}^{x+2} y^2\,dy\,dx = k\int_{-1}^{2}\left[y^3/3\right]_{x^2}^{x+2} dx$$

$$= \frac{k}{3}\int_{-1}^{2}(x^3+6x^2+12x+8-x^6)\,dx = \frac{k}{3}\left[\frac{x^4}{4}+2x^3+6x^2+8x-\frac{x^7}{7}\right]_{-1}^{2}$$

$$= \frac{k}{3}\left[\frac{1}{4}(15)+2(9)+6(3)+8(3)-\frac{1}{7}(128+1)\right] = \frac{k}{3}\cdot\left(\frac{1269}{28}\right) = \frac{423}{28}k.$$

Moment of inertia about the y-axis is given by

$$I_y = \iint_R x^2\,\rho(x,y)\,dy\,dx = k\int_{-1}^{2}\int_{x^2}^{x+2} x^2\,dy\,dx = k\int_{-1}^{2}\left[x^2 y\right]_{x^2}^{x+2} dx$$

$$= k\int_{-1}^{2}(x^3+2x^2-x^4)\,dx = k\left[\frac{x^4}{4}+\frac{2x^3}{3}-\frac{x^5}{5}\right]_{-1}^{2} = k\left[\frac{1}{4}(15)+\frac{2}{3}(9)-\frac{1}{5}(33)\right] = \frac{63}{20}k.$$

Moment of inertia about the origin (polar moment) is given by

$$I_o = I_x + I_y = \frac{423}{28}k + \frac{63}{20}k = \frac{639}{35}k.$$

Remark To determine $I_o = \iint_R (x^2+y^2)\,\rho(x,y)\,dy\,dx$, we donot need to evaluate an integral, since we know I_x and I_y.

4.5.7 Average Value or Mean Value

The average value or mean value of a continuous function $f(x, y)$ of two variables defined on a closed and bounded region R is the integral over the region divided by the area of the region. Thus,

$$\text{Average value of } f(x,y) \text{ over } R = \frac{1}{\text{area of the region R}}\iint_R f(x,y)\,dA.$$

EXAMPLE 4.26 Find the average value of $f(x, y) = x \cos xy$ over the rectangle

$$R: 0 \le x \le \pi,\ 0 \le y \le 1.$$

Solution The value of the integral of $f(x, y)$ over R is

$$\int_{x=0}^{\pi}\int_{y=0}^{1} x\cos xy\,dy\,dx = \int_0^{\pi} dx\left[\sin xy\right]_{y=0}^{1} = \int_0^{\pi}(\sin x)\,dx$$

$$= \left[-\cos x\right]_0^{\pi} = 1+1=2.$$

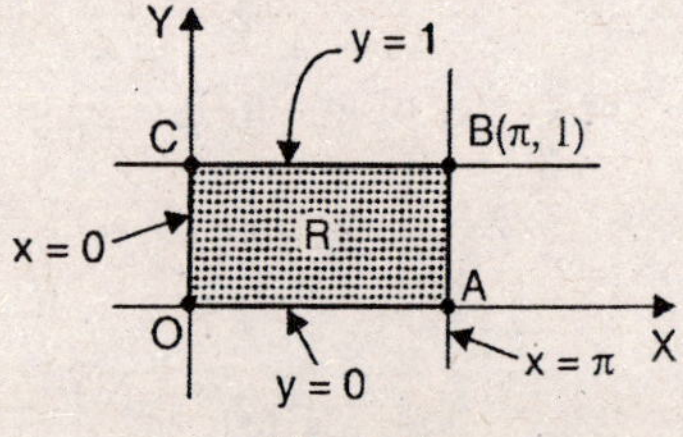

Fig. 4.41

The area of Region $R = \pi$. Thus, the average value of f over R is $2/\pi$.

EXAMPLE 4.27 Suppose that the temperture in degree celsius at a point (x, y) on a flat metal plate in $T(x, y) = 10 - 8x^2 - 2y^2$, where x and y are in meters. Find the average temperature of the rectangular portion of the plate for which $0 \le x \le 1$ and $0 \le y \le 2$.

Solution $$T_{ave} = \frac{1}{\text{Area of Region R}} \iint_R T(x, y)\, dy\, dx$$

$$= \frac{1}{2 \times 1} \int_{x=0}^{1} \int_{y=0}^{2} (10 - 8x^2 - 2y^2)\, dy\, dx = \frac{1}{2}\int_{x=0}^{1} \left[10\,y - 8x^2 y - \frac{2y^3}{3}\right]_0^2 dx$$

$$= \frac{1}{2}\int_0^1 \left(20 - 16x^2 - \frac{16}{3}\right) dx = \frac{1}{2}\left[20x - \frac{16x^3}{3} - \frac{16x}{3}\right]_0^1 = \frac{1}{2}\cdot\frac{28}{3} = (14/3)°C.$$

ILLUSTRATIVE EXAMPLES

EXAMPLE 4.28 Find by double integration

(*i*) the area lying between the parabola $y = 4x - x^2$ and the line $y = x$, *(DCE, 2004; MDU 2003)*

(*ii*) the area bounded by the parabolas $y^2 = 4 - x$ and $y^2 = 4 - 4x$,

(*iii*) the area included between the curves $y = x^2 - 6x + 3$ and $y = 2x - 9$, *(AMIETE; June 2001)*

(*iv*) the area included between the curves $y^2 = 4a(x + a)$ and $y^2 = 4b(b - x)$. *(AMIETE, June 2001)*

Solution (*i*) The two curves intersect at points whose abscissae are given by $4x - x^2 = x$, that is, $x = 0$ or 3.

The area under consideration lies between the curves $y = x$, $y = 4x - x^2$, $x = 0$ and $x = 3$. Using vertical strips, required area

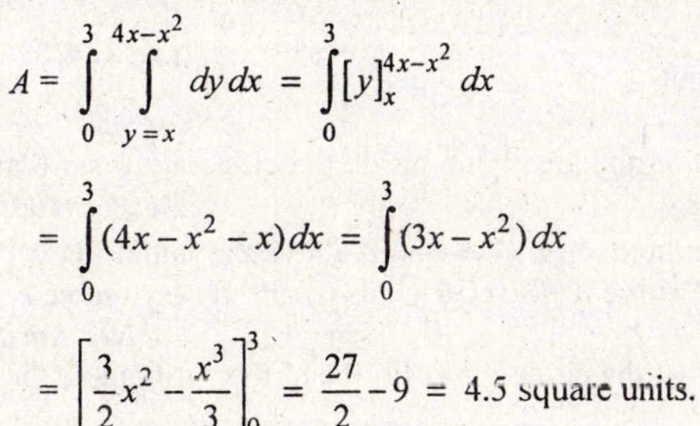

$$A = \int_0^3 \int_{y=x}^{4x-x^2} dy\, dx = \int_0^3 [y]_x^{4x-x^2} dx$$

$$= \int_0^3 (4x - x^2 - x)\, dx = \int_0^3 (3x - x^2)\, dx$$

$$= \left[\frac{3}{2}x^2 - \frac{x^3}{3}\right]_0^3 = \frac{27}{2} - 9 = 4.5 \text{ square units.}$$

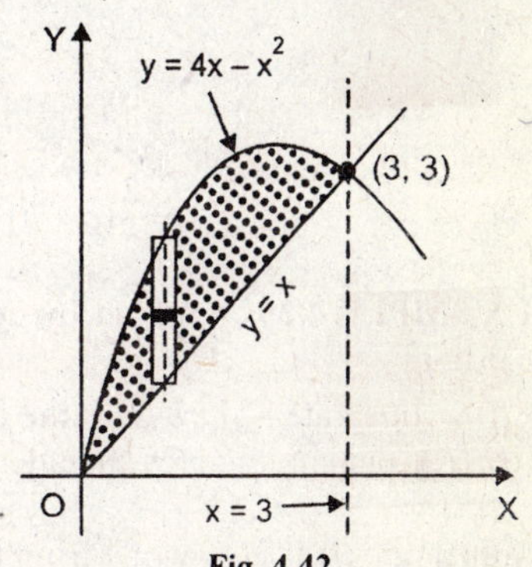

Fig. 4.42

(*ii*) Using horizontal strips and taking advantage of symmetry, we have

$$A = 2\int_0^2 \int_{1-(y^2/4)}^{4-y^2} dx\, dy = 2\int_0^2 \left[(4 - y^2) - \left(1 - \frac{y^2}{4}\right)\right] dy$$

$$= 6\int_0^2 \left(1 - \frac{y^2}{4}\right) dy = 6\left[y - \frac{y^3}{12}\right]_0^2 = 8 \text{ sq. units.}$$

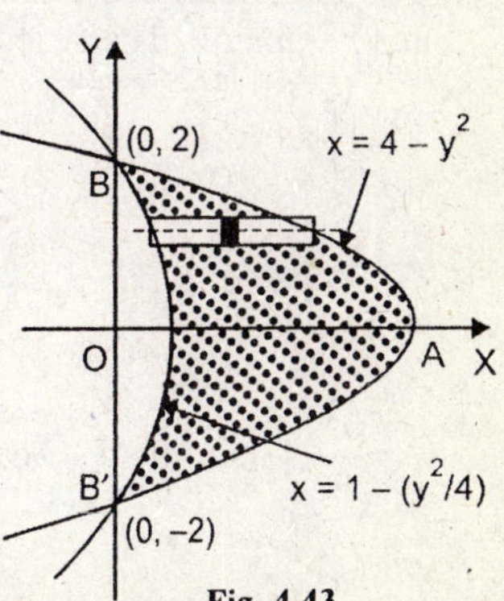

Fig. 4.43

(*iii*) Required area, $A = \displaystyle\int_{x=2}^{6} \int_{y=x^2-6x+3}^{2x-9} dy\, dx$ (Refer Fig. 4.44)

$$= \int_2^6 (2x - 9 - x^2 + 6x - 3)\, dx = -\int_2^6 (x^2 - 8x + 12)\, dx$$

$$= \int_6^2 (x^2 - 8x + 12)\,dx = \left[\frac{x^3}{3} - 4x^2 + 12x\right]_6^2$$

$$= (8 - 216)/3 - 4(4 - 36) + 12\,(2 - 6)$$

$$= -(208)/3 + 128 - 48 = 32/3 \text{ square units.}$$

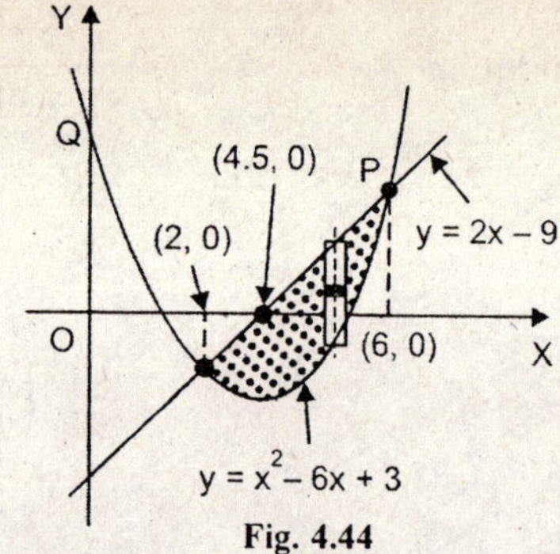

Fig. 4.44

(*iv*) The region of integration is shown in Fig. 4.45. We imagine a horizontal line passing from left to right through the region. It enters at $x = (y^2/4a) - a$ and leaves at $x = b - (y^2/4b)$. To include the area between the curves, we let y run from $y = -2\sqrt{ab}$ to $y = 2\sqrt{ab}$.

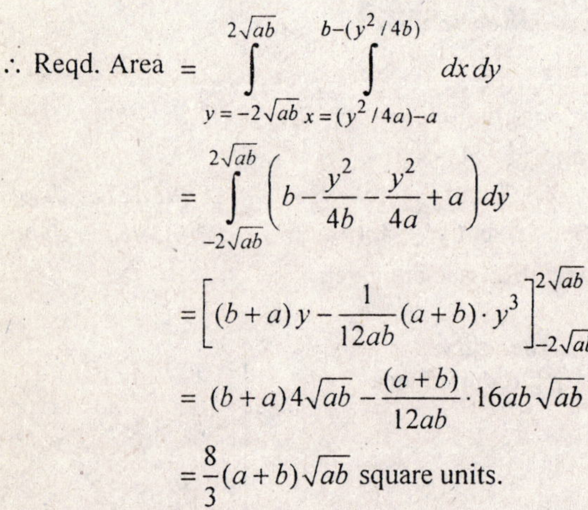

$$\therefore \text{ Reqd. Area } = \int_{y=-2\sqrt{ab}}^{2\sqrt{ab}} \int_{x=(y^2/4a)-a}^{b-(y^2/4b)} dx\,dy$$

$$= \int_{-2\sqrt{ab}}^{2\sqrt{ab}} \left(b - \frac{y^2}{4b} - \frac{y^2}{4a} + a\right) dy$$

$$= \left[(b+a)\,y - \frac{1}{12ab}(a+b)\cdot y^3\right]_{-2\sqrt{ab}}^{2\sqrt{ab}}$$

$$= (b+a)4\sqrt{ab} - \frac{(a+b)}{12ab}\cdot 16ab\sqrt{ab}$$

$$= \frac{8}{3}(a+b)\sqrt{ab} \text{ square units.}$$

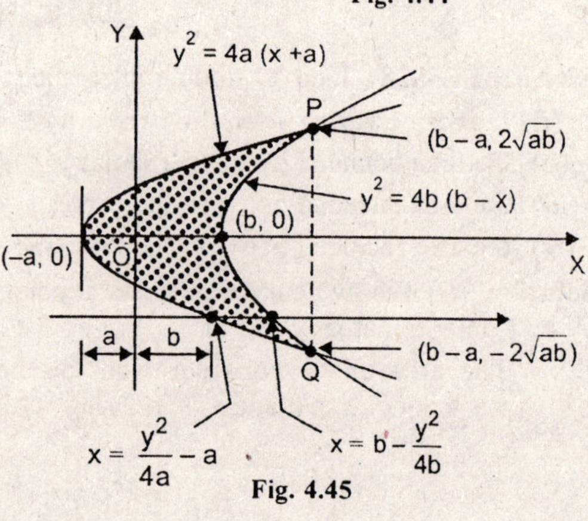

Fig. 4.45

EXAMPLE 4.29 (*i*) Find by double integration the area lying inside the circle $r = a \sin\theta$ and outside the cardioid $r = a\,(1 - \cos\theta)$. *(DCE 2005, MDU 2004)*

(*ii*) Integrate $r \sin\theta$ over the area of the cardioid $r = a\,(1 + \cos\theta)$ above the initial line. *(MDU, 2004)*

(*iii*) Compute the area included between the curve $r = a(\sec\theta + \cos\theta)$ and its asymptote $r = a\sec\theta$. *(NIT Kurukshetra 2007)*

Solution (*i*) It is obvious from the diagram that the circle $r = a\sin\theta$ and the cardioid $r = a(1 - \cos\theta)$ cut at $\theta = 0$ and $\theta = \pi/2$.

The area A between the two curves can be covered by radial strips whose ends are at $r = a(1 - \cos\theta)$ and $r = a\sin\theta$. The strips start at $\theta = 0$ and end at $\theta = \pi/2$. Hence the desired area

$$= \iint_A dA = \int_0^{\pi/2} \int_{a(1-\cos\theta)}^{a\sin\theta} r\,dr\,d\theta = \int_0^{\pi/2} \left[\frac{r^2}{2}\right]_{a(1-\cos\theta)}^{a\sin\theta} d\theta$$

$$= \frac{1}{2}\int_0^{\pi/2} \left[a^2\sin^2\theta - a^2(1-\cos\theta)^2\right] d\theta$$

$$= \frac{a^2}{2}\int_0^{\pi/2} \left[\sin^2\theta - 1 + 2\cos\theta - \cos^2\theta\right] d\theta$$

$$= \frac{a^2}{2}\int_0^{\pi/2} [2\cos\theta - 1 - \cos 2\theta]\,d\theta = \frac{a^2}{2}\left[2\sin\theta - \theta - \frac{\sin 2\theta}{2}\right]_0^{\pi/2}$$

$$= \frac{a^2}{2}\left[2 - \frac{\pi}{2} - \frac{1}{2}\sin\pi\right] = a^2\left(1 - \frac{\pi}{4}\right).$$

Y
θ = π/2
r = a (1 – cos θ)
r = a sin θ
O
θ = 0
X

Fig. 4.46

(*ii*) Here the region of integration A can be covered by radial strips, whose ends are at $r = 0$ and $r = a(1 + \cos\theta)$. The strips start from $\theta = 0$ and end at $\theta = \pi$. Therefore the required integral is

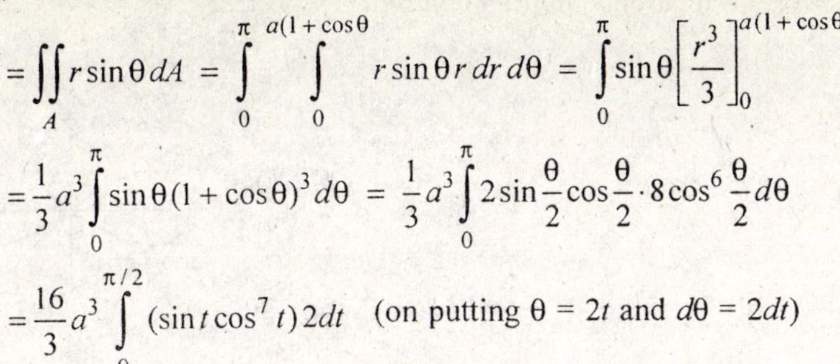

$$= \iint_A r\sin\theta\, dA = \int_0^{\pi}\int_0^{a(1+\cos\theta)} r\sin\theta\, r\, dr\, d\theta = \int_0^{\pi} \sin\theta\left[\frac{r^3}{3}\right]_0^{a(1+\cos\theta)} d\theta$$

$$= \frac{1}{3}a^3\int_0^{\pi}\sin\theta(1+\cos\theta)^3 d\theta = \frac{1}{3}a^3\int_0^{\pi} 2\sin\frac{\theta}{2}\cos\frac{\theta}{2}\cdot 8\cos^6\frac{\theta}{2}\, d\theta$$

$$= \frac{16}{3}a^3\int_0^{\pi/2}(\sin t\cos^7 t)\, 2dt \quad \text{(on putting } \theta = 2t \text{ and } d\theta = 2dt\text{)}$$

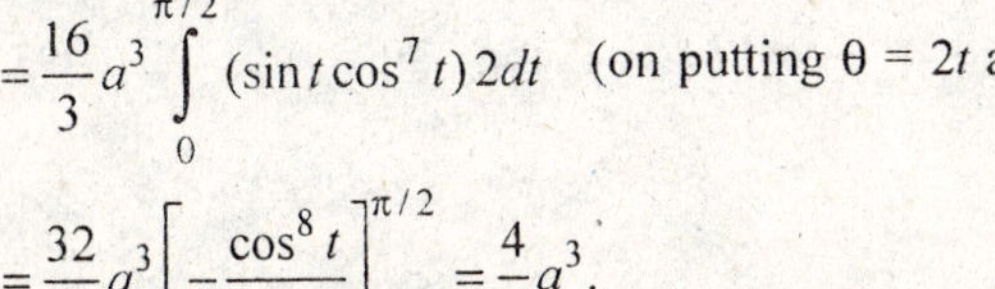

$$= \frac{32}{3}a^3\left[-\frac{\cos^8 t}{8}\right]_0^{\pi/2} = \frac{4}{3}a^3.$$

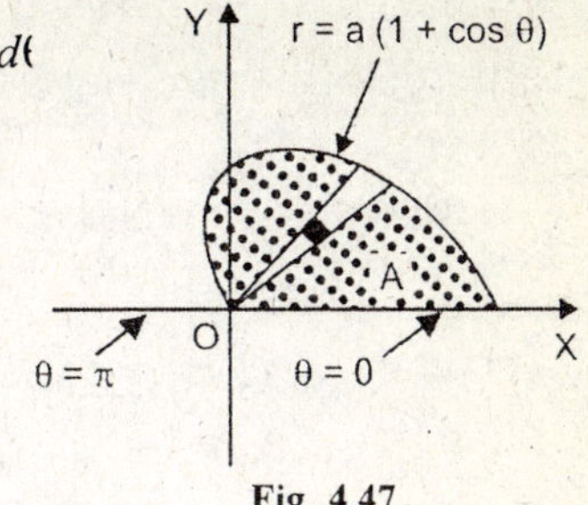

Fig. 4.47

(*iii*) The curve is symmetrical about the initial line and has an asymptote $r = a\sec\theta$. Taking any line OB cutting the curve at B and its asymptote at B′. Along this line, θ is constant and r varies from $a\sec\theta$ at B′ to $a(\sec\theta + \cos\theta)$ at B. To get the area between the asymptote and the curve, θ varies from $-\pi/2$ to $+\pi/2$. By symmetry, the required area

$$= 2\int_0^{\pi/2}\int_{a\sec\theta}^{a(\sec\theta+\cos\theta)} r\, dr\, d\theta = 2\int_0^{\pi/2}\left[\frac{r^2}{2}\right]_{a\sec\theta}^{a(\sec\theta+\cos\theta)} d\theta$$

$$= a^2\int_0^{\pi/2}(2+\cos^2\theta)\, d\theta = a^2\left[\pi + \frac{\pi}{4}\right] = \frac{5\pi a^2}{4} \text{ square units.}$$

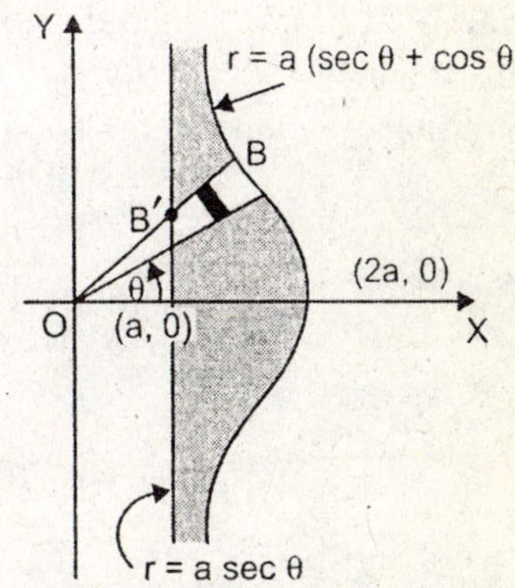

Fig. 4.48

EXAMPLE 4.30 (*i*) Find the mass of a circular plate of radius r if the density varies as the square of the distance from a point on the circumference.

(*ii*) Find the mass of the area between $y = x^3$ and $x = y^2$, if $\rho = k(x^2 + y^2)$.

(*iii*) Find the mass of a plate in the form of one loop of the lemniscate $r^2 = a^2\cos 2\theta$, if the density varies as the square of the distance from the pole.

Solution (*i*) Let $A(r, 0)$ be the fixed point on the circumference of a circle.

Then density at $P(x, y) = k[(x - r)^2 + y^2]$ and therefore,

$$M = \iint_A \rho(x, y)\, dA$$

$$= 2\int_{-r}^{r}\int_0^{\sqrt{r^2-x^2}} k\left[(x-r)^2 + y^2\right] dy\, dx = \frac{3}{2}k\pi r^4 \text{ units.}$$

(*ii*) The curves $y = x^3$ and $x = y^2$ cut at $x = 0$ and $x = 1$.

Hence the region of integration can be considered to lie between the curves $y = x^3$, $y = \sqrt{x}$, $x = 0$ and $x = 1$.

Hence the mass of the area between the curves

$$= \iint_A \rho(x, y)\, dx\, dy = \iint_A k(x^2 + y^2)\, dx\, dy$$

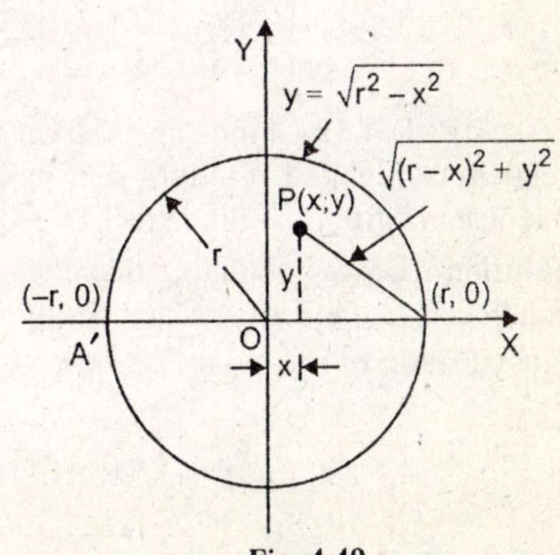

Fig. 4.49

$$= \int_0^1 \int_{x^3}^{\sqrt{x}} k(x^2+y^2)\,dy\,dx, \text{ which on simplification reduces to } \frac{23k}{105}.$$

(*iii*) At the point $P(r, \theta)$ the density $\rho = kr^2$.

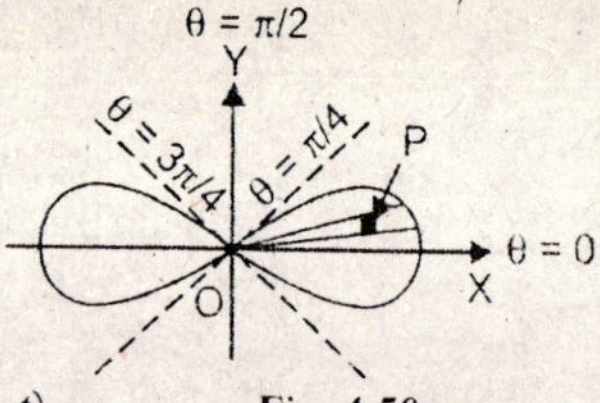

$$\text{Therefore, } M = 2\int_0^{\pi/4}\int_0^{a\sqrt{\cos 2\theta}} k\,r^2 \cdot r\,dr\,d\theta = 2k\int_0^{\pi/4}\int_0^{a\sqrt{\cos 2\theta}} r^3 dr\,d\theta$$

$$= \frac{2a^4}{4}k\int_0^{\pi/4}\cos^2 2\theta\,d\theta = \frac{a^4 k}{2}\int_0^{\pi/2}\frac{1}{2}\cos^2 t\,dt \quad (\text{Putting } 2\theta = t)$$

$$= \frac{1}{4}a^4 k\frac{1}{2}\cdot\frac{\pi}{2} = \frac{a^4 k\pi}{16}.$$

Fig. 4.50

EXAMPLE 4.31 Find the volume generated by reolving one loop of the curve $r = a\cos 2\theta$ about the initial line by double integration.

Solution Volume of the solid generated by the revolution of one loop of the curve $r = a\cos 2\theta$ about the x-axis

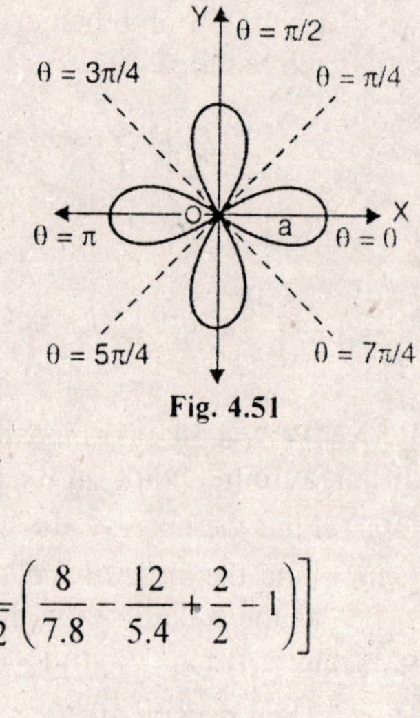

$$= 2\pi\int_{\theta=0}^{\pi/4}\sin\theta\,d\theta\int_{r=0}^{r=a\cos 2\theta} r^2\,dr = 2\pi\int_0^{\pi/4}\left(\frac{r^3}{3}\right)_0^{a\cos 2\theta}\sin\theta\,d\theta$$

$$= \frac{2}{3}\pi a^3\int_0^{\pi/4}\cos^3 2\theta\sin\theta\,d\theta$$

$$= \frac{2}{3}\pi a^3\int_0^{\pi/4}(2\cos^2\theta - 1)^3\sin\theta\,d\theta, \quad \text{put } \cos\theta = z$$

$$= \frac{2}{3}\pi a^3\int_{z=1/\sqrt{2}}^{1}(2z^2-1^3)dz = \frac{2}{3}\pi a^3\int_{1/\sqrt{2}}^{1}(8z^6-12z^4+6z^2-1)\,dz$$

$$= \frac{2}{3}\pi a^3\left[\frac{8}{7}z^7 - \frac{12}{5}z^5 + 2z^3 - z\right]_{1/\sqrt{2}}^{1} = \frac{2}{3}\pi a^3\left[\left(\frac{8}{7}-\frac{12}{5}+2-1\right)-\frac{1}{\sqrt{2}}\left(\frac{8}{7.8}-\frac{12}{5.4}+\frac{2}{2}-1\right)\right]$$

$$= \frac{2}{3}\pi a^3\left[-\frac{9}{35}+\frac{16}{35}\cdot\frac{1}{\sqrt{2}}\right] = \frac{2\pi a^3}{105}(8\sqrt{2}-9) \text{ cubic units.}$$

Fig. 4.51

EXAMPLE 4.32 Find the position of the centre of gravity of a semicircular lamina of radius '*a*' if its density varies as the square of the distance from the diameter.*(AMIETE, Dec 2010)*

Solution Let the bounding diameter be as the x-axis and line perpendicular to the diameter and passing through the center is y-axis. The equation of the circle is $x^2 + y^2 = a^2$. By symmetry $\bar{x} = 0$.

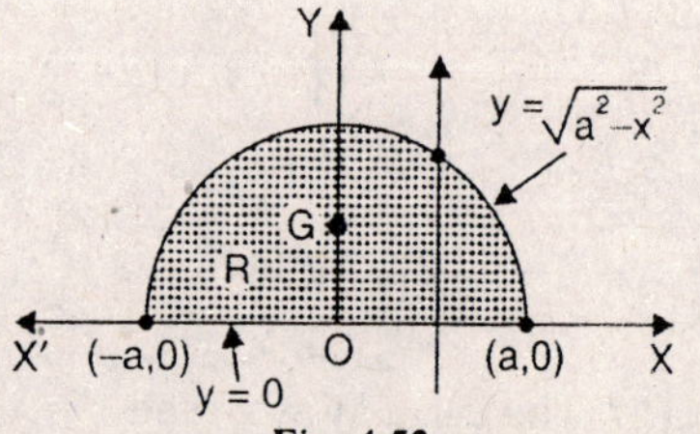

Fig. 4.52

$$\bar{y} = \frac{\iint_R y\,\rho(x,y)\,dy\,dx}{\iint_R \rho(x,y)\,dy\,dx} = \frac{\iint_R y(ky^2)\,dy\,dx}{\iint_R ky^2\,dy\,dx} = \frac{k\int_{-a}^{a}\int_0^{\sqrt{a^2-x^2}} y^3\,dy\,dx}{k\int_{-a}^{a}\int_0^{\sqrt{a^2-x^2}} y^2\,dy\,dx}$$

$$= \frac{\int_{-a}^{a}\left[y^4/4\right]_0^{\sqrt{a^2-x^2}} dx}{\int_{-a}^{a}\left[y^3/3\right]_0^{\sqrt{a^2-x^2}} dx} = \frac{3}{4}\cdot\frac{\int_{-a}^{a}(a^2-x^2)^2\,dx}{\int_{-a}^{a}(a^2-x^2)^{3/2}\,dx}$$

$\left(\begin{array}{l}\text{Put } x = a\sin\theta, \\ dx = a\cos\theta\, d\theta \\ \text{and also change the limits}\end{array}\right)$

$$= \frac{3}{4}\cdot\frac{\int_{-\pi/2}^{\pi/2}(a^4\cos^4\theta)\, a\cos\theta\, d\theta}{\int_{-\pi/2}^{\pi/2}(a^3\cos^3\theta)\, a\cos\theta\, d\theta} = \frac{3}{4}\cdot\frac{\int_{-\pi/2}^{\pi/2} a^5\cos^5\theta\, d\theta}{\int_{-\pi/2}^{\pi/2} a^4\cos^4\theta\, d\theta} = \frac{3}{4}a\,\frac{(4.2)/(5.3)}{(3.1)\frac{\pi}{2}/(4.2)} = \frac{32a}{15\pi}.$$

Thus the coordinates of the center of gravity are $\bar{x} = 0, \bar{y} = 32a/15\pi$.

EXAMPLE 4.33 Find the center of mass of a plate in the form of the upper half of the cardioid $r = a(1 + \cos\theta)$ if the density varies as the distance from the pole. *(AMIETE, Dec 2010)*

Solution Refer Fig. 4.53.

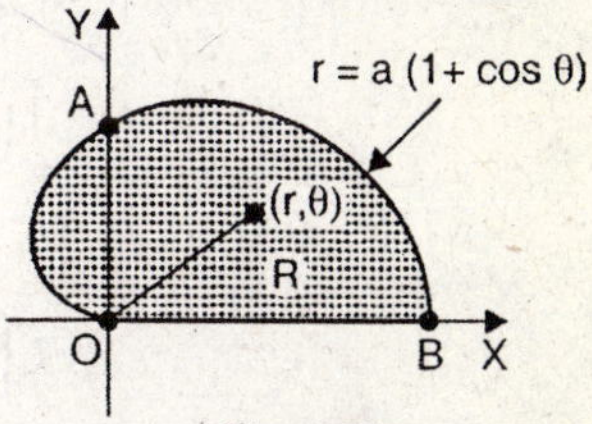

Mass, $$M = \iint_R \rho(r,\theta)\, dA = \int_{\theta=0}^{\pi}\int_{r=0}^{a(1+\cos\theta)} (kr)\, r\, dr\, d\theta$$

$$= \frac{a^3k}{3}\int_0^{\pi}(1+\cos\theta)^3\, d\theta = \frac{5}{6}k\pi a^3.$$

$$M_x = \iint_R \rho(r,\theta)\, y\, dA = \int_{\theta=0}^{\pi}\int_{r=0}^{a(1+\cos\theta)} (kr)(r\sin\theta)\, r\, dr\, d\theta$$

$$= \frac{a^4k}{4}\int_0^{\pi}(1+\cos\theta)^4\sin\theta\, d\theta = \frac{8}{5}k\, a^4.$$

$$M_y = \iint_R \rho(r,\theta)\, x\, dA = \int_0^{\pi}\int_0^{a(1+\cos\theta)} (kr)(r\cos\theta)\, r\, dr\, d\theta = \frac{7}{8}k\,\pi\, a^4.$$

Then $$\bar{x} = \frac{M_y}{M} = \frac{7k\pi a^4}{8(5k\pi a^3)/6} = \frac{21a}{20}, \quad \bar{y} = \frac{M_x}{M} = \frac{8ka^4}{5(5k\pi a^3)/6} = \frac{48a}{25\pi},$$

and the center of mass has coordinates $\left(\frac{21a}{20}, \frac{48a}{25\pi}\right)$.

Fig. 4.53

EXAMPLE 4.34 Find the moment of inertia of a quadrant of the ellipse $x^2/a^2 + y^2/b^2 = 1$ of mass M about the z-axis, if the density at a point is proportional to xy.

Solution Refer Fig. 4.54. The quadrant of the ellipse $x^2/a^2 + y^2/b^2 = 1$ is bounded by the curves: $y = 0$, $y = b\sqrt{1-(x^2/a^2)}$, $x = 0$, $x = a$.

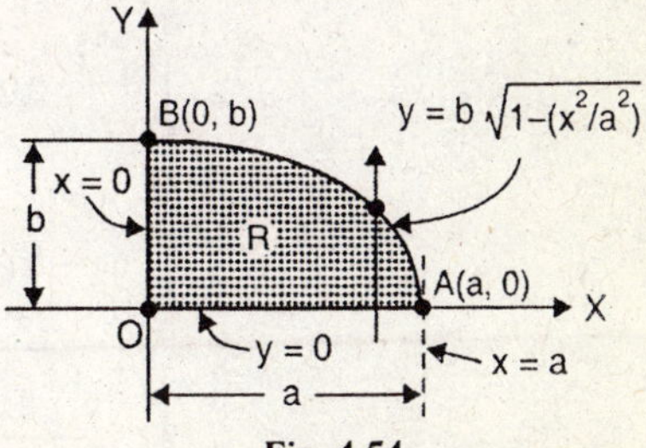

Moment of inertia about z-axis is given by

$$I_z = \iint_R (x^2+y^2)\,\rho(x,y)\, dy\, dx$$

$$= \int_{x=0}^{a}\int_{y=0}^{b\sqrt{1-(x^2/a^2)}} (x^2+y^2)(k\,xy)\, dy\, dx$$

Fig. 4.54

$$= k\int_0^a dx\left[x^3\frac{y^2}{2} + x\frac{y^4}{4}\right]_0^{b\sqrt{1-(x^2/a^2)}} = k\int_0^a\left[\frac{x^3}{2}b^2\left(1-\frac{x^2}{a^2}\right) + \frac{x}{4}b^4\left(1-\frac{x^2}{a^2}\right)^2\right]dx$$

$$= k\int_0^a \left(\frac{b^2}{2}x^3 - \frac{b^2x^5}{2a^2} + \frac{b^4x}{4} + \frac{b^4x^5}{4a^4} - \frac{b^4x^3}{2a^2}\right) dx$$

$$= k\int_0^a \left[\left(\frac{b^4 - 2a^2b^2}{4a^4}\right)x^5 + \left(\frac{a^2b^2 - b^4}{2a^2}\right)x^3 + \frac{b^4}{4}x\right] dx$$

$$= k\left[\left(\frac{b^4 - 2a^2b^2}{24a^4}\right)x^6 + \left(\frac{a^2b^2 - b^4}{8a^2}\right)x^4 + \frac{b^4}{8}x^2\right]_0^a$$

$$= k\left[\frac{(b^4 - 2a^2b^2)a^2}{24} + \frac{(a^2b^2 - b^4)a^2}{8} + \frac{a^2b^4}{8}\right] = \frac{k}{24}a^2\, b^2\, (a^2 + b^2).$$

The mass of the quadrant is given by

$$M = \iint_R \rho\,(x, y)\, dy\, dx = \int_{x=0}^{a}\int_{y=0}^{b\sqrt{1-(x^2/a^2)}} k\, xy\, dy\, dx$$

$$= k\int_0^a dx\left[\frac{xy^2}{2}\right]_0^{b\sqrt{1-(x^2/a^2)}} = k\int_0^a \frac{x}{2}b^2\left(1 - \frac{x^2}{a^2}\right)dx$$

$$= \frac{b^2k}{2a^2}\int_0^a (a^2x - x^3)\, dx = \frac{kb^2}{2a^2}\left[a^2\frac{x^2}{2} - \frac{x^4}{4}\right]_0^a = \frac{kb^2}{2a^2}\cdot\frac{a^4}{4} = \frac{ka^2b^2}{8}.$$

Therefore, $$I_z = \frac{k}{24}a^2\, b^2\, (a^2 + b^2) = \frac{1}{3}\left(\frac{k}{8}a^2\, b^2\right)(a^2 + b^2) = \frac{M}{3}(a^2 + b^2).$$

EXAMPLE 4.35 (*i*) Find the volume under the plane $x + y + z = 6$ and above the triangle in the xy-plane bounded by $2x = 3y$, $y = 0$, $x = 3$.

(*ii*) Find the volume bounded by the plane $z = 0$, surface $z = x^2 + y^2 + 2$ and the cylinder $x^2 + y^2 = 4$.

(*iii*) Find the volume in the first octant bounded by the circular cylinder $x^2 + y^2 = 2$ and the planes $z = x + y$, $y = x$, $z = 0$ and $x = 0$.

(*iv*) Find the volume cut off from the paraboloid $x^2 + (y^2/4) + z = 1$ by the plane $z = 0$.

Solution (*i*) The required volume, $V = \iint_A z\, dy\, dx = \iint_A (6 - x - y)\, dy\, dx$,

where A is the region, shown in Fig. 4.55 bounded by the curves $y = 0$, $y = 2x/3$, $x = 0$ and $x = 3$.

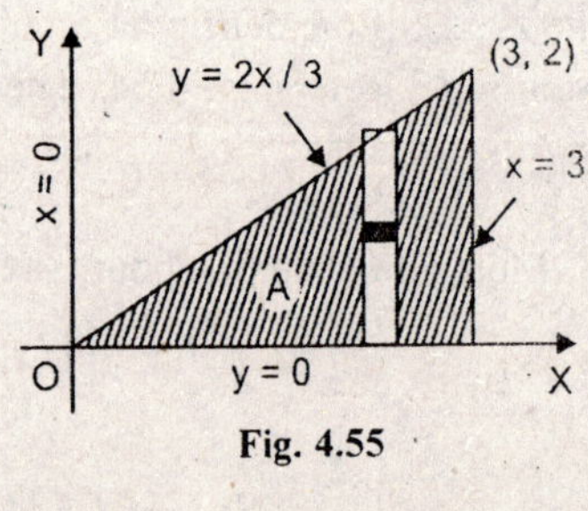

Fig. 4.55

$$\therefore \quad V = \int_0^3\int_0^{2x/3} (6 - x - y)\, dy\, dx$$

$$= \int_0^3\left[6y - xy - \frac{y^2}{2}\right]_0^{2x/3} dx = \int_0^3\left(4x - \frac{2}{3}x^2 - \frac{2}{9}x^2\right)dx$$

$$= \int_0^3\left(4x - \frac{8x^2}{9}\right)dx = \left[2x^2 - \frac{8}{27}x^3\right]_0^3 = 10 \text{ cubic units.}$$

(*ii*) The region A in the positive quadrant of XOY plane is bounded by the curves $y = 0$, $y = \sqrt{(4 - x^2)}$, $x = 0$ and $x = 2$.

Required volume = 4 times the volume lying in the first octant

$$= 4\int_0^2 \int_0^{\sqrt{(4-x^2)}} (x^2+y^2+2)\,dy\,dx = 4\int_0^2 \left[x^2 y + (y^3/3) + 2y\right]_0^{\sqrt{(4-x^2)}} dx$$

$$= 4\int_0^2 \left[x^2\sqrt{(4-x^2)} + (1/3)(4-x^2)^{3/2} + 2\sqrt{(4-x^2)}\right] dx$$

$$= 4\int_0^{\pi/2} \left[4\sin^2\theta \cdot 2\cos\theta + (1/3)8\cos^3\theta + 2\cdot 2\cos\theta\right] 2\cos\theta\, d\theta$$

(on putting $x = 2\sin\theta$, $\sqrt{(4-x^2)} = \sqrt{\{4(1-\sin^2\theta)\}} = 2\cos\theta$ and $dx = 2\cos\theta\, d\theta$.)

$$= 32\int_0^{\pi/2} \left[2\sin^2\theta\cos^2\theta + \frac{2}{3}\cos^4\theta + \cos^2\theta\right] d\theta$$

$$= 32\left[2\cdot\frac{1}{4}\cdot\frac{1}{2}\cdot\frac{\pi}{2} + \frac{2}{3}\cdot\frac{3\cdot 1}{4\cdot 2}\cdot\frac{\pi}{2} + \frac{1}{2}\cdot\frac{\pi}{2}\right]$$

$$= 32\left(\frac{\pi}{8} + \frac{\pi}{8} + \frac{\pi}{4}\right) = 16\pi.$$

(*iii*) The region of integration A is shown by shaded lines in the diagram, which is bounded by the curves

$$y = x,\ y = \sqrt{(2-x^2)},\ x = 0 \text{ and } x = 1.$$

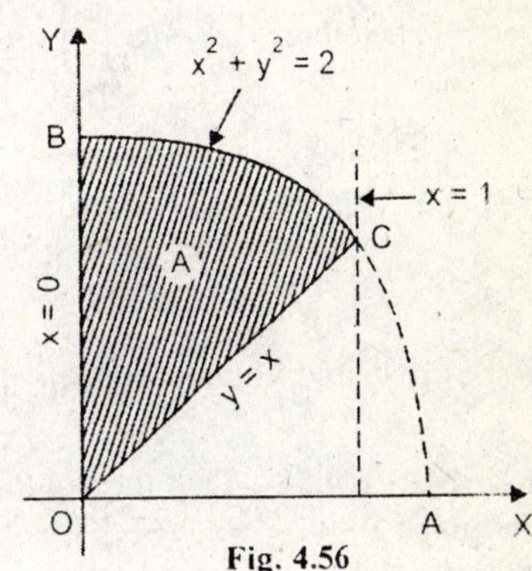

Fig. 4.56

Hence the required volume, $V = \int_0^1 \int_x^{\sqrt{(2-x^2)}} (x+y)\,dy\,dx,$

which on simplification reduces to $2\sqrt{2}/3$.

(iv) The equation of the paraboloid is $z = 1 - x^2 - \dfrac{y^2}{4}$ and it cuts the plane $z = 0$ in the ellipes $x^2 + \dfrac{y^2}{4} = 1$.
Hence, the volume of the paraboloid above the XOY plane

$$= \iint \left(1 - x^2 - \frac{y^2}{4}\right) dx\,dy \text{ taken over the area of the ellipse.}$$

Therefore, the required volume $= \displaystyle\int_{x=-1}^{+1} dx \int_{y=-2\sqrt{1-x^2}}^{+2\sqrt{1-x^2}} \left(1 - x^2 + \frac{y^2}{4}\right) dy$

$$= \int_{-1}^{1} \left[(1-x^2)\,y - \frac{y^3}{12}\right]_{-2\sqrt{1-x^2}}^{+2\sqrt{1-x^2}} dx = \int_{-1}^{1} \frac{8}{3}(1-x^2)^{3/2}\,dx = \frac{8}{3}\int_{-\pi/2}^{\pi/2} \cos^4\theta\, d\theta, \text{ (where } x = \sin\theta)$$

$$= \frac{8}{3}\cdot 2\int_0^{\pi/2} \cos^4\theta\, d\theta = \frac{16}{3}\cdot\frac{3}{4}\cdot\frac{1}{2}\cdot\frac{\pi}{2} = \pi \text{ cubic units.}$$

EXAMPLE 4.36 (*i*) Find the volume common to the cylinders $x^2 + y^2 = a^2$, $x^2 + z^2 = a^2$. (*SVTU, 2006*)

(*ii*) Prove that the volume enclosed between the cylinders $x^2 + y^2 = 2ax$ and $z^2 = 2ax$ is $128a^3/15$.

Solution (*i*) Figure 4.57 shows one-eighth of the volume common to the cylinders.

Required volume = 8 times volume of region shown in Fig. 4.57.

$$= 8\int_{x=0}^{a}\int_{y=0}^{\sqrt{(a^2-x^2)}} z\,dy\,dx = 8\int_{x=0}^{a}\int_{y=0}^{\sqrt{(a^2-x^2)}} \sqrt{a^2-x^2)}\,dy\,dx$$

$$= 8\int_0^a \sqrt{a^2-x^2}\,[y]_0^{\sqrt{(a^2-x^2)}}\,dx = 8\int_0^a (a^2-x^2)\,dx = 8\left[a^2x-\frac{x^3}{3}\right]_0^a$$

$$= 8\left(a^3-\frac{a^3}{3}\right) = \frac{16a^3}{3} \text{ cubic units.}$$

Fig. 4.57

(*ii*) Here $z = \pm\sqrt{2ax}$, Thus half of the volume lies above the *xy*-plane and half below it.

Therefore, $$V = 2\int_0^{2a}\int_{-\sqrt{(2ax-x^2)}}^{\sqrt{(2ax-x^2)}} \sqrt{2ax}\,dy\,dx = 4\int_0^{2a}\int_0^{\sqrt{(2ax-x^2)}} \sqrt{2ax}\,dy\,dx$$

$$= 4\int_0^{2a} \sqrt{2ax}\cdot\sqrt{(2ax-x^2)}\,dx = 4\int_0^{\pi/2} 2a\sin\theta\sqrt{\{2a\sin^2\theta(2a-2a\sin^2\theta)\}}\cdot 4a\sin\theta\cos\theta\,d\theta$$

(on putting $x = 2a\sin^2\theta$, $dx = 4a\sin\theta\cos\theta\,d\theta$).

$$\text{Required volume} = 64a^3\int_0^{\pi/2}\sin^3\theta\cos^2\theta\,d\theta = 64a^3\frac{2}{5\cdot 3\cdot 1} = \frac{128a^3}{15}.$$

EXAMPLE 4.37 Find the volume bounded by the paraboloid $x^2 + y^2 = a\,z$, the cylinder $x^2 + y^2 = 2$ ay and the plane $z = 0$.

Solution The required volume is obtained by integrating $z = (x^2 + y^2)/a$ over the circle $x^2 + y^2 - 2ay = 0$. Changing to polar coordinates in the *xy*-plane, we have $x = r\cos\theta$ $y = r\sin\theta$ so that $z = r^2/a$ and the polar equation of the circle is $r = 2a\sin\theta$. To cover the circle, r varies from 0 to $2a\sin\theta$ and θ varies from 0 to π. (Refer Fig. 4.58).

Hence the required volume, $V = \iint_R z\,dA$

$$= \int_{\theta=0}^{\pi}\int_{r=0}^{2a\sin\theta} z\cdot r\,dr\,d\theta = \frac{1}{a}\int_0^{\pi} d\theta\int_0^{2a\sin\theta} r^3\,dr$$

$$= \frac{1}{a}\int_0^{\pi}\left[\frac{r^4}{4}\right]_0^{2a\sin\theta} d\theta = 4a^3\int_0^{\pi}\sin^4\theta\,d\theta = \frac{3\pi a^3}{2} \text{ cubic units.}$$

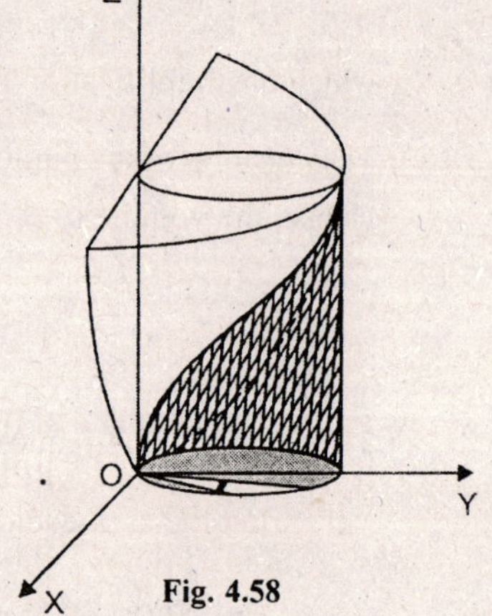

Fig. 4.58

EXAMPLE 4.38 Find the volume of the ellipsoid $\frac{x^2}{a^2}+\frac{y^2}{b^2}+\frac{z^2}{c^2}=1.$

(*PTU 2006; Kottayam, 2005; DCE 2004; MDU 2001*)

Solution On account of symmetry, the required volume is 8 times the volume of the ellipsoid in the first octant. The projection of the surface $z = c\sqrt{1-(x^2/a^2)-(y^2/b^2)}$ in the *xy*-plane is the region in the first quadrant of the ellipse $(x^2/a^2)+(y^2/b^2) = 1$.

Therefore, $V = 8\int_0^a dx \int_0^{b\sqrt{1-(x^2/a^2)}} c\sqrt{1-\frac{x^2}{a^2}-\frac{y^2}{b^2}}\,dy = 8c\int_0^a dx\int_0^{bk}\sqrt{k^2-\frac{y^2}{b^2}}\,dy$,

where $k^2 = 1 - (x^2/a^2)$. Putting $y = bk \sin\theta$ and $dy = bk\cos\theta\, d\theta$, we obtain

$$V = 8c\int_0^a dx\int_0^{\pi/2}\sqrt{k^2-k^2\sin^2\theta}\,(bk\cos\theta)\,d\theta = 8bc\int_0^a dx\int_0^{\pi/2} k^2\cos^2\theta\, d\theta$$

$$= 8bc\cdot\frac{1}{2}\cdot\frac{\pi}{2}\int_0^a\left(1-\frac{x^2}{a^2}\right)dx = \frac{2\pi bc}{a^2}\int_0^a (a^2-x^2)\,dx = \frac{2\pi bc}{a^2}\left[a^2x-\frac{x^3}{3}\right]_0^a = \frac{4\pi abc}{3}\text{ units}^3.$$

EXAMPLE 4.39 Find the volume of the solid bounded by the conical surface

$$(z-2)^2 = \frac{x^2}{3}+\frac{y^2}{2}\text{ and the plane } z = 0.$$

Solution By symmetry, volume required is given by

$V = 4\iint_R\left(2+\sqrt{\frac{x^2}{3}+\frac{y^2}{2}}\right)dy\,dx$, where R is the part of the ellipse $\frac{x^2}{12}+\frac{y^2}{8} = 1$ in the first quadrant of the xy plane. Putting $x = u\sqrt{3},\ y = v\sqrt{2}$, we obtain

$V = 4\iint_{R_1}(2+\sqrt{u^2+v^2})\sqrt{6}\,du\,dv$, where R_1 is the quadrant of the circle $u^2+v^2 = 4$.

Again putting $u = r\cos\theta$, $v = r\sin\theta$, we obtain

$$V = 4\sqrt{6}\int_0^{\pi/2}\int_0^2 (2+r)\,r\,dr\,d\theta = 4\sqrt{6}\int_0^{\pi/2}\left[r^2+\frac{r^3}{3}\right]_0^2 d\theta = 4\sqrt{6}\cdot\frac{20}{3}\int_0^{\pi/2} 1\,d\theta$$

$$= 4\sqrt{6}\times\frac{20}{3}\times\frac{\pi}{2} = \frac{40}{3}\pi\sqrt{6}\text{ cubic units.}$$

EXERCISE 4.2

1. Find the area bounded by the curves $y = x$, $y = 2 - x^2$. **Ans.** 4.5 units^2.
2. (*i*) Find the smaller of the areas bounded by $y = 2 - x$ and $x^2 + y^2 = 4$. **Ans.** $(\pi - 4)$ unit^2.

   (*ii*) Find the area enclosed by the ellipse $x^2/a^2 + y^2/b^2 = 1$.
   (*VTU 2001; Osmania 2000S; MDU 2000*) **Ans.** πab sq. units.

   (*iii*) Evaluate $\iint xy\,dx\,dy$ over the area between $y = x^2$ and $y = x$. (*AMIE, S-2009*) **Ans.** 1/24.
3. Find the area in the first quadrant bounded by the x-axis and the curves $x^2 + y^2 = 10$ and $y^2 = 9x$.
   Ans. 6.75 square units.
4. Find the area of the region enclosed by the curves $\sqrt{x}+\sqrt{y}=\sqrt{a}$ and $x + y = a$.
   Ans. $a^2/3$ square units.
5. Find the area enclosed by the curves $y^2 = 4ax$ and $x^2 = 4ay$
   (*NIT Kurukshetra 2010; Kerala 2005, MDU 2003*) **Ans.** $16a^2/3$ square units.
6. Find the area of a figure bounded by the curves $y = \sin x$, $y = \cos x$, $x = 0$. **Ans.** $(\sqrt{2}-1)$ square units.
7. Find the area enclosed between the parabola $y = x(4 - x)$ and the axis of x. **Ans.** 10.667 square units.

8. Evaluate the integral $\iint\limits_R \sqrt{x^2+y^2}\,dx\,dy$ by changing to polar coordinates, where R is the region in the xy-plane bounded by the circles $x^2+y^2=4$ and $x^2+y^2=9$. (*AMIETE Dec. 2009, June 2007*)
[**Hint:** Using $x = r\cos\theta$, $y = r\sin\theta$, $dxdy = r\,dr\,d\theta$, we obtain

$$I = \int_0^{2\pi}\int_2^3 r(r\,dr\,d\theta) = \int_0^{2\pi}\left[\frac{r^3}{3}\right]_2^3 d\theta = \frac{19}{3}\int_0^{2\pi} d\theta = \frac{38\pi}{3}.\Bigg]$$

9. Evaluate $\iint\limits_A r^3\,dr\,d\theta$, where A is the area included between the circles $r = 2\sin\theta$ and $r = 4\sin\theta$.

[**Hint:** The given integral $= \int_{\theta=0}^{\pi} d\theta \int_{r=2\sin\theta}^{4\sin\theta} r^3\,dr$] (*MDU-2003*) **Ans.** $22.5\,\pi$

10. Find the area in the xy-plane bounded by the lemniscate $r^2 = a^2\cos 2\theta$.
(*Madras, 2000S*) **Ans.** a^2 square units

11. (*i*) Find the area inside the circle $r = 2a\cos\theta$ and outside the circle $r = a$.
Ans. $2a^2(\pi/6+\sqrt{3}/4)$ square units
(*ii*) Find the area lying inside a cardioide $r = 1+\cos\theta$ and outside the parabola $r(1+\cos\theta) = 1$.
(*DCE 2005*) **Ans.** $(9\pi+16)/12$.

12. Find the area common to the cardioids $r = a(1-\cos\theta)$ and $r = a(1+\cos\theta)$.
Ans. $a^2(3\pi-8)/2$ square units.

13. Find the area inside the circle $r = 4\sin\theta$ and outside the lemniscate $r^2 = 8\cos 2\theta$.

[**Hint:** Area $= 2\int_{\pi/6}^{\pi/4}\int_{2\sqrt{2\cos 2\theta}}^{4\sin\theta} r\,dr\,d\theta + 2\int_{\pi/4}^{\pi/2}\int_0^{4\sin\theta} r\,dr\,d\theta$.] **Ans.** $(8\pi+12\sqrt{3}-12)/3$ square units.

14. Find the mass of a plate in the form of a right triangle with legs 'a' and 'b', if the density varies as the sum of distances from the legs. **Ans.** $kab(a+b)/6$ units.

15. Find the mass of a plate in the form of an ellipse $b^2x^2 + a^2y^2 = a^2b^2$, if the density varies as the sum of the distances from the axes. **Ans.** $4kab(a+b)/3$ units.

16. Find the mass of a plate in the form of a quadrant of an ellipse $x^2/a^2 + y^2/b^2 = 1$, whose density per unit area is given by $\rho = kxy$. **Ans.** $ka^2b^2/8$ units.
[**Hint:** The quadrant of the ellipse $x^2/a^2 + y^2/b^2 = 1$ is enclosed by the curves $y = 0$, $y = b\sqrt{(1-x^2/a^2)}$, $x = 0$ and $x = a$. This gives the region of integration R].

17. A plate has its edge the curve $y = e^x$, the line $x = 1$ and the coordinate axes. If the density varies as the square of the distance from the origin, find the mass of the plate. **Ans.** 2.84 k units.

[**Hint:** $M = \int_0^1\int_0^{e^x} k(x^2+y^2)\,dy\,dx$]

18. A plate is in the form of a parabolic segment bounded by parabola $y^2 = 8x$ and its latus rectum $x = 2$. If the density varies as the distance from the latus rectum, find the mass of the plate.
Ans. $64k/15$ units.

19. Determine the mass of a circular plate of radius r if the surface density $f(x, y)$ of the material at each point $P(x, y)$ is proportional to the distance of the point (x, y) from the centre of the circle.
Ans. $2k\pi r^3/3$ units.

20. Find the mass of a plate in the form of a cardioid $r = a(1 + \cos\theta)$, if the density of the mass varies as the distance from the pole. **Ans.** $5\pi k a^3/3$.

[**Hint:** $M = 2\int_{\theta=0}^{\pi}\int_{r=0}^{a(1+\cos\theta)} kr\,(r\,dr\,d\theta) = 5\pi k a^3/3$]

21. Find the mass and center of gravity of the lamina bounded by the x-axis, the line $x = 1$, and the curve $y = \sqrt{x}$. The density function $\rho(x, y) = x + y$. **Ans.** $M = 13/20$, Center of gravity (190/273, 6/13).

22. Find the centroid of the semicircular region in Fig 4.59.

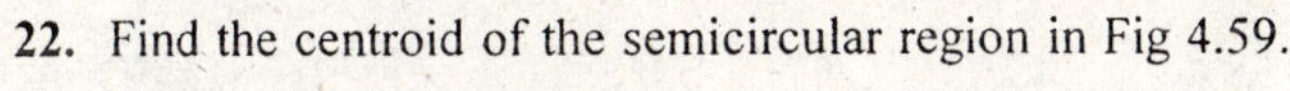

[**Hint:** By symmetry $\bar{x} = 0$,

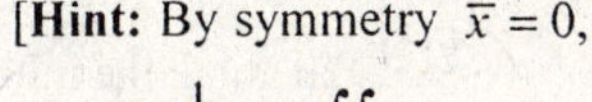

$$\bar{y} = \frac{1}{\text{area of } R}\iint_R y\,dy\,dx = \frac{1}{(\pi a^2/2)}\iint_R y\,dy\,dx$$

$$\frac{1}{(\pi a^2/2)}\int_0^{\pi}\int_0^{a}(r\sin\theta)\,r\,dr\,d\theta = 4a/3\pi.$$ The centroid is $(0, 4a/3\pi)$.]

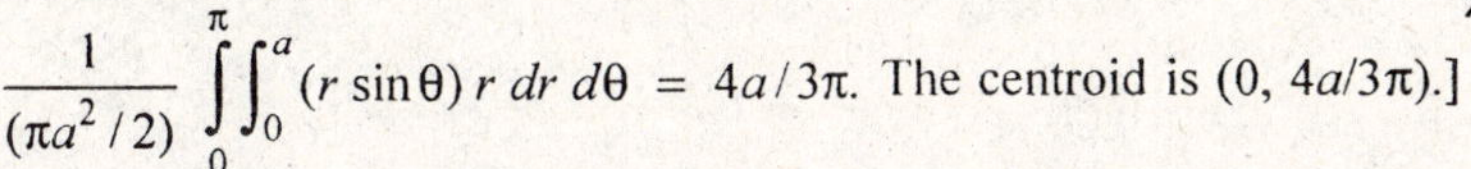

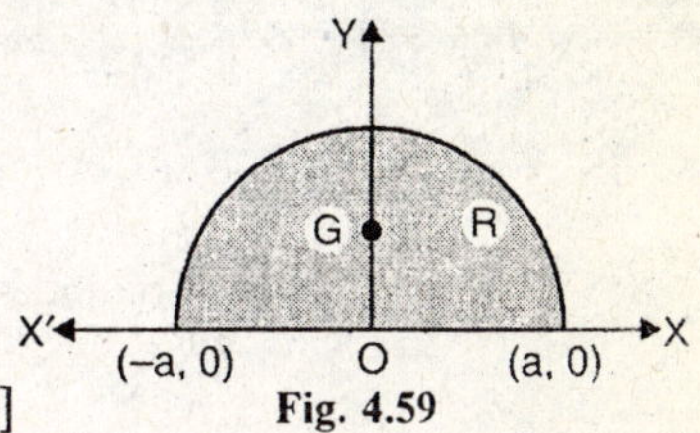

Fig. 4.59

23. Find the centroid of the region in the first quadrant that is bounded above by the line $y = x$ and below by the parabola $y = x^2$. **Ans.** (1/2, 2/5)

24. Find the centroid of the area in the first quadrant bouned by the semi-cubical parabola $y^2 = x^3$ and the line $y = x$. **Ans.** G (5/12, 10/21).

25. Find the center of gravity of a lamina in the shape of a quadrant of the curve $(x/a)^{2/3} + (y/b)^{2/3} = 1$, the density being $\rho = k\,xy$, where k is a constant. **Ans.** $G\left(\frac{128a}{429}, \frac{128b}{429}\right)$.

26. Determine the coordinates of the centre of gravity of a quarter of the ellipse $x^2/a^2 + y^2/b^2 = 1$, assuming that the $\rho(x, y) = 1$. **Ans.** $\left(\frac{4a}{3\pi}, \frac{4b}{3\pi}\right)$.

27. Find the total mass and the center of gravity of the region bounded by $x^{2/3} + y^{2/3} = a^{2/3}$, $x \geq 0$, $y \geq 0$, when the density is constant k. **Ans.** $M = \frac{3\pi k a^2}{32}$, $\bar{x} = \bar{y}\ \frac{8ka^3}{105M}$.

28. Find the center of gravity of a plate bounded by the cardioide $r = 1 - \cos\theta$, if the surface density $\rho(x, y)$ is inversely proportional to r. **Ans.** (–1/2, 0)

[**Hint:** $\rho(x, y) = k/r$, where k is constant of proportionality.

$$\bar{x} = \int_0^{2\pi}\int_0^{(1-\cos\theta)}(k/r)(r\cos\theta)\cdot r\,dr\,d\theta \Bigg/ \int_0^{2\pi}\int_0^{(1-\cos\theta)}(k/r)\cdot r\,dr\,d\theta$$

$$= 2\int_0^{\pi}\int_0^{1-\cos\theta} kr\cos\theta\,dr\,d\theta \Bigg/ 2\int_0^{\pi}\int_0^{1-\cos\theta} k\,dr\,d\theta = -1/2.\quad \bar{y} = 0, \text{ from symmetry.}]$$

29. Find by double integration, the centroid of the area of the cardioid $r = a(1 + \cos\theta)$.

Ans. The centroid of the cardioid is at $(5a/6, 0)$.

30. Determine the moment of inertia of a circular area of radius r with respect to a diametral axis.

Ans. $I_x = \pi r^4/4$.

31. Find the center of mass and the moment of inertia about the y-axis of a thin rectangular plate cut from the first quadrant by the lines $x = 6$ and $y = 1$ if $\rho(x, y) = x + y + 1$.

Ans. $\bar{x} = 11/3$, $\bar{y} = 14/27$, $I_y = 432$.

32. Find the average height of the paraboloid $z = x^2 + y^2$ over the square $0 \le x \le 2,\ 0 \le y \le 2$. **Ans.** 8/3.

33. Find the average value of $1/(1 + x^2)$ over the triangular region with vertices (0, 0), (1, 1) and (0, 1).

Ans. $\left(\frac{\pi}{2} - \log 2\right)$.

34. Find the average value of $f(x, y) = y \sin xy$ over the rectangle $[0, 1] \times [0, \pi/2]$.

[**Hint:** We may express the rectangle $\{(x, y) : a \le x \le b,\ c \le y \le d\}$ as $[a, b] \times [c, d]$ for simplicity.

$$f_{ave} = \left(\frac{\pi}{2} \times 1\right)^{-1} \int_{y=0}^{\pi/2} \int_{x=0}^{1} y \sin xy \, dx\, dy = 1 - \frac{2}{\pi}.]$$

35. Find the volume of solid obtained by the revolution of the lemniscate $r^2 = a^2 \cos 2\theta$ about the initial line.

[**Hint:** Required volume $= 2\int_0^{\pi/4} \frac{2}{3} \pi r^3 \cos\theta \, d\theta = \frac{4}{3} \pi a^3 \int_0^{\pi/4} (\cos 2\theta)^{3/2} \cos\theta \, d\theta$

$$= \frac{4}{3} \pi a^3 \int_0^{\pi/4} (1 - 2\sin^2\theta)^{3/2} \cos\theta \, d\theta$$

Putting $\sqrt{2} \sin\theta = \sin\alpha$, $\sqrt{2}\cos\theta\, d\theta = \cos\alpha\, d\alpha$.
When $\theta = 0$, $\alpha = 0$; when $\theta = \pi/4$, $\alpha = \pi/2$.

Therefore, required volume $= \frac{4\pi a^3}{3\sqrt{2}} \int_0^{\pi/2} \cos^4\alpha \, d\alpha = \frac{\pi^2 a^3}{4\sqrt{2}}$ cubic units.]

Use double integration to find the volume of each solid.

36. The solid bounded by the elliptic paraboloid $4z = 16 - 4x^2 - y^2$ and the XOY-plane. **Ans.** 16π cubic units.

37. The solid bounded by the cylinder $x^2 + y^2 = 9$ and the planes $y + z = 9$ and $z = 0$. **Ans.** 81π cubic units.

38. The solid enclosed by the surfaces $y^2 + z^2 = 4x$ and $x = 5$. **Ans.** 50π units^3.

[**Hint:** $V = 4\int_0^5 \int_0^{2\sqrt{x}} \sqrt{4x - y^2}\, dy\, dx.$]

39. The solid bounded under $z = 3x$ and above the first quadrant area enclosed by $x = 0$, $y = 0$, $x = 4$ and $x^2 + y^2 = 25$. **Ans.** 98 cubic units

40. The solid bounded under the plane $z = x + y$ and above the area cut from the first quadrant by the ellipse $4x^2 + 9y^2 = 36$. **Ans.** 10 cubic units.

[**Hint:** Reqd. Volume $= \int_0^3 \int_0^{2\sqrt{(9-x^2)/3}} (x + y)\, dy\, dx$].

41. The solid bounded by the cylinders $y = x^2$, $y^2 = x$ and the planes $z = 0$ and $x + y + z = 2$.

[**Hint:** Volume $= \int_0^1 \int_{x^2}^{\sqrt{x}} (2 - x - y)\, dy\, dx$ **Ans.** 11/30 cubic units.

42. The solid bounded under the plane $x + z = 2$, above the plane $z = 0$ and within the cylinder $x^2 + y^2 = 4$. **Ans.** 8π cubic units.

43. The solid bounded by the co-ordinate planes and that portion of the plane which lie in the first quadrant.

(*a*) $lx + my + nz = 1$ (*AMIETE, Dec. 2001*) **Ans.** $1/6lmn$ cubic units

(*b*) $x + 2y + 3z = 4$ **Ans.** 16/9 cubic units

[**Hint:** (a) Required Volume $= \int_0^{1/l} \int_0^{(1-lx)/m} \frac{(1 - lx - my)}{n}\, dy\, dx = \frac{1}{6\, lmn}$ cubic units.]

44. The solid (tetrahedron) bounded by the coordinate planes $x = 0$, $y = 0$, $z = 0$ and the plane
$x/a + y/b + z/c = 1$. *(AMIE, S-2004)*

[**Hint:** Volume $= \int_0^a \int_0^{b\{1-(x/a)\}} c\left(1 - \dfrac{x}{a} - \dfrac{y}{b}\right) dy\, dx$

$$= c\int_0^a \int_0^{b(a-x)/y} \left[\frac{1}{a}(a-x) - \frac{y}{b}\right] dy\, dx = c\int_0^a \left[\frac{1}{a}(a-x)y - \frac{1}{b}\cdot\frac{y^2}{2}\right]_0^{b(a-x)/a} dx$$

$$= c\int_0^a \left[\frac{b}{a^2}(a-x)^2 - \frac{b(a-x)^2}{2a^2}\right] dx = \frac{bc}{2a^2}\int_0^a (a-x)^2\, dx = \frac{abc}{6}\text{cubic units}].$$

45. A triangular prism is formed by the planes, whose equations are $ay = bx$, $y = 0$ and $x = a$. Show that the volume of this prism between the plane $z = 0$ and the surface $z = c + xy$ is $ab\,(4c + ab)/8$.

[**Hint:** $V = \int_0^a \int_0^{bx/a} (c + xy)\, dy\, dx$].

46. Show that the volume common to the sphere $x^2 + y^2 + z^2 = a^2$ and the cylinder $x^2 + y^2 = ay$ is $2a^3(3\pi - 4)/9$ cubic units. *(DCE, 2004)*

[**Hint:** The required volume is twice the volume obtained by integrating $z = \sqrt{a^2 - x^2 - y^2}$ over the circle $x^2 + y^2 = ay$ in the xy-plane.

$$\text{Required volume } = 2\int_0^{\pi} \int_0^{a\sin\theta} \sqrt{a^2 - r^2}\cdot r\, dr\, d\theta.]$$

47. Show that the volume cut off the sphere $x^2 + y^2 + z^2 = a^2$ by the cone $x^2 + y^2 = z^2$ is $\pi(2 - \sqrt{2})\, a^3/3$ cubic units.

48. Find the volume bounded by the surface $z = c\left(1 - \dfrac{x}{a}\right)\left(1 - \dfrac{y}{b}\right)$ and the positive quadrant of the elliptic cylinder $x^2/a^2 + y^2/b^2 = 1$, $z = 0$. *(AMIETE, Dec. 2005)* **Ans.** $\dfrac{abc}{4}\left(\pi - \dfrac{13}{16}\right)$ cubic units.

49. Find the volume of the paraboloid of revolution $x^2 + y^2 = 4z$ cut off by the plane $z = 4$. *(DCE 2005)*

[**Hint:** $V = 4\int_0^4 \int_0^{2\sqrt{z}} \sqrt{4z - y^2}\, dy\, dz$

$$= 4\int_0^4 \left[\frac{y}{2}\sqrt{4z - y^2} + \frac{4z}{2}\sin^{-1}\frac{y}{2\sqrt{z}}\right]_0^{2\sqrt{z}} dz$$

$$= 4\int_0^4 \left[0 + 2z\cdot\frac{\pi}{2}\right] dz = 4\pi\int_0^4 z\, dz = 32\pi \text{ cubic units.}]$$

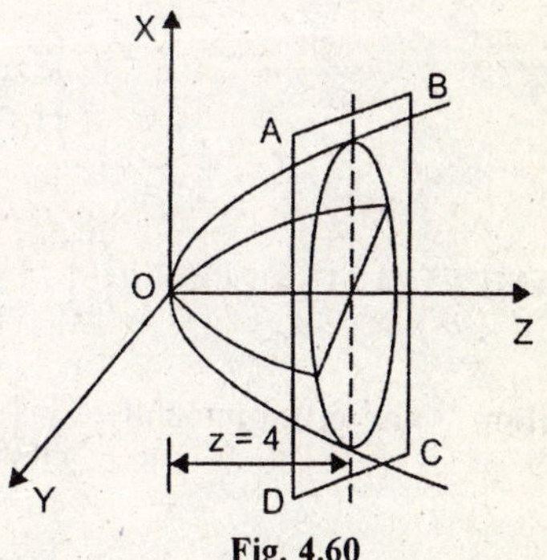

Fig. 4.60

50. Find the volume of the cylindrical column standing on the area common to the parabolas $y^2 = x$, $x^2 = y$ and cut off by the surface $z = 12 + y - x^2$. *(DCE-2003; UPTU 2001)* **Ans.** 49/140.

4.6 TRIPLE INTEGRALS

Let $f(x, y, z)$ be a continuous function on a three dimeneional region R. The definition of the double intègral can be extended in an obvious way to obtain the definition of the triple integral $\iiint_R f(x, y, z)\, dv$.

If $f(x, y, z) = 1$, then may be intepreced as measuring the volume of the region R.

Just as a double-integral can be evaluated by two successive single integrations, so a triple integral can be evaluated by three successive integrations.

Notation: The symbol $\int_a^b \int_c^d \int_h^k f(x, y, z)\, dz\, dy\, dx$ is called a *triple integral.*

We integrate $f(x, y, z)$ with respect to z (treating x and y as constants), between the limits h and k. The resulting expression is then integrated with respect to y (treating x as a constant), between the limits c and d. The result is integrated with respect to x between the limits a and b.

Thus $$\int_a^b \int_c^d \int_h^k f(x, y, z)\, dz\, dy\, dx = \int_a^b \left[\int_c^d \left(\boxed{\int_h^k f(x, y, z)\, dz} \right) dy \right] dx$$

or $$\int_a^b dx \int_c^d dy \int_h^k f(x, y, z)\, dz .$$

Remark h, k may be constants or functions of x and y. Similarly c, d may be constants or functions of x.

EXAMPLE 4.40 Evaluate $\int_0^1 \int_0^1 \int_0^x (x - 2y + z)\, dz\, dy\, dx$.

Solution The integral $= \int_0^1 \int_0^1 \left[(x-2y)z + \dfrac{z^2}{2} \right]_0^x dy\, dx = \int_0^1 \int_0^1 \left[(x-2y)x + \dfrac{x^2}{2} \right] dy\, dx = \int_0^1 \int_0^1 \left(\dfrac{3x^2}{2} - 2xy \right) dy\, dx$

$$= \int_0^1 \left[\frac{3x^2}{2} \cdot y - 2x \frac{y^2}{2} \right]_0^1 dx = \int_0^1 \left[\frac{3x^2}{2} \cdot 1 - 2x \cdot \frac{1}{2} \right] dx$$

$$= \int_0^1 \left(\frac{3x^2}{2} - x \right) dx = \left[\frac{3}{2} \cdot \frac{x^3}{3} - \frac{x^2}{2} \right]_0^1 = 0.$$

EXAMPLE 4.41 Evaluate $\int_0^1 \int_0^{1-x} \int_0^{1-x^2-y^2} dz\, dy\, dx$.

Solution The given integral $= \int_0^1 \int_0^{1-x} (1 - x^2 - y^2)\, dy\, dx$.

$$= \int_0^1 \left[y - x^2 y - \frac{y^3}{3} \right]_0^{1-x} dx = \int_0^1 \left\{ (1-x) - x^2(1-x) - \frac{1}{3}(1-x)^3 \right\} dx$$

$$= \left[x - \frac{x^2}{2} - \frac{x^3}{3} + \frac{x^4}{4} + \frac{1}{3} \frac{(1-x)^4}{4} \right]_0^1 = 1 - \frac{1}{2} - \frac{1}{3} + \frac{1}{4} - \frac{1}{12} = \frac{1}{3}.$$

EXAMPLE 4.42 Evaluate $\int_0^1 \int_0^{\sqrt{1-x^2}} \int_0^{\sqrt{1-x^2-y^2}} \frac{1}{\sqrt{1-x^2-y^2-z^2}}\, dz\, dy\, dx.$

Solution $I = \int_0^1 \int_0^{\sqrt{1-x^2}} \int_0^{\sqrt{1-x^2-y^2}} \frac{1}{\sqrt{(1-x^2-y^2)-z^2}}\, dz\, dy\, dx$

$$= \int_0^1 \int_0^{\sqrt{1-x^2}} \left[\sin^{-1}(z)/\sqrt{1-x^2-y^2}\right]_0^{\sqrt{1-x^2-y^2}} dy\, dx = \int_0^1 \int_0^{\sqrt{1-x^2}} (\sin^{-1}1 - \sin^{-1}0)\, dy\, dx$$

$$= \int_0^1 \int_0^{\sqrt{1-x^2}} \frac{\pi}{2}\, dy\, dx = \frac{\pi}{2}\int_0^1 [y]_0^{\sqrt{1-x^2}} dx = \frac{\pi}{2}\int_0^1 \sqrt{1-x^2}\, dx$$

$$= \frac{\pi}{2}\left[\frac{x}{2}\sqrt{1-x^2} + \frac{1}{2}\sin^{-1}x\right]_0^1 = \frac{\pi}{2}\cdot\frac{1}{2}\sin^{-1}1 = \frac{\pi}{2}\cdot\frac{\pi}{4} = \frac{\pi^2}{8}.$$

EXAMPLE 4.43 Evaluate the integral $\int_0^{\log 2} \int_0^x \int_0^{x+\log y} e^{x+y+z}\, dz\, dy\, dx.$ *(NIT Kurukshetra, 2008)*

Solution $I = \int_0^{\log 2} \int_0^x \int_0^{x+\log y} e^{x+y+z}\, dz\, dy\, dx$

$$= \int_0^{\log 2} e^x dx \int_0^x e^y dy \int_0^{x+\log y} e^z dz = \int_0^{\log 2} e^x dx \int_0^x e^y dy\, [e^z]_0^{x+\log y}$$

$$= \int_0^{\log 2} e^x dx \int_0^x e^y dy\, (e^{x+\log y} - 1) = \int_0^{\log 2} e^x dx \int_0^x e^y dy\, (e^x \cdot e^{\log y} - 1)$$

$$= \int_0^{\log 2} e^x dx \int_0^x e^y dy\, (ye^x - 1) = \int_0^{\log 2} e^x dx \int_0^x \underset{\text{I}}{(ye^x - 1)}\, \underset{\text{II}}{e^y}\, dy$$

$$= \int_0^{\log 2} e^x dx \left[(ye^x - 1)\, e^y - \int e^x \cdot e^y\, dy\right]_0^x = \int_0^{\log 2} e^x dx \left[(ye^x - 1)\, e^y - e^{x+y}\right]_0^x$$

$$= \int_0^{\log 2} e^x dx \left[(xe^x - 1)\, e^x - e^{2x} + 1 + e^x\right] = \int_0^{\log 2} e^x dx \left[(xe^{2x} - e^x - e^{2x} + 1 + e^x\right]$$

$$= \int_0^{\log 2} (\underset{\text{I}}{x}\underset{\text{II}}{e^{3x}} - e^{3x} + e^x)\, dx = \left[x\frac{e^{3x}}{3} - \int 1\cdot\frac{e^{3x}}{3}\, dx - \frac{e^{3x}}{3} + e^x\right]_0^{\log 2}$$

$$= \left[\frac{x}{3}e^{3x} - \frac{e^{3x}}{9} - \frac{e^{3x}}{3} + e^x\right]_0^{\log 2} = \frac{\log 2}{3}e^{3\log 2} - \frac{e^{3\log 2}}{9} - \frac{e^{3\log 2}}{3} + e^{\log 2} + \frac{1}{9} + \frac{1}{3} - 1$$

$$= \frac{8}{3}\log 2 - \frac{8}{9} - \frac{8}{3} + 2 + \frac{1}{9} + \frac{1}{3} - 1 = \frac{8}{3}\log 2 - \frac{19}{9}.$$

EXAMPLE 4.44 Evaluate $\int_0^{\pi/2} \int_0^{a\sin\theta} \int_0^{(a^2-r^2)/a} r\, dz\, dr\, d\theta.$ *(NIT Kurukshetra, 2010; VTU, 2007)*

Solution The given integral $= \int_0^{\pi/2}\left[\int_0^{a\sin\theta}\left(\int_0^{(a^2-r^2)/a} r\,dz\right)dr\right]d\theta$

$$= \int_0^{\pi/2}\left[\int_0^{a\sin\theta} r[z]_0^{(a^2-r^2)/a}\,dr\right]d\theta = \int_0^{\pi/2}\left[\int_0^{a\sin\theta} r\left(\frac{a^2-r^2}{a}\right)dr\right]d\theta$$

$$= \frac{1}{a}\int_0^{\pi/2}\left[\frac{a^2r^2}{2}-\frac{r^4}{4}\right]_0^{a\sin\theta} d\theta = \frac{1}{a}\int_0^{\pi/2}\frac{a^4}{4}(2\sin^2\theta-\sin^4\theta)\,d\theta$$

$$= \frac{a^3}{4}\left[2\cdot\frac{1}{2}\cdot\frac{\pi}{2}-\frac{3}{4}\cdot\frac{1}{2}\cdot\frac{\pi}{2}\right] = \frac{5\pi a^3}{64}.$$

4.7 VOLUME UNDER A SURFACE (TRIPLE INTEGRATION)

The volume of a vertical column (that is, a column whose lateral surface is generated by lines parallel to z-axis) whose upper end is in the surface $z = f(x, y)$ and whose lower end is in the XOY plane is also given by the triple integral

$\iint_R\left[\int_0^{f(x,y)} dz\right]dy\,dx$, where R is the lower base of the column. Obviously R is the orthogonal projection of the upper end S of the column in XOY plane.

If the upper end of the vertical column lies in the surface $z = f_2(x, y)$ and lower end in $z = f_1(x, y)$ then the volume is given by $V = \iint_R\left[\int_{f_1}^{f_2} dz\right]dy\,dx$, where R is the projection of the upper end (or lower end) of the column in the XOY plane.

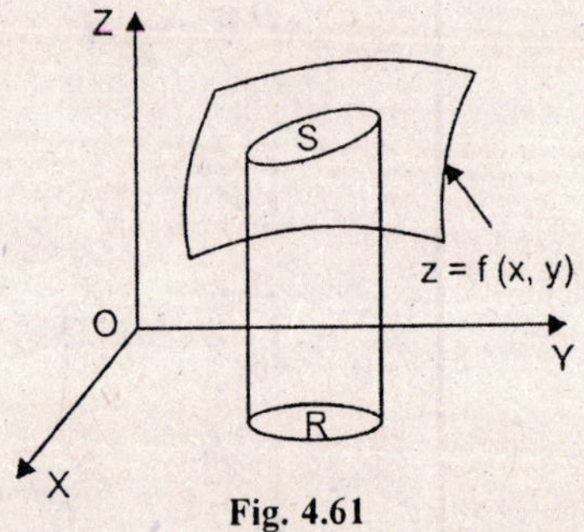

Fig. 4.61

NOTE: In case the lateral surface of the column is generated by lines parallel to x-axis and one end lies in the surface $x = \phi(y, z)$ and base R in the YOZ plane then

$$V = \iint_R\left[\int_0^{\phi(y,z)} dx\right]dy\,dz.$$

When the base of the column lies in XOZ plane and the other end in the surface $y = F(x, z)$ then $V = \iint_R\left[\int_0^{F(x,z)} dy\right]dx\,dz$.

4.8 CYLINDRICAL AND SPHERICAL COORDINATES

In this section we will discuss two new types of coordinate systems in 3-space that are often more useful than rectangular coordinate systems. These are: the cylindrical coordinate system and the spherical coordinate system. Cylindrical coordinates simplify the equations of cylinders. Spherical coordinates simplify the equations of spheres and cones.

4.8.1 Cylindrical Coordinates (r, θ, z)

See Fig. 4.62. Cylindrical coordinates represent a point P in space by ordered triples (r, θ, z) in which r and θ are polar coordinates for the vertical projection of P on the xy-plane, and z is the rectangular vertical coordinate.

The values of x, y, r and θ in rectangular and cylindrical coordinates are related by the following equations.

$x = r\cos\theta$, $y = r\sin\theta$, $z = z$, $r^2 = x^2 + y^2$, $\tan\theta = y/x$.

In cylindrical co-ordinates, the equation $r = r_0$ is a right circular cylinder of radius r_0 centered on the z-axis, where $0 \le \theta < 2\pi$.

Transformation equations: $x = r\cos\theta$, $y = r\sin\theta$, $z = z$;

Element of Volume: $dV = r\,dr\,d\theta\,dz$.

Sometimes a triple integral that is difficult to integrate in rectangular coordinates, can be evaluated more easily by making the substitution $x = r\cos\theta$, $y = r\sin\theta$, $z = z$ to convert it to an integral in cylindrical coordinates.

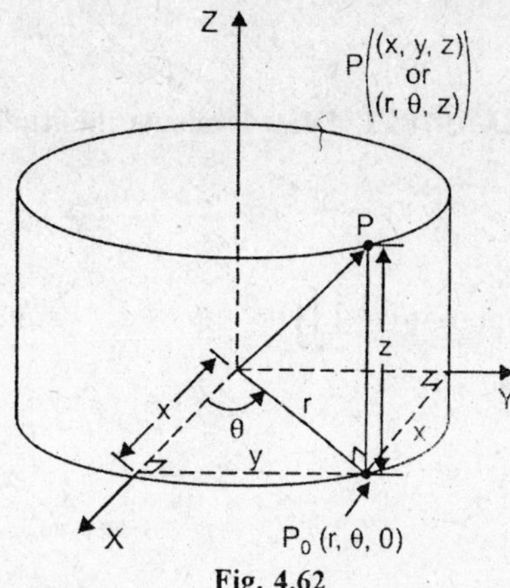

Fig. 4.62

Under such a substitution, a rectangular triple integral can be expressed as an iterated integral in cylindrical coordinates as

$$\iiint_R f(x, y, z)\,dV = \iiint_{\substack{\text{appropriate}\\ \text{limits}}} f(r\cos\theta, r\sin\theta, z)\,dz \cdot r\,dr\,d\theta$$

4.8.2 Spherical Coordinates (ρ, ϕ, θ)

See Fig. 4.63. Spherical coordinates represent a point P in space by an ordered triples (ρ, ϕ, θ) in which the first coordinate, ρ is the distance from P to the origin, the second coordinate, ϕ, is the angle $\overline{OP}$ makes with the positive z-axis ($0 \le \phi \le \pi$), and the third coordinate is the angle θ as measured in cylindrical coordinates.

The equation $\rho = \rho_0$ describes the sphere of radius ρ_0 centered at the origin. The equation $\phi = \phi_0$ describes a single cone whose vertex lies at the origin and whose axis lies along the z-axis. If ϕ_0 is greater than $\pi/2$, the cone $\phi = \phi_0$ opens downward.

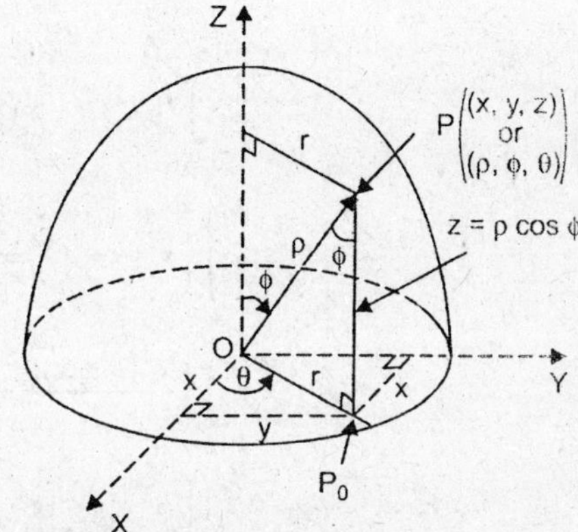

Fig. 4.63. Spherical coordinates are measured with a distance and two angle ($0 \le \theta \le 2\pi$, $0 \le \phi \le \pi$)

4.8.3 Equations Relating Spherical Coordinates to Cylindrical Coordinates and Cylindrical Coordinates to Rectangular Coordinates

$r = \rho\sin\phi$, $\theta = \theta$, $z = \rho\cos\phi$

$x = r\cos\theta = \rho\sin\phi\cos\theta$, $y = r\sin\theta = \rho\sin\phi\sin\theta$

$$\rho = \sqrt{x^2 + y^2 + z^2} = \sqrt{r^2 + z^2}.$$

Transformation equations: $x = \rho\sin\phi\cos\theta$, $y = \rho\sin\phi\sin\theta$, $z = \rho\cos\phi$.

Element of volume: $dV = \rho^2\sin\phi\,d\rho\,d\phi\,d\theta$.

A triple integrals in spherical coordinates can be evaluated as an iterated integral of the form

$$\iiint_D f(\rho, \phi, \theta)\,dV = \iiint_{\substack{\text{appropriate}\\ \text{limits}}} f(\rho, \phi, \theta)\rho^2\sin\phi\,d\rho\,d\phi\,d\theta$$

To evaluatc $\iiint_D f(\rho, \phi, \theta)\,dV$ over a region D in space in spherical coordinates, integrating first with respect to ρ, then with respect to ϕ, and finally with respect to θ.

ILLUSTRATIVE EXAMPLE

EXAMPLE 4.45 Evaluate the triple integral $\iiint_R (x^2+y^2+z^2)\,dx\,dy\,dz$, the region R bounded by

$x+y+z=a\ (a>0),\ x=0,\ y=0,\ z=0.$ *(AMIE, W-2002)*

Solution
$$\iiint_R (x^2+y^2+z^2)\,dz\,dy\,dx = \int_{x=0}^{a}\int_{y=0}^{a-x}\int_{z=0}^{a-x-y}(x^2+y^2+z^2)\,dz\,dy\,dx$$

$$= \int_{x=0}^{a}\int_{y=0}^{a-x}\left[x^2z+y^2z+\frac{z^3}{3}\right]_{z=0}^{a-x-y} dy\,dx$$

$$= \int_{x=0}^{a}\int_{y=0}^{a-x}\left\{x^2(a-x)-x^2y+(a-x)y^2-y^3+\frac{(a-x-y)^3}{3}\right\}dy\,dx$$

$$= \int_{x=0}^{a}\left[x^2(a-x)y-\frac{x^2y^2}{2}+\frac{(a-x)y^3}{3}-\frac{y^4}{4}-\frac{(a-x-y)^4}{12}\right]_{y=0}^{a-x} dx$$

$$= \int_0^a\left\{x^2(a-x)^2-\frac{x^2(a-x)^2}{2}+\frac{(a-x)^4}{3}-\frac{(a-x)^4}{4}+\frac{(a-x)^4}{12}\right\}dx$$

$$= \int_0^a\left\{\frac{x^2(a-x)^2}{2}+\frac{(a-x)^4}{6}\right\}dx = \frac{1}{2}\int_0^a (a^2x^2-2ax^3+x^4)\,dx+\frac{1}{6}\int_0^a (a-x)^4\,dx$$

$$= \frac{1}{2}\left[\frac{a^2x^3}{3}-\frac{ax^4}{2}+\frac{x^5}{5}\right]_0^a-\frac{1}{6}\left[\frac{(a-x)^5}{5}\right]_0^a = \frac{a^5}{6}-\frac{a^5}{4}+\frac{a^5}{10}+\frac{a^5}{30} = \frac{a^5}{20}.$$

EXAMPLE 4.46 Evaluate the integral $\iiint_T z\,dx\,dy\,dz$, where T is the region bounded by the cone $x^2\tan^2\alpha+y^2\tan^2\beta=z^2$ and the planes $z=0$ to $z=h$ in the first octant. *(AMIETE, Dec. 2009, Dec. 2007)*

Solution The required region can be written as

$$0\le z\le\sqrt{x^2\tan^2\alpha+y^2\tan^2\beta},\ 0\le y\le\left(\sqrt{h^2-x^2\tan^2\alpha}\right)\cot\beta,\ 0\le x\le h\cot\alpha.$$

Therefore, $I=\int_0^{h\cot\alpha}dx\left[\int_0^{\left(\sqrt{h^2-x^2\tan^2\alpha}\right)\cot\beta}\frac{1}{2}(x^2\tan^2\alpha+y^2\tan^2\beta)\,dy\right.$

$$=\frac{1}{2}\int_0^{h\cot\alpha}\left[x^2\sqrt{h^2-x^2\tan^2\alpha}\,\tan^2\alpha+\frac{1}{3}(h^2-x^2\tan^2\alpha)^{3/2}\right]\cot\beta\,dx.$$

Substituting $x\tan\alpha=h\sin t$, $dx=h\cot\alpha\cos t\,dt$. When $x=0$, $t=0$ and when $x=h\cot\alpha$, $t=\pi/2$. We obtain

$$I=\frac{\cot\beta}{2}\int_{t=0}^{\pi/2}\left[h^2\sin^2 t(h\cos t)+\frac{1}{3}(h^3\cos^3 t)\right]h\cot\alpha\cos t\,dt$$

$$=\frac{h^4}{2}\cot\alpha\cot\beta\left[\int_0^{\pi/2}\left(\sin^2 t\cos^2 t+\frac{1}{3}\cos^4 t\right)dt\right]$$

$$= \frac{h^4}{2}\cot\alpha\cot\beta\left[\int_0^{\pi/2}\left(\sin^2 t - \sin^4 t + \frac{1}{3}\cos^4 t\right)dt\right]$$

$$= \frac{h^4}{2}\cot\alpha\cot\beta\left[\frac{1}{2}\cdot\frac{\pi}{2} - \frac{3\cdot 1}{4\cdot 2}\cdot\frac{\pi}{2} + \frac{1}{3}\cdot\frac{3\cdot 1}{4\cdot 2}\cdot\frac{\pi}{2}\right] = \frac{h^4}{2}\cot\alpha\cot\beta\left(\frac{\pi}{4} - \frac{3\pi}{16} + \frac{\pi}{16}\right) = \frac{\pi h^4}{16}\cot\alpha\cot\beta.$$

EXAMPLE 4.47 Using triple integral, find the

(*i*) volume of the region R bounded by the parabolic cylinder $z = 4 - x^2$ and the plane $x = 0$, $y = 0$, $y = 6$, $z = 0$.

(*ii*) volume in the first octant bounded by the co-ordinate planes and the plane $x + 2y + 3z = 4$.

(*iii*) volume of the sphere $x^2 + y^2 + z^2 = a^2$. *(AMIETE, June 2009; Madras, 2006; VTU, 2003S)*

Solution (*i*) The region R is shown in Fig. 4.64.

Required Volume $\iiint_R dz\,dy\,dx$

$$= \int_{x=0}^{2}\int_{y=0}^{6}\int_{z=0}^{4-x^2} dz\,dy\,dx = \int_0^2\int_0^6 (4-x^2)\,dy\,dx$$

$$= \int_0^2 \left[4y - x^2 y\right]_0^6 dx = \int_0^2 (24 - 6x^2)\,dx$$

$$= \left[24x - 2x^3\right]_0^2 = 48 - 16 = 32 \text{ cubic units.}$$

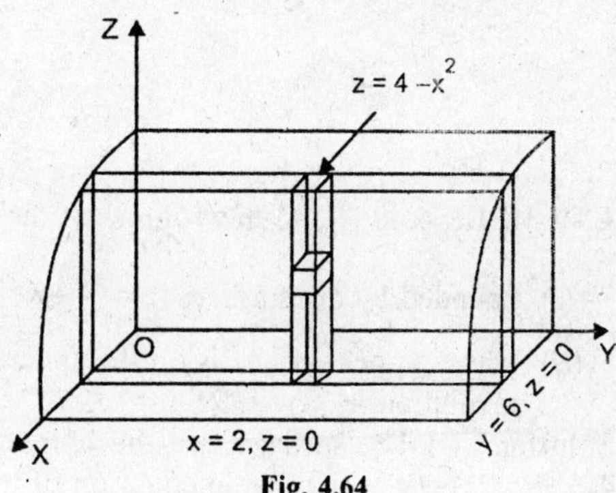

Fig. 4.64

(*ii*) Here $z = (4 - x - 2y)/3$ and region R is the triangle OAB.

$$\therefore \text{ Volume required } = \iint_R\left[\int_0^z dz\right]dy\,dx$$

$$= \iint_R z\,dy\,dx = \frac{1}{3}\int_0^4\int_0^{(4-x)/2}(4 - x - 2y)\,dy\,dx$$

$$= \frac{1}{3}\int_0^4\left[(4-x)y - y^2\right]_0^{(4-x)/2} dx$$

$$= \frac{1}{3}\int_0^4\left[\frac{1}{2}(4-x)^2 - \frac{1}{4}(4-x)^2\right]dx$$

$$= \frac{1}{12}\int_0^4 (4-x)^2\,dx = -\frac{1}{12}\left[\frac{(4-x)^3}{3}\right]_0^4 = \frac{16}{9} \text{ cubic units.}$$

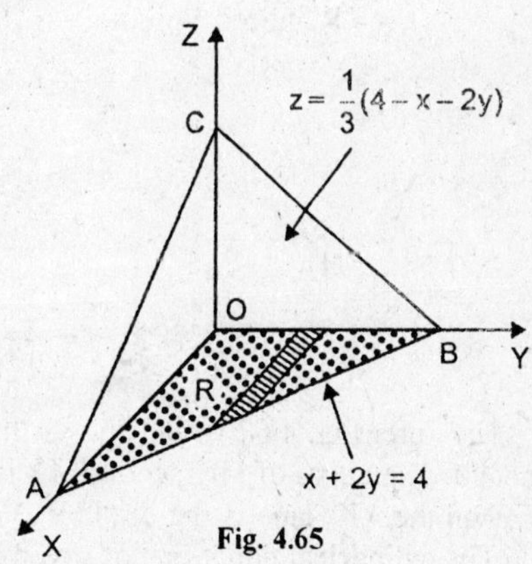

Fig. 4.65

(*iii*) Let the equation of the sphere be $x^2 + y^2 + z^2 = a^2$.

By symmetry the volume of the sphere is 8 times the volume in the first octant.

$$\therefore \text{ Volume } = 8\int_0^a\int_0^{\sqrt{a^2-x^2}}\int_0^{\sqrt{a^2-x^2-y^2}} dz\,dy\,dx$$

(Here region, R is the quadrant of the circle $x^2 + y^2 = a^2$ in the XY plane)

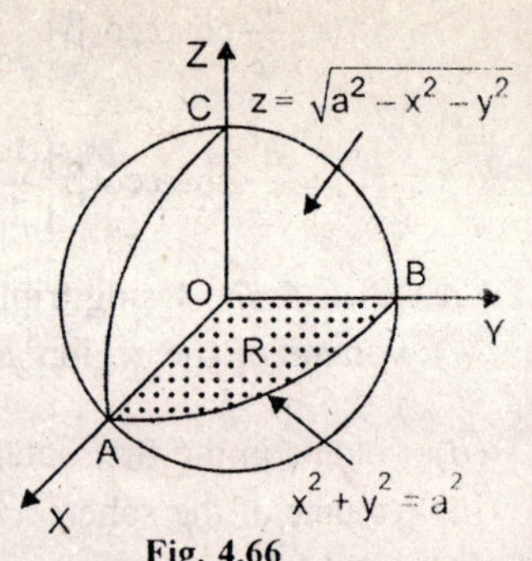

Fig. 4.66

$$= 8\int_0^a \int_0^{\sqrt{a^2-x^2}} \sqrt{a^2 - x^2 - y^2}\, dy\, dx$$

$$= 8\int_0^a \int_0^t \sqrt{t^2 - y^2}\, dy\, dx, \text{ where } t = \sqrt{a^2 - x^2} \qquad \ldots(i)$$

$$= 8\int_0^a \left[\frac{t}{2}\sqrt{t^2 - y^2} + \frac{t^2}{2}\sin^{-1}\frac{y}{t}\right]_0^t dx$$

$$= 8\int_0^a \left(\frac{t^2}{2}\sin^{-1}1\right) dx = 4\int_0^a \left(t^2 \cdot \frac{\pi}{2}\right) dx$$

$$= 2\pi\int_0^a (a^2 - x^2)\, dx = 2\pi\left[a^2 x - \frac{x^3}{3}\right]_0^a = \frac{4\pi a^3}{3}.$$

EXAMPLE 4.48 Find the volume of the solid

(*i*) bounded by the surfaces $z = 4 - x^2 - \frac{1}{4}y^2,\ z = 3x^2 + \frac{1}{4}y^2$.

(*ii*) whose upper surface is on the sphere $x^2 + y^2 + z^2 = 8$ and whose lower surface is on the paraboloid of revolution $x^2 + y^2 = 2z$.

Solution (*i*) The surfaces are the elliptic paraboloids. Eliminating z between these equations, we obtain $4x^2 + (y^2/2) = 4$ which is the equation of the cylinder having generator parallel to z-axis.

$$\text{Required Volume} = \int_{x=-1}^{1} \int_{y=-2\sqrt{2(1-x^2)}}^{2\sqrt{2(1-x^2)}} \int_{z=3x^2+\frac{1}{4}y^2}^{4-x^2-\frac{1}{4}y^2} dz\, dy\, dx$$

$$= 4\int_{x=0}^{1} \int_{y=0}^{2\sqrt{2(1-x^2)}} \int_{z=3x^2+\frac{1}{4}y^2}^{4-x^2-\frac{1}{4}y^2} dz\, dy\, dx$$

$$= 4\int_0^1 \int_0^{2\sqrt{2(1-x^2)}} \left(4 - 4x^2 - \frac{1}{2}y^2\right) dy\, dx = 4\pi\sqrt{2} \text{ cubic units.}$$

(*ii*) Refer Fig. 4.67 which shows the sphere and the paraboloid in the first octant. The curve of intersection AB lies in the plane $z = 2$ and its projection DE on the XY plane is the circle $x^2 + y^2 = 4$.

The cylindrical equations of given sphere and paraboloid are respectively $r^2 + z^2 = 8$ and $r^2 = 2z$. The polar equation of the circle, $x^2 + y^2 = 4$ is $r = 2$.

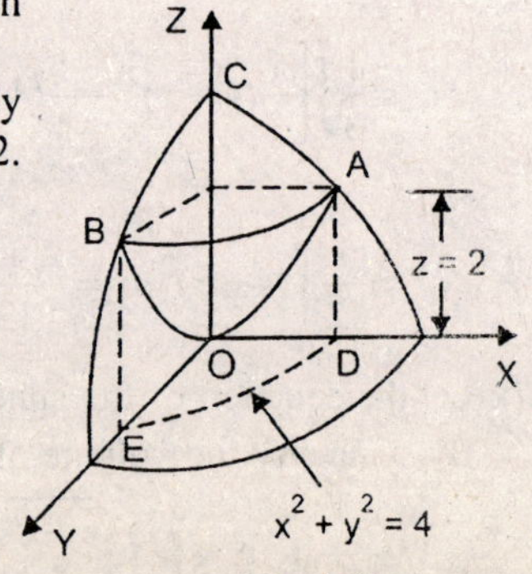

Fig. 4.67

$$\therefore \text{ Required volume, } V = \int_{\theta=0}^{2\pi} \int_{r=0}^{2} \int_{z=r^2/2}^{\sqrt{8-r^2}} dz \cdot r\, dr\, d\theta$$

$$= \int_0^{2\pi} \int_0^2 [z]_{r^2/a}^{\sqrt{(8-r^2)}} r\, dr\, d\theta$$

$$= \int_0^{2\pi}\int_0^{2}\left[\sqrt{(8-r^2)} - \frac{r^2}{2}\right] r\,dr\,d\theta = \int_0^{2\pi}\left[\frac{(8-r^2)^{3/2}}{-2\cdot 3/2} - \frac{r^4}{8}\right]_0^2 d\theta$$

$$= \int_0^{2\pi}\left[-\frac{1}{3}(8) + \frac{16\sqrt{2}}{3} - 2\right] d\theta = \int_0^{2\pi}\frac{2}{3}(8\sqrt{2}-7)\,d\theta = \frac{4\pi}{3}(8\sqrt{2}-7) \text{ cubic units.}$$

EXAMPLE 4.49 Use triple integration in spherical coordinates to find the volume of the solid G bounded above by the sphere $x^2 + y^2 + z^2 = 16$ and below by the cone $z = \sqrt{x^2 + y^2}$.

Solution The solid G is shown in Fig. 4.68. In spherical coordinates, the equation of the sphere $x^2 + y^2 + z^2 = 16$ is $\rho = 4$ and the equation of the cone $z = \sqrt{x^2 + y^2}$ is $\rho\cos\phi = \sqrt{\rho^2\sin^2\phi\cos^2\theta + \rho^2\sin^2\phi\sin^2\theta}$, which simplifies to $\rho \cos \phi = \rho \sin \phi$. On dividing both sides by $\rho \cos \phi$, we get $\tan \phi = 1 = \tan \pi/4$.

Thus $\phi = \pi/4$. The volume of G is

$$V = \iiint_G dV = \int_{\theta=0}^{2\pi}\int_{\phi=0}^{\pi/4}\int_{\rho=0}^{4} \rho^2 \sin\phi\, d\rho\, d\phi\, d\theta$$

$$= \int_0^{2\pi}\int_0^{\pi/4}\left[\frac{\rho^3}{3}\sin\phi\right]_0^4 d\phi\, d\theta = \int_0^{2\pi}\int_0^{\pi/4}\frac{64}{3}\sin\phi\, d\phi\, d\theta$$

$$= \frac{64}{3}\int_0^{2\pi}[-\cos\phi]_{\phi=0}^{\pi/4}\, d\theta = \frac{64}{3}\int_0^{2\pi}\left(1 - \frac{\sqrt{2}}{2}\right)d\theta$$

$$= \frac{64\pi}{3}(2-\sqrt{2}) \text{cubic units.}$$

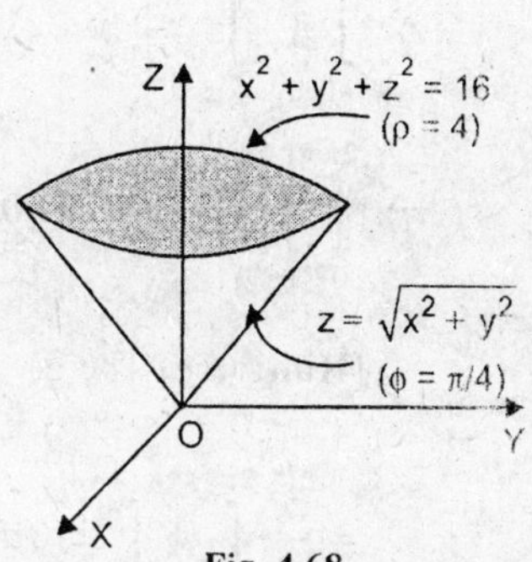

Fig. 4.68

EXERCISE 4.3

1. Evaluate the following triple integrals:

(*i*) $\int_2^3\int_1^2\int_2^5 xy^2\, dz\, dy\, dx$

(*ii*) $\int_1^2\int_0^1\int_{-1}^1 (x^2+y^2+z^2)\, dx\, dy\, dz$,

(*iii*) $\int_{x=0}^{1}\int_{y=0}^{1}\int_{z=\sqrt{x^2+y^2}}^{2} xyz\, dz\, dy\, dx$,

(*iv*) $\int_{-1}^{1} dz \int_0^z dx \int_{x-z}^{x+z} (x+y+z)\, dy$,

(AMIETE, Dec,' 08; JNTU, 2006)

(*v*) $\int_0^4\int_0^{2\sqrt{z}}\int_0^{\sqrt{4z-x^2}} dy\, dx\, dz$,

(*vi*) $\int_0^a\int_0^x\int_0^{x+y} e^{x+y+z}\, dz\, dy\, dx$, *(JNTU, 2005)*

(*vii*) $\iiint_R (x-2y+z)\, dz\, dy\, dx$, where R is the region determined by $0 \le x \le 1$, $0 \le y \le x^2$, $0 \le z \le x + y$

[Hint: (*vii*) $\iiint_R f(x, y, z)\, dz\, dy\, dx = \int_0^1\int_0^{x^2}\int_0^{x+y} (x-2y+z)\, dz\, dy\, dx$].

(viii) $\int_0^1 \int_{y^2}^1 \int_0^{1-x} x\, dz\, dx\, dy$, (ix) $\int_0^{\pi/2} \int_0^{\cos\theta} \int_{r^2}^{r\cos\theta} r\, dz\, dr\, d\theta$,

(x) $\int_0^a \int_0^x \int_0^y xyz\, dz\, dy\, dx$, (xi) $\int_0^1 \int_0^{\sqrt{1-x^2}} \int_0^{\sqrt{1-x^2-y^2}} xyz\, dz\, dy\, dx$, (VTU, 2003S)

(xii) $\int_0^1 \int_0^{1-y} \int_0^{1-x^2-y^2} dz\, dy\, dx$, (xiii) $\int_1^e \int_1^{\log y} \int_1^{e^x} \log z\, dz\, dx\, dy$, (Rohtak, 2005)

(xiv) $\int_0^2 \int_1^3 \int_1^2 xy^2 z\, dz\, dy\, dx$, (AMIETE, June 2009, Dec 2007)

(xv) $\int_{-1}^1 \int_0^z \int_{x-z}^{x+z} (x+y+z)\, dy\, dx\, dz$, (AMIETE, 2008, Cochin, 2005)

(xvi) $\int_0^{2\pi} \int_0^{\pi/4} \int_0^a r^2 \sin\theta\, dr\, d\theta\, d\phi$, where a = constant.

[**Hint:** (xvi) The given integral $= \int_0^{2\pi} d\phi \int_0^{\pi/4} d\theta \int_0^a r^2 \sin\theta\, dr$

$$= \frac{a^3}{3}\int_0^{2\pi} d\phi \int_0^{\pi/4} \sin\theta\, d\theta = \frac{a^3}{3}\int_0^{2\pi}\left(-\frac{1}{\sqrt{2}}+1\right)d\phi = \frac{a^3}{3}\cdot\frac{\sqrt{2}-1}{\sqrt{2}}\cdot 2\pi = \frac{\pi a^3}{3}(2-\sqrt{2})].$$

Ans. (i) 35/2, (ii) 6, (iii) 3/8, (iv) 0, (v) 8π, (vi) $\frac{e^{4a}}{8} - \frac{3}{4}e^{2a} + e^a - \frac{3}{8}$, (vii) 8/35, (viii) 4/35,

(ix) $\pi/64$, (x) $a^6/48$, (xi) 1/48, (xii) 1/3, (xiii) $\frac{1}{4}(e^2 - 8e + 13)$, (xv) 0.

2. Find the volume of the solid cut off by the surface $z = (x + y)^2$ from the right prism whose base, in the plane $z = 0$, is the triangle bounded by the lines $x = 0$, $y = 0$, $x + y = 1$. **Ans.** 1/4 cubic units.

[**Hint:** Required volume $= \int_{x=0}^{1} \int_{y=0}^{1-x} \int_{z=0}^{(x+y)^2} dz\, dy\, dx$.]

3. Find the volume bounded by the paraboloid $x^2 + y^2 = 1 + z$ and $z = 0$. **Ans.** $\pi/2$ unit^3.

[**Hint:** Required volume $= \int_{x=-1}^{1} \int_{y=-\sqrt{(1-x^2)}}^{\sqrt{(1-x^2)}} \int_{z=0}^{x^2+y^2-1} dz\, dy\, dx$]

4. Evaluate the triple integral of the function $f(x, y, z) = x^2$ over the region V enclosed by the planes $x = 0$, $y = 0$, $z = 0$, and $x + y + z = a$. **Ans.** $a^5/60$

5. Evaluate $\iiint_R xyz\, dz\, dy\, dx$ (i) R: Region bounded by $x + y + z = 1$ and the coordinate planes.

(ii) Over the positive octant of the sphere $x^2 + y^2 + z^2 = a^2$. **Ans.** (i) 1/720, (ii) $a^6/48$

6. Compute $\iiint_D \frac{dz\,dy\,dx}{(x+y+z+1)^3}$ if the domain D, of integration is bounded by the coordinate planes and the plane $x + y + z = 1$. **Ans.** $\frac{\log 2}{2} - \frac{5}{16}$. (*AMIE; S-2003; AMIETE, Dec. 2003*)

[**Hint:** We have to evaluate $I = \int_0^1 \int_0^{1-x} \int_0^{1-x-y} (x+y+z+1)^{-3}\, dz\, dy\, dx$

$$= \int_0^1 \int_0^{1-x} \left[-\frac{1}{2}(x+y+z+1)^{-2} \right]_0^{1-x-y} dy\,dx = -\frac{1}{2}\int_0^1 \int_0^{1-x} \left[(2)^{-2} - (x+y+1)^{-2} \right] dy\,dx$$

$$= -\frac{1}{2}\int_0^1 \left[\frac{y}{4} + \frac{1}{x+y+1} \right]_0^{1-x} dx = -\frac{1}{2}\int_0^1 \left[\left(\frac{1-x}{4} + \frac{1}{2} \right) - \frac{1}{1+x} \right] dx$$

$$= -\frac{1}{2}\int_0^1 \left(\frac{3}{4} - \frac{x}{4} - \frac{1}{1+x} \right) dx = -\frac{1}{2}\int_0^1 \left(\frac{3}{4}x - \frac{x^2}{8} - \log(1+x) \right)_0^1 = \frac{1}{16}(8 \log 2 - 5).$$

7. Find the volume of the solid within the cylinder $x^2 + y^2 = 9$ and between the planes $z = 1$ and $x + z = 5$. **Ans.** 36π cubic units.

8. (*a*) Prove that $\iiint_V (1+x+y+z)^2\, dz\, dy\, dx = \frac{31}{60}$, where V is the region of the tetrahedron bounded by $x = 0$, $y = 0$, $z = 0$ and $x + y + z = 1$. (*A.M.I.E., W-2003*)

(*b*) Evaluate the integral $\iiint_T y\, dx\, dy\, dz$ where T is the region bounded by the surfaces $x = y^2$, $x = y + 2$, $4z = x^2 + y^2$ and $z = y + 3$. (*AMIETE Dec. 2008*)

[**Hint:** Limits of z are from $(x^2 + y^2)/4$ to $y + 3$, limits of x from y^2 to $y + 2$ and limits of y from -1 to 2.

$$\therefore \iiint_T y\, dx\, dy\, dz = \int_{-1}^{2} \left[\int_{y^2}^{(y+2)} \left[\int_{(x^2+y^2)/4}^{(y+3)} y\, dz \right] dx \right] dy = \frac{837}{160}.]$$

9. Evaluate $\iiint_D x\, dx\, dy\, dz$, where D is the domain bounded by the tetrahedron with faces,

$x/a + y/b + z/c = 1$, $x = 0$, $y = 0$, $z = 0$. (*AMIE, S-2004*) **Ans.** $a^2bc/24$.

10. Find the volume of that portion of the ellipsoid $x^2/a^2 + y^2/b^2 + z^2/c^2 = 1$ which lies in the first octant.

[**Hint:** $V = \int_0^a \int_0^{b\sqrt{1-x^2/a^2}} \int_0^{c\sqrt{1-x^2/a^2-y^2/b^2}} dz\, dy\, dx.$] **Ans.** $\pi abc/6$ cubic units.

11. Find the volume of the tetrahedron bounded by the coordinate planes $x = 0$, $y = 0$, $z = 0$ and the plane $x/a + y/b + z/c = 1$. (*AMIE, S-2004*)

[**Hint:** Volume, $V = \int_0^a \int_0^{b(1-x/a)} \int_0^{c\{1-(x/a)-(y/b)\}} dz\, dy\, dx$

$$= \int_0^a \int_0^{b(1-x/a)} c(1 - x/a - y/b)\, dy\, dx = c\int_0^a \int_0^{b(a-x)/a} \left\{ \frac{1}{a}(a-x) - \frac{y}{b} \right\} dy\, dx$$

$$= c\int_0^a \left[\frac{1}{a}(a-x)y - \frac{1}{b}\frac{y^2}{2} \right]_0^{b(a-x)/a} dx = \frac{bc}{2a^2}\int_0^a (a-x)^2\, dx = abc/6 \text{ cubic units.}]$$

12. Evaluate $\iiint \frac{dx\,dy\,dz}{\sqrt{1-x^2-y^2-z^2}}$ throughout the volume of the sphere $x^2+y^2+z^2=1$. **Ans.** π^2.

13. Find the volume of the region above the xy-plane bounded by the paraboloid $z = x^2 + y^2$ and the cylinder $x^2 + y^2 = a^2$. **Ans.** $\pi a^4/2$ cubic units

[**Hint:** The volume is most easily found by using cylindrical co-ordinates. The equations of the paraboloid and cylinder in cylindrical co-ordinates are respectively $z = r^2$ and $r = a$. Required volume = 4 times volume in the first octant

$$= 4\int_{\theta=0}^{\pi/2}\int_{r=0}^{a}\int_{z=0}^{r^2} r\,dz\,dr\,d\phi = \pi a^4/2.]$$

14. Find the volume of the region bounded by $z = x^2 + y^2$ and $z = 2x$. **Ans.** $\pi/2$.

15. Find the volume of the paraboloid of revolution $x^2 + y^2 = 4z$ cut off by the plane $z = 4$. **Ans.** 32π cubic units.

[**Hint:** Required volume $= \int_{x=-4}^{4}\int_{y=-\sqrt{16-x^2}}^{\sqrt{16-x^2}}\int_{z=(x^2+y^2)/4}^{z=4} dz\,dy\,dx = 4\int_0^4\int_0^{\sqrt{16-x^2}}\int_{(x^2+y^2)/4}^{4} dz\,dy\,dx.$]

16. Find the volume of the region bounded by the paraboloids $z = x^2 + y^2$ and $z = 6 - \frac{x^2+y^2}{2}$.

[**Hint:** $V = \int_{x=-2}^{2}\int_{y=-\sqrt{4-x^2}}^{\sqrt{4-x^2}}\int_{x^2+y^2}^{6-\frac{1}{2}(x^2+y^2)} dz\,dy\,dx = \int_{-2}^{2}\int_{-\sqrt{4-x^2}}^{\sqrt{4-x^2}}\left[6-\frac{1}{2}(x^2+y^2)-(x^2+y^2)\right]dy\,dx$

$= 6\int_0^2\int_0^{\sqrt{4-x^2}}\left[4-(x^2+y^2)\right]dy\,dx = 6\int_0^{\pi/2}\int_0^2(4-r^2)r\,dr\,d\theta = 12\pi$]. **Ans.** 12π cubic units.

17. Evaluate $\iiint (x+y+z)\,dz\,dy\,dx$ over the tetrahedron bounded by the planes $x = 0$, $y = 0$, $z = 0$ and $x + y + z = 1$. **Ans.** 1/4.

18. Evaluate $\iiint z\,dx\,dy\,dz$, over the volume enclosed between the cone $x^2 + y^2 = z^2$ and the sphere $x^2 + y^2 + z^2 = 1$ on the positive side of the xy-plane. [**Hint.** Integrate w.r.t. z and then change to polars]. **Ans.** $\pi/8$.

19. Find the volume of the solid bounded by the sphere $x^2 + y^2 + z^2 = 4$ and the surface of the paraboloid $x^2 + y^2 = 3z$. **Ans.** $19\pi/6$ cubic units

[**Hint:** The two surfaces intersect at $z = 1$.

$$V = \iiint_D dz\,dy\,dx = \iint_R dy\,dx\int_{z=\frac{1}{3}(x^2+y^2)}^{\sqrt{(4-x^2-y^2)}} dz = \iint_R\left[\sqrt{4-x^2-y^2}-\frac{1}{3}(x^2+y^2)\right]dy\,dx.$$

Changing to polars, R being the circle $x^2 + y^2 \le 3$.

Therefore, $V = \int_0^{2\pi} d\theta\int_0^{\sqrt{3}}\left(\sqrt{4-r^2}-\frac{r^2}{3}\right)r\,dr = \frac{19\pi}{6}$.]

20. Find the volume of the solid bounded by the surfaces $z = 0$, $3z = x^2 + y^2$ and $x^2 + y^2 = 9$.
(*AMIETE June 2005*) **Ans.** $27\pi/2$.

4.9 CHANGE OF VARIABLES IN MULTIPLE INTEGRALS; JACOBIAN

Sometimes it may not be possible to evaluate a double integral with the help of methods discussed so far. Those methods fail because the integrand and/or, the region of integration happen to be complicated. In such cases certain substitutions of the type $x = f_1(u, v)$, $y = f_2(u, v)$ may

(*i*) reduce the integrand to a simple form.

(*ii*) map a complicated region R in xy-plane into a simpler region R' (say a rectangle) in *uv*-plane, over which the integral can be easily evaluated.

The procedure for evaluation of double integral and triple integral without proof, is given below:

4.9.1 Double Integrals

Consider the double integral $\iint\limits_R f(x, y)\,dx\,dy$ ...(4.8)

and let the substitutions be $x = f_1(u, v)$, $y = f_2(u, v)$. ...(4.9)

Then $dx\,dy$ is given by $dx\,dy = \left|\dfrac{\partial(x, y)}{\partial(u, v}\right|\,du\,dv$.

Thus, under the transformation (4.9), (4.8) is transformed as

$$\iint\limits_R f(x, y)\,dx\,dy = \iint\limits_{R'} F(u, v)\left|\frac{\partial(x, y)}{\partial(u, v)}\right|du\,dv,$$

that is, the integrand is expressed in terms of u and v, and $dx\,dy$ is replaced by $du\,dv$ times the absolute value of the **Jacobian**.

Remark: Sometimes we use the transformation $u = \phi_1(x, y)$, $v = \phi_2(x, y)$.

In some cases it may not be possible to obtain the inverse transformation that is, $x = x(u, v)$, $y = y(u, v)$. In such cases, we get $\dfrac{\partial(u, v)}{\partial(x, y)}$. Then with the help of the property $\dfrac{\partial(x, y)}{\partial(u, v)} \bullet \dfrac{\partial(u, v)}{\partial(x, y)} = 1$, we obtain

$$\frac{\partial(x, y)}{\partial(u, v)} = 1\Big/\frac{\partial(u, v)}{\partial(x, y)}.$$

4.9.2 Triple Integrals

Consider the triple integral $\iiint\limits_V f(x, y, z)\,dx\,dy\,dz$. Let the transformation be

$$x = f_1(u, v, w),\ y = f_2(u, v, w),\ z = f_3(u, v, w).$$

Then $\quad dx\,dy\,dz = \left|\dfrac{\partial(x, y, z)}{\partial(u, v, w)}\right|du\,dv\,dz$,

and $\quad \iiint\limits_V f(x, y, z)\,dx\,dy\,dz = \iiint\limits_{V'} F(u, v, w)\left|\dfrac{\partial(x, y, z)}{\partial(u, v, w)}\right|du\,dv\,dw,$

where V' is the new region of integration.

***NOTE*:** The remark after case (*A*) above is applicable in this case also that is, if we find convenient to use the transformation

$u = \phi_1(x, y, z)$, $v = \phi_2(x, y, z)$, $w = \phi_3(x, y, z)$, then we get $\dfrac{\partial(u, v, w)}{\partial(x, y, z)}$.

The value of $\dfrac{\partial(x, y, z)}{\partial(u, v, w)}$ is given by $\dfrac{\partial(x, y, z)}{\partial(u, v, w)} = 1\Big/\dfrac{\partial(u, v, w)}{\partial(x, y, z)}$.

ILLUSTRATIVE EXAMPLES

EXAMPLE 4.50 Evaluate $\iint xy\,dy\,dx$ over the area bounded by $y^2 = 4x$, $y^2 = 8x$, $x^2 = 4y$, $x^2 = 8y$.

Solution Obviously the region of integration R in the xy-plane is very complicated.

By the substitutions $y^2/x = u$, $x^2/y = v$, the region R maps into region R' of uv-plane which is bounded by the lines $u = 4$, $u = 8$ and $v = 4$ and $v = 8$. Thus R' is a square.

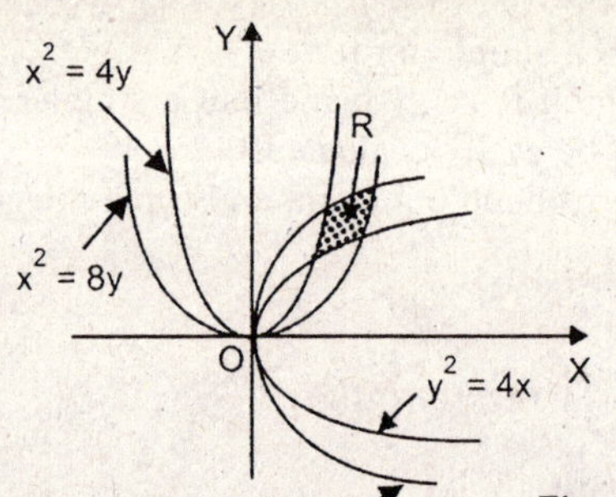

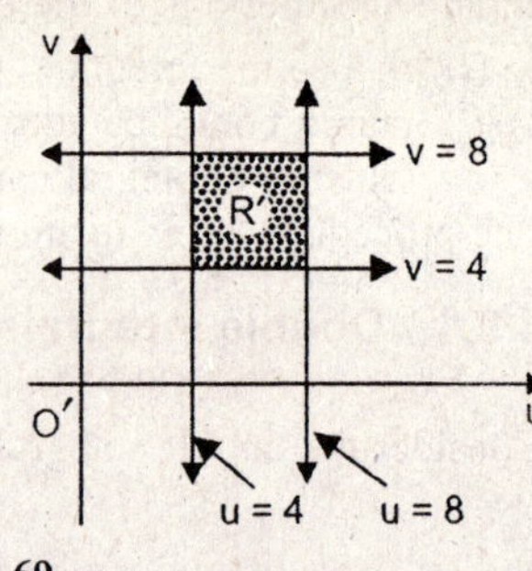

Fig. 4.69

Also $$xy = \left(y^2/x\right)\left(x^2/y\right) = uv$$

and $$\frac{\partial(u, v)}{\partial(x, y)} = \begin{vmatrix} -y^2/x^2 & 2y/x \\ 2x/y & -x^2/y^2 \end{vmatrix} = 1 - 4 = -3.$$

Therefore, $$\left|\frac{\partial(u, v)}{\partial(x, y)}\right| = 3 \Rightarrow \left|\frac{\partial(x, y)}{\partial(u, v)}\right| = \frac{1}{3}.\ \text{Hence, } dx\,dy = \frac{1}{3}du\,dv.$$

Thus $$\iint_R xy\,dy\,dx = \int_4^8\int_4^8 uv\frac{1}{3}du\,dv = \frac{1}{3}\int_4^8\left[\frac{u^2}{2}\right]_4^8 v\,dv = \frac{1}{3}\int_4^8\frac{1}{2}\cdot 48v\,dv = 4\left[v^2\right]_4^8 = 192.$$

EXAMPLE 4.51 Use the transformation

$$x = \frac{u}{2}(1+v),\ y = \frac{u}{2}(1-v) \text{ to evaluate } \int_0^\infty\int_0^\infty e^{-(x+y)}\sqrt{xy}\,dy\,dx.$$

Solution We have $x + y = u$, $\sqrt{xy} = u\sqrt{1-v^2}/2$.

Also $$\frac{\partial(x, y)}{\partial(u, v)} = \begin{vmatrix} \frac{1+v}{2} & \frac{u}{2} \\ \frac{1-v}{2} & \frac{-u}{2} \end{vmatrix} = -\frac{u}{2}.\ \text{Therefore, } dy\,dx = \left|\frac{\partial(x, y)}{\partial(u, v)}\right| dv\,du = \frac{u}{2}dv\,du.$$

Obviously, the region of integration R in xy-plane is in the first quadrant.

The line $x = 0$ transforms into $u = 0$, $v = -1$. Also the transform of $y = 0$ is $u = 0$, $v = 1$.

As $x \to \infty$, $y \to \infty$ we find $u \to \infty$, while $-1 < v < 1$.

Fig. 4.70

Hence $$\int_0^\infty\int_0^\infty e^{-(x+y)}\sqrt{xy}\,dy\,dx = \int_0^\infty\int_{-1}^1 e^{-u}\cdot\frac{u}{2}\sqrt{1-v^2}\,\frac{u}{2}dv\,du$$

$$= \frac{1}{4}\int_0^\infty u^2e^{-u}du\int_{-1}^1\sqrt{1-v^2}\,dv = \frac{1}{4}\left[u^2(-e^{-u}) - (2u)(e^{-u}) + 2(-e^{-u})\right]_0^\infty$$

$$= 2\left[\frac{v}{2}\sqrt{1-v^2} + \frac{1}{2}\sin^{-1}v\right]_0^1 = \frac{1}{4}(2)\cdot 2\left[\frac{1}{2}\cdot\frac{\pi}{2}\right] = \frac{\pi}{4}.$$

EXAMPLE 4.52 Evaluate $\int_{y=0}^{2}\int_{x=0}^{y}\left[(x-y)^2+2(x+y)+1\right]^{-1/2}dx\,dy$ using the substitutions $x = u(1 + v)$, $y = v(1 + u)$, $u \geq 0$, $v \geq 0$.

Solution Substituting $x = u(1 + v)$, $y = v(1 + u)$ in the integrand, we get

$$\left[(x-y)^2+2(x+y)+1\right]^{-1/2}=\left[(u-v)^2+2(u+v+2uv)+1\right]^{-1/2}=\left[(u+v+1)^2\right]^{-1/2}=1/(u+v+1).$$

Also $dx\,dy = \left|\dfrac{\partial(x, y)}{\partial(u, v)}\right| dv\,du$. Now $\dfrac{\partial(x, y)}{\partial(u, v)} = \begin{vmatrix} 1+v & u \\ v & 1+u \end{vmatrix} = 1+u+v.$

The region of integration R of the xy-plane is bounded by the curves $x = 0$, $x = y$, $y = 0$ and $y = 2$. When $x = 0$, $u = 0$ or $v = -1$. But $v \geq 0$, therefore, $x = 0 \Rightarrow u = 0$.

When $x = y$, we have $u = v$.

Also $y = 0 \Rightarrow v = 0$ and $y = 2 \Rightarrow v(1 + u) = 2$, which is a rectangular hyperbola having $v = 0$ and $u = -1$ as asymptotes.

The region R' into which R is mapped under the given transformation is as shown.

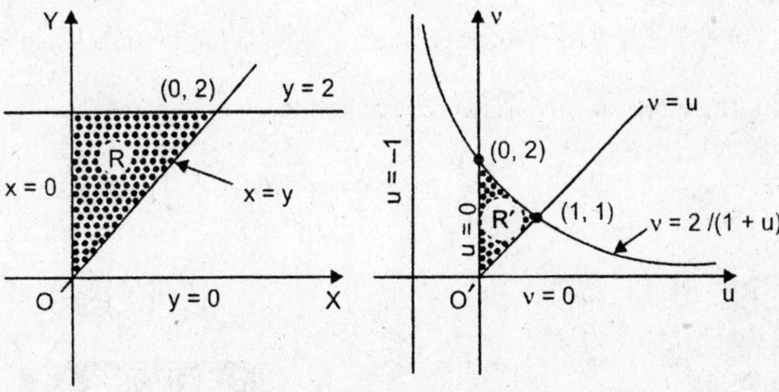

Fig. 4.71

$$\int_0^2\int_0^y\left[(x-y)^2+2(x+y)+1\right]^{-1/2}dx\,dy=\int_{u=0}^{1}\int_{v=u}^{2/(1+u)}\frac{1}{u+v+1}(u+v+1)\,dv\,du$$

$$=\int_0^1\int_u^{2/(1+u)}1\,dv\,du=\int_0^1\left(\frac{2}{1+u}-u\right)du=\left[2\log(1+u)-\frac{u^2}{2}\right]_0^1=2\log 2-\frac{1}{2}.$$

EXAMPLE 4.53 Using the transformation $x + y = u$, $y = uv$, show that $\int_{x=0}^{1}\int_{y=0}^{1-x}e^{y/(x+y)}dy\,dx=\frac{1}{2}(e-1)$.

(*PTU, 2003*)

Solution The region R of integration of xy-plane is as shown in Fig. 4.72(a). It is bounded by the curves $y = 0$, $y = 1 - x$, $x = 0$, $x = 1$. We have $e^{y/(x+y)} = e^{uv/u} = e^v$.

Since $y = uv$, $x = u - y = u - uv$ that is, $x = u(1 - v)$.

Now, $\dfrac{\partial(x, y)}{\partial(u, v)} = \begin{vmatrix} 1-v & -u \\ v & u \end{vmatrix} = u.$

Therefore, $dydx = ududv$.

When $y = 0$, $u = 0$, $v = 0$. When $y = 1 - x$ that is, $x + y = 1$, we have $u = 1$.

$x = 0 \Rightarrow u = 0$, $1 - v = 0$ or $v = 1$.

$x = 1 \Rightarrow u(1 - v) = 1$, a rectangular hyperbola.

The R' in the uv-plane is bounded by $u = 0$, $v = 0$, $u = 1$, $v = 1$ and $u(1 - v) = 1$. It is a square as shown in the figure 4.72(b).

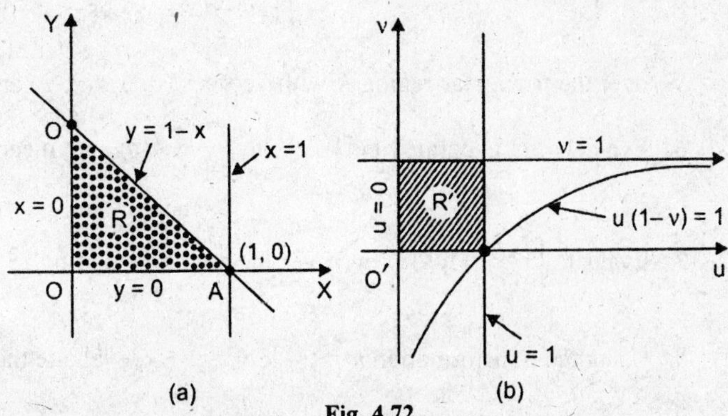

Fig. 4.72

$$\therefore \int_{x=0}^{1}\int_{y=0}^{1-x} e^{y/(x+y)}dy\,dx = \int_{v=0}^{1}\int_{u=0}^{1} e^{v}u\,du\,dv = \int_0^1 e^{v}\left[\frac{u^2}{2}\right]_0^1 dv = \frac{1}{2}\int_0^1 e^{v}dv = \frac{1}{2}(e-1).$$

EXAMPLE 4.54 Evaluate $\iiint_R \frac{dx\,dy\,dz}{(x^2+y^2+z^2)^{3/2}}$, where R is the region bounded by the spheres

$$x^2 + y^2 + z^2 = a^2 \text{ and } x^2 + y^2 + z^2 = b^2 \ (a > b > 0).$$

Solution Transforming to spherical polar co-ordinates (ρ, ϕ, θ) that is, using the substitutions $x = \rho \sin\phi \cos\theta$, $y = \rho \sin \phi \sin \theta$, and $z = \rho \cos \phi$.

We have by Ex. 1.70, Page 1.56, $\frac{\partial(x, y, z)}{\partial(\rho, \phi, \theta)} = \rho^2 \sin\phi$. Also $x^2 + y^2 + z^2 = \rho^2$.

Therefore, the given integral reduces to

$$\int_{\theta=0}^{2\pi}\int_{\phi=0}^{\pi}\int_{\rho=b}^{a} \frac{1}{\rho^3}\rho^2 \sin\phi\, d\rho\, d\phi\, d\theta = \int_0^{2\pi}\int_0^{\pi}[\log\rho]_b^a \sin\phi\, d\phi\, d\theta = \log\frac{a}{b}\int_0^{2\pi}[-\cos\phi]_0^{\pi}\, d\theta$$

$$= \log\left(\frac{a}{b}\right)2\cdot 2\pi = 4\pi\log\left(\frac{a}{b}\right).$$

EXERCISE 4.4

Evaluate the following problems 1 to 3.

1. $\iint (x^4 - y^4)\,dx\,dy$ in the first quadrant in which $1 \le x^2 - y^2 \le 2$; $1 \le xy \le 2$.

[**Hint:** Let $x^2 - y^2 = u$, $xy = v$]. **Ans.** 3/4.

2. $\iint_R \cos\left(\frac{x-y}{x+y}\right)dx\,dy$, where R be the region bounded by $x + y = 1$, $x = 0$, $y = 0$.

[**Hint:** Let $x - y = u$, $x + y = v$] **Ans.** (sin1)/2.

3. $\iint y^3\,dx\,dy$ over the area enclosed by $y^2 = x$, $y^2 = 2x$, $x^2 = y$, $x^2 = 3y$ **Ans.** 28/9.

4. Using the transformation $x + y = u$, $y = uv$, show that $\iint \sqrt{xy(1-x-y)}\,dx\,dy = \frac{2\pi}{105}$, integration being taken over the area of the triangle bounded by the lines $x = 0$, $y = 0$, $x + y = 1$.

(AMIETE, June 2011, Dec. 2009)

5. Using the transformation $u = (x + y)/2$, $v = (x - y)/2$, evaluate the integral

$$\iint_R \sin\frac{1}{2}(x+y)\cos\frac{1}{2}(x-y)\,dx\,dy,$$

over the triangular region R with vertices (0, 0), (2, 0) and (1, 1). **Ans.** (2 – sin 2)/2.

6. Evaluate the integral $\int_0^{\infty}\int_0^{\infty} e^{-(x+y)}\sin\frac{\pi y}{x+y}\,dx\,dy$, by means of transformation $u = x + y$ and $v = y$.

Ans. $2/\pi$.

7. Evaluate $\iint (x^2 + y^2)\,dx\,dy$, over the area bounded by $x^2 - y^2 = 1$, $x^2 - y^2 = 9$, $xy = 2$ and $xy = 4$.

Ans. 8.

8. Using the transformation $u = y - x$, $v = y + x$, evaluate the integral $\iint_S e^{(y-x)/(y+x)}dx\,dy$, where S is the triangle bounded by the line $x + y = 2$ and the two coordinate axes. **Ans.** $(e - e^{-1})$.

[**Hint:** Solving for x and y, we find $x = (v - u)/2$ and $y = (v + u)/2$.

The Jacobian determinant is $J(u, v) = -1/2$, $|J| = 1/2$.

$$\iint_S e^{(y-x)/(y+x)} dx\, dy = \frac{1}{2} \iint_T e^{u/v} du\, dv$$

$$= \frac{1}{2} \int_{v=0}^{2} \int_{u=-v}^{v} e^{u/v}\, du\, dv = e - e^{-1}.]$$

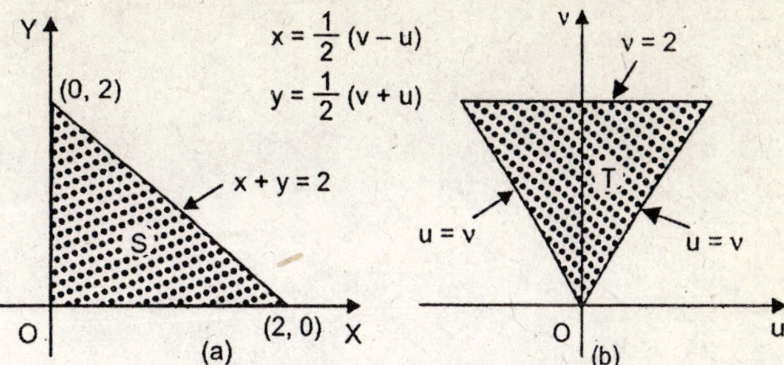

Fig. 4.73: Mapping by a linear transformation

9. Use a suitable linear transformation to evaluate the double integral $\iint_S (x-y)^2 \cos^2(x+y)\, dx\, dy$, where S is the rhombus with successive vertices at $(\pi, 0)$, $(2\pi, \pi)$, $(\pi, 2\pi)$, $(0, \pi)$. *(AMIETE, Dec 2006)* **Ans.** $\pi^4/3$.

[**Hint:** The region S is given in Fig. 4.74. The equations of the sides AB, BC, CD and DA are respectively $x - y = \pi$, $x + y = 3\pi$, $x - y = -\pi$ and $x + y = \pi$.

Substitute $y - x = u$ and $y + x = v$. Then $-\pi \le u \le \pi$ and $\pi \le v \le 3\pi$. We obtain $x = (v - u)/2$, $y = (v + u)/2$ and

$$J = \frac{\partial(x, y)}{\partial(u, v)} = -\frac{1}{2},\ |J| = \frac{1}{2}.$$

Therefore,

$$I = \iint_S (x-y)^2 \cos^2(x+y)\, dx\, dy = \frac{1}{2} \int_{\pi}^{3\pi} \int_{-\pi}^{\pi} u^2 \cos^2 v\, du\, dv = \frac{\pi^4}{3}.]$$

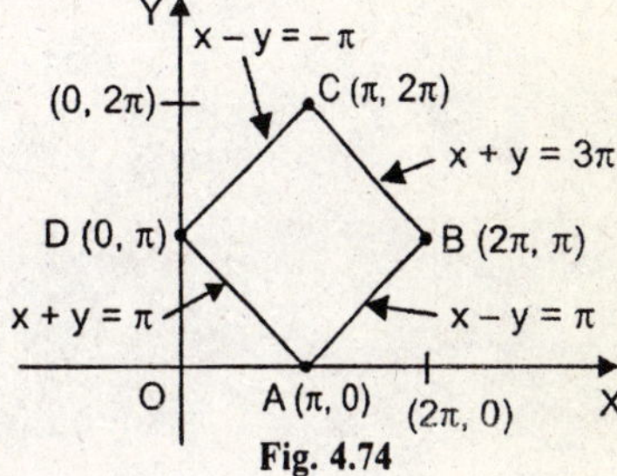

Fig. 4.74

10. Compute the double integral $\iint_D (y-x)\, dx\, dy$ over the region D in the xy-plane bounded by the straight lines $y = x + 1$, $y = x - 3$, $y = (-x/3) + (7/3)$, $y = (-x/3) + 5$. **Ans.** –8.

[**Hint:** Put $u = y - x$, $v = y + (x/3)$, $|J| = 3/4$. Therefore, $\iint_D (y-x)\, dx\, dy = \int_{7/3}^{5} \int_{-3}^{1} \frac{3}{4} u\, du\, dv = -8$].

11. Evaluate $\int_{y=0}^{4} \int_{x=y/2}^{(y/2)+1} \frac{2x-y}{2} dx\, dy$ by applying the transformation $u = (2x - y)/2$, $v = y/2$. **Ans.** 2.

12. Evaluate $\int_0^1 \int_0^{1-x} \sqrt{x+y}\,(y-2x)^2\, dy\, dx$, by applying the transformation $u = x + y$ and $v = y - 2x$. **Ans.** 2/9.

13. Evaluate $\iint_R e^{xy} dy\, dx$, where R is the region enclosed by the lines $y = x/2$, $y = x$ and the hyperbolas $y = 1/x$, $y = 2/x$. [**Hint:** Use $u = y/x$, $v = xy$.] **Ans.** $[(e^2 - e)\log 2]/2$.

14. Using the transformation $u = x + y$ and $v = x - 2y$, evaluate $\iint_R (x+y)^2 dx\, dy$, where R is the parallelogram in the xy-plane with vertices (1, 0), (3, 1), (2, 2) and (0, 1). *(UPTU, 2004)* **Ans.** 21.

15. Evaluate the integral $\iiint (x^2 + y^2 + z^2)\, dx\, dy\, dz$ taken over the volume enclosed by the sphere $x^2 + y^2 + z^2 = 1$. [**Hint:** Changing to spherical polar coordinates]. **Ans.** $4\pi/5$.

16. Evaluate $\iiint_V dz\, dy\, dx$, where V is the volume bounded by the ellipsoid $(2x + y + z)^2 + (3x - y + z)^2 + 4(x + 2y)^2 = 16$ **Ans.** $32\pi/3$.

CHAPTER 5

Linear Algebra: Matrices*, Vectors, Determinants, Linear Systems of Equations

5.1 INTRODUCTION

Linear algebra includes the theory and applications of linear systems of equations, linear transformations and eigenvalue problems. Linear algebra makes systematic use of vectors and matrices and, to a lesser extent, determinants. These days, matrix theory occupies an important place and has applications in almost all branches of Engineering and physical sciences. Certain mathematical operations and results can be expressed in elegant and compact form by using matrix algebra.

5.2 DEFINITION

A **matrix** is a rectangular array of numbers (real or complex) or functions enclosed within square brackets. These numbers (real or complex) or functions in the array are called **entries** or **elements** of the matrix. Each element or entry of the matrix can be a real or a complex number or a function of one or more variables or any other object. For example $\mathbf{A} = \begin{bmatrix} 4 & -7 & 5 \\ 8 & 1 & 6 \end{bmatrix}$ is a 2×3 matrix or matrix of order 2×3 (read as 2 by 3 matrix). It has two *"rows"* (horizontal lines) and three *columns* (vertical lines).

If a matrix has m rows and n columns, we say that its **size** is m by n (written $m \times n$).

5.2.1 General Notations and Concepts

The matrices are usually denoted by single capital boldface letters **A, B, C**,... etc. Thus, in general

$$\mathbf{A} = \begin{bmatrix} a_{11} & a_{12} & a_{13} & \cdots & a_{1n} \\ a_{21} & a_{22} & a_{23} & \cdots & a_{2n} \\ \vdots & & & & \vdots \\ a_{m1} & a_{m2} & a_{m3} & \cdots & a_{mn} \end{bmatrix} \quad \text{...(5.1)}$$

Minor diagonal (a_{m1} to a_{1n}); Principal diagonal (a_{11} to a_{mn})

is an $m \times n$ matrix.

Here each element has two subscripts. The first subscript indicates the row and the second indicates the column in which the element is located. For example a_{12} is the element in the first row and the second column and a_{ij} is the element in the ith row and jth column. This element is also known as the (i, j)th element. All elements in the second row have 2 as first subscript and each element in the 3rd column has 3 as second subscript. An $m \times n$ matrix **A** above is abbreviated as $\mathbf{A} = (a_{ij})_{m \times n}$. Sometimes double bars $\| \ \|$ or pair of parentheses ( ) are used in place of square brackets in indicating a matrix.

If all the elements of a matrix are real, it is called a **real matrix**, whereas if one or more elements of a matrix are complex, it is called a **complex matrix**.

Remark A matrix, by definition, is simply an arrangement of numbers and has no numerical value.

Some physical examples of matrices

(*a*) The coordinates of a point can be written in the form of a 1×3 matrix $[x_1 \ y_1 \ z_1]$.

(*b*) The direction-ratios of a line can be written in the form of a matrix $[a \ b \ c]$.

---

* The theory of matrices was the invention of the eminent English mathematician Arthur Cayley (1821-1895).

(*c*) The co-efficients of *x, y, z* in the equations:

$$a_1x + b_1y + c_1z = d_1,\ a_2x + b_2y + c_2z = d_2,\ a_3x + b_3y + c_3z = d_3$$

can be written in the form of a 3 × 3 matrix $\begin{bmatrix} a_1 & b_1 & c_1 \\ a_2 & b_2 & c_2 \\ a_3 & b_3 & c_3 \end{bmatrix}$, which is called coefficient matrix.

5.3 SOME TYPES OF MATRICES

Row vector A matrix of order 1 × *n*, that is, it has exactly one row and *n* columns is called a *row vector* or a *row matrix* of order *n* and is written as

$[a_{11}\ a_{12}\ a_{13} \ldots\ a_{1n}]$, or $[a_1\ a_2\ a_3 \ldots\ a_n]$ in which a_{1j} (or a_j) is the *j*th element.

Column vector A matrix of order *m* × 1, that is, it has *m* rows and exactly one column is called a *column vector* or a *column matrix* of order *m* and is written as

$$\begin{bmatrix} b_{11} \\ b_{21} \\ \vdots \\ b_{m1} \end{bmatrix}, \text{ or } \begin{bmatrix} b_1 \\ b_2 \\ \vdots \\ b_m \end{bmatrix}$$ in which b_{j1} (or b_j) is the *j*th element.

The number of elements in a row/column vector is called its *order*. The vectors are usually denoted by boldface lower case letters such as **a, b, c**,... etc.

Rectangular matrix A matrix **A** of order $m \times n,\ m \neq n$ is called a *rectangular matrix*.

Square matrix A matrix **A** of order $m \times n$ in which $m = n$ that is, number of rows is equal to the number of columns is called a *square matrix* of order *n*.

For example $\mathbf{B} = \begin{bmatrix} 4 & 3 \\ 1 & 2 \end{bmatrix}$ and $\mathbf{C} = \begin{bmatrix} a & b & c \\ d & e & f \\ g & h & i \end{bmatrix}$ are square matrices of order 2 and 3 respectively.

The elements a_{ii}, that is the elements $a_{11}, a_{22}, \ldots, a_{nn}$ are called the *diagonal elements* and the line on which these elements lie is called the **Principal diagonal** or the **main** or **leading** diagonal of the matrix. The sum of the diagonal elements of a square matrix is called the **Trace** or **Spur** of the matrix. The *trace* of A, written *tr* (**A**), is the sum of the diagonal elements. Namely $tr(\mathbf{A}) = a_{11} + a_{22} + a_{33} + \ldots + a_{nn}$. In the above examples $tr(\mathbf{B}) = 4 + 2 = 6$, $tr(\mathbf{C}) = a + e + i$.

Null matrix A matrix **A** of order *m* × *n* in which all the elements are zero is called a *null matrix* or a *zero matrix* and is denoted by **O**. For example, $\mathbf{O} = \begin{bmatrix} 0 & 0 & 0 \\ 0 & 0 & 0 \end{bmatrix}$ is a 2 × 3 null matrix, $\mathbf{O} = \begin{bmatrix} 0 & 0 \\ 0 & 0 \\ 0 & 0 \end{bmatrix}$ is a 3 × 2 null matrix, etc.

Diagonal matrix A square matrix **A** having non-zero elements in the principal diagonal positions and zero in all other positions, is called a *diagonal matrix*. A diagonal matrix of order *n* with diagonal elements $a_{11}, a_{22}, \ldots, a_{nn}$ is denoted by $\mathbf{D} = \text{diag}\ [a_{11}, a_{22}, a_{33} \ldots a_{nn}]$.

If all the elements of a diagonal matrix of order *n* are equal, that is $a_{ii} = k$ for all *i*, (*k* being some number) then the matrix

$$\mathbf{A} = \begin{bmatrix} k & 0 & \cdots & 0 \\ 0 & k & & 0 \\ & & \ddots & \\ 0 & 0 & & k \end{bmatrix}$$ is called a **scalar matrix** of order *n*.

If all the elements of a diagonal matrix of order *n* are 1, then the matrix

$\mathbf{A} = \begin{bmatrix} 1 & & & 0 \\ & 1 & & \\ & & \ddots & \\ 0 & & & 1 \end{bmatrix}$ is called an **unit matrix** or an **identity matrix** of order n. An identity matrix of order n is denoted by $\mathbf{I}_n$ or more simply by **I**.

For example, $\begin{bmatrix} 1 & 0 \\ 0 & 1 \end{bmatrix}$ is a unit matrix of order 2 and is denoted by $\mathbf{I}_2$, whereas $\begin{bmatrix} 1 & 0 & 0 \\ 0 & 1 & 0 \\ 0 & 0 & 1 \end{bmatrix}$ is a unit matrix of order 3 and is denoted by $\mathbf{I}_3$.

Triangular matrices: A square matrix $\mathbf{A} = (a_{ij})$ is called a *lower triangular matrix* if $a_{ij} = 0$ for $i < j$, that is all elements above the principal diagonal are zero and an *upper triangular matrix* if $a_{ij} = 0$ for $i > j$, that is all the elements below the principal diagonal are zero.

Examples of lower and upper triangular matrices are given below:

Lower triangular matrices $\begin{pmatrix} 4 & 0 \\ -2 & 3 \end{pmatrix}, \begin{pmatrix} 3 & 0 & 0 \\ 8 & -2 & 0 \\ 7 & 5 & 4 \end{pmatrix}, \begin{pmatrix} 3 & 0 & 0 & 0 \\ 8 & -4 & 0 & 0 \\ 3 & 0 & 2 & 0 \\ 2 & 4 & 3 & 6 \end{pmatrix}$

Upper triangular matrices $\begin{pmatrix} 1 & 4 \\ 0 & 3 \end{pmatrix}, \begin{pmatrix} 1 & 2 & 8 \\ 0 & 4 & 2 \\ 0 & 0 & 5 \end{pmatrix}, \begin{pmatrix} 4 & 2 & 2 & 0 \\ 0 & -5 & 4 & 1 \\ 0 & 0 & 3 & -6 \\ 0 & 0 & 0 & -7 \end{pmatrix}$

It is to be noted that there is no restriction on the elements on or below the principal diagonal for an lower triangular matrix; they may be any number, even zero. The elements above the principal diagonal must be zero. Similarly, for upper triangular matrix, there is no restriction on the elements on or above the principal diagonal. The null matrix, the diagonal matrix, and the identity matrix, defined above, are in the upper triangular form. The identity matrix is both upper as well as lower triangular.

Submatrix: If a finite number of rows or columns are deleted in a matrix the resulting matrix is known as a *submatrix*.

Consider the matrix $\mathbf{A} = \begin{bmatrix} 6 & 7 & 8 & 5 \\ 4 & 3 & 2 & -1 \\ 1 & 0 & 7 & 5 \end{bmatrix}$. **A** is a 3 × 4 matrix. If we delete the first column, we get $\begin{bmatrix} 7 & 8 & 5 \\ 3 & 2 & -1 \\ 0 & 7 & 5 \end{bmatrix}$, which is a 3 × 3 matrix. If we delete the second row, we get $\begin{bmatrix} 6 & 7 & 8 & 5 \\ 1 & 0 & 7 & 5 \end{bmatrix}$, which is a 2 × 4 matrix. If we delete the second column and third row, we get $\begin{bmatrix} 6 & 8 & 5 \\ 4 & 2 & -1 \end{bmatrix}$ which is a 2 × 3 matrix. Thus by deleting some rows and columns of a given matrix, we get different matrices. Each of these is called a *sub-matrix* of **A**.

Remark The matrix **A** itself is also a submatrix of **A** obtained by omitting no rows or columns.

Equal matrices Two matrices $\mathbf{A} = (a_{ij})_{m \times n}$ and $\mathbf{B} = (b_{ij})_{p \times q}$ are said to be equal, when

(*i*) they are of the same order, that is $m = p$, $n = q$ and

(*ii*) their corresponding elements are equal, that is $a_{ij} = b_{ij}$ for each i, and j. Matrices of different sizes cannot be equal.

Two matrices are said to be *conformable* if they are of the same order, that is if they have the same number of rows and the same number of columns.

EXAMPLE Find the values of x, y, z, t which satisfy the matrix equation. (*AMIE, W 2005*)

$$\begin{bmatrix} x+3 & 2y+x \\ z-1 & 4t-6 \end{bmatrix} = \begin{bmatrix} 0 & -7 \\ 3 & 2t \end{bmatrix}.$$

Solution By the definition of equality of matrices, the four corresponding entries must be equal. Thus: $x + 3 = 0$, $2y + x = -7$, $z - 1 = 3$, and $4t - 6 = 2t$.

Solving the above system of equations yields $x = -3$, $y = -2$, $z = 4$, and $t = 3$.

5.4 MATRIX ALGEBRA—OPERATIONS ON MATRICES

5.4.1 Scalar Multiplication (Multiplication of a Matrix by a Number)

If $\mathbf{A} = (a_{ij})$ be a given matrix of order $m \times n$ and k be a scalar (real or complex number), then **B** is the matrix $k\mathbf{A}$ of order $m \times n = (ka_{ij})$, obtained by multiplying each entry in **A** by k. The order of the new matrix **B** is same as that of the matrix **A**.

We also define $-\mathbf{A} = (-1)\mathbf{A}$. Similarly $(-k)$ **A** is written $-k\mathbf{A}$. The matrix $-\mathbf{A}$ is called the *negative* of the matrix **A**.

5.4.2 Addition/Subtraction of Two Matrices

If **A** and **B** are two $m \times n$ matrices then the sum of **A** and **B**, written **A + B**, is defined as an $m \times n$ matrix **C** such that each element of **C** is the sum of the corresponding elements of **A** and **B**. The sum of matrices with different sizes is not defined.

Thus, if $\mathbf{A} = \begin{bmatrix} 6 & 5 & 4 \\ -2 & 3 & 7 \end{bmatrix}$, $\mathbf{B} = \begin{bmatrix} 4 & 1 & -5 \\ 3 & 9 & 2 \end{bmatrix}$

then $$\mathbf{C} = \mathbf{A} + \mathbf{B} = \begin{bmatrix} 6+4 & 5+1 & 4-5 \\ -2+3 & 3+9 & 7+2 \end{bmatrix} = \begin{bmatrix} 10 & 6 & -1 \\ 1 & 12 & 9 \end{bmatrix}.$$

Simlarly the difference **A–B** of the matrices **A** and **B** is a matrix **C** each element of which is obtained by subtracting the elements of **B** from the corresponding elements of **A**.

Thus if $\mathbf{A} = \begin{bmatrix} 6 & 2 \\ 7 & -5 \end{bmatrix}, \mathbf{B} = \begin{bmatrix} 8 & 1 \\ 3 & 4 \end{bmatrix}$

then $$\mathbf{C} = \mathbf{A} - \mathbf{B} = \begin{bmatrix} 6 & 2 \\ 7 & -5 \end{bmatrix} - \begin{bmatrix} 8 & 1 \\ 3 & 4 \end{bmatrix} = \begin{bmatrix} 6-8 & 2-1 \\ 7-3 & -5-4 \end{bmatrix} = \begin{bmatrix} -2 & 1 \\ 4 & -9 \end{bmatrix}.$$

Matrices of the same order are said to be *conformable* for addition/subtraction.

Properties of the matrix addition and scalar multiplication

Let **A, B, C** be the matrices which are conformable for addition and k, λ be scalars we state:

1. $\mathbf{A} + \mathbf{B} = \mathbf{B} + \mathbf{A}$ (*i.e.*, Commutative law holds good)
2. $(\mathbf{A} + \mathbf{B}) + \mathbf{C} = \mathbf{A} + (\mathbf{B} + \mathbf{C})$ (Associative law holds good)
3. $\mathbf{A} + \mathbf{O} = \mathbf{A}$ (**O** is the null matrix of the same order as **A**)
4. $\mathbf{A} + (-\mathbf{A}) = \mathbf{O}$
5. $k(\mathbf{A} + \mathbf{B}) = k\mathbf{A} + k\mathbf{B} = (\mathbf{A} + \mathbf{B})k$ (*i.e.*, Distributive law holds good)
6. $(k + \lambda)\mathbf{A} = k\mathbf{A} + \lambda\mathbf{A}$.
7. $k(\lambda\mathbf{A}) = (k\lambda)\mathbf{A} = k\lambda\mathbf{A}$. (Associative law for multiplication by scalars)
8. $1 \times \mathbf{A} = \mathbf{A}$ and $0 \times \mathbf{A} = \mathbf{O}$

5.4.3 Multiplication of Two Matrices

The product **AB** of two matrices **A** and **B** is defined only when the number of columns in **A** is equal to the number of rows in **B**. Such matrices are said to be *conformable* for multiplication.

Let $\mathbf{A} = (a_{ij})$ be an $m \times p$ matrix and $\mathbf{B} = (b_{ij})$ be an $p \times n$ matrix.

Then the product matrix $\mathbf{C} = (c_{ij}) = \mathbf{AB}$ is the $m \times n$ matrix whose *ij*-entry is obtained by multiplying the *i*th rows of **A** by the *j*th column of **B**. Figure 5.1 illustrates this.

$$\underset{m\times p}{\begin{bmatrix} a_{11} & a_{12} & \cdots & a_{1p} \\ a_{21} & a_{22} & \cdots & a_{2p} \\ \vdots & & & \\ a_{i1} & a_{i2} & \cdots & a_{ip} \\ \vdots & & & \\ a_{m1} & a_{m2} & \cdots & a_{mp} \end{bmatrix}} \underset{p\times n}{\begin{bmatrix} b_{11} & b_{12} & \cdots & b_{1j} & \cdots & b_{1n} \\ b_{21} & b_{22} & \cdots & b_{2j} & \cdots & b_{2n} \\ \vdots & & & & & \\ b_{p1} & b_{p2} & \cdots & b_{pj} & \cdots & b_{pn} \end{bmatrix}} = \underset{m\times n}{\begin{bmatrix} c_{11} & c_{12} & \cdots & c_{1n} \\ c_{21} & c_{22} & \cdots & c_{2n} \\ \vdots & & \cdots & \vdots \\ \vdots & & c_{ij} & \vdots \\ \vdots & & \cdots & \vdots \\ c_{m1} & c_{m2} & \cdots & c_{mn} \end{bmatrix}}$$

(A; B; = C; *i*th row; *j*th column)

Fig. 5.1: Matrix multiplication **AB** = **C**

$$\text{where } c_{ij} = \left(\sum_{k=1}^{p} a_{ik} b_{kj}\right)_{m\times n} = a_{i1}b_{1j} + a_{i2}b_{2j} + \cdots + a_{ip}b_{pj}. \qquad \text{...(5.2)}$$

The product **AB** is not defined if **A** is an $m \times p$ matrix and **B** is a $q \times n$ matrix, where $p \neq q$.

In the product **AB**, the matrix **B** is said to be pre-multiplied by **A** or **A** is said to be post-multiplied by **B**.

The product **AB** of a row matrix **A** of order $1 \times n$ and column matrix **B** of order $n \times 1$ is a matrix of order 1×1, that is a single element, and **BA** is a matrix of order $n \times n$.

Important Notes

1. It is possible that for two given matrices **A** and **B**, the product matrices **AB** and **BA** may or may not exist and may be equal or different.
   For example, if **A** is a $m \times n$ matrix and **B** is $n \times k$ matrix then the product matrix **AB** is defined and is a matrix of order $m \times k$, whereas the product matrix **BA** is not defined.
2. If both the product matrices **AB** and **BA** are defined, then both the matrices **AB** and **BA** are square matrices. In general **AB** ≠ **BA**. Thus, the matrix product is not commutative.
   If **AB** = **BA**, then the matrices **A** and **B** are said to *commute* with each other.
3. If **AB** = **O**, then it does not necessarily imply that either **A** = **O** or **B** = **O** or **BA** = **O**.
4. If **AB** = **AC** it does not necessarily imply that **B** = **C**.
5. If $\mathbf{A}^2 = \mathbf{A}$, then **A** is called an *idempotent matrix*.
6. If $\mathbf{A}^k = \mathbf{O}$ for some positive integer k, then matrix **A** is termed *nilpotent*. The smallest value of k for which $\mathbf{A}^k = \mathbf{O}$ is called the *index of nilpotency* of the matrix **A**.
7. A matrix **A** will be called an involuntary matrix, if $\mathbf{A}^2 = \mathbf{I}$ (unit matrix). Since $\mathbf{I}^2$ always is equal to **I**, therefore unit matrix is involutary.

Properties of matrix multiplication

1. If **A, B, C** are matrices of orders $m \times n$, $n \times p$ and $p \times q$ respectively, then
   (**AB**)**C** = **A**(**BC**) is a matrix of order $m \times q$. (associative law for product hold good)
2. **A**(**B** + **C**) = **AB** + **AC** (left distributive law for product hold good)
3. (**A** + **B**)**C** = **AC** + **BC** (right distributive law for product hold good)
   (provided the products and sums of the matrices are defined).
4. k(**AB**) = **A**(k**B**) = (k**A**)**B**, where k is a scalar.

It may be noted that there is no concept of dividing a matrix by a matrix. Therefore, the operation **A/B** where **A** and **B** are matrices is not defined.

ILLUSTRATIVE EXAMPLES

EXAMPLE 5.1 (*i*) If $\mathbf{A}=\begin{bmatrix}10\\3\\12\end{bmatrix}$, $\mathbf{B}=\begin{bmatrix}5\\-7\\9\end{bmatrix}$ and $\mathbf{C}=\begin{bmatrix}-9\\4\\-8\end{bmatrix}$ find **A – B – C.**

(*ii*) If $\mathbf{A}=\begin{bmatrix}1 & 2 & -1\\0 & -1 & 3\end{bmatrix}$ and $\mathbf{B}=\begin{bmatrix}2 & 3 & 1\\0 & -1 & 5\end{bmatrix}$, find **2A – 3B**.

Solution (*i*) We have $\mathbf{A}-\mathbf{B}-\mathbf{C}=\begin{bmatrix}10\\3\\12\end{bmatrix}-\begin{bmatrix}5\\-7\\9\end{bmatrix}-\begin{bmatrix}-9\\4\\-8\end{bmatrix}$.

First perform the scalar multiplication and then a matrix addition.

$$\mathbf{A}-\mathbf{B}-\mathbf{C} = \mathbf{A}+(-\mathbf{B})+(-\mathbf{C}) = \begin{bmatrix}10\\3\\12\end{bmatrix}+\begin{bmatrix}-5\\7\\-9\end{bmatrix}+\begin{bmatrix}9\\-4\\8\end{bmatrix}=\begin{bmatrix}10-5+9\\3+7-4\\12-9+8\end{bmatrix}=\begin{bmatrix}14\\6\\11\end{bmatrix}.$$

(*ii*) First perform the scalar multiplication and then a matrix addition.

$$2\mathbf{A}-3\mathbf{B} = 2\mathbf{A}+(-3\mathbf{B}) = \begin{bmatrix}2 & 4 & -2\\0 & -2 & 6\end{bmatrix}+\begin{bmatrix}-6 & -9 & -3\\0 & 3 & -15\end{bmatrix}=\begin{bmatrix}-4 & -5 & -5\\0 & 1 & -9\end{bmatrix}.$$

NOTE The matrix **2A – 3B** that is, **2A + (–3B)** is called a *linear combination* of matrices **A** and **B**.

EXAMPLE 5.2 If $\mathbf{P}=\begin{bmatrix}3 & 5 & 7\\9 & 3 & 2\\1 & 2 & 5\end{bmatrix}$, and $\mathbf{Q}=\begin{bmatrix}3 & 5 & 4\\4 & 7 & 8\\3 & 2 & 1\end{bmatrix}$, find the matrix **R** such that **3P + 2Q – 2R** is a null matrix.

Solution We have $3\mathbf{P}+2\mathbf{Q}-2\mathbf{R} = \begin{bmatrix}9 & 15 & 21\\27 & 9 & 6\\3 & 6 & 15\end{bmatrix}+\begin{bmatrix}6 & 10 & 8\\8 & 14 & 16\\6 & 4 & 2\end{bmatrix}-2\mathbf{R} = \begin{bmatrix}0 & 0 & 0\\0 & 0 & 0\\0 & 0 & 0\end{bmatrix}$

$$\Rightarrow \quad 2\mathrm{R} = \begin{bmatrix}15 & 25 & 29\\35 & 23 & 22\\9 & 10 & 17\end{bmatrix} \text{ or } \mathrm{R} = 1/2\begin{bmatrix}15 & 25 & 29\\35 & 23 & 22\\9 & 10 & 17\end{bmatrix}.$$

EXAMPLE 5.3 Find x, y, z, w where $3\begin{bmatrix}x & y\\z & w\end{bmatrix}=\begin{bmatrix}x & 6\\-1 & 2w\end{bmatrix}+\begin{bmatrix}4 & x+y\\z+w & 3\end{bmatrix}$.

Solution Write each side as a single equation we have $\begin{bmatrix}3x & 3y\\3z & 3w\end{bmatrix}=\begin{bmatrix}x+4 & x+y+6\\z+w-1 & 2w+3\end{bmatrix}$.

By the definition of equality of matrices, the four corresponding entries must be equal. Thus, we obtain the following four equations: $3x = x + 4$, $3y = x + y + 6$, $3z = z + w - 1$, $3w = 2w + 3$.

Solving the above system of equations yields $x = 2$, $y = 4$, $z = 1$, $w = 3$.

EXAMPLE 5.4 Let $\mathbf{A}=[1 \;\; 2 \;\; 3]$, $\mathbf{B}=\begin{bmatrix}1 & 2\\-1 & 0\\1 & 3\end{bmatrix}$ and $\mathbf{C}=\begin{bmatrix}3\\1\\5\end{bmatrix}$. Compute the ones which are defined

CA, AC + B, AB + C.

Solution Since **C** is a matrix of order 3 × 1 and **A** is a matrix of order 1 × 3, **CA** is defined and is a matrix of order 3 × 3.

$$\mathbf{CA} = \begin{bmatrix} 3 \\ 1 \\ 5 \end{bmatrix} \begin{bmatrix} 1 & 2 & 3 \end{bmatrix} = \begin{bmatrix} 3 & 6 & 9 \\ 1 & 2 & 3 \\ 5 & 10 & 15 \end{bmatrix}.$$

A is a 1 × 3 matrix and **C** is a 3 × 1 matrix, **AC** is defined and is a 1 × 1 matrix. But **B** is 3 × 2 matrix, therefore **AC** + **B** is not defined.

Further, **AB** is 1 × 2 matrix but **C** is 3 × 1 matrix, therefore **AB** + **C** is not defined.

EXAMPLE 5.5 If $\mathbf{A} = \begin{bmatrix} 1 & 2 \\ 3 & 4 \end{bmatrix}$, show that $\mathbf{A}^2 - 5\mathbf{A} = 2\mathbf{I}$, where **I** is the unit matrix of order 2. Hence, or otherwise determine $\mathbf{A}^4$.

Solution We have $\mathbf{A}^2 = \mathbf{AA} = \begin{bmatrix} 1 & 2 \\ 3 & 4 \end{bmatrix}\begin{bmatrix} 1 & 2 \\ 3 & 4 \end{bmatrix} = \begin{bmatrix} (1)(1)+(2)(3) & (1)(2)+(2)(4) \\ (3)(1)+(4)(3) & (3)(2)+(4)(4) \end{bmatrix} = \begin{bmatrix} 7 & 10 \\ 15 & 22 \end{bmatrix}$.

$$\therefore \quad \mathbf{A}^2 - 5\mathbf{A} = \mathbf{A}^2 + (-5\mathbf{A}) = \begin{bmatrix} 7 & 10 \\ 15 & 22 \end{bmatrix} + \begin{bmatrix} -5 & -10 \\ -15 & -20 \end{bmatrix} = \begin{bmatrix} 2 & 0 \\ 0 & 2 \end{bmatrix} = 2\begin{bmatrix} 1 & 0 \\ 0 & 1 \end{bmatrix} = 2\mathbf{I}$$

We know $\mathbf{A}^2 = 5\mathbf{A} + 2\mathbf{I}$. Hence $\mathbf{A}^4 = \mathbf{A}^2\mathbf{A}^2 = (5\mathbf{A} + 2\mathbf{I})(5\mathbf{A} + 2\mathbf{I})$

$$= 25\mathbf{A}^2 + 10\mathbf{AI} + 10\mathbf{IA} + 4\mathbf{I}^2$$

$$= 25(5\mathbf{A} + 2\mathbf{I}) + 20\mathbf{A} + 4\mathbf{I}, \text{ since } \mathbf{AI} = \mathbf{IA} = \mathbf{A} \text{ and } \mathbf{I}^2 = \mathbf{I}$$

$$= 145\mathbf{A} + 54\mathbf{I} = 145\begin{bmatrix} 1 & 2 \\ 3 & 4 \end{bmatrix} + 54\begin{bmatrix} 1 & 0 \\ 0 & 1 \end{bmatrix} = \begin{bmatrix} 199 & 290 \\ 435 & 634 \end{bmatrix}.$$

EXAMPLE 5.6 If $\mathbf{A} = \begin{bmatrix} 2 & -1 \\ 1 & 0 \\ -3 & 4 \end{bmatrix}$ and $\mathbf{B} = \begin{bmatrix} 1 & -2 & -5 \\ 3 & 4 & 0 \end{bmatrix}$, evaluate **AB** and **BA**. *(AMIE S-2001)*

Solution Since **A** is a 3 × 2 matrix and **B** is 2 × 3 matrix, the product **AB** is defined and is a 3 × 3 matrix. Multiply the rows of **A** by the columns of **B** to obtain.

$$\mathbf{AB} = \begin{bmatrix} 2 & -1 \\ 1 & 0 \\ -3 & 4 \end{bmatrix}\begin{bmatrix} 1 & -2 & -5 \\ 3 & 4 & 0 \end{bmatrix} = \begin{bmatrix} 2(1)+(-1)3 & 2(-2)+(-1)4 & 2(-5)+0 \\ 1(1)+0 & 1(-2)+0 & 1(-5)+0 \\ (-3)(1)+4(3) & (-3)(-2)+4(4) & (-3)(-5)+0 \end{bmatrix}$$

$$= \begin{bmatrix} 2-3 & -4-4 & -10 \\ 1 & -2 & -5 \\ -3+12 & 6+16 & 15 \end{bmatrix} = \begin{bmatrix} -1 & -8 & -10 \\ 1 & -2 & -5 \\ 9 & 22 & 15 \end{bmatrix}.$$

Further, product **BA** is defined and is a 2 × 2 matrix. Multiply the rows of **B** by the columns of **A** to obtain

$$\mathbf{BA} = \begin{bmatrix} 1 & -2 & -5 \\ 3 & 4 & 0 \end{bmatrix}\begin{bmatrix} 2 & -1 \\ 1 & 0 \\ -3 & 4 \end{bmatrix} = \begin{bmatrix} 2-2+15 & -1+0-20 \\ 6+4+0 & -3+0+0 \end{bmatrix} = \begin{bmatrix} 15 & -21 \\ 10 & -3 \end{bmatrix}.$$

Note: AB ≠ BA.

EXAMPLE 5.7 Evaluate $\mathbf{A}^2 - 3\mathbf{A} + 9\mathbf{I}$ if $\mathbf{I}$ is the unit matrix of order 3 and $\mathbf{A} = \begin{bmatrix} 1 & -2 & 3 \\ 2 & 3 & -1 \\ -3 & 1 & 2 \end{bmatrix}$.

Solution We have $\mathbf{A}^2 = \mathbf{AA} = \begin{bmatrix} 1 & -2 & 3 \\ 2 & 3 & -1 \\ -3 & 1 & 2 \end{bmatrix}\begin{bmatrix} 1 & -2 & 3 \\ 2 & 3 & -1 \\ -3 & 1 & 2 \end{bmatrix}$

$$= \begin{bmatrix} 1(1)+(-2)2+3(-3) & 1(-2)+(-2)3+3(1) & 1(3)+(-2)(-1)+3(2) \\ 2(1)+3(2)+(-1)(-3) & 2(-2)+3(3)+(-1)1 & 2(3)+3(-1)+(-1)2 \\ (-3)1+1(2)+2(-3) & (-3)(-2)+1(3)+2(1) & (-3)3+1(-1)+2(2) \end{bmatrix}$$

$$= \begin{bmatrix} 1-4-9 & -2-6+3 & 3+2+6 \\ 2+6+3 & -4+9-1 & 6-3-2 \\ -3+2-6 & 6+3+2 & -9-1+4 \end{bmatrix} = \begin{bmatrix} -12 & -5 & 11 \\ 11 & 4 & 1 \\ -7 & 11 & -6 \end{bmatrix}.$$

Therefore, $$\mathbf{A}^2 - 3\mathbf{A} + 9\mathbf{I} = \begin{bmatrix} -12 & -5 & 11 \\ 11 & 4 & 1 \\ -7 & 11 & -6 \end{bmatrix} + \begin{bmatrix} -3 & 6 & -9 \\ -6 & -9 & 3 \\ 9 & -3 & -6 \end{bmatrix} + \begin{bmatrix} 9 & 0 & 0 \\ 0 & 9 & 0 \\ 0 & 0 & 9 \end{bmatrix} = \begin{bmatrix} -6 & 1 & 2 \\ 5 & 4 & 4 \\ 2 & 8 & -3 \end{bmatrix}.$$

EXAMPLE 5.8 Let $\mathbf{A} = \begin{bmatrix} 2 & -5 \\ 3 & 1 \end{bmatrix}$ and let $f(x) = x^3 - 2x^2 - 5$. Find $f(\mathbf{A})$.

Solution $$\mathbf{A}^2 = \mathbf{AA} = \begin{bmatrix} 2 & -5 \\ 3 & 1 \end{bmatrix}\begin{bmatrix} 2 & -5 \\ 3 & 1 \end{bmatrix} = \begin{bmatrix} 4-15 & -10-5 \\ 6+3 & -15+1 \end{bmatrix} = \begin{bmatrix} -11 & -15 \\ 9 & -14 \end{bmatrix}.$$

$$\mathbf{A}^3 = \mathbf{A}^2\mathbf{A} = \begin{bmatrix} -11 & -15 \\ 9 & -14 \end{bmatrix}\begin{bmatrix} 2 & -5 \\ 3 & 1 \end{bmatrix} = \begin{bmatrix} -22-45 & 55-15 \\ 18-42 & -45-14 \end{bmatrix} = \begin{bmatrix} -67 & 40 \\ -24 & -59 \end{bmatrix}.$$

Substitute $\mathbf{A}$ for x and $5\mathbf{I}$ for constant term in $f(x)$, we obtain

$$f(\mathbf{A}) = \mathbf{A}^3 - 2\mathbf{A}^2 - 5\mathbf{I} = \begin{bmatrix} -67 & 40 \\ -24 & -59 \end{bmatrix} + \begin{bmatrix} 22 & 30 \\ -18 & 28 \end{bmatrix} + \begin{bmatrix} -5 & 0 \\ 0 & -5 \end{bmatrix} = \begin{bmatrix} -50 & 70 \\ -42 & -36 \end{bmatrix}.$$

EXAMPLE 5.9 Find the diagonal and trace of each matrix

(*i*) $\mathbf{P} = \begin{bmatrix} 2 & -5 & 8 \\ 3 & -6 & -7 \\ 4 & 0 & -1 \end{bmatrix}$, (*ii*) $\mathbf{Q} = \begin{bmatrix} 4 & 3 & -6 \\ 2 & -5 & 0 \end{bmatrix}$.

Solution (*i*) The diagonal elements of matrix **P** are 2, –6 and –1. The trace of **P** is the sum of these diagonal elements Thus $tr(\mathbf{P}) = 2 - 6 - 1 = -5$.

(*ii*) The diagonal and trace are only defined for square matrices, whereas the given matrix is rectangle of order 2 × 3.

EXAMPLE 5.10 If $\begin{bmatrix} 4 \\ 1 \\ 3 \end{bmatrix}\mathbf{X} = \begin{bmatrix} -4 & 8 & 4 \\ -1 & 2 & 1 \\ -3 & 6 & 3 \end{bmatrix}$, find **X**.

Solution Let $\mathbf{X} = [a \quad b \quad c]$.

We have
$$\begin{bmatrix} 4 \\ 1 \\ 3 \end{bmatrix} [a \quad b \quad c] = \begin{bmatrix} 4a & 4b & 4c \\ a & b & c \\ 3a & 3b & 3c \end{bmatrix} = \begin{bmatrix} -4 & 8 & 4 \\ -1 & 2 & 1 \\ -3 & 6 & 3 \end{bmatrix}. \qquad \text{(given)}$$

By the definition of equality of matrices, we obtain $a = -1$, $b = 2$ and $c = 1$.
Therefore, $\mathbf{X} = [-1 \ 2 \ 1]$.

EXAMPLE 5.11 If $e^{\mathbf{A}}$ is defined as $\mathbf{I} + \mathbf{A} + \dfrac{\mathbf{A}^2}{2!} + \dfrac{\mathbf{A}^3}{3!} + \cdots$ show that

$$e^{\mathbf{A}} = e^x \begin{bmatrix} \cosh x & \sinh x \\ \sinh x & \cosh x \end{bmatrix} \text{ when } \mathbf{A} = \begin{bmatrix} x & x \\ x & x \end{bmatrix}. \qquad \text{(AMIE, W-2002)}$$

Solution Here $\mathbf{A} = \begin{bmatrix} x & x \\ x & x \end{bmatrix}$. Therefore, $\mathbf{A}^2 = \mathbf{AA} = \begin{bmatrix} x & x \\ x & x \end{bmatrix}\begin{bmatrix} x & x \\ x & x \end{bmatrix} = \begin{bmatrix} 2x^2 & 2x^2 \\ 2x^2 & 2x^2 \end{bmatrix}$ and

$$\mathbf{A}^3 = \mathbf{A}^2\mathbf{A} = \begin{bmatrix} 2x^2 & 2x^2 \\ 2x^2 & 2x^2 \end{bmatrix}\begin{bmatrix} x & x \\ x & x \end{bmatrix} = \begin{bmatrix} 4x^3 & 4x^3 \\ 4x^3 & 4x^3 \end{bmatrix} \text{etc.}$$

In general
$$\mathbf{A}^n = \begin{bmatrix} 2^{n-1}x^n & 2^{n-1}x^n \\ 2^{n-1}x^n & 2^{n-1}x^n \end{bmatrix} = \frac{1}{2}\begin{bmatrix} (2x)^n & (2x)^n \\ (2x)^n & (2x)^n \end{bmatrix}.$$

Hence
$$e^{\mathbf{A}} = \mathbf{I} + \mathbf{A} + \frac{\mathbf{A}^2}{2!} + \frac{\mathbf{A}^3}{3!} + \cdots$$

$$= \begin{bmatrix} 1 & 0 \\ 0 & 1 \end{bmatrix} + \begin{bmatrix} x & x \\ x & x \end{bmatrix} + \frac{1}{2!}\begin{bmatrix} (2x)^2 & (2x)^2 \\ (2x)^2 & (2x)^2 \end{bmatrix} + \frac{1}{3!}\begin{bmatrix} 4x^3 & 4x^3 \\ 4x^3 & 4x^3 \end{bmatrix} + \cdots$$

$$= \frac{1}{2}\begin{bmatrix} 2 + 2x + \dfrac{(2x)^2}{2!} + \dfrac{(2x)^3}{3!} + \cdots & 2x + \dfrac{(2x)^2}{2!} + \dfrac{(2x)^3}{3!} + \cdots \\ 2x + \dfrac{(2x)^2}{2!} + \dfrac{(2x)^3}{3!} + \cdots & 2 + 2x + \dfrac{(2x)^2}{2!} + \dfrac{(2x)^3}{3!} + \cdots \end{bmatrix}$$

$$= \frac{1}{2}\begin{bmatrix} 1 + e^{2x} & e^{2x} - 1 \\ e^{2x} - 1 & 1 + e^{2x} \end{bmatrix} = e^x \begin{bmatrix} \dfrac{e^x + e^{-x}}{2} & \dfrac{e^x - e^{-x}}{2} \\ \dfrac{e^x - e^{-x}}{2} & \dfrac{e^x + e^{-x}}{2} \end{bmatrix} = e^x \begin{bmatrix} \cosh x & \sinh x \\ \sinh x & \cosh x \end{bmatrix}.$$

EXERCISE 5.1

1. (*i*) Evaluate $3\begin{bmatrix} 2 & 9 & -6 \\ 4 & -5 & 3 \end{bmatrix} - 2\begin{bmatrix} 1 & 6 & 2 \\ 4 & 3 & -5 \end{bmatrix}$.

(*ii*) Given $\mathbf{A}=\begin{bmatrix}1 & 2 & -3\\ 5 & 0 & 2\\ 1 & -1 & 1\end{bmatrix}$, $\mathbf{B}=\begin{bmatrix}3 & -1 & 2\\ 4 & 2 & 5\\ 2 & 0 & 3\end{bmatrix}$ and $\mathbf{C}=\begin{bmatrix}4 & 1 & 2\\ 0 & 3 & 2\\ 1 & -2 & 3\end{bmatrix}$,

compute (*a*) **A** + **B** and **A** – **C** (*b*) verify **A** + (**B** – **C**) = (**A** + **B**) – **C**.

Ans. (*i*) $\begin{bmatrix}4 & 15 & -22\\ 4 & -21 & 19\end{bmatrix}$ (*ii*) (*a*) $\begin{bmatrix}4 & 1 & -1\\ 9 & 2 & 7\\ 3 & -1 & 4\end{bmatrix}$; $\begin{bmatrix}-3 & 1 & -5\\ 5 & -3 & 0\\ 0 & 1 & -2\end{bmatrix}$.

2. (*i*) If $\mathbf{A}=\begin{bmatrix}2 & -1\\ 4 & 3\end{bmatrix}$, $\mathbf{B}=\begin{bmatrix}-1 & 1\\ 2 & -4\end{bmatrix}$, and $\mathbf{C}=\begin{bmatrix}1 & 4\\ -2 & -1\end{bmatrix}$, find 3**A** + 2**B** – 4**C**.

(*ii*) Find *x*, *y*, *z*, *t* such that $\begin{bmatrix}x+2 & 2y+x\\ z-1 & 4t+1\end{bmatrix}=\begin{bmatrix}0 & -6\\ 3 & 2t\end{bmatrix}$.

Ans. (*i*) $\begin{bmatrix}0 & -17\\ 24 & 5\end{bmatrix}$ (*ii*) $x=-2$, $y=-2$, $z=4$, $t=-1/2$.

3. If $\mathbf{A}=\begin{bmatrix}1 & 2\\ 3 & 4\\ 5 & 6\end{bmatrix}$, $\mathbf{B}=\begin{bmatrix}-3 & -2\\ 1 & -5\\ 4 & 3\end{bmatrix}$, find $\mathbf{D}=\begin{bmatrix}p & q\\ r & s\\ t & u\end{bmatrix}$ such that **A** + **B** – **D** = **O**. **Ans.** $\begin{bmatrix}-2 & 0\\ 4 & -1\\ 9 & 9\end{bmatrix}$.

4. If $\mathbf{A}=\begin{bmatrix}0 & 1 & 2\\ 1 & 2 & 3\\ 2 & 3 & 4\end{bmatrix}$, and $\mathbf{B}=\begin{bmatrix}1 & -2\\ -1 & 0\\ 2 & -1\end{bmatrix}$, obtain the product **AB**. Explain why **BA** is not defined?

Ans. $\begin{bmatrix}3 & -2\\ 5 & -5\\ 7 & -8\end{bmatrix}$.

5. If $\mathbf{A}=\begin{bmatrix}1 & 0 & 0\\ 0 & -2 & 3\\ 2 & 1 & -1\end{bmatrix}$, $\mathbf{B}=\begin{bmatrix}0 & 2 & -1\\ 3 & 1 & 0\\ 0 & -1 & 2\end{bmatrix}$ and $\mathbf{C}=\begin{bmatrix}1 & 2\\ -4 & -2\\ 0 & 1\end{bmatrix}$, find (*i*) **AB**, (*ii*) **BA**, (*iii*) **BC**, (*iv*) **CB**, (*v*) **AC**, (*vi*) **CA**.

Ans. (*i*) $\begin{bmatrix}0 & 2 & -1\\ -6 & -5 & 6\\ 3 & 6 & -4\end{bmatrix}$, (*ii*) $\begin{bmatrix}-2 & -5 & 7\\ 3 & -2 & 3\\ 4 & 4 & -5\end{bmatrix}$, (*iii*) $\begin{bmatrix}-8 & -5\\ -1 & 4\\ 4 & 4\end{bmatrix}$, (*iv*) Not defined, (*v*) $\begin{bmatrix}1 & 2\\ 8 & 7\\ -2 & 1\end{bmatrix}$ (*vi*) Not defined.

6. (*i*) If $\mathbf{A}=\begin{bmatrix}1 & -2 & 3\\ 2 & 3 & -1\\ -3 & 1 & 2\end{bmatrix}$ and $\mathbf{B}=\begin{bmatrix}1 & 0 & 2\\ 0 & 1 & 2\\ 1 & 2 & 0\end{bmatrix}$, obtain the products **AB** and **BA** and show that **AB** ≠ **BA**.

(*ii*) Evaluate $\begin{bmatrix}1\\ -2\\ 3\end{bmatrix}\times\begin{bmatrix}4 & 5 & 2\end{bmatrix}\times\begin{bmatrix}2\\ -3\\ 5\end{bmatrix}\times\begin{bmatrix}3 & 2\end{bmatrix}$.

Ans. (*i*) $\mathbf{AB}=\begin{bmatrix}4 & 4 & -2\\ 1 & 1 & 10\\ -1 & 5 & -4\end{bmatrix}$, $\mathbf{BA}=\begin{bmatrix}-5 & 0 & 7\\ -4 & 5 & 3\\ 5 & 4 & 1\end{bmatrix}$; (*ii*) $\begin{bmatrix}9 & 6\\ -18 & 2\\ 27 & 18\end{bmatrix}$.

7. (*i*) If $\mathbf{A} = \begin{bmatrix} 2 & 1 & 1 \\ -1 & 1 & 4 \end{bmatrix}$, $\mathbf{B} = \begin{bmatrix} 1 & 5 \\ 7 & 8 \\ -1 & 0 \end{bmatrix}$ and $\mathbf{C} = \begin{bmatrix} -1 & 1 \\ 2 & -1 \end{bmatrix}$, verify that **A(B + C) = AB + AC** and **(AB)C = A (BC).**

(*ii*) If $\mathbf{E} = \begin{bmatrix} 0 & 1 & 0 \\ 0 & 0 & 1 \\ 0 & 0 & 0 \end{bmatrix}$ and $\mathbf{F} = \begin{bmatrix} 0 & 0 & 0 \\ 1 & 0 & 0 \\ 0 & 1 & 0 \end{bmatrix}$, calculate the matrices products **EF** and **FE** and show that $\mathbf{E}^2\mathbf{F} + \mathbf{F}\mathbf{E}^2 = \mathbf{E}$.

Ans. (*i*) Each is $\begin{bmatrix} 28 & -10 \\ 4 & -1 \end{bmatrix}$ (*ii*) $\mathbf{EF} = \begin{bmatrix} 1 & 0 & 0 \\ 0 & 1 & 0 \\ 0 & 0 & 0 \end{bmatrix}$ and $\mathbf{FE} = \begin{bmatrix} 0 & 0 & 0 \\ 0 & 1 & 1 \\ 0 & 0 & 0 \end{bmatrix}$.

8. If $\mathbf{A} = \begin{bmatrix} 2 & 1 & -1 \\ 1 & -2 & 3 \\ -2 & 1 & 2 \end{bmatrix}$ and $\mathbf{B} = \begin{bmatrix} 1 & -1 & 2 \\ -2 & 1 & 3 \\ 2 & -1 & 1 \end{bmatrix}$, show that $(\mathbf{A} + \mathbf{B})^2 = \mathbf{A}^2 + \mathbf{AB} + \mathbf{BA} + \mathbf{B}^2$.

9. If $\mathbf{A} = \begin{bmatrix} 3 & -2 \\ 1 & -1 \end{bmatrix}$ and $\mathbf{B} = \begin{bmatrix} 1 & 2 \\ -1 & 1 \end{bmatrix}$, verify that $(\mathbf{A} + \mathbf{B})(\mathbf{A} - \mathbf{B}) = \mathbf{A}^2 + \mathbf{BA} - \mathbf{AB} - \mathbf{B}^2$.

10. Find x and y such that $\begin{bmatrix} 2 & -1 \\ -3 & 4 \end{bmatrix}\begin{bmatrix} x \\ y \end{bmatrix} + \begin{bmatrix} 8 \\ 1 \end{bmatrix} = \begin{bmatrix} 0 \\ 0 \end{bmatrix}$. **Ans.** $x = -33/5$, $y = -26/5$.

11. If $\mathbf{A} = \begin{bmatrix} 1 & -2 & 3 \\ 2 & 3 & -1 \\ -3 & 1 & 2 \end{bmatrix}$, $\mathbf{B} = \begin{bmatrix} 6 & -1 & -2 \\ -5 & -4 & -4 \\ -2 & -8 & 3 \end{bmatrix}$ and **I** is a unit matrix of order 3, evaluate $\mathbf{A}^2 - 3\mathbf{A} + \mathbf{I} + \mathbf{B}$.

Ans. $\begin{bmatrix} -8 & 0 & 0 \\ 0 & -8 & 0 \\ 0 & 0 & -8 \end{bmatrix}$.

12. Find matrix **A**, such that $\mathbf{A}\begin{bmatrix} 0 & 1 \\ 2 & -1 \end{bmatrix} = \begin{bmatrix} 2 & 1 \\ -1 & 0 \end{bmatrix}$. (*AMIETE, June 2008*) **Ans.** $\begin{bmatrix} 2 & 1 \\ -1/2 & -1/2 \end{bmatrix}$.

13. Prove that the product of the two matrices $\begin{bmatrix} \cos^2\theta & \cos\theta\sin\theta \\ \cos\theta\sin\theta & \sin^2\theta \end{bmatrix}$ and $\begin{bmatrix} \cos^2\phi & \cos\phi\sin\phi \\ \cos\phi\sin\phi & \sin^2\phi \end{bmatrix}$ is a null matrix when θ and ϕ differ by an odd multiple of $\pi/2$.

14. (*i*) Given $\mathbf{A} = \begin{bmatrix} 3 & -4 \\ 1 & -1 \end{bmatrix}$ show that $\mathbf{A}^n = \begin{bmatrix} 1+2n & -4n \\ n & 1-2n \end{bmatrix}$.

(*ii*) If $\mathbf{A} = \begin{bmatrix} k & 1 \\ 0 & k \end{bmatrix}$, prove that $\mathbf{A}^n = \begin{bmatrix} k^n & nk^{n-1} \\ 0 & k^n \end{bmatrix}$ for every positive integer n.

[**Hint:** Use the method of mathematical Induction] (*AMIE, S-2001*)

15. Given $\mathbf{A} = \begin{bmatrix} \cos\theta & \sin\theta \\ -\sin\theta & \cos\theta \end{bmatrix}$, prove that $\mathbf{A}^2 = \begin{bmatrix} \cos 2\theta & \sin 2\theta \\ -\sin 2\theta & \cos 2\theta \end{bmatrix}$. Also find the value of $\mathbf{A}^n$.

(*AMIE, S-2002*) **Ans.** $\begin{bmatrix} \cos n\theta & \sin n\theta \\ -\sin n\theta & \cos n\theta \end{bmatrix}$.

16. If $\mathbf{A} = \begin{bmatrix} 0 & 1 & 0 \\ 0 & 0 & 1 \\ p & q & r \end{bmatrix}$ and **I** is a unit matrix of order 3, show that $\mathbf{A}^3 = p\mathbf{I} + q\mathbf{A} + r\mathbf{A}^2$.

17. If $\mathbf{A} = \begin{bmatrix} 0 & 1 \\ -1 & 0 \end{bmatrix}$, choose α and β so that $(\alpha\mathbf{I} + \beta\mathbf{A})^2 = \mathbf{A}$. **Ans.** $\alpha = \beta = 1/\sqrt{2}$.

18. Show that the matrix $\mathbf{A} = \begin{bmatrix} 1 & 3 & 2 \\ 2 & 0 & -1 \\ 1 & 2 & 3 \end{bmatrix}$ satisfies the matrix equation $\mathbf{A}^3 - 4\mathbf{A}^2 - 3\mathbf{A} + 11\mathbf{I} = \mathbf{O}$, where **I** is an identity matrix of order 3.

19. Let **A**, **B** be the indicated matrices. Show that $tr(\mathbf{AB}) = tr(\mathbf{BA})$.

(*i*) $\mathbf{A} = \begin{pmatrix} 1 & -1 & 1 \\ 2 & 4 & 1 \\ 3 & 0 & 1 \end{pmatrix}$, $\mathbf{B} = \begin{pmatrix} 3 & 1 & 2 \\ 1 & 1 & 0 \\ -1 & 2 & 1 \end{pmatrix}$.

(*ii*) $\mathbf{A} = \begin{pmatrix} 1 & -1 & 1 \\ 2 & 4 & 1 \\ 3 & 0 & 1 \end{pmatrix}$, $\mathbf{B} = \begin{pmatrix} 3 & 1 & 2 \\ 1 & 1 & 0 \\ -1 & 2 & 1 \end{pmatrix}$.

20. If $\mathbf{A} = \begin{bmatrix} 1 & -4 & 2 \\ -1 & 4 & -2 \end{bmatrix}$, $\mathbf{B} = \begin{bmatrix} 1 & 2 \\ -1 & 3 \\ 5 & -2 \end{bmatrix}$, $\mathbf{C} = \begin{bmatrix} 2 & 2 \\ 1 & -1 \\ 1 & -3 \end{bmatrix}$, compute $\mathbf{A}(2\mathbf{B} - 3\mathbf{C})$. **Ans.** $\begin{bmatrix} 30 & -28 \\ -30 & 28 \end{bmatrix}$

21. Calculate **AB** – **BA** in each case

(*i*) $\mathbf{A} = \begin{bmatrix} 1 & 2 & 2 \\ 2 & 1 & 2 \\ 1 & 2 & 3 \end{bmatrix}$, $\mathbf{B} = \begin{bmatrix} 4 & 1 & 1 \\ -4 & 2 & 0 \\ 1 & 2 & 1 \end{bmatrix}$. (*ii*) $\mathbf{A} = \begin{pmatrix} 2 & 0 & 0 \\ 1 & 1 & 2 \\ -1 & 2 & 1 \end{pmatrix}$, $\mathbf{B} = \begin{pmatrix} 3 & 1 & -2 \\ 3 & -2 & 4 \\ -3 & 5 & 11 \end{pmatrix}$.

Ans. (*i*) $\begin{pmatrix} -9 & -2 & -10 \\ 6 & 14 & 8 \\ -7 & 5 & -5 \end{pmatrix}$, (*ii*) $\begin{pmatrix} -3 & 5 & -4 \\ 0 & 3 & 24 \\ 12 & -27 & 0 \end{pmatrix}$.

5.5 TRANSPOSE OF A MATRIX

A matrix obtained by interchanging the corresponding rows and columns of a given matrix **A** is called the *transpose of matrix* **A** and is denoted by $\mathbf{A}^T$ or $\mathbf{A}'$. That is, if

$$\mathbf{A} = \begin{bmatrix} a_{11} & a_{12} & \cdots & a_{1n} \\ a_{21} & a_{22} & \cdots & a_{2n} \\ \vdots & & & \\ a_{m1} & a_{m2} & \cdots & a_{mn} \end{bmatrix}, \text{ then } \mathbf{A}^T \text{ or } \mathbf{A}' = \begin{bmatrix} a_{11} & a_{21} & \cdots & a_{m1} \\ a_{12} & a_{22} & & a_{m2} \\ \vdots & \vdots & & \vdots \\ a_{1n} & a_{2n} & & a_{mn} \end{bmatrix}.$$

If **A** is an $m \times n$ matrix, then $\mathbf{A}^T$ is an $n \times m$ matrix.

For example, if $\mathbf{A} = \begin{bmatrix} 6 & 5 & 4 \\ 3 & 2 & -7 \end{bmatrix}$ then $\mathbf{A}^T = \begin{bmatrix} 6 & 3 \\ 5 & 2 \\ 4 & -7 \end{bmatrix}$.

Also, both the product matrices $\mathbf{A}^T\mathbf{A}$ and $\mathbf{A}\mathbf{A}^T$ are defined and $\mathbf{A}^T\mathbf{A} = (n \times m)(m \times n)$ is an $n \times n$ square matrix and $\mathbf{A}\mathbf{A}^T = (m \times n)(n \times m)$ is an $m \times m$ square matrix. A row vector **a** can also be written as $\begin{bmatrix} a_1 \\ a_2 \\ \vdots \\ a_n \end{bmatrix}^T$.

The transpose of a row matrix is a column matrix and the transpose of a column matrix is a row matrix.

5.5.1 Properties of Transpose

If $\mathbf{A}^T$ and $\mathbf{B}^T$ be the transposes respectively of matrices **A** and **B** and if k a scalar then whenever the sum and product defined.

1. $(\mathbf{A}^T)^T = \mathbf{A}$ that is, the transpose of the transpose of a matrix is the matrix itself.
2. $(\mathbf{kA})^T = \mathbf{kA}^T$
3. $(\mathbf{A} + \mathbf{B})^T = \mathbf{A}^T + \mathbf{B}^T$ that is, the transpose of the sum of two matrices is equal to the sum of their transposes.
4. $(\mathbf{AB})^T = \mathbf{B}^T\mathbf{A}^T$ that is, the transpose of a product of two matrices is equal to the product of the transposes taken in the reverse order.

Proof

1. Let $\mathbf{A} = (a_{ij})$ is an m × n matrix, then $\mathbf{A}^T$, is an n × m matrix and $(\mathbf{A}^T)^T$, the transpose of $\mathbf{A}^T$, is an m × n matrix. Thus the matrices $(\mathbf{A}^T)^T$ and **A** are of the same order.
Further, the (i, j)-th element of $(\mathbf{A}^T)^T = (j, i)$-th element of $\mathbf{A}^T = (i, j)$-th element of **A**.
Since $(\mathbf{A}^T)^T$ and **A** are comparable and their corresponding elements are equal, $(\mathbf{A}^T)^T = \mathbf{A}$
2. Let $\mathbf{A} = (a_{ij})$ is an m × n matrix. Thus **kA** is an m × n matrix, so that $(\mathbf{kA})^T$ is an $n \times m$ matrix.
Further $\mathbf{A}^T$ is an $n \times m$ matrix and, therefore, $\mathbf{kA}^T$ is also an n × m matrix.
Thus both the matrices $(\mathbf{kA})^T$ and $\mathbf{kA}^T$ are of the same order.
Also (i, j)-th element of $(\mathbf{kA})^T = (j, i)$-th element of $\mathbf{kA} = k\,(j, i)$-th element of $\mathbf{A} = k\,(i, j)$-th element of $\mathbf{A}^T = (i, j)$-th element of $(\mathbf{kA}^T)$.
Since the matrices $(k\mathbf{A})^T$ and $k\mathbf{A}^T$ are of the same order and their corresponding elements are equal, they are equal.
3. Let $\mathbf{A} = (a_{ij})$, $\mathbf{B} = (b_{ij})$ be $m \times n$ matrices.
Since each of the matrices **A** and **B** is $m \times n$ matrix, $(\mathbf{A} + \mathbf{B})$ is an $m \times n$ matrix and, therefore, $(\mathbf{A} + \mathbf{B})^T$ is an $n \times m$ matrix.
Also $\mathbf{A}^T$ and $\mathbf{B}^T$ are both $n \times m$ matrices. So that $(\mathbf{A}^T + \mathbf{B}^T)$ is an $n \times m$ matrix. Thus $(\mathbf{A} + \mathbf{B})^T$ and $(\mathbf{A}^T + \mathbf{B}^T)$ are of the same order.
Further (i, j)-th element of $(\mathbf{A} + \mathbf{B})^T = (j, i)$-th element of $(\mathbf{A} + \mathbf{B}) = a_{ji} + b_{ji} = (j, i)$-th element of **A** + (j, i)-th element of $\mathbf{B} = (i, j)$-th element of $\mathbf{A}^T + (i, j)$-th element of $\mathbf{B}^T = (i, j)$-th element of $(\mathbf{A}^T + \mathbf{B}^T)$.
Since $(\mathbf{A} + \mathbf{B})^T$ and $\mathbf{A}^T + \mathbf{B}^T$ are of the same order and their corresponding elements are equal.
Thus, $(\mathbf{A} + \mathbf{B})^T + \mathbf{A}^T + \mathbf{B}^T$.
4. Let $\mathbf{A} = (a_{ij})$ is an $m \times n$ matrix and $\mathbf{B} = (b_{ij})$ is an $n \times p$ matrix.
Since **AB** is an $m \times p$ matrix, $(\mathbf{AB})^T$ is a $p \times m$ matrix.
Further $\mathbf{B}^T$ is $p \times n$ matrix and $\mathbf{A}^T$ is an $n \times m$ matrix and therefore $\mathbf{B}^T\mathbf{A}^T$ is a $p \times m$ matrix.
Thus $(\mathbf{AB})^T$ and $\mathbf{B}^T\mathbf{A}^T$ are matrices of the same order.

Now the (j, i)-th element of $(\mathbf{AB})^T = (i, j)$-th element of $\mathbf{AB} = \sum_{k=1}^{n} a_{ik} b_{kj}$ ...(5.3)

Also the jth row of $\mathbf{B}^T$ is $b_{1j}, b_{2j}, \ldots, b_{nj}$ and ith column of $\mathbf{A}^T$ is $a_{i1}, a_{i2}, \ldots, a_{in}$.

$\therefore$ (j, i)-th element of $\mathrm{B}^T\mathrm{A}^T = \sum_{k=1}^{n} a_{kj} b_{ik}$ ...(5.4)

From (5.3) and (5.4), we have

$\therefore$ (j, i)-th element of $(\mathbf{AB})^T$ = (j, i)-th element of $\mathbf{B}^T\mathbf{A}^T$.

Since the matrices $(\mathbf{AB})^T$ and $\mathbf{B}^T\mathbf{A}^T$ are of the same order and their corresponding elements are equal, we have $(\mathbf{AB})^T = \mathbf{B}^T\mathbf{A}^T$.

5.6 SYMMETRIC AND SKEW-SYMMETRIC MATRICES

Def. A real square matrix $\mathbf{A} = (a_{ij})$ is said to be *symmetric*, if $a_{ij} = a_{ji}$ for all $i\,j$, that is $\mathbf{A}^T = \mathbf{A}$. This means that the entries in a symmetric matrix are symmetric with respect to the main diagonal of the matrix.

Skew-symmetric, if $a_{ij} = -a_{ji}$ for all i and j, that is $\mathbf{A}^T = -\mathbf{A}$.

$$\begin{pmatrix} 4 & 3 & 1 \\ 3 & 7 & 2 \\ 1 & 2 & 5 \end{pmatrix} \qquad \begin{pmatrix} 0 & 2 & 5 \\ -2 & 0 & 1 \\ -5 & -1 & 0 \end{pmatrix}$$

symmetric matrix — skew-symmetric matrix

5.6.1 Properties of Symmetric and Skew-symmetric Matrices

1. In a skew-symmetric matrix $\mathbf{A} = (a_{ij})$, all the elements in the principal diagonal are zero.
2. If $\mathbf{A}$ is a real square matrix, then matrix $\mathbf{A} + \mathbf{A}^T$ is symmetric and matrix $\mathbf{A} - \mathbf{A}^T$ is skew-symmetric or anti-symmetric matrix. Therefore any real square matrix can always be expressed as the sum of a real symmetric matrix and a real skew-symmetric matrix. That is,

$$\mathbf{A} = \frac{1}{2}(\mathbf{A} + \mathbf{A}^T) + \frac{1}{2}(\mathbf{A} - \mathbf{A}^T)$$

Proof: $(\mathbf{A} + \mathbf{A}^T)^T = \mathbf{A}^T + \mathbf{A} = \mathbf{A} + \mathbf{A}^T$.

$\therefore$ $\mathbf{A} + \mathbf{A}^T$ and hence $\frac{1}{2}(\mathbf{A} + \mathbf{A}^T)$ is symmetric.

Also, since $(\mathbf{A} - \mathbf{A}^T)^T = \mathbf{A}^T - \mathbf{A} = -(\mathbf{A} - \mathbf{A}^T)$, it follows that $\frac{1}{2}(\mathbf{A} - \mathbf{A}^T)$ is skew-symmetric. The result is thus proved.

3. If **A** and **B** be two symmetric matrices of the same order, then the product matrix **AB** is symmetric if and only if **AB** = **BA**, that is the matrices **A** and **B** commute.

Proof. Since the matrices **A** and **B** are symmetric we have $\mathbf{A}^T = \mathbf{A}$ **and** $\mathbf{B}^T = \mathbf{B}$.

Let **AB** be symmetric. Then $(\mathbf{AB})^T = \mathbf{AB}$, or $\mathbf{B}^T\mathbf{A}^T = \mathbf{AB}$, $\Rightarrow \mathbf{BA} = \mathbf{AB}$.

Now, let **AB** = **BA**. Taking transpose on both sides, we obtain

$(\mathbf{AB})^T = (\mathbf{BA})^T = \mathbf{A}^T\mathbf{B}^T = \mathbf{AB}$. Hence, the result.

EXAMPLE 5.12 Given $\mathbf{A} = \begin{bmatrix} 1 & 2 & 3 \\ 4 & 5 & 6 \\ 7 & 8 & 9 \end{bmatrix}$ and $\mathbf{B} = \begin{bmatrix} 1 & 1 & 1 \\ 1 & 1 & 1 \\ 1 & 1 & 1 \end{bmatrix}$. Show that $(\mathbf{AB})^T = \mathbf{B}^T\mathbf{A}^T$. *(AMIE, S-2002)*

Solution We have
$$\mathbf{AB} = \begin{bmatrix} 1 & 2 & 3 \\ 4 & 5 & 6 \\ 7 & 8 & 9 \end{bmatrix}\begin{bmatrix} 1 & 1 & 1 \\ 1 & 1 & 1 \\ 1 & 1 & 1 \end{bmatrix} = \begin{bmatrix} 1+2+3 & 1+2+3 & 1+2+3 \\ 4+5+6 & 4+5+6 & 4+5+6 \\ 7+8+9 & 7+8+9 & 7+8+9 \end{bmatrix} = \begin{bmatrix} 6 & 6 & 6 \\ 15 & 15 & 15 \\ 24 & 24 & 24 \end{bmatrix}$$

$$\therefore \quad (\mathbf{AB})^T = \begin{bmatrix} 6 & 6 & 6 \\ 15 & 15 & 15 \\ 24 & 24 & 24 \end{bmatrix}^T = \begin{bmatrix} 6 & 15 & 24 \\ 6 & 15 & 24 \\ 6 & 15 & 24 \end{bmatrix}.$$

Now
$$\mathbf{B}^T\mathbf{A}^T = \begin{bmatrix} 1 & 1 & 1 \\ 1 & 1 & 1 \\ 1 & 1 & 1 \end{bmatrix}\begin{bmatrix} 1 & 4 & 7 \\ 2 & 5 & 8 \\ 3 & 6 & 9 \end{bmatrix} = \begin{bmatrix} 1+2+3 & 4+5+6 & 7+8+9 \\ 1+2+3 & 4+5+6 & 7+8+9 \\ 1+2+3 & 4+5+6 & 7+8+9 \end{bmatrix} = \begin{bmatrix} 6 & 15 & 24 \\ 6 & 15 & 24 \\ 6 & 15 & 24 \end{bmatrix}.$$
Hence $(\mathbf{AB})^T = \mathbf{B}^T\mathbf{A}^T$.

EXAMPLE 5.13 Express the following matrix **A** as a sum of a symmetric and skew-symmetric matrix

$$\mathbf{A} = \begin{bmatrix} -1 & 7 & 1 \\ 2 & 3 & 4 \\ 5 & 0 & 5 \end{bmatrix}$$

Solution We have $\mathbf{A}^T = \begin{bmatrix} -1 & 2 & 5 \\ 7 & 3 & 0 \\ 1 & 4 & 5 \end{bmatrix}$. Then $\mathbf{A} + \mathbf{A}^T = \begin{bmatrix} -1 & 7 & 1 \\ 2 & 3 & 4 \\ 5 & 0 & 5 \end{bmatrix} + \begin{bmatrix} -1 & 2 & 5 \\ 7 & 3 & 0 \\ 1 & 4 & 5 \end{bmatrix} = \begin{bmatrix} -2 & 9 & 6 \\ 9 & 6 & 4 \\ 6 & 4 & 10 \end{bmatrix}$.

Therefore, $\frac{1}{2}(\mathbf{A} + \mathbf{A}^T) = \begin{bmatrix} -1 & 9/2 & 3 \\ 9/2 & 3 & 2 \\ 3 & 2 & 5 \end{bmatrix}$. Similarly, $\frac{1}{2}(\mathbf{A} - \mathbf{A}^T) = \begin{bmatrix} 0 & 5/2 & -2 \\ -5/2 & 0 & 2 \\ 2 & -2 & 0 \end{bmatrix}$.

$$\therefore \quad \mathbf{A} = \frac{1}{2}(\mathbf{A} + \mathbf{A}^T) + \frac{1}{2}(\mathbf{A} - \mathbf{A}^T) = \underbrace{\begin{bmatrix} -1 & 9/2 & 3 \\ 9/2 & 3 & 2 \\ 3 & 2 & 5 \end{bmatrix}}_{\text{symmetrical matrix}} + \underbrace{\begin{bmatrix} 0 & 5/2 & -2 \\ -5/2 & 0 & 2 \\ 2 & -2 & 0 \end{bmatrix}}_{\text{skew-symmetric matrix}}.$$

EXAMPLE 5.14 If $2\mathbf{A} + \mathbf{B}^T = \begin{bmatrix} 1 & 2 \\ 0 & 3 \end{bmatrix}$ ...(*i*) and $-\mathbf{A} + 2\mathbf{B} = \begin{bmatrix} 4 & 3 \\ 4 & 6 \end{bmatrix}$, ...(*ii*) find the matrices **A** and **B**.

(*AMIE, S-2003*)

Solution Multiplying (*ii*) by 2 and adding, we get

$4\mathbf{B} + \mathbf{B}^T = \begin{bmatrix} 9 & 8 \\ 8 & 15 \end{bmatrix}$ ...(*iii*) Let $\mathbf{B} = \begin{bmatrix} p & q \\ r & s \end{bmatrix}$. Therefore, $\mathbf{B}^T = \begin{bmatrix} p & r \\ q & s \end{bmatrix}$.

Matrix equation (*iii*) may be re-written as $\begin{bmatrix} 5p & 4q + r \\ 4r + q & 5s \end{bmatrix} = \begin{bmatrix} 9 & 8 \\ 8 & 15 \end{bmatrix}$.

By the definition of equality of matrices, we have

$5p = 9,\ 4q + r = 8,\ 4r + q = 8,\ 5s = 15.$

On solving these equations, we obtain $p = 9/5$, $q = 8/5$, $r = 8/5$, $s = 3$.

$$\therefore\ \mathbf{B} = \begin{bmatrix} 9/5 & 8/5 \\ 8/5 & 3 \end{bmatrix} \text{ and } \mathbf{A} = 2\mathbf{B} - \begin{bmatrix} 4 & 3 \\ 4 & 6 \end{bmatrix} = \begin{bmatrix} 18/5 & 16/5 \\ 16/5 & 6 \end{bmatrix} + \begin{bmatrix} -4 & -3 \\ -4 & -6 \end{bmatrix} = \begin{bmatrix} -2/5 & 1/5 \\ -4/5 & 0 \end{bmatrix}.$$

EXERCISE 5.2

1. If $\mathbf{A} = \begin{bmatrix} 1 & 2 & 0 \\ 3 & -1 & 4 \end{bmatrix}$, find $\mathbf{AA}^T$ and $\mathbf{A}^T\mathbf{A}$. **Ans.** $\begin{bmatrix} 5 & 1 \\ 1 & 26 \end{bmatrix}, \begin{bmatrix} 10 & -1 & 12 \\ -1 & 5 & -4 \\ 12 & -4 & 16 \end{bmatrix}$.

2. If $\mathbf{A} = \begin{bmatrix} 1 & 2 & -1 \\ 3 & 0 & 2 \\ 4 & 5 & 0 \end{bmatrix}$, $\mathbf{B} = \begin{bmatrix} 1 & 0 & 0 \\ 2 & 1 & 0 \\ 0 & 1 & 3 \end{bmatrix}$. Verify that $(\mathbf{AB})^T = \mathbf{B}^T\mathbf{A}^T$.

3. If $\mathbf{A}=\begin{bmatrix}4\\8\\-10\end{bmatrix}$ and $\mathbf{B} = [2\ 4\ 5]$, find (*i*) $\mathbf{A}^T\mathbf{A}$, (*ii*) $\mathbf{B}^T\mathbf{B}$, (*iii*) $\mathbf{A}+\mathbf{B}^T$.

Ans. (*i*) 180; (*ii*) $\begin{bmatrix}4 & 8 & 10\\8 & 16 & 20\\10 & 20 & 25\end{bmatrix}$; (*iii*) $\begin{bmatrix}6\\12\\-5\end{bmatrix}$.

4. Find (*i*) $\mathbf{A}^T\mathbf{B}$, (*ii*) $\mathbf{A}^T\mathbf{C}$, where $\mathbf{A}=\begin{bmatrix}1 & -1 & 2\\0 & 3 & 4\end{bmatrix}$, $\mathbf{B}=\begin{bmatrix}4 & 0 & -3\\-1 & -2 & 3\end{bmatrix}$, $\mathbf{C}=\begin{bmatrix}2 & -3 & 0 & 1\\5 & -1 & -4 & 2\\-1 & 0 & 0 & 3\end{bmatrix}$.

Ans. (*i*) $\begin{bmatrix}4 & 0 & -3\\-7 & -6 & 12\\4 & -8 & 6\end{bmatrix}$; (*ii*) not defined.

5. (*i*) Express the matrix $\begin{bmatrix}4 & 5 & 6\\-1 & 0 & 1\\2 & 1 & 2\end{bmatrix}$ as the sum of a symmetric and a skew-symmetric matrix.

(*ii*) If $\mathbf{A}=\begin{pmatrix}3 & 1\\-4 & -1\end{pmatrix}$, represent it as $\mathbf{A}=\mathbf{B}+\mathbf{C}$, where $\mathbf{B}$ is a symmetric and $\mathbf{C}$ is a skew-symmetric matrix.

Ans. (*i*) $\begin{bmatrix}4 & 2 & 4\\2 & 0 & 1\\4 & 1 & 2\end{bmatrix}+\begin{bmatrix}0 & 3 & 2\\-3 & 0 & 0\\-2 & 0 & 0\end{bmatrix}$; (*ii*) $\mathbf{B}=\begin{bmatrix}3 & -3/2\\-3/2 & -1\end{bmatrix}$; $\mathbf{C}=\begin{bmatrix}0 & 5/2\\-5/2 & 0\end{bmatrix}$.

6. If **A** and **B** are symmetric matrices, prove that **AB** – **BA** is a skew-symmetric matrix.

5.7 DETERMINANTS

Corresponding to every square matrix **A** of order n, that is, for every $n \times n$ matrix there is associated an expression, called the *determinant* of **A** denoted by either det (**A**) or |**A**|. The determinant has a value and this value is real if the matrix is real and may be real or complex, if the matrix is complex. A determinant of order n is defined as

$$\det(\mathbf{A}) \text{ or } |\mathbf{A}| = \begin{vmatrix}a_{11} & a_{12} & \cdots & a_{1n}\\a_{21} & a_{22} & \cdots & a_{2n}\\\vdots & & & \vdots\\a_{n1} & a_{n2} & \cdots & a_{nn}\end{vmatrix}. \qquad ...(5.5)$$

The n^2 quantities $a_{11}, a_{12}, ..., a_{nn}$ are called **elements** of the determinant. The horizontal lines of elements are called **rows** and the vertical lines of elements are called **columns.** A determinant of nth order has n rows and n columns.

We now discuss methods to find the value of a determinant. For a, 1×1 matrix $\mathbf{A} = [a]$, we have det (**A**) = $|a| = a$.

A determinant of 2×2 matrix is written as $|\mathbf{A}|=\begin{vmatrix}a_{11} & a_{12}\\a_{21} & a_{22}\end{vmatrix}$ and its value is given by $|\mathbf{A}| = a_{11}a_{22} - a_{12}a_{21}$.

We evaluate higher order determinants using minors and co-factors.

5.7.1 Minors, Co-factors and Laplace Expansion

Minor Let a_{ij} be the general element of a determinant. If we delete the i-th row and the j-th column from the determinant, we obtain a new determinant of order $(n - 1)$ which is called the *minor* of the element a_{ij}. We denote the minor by M_{ij}.

For example, the minor corresponding to the element 3 in the 3rd row and 2nd column of the fourth order

$$\det(\mathbf{B}) = \begin{vmatrix} 2 & 1 & -1 & 4 \\ -3 & 2 & 5 & 6 \\ 1 & 3 & -2 & 2 \\ 4 & 5 & -3 & 1 \end{vmatrix} \text{ is } \mathbf{M}_{32} = \begin{vmatrix} 2 & -1 & 4 \\ -3 & 5 & 6 \\ 4 & -3 & 1 \end{vmatrix}$$ which is obtained by deleting the elements shown in the blocks.

Co-factor If we multiply the minor of a_{ij} by $(-1)^{i+j}$, the result is called the co-factor of a_{ij} and is denoted by A_{ij}, that is $A_{ij} = (-1)^{i+j} \cdot M_{ij}$. ...(5.6)

For example the co-factor A_{32} corresponding to the element 3 in the det (**B**) is $(-1)^{3+2}$ times its minor, that

is $A_{32} = -\begin{vmatrix} 2 & -1 & 4 \\ -3 & 5 & 6 \\ 4 & -3 & 1 \end{vmatrix}$. We can expand a determinant of order n through the elements of any row or any column.

The Laplace expansion The value of a determinant is defined as the sum of the products of the elements of the ith row (or jth column) by their corresponding cofactors and is called the *Laplace expansion.*

Thus, we have $\det(\mathbf{A})$ or $|\mathbf{A}| = \sum_{j=1}^{n} (-1)^{i+j} a_{ij} M_{ij} = \sum_{j=1}^{n} a_{ij} A_{ij}$, ...(5.7a)

when we expand through the elements of the ith row

or $$\det(\mathbf{A}) \text{ or } |\mathbf{A}| = \sum_{i=1}^{n} (-1)^{i+j} a_{ij} M_{ij} = \sum_{i=1}^{n} a_{ij} A_{ij}$$ ...(5.7b)

when we expand through the elements of the jth column.

Remark Note that, the sum of the products formed by multiplying the elements of any row (or column) of an n-squared matrix **A** by their corresponding co-factors gives the value of the determinant. However, the sum of the products of the elements of any row (or column) of an n-square matrix **A** by the corresponding co-factors of **any other** row (or column) is zero.

A determinant of third order can by defined by

$$\det(\mathbf{A}) = \begin{vmatrix} a_{11} & a_{12} & a_{13} \\ a_{21} & a_{22} & a_{23} \\ a_{31} & a_{32} & a_{33} \end{vmatrix} = a_{11}\begin{vmatrix} a_{22} & a_{23} \\ a_{32} & a_{33} \end{vmatrix} - a_{12}\begin{vmatrix} a_{21} & a_{23} \\ a_{31} & a_{33} \end{vmatrix} + a_{13}\begin{vmatrix} a_{21} & a_{22} \\ a_{31} & a_{32} \end{vmatrix}$$

$$= a_{11}\begin{vmatrix} a_{22} & a_{23} \\ a_{32} & a_{33} \end{vmatrix} + a_{12}\left(-\begin{vmatrix} a_{21} & a_{23} \\ a_{31} & a_{33} \end{vmatrix}\right) + a_{13}\begin{vmatrix} a_{21} & a_{22} \\ a_{31} & a_{32} \end{vmatrix}$$

$$= a_{11}A_{11} + a_{12}A_{12} + a_{13}A_{13}$$ [Co-factor expansion of |**A**| along the first row].

A 3 × 3 matrix has nine co-factors

$$A_{11} = M_{11} \qquad A_{12} = -M_{12} \qquad A_{13} = M_{13}$$
$$A_{21} = -M_{21} \qquad A_{22} = M_{22} \qquad A_{23} = -M_{23}$$
$$A_{31} = M_{31} \qquad A_{32} = -M_{32} \qquad A_{33} = M_{33}$$

Similarly, it can be shown that

det (**A**) = $a_{12} A_{12} + a_{22} A_{22} + a_{32} A_{32}$, which is the co-factor expansion of |**A**| along the second column.

Generally, we expand a determinant through that row or column which has a number of zeroes.

Note that determinants are defined only for square matrices.

EXAMPLE 5.15 Evaluate the determinant of $\mathbf{A} = \begin{bmatrix} 3 & -2 & 2 \\ 1 & 2 & -3 \\ 4 & 1 & 2 \end{bmatrix}$, taking expansion along the first row.

Solution $\det(\mathbf{A}) = \begin{vmatrix} 3 & -2 & 2 \\ 1 & 2 & -3 \\ 4 & 1 & 2 \end{vmatrix} = 3A_{11} + (-2)A_{12} + 2A_{13}.$

Now, the co-factors of the entries/elements in the first row of **A** are

$$A_{11} = (-1)^{1+1}\begin{vmatrix} 3 & -2 & 2 \\ 1 & 2 & -3 \\ 4 & 1 & 2 \end{vmatrix} = (-1)^2\begin{vmatrix} 2 & -3 \\ 1 & 2 \end{vmatrix} = 4 + 3 = 7$$

$$A_{12} = (-1)^{1+2}\begin{vmatrix} 3 & -2 & 2 \\ 1 & 2 & -3 \\ 4 & 1 & 2 \end{vmatrix} = (-1)^3\begin{vmatrix} 1 & -3 \\ 4 & 2 \end{vmatrix} = -(2+12) = -14,$$

$$A_{13} = (-1)^{1+3}\begin{vmatrix} 3 & -2 & 2 \\ 1 & 2 & -3 \\ 4 & 1 & 2 \end{vmatrix} = (-1)^4\begin{vmatrix} 1 & 2 \\ 4 & 1 \end{vmatrix} = (1-8) = -7.$$

Thus det (**A**) = 3(7) + (–2) (–14) + 2(–7) = 21 + 28 – 14 = 35.

NOTE In place of first row we could expand along any other row or column, the value of the determinant remain same. Let us take co-factor expansion along the third column. The co-factors of elements of third column that is, 2, –3 and 2 are respectively $(-1)^{1+3}\begin{vmatrix} 1 & 2 \\ 4 & 1 \end{vmatrix}$, $(-1)^{2+3}\begin{vmatrix} 3 & -2 \\ 4 & 1 \end{vmatrix}$, $(-1)^{3+3}\begin{vmatrix} 3 & -2 \\ 1 & 2 \end{vmatrix}$.

That is, –7, –11 and 8. Therefore expansion along third column gives det (**A**) = 2(–7) –3 (–11) +2 (8) = –14 + 33 + 16 = 35, which is same as above.

EXAMPLE 5.16 Solve the equation $\begin{vmatrix} x-1 & -6 & 2 \\ -6 & x-2 & -4 \\ 2 & -4 & x-6 \end{vmatrix} = 0.$

Solution Expansion along the first row gives

$$(x-1)\begin{vmatrix} x-2 & -4 \\ -4 & x-6 \end{vmatrix} - (-6)\begin{vmatrix} -6 & -4 \\ 2 & x-6 \end{vmatrix} + 2\begin{vmatrix} -6 & x-2 \\ 2 & -4 \end{vmatrix} = 0$$

or $(x-1)(x^2 - 8x + 12 - 16) + 6(-6x + 36 + 8) + 2(24 - 2x + 4) = 0.$

Simplifying, we obtain $x^3 - 9x^2 - 36x + 324 = 0$ ...(*i*)

By Remainder Theorem, $x = 6$ is a root of this equation. Hence we can write (*i*) as

$(x-6)(x^2 - 3x - 54) = 0$ or $(x-6)(x+6)(x-9) = 0.$

$\therefore$ $x = 6, -6, 9$ are the required roots.

EXERCISE 5.3

1. Find the co-factors of the elements of the second row and third column of $|\mathbf{A}| = \begin{vmatrix} 1 & 2 & 3 \\ 2 & 3 & 2 \\ 1 & 2 & 2 \end{vmatrix}$.

Ans. 2, –1, 0; 1, 0, –1.

2. Show that the co-factor of each element of $\begin{vmatrix} -1/3 & -2/3 & -2/3 \\ 2/3 & 1/3 & -2/3 \\ 2/3 & -2/3 & 1/3 \end{vmatrix}$ is that element itself.

3. Show that the co-factor of an element of any row of $\begin{vmatrix} -4 & -3 & -3 \\ 1 & 0 & 1 \\ 4 & 4 & 3 \end{vmatrix}$ is the corresponding element of the same numbered column that is, $A_{ij} = a_{ji}$.

4. Compute the determinant of each of the following matrices

$$(i)\ \begin{bmatrix} 2 & -1 & 1 \\ 3 & 2 & 4 \\ -1 & 0 & 3 \end{bmatrix}, \quad (ii)\ \begin{bmatrix} 3 & 4 & 5 \\ 5 & 3 & 4 \\ 4 & 5 & 3 \end{bmatrix}, \quad (iii)\ \begin{bmatrix} 3 & -4 & -3 \\ 2 & 7 & -31 \\ 5 & -9 & 2 \end{bmatrix}, \quad (iv)\ \begin{bmatrix} -2 & -1 & 4 \\ 6 & -3 & -2 \\ 4 & 1 & 2 \end{bmatrix}.$$

Ans. (*i*) 27, (*ii*) 36, (*iii*) zero, (*iv*) 100.

5. Solve the following equations:

$$(i)\ \begin{vmatrix} 2x & 5 \\ 9 & x+3 \end{vmatrix} = \begin{vmatrix} 5 & 4 \\ 13 & 3x \end{vmatrix}, \qquad (ii)\ \begin{vmatrix} 2x^2 & -3 & -16 \\ x & 5 & 5 \\ 11 & 20 & -15 \end{vmatrix} = 0,$$

$$(iii)\ \begin{vmatrix} x & 2 & 3 \\ 6 & x+4 & 4 \\ 7 & 8 & x+8 \end{vmatrix} = 0, \qquad (iv)\ \begin{vmatrix} x-4 & 3 \\ 2 & x-9 \end{vmatrix} = 0.$$

Ans. (*i*) 3.5, 1 (*ii*) 1, $-2\frac{3}{70}$ (*iii*) 1, 1.39, −14.39 (*iv*) 3, 10.

6. Find values of λ for which $\begin{vmatrix} 3-\lambda & 0 & 0 \\ 0 & 4-\lambda & \sqrt{3} \\ 0 & \sqrt{3} & 6-\lambda \end{vmatrix} = 0.$ **Ans.** 3, 7.

5.8 PROPERTIES OF DETERMINANTS

In applied mathematics very often we are required to evaluate determinants whose elements are numbers. Such determinants can be evaluated without much difficulty by the method of Laplace expansion when the determinants are of second or third order. In case the determinants are of higher order, this process becomes very cumbersome. The expansion in such cases can be effected very conveniently with the help of the following properties of determinants.

1. If all the elements in a row (or column) of a determinant are zero, then the value of the determinant is zero.
2. The value of a determinant is not changed if its corresponding rows and columns are interchanged, that is, $|\mathbf{A}| = |\mathbf{A}^T|$.
3. If any two rows (or columns) of a determinant are interchanged, then the value of the determinant is multiplied by (−1).
4. If the corresponding elements of two rows (or columns) are same, that is two rows (or columns) are identical, then the value of the determinant is zero.
5. If the corresponding elements of two rows (or columns) are proportional to each other, then the value of the determinant is zero.
6. If each element of a row (or column) of a determinant is multiplied by a scalar k, then the determinant is multiplied by the scalar k.

7. If each element of any row (or column) of a determinant can be expressed as the sum of two (or more) terms, then the determinant can be expressed as sum of two (or more) determinants.
8. If we multiply the elements of any row (or column) by a non-zero number and add to corresponding elements of any other row (or column), then the value of the determinant remains the same.
9. The value of the determinant of a diagonal or a lower triangular or an upper triangular matrix is the product of its diagonal elements.

Remark When the elements of the jth row are multiplied by a non-zero constant k and added to the corresponding elements of the ith row, we denote this operation as $R_i \to R_i + kR_j$, where R_i is the ith row of $|\mathbf{A}|$. The elements of the jth row remain unchanged whereas the elements of the jth row get changed. This operation is called an *elementary row operation.* Similarly, the operation $C_i \to C_i + kC_j$, where C_i is the ith column of $|\mathbf{A}|$, is called the *elementary column operation.* Therefore, under elementary row (or column) operations, the value of the determinant is unchanged.

ILLUSTRATIVE EXAMPLES

EXAMPLE 5.17 Evaluate: (*i*) $|\mathbf{A}| = \begin{vmatrix} 28 & 25 & 38 \\ 42 & 38 & 65 \\ 56 & 47 & 83 \end{vmatrix}$, (*ii*) $|\mathbf{A}| = \begin{vmatrix} 1 & a & b+c \\ 1 & b & c+a \\ 1 & c & a+b \end{vmatrix}$.

Solution (*i*) Taking 14 as common factor, we obtain

$$|\mathbf{A}| = 14\begin{vmatrix} 2 & 25 & 38 \\ 3 & 38 & 65 \\ 4 & 47 & 83 \end{vmatrix} \quad (\text{operate } C_2 \to C_2 - 12C_1 \text{ and } C_3 \to C_3 - 19C_1)$$

$$= 14\begin{vmatrix} 2 & 1 & 0 \\ 3 & 2 & 8 \\ 4 & -1 & 7 \end{vmatrix} \quad (\text{operate } C_1 \to C_1 - 2C_2) = 14\begin{vmatrix} 0 & 1 & 0 \\ -1 & 2 & 8 \\ 6 & -1 & 7 \end{vmatrix}.$$

Expanding the determinant by using the first row, we have $|\mathbf{A}| = 14\,(-1)\begin{vmatrix} -1 & 8 \\ 6 & 7 \end{vmatrix} = 14 \times 55 = 770.$

(*ii*) Operating $C_3 \to C_3 + C_2$, the given determinant becomes

$$|\mathbf{A}| = \begin{vmatrix} 1 & a & a+b+c \\ 1 & b & a+b+c \\ 1 & c & a+b+c \end{vmatrix} = (a+b+c)\begin{vmatrix} 1 & a & 1 \\ 1 & b & 1 \\ 1 & c & 1 \end{vmatrix} = 0,$$

because two columns are identical.

EXAMPLE 5.18 Prove, without expanding, that each of the following determinants vanishes.

(*i*) $|\mathbf{A}| = \begin{vmatrix} 1 & 15 & 14 & 4 \\ 12 & 6 & 7 & 9 \\ 8 & 10 & 11 & 5 \\ 13 & 3 & 2 & 16 \end{vmatrix}$, (*ii*) $|\mathbf{A}| = \begin{vmatrix} 1 & a & a^2 & a^3+bcd \\ 1 & b & b^2 & b^3+cda \\ 1 & c & c^2 & c^3+dab \\ 1 & d & d^2 & d^3+abc \end{vmatrix}$.

Solution (*i*) Operating $C_1 \to C_1 + C_2 + C_3 + C_4$ and taking out 34, the common factor from C_1, the given determinant

$$|\mathbf{A}| = 34\begin{vmatrix} 1 & 15 & 14 & 4 \\ 1 & 6 & 7 & 9 \\ 1 & 10 & 11 & 5 \\ 1 & 3 & 2 & 16 \end{vmatrix} \quad \begin{pmatrix}\text{operate } R_4 \to R_4 - R_1, \\ R_3 \to R_3 - R_1, \\ \text{and } R_2 \to R_2 - R_1\end{pmatrix} = 34\begin{vmatrix} 1 & 15 & 14 & 4 \\ 0 & -9 & -7 & 5 \\ 0 & -5 & -3 & 1 \\ 0 & -12 & -12 & 12 \end{vmatrix} \quad \begin{pmatrix}\text{taking out 12} \\ \text{as common} \\ \text{factor from } R_4\end{pmatrix}$$

$$= 408\begin{vmatrix} 1 & 15 & 14 & 4 \\ 0 & -9 & -7 & 5 \\ 0 & -5 & -3 & 1 \\ 0 & -1 & -1 & 1 \end{vmatrix}$$. Expanding the determinant by using the first column, we have

$$|\mathbf{A}| = 408\begin{vmatrix} -9 & -7 & 5 \\ -5 & -3 & 1 \\ -1 & -1 & 1 \end{vmatrix} \begin{matrix}\text{(operate } C_3 \to C_3 + C_2 \\ \text{and } C_2 \to C_2 - C_1)\end{matrix} = 408\begin{vmatrix} -9 & 2 & -2 \\ -5 & 2 & -2 \\ -1 & 0 & 0 \end{vmatrix} = 0. \quad \begin{matrix}\text{(because } C_2 \text{ and } C_3 \\ \text{are proportional).}\end{matrix}$$

(*ii*) By property 7, the given determinant

$$|\mathbf{A}| = \begin{vmatrix} 1 & a & a^2 & a^3 \\ 1 & b & b^2 & b^3 \\ 1 & c & c^2 & c^3 \\ 1 & d & d^2 & d^3 \end{vmatrix} + \begin{vmatrix} 1 & a & a^2 & bcd \\ 1 & b & b^2 & cda \\ 1 & c & c^2 & dab \\ 1 & d & d^2 & abc \end{vmatrix} = |\mathbf{B}| + |\mathbf{C}| \text{ (say).}$$

Taking $a^{-1}, b^{-1}, c^{-1}, d^{-1}$ common from R_1, R_2, R_3 and R_4 respectively in $|\mathbf{C}|$, we have

$$|\mathbf{C}| = \frac{1}{abcd}\begin{vmatrix} a & a^2 & a^3 & abcd \\ b & b^2 & b^3 & bcda \\ c & c^2 & c^3 & cdab \\ d & d^2 & d^3 & dabc \end{vmatrix} \quad \text{(taking } abcd \text{ common from } C_4)$$

$$= \begin{vmatrix} a & a^2 & a^3 & 1 \\ b & b^2 & b^3 & 1 \\ c & c^2 & c^3 & 1 \\ d & d^2 & d^3 & 1 \end{vmatrix} \begin{matrix}\text{(passing } C_4 \text{ over} \\ \text{first three columns)}\end{matrix} = (-1)^3\begin{vmatrix} 1 & a & a^2 & a^3 \\ 1 & b & b^2 & b^3 \\ 1 & c & c^2 & c^3 \\ 1 & d & d^2 & d^3 \end{vmatrix} = -|\mathbf{B}|.$$

$\therefore$ $|\mathbf{A}| = |\mathbf{B}| - |\mathbf{B}| = 0$. Hence the result.

EXAMPLE 5.19 Prove that

$$|\mathbf{A}| = \begin{vmatrix} 1+a & 1 & 1 & 1 \\ 1 & 1+b & 1 & 1 \\ 1 & 1 & 1+c & 1 \\ 1 & 1 & 1 & 1+d \end{vmatrix} = abcd(1 + a^{-1} + b^{-1} + c^{-1} + d^{-1}).$$

Solution Taking a, b, c, d common from R_1, R_2, R_3, R_4 respectively, we obtain

$$|\mathbf{A}| = abcd\begin{vmatrix} a^{-1}+1 & a^{-1} & a^{-1} & a^{-1} \\ b^{-1} & b^{-1}+1 & b^{-1} & b^{-1} \\ c^{-1} & c^{-1} & c^{-1}+1 & c^{-1} \\ d^{-1} & d^{-1} & d^{-1} & d^{-1}+1 \end{vmatrix}$$

(operating $R_1 \to R_1 + (R_2 + R_3 + R_4)$ and taking out the common factor from R_1)

$$= abcd\,(1+a^{-1}+b^{-1}+c^{-1}+d^{-1})\begin{vmatrix} 1 & 1 & 1 & 1 \\ b^{-1} & b^{-1}+1 & b^{-1} & b^{-1} \\ c^{-1} & c^{-1} & c^{-1}+1 & c^{-1} \\ d^{-1} & d^{-1} & d^{-1} & d^{-1}+1 \end{vmatrix}$$

(operating $C_2 \to C_2 - C_1$, $C_3 \to C_3 - C_1$ and $C_4 \to C_4 - C_1$)

$$= abcd\,(1+a^{-1}+b^{-1}+c^{-1}+d^{-1})\begin{vmatrix} 1 & 0 & 0 & 0 \\ b^{-1} & 1 & 0 & 0 \\ c^{-1} & 0 & 1 & 0 \\ d^{-1} & 0 & 0 & 1 \end{vmatrix}.$$

Expanding the determinant by using the first row we obtain

$$|\mathbf{A}| = abcd\,(1 + a^{-1} + b^{-1} + c^{-1} + d^{-1}).$$

EXAMPLE 5.20 Show that $\begin{vmatrix} a-b-c & 2b & 2c \\ 2a & b-c-a & 2c \\ 2a & 2b & c-a-b \end{vmatrix} = (a+b+c)^3.$

Solution Let $|\mathbf{A}|$ be the given determinant. Operating $C_1 \to C_1 + (C_2 + C_3)$, we obtain

$$|\mathbf{A}| = \begin{vmatrix} a+b+c & 2b & 2c \\ a+b+c & b-c-a & 2c \\ a+b+c & 2b & c-a-b \end{vmatrix} = (a+b+c)\begin{vmatrix} 1 & 2b & 2c \\ 1 & b-c-a & 2c \\ 1 & 2b & c-a-b \end{vmatrix}$$

(operate $R_2 \to R_2 - R_1$ and $R_3 \to R_3 - R_1$)

$$= (a+b+c)\begin{vmatrix} 1 & 2b & 2c \\ 0 & -b-c-a & 0 \\ 0 & 0 & -c-a-b \end{vmatrix}.$$

Expanding the determinant by using the first column, we have

$$|\mathbf{A}| = (a + b + c)\,(a + b + c)^2 = (a + b + c)^3.$$

EXAMPLE 5.21 Prove that $|\mathbf{A}| = \begin{vmatrix} 1+a^2+b^2 & 2ab & -2b \\ 2ab & 1+a^2+b^2 & 2a \\ 2b & -2a & 1-a^2-b^2 \end{vmatrix} = (1+a^2+b^2)^3.$

Solution Operating $C_1 \to C_1 - bC_3$ and $C_2 \to C_2 + aC_3$, the

$$\det(\mathbf{A}) = \begin{vmatrix} 1+a^2+b^2 & 0 & -2b \\ 0 & 1+a^2+b^2 & 2a \\ b+a^2b+b^3 & -a-a^3-ab^2 & 1-a^2-b^2 \end{vmatrix}$$

[taking out $(1+a^2+b^2)$ common from C_1 and C_2]

$$= (1+a^2+b^2)^2\begin{vmatrix} 1 & 0 & -2b \\ 0 & 1 & 2a \\ b & -a & 1-a^2-b^2 \end{vmatrix}$$

(operate $C_3 \to C_3 + 2b\,C_1$

$$= (1+a^2+b^2)^2\begin{vmatrix} 1 & 0 & 0 \\ 0 & 1 & 2a \\ b & -a & 1-a^2+b^2 \end{vmatrix}.$$

Expanding the determinant by using the first row, we have

$$|\mathbf{A}| = (1+a^2+b^2)^2 \begin{vmatrix} 1 & 2a \\ -a & 1-a^2+b^2 \end{vmatrix} = (1 + a^2 + b^2)^2[1 - a^2 + b^2 + 2a^2] = (1 + a^2 + b^2)^3.$$

EXAMPLE 5.22 Show that

$$|\mathbf{A}| = \begin{vmatrix} x^2+1 & xy & xz & xu \\ xy & y^2+1 & yz & yu \\ zx & zy & z^2+1 & zu \\ ux & uy & uz & u^2+1 \end{vmatrix} = x^2+y^2+z^2+u^2+1.$$

Solution Multiplying C_1, C_2, C_3, C_4 by x, y, z, u respectively and dividing the determinant by $xyzu$, we obtain

$$|\mathbf{A}| = \frac{1}{xyzu}\begin{vmatrix} x(x^2+1) & xy^2 & xz^2 & xu^2 \\ x^2y & y(y^2+1) & yz^2 & yu^2 \\ x^2z & zy^2 & z(z^2+1) & zu^2 \\ x^2u & uy^2 & uz^2 & u(u^2+1) \end{vmatrix}$$ (taking x, y, z, u common from R_1, R_2, R_3, R_4 respectively)

$$= \frac{xyzu}{xyzu}\begin{vmatrix} x^2+1 & y^2 & z^2 & u^2 \\ x^2 & y^2+1 & z^2 & u^2 \\ x^2 & y^2 & z^2+1 & u^2 \\ x^2 & y^2 & z^2 & u^2+1 \end{vmatrix}$$ [operating $C_1 \to C_1 + (C_2 + C_3 + C_4)$ and then taking $x^2+y^2+z^2+u^2+1$ common from C_1)]

$$= (x^2+y^2+z^2+u^2+1)\begin{vmatrix} 1 & y^2 & z^2 & u^2 \\ 1 & y^2+1 & z^2 & u^2 \\ 1 & y^2 & z^2+1 & u^2 \\ 1 & y^2 & z^2 & u^2+1 \end{vmatrix}$$ (operate $R_2 \to R_2 - R_1$, $R_3 \to R_3 - R_1$ and $R_4 \to R_4 - R_1$)

$$= (x^2+y^2+z^2+u^2+1)\begin{vmatrix} 1 & y^2 & z^2 & u^2 \\ 0 & 1 & 0 & 0 \\ 0 & 0 & 1 & 0 \\ 0 & 0 & 0 & 1 \end{vmatrix} = x^2 + y^2 + z^2 + u^2 + 1.$$

EXAMPLE 5.23 If $|\mathbf{A}| = \begin{vmatrix} a & a^2 & a^3-1 \\ b & b^2 & b^3-1 \\ c & c^2 & c^3-1 \end{vmatrix} = 0$ in which a, b, c are all different, show that $abc = 1$.

Solution Since each term of C_3 in the given det (**A**) consists of two terms, the given determinant can be expressed as a sum of two determinants. (Property 7).

Therefore, $$\begin{vmatrix} a & a^2 & a^3-1 \\ b & b^2 & b^3-1 \\ c & c^2 & c^3-1 \end{vmatrix} = \begin{vmatrix} a & a^2 & a^3 \\ b & b^2 & b^3 \\ c & c^2 & c^3 \end{vmatrix} + \begin{vmatrix} a & a^2 & -1 \\ b & b^2 & -1 \\ c & c^2 & -1 \end{vmatrix}$$

(taking a, b, c common from R_1, R_2, R_3 respectively of the first determinant and –1 from C_3 of the second determinant). The given det (**A**)

$$= abc\begin{vmatrix} 1 & a & a^2 \\ 1 & b & b^2 \\ 1 & c & c^2 \end{vmatrix} - \begin{vmatrix} a & a^2 & 1 \\ b & b^2 & 1 \\ c & c^2 & 1 \end{vmatrix} \quad \text{(passing } C_3 \text{ over } C_2 \text{ and } C_1 \text{ in the second determinant)}$$

$$= abc\begin{vmatrix} 1 & a & a^2 \\ 1 & b & b^2 \\ 1 & c & c^2 \end{vmatrix} - (-1)^2\begin{vmatrix} 1 & a & a^2 \\ 1 & b & b^2 \\ 1 & c & c^2 \end{vmatrix} = (abc-1)\begin{vmatrix} 1 & a & a^2 \\ 1 & b & b^2 \\ 1 & c & c^2 \end{vmatrix} = 0. \quad \text{(as the given determinant is equal to zero).}$$

Since a, b, c are all different, so $\begin{vmatrix} 1 & a & a^2 \\ 1 & b & b^2 \\ 1 & c & c^2 \end{vmatrix} \neq 0$. Hence $abc - 1 = 0$ or $abc = 1$.

EXAMPLE 5.24 Solve the equation $\begin{vmatrix} a+x & b+x & c+x \\ b+x & c+x & a+x \\ c+x & a+x & b+x \end{vmatrix} = 0.$

Solution Operating $C_1 \to C_1 + (C_2 + C_3)$ and taking $(a + b + c + 3x)$ as common from first column, we obtain

$$(a+b+c+3x)\begin{vmatrix} 1 & b+x & c+x \\ 1 & c+x & a+x \\ 1 & a+x & b+x \end{vmatrix} = 0.$$ Operating $R_2 \to R_2 - R_1$ and $R_3 \to R_3 - R_1$. We obtain,

$$(a+b+c+3x)\begin{vmatrix} 1 & b+x & c+x \\ 0 & c-b & a-c \\ 0 & a-b & b-c \end{vmatrix} = 0.$$ That is, $(a + b + c + 3x)\,[-(b - c)^2 + (c - a)(a - b)] = 0.$

Therefore, $x = -(a + b + c)/3$, provided that $(b - c)^2 \neq (a - b)(c - a)$.

EXAMPLE 5.25 Solve the equation $\begin{vmatrix} x+2 & 2x+3 & 3x+4 \\ 2x+3 & 3x+4 & 4x+5 \\ 3x+5 & 5x+8 & 10x+17 \end{vmatrix} = 0.$

Solution Operating $R_3 \to R_3 - (R_1 + R_2)$, we obtain

$$\begin{vmatrix} x+2 & 2x+3 & 3x+4 \\ 2x+3 & 3x+4 & 4x+5 \\ 0 & 1 & 3x+8 \end{vmatrix} = 0 \quad \text{(operate } R_2 \to R_2 - R_1, \text{ and } R_1 \to R_1 + R_3) \quad \text{or} \quad \begin{vmatrix} x+2 & 2x+4 & 6x+12 \\ x+1 & x+1 & x+1 \\ 0 & 1 & 3x+8 \end{vmatrix} = 0$$

or $$(x+1)(x+2)\begin{vmatrix} 1 & 2 & 6 \\ 1 & 1 & 1 \\ 0 & 1 & 3x+8 \end{vmatrix} = 0 \quad \text{(operate } R_1 \to R_1 - R_2) \text{ or } (x+1)(x+2)\begin{vmatrix} 0 & 1 & 5 \\ 1 & 1 & 1 \\ 0 & 1 & 3x+8 \end{vmatrix} = 0.$$

Expanding the determinant by using the first column, we have $-(x + 1)(x + 2)(3x + 8 - 5) = 0$ or $(x + 1)(x + 2)(x + 1) = 0$. Therefore, $x = -1, -1, -2$.

EXERCISE 5.4

1. Compute the determinant of each of the following matrices:

(i) $\begin{bmatrix} 2 & 3 & -2 & 4 \\ 3 & -2 & 1 & 2 \\ 3 & 2 & 3 & 4 \\ -2 & 4 & 0 & 5 \end{bmatrix}$, (ii) $\begin{bmatrix} 3 & 5 & 8 & 3 \\ 2 & -3 & -5 & -10 \\ 4 & 1 & 6 & 7 \\ 5 & -4 & -2 & -4 \end{bmatrix}$,

(iii) $\begin{bmatrix} 2 & -5 & -7 & -5 \\ 8 & 3 & -11 & -27 \\ 3 & -9 & 4 & 2 \\ -5 & 2 & 3 & 13 \end{bmatrix}$, (iv) $\begin{bmatrix} 5 & -3 & 7 & 2 \\ 3 & 8 & -6 & -9 \\ 8 & -1 & 1 & -10 \\ 1 & 3 & 5 & 12 \end{bmatrix}$.

Ans. (i) –286, (ii) 0, (iii) –868, (iv) –1368.

2. For the matrix $\mathbf{A} = \begin{bmatrix} 3 & 9 & 5 \\ 4 & 6 & 0 \\ -1 & -3 & 2 \end{bmatrix}$, show that $\det(\mathbf{A}^T) = \det(\mathbf{A})$.

3. Evaluate the following determinants:

(i) $\begin{vmatrix} a+b & a & b \\ a & c+a & c \\ b & c & b+c \end{vmatrix}$, (ii) $\begin{vmatrix} 1+i & 1-i & i \\ 1-i & i & 1+i \\ i & 1+i & 1-i \end{vmatrix}$ where $i = \sqrt{-1}$.

Ans. (i) 4 *abc*, (ii) 4 + 7*i*.

4. Prove, without expanding, that each of the following determinants vanishes:

(i), $\begin{vmatrix} -3 & 1 & 1 & 1 \\ 1 & -3 & 1 & 1 \\ 1 & 1 & -3 & 3 \\ 1 & 1 & 1 & -3 \end{vmatrix}$, (ii) $\begin{vmatrix} 1 & a & a^2-bc \\ 1 & b & b^2-ca \\ 1 & c & c^2-ab \end{vmatrix}$, (iii) $\begin{vmatrix} a-b & b-c & c-a \\ b-c & c-a & a-b \\ c-a & a-b & b-c \end{vmatrix}$,

(iv) $\begin{vmatrix} 1^2 & 2^2 & 3^2 & 4^2 \\ 2^2 & 3^2 & 4^2 & 5^2 \\ 3^2 & 4^2 & 5^2 & 6^2 \\ 4^2 & 5^2 & 6^2 & 7^2 \end{vmatrix}$, (v) $\begin{vmatrix} 3 & 5 & 8 & 3 \\ 2 & -3 & -5 & -10 \\ 4 & 1 & 6 & 7 \\ 5 & -4 & -2 & -4 \end{vmatrix}$, (vi) $\begin{vmatrix} 1 & bc & a(b+c) \\ 1 & ca & b(c+a) \\ 1 & ab & c(a+b) \end{vmatrix}$.

Prove the following:

5. $\begin{vmatrix} a & b & c \\ l & m & n \\ p & q & r \end{vmatrix} = \begin{vmatrix} m & b & q \\ l & a & p \\ n & c & r \end{vmatrix} = \begin{vmatrix} l & m & n \\ p & q & r \\ a & b & c \end{vmatrix}$.

6. $\begin{vmatrix} a+b & b+c & c+a \\ b+c & c+a & a+b \\ c+a & a+b & b+c \end{vmatrix} = 2\begin{vmatrix} a & b & c \\ b & c & a \\ c & a & b \end{vmatrix}$.

7. $\begin{vmatrix} a & a+b & a+b+c \\ 2a & 3a+2b & 4a+3b+2c \\ 3a & 6a+3b & 10a+6b+3c \end{vmatrix} = a^3$.

8. $\begin{vmatrix} b+c & a-c & a-b \\ b-c & c+a & b-a \\ c-b & c-a & a+b \end{vmatrix} = 8abc$

9. $\begin{vmatrix} a & b & c \\ a-b & b-c & c-a \\ b+c & c+a & a+b \end{vmatrix} = a^3+b^3+c^3-3abc$

10. $\begin{vmatrix} 1 & 1 & 1 \\ \alpha & \beta & r \\ \alpha^2 & \beta^2 & r^2 \end{vmatrix} = (\alpha-\beta)(\beta-\gamma)(\gamma-\alpha)$.

11. $\begin{vmatrix} (b+c)^2 & a^2 & a^2 \\ b^2 & (c+a)^2 & b^2 \\ c^2 & c^2 & (a+b)^2 \end{vmatrix} = 2abc(a+b+c)^3.$

12. $\begin{vmatrix} 1 & \alpha & \beta\gamma \\ 1 & \beta & \gamma\alpha \\ 1 & \gamma & \alpha\beta \end{vmatrix} = (\alpha-\beta)(\beta-\gamma)(\gamma-\alpha).$

13. $\begin{vmatrix} 1 & 1 & 1 \\ a & b & c \\ a^3 & b^3 & c^3 \end{vmatrix} = (a-b)(b-c)(c-a)(a+b+c).$

14. $\begin{vmatrix} a & a^2 & b+c \\ b & b^2 & c+a \\ c & c^2 & a+b \end{vmatrix} = (a-b)(b-c)(c-a)(a+b+c).$

15. $\begin{vmatrix} a+b+c & -c & -b \\ -c & a+b+c & -a \\ -b & -a & a+b+c \end{vmatrix} = 2(b+c)(c+a)(a+b).$

16. $\begin{vmatrix} a & a^2 & bc \\ b & b^2 & ca \\ c & c^2 & ab \end{vmatrix} = \begin{vmatrix} 1 & a^2 & a^3 \\ 1 & b^2 & b^3 \\ 1 & c^2 & c^3 \end{vmatrix} = (a-b)(b-c)(c-a)(ab+bc+ca).$

17. $\begin{vmatrix} 1+a & 2 & 3 \\ 1 & 2+a & 3 \\ 1 & 2 & 3+a \end{vmatrix} = a^2(a+6).$

18. $\begin{vmatrix} 1+a & 1 & 1 & 1 \\ 1 & 1+a & 1 & 1 \\ 1 & 1 & 1+a & 1 \\ 1 & 1 & 1 & 1+a \end{vmatrix} = a^4 + 4a^3.$

19. $\begin{vmatrix} a^2+\lambda & ab & ac & ad \\ ab & b^2+\lambda & bc & bd \\ ac & bc & c^2+\lambda & cd \\ ad & bd & cd & d^2+\lambda \end{vmatrix} = \lambda^3(a^2+b^2+c^2+d^2+\lambda).$

[**Hint:** Operate $C_1 \to aC_1$, $C_2 \to bC_2$, $C_3 \to cC_3$, and $C_4 \to dC_4$, divide the determinant, by $abcd$. Now take a, b, c, d common from R_1, R_2, R_3, R_4 respectively. Next operate $C_1 \to C_1 + (C_2 + C_3 + C_4)$ and take out $(\Sigma a^2 + \lambda)$ as a common factor from C_1. Finally, operating $R_2 \to R_2 - R_1$, $R_3 \to R_3 - R_1$ and $R_4 \to R_4 - R_1$, we obtain the desired result].

20. $\begin{vmatrix} a^3 & 3a^2 & 3a & 1 \\ a^2 & a^2+2a & 2a+1 & 1 \\ a & 2a+1 & a+2 & 1 \\ 1 & 3 & 3 & 1 \end{vmatrix} = (a-1)^6.$

Solve the following equations:

21. $\begin{vmatrix} x-3 & x+2 & x-1 \\ x+2 & x-4 & x+4 \\ x-1 & x+4 & x-5 \end{vmatrix} = 0.$ **Ans.** 2/33.

22. $\begin{vmatrix} 4x+5 & 4x+7 & 4x+9 \\ 4x+9 & 4x+5 & 4x+7 \\ 4x+7 & 4x+9 & 4x+5 \end{vmatrix} = 0.$ **Ans.** $x = -1.75$

23. $\begin{vmatrix} x+1 & 2x+1 & 3x+1 \\ 2x & 4x+3 & 6x+3 \\ 4x+1 & 6x+4 & 8x+4 \end{vmatrix} = 0.$ **Ans.** $x = 0, -1/2$.

5.9 PRODUCT OF TWO DETERMINANTS

The determinant of a product of two square matrices of the same order **A** and **B** is the product of their determinants, that is det (**AB**) = det (**A**) det (**B**) or |**AB**| = |**A**||**B**|.

Since $|\mathbf{A}| = |\mathbf{A}^T|$, we can multiply two determinants in any one of the following ways:

(*i*) row by row, (*ii*) column by column, (*iii*) row by column, (*iv*) column by row. The value of the determinant is same in each case.

ILLUSTRATIVE EXAMPLES

EXAMPLE 5.26 Given the matrices $\mathbf{A} = \begin{bmatrix} 2 & 3 & 1 \\ 3 & 2 & -2 \\ 4 & -4 & 3 \end{bmatrix}$, $\mathbf{B} = \begin{bmatrix} 1 & 3 & -2 \\ 2 & 1 & 3 \\ -1 & 2 & 2 \end{bmatrix}$. Verify that |**AB**| = |**A**||**B**|.

Solution We have

$$\mathbf{AB} = \begin{bmatrix} 2 & 3 & 1 \\ 3 & 2 & -2 \\ 4 & -4 & 3 \end{bmatrix}\begin{bmatrix} 1 & 3 & -2 \\ 2 & 1 & 3 \\ -1 & 2 & 2 \end{bmatrix} = \begin{bmatrix} 2+6-1 & 6+3+2 & -4+9+2 \\ 3+4+2 & 9+2-4 & -6+6-4 \\ 4-8-3 & 12-4+6 & -8-12+6 \end{bmatrix} = \begin{bmatrix} 7 & 11 & 7 \\ 9 & 7 & -4 \\ -7 & 14 & -14 \end{bmatrix}.$$

Therefore, $|\mathbf{AB}| = \begin{vmatrix} 7 & 11 & 7 \\ 9 & 7 & -4 \\ -7 & 14 & -14 \end{vmatrix} = 7(-98+56) - 11(-126-28) + 7(126+49)$

$$= 7(-42) - 11\,(-154) + 7(175) = -294 + 1694 + 1225 = 2625.$$

The product $|\mathbf{A}||\mathbf{B}| = \begin{vmatrix} 2 & 3 & 1 \\ 3 & 2 & -2 \\ 4 & -4 & 3 \end{vmatrix}\begin{vmatrix} 1 & 3 & -2 \\ 2 & 1 & 3 \\ -1 & 2 & 2 \end{vmatrix} = \begin{vmatrix} 2+9-2 & 4+3+3 & -2+6+2 \\ 3+6+4 & 6+2-6 & -3+4-4 \\ 4-12-6 & 8-4+9 & -4-8+6 \end{vmatrix}$ (multiplying determinants row by row)

$$= \begin{vmatrix} 9 & 10 & 6 \\ 13 & 2 & -3 \\ -14 & 13 & -6 \end{vmatrix} \begin{matrix} (\text{Operate } R_1 \to R_1 + 2R_2) \\ \\ (\text{Operate } R_3 \to R_3 - 2R_2) \end{matrix} = \begin{vmatrix} 35 & 14 & 0 \\ 13 & 2 & -3 \\ -40 & 9 & 0 \end{vmatrix}.$$

Expanding the determinant by using the third column, we have $|\mathbf{A}||\mathbf{B}| = 3\begin{vmatrix} 35 & 14 \\ -40 & 9 \end{vmatrix} = 3(315+560) = 2625$,

which is same as |**AB**|.

NOTE We can also find |**A**| and |**B**| and then multiply.

EXAMPLE 5.27 Express $\begin{vmatrix} 0 & z & y \\ z & 0 & x \\ y & x & 0 \end{vmatrix}^2$ as a determinant. Hence evaluate the determinant.

Solution

$$\begin{vmatrix} 0 & z & y \\ z & 0 & x \\ y & x & 0 \end{vmatrix}^2 = \begin{vmatrix} 0 & z & y \\ z & 0 & x \\ y & x & 0 \end{vmatrix}\begin{vmatrix} 0 & z & y \\ z & 0 & x \\ y & x & 0 \end{vmatrix}$$

$$= \begin{vmatrix} 0+z^2+y^2 & 0+0+xy & 0+zx+0 \\ 0+0+xy & z^2+0+x^2 & zy+0+0 \\ 0+xz+0 & yz+0+0 & y^2+x^2+0 \end{vmatrix} \quad \text{(multiplying determinants row by row)}$$

$$= \begin{vmatrix} y^2+z^2 & xy & zx \\ xy & z^2+x^2 & yz \\ zx & yz & x^2+y^2 \end{vmatrix}.$$

Since $\begin{vmatrix} 0 & z & y \\ z & 0 & x \\ y & x & 0 \end{vmatrix} = -z\begin{vmatrix} z & x \\ y & 0 \end{vmatrix} + y\begin{vmatrix} z & 0 \\ y & x \end{vmatrix} = 2xyz$. Therefore, $\begin{vmatrix} 0 & z & y \\ z & 0 & x \\ y & x & 0 \end{vmatrix}^2 = 4x^2y^2z^2$.

EXAMPLE 5.28 Show that the

$$\det(\mathbf{A}) = \begin{vmatrix} 2b_1+c_1 & c_1+3a_1 & 2a_1+3b_1 \\ 2b_2+c_2 & c_2+3a_2 & 2a_2+3b_2 \\ 2b_3+c_3 & c_3+3a_3 & 2a_3+3b_3 \end{vmatrix} \text{ is a multiple of the determinant } \begin{vmatrix} a_1 & b_1 & c_1 \\ a_2 & b_2 & c_2 \\ a_3 & b_3 & c_3 \end{vmatrix}.$$

Solution The given determinant can be written in the form

$$|\mathbf{A}| = \begin{vmatrix} a_1.0+b_1.2+c_1.1 & a_1.3+b_1.0+c_1.1 & a_1.2+b_1.3+c_1.0 \\ a_2.0+b_2.2+c_2.1 & a_2.3+b_2.0+c_2.1 & a_2.2+b_2.3+c_2.0 \\ a_3.0+b_3.2+c_3.1 & a_3.3+b_3.0+c_3.1 & a_3.2+b_3.3+c_3.0 \end{vmatrix}.$$

From the rule for writing the elements of the product determinant, we can write

$$|\mathbf{A}| = \begin{vmatrix} a_1 & b_1 & c_1 \\ a_2 & b_2 & c_2 \\ a_3 & b_3 & c_3 \end{vmatrix}\begin{vmatrix} 0 & 2 & 1 \\ 3 & 0 & 1 \\ 2 & 3 & 0 \end{vmatrix}, \text{ which proves the result.}$$

EXAMPLE 5.29 Prove that $\begin{vmatrix} 2ll' & lm'+ml' & ln'+nl' \\ lm'+ml' & 2mm' & mn'+nm' \\ ln'+nl' & mn'+nm' & 2nn' \end{vmatrix} = 0.$

Solution The given determinant can be written as

$$\begin{vmatrix} ll'+l'l+00 & lm'+l'm+00 & ln'+l'n+00 \\ ml'+m'l+00 & mm'+m'm+00 & mn'+m'n+00 \\ nl'+n'l+00 & nm'+n'm+00 & nn'+n'n+00 \end{vmatrix} = \begin{vmatrix} l & l' & 0 \\ m & m' & 0 \\ n & n' & 0 \end{vmatrix}\begin{vmatrix} l' & l & 0 \\ m' & m & 0 \\ n' & n & 0 \end{vmatrix} = 0,$$

because each of the determinants on the left is a zero-determinant.

EXAMPLE 5.30 Show that $\begin{vmatrix} A_1 & B_1 & C_1 \\ A_2 & B_2 & C_2 \\ A_3 & B_3 & C_3 \end{vmatrix} = \begin{vmatrix} a_1 & b_1 & c_1 \\ a_2 & b_2 & c_2 \\ a_3 & b_3 & c_3 \end{vmatrix}^2$ where capital letters denote the co-factors of the corresponding small letters of the determinant on the right.

Solution Let $|\mathbf{P}| = \begin{vmatrix} a_1 & b_1 & c_1 \\ a_2 & b_2 & c_2 \\ a_3 & b_3 & c_3 \end{vmatrix}$ and $|\mathbf{Q}| = \begin{vmatrix} A_1 & B_1 & C_1 \\ A_2 & B_2 & C_2 \\ A_3 & B_3 & C_3 \end{vmatrix}$.

Then multiplication of determinants row by row gives

$$|\mathbf{P}||\mathbf{Q}| = \begin{vmatrix} a_1A_1 + b_1B_1 + c_1C_1 & a_1A_2 + b_1B_2 + c_1C_2 & a_1A_3 + b_1B_3 + c_1C_3 \\ a_2A_1 + b_2B_1 + c_2C_1 & a_2A_2 + b_2B_2 + c_2C_2 & a_2A_3 + b_2B_3 + c_2C_3 \\ a_3A_1 + b_3B_1 + c_3C_1 & a_3A_2 + b_3B_2 + c_3C_2 & a_3A_3 + b_3B_3 + c_3C_3 \end{vmatrix} = \begin{vmatrix} |\mathbf{P}| & 0 & 0 \\ 0 & |\mathbf{P}| & 0 \\ 0 & 0 & |\mathbf{P}| \end{vmatrix} = \{|\mathbf{P}|\}^3.$$

Therefore, $|\mathbf{Q}| = \{|\mathbf{P}|\}^2$, if $|\mathbf{P}| \neq 0$.

Note. $|\mathbf{Q}|$ is called the *adjugate* or *reciprocal* determinant of **P**.

EXAMPLE 5.31 Show that the $\det(\mathbf{D}) = \begin{vmatrix} 1 & \cos(\beta-\alpha) & \cos(\gamma-\alpha) \\ \cos(\alpha-\beta) & 1 & \cos(\gamma-\beta) \\ \cos(\alpha-\gamma) & \cos(\beta-\gamma) & 1 \end{vmatrix}$ is a perfect square (of a determinant) and find its value.

Solution The give det (**D**)= $\begin{vmatrix} \cos\alpha\cos\alpha + \sin\alpha\sin\alpha & \cos\alpha\cos\beta + \sin\alpha\sin\beta & \cos\alpha\cos\gamma + \sin\alpha\sin\gamma \\ \cos\beta\cos\alpha + \sin\beta\sin\alpha & \cos\beta\cos\beta + \sin\beta\sin\beta & \cos\beta\cos\gamma + \sin\beta\sin\gamma \\ \cos\gamma\cos\alpha + \sin\gamma\sin\alpha & \cos\gamma\cos\beta + \sin\gamma\sin\beta & \cos\gamma\cos\gamma + \sin\gamma\sin\gamma \end{vmatrix}$

$$= \begin{vmatrix} \cos\alpha & \sin\alpha & 0 \\ \cos\beta & \sin\beta & 0 \\ \cos\gamma & \sin\gamma & 0 \end{vmatrix}\begin{vmatrix} \cos\alpha & \sin\alpha & 0 \\ \cos\beta & \sin\beta & 0 \\ \cos\gamma & \sin\gamma & 0 \end{vmatrix} = \begin{vmatrix} \cos\alpha & \sin\alpha & 0 \\ \cos\beta & \sin\beta & 0 \\ \cos\gamma & \sin\gamma & 0 \end{vmatrix}^2 = 0 \times 0 = 0.$$

EXERCISE 5.5

1. Express $\begin{vmatrix} 1 & 4 & -3 \\ -3 & -2 & 1 \\ 1 & 3 & 5 \end{vmatrix}\begin{vmatrix} 3 & 4 & 2 \\ 2 & 5 & 4 \\ -1 & 6 & 1 \end{vmatrix}$ as a single determinant and hence find its value. **Ans.** –3384

2. Find the value of the det (**M**) if $|\mathbf{M}| = 3\{|\mathbf{A}|\}^2 + |\mathbf{AB}| + \{|\mathbf{B}|\}^2$, where $|\mathbf{A}| = \begin{vmatrix} 2 & 1 & 1 \\ 1 & 2 & 1 \\ 0 & -1 & 0 \end{vmatrix}$, $|\mathbf{B}| = \begin{vmatrix} 1/2 & 0 & -1 \\ -1 & 0 & 1 \\ 1 & 2 & 3 \end{vmatrix}$, without evaluating $|\mathbf{A}|$ and $|\mathbf{B}|$ independently. **Ans.** 5

3. Show that $\begin{vmatrix} (x-a)^2 & (y-a)^2 & (z-a)^2 \\ (x-b)^2 & (y-b)^2 & (z-b)^2 \\ (x-c)^2 & (y-c)^2 & (z-c)^2 \end{vmatrix} = 2\begin{vmatrix} x^2 & x & 1 \\ y^2 & y & 1 \\ z^2 & z & 1 \end{vmatrix} \times \begin{vmatrix} a^2 & a & 1 \\ b^2 & b & 1 \\ c^2 & c & 1 \end{vmatrix}$

$= 2(x-y)(y-z)(z-x)(a-b)(b-c)(c-a).$

$$\left[\textbf{Hint: } \text{R.H.S.} = -2\begin{vmatrix} x^2 & x & 1 \\ y^2 & y & 1 \\ z^2 & z & 1 \end{vmatrix}\begin{vmatrix} 1 & a & a^2 \\ 1 & b & b^2 \\ 1 & c & c^2 \end{vmatrix} = \begin{vmatrix} x^2 & x & 1 \\ y^2 & y & 1 \\ z^2 & z & 1 \end{vmatrix}\begin{vmatrix} 1 & -2a & a^2 \\ 1 & -2b & b^2 \\ 1 & -2c & c^2 \end{vmatrix}\right]$$

4. Evaluate $\begin{vmatrix} \lambda & c & -b \\ -c & \lambda & a \\ b & -a & \lambda \end{vmatrix}\begin{vmatrix} a^2+\lambda^2 & ab+\lambda c & ca-\lambda b \\ ab-\lambda c & b^2+\lambda^2 & bc+\lambda a \\ ca+\lambda b & bc-\lambda a & c^2+\lambda^2 \end{vmatrix}$.

[**Hint:** The second determinant is adjugate of first.] **Ans.** $\lambda^3(a^2+b^2+c^2+\lambda^2)^3$.

5. Express $\begin{vmatrix} 2bc-a^2 & c^2 & b^2 \\ c^2 & 2ca-b^2 & a^2 \\ b^2 & a^2 & 2ab-c^2 \end{vmatrix}$ as the square of a determinant and hence find its value.

[**Hint:** Proceeding as in examples 5.28 and 5.29 the given determinant

$$= \begin{vmatrix} a & b & c \\ b & c & a \\ c & a & b \end{vmatrix}\begin{vmatrix} -a & c & b \\ -b & a & c \\ -c & b & a \end{vmatrix} = \begin{vmatrix} a & b & c \\ b & c & a \\ c & a & b \end{vmatrix}^2\Bigg].$$

Ans. $(a^3+b^3+c^3-3\,abc)^2$.

6. Show that $\begin{vmatrix} b^2+c^2 & ab & ca \\ ab & c^2+a^2 & bc \\ ca & bc & a^2+b^2 \end{vmatrix} = \begin{vmatrix} 0 & c & b \\ c & 0 & a \\ b & a & 0 \end{vmatrix}^2 = 4a^2b^2c^2$.

7. If $|\mathbf{P}| = \begin{vmatrix} a & b & c \\ b & c & a \\ c & a & b \end{vmatrix}$, show that $|\mathbf{Q}| = \begin{vmatrix} bc-a^2 & ca-b^2 & ab-c^2 \\ ca-b^2 & ab-c^2 & bc-a^2 \\ ab-c^2 & bc-a^2 & ca-b^2 \end{vmatrix} = \{|\mathbf{P}|\}^2$. [**Hint:** $|\mathbf{Q}| = adj|\mathbf{P}|$].

8. Show that $\begin{vmatrix} \sin(A+P) & \sin(B+P) & \sin(C+P) \\ \sin(A+Q) & \sin(B+Q) & \sin(C+Q) \\ \sin(A+R) & \sin(B+R) & \sin(C+R) \end{vmatrix} = 0$.

5.10 RANK OF A MATRIX

Def. *A matrix* **A** *is said to be of rank r when it has at least one* non-zero *minor of* |**A**| *of order r,* all *minors of order* (*r* + 1) *are zero.*

The rank of a matrix **A** is denoted symbolically by r or $r(\mathbf{A})$ or $\rho(\mathbf{A})$.

We know that any determinant can be expressed in terms of determinants of next lower order. As such if all minors of order $(r+1)$ are zero then all minors of order $(r+2)$ and higher orders (if any) are also zero. *Thus the rank of matrix* **A** *is equal to the order of its largest non-zero minor of* |**A**|.

Since as a result of elementary operations the zero or non-zero character of a minor is not changed, the rank of a matrix remains unaltered by these operations.

From the definition of the rank of a matrix it follows that:

(*i*) The rank of a non-singular matrix of order n is n. If the matrix is singular its rank is less than n.

(*ii*) The rank of an $m \times n$ matrix can at most be equal to the smaller of numbers m and n but it may be less.

(*iii*) If there is a non-zero minor of order r, then rank is $\geq r$.

(*iv*) The rank of a null matrix is zero and if the rank of a matrix is zero, then it must be a null matrix. The rank of a non-zero matrix is ≥ 1.

(*v*) The rank of $I_n = n$.

In practice, while determining the rank of a matrix, we may start with the largest order minor or minors. If they are all zero we go to the minors of next lower order and so on till we get a non-zero minor. The order of that minor is the rank of the given matrix.

EXAMPLE 5.32 Determine the rank of the following matrices:

(*i*) $\begin{bmatrix} 1 & 2 & 3 \\ -4 & 0 & 5 \end{bmatrix}$, (*ii*) $\begin{bmatrix} 0 & 2 & 3 \\ 0 & 4 & 6 \\ 0 & 6 & 9 \end{bmatrix}$, (*iii*) $\begin{bmatrix} 1 & 2 & 3 \\ 2 & 4 & 7 \\ 3 & 6 & 10 \end{bmatrix}$, (*iv*) $\begin{bmatrix} 1 & 2 & 3 \\ 1 & 4 & 2 \\ 2 & 6 & 5 \end{bmatrix}$. (*AMIETE, June 2009*)

Solution (*i*) It is a 2 × 3 matrix, therefore the rank cannot exceed 2. Since $\begin{vmatrix} 1 & 2 \\ -4 & 0 \end{vmatrix} \neq 0$, the rank is 2.

(*ii*) Obviously the given matrix is singular. Therefore its rank is less than 3. Also each of the nine minors of order 2 is zero. Hence the rank is less than 2. Since it is non-zero matrix, the rank is 1.

(*iii*) Let the given matrix be denoted by **A**.

$$\therefore |\mathbf{A}| = \begin{vmatrix} 1 & 2 & 3 \\ 2 & 4 & 7 \\ 3 & 6 & 10 \end{vmatrix} \text{ (Operate } R_2 \to R_2 - 2R_1 \text{ and } R_3 \to R_3 - 3R_1) = \begin{vmatrix} 1 & 2 & 3 \\ 0 & 0 & 1 \\ 0 & 0 & 1 \end{vmatrix} = 0.$$

Hence the matrix **A** is singular and therefore, the rank is less than 3.

Since $\begin{vmatrix} 2 & 3 \\ 4 & 7 \end{vmatrix} = 2 \neq 0$, the rank of the given matrix is 2.

(*iv*) Let the given matrix be denoted by **A**.

$$\therefore |\mathbf{A}| = \begin{vmatrix} 1 & 2 & 3 \\ 1 & 4 & 2 \\ 2 & 6 & 5 \end{vmatrix} \text{ (operate } R_2 \to R_2 - R_1 \text{ and } R_3 \to R_3 - 2R_1) = \begin{vmatrix} 1 & 2 & 3 \\ 0 & 2 & -1 \\ 0 & 2 & -1 \end{vmatrix} = 0.$$

Matrix **A** is singular and therefore the rank is less than 3.

Since $\begin{vmatrix} 1 & 2 \\ 1 & 4 \end{vmatrix} \neq 0$, the rank is 2.

EXAMPLE 5.33 Find all values of μ for which rank of the matrix

$$\mathbf{A} = \begin{bmatrix} \mu & -1 & 0 & 0 \\ 0 & \mu & -1 & 0 \\ 0 & 0 & \mu & -1 \\ -6 & 11 & -6 & 1 \end{bmatrix} \text{ is equal to 3.}$$

(*AMIETE, Dec. 2005*)

Solution The matrix **A** is of order 4. Rank (**A**) ≤ 4. Now Rank (**A**) = 3, if |**A**| = 0 and there is atleast one submatrix of order 3 whose determinant is not zero. Expanding the determinant through the elements of first row, we get

$$|\mathbf{A}| = \mu \begin{vmatrix} \mu & -1 & 0 \\ 0 & \mu & -1 \\ 11 & -6 & 1 \end{vmatrix} + 1 \begin{vmatrix} 0 & -1 & 0 \\ 0 & \mu & -1 \\ -6 & -6 & 1 \end{vmatrix} = \mu[\mu(\mu - 6) + 1(11)] - 6$$

$$= \mu^3 - 6\mu^2 + 11\mu - 6 = (\mu - 1)(\mu - 2)(\mu - 3).$$

Setting |**A**| = 0, we obtain μ = 1, 2, 3. The given matrix **A** possesses a minor of order 3 *viz* $\begin{vmatrix} \mu & -1 & 0 \\ 0 & \mu & -1 \\ 0 & 0 & \mu \end{vmatrix} = \mu^3 \neq 0.$

(For μ = 1, 2, 3.)

Hence r (**A**) = 3, when μ = 1 or 2 or 3. For other values of μ, r (**A**) = 4.

5.11 ADJOINT OF A SQUARE MATRIX

Definition: Let $\mathbf{A} = (a_{ij})$ be a square matrix of order n and let A_{ij} be the cofactor of elements a_{ij} in the det(**A**). Then the transpose of the cofactor matrix (A_{ij}) is defined as the adjoint of **A**, and is denoted by *adj***A**.

Thus if $\mathbf{A} = \begin{bmatrix} a_{11} & a_{12} & a_{13} \\ a_{21} & a_{22} & a_{23} \\ a_{31} & a_{32} & a_{33} \end{bmatrix}$ is a square matrix of order 3, then

$$adj\ \mathbf{A} = \begin{bmatrix} A_{11} & A_{12} & A_{13} \\ A_{21} & A_{22} & A_{23} \\ A_{31} & A_{32} & A_{33} \end{bmatrix}^T = \begin{bmatrix} A_{11} & A_{21} & A_{31} \\ A_{12} & A_{22} & A_{32} \\ A_{13} & A_{23} & A_{33} \end{bmatrix}.$$

If **A** is a square matrix of order 2, then adjoint **A** can be obtained directly by simply interchanging the diagonal elements and changing the signs of non-diagonal elements as is illustrated below:

If $\mathbf{A} = \begin{bmatrix} 1 & 4 \\ 2 & 10 \end{bmatrix}$, then $adj\ \mathbf{A} = \begin{bmatrix} 10 & -4 \\ -2 & 1 \end{bmatrix}$.

5.11.1 Properties of the Adjoint of a Matrix

1. If **A** is a square matrix of order n, then $\mathbf{A}(adj\mathbf{A}) = (adj\mathbf{A})\mathbf{A} = |\mathbf{A}|\mathbf{I}_n$, where $\mathbf{I}_n$ is a unit or identity matrix of order n.
2. If **A** is a square matrix of order n, then $adj(\mathbf{A}^T) = (adj\mathbf{A})^T$.
3. If **A** and **B** are two square matrices of the same order, then $adj(\mathbf{AB}) = adj(\mathbf{B})\ adj(\mathbf{A})$.

We prove the property **1** when $n = 3$.

$$\mathbf{A}(adj\ \mathbf{A}) = \begin{bmatrix} a_{11} & a_{12} & a_{13} \\ a_{21} & a_{22} & a_{23} \\ a_{31} & a_{32} & a_{33} \end{bmatrix}\begin{bmatrix} A_{11} & A_{21} & A_{31} \\ A_{12} & A_{22} & A_{32} \\ A_{13} & A_{23} & A_{33} \end{bmatrix}$$

$$= \begin{bmatrix} a_{11}A_{11}+a_{12}A_{12}+a_{13}A_{13} & a_{11}A_{21}+a_{12}A_{22}+a_{13}A_{23} & a_{11}A_{31}+a_{12}A_{32}+a_{13}A_{33} \\ a_{21}A_{11}+a_{22}A_{12}+a_{23}A_{13} & a_{21}A_{21}+a_{22}A_{22}+a_{23}A_{23} & a_{21}A_{31}+a_{22}A_{32}+a_{23}A_{33} \\ a_{31}A_{11}+a_{32}A_{12}+a_{33}A_{13} & a_{31}A_{21}+a_{32}A_{22}+a_{33}A_{23} & a_{31}A_{31}+a_{32}A_{32}+a_{33}A_{33} \end{bmatrix}$$

$$= \begin{bmatrix} \det(\mathbf{A}) & 0 & 0 \\ 0 & \det(\mathbf{A}) & 0 \\ 0 & 0 & \det(\mathbf{A}) \end{bmatrix} = \det(\mathbf{A})\begin{bmatrix} 1 & 0 & 0 \\ 0 & 1 & 0 \\ 0 & 0 & 1 \end{bmatrix} = |\mathbf{A}|I_3 = |\mathbf{A}|I. \qquad ...(5.8)$$

Similarly, it can be shown in exactly the same manner that $(adj\mathbf{A})\mathbf{A} = |\mathbf{A}|\mathbf{I}$. ...(5.9)

EXAMPLE 5.34 If $\mathbf{A} = \begin{bmatrix} 1 & 1 & 1 \\ 1 & 2 & -3 \\ 2 & -1 & 3 \end{bmatrix}$, verify that $\mathbf{A}(adj\mathbf{A}) = |\mathbf{A}|\mathbf{I}_3 = (adj\mathbf{A})\mathbf{A}$.

Solution We have $|\mathbf{A}| = \begin{vmatrix} 1 & 1 & 1 \\ 1 & 2 & -3 \\ 2 & -1 & 3 \end{vmatrix}$. The cofactors corresponding to the elements in $|\mathbf{A}|$ are

$$A_{11} = \begin{vmatrix} 2 & -3 \\ -1 & 3 \end{vmatrix} = 3,\ A_{12} = -\begin{vmatrix} 1 & -3 \\ 2 & 3 \end{vmatrix} = -9,\ A_{13} = \begin{vmatrix} 1 & 2 \\ 2 & -1 \end{vmatrix} = -5,$$

$$A_{21} = -\begin{vmatrix} 1 & 1 \\ -1 & 3 \end{vmatrix} = -4,\ A_{22} = \begin{vmatrix} 1 & 1 \\ 2 & 3 \end{vmatrix} = 1,\ A_{23} = -\begin{vmatrix} 1 & 1 \\ 2 & -1 \end{vmatrix} = 3,$$

$$A_{31} = \begin{vmatrix} 1 & 1 \\ 2 & -3 \end{vmatrix} = -5,\ A_{32} = -\begin{vmatrix} 1 & 1 \\ 1 & -3 \end{vmatrix} = 4,\ A_{33} = \begin{vmatrix} 1 & 1 \\ 1 & 2 \end{vmatrix} = 1.$$

$\therefore$ $|\mathbf{A}| = (1)(3) + (1)(-9) + (1)(-5) = -11.$

The matrix of cofactors $= \begin{bmatrix} 3 & -9 & -5 \\ -4 & 1 & 3 \\ -5 & 4 & 1 \end{bmatrix}$. Therefore, $adj\ \mathbf{A} = \begin{bmatrix} 3 & -9 & -5 \\ -4 & 1 & 3 \\ -5 & 4 & 1 \end{bmatrix}^T = \begin{bmatrix} 3 & -4 & -5 \\ -9 & 1 & 4 \\ -5 & 3 & 1 \end{bmatrix}$.

Hence
$$\mathbf{A}(adj\ \mathbf{A}) = \begin{bmatrix} 1 & 1 & 1 \\ 1 & 2 & -3 \\ 2 & -1 & 3 \end{bmatrix}\begin{bmatrix} 3 & -4 & -5 \\ -9 & 1 & 4 \\ -5 & 3 & 1 \end{bmatrix} = \begin{bmatrix} 3-9-5 & -4+1+3 & -5+4+1 \\ 3-18+15 & -4+2-9 & -5+8-3 \\ 6+9-15 & -8-1+9 & -10-4+3 \end{bmatrix}$$

$$= \begin{bmatrix} -11 & 0 & 0 \\ 0 & -11 & 0 \\ 0 & 0 & -11 \end{bmatrix} = -11\begin{bmatrix} 1 & 0 & 0 \\ 0 & 1 & 0 \\ 0 & 0 & 1 \end{bmatrix} = |\mathbf{A}|\, I_3$$

Similarly, it can be shown that $(adj\mathbf{A})\mathbf{A} = |\mathbf{A}|\mathbf{I}_3$. Hence the result.

5.12 SINGULAR AND THE NON-SINGULAR MATRICES

A square matrix **A** of order n is said to be **singular** if $|\mathbf{A}| = 0$, and it is called **non-singular** if $|\mathbf{A}| \neq 0$. In other words, a square matrix of order n is singular if its rank $\mathbf{A} < n$ and non-singular if its rank $\mathbf{A} = n$.

(Refer Article 5.10).

5.13 INVERSE (OR RECIPROCAL) OF A SQUARE MATRIX

Def. If **A** and **B** are square matrices of order n such that $\mathbf{AB} = \mathbf{BA} = \mathbf{I}$, ...(5.10)
than **B** is called the inverse of **A** and is denoted by $\mathbf{A}^{-1}$, that is $\mathbf{B} = \mathbf{A}^{-1}$.

From the symmetry of (5.10) it follows that if **B** is inverse of **A** then **A** is inverse of **B**.

Thus, it follows that $\mathbf{A} = \mathbf{B}^{-1} = (\mathbf{A}^{-1})^{-1}$

If $|\mathbf{A}| \neq 0$, we have from results (5.8) and (5.9) of Art. 5.14.

$$\mathbf{A} \cdot \frac{adj\ \mathbf{A}}{|\mathbf{A}|} = \frac{adj\ \mathbf{A}}{|\mathbf{A}|} \cdot \mathbf{A} = \mathbf{I}.$$

From the definition of inverse of **A** it follows that

$$\mathbf{A}^{-1} = \frac{adj\ \mathbf{A}}{|\mathbf{A}|}. \qquad ...(5.11)$$

Note. If a matrix has an inverse, both the matrix and its inverse must be non-singular. When a matrix **A** is singular, that is $|\mathbf{A}| = 0$, its inverse is not defined, that is, a singular matrix has no inverse.

5.13.1 Properties of the Inverse of a Matrix

1. The inverse of a non-sigular matrix, is unique.
2. The inverse of the inverse is the original matrix itself, that is $(\mathbf{A}^{-1})^{-1} = \mathbf{A}$.
3. The inverse of the transpose of a matrix is the transpose of its inverse, that is $(\mathbf{A}^T)^{-1} = (\mathbf{A}^{-1})^T$.
4. Let $\mathbf{D} = diag(d_{11}, d_{22}, ..., d_{nn})$, $d_{ii} \neq 0$. Then $\mathbf{D}^{-1} = diag(1/d_{11}, 1/d_{22}, ..., 1/d_{nn})$.
5. If **A** and **B** are two invertible matrices of the same order, then
   $(\mathbf{AB})^{-1} = \mathbf{B}^{-1}\,\mathbf{A}^{-1}$.

   Proof: We have $(\mathbf{AB})(\mathbf{AB})^{-1} = \mathbf{I}$.

   Premultiplying both sides first by $\mathbf{A}^{-1}$ and then by $\mathbf{B}^{-1}$, we obtain

$B^{-1}A^{-1}(AB)(AB)^{-1} = B^{-1}A^{-1}$

or $\quad B^{-1}(A^{-1}A)B(AB)^{-1} = B^{-1}A^{-1}$ or $(AB)^{-1} = B^{-1}A^{-1}$.

Hence for more than two factors,

$(AB \ldots PQ)^{-1} = Q^{-1}P^{-1} \ldots B^{-1}A^{-1}$.

6. If a matrix **A** is symmetric and invertible, then its inverse matrix, that is A^{-1} is symmetric matrix.

ILLUSTRATIVE EXAMPLES

EXAMPLE 5.35 Find the inverse of the matrix $\mathbf{A} = \begin{bmatrix} -1 & 1 & 2 \\ 3 & -1 & 1 \\ -1 & 3 & 4 \end{bmatrix}$ and check the answer by direct multiplication.

Solution $\det(\mathbf{A}) = -1\begin{vmatrix} -1 & 1 \\ 3 & 4 \end{vmatrix} - 1\begin{vmatrix} 3 & 1 \\ -1 & 4 \end{vmatrix} + 2\begin{vmatrix} 3 & -1 \\ -1 & 3 \end{vmatrix} = -1(-4-3) - 1(12+1) + 2(9-1) = 10 \neq 0.$

Thus A^{-1} exists and is given by $\mathbf{A}^{-1} = \dfrac{1}{|\mathbf{A}|}\, adj\, \mathbf{A}$.

The cofactors corresponding to the elements in det (**A**) are

$$A_{11} = \begin{vmatrix} -1 & 1 \\ 3 & 4 \end{vmatrix} = -7,\ A_{12} = -\begin{vmatrix} 3 & 1 \\ -1 & 4 \end{vmatrix} = -13,\ A_{13} = \begin{vmatrix} 3 & -1 \\ -1 & 3 \end{vmatrix} = 8,$$

$$A_{21} = -\begin{vmatrix} 1 & 2 \\ 3 & 4 \end{vmatrix} = 2,\ A_{22} = \begin{vmatrix} -1 & 2 \\ -1 & 4 \end{vmatrix} = -2,\ A_{23} = -\begin{vmatrix} -1 & 1 \\ -1 & 3 \end{vmatrix} = 2,$$

$$A_{31} = \begin{vmatrix} 1 & 2 \\ -1 & 1 \end{vmatrix} = 3,\ A_{32} = -\begin{vmatrix} -1 & 2 \\ 3 & 1 \end{vmatrix} = 7,\ A_{33} = \begin{vmatrix} -1 & 1 \\ 3 & -1 \end{vmatrix} = -2.$$

Therefore, $$adj\, \mathbf{A} = \begin{bmatrix} -7 & -13 & 8 \\ 2 & -2 & 2 \\ 3 & 7 & -2 \end{bmatrix}^T = \begin{bmatrix} -7 & 2 & 3 \\ -13 & -2 & 7 \\ 8 & 2 & -2 \end{bmatrix}.$$

Hence, $$\mathbf{A}^{-1} = \frac{1}{|\mathbf{A}|} adj\, \mathbf{A} = \frac{1}{10}\begin{bmatrix} -7 & 2 & 3 \\ -13 & -2 & 7 \\ 8 & 2 & -2 \end{bmatrix} = \begin{bmatrix} -0.7 & 0.2 & 0.3 \\ -1.3 & -0.2 & 0.7 \\ 0.8 & 0.2 & -0.2 \end{bmatrix}.$$

$$\mathbf{AA}^{-1} = \begin{bmatrix} -1 & 1 & 2 \\ 3 & -1 & 1 \\ -1 & 3 & 4 \end{bmatrix}\begin{bmatrix} -0.7 & 0.2 & 0.3 \\ -1.3 & -0.2 & 0.7 \\ 0.8 & 0.2 & -0.2 \end{bmatrix}$$

$$= \begin{bmatrix} 0.7-1.3+1.6 & -0.2-0.2+0.4 & -0.3+0.7-0.4 \\ -2.1+1.3+0.8 & 0.6+0.2+0.2 & 0.9-0.7-0.2 \\ 0.7-3.9+3.2 & -0.2-0.6+0.8 & -0.3+2.1-0.8 \end{bmatrix} = \begin{bmatrix} 1 & 0 & 0 \\ 0 & 1 & 0 \\ 0 & 0 & 1 \end{bmatrix} = \mathbf{I}.$$

We can also show that $\mathbf{A}^{-1}\mathbf{A} = \mathbf{I}$. This gives the required check.

EXAMPLE 5.36 Show that the matrix $\mathbf{A} = \begin{bmatrix} 2 & -1 & 1 \\ -1 & 2 & -1 \\ 1 & -1 & 2 \end{bmatrix}$ satisfies the matrix equation $\mathbf{A}^3 - 6\mathbf{A}^2 + 9\mathbf{A} - 4\mathbf{I} = \mathbf{O}$ where **I** is an identity matrix of order 3. Hence, find the matrix (*i*) $\mathbf{A}^{-1}$ and (*ii*) $\mathbf{A}^{-2}$.

Solution We have

$$\mathbf{A}^2 = \mathbf{AA} = \begin{bmatrix} 2 & -1 & 1 \\ -1 & 2 & -1 \\ 1 & -1 & 2 \end{bmatrix}\begin{bmatrix} 2 & -1 & 1 \\ -1 & 2 & -1 \\ 1 & -1 & 2 \end{bmatrix} = \begin{bmatrix} 4+1+1 & -2-2-1 & 2+1+2 \\ -2-2-1 & 1+4+1 & -1-2-2 \\ 2+1+2 & -1-2-2 & 1+1+4 \end{bmatrix} = \begin{bmatrix} 6 & -5 & 5 \\ -5 & 6 & -5 \\ 5 & -5 & 6 \end{bmatrix}.$$

$$\mathbf{A}^3 = \mathbf{A}^2\mathbf{A} = \begin{bmatrix} 6 & -5 & 5 \\ -5 & 6 & -5 \\ 5 & -5 & 6 \end{bmatrix}\begin{bmatrix} 2 & -1 & 1 \\ -1 & 2 & -1 \\ 1 & -1 & 2 \end{bmatrix} = \begin{bmatrix} 12+5+5 & -6-10-5 & 6+5+10 \\ -10-6-5 & 5+12+5 & -5-6-10 \\ 10+5+6 & -5-10-6 & 5+5+12 \end{bmatrix} = \begin{bmatrix} 22 & -21 & 21 \\ -21 & 22 & -21 \\ 21 & -21 & 22 \end{bmatrix}.$$

Substituting in $\mathbf{C} = \mathbf{A}^3 - 6\mathbf{A}^2 + 9\mathbf{A} - 4\mathbf{I}$, we get

$$\mathbf{C} = \begin{bmatrix} 22 & -21 & 21 \\ -21 & 22 & -21 \\ 21 & -21 & 22 \end{bmatrix} + \begin{bmatrix} -36 & 30 & -30 \\ 30 & -36 & 30 \\ -30 & 30 & -36 \end{bmatrix} + \begin{bmatrix} 18 & -9 & 9 \\ -9 & 18 & -9 \\ 9 & -9 & 18 \end{bmatrix} + \begin{bmatrix} -4 & 0 & 0 \\ 0 & -4 & 0 \\ 0 & 0 & -4 \end{bmatrix} = \begin{bmatrix} 0 & 0 & 0 \\ 0 & 0 & 0 \\ 0 & 0 & 0 \end{bmatrix} = \mathbf{O}.$$

(*i*) Premultiplying $\mathbf{A}^3 - 6\mathbf{A}^2 + 9\mathbf{A} - 4\mathbf{I} = \mathbf{O}$ by $\mathbf{A}^{-1}$, we get

$$\mathbf{A}^{-1}\mathbf{A}^3 - 6\mathbf{A}^{-1}\mathbf{A}^2 + 9\mathbf{A}^{-1}\mathbf{A} - 4\mathbf{A}^{-1} = \mathbf{O} \text{ or } 4\mathbf{A}^{-1} = \mathbf{A}^2 - 6\mathbf{A} + 9\mathbf{I}$$

$$= \begin{bmatrix} 6 & -5 & 5 \\ -5 & 6 & -5 \\ 5 & -5 & 6 \end{bmatrix} + \begin{bmatrix} -12 & 6 & -6 \\ 6 & -12 & 6 \\ -6 & 6 & -12 \end{bmatrix} + \begin{bmatrix} 9 & 0 & 0 \\ 0 & 9 & 0 \\ 0 & 0 & 9 \end{bmatrix} = \begin{bmatrix} 3 & 1 & -1 \\ 1 & 3 & 1 \\ -1 & 1 & 3 \end{bmatrix}.$$

Hence $\mathbf{A}^{-1} = 1/4\begin{bmatrix} 3 & 1 & -1 \\ 1 & 3 & 1 \\ -1 & 1 & 3 \end{bmatrix}.$

(*ii*) $$\mathbf{A}^{-2} = (\mathbf{A}^{-1})^2 = \frac{1}{16}\begin{bmatrix} 3 & 1 & -1 \\ 1 & 3 & 1 \\ -1 & 1 & 3 \end{bmatrix}\begin{bmatrix} 3 & 1 & -1 \\ 1 & 3 & 1 \\ -1 & 1 & 3 \end{bmatrix}$$

$$= \frac{1}{16}\begin{bmatrix} 9+1+1 & 3+3-1 & -3+1-3 \\ 3+3-1 & 1+9+1 & -1+3+3 \\ -3+1-3 & -1+3+3 & 1+1+9 \end{bmatrix} = \frac{1}{16}\begin{bmatrix} 11 & 5 & -5 \\ 5 & 11 & 5 \\ -5 & 5 & 11 \end{bmatrix}.$$

EXAMPLE 5.37 If $\mathbf{A} = \begin{bmatrix} 1 & 1 & -2 \\ 1 & 9 & -11 \\ 1 & -1 & 2 \end{bmatrix}$ and $\mathbf{B} = \begin{bmatrix} 1 & 2 & 0 \\ 2 & 3 & -1 \\ 1 & 1 & -3 \end{bmatrix}$, find $(\mathbf{AB})^{-1}$.

Solution We have $\mathbf{AB} = \begin{bmatrix} 1 & 1 & -2 \\ 1 & 9 & -11 \\ 1 & -1 & 2 \end{bmatrix}\begin{bmatrix} 1 & 2 & 0 \\ 2 & 3 & -1 \\ 1 & 1 & -3 \end{bmatrix} = \begin{bmatrix} 1 & 3 & 5 \\ 8 & 18 & 24 \\ 1 & 1 & -5 \end{bmatrix}.$

Therefore, $$|\mathbf{AB}| = \begin{vmatrix} 1 & 3 & 5 \\ 8 & 18 & 24 \\ 1 & 1 & -5 \end{vmatrix} = 1(-90-24) - 3(-40-24) + 5(8-18) = -114 + 192 - 50 = 28 \neq 0.$$

Hence $(\mathbf{AB})^{-1}$ exists and is given by $(\mathbf{AB})^{-1} = \dfrac{1}{|\mathbf{AB}|} adj(\mathbf{AB}).$

It can be easily seen that $adj(\mathbf{AB}) = \begin{bmatrix} -114 & 20 & -18 \\ 64 & -10 & 16 \\ -10 & 2 & -6 \end{bmatrix}$.

Therefore, $$(\mathbf{AB})^{-1} = \frac{1}{28}\begin{bmatrix} -114 & 20 & -18 \\ 64 & -10 & 16 \\ -10 & 2 & -6 \end{bmatrix} = \begin{bmatrix} -57/14 & 5/7 & -9/14 \\ 16/7 & -5/14 & 4/7 \\ -5/14 & 1/14 & -3/14 \end{bmatrix}.$$

EXAMPLE 5.38 Find matrix **A** such that $\begin{bmatrix} 2 & 1 \\ 3 & 2 \end{bmatrix}\mathbf{A}\begin{bmatrix} -3 & 2 \\ 5 & -3 \end{bmatrix} = \begin{bmatrix} -2 & 4 \\ 3 & -1 \end{bmatrix}$.

Solution Let $\mathbf{B} = \begin{bmatrix} 2 & 1 \\ 3 & 2 \end{bmatrix}$, $\mathbf{C} = \begin{bmatrix} -3 & 2 \\ 5 & -3 \end{bmatrix}$ and $\mathbf{D} = \begin{bmatrix} -2 & 4 \\ 3 & -1 \end{bmatrix}$. Then the given matrix equation is

$$\mathbf{BAC} = \mathbf{D}. \qquad ...(i)$$

Now $|\mathbf{B}| = 4 - 3 = 1 \neq 0$. Thus **B** is non-singular and hence $\mathbf{B}^{-1}$ exists. Premultiplying both sides of equation (*i*) by $\mathbf{B}^{-1}$ gives $\mathbf{AC} = \mathbf{B}^{-1}\mathbf{D}$.

But $$\mathbf{B}^{-1} = \frac{1}{|\mathbf{B}|} adj\ \mathbf{B} = \begin{bmatrix} 2 & -1 \\ -3 & 2 \end{bmatrix}.$$

Therefore, $$\mathbf{AC} = \mathbf{B}^{-1}\mathbf{D} = \begin{bmatrix} 2 & -1 \\ -3 & 2 \end{bmatrix}\begin{bmatrix} -2 & 4 \\ 3 & -1 \end{bmatrix} = \begin{bmatrix} -7 & 9 \\ 12 & -14 \end{bmatrix}. \qquad ...(ii)$$

Now $|\mathbf{C}| = 9 - 10 = -1 \neq 0$. Thus **C** is non-singular and hence $\mathbf{C}^{-1}$ exists.

But $$\mathbf{C}^{-1} = \frac{1}{|\mathbf{C}|} adj \mathbf{C} = -\begin{bmatrix} -3 & -2 \\ -5 & -3 \end{bmatrix} = \begin{bmatrix} 3 & 2 \\ 5 & 3 \end{bmatrix}.$$

Now post-multiplying both sides of equation (*ii*) by $\mathbf{C}^{-1}$ gives

$$\mathbf{A} = \begin{bmatrix} -7 & 9 \\ 12 & -14 \end{bmatrix}\begin{bmatrix} 3 & 2 \\ 5 & 3 \end{bmatrix} = \begin{bmatrix} 24 & 13 \\ -34 & -18 \end{bmatrix}, \text{ the required result.}$$

EXAMPLE 5.39 Given the inverse of the matrix **A** as

$$\mathbf{A}^{-1} = \begin{bmatrix} 2/21 & 1/7 & -13/21 \\ -1/7 & 2/7 & 3/7 \\ 5/21 & -1/7 & -1/21 \end{bmatrix}, \text{ obtain } \mathbf{A}^2 + 2\mathbf{A}.$$

Solution We have

$$|\mathbf{A}^{-1}| = \begin{vmatrix} 2/21 & 1/7 & -13/21 \\ -1/7 & 2/7 & 3/7 \\ 5/21 & -1/7 & -1/21 \end{vmatrix} \quad \text{(Operating } R_1 \to 2R_1 - R_2 \text{ and } R_3 \to 2R_3 + R_2\text{)}$$

$$= 1/4\begin{vmatrix} 1/3 & 0 & -35/21 \\ -1/7 & 2/7 & 3/7 \\ 1/3 & 0 & 1/3 \end{vmatrix} = -\frac{1}{4 \times 63}\begin{vmatrix} 0 & 1 & -35/7 \\ 2 & -1 & 3 \\ 0 & 1 & 1 \end{vmatrix} = \frac{1 \times 2}{4 \times 63}\left(1 + \frac{35}{7}\right) = \frac{1}{21} \neq 0.$$

It can be easily seen that $adj(\mathbf{A}^{-1}) = \begin{bmatrix} 1/21 & 2/21 & 5/21 \\ 2/21 & 3/21 & 1/21 \\ -1/21 & 1/21 & 1/21 \end{bmatrix}$.

$$\therefore \quad \mathbf{A} = (\mathbf{A}^{-1})^{-1} = \frac{1}{|\mathbf{A}^{-1}|} adj\,(\mathbf{A}^{-1}) = 21 \begin{bmatrix} 1/21 & 2/21 & 5/21 \\ 2/21 & 3/21 & 1/21 \\ -1/21 & 1/21 & 1/21 \end{bmatrix} = \begin{bmatrix} 1 & 2 & 5 \\ 2 & 3 & 1 \\ -1 & 1 & 1 \end{bmatrix}.$$

Now $$\mathbf{A}^2 = \mathbf{AA} = \begin{bmatrix} 1 & 2 & 5 \\ 2 & 3 & 1 \\ -1 & 1 & 1 \end{bmatrix}\begin{bmatrix} 1 & 2 & 5 \\ 2 & 3 & 1 \\ -1 & 1 & 1 \end{bmatrix} = \begin{bmatrix} 0 & 13 & 12 \\ 7 & 14 & 14 \\ 0 & 2 & -3 \end{bmatrix},$$

and hence, $$\mathbf{A}^2 + 2\mathbf{A} = \begin{bmatrix} 0 & 13 & 12 \\ 7 & 14 & 14 \\ 0 & 2 & -3 \end{bmatrix} + \begin{bmatrix} 2 & 4 & 10 \\ 4 & 6 & 2 \\ -2 & 2 & 2 \end{bmatrix} = \begin{bmatrix} 2 & 17 & 22 \\ 11 & 20 & 16 \\ -2 & 4 & -1 \end{bmatrix}.$$

5.14 SYSTEMS OF *n* × *n* LINEAR EQUATIONS AND THEIR SOLUTION

Consider a system of n equations in the n unknowns $x_1, x_2, ...x_n$

$$\left.\begin{aligned} a_{11}x_1 + a_{12}x_2 + \cdots\cdots + a_{1n}x_n &= b_1 \\ a_{21}x_1 + a_{22}x_2 + \cdots\cdots + a_{2n}x_n &= b_2 \\ \vdots \qquad\qquad & \vdots \\ a_{n1}x_1 + a_{n2}x_2 + \cdots\cdots + a_{nn}x_n &= b_n \end{aligned}\right\}. \qquad ...(5.12)$$

The coefficients of the unknowns in the linear system (5.12) can be abbreviated as a_{ij}, where i denotes the row and j denotes the column in which the coefficient appears. Thus, $i = j = 1, 2, 3, ..., n$. The numbers $b_1, b_2, ..., b_n$ are called the **constants** of the system.

5.14.1 Matrix Representation of a System of Equations

In matrix form, we can write the system of equations (5.12) as

$$\mathbf{Ax} = \mathbf{b} \qquad ...(5.13)$$

where $\mathbf{A} = \begin{bmatrix} a_{11} & a_{12} & \cdots\cdots & a_{1n} \\ a_{21} & a_{22} & \cdots\cdots & a_{2n} \\ \vdots & & & \vdots \\ a_{n1} & a_{n2} & \cdots\cdots & a_{nn} \end{bmatrix}$, $\mathbf{b} = \begin{bmatrix} b_1 \\ b_2 \\ \vdots \\ b_n \end{bmatrix}$, $\mathbf{x} = \begin{bmatrix} x_1 \\ x_2 \\ \vdots \\ x_n \end{bmatrix}$

and **A, b, x** are respectively called the *coefficient matrix*, the right hand side column vector and the solution vector.

If $\mathbf{b} = \mathbf{0}$, that is all elements $b_1, b_2, ..., b_n$ is 0, the system (5.12) is said to be **homogeneous**. If $b \neq 0$, that is, atleast one of the elements $b_1, b_2, ..., b_n$ is not zero, then the system is said to be **non homogeneous**.

A **solution** of a linear system (5.12) is a set of n numbers $x_1, x_2, ..., x_n$ that satisfies each equation in the system.

The system of equations is said to be **consistent** if it has atleast one solution and **inconsistent**, if it has no solution. If a linear system is consistent, it has either a unique solution (that is, precisely one solution), or infinitely many solutions.

5.14.2 Non-homogeneous System of Equations

The non-homogeneous system of equation **Ax** = **b** can be solved by the following methods.

Matrix method

Let coefficient matrix **A** be non-singular, that is $|\mathbf{A}| \neq 0$, or the rank of the matrix **A** is n so that $\mathbf{A}^{-1}$ exists. Premultiplying **Ax** = **b** by $\mathbf{A}^{-1}$, we obtain

$$\mathbf{x} = \mathbf{A}^{-1}\mathbf{b}. \qquad \text{...(5.14)}$$

The system of equations is consistent and has a unique solution. If **b** = **O**, then **x** = **O** (trivial solution) is the only solution.

Cramer's rule

Let **A** be non-singular matrix. The Cramer's rule for the solution of **Ax** = **b** is given by

$$x_k = \frac{|\mathbf{A}_k|}{|\mathbf{A}|}, k = 1, 2, \ldots, n \qquad \text{...(5.15)}$$

where $|\mathbf{A}_k|$ is the determinant of the matrix $\mathbf{A}_k$ obtained by replacing the kth column of **A** by the right hand side column vector **b**. We discuss the following cases.

Case 1: When $|\mathbf{A}| \neq 0$, the system of equations is consistent and the unique solution is obtained by using equation (5.15).

Case 2: When $|\mathbf{A}| = 0$ and all $|\mathbf{A}_k| = 0$, $k = 1, 2, \ldots, n$, then the system of equations is consistent and has infinite number of solutions.

Case 3: When $|\mathbf{A}| = 0$, and one or more of $|\mathbf{A}_k|$, $k = 1, 2, \ldots n$, are not zero, then the system of equations has no solution, that is the system is inconsistent.

5.14.3 Homogeneous System of Equations

Consider the homogeneous system of equations **Ax** = **O**. ...(5.16)

Trivial solution **x** = **O** (i.e., $x_1 = 0, x_2 = 0, \ldots, x_n = 0$) is alsways a solution of this system If **A** is non-singular, then again $\mathbf{x} = \mathbf{A}^{-1}\mathbf{O} = \mathbf{O}$ is the solution. Therefore, a homogeneous sysem of equations is always consistent.

We conclude that non-trivial solution for **Ax** = **O** exist if and only if **A** is singular.

In this case, the homogeneous system of equations has infinite number of solutions.

EXAMPLE 5.40 Solve the following system of equations with the help of matrix inversion.

$$x + y + z = 3,\ x + 2y + 3z = 4,\ x + 4y + 9z = 6.$$

Solution The given system can be written in the matrix form as **Ax** = **b**,

where $\mathbf{A} = \begin{bmatrix} 1 & 1 & 1 \\ 1 & 2 & 3 \\ 1 & 4 & 9 \end{bmatrix}$, $\mathbf{x} = \begin{bmatrix} x \\ y \\ z \end{bmatrix}$, and $\mathbf{b} = \begin{bmatrix} 3 \\ 4 \\ 6 \end{bmatrix}$.

We find that $|\mathbf{A}| = \begin{vmatrix} 1 & 1 & 1 \\ 1 & 2 & 3 \\ 1 & 4 & 9 \end{vmatrix} = 1(18-12) - 1(9-3) + 1(4-2) = 2 \neq 0.$

Therefore, the coefficient matrix **A** is non-singular and the given system of equations has a unique solution.

We find $adj\ \mathbf{A} = \begin{bmatrix} 6 & -5 & 1 \\ -6 & 8 & -2 \\ 2 & -3 & 1 \end{bmatrix}$ and obtain $\mathbf{A}^{-1} = \frac{1}{|\mathbf{A}|} adj\ \mathbf{A} = \frac{1}{2}\begin{bmatrix} 6 & -5 & 1 \\ -6 & 8 & -2 \\ 2 & -3 & 1 \end{bmatrix}$.

Therefore, $\mathbf{x} = \mathbf{A}^{-1}\mathbf{b} = \frac{1}{2}\begin{bmatrix} 6 & -5 & 1 \\ -6 & 8 & -2 \\ 2 & -3 & 1 \end{bmatrix}\begin{bmatrix} 3 \\ 4 \\ 6 \end{bmatrix} = \frac{1}{2}\begin{bmatrix} 4 \\ 2 \\ 0 \end{bmatrix} = \begin{bmatrix} 2 \\ 1 \\ 0 \end{bmatrix}$.

Hence, $x = 2$, $y = 1$ and $z = 0$.

EXAMPLE 5.41 If $\mathbf{A} = \begin{bmatrix} 3 & -3 & 4 \\ 2 & -3 & 4 \\ 0 & -1 & 1 \end{bmatrix}$, find $\mathbf{A}^{-1}$ and hence solve the following system of linear equations:

$x - y = 0, -2x + 3y - 4z = 9, -2x + 3y - 3z = -5.$ (*AMIE, S, 2003*)

Solution We find $|\mathbf{A}| = \begin{vmatrix} 3 & -3 & 4 \\ 2 & -3 & 4 \\ 0 & -1 & 1 \end{vmatrix} = 3(-3+4) + 3(2) + 4(-2) = 1 \neq 0.$

Therefore $\mathbf{A}^{-1}$ exists and is given by $\mathbf{A}^{-1} = \dfrac{1}{|\mathbf{A}|} \, adj\ \mathbf{A}.$

The Cofactors A_{ij} of **A** are:

$$A_{11} = \begin{vmatrix} -3 & 4 \\ -1 & 1 \end{vmatrix} = 1,\ A_{12} = -\begin{vmatrix} 2 & 4 \\ 0 & 1 \end{vmatrix} = -2,\ A_{13} = \begin{vmatrix} 2 & -3 \\ 0 & -1 \end{vmatrix} = -2,$$

$$A_{21} = -\begin{vmatrix} -3 & 4 \\ -1 & 1 \end{vmatrix} = -1,\ A_{22} = \begin{vmatrix} 3 & 4 \\ 0 & 1 \end{vmatrix} = 3,\ A_{23} = -\begin{vmatrix} 3 & -3 \\ 0 & -1 \end{vmatrix} = 3,$$

$$A_{31} = \begin{vmatrix} -3 & 4 \\ -3 & 4 \end{vmatrix} = 0,\ A_{32} = -\begin{vmatrix} 3 & 4 \\ 2 & 4 \end{vmatrix} = -4,\ A_{33} = \begin{vmatrix} 3 & -3 \\ 2 & -3 \end{vmatrix} = -3.$$

$$adj\ \mathbf{A} = \begin{bmatrix} 1 & -2 & -2 \\ -1 & 3 & 3 \\ 0 & -4 & -3 \end{bmatrix}^T = \begin{bmatrix} 1 & -1 & 0 \\ -2 & 3 & -4 \\ -2 & 3 & -3 \end{bmatrix}.$$

We obtain $\mathbf{A}^{-1} = \dfrac{adj\ \mathbf{A}}{|\mathbf{A}|} = \begin{bmatrix} 1 & -1 & 0 \\ -2 & 3 & -4 \\ -2 & 3 & -3 \end{bmatrix}.$

Now the given system of equations can be written in the matrix form as $\begin{bmatrix} 1 & -1 & 0 \\ -2 & 3 & -4 \\ -2 & 3 & -3 \end{bmatrix} \begin{bmatrix} x \\ y \\ z \end{bmatrix} = \begin{bmatrix} 0 \\ 9 \\ -5 \end{bmatrix},$

that is, $\mathbf{A}^{-1}\mathbf{x} = \mathbf{b} \Rightarrow \mathbf{x} = \mathbf{A}\mathbf{b}$, where $\mathbf{x} = [x\ y\ z]^T$ and $\mathbf{b} = [0 \quad 9 \quad 5]^T$.

Therefore, $\mathbf{x} = \begin{bmatrix} 3 & -3 & 4 \\ 2 & -3 & 4 \\ 0 & -1 & 1 \end{bmatrix} \begin{bmatrix} 0 \\ 9 \\ -5 \end{bmatrix} = \begin{bmatrix} -47 \\ -47 \\ -14 \end{bmatrix}$. Hence $x = -47$, $y = -47$, and $z = -14$.

EXAMPLE 5.42 Solve the homogeneous system of equations

$$x + y - z = 0,\ x - 2y + z = 0,\ 3x + 6y - 5z = 0.$$

Solution In matrix form, we can write the given system of equations as $\mathbf{Ax} = \mathbf{O}$.

$$\begin{bmatrix} 1 & 1 & -1 \\ 1 & -2 & 1 \\ 3 & 6 & -5 \end{bmatrix} \begin{bmatrix} x \\ y \\ z \end{bmatrix} = \begin{bmatrix} 0 \\ 0 \\ 0 \end{bmatrix}.$$

We find that $|\mathbf{A}| = \begin{vmatrix} 1 & 1 & -1 \\ 1 & -2 & 1 \\ 3 & 6 & -5 \end{vmatrix} = 1(10-6) - 1(-5-3) - 1(6+6) = 0$. The system possesses non-trivial solutions. Hence, the given system has infinite number of solutions. Solving the first two equations

$$\begin{bmatrix}1 & 1\\ 1 & -2\end{bmatrix}\begin{bmatrix}x\\ y\end{bmatrix}=\begin{bmatrix}z\\ -z\end{bmatrix} \Rightarrow \begin{bmatrix}x\\ y\end{bmatrix}=\begin{bmatrix}1 & 1\\ 1 & -2\end{bmatrix}^{-1}\begin{bmatrix}z\\ -z\end{bmatrix}=-1/3\begin{bmatrix}-2 & -1\\ -1 & 1\end{bmatrix}\begin{bmatrix}-z\\ -z\end{bmatrix}=-1/3\begin{bmatrix}-z\\ -2z\end{bmatrix}=\begin{bmatrix}z/3\\ 2z/3\end{bmatrix}.$$

We obtain $x = z/3$, $y = 2z/3$, where z is arbitrary. This solution satisfies the third equation.

EXAMPLE 5.43 Find the value (s) of λ for which the equations

$$(\lambda - 1)\, x + (3\lambda + 1)\, y + 2\lambda z = 0$$
$$(\lambda - 1)\, x + 2(2\lambda - 1)\, y + (\lambda + 3)z = 0$$
$$2x + (3\lambda + 1)\, y + 3(\lambda - 1)z = 0$$

have non-trivial solution(s). Find the ratios $x : y : z$ when λ has the smallest of these values. What happens when λ has the greatest of these values? *(AMIE, Dec, 2010; AMIETE, Dec 2010)*

Solution For the given system of equations to have a non-trivial solution, the determinant of the coefficient matrix should be zero, that is

$$\begin{vmatrix}\lambda-1 & 3\lambda+1 & 2\lambda\\ \lambda-1 & 4\lambda-2 & \lambda+3\\ 2 & 3\lambda+1 & 3(\lambda-1)\end{vmatrix}=0 \text{ (Operate } R_2 \to R_2 - R_1) \text{ or } \begin{vmatrix}\lambda-1 & 3\lambda+1 & 2\lambda\\ 0 & \lambda-3 & 3-\lambda\\ 2 & 3\lambda+1 & 3(\lambda-1)\end{vmatrix}=0 \text{ (Operate } C_3 \to C_3 + C_2)$$

or
$$\begin{vmatrix}\lambda-1 & 3\lambda+1 & 5\lambda+1\\ 0 & \lambda-3 & 0\\ 2 & 3\lambda+1 & 2(3\lambda-1)\end{vmatrix}=0.$$

Expanding along R_2, we obtain

$$(\lambda-3)\begin{vmatrix}\lambda-1 & 5\lambda+1\\ 2 & 2(3\lambda-1)\end{vmatrix}=0$$

or $\quad 2(\lambda - 3)\,[(\lambda - 1)\,(3\lambda - 1) - (5\lambda + 1)] = 0 \quad$ or $\quad 6\lambda\,(\lambda - 3)^2 = 0 \quad$ or $\quad \lambda = 0, 3.$

(*i*) When $\lambda = 0$, equations becomes $-x + y = 0$, $\;-x - 2y + 3z = 0$, $\;2x + y - 3z = 0$.
On solving these equations, we get $x = y = z$. Thus, $x : y : z = 1 : 1 : 1$.

(ii) When $\lambda = 3$, equations becomes identical.

EXAMPLE 5.44 Show that the system of equations $2x + y + 4z = 2$, $x + 3y = 2z + 7$, $5z - 8 = 5x + 3y$ has a unique solution. Solve this system using Cramer's rule.

Solution We have $|\mathbf{A}| = \begin{vmatrix}2 & 1 & 4\\ 1 & 3 & -2\\ 5 & 3 & -5\end{vmatrix} = 2(-15+6)-1(-5+10)+4(3-15) = -71 \neq 0$. Therefore, the coefficient matrix **A** is non-singular and the given system of equations has a unique solution.

We find that $\quad |\mathbf{A}_1| = \begin{vmatrix}2 & 1 & 4\\ 7 & 3 & -2\\ -8 & 3 & -5\end{vmatrix} = 2(-15+6)-1(-35-16)+4(21+24) = 213.$

$$|\mathbf{A}_2| = \begin{vmatrix}2 & 2 & 4\\ 1 & 7 & -2\\ 5 & -8 & -5\end{vmatrix} = 2(-35-16)-2(-5+10)+4(-8-35) = -284.$$

$$|\mathbf{A}_3| = \begin{vmatrix}2 & 1 & 2\\ 1 & 3 & 7\\ 5 & 3 & -8\end{vmatrix} = 2(-24-21)-1(-8-35)+2(3-15) = -71.$$

Therefore, $x = \frac{|\mathbf{A}_1|}{|\mathbf{A}|} = -3,\ y = \frac{|\mathbf{A}_2|}{|\mathbf{A}|} = 4,\ z = \frac{|\mathbf{A}_3|}{|\mathbf{A}|} = 1.$

Hence $x = -3$, $y = 4$ and $z = 1$ is the required solution.

EXAMPLE 5.45 Show that the system of equations $2x - y + 3z = 4$, $x + y - 3z = -1$, $5x - y + 3z = 7$ has infinite number of solution. Hence, find the solutions.

Solution We find that

$$|\mathbf{A}| = \begin{vmatrix} 2 & -1 & 3 \\ 1 & 1 & -3 \\ 5 & -1 & 3 \end{vmatrix} = 2(3-3) + 1(3+15) + 3(-1-5) = 0,$$

$$|\mathbf{A}_1| = \begin{vmatrix} 4 & -1 & 3 \\ -1 & 1 & -3 \\ 7 & -1 & 3 \end{vmatrix} = 4(3-3) + 1(-3+21) + 3(1-7) = 0,$$

$$|\mathbf{A}_2| = \begin{vmatrix} 2 & 4 & 3 \\ 1 & -1 & -3 \\ 5 & 7 & 3 \end{vmatrix} = 2(-3+21) - 4(3+15) + 3(7+5) = 0,$$

$$|\mathbf{A}_3| = \begin{vmatrix} 2 & -1 & 4 \\ 1 & 1 & -1 \\ 5 & -1 & 7 \end{vmatrix} = 2(7-1) + 1(7+5) + 4(-1-5) = 0.$$

The system is consistent and has infinite number of solution. Taking the first two equations.

$2x - y = 4 - 3z$, $x + y = -1 + 3z$ and solving, we obtain $x = 1$ and $y = 3z - 2$ where z is arbitrary. This solution satisfies the third equation.

EXAMPLE 5.46 Show that the system of equations $x + 2y + z = 6$, $x + 4y = 10 - 3z$, $x + 4y + 3z = 8$ is inconsistent.

Solution We find that $|\mathbf{A}| = \begin{vmatrix} 1 & 2 & 1 \\ 1 & 4 & 3 \\ 1 & 4 & 3 \end{vmatrix} = 0,\quad |\mathbf{A}_1| = \begin{vmatrix} 6 & 2 & 1 \\ 10 & 4 & 3 \\ 8 & 4 & 3 \end{vmatrix}$

$$= 6(12-12) - 2(30-24) + 1(40-32) = -12 + 8 = -4 \neq 0.$$

Since $|\mathbf{A}| = 0$ and $|\mathbf{A}_1| = -4 \neq 0$, the system of equations is inconsistent, that is the system has no solution.

EXERCISE 5.6

1. Find the adjoint of the matrix $\mathbf{A} = \begin{bmatrix} 1 & 2 & 0 \\ -1 & 3 & 4 \\ -2 & 5 & 6 \end{bmatrix}$. **Ans.** $\begin{bmatrix} -2 & -12 & 8 \\ -2 & 6 & -4 \\ 1 & -9 & 5 \end{bmatrix}$.

2. Find the adjoint of the matrix $\mathbf{A} = \begin{bmatrix} 1 & 1 & 1 \\ 2 & 2 & 3 \\ 1 & 4 & 9 \end{bmatrix}$, verify that $\mathbf{A}(adj\ \mathbf{A}) = |\mathbf{A}|\mathbf{I}_3 = (adj\ \mathbf{A})\mathbf{A}$.

   Ans. $\begin{bmatrix} 6 & -5 & 1 \\ -15 & 8 & -1 \\ 6 & -3 & 0 \end{bmatrix}$.

3. If matrix $\mathbf{C} = \begin{bmatrix} -1 & -2 & -2 \\ 2 & 1 & -2 \\ 2 & -2 & 1 \end{bmatrix}$, show that $adj\mathbf{C} = 3\mathbf{C}^T$.

4. Given two matrices $\mathbf{A}=\begin{bmatrix}2 & 3 & 4\\ 1 & 2 & 3\\ -4 & -5 & 7\end{bmatrix}$, and $\mathbf{B}=\begin{bmatrix}2 & 1 & 3\\ 0 & -1 & 3\\ 4 & 2 & 1\end{bmatrix}$. Prove that $adj(\mathbf{AB}) = adj\mathbf{B}\cdot adj\mathbf{A}$

5. For what values of x is the matrix $\begin{bmatrix}3-x & 2 & 2\\ 2 & 4-x & 2\\ -2 & -4 & -1-x\end{bmatrix}$ singular? (*AMIETE, W-2010*) **Ans.** 0 and 3.

6. Find the inverse of each of the following matrices:

(*i*) $\begin{bmatrix}5 & 1\\ -3 & 4\end{bmatrix}$, (*ii*) $\begin{bmatrix}3 & 5\\ 7 & -11\end{bmatrix}$, (*iii*) $\begin{bmatrix}3 & 2 & 2\\ 2 & 5 & 3\\ 1 & 2 & 1\end{bmatrix}$, (*iv*) $\begin{bmatrix}1 & 3 & 7\\ 4 & 2 & 3\\ 1 & 2 & 1\end{bmatrix}$,

(*v*) $\begin{bmatrix}1 & 2 & 5\\ 3 & 1 & 4\\ 1 & 1 & 2\end{bmatrix}$, (*vi*) $\begin{bmatrix}1 & 0 & -1\\ 3 & 4 & 5\\ 0 & -6 & -7\end{bmatrix}$, (*vii*) $\begin{bmatrix}2 & 5 & 3\\ 3 & 1 & 2\\ 1 & 2 & 1\end{bmatrix}$, (*viii*) $\begin{bmatrix}3 & -3 & 4\\ 2 & -3 & 4\\ 0 & -1 & 1\end{bmatrix}$,

(*ix*) $\begin{bmatrix}1 & -2 & 3\\ 0 & 2 & -1\\ -4 & 5 & 2\end{bmatrix}$, (*x*) $\begin{bmatrix}1 & 0 & 2\\ 2 & -1 & 3\\ 4 & 1 & 8\end{bmatrix}$, (*xi*) $\begin{bmatrix}2 & 4 & 3\\ 0 & 1 & 1\\ 2 & 2 & -1\end{bmatrix}$. (*AMIETE, Dec. 2001*)

Ans. (*i*) $\begin{bmatrix}4/23 & -1/23\\ 3/23 & 5/23\end{bmatrix}$, (*ii*) $\begin{bmatrix}11/68 & 5/68\\ 7/68 & -3/68\end{bmatrix}$, (*iii*) $1/3\begin{bmatrix}1 & -2 & 4\\ -1 & -1 & 5\\ 1 & 4 & -11\end{bmatrix}$, (*iv*) $1/35\begin{bmatrix}-4 & 11 & -5\\ -1 & -6 & 25\\ 6 & 1 & -10\end{bmatrix}$

(*v*) $1/4\begin{bmatrix}-2 & 1 & 3\\ -2 & -3 & 11\\ 2 & 1 & -5\end{bmatrix}$, (*vi*) $1/20\begin{bmatrix}2 & 6 & 4\\ 21 & -7 & -8\\ -18 & 6 & 4\end{bmatrix}$, (*vii*) $1/4\begin{bmatrix}-3 & 1 & 7\\ -1 & -1 & 5\\ 5 & 1 & -13\end{bmatrix}$, (*viii*) $\begin{bmatrix}1 & -1 & 0\\ -2 & 3 & -4\\ -2 & 3 & -3\end{bmatrix}$,

(*ix*) $1/25\begin{bmatrix}9 & 19 & -4\\ 4 & 14 & 1\\ 8 & 3 & 2\end{bmatrix}$, (*x*) $\begin{bmatrix}-11 & 2 & 2\\ -4 & 0 & 1\\ 6 & -1 & -1\end{bmatrix}$, (*xi*) $1/4\begin{bmatrix}3 & -10 & -1\\ -2 & 8 & 2\\ 2 & -4 & -2\end{bmatrix}$.

7. For any three non-singular matrices **A, B, C** each of order n, show that $(\mathbf{ABC})^{-1} = \mathbf{C}^{-1}\mathbf{B}^{-1}\mathbf{A}^{-1}$. (*AMIE, W-2005*)

8. Does the inverse of an identity matrix exist? If so, what is it?

If $\mathbf{A} = 1/9\begin{bmatrix}-8 & 1 & 4\\ 4 & 4 & 7\\ 1 & -8 & 4\end{bmatrix}$, prove that $\mathbf{A}^{-1} = \mathbf{A}^{T}$. **Ans.** $\mathbf{I}_3$

9. The inverse of $\begin{bmatrix}1 & 2 & -2\\ -1 & 3 & 0\\ 0 & -2 & 1\end{bmatrix}$ is $\begin{bmatrix}3 & 2 & 6\\ 1 & 1 & k\\ 2 & 2 & 5\end{bmatrix}$, then the value of k is... **Ans.** 2.

10. For the matrix $\mathbf{A}=\begin{bmatrix}3 & -3 & 4\\ 2 & -3 & 4\\ 0 & -1 & 1\end{bmatrix}$, verify that $\mathbf{A}^3 = \mathbf{A}^{-1}$.

11. Show that $\begin{bmatrix} \cos\theta & -\sin\theta \\ \sin\theta & \cos\theta \end{bmatrix} = \begin{bmatrix} 1 & -\tan\theta/2 \\ \tan\theta/2 & 1 \end{bmatrix}\begin{bmatrix} 1 & \tan\theta/2 \\ -\tan\theta/2 & 1 \end{bmatrix}^{-1}$.

12. If $\mathbf{A}^T$ denotes the transpose of matrix **A** and $\mathbf{A} = \begin{bmatrix} 1 & -2 & 3 \\ 0 & -1 & 4 \\ -2 & 2 & 1 \end{bmatrix}$, obtain $(\mathbf{A}^T)^{-1}$. *(AMIE, S-2003)*

Ans. $\begin{bmatrix} -9 & -8 & -2 \\ 8 & 7 & 2 \\ -5 & -4 & -1 \end{bmatrix}$.

13. Find the matrix **A** satisfying the matrix equation $\begin{bmatrix} 5 & 4 \\ 1 & 1 \end{bmatrix}\mathbf{A} = \begin{bmatrix} 1 & -2 \\ 1 & 3 \end{bmatrix}$. **Ans.** $\begin{bmatrix} -3 & -14 \\ 4 & 17 \end{bmatrix}$.

14. Find matrix **X** such that $\mathbf{X}\begin{bmatrix} 1 & -2 & 3 \\ 0 & 2 & -1 \\ -4 & 5 & 2 \end{bmatrix} = \begin{bmatrix} 1 & -2 & 3 \\ -4 & 5 & 2 \\ 0 & 5 & -1 \end{bmatrix}$. **Ans.** $\begin{bmatrix} 1 & 0 & 0 \\ 0 & 0 & 1 \\ 0 & 1 & 0 \end{bmatrix}$.

15. If a matrix **A** satisfies a relation $\mathbf{A}^2 + \mathbf{A} - \mathbf{I} = \mathbf{O}$, prove that $\mathbf{A}^{-1}$ exists and that $\mathbf{A}^{-1} = \mathbf{I} + \mathbf{A}$, **I** being an identity matrix. *(AMIE, W-2003)*

[**Hint:** If **A** and **B** are square matrices of the same order, then $|\mathbf{AB}| = |\mathbf{A}||\mathbf{B}|$.

We have, $\mathbf{A}^2 + \mathbf{A} = \mathbf{I}$ or $\mathbf{A}(\mathbf{A} + \mathbf{I}) = \mathbf{I}$. ...(*i*)

$\therefore$ $|\mathbf{A}||\mathbf{A} + \mathbf{I}| = |\mathbf{I}| = 1$. Thus $|\mathbf{A}| \neq 0$ and therefore $\mathbf{A}^{-1}$ exists.

Pre- multiplying (*i*) by $\mathbf{A}^{-1}$, we obtain $\mathbf{A} + \mathbf{I} = \mathbf{A}^{-1}$].

16. Show that the matrix $\mathbf{A} = \begin{bmatrix} 1 & 0 & 0 \\ 2 & 1 & 0 \\ 3 & 2 & 1 \end{bmatrix}$ satisfies the equation $\mathbf{A}^3 - 3\mathbf{A}^2 + 3\mathbf{A} - \mathbf{I} = \mathbf{O}$. Hence obtain $\mathbf{A}^{-1}$.

(AMIE, S-2001) **Ans.** $\begin{bmatrix} 1 & 0 & 0 \\ -2 & 1 & 0 \\ 1 & -2 & 1 \end{bmatrix}$.

17. If $\mathbf{A} = \begin{bmatrix} 1 & 1 & -2 \\ -1 & 2 & 1 \\ 0 & 1 & -1 \end{bmatrix}$ and $\mathbf{B} = \begin{bmatrix} 1 & 1 & 3 \\ 0 & 3 & 2 \\ 1 & 1 & 1 \end{bmatrix}$ be two square matrices of order 3, verify that

$\mathbf{B}^{-1}\mathbf{AB} = \begin{bmatrix} -1 & 0 & 0 \\ 0 & 2 & 0 \\ 0 & 0 & 1 \end{bmatrix}$.

18. Given two matrixes $\mathbf{A} = \begin{bmatrix} 1 & 1 & 1 \\ 2 & 4 & 1 \\ 2 & 3 & 1 \end{bmatrix}$ and $\mathbf{B} = \begin{bmatrix} 2 & 3 \\ 3 & 4 \end{bmatrix}$. Find **X** such that $\mathbf{BXA} = \begin{bmatrix} 1 & 0 & 1 \\ 0 & 1 & 0 \end{bmatrix}$.

Ans. $\begin{bmatrix} -4 & 7 & -7 \\ 3 & -5 & 5 \end{bmatrix}$.

19. If $\mathbf{A} = \begin{bmatrix} 1 & 1 & 1 \\ 1 & 2 & 3 \\ 1 & 4 & 9 \end{bmatrix}$ and $\mathbf{B} = \begin{bmatrix} 2 & 5 & 3 \\ 3 & 1 & 2 \\ 1 & 2 & 1 \end{bmatrix}$ be two square matrices of order 3, verify that $(\mathbf{AB})^{-1} = \mathbf{B}^{-1}\mathbf{A}^{-1}$.

20. Given $\mathbf{A}^{-1} = \begin{bmatrix} 5/7 & 1/7 \\ 3/7 & 2/7 \end{bmatrix}$, evaluate $\mathbf{A}^2 + 2\mathbf{A}$. **Ans.** $\begin{bmatrix} 11 & -9 \\ -27 & 38 \end{bmatrix}$.

21. If $\mathbf{A}=\begin{bmatrix}1 & 2 & 3\\ 2 & 3 & 1\\ 3 & 1 & 2\end{bmatrix}$, verify that $(\mathbf{A}^2)^{-1} = (\mathbf{A}^{-1})^2$.

22. If $\mathbf{A}=\begin{bmatrix}2 & -1\\ -2 & 3\end{bmatrix}$ and $\mathbf{B}=\begin{bmatrix}7 & 6\\ 9 & 8\end{bmatrix}$, find 2×2 matrices **C** and **D** such that **AC** = **B** and **DA** = **B**.

[Hint: $\mathbf{C} = \mathbf{A}^{-1}\mathbf{B}, \mathbf{D} = \mathbf{B}\mathbf{A}^{-1}$]. **Ans.** $\mathbf{C}=\begin{bmatrix}15/2 & 13/2\\ 8 & 7\end{bmatrix}, \mathbf{D}=\begin{bmatrix}33/4 & 19/4\\ 43/4 & 25/4\end{bmatrix}$.

23. If $\mathbf{A}=\begin{bmatrix}1 & 2 & -1\\ 3 & 8 & 2\\ 4 & 9 & -1\end{bmatrix}$, find $\mathbf{A}^{-1}$. Using $\mathbf{A}^{-1}$, solve the following system of linear equations: $x + 2y - z = 2$, $3x + 8y + 2z = 10$, $4x + 9y - z = 12$.

Ans. $\begin{bmatrix}-26 & -7 & 12\\ 11 & 3 & -5\\ -5 & -1 & 2\end{bmatrix}$; 22, –8, 4.

24. Solve the following system of equations by (*i*) matrix method (*ii*) Cramer's rule

(*a*) $x + 2y + 3z = 1$, $2x + 3y + 2z = 2$, $3x + 3y + 4z = 1$.

(*b*) $3x - 2y + 2z = 10$, $x + 2y - 3z = -1$, $4x + y + 2z = 3$.

(*c*) $x - y + z = 4$, $2x + y = 3z$, $x + y + z = 2$.

(*d*) $x + y + z = 3$, $x + 2y + 3z = 4$, $x + 4y + 9z = 6$.

(*e*) $x + 2y + 3y = 2$, $2x + 3y + 5z = 3$, $3x + 5y + z = 12$. (*AMIE, W-2002*)

Ans. (*a*) –3/7, 8/7, –2/7; (*b*) 2, –3, –1; (*c*) 2, –1, 1; (*d*) 2, 1, 0; (*e*) 1, 2, –1.

25. The currents I_1, I_2, I_3 and I_4 in an electric network satisfy the system of equations $3I_1 + 2I_3 - I_4 = 60$, $2I_1 - I_2 + 4I_3 = 160$, $4I_2 + I_3 - 2I_4 = 20$, $5I_1 - I_2 - 2I_3 + I_4 = 0$. Find I_3. **Ans.** 40.

26. Find the value of λ for which the equations

$$(2 - \lambda)x + 2y + 3 = 0,\quad 2x + (4 - \lambda)y + 7 = 0,\quad 2x + 5y + (6 - \lambda) = 0$$

are consistent and find the values of x and y corresponding to each of these value of λ.

Ans. $\lambda = -1$, (–1/11, –15/11); 1, (–5, 1); 12, (1/2, 1).

27. Investigate the following system of equations. Find the solution if it exists.

$$\begin{bmatrix}2 & 1 & -1\\ 1 & -2 & 1\\ 4 & -3 & 1\end{bmatrix}\begin{bmatrix}x\\ y\\ z\end{bmatrix}=\begin{bmatrix}1\\ 3\\ 5\end{bmatrix}.$$

Ans. Inconsistent.

28. Find the values of k for which the system of equations

$$(3k - 8)x + 3y + 3z = 0,\quad 3x + (3k - 8)y + 3z = 0,\quad 3x + 3y + (3k - 8)z = 0.$$

has a non-trivial solution. (*AMIETE, June 2010; UPTU-2006*)

[Hint: For the given system of equations to have non-trivial solution, the determinant of the coefficient matrix is zero.

That is,
$$\begin{vmatrix}(3k-8) & 3 & 3\\ 3 & (3k-8) & 3\\ 3 & 3 & (3k-8)\end{vmatrix} = 0, \quad [\text{Operating } C_1 \to C_1 + (C_2 + C_3)].$$

or
$$(3k-2)\begin{vmatrix}1 & 3 & 3\\ 1 & 3k-8 & 3\\ 1 & 3 & 3k-5\end{vmatrix} = 0, \quad [\text{Operating } R_2 \to R_2 - R_1, R_3 \to R_3 - R_1].$$

$$\text{or} \qquad (3k-2)\begin{vmatrix} 1 & 3 & 3 \\ 0 & (3k-11) & 0 \\ 0 & 0 & (3k-11) \end{vmatrix} = 0$$

or $(3k - 2)(3k - 11)^2 = 0$, whence $k = 2/3,\ 11/3$.

29. If the following system

$$ax + by + cz = 0, \quad bx + cy + az = 0, \quad cx + ay + bz = 0.$$

has non-trivial solution. Prove that $a + b + c = 0$ or $a = b = c$.

5.15 VECTOR SPACES

Let V denote a non-empty set of certain objects, which may be vectors, matrices, functions or some other objects. Each object is an element of V and is called a vector. The elements of V are denoted by lowercase bold letters **a, b, c, u, v**. etc.

Definition of a Vector Space

A vector space V is a non-empty set of objects which can be added and multiplied by numbers, in such a way that the sum of two elements of V is again an element of V, the product of an element of V by a number is an element of V and the following ten properties, (axioms) are satisfied.

Properties (axioms) for Vector Addition

1. If **a** and **b** are in V, then **a** + **b** is in V.
2. For any two vectors **a** and **b** of V, **a** + **b** = **b** + **a**. (Commutative law)
3. For any three vectors **a, b, c** of V, **a** + (**b** + **c**) = (**a** + **b**) + **c** (associative law)
4. There is a unique vector **0** in V such that **a** + **0** = **0** + **a** = **a** for all **a** in V (existence of zero vector)
5. For every **a** in V, there exists a vector –**a** such that
   a + (– **a**) = (–**a**) + **a** = **0** (existence of additive inverse or negative vector in V)

Properties (axioms) for Scalar Multiplication

6. If k_1 is any scalar and **a** is in V, then k_1**a** is in V.
7. $k_1(\mathbf{a} + \mathbf{b}) = k_1\mathbf{a} + k_1\mathbf{b}$ (right distributive law)
8. $(k_1 + k_2)\,\mathbf{a} = k_1\mathbf{a} + k_2\mathbf{a}$ (left distributive law)
9. $k_1(k_2\mathbf{a}) = (k_1k_2)\,\mathbf{a}$.
10. 1**a** = **a** (existence of multiplicative inverse)

The properties defined in **1** and **6** are called the *closure* properties. When these two properties are satisfied, we say that the vector space V is closed under the vector addition and scalar multiplication.

If the elements of V are real, then it is called a *real vector space* when the scalars k_1, k_2 are real numbers. If *real number* is replaced by *complex number* in properties **6, 7, 8, 9,** the resulting structure is called a *complex vector space*. When we use the term vector space without further designation, it is to be understood that the space can be real or complex.

The following are some examples of vector spaces under the usual operations of vector addition and scalar multiplication.

1. The set V of real or complex numbers.
2. The set V of n-tuples of numbers in R^n* or $\mathbb{C}^n$**.
3. The set of polynomials P_n of degree less than or equal to n.
4. The set of real valued continuous functions f on any closed interval $[a, b]$.
5. The set V of all $m \times n$ matrices.

---

* The set of all n-tuples of real numbers, denoted by R^n *is called* n-*space*.

** The set of all n-tuples of complex numbers, denoted by $\mathbb{C}^n$, *is called* complex n-*space*.

EXAMPLE 5.47 Let V be the set of all ordered pairs (x, y), where x, y are real numbers. Let $\mathbf{u} = (x_1, y_1)$ and $\mathbf{v} = (x_2, y_2)$ be two elements in V. Define the addition as

$$\mathbf{u} + \mathbf{v} = (x_1, y_1) + (x_2, y_2) = (2x_1 - 3x_2, y_1 - y_2)$$

and the scalar multiplication as

$$k(x_1, y_1) = (kx_1/3, ky_1/3).$$

Show that V is not a vector space.

Solution We shall go through properties that are not satisfied.

(*i*) $$(x_2, y_2) + (x_1, y_1) = (2x_2 - 3x_1, y_2 - y_1) \neq (x_1, y_1) + (x_2, y_2).$$

Therefore, **property 2** (commutative law) does not hold.

(*ii*) $$((x_1, y_1) + (x_2, y_2)) + (x_3, y_3) = (2x_1 - 3x_2, y_1 - y_2) + (x_3, y_3) = (4x_1 - 6x_2 - 3x_3, y_1 - y_2 - y_3).$$
$$(x_1, y_1) + ((x_2, y_2) + (x_3, y_3)) = (x_1, y_1) + (2x_2 - 3x_3, y_2 - y_3) = 2x_1 - 6x_2 + 9x_3, y_1 - y_2 + y_3).$$

Therefore, **property 3** (associative law) is not satisfied.

(*iii*) $$1(x_1, y_1) = (x_1/3, y_1/3) \neq (x_1, y_1).$$

Therefore, **property 10.** (existence of multiplicative inverse) is not satisfied.

Hence, V is not a vector space.

EXAMPLE 5.48 Let V be the set of all positive real numbers where addition is defined by $\mathbf{x} + \mathbf{y} = xy$ and scalar multiplication is defined by $k\mathbf{x} = x^k$. Determine whether V is a vector space.

Solution We shall go through all ten properties.

1. For $\mathbf{x} = x > 0$ and $\mathbf{y} = y > 0$, $\mathbf{x} + \mathbf{y} = xy > 0$. Thus, the sum $\mathbf{x} + \mathbf{y}$ is in V; V is closed under addition.
2. For all $\mathbf{x} = x$ and $\mathbf{y} = y$ in V, $\mathbf{x} + \mathbf{y} = xy = yx = \mathbf{y} + \mathbf{x}$. Thus addition is commutative.
3. For all $\mathbf{x} = x$, $\mathbf{y} = y$, $\mathbf{z} = z$ in V, $\mathbf{x} + (\mathbf{y} + \mathbf{z}) = x(yz) = (xy)z = (\mathbf{x} + \mathbf{y}) + \mathbf{z}$. Thus addition is associative.
4. Since $\mathbf{1} + \mathbf{x} = 1x = x = \mathbf{x}$ and $\mathbf{x} + \mathbf{1} = x1 = x = \mathbf{x}$, the zero vector $\mathbf{O}$ is $\mathbf{1} = 1$.
5. If we define $-\mathbf{x} = 1/x$, then $\mathbf{x} + (-\mathbf{x}) = x\frac{1}{x} = 1 = \mathbf{1} = \mathbf{0}$ and $(-\mathbf{x}) + \mathbf{x} = \frac{1}{x}x = 1 = \mathbf{1} = \mathbf{0}$. The negative of a vector is its reciprocal.
6. If k is any scalar and $\mathbf{x} = x > 0$ is any vector, then $k\mathbf{x} = x^k > 0$. Hence V is closed under scalar multiplication.
7. If k is any scalar, then $k(\mathbf{x} + \mathbf{y}) = (xy)^k = x^k y^k = k\mathbf{x} + k\mathbf{y}$.
8. For scalars k_1 and k_2, $(k_1 + k_2)\mathbf{x} = x^{(k_1 + k_2)} = x^{k_1} x^{k_2} = k_1\mathbf{x} + k_2\mathbf{x}$.
9. For scalars k_1 and $k_2, k_1(k_2\mathbf{x}) = (x^{k_2})^{k_1} = x^{k_1 k_2} = (k_1 k_2)\mathbf{x}$.
10. $1\mathbf{x} = x^1 = x = \mathbf{x}$.

Since all the properties of vector space are satisfied, we conclude that V is a vector space.

5.15.1 Sub space of a vector space

If a subset W of a vector space V is itself a vector space under the operations of vector addition and scalar multiplication defined on V, then W is called a **subspace** of V. Every vector space V has atleast two subspaces: V itself and the zero subspace $\{0\}$; $\{0\}$ is a subspace since the zero vector must be an element in every vector space. The vector space V is also taken as a subspace of V.

To show that a subset W of a vector space V is a subspace, it is not necessary to show that all the ten properties are satisifed.

Criteria for a subspace

A non empty subset W of a vector space V is a subspace of V if and only if W is closed under vector addition and scalar multiplication defined on V:

(*i*) If $\mathbf{a}$ and $\mathbf{b}$ are in W, then $\mathbf{a} + \mathbf{b}$ is in W.

(*ii*) If $\mathbf{a}$ is in W and k is any scalar, then $k\mathbf{a}$ is W.

Example 1 Let $V = R^n$ and let W be the set of vectors in V whose last coordinate is equal to 0. Then W is a subspace of V, which we could identify with R^{n-1}.

Example 2 Let U and W be subspaces of a vector space V. We denote by $U \cap W$ the intersection of U and W, that is, the set of elements which lie both in U and W. Then $U \cap W$ is a subspace of V.

Linear Combinations, Spanning Set

Let V be a vector space and let $\mathbf{v}_1, \mathbf{v}_2, ..., \mathbf{v}_m$ be any m elements in V under usual vector addition and scalar multiplication. Then, the set of all linear combinations of these elements, that is the set of all elements of the form

$$k_1\mathbf{v}_1 + k_2\mathbf{v}_2 + ...+ k_m\mathbf{v}_m \quad ...(5.17)$$

is a subspace of V, where $k_1, k_2 ..., k_m$ are scalars.

EXAMPLE 5.49 Express $\mathbf{x} = (2, -5, 3)$ in R^3 as a linear combination of the vectors

$$\mathbf{u} = (1, -3, 2),\ \mathbf{v} = (2, -4, -1) \text{ and } \mathbf{w} = (1, -5, 7).$$

Solution Let k_1, k_2, k_3 be scalars, such that $\mathbf{x} = k_1\mathbf{u} + k_2\mathbf{v} + k_3\mathbf{w}$. Thus, we have

$$\begin{pmatrix} 2 \\ -5 \\ 3 \end{pmatrix} = k_1\begin{pmatrix} 1 \\ -3 \\ 2 \end{pmatrix} + k_2\begin{pmatrix} 2 \\ -4 \\ -1 \end{pmatrix} + k_3\begin{pmatrix} 1 \\ -5 \\ 7 \end{pmatrix} \quad \text{or} \quad \begin{aligned} k_1 + 2k_2 + k_3 &= 2, \\ -3k_1 - 4k_2 - 5k_3 &= -5, \\ 2k_1 - k_2 + 7k_3 &= 3. \end{aligned}$$

We write the augmented matrix and reduce it to *row-echelon form by applying elementary row-operations.

$$(\mathbf{A} \mid \mathbf{b}) = \left[\begin{array}{ccc|c} 1 & 2 & 1 & 2 \\ -3 & -4 & -5 & -5 \\ 2 & -1 & 7 & 3 \end{array}\right] \sim \left[\begin{array}{ccc|c} 1 & 2 & 1 & 2 \\ 0 & 2 & -2 & 1 \\ 0 & -5 & 5 & -1 \end{array}\right] \sim \left[\begin{array}{ccc|c} 1 & 2 & 1 & 2 \\ 0 & 2 & -2 & 1 \\ 0 & 0 & 0 & 3/2 \end{array}\right].$$

Using the back substitution method, we obtain.

$$k_1 + 2k_2 + k_3 = 2,\ 2k_2 - 2k_3 = 1,\ 0 = 3/2.$$

The system is inconsistent and has no solution. Thus $\mathbf{x}$ cannot be written as a linear combination of $\mathbf{u}, \mathbf{v}, \mathbf{w}$.

EXAMPLE 5.50 Express $\mathbf{x} = (4, 3, 10)$ in R^3 as a linear combination of the vectors

$$\mathbf{u} = (1, 2, -1),\ \mathbf{v} = (2, 3, 4) \text{ and } \mathbf{w} = (1, 5, -3).$$

Solution Let k_1, k_2, k_3 be scalars, such that $\mathbf{x} = k_1\mathbf{u} + k_2\mathbf{v} + k_3\mathbf{w}$; that is

$$\begin{bmatrix} 4 \\ 3 \\ 10 \end{bmatrix} = k_1\begin{bmatrix} 1 \\ 2 \\ -1 \end{bmatrix} + k_2\begin{bmatrix} 2 \\ 3 \\ 4 \end{bmatrix} + k_3\begin{bmatrix} 1 \\ 5 \\ -3 \end{bmatrix} \quad \text{or} \quad \begin{aligned} k_1 + 2k_2 + k_3 &= 4, \\ 2k_1 + 3k_2 + 5k_3 &= 3, \\ -k_1 + 4k_2 - 3k_3 &= 10. \end{aligned}$$

We write the augmented matrix and reduce it to row-echelon form by applying elementary row-operations.

$$(\mathbf{A} \mid \mathbf{b}) = \left[\begin{array}{ccc|c} 1 & 2 & 1 & 4 \\ 2 & 3 & 5 & 3 \\ -1 & 4 & -3 & 10 \end{array}\right] \quad (\text{operate } R_2 \to R_2 - 2R_1 \text{ and } R_3 \to R_3 + R_1)$$

$$\sim \left[\begin{array}{ccc|c} 1 & 2 & 1 & 4 \\ 0 & -1 & 3 & -5 \\ 0 & 6 & -2 & 14 \end{array}\right] \quad (\text{operate } R_3 \to R_3 + 6R_2) \quad \sim \left[\begin{array}{ccc|c} 1 & 2 & 1 & 4 \\ 0 & -1 & 3 & -5 \\ 0 & 0 & 16 & -16 \end{array}\right].$$

* Refer Article 5.17, Example 5.68.

Using the back substitution method, we obtain the solution

$$16k_3 = -16 \text{ or } k_3 = -1,$$
$$-k_2 + 3k_3 = -5 \text{ or } k_2 = 2,$$
$$k_1 + 2k_2 + k_3 = 4 \text{ or } k_1 = 1. \text{ Thus } \mathbf{x} = \mathbf{u} + 2\mathbf{v} - \mathbf{w}.$$

Spanning Set: Let S be a subset of a vector space V and suppose that every element in V can be obtained as a linear combination of the elements taken from S, then S is said to be the *spanning set* for V. We may also say that S spans V.

EXAMPLE 5.51 Let V be the set of all 2×2 real matrices. Show that the set

$$S = \left\{ \begin{pmatrix} 2 & 1 \\ 1 & -2 \end{pmatrix}, \begin{pmatrix} 1 & 1 \\ 1 & 0 \end{pmatrix}, \begin{pmatrix} 0 & 0 \\ 1 & 1 \end{pmatrix}, \begin{pmatrix} 0 & 2 \\ 0 & -1 \end{pmatrix} \right\} \text{ Spans } V.$$

Solution Let $\mathbf{Y} = \begin{bmatrix} a & b \\ c & d \end{bmatrix}$ be an arbitrary element of V. We are required to determine the scalars α_1, α_2, α_3, α_4 so that

$$\begin{bmatrix} a & b \\ c & d \end{bmatrix} = \alpha_1 \begin{bmatrix} 2 & 1 \\ 1 & -2 \end{bmatrix} + \alpha_2 \begin{bmatrix} 1 & 1 \\ 1 & 0 \end{bmatrix} + \alpha_3 \begin{bmatrix} 0 & 0 \\ 1 & 1 \end{bmatrix} + \alpha_4 \begin{bmatrix} 0 & 2 \\ 0 & -1 \end{bmatrix}.$$

Equating the corresponding elements, we obtain the system of equations

$$2\alpha_1 + \alpha_2 = a, \qquad \alpha_1 + \alpha_2 + 2\alpha_4 = b,$$
$$\alpha_1 + \alpha_2 + \alpha_3 = c, \qquad -2\alpha_1 + \alpha_3 - \alpha_4 = d.$$

We write the augmented matrix and reduce it to row-echelon form by applying elementary row-operations.

$$(\mathbf{A}|\mathbf{b}) = \left[\begin{array}{cccc|c} 2 & 1 & 0 & 0 & a \\ 1 & 1 & 0 & 2 & b \\ 1 & 1 & 1 & 0 & c \\ -2 & 0 & 1 & -1 & d \end{array}\right] \sim \left[\begin{array}{cccc|c} 2 & 1 & 0 & 0 & a \\ 0 & 1 & 0 & 4 & 2b-a \\ 0 & 0 & 1 & -2 & c-b \\ 0 & 1 & 1 & -1 & d+a \end{array}\right]$$

$$\sim \left[\begin{array}{cccc|c} 2 & 1 & 0 & 0 & a \\ 0 & 1 & 0 & 4 & 2b-a \\ 0 & 0 & 1 & -2 & c-b \\ 0 & 0 & 1 & -5 & 2a-2b+d \end{array}\right] \sim \left[\begin{array}{cccc|c} 2 & 1 & 0 & 0 & a \\ 0 & 1 & 0 & 4 & 2b-a \\ 0 & 0 & 1 & -2 & c-b \\ 0 & 0 & 0 & -3 & 2a-b-c+d \end{array}\right].$$

Using the back substitution method, we obtain the solution

$$-3\alpha_4 = 2a - b - c + d \text{ or } \alpha_4 = (-2a + b + c - d)/3$$

$$\alpha_3 - 2\alpha_4 = c - b \text{ or } \alpha_3 = c - b + \frac{2b + 2c - 4a - 2d}{3} = (-4a - b + 5c - 2d)/3$$

$$\alpha_2 + 4\alpha_4 = 2b - a \text{ or } \alpha_2 = 2b - a + (8a - 4b - 4c + 4d)/3 = (5a + 2b - 4c + 4d)/3$$

$$2\alpha_1 + \alpha_2 = a \text{ or } \alpha_1 = \frac{a}{2} + \frac{4c - 4d - 5a - 2b}{6} = \frac{(-a - b + 2c - 2d)}{3}.$$

Therefore, we can write

$$\begin{bmatrix} a & b \\ c & d \end{bmatrix} = \frac{(-a-b+2c-2d)}{3} \begin{bmatrix} 2 & 1 \\ 1 & -2 \end{bmatrix} + \frac{(5a+2b-4c+4d)}{3} \begin{bmatrix} 1 & 1 \\ 1 & 0 \end{bmatrix}$$
$$+ \frac{(-4a-b+5c-2d)}{3} \begin{bmatrix} 0 & 0 \\ 1 & 1 \end{bmatrix} + \frac{(-2a+b+c-d)}{3} \begin{bmatrix} 0 & 2 \\ 0 & -1 \end{bmatrix}.$$

Since every element of V can be written as a linear combination of the element of S, the set S spans the vector space V.

EXAMPLE 5.52 Let V be the set of all 3×1 real matrices. Show that the set

$$S = \left\{ \begin{pmatrix} 1 \\ 1 \\ 0 \end{pmatrix}, \begin{pmatrix} 1 \\ -1 \\ 0 \end{pmatrix}, \begin{pmatrix} 0 \\ 0 \\ 1 \end{pmatrix} \right\} \text{ spans } V.$$

Solution Let $\mathbf{X} = [a\ b\ c]^T$ be an arbitrary element of V. We are required to determine the scalars $\alpha_1, \alpha_2, \alpha_3$ so that

$$\begin{bmatrix} a \\ b \\ c \end{bmatrix} = \alpha_1 \begin{bmatrix} 1 \\ 1 \\ 0 \end{bmatrix} + \alpha_2 \begin{bmatrix} 1 \\ -1 \\ 0 \end{bmatrix} + \alpha_3 \begin{bmatrix} 0 \\ 0 \\ 1 \end{bmatrix}.$$

Equating the corresponding elements, we obtain the system of equations

$$\alpha_1 + \alpha_2 = a, \quad \alpha_1 - \alpha_2 = b, \quad \alpha_3 = c.$$

The solution of this system of equations is

$$\alpha_1 = (a + b)/2, \ \alpha_2 = (a - b)/2, \text{ and } \alpha_3 = c.$$

Therefore, we can write $\begin{bmatrix} a \\ b \\ c \end{bmatrix} = \dfrac{(a+b)}{2} \begin{bmatrix} 1 \\ 1 \\ 0 \end{bmatrix} + \dfrac{(a-b)}{2} \begin{bmatrix} 1 \\ -1 \\ 0 \end{bmatrix} + c \begin{bmatrix} 0 \\ 0 \\ 1 \end{bmatrix}.$

Since every element of V can be written as a linear combination of the elements of S, the set S spans the vector space V.

5.15.2 Linear Independence of vectors

Let V be a vector space. A finite set $\{\mathbf{v}_1, \mathbf{v}_2, ..., \mathbf{v}_n\}$ of the elements of V is said to be *linearly dependent* if there exists scalars $k_1, k_2, ..., k_n$, not all of them zero, such that

$$k_1\mathbf{v}_1 + k_2\mathbf{v}_2 + ... + k_n\mathbf{v}_n = \mathbf{O}, \qquad ...(5.18)$$

where $\mathbf{O}$ on the right hand side denotes a null vector. Otherwise, we say that the vectors are *linearly independent*. In this case the equation (5.18) will be satisfied only when

$$k_1 = k_2 = ... = k_n = 0.$$

The above definition of linear dependence of $\mathbf{v}_1, \mathbf{v}_2, ..., \mathbf{v}_n$ can be written alternately as follows.

Theorem: The set of vectors $\{\mathbf{v}_1, \mathbf{v}_2, ..., \mathbf{v}_n\}$ is linearly dependent if and only if at least one element of the set is a linear combination of the remaining elements.

Proof: Let the elements $\mathbf{v}_1, \mathbf{v}_2, ..., \mathbf{v}_n$ be linearly dependent. Then, there exists scalars $k_1, k_2, ..., k_n$, not all zero such that

$$k_1\mathbf{v}_1 + k_2\mathbf{v}_2 + ...k_{i-1}\mathbf{v}_{i-1} + k_i\mathbf{v}_i + k_{i+1}\mathbf{v}_{i+1} ...+ k_n\mathbf{v}_n = \mathbf{0}.$$

Let $k_i \neq 0$. Then, we can write

or $$\mathbf{v}_i = -\left(\frac{k_1}{k_i}\right)\mathbf{v}_1 - \left(\frac{k_2}{k_i}\right)\mathbf{v}_2 - ...... - \left(\frac{k_{i-1}}{k_i}\right)\mathbf{v}_{i-1} - \left(\frac{k_{i+1}}{k_i}\right)\mathbf{v}_{i+1} - ...... \left(\frac{k_n}{k_i}\right)\mathbf{v}_n$$

$$= c_1\mathbf{v}_1 + c_2\mathbf{v}_2 + c_{i-1}\mathbf{v}_{i-1} + c_{i+1}\mathbf{v}_{i+1} + ... + c_n\mathbf{v}_n, \quad \text{where } c_1, c_2, ..., c_n \text{ are some scalars.}$$

Hence, the vector $\mathbf{v}_i$ is a linear combination of the vectors $\mathbf{v}_1, \mathbf{v}_2, ..., \mathbf{v}_{i-1}, \mathbf{v}_{i+1}, ..., \mathbf{v}_n$.

Now let $\mathbf{v}_i$ be a linear combination of $\mathbf{v}_1, \mathbf{v}_2, ..., \mathbf{v}_{i-1}, \mathbf{v}_{i+1}, ..., \mathbf{v}_n$. Therefore, we have

$$\mathbf{v}_i = a_1\mathbf{v}_1 + a_2\mathbf{v}_2 + ...+ a_{i-1}\mathbf{v}_{i-1} + a_{i+1}\mathbf{v}_{i+1} + ... + a_n\mathbf{v}_n$$

where a_i's are scalars. Then $a_1\mathbf{v}_1 + a_2\mathbf{v}_2 + ...+a_{i-1}\mathbf{v}_{i-1} + (-1)\mathbf{v}_i + a_{i+1}\mathbf{v}_{i+1} + ...+ a_n\mathbf{v}_n = \mathbf{0}.$

Since the coefficient of $\mathbf{v}_i$ is not zero, the elements are linearly dependent.

Remark Equation (5.18) gives a homogeneous system of algebraic equations. Non-trivial solution exist if det (coefficient matrix) = 0, that is the vectors are linearly dependent in this case. If the det (coefficient matrix) $\neq 0$, then by Cramer's rule, $k_1 = k_2 = ..., = k_n = 0$ and the vectors are linearly independent.

Example Let $V = \mathbb{R}^n$ and consider the vectors

$$\mathbf{v}_1 = (1, 0, ..., 0), \mathbf{v}_2 = (0, 1, ..., 0), ..., \mathbf{v}_n = (0, 0, ..., 1),$$ then $\mathbf{v}_1, \mathbf{v}_2, ..., \mathbf{v}_n$ are linearly independent.

Solution Let $k_1, k_2, ..., k_n$ be scalars. Consider the vector equation

$$k_1\mathbf{v}_1 + k_2\mathbf{v}_2 + ... + k_n\mathbf{v}_n = \mathbf{0}.$$

Substituting for $\mathbf{v}_1, \mathbf{v}_2, ..., \mathbf{v}_n$ we obtain

$$k_1(1, 0, ..., 0) + k_2(0, 1, ..., 0) + ... k_n(0, 0, ..., 1) = \mathbf{0} \quad \text{or} \quad (k_1, k_2, ..., k_n) = \mathbf{0}.$$

Comparing, we obtain $k_i = 0$. Therefore, the given set of vectors is linearly independent.

Alternative

$$\det(\mathbf{v}_1, \mathbf{v}_2, \cdots \mathbf{v}_n) = \begin{vmatrix} 1 & 0 & 0 & \cdots & 0 \\ 0 & 1 & 0 & \cdots & 0 \\ \cdots & \cdots & \cdots & \cdots & \cdots \\ \cdots & \cdots & \cdots & \cdots & \cdots \\ 0 & 0 & 0 & \cdots & 1 \end{vmatrix} = 1 \neq 0.$$

Therefore, the given vectors are linearly independent.

Rule for testing linear dependence: Let the number of vectors be m.

(*i*) Write the given vectors as row vectors.

(*ii*) Add such multiples of one vector to the other vectors so that the resulting vectors have zero as first component. In this way, we are left with $(m - 1)$ vectors each with first component zero.

(*iii*) Again add suitable multiples of one from the $(m - 1)$ vectors to the remaining vectors so that the resulting vectors have their second component zero. We now have $(m - 2)$ vectors with first two components zero.

(*iv*) Repeat this process till we are left with only one vector.

If all the components of this vector are zero, the given m vectors are linearly dependent, otherwise they are linearly independent.

EXAMPLE 5.53 Investigate the linear dependence or independence of the vectors.

(*i*) $\mathbf{v}_1 = (2, -1, 3, 2)$, $\mathbf{v}_2 = (1, 3, 4, 2)$ and $\mathbf{v}_3 = (3, -5, 2, 2)$. *(MDU 2001)*

(*ii*) $\mathbf{v}_1 = (1, 1, 1)$, $\mathbf{v}_2 = (1, 2, 3)$ and $\mathbf{v}_3 = (2, 3, 8)$.

(*iii*) $\mathbf{v}_1 = (1, -1, 0)$, $\mathbf{v}_2 = (0, 1, -1)$, $\mathbf{v}_3 = (0, 2, 1)$ and $\mathbf{v}_4 = (1, 0, 3)$ be elements of $\mathbb{R}^3$. *(AMIE, S-2008)*

Solution (*i*) Using $\mathbf{v}_2$ to reduce the first component of resulting vectors to zero, we have

$$\mathbf{v}_1 - 2\mathbf{v}_2 = (0, -7, -5, -2) \quad ...(1)$$

and

$$\mathbf{v}_3 - 3\mathbf{v}_2 = (0, -14, -10, -4). \quad ...(2)$$

Multiplying (1) by 2 and subtracting from (2), we get

$$(\mathbf{v}_3 - 3\mathbf{v}_2) - 2(\mathbf{v}_1 - 2\mathbf{v}_2) = (0, 0, 0, 0) \text{ or } 2\mathbf{v}_1 - \mathbf{v}_2 - \mathbf{v}_3 = \mathbf{0}.$$

Therefore, the given vectors are linearly dependent.

(*ii*) We have

$$\mathbf{v}_2 - \mathbf{v}_1 = (0, 1, 2), \mathbf{v}_3 - 2\mathbf{v}_1 = (0, 1, 6).$$

$$(\mathbf{v}_3 - 2\mathbf{v}_1) - (\mathbf{v}_2 - \mathbf{v}_1) = (0, 0, 4) \text{ or } \mathbf{v}_3 - \mathbf{v}_1 - \mathbf{v}_2 = (0, 0, 4) \neq \mathbf{0}.$$

Since the resulting vector is not a null vector, so the given vectors are linearly independent.

(*iii*) We have $\mathbf{v}_1 - \mathbf{v}_4 = (0, -1, -3)$, $\mathbf{v}_2 = (0, 1, -1)$ and $\mathbf{v}_3 = (0, 2, 1)$.

$$(\mathbf{v}_1 - \mathbf{v}_4) + \mathbf{v}_2 = (0, 0, -4), \quad ...(3)$$

and

$$2\mathbf{v}_2 - \mathbf{v}_3 = (0, 0, -3). \quad ...(4)$$

Multiplying (3) by 3, (4) by –4 and adding, we obtain

$$3(\mathbf{v}_1 - \mathbf{v}_4) + 3\mathbf{v}_2 - 8\mathbf{v}_2 + 4\mathbf{v}_3 = (0, 0, 0) \text{ or } \mathbf{v}_1 - \frac{5}{3}\mathbf{v}_2 + \frac{4}{3}\mathbf{v}_3 - \mathbf{v}_4 = 0.$$

Therefore, the set of vectors $\{\mathbf{v}_1, \mathbf{v}_2, \mathbf{v}_3, \mathbf{v}_4\}$ is linearly dependent.

EXAMPLE 5.54 Are the following vectors linearly dependent

$$\mathbf{v}_1 = (1, 2, 4), \mathbf{v}_2 = (2, -1, 3), \mathbf{v}_3 = (0, 1, 2) \text{ and } \mathbf{v}_4 = (-3, 7, 2)?$$

If so, find the relation between them. (*AMIE, S-2007*)

Solution Using $\mathbf{v}_1$ to reduce the first component to zero we obtain

$$\mathbf{v}_2 - 2\mathbf{v}_1 = (0, -5, -5), \mathbf{v}_3 = (0, 1, 2), \mathbf{v}_4 + 3\mathbf{v}_1 = (0, 13, 14).$$

Using $\mathbf{v}_3$ (the simplest of the above three vectors) to reduce the second component to zero, we obtain

$$(\mathbf{v}_2 - 2\mathbf{v}_1) + 5\mathbf{v}_3 = (0, 0, 5), (\mathbf{v}_4 + 3\mathbf{v}_1) - 13\mathbf{v}_3 = (0, 0, -12).$$

Multiplying the first equation by 12 and the second by 5, and adding, we obtain

$$12(\mathbf{v}_2 - 2\mathbf{v}_1 + 5\mathbf{v}_3) + 5(\mathbf{v}_4 + 3\mathbf{v}_1 - 13\mathbf{v}_3) = \mathbf{0}$$

or

$$-9\mathbf{v}_1 + 12\mathbf{v}_2 - 5\mathbf{v}_3 + 5\mathbf{v}_4 = \mathbf{0}. \qquad ...(i)$$

Therefore, the given vectors are linearly dependent and are connected by relation (*i*).

EXAMPLE 5.55 Let $\mathbf{v}_1 = (2, 2, 1)$, $\mathbf{v}_2 = (1, -1, 1)$ and $\mathbf{v}_3 = (1, 0, 1)$ be elements in R^3. Show that the set of vectors $\{\mathbf{v}_1, \mathbf{v}_2, \mathbf{v}_3\}$ is linearly independent.

Solution Consider the vector equation $k_1\mathbf{v}_1 + k_2\mathbf{v}_2 + k_3\mathbf{v}_3 = \mathbf{0}$.

Substituting for $\mathbf{v}_1, \mathbf{v}_2, \mathbf{v}_3$, we obtain

$$k_1(2, 2, 1) + k_2(1, -1, 1) + k_3(1, 0, 1) = \mathbf{0},$$

or

$$(2k_1 + k_2 + k_3;\ \ 2k_1 - k_2, k_1 + k_2 + k_3) = \mathbf{0}.$$

Comparing, we obtain $2k_1 + k_2 + k_3 = 0$, $2k_1 - k_2 = 0$ and $k_1 + k_2 + k_3 = 0$. The solution of these equations is $k_1 = k_2 = k_3 = 0$. Therefore, the given set of vectors is linearly independent.

Alternative

$$\det(\mathbf{v}_1, \mathbf{v}_2, \mathbf{v}_3) = \begin{vmatrix} 2 & 2 & 1 \\ 1 & -1 & 1 \\ 1 & 0 & 1 \end{vmatrix} = 2(-1) - 2(1-1) + 1(1) = -2+1 = -1 \neq 0.$$

Therefore, the given vectors are linearly independent.

5.15.3 Dimension and Basis

If in a vector space V there exists a set S of n linearly independent elements of V and if every set of $n + 1$ or more elements in V is linearly dependent, then V is said to have *dimension n*. Thus, the maximum number of linearly independent elements in V is the *dimension* of V and is denoted by dim (V). We write dim (V) $= n$. The vector space $\{0\}$ is defined to have dimension zero.

A linearly independent set in V consisting of a maximum possible number of elements in V is called a *basis* for V. Thus the number of elements of a basis for V equals dim (V).

Some of the standard basis are listed below:

1. If V consists of n-tuples in R^n, then
   $\mathbf{e}_1 = (1, 0, 0, ..., 0)$, $\mathbf{e}_2 = (0, 1, 0, ..., 0)$, ..., $\mathbf{e}_n = (0, 0, ..., 0, 1)$ is called a standard basis in R^n.
2. If V consists of all polynomials $P(t)$ of degree $\leq n$, then $\{1, t, t^2, ..., t^n\}$ is taken as its standard basis.
3. The real 2×2 matrices form a four-dimensional real vector space. The following four matrices form a basis:

$$\mathbf{E}_{11} = \begin{bmatrix} 1 & 0 \\ 0 & 0 \end{bmatrix}, \mathbf{E}_{12} = \begin{bmatrix} 0 & 1 \\ 0 & 0 \end{bmatrix}, \mathbf{E}_{21} = \begin{bmatrix} 0 & 0 \\ 1 & 0 \end{bmatrix}, \mathbf{E}_{22} = \begin{bmatrix} 0 & 0 \\ 0 & 1 \end{bmatrix}.$$

Any matrix $\begin{bmatrix} l & m \\ p & q \end{bmatrix}$ in V can be written as $\begin{bmatrix} l & m \\ p & q \end{bmatrix} = l\mathbf{E}_{11} + m\mathbf{E}_{12} + p\mathbf{E}_{21} + q\mathbf{E}_{22}$.

EXAMPLE 5.56 Determine whether or not each of the following set of vectors $\{\mathbf{u}, \mathbf{v}, \mathbf{w}\}$ forms a basis in $\mathbf{R}^3$, where

(*i*) $\mathbf{u} = (1, 1, 1)$, $\mathbf{v} = (1, 2, 3)$, $\mathbf{w} = (2, -1, 1)$,

(*ii*) $\mathbf{u} = (1, 2, 5)$, $\mathbf{v} = (2, 5, 1)$, $\mathbf{w} = (1, 5, 2)$.

Solution If the set {**u, v, w**} forms a basis in $\mathbf{R}^3$, then **u, v, w** must be linearly independent. Let k_1, k_2, k_3 be scalars. Then, the only solution of the equation

$$k_1\mathbf{u} + k_2\mathbf{v} + k_3\mathbf{w} = \mathbf{0} \quad ...(1) \quad \text{must be} \quad k_1 = k_2 = k_3 = 0.$$

(*i*) Using equation (1), we obtain the system of equations

$$k_1 + k_2 + 2k_3 = 0,\; k_1 + 2k_2 - k_3 = 0 \text{ and } k_1 + 3k_2 + k_3 = 0.$$

The solution of this system of equations is $k_1 = k_2 = k_3 = 0$. Therefore, **u, v, w** are linearly independent and they form a basis in $\mathbf{R}^3$.

(*ii*) Using equation (1), we obtain the system of equations

$$k_1 + 2k_2 + k_3 = 0,\; 2k_1 + 5k_2 + 5k_3 = 0 \text{ and } 5k_1 + k_2 + 2k_3 = 0.$$

The solution of this system of equations is $k_1 = k_2 = k_3 = 0$. Therefore, **u, v, w** are linearly independent and they form a basis in $\mathbf{R}^3$.

EXAMPLE 5.57 Find the value(s) of k for whcih the vectors $(1, k, 5)$, $(1 - 3, 2)$, $(2, -1, 1)$ will form a basis in R^3. (*AMIE, W-2010*)

Solution The three vectors form a basis if and only if they are linearly independent. Thus form the matrix whose rows are the given vectors, and row reduce the matrix to echelon form:

$$\begin{bmatrix} 1 & -3 & 2 \\ 2 & -1 & 1 \\ 1 & k & 5 \end{bmatrix} \sim \begin{bmatrix} 1 & -3 & 2 \\ 0 & 5 & -3 \\ 0 & k+3 & 3 \end{bmatrix} \sim \begin{bmatrix} 1 & -3 & 2 \\ 0 & 5 & -3 \\ 0 & k+8 & 0 \end{bmatrix}.$$

If the value of $k \neq -8$, then the echelon matrix has no zero rows. Hence the three vectors are linearly independent, and so they do form a basis of R^3.

5.15.4 Linear Transformations

First we introduce some notation and terminology concerning arbitrary functions. A function is a rule which assigns to each number some number.

Let A and B be two arbitrary non-empty sets. Suppose to each element in A there is assigned a unique element of B; the collection f of such assignments is called a *mapping* or a transformation from A into B. Thus, a transformation maps the elements of A into the elements of B. The set A is called the *domain* of the transformation, and B is called the *target set.* We denote transformation by capital letters T, S etc. If T is the transformation from A into B, we write

$$T : A \to B. \qquad ...(5.19)$$

For each element **a** in A, the element $T(\mathbf{a})$ in **b** is called the *image of* **a** under the mapping and we say that T maps **a** onto $T(\mathbf{a})$. The collection of all such images in B is called the *range* or the image set of the transformation T.

Now we assume that V and W be two vector spaces, both real or complex, over the same field F of scalars. Let T be a mapping from V into W. The mapping T is said to be a *Linear transformation* or a *linear mapping*, if it satisfies the following two properties:

(*i*) For any two elements $\mathbf{v}_1, \mathbf{v}_2$ in V

$$T(\mathbf{v}_1 + \mathbf{v}_2) = T(\mathbf{v}_1) + T(\mathbf{v}_2) \qquad ...(5.20)$$

(*ii*) For every scalar k and every element **v** in V

$$T(k\mathbf{v}) = kT(\mathbf{v}) \qquad ...(5.21)$$

The two properties can be combined into one formula which states that

$$T(k_1\mathbf{v}_1 + k_2\mathbf{v}_2) = T(k_1\mathbf{v}_1) + T(k_2\mathbf{v}_2) = k_1T(\mathbf{v}_1) + k_2T(\mathbf{v}_2)$$

for $\mathbf{v}_1$ and $\mathbf{v}_2$ in V and any scalars k_1 and k_2.

Matrix Mappings

Let **A** be an $m \times n$ real (or complex) matrix. Let the rows of **A** represent the elements in R^n (or $\mathbb{C}^n$) and the columns of **A** represent the elements in R^m (or $\mathbb{C}^m$). If **x** is in R^n, then **Ax** is in R^m.

Thus, an $m \times n$ matrix maps the element in R^n into the element in R^m. We write

$T = \mathbf{A} : \mathrm{R}^n \to \mathrm{R}^m$, and $\mathrm{T}\mathbf{x} = \mathbf{A}\mathbf{x}$. The mapping **A** is a linear transformation.

EXAMPLE 5.58 Let L: $R^2 \to R^2$ be a linear map, having the following effect on the indicated vectors: $L(1, 1) = (1, 4)$ and $L(2, -1) = (-2, 3)$. Compute $L(3, -1)$.

Solution We write $(3, -1)$ as a linear combination of $(1, 1)$ and $(2, -1)$. Thus, we have

$$(3, -1) = k_1 (1, 1) + k_2 (2, -1).$$

That is, $k_1 + 2k_2 = 3$, $k_1 - k_2 = -1$. The solution is $k_1 = 1/3$, $k_2 = 4/3$. Hence

$$L(3, -1) = k_1 L(1, 1) + k_2 L(2, -1) = \frac{1}{3}(1,4) + \frac{4}{3}(-2,3) = \left(-\frac{7}{3}, \frac{16}{3}\right).$$

EXAMPLE 5.59 Let T be a linear transformation from R^3 into R^2 defined by the relations

$$T\mathbf{X} = \mathbf{AX},\ \mathbf{A} = \begin{bmatrix} 3 & 4 & 5 \\ 6 & 7 & 8 \end{bmatrix}. \quad \text{Find } T\,\mathbf{X} \text{ when } \mathbf{X} \text{ is given by } [5\ 6\ 7]^T.$$

Solution We have

$$T\mathbf{X} = \mathbf{AX} = \begin{bmatrix} 3 & 4 & 5 \\ 6 & 7 & 8 \end{bmatrix}\begin{bmatrix} 5 \\ 6 \\ 7 \end{bmatrix} = \begin{bmatrix} 15 + 24 + 35 \\ 30 + 42 + 56 \end{bmatrix} = \begin{bmatrix} 74 \\ 128 \end{bmatrix}.$$

EXAMPLE 5.60 Let T be a linear transformation defined by

$$T\left[\begin{pmatrix} 1 & 1 \\ 1 & 1 \end{pmatrix}\right] = \begin{bmatrix} 1 \\ 2 \\ 3 \end{bmatrix}, T\left[\begin{pmatrix} 0 & 1 \\ 1 & 1 \end{pmatrix}\right] = \begin{bmatrix} 1 \\ -2 \\ 3 \end{bmatrix}, T\left[\begin{pmatrix} 0 & 0 \\ 1 & 1 \end{pmatrix}\right] = \begin{bmatrix} 1 \\ -2 \\ -3 \end{bmatrix}, T\left[\begin{pmatrix} 0 & 0 \\ 0 & 1 \end{pmatrix}\right] = \begin{bmatrix} -1 \\ 2 \\ 3 \end{bmatrix}. \text{ Find } T\left[\begin{pmatrix} 4 & 5 \\ 3 & 8 \end{pmatrix}\right].$$

(AMIETE, W, 2005)

Solution The matrices $\begin{bmatrix} 1 & 1 \\ 1 & 1 \end{bmatrix}, \begin{bmatrix} 0 & 1 \\ 1 & 1 \end{bmatrix}, \begin{bmatrix} 0 & 0 \\ 1 & 1 \end{bmatrix}, \begin{bmatrix} 0 & 0 \\ 0 & 1 \end{bmatrix}$ are linearly independent and hence form a basis in the space of 2×2 matrices. We write for any scalars k_1, k_2, k_3, k_4, not all zero

$$\begin{bmatrix} 4 & 5 \\ 3 & 8 \end{bmatrix} = k_1\begin{bmatrix} 1 & 1 \\ 1 & 1 \end{bmatrix} + k_2\begin{bmatrix} 0 & 1 \\ 1 & 1 \end{bmatrix} + k_3\begin{bmatrix} 0 & 0 \\ 1 & 1 \end{bmatrix} + k_4\begin{bmatrix} 0 & 0 \\ 0 & 1 \end{bmatrix} = \begin{bmatrix} k_1 & k_1 + k_2 \\ k_1 + k_2 + k_3 & k_1 + k_2 + k_3 + k_4 \end{bmatrix}.$$

Comparing the elements and solving the resulting system of equations, we obtain $k_1 = 4$, $k_2 = 1$, $k_3 = -2$, $k_4 = 5$. Since T is a linear transformation, we get

$$T\left[\begin{pmatrix} 4 & 5 \\ 3 & 8 \end{pmatrix}\right] = k_1 T\left[\begin{pmatrix} 1 & 1 \\ 1 & 1 \end{pmatrix}\right] + k_2 T\left[\begin{pmatrix} 0 & 1 \\ 1 & 1 \end{pmatrix}\right] + k_3 T\left[\begin{pmatrix} 0 & 0 \\ 1 & 1 \end{pmatrix}\right] + k_4 T\left[\begin{pmatrix} 0 & 0 \\ 1 & 1 \end{pmatrix}\right]$$

$$= 4\begin{bmatrix} 1 \\ 2 \\ 3 \end{bmatrix} + 1\begin{bmatrix} 1 \\ -2 \\ 3 \end{bmatrix} + (-2)\begin{bmatrix} 1 \\ -2 \\ -3 \end{bmatrix} + 5\begin{bmatrix} -1 \\ 2 \\ 3 \end{bmatrix} = \begin{bmatrix} 4 + 1 - 2 - 5 \\ 8 - 2 + 4 + 10 \\ 12 + 3 + 6 + 15 \end{bmatrix} = \begin{bmatrix} -2 \\ 20 \\ 36 \end{bmatrix}.$$

EXAMPLE 5.61 The set of vectors $\{\mathbf{x}_1, \mathbf{x}_2\}$, where $\mathbf{x}_1 = (1, 3)^T$, $\mathbf{x}_2 = (4, 6)^T$ is a basis in R^2. Find a linear transformation $T : R^2 \to R^3$, such that

$$T\mathbf{x}_1 = (-2, 2, -7)^T \text{ and } T\mathbf{x}_2 = (-2, -4, -10)^T.$$ *(AMIETE, W, 2007)*

Solution The transformation T maps column vectors in R^2 into column vectors in R^3. Therefore T must be a matrix **A** of order 3×2. Let $\mathbf{A} = \begin{bmatrix} a_1 & b_1 \\ a_2 & b_2 \\ a_3 & b_3 \end{bmatrix}$.

Therefore, we have $\underset{3\times 2}{\begin{bmatrix} a_1 & b_1 \\ a_2 & b_2 \\ a_3 & b_3 \end{bmatrix}} \underset{2\times 1}{\begin{bmatrix} 1 \\ 3 \end{bmatrix}} = \underset{3\times 1}{\begin{bmatrix} -2 \\ 2 \\ -7 \end{bmatrix}}$ and $\begin{bmatrix} a_1 & b_1 \\ a_2 & b_2 \\ a_3 & b_3 \end{bmatrix} \begin{bmatrix} 4 \\ 6 \end{bmatrix} = \begin{bmatrix} -2 \\ -4 \\ -10 \end{bmatrix}$.

Multiplying and comparing the corresponding elements, we get

$$a_1 + 3b_1 = -2, \qquad 4a_1 + 6b_1 = -2,$$
$$a_2 + 3b_2 = 2, \qquad 4a_2 + 6b_2 = -4,$$
$$a_3 + 3b_3 = -7, \qquad 4a_3 + 6b_3 = -10.$$

Solving these equations, we obtain $\mathbf{A} = \begin{bmatrix} 1 & -1 \\ -4 & 2 \\ 2 & -3 \end{bmatrix}$.

EXAMPLE 5.62 Find a linear transformation T from R^3 into R^3 such that

$$T\begin{bmatrix} 1 \\ 1 \\ 1 \end{bmatrix} = \begin{bmatrix} 6 \\ 2 \\ 4 \end{bmatrix}, \; T\begin{bmatrix} 1 \\ -1 \\ 1 \end{bmatrix} = \begin{bmatrix} 2 \\ -4 \\ 2 \end{bmatrix}, \; T\begin{bmatrix} 1 \\ -2 \\ 3 \end{bmatrix} = \begin{bmatrix} 6 \\ 6 \\ 5 \end{bmatrix}.$$

(AMIETE, June 2007)

Solution The transformation T maps elements in R^3 into R^3. Therefore, the transformation is a matrix of order 3×3. Let this matrix be written as

$T = \mathbf{A} = \begin{bmatrix} a_1 & b_1 & c_1 \\ a_2 & b_2 & c_2 \\ a_3 & b_3 & c_3 \end{bmatrix}$. We determine the elements of the matrix **A** such as

$$\begin{bmatrix} a_1 & b_1 & c_1 \\ a_2 & b_2 & c_2 \\ a_3 & b_3 & c_3 \end{bmatrix}\begin{bmatrix} 1 \\ 1 \\ 1 \end{bmatrix} = \begin{bmatrix} 6 \\ 2 \\ 4 \end{bmatrix}, \begin{bmatrix} a_1 & b_1 & c_1 \\ a_2 & b_2 & c_2 \\ a_3 & b_3 & c_3 \end{bmatrix}\begin{bmatrix} 1 \\ -1 \\ 1 \end{bmatrix} = \begin{bmatrix} 2 \\ -4 \\ 2 \end{bmatrix}, \begin{bmatrix} a_1 & b_1 & c_1 \\ a_2 & b_2 & c_2 \\ a_3 & b_3 & c_3 \end{bmatrix}\begin{bmatrix} 1 \\ -2 \\ 3 \end{bmatrix} = \begin{bmatrix} 6 \\ 6 \\ 5 \end{bmatrix}.$$

Multiplying and comparing the corresponding elements, we obtain

$$a_1 + b_1 + c_1 = 6 \qquad a_1 - b_1 + c_1 = 2 \qquad a_1 - 2b_1 + 3c_1 = 6$$
$$a_2 + b_2 + c_2 = 2 \qquad a_2 - b_2 + c_2 = -4 \qquad a_2 - 2b_2 + 3c_2 = 6$$
$$a_3 + b_3 + c_3 = 4 \qquad a_3 - b_3 + c_3 = 2 \qquad a_3 - 2b_3 + 3c_3 = 5$$

Solving these equations, we obtain $\mathbf{A} = \begin{bmatrix} 1 & 2 & 3 \\ -15/2 & 3 & 13/2 \\ 1 & 1 & 2 \end{bmatrix}$.

Matrix Representation of a Linear Transformation

Let T be a linear transformation from a n dimensional vector space V into another m-dimensional vector space W over the same field F, such that $T: V \to W$. Let

$$\mathbf{X} = \{\mathbf{v}_1, \mathbf{v}_2, ..., \mathbf{v}_n\}, \mathbf{Y} = \{\mathbf{w}_1, \mathbf{w}_2, ..., \mathbf{w}_m\}$$

be the ordered basis of V and W respectively. Since T has values in W, each element $T(\mathbf{v}_i)$ can be expressed uniquely as a linear combination of the basis elements $\mathbf{w}_1, \mathbf{w}_2, ..., \mathbf{w}_m$ in W. That is, there exist scalars a_{ij}, $i = 1, 2 ..., n, j = 1, 2, ..., m$ not all zero, such that

$$T(\mathbf{v}_i) = a_{1i}\mathbf{w}_1 + a_{2i}\,\mathbf{w}_2 + ... + a_{mi}\,\mathbf{w}_m = [\mathbf{w}_1, \mathbf{w}_2, ..., \mathbf{w}_m]\;[a_{1i}, a_{2i}, ...a_{mi}] \quad i = 1, 2, ..., n$$

where $a_{1i}, a_{2i}, ..., a_{mi}$ are the components of $T(\mathbf{v}_i)$ relative of the ordered basis $(\mathbf{w}_1, \mathbf{w}_2, ..., \mathbf{w}_m)$.

Hence, we can write

$$T[\mathbf{v}_1, \mathbf{v}_2, \cdots, \mathbf{v}_n] = [\mathbf{w}_1, \mathbf{w}_2, \cdots, \mathbf{w}_m]\begin{bmatrix} a_{11} & a_{12} & \cdots & a_{1n} \\ a_{21} & a_{22} & \cdots & a_{2n} \\ \vdots & & & \vdots \\ a_{m1} & a_{m2} & \cdots & a_{mn} \end{bmatrix} \qquad ...(5.22)$$

or $\qquad T\mathbf{X} = \mathbf{YA}$ where $\mathbf{A}$ is the m by n matrix.

Thus, every linear transformation T of an n-dimensional space V into an m-dimensional space W give rise to an $m \times n$ matrix, $\mathbf{A} = (a_{ij})$ whose columns consist of the components of $T(\mathbf{v}_1)$, $T(\mathbf{v}_2)$, ..., $T(\mathbf{v}_n)$ relative to the basis $(\mathbf{w}_1, \mathbf{w}_2, ..., \mathbf{w}_m)$. We call this the matrix representation of T relative to the given choice of ordered basis $(\mathbf{v}_1, \mathbf{v}_2, ..., \mathbf{v}_n)$ for V and $(\mathbf{w}_1, \mathbf{w}_2, ..., \mathbf{w}_m)$ for W.

EXAMPLE 5.63 Let $T : \mathrm{R}^3 \to \mathrm{R}^2$ be a linear transformation defined by $T\begin{bmatrix} x \\ y \\ z \end{bmatrix} = \begin{pmatrix} x+y \\ x-z \end{pmatrix}$.

Find the matrix representation of T with respect to the ordered basis.

(*i*) $\mathbf{X} = \left\{ \begin{pmatrix} 1 \\ 0 \\ 1 \end{pmatrix}, \begin{pmatrix} 1 \\ 1 \\ 0 \end{pmatrix}, \begin{pmatrix} 0 \\ 1 \\ 1 \end{pmatrix} \right\}$ in R^3 and $\mathbf{Y} = \left\{ \begin{pmatrix} 1 \\ 0 \end{pmatrix}, \begin{pmatrix} 0 \\ 1 \end{pmatrix} \right\}$ in R^2.

(*ii*) $\mathbf{X} = \left\{ \begin{pmatrix} 1 \\ 0 \\ 1 \end{pmatrix}, \begin{pmatrix} 1 \\ 1 \\ 0 \end{pmatrix}, \begin{pmatrix} 0 \\ 1 \\ 1 \end{pmatrix} \right\}$ in R^3 and $\mathbf{Y} = \left\{ \begin{pmatrix} 1 \\ 3 \end{pmatrix}, \begin{pmatrix} 2 \\ 5 \end{pmatrix} \right\}$ in R^2.

Solution Let $V = \mathrm{R}^3$, $W = \mathrm{R}^2$. Let $\mathbf{X} = \{\mathbf{v}_1, \mathbf{v}_2, \mathbf{v}_3\}$, $\mathbf{Y} = \{\mathbf{w}_1, \mathbf{w}_2\}$.

(*i*) We have $\mathbf{v}_1 = \begin{bmatrix} 1 \\ 0 \\ 1 \end{bmatrix}$, $\mathbf{v}_2 = \begin{bmatrix} 1 \\ 1 \\ 0 \end{bmatrix}$, $\mathbf{v}_3 = \begin{bmatrix} 0 \\ 1 \\ 1 \end{bmatrix}$, $\mathbf{w}_1 = \begin{bmatrix} 1 \\ 0 \end{bmatrix}$, $\mathbf{w}_2 = \begin{bmatrix} 0 \\ 1 \end{bmatrix}$.

We obtain $T\begin{bmatrix} 1 \\ 0 \\ 1 \end{bmatrix} = \begin{bmatrix} 1 \\ 0 \end{bmatrix} = \begin{bmatrix} 1 \\ 0 \end{bmatrix}(1) + \begin{bmatrix} 0 \\ 1 \end{bmatrix}(0)$, $T\begin{bmatrix} 1 \\ 1 \\ 0 \end{bmatrix} = \begin{bmatrix} 2 \\ 1 \end{bmatrix} = \begin{bmatrix} 1 \\ 0 \end{bmatrix}(2) + \begin{bmatrix} 0 \\ 1 \end{bmatrix}(1)$,

$$T\begin{bmatrix} 0 \\ 1 \\ 1 \end{bmatrix} = \begin{bmatrix} 1 \\ -1 \end{bmatrix} = \begin{bmatrix} 1 \\ 0 \end{bmatrix}(1) + \begin{bmatrix} 0 \\ 1 \end{bmatrix}(-1).$$

Using the notation $T\mathbf{X} = \mathbf{YA}$, we write

$$T[\mathbf{v}_1, \mathbf{v}_2, \mathbf{v}_3] = [\mathbf{w}_1, \mathbf{w}_2]\begin{bmatrix} 1 & 2 & 1 \\ 0 & 1 & -1 \end{bmatrix}$$

or

$$T\left[\begin{pmatrix} 1 \\ 0 \\ 1 \end{pmatrix}, \begin{pmatrix} 1 \\ 1 \\ 0 \end{pmatrix}, \begin{pmatrix} 0 \\ 1 \\ 1 \end{pmatrix} \right] = \left[\begin{pmatrix} 1 \\ 0 \end{pmatrix}, \begin{pmatrix} 0 \\ 1 \end{pmatrix} \right]\begin{bmatrix} 1 & 2 & 1 \\ 0 & 1 & -1 \end{bmatrix}.$$

Therefore, the matrix of the linear transformation T with respect to the given basis vectors is given by

$$\mathbf{A} = \begin{bmatrix} 1 & 2 & 1 \\ 0 & 1 & -1 \end{bmatrix}.$$

(*ii*) We have $\mathbf{v}_1 = \begin{bmatrix} 1 \\ 0 \\ 1 \end{bmatrix}, \mathbf{v}_2 = \begin{bmatrix} 1 \\ 1 \\ 0 \end{bmatrix}, \mathbf{v}_3 = \begin{bmatrix} 0 \\ 1 \\ 1 \end{bmatrix}, \mathbf{w}_1 = \begin{bmatrix} 1 \\ 3 \end{bmatrix}, \mathbf{w}_2 = \begin{bmatrix} 2 \\ 5 \end{bmatrix}.$

We obtain $T\begin{bmatrix} 1 \\ 0 \\ 1 \end{bmatrix} = \begin{bmatrix} 1 \\ 0 \end{bmatrix} = \begin{bmatrix} 1 \\ 3 \end{bmatrix}(-5) + \begin{bmatrix} 2 \\ 5 \end{bmatrix}(3),$

$$T\begin{bmatrix} 1 \\ 1 \\ 0 \end{bmatrix} = \begin{bmatrix} 2 \\ 1 \end{bmatrix} = \begin{bmatrix} 1 \\ 3 \end{bmatrix}(-8) + \begin{bmatrix} 2 \\ 5 \end{bmatrix}(5), \; T\begin{bmatrix} 0 \\ 1 \\ 1 \end{bmatrix} = \begin{bmatrix} 1 \\ -1 \end{bmatrix} = \begin{bmatrix} 1 \\ 3 \end{bmatrix}(-7) + \begin{bmatrix} 2 \\ 5 \end{bmatrix}(4).$$

Using $T\mathbf{X} = \mathbf{YA}$, we write

$$T\left[\begin{pmatrix} 1 \\ 0 \\ 1 \end{pmatrix}, \begin{pmatrix} 1 \\ 1 \\ 0 \end{pmatrix}, \begin{pmatrix} 0 \\ 1 \\ 1 \end{pmatrix}\right] = \left[\begin{pmatrix} 1 \\ 3 \end{pmatrix}, \begin{pmatrix} 2 \\ 5 \end{pmatrix}\right]\begin{bmatrix} -5 & -8 & -7 \\ 3 & 5 & 4 \end{bmatrix}.$$

Therefore, the matrix of the linear transformation T with respect to the given basis vectors is given by

$$\mathbf{A} = \begin{bmatrix} -5 & -8 & -7 \\ 3 & 5 & 4 \end{bmatrix}.$$

EXAMPLE 5.64 Let V and W be two vector spaces in R^3. Let $T : V \to W$ be a linear transformation defined by

$$T\begin{bmatrix} x \\ y \\ z \end{bmatrix} = \begin{bmatrix} x+z \\ x+y \\ x+y+z \end{bmatrix}.$$

Find the matrix representation of T with respect to the ordered basis

$$\mathbf{X} = \left\{\begin{bmatrix} -1 \\ 1 \\ 1 \end{bmatrix}, \begin{bmatrix} 1 \\ -1 \\ 1 \end{bmatrix}, \begin{bmatrix} 1 \\ 1 \\ -1 \end{bmatrix}\right\} \text{ in } V \text{ and } \mathbf{Y} = \left\{\begin{bmatrix} 1 \\ -1 \\ -1 \end{bmatrix}, \begin{bmatrix} -1 \\ 1 \\ -1 \end{bmatrix}, \begin{bmatrix} -1 \\ -1 \\ 1 \end{bmatrix}\right\} \text{ in } W.$$

Solution Let $\mathbf{X} = \{\mathbf{v}_1, \mathbf{v}_2, \mathbf{v}_3\}$, $\mathbf{Y} = \{\mathbf{w}_1, \mathbf{w}_2, \mathbf{w}_3\}$.

We have $\mathbf{v}_1 = \begin{bmatrix} -1 \\ 1 \\ 1 \end{bmatrix}, \mathbf{v}_2 = \begin{bmatrix} 1 \\ -1 \\ 1 \end{bmatrix}, \mathbf{v}_3 = \begin{bmatrix} 1 \\ 1 \\ -1 \end{bmatrix}, \mathbf{w}_1 = \begin{bmatrix} 1 \\ -1 \\ -1 \end{bmatrix}, \mathbf{w}_2 = \begin{bmatrix} -1 \\ 1 \\ -1 \end{bmatrix}, \mathbf{w}_3 = \begin{bmatrix} -1 \\ -1 \\ 1 \end{bmatrix}.$

We obtain $T\begin{bmatrix} -1 \\ 1 \\ 1 \end{bmatrix} = \begin{bmatrix} 0 \\ 0 \\ 1 \end{bmatrix} = \begin{bmatrix} 1 \\ -1 \\ -1 \end{bmatrix}\left(-\frac{1}{2}\right) + \begin{bmatrix} -1 \\ 1 \\ -1 \end{bmatrix}\left(-\frac{1}{2}\right) + \begin{bmatrix} -1 \\ -1 \\ 1 \end{bmatrix}(0),$

$$T\begin{bmatrix} 1 \\ -1 \\ 1 \end{bmatrix} = \begin{bmatrix} 2 \\ 0 \\ 1 \end{bmatrix} = \begin{bmatrix} 1 \\ -1 \\ -1 \end{bmatrix}\left(-\frac{1}{2}\right) + \begin{bmatrix} -1 \\ 1 \\ -1 \end{bmatrix}\left(-\frac{3}{2}\right) + \begin{bmatrix} -1 \\ -1 \\ 1 \end{bmatrix}(-1),$$

$$T\begin{bmatrix} 1 \\ 1 \\ -1 \end{bmatrix} = \begin{bmatrix} 0 \\ 2 \\ 1 \end{bmatrix} = \begin{bmatrix} 1 \\ -1 \\ -1 \end{bmatrix}\left(-\frac{3}{2}\right) + \begin{bmatrix} -1 \\ 1 \\ -1 \end{bmatrix}\left(-\frac{1}{2}\right) + \begin{bmatrix} -1 \\ -1 \\ 1 \end{bmatrix}(-1).$$

Using $TX = YA$, we write

$$T\left[\begin{bmatrix}-1\\1\\1\end{bmatrix}\begin{bmatrix}1\\-1\\1\end{bmatrix}\begin{bmatrix}1\\1\\-1\end{bmatrix}\right]=\left[\begin{bmatrix}1\\-1\\-1\end{bmatrix}\begin{bmatrix}-1\\1\\-1\end{bmatrix}\begin{bmatrix}-1\\-1\\1\end{bmatrix}\right]\begin{bmatrix}-1/2 & -1/2 & -3/2\\-1/2 & -3/2 & -1/2\\0 & -1 & -1\end{bmatrix}.$$

Therefore, the matrix of the linear transformation T with respect to the given basis vectors is given by

$$\mathbf{A}=\begin{bmatrix}-1/2 & -1/2 & -3/2\\-1/2 & -3/2 & -1/2\\0 & -1 & -1\end{bmatrix}.$$

EXAMPLE 5.65 Let $T: R^3 \to R^2$ be a linear transformation. Let $\mathbf{A}=\begin{bmatrix}1 & 2 & 1\\2 & -3 & -4\end{bmatrix}$ be the matrix representation of the linear transformation with respect to the ordered basis vectors $\mathbf{v}_1 = [1, -1, 1]^T$, $\mathbf{v}_2 = [2, 3, -1]^T$, $\mathbf{v}_3 = [1, 1, 1]^T$ in R^3 and $\mathbf{w}_1 = [1, 1]^T$, $\mathbf{w}_2 = [2, 3]^T$ in R^2. Then determine the linear transformation T.

Solution We have $T[\mathbf{v}_1, \mathbf{v}_2, \mathbf{v}_3]=[\mathbf{w}_1, \mathbf{w}_2]\,\mathbf{A}=\begin{bmatrix}1 & 2\\1 & 3\end{bmatrix}\begin{bmatrix}1 & 2 & 1\\2 & -3 & -4\end{bmatrix}=\begin{bmatrix}5 & -4 & -7\\7 & -7 & -11\end{bmatrix}.$

Now, any vector $\mathbf{x} = (x_1, x_2, x_3)^T$ in R^3 with respect to the given basis can be written as

$$\begin{bmatrix}x_1\\x_2\\x_3\end{bmatrix}=k_1\begin{bmatrix}1\\-1\\1\end{bmatrix}+k_2\begin{bmatrix}2\\3\\-1\end{bmatrix}+k_3\begin{bmatrix}1\\1\\-1\end{bmatrix}.$$

We obtain $k_1=\dfrac{2x_1-x_2+x_3}{4},\ k_2=\dfrac{x_2+x_3}{2},\ k_3=\dfrac{2x_1-3x_2-5x_3}{4}$. Hence, we have

$$Tx = k_1Tv_1 + k_2Tv_2 + k_3Tv_3$$

or

$$T\mathbf{x}=k_1\begin{bmatrix}5\\7\end{bmatrix}+k_2\begin{bmatrix}-4\\-7\end{bmatrix}+k_3\begin{bmatrix}-7\\-11\end{bmatrix}=\begin{bmatrix}5k_1-4k_2-7k_3\\7k_1-7k_2-11k_3\end{bmatrix}=\begin{bmatrix}-x_1+2x_2+8x_3\\-2x_1+3x_2+12x_3\end{bmatrix}.$$

EXERCISE 5.7

1. Let V be the set of all positive real numbers where addition is defined by $\mathbf{x} + \mathbf{y} = xy$ and scalar multiplication is defined by $k\mathbf{x} = x$. Determine whether V is a vector space. **Ans.** yes.
2. The vectors $\mathbf{u}_1 = (1, 0, 0)$, $\mathbf{u}_2 = (1, 1, 0)$ and $\mathbf{u}_3 = (1, 1, 1)$ form a basis for the vector space R^3. Express the vector $\mathbf{x} = (3, -4, 8)$ as a linear combination of $\mathbf{u}_1$, $\mathbf{u}_2$ and $\mathbf{u}_3$. **Ans.** $\mathbf{x} = 7\mathbf{u}_1 - 12\mathbf{u}_2 + 8\mathbf{u}_3$.
3. Express **M** as a linear combination of the matrices **A, B, C** where
$$\mathbf{M}=\begin{bmatrix}4 & 7\\7 & 9\end{bmatrix},\ \text{and } \mathbf{A}=\begin{bmatrix}1 & 1\\1 & 1\end{bmatrix},\ \mathbf{B}=\begin{bmatrix}1 & 2\\3 & 4\end{bmatrix},\ \mathbf{C}=\begin{bmatrix}1 & 1\\4 & 5\end{bmatrix}.$$
Ans. $\mathbf{M} = 2\mathbf{A} + 3\mathbf{B} - \mathbf{C}$.
4. Let $\mathbf{u} = (1, 2, -1)$, $\mathbf{v} = (2, 3, 4)$ and $\mathbf{w} = (1, 5, -3)$. Determine whether or not $\mathbf{x}$ is a linear combinations of **u, v, w** where $\mathbf{x} = (3, 2, 5)$ **Ans.** $2\mathbf{u} + \mathbf{v} - \mathbf{w}$
5. Investigate the linear dependence or independence of the following set of vectors:
   (*i*) $\mathbf{v}_1 = (1, 2, 1)$, $\mathbf{v}_2 = (4, 1, 2)$, $\mathbf{v}_3 = (6, 5, 4)$ and $\mathbf{v}_4 = (-3, 8, 1)$.
   (*ii*) $\mathbf{v}_1 = (2, -1, 4)$, $\mathbf{v}_2 = (0, 1, 2)$, $\mathbf{v}_3 = (6, -1, 16)$, $v_4 = (4, 0, 12)$.
   (*iii*) $\mathbf{v}_1 = (1, 2, -1, 0)$, $\mathbf{v}_2 = (1, 3, 1, 2)$, $\mathbf{v}_3 = (6, 1, 0, 1)$ and $\mathbf{v}_4 = (4, 1, 2, 0)$.
   (*iv*) $\mathbf{v}_1 = (1, 2, 2)$, $\mathbf{v}_2 = (2, 1, -2)$ and $\mathbf{v}_3 = (2, -2, 1)$.

Ans. (*i*) Dependent, (*ii*) Dependent, (*iii*) Indepe ndent, (*iv*) Independent.

6. Are the following vectors linearly dependent? If so, find the relation between them.
   (*i*) $\mathbf{v}_1 = (1, 2, 1)$, $\mathbf{v}_2 = (2, 1, 4)$, $\mathbf{v}_3 = (4, 5, 6)$, and $\mathbf{v}_4 = (1, 8, -3)$.
   (*ii*) $\mathbf{v}_1 = (3, 1, -4)$, $\mathbf{v}_2 = (2, 2, -3)$, $\mathbf{v}_3 = (0, -4, 1)$ and $\mathbf{v}_4 = (-4, -4, 6)$.

Ans. (*i*) yes, $\mathbf{v}_3 = 2\mathbf{v}_1 + \mathbf{v}_2$ and $\mathbf{v}_4 = 5\mathbf{v}_1 - 2\mathbf{v}_2$; (*ii*) yes, $2\mathbf{v}_1 - \mathbf{v}_2 - \mathbf{v}_3 + \mathbf{v}_4 = 0$.

7. Let $\mathbf{v}_1 = (1, -1, 0)$, $\mathbf{v}_2 = (0, 1, -1)$ and $\mathbf{v}_3 = (0, 0, 1)$ be elements of R^3. Show that the set of vectors $\{\mathbf{v}_1, \mathbf{v}_2, \mathbf{v}_3\}$ is linearly independent. *(AMIETE, Dec. 2005)*

8. Verify whether the vector $\mathbf{x}_1 = (1\ 3\ 4\ 2)$, $\mathbf{x}_2 = (3\ -5\ 2\ 2)$ and $\mathbf{x}_3 = (2\ -1\ 3\ 2)$ are linearly dependent. If so, express one of these vectors as a linear combination of others. *(AMIE, W-2010)*

Ans. $\mathbf{x}_1 = 2\,\mathbf{x}_3 - \mathbf{x}_2$.

9. Examine whether the transformation T given is linear or not. If not linear, state why?

$$T: R^2 \to R^2;\ T\begin{bmatrix} x \\ y \end{bmatrix} = \begin{bmatrix} x+y \\ x \end{bmatrix}.$$

Ans. Linear.

10. Let T be a transformation from R^3 into R^2 defined by $T(x_1, x_2, x_3) = x_1^2 + x_2^2 + x_3^2$. Show that T is not a linear transformation. *(PTU, 2005)*

[**Hint:** Let $\mathbf{x} = (x_1, x_2, x_3)$ and $\mathbf{y} = (y_1, y_2, y_3)$ be any two elements in R^3.

Then $\mathbf{x} + \mathbf{y} = (x_1 + y_1, x_2 + y_2, x_3 + y_3)$.

$$T(\mathbf{x}) = x_1^2 + x_2^2 + x_3^2,\ T(\mathbf{y}) = y_1^2 + y_2^2 + y_3^2$$

$$T(\mathbf{x}+\mathbf{y}) = (x_1 + y_1)^2 + (x_2 + y_2)^2 + (x_3 + y_3)^2 \neq T(\mathbf{x}) + T(\mathbf{y}).$$

Therefore, T is not a linear transformation.]

11. Let T be a linear transformation from R^3 into R^2 defined by the relations $T\mathbf{x} = \mathbf{A}\mathbf{x}$, $\mathbf{A} = \begin{bmatrix} 1 & 2 & 3 \\ 4 & 5 & 6 \end{bmatrix}$.

Find $T\mathbf{x}$, when $\mathbf{x}$ is given by $[3\ \ 4\ \ 5]^T$. *(AMIETE, Dec 2009)* **Ans.** $\begin{bmatrix} 26 \\ 62 \end{bmatrix}$.

12. Let $T: R^3 \to R^2$ be a linear transformation defined by $T\begin{bmatrix} x \\ y \\ z \end{bmatrix} = \begin{pmatrix} y+z \\ y-z \end{pmatrix}$.

Taking $\left\{\begin{bmatrix} 1 \\ 1 \\ 0 \end{bmatrix}, \begin{bmatrix} 0 \\ 1 \\ 1 \end{bmatrix}, \begin{bmatrix} 1 \\ 0 \\ 1 \end{bmatrix}\right\}$ as a basis in R^3 and $\left\{\begin{pmatrix} 1 \\ 0 \end{pmatrix}, \begin{pmatrix} 0 \\ 1 \end{pmatrix}\right\}$ as a basis in R^2, determine the matrix of linear transformation. *(AMIETE, Dec. 2006)* **Ans.** $\begin{bmatrix} 1 & 2 & 1 \\ 1 & 0 & -1 \end{bmatrix}$.

13. Let $T: R^3 \to R^2$ be a linear transformation defined by $T\begin{bmatrix} x \\ y \\ z \end{bmatrix} = \begin{bmatrix} 2x+3y-z \\ 4x-y+2z \end{bmatrix}$.

Find the matrix representation of T with respect to the ordered basis

$$\mathbf{X} = \left\{\begin{bmatrix} 1 \\ 1 \\ 0 \end{bmatrix}, \begin{bmatrix} 1 \\ 2 \\ 3 \end{bmatrix}, \begin{bmatrix} 1 \\ 3 \\ 5 \end{bmatrix}\right\} \text{ in } R^3 \text{ and } \mathbf{Y} = \left\{\begin{bmatrix} 1 \\ 2 \end{bmatrix}, \begin{bmatrix} 2 \\ 3 \end{bmatrix}\right\} \text{ in } R^2.$$

Ans. $\begin{bmatrix} -9 & 1 & 4 \\ 7 & 2 & 1 \end{bmatrix}$.

5.16 ELEMENTARY TRANSFORMATION (OR OPERATION) OF A MATRIX

Any one of the following three operations applied on the rows (or columns) of a matrix is called an *elementary transformation* (or *operation*).

(*i*) The interchange of any two rows (or columns).

(*ii*) The multiplication of each element of any row (or column) by a non-zero constant.

(*iii*) The addition to the elements of any row (or column), the constant times the corresponding elements of any other row (or column).

Notation: *The following symbols are used for elementary transformations:*

(*a*) $R_i \leftrightarrow R_j$ *for the interchange of* i^{th} *row with* j^{th} *row.*
$C_i \leftrightarrow C_j$ *for the interchange of ith column with* j^{th} *column.*

(*b*) $R_i \to kR_i$ *for the multiplication of* i^{th} *row by non-zero constant k.*
$C_i \to kC_i$ *for the multiplication of* i^{th} *column by non-zero constant k.*

(*c*) $R_i \to R_i + kR_j$ *for addition to* i^{th} *row, k times the elements of the* j^{th} *row.*
$C_i \to C_i + kC_j$ *for addition to* i^{th} *column, k times the elements of the* j^{th} *column.*

Two matrices are said to be equivalent if one is obtained from the other by elementary transformations. The symbol ~ is used the **equivalence**.

Any elementary row transformation of a matrix **A** can be accomplished by **premultiplying A** by a unit matrix on whose rows the same elementary transformation has been performed. Similarly, any elementary column transformation of a matrix **A** can be accomplished by **post multiplying A** by a unit matrix on whose columns the same elementary transformation has been performed.

5.16.1 Inverse of a matrix by elementary operations

If **A** is a matrix such that $\mathbf{A}^{-1}$ exists, then to find $\mathbf{A}^{-1}$ using elementary row operations, write **A** = **IA** till we get **I** = **BA**. The matrix **B** will be the inverse of **A**. Similarly, if we wish to find $\mathbf{A}^{-1}$ using column operations, then write **A** = **AI** and apply a sequence of column operations on **A** = **AI** till we get, **I** = **AB**.

Remark In case, after applying one or more elementary row (column) operations on **A** = **IA** (**A** = **AI**), if we obtain all zeros in one or more rows of the matrix **A** on L.H.S., then $\mathbf{A}^{-1}$ does not exist.

NOTE A matrix obtained from a unit matrix by subjecting it to any of the elementary transformations, is called an **elementary matrix.**

EXAMPLE 5.66 Obtain the inverse of the matrix $\mathbf{A} = \begin{bmatrix} 3 & -3 & 4 \\ 2 & -3 & 4 \\ 0 & -1 & 1 \end{bmatrix}$ using elementary row transformations.

Solution Write **A** = **IA**, that is

$$\begin{bmatrix} 3 & -3 & 4 \\ 2 & -3 & 4 \\ 0 & -1 & 1 \end{bmatrix} = \begin{bmatrix} 1 & 0 & 0 \\ 0 & 1 & 0 \\ 0 & 0 & 1 \end{bmatrix} \mathbf{A}$$

Applying $R_1 \to R_1/3$, we obtain

$$\begin{bmatrix} 1 & -1 & 4/3 \\ 2 & -3 & 4 \\ 0 & -1 & 1 \end{bmatrix} = \begin{bmatrix} 1/3 & 0 & 0 \\ 0 & 1 & 0 \\ 0 & 0 & 1 \end{bmatrix} \mathbf{A} \quad \begin{pmatrix}\text{Operate} \\ R_2 \to R_2 - 2R_1\end{pmatrix} \quad \text{or} \quad \begin{bmatrix} 1 & -1 & 4/3 \\ 0 & -1 & 4/3 \\ 0 & -1 & 1 \end{bmatrix} = \begin{bmatrix} 1/3 & 0 & 0 \\ -2/3 & 1 & 0 \\ 0 & 0 & 1 \end{bmatrix} \mathbf{A} \quad \begin{pmatrix}\text{Operate} \\ R_2 \to -R_2\end{pmatrix}$$

$$\text{or} \begin{bmatrix} 1 & -1 & 4/3 \\ 0 & 1 & -4/3 \\ 0 & -1 & 1 \end{bmatrix} = \begin{bmatrix} 1/3 & 0 & 0 \\ 2/3 & -1 & 0 \\ 0 & 0 & 1 \end{bmatrix} \mathbf{A} \quad \begin{pmatrix}\text{Performing} \\ R_1 \to R_1 + R_2, \\ \text{and } R_3 \to R_3 + R_2\end{pmatrix} \quad \text{or} \begin{bmatrix} 1 & 0 & 0 \\ 0 & 1 & -4/3 \\ 0 & 0 & -1/3 \end{bmatrix} = \begin{bmatrix} 1 & -1 & 0 \\ 2/3 & -1 & 0 \\ 2/3 & -1 & 1 \end{bmatrix} \mathbf{A} \quad \begin{pmatrix}\text{Operate} \\ R_3 \to -3R_3\end{pmatrix}$$

$$\text{or} \quad \begin{bmatrix} 1 & 0 & 0 \\ 0 & 1 & -4/3 \\ 0 & 0 & 1 \end{bmatrix} = \begin{bmatrix} 1 & -1 & 0 \\ 2/3 & -1 & 0 \\ -2 & 3 & -3 \end{bmatrix} \mathbf{A} \qquad [\text{Operate } R_2 \to R_2 + (4/3)R_3]$$

or $$\begin{bmatrix} 1 & 0 & 0 \\ 0 & 1 & 0 \\ 0 & 0 & 1 \end{bmatrix} = \begin{bmatrix} 1 & -1 & 0 \\ -2 & 3 & -4 \\ -2 & 3 & -3 \end{bmatrix} \mathbf{A}. \quad \text{Hence, } \mathbf{A}^{-1} = \begin{bmatrix} 1 & -1 & 0 \\ -2 & 3 & -4 \\ -2 & 3 & -3 \end{bmatrix}.$$

EXAMPLE 5.67 Obtain the inverse of the following matrix using elementary column operations

$$\mathbf{A} = \begin{bmatrix} 0 & 1 & 2 \\ 1 & 2 & 3 \\ 3 & 1 & 1 \end{bmatrix}.$$

Solution Write **A** = **IA**, that is

$$\begin{bmatrix} 0 & 1 & 2 \\ 1 & 2 & 3 \\ 3 & 1 & 1 \end{bmatrix} = \mathbf{A} \begin{bmatrix} 1 & 0 & 0 \\ 0 & 1 & 0 \\ 0 & 0 & 1 \end{bmatrix}.$$ Applying $C_1 \leftrightarrow C_2$, we obtain

$$\begin{bmatrix} 1 & 0 & 2 \\ 2 & 1 & 3 \\ 1 & 3 & 1 \end{bmatrix} = \mathbf{A} \begin{bmatrix} 0 & 1 & 0 \\ 1 & 0 & 0 \\ 0 & 0 & 1 \end{bmatrix}, \text{ (Operate } C_3 \to C_3 - 2C_1) \text{ or } \begin{bmatrix} 1 & 0 & 0 \\ 2 & 1 & -1 \\ 1 & 3 & -1 \end{bmatrix} = \mathbf{A} \begin{bmatrix} 0 & 1 & 0 \\ 1 & 0 & -2 \\ 0 & 0 & 1 \end{bmatrix} \text{ (Performing } C_3 \to C_3 + C_2)$$

$$\text{or } \begin{bmatrix} 1 & 0 & 0 \\ 2 & 1 & 0 \\ 1 & 3 & 2 \end{bmatrix} = \mathbf{A} \begin{bmatrix} 0 & 1 & 1 \\ 1 & 0 & -2 \\ 0 & 0 & 1 \end{bmatrix} \text{ (Operate } C_3 \to C_3/2) \text{ or } \begin{bmatrix} 1 & 0 & 0 \\ 2 & 1 & 0 \\ 1 & 3 & 1 \end{bmatrix} = \mathbf{A} \begin{bmatrix} 0 & 1 & 1/2 \\ 1 & 0 & -1 \\ 0 & 0 & 1/2 \end{bmatrix} \text{ (Operate } C_1 \to C_1 - 2C_2)$$

$$\text{or } \begin{bmatrix} 1 & 0 & 0 \\ 0 & 1 & 0 \\ -5 & 3 & 1 \end{bmatrix} = \mathbf{A} \begin{bmatrix} -2 & 1 & 1/2 \\ 1 & 0 & -1 \\ 0 & 0 & 1/2 \end{bmatrix} \text{ (Operate } C_1 \to C_1 + 5C_3)$$

$$\text{or } \begin{bmatrix} 1 & 0 & 0 \\ 0 & 1 & 0 \\ 0 & 3 & 1 \end{bmatrix} = \mathbf{A} \begin{bmatrix} 1/2 & 1 & 1/2 \\ -4 & 0 & -1 \\ 5/2 & 0 & 1/2 \end{bmatrix} \text{ (Operate } C_2 \to C_2 - 3C_3) \text{ or } \begin{bmatrix} 1 & 0 & 0 \\ 0 & 1 & 0 \\ 0 & 0 & 1 \end{bmatrix} = \mathbf{A} \begin{bmatrix} 1/2 & -1/2 & 1/2 \\ -4 & 3 & -1 \\ 5/2 & -3/2 & 1/2 \end{bmatrix}$$

$$\text{Hence, } \mathbf{A}^{-1} = \begin{bmatrix} 1/2 & -1/2 & 1/2 \\ -4 & 3 & -1 \\ 5/2 & -3/2 & 1/2 \end{bmatrix} = \frac{1}{2} \begin{bmatrix} 1 & -1 & 1 \\ -8 & 6 & -2 \\ 5 & -3 & 1 \end{bmatrix}.$$

5.17 ECHELON FORM OF A MATRIX

An $m \times n$ matrix **A** is called a row *echelon matrix*, or in row *echelon form* if the number of zeros preceeding the first non-zero element of a row increases row by row until a row having all zero elements is obtained. Therefore, a matrix is in row echelon form if the following are satisfied:

(*i*) if the *i*th row contains all zero, it is true for all subsequent rows.

(*ii*) If a column contains a non-zero element of any row, then every subsequent entry in this column is zero.

(*iii*) Rows containing all zeros occur only after all non-zero rows.

For example, the following matrices are in row echelon form whose pivots have been circled.

$$\begin{bmatrix} \textcircled{1} & 3 & 7 & 9 \\ 0 & \textcircled{5} & 4 & 1 \\ 0 & 0 & 0 & 6 \end{bmatrix} \quad \begin{bmatrix} \textcircled{1} & -4 & 5 & 6 \\ 0 & 0 & 3 & 5 \\ 0 & 0 & 0 & 0 \end{bmatrix} \quad \begin{bmatrix} \textcircled{1} & -3 & 4 & 6 \\ 0 & \textcircled{2} & -3 & -3 \\ 0 & 0 & \textcircled{6} & 0 \\ 0 & 0 & 0 & 0 \end{bmatrix}$$

Remarks

1. If **A** is a square matrix, then the row-echelon form is an upper triangular matrix and column echelon form is a lower-triangular matrix.
2. From echelon form of a matrix, we examine whether a given set of vectors are linearly independent or not. We form the matrix with each vector as its row (or column) and reduce it to the row (column) echelon form. The given vectors are *linearly independent*, if the row echelon form has *no row with all its elements as zeros.* The number of non-zero rows is the *dimension* of the given set of vectors and the set of vectors consisting of the non-zero rows is the *basis*.

Rank of a matrix A To find the rank of a matrix from definition, lot of computational work is involved because we have to evaluate several determinants. The given matrix reduces to row-echelon form give us the rank, as per following norms.

Let $\mathbf{A} = (a_{ij})$ be a given $m \times n$ matrix. Assume that $a_{11} \neq 0$. If $a_{11} = 0$, we interchange the first row with some other row to make the element in the (1, 1) position as non-zero. Using elementary row operations, we reduce the matrix **A** to its row-echelon form, elements of first column below a_{11} are made zero, then elements in the second column below a_{22} are made zero and so on. The number of non-zero rows in the row echelon form of matrix **A** gives the *rank of the matrix* **A.**

The following examples will make the procedure clear.

EXAMPLE 5.68 Find the rank of the following matrices:

$$(i)\ \mathbf{A} = \begin{bmatrix} 1 & 3 & 4 & 5 \\ 1 & 2 & 6 & 7 \\ 1 & 5 & 0 & 10 \end{bmatrix},\quad (ii)\ \mathbf{A} = \begin{bmatrix} 1 & 2 & 3 & 0 \\ 2 & 4 & 3 & 2 \\ 3 & 2 & 1 & 3 \\ 6 & 8 & 7 & 5 \end{bmatrix},\quad (iii)\ \mathbf{A} = \begin{bmatrix} 6 & 1 & 3 & 8 \\ 4 & 2 & 6 & -1 \\ 10 & 3 & 9 & 7 \\ 16 & 4 & 12 & 15 \end{bmatrix}.$$

(*Kottayam*, 2005; *Osmania* 2003; *MDU* 2003)

Solution (*i*) Since it is a 3×4 matrix, the rank is ≤ 3.

Applying the operations $R_2 \to R_2 - R_1$, and $R_3 \to R_3 - R_1$, we obtain

$$\mathbf{A} \approx \begin{bmatrix} 1 & 3 & 4 & 5 \\ 0 & -1 & 2 & 2 \\ 0 & 2 & -4 & 5 \end{bmatrix} \text{(operate } R_3 \to R_3 + 2R_2) \approx \begin{bmatrix} 1 & 3 & 4 & 5 \\ 0 & -1 & 2 & 2 \\ 0 & 0 & 0 & 9 \end{bmatrix}.$$

This is the row echelon form of **A.** Since the number of non-zero rows in the row echelon form is 3, we get rank (**A**) = 3.

(*ii*) Applying the operations $R_2 \to R_2 - 2R_1$, $R_3 \to R_3 - 3R_1$ and $R_4 \to R_4 - 6R_1$, we obtain

$$\mathbf{A} \approx \begin{bmatrix} 1 & 2 & 3 & 0 \\ 0 & 0 & -3 & 2 \\ 0 & -4 & -8 & 3 \\ 0 & -4 & -11 & 5 \end{bmatrix} \text{(operate } R_3 \to -R_3/4) \approx \begin{bmatrix} 1 & 2 & 3 & 0 \\ 0 & 0 & -3 & 2 \\ 0 & 1 & 2 & -3/4 \\ 0 & -4 & -11 & 5 \end{bmatrix} \text{(operate } R_4 \to R_4 + 4R_3)$$

$$\approx \begin{bmatrix} 1 & 2 & 3 & 0 \\ 0 & 0 & -3 & 2 \\ 0 & 1 & 2 & -3/4 \\ 0 & 0 & -3 & 2 \end{bmatrix} \text{(performing } R_2 \leftrightarrow R_3) \approx \begin{bmatrix} 1 & 2 & 3 & 0 \\ 0 & 1 & 2 & -3/4 \\ 0 & 0 & -3 & 2 \\ 0 & 0 & -3 & 2 \end{bmatrix} \text{(operate } R_4 \to R_4 - R_3)$$

$$\approx \begin{bmatrix} 1 & 2 & 3 & 0 \\ 0 & 1 & 2 & -3/4 \\ 0 & 0 & -3 & 2 \\ 0 & 0 & 0 & 0 \end{bmatrix}$$

Since the number of non-zero rows in the row echelon form of **A** is 3, we get rank (**A**) = 3.

(*iii*) $$\mathbf{A} = \begin{bmatrix} 6 & 1 & 3 & 8 \\ 4 & 2 & 6 & -1 \\ 10 & 3 & 9 & 7 \\ 16 & 4 & 12 & 15 \end{bmatrix} R_1 \leftrightarrow R_2 \approx \begin{bmatrix} 4 & 2 & 6 & -1 \\ 6 & 1 & 3 & 8 \\ 10 & 3 & 9 & 7 \\ 16 & 4 & 12 & 15 \end{bmatrix} \begin{matrix} R_2 \to R_2 - \frac{3}{2}R_1 \\ R_3 \to R_3 - \frac{5}{2}R_1 \\ R_4 \to R_4 - 4R_1 \end{matrix}$$

$$\approx \begin{bmatrix} 4 & 2 & 6 & -1 \\ 0 & -2 & -6 & 19/2 \\ 0 & -2 & -6 & 19/2 \\ 0 & -4 & -12 & 19 \end{bmatrix} \begin{matrix} \\ \\ R_3 \to R_3 - R_2 \\ R_4 \to R_4 - 2R_2 \end{matrix} \approx \begin{bmatrix} 4 & 2 & 6 & -1 \\ 0 & -2 & -6 & 19/2 \\ 0 & 0 & 0 & 0 \\ 0 & 0 & 0 & 0 \end{bmatrix}.$$

Since the number of non-zero rows in the row echelon form of **A** is 2, we get rank (**A**) = 2.

EXAMPLE 5.69 Find the rank of following matrices:

(*i*) $$\mathbf{A} = \begin{bmatrix} 2 & 3 & -1 & -1 \\ 1 & -1 & -2 & -4 \\ 3 & 1 & 3 & -2 \\ 6 & 3 & 0 & -7 \end{bmatrix},$$ (*PTU*, 2007; *NIT Kurukshetra*, 2005; *UPTU 2005, MDU 2001*)

(*ii*) $$\mathbf{B} = \begin{bmatrix} 1 & 2 & -2 & 3 & 1 \\ 1 & 3 & -2 & 3 & 0 \\ 2 & 4 & -3 & 6 & 4 \\ 1 & 1 & -1 & 4 & 6 \end{bmatrix}$$

Solution (*i*) Performing the operation $R_1 \leftrightarrow R_2$, we get

$$\mathbf{A} \approx \begin{bmatrix} 1 & -1 & -2 & -4 \\ 2 & 3 & -1 & -1 \\ 3 & 1 & 3 & -2 \\ 6 & 3 & 0 & -7 \end{bmatrix}$$ (applying the operations $R_2 \to R_2 - 2R_1$, $R_3 \to R_3 - 3R_1$ and $R_4 \to R_4 - 6R_1$)

$$\approx \begin{bmatrix} 1 & -1 & -2 & -4 \\ 0 & 5 & 3 & 7 \\ 0 & 4 & 9 & 10 \\ 0 & 9 & 12 & 17 \end{bmatrix}$$ (applying the operations $R_3 \to R_3 - \frac{4}{5}R_2$ and $R_4 \to R_4 - \frac{9}{5}R_2$)

$$\approx \begin{bmatrix} 1 & -1 & -2 & -4 \\ 0 & 5 & 3 & 7 \\ 0 & 0 & 33/5 & 22/5 \\ 0 & 0 & 33/5 & 22/5 \end{bmatrix} \text{(applying } R_4 \to R_4 - R_3) \approx \begin{bmatrix} 1 & -1 & -2 & -4 \\ 0 & 5 & 3 & 7 \\ 0 & 0 & 33/5 & 22/5 \\ 0 & 0 & 0 & 0 \end{bmatrix}$$

This is the row echelon form of **A**. Since the number of non-zero rows in the row echelon form of **A** is 3. We get rank (**A**) = 3.

(*ii*) By applying the operations $R_2 \to R_2 - R_1$, $R_3 \to R_3 - 2R_1$, and $R_4 \to R_4 - R_1$, we obtain

$$\mathbf{B} \approx \begin{bmatrix} 1 & 2 & -2 & 3 & 1 \\ 0 & 1 & 0 & 0 & -1 \\ 0 & 0 & 1 & 0 & 2 \\ 0 & -1 & 1 & 1 & 5 \end{bmatrix}$$ (operate $R_4 \to R_4 + R_2$)

$$\approx \begin{bmatrix} 1 & 2 & -2 & 3 & 1 \\ 0 & 1 & 0 & 0 & -1 \\ 0 & 0 & 1 & 0 & 2 \\ 0 & 0 & 1 & 1 & 4 \end{bmatrix} \text{ (operate } R_4 \to R_4 - R_3) \approx \begin{bmatrix} 1 & 2 & -2 & 3 & 1 \\ 0 & 1 & 0 & 0 & -1 \\ 0 & 0 & 1 & 0 & 2 \\ 0 & 0 & 0 & 1 & 2 \end{bmatrix}.$$

This is the row echelon form of **B**. Since the number of non-zero rows in the row echelon form of **B** is 4, we get rank (**B**) = 4.

5.17.1 Normal form of a Matrix

Every non-zero matrix **A** of rank r, can be reduced by a sequence of elementary transformations to one of the following four forms, called the **normal form of matrix A.**

(*i*) $[\mathbf{I}_r]$, (*ii*) $[\mathbf{I}_r \ \ \mathbf{O}]$, (*iii*) $\begin{bmatrix} \mathbf{I}_r \\ \mathbf{O} \end{bmatrix}$, (*iv*) $\left[\begin{array}{c|c} \mathbf{I}_r & \mathbf{O} \\ \hline \mathbf{O} & \mathbf{O} \end{array}\right]$, where $\mathbf{I}_r$ is $r \times r$ identity matrix and **O** is null matrix of any order.

By a combination of rows transformations followed by column transformations a non-zero matrix **A** can be easily reduced to normal form which gives the rank of the matrix **A**.

EXAMPLE 5.70 Find the ranks of the following matrices by reducing them to the normal form:

(*i*) $\mathbf{A} = \begin{bmatrix} 3 & 2 & 5 & 7 & 12 \\ 1 & 1 & 2 & 3 & 5 \\ 3 & 3 & 6 & 9 & 15 \end{bmatrix}$, (*ii*) $\mathbf{B} = \begin{bmatrix} 0 & 1 & -3 & -1 \\ 1 & 0 & 1 & 1 \\ 3 & 1 & 0 & 2 \\ 1 & 1 & -2 & 0 \end{bmatrix}$. *(AMIE, W- 2003)*

Solution (*i*) By performing the operation $R_1 \leftrightarrow R_2$, the given matrix

$$\mathbf{A} \approx \begin{bmatrix} 1 & 1 & 2 & 3 & 5 \\ 3 & 2 & 5 & 7 & 12 \\ 3 & 3 & 6 & 9 & 15 \end{bmatrix} \quad \text{(perform the operations } (R_2 \to R_2 - 3R_1 \text{ and } R_3 \to R_3 - 3R_1)$$

$$\approx \begin{bmatrix} 1 & 1 & 2 & 3 & 5 \\ 0 & -1 & -1 & -2 & -3 \\ 0 & 0 & 0 & 0 & 0 \end{bmatrix} \quad \text{(perform the operations } C_2 \to C_2 - C_1, C_3 \to C_3 - 2C_1 \text{ and } C_4 \to C_4 - 3C_1 \text{ and } C_5 \to C_5 - 5C_1)$$

$$\approx \begin{bmatrix} 1 & 0 & 0 & 0 & 0 \\ 0 & -1 & -1 & -2 & -3 \\ 0 & 0 & 0 & 0 & 0 \end{bmatrix} \quad \text{(perform the operation } R_2 \to -R_2)$$

$$\approx \begin{bmatrix} 1 & 0 & 0 & 0 & 0 \\ 0 & 1 & 1 & 2 & 3 \\ 0 & 0 & 0 & 0 & 0 \end{bmatrix} \quad \text{(perform the operations } (C_3 \to C_3 - C_2, C_4 \to C_4 - 2C_2, \text{ and } C_5 \to C_5 - 3C_2)$$

$$\approx \begin{bmatrix} 1 & 0 & 0 & 0 & 0 \\ 0 & 1 & 0 & 0 & 0 \\ 0 & 0 & 0 & 0 & 0 \end{bmatrix} \approx \left[\begin{array}{cc|ccc} 1 & 0 & 0 & 0 & 0 \\ 0 & 1 & 0 & 0 & 0 \\ \hline 0 & 0 & 0 & 0 & 0 \end{array}\right] \approx \begin{bmatrix} \mathbf{I}_2 & \mathbf{O} \\ \mathbf{O} & \mathbf{O} \end{bmatrix},$$

which is the required normal form. Hence the rank (**A**) = 2.

(*ii*) Applying the operations $C_3 \to C_3 - C_1$ and $C_4 \to C_4 - C_1$, we get

$$\mathbf{B} \approx \begin{bmatrix} 0 & 1 & -3 & -1 \\ 1 & 0 & 0 & 0 \\ 3 & 1 & -3 & -1 \\ 1 & 1 & -3 & -1 \end{bmatrix} \begin{matrix} R_3 \to R_3 - R_1 \\ R_4 \to R_4 - R_1 \end{matrix} \approx \begin{bmatrix} 0 & 1 & -3 & -1 \\ 1 & 0 & 0 & 0 \\ 3 & 0 & 0 & 0 \\ 1 & 0 & 0 & 0 \end{bmatrix} \begin{matrix} \text{(Operating } C_3 \to C_3 + 3C_2 \\ \text{and } C_4 \to C_4 + C_2) \end{matrix}$$

$$\approx \begin{bmatrix} 0 & 1 & 0 & 0 \\ 1 & 0 & 0 & 0 \\ 3 & 0 & 0 & 0 \\ 1 & 0 & 0 & 0 \end{bmatrix} \begin{matrix} (R_3 \to R_3 - 3R_2 \\ R_4 \to R_4 - R_2) \end{matrix} \approx \begin{bmatrix} 0 & 1 & 0 & 0 \\ 1 & 0 & 0 & 0 \\ 0 & 0 & 0 & 0 \\ 0 & 0 & 0 & 0 \end{bmatrix} \text{(operating } R_1 \leftrightarrow R_2)$$

$$\approx \left[\begin{array}{cc|cc} 1 & 0 & 0 & 0 \\ 0 & 1 & 0 & 0 \\ \hline 0 & 0 & 0 & 0 \\ 0 & 0 & 0 & 0 \end{array}\right] \approx \begin{bmatrix} \mathbf{I}_2 & \mathbf{O} \\ \mathbf{O} & \mathbf{O} \end{bmatrix}.$$ Hence Rank of matrix **B** = 2.

EXAMPLE 5.71 For the matrix $\mathbf{A} = \begin{bmatrix} 1 & 2 & 3 \\ 3 & 1 & 2 \end{bmatrix}$, find non-singular matrices **P** and **Q** such that **PAQ** is in the normal form.

Solution The given matrix **A** is of order 2 × 3. we start with $\mathbf{I}_2\mathbf{A}\mathbf{I}_3$, while $\mathbf{I}_2$ is meant for row transformations only, $\mathbf{I}_3$ is meant for column transformations only, which are performed on **A** to reduce it to normal form. The procedure is as follows.

$$\mathbf{I}_2\mathbf{A}\mathbf{I}_3 = \begin{bmatrix} 1 & 0 \\ 0 & 1 \end{bmatrix}\begin{bmatrix} 1 & 2 & 3 \\ 3 & 1 & 2 \end{bmatrix}\begin{bmatrix} 1 & 0 & 0 \\ 0 & 1 & 0 \\ 0 & 0 & 1 \end{bmatrix} \text{(Applying the operation } R_2 \to R_2 - 3R_1)$$

$$\approx \begin{bmatrix} 1 & 0 \\ -3 & 1 \end{bmatrix}\begin{bmatrix} 1 & 2 & 3 \\ 0 & -5 & -7 \end{bmatrix}\begin{bmatrix} 1 & 0 & 0 \\ 0 & 1 & 0 \\ 0 & 0 & 1 \end{bmatrix} \begin{matrix} \text{(Applying the operations } C_2 \to C_2 - 2C_1 \\ \text{and } C_3 \to C_3 - 3C_1) \end{matrix}$$

$$\approx \begin{bmatrix} 1 & 0 \\ -3 & 1 \end{bmatrix}\begin{bmatrix} 1 & 0 & 0 \\ 0 & -5 & -7 \end{bmatrix}\begin{bmatrix} 1 & -2 & -3 \\ 0 & 1 & 0 \\ 0 & 0 & 1 \end{bmatrix} \text{(Applying the operation } C_2 \to -C_2/5)$$

$$\approx \begin{bmatrix} 1 & 0 \\ -3 & 1 \end{bmatrix}\begin{bmatrix} 1 & 0 & 0 \\ 0 & 1 & -7 \end{bmatrix}\begin{bmatrix} 1 & 2/5 & -3 \\ 0 & -1/5 & 0 \\ 0 & 0 & 1 \end{bmatrix} \text{(Applying the operation } C_3 \to C_3 + 7C_2)$$

$$\approx \begin{bmatrix} 1 & 0 \\ -3 & 1 \end{bmatrix}\left[\begin{array}{cc|c} 1 & 0 & 0 \\ 0 & 1 & 0 \end{array}\right]\begin{bmatrix} 1 & 2/5 & -1/5 \\ 0 & -1/5 & -7/5 \\ 0 & 0 & 1 \end{bmatrix}$$ which may be written as

$$\begin{bmatrix} 1 & 0 \\ -3 & 1 \end{bmatrix}[I_2 \quad 0]\begin{bmatrix} 1 & 2/5 & -1/5 \\ 0 & -1/5 & -7/5 \\ 0 & 0 & 1 \end{bmatrix}.$$ Hence $\mathbf{P} = \begin{bmatrix} 1 & 0 \\ -3 & 1 \end{bmatrix}$ and $\mathbf{Q} = \frac{1}{5}\begin{bmatrix} 5 & 2 & -1 \\ 0 & -1 & -7 \\ 0 & 0 & 5 \end{bmatrix}$.

NOTES 1. It is clear that **P** and **Q** are not unique.

2. As is clear from the procedure, we start with unit matrices of suitable order. For example if **A** is $m \times n$ then we start with $\mathbf{I}_m \mathbf{A} \mathbf{I}_n$, while $\mathbf{I}_m$ is meant for row transformations only, $\mathbf{I}_n$ is meant for column transformations only, which are performed on **A** to reduce it to normal form.

EXAMPLE 5.72 Examine whether the following set of vectors is linearly independent. Find the dimension and the basis of the given set of vectors.

(*i*) {(1, 1, 0, 1), (1, 1, 1, 1), (4, 4, 1, 1), (1, 0, 0, 1)}. *(AMIETE, June 2006)*

(*ii*) {(2, 2, 0, 2), (4, 1, 4, 1), (3, 0, 4, 0)}.

(*iii*) {(2, 3, 6, –3, 4), (4, 2, 12, –3, 6), (4, 10, 12, –9, 10)}.

Solution Let each given vector represent a row of a matrix **A**. We reduce **A** to row echelon form. If all the rows of the row echelon form have some non-zero elements, then the given set of vectors are linearly independent.

$$(i)\quad \mathbf{A} = \begin{bmatrix} 1 & 1 & 0 & 1 \\ 1 & 1 & 1 & 1 \\ 4 & 4 & 1 & 1 \\ 1 & 0 & 0 & 1 \end{bmatrix} \begin{matrix} R_2 \to R_2 - R_1 \\ R_3 \to R_3 - 4R_1 \\ R_4 \to R_4 - R_1 \end{matrix} \approx \begin{bmatrix} 1 & 1 & 0 & 1 \\ 0 & 0 & 1 & 0 \\ 0 & 0 & 1 & -3 \\ 0 & -1 & 0 & 0 \end{bmatrix} \text{(operate } R_2 \leftrightarrow R_4)$$

$$\approx \begin{bmatrix} 1 & 1 & 0 & 1 \\ 0 & -1 & 0 & 0 \\ 0 & 0 & 1 & -3 \\ 0 & 0 & 1 & 0 \end{bmatrix} (R_4 \to R_4 - R_3) \approx \begin{bmatrix} 1 & 1 & 0 & 1 \\ 0 & -1 & 0 & 0 \\ 0 & 0 & 1 & -3 \\ 0 & 0 & 0 & 3 \end{bmatrix}.$$

Since all the rows in the row echelon form of **A** are non-zero, the given set of vectors are linearly independent and the dimension of the given set of vectors is 4. The set of vectors [(1, 1, 0, 1), (0, –1, 0, 0), (0, 0, 1, –3), (0, 0, 0, 3) or the given set itself forms the basis.

$$(ii)\quad \mathbf{A} = \begin{bmatrix} 2 & 2 & 0 & 2 \\ 4 & 1 & 4 & 1 \\ 3 & 0 & 4 & 0 \end{bmatrix} \begin{matrix} \text{operate } R_2 \to R_2 - 2R_1 \\ \text{and } R_3 \to R_3 - \dfrac{3}{2}R_1 \end{matrix} \approx \begin{bmatrix} 2 & 2 & 0 & 2 \\ 0 & -3 & 4 & -3 \\ 0 & -3 & 4 & -3 \end{bmatrix} \text{(operate } R_3 \to R_3 - R_2) \approx \begin{bmatrix} 2 & 2 & 0 & 2 \\ 0 & -3 & 4 & -3 \\ 0 & 0 & 0 & 0 \end{bmatrix}.$$

Since all the rows in the row echelon form of **A** are not non-zero, the given set of vectors are linearly dependent. Since the number of non-zero rows is 2, the dimension of the given set of vectors is 2 and its basis can be taken as the set {(2, 2, 0, 2), (0, –3, 4, –3)}

$$(iii)\quad \mathbf{A} = \begin{bmatrix} 2 & 3 & 6 & -3 & 4 \\ 4 & 2 & 12 & -3 & 6 \\ 4 & 10 & 12 & -9 & 10 \end{bmatrix} \text{ (operate } R_2 \to R_2 - 2R_1 \text{ and } R_3 \to R_3 - 2R_1)$$

$$\approx \begin{bmatrix} 2 & 3 & 6 & -3 & 4 \\ 0 & -4 & 0 & 3 & -2 \\ 0 & 4 & 0 & -3 & 2 \end{bmatrix} R_3 \to R_3 + R_2 \approx \begin{bmatrix} 2 & 3 & 6 & -3 & 4 \\ 0 & -4 & 0 & 3 & -2 \\ 0 & 0 & 0 & 0 & 0 \end{bmatrix}.$$

Since all the rows in the row echelon form of **A** are not non-zero, the given set of vectors are linearly dependent. Since the number of non-zero rows is 2, the dimension of the given set of vectors is 2 and its basis can be taken as the set {(2, 3, 6, –3, 4), (0, –4, 0, 3, –2)}.

5.18 SOLUTION OF GENERAL LINEAR SYSTEM OF EQUATIONS

In Article 5.14, we have discussed the matrix method and the Cramer's rule for solving a system of n linear equations in n unknowns, $\mathbf{Ax} = \mathbf{b}$. We assumed that the coefficient matrix **A** is non-singular, that is $|\mathbf{A}| \neq 0$, or the rank of the matrix **A** is n. The matrix method requires evaluation of n^2 determinants each of order $(n - 1)$ to generate the cofactor matrix, and one determinant of order n, whereas the Cramer's rule requires evaluation of $(n + 1)$ determinants each of order n. Since the evaluation of high order determinants is very time consuming, these methods are not used for large value of n , say greater than four.

A system of m linear equations in n unknowns $x_1, x_2, ..., x_n$ has the general form

$$\left.\begin{aligned} a_{11}x_1 + a_{12}x_2 + \cdots\cdots + a_{1n}x_n &= b_1 \\ a_{21}x_1 + a_{22}x_2 + \cdots\cdots + a_{2n}x_n &= b_2 \\ \vdots \\ a_{m1}x_1 + a_{m2}x_2 + \cdots\cdots + a_{mn}x_n &= b_m \end{aligned}\right\} \qquad ...(5.23)$$

can be written compactly as a matrix equation $\mathbf{Ax} = \mathbf{b}$ ...(5.24)

where
$$\mathbf{A} = \begin{bmatrix} a_{11} & a_{12} & \cdots & a_{1n} \\ a_{21} & a_{22} & \cdots & a_{2n} \\ \vdots & & & \vdots \\ a_{m1} & a_{m2} & \cdots & a_{mn} \end{bmatrix}, \quad \mathbf{b} = \begin{bmatrix} b_1 \\ b_2 \\ \vdots \\ b_m \end{bmatrix}, \quad \mathbf{x} = \begin{bmatrix} x_1 \\ x_2 \\ \vdots \\ x_n \end{bmatrix}$$

are respectively called the *coefficient matrix, right hand side column vector* and the *solution vector.* The orders of the matrices **A, b, x** are respectively $m \times n$, $m \times 1$ and $n \times 1$. The matrix

$$(\mathbf{A}|\mathbf{b}) = \left[\begin{array}{cccccc|c} a_{11} & a_{12} & \cdots & \cdots & a_{1n} & b_1 \\ a_{21} & a_{22} & \cdots & \cdots & a_{2n} & b_2 \\ \vdots & & & & \vdots & \vdots \\ \vdots & & & & \vdots & \vdots \\ a_{m1} & a_{m2} & \cdots & \cdots & a_{mn} & b_m \end{array}\right] \qquad ...(5.25)$$

is called the *augmented matrix* of the given system of equations and has m rows and $(n + 1)$ columns that is, of order $m \times (n + 1)$.

To solve a system of m linear equations in n unknowns using an augmented matrix $(\mathbf{A}|\mathbf{b})$, we may use either **Gauss elimination method** or the **Gauss-Jordan method.**

5.18.1 Gauss Elimination Method for Non-homogeneous System

Consider a non-homogeneous system of m equations in n unknown

$$\mathbf{Ax} = \mathbf{b}. \qquad \text{(Refer equation 5.24)}$$

We write the augmented matrix $(\mathbf{A}|\mathbf{b})$ of order $m \times (n + 1)$ as

$$\mathbf{A}|\mathbf{b} = \left[\begin{array}{cccc|c} a_{11} & a_{12} & \cdots & a_{1n} & b_1 \\ a_{21} & a_{22} & \cdots & a_{2n} & b_2 \\ \vdots & \vdots & & & \vdots \\ a_{m1} & a_{m2} & \cdots & a_{mn} & b_m \end{array}\right]$$

and reduce it to the row echelon form by using elementary row operations. We need a maximum of $(m - 1)$ stages of elimination to reduce the given augmented matrix to the equivalent row echelon form. This process may terminate at an earlier stage. We then obtained an equivalent system of the form

$$(\mathbf{A}|\mathbf{b}) = \left[\begin{array}{cccccccc|c} a_{11} & a_{12} & \cdots & \cdots & a_{1r} & \cdots & a_{1n} & b_1 \\ 0 & \bar{a}_{22} & \cdots & \cdots & \bar{a}_{2r} & \cdots & \bar{a}_{2n} & \bar{b}_2 \\ \vdots & & & & & & & \\ 0 & 0 & \cdots & \cdots & a^*_{rr} & \cdots & a^*_{rn} & b^*_r \\ 0 & 0 & \cdots & \cdots & 0 & \cdots & 0 & b^*_{r+1} \\ \vdots & & & & & & & \vdots \\ 0 & 0 & \cdots & \cdots & 0 & \cdots & 0 & b^*_m \end{array}\right]$$

where $r \le m$ and $a_{11} \ne 0$, $\bar{a}_{22} \ne 0$, ..., $a^*_{rr} \ne 0$ are called pivots. We have the following cases:

(*i*) **No solution.** If $r < m$ and one or more of the elements $b^*_{r+1}, b^*_{r+2}, \cdots\cdots, b^*_m$ are not zero. Then rank (**A**) ≠ rank (**A**|**b**).

(*ii*) **Unique solution**, that is **precisely one solution.** Let $m \geq n$ and $r = n$ and $b^*_{r+1}, b^*_{r+2}, \cdots\cdots, b^*_m$ are all zeros. In this case rank (**A**) = rank (**A**|**b**) = *n*. We solve the *n*th equations for x_n, the $(n-1)^{th}$ equation for x_{n-1} and so on. This procedure is called the *back substitution method.*

(*iii*) **Infinitely many solutions.** If $r < n$ and $b^*_{r+1}, b^*_{r+2}, \cdots\cdots, b^*_m$ are all zeros. In this case, *r* unknown, $x_1, x_2, \ldots\ldots, x_r$ can be obtained in terms of the remaining $(n - r)$ unknown $x_{r+1}, x_{r+2}, \ldots\ldots, x_r$ by solving the *r*th equation for x_r, $(r - 1)$ th equation for x_{r-1} and so on. In this case, the solution of the system contains $(n - r)$ parameters.

Remarks

1. We donot, normally use column elementary operations in solving the linear system of equations.
2. Gauss elimination method may be written as $(\mathbf{A}|\mathbf{b}) \xrightarrow[\text{row operations}]{\text{Elementary}} (\mathbf{B}|\mathbf{c})$.
   The matrix **B** is the row echelon form of the matrix **A** and **c** is the new right hand side column vector. We obtain the solution vector (if it exists) using the back substitution method.

5.18.2 Gauss-Jordon Method to Solve System of Equations

In the Gauss-Jordan method, we reduce the augmented matrix (**A**|**b**) of the given system of equations to the form (**I**|**d**) by using elementary row transformations only, where **I** is the unit matrix and matrix **d** is the new right hand side column vector. The solution of the system of equations is given by column vector **d**. This method does not require back-substitution. The solution of the system will be apparent by inspection of the final matrix.

Gauss-Jordan method involves the following sequence of elementary row operations:

(*a*) In matrix **A**, make the element in the first row first column, pivot a_{11}, to unity by some suitable elementary row operation.

(*b*) Reduce all other elements in the first column to zero with the help of unity obtained in first step.

(*c*) Reduce the element in the second row second column, pivot $\bar{a}_{22}$, to unity by suitable elementary row operations.

(*d*) Reduce all other elements in the second column to zero with the help of unity obtained in third step.

(*e*) At the *k*th step, all the elements above and below the pivot a^*_{kk} are made zero. The pivot in the (k, k) position be made 1. This process is continued until an identity matrix is obtained.

Remark

1. Whenever a row of zeros exists to the left of the vertical line and a non-zero number appears to the right, the system is inconsistent, that is, it has no solution.
2. Whenever a complete row of zeros occurs (that is zeros on both sides of the vertical line), the system is dependent, that is it has infinite number of solutions.
3. Otherwise, the given system has a unique solution given in the column to the right of the vertical line.

EXAMPLE 5.73 Solve the following systems of equations (if possible) using Gauss elimination methods.

(*i*) $x - y + z = 0, -x + y - z = 0, 10y + 25z = 90, 20x + 10y = 80$.

(*ii*) $2x_1 + x_2 + 2x_3 + x_4 = 6, 6x_1 - 6x_2 + 6x_3 + 12x_4 = 36, 4x_1 + 3x_2 + 3x_3 - 3x_4 = -1, 2x_1 + 2x_2 - x_3 + x_4 = 10$.

(*iii*) $4y + 3z = 8, 2x - z = 2, 3x + 2y = 5$.

(*iv*) $5x + 5y - 10z = 0, 2w - 3x - 3y + 6z = 2, 4w + x + y - 2z = 4$.

Solution We write the augmented matrix and reduce it to row echelon form by applying elementary row operations.

$$(i)\ (\mathbf{A}|\mathbf{b}) = \left[\begin{array}{ccc|c} 1 & -1 & 1 & 0 \\ -1 & 1 & -1 & 0 \\ 0 & 10 & 25 & 90 \\ 20 & 10 & 0 & 80 \end{array}\right] \begin{array}{l} R_2 \to R_2 + R_1 \\ R_4 \to R_4 - 20R_1 \end{array} \approx \begin{bmatrix} 1 & -1 & 1 & 0 \\ 0 & 0 & 0 & 0 \\ 0 & 10 & 25 & 90 \\ 0 & 30 & -20 & 80 \end{bmatrix} (R_2 \leftrightarrow R_4)$$

$$\approx\left[\begin{array}{rrr|r} 1 & -1 & 1 & 0 \\ 0 & 30 & -20 & 80 \\ 0 & 10 & 25 & 90 \\ 0 & 0 & 0 & 0 \end{array}\right] (R_3 \to R_3 - R_2/3) \quad \approx\left[\begin{array}{rrr|r} 1 & -1 & 1 & 0 \\ 0 & 30 & -20 & 80 \\ 0 & 0 & 95/3 & 190/3 \\ 0 & 0 & 0 & 0 \end{array}\right].$$

Using the back substitution method, we obtain the solution as

$$95z/3 = 190/3, \text{ or } z = 2,$$
$$30y - 20z = 80, \text{ or } y = (2z + 8)/3 = 4,$$
$$x - y + z = 0, \text{ or } x = y - z = 2.$$

Therefore, the system of equations has the unique solution $x = 2, y = 4, z = 2$.

(*ii*) We have

$$(\mathbf{A}|\mathbf{b}) = \left[\begin{array}{rrrr|r} 2 & 1 & 2 & 1 & 6 \\ 6 & -6 & 6 & 12 & 36 \\ 4 & 3 & 3 & -3 & -1 \\ 2 & 2 & -1 & 1 & 10 \end{array}\right] \quad \begin{array}{l}(\text{operating } R_2 \to R_2 - 3R_1, R_3 \to R_3 - 2R_1 \\ \text{and } R_4 \to R_4 - R_1)\end{array}$$

$$\approx\left[\begin{array}{rrrr|r} 2 & 1 & 2 & 1 & 6 \\ 0 & -9 & 0 & 9 & 18 \\ 0 & 1 & -1 & -5 & -13 \\ 0 & 1 & -3 & 0 & 4 \end{array}\right] R_2 \to -R_2/9 \quad \approx\left[\begin{array}{rrrr|r} 2 & 1 & 2 & 1 & 6 \\ 0 & 1 & 0 & -1 & -2 \\ 0 & 1 & -1 & -5 & -13 \\ 0 & 1 & -3 & 0 & 4 \end{array}\right] \begin{array}{l} R_3 \to R_3 - R_2 \\ R_4 \to R_4 - R_2 \end{array}$$

$$\approx\left[\begin{array}{rrrr|r} 2 & 1 & 2 & 1 & 6 \\ 0 & 1 & 0 & -1 & -2 \\ 0 & 0 & -1 & -4 & -11 \\ 0 & 0 & -3 & 1 & 6 \end{array}\right] R_4 \to R_4 - 3R_3 \quad \approx\left[\begin{array}{rrrr|r} 2 & 1 & 2 & 1 & 6 \\ 0 & 1 & 0 & -1 & -2 \\ 0 & 0 & -1 & -4 & -11 \\ 0 & 0 & 0 & 13 & 39 \end{array}\right].$$

Using the back substitution method, we obtain the solution as

$$13x_4 = 39, \text{ or } x_4 = 3,$$
$$-x_3 - 4x_4 = -11, \text{ or } x_3 = -1,$$
$$x_2 - x_4 = -2, \text{ or } x_2 = 1,$$
$$2x_1 + x_2 + 2x_3 + x_4 = 6, \text{ or } x_1 = 2.$$

Therefore, the system of equations has the unique solution $x_1 = 2, x_2 = 1, x_3 = -1, x_4 = 3$.

(*iii*)

$$(\mathbf{A}|\mathbf{b}) = \left[\begin{array}{rrr|r} 0 & 4 & 3 & 8 \\ 2 & 0 & -1 & 2 \\ 3 & 2 & 0 & 5 \end{array}\right] (\text{perform } R_1 \leftrightarrow R_3)$$

$$\approx\left[\begin{array}{rrr|r} 3 & 2 & 0 & 5 \\ 2 & 0 & -1 & 2 \\ 0 & 4 & 3 & 8 \end{array}\right] \left(\text{operate } R_2 \to R_2 - \frac{2}{3}R_1\right)$$

$$\approx\left[\begin{array}{rrr|r} 3 & 2 & 0 & 5 \\ 0 & -4/3 & -1 & -4/3 \\ 0 & 4 & 3 & 8 \end{array}\right] (\text{operate } R_3 \to R_3 + 3R_2) \approx\left[\begin{array}{rrr|r} 3 & 2 & 0 & 5 \\ 0 & -4/3 & -1 & -4/3 \\ 0 & 0 & 0 & 4 \end{array}\right].$$

The third row of the last augmented matrix means $0x + 0y + 0z = 4$ (or $0 = 4$). Since no numbers x, y and z can satisfy this equation, we conclude that the system has no solution—inconsistent system.

Alternately We find that rank (**A**) = 2 and rank (**A**|**b**) = 3. Therefore, the system of equations has no solution.

$$(iv)\quad (\mathbf{A}|\mathbf{b}) = \left[\begin{array}{cccc|c} 5 & 5 & -10 & 0 & 0 \\ -3 & -3 & 6 & 2 & 2 \\ 1 & 1 & -2 & 4 & 4 \end{array}\right] \quad \text{(perform } (R_1 \leftrightarrow R_3)\text{)}$$

$$\approx \left[\begin{array}{cccc|c} 1 & 1 & -2 & 4 & 4 \\ -3 & -3 & 6 & 2 & 2 \\ 5 & 5 & -10 & 0 & 0 \end{array}\right] \quad \text{(operate } R_2 \to R_2 + 3R_1 \text{ and } R_3 \to R_3 - 5R_1\text{)}$$

$$\approx \left[\begin{array}{cccc|c} 1 & 1 & -2 & 4 & 4 \\ 0 & 0 & 0 & 14 & 14 \\ 0 & 0 & 0 & -20 & -20 \end{array}\right] \left(\text{operate } R_3 \to R_3 + \frac{10}{7}R_2\right) \approx \left[\begin{array}{cccc|c} 1 & 1 & -2 & 4 & 4 \\ 0 & 0 & 0 & 14 & 14 \\ 0 & 0 & 0 & 0 & 0 \end{array}\right].$$

The system is consistent and has infinite number of solutions. From the second equation, we obtain $w = 1$. From the first equation, we obtain $x = 4 - y + 2z - 4w = 2z - y$.

Thus, we obtain a two parameter family of solutions:

$x = 2z - y$ and $w = 1$, where y and z are arbitrary.

EXAMPLE 5.74 Use Gauss-Jordon elimination to solve the given system or show that no solution exists.

(*i*) $x - y - z = -3,\ 2x + 3y + 5z = 7,\ x - 2y + 3z = -11.$

(*ii*) $x - y - z = 8,\ x - y + z = 3,\ -x + y + z = 4.$

(*iii*) $x + 3y - 2z = -7,\ 4x + y + 3z = 5,\ 2x - 5y + 7z = 19.$

Solution (*i*) We have

$$(\mathbf{A}|\mathbf{b}) = \left[\begin{array}{ccc|c} 1 & -1 & -1 & -3 \\ 2 & 3 & 5 & 7 \\ 1 & -2 & 3 & -11 \end{array}\right] \quad \text{(operate } R_2 \to R_2 - 2R_1 \text{ and } R_3 \to R_3 - R_1\text{)}$$

$$\approx \left[\begin{array}{ccc|c} 1 & -1 & -1 & -3 \\ 0 & 5 & 7 & 13 \\ 0 & -1 & 4 & -8 \end{array}\right] \quad \text{(operate } R_2 \to R_2/5\text{)} \approx \left[\begin{array}{ccc|c} 1 & -1 & -1 & -3 \\ 0 & 1 & 7/5 & 13/5 \\ 0 & -1 & 4 & -8 \end{array}\right] \quad \begin{array}{l}\text{(operate } R_1 \to R_1 + R_2 \\ \text{and } R_3 \to R_3 + R_2\text{)}\end{array}$$

$$\approx \left[\begin{array}{ccc|c} 1 & 0 & 2/5 & -2/5 \\ 0 & 1 & 7/5 & 13/5 \\ 0 & 0 & 27/5 & -27/5 \end{array}\right] \left(\text{operate } R_3 \to \frac{5R_3}{27}\right)$$

$$\approx \left[\begin{array}{ccc|c} 1 & 0 & 2/5 & -2/5 \\ 0 & 1 & 7/5 & 13/5 \\ 0 & 0 & 1 & -1 \end{array}\right] \left(\text{operate } R_1 \to R_1 - \frac{2}{5}R_3 \text{ and } R_2 \to R_2 - \frac{7}{5}R_3\right) \approx \left[\begin{array}{ccc|c} 1 & 0 & 0 & 0 \\ 0 & 1 & 0 & 4 \\ 0 & 0 & 1 & -1 \end{array}\right].$$

The last matrix is in reduced row-echelon form. We see that the solution of the system is

$$x = 0, \quad y = 4, \quad z = -1.$$

$$(ii)\quad (\mathbf{A}|\mathbf{b}) = \left[\begin{array}{ccc|c} 1 & -1 & -1 & 8 \\ 1 & -1 & 1 & 3 \\ -1 & 1 & 1 & 4 \end{array}\right] \text{(operate } R_2 \to R_2 - R_1 \text{ and } R_3 \to R_3 + R_1\text{)} \approx \left[\begin{array}{ccc|c} 1 & -1 & -1 & 8 \\ 0 & 0 & 2 & -5 \\ 0 & 0 & 0 & 12 \end{array}\right].$$

This shows that the system has no solution-inconsistent.

$$(iii)\quad (\mathbf{A}|\mathbf{b}) = \left[\begin{array}{ccc|c} 1 & 3 & -2 & -7 \\ 4 & 1 & 3 & 5 \\ 2 & -5 & 7 & 19 \end{array}\right] \text{(operate } R_2 \to R_2 - R_1 \text{ and } R_3 \to R_3 - 2R_1\text{)}$$

$$\approx \left[\begin{array}{ccc|c} 1 & 3 & -2 & -7 \\ 0 & -11 & 11 & 33 \\ 0 & -11 & 11 & 33 \end{array}\right] \text{(operate } R_2 \to -R_2/11 \text{ and } R_3 \to -R_3/11)$$

$$\approx \left[\begin{array}{ccc|c} 1 & 3 & -2 & -7 \\ 0 & 1 & -1 & -3 \\ 0 & 1 & -1 & -3 \end{array}\right] \begin{array}{l} R_1 \to R_1 - 3R_2 \\ R_3 \to R_3 - R_2 \end{array} \approx \left[\begin{array}{ccc|c} 1 & 0 & 1 & 2 \\ 0 & 1 & -1 & -3 \\ 0 & 0 & 0 & 0 \end{array}\right].$$

The system of equations is consistent and has infinite number of solutions. Choose z as arbitrary. From the second equation, we obtain $y - z = -3$ or $y = z - 3$.
From the first equation, we obtain $x + z = 2$ or $x = 2 - z$.

EXAMPLE 5.75 Test for consistency and solve the system of equation (if possible) using Gauss elimination method $x - 2y + 3z = 2$, $2x - 3z = 3$, $x + y + z = 0$.

Solution We write the augmented matrix of the system and reduce it to row echelon form by applying elementary row operations.

$$(\mathbf{A}|\mathbf{b}) = \left[\begin{array}{ccc|c} 1 & -2 & 3 & 2 \\ 2 & 0 & -3 & 3 \\ 1 & 1 & 1 & 0 \end{array}\right] \text{(operate } R_2 \to R_2 - 2R_1 \text{ and } R_3 \to R_3 - R_1)$$

$$\approx \left[\begin{array}{ccc|c} 1 & -2 & 3 & 2 \\ 0 & 4 & -9 & -1 \\ 0 & 3 & -2 & -2 \end{array}\right] R_2 \to R_2/4 \approx \left[\begin{array}{ccc|c} 1 & -2 & 3 & 2 \\ 0 & 1 & -9/4 & -1/4 \\ 0 & 3 & -2 & -2 \end{array}\right] R_3 \to R_3 - 3R_2$$

$$\approx \left[\begin{array}{ccc|c} 1 & -2 & 3 & 2 \\ 0 & 1 & -9/4 & -1/4 \\ 0 & 0 & 19/4 & -5/4 \end{array}\right].$$ We find that rank $(\mathbf{A})$ = rank $(\mathbf{A}|\mathbf{b}) = 3 = n$.

Therefore, the system of equations is consistent.
Using the back substitution method, we obtain the solution as

$$\frac{19}{4}z = \frac{-5}{4}, \text{ or } z = \frac{-5}{19}, \quad y - \frac{9}{4}z = -\frac{1}{4}, \text{ or } y = \frac{-16}{19}, \quad x - 2y + 3z = 2, \text{ or } x = \frac{21}{19}.$$

Therefore, the system of equation has the unique solution

$$x = 21/19,\ y = -16/19,\ z = -5/19.$$

EXAMPLE 5.76 Show that the system of equations

$2x - y + z = 4$, $3x - y + z = 6$, $4x - y + 2z = 7$, $-x + y - z = 9$ is inconsistent.

Solution We write the augmented matrix and reduce it to row echelon form by applying elementary row operations.

$$(\mathbf{A}|\mathbf{b}) = \left[\begin{array}{ccc|c} 2 & -1 & 1 & 4 \\ 3 & -1 & 1 & 6 \\ 4 & -1 & 2 & 7 \\ -1 & 1 & -1 & 9 \end{array}\right] \text{(operate } R_1 \leftrightarrow R_4)$$

$$\approx \left[\begin{array}{ccc|c} -1 & 1 & -1 & 9 \\ 3 & -1 & 1 & 6 \\ 4 & -1 & 2 & 7 \\ 2 & -1 & 1 & 4 \end{array}\right]. \begin{array}{l} R_2 \to R_2 + 3R_1 \\ R_3 \to R_3 + 4R_1 \\ R_4 \to R_4 + 2R_1 \end{array} \approx \left[\begin{array}{ccc|c} -1 & 1 & -1 & 9 \\ 0 & 2 & -2 & 33 \\ 0 & 3 & -2 & 43 \\ 0 & 1 & -1 & 22 \end{array}\right] \text{(operate } R_2 \leftrightarrow R_4)$$

$$\approx \begin{bmatrix} -1 & 1 & -1 & 9 \\ 0 & 1 & -1 & 22 \\ 0 & 3 & -2 & 43 \\ 0 & 2 & -2 & 33 \end{bmatrix} \begin{matrix} R_3 \to R_3 - 3R_2 \\ R_4 \to R_4 - 2R_2 \end{matrix} \approx \begin{bmatrix} -1 & 1 & -1 & 9 \\ 0 & 1 & -1 & 22 \\ 0 & 0 & 1 & -23 \\ 0 & 0 & 0 & -11 \end{bmatrix}.$$

We find that rank (**A**) = 3 and rank (**A**|**b**) = 4. Therefore, the given system of equations is inconsistent and has no solution.

EXAMPLE 5.77 Test the following system of equations for consistent and find the solution for the consistent system.

$$\begin{bmatrix} 1 & 4 & 7 \\ 2 & 5 & 8 \\ 1 & 2 & 3 \end{bmatrix} \begin{bmatrix} x \\ y \\ z \end{bmatrix} = \begin{bmatrix} 1 \\ 2 \\ 1 \end{bmatrix}.$$

Solution We write the augmented matrix and reduce it to row echelon form by applying elementary row operations.

$$(\mathbf{A}|\mathbf{b}) = \begin{bmatrix} 1 & 4 & 7 & 1 \\ 2 & 5 & 8 & 2 \\ 1 & 2 & 3 & 1 \end{bmatrix} \quad (\text{operate } R_2 \to R_2 - 2R_1 \text{ and } R_3 \to R_3 - R_1)$$

$$\approx \begin{bmatrix} 1 & 4 & 7 & 1 \\ 0 & -3 & -6 & 0 \\ 0 & -2 & -4 & 0 \end{bmatrix} R_3 \to R_3 - 2R_2/3 \approx \begin{bmatrix} 1 & 4 & 7 & 1 \\ 0 & -3 & -6 & 0 \\ 0 & 0 & 0 & 0 \end{bmatrix}.$$

We find that rank (**A**) = rank (**A**|**b**) = 2 < number of unknowns, the system is consistent and has infinite number of solutions. We find that the last equation is satisfied for all values of x, y, z.

From the second equation, we get $y = -2z$. From the first equation, we get $x + 4y + 7z = 1$. or $x = 1 + z$. Therefore, we obtain the solution $x = 1 + z$, $y = -2z$ and z is arbitrary.

EXAMPLE 5.78 Determine for what values of λ and μ the following equations have (*i*) a unique solution, (*ii*) infinite number of solutions and (*iii*) no solution.

$$x + y + z = 6,\ x + 2y + 3z = 10,\ x + 2y + \lambda z = \mu$$

(AMIE, W-2007; PTU, 2007; UPTU 2006; AMIETE, June 2002)

Solution We write the augmented matrix and reduce it to row echelon form by applying elementary row operation.

$$(\mathbf{A}|\mathbf{b}) = \begin{bmatrix} 1 & 1 & 1 & 6 \\ 1 & 2 & 3 & 10 \\ 1 & 2 & \lambda & \mu \end{bmatrix} \quad (\text{operate } R_2 \to R_2 - R_1 \text{ and } R_3 \to R_3 - R_1)$$

$$\approx \begin{bmatrix} 1 & 1 & 1 & 6 \\ 0 & 1 & 2 & 4 \\ 0 & 1 & \lambda - 1 & \mu - 6 \end{bmatrix} R_3 \to R_3 - R_2 \approx \begin{bmatrix} 1 & 1 & 1 & 6 \\ 0 & 1 & 2 & 4 \\ 0 & 0 & \lambda - 3 & \mu - 10 \end{bmatrix}.$$

(*i*) There is a unique solution if rank (**A**) = rank (**A**|**b**) = 3 that is, if $\lambda \neq 3$, μ may have any value.

(*ii*) There are infinite number of solutions if rank (**A**) = rank (**A**|**b**) < 3 that is $\lambda = 3$, $\mu = 10$.

(*iii*) There is no solution if rank (**A**) ≠ rank (**A**|**b**) that is, $\lambda = 3$, $\mu \neq 10$.

.18.3 Homogeneous System of *m* × *n* linear equations

onsider the homogeneous system of equations **Ax** = **O** ...(5.26) where **A** is an $m \times n$ matrix. The homogeneous system is always *consistent*, since $x_1 = 0, x_2 = 0, ...x_n = 0$ will satisfy each equation in the system. The solution consisting of all zeros is called the **trivial solution**. In this case, rank(**A**) = rank (**A**|**O**) = number of unknowns.

Non-trivial solution exists if and only if rank (**A**) $< n$. If rank **A** $= r < n$, we obtain an $(n - r)$ parameter family of solutions which form a vector space of dimension $(n - r)$ called the **solution space** of homogeneous system. The solution space of homogeneous system is called the **null space** of **A** because **Ax** = **O** for every **x** in the solution space and its dimension is called the **nullity** of **A**. In view of above, we obtain the result

$$\text{rank }(\mathbf{A}) + \text{nullity }(\mathbf{A}) = n.$$

Remarks

1. A homogeneous linear system with fewer equations than unknowns always possesses non-trivial solution.
2. The following diagram 5.2 outlines the connection between the concept of rank of a matrix and the solution of m linear equations in n unknown **Ax** = **b**. Let rank (**A**) = r.

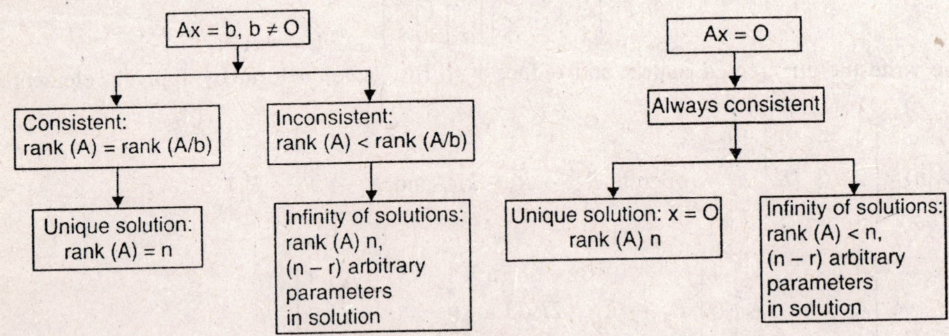

Fig. 5.2

EXAMPLE 5.79 Solve the following homogeneous system of equations **Ax** = **O**, where **A** is given by

(i) $\begin{bmatrix} 2 & 1 \\ 1 & -1 \\ 3 & 2 \end{bmatrix}$, (ii) $\begin{bmatrix} 1 & 2 & 3 \\ 3 & 4 & 4 \\ 7 & 10 & 12 \end{bmatrix}$, (iii) $\begin{bmatrix} 2 & -4 & 3 \\ 1 & 1 & -2 \end{bmatrix}$, (iv) $\begin{bmatrix} 4 & 2 & 1 & 3 \\ 6 & 3 & 4 & 7 \\ 2 & 1 & 0 & 1 \end{bmatrix}$.

Find the rank (**A**) and nullity (**A**).

Solution We write the augmented matrix (**A**|**O**) and reduce it to row echelon form.

(i) $(\mathbf{A}|\mathbf{O}) = \left[\begin{array}{cc|c} 2 & 1 & 0 \\ 1 & -1 & 0 \\ 3 & 2 & 0 \end{array}\right]$ (perform $R_2 \to R_2 - R_1/2$ and $R_3 \to R_3 - 3R_1/2$) $\approx \left[\begin{array}{cc|c} 2 & 1 & 0 \\ 0 & -3/2 & 0 \\ 0 & 1/2 & 0 \end{array}\right]$ $R_3 \to R_3 + R_2/3$

$\approx \left[\begin{array}{cc|c} 2 & 1 & 0 \\ 0 & -3/2 & 0 \\ 0 & 0 & 0 \end{array}\right]$. Since, rank (**A**) = 2 = number of unknown, the system has only a trivial solution.

Hence nullity (**A**) = **0**.

(ii) $(\mathbf{A}|\mathbf{O}) = \left[\begin{array}{ccc|c} 1 & 2 & 3 & 0 \\ 3 & 4 & 4 & 0 \\ 7 & 10 & 12 & 0 \end{array}\right]$ (operate $R_2 \to R_2 - 3R_1$ and $R_3 \to R_3 - 7R_1$) $\approx \left[\begin{array}{ccc|c} 1 & 2 & 3 & 0 \\ 0 & -2 & -5 & 0 \\ 0 & -4 & -9 & 0 \end{array}\right]$ $R_3 \to R_3 - 2R_2$

$\approx \left[\begin{array}{ccc|c} 1 & 2 & 3 & 0 \\ 0 & -2 & -5 & 0 \\ 0 & 0 & 1 & 0 \end{array}\right]$. Since rank (**A**) = 3 = number of unknown, the homogeneous system has only a trivial solution. Therefore, nullity (**A**) = **0**.

(*iii*) Since the number of equations is less than the number of unknowns, we know that the given system has non trivial solution (See Remark of Art. 5.18.3). Using Gauss-Jordan elimination, we find

$$(\mathbf{A}|\mathbf{O}) = \left[\begin{array}{ccc|c} 2 & -4 & 3 & 0 \\ 1 & 1 & -2 & 0 \end{array}\right] R_1 \leftrightarrow R_2 \approx \left[\begin{array}{ccc|c} 1 & 1 & -2 & 0 \\ 2 & -4 & 3 & 0 \end{array}\right] R_2 \to R_2 - 2R_1$$

$$\approx \left[\begin{array}{ccc|c} 1 & 1 & -2 & 0 \\ 0 & -6 & 7 & 0 \end{array}\right] (\text{perform } R_2 \to -R_2/6) \approx \left[\begin{array}{ccc|c} 1 & 1 & -2 & 0 \\ 0 & 1 & -7/6 & 0 \end{array}\right] R_1 \to R_1 - R_2 \approx \left[\begin{array}{ccc|c} 1 & 0 & -5/6 & 0 \\ 0 & 1 & -7/6 & 0 \end{array}\right].$$

Here, rank (**A**) = 2 and the number of unknowns is 3, we obtain a one parameter family of solutions as $x = 5z/6$, $y = 7z/6$, where z is arbitrary. Therefore, nullity (**A**) = 1.

$$(iv)\quad (\mathbf{A}|\mathbf{O}) = \left[\begin{array}{cccc|c} 4 & 2 & 1 & 3 & 0 \\ 6 & 3 & 4 & 7 & 0 \\ 2 & 1 & 0 & 1 & 0 \end{array}\right] \begin{array}{l} R_2 \to R_2 - 3R_1/2 \\ R_3 \to R_3 - R_1/2 \end{array}$$

$$\approx \left[\begin{array}{cccc|c} 4 & 2 & 1 & 3 & 0 \\ 0 & 0 & 5/2 & 5/2 & 0 \\ 0 & 0 & -1/2 & -1/2 & 0 \end{array}\right] (\text{operate } R_3 \to R_3 + R_2/5) \approx \left[\begin{array}{cccc|c} 4 & 2 & 1 & 3 & 0 \\ 0 & 0 & 5/2 & 5/2 & 0 \\ 0 & 0 & 0 & 0 & 0 \end{array}\right].$$

Rank (**A**) = 2 and the number of unknowns is 4.

Hence, we obtain a two parameter family of solutions as $x_3 = -x_4$, $x_2 = (-4x_1 - x_3 - 3x_4)/2 = -(2x_1 + x_4)$, where x_1 and x_4 are arbitrary. Therefore, nullity (**A**) = 2.

EXAMPLE 5.80 Given the following system of equations:

$$\left.\begin{array}{l} 2x - 2y + 5z + 3w = 0 \\ 4x - y + z + w = 0 \\ 3x - 2y + 3z + 4w = 0 \\ x - 3y + 7z + 6w = 0 \end{array}\right\} \begin{array}{l} \text{or } \mathbf{AX} = \mathbf{O}, \\ \text{where } \mathbf{A} \text{ is} \\ \text{the coefficient} \\ \text{matrix.} \end{array}$$

Reduce the coefficient matrix **A** into Echelon form and find the rank. Utilising the property of the rank, test the given system of equations for consistency and if possible, find the solution of the given system.

(AMIETE, June 2001)

Solution In matrix form, we can write the system of equations as **AX** = **O**.

$$\begin{bmatrix} 2 & -2 & 5 & 3 \\ 4 & -1 & 1 & 1 \\ 3 & -2 & 3 & 4 \\ 1 & -3 & 7 & 6 \end{bmatrix} \begin{bmatrix} x \\ y \\ z \\ w \end{bmatrix} = \begin{bmatrix} 0 \\ 0 \\ 0 \\ 0 \end{bmatrix}.$$

$$\text{The coefficient matrix A} = \begin{bmatrix} 2 & -2 & 5 & 3 \\ 4 & -1 & 1 & 1 \\ 3 & -2 & 3 & 4 \\ 1 & -3 & 7 & 6 \end{bmatrix} \begin{array}{l} (R_1 \to R_1 - 2R_4) \\ (R_2 \to R_2 - 4R_4) \\ (R_3 \to R_3 - 3R_4) \end{array}$$

$$\approx \begin{bmatrix} 0 & 4 & -9 & -9 \\ 0 & 11 & -27 & -23 \\ 0 & 7 & -18 & -14 \\ 1 & -3 & 7 & 6 \end{bmatrix} (R_1 \leftrightarrow R_4) \qquad \approx \begin{bmatrix} 1 & -3 & 7 & 6 \\ 0 & 11 & -27 & -23 \\ 0 & 7 & -18 & -14 \\ 0 & 4 & -9 & -9 \end{bmatrix} (R_2 \to R_2 - R_3)$$

$$\approx \begin{bmatrix} 1 & -3 & 7 & 6 \\ 0 & 4 & -9 & -9 \\ 0 & 7 & -18 & -14 \\ 0 & 4 & -9 & -9 \end{bmatrix} \quad (R_4 \to R_4 - R_2) \quad \approx \begin{bmatrix} 1 & -3 & 7 & 6 \\ 0 & 4 & -9 & -9 \\ 0 & 7 & -18 & -14 \\ 0 & 0 & 0 & 0 \end{bmatrix} \quad (C_2 \to C_2 + 3C_1)$$

$$\approx \begin{bmatrix} 1 & 0 & 7 & 6 \\ 0 & 4 & -9 & -9 \\ 0 & 7 & -18 & -14 \\ 0 & 0 & 0 & 0 \end{bmatrix} \quad (R_3 \to R_3 - \frac{7}{4} R_2)$$

$$\approx \begin{bmatrix} 1 & 0 & 7 & 6 \\ 0 & 4 & -9 & -9 \\ 0 & 0 & -9/4 & 7/4 \\ 0 & 0 & 0 & 0 \end{bmatrix} \quad (R_2 \to R_2 / 4, \; R_3 \to -\frac{4}{9} R_3) \quad \approx \begin{bmatrix} 1 & 0 & 7 & 6 \\ 0 & 1 & -9/4 & -9/4 \\ 0 & 0 & 1 & -7/9 \\ 0 & 0 & 0 & 0 \end{bmatrix}.$$

This is a matrix in the reduced Echelon form having three non zero rows, hence the rank (**A**) = 3 and is less than the number of unknown viz 4.

Now the matrix form of the given equations is

$$\begin{bmatrix} 1 & 0 & 7 & 6 \\ 0 & 1 & -9/4 & -9/4 \\ 0 & 0 & 1 & -7/9 \\ 0 & 0 & 0 & 0 \end{bmatrix} \begin{bmatrix} x \\ y \\ z \\ w \end{bmatrix} = \begin{bmatrix} 0 \\ 0 \\ 0 \\ 0 \end{bmatrix}$$

or $\qquad x + 7z + 6w = 0 \qquad$ ...(*i*)

$$y - \frac{9}{4} z - \frac{9}{4} w = 0 \quad ...(ii) \qquad z - \frac{7}{9} w = 0 \quad ...(iii)$$

The system is consistent and has infinite number of solutions.

From (iii), we get $z = \frac{7}{9} w$. From (i), $x = -7z - 6w = \left(\frac{-49}{9} - 6 \right) w = \frac{-103}{9} w$

and $\qquad y = \frac{9}{4} z + \frac{9}{4} w = \left(\frac{7}{4} + \frac{9}{4} \right) w = 4w.$

Thus we find that the three of the unknown viz x, y and z are expressed in terms of the 4th unknown viz w. An infinite number of solutions of the given equations can be obtained by assigning arbitrary values to w. Also we know that the system has $n - r$, that is, 4 – 3 that is, 1 linearly independent solutions.

5.19 GAUSS-JORDON METHOD TO FIND THE INVERSE OF A MATRIX

Any non-singular matrix can be reduced to a unit matrix by successive-elementary row (or column) transformations. Assuming this result, we have the following theorem.

Theorem: *The elementary row transformations which reduce a given matrix* **A** *to a unit matrix, when applied to the unit matrix give the inverse matrix* $\mathbf{A}^{-1}$.

Let **A** be a non-singular matrix of order n. We may write **A** = **IA** ...(5.27).

Now we apply elementary row operations to **A** on the left hand side (5.27) so that **A** is reduced to **I**.

Simultaneously, apply the same elementary row operations to the prefactor **I** on the right side. Let **I** reduce to **B**, so that (5.27) becomes **I** = **BA.**

Post multiplying by $\mathbf{A}^{-1}$, we obtain $\mathbf{IA}^{-1} = \mathbf{BAA}^{-1} = \mathbf{BI} = \mathbf{B}$

$\Rightarrow \qquad \mathbf{A}^{-1} = \mathbf{B}$. Thus **B** is the inverse of **A**.

Similarly, the elementary column transformations that reduce a given matrix **A** to a unit matrix, when applied to the unit matrix give inverse matrix $\mathbf{A}^{-1}$.

In practice the two matrices **A** and **I** are written side by side and the same row operations are performed on them. The aim is to reduce **A** to **I**. As soon as this is achieved, the other matrix gives $\mathbf{A}^{-1}$.

$$(\mathbf{A}|\mathbf{I}) \xrightarrow[\text{row operations}]{\text{Elementary}} (\mathbf{I}|\mathbf{B}). \text{ Hence } \mathbf{B} = \mathbf{A}^{-1}.$$

This method is called the *Gauss-Jordan* method. In the first step, all the elements below the pivot a_{11} are made zero. In the second step, all the elements above and below the second pivot $\bar{a}_{22}$ are made zero. At the kth step, all the elements above and below a_{kk}^* are made zero. The pivot in the (i, i) position can be made 1 at every step or when the eliminetion is completed.

EXAMPLE 5.81 Using the Gauss-Jordan method, find the inverses of the following matrices:

$$(i)\ \mathbf{A} = \begin{bmatrix} 2 & 3 & 4 \\ 4 & 3 & 1 \\ 1 & 2 & 4 \end{bmatrix}, \qquad (ii)\ \mathbf{A} = \begin{bmatrix} 4 & 2 & -1 \\ 3 & -2 & 1 \\ 2 & 4 & -3 \end{bmatrix}, \qquad (iii)\ \mathbf{A} = \begin{bmatrix} 1 & -1 & 1 \\ -3 & 2 & -1 \\ -2 & 1 & -1 \end{bmatrix}.$$

Solution (*i*) Writing the given matrix side by side with unit matrix of order 3, that is **A|I** we have

$$\mathbf{A}|\mathbf{I} = \left[\begin{array}{ccc|ccc} 2 & 3 & 4 & 1 & 0 & 0 \\ 4 & 3 & 1 & 0 & 1 & 0 \\ 1 & 2 & 4 & 0 & 0 & 1 \end{array}\right] (\text{operate } R_1 \leftrightarrow R_3) \approx \left[\begin{array}{ccc|ccc} 1 & 2 & 4 & 0 & 0 & 1 \\ 4 & 3 & 1 & 0 & 1 & 0 \\ 2 & 3 & 4 & 1 & 0 & 0 \end{array}\right] \begin{pmatrix}\text{operate } R_2 \to R_2 - 4R_1 \\ \text{and } R_3 \to R_3 - 2R_1\end{pmatrix}$$

$$\approx \left[\begin{array}{ccc|ccc} 1 & 2 & 4 & 0 & 0 & 1 \\ 0 & -5 & -15 & 0 & 1 & -4 \\ 0 & -1 & -4 & 1 & 0 & -2 \end{array}\right] \begin{pmatrix}\text{operate } R_2 \to \dfrac{-R_2}{5} \\ \text{and } R_3 \to -R_3\end{pmatrix} \approx \left[\begin{array}{ccc|ccc} 1 & 2 & 4 & 0 & 0 & 1 \\ 0 & 1 & 3 & 0 & -1/5 & 4/5 \\ 0 & 1 & 4 & -1 & 0 & 2 \end{array}\right] \begin{pmatrix}\text{operate } R_1 \to R_1 - 2R_2 \\ \text{and } R_3 \to R_3 - R_2\end{pmatrix}$$

$$\approx \left[\begin{array}{ccc|ccc} 1 & 0 & -2 & 0 & 2/5 & -3/5 \\ 0 & 1 & 3 & 0 & -1/5 & 4/5 \\ 0 & 0 & 1 & -1 & 1/5 & 6/5 \end{array}\right] \qquad (\text{operate } R_1 \to R_1 + 2R_3 \text{ and } R_2 \to R_2 - 3R_3)$$

$$\approx \left[\begin{array}{ccc|ccc} 1 & 0 & 0 & -2 & 4/5 & 9/5 \\ 0 & 1 & 0 & 3 & -4/5 & -14/5 \\ 0 & 0 & 1 & -1 & 1/5 & 6/5 \end{array}\right]. \qquad \text{Hence, } \mathbf{A}^{-1} = \frac{1}{5}\begin{bmatrix} -10 & 4 & 9 \\ 15 & -4 & -14 \\ -5 & 1 & 6 \end{bmatrix}.$$

(*ii*) We have

$$\mathbf{A}|\mathbf{I} = \left[\begin{array}{ccc|ccc} 4 & 2 & -1 & 1 & 0 & 0 \\ 3 & -2 & 1 & 0 & 1 & 0 \\ 2 & 4 & -3 & 0 & 0 & 1 \end{array}\right] \left(\text{operate } R_1 \to \frac{R_1}{4}\right)$$

$$\approx \left[\begin{array}{ccc|ccc} 1 & 1/2 & -1/4 & 1/4 & 0 & 0 \\ 3 & -2 & 1 & 0 & 1 & 0 \\ 2 & 4 & -3 & 0 & 0 & 1 \end{array}\right] \qquad (\text{operate } R_2 \to R_2 - 3R_1 \text{ and } R_3 \to R_3 - 2R_1)$$

$$\approx \left[\begin{array}{ccc|ccc} 1 & 1/2 & -1/4 & 1/4 & 0 & 0 \\ 0 & -7/2 & 7/4 & -3/4 & 1 & 0 \\ 0 & 3 & -5/2 & -1/2 & 0 & 1 \end{array}\right] \qquad (\text{operate } R_2 \leftrightarrow R_3)$$

$$\approx \left[\begin{array}{ccc|ccc} 1 & 1/2 & -1/4 & 1/4 & 0 & 0 \\ 0 & 3 & -5/2 & -1/2 & 0 & 1 \\ 0 & -7/2 & 7/4 & -3/4 & 1 & 0 \end{array}\right] \quad \left(\text{operate } R_2 \to \frac{1}{3}R_2\right)$$

$$\approx \left[\begin{array}{ccc|ccc} 1 & 1/2 & -1/4 & 1/4 & 0 & 0 \\ 0 & 1 & -5/6 & -1/6 & 0 & 1/3 \\ 0 & -7/2 & 7/4 & -3/4 & 1 & 0 \end{array}\right] \quad \left(\begin{array}{l}\text{operate } R_1 \to R_1 - \frac{1}{2}R_2 \\ \text{and } R_3 \to R_3 + \frac{7}{2}R_2\end{array}\right)$$

$$\approx \left[\begin{array}{ccc|ccc} 1 & 0 & 1/6 & 1/3 & 0 & -1/6 \\ 0 & 1 & -5/6 & -1/6 & 0 & 1/3 \\ 0 & 0 & -7/6 & -4/3 & 1 & 7/6 \end{array}\right] \quad \left(\text{operate } R_3 \to -\frac{6}{7}R_3\right)$$

$$\approx \left[\begin{array}{ccc|ccc} 1 & 0 & 1/6 & 1/3 & 0 & -1/6 \\ 0 & 1 & -5/6 & -1/6 & 0 & 1/3 \\ 0 & 0 & 1 & 8/7 & -6/7 & -1 \end{array}\right] \quad \left(\begin{array}{l}\text{operate } R_1 \to R_1 - \frac{1}{6}R_3 \\ \text{and } R_2 \to R_2 + \frac{5}{6}R_3\end{array}\right)$$

$$\approx \left[\begin{array}{ccc|ccc} 1 & 0 & 0 & 1/7 & 1/7 & 0 \\ 0 & 1 & 0 & 11/14 & -5/7 & -1/2 \\ 0 & 0 & 1 & 8/7 & -6/7 & -1 \end{array}\right]. \qquad \text{Hence } \mathbf{A}^{-1} = \frac{1}{14}\begin{bmatrix} 2 & 2 & 0 \\ 11 & -10 & -7 \\ 16 & -12 & -14 \end{bmatrix}.$$

(*iii*) We have $\mathbf{A}|\mathbf{I} = \left[\begin{array}{ccc|ccc} 1 & -1 & 1 & 1 & 0 & 0 \\ -3 & 2 & -1 & 0 & 1 & 0 \\ -2 & 1 & -1 & 0 & 0 & 1 \end{array}\right]$ (operating $R_2 \to R_2 + 3R_1$, and $R_3 \to R_3 + 2R_1$)

$$\approx \left[\begin{array}{ccc|ccc} 1 & -1 & 1 & 1 & 0 & 0 \\ 0 & -1 & 2 & 3 & 1 & 0 \\ 0 & -1 & 1 & 2 & 0 & 1 \end{array}\right] \quad (\text{operate } R_2 \to -R_2 \text{ and } R_3 \to -R_3)$$

$$\approx \left[\begin{array}{ccc|ccc} 1 & -1 & 1 & 1 & 0 & 0 \\ 0 & 1 & -2 & -3 & -1 & 0 \\ 0 & 1 & -1 & -2 & 0 & -1 \end{array}\right] \quad \begin{array}{l} R_1 \to R_1 + R_2 \\ R_3 \to R_3 - R_2 \end{array}$$

$$\approx \left[\begin{array}{ccc|ccc} 1 & 0 & -1 & -2 & -1 & 0 \\ 0 & 1 & -2 & -3 & -1 & 0 \\ 0 & 0 & 1 & 1 & 1 & -1 \end{array}\right] \quad \begin{array}{l} R_1 \to R_1 + R_3 \\ R_2 \to R_2 + 2R_3 \end{array}$$

$$\approx \left[\begin{array}{ccc|ccc} 1 & 0 & 0 & -1 & 0 & -1 \\ 0 & 1 & 0 & -1 & 1 & -2 \\ 0 & 0 & 1 & 1 & 1 & -1 \end{array}\right]. \qquad \text{Hence, } \mathbf{A}^{-1} = \begin{bmatrix} -1 & 0 & -1 \\ -1 & 1 & -2 \\ 1 & 1 & -1 \end{bmatrix}.$$

5.20 INVERSE TRANSFORMATION

The relation

$$\left.\begin{array}{l} y_1 = a_{11}x_1 + a_{12}x_2 + a_{13}x_3 \\ y_2 = a_{21}x_1 + a_{22}x_2 + a_{23}x_3 \\ y_3 = a_{31}x_1 + a_{32}x_2 + a_{33}x_3 \end{array}\right\} \quad \text{...(5.28)}$$

is a linear transformation in three dimensions which sends the point with coordinates (x_1, x_2, x_3) into a point with coordinates (y_1, y_2, y_3).

The relation (5.28) may be written in the matrix form $\mathbf{y} = \mathbf{A}\mathbf{x}$

where $$\mathbf{y} = \begin{pmatrix} y_1 \\ y_2 \\ y_3 \end{pmatrix}, \mathbf{x} = \begin{pmatrix} x_1 \\ x_2 \\ x_3 \end{pmatrix} \text{ and } \mathbf{A} = \begin{pmatrix} a_{11} & a_{12} & a_{13} \\ a_{21} & a_{22} & a_{23} \\ a_{31} & a_{32} & a_{33} \end{pmatrix}.$$

Here $\mathbf{A}$ is the matrix of transformation.

In general $\mathbf{y} = \mathbf{A}\mathbf{x}$, ...(5.29)

where $$\mathbf{y} = \begin{pmatrix} y_1 \\ y_2 \\ \vdots \\ y_n \end{pmatrix}, \mathbf{A} = \begin{pmatrix} a_{11} & a_{12} & \cdots & a_{1n} \\ a_{21} & a_{22} & \cdots & a_{2n} \\ \cdots & \cdots & \cdots & \cdots \\ a_{n1} & a_{n2} & \cdots & a_{nn} \end{pmatrix} \text{ and } \mathbf{x} = \begin{pmatrix} x_1 \\ x_2 \\ \vdots \\ x_n \end{pmatrix}$$

is a transformation of n-component vector $\mathbf{x}$ into n-component vector $\mathbf{y}$ by the matrix of transformation $\mathbf{A}$.

The determinant $|\mathbf{A}|$ is called the *modulus of transformation.*

If $|\mathbf{A}| = 0$, the transformation matrix $\mathbf{A}$ is called *singular* otherwise *non-singular or regular.*

If $\mathbf{A}$ is a non-singular matrix, $\mathbf{A}^{-1}$ exists. Then from (5.29), we have

$\mathbf{A}^{-1}\mathbf{y} = \mathbf{A}^{-1}\mathbf{A}\mathbf{x} = \mathbf{x}$ or $\mathbf{x} = \mathbf{A}^{-1}\mathbf{y}$ which gives the inverse transformation.

Remark If a transformation from $(x_1, x_2, ..., x_n)$ to $(y_1, y_2, ..., y_n)$ is given by $\mathbf{y} = \mathbf{A}\mathbf{x}$ and another transformation from $(y_1, y_2, ..., y_n)$ to $(z_1, z_2, ..., z_n)$ is given by $\mathbf{z} = \mathbf{B}\mathbf{y}$, then the transformation from $(x_1, x_2, ..., x_n)$ to $(z_1, z_2, ..., z_n)$ is given by $\mathbf{z} = \mathbf{B}\mathbf{y} = \mathbf{B}(\mathbf{A}\mathbf{x}) = (\mathbf{B}\mathbf{A})\,\mathbf{x}$.

EXAMPLE 5.82 Show that the transformation

$$y_1 = x_1 - x_2 + x_3,\ y_2 = 3x_1 - x_2 + 2x_3,\ y_3 = 2x_1 - 2x_2 + 3x_3$$

is non-singular. Find the inverse transformation. *(AMIETE, June 2008)*

Solution In matrix notation, the given transformation is $\mathbf{y} = \mathbf{A}\mathbf{x}$,

where $$\mathbf{y} = \begin{pmatrix} y_1 \\ y_2 \\ y_3 \end{pmatrix}, \mathbf{A} = \begin{pmatrix} 1 & -1 & 1 \\ 3 & -1 & 2 \\ 2 & -2 & 3 \end{pmatrix}, \mathbf{x} = \begin{pmatrix} x_1 \\ x_2 \\ x_3 \end{pmatrix}.$$

Since $$|\mathbf{A}| = \begin{vmatrix} 1 & -1 & 1 \\ 3 & -1 & 2 \\ 2 & -2 & 3 \end{vmatrix} \begin{matrix} R_2 \to R_2 - 3R_1 \\ R_3 \to R_3 - 2R_1 \end{matrix} = \begin{vmatrix} 1 & -1 & 1 \\ 0 & 2 & -1 \\ 0 & 0 & 1 \end{vmatrix} = 1\begin{vmatrix} 2 & -1 \\ 0 & 1 \end{vmatrix} = 2 \neq 0,$$

the matrix $\mathbf{A}$ is non-singular and hence, the given transformation is non-singular or regular.

We compute $\mathbf{A}^{-1}$ using the Gauss-Jordan method. We have

$$\mathbf{A}|\mathbf{I} = \left[\begin{array}{ccc|ccc} 1 & -1 & 1 & 1 & 0 & 0 \\ 3 & -1 & 2 & 0 & 1 & 0 \\ 2 & -2 & 3 & 0 & 0 & 1 \end{array}\right] \begin{matrix} R_2 \to R_2 - 3R_1 \\ R_3 \to R_3 - 2R_1 \end{matrix} \approx \left[\begin{array}{ccc|ccc} 1 & -1 & 1 & 1 & 0 & 0 \\ 0 & 2 & -1 & -3 & 1 & 0 \\ 0 & 0 & 1 & -2 & 0 & 1 \end{array}\right] R_2 \to R_2/2$$

$$\approx \left[\begin{array}{ccc|ccc} 1 & -1 & 1 & 1 & 0 & 0 \\ 0 & 1 & -1/2 & -3/2 & 1/2 & 0 \\ 0 & 0 & 1 & -2 & 0 & 1 \end{array}\right] R_1 \to R_1 + R_2$$

$$\approx \left[\begin{array}{ccc|ccc} 1 & 0 & 1/2 & -1/2 & 1/2 & 0 \\ 0 & 1 & -1/2 & -3/2 & 1/2 & 0 \\ 0 & 0 & 1 & -2 & 0 & 1 \end{array}\right] \begin{matrix} R_1 \to R_1 - (R_3/2) \\ R_2 \to R_2 + (R_3/2) \end{matrix}$$

$$\approx\left[\begin{array}{ccc|ccc}1 & 0 & 0 & 1/2 & 1/2 & -1/2\\ 0 & 1 & 0 & -5/2 & 1/2 & 1/2\\ 0 & 0 & 1 & -2 & 0 & 1\end{array}\right].\quad \text{Hence } \mathbf{A}^{-1}=\frac{1}{2}\begin{pmatrix}1 & 1 & -1\\ -5 & 1 & 1\\ -4 & 0 & 2\end{pmatrix}.$$

The inverse transformation is given by $\mathbf{x} = \mathbf{A}^{-1}\mathbf{y}$

That is, $$\begin{pmatrix}x_1\\ x_2\\ x_3\end{pmatrix}=\frac{1}{2}\begin{pmatrix}1 & 1 & -1\\ -5 & 1 & 1\\ -4 & 0 & 2\end{pmatrix}\begin{pmatrix}y_1\\ y_2\\ y_3\end{pmatrix}$$

$$\Rightarrow \quad x_1=\frac{1}{2}(y_1+y_2-y_3),\ x_2=\frac{1}{2}(-5y_1+y_2+y_3),\ x_3=-2y_1+y_3.$$

EXAMPLE 5.83 Let

$$y_1 = 5x_1 + 3x_2 + 3x_3$$
$$y_2 = 3x_1 + 2x_2 - 2x_3$$
$$y_3 = 2x_1 - x_2 + 2x_3$$

be a linear transformation from (x_1, x_2, x_3) to (y_1, y_2, y_3) and $z_1 = 4x_1 + 2x_3,\ z_2 = x_2 + 4x_3,\ z_3 = 5x_3$ be a linear transformation from (x_1, x_2, x_3) to (z_1, z_2, z_3). Find the linear transformation from (z_1, z_2, z_3) to (y_1, y_2, y_3) by inverting appropriate matrix and matrix multiplication. *(AMIETE, Dec. 2004)*

Solution The transformation from the variables x_1, x_2, x_3 to y_1, y_2, y_3 is given by

$$\mathbf{y} = \mathbf{A}\mathbf{x}, \text{ where } \mathbf{A}=\begin{bmatrix}5 & 3 & 3\\ 3 & 2 & -2\\ 2 & -1 & 2\end{bmatrix}.$$

Another transformation from x_1, x_2, x_3 to z_1, z_2, z_3 is given by $\mathbf{z} = \mathbf{B}\mathbf{x}$, where $\mathbf{B}=\begin{bmatrix}4 & 0 & 2\\ 0 & 1 & 4\\ 0 & 0 & 5\end{bmatrix}$.

The required transformation is $\mathbf{y} = \mathbf{A}\mathbf{x} = (\mathbf{A}\mathbf{B}^{-1})\mathbf{z}$. We now determine $\mathbf{B}^{-1}$ by using Gauss-Jordon elimination method. We have

$$\mathbf{B}|\mathbf{I}=\left[\begin{array}{ccc|ccc}4 & 0 & 2 & 1 & 0 & 0\\ 0 & 1 & 4 & 0 & 1 & 0\\ 0 & 0 & 5 & 0 & 0 & 1\end{array}\right] R_1\to R_1/4 \approx\left[\begin{array}{ccc|ccc}1 & 0 & 1/2 & 1/4 & 0 & 0\\ 0 & 1 & 4 & 0 & 1 & 0\\ 0 & 0 & 5 & 0 & 0 & 1\end{array}\right] R_3\to R_3/5$$

$$\approx\left[\begin{array}{ccc|ccc}1 & 0 & 1/2 & 1/4 & 0 & 0\\ 0 & 1 & 4 & 0 & 1 & 0\\ 0 & 0 & 1 & 0 & 0 & 1/5\end{array}\right]\begin{matrix}R_1\to R_1-\frac{1}{2}R_3\\ R_2\to R_2-4R_3\end{matrix}\approx\left[\begin{array}{ccc|ccc}1 & 0 & 0 & 1/4 & 0 & -1/10\\ 0 & 1 & 0 & 0 & 1 & -4/5\\ 0 & 0 & 1 & 0 & 0 & 1/5\end{array}\right]. \text{ Therefore, } \mathbf{B}^{-1}=\begin{bmatrix}1/4 & 0 & -1/10\\ 0 & 1 & -4/5\\ 0 & 0 & 1/5\end{bmatrix}.$$

Hence $$\mathbf{A}\mathbf{B}^{-1}=\begin{bmatrix}5 & 3 & 3\\ 3 & 2 & -2\\ 2 & -1 & 2\end{bmatrix}\begin{bmatrix}1/4 & 0 & -1/10\\ 0 & 1 & -4/5\\ 0 & 0 & 1/5\end{bmatrix}=\begin{bmatrix}5/4 & 3 & -1/2-12/5+3/5\\ 3/4 & 2 & -3/10-8/5-2/5\\ 1/2 & -1 & -2/10+4/5+2/5\end{bmatrix}$$

$$=\begin{bmatrix}5/4 & 3 & -23/10\\ 3/4 & 2 & -23/10\\ 1/2 & -1 & 1\end{bmatrix}=\frac{1}{20}\begin{bmatrix}25 & 60 & -46\\ 15 & 40 & -46\\ 10 & -20 & 20\end{bmatrix}.$$

Thus $y_1=\frac{1}{20}(25z_1+60z_2-46z_3)$; $y_2=\frac{1}{20}(15z_1+40z_2-46z_3)$; $y_3=\frac{1}{2}(z_1-2z_2+2z_3)$ is the required transformation.

EXERCISE 5.8

1. Using the elementary row operations, determine the ranks of the following matrices.

(*i*) $\begin{bmatrix} 1 & -2 & 3 \\ -2 & 4 & -1 \\ -1 & 2 & 7 \end{bmatrix}$, (*ii*) $\begin{bmatrix} 3 & -1 & 2 \\ -6 & 2 & 4 \\ -3 & 1 & 2 \end{bmatrix}$, (*MDU* 2005, *Andhra*, 2000) (*iii*) $\begin{bmatrix} 0 & 1 & 2 & -2 \\ 4 & 0 & 2 & 6 \\ 2 & 1 & 3 & 1 \end{bmatrix}$,

(*iv*) $\begin{bmatrix} 1 & -1 & 2 & -3 \\ 4 & 1 & 0 & 2 \\ 0 & 3 & 0 & 4 \\ 0 & 1 & 0 & 2 \end{bmatrix}$, (*v*) $\begin{bmatrix} 1 & -1 & 1 & -1 \\ 4 & 2 & -1 & 2 \\ 2 & 2 & -2 & 2 \end{bmatrix}$, (*vi*) $\begin{bmatrix} 1 & 3 & 4 & 3 \\ 3 & 9 & 12 & 9 \\ -1 & -3 & -4 & -3 \end{bmatrix}$,

(*vii*) $\begin{bmatrix} 1 & 3 & 4 & 7 \\ 2 & 4 & 5 & 8 \\ 3 & 1 & 2 & 4 \end{bmatrix}$, (*viii*) $\begin{bmatrix} 1 & 0 & 2 & 1 \\ 0 & 1 & -2 & 1 \\ 1 & -1 & 4 & 0 \\ -2 & 2 & 8 & 0 \end{bmatrix}$, (*PTU, 2004*) (*ix*) $\begin{bmatrix} 9 & 3 & 1 & 0 \\ 3 & 0 & 1 & -6 \\ 1 & 1 & 1 & 1 \\ 0 & -6 & 1 & 9 \end{bmatrix}$ (*AMIETE Dec*. 2004),

(*x*) $\begin{bmatrix} 1 & 2 & 3 & 4 \\ 2 & 3 & 4 & 5 \\ 3 & 4 & 5 & 6 \\ 4 & 5 & 6 & 7 \end{bmatrix}$ (*AMIETE June*. 2002), (*xi*) $\begin{bmatrix} 1 & 2 & 3 & 2 \\ 1 & 4 & 2 & 5 \\ 2 & 6 & 5 & 7 \end{bmatrix}$ (*AMIETE Dec*. 2010).

Ans. (*i*) 2, (*ii*) 2, (*iii*) 2, (*iv*) 4, (*v*) 3; (*vi*) 1, (*vii*) 3, (*viii*) 3, (*ix*) 4, (*x*) 2, (*xi*) 2.

2. For which value of b the rank of the matrix

$$\mathbf{A} = \begin{bmatrix} 1 & 5 & 4 \\ 0 & 3 & 2 \\ b & 13 & 10 \end{bmatrix} \text{ is 2.}$$

(*AMIETE, Dec 2010)* **Ans.** 2

3. Reduce each of the following matrices to normal form and hence, find their ranks.

(*i*) $\begin{bmatrix} 8 & 1 & 3 & 6 \\ 0 & 3 & 2 & 2 \\ -8 & -1 & -3 & 4 \end{bmatrix}$ (*JNTU-2000*), (*ii*) $\begin{bmatrix} 2 & 1 & -3 & -6 \\ 3 & -3 & 1 & 2 \\ 1 & 1 & 1 & 2 \end{bmatrix}$ (*UPTU-2003*),

(*iii*) $\begin{bmatrix} 2 & 3 & -1 & -1 \\ 1 & -1 & -2 & -4 \\ 3 & 1 & 3 & -2 \\ 6 & 3 & 0 & -7 \end{bmatrix}$, (*AMIE - W 2007; UPTU - 2005; MDU 2000*) (*iv*) $\begin{bmatrix} 1 & 2 & -1 & 4 \\ 2 & 4 & 3 & 4 \\ 1 & 2 & 3 & 4 \\ -1 & -2 & 6 & -7 \end{bmatrix}$ (*UPTU* – 2005, 2002).

Ans. (*i*) 3, (*ii*) 3, (*iii*) 3, (*iv*) 3.

4. Find the non-singular matrices **P** and **Q** such that **PAQ** is in the normal form, where $\mathbf{A} = \begin{bmatrix} 1 & 1 & 1 \\ 1 & -1 & -1 \\ 3 & 1 & 1 \end{bmatrix}$

Ans. $\mathbf{P} = \begin{bmatrix} 1/2 & 1/2 & 0 \\ 1/2 & -1/2 & 0 \\ -2 & -1 & 1 \end{bmatrix}, \mathbf{Q} = \begin{bmatrix} 1 & 0 & 0 \\ 0 & 1 & -1 \\ 0 & 0 & 1 \end{bmatrix}$.

5. If $\mathbf{A}=\begin{bmatrix}3 & -3 & 4\\ 2 & -3 & 4\\ 0 & -1 & 1\end{bmatrix}$, find $\mathbf{A}^{-1}$. Also find non-singular matrices **P** and **Q** such that **PAQ = I**, where **I** is a unit matrix and verify that $\mathbf{A}^{-1}=\mathbf{QP}$.

Ans. $\mathbf{P}=\begin{bmatrix}1 & -1 & 0\\ 0 & 0 & 1\\ -2 & 3 & -3\end{bmatrix}, \mathbf{Q}=\begin{bmatrix}1 & 0 & 0\\ 0 & -1 & 1\\ 0 & 0 & 1\end{bmatrix}.$

6. Find the non-singular matrices **R** and **S**, such that **RAS** is in the normal form, where $\mathbf{A}=\begin{bmatrix}2 & 2 & -6\\ -1 & 2 & 2\end{bmatrix}$.

Ans. $\mathbf{R}=\begin{bmatrix}1/2 & 0\\ 1/2 & 1\end{bmatrix}, \mathbf{S}=\begin{bmatrix}1 & -1/3 & 8/3\\ 0 & 1/3 & 1/3\\ 0 & 0 & 1\end{bmatrix}.$

7. Find the non-singular matrices **P** and **Q** such that **PAQ** is in the normal form where

(*i*) $\mathbf{A}=\begin{bmatrix}1 & -1 & -1\\ 1 & 1 & 1\\ 3 & 1 & 1\end{bmatrix}$, (*ii*) $\mathbf{A}=\begin{bmatrix}1 & 1 & 2\\ 1 & 2 & 3\\ 0 & -1 & -1\end{bmatrix}$. (*UPTU 2001, MDU 2001*)

Ans. (*i*) $\mathbf{P}=\begin{bmatrix}1 & 0 & 0\\ -1/2 & 1/2 & 0\\ 1/2 & 1 & -1/2\end{bmatrix}, \quad \mathbf{Q}=\begin{bmatrix}1 & 1 & 0\\ 0 & 1 & -1\\ 0 & 0 & 1\end{bmatrix}.$ (*ii*) $\mathbf{P}=\begin{bmatrix}1 & 0 & 0\\ -1 & 1 & 0\\ -1 & 1 & 0\end{bmatrix}, \quad \mathbf{Q}=\begin{bmatrix}1 & -1 & -1\\ 0 & 1 & -1\\ 0 & 0 & 1\end{bmatrix}.$

8. Test for consistency and hence solve:

(*i*) $x+y+z=6,\ x-y+2z=5,\ 3x+y+z=8,\ 2x-2y+3z=7.$

(*ii*) $5x+3y+7z=4,\ 3x+26y+2z=9,\ 7x+2y+10z=5.$

(*AMIETE June 2009, 2008; PTU 2005; VTU 2004*)

(*iii*) $2x-3y+7z=5,\ 3x+y-3z=13,\ 2x+19y-47z=32.$

(*GGSIPU 2005; MDU 2003, D.U. 2002, AMIE, S-2002, Andhra, 2000*)

Ans. (*i*) 1, 2, 3, (*ii*) $x=(7-16z)/11,\ y=(z+3)/11$, (*iii*) inconsistent and possesses no solution.

9. Investigate for consistency of the following systems and find the solution if it exists.

(*i*) $4x+7y-5z+3=0,\ 9x-11y+5z+1=0,\ 11x-9y+7z-5=0,\ 3y+10z-23=0.$

(*ii*) $3x+3y+2z=1,\ x+2y=4,\ 10y+3z=-2.$ (*AMIETE, June 2011*)

(*iii*) $4x-2y+6z=8, x+y-3z=-1, 15x-3y+9z=21.$ (*AMIE, June 2011; AMIETE, June 2009*)

Ans. (*i*) 0, 1, 2; (*ii*) 2, 1, –4. (*iii*) Consistent; $x=1,\ y=3k-2,\ z=k$ for all k.

10. Show that the equations

$$3x+4y+5z=a,\ 4x+5y+6z=b,\ 5x+6y+7z=c$$

do not have a solution unless $a+c=2b$. Solve the equations when $a=b=c=-1$.

(*KUK, 2006; UPTU, 2004*)

Ans. $x=1+z,\ y=-(1+2z)$ and z is arbitrary.

11. Investigate the values of μ and λ, so that the equations

$$2x+3y+5z=9,\quad 7x+3y-2z=8,\quad 2x+3y+\lambda z=\mu \text{ has}$$

(*i*) a unique solution; (*ii*) an infinite number of solutions (*iii*) no solution.

(*AMIETE, June 2007; Ranchi 2000*)

Ans. (*i*) $\lambda\neq 5$, μ may have any value; (*ii*) $\lambda=5,\ \mu=9$; (*iii*) $\lambda=5$ and $\mu\neq 9$.

12. For what value of λ the equations $x + y + z = 1$, $x + 2y + 4z = \lambda$, $x + 4y + 10z = \lambda^2$, $3x + 7y + 15z = 3$ are consistent. Solve the set of equations for such a value of λ. *(AMIE, S-2003)*

Ans. $\lambda = 1$, $x = 1 + 2z$, $y = -3z$, and z is arbitrary.

13. Find the values of a and b for which the equations

$$x + ay + z = 3, \quad x + 2y + 2z = b, \quad x + 5y + 3z = 9$$

are consistent. When will these equations have a unique solution? *(AMIETE, Dec 2010)*

Ans. For a unique solution $a \neq -1$ and b can have any value. For $a = -1$, $b = 6$ the system has infinite solutions.

14. Find the values of the parameter λ for which the system of equations

$$x + y + 4z = 1, \quad x + 2y - 2z = 1, \quad \lambda x + y + z = 1$$

will have (*i*) a unique solution (*ii*) no solution. Also find the solution for $\lambda = 1/2$ by applying elementary row transformations to the augmented matrix. *(AMIE, W-2000)*

Ans. (*i*) $\lambda \neq 7/10$, (*ii*) $\lambda = 7/10$; $[-3/2, 3/2, 1/4]$.

15. For what values of k the equations $x + y + z = 1$, $2x + y + 4z = k$, $4x + y + 10z = k^2$ have a solution and solve them completely in each case. *(KUK, 2005; PTU 2005; VTU 2004)*

Ans. For $k = 1$, $x = -3\lambda$, $y = 1 + 2\lambda$, $z = \lambda$; and for $k = 2$, $x = 1 - 3\lambda$, $y = 2\lambda$, $z = \lambda$.

16. Determine the values of λ for which the system of equations

$$x - \lambda y + z = 0, \quad \lambda x + 3y - \lambda z = 0, \quad 3x + y - z = 0$$

has (*i*) only trivial solution, (*ii*) non-trivial solution.

Ans. (*i*) $\lambda \neq 2$ and $\lambda \neq -3$, (*ii*) $\lambda = 2$, or $\lambda = -3$.

17. For what values of k will be system

$$2x + ky + z = 0, \quad (k - 1)x - y + 2z = 0, \quad 4x + y + 4z = 0$$

have non-trivial solutions? **Ans.** $k = 1, 9/4$.

18. Consider the system of equations:

$$x + 2y + z = 3, \quad \lambda y + 5z = 10, \quad 2x + 7y + \lambda z = \mu.$$

Find (*a*) those values of λ for which the system has a unique solution (*b*) those pairs of values (λ, μ) for which the system has more than one solution.

Ans. (*a*) $\lambda \neq 5$ and $\lambda \neq -3$; (*b*) (5, 12) and (–3, –4).

19. Discuss the consistency of the following system of equations for various values of λ:
$2x_1 - 3x_2 + 6x_3 - 5x_4 = 3$, $x_2 - 4x_3 + x_4 = 1$, $4x_1 - 5x_2 + 8x_3 - 9x_4 = \lambda$ and if consistent, solve it. *(AMIETE, Dec. 2006)*

Ans. $\lambda = 7$; two parameter family of solutions $x_1 = 3 + 3x_3 + x_4$ and $x_2 = 1 + 4x_3 - x_4$, where x_3 and x_4 are arbitrary.

20. Test for consistency the following system of equations, and if consistent, solve them.
$x_1 + 2x_2 - x_3 = 3$; $3x_1 - x_2 + 2x_3 = 1$; $2x_1 - 2x_2 + 3x_3 = 2$; $x_1 - x_2 + x_3 = -1$. *(AMIETE, Dec. 2009)*

Ans. The system of equations has a unique solution $x_1 = -1$, $x_2 = 4$ and $x_3 = 4$.

21. (*a*) Solve system of equations by matrix method

$$5x + 3y + 14z = 4, \quad y + 2z = 1, \quad 2x + y + 6z = 2, \quad x + y + 2z = 0.$$

(AMIETE, June 2006) **Ans.** $[-1, -1/2, 3/4]$.

(*b*) Solve the following system of linear equations by Gauss-elimination method.

$$2x + y + z = 10, \quad 3x + 2y + 3z = 18, \quad x + 4y + 9z = 16.$$

(AMIE, S-2009, W-2008) **Ans.** $x = 7$, $y = -9$, $z = 5$.

22. Examine for non-trivial solution the system:

$$7x_1 + x_2 - 2x_3 = 0, \quad x_1 + 5x_2 - 4x_3 = 0, \quad 3x_1 - 2x_2 + x_3 = 0.$$

Ans. $x_1 = 3t$, $x_2 = 13t$, $x_3 = 17t$, where t is arbitrary.

23. Investigate the values of λ and μ so that the equations

$$x + 2y + 3z = 6, \quad 3x + 4y - 10z = 1, \quad 3x + 4y + \lambda z = \mu$$

have (*i*) unique solution, (*ii*) infinite number of solutions (*iii*) no solution.

Ans. (*i*) $\lambda \neq -10$, μ may have any value; (*ii*) $\lambda = -10, \mu = 1$; (*iii*) $\lambda = -10, \lambda \neq 1$.

24. Determine the condition so that the equations

$$x + y + z = 1, \quad x + 2y + 3z = 2, \quad 2x + 3y + \lambda z = \mu$$

have (*i*) unique solution, (*ii*) infinite many solutions, (*iii*) no solution.

Ans. (*i*) $\lambda \neq 4$, μ may have any value, (*ii*) $\lambda = 4, \mu = 3$, (*iii*) $\lambda = 4, \mu \neq 3$.

25. Using the Gauss-Jordan elimination method, to find the inverses of the following matrices.

(*i*) $\begin{bmatrix} 0 & 1 & 2 \\ 1 & 2 & 3 \\ 3 & 1 & 1 \end{bmatrix}$, (*UPTU – 2001*) (*ii*) $\begin{bmatrix} 1 & 1 & 3 \\ 1 & 3 & -3 \\ -2 & -4 & -4 \end{bmatrix}$, (*Kurukshetra, 2006*) (*iii*) $\begin{bmatrix} 2 & 0 & 1 \\ -2 & 3 & 4 \\ -5 & 5 & 6 \end{bmatrix}$,

(*iv*) $\begin{bmatrix} -1 & 1 & 2 \\ 3 & -1 & 1 \\ -1 & 3 & 4 \end{bmatrix}$, (*v*) $\begin{bmatrix} 1 & 1 & 1 & 1 \\ -2 & 1 & -1 & -2 \\ -4 & -2 & -3 & 1 \\ 1 & 1 & 1 & 0 \end{bmatrix}$, (*vi*) $\begin{bmatrix} 8 & 4 & -3 \\ 2 & 1 & 1 \\ 1 & 2 & 1 \end{bmatrix}$, (*GGSIPU -2005*)

(*vii*) $\begin{bmatrix} 1 & 2 & 3 \\ 2 & 4 & 5 \\ 3 & 5 & 6 \end{bmatrix}$, (*AMIETE, June 2009*) (*viii*) $\begin{bmatrix} 2 & 1 & -1 \\ 0 & 2 & 1 \\ 5 & 2 & -3 \end{bmatrix}$, (*AMIETE, Dec 2010; NIT Kurushetra, 2008*)

Ans. (*i*) $\frac{1}{2}\begin{bmatrix} 1 & -1 & 1 \\ -8 & 6 & -2 \\ 5 & -3 & 1 \end{bmatrix}$, (*ii*) $\frac{1}{4}\begin{bmatrix} 12 & 4 & 6 \\ -5 & -1 & -3 \\ -1 & -1 & -1 \end{bmatrix}$, (*iii*) $\begin{bmatrix} -2 & 5 & -3 \\ -8 & 17 & -10 \\ -5 & -10 & 6 \end{bmatrix}$,

(*iv*) $\frac{1}{10}\begin{bmatrix} -7 & 2 & 3 \\ -13 & -2 & 7 \\ 8 & 2 & -2 \end{bmatrix}$, (*v*) $\begin{bmatrix} 4 & 1 & -2 & -9 \\ 3 & 1 & -1 & -5 \\ -7 & -2 & 3 & 15 \\ -1 & 0 & 0 & -1 \end{bmatrix}$, (*vi*) $\frac{1}{21}\begin{bmatrix} 1 & 10 & -7 \\ 1 & -11 & 14 \\ -3 & 12 & 0 \end{bmatrix}$,

(*vii*) $\begin{bmatrix} 1 & -3 & 2 \\ -3 & 3 & -1 \\ 2 & -1 & 0 \end{bmatrix}$ (*viii*) $\begin{bmatrix} 8 & -1 & -3 \\ -5 & 1 & 2 \\ 10 & -1 & -4 \end{bmatrix}$.

26. Using matrix method, show that the equations

$$3x_1 + 2x_2 + 2x_3 - 5x_4 = 8, \quad 2x_1 + 5x_2 + 5x_3 - 18x_4 = 9, \quad 4x_1 - x_2 - x_3 + 8x_4 = 7.$$

are consistent and hence solve them. (*AMIE, S-2009*)

[**Hint:** The augmented matrix, (**A/b**) = $\left[\begin{array}{cccc|c} 3 & 2 & 2 & -5 & 8 \\ 2 & 5 & 5 & -18 & 9 \\ 4 & -1 & -1 & 8 & 7 \end{array}\right]$.

Applying the row operations. Finally we arrive at the augmented matrix (*i*).

$$\begin{bmatrix} 3 & 2 & 2 & -5 & 8 \\ 2 & 5 & 5 & -18 & 9 \\ 4 & -1 & -1 & 8 & 7 \end{bmatrix} \approx \begin{bmatrix} 1 & 2/3 & 2/3 & -5/3 & 8/3 \\ 2 & 5 & 5 & -18 & 9 \\ 4 & -1 & -1 & 8 & 7 \end{bmatrix} \approx \begin{bmatrix} 1 & 2/3 & 2/3 & -5/3 & 8/3 \\ 0 & 11/3 & 11/3 & -44/3 & 11/3 \\ 0 & -11/3 & -11/3 & 44/3 & -11/3 \end{bmatrix}$$

$$\approx \begin{bmatrix} 1 & 2/3 & 2/3 & -5/3 & 8/3 \\ 0 & 1 & 1 & -4 & 1 \\ 0 & -1 & -1 & 4 & -1 \end{bmatrix} \approx \begin{bmatrix} 1 & 2/3 & 2/3 & -5/3 & 8/3 \\ 0 & 1 & 1 & -4 & 1 \\ 0 & 0 & 0 & 0 & 0 \end{bmatrix} \approx \begin{bmatrix} 1 & 0 & 0 & 1 & 2 \\ 0 & 1 & 1 & -4 & 1 \\ 0 & 0 & 0 & 0 & 0 \end{bmatrix} \cdots\cdots (i)$$

The corresponding system of equations can be solved for x_1, x_2 in terms of x_3 and x_4, giving us

$$x_1 = 2 - x_4, \quad x_2 = 1 - x_3 + 4x_4.$$

If we let $x_3 = t_1$ and $x_4 = t_2$, where t_1 and t_2 are arbitrary real numbers, the vector (x_1, x_2, x_3, x_4) in V_4 is given by

$$(x_1, x_2, x_3, x_4) = (2 - t_2, 1 - t_1 + 4t_2, t_1, t_2)$$

is a solution. By separating the parts involving t_1 and t_2, we can rewrite this as follows:

$$(x_1, x_2, x_3, x_4) = (2, 1, 0, 0) + t_1 (0 - 1, 1, 0) + t_2 (-1, 4, 0, 1).$$

This equation gives the general solution of the system. The vector (2, 1, 0, 0) is a particular solution of the non-homogeneous system. The two vectors (0, –1, 1, 0) and (–1, 4, 0, 1) are solution of the corresponding homogeneous system.]

27. Show that the transformation

$$y_1 = 2x_1 + x_2 + x_3, \quad y_2 = x_1 + x_2 + 2x_3, \quad y_3 = x_1 - 2x_3$$

is non-singular. Write down the inverse transformation.(*AMIETE, Dec. 2008*)

Ans. $x_1 = 2y_1 - 2y_2 - y_3$; $x_2 = -4y_1 + 5y_2 + 3y_3$; $x_3 = y_1 - y_2 - y_3$.

28. A transformation from the variables x_1, x_2, x_3 to y_1, y_2, y_3 is given by **Y** = **AX**, and another transformation from y_1, y_2, y_3 to z_1, z_2, z_3 is given by **Z** = **BY**, where

$$\mathbf{A} = \begin{bmatrix} 2 & 1 & 0 \\ 0 & 1 & -2 \\ -1 & 2 & 1 \end{bmatrix}, \ \mathbf{B} = \begin{bmatrix} 1 & 1 & 1 \\ 1 & 2 & 3 \\ 1 & 3 & 5 \end{bmatrix}$$

obtain the transformation from x_1, x_2, x_3 to z_1, z_2, z_3.

[**Hint:** The required transformation is **Z** = **BY** = **B** (**AX**) = (**BA**) **X**,

$$\text{where } \mathbf{BA} = \begin{bmatrix} 1 & 1 & 1 \\ 1 & 2 & 3 \\ 1 & 3 & 5 \end{bmatrix} \begin{bmatrix} 2 & 1 & 0 \\ 0 & 1 & -2 \\ -1 & 2 & 1 \end{bmatrix} = \begin{bmatrix} 1 & 4 & -1 \\ -1 & 9 & -1 \\ -3 & 14 & -1 \end{bmatrix}.$$

Ans. $z_1 = x_1 + 4x_2 - x_3$; $z_2 = -x_1 + 9x_2 - x_3$; $z_3 = -3x_1 + 14x_2 - x_3$.

29. Represent each of the transformations

$$x_1 = 3y_1 + 2y_2; \quad y_1 = z_1 + 2z_2 \quad \text{and} \quad x_2 = -y_1 + 4y_2, \quad y_3 = 3z_1$$

by the use of matrices and find the composite transformation which expresses x_1, x_2 in terms of z_1, z_2.

$$\textbf{Ans.} \begin{pmatrix} x_1 \\ x_2 \end{pmatrix} = \begin{pmatrix} 3 & 2 \\ -1 & 4 \end{pmatrix} \begin{pmatrix} y_1 \\ y_2 \end{pmatrix}; \begin{pmatrix} y_1 \\ y_2 \end{pmatrix} = \begin{pmatrix} 1 & 2 \\ 3 & 0 \end{pmatrix} \begin{pmatrix} z_1 \\ z_2 \end{pmatrix}; \begin{pmatrix} x_1 \\ x_2 \end{pmatrix} = \begin{pmatrix} 9 & 6 \\ 11 & -2 \end{pmatrix} \begin{pmatrix} z_1 \\ z_2 \end{pmatrix}.$$

CHAPTER

6

Linear Algebra: Matrix Eigenvalue Problems

6.1 EIGENVALUES AND EIGENVECTORS

Let $A = [a_{ij}]$ be a given square matrix of order n. Consider the vector equation

$$\mathbf{Ax} = \lambda\mathbf{x} \text{ or in matrix notation, } (\mathbf{A} - \lambda\mathbf{I})\,\mathbf{x} = \mathbf{0} \qquad ...(6.1)$$

where λ is a scalar, **I** is an identity matrix of order n, **x** is an solution vector. If we let $\mathbf{x} = [x_1, x_2, ..x_n]^T$ then (6.1) is the same as

$$\left.\begin{array}{l}(a_{11}-\lambda)x_1 + a_{12}x_2 + ... + a_{1n}x_n = 0 \\ a_{21}x_1 + (a_{22}-\lambda)x_2 + ... + a_{2n}x_n = 0 \\ \dots\dots\dots\dots\dots\dots\dots\dots\dots\dots \\ a_{n1}x_1 + a_{n2}x_2 + ... + (a_{nn}-\lambda)x_n = 0\end{array}\right\} \qquad ...(6.2)$$

Although an obvious solution of homogeneous system (6.2) is $x_1 = 0$, $x_2 = 0$, ...$x_n = 0$, for any value of λ. This is of no practical interest. We are seeking only non-trivial solutions. We know that a homogeneous system of n linear equations in n unknowns has a nontrivial solution if and only if the determinant of the coefficient matrix is equal to zero. Thus to find a non zero solution of **x** for (6.1), we must have

$$|\mathbf{A} - \lambda\mathbf{I}| = 0. \qquad ...(6.3)$$

The expansion of $|\mathbf{A} - \lambda\mathbf{I}|$ by cofactors results in an nth degree polynomial in λ, which is of the form $\lambda^n - k_1\lambda^{n-1} + k_2\lambda^{n-2} - ... + (-1)^n k_n = 0$, where $k_1, k_2, ..., k_n$ are functions of the a's. The equation (6.3) is called **characteristic equation** of the matrix **A**. The polynomial equation of degree n in λ will have n roots ($\lambda_1, \lambda_2, ...\lambda_n$) which can be real or complex, simple or repeated. Thus, the **eigenvalues** or the **characteristic values** (or *latent roots*) of the matrix **A** are the roots of the **characteristic equation.** The corresponding nontrivial solution vectors **x** are called the eigenvectors or the characteristic vectors of **A** corresponding to that eigenvalue λ.

If **x** is a nontrivial solution of the homogeneous system (6.2), then k**x**, where k is any constant is also a solution of the homogeneous system. Hence, an eigenvector is unique only upto a constant multiple. Thus the eigenvector corresponding to an eigenvalue is not unique.

The set of the eigenvalues is called the **spectrum** of **A**. The largest of the absolute values of the eigenvalues of **A** is called the **spectral radius** of **A** and is denoted by $\rho(\mathbf{A})$. After determining the eigenvalues λ_i's, we solve the homogeneous system $(\mathbf{A} - \lambda_i\mathbf{I})\,\mathbf{x} = \mathbf{0}$ for each λ_i, $i = 1, 2, ..., n$ to obtain the corresponding eigenvectors.

The set of all eigenvectors corresponding to an eigenvalue of **A**, together with **O**, forms a vector space called the **eigenspace** of **A** corresponding to this eigenvalue. The problem of determining the eigenvalues and eigenvectors of a matrix is called an *eigenvalue problem.*

Remark: A square matrix **A** of order n has always n linearly independent eigenvectors when its eigenvalues are distinct. The matrix **A** may also have n linearly independent eigenvectors even when some eigenvalues are repeated.

6.1.1 Properties of Eigenvalues

We give below, without proof, some important properties of eigenvalues.

1. The sum of the eigenvalues of a matrix **A** is equal to the *trace* of the matrix A.
2. The product of the eigenvalues of a matrix **A** is equal to the determinant of **A**. Note that matrix **A** is singular if and only if at least one of its eigenvalues is zero.

3. If **A** is a square matrix, **A** and $\mathbf{A}^T$ have the same eigenvalues, since a determinant can be expanded by rows or columns.
4. The eigenvalues of an upper triangular, lower triangular, or diagonal matrix are the main diagonal entries.
5. If $\lambda_1, \lambda_2, ..., \lambda_n$ are the eigenvalues of **A**, then the eigenvalues of

   (*i*) $k\mathbf{A}$ are $k\lambda_1, k\lambda_2, ..., k\lambda_n$. (*ii*) $\mathbf{A}^m$ are $\lambda_1^m, \lambda_2^m, ..., \lambda_n^m$

   (*iii*) $\mathbf{A}^{-1}$ (if it exists) are $1/\lambda_1, 1/\lambda_2, ..., 1/\lambda_n$.
6. $\mathbf{A} - k\mathbf{I}$ has the eigenvalue $\lambda_1 - k, ..., \lambda_n - k$, for any scalar k.
7. $(\mathbf{A} - k\mathbf{I})^{-1}$ has the eigenvalue $1/(\lambda_1 - k), ..., 1/(\lambda_n - k)$.
8. A real matrix **A** may have complex eigenvalues. That is if $\alpha + i\beta$ is an eigenvalue, then its conjugate $\alpha - i\beta$ is also an eigenvalue.
9. The eigenvalues of a symmetric matrix are real.
10. The eigenvalues of a skew-symmetric matrix are pure imaginary or zero.

6.2 CAYLEY-HAMILTON THEOREM

Statement: Every square matrix **A** satisfies its own characteristic equation.

Proof: Let
$$\mathbf{A} = \begin{bmatrix} a_{11} & a_{12} & \cdots & a_{1n} \\ a_{21} & a_{22} & \cdots & a_{2n} \\ \cdots & \cdots & \cdots & \cdots \\ a_{n1} & a_{n2} & \cdots & a_{nn} \end{bmatrix}.$$

Then characteristic equation of **A** is
$$|\mathbf{A} - \lambda\mathbf{I}| = 0.$$

Let $|\mathbf{A} - \lambda\mathbf{I}| = p_0 + p_1\lambda + p_2\lambda^2 + ... + p_n\lambda^n$ where $p_n = (-1)^n$.

$\therefore$ Characteristic equation is
$$p_0 + p_1\lambda + p_2\lambda^2 + ... + p_n\lambda^n = 0.$$

Now we have to prove that
$$p_0\mathbf{I} + p_1\mathbf{A} + p_2\mathbf{A}^2 + ... + p_n\mathbf{A}^n = \mathbf{0}.$$

Let
$$\mathbf{B} = Adj(\mathbf{A} - \lambda\mathbf{I}).$$

Since co-factors of elements in $|\mathbf{A} - \lambda\mathbf{I}|$ are polynomials in λ of degree $(n - 1)$ or less, the elements of **B** are also such polynomials.

We can, therefore break up **B** into a number of matrices each containing the same power of λ and write
$$\mathbf{B} = \mathbf{C}_0 + \mathbf{C}_1\lambda + \mathbf{C}_2\lambda^2 + ... + \mathbf{C}_{n-1}\lambda^{n-1}$$
where $\mathbf{C}_0, \mathbf{C}_1, \mathbf{C}_2, ..., \mathbf{C}_{n-1}$ are square matrices of order n whose elements are functions of the elements of **A**.

Then
$$(\mathbf{A} - \lambda\mathbf{I})\mathbf{B} = |\mathbf{A} - \lambda\mathbf{I}|\mathbf{I}$$
$$\Rightarrow (\mathbf{A} - \lambda\mathbf{I})(\mathbf{C}_0 + \mathbf{C}_1\lambda + \mathbf{C}_2\lambda^2 + ... + \mathbf{C}_{n-1}\lambda^{n-1}) = (p_0 + p_1\lambda + p_2\lambda^2 + ... + p_n\lambda^n)\mathbf{I}.$$

Equating Coeff. of different powers of λ, we get
$$\mathbf{A}\mathbf{C}_0 = p_0\mathbf{I}$$
$$\mathbf{A}\mathbf{C}_1 - \mathbf{I}\mathbf{C}_0 = p_1\mathbf{I}$$
$$\mathbf{A}\mathbf{C}_2 - \mathbf{I}\mathbf{C}_1 = p_2\mathbf{I}$$
$$\cdots \quad \cdots \quad \cdots \quad \cdots \quad \cdots$$
$$\mathbf{A}\mathbf{C}_{n-1} - \mathbf{I}\mathbf{C}_{n-2} = p_{n-1}\mathbf{I}$$
$$-\mathbf{I}\mathbf{C}_{n-1} = p_n\mathbf{I}.$$

Pre-multiplying the above equations by 1, **A**, $\mathbf{A}^2$, ...$\mathbf{A}^n$ respectively and adding, we get
$$p_0\mathbf{I} + p_1\mathbf{A} + p_2\mathbf{A}^2 + ... + p_n\mathbf{A}^n$$
$$= \mathbf{A}\mathbf{C}_0 + (\mathbf{A}^2\mathbf{C}_1 - \mathbf{A}\mathbf{C}_0) + (\mathbf{A}^3\mathbf{C}_2 - \mathbf{A}^2\mathbf{C}_1) + ... + (\mathbf{A}^n\mathbf{C}_{n-1} - \mathbf{A}^{n-1}\mathbf{C}_{n-2}) - \mathbf{A}^n\mathbf{C}_{n-1} = \mathbf{0}$$

that is, $$p_0\mathbf{I} + p_1\mathbf{A} + p_2\mathbf{A}^2 + \dots + p_n\mathbf{A}^n = \mathbf{0} \qquad \dots(6.4)$$

which proves the theorem.

Note: We can use (6.4) to find $\mathbf{A}^{-1}$ (if it exists) in terms of the powers of the matrix **A**. Pre-multiplying equation (6.4) by $\mathbf{A}^{-1}$, we have

$$p_0\mathbf{A}^{-1} + p_1\mathbf{I} + p_2\mathbf{A} + \dots + p_n\mathbf{A}^{n-1} = \mathbf{0} \Rightarrow \mathbf{A}^{-1} = \frac{1}{p_0}(-p_1\mathbf{I} - p_2\mathbf{A} \dots\dots - p_n\mathbf{A}^{n-1}).$$

ILLUSTRATIVE EXAMPLES

EXAMPLE 6.1 Verify that $\mathbf{x} = [1, -1, 1]^T$ is an eigenvector of the matrix.

$$\mathbf{A} = \begin{bmatrix} 0 & -1 & -3 \\ 2 & 3 & 3 \\ -2 & 1 & 1 \end{bmatrix}.$$

Solution By carrying out the multiplication **Ax**, we obtain

$$\mathbf{Ax} = \begin{bmatrix} 0 & -1 & -3 \\ 2 & 3 & 3 \\ -2 & 1 & 1 \end{bmatrix}\begin{bmatrix} 1 \\ -1 \\ 1 \end{bmatrix} = \begin{bmatrix} 1-3 \\ 2-3+3 \\ -2-1+1 \end{bmatrix} = \begin{bmatrix} -2 \\ 2 \\ -2 \end{bmatrix} = (-2)\begin{bmatrix} 1 \\ -1 \\ 1 \end{bmatrix} = (-2)\mathbf{x}.$$

(eigenvalue ↓)

We see from the definition that $\lambda = -2$ is an eigenvalue of **A**.

EXAMPLE 6.2 Verify Cayley-Hamilton theorem for the matrix

$$\mathbf{A} = \begin{bmatrix} 7 & -1 & 3 \\ 6 & 1 & 4 \\ 2 & 4 & 8 \end{bmatrix}.$$ Also obtain $\mathbf{A}^{-1}$ and $\mathbf{A}^3$. *(MDU 2003)*

Solution The characteristic equation of **A** is given by

$$|\mathbf{A} - \lambda\mathbf{I}| = \begin{vmatrix} 7-\lambda & -1 & 3 \\ 6 & 1-\lambda & 4 \\ 2 & 4 & 8-\lambda \end{vmatrix} = (7-\lambda)[(1-\lambda)(8-\lambda) - 16] + 1[6(8-\lambda) - 8] + 3[24 - 2(1-\lambda)]$$

$$= (7-\lambda)[\lambda^2 - 9\lambda - 8] + (40 - 6\lambda) + 3(22 + 2\lambda)$$
$$= -\lambda^3 + 16\lambda^2 - 55\lambda + 50 = 0.$$

$$\text{Now } \mathbf{A}^2 = \begin{bmatrix} 7 & -1 & 3 \\ 6 & 1 & 4 \\ 2 & 4 & 8 \end{bmatrix}\begin{bmatrix} 7 & -1 & 3 \\ 6 & 1 & 4 \\ 2 & 4 & 8 \end{bmatrix} = \begin{bmatrix} 49 & 4 & 41 \\ 56 & 11 & 54 \\ 54 & 34 & 86 \end{bmatrix}.$$

$$\mathbf{A}^3 = \mathbf{A}^2\mathbf{A} = \begin{bmatrix} 49 & 4 & 41 \\ 56 & 11 & 54 \\ 54 & 34 & 86 \end{bmatrix}\begin{bmatrix} 7 & -1 & 3 \\ 6 & 1 & 4 \\ 2 & 4 & 8 \end{bmatrix} = \begin{bmatrix} 449 & 119 & 491 \\ 566 & 171 & 644 \\ 754 & 324 & 986 \end{bmatrix}.$$

$$\text{We have } -\mathbf{A}^3 + 16\mathbf{A}^2 - 55\mathbf{A} + 50\mathbf{I} = -\begin{bmatrix} 449 & 119 & 491 \\ 566 & 171 & 644 \\ 754 & 324 & 986 \end{bmatrix} + 16\begin{bmatrix} 49 & 4 & 41 \\ 56 & 11 & 54 \\ 54 & 34 & 86 \end{bmatrix}$$

$$-55\begin{bmatrix}7 & -1 & 3\\ 6 & 1 & 4\\ 2 & 4 & 8\end{bmatrix}+50\begin{bmatrix}1 & 0 & 0\\ 0 & 1 & 0\\ 0 & 0 & 1\end{bmatrix}=\begin{bmatrix}0 & 0 & 0\\ 0 & 0 & 0\\ 0 & 0 & 0\end{bmatrix}=\mathbf{0}. \quad ...(i)$$

Hence, **A** satisfies the characteristic equation $-\lambda^3 + 16\lambda^2 - 55\lambda + 50 = 0$.
From equation (*i*), we obtain

$$\mathbf{A}^{-1}=\frac{1}{50}[\mathbf{A}^2-16\mathbf{A}+55\mathbf{I}]=\frac{1}{50}\left[\begin{pmatrix}49 & 4 & 41\\ 56 & 11 & 54\\ 54 & 34 & 86\end{pmatrix}-\begin{pmatrix}112 & -16 & 48\\ 96 & 16 & 64\\ 32 & 64 & 128\end{pmatrix}+\begin{pmatrix}55 & 0 & 0\\ 0 & 55 & 0\\ 0 & 0 & 55\end{pmatrix}\right]$$

$$=\frac{1}{50}\begin{bmatrix}-8 & 20 & -7\\ -40 & 50 & -10\\ 22 & -30 & 13\end{bmatrix}.$$

From equation (*i*), we get

$$\mathbf{A}^3=16\mathbf{A}^2-55\mathbf{A}+50\mathbf{I}=\begin{bmatrix}784 & 64 & 656\\ 896 & 176 & 864\\ 864 & 544 & 1376\end{bmatrix}-\begin{bmatrix}385 & -55 & 165\\ 330 & 55 & 220\\ 110 & 220 & 440\end{bmatrix}+\begin{bmatrix}50 & 0 & 0\\ 0 & 50 & 0\\ 0 & 0 & 50\end{bmatrix}=\begin{bmatrix}449 & 119 & 491\\ 566 & 171 & 644\\ 754 & 324 & 986\end{bmatrix}.$$

EXAMPLE 6.3 If $\mathbf{A}=\begin{bmatrix}2 & 2 & 0\\ 2 & 1 & 1\\ -7 & 2 & -3\end{bmatrix}$, find the eigenvalues of the matrix **A**, $\mathbf{A}^2$ and verify that eigenvalues of $\mathbf{A}^2$ are squares of those of **A**. Also determine the spectral radius of **A**.

Solution The characteristic equation of **A** is given by

$$|\mathbf{A}-\lambda\mathbf{I}|=\begin{vmatrix}2-\lambda & 2 & 0\\ 2 & 1-\lambda & 1\\ -7 & 2 & -3-\lambda\end{vmatrix}=(2-\lambda)[(1-\lambda)(-3-\lambda)-2]-2[2(-3-\lambda+7)]=0$$

or $\quad (2-\lambda)(\lambda^2+2\lambda-5)-2(1-2\lambda)=0$ or $-\lambda^3+13\lambda-12=0$.

We have $\quad \mathbf{A}^2=\begin{bmatrix}2 & 2 & 0\\ 2 & 1 & 1\\ -7 & 2 & -3\end{bmatrix}\begin{bmatrix}2 & 2 & 0\\ 2 & 1 & 1\\ -7 & 2 & -3\end{bmatrix}=\begin{bmatrix}8 & 6 & 2\\ -1 & 7 & -2\\ 11 & -18 & 11\end{bmatrix}.$

Eigenvalues of **A** are the roots of $\lambda^3 - 13\lambda + 12 = 0$. ...(*i*)
By trial $\lambda = 1$ is a root of this equation. Now (*i*) can be written as
$(\lambda - 1)(\lambda^2 + \lambda - 12) = 0$ or $(\lambda - 1)(\lambda - 3)(\lambda + 4) = 0$ or $\lambda = 1, 3, -4$.

The characteristic equation of $\mathbf{A}^2$ is given by $\begin{vmatrix}8-\lambda & 6 & 2\\ -1 & 7-\lambda & -2\\ 11 & -18 & 11-\lambda\end{vmatrix}$

$$=(8-\lambda)[(7-\lambda)(11-\lambda)-36]-6[\lambda-11+22]+2[18-11(7-\lambda)]=0$$

or $\quad (8-\lambda)(\lambda^2-18\lambda+41)-6\lambda-66+22\lambda-118=0$

or $\quad \lambda^3-26\lambda^2+169\lambda-144=0$ or $(\lambda-1)(\lambda^2-25\lambda+144)=0$ or $(\lambda-1)(\lambda-9)(\lambda-16)=0$.

The eigenvalues of $\mathbf{A}^2$ are 1, 9, 16 which are the squares of the eigenvalues of **A**.
The spectral radius of **A** is given by
$\rho(\mathbf{A})$ = largest eigenvalue in magnitude = $\max|\lambda_i| = |-4| = 4$.

EXAMPLE 6.4 If $\mathbf{A}=\begin{bmatrix}1 & 0 & 0\\ 1 & 0 & 1\\ 0 & 1 & 0\end{bmatrix}$, then show that $\mathbf{A}^n = \mathbf{A}^{n-2} + \mathbf{A}^2 - \mathbf{I}$ for $n \geq 3$. Hence, find $\mathbf{A}^{50}$.

(AMIETE, W-2006)

Solution The characteristic equation of **A** is given by

$$|\mathbf{A}-\lambda\mathbf{I}| = \begin{vmatrix}1-\lambda & 0 & 0\\ 1 & -\lambda & 1\\ 0 & 1 & -\lambda\end{vmatrix} = (1-\lambda)\,(\lambda^2-1)=0, \text{ or } \lambda^3-\lambda^2-\lambda+1=0$$

and thus by the Cayley-Hamilton theorem, we get

$$\mathbf{A}^3 - \mathbf{A}^2 - \mathbf{A} + \mathbf{I} = \mathbf{0} \text{ or } \mathbf{A}^3 - \mathbf{A}^2 = \mathbf{A} - \mathbf{I}.$$

Premultiplying both sides successively by **A**, we obtain

$$\mathbf{A}^4 \quad \mathbf{A}^3 = \mathbf{A}^2 \quad \mathbf{A}$$

....

....

$$\mathbf{A}^{n-1} - \mathbf{A}^{n-2} = \mathbf{A}^{n-3} - \mathbf{A}^{n-4}$$
$$\mathbf{A}^{n} - \mathbf{A}^{n-1} = \mathbf{A}^{n-2} - \mathbf{A}^{n-3}.$$

Adding these equations, we get

$$\mathbf{A}^n - \mathbf{A}^2 = \mathbf{A}^{n-2} - \mathbf{I}, \text{ or } \mathbf{A}^n = \mathbf{A}^{n-2} + \mathbf{A}^2 - \mathbf{I},\ n \geq 3. \qquad ...(i)$$

Replacing n by $n-2$, $n-4$, etc., we obtain

$$\mathbf{A}^{n-2} = \mathbf{A}^{n-4} + \mathbf{A}^2 - \mathbf{I}, \quad \mathbf{A}^{n-4} = \mathbf{A}^{n-6} + \mathbf{A}^2 - \mathbf{I}$$

Now (*i*) may be re-written as

$$\mathbf{A}^n = (\mathbf{A}^{n-4} + \mathbf{A}^2 - \mathbf{I}) + \mathbf{A}^2 - \mathbf{I} = \mathbf{A}^{n-4} + 2(\mathbf{A}^2 - \mathbf{I})$$
$$= (\mathbf{A}^{n-6} + \mathbf{A}^2 - \mathbf{I}) + 2(\mathbf{A}^2 - \mathbf{I}) = \mathbf{A}^{n-6} + 3(\mathbf{A}^2 - \mathbf{I})$$

....

....

$$= \mathrm{A}^{n-(n-2)} + \frac{(n-2)}{2}(\mathrm{A}^2 - \mathrm{I}) = \frac{n}{2}\mathrm{A}^2 - \left(\frac{n-2}{2}\right)\mathbf{I}$$

Substituting $n = 50$, we get

$$\mathbf{A}^{50} = 25\mathbf{A}^2 \quad 24\mathbf{I} = 25\begin{bmatrix}1 & 0 & 0\\ 1 & 1 & 0\\ 1 & 0 & 1\end{bmatrix} - 24\begin{bmatrix}1 & 0 & 0\\ 0 & 1 & 0\\ 0 & 0 & 1\end{bmatrix} = \begin{bmatrix}1 & 0 & 0\\ 25 & 1 & 0\\ 25 & 0 & 1\end{bmatrix}.$$

EXAMPLE 6.5 If $\mathbf{A}=\begin{bmatrix}1 & -1\\ 2 & 3\end{bmatrix}$. Find $\mathbf{A}^8$.

Solution The characteristic equation of **A** is given by

$$|\mathbf{A}-\lambda\mathbf{I}| = \begin{vmatrix}1-\lambda & -1\\ 2 & 3-\lambda\end{vmatrix} = (1-\lambda)(3-\lambda)+2 = \lambda^2 - 4\lambda + 5 = 0$$

and thus by the Cayley-Hamilton theorem, we get

$$\mathbf{A}^2 - 4\mathbf{A} + 5\mathbf{I} = \mathbf{0} \Rightarrow \mathbf{A}^2 = 4\mathbf{A} - 5\mathbf{I}.$$

$$\mathbf{A}^4 = 16\mathbf{A}^2 - 40\mathbf{A} + 25\mathbf{I}, \text{ using the Commutative rule,}$$
$$= 16(4\mathbf{A} - 5\mathbf{I}) - 40\mathbf{A} + 25\mathbf{I} = 24\mathbf{A} - 55\mathbf{I};$$
$$\mathbf{A}^8 = 24^2\,\mathbf{A}^2 - (48)\,(55)\,\mathbf{A} + 55^2\mathbf{I} = 24^2(4\mathbf{A} - 5\mathbf{I}) - (48)\,(55)\,\mathbf{A} + 55^2\mathbf{I}$$
$$= -336\mathbf{A} + 145\mathbf{I} = -336\begin{bmatrix}1 & -1\\ 2 & 3\end{bmatrix} + 145\begin{bmatrix}1 & 0\\ 0 & 1\end{bmatrix} = \begin{bmatrix}-191 & 336\\ -672 & -863\end{bmatrix}.$$

EXAMPLE 6.6 Verify Cayley-Hamilton theorem for the matrix $\mathbf{A} = \begin{bmatrix} 1 & 4 \\ 2 & 3 \end{bmatrix}$ and find its inverse. Also express $\mathbf{A}^5 - 4\mathbf{A}^4 - 7\mathbf{A}^3 + 11\mathbf{A}^2 - \mathbf{A} - 10\mathbf{I}$ as a linear polynomial in **A**. *(MDU - 2003; AMIETE, June 2003)*

Solution The characteristic equation of **A** is given by

$$|\mathbf{A} - \lambda\mathbf{I}| = \begin{vmatrix} 1-\lambda & 4 \\ 2 & 3-\lambda \end{vmatrix} = (1-\lambda)(3-\lambda) - 8 = \lambda^2 - 4\lambda - 5 = 0 \qquad ...(i)$$

Now $$\mathbf{A}^2 = \begin{bmatrix} 1 & 4 \\ 2 & 3 \end{bmatrix}\begin{bmatrix} 1 & 4 \\ 2 & 3 \end{bmatrix} = \begin{bmatrix} 9 & 16 \\ 8 & 17 \end{bmatrix}.$$

We have $$\mathbf{A}^2 - 4\mathbf{A} - 5\mathbf{I} = \begin{bmatrix} 9 & 16 \\ 8 & 17 \end{bmatrix} - \begin{bmatrix} 4 & 16 \\ 8 & 12 \end{bmatrix} - \begin{bmatrix} 5 & 0 \\ 0 & 5 \end{bmatrix} = \begin{bmatrix} 0 & 0 \\ 0 & 0 \end{bmatrix} = \mathbf{0}. \qquad ...(ii)$$

Hence, **A** satisfies the characteristic equation $\lambda^2 - 4\lambda + 5 = 0$.
From equation (*ii*), we get

$$\mathbf{A}^{-1} = \frac{1}{5}[\mathbf{A} - 4\mathbf{I}] = \frac{1}{5}\left[\begin{pmatrix} 1 & 4 \\ 2 & 3 \end{pmatrix} - \begin{pmatrix} 4 & 0 \\ 0 & 4 \end{pmatrix}\right] = \frac{1}{5}\begin{bmatrix} -3 & 4 \\ 2 & -1 \end{bmatrix}.$$

Now dividing the polynomial $\lambda^5 - 4\lambda^4 - 7\lambda^3 + 11\lambda^2 - \lambda - 10$ by the polynomial $\lambda^2 - 4\lambda - 5$, we obtain

$$\lambda^5 - 4\lambda^4 - 7\lambda^3 + 11\lambda^2 - \lambda - 10 = (\lambda^2 - 4\lambda - 5)(\lambda^3 - 2\lambda + 3) + \lambda + 5 = \lambda + 5 \text{ [by (i)]}$$

Hence $\mathbf{A}^5 - 4\mathbf{A}^4 - 7\mathbf{A}^3 + 11\mathbf{A}^2 - \mathbf{A} - 10\mathbf{I} = \mathbf{A} + 5\mathbf{I}$, which is a linear polynomial in **A**.

EXAMPLE 6.7: Find the eigenvalues and the corresponding eigenvectors of the following matrices

(*i*) $\mathbf{A} = \begin{bmatrix} 4 & 2 \\ 6 & 5 \end{bmatrix}$, (*ii*) $\mathbf{A} = \begin{bmatrix} 6 & -1 \\ 5 & 4 \end{bmatrix}$.

Solution (*i*) The characteristic equation of **A** is given by

$$|\mathbf{A} - \lambda\mathbf{I}| = \begin{vmatrix} 4-\lambda & 2 \\ 6 & 5-\lambda \end{vmatrix} = 0 \text{ or } (4-\lambda)(5-\lambda) - 12 = 0$$

or $$\lambda^2 - 9\lambda + 8 = 0 \text{ or } \lambda = 1, 8.$$

Corresponding to the eigenvalue $\lambda = 1$, we have

$$(\mathbf{A} - \mathbf{I})\mathbf{x} = \begin{bmatrix} 3 & 2 \\ 6 & 4 \end{bmatrix}\begin{bmatrix} x_1 \\ x_2 \end{bmatrix} = \begin{bmatrix} 0 \\ 0 \end{bmatrix} \text{ or } \left.\begin{matrix} 3x_1 + 2x_2 = 0 \\ 6x_1 + 4x_2 = 0 \end{matrix}\right\} \begin{matrix} \text{or } 3x_1 + 2x_2 = 0, \\ \text{or } x_1 = -2x_2/3. \end{matrix}$$

Hence, the eigenvector **x** is given by

$$\mathbf{x} = \begin{bmatrix} x_1 \\ x_2 \end{bmatrix} = \begin{bmatrix} -2x_2/3 \\ x_2 \end{bmatrix} = x_2\begin{bmatrix} -2/3 \\ 1 \end{bmatrix}.$$

Since an eigenvector is unique upto a constant multiple, we can take the eigenvector as $[-2, 3]^T$.
Corresponding to the eigenvalue $\lambda = 8$, we have

$$(\mathbf{A} - 8\mathbf{I})\mathbf{x} = \begin{bmatrix} -4 & 2 \\ 6 & -3 \end{bmatrix}\begin{bmatrix} x_1 \\ x_2 \end{bmatrix} = \mathbf{0} \text{ or } \left.\begin{matrix} -4x_1 + 2x_2 = 0 \\ 6x_1 - 3x_2 = 0 \end{matrix}\right\} \text{ or } x_1 = \frac{x_2}{2}.$$

Choose $x_2 = 2$, $x_1 = 1$. Therefore, the eigenvector is given by

$$\mathbf{x} = (x_1, x_2)^T = x_1(1, 2)^T \text{ or simply } (1, 2)^T.$$

(*ii*) The characteristic equation of **A** is given by

$$|\mathbf{A} - \lambda\mathbf{I}| = \begin{vmatrix} 6-\lambda & -1 \\ 5 & 4-\lambda \end{vmatrix} = 0 \text{ or } \lambda^2 - 10\lambda + 29 = 0, \text{ or } \lambda = 5 \pm 2i.$$

Corresponding to the eigenvalue $\lambda = 5 + 2i$, we have

$$[\mathbf{A}-(5+2i)\mathbf{I}]\mathbf{x} = \begin{bmatrix} 1-2i & -1 \\ 5 & -1-2i \end{bmatrix}\begin{bmatrix} x_1 \\ x_2 \end{bmatrix} = \begin{bmatrix} 0 \\ 0 \end{bmatrix}$$

or $(1 - 2i)x_1 - x_2 = 0$ and $5x_1 - (1 + 2i)x_2 = 0$.

On solving, we get $x_2 = (1 - 2i)x_1$. Choosing $x_1 = 1$, we get $x_2 = 1 - 2i$.

Therefore, the eigenvector is $\mathbf{x} = [1, 1 - 2i]^T$.

Corresponding to the eigenvalue $\lambda = 5 - 2i$, we have

$$[\mathbf{A}-(5-2i)\mathbf{I}]\mathbf{x} = \begin{bmatrix} 1+2i & -1 \\ 5 & -1+2i \end{bmatrix}\begin{bmatrix} x_1 \\ x_2 \end{bmatrix} = \begin{bmatrix} 0 \\ 0 \end{bmatrix}$$

or $(1 + 2i)\, x_1 - x_2 = 0$ and $5x_1 + x_2\,(-1 + 2i) = 0$.

On solving, we get $x_2 = (1 + 2i)x_1$. Choosing $x_1 = 1$, we get $x_2 = (1 + 2i)$.

Therefore, the eigenvector is $\mathbf{x} = [1, 1 + 2i]^T$.

Remark For a real matrix **A**, the eigenvalues and the corresponding eigenvectors can be complex.

EXAMPLE 6.8 Find the eigenvalues and the corresponding eigenvectors of the matrix

$$\mathbf{A} = \begin{bmatrix} 1 & 1 & 0 \\ 0 & 1 & 0 \\ 0 & 0 & 1 \end{bmatrix}$$

(AMIE, W-2010)

Solution The characteristic equation of **A** is given by

$$|\mathbf{A}-\lambda\mathbf{I}| = \begin{vmatrix} 1-\lambda & 1 & 0 \\ 0 & 1-\lambda & 0 \\ 0 & 0 & 1-\lambda \end{vmatrix} = 0, \text{ or } (1-\lambda)\,[(1-\lambda)^2 - 0] = 0 \quad \text{or} \quad (1-\lambda)^3 = 0.$$

Therefore, the eigenvalues are $\lambda = 1, 1, 1$, a repeated value.

Corresponding to the eigenvalue $\lambda = 1$, we obtain the following eigen vectors.

$$(\mathbf{A}-\mathbf{I})\mathbf{x} = \begin{bmatrix} 0 & 1 & 0 \\ 0 & 0 & 0 \\ 0 & 0 & 0 \end{bmatrix}\begin{bmatrix} x_1 \\ x_2 \\ x_3 \end{bmatrix} = \begin{bmatrix} 0 \\ 0 \\ 0 \end{bmatrix}, \text{ or } \begin{cases} x_2 = 0 \\ x_1,\ x_3 \text{ arbitrary.} \end{cases}$$

Taking $x_1 = 0, x_3 = 1$ and $x_1 = 1, x_3 = 0$, we obtain two linearly independent solutions

$$\mathbf{x}_1 = [0,\ 0,\ 1]^T, \quad \mathbf{x}_2 = [1,\ 0,\ 0]^T.$$

Thus A has two linearly independent eigenvectors.

EXAMPLE 6.9 Find the eigenvalues and the corresponding eigenvectors of the matrix

$$\mathbf{A} = \begin{bmatrix} 3 & -4 & 4 \\ 1 & -2 & 4 \\ 1 & -1 & 3 \end{bmatrix}$$

(AMIE, S-2005)

Solution The characteristic equation of **A** is given by

$$|\mathbf{A}-\lambda\mathbf{I}| = \begin{vmatrix} 3-\lambda & -4 & 4 \\ 1 & -2-\lambda & 4 \\ 1 & -1 & 3-\lambda \end{vmatrix} = 0.$$

Operating $C_2 \to C_2 + C_3$, we get

$$|\mathbf{A}-\lambda\mathbf{I}|=\begin{vmatrix}3-\lambda & 0 & 4\\ 1 & 2-\lambda & 4\\ 1 & 2-\lambda & 3-\lambda\end{vmatrix}=0 \text{ or } (2-\lambda)\begin{vmatrix}3-\lambda & 0 & 4\\ 1 & 1 & 4\\ 1 & 1 & 3-\lambda\end{vmatrix}=0.$$

Expanding the determinant along R_1, we obtain

$$(2-\lambda)[(3-\lambda)(3-\lambda-4)]+4(1-1)=0 \quad \text{or} \quad (2-\lambda)(3-\lambda)(-\lambda-1)=0 \text{ or } \lambda=2, 3, -1.$$

The eigenvalues are 2, 3, –1.

We have the $(\mathbf{A}-\lambda\mathbf{I})\,\mathbf{x}=\mathbf{0}$,

that is,
$$\begin{bmatrix}3-\lambda & -4 & 4\\ 1 & -2-\lambda & 4\\ 1 & -1 & 3-\lambda\end{bmatrix}\begin{bmatrix}x_1\\ x_2\\ x_3\end{bmatrix}=\begin{bmatrix}0\\0\\0\end{bmatrix}$$

that is,
$$\left.\begin{aligned}(3-\lambda)x_1-4x_2+4x_3&=0\\ x_1-(2+\lambda)x_2+4x_3&=0\\ x_1-x_2+(3-\lambda)x_3&=0\end{aligned}\right\}. \qquad \text{...}(i)$$

Putting the different values of λ (*i.e.*, eigenvalues) in (*i*) we get the corresponding eigenvectors which are non-zero solutions of (*i*).

(*a*) When $\lambda = 2$, the system of equations (*i*) is

$$\left.\begin{aligned}x_1-4x_2+4x_3&=0\\ x_1-4x_2+4x_3&=0\\ x_1-x_2+x_3&=0\end{aligned}\right\} \text{ or } \left.\begin{aligned}x_1-4x_2+4x_3&=0,\\ x_1-x_2+x_3&=0.\end{aligned}\right\}$$

Obviously only two equations are independent. Solving the last two equations, we get

$$\frac{x_1}{-4+4}=\frac{x_2}{4-1}=\frac{x_3}{-1+4} \text{ or } \frac{x_1}{0}=\frac{x_2}{3}=\frac{x_3}{3} \text{ or } \frac{x_1}{0}=\frac{x_2}{1}=\frac{x_3}{1}.$$

Hence the corresponding eigenvector is $(0, 1, 1)^T$.

(*b*) When $\lambda = 3$, equations (*i*) are

$$\begin{aligned}0\cdot x_1-4x_2+4x_3&=0,\\ x_1-5x_2+4x_3&=0,\\ x_1-x_2+0\cdot x_3&=0.\end{aligned}$$

Since only two equations are independent we can omit one of them.

From first two equations, we obtain

$$\frac{x_1}{-16+20}=\frac{x_2}{4-0}=\frac{x_3}{0+4} \text{ or } \frac{x_1}{1}=\frac{x_2}{1}=\frac{x_3}{1}.$$

Thus the corresponding eigenvector is $(1, 1, 1)^T$.

(*c*) When $\lambda = -1$, equations (*i*) are

$$\left.\begin{aligned}4x_1-4x_2+4x_3&=0\\ x_1-x_2+4x_3&=0\\ x_1-x_2+4x_3&=0\end{aligned}\right\} \text{ or } \begin{aligned}x_1-x_2+x_3&=0\\ x_1-x_2+4x_3&=0\end{aligned}$$

Since only two equations are independent, from first two equations we have

$$\frac{x_1}{-4+1}=\frac{x_2}{1-4}=\frac{x_3}{-1+1} \text{ or } \frac{x_1}{1}=\frac{x_2}{1}=\frac{x_3}{0}.$$

The corresponding eigenvector is $(1, 1, 0)^T$.

Hence the eigenvectors of a 3 × 3 matrix **A** corresponding to the eigenvalues 2, 3, –1 are $[0, 1, 1]^T$, $[1, 1, 1]^T$ and $[1, 1, 0]^T$ respectively.

EXAMPLE 6.10 Find the eigenvalues and eigenvectors of the matrix

$$\mathbf{A} = \begin{bmatrix} 6 & -2 & 2 \\ -2 & 3 & -1 \\ 2 & -1 & 3 \end{bmatrix}.$$ *(Kurukshetra, 2006; DU 2002; MDU 2002; Madras 2002)*

Solution The characteristic equation of **A** is given by

$$|\mathbf{A} - \lambda \mathbf{I}| = \begin{vmatrix} 6-\lambda & -2 & 2 \\ -2 & 3-\lambda & -1 \\ 2 & -1 & 3-\lambda \end{vmatrix} = 0$$

Operating $C_2 \to C_2 + C_3$, we get

$$\begin{vmatrix} 6-\lambda & 0 & 2 \\ -2 & 2-\lambda & -1 \\ 2 & 2-\lambda & 3-\lambda \end{vmatrix} = 0 \text{ or } (2-\lambda)\begin{vmatrix} 6-\lambda & 0 & 2 \\ -2 & 1 & -1 \\ 2 & 1 & 3-\lambda \end{vmatrix} = 0$$

or $(2 - \lambda)\,[(6 - \lambda)(3 - \lambda + 1) + 2(-2 -2)] = 0$

or $(2 - \lambda)\,(\lambda^2 - 10\lambda + 16) = 0$ that is, $(\lambda - 2)(\lambda - 2)(\lambda - 8) = 0$ or $\lambda = 2, 2, 8$.

We have $(\mathbf{A} - \lambda\mathbf{I})\mathbf{x} = \mathbf{0}$,

$$\text{that is, } \left.\begin{aligned} (6-\lambda)x_1 - 2x_2 + 2x_3 &= 0 \\ -2x_1 + (3-\lambda)x_2 - x_3 &= 0 \\ 2x_1 - x_2 + (3-\lambda)x_3 &= 0 \end{aligned}\right\} \qquad \text{...}(i)$$

(*a*) When $\lambda = 8$, the equations (*i*) are

$$\left.\begin{aligned} -2x_1 - 2x_2 + 2x_3 &= 0 \\ -2x_1 - 5x_2 - x_3 &= 0 \\ 2x_1 - x_2 - 5x_3 &= 0 \end{aligned}\right\} \quad \text{or} \quad \begin{aligned} x_1 + x_2 - x_3 &= 0, \\ x_2 + x_3 &= 0. \end{aligned}$$

Choosing $x_3 = 1$, we get $x_2 = -1$, $x_1 = 2$.

Hence $(2, -1, 1)^T$ or any multiple is a characteristic vector.

(*b*) When $\lambda = 2$, (the equal root) the equations (*i*) reduces to

$$\left.\begin{aligned} 4x_1 - 2x_2 + 2x_3 &= 0 \\ -2x_1 + x_2 - x_3 &= 0 \\ 2x_1 - x_2 + x_3 &= 0 \end{aligned}\right\} \quad \text{or} \quad 2x_1 - x_2 + x_3 = 0.$$

Clearly only one equation is independent, the other two being multiples of that. Here we can get any number of eigenvectors by giving arbitrary values to any two of the three variables. To get two linearly independent vectors we put $x_2 = 0$ and $x_3 = 0$ successively in any one equation.

For $x_2 = 0$, $x_1 = 1$, we get $x_3 = -2$. The eigenvector is $(1, 0, -2)^T$.

For $x_3 = 0$, $x_1 = 1$, we get $x_2 = 2$. The eigenvector is $(1, 2, 0)^T$.

It can be verified that any other eigenvector corresponding to $\lambda = 2$ is a linear combination of these two.

Hence the three eigenvectors are $(2, -1, 1)^T$, $(1, 0, -2)^T$ and $(1, 2, 0)^T$.

EXAMPLE 6.11 Find the eigenvalues and eigenvectors of the matrix $\mathbf{A} = \begin{bmatrix} 2 & -2 & 2 \\ 1 & 1 & 1 \\ 1 & 3 & -1 \end{bmatrix}$. *(AMIE, W-2009)*

Solution The characteristic equation of **A** is given by

$$|\mathbf{A}-\lambda\mathbf{I}| = \begin{vmatrix} 2-\lambda & -2 & 2 \\ 1 & 1-\lambda & 1 \\ 1 & 3 & -1-\lambda \end{vmatrix} = 0.$$

On expanding the determinant, we obtain

$$(2-\lambda)\,[(1-\lambda)(-1-\lambda)-3] + 2\,[-1-\lambda-1] + 2\,[3-(1-\lambda)] = 0$$

or $(2-\lambda)\,[\lambda^2-4] - 2\lambda - 4 + 2\lambda + 4 = 0$.

or $(2-\lambda)(\lambda-2)(\lambda+2) = 0$ or $\lambda = -2, 2, 2$. The eigenvalues are $-2, 2, 2$. We have $(\mathbf{A}-\lambda\mathbf{I})\,\mathbf{x} = \mathbf{0}$,

$$\left.\begin{aligned} \text{that is, } (2-\lambda)x_1 - 2x_2 + 2x_3 &= 0 \\ x_1 + (1-\lambda)x_2 + x_3 &= 0 \\ x_1 + 3x_2 - (1+\lambda)x_3 &= 0 \end{aligned}\right\} \qquad \text{...}(i)$$

(*a*) When $\lambda = -2$, equations (*i*) are

$$\left.\begin{aligned} 4x_1 - 2x_2 + 2x_3 &= 0, \\ x_1 + 3x_2 + x_3 &= 0, \\ x_1 + 3x_2 + x_3 &= 0. \end{aligned}\right\} \text{ or } \left.\begin{aligned} 2x_1 - x_2 + x_3 &= 0, \\ x_1 + 3x_2 + x_3 &= 0. \end{aligned}\right\}$$

Obviously only first two equations are independent. Solving these equations

$$\frac{x_1}{-1-3} = \frac{x_2}{1-2} = \frac{x_3}{6+1} \text{ that is, } \frac{x_1}{-4} = \frac{x_2}{-1} = \frac{x_3}{7}.$$

$\therefore$ Corresponding eigenvector is $(-4, -1, 7)^T$.

(*b*) When $\lambda = 2$ (the equal root) equations (*i*) become

$$\left.\begin{aligned} 0\cdot x_1 - 2x_2 + 2x_3 &= 0 \\ x_1 - x_2 + x_3 &= 0 \\ x_1 + 3x_2 - 3x_3 &= 0 \end{aligned}\right\} \text{ or } \left.\begin{aligned} x_2 - x_3 &= 0, \\ x_1 - x_2 + x_3 &= 0, \\ x_1 + 3x_2 - 3x_3 &= 0. \end{aligned}\right\}$$

Here only two equations are independent. Solving first two equations, we get

$$\frac{x_1}{-1+1} = \frac{x_2}{1} = \frac{x_3}{1} \text{ that is, } \frac{x_1}{0} = \frac{x_2}{1} = \frac{x_3}{1}.$$

The eigenvector corresponding to the eigenvalues 2 is $(0, 1, 1)^T$.
In this case we have only two independent eigenvectors.

Remarks

(1) Eigenvectors corresponding to distinct eigenvalues are linearly independent.

(2) If λ is an eigenvalue of multiplicity m of a square matrix **A** of order n, then the number of linearly independent eigenvectors associated with λ is given by

$$p = n - r, \text{ where } r = \text{rank } (\mathbf{A}-\lambda\mathbf{I}),\ 1 \le p \le m.$$

For example let a 3×3 matrix has the eigenvalue $\lambda = 1$ of multiplicity 3.

(*a*) If the rank of the matrix **A** – **I** is 2, we obtain one linearly independent eigenvector.

(*b*) If the rank of matrix **A** – **I** is 1, we obtain two linearly independent eigenvectors.

(*c*) If the rank of matrix **A** – **I** is 0, we obtain three linearly independent eigenvectors.

EXAMPLE 6.12 Let a 4×4 matrix **A** have eigenvalues $1, -1, 2, -2$. Find the value of the determinant of the matrix $\mathbf{B} = 2\mathbf{A} + \mathbf{A}^{-1} - \mathbf{I}$ and trace of matrix **B**. (*AMIETE, June 2005*)

Solution Eigenvalues of **B** are $2\lambda_i + (1/\lambda_i) - 1$, $i = 1, 2, 3, 4$.

That is, $2(1) + 1 - 1$, $2(-1) - 1 - 1$, $2\,(2) + (1/2) - 1$, $2\,(-2) - (1/2) - 1$ or $2, -4, 7/2, -11/2$.

$|\mathbf{B}|$ = Product of eigenvalues of $\mathbf{B} = 2(-4)(7/2)(-11/2) = 154$.

Trace of **B** = Sum of eigenvalues of $\mathbf{B} = 2 - 4 + (7/2) - (11/2) = -4$.

EXAMPLE 6.13 Let a 3 × 3 matrix **A** have eigenvalues 1, 2, –1. Find the trace of the matrix $\mathbf{B} = \mathbf{A} - \mathbf{A}^{-1} + \mathbf{A}^2$.

Solution Eigenvalues of **B** are $\lambda_i - (1/\lambda_i) + \lambda_i^2$, $i = 1, 2, 3$.

That is, $1 - 1 + 1$, $2 - (1/2) + 2^2$, $-1 + 1 + 1$ or 1, 11/2, 1.

Trace of **B** = sum of eigenvalues of **B** = 1 + (11/2) + 1 = 15/2.

6.3 SIMILAR MATRICES

Suppose **A** and **B** are square matrices of the same order. The matrix **B** is said to be similar to the matrix **A** if there exists an invertible matrix **P** such that

$$\mathbf{B} = \mathbf{P}^{-1}\mathbf{A}\mathbf{P} \text{ or } \mathbf{P}\mathbf{B} = \mathbf{A}\mathbf{P}. \qquad ...(6.5)$$

That is, **B** is said to be obtained from **A** by similarity transformation. Post multiplying both sides in (6.5) by $\mathbf{P}^{-1}$, we get

$$\mathbf{P}\mathbf{B}\mathbf{P}^{-1} = \mathbf{A}$$

Thus, **B** is similar to **A**, if and only if **A** is similar to **B**. The matrix **P** is called the *similarity matrix.*

Important Results

1. Two similar matrices have the same characteristic equation and hence the same eigenvalues. However, the converse is not true. Two matrices which have the same characteristic equation need not always be similar.
2. If **x** is an eigenvector of **A** corresponding to the eigenvalue λ, then $\mathbf{y} = \mathbf{P}^{-1}\mathbf{x}$ is an eigenvector of **B** corresponding to the eigenvalue λ, where **P** is the similarity matrix. Thus similarity transformation retains same eigenvalues but changed eigenvectors.
3. If **A** is similar to **B** and **B** is similar to **C**, then **A** is similar to **C**.
   In practice, it is usually difficult to obtain a non-singular matrix **P** which satisfies the equation $\mathbf{B} = \mathbf{P}^{-1}\mathbf{A}\mathbf{P}$ for any two matrices **A** and **B**. However, it is possible to obtain the matrix **P** when **A** or **B** is a diagonal matrix.

EXAMPLE 6.14 Examine whether **A** is similar to **B**, where

(*i*) $\mathbf{A} = \begin{bmatrix} 1 & 1 \\ 2 & -3 \end{bmatrix}$ and $\mathbf{B} = \begin{bmatrix} -34 & 57 \\ -19 & 32 \end{bmatrix}$, (*ii*) $\mathbf{A} = \begin{bmatrix} 1 & 0 \\ 0 & 1 \end{bmatrix}$ and $\mathbf{B} = \begin{bmatrix} 1 & 1 \\ 0 & 1 \end{bmatrix}$.

Solution The given matrices are similar if there exists a non-singular matrix **P** such that $\mathbf{B} = \mathbf{P}^{-1}\mathbf{A}\mathbf{P}$ or $\mathbf{P}\mathbf{B} = \mathbf{A}\mathbf{P}$.

Let $\mathbf{P} = \begin{bmatrix} a & b \\ c & d \end{bmatrix}$. We shall find *a, b, c, d* such that **PB** = **AP** and then check whether **P** is non-singular.

(*i*)
$$\begin{bmatrix} a & b \\ c & d \end{bmatrix}\begin{bmatrix} -34 & 57 \\ -19 & 32 \end{bmatrix} = \begin{bmatrix} 1 & 1 \\ 2 & -3 \end{bmatrix}\begin{bmatrix} a & b \\ c & d \end{bmatrix}$$

or
$$\begin{bmatrix} -34a-19b & 57a+32b \\ -34c-19d & 57c+32d \end{bmatrix} = \begin{bmatrix} a+c & b+d \\ 2a-3c & 2b-3d \end{bmatrix}.$$

Equating the corresponding elements, we obtain the system of equations

$$\begin{aligned} -34a - 19b &= a + c, &\quad \text{or} \quad & 35a + 19b + c = 0 \\ 57a + 32b &= b + d, & \text{or} \quad & 57a + 31b - d = 0 \\ -34c - 19d &= 2a - 3c, & \text{or} \quad & 2a + 31c + 19d = 0 \\ 57c + 32d &= 2b - 3d, & \text{or} \quad & 2b - 57c - 35d = 0 \end{aligned}$$

On solving these equations, we obtain $a = 1$, $b = -2$, $c = 3$, $d = -5$.

Therefore, we get $\mathbf{P} = \begin{bmatrix} 1 & -2 \\ 3 & -5 \end{bmatrix}$, which is a non-singular matrix. Hence the matrices **A** and **B** are similar.

(*ii*) $\begin{bmatrix} a & b \\ c & d \end{bmatrix}\begin{bmatrix} 1 & 1 \\ 0 & 1 \end{bmatrix} = \begin{bmatrix} 1 & 0 \\ 0 & 1 \end{bmatrix}\begin{bmatrix} a & b \\ c & d \end{bmatrix}$, or $\begin{bmatrix} a & a+b \\ c & c+d \end{bmatrix} = \begin{bmatrix} a & b \\ c & d \end{bmatrix}$.

Equating the corresponding elements, we obtain

$a + b = b$, $c + d = d$ or $a = c = 0$. Therefore, $\mathbf{P} = \begin{bmatrix} 0 & b \\ 0 & d \end{bmatrix}$, which is singular matrix. Since an invertible matrix **P** does not exist, the matrices **A** and **B** are not similar.

6.4 DIAGONALIZABLE MATRICES

A matrix **A** is diagonalizable if it is similar to a diagonal matrix, that is there exists an invertible matrix **P** such that $\mathbf{P}^{-1}\mathbf{AP} = \mathbf{D}$, where **D** is a diagonal matrix. Since, similar matrices have the same eigenvalues, the diagonal elements of matrix **D** are the eigenvalues of **A**.

Theorem. A square matrix **A** of order n is diagonalizable if and only if it has n linearly independent eigenvectors.

Proof: Let $\mathbf{x}_1, \mathbf{x}_2, ..., \mathbf{x}_n$ be n linearly independent eigenvectors, corresponding to the eigenvalues $\lambda_1, \lambda_2, ..., \lambda_n$ of the matrix **A** in the same order, that is the eigenvector x_i corresponds to the eigenvalues λ_i, $i = 1, 2, ..., n$. Then $\mathbf{Ax}_i = \lambda_i\, \mathbf{x}_i$.

Denote by **P** the square matrix whose columns are $\mathbf{x}_1, \mathbf{x}_2, ..., \mathbf{x}_n$. We write

$$\mathbf{P} = [\mathbf{x}_1, \mathbf{x}_2, ..., \mathbf{x}_n]. \text{ Then}$$

$$\mathbf{AP} = \mathbf{A}[\mathbf{x}_1, \mathbf{x}_2, ..., \mathbf{x}_n] = [\mathbf{Ax}_1, \mathbf{Ax}_2, ..., \mathbf{Ax}_n]$$
$$= [\lambda_1 \mathbf{x}_1, \lambda_2 \mathbf{x}_2, ..., \lambda_n \mathbf{x}_n] = [\mathbf{x}_1, \mathbf{x}_2, ..., \mathbf{x}_n]\, \mathbf{D} = \mathbf{PD}, \qquad ...(6.6)$$

where $\mathbf{D} = \begin{bmatrix} \lambda_1 & 0 & 0 & \cdots & 0 \\ 0 & \lambda_2 & 0 & \cdots & 0 \\ \vdots & \vdots & \vdots & \vdots & \vdots \\ 0 & 0 & 0 & \cdots & \lambda_n \end{bmatrix}$.

Premultiplying both sides of (6.6) by $\mathbf{P}^{-1}$, we obtain

$\mathbf{P}^{-1}\mathbf{AP} = \boxed{\mathbf{P}^{-1}\mathbf{P}}\,\mathbf{D} = \mathbf{D}$, which implies that **A** is similar to **D**.

The matrix of eigenvectors **P** reduces a matrix **A** to its diagonal form. The matrix **P** is called a *modal matrix* of **A** and **D** is called the *spectral matrix* of **A**.

Post multiplying both sides of (6.6) by $\mathbf{P}^{-1}$, we obtain

$$\mathbf{A} = \mathbf{PDP}^{-1}. \qquad ...(6.7)$$

Remarks

1. The reduction to diagonal form enables us to calculate powers of **A**. We have $\mathbf{P}^{-1}\,\mathbf{AP} = \mathbf{D}$ ...(*i*)

   Premultiplying both sides of (*i*) by **P**, we obtain $\mathbf{AP} = \mathbf{PD}$ ...(*ii*)

   Post multiplying both sides of (*ii*) by $\mathbf{P}^{-1}$, we obtain $\mathbf{A} = \mathbf{PDP}^{-1}$.

   Therefore, $\mathbf{A}^2 = \mathbf{AA} = (\mathbf{PDP}^{-1})(\mathbf{PDP}^{-1}) = \mathbf{PD}(\mathbf{P}^{-1}\mathbf{P})\mathbf{DP}^{-1} = \mathbf{PD}^2\mathbf{P}^{-1}$.

   Repeating the pre-multiplication (post-multiplication) m times, we obtain

   $\mathbf{A}^m = \mathbf{PD}^m\mathbf{P}^{-1}$, for any positive integer m.

2. If **D** is a diagonal matrix of order n, and

$$\mathbf{D} = \begin{bmatrix} \lambda_1 & 0 & 0 \\ 0 & \lambda_2 & 0 \\ \vdots & \vdots & \vdots \\ 0 & 0 & \lambda_n \end{bmatrix}, \text{ then } \mathbf{D}^m = \begin{bmatrix} \lambda_1^m & 0 & 0 \\ 0 & \lambda_2^m & 0 \\ \vdots & \vdots & \vdots \\ 0 & 0 & \lambda_n^m \end{bmatrix}$$

for any positive integer m. If $f(\mathbf{D})$ is a polynomial in $\mathbf{D}$, then we obtain

$$f(D)=\begin{bmatrix} f(\lambda_1) & 0 & \cdots & 0 \\ 0 & f(\lambda_2) & \cdots & 0 \\ \cdots & \cdots & \ddots & \cdots \\ 0 & 0 & \cdots & f(\lambda_n) \end{bmatrix}.$$

Let a matrix $\mathbf{A}$ be diagonalizable. Then we have $\mathbf{A} = \mathbf{PDP}^{-1}$ and $\mathbf{A}^m = \mathbf{PD}^m\mathbf{P}^{-1}$, for any positive integer m. Hence, we obtain $f(\mathbf{A}) = f(\mathbf{PDP}^{-1}) = \mathbf{P}f(\mathbf{D})\mathbf{P}^{-1}$, for any matrix polynomial $f(\mathbf{A})$.

LLUSTRATIVE EXAMPLES

EXAMPLE 6.15 Show that the matrix

$$\mathbf{A}=\begin{bmatrix} 2 & 2 & 1 \\ 1 & 3 & 1 \\ 1 & 2 & 2 \end{bmatrix}$$

(AMIE S-2005, AMIETE., Dec. 2003)

is diagonalizable. Hence, find $\mathbf{P}$ such that $\mathbf{P}^{-1}\mathbf{AP}$ is a diagonal matrix.

Solution The characteristic equation of $\mathbf{A}$ is given by

$$|\mathbf{A}-\lambda\mathbf{I}|=\begin{vmatrix} 2-\lambda & 2 & 1 \\ 1 & 3-\lambda & 1 \\ 1 & 2 & 2-\lambda \end{vmatrix} = (2-\lambda)\,[(3-\lambda)\,(2-\lambda)-2] - 2\,(2-\lambda-1) + 1\,(2-3+\lambda)$$

$= (2-\lambda)\,(\lambda^2 - 5\lambda + 4) - 2 + 2\lambda + \lambda - 1 = \lambda^3 - 7\lambda^2 + 11\lambda - 5 = 0$ or $(\lambda - 5)\,(\lambda - 1)^2 = 0$ or $\lambda = 5, 1, 1$.

We first find the eigenvectors corresponding to the repeated eigenvalue $\lambda = 1$. We have the system

$$(\mathbf{A}-\mathbf{I})\mathbf{x}=\begin{bmatrix} 1 & 2 & 1 \\ 1 & 2 & 1 \\ 1 & 2 & 1 \end{bmatrix}\begin{bmatrix} x_1 \\ x_2 \\ x_3 \end{bmatrix}=\begin{bmatrix} 0 \\ 0 \\ 0 \end{bmatrix} \text{ or } x_1+2x_2+x_3=0$$

The rank of the coefficient matrix is 1. Therefore, the system has two linearly independent solutions. We use the equation $x_1 + 2x_2 + x_3 = 0$ to find two linearly independent eigenvectors. Taking $x_3 = 0$, $x_2 = -1$, we obtain the eigenvector $[2, -1, 0]^T$ and taking $x_2 = 0$, $x_3 = -1$, we obtain the eigenvector $[1, 0, -1]^T$.

Eigenvector corresponding to the eigenvalue $\lambda = 5$ is the solution of the system.

$$(\mathbf{A}-5\mathbf{I})\mathbf{x}=\begin{bmatrix} -3 & 2 & 1 \\ 1 & -2 & 1 \\ 1 & 2 & -3 \end{bmatrix}\begin{bmatrix} x_1 \\ x_2 \\ x_3 \end{bmatrix}=\begin{bmatrix} 0 \\ 0 \\ 0 \end{bmatrix}.$$

A solution of this system is $[1, 1, 1]^T$.

The given 3×3 matrix has three linearly independent eigenvectors. Therefore, the matrix $\mathbf{A}$ is diagonalizable. The modal matrix $\mathbf{P}$ is given by

$$\mathbf{P}=\begin{bmatrix} 1 & 2 & 1 \\ 1 & -1 & 0 \\ 1 & 0 & -1 \end{bmatrix} \text{ and } \mathbf{P}^{-1}=\frac{1}{4}\begin{bmatrix} 1 & 2 & 1 \\ 1 & -2 & 1 \\ 1 & 2 & -3 \end{bmatrix}.$$

We shall now verify that $\mathbf{P}^{-1}\mathbf{AP} = \mathbf{D} = \text{diag}\,(5, 1, 1)$.

$$\mathbf{P}^{-1}\mathbf{AP} = \frac{1}{4}\begin{bmatrix} 1 & 2 & 1 \\ 1 & -2 & 1 \\ 1 & 2 & -3 \end{bmatrix}\begin{bmatrix} 2 & 2 & 1 \\ 1 & 3 & 1 \\ 1 & 2 & 2 \end{bmatrix}\begin{bmatrix} 1 & 2 & 1 \\ 1 & -1 & 0 \\ 1 & 0 & -1 \end{bmatrix}$$

$$= \frac{1}{4}\begin{bmatrix} 5 & 10 & 5 \\ 1 & -2 & 1 \\ 1 & 2 & -3 \end{bmatrix}\begin{bmatrix} 1 & 2 & 1 \\ 1 & -1 & 0 \\ 1 & 0 & -1 \end{bmatrix} = \frac{1}{4}\begin{bmatrix} 20 & 0 & 0 \\ 0 & 4 & 0 \\ 0 & 0 & 4 \end{bmatrix} = \begin{bmatrix} 5 & 0 & 0 \\ 0 & 1 & 0 \\ 0 & 0 & 1 \end{bmatrix} = \text{diag}(5, 1, 1).$$

EXAMPLE 6.16 Examine whether the matrix **A**, where **A** is given by

(*i*) $\mathbf{A} = \begin{bmatrix} 1 & 3 & 3 \\ 1 & 4 & 3 \\ -1 & 3 & 4 \end{bmatrix}$, (*ii*) $\mathbf{A} = \begin{bmatrix} -2 & 2 & -3 \\ 2 & 1 & -6 \\ -1 & -2 & 0 \end{bmatrix}$, (*AMIETE, Dec. 2008, June 2007*) (*iii*) $\mathbf{A} = \begin{bmatrix} 3 & 2 & 1 \\ 0 & 2 & 0 \\ 1 & 2 & 3 \end{bmatrix}$. (*AMIETE, June 2006*)

is diagonalizable. If so, obtain the matrix **P** such that $\mathbf{P}^{-1}\mathbf{AP}$ is a diagonal matrix.

Solution (*i*) The characteristic equation of the matrix **A** is given by

$$|\mathbf{A} - \lambda\mathbf{I}| = \begin{vmatrix} 1-\lambda & 3 & 3 \\ 1 & 4-\lambda & 3 \\ -1 & 3 & 4-\lambda \end{vmatrix}$$

$$= (1 - \lambda)\,[(4 - \lambda)^2 - 9] - 3(4 - \lambda + 3) + 3(3 + 4 - \lambda)$$

$$= (\lambda - 1)^2\,(\lambda - 7) = 0, \text{ or } \lambda = 1, 1, 7.$$

We first find the eigenvectors corresponding to the repeated eigenvalue $\lambda = 1$. We have the system

$$(\mathbf{A} - \mathbf{I})\mathbf{x} = \begin{bmatrix} 0 & 3 & 3 \\ 1 & 3 & 3 \\ -1 & 3 & 3 \end{bmatrix}\begin{bmatrix} x_1 \\ x_2 \\ x_3 \end{bmatrix} = \begin{bmatrix} 0 \\ 0 \\ 0 \end{bmatrix}.$$

Since the rank of the coefficient matrix is 2, it has one linearly independent eigenvector. We obtain another linearly independent eigenvector corresponding to the eigenvalue $\lambda = 7$. Since the matrix **A** has only two linearly independent eigen-vectors, the matrix is not diagonalizable.

(*ii*) The characteristic equation of the matrix **A** is given by

$$|\mathbf{A} - \lambda\mathbf{I}| = \begin{vmatrix} -2-\lambda & 2 & -3 \\ 2 & 1-\lambda & -6 \\ -1 & -2 & 0-\lambda \end{vmatrix} = 0 \quad \text{or} \quad \lambda^3 + \lambda^2 - 21\lambda - 45 = 0$$

or $(\lambda - 5)(\lambda + 3)^2 = 0$ or $\lambda = 5, -3, -3$.

Eigenvector corresponding to the eigenvalue $\lambda = 5$ is the solution of the system

$$(\mathbf{A} - 5\mathbf{I})\mathbf{x} = \begin{bmatrix} -7 & 2 & -3 \\ 2 & -4 & -6 \\ -1 & -2 & -5 \end{bmatrix}\begin{bmatrix} x_1 \\ x_2 \\ x_3 \end{bmatrix} = \begin{bmatrix} 0 \\ 0 \\ 0 \end{bmatrix}.$$

A solution of this system is $[1, 2, -1]^T$.

Eigenvectors corresponding to $\lambda = -3$ are the solutions of the system

$$(\mathbf{A} + 3\mathbf{I})\mathbf{x} = \begin{bmatrix} 1 & 2 & -3 \\ 2 & 4 & -6 \\ -1 & -2 & 3 \end{bmatrix}\begin{bmatrix} x_1 \\ x_2 \\ x_3 \end{bmatrix} = \begin{bmatrix} 0 \\ 0 \\ 0 \end{bmatrix} \quad \text{or} \quad x_1 + 2x_2 - 3x_3 = 0.$$

The rank of the coefficient matrix is 1. Therefore, the system has two linearly independent solutions. We use the equation $x_1 + 2x_2 - 3x_3 = 0$ to find two linearly independent eigenvectors. Choosing $x_3 = 0$, $x_2 = -1$, we obtain the eigenvector $[2, -1, 0]^T$ and taking $x_2 = 0$, $x_3 = 1$, we obtain the eigenvector

$[3, 0, 1]^T$. The given 3 × 3 matrix has three linearly independent eigenvectors. Therefore, the matrix **A** is diagonalizable. The modal matrix **P** is given by

$$\mathbf{P} = \begin{bmatrix} 1 & 2 & 3 \\ 2 & -1 & 0 \\ -1 & 0 & 1 \end{bmatrix} \text{ and } \mathbf{P}^{-1} = \frac{1}{8}\begin{bmatrix} 1 & 2 & -3 \\ 2 & -4 & -6 \\ 1 & 2 & 5 \end{bmatrix}.$$

Now we have to verified that $\mathbf{P}^{-1}\mathbf{AP}$ = diag [5, –3, –3].

$$\mathbf{P}^{-1}\mathbf{AP} = \frac{1}{8}\begin{bmatrix} 1 & 2 & -3 \\ 2 & -4 & -6 \\ 1 & 2 & 5 \end{bmatrix}\begin{bmatrix} -2 & 2 & -3 \\ 2 & 1 & -6 \\ -1 & -2 & 0 \end{bmatrix}\begin{bmatrix} 1 & 2 & 3 \\ 2 & -1 & 0 \\ -1 & 0 & 1 \end{bmatrix}$$

$$= \frac{1}{8}\begin{bmatrix} 5 & 10 & -15 \\ -6 & 12 & 18 \\ -3 & -6 & -15 \end{bmatrix}\begin{bmatrix} 1 & 2 & 3 \\ 2 & -1 & 0 \\ -1 & 0 & 1 \end{bmatrix} = \frac{1}{8}\begin{bmatrix} 40 & 0 & 0 \\ 0 & -24 & 0 \\ 0 & 0 & -24 \end{bmatrix} = \text{diag}[5, -3, -3].$$

(*iii*) The characteristic equation of the matrix **A** is given by

$$|\mathbf{A} - \lambda \mathbf{I}| = \begin{vmatrix} 3-\lambda & 2 & 1 \\ 0 & 2-\lambda & 0 \\ 1 & 2 & 3-\lambda \end{vmatrix} = (3-\lambda)^2(2-\lambda) + 1(\lambda - 2) = (\lambda-2)^2(\lambda-4) = 0 \text{ or } \lambda = 2, 2, 4.$$

Eignvector corresponding to the eigenvalue $\lambda = 2$ is the solution of the system

$$(\mathbf{A} - 2\mathbf{I})\mathbf{x} = \begin{bmatrix} 1 & 2 & 1 \\ 0 & 0 & 0 \\ 1 & 2 & 1 \end{bmatrix}\begin{bmatrix} x_1 \\ x_2 \\ x_3 \end{bmatrix} = \begin{bmatrix} 0 \\ 0 \\ 0 \end{bmatrix} \text{ or } x_1 + 2x_2 + x_3 = 0.$$

The rank of the coefficient matrix is 1. Therefore, the system has two linearly independent solutions. We use the equation $x_1 + 2x_2 + x_3 = 0$ to find two linearly independent eigenvectors. Choosing $x_2 = 0$, $x_3 = -1$, we obtain the eigenvector $[1, 0, -1]^T$ and taking $x_3 = 0$, $x_2 = -1$, we obtain the eigen vector $[2, -1, 0]^T$.

Eigenvector corresponding to the eigenvalue $\lambda = 4$ is the solution of the system

$$(\mathbf{A} - 4\mathbf{I})\mathbf{x} = \begin{bmatrix} -1 & 2 & 1 \\ 0 & -2 & 0 \\ 1 & 2 & -1 \end{bmatrix}\begin{bmatrix} x_1 \\ x_2 \\ x_3 \end{bmatrix} = \begin{bmatrix} 0 \\ 0 \\ 0 \end{bmatrix}.$$

The solution of this system is $[1, 0, 1]^T$.

The given 3 × 3 matrix has three linearly independent eigenvectors.

Therefore, the matrix **A** is diagonalizable.

The modal matrix **P** is given by

$$\mathbf{P} = \begin{bmatrix} 1 & 2 & 1 \\ 0 & -1 & 0 \\ -1 & 0 & 1 \end{bmatrix} \text{ and } \mathbf{P}^{-1} = \frac{1}{2}\begin{bmatrix} 1 & 2 & -1 \\ 0 & -2 & 0 \\ 1 & 2 & 1 \end{bmatrix}.$$

It can be verified that $\mathbf{P}^{-1}\mathbf{AP} = \frac{1}{2}\begin{bmatrix} 1 & 2 & -1 \\ 0 & -2 & 0 \\ 1 & 2 & 1 \end{bmatrix}\begin{bmatrix} 3 & 2 & 1 \\ 0 & 2 & 0 \\ 1 & 2 & 3 \end{bmatrix}\begin{bmatrix} 1 & 2 & 1 \\ 0 & -1 & 0 \\ -1 & 0 & 1 \end{bmatrix}$

$$= \frac{1}{2}\begin{bmatrix} 2 & 4 & -2 \\ 0 & -4 & 0 \\ 4 & 8 & 4 \end{bmatrix}\begin{bmatrix} 1 & 2 & 1 \\ 0 & -1 & 0 \\ -1 & 0 & 1 \end{bmatrix} = \frac{1}{2}\begin{bmatrix} 4 & 0 & 0 \\ 0 & 4 & 0 \\ 0 & 0 & 8 \end{bmatrix} = \begin{bmatrix} 2 & 0 & 0 \\ 0 & 2 & 0 \\ 0 & 0 & 4 \end{bmatrix} = \text{diag}(2, 2, 4).$$

EXAMPLE 6.17 The eigenvectors of a 3 × 3 matrix **A** corresponding to the eigen-values 1, 2, 3 are $[1, 2, 1]^T$, $[2, 3, 4]^T$, $[1, 4, 9]^T$ respectively. Find the matrix **A**.

Solution We have

$$\text{modal matrix } \mathbf{P} = \begin{bmatrix} 1 & 2 & 1 \\ 2 & 3 & 4 \\ 1 & 4 & 9 \end{bmatrix} \text{ and the spectral matrix } \mathbf{D} = \begin{bmatrix} 1 & 0 & 0 \\ 0 & 2 & 0 \\ 0 & 0 & 3 \end{bmatrix}.$$

$$\text{We find that } \mathbf{P}^{-1} = \frac{1}{12}\begin{bmatrix} -11 & 14 & -5 \\ 14 & -8 & 2 \\ -5 & 2 & 1 \end{bmatrix}.$$

$$\text{Therefore, } \mathbf{A} = \mathbf{PDP}^{-1} = \frac{1}{12}\begin{bmatrix} 1 & 2 & 1 \\ 2 & 3 & 4 \\ 1 & 4 & 9 \end{bmatrix}\begin{bmatrix} 1 & 0 & 0 \\ 0 & 2 & 0 \\ 0 & 0 & 3 \end{bmatrix}\begin{bmatrix} -11 & 14 & -5 \\ 14 & -8 & 2 \\ -5 & 2 & 1 \end{bmatrix}$$

$$= \frac{1}{12}\begin{bmatrix} 1 & 4 & 3 \\ 2 & 6 & 12 \\ 1 & 8 & 27 \end{bmatrix}\begin{bmatrix} -11 & 14 & -5 \\ 14 & -8 & 2 \\ -5 & 2 & 1 \end{bmatrix} = \frac{1}{12}\begin{bmatrix} 30 & -12 & 6 \\ 2 & 4 & 14 \\ -34 & 4 & 38 \end{bmatrix}.$$

EXAMPLE 6.18 Show that the matrix

$$\mathbf{A} = \begin{bmatrix} 1 & 2 & 0 \\ 2 & 1 & -6 \\ 2 & -2 & 3 \end{bmatrix}$$

is diagonalizable. Hence, find **P** such that $\mathbf{P}^{-1}\mathbf{AP}$ is a diagonal matrix. Then, obtain the matrix $\mathbf{B} = \mathbf{A}^2 + 6\mathbf{A} + 4\mathbf{I}$.

Solution The characteristic equation of **A** is given by

$$|\mathbf{A} - \lambda\mathbf{I}| = \begin{vmatrix} 1-\lambda & 2 & 0 \\ 2 & 1-\lambda & -6 \\ 2 & -2 & 3-\lambda \end{vmatrix} \begin{aligned} &= \lambda^3 - 5\lambda^2 - 9\lambda + 45 \\ &= (\lambda-5)(\lambda-3)(\lambda+3) = 0, \text{ or } \lambda = 5, 3, -3. \end{aligned}$$

Since the matrix **A** has three distinct eigenvalues, it has three linearly independent eigenvectors and hence it is diagonalizable.

The eigenvector corresponding to the eigenvalue $\lambda = 5$ is the solution of the system

$$(\mathbf{A} - 5\mathbf{I})\mathbf{x} = \begin{bmatrix} -4 & 2 & 0 \\ 2 & -4 & -6 \\ 2 & -2 & -2 \end{bmatrix}\begin{bmatrix} x_1 \\ x_2 \\ x_3 \end{bmatrix} = \begin{bmatrix} 0 \\ 0 \\ 0 \end{bmatrix}. \text{ The solution of } \mathbf{x}_1 = \begin{bmatrix} 1 \\ 2 \\ -1 \end{bmatrix}.$$

The eigenvector corresponding to the eigenvalue $\lambda = 3$ is the solution of the system

$$(\mathbf{A} - 3\mathbf{I})\mathbf{x} = \begin{bmatrix} -2 & 2 & 0 \\ 2 & -2 & -6 \\ 2 & -2 & 0 \end{bmatrix}\begin{bmatrix} x_1 \\ x_2 \\ x_3 \end{bmatrix} = \begin{bmatrix} 0 \\ 0 \\ 0 \end{bmatrix}. \text{ The solution of } \mathbf{x}_2 = \begin{bmatrix} 1 \\ 1 \\ 0 \end{bmatrix}.$$

The eigenvector corresponding to the eigenvalue $\lambda = -3$ is the solution of the system

$$(\mathbf{A} + 3\mathbf{I})\mathbf{x} = \begin{bmatrix} 4 & 2 & 0 \\ 2 & 4 & -6 \\ 2 & -2 & 6 \end{bmatrix}\begin{bmatrix} x_1 \\ x_2 \\ x_3 \end{bmatrix} = \begin{bmatrix} 0 \\ 0 \\ 0 \end{bmatrix}. \text{ The solution of } \mathbf{x}_3 = \begin{bmatrix} -1 \\ 2 \\ 1 \end{bmatrix}.$$

Hence, the modal matrix is given by

$$\mathbf{P} = [\mathbf{x}_1, \mathbf{x}_2, \mathbf{x}_3] = \begin{bmatrix} 1 & 1 & -1 \\ 2 & 1 & 2 \\ -1 & 0 & 1 \end{bmatrix}.$$

We shall verify that $\mathbf{P}^{-1}\mathbf{AP} = \mathbf{D} = \text{diag}\ (5, 3, -3)$.

The characteristic equation of $\mathbf{P}$ is $\lambda^3 - 3\lambda^2 + 4 = 0$.

Hence, by the Cayley-Hamilton theorem:

$\mathbf{P}^3 - 3\mathbf{P}^2 + 4\mathbf{I} = \mathbf{0}$, where $\mathbf{I}$ is the unit matrix of order 3. Premultiplying by $\mathbf{P}^{-1}$, we obtain

$\mathbf{P}^2 - 3\mathbf{P} + 4\mathbf{P}^{-1} = \mathbf{0}$ or $\mathbf{P}^{-1} = (-\mathbf{P}^2 + 3\mathbf{P})/4$

$$= -\frac{1}{4}\begin{bmatrix} 1 & 1 & -1 \\ 2 & 1 & 2 \\ -1 & 0 & 1 \end{bmatrix}\begin{bmatrix} 1 & 1 & -1 \\ 2 & 1 & 2 \\ -1 & 0 & 1 \end{bmatrix} + \frac{3}{4}\begin{bmatrix} 1 & 1 & -1 \\ 2 & 1 & 2 \\ -1 & 0 & 1 \end{bmatrix}$$

$$= -\frac{1}{4}\begin{bmatrix} 4 & 2 & 0 \\ 2 & 3 & 2 \\ -2 & -1 & 2 \end{bmatrix} + \frac{3}{4}\begin{bmatrix} 1 & 1 & -1 \\ 2 & 1 & 2 \\ -1 & 0 & 1 \end{bmatrix} = \begin{bmatrix} -1/4 & 1/4 & -3/4 \\ 1 & 0 & 1 \\ -1/4 & 1/4 & 1/4 \end{bmatrix}.$$

$$\mathbf{AP} = \begin{bmatrix} 1 & 2 & 0 \\ 2 & 1 & -6 \\ 2 & -2 & 3 \end{bmatrix}\begin{bmatrix} 1 & 1 & -1 \\ 2 & 1 & 2 \\ -1 & 0 & 1 \end{bmatrix} = \begin{bmatrix} 5 & 3 & 3 \\ 10 & 3 & -6 \\ -5 & 0 & -3 \end{bmatrix}.$$

Therefore,
$$\mathbf{P}^{-1}\mathbf{AP} = \begin{bmatrix} -1/4 & 1/4 & -3/4 \\ 1 & 0 & 1 \\ -1/4 & 1/4 & 1/4 \end{bmatrix}\begin{bmatrix} 5 & 3 & 3 \\ 10 & 3 & -6 \\ -5 & 0 & -3 \end{bmatrix} = \begin{bmatrix} 5 & 0 & 0 \\ 0 & 3 & 0 \\ 0 & 0 & -3 \end{bmatrix} = \text{diag}\ (5, 3, -3).$$

We have $\mathbf{D} = \text{diag}\ (5, 3, -3)$, $\mathbf{D}^2 = \text{diag}\ (25, 9, 9)$.

Therefore, $\quad \mathbf{A}^2 + 6\mathbf{A} + 4\mathbf{I} = \mathbf{P}(\mathbf{D}^2 + 6\mathbf{D} + 4\mathbf{I})\mathbf{P}^{-1}$.

Now
$$\mathbf{D}^2 + 6\mathbf{D} + 4\mathbf{I} = \begin{bmatrix} 25 & 0 & 0 \\ 0 & 9 & 0 \\ 0 & 0 & 9 \end{bmatrix} + \begin{bmatrix} 30 & 0 & 0 \\ 0 & 18 & 0 \\ 0 & 0 & -18 \end{bmatrix} + \begin{bmatrix} 4 & 0 & 0 \\ 0 & 4 & 0 \\ 0 & 0 & 4 \end{bmatrix} = \begin{bmatrix} 59 & 0 & 0 \\ 0 & 31 & 0 \\ 0 & 0 & -5 \end{bmatrix}.$$

Hence, we obtain

$$\mathbf{A}^2 + 6\mathbf{A} + 4\mathbf{I} = \begin{bmatrix} 1 & 1 & -1 \\ 2 & 1 & 2 \\ -1 & 0 & 1 \end{bmatrix}\begin{bmatrix} 59 & 0 & 0 \\ 0 & 31 & 0 \\ 0 & 0 & -5 \end{bmatrix}\begin{bmatrix} -1/4 & 1/4 & -3/4 \\ 1 & 0 & 1 \\ -1/4 & 1/4 & 1/4 \end{bmatrix} = \begin{bmatrix} 15 & 16 & -12 \\ 4 & 27 & -60 \\ 16 & -16 & 43 \end{bmatrix}.$$

EXAMPLE 6.19 Find $\mathbf{A}^n$, given $\mathbf{A} = \begin{bmatrix} -1 & 2 \\ -10 & 8 \end{bmatrix}$.

Solution The characteristic equation of $\mathbf{A}$ is given by

$$|\mathbf{A} - \lambda\mathbf{I}| = \begin{vmatrix} -1-\lambda & 2 \\ -10 & 8-\lambda \end{vmatrix} = 0 \text{ or } (-1-\lambda)(8-\lambda) + 20 = 0 \text{ or } \lambda^2 - 7\lambda + 12 = 0, \text{ or } \lambda = 3, 4.$$

Corresponding to the eigenvalue $\lambda = 3$, we have

$$(\mathbf{A} - 3\mathbf{I})\mathbf{x} = \begin{bmatrix} -4 & 2 \\ -10 & 5 \end{bmatrix}\begin{bmatrix} x_1 \\ x_2 \end{bmatrix} = \begin{bmatrix} 0 \\ 0 \end{bmatrix} \text{ or } 2x_1 - x_2 = 0 \text{ or } x_1 = \frac{x_2}{2}.$$

Hence, the eigenvector $\mathbf{x}$ is given by

$$\mathbf{x}=\begin{bmatrix}x_1\\x_2\end{bmatrix}=\begin{bmatrix}x_2/2\\x_2\end{bmatrix}=x_2\begin{bmatrix}1/2\\1\end{bmatrix}.$$

Since an eigenvector is unique upto a constant multiple, we can take the eigenvector as $[1, 2]^T$. Corresponding to the eigenvalue $\lambda = 4$, we have

$$(\mathbf{A}-4\mathbf{I})\,\mathbf{x}=\begin{bmatrix}-5&2\\-10&4\end{bmatrix}\begin{bmatrix}x_1\\x_2\end{bmatrix}=\begin{bmatrix}0\\0\end{bmatrix}\text{ or }5x_1-2x_2=0\text{ or }x_1=\frac{2}{5}x_2.$$

Therefore, the eigenvector is given by $\mathbf{x} = (x_1, x_2)^T = x_2(2, 5)^T$ or simply $(2, 5)^T$
Hence, the modal matrix is given by

$$\mathbf{P}=\begin{bmatrix}1&2\\2&5\end{bmatrix}\text{ and }\mathbf{P}^{-1}=\begin{bmatrix}5&-2\\-2&1\end{bmatrix}.$$

Now $\quad \mathbf{AP}=\begin{bmatrix}-1&2\\-10&8\end{bmatrix}\begin{bmatrix}1&2\\2&5\end{bmatrix}=\begin{bmatrix}3&8\\6&20\end{bmatrix}$ and $\mathbf{P}^{-1}\mathbf{AP}=\begin{bmatrix}5&-2\\-2&1\end{bmatrix}\begin{bmatrix}3&8\\6&20\end{bmatrix}=\begin{bmatrix}3&0\\0&4\end{bmatrix}.$

The reduction to diagonal form enables us to calculate powers of **A**. For example,

$$\begin{bmatrix}3^3&0\\0&4^3\end{bmatrix}=(\mathbf{P}^{-1}\mathbf{AP})^3=(\mathbf{P}^{-1}\mathbf{AP})(\mathbf{P}^{-1}\mathbf{AP})(\mathbf{P}^{-1}\mathbf{AP})=\mathbf{P}^{-1}\mathbf{A}(\mathbf{PP}^{-1})\mathbf{A}(\mathbf{PP}^{-1})\mathbf{AP}=\mathbf{P}^{-1}\mathbf{A}^3\mathbf{P}.$$

Therefore, $\mathbf{A}^3=\mathbf{P}\begin{bmatrix}3^3&0\\0&4^3\end{bmatrix}\mathbf{P}^{-1}.$

More generally, we have $\mathbf{A}^n=\begin{bmatrix}1&2\\2&5\end{bmatrix}\begin{bmatrix}3^n&0\\0&4^n\end{bmatrix}\begin{bmatrix}5&-2\\-2&1\end{bmatrix}$

$$=\begin{bmatrix}3^n&2\cdot4^n\\2\cdot3^n&5\cdot4^n\end{bmatrix}\begin{bmatrix}5&-2\\-2&1\end{bmatrix}=\begin{bmatrix}3^n\cdot5-2^2\cdot4^n&-2\cdot3^n+2\cdot4^n\\10\cdot3^n-5\cdot4^n\cdot2&-4\cdot3^n+5\cdot4^n\end{bmatrix}=\begin{bmatrix}5\cdot3^n-2^{2n+2}&-2\cdot3^n+2^{2n+1}\\10\cdot3^n-5\cdot2^{2n+1}&-4\cdot3^n+5\cdot2^{2n}\end{bmatrix}.$$

6.5 INNER PRODUCT (DOT PRODUCT) OF VECTORS

Let $\mathbf{x} = (x_1, x_2, ..., x_n)^T$ and $\mathbf{y} = (y_1, y_2, ..., y_n)^T$ are two vectors of dimensions n in R^n or $\mathbb{C}^n$. We define the following:

Inner Product or dot product of two vectors **x** and **y** in R^n is denoted and defined by

$$\mathbf{x}\cdot\mathbf{y}=\mathbf{x}^T\mathbf{y}=x_1y_1+x_2y_2+........+x_ny_n=\sum_{i=1}^{n}x_iy_i \qquad ...(6.8)$$

and is a scalar. The inner product is also denoted by ⟨x, y⟩. In this case $\mathbf{x}\cdot\mathbf{y}=\mathbf{y}\cdot\mathbf{x}$. Note that

$$\left.\begin{array}{l}\mathbf{x}\cdot\mathbf{x}\ge0\\ \mathbf{x}\cdot\mathbf{x}=0\text{ if and only if }\mathbf{x}=\mathbf{0}\end{array}\right\}.$$

EXAMPLE 6.20 Consider vectors $\mathbf{x} = (2, 3, -4)^T$ and $\mathbf{y} = (3, -1, -2)^T$. find ⟨**x**, **y**⟩.

Solution ⟨**x**, **y**⟩ = $\mathbf{x}^T\mathbf{y}$ = (2) (3) + (3) (–1) + (–4) (–2) = 11.

If **x** and **y** are in $\mathbb{C}^n$, then the inner product of these vectors is defined as

$$\mathbf{x}\cdot\mathbf{y}=\mathbf{x}^T\overline{\mathbf{y}}=\sum_{i=1}^{n}x_i\,\overline{y}_i\text{ and }\mathbf{y}\cdot\mathbf{x}=\mathbf{y}^T\overline{\mathbf{x}}=\sum_{i=1}^{n}y_i\,\overline{x}_i, \qquad ...(6.9)$$

where $\overline{\mathbf{x}}$ and $\overline{\mathbf{y}}$ are complex conjugate vectors of **x** and **y** respectively. Note that $\mathbf{x}\cdot\mathbf{y}=\overline{\mathbf{y}\cdot\mathbf{x}}$. ...(6.10)

EXAMPLE 6.21 Consider vectors $\mathbf{x} = (5+i, 2-3i, 7+2i)^T$ and $\mathbf{y} = (3-2i, 4i, 1+6i)^T$, find $\langle \mathbf{x}, \mathbf{y} \rangle$.

Solution We have
$$\mathbf{x}\cdot\mathbf{y} = \mathbf{x}^T\bar{\mathbf{y}} = (5+i)\overline{(3-2i)} + (2-3i)\overline{(4i)} + (7+2i)\overline{(1+6i)}$$
$$= (5+i)(3+2i) + (2-3i)(-4i) + (7+2i)(1-6i)$$
$$= (13+13i) + (-12-8i) + (19-40i) = 20-35i.$$

NOTE It can be easily verified that for any vectors **x**, **y**, **z** and scalars k_1 and k_2
$$\langle k_1\mathbf{x} + k_2\mathbf{y}, \mathbf{z}\rangle = k_1\langle \mathbf{x}, \mathbf{z}\rangle + k_2\langle \mathbf{y}, \mathbf{z}\rangle = k_1(\mathbf{x}\cdot\mathbf{z}) + k_2(\mathbf{y}\cdot\mathbf{z})$$

6.5.1 Problems based on Inner Products

EXAMPLE 6.22 Expand:

(*i*) $\langle 3\mathbf{u}_1 - 4\mathbf{u}_2, 2\mathbf{v}_1 - 5\mathbf{v}_2 + 6\mathbf{v}_3\rangle$, (*ii*) $\langle 3\mathbf{u} + 5\mathbf{v}, 4\mathbf{u} - 6\mathbf{v}\rangle$.

Solution (*i*) $\langle 3\mathbf{u}_1 - 4\mathbf{u}_2, 2\mathbf{v}_1 - 5\mathbf{v}_2 + 6\mathbf{v}_3\rangle = 6\langle \mathbf{u}_1, \mathbf{v}_1\rangle - 15\langle \mathbf{u}_1, \mathbf{v}_2\rangle + 18\langle \mathbf{u}_1, \mathbf{v}_3\rangle - 8\langle \mathbf{u}_2, \mathbf{v}_1\rangle + 20\langle \mathbf{u}_2, \mathbf{v}_2\rangle - 24\langle \mathbf{u}_2, \mathbf{v}_3\rangle$.

(*ii*) $\langle 3\mathbf{u} + 5\mathbf{v}, 4\mathbf{u} - 6\mathbf{v}\rangle = 12\langle \mathbf{u}, \mathbf{u}\rangle - 18\langle \mathbf{u}, \mathbf{v}\rangle + 20\langle \mathbf{v}, \mathbf{u}\rangle - 30\langle \mathbf{v}, \mathbf{v}\rangle$
$= 12\langle \mathbf{u}, \mathbf{u}\rangle + 2\langle \mathbf{u}, \mathbf{v}\rangle - 30\langle \mathbf{v}, \mathbf{v}\rangle$, [use symmetry, $\langle \mathbf{u}, \mathbf{v}\rangle = \langle \mathbf{v}, \mathbf{u}\rangle$].

EXAMPLE 6.23 Let $\langle \mathbf{u}, \mathbf{v}\rangle = 3 + 2i$ in a complex inner product space **v**. Find:

(*i*) $\langle (4-6i)\mathbf{u}, \mathbf{v}\rangle$, (*ii*) $\langle \mathbf{u}, (8+3i)\mathbf{v}\rangle$, (*iii*) $\langle (3-6i)\mathbf{u}, (5-2i)\mathbf{v}\rangle$.

Solution (*i*) $\langle (4-6i)\mathbf{u}, \mathbf{v}\rangle = (4-6i)\langle \mathbf{u}, \mathbf{v}\rangle = (4-6i)(3+2i) = 24-10i$

(*ii*) $\langle \mathbf{u}, (8+3i)\mathbf{v}\rangle = \overline{(8+3i)}\langle \mathbf{u}, \mathbf{v}\rangle = (8-3i)(3+2i) = 30+7i.$

(*iii*) $\langle (3-6i)\mathbf{u}, (5-2i)\mathbf{v}\rangle = (3-6i)\overline{(5-2i)}\langle \mathbf{u}, \mathbf{v}\rangle = (3-6i)(5+2i)(3+2i)$
$= (27-24i)(3+2i) = 81+48-18i = 129-18i.$

EXAMPLE 6.24 Let $\mathbf{u} = (1, 3, -4, 2)$, $\mathbf{v} = (4, -2, 2, 1)$, $\mathbf{w} = (5, -1, -2, 6)$ in R^4. Show that
$$\langle 3\mathbf{u} - 2\mathbf{v}, \mathbf{w}\rangle = 3\langle \mathbf{u}, \mathbf{w}\rangle - 2\langle \mathbf{v}, \mathbf{w}\rangle.$$

Solution We have
$$3\mathbf{u} - 2\mathbf{v} = (3, 9, -12, 6) - (8, -4, 4, 2) = (-5, 13, -16, 4).$$

Therefore, $\langle 3\mathbf{u} - 2\mathbf{v}, \mathbf{w}\rangle = (-5)(5) + (13)(-1) + (-16)(-2) + 4(6)$
$= -25 - 13 + 32 + 24 = 18.$

$\langle \mathbf{u}, \mathbf{w}\rangle = 5 - 3 + 8 + 12 = 22$ and $\langle \mathbf{v}, \mathbf{w}\rangle = 20 + 2 - 4 + 6 = 24$

$\therefore$ $3\langle \mathbf{u}, \mathbf{w}\rangle - 2\langle \mathbf{v}, \mathbf{w}\rangle = 3(22) - 2(24) = 18 = \langle 3\mathbf{u} - 2\mathbf{v}, \mathbf{w}\rangle.$

EXAMPLE 6.25 Consider $\mathbf{u} = (1+i, 3, 4-i)$ and $\mathbf{v} = (3-4i, 1+i, 2i)$ in $\mathbb{C}^3$. Find (*i*) $\langle \mathbf{u}, \mathbf{v}\rangle$, (*ii*) $\langle \mathbf{v}, \mathbf{u}\rangle$.

Solution (*i*)
$$\langle \mathbf{u}, \mathbf{v}\rangle = (1+i)\overline{(3-4i)} + 3\overline{(1+i)} + (4-i)\overline{(2i)}$$
$$= (1+i)(3+4i) + 3(1-i) + (4-i)(-2i)$$
$$= -1 + 7i + 3 - 3i - 8i - 2 = -4i.$$

(*ii*)
$$\langle \mathbf{v}, \mathbf{u}\rangle = (3-4i)\overline{(1+i)} + (1+i)(3) + (2i)\overline{(4-i)}$$
$$= (3-4i)(1-i) + 3 + 3i + 2i(4+i)$$
$$= -1 - 7i + 3 + 3i + 8i - 2 = 4i.$$

Alternatively $\langle \mathbf{v}, \mathbf{u}\rangle = \overline{\langle \mathbf{u}, \mathbf{v}\rangle} = -\overline{4i} = 4i.$

6.5.2 Norm (Length) of a Vector

The *norm* or *length* of a vector **x** in R^n or $\mathbb{C}^n$, denoted by $\|\mathbf{x}\|$, is defined to be the non-negative square root of $\mathbf{x}\cdot\mathbf{x}$. Thus
$$\|\mathbf{x}\| = \sqrt{\mathbf{x}\cdot\mathbf{x}} = \sqrt{\mathbf{x}^T\mathbf{x}} = \sqrt{x_1^2 + x_2^2 + \ldots\ldots + x_n^2} \qquad \ldots(6.11)$$
is called the *length* or the *norm* of the vector **x**.

EXAMPLE 6.26 Find ||(3, –4, 12)||.

Solution $\|3,-4,12\| = \sqrt{(3)^2+(-4)^2+12^2} = \sqrt{9+16+144} = \sqrt{169} = 13.$

EXAMPLE 6.27 Prove $\|k\mathbf{x}\| = |k|\,\|\mathbf{x}\|$, for real number k.

Solution $\|k\mathbf{x}\| = \|k(x_1, x_2, \ldots, x_n)\| = \|(kx_1, kx_2, \ldots\ldots, kx_n)\|$

$$= \sqrt{(kx_1)^2+(kx_2)^2+\ldots..+(kx_n)^2} = \sqrt{k^2(x_1^2+x_2^2+\ldots+x_n^2)} = \sqrt{k^2}\,\sqrt{x_1^2+x_2^2+\ldots+x_n^2} = |k|\,\|\mathbf{x}\|.$$

Unit Vector

The vector **x** is called a unit vector if $\|\mathbf{x}| = 1$. For any non zero vector **x** in R^n, the vector $\mathbf{x}/\|\mathbf{x}\|$ is always a unit vector. Vectors of unit length are called **unit or normalized vectors.**

Orthogonal Vectors

The vectors **x** and **y** are said to be *orthogonal vectors* if and only if $\mathbf{x}\cdot\mathbf{y} = 0$. That is, inner product of these vectors is zero.

A non empty set of mutually orthogonal nonzero vector is called an **orthogonal set** of vectors.

EXAMPLE 6.28 Show that vectors $\mathbf{x} = [1/3, -2/3, -2/3]^T$ and $\mathbf{y} = [2/3, -1/3, 2/3]^T$ are orthogonal.

Solution $$\mathbf{x}\cdot\mathbf{y} = \mathbf{x}^{\mathrm{T}}\mathbf{y} = [1/3, -2/3, -2/3]\begin{bmatrix} 2/3 \\ -1/3 \\ 2/3 \end{bmatrix} = (1/3)(2/3) + (-2/3)(-1/3) + (-2/3)(2/3)$$

$= 2/9 + 2/9 - 4/9 = \mathbf{0}$ and the vectors are orthogonal.

EXAMPLE 6.29 Consider the vectors $\mathbf{x} = (1, 1, 1)$, $\mathbf{y} = (1, 2, -3)$, $\mathbf{z} = (1, -4, 3)$ in R^3. Show that **x** is orthogonal to both **y** and **z**. But **y** and **z** are not orthogonal.

Solution $\langle \mathbf{x}, \mathbf{y}\rangle = (1)(1) + (1)(2) + (1)(-3) = 1 + 2 - 3 = \mathbf{0},$

$\langle \mathbf{x}, \mathbf{z}\rangle = (1)(1) + (1)(-4) + (1)(3) = 1 - 4 + 3 = \mathbf{0},$

$\langle \mathbf{y}, \mathbf{z}\rangle = (1)(1) + (2)(-4) + (-3)(3) = 1 - 8 - 9 = -16.$

Thus **x** is orthogonal to vectors **y** and **z**, but vectors **y** and **z** are not orthogonal.

Orthonormal Vectors: The vectors **x** and **y** for which $\mathbf{x}\cdot\mathbf{y} = 0$ and $\|\mathbf{x}\| = 1$, $\|\mathbf{y}\| = 1$ are called *orthonormal* vectors. If **x**, **y** are any vector and $\mathbf{x}\cdot\mathbf{y} = 0$, then $\mathbf{x}/\|\mathbf{x}\|$, $\mathbf{y}/\|\mathbf{y}\|$ are orthonormal.

Orthonormal System of Vectors: Let $\mathbf{x}_1, \mathbf{x}_2, \ldots, \mathbf{x}_m$ be m vectors in R^n. Then this set of vectors forms an orthonormal system of vectors, if

$$\mathbf{x}_i\cdot\mathbf{x}_j = \mathbf{x}_i^T\mathbf{x}_j = \begin{cases} 0, & i \neq j \\ 1, & i = j. \end{cases}$$

6.6 ORTHOGONAL MATRIX

A real matrix **A** is called an orthogonal matrix if **A** is non-singular and its transpose is the same as its inverse. That is, $\mathbf{A}^T = \mathbf{A}^{-1}$ or $\mathbf{A}^T\mathbf{A} = \mathbf{A}\mathbf{A}^T = \mathbf{I}$.

A simple example is

$$\mathbf{A} = \begin{bmatrix} \sin\theta & \cos\theta \\ -\cos\theta & \sin\theta \end{bmatrix}.$$

∴ Consider an orthogonal matrix $$\mathbf{B} = \begin{bmatrix} 1/\sqrt{3} & 1/\sqrt{6} & -1/\sqrt{2} \\ 1/\sqrt{3} & -2/\sqrt{6} & 0 \\ 1/\sqrt{3} & 1/\sqrt{6} & 1/\sqrt{2} \end{bmatrix}.$$

The column vectors and also the (row vectors) of an orthogonal matrix form an orthonormal system of vectors.

Remarks:

1. The product of two or more orthogonal matrices is orthogonal.
2. The inverse and the transpose of an orthogonal matrix are orthogonal.
3. The determinant of an orthogonal matrix has the value ±1.
4. The eigenvalues of an orthogonal matrix are real or complex conjugates in pairs and have absolute value 1.

EXAMPLE 6.30 Verify that the following matrices are orthogonal

$$(i)\ \mathbf{A} = \begin{bmatrix} 1/\sqrt{3} & 0 & -2/\sqrt{6} \\ 1/\sqrt{3} & 1/\sqrt{2} & 1/\sqrt{6} \\ 1/\sqrt{3} & -1/\sqrt{2} & 1/\sqrt{6} \end{bmatrix}, \qquad (ii)\ \mathbf{A} = \frac{1}{3}\begin{bmatrix} -1 & 2 & -2 \\ -2 & 1 & 2 \\ 2 & 2 & 1 \end{bmatrix} \qquad (AMIE,\ W\text{-}2003)$$

Solution (*i*) We need only verify that $\mathbf{A}^{\mathrm{T}}\mathbf{A} = \mathbf{I}$

$$\mathbf{A}^T\mathbf{A} = \begin{bmatrix} 1/\sqrt{3} & 1/\sqrt{3} & 1/\sqrt{3} \\ 0 & 1/\sqrt{2} & -1/\sqrt{2} \\ -2/\sqrt{6} & 1/\sqrt{6} & 1/\sqrt{6} \end{bmatrix}\begin{bmatrix} 1/\sqrt{3} & 0 & -2/\sqrt{6} \\ 1/\sqrt{3} & 1/\sqrt{2} & 1/\sqrt{6} \\ 1/\sqrt{3} & -1/\sqrt{2} & 1/\sqrt{6} \end{bmatrix}$$

$$= \begin{bmatrix} \frac{1}{3}+\frac{1}{3}+\frac{1}{3} & \frac{1}{\sqrt{6}}-\frac{1}{\sqrt{6}} & -\frac{2}{3\sqrt{2}}+\frac{1}{3\sqrt{2}}+\frac{1}{3\sqrt{2}} \\ 0+\frac{1}{\sqrt{6}}-\frac{1}{\sqrt{6}} & \frac{1}{2}+\frac{1}{2} & 0+\frac{1}{2\sqrt{3}}-\frac{1}{2\sqrt{3}} \\ -\frac{2}{3\sqrt{2}}+\frac{1}{3\sqrt{2}}+\frac{1}{3\sqrt{2}} & \frac{1}{2\sqrt{3}}-\frac{1}{2\sqrt{3}} & \frac{2}{3}+\frac{1}{6}+\frac{1}{6} \end{bmatrix} = \begin{bmatrix} 1 & 0 & 0 \\ 0 & 1 & 0 \\ 0 & 0 & 1 \end{bmatrix} = \mathbf{I}.$$

$$(ii)\qquad \mathbf{A}^T\mathbf{A} = \frac{1}{3}\cdot\frac{1}{3}\begin{bmatrix} -1 & -2 & 2 \\ 2 & 1 & 2 \\ -2 & 2 & 1 \end{bmatrix}\begin{bmatrix} -1 & 2 & -2 \\ -2 & 1 & 2 \\ 2 & 2 & 1 \end{bmatrix} = \frac{1}{9}\begin{bmatrix} 9 & 0 & 0 \\ 0 & 9 & 0 \\ 0 & 0 & 9 \end{bmatrix} = \begin{bmatrix} 1 & 0 & 0 \\ 0 & 1 & 0 \\ 0 & 0 & 1 \end{bmatrix} = \mathbf{I}.$$

EXAMPLE 6.31 Show that the vectors $\mathbf{x}_1 = [\cos\theta, \sin\theta, 0]^T$, $\mathbf{x}_2 = [-\sin\theta, \cos\theta, 0]^T$, $\mathbf{x}_3 = [0, 0, 1]^T$ form an orthogona! set or system of vectors.

Solution Since the vectors are real, we must show that

$$\mathbf{x}_i \cdot \mathbf{x}_j = \mathbf{x}_i^T\mathbf{x}_j = \begin{cases} 0, & i \neq j \\ 1, & i = j. \end{cases}$$

If $i = j = 1$, we have

$$\mathbf{x}_1^T\mathbf{x}_1 = [\cos\theta, \sin\theta, 0]\begin{bmatrix} \cos\theta \\ \sin\theta \\ 0 \end{bmatrix} = \cos^2\theta + \sin^2\theta = 1.$$

Similarly, we find if $i = j = 2$ and $i = j = 3$, $\mathbf{x}_2^T\mathbf{x}_2 = 1$, $\mathbf{x}_3^T\mathbf{x}_3 = 1$.

Thus $\mathbf{x}_1$, $\mathbf{x}_2$, $\mathbf{x}_3$ are unit vectors.

Take $i = 1, j = 2$, the orthogonality of any two of the vectors.

$$\mathbf{x}_1^T\mathbf{x}_2 = [\cos\theta \quad \sin\theta \quad 0]\begin{bmatrix} -\sin\theta \\ \cos\theta \\ 0 \end{bmatrix} = 0.$$

Similarly $\mathbf{x}_2^T\mathbf{x}_3 = 0$, $\mathbf{x}_1^T\mathbf{x}_3 = 0$. The vectors are mutually orthogonal.

Thus the vectors form an orthonormal system.

EXAMPLE 6.32 Let $\mathbf{A} = \begin{bmatrix} 1 & 1 & -1 \\ 1 & 3 & 4 \\ 7 & -5 & 2 \end{bmatrix}$. Determine whether or not

(*i*) the rows of **A** are orthogonal;
(*ii*) **A** is an orthogonal matrix;
(*iii*) the columns of **A** are orthogonal;
(*iv*) normalizing each row of matrix **A** and find matrix **B**;
(*v*) Is **B** an orthogonal matrix?
(*vi*) Are the columns of matrix **B** orthogonal?

Solution (*i*) yes, since $(1, 1, -1) \cdot (1, 3, 4) = (1)(1) + (1)(3) + (-1)(4) = 0,$

$$(1, 3, 4) \cdot (7, -5, 2) = 7 - 15 + 8 = 0,$$

and $$(1, 1, -1) \cdot (7, -5, 2) = 7 - 5 - 2 = 0.$$

(*ii*) No, since the rows of **A** are not unit vector.

$$\|(1, 1, -1)\| = \sqrt{1^2 + 1^2 + (-1)^2} = \sqrt{3}.$$

(*iii*) No, for example $(1, 1, 7) \cdot (1, 3, -5) = 1 + 3 - 35 = -31 \neq 0.$

(*iv*) $$\|(1, 1, -1)\| = \sqrt{1+1+1} = \sqrt{3},\ \|(1, 3, 4)\| = \sqrt{1+9+16} = \sqrt{26},$$

$$\|(7, -5, 2)\| = \sqrt{49+25+4} = \sqrt{78}.$$

Thus $$\mathbf{B} = \begin{bmatrix} 1/\sqrt{3} & 1/\sqrt{3} & -1/\sqrt{3} \\ 1/\sqrt{26} & 3/\sqrt{26} & 4/\sqrt{26} \\ 7/\sqrt{78} & -5/\sqrt{78} & 2/\sqrt{78} \end{bmatrix}.$$

(*v*) Yes, since the rows of matrix **B** are still orthogonal and are now unit vectors.
(*vi*) Yes, since the rows of matrix **B** form an orthonormal set of vectors, the columns of **B** must automatically form an orthonormal set.

EXAMPLE 6.33 Show that the matrices **A** and $\mathbf{A^T}$ have the same eigenvalues. Further if **λ, μ** are two distinct eigenvalues, then show that the eigenvector corresponding to λ for **A** is orthogonal to eigenvector corresponding to **μ** for $\mathbf{A^T}$. *[AMIETE, Dec. 2005]*

Solution We have $|\mathbf{A} - \lambda\mathbf{I}| = |(\mathbf{A}^T)^T - \lambda\mathbf{I}^T| = |[\mathbf{A}^T - \lambda\mathbf{I}]^T| = |\mathbf{A}^T - \lambda\mathbf{I}|.$

Since **A** and $\mathbf{A}^T$ have the same characteristic equation, they have the same eigenvalues.

Let **x** be the eigenvector corresponding to the eigenvalue λ for **A** and **y** be the eigenvector corresponding to the eigenvalue μ for $\mathbf{A}^T$. We have $\mathbf{Ax} = \lambda\mathbf{x}$.

Premultiplying by $\mathbf{y}^T$, we obtain $\mathbf{y}^T\mathbf{Ax} = \lambda\mathbf{y}^T\mathbf{x}$ ...(*i*)

we also have $\mathbf{A}^T\mathbf{y} = \mu\mathbf{y}$, or $(\mathbf{A}^T\mathbf{y})^T = (\mu\mathbf{y})^T$ or $\mathbf{y}^T\mathbf{A} = \mu\mathbf{y}^T$.

Postmultiplying by **x**, we get $\mathbf{y}^T\mathbf{Ax} = m\mathbf{y}^T\mathbf{x}$. ...(*ii*)

From (*i*) and (*ii*), we obtain $(\lambda - \mu)\mathbf{y}^T\mathbf{x} = 0$.

Since $\lambda \neq \mu$, we obtain $\mathbf{y}^T\mathbf{x} = \mathbf{0}$. Therefore, the vectors **x** and **y** are orthogonal.

EXERCISE 6.1

1. (*a*) Find the sum and product of the eigenvalues of the following matrices:

(*i*) $\mathbf{A} = \begin{bmatrix} 2 & 2 & 1 \\ 1 & 3 & 1 \\ 1 & 2 & 2 \end{bmatrix}$, (*ii*) $\mathbf{B} = \begin{bmatrix} 1 & 2 & 2 \\ 0 & 2 & 1 \\ -1 & 2 & 2 \end{bmatrix}$. *(AMIETE, June 2010)* **Ans.** (*i*) (7,5), (*ii*) (5, 4)

(b) If $\mathbf{A}=\begin{bmatrix}-1 & 0 & 0\\ 2 & -3 & 0\\ 1 & 4 & 2\end{bmatrix}$, find the eigenvalues of $\mathbf{A}^2$. *(AMIETE, Dec. 2008)* **Ans.** 1, 9, 4.

Verify the Cayley-Hamilton theorem for the matrix **A**. Find $\mathbf{A}^{-1}$, if it exists, where **A** is as given in Problems 2 (*i*) to (*viii*).

2. (*i*) $\begin{bmatrix}2 & 3\\ 3 & 5\end{bmatrix}$, (*ii*) $\begin{bmatrix}2 & -1 & 1\\ -1 & 2 & -1\\ 1 & -1 & 2\end{bmatrix}$, *(Kuk, 2008; Madras, 2006; UPTU, 2005; AMIE, W - 2005)*

(*iii*) $\begin{bmatrix}1 & -2 & 2\\ 1 & 2 & 3\\ 0 & -1 & 2\end{bmatrix}$, (*iv*) $\begin{bmatrix}1 & 2 & 1\\ -1 & 0 & 3\\ 2 & -1 & 1\end{bmatrix}$,

(*v*) $\begin{bmatrix}1 & 3 & 7\\ 4 & 2 & 3\\ 1 & 2 & 1\end{bmatrix}$, *(Madras, 2003; MDU 2002)* (*vi*) $\begin{bmatrix}1 & 1 & 3\\ 1 & 3 & -3\\ -2 & -4 & -4\end{bmatrix}$, *(AMIETE, June 2002)*

(*vii*) $\begin{bmatrix}1 & 0 & 3\\ 2 & 1 & -1\\ 1 & -1 & 1\end{bmatrix}$ *(Osmania, 2000S)* (*viii*) $\begin{bmatrix}7 & 2 & -2\\ -6 & -1 & 2\\ 6 & 2 & -1\end{bmatrix}$. *(Coimbatore 2001; Madras 2000S)*

Ans. (*i*) $\begin{bmatrix}5 & -3\\ -3 & 2\end{bmatrix}$, (*ii*) $\dfrac{1}{4}\begin{bmatrix}3 & 1 & -1\\ 1 & 3 & 1\\ -1 & 1 & 3\end{bmatrix}$, (*iii*) $\dfrac{1}{9}\begin{bmatrix}7 & 2 & -10\\ -2 & 2 & -1\\ -1 & 1 & 4\end{bmatrix}$,

(*iv*) $\dfrac{1}{18}\begin{bmatrix}3 & -3 & 6\\ 7 & -1 & -4\\ 1 & 5 & 2\end{bmatrix}$, (*v*) $\dfrac{1}{35}\begin{bmatrix}-4 & 11 & -5\\ -1 & -6 & 25\\ 6 & 1 & -10\end{bmatrix}$, (*vi*) $\dfrac{1}{4}\begin{bmatrix}12 & 4 & 6\\ -5 & -1 & -3\\ -1 & -1 & -1\end{bmatrix}$,

(*vii*) $\dfrac{1}{9}\begin{bmatrix}0 & 3 & 3\\ 3 & 2 & -7\\ 3 & -1 & -1\end{bmatrix}$, (*viii*) $\dfrac{1}{3}\begin{bmatrix}-3 & -2 & 2\\ 6 & 5 & -2\\ -6 & -2 & 5\end{bmatrix}$.

3. If $\mathbf{A}=\begin{bmatrix}3 & 1\\ -1 & 2\end{bmatrix}$, express $2\mathbf{A}^5 - 3\mathbf{A}^4 + \mathbf{A}^2 - 4\mathbf{I}$ as a linear polynomial in **A**. **Ans.** $138\mathbf{A} - 403\mathbf{I}$

4. Evaluate the matrix $\mathbf{A}^5 - 27\mathbf{A}^3 + 65\mathbf{A}^2$, where $\mathbf{A}=\begin{bmatrix}0 & 0 & 1\\ 3 & 1 & 0\\ -2 & 1 & 4\end{bmatrix}$. **Ans.** $\begin{bmatrix}-40 & 0 & 73\\ 219 & 33 & 0\\ -146 & 73 & 252\end{bmatrix}$.

5. Given $\mathbf{A}=\begin{bmatrix}1 & 1 & 2\\ 3 & 1 & 1\\ 2 & 3 & 1\end{bmatrix}$. Use the fact that **A** is satisfies its characteristic equation to compute $\mathbf{A}^3$ and $\mathbf{A}^4$.

Also compute $\mathbf{A}^{-1}$ and $\mathbf{A}^{-2}$ since **A** is non-singular matrix.

[**Hint:** The characteristic equation of **A** is $\lambda^3 - 3\lambda^2 - 7\lambda - 11 = 0$.

$$A^3 = 3A^2 + 7A + 11I = \begin{bmatrix} 42 & 31 & 29 \\ 45 & 39 & 31 \\ 53 & 45 & 42 \end{bmatrix}.$$

$$A^4 = 3A^3 + 7A^2 + 11A = \begin{bmatrix} 193 & 160 & 144 \\ 224 & 177 & 160 \\ 272 & 224 & 193 \end{bmatrix}.$$

$$A^{-1} = \frac{1}{11}(A^2 - 3A - 7I) = \frac{1}{11}\begin{bmatrix} -2 & 5 & -1 \\ -1 & -3 & 5 \\ 7 & -1 & -2 \end{bmatrix}.$$

$$A^{-2} = \frac{1}{11}[A - 3I - 7A^{-1}] = \frac{1}{121}\begin{bmatrix} -8 & -24 & 29 \\ 40 & -1 & -24 \\ -27 & 40 & -8 \end{bmatrix}].$$

6. (*i*) Find the characteristic equation of the matrix $A = \begin{bmatrix} 4 & 3 & 1 \\ 2 & 1 & -2 \\ 1 & 2 & 1 \end{bmatrix}$. Hence find A^{-1}. (*UPTU-2001*)

Ans. $\lambda^3 - 6\lambda^2 + 6\lambda - 11 = 0,\ A^{-1} = 1/11\begin{bmatrix} 5 & -1 & -7 \\ -4 & 3 & 10 \\ 3 & -5 & -2 \end{bmatrix}$.

(*ii*) Show that the characteristic equation of the matrix $A = \begin{bmatrix} 5 & 7 & 3 \\ 1 & 5 & 2 \\ 3 & 2 & 1 \end{bmatrix}$ is $\lambda^3 - 11\lambda^2 + 15\lambda - 1 = 0$.

Deduce that A is non-singular and that $A^{-1} = A^2 - 11A + 15I$.

7. If $A = \begin{bmatrix} 3 & -7 \\ -4 & 1 \end{bmatrix}$, find $A^3 + 3A^2 + 12A$. **Ans.** $\begin{bmatrix} 370 & -455 \\ -260 & 240 \end{bmatrix}$.

8. Find $A^4 + A^2 + I$ if $A = \begin{bmatrix} 3 & 5 \\ -1 & -2 \end{bmatrix}$. **Ans.** $\begin{bmatrix} 16 & 20 \\ -4 & -4 \end{bmatrix}$.

9. If $A = \begin{bmatrix} 1 & 1 & 2 \\ 0 & 2 & 1 \\ 1 & 0 & 2 \end{bmatrix}$, show that $A^3 = (5A - I)(A - I)$ and $A^{-1} = (A - 3I)(A - 2I)$. Deduce explicit forms for A^3 and A^{-1}.

Ans. $A^3 = \begin{bmatrix} 10 & 9 & 23 \\ 5 & 9 & 14 \\ 9 & 5 & 19 \end{bmatrix},\ A^{-1} = \begin{bmatrix} 4 & -2 & -3 \\ 1 & 0 & -1 \\ -2 & 1 & 2 \end{bmatrix}$.

10. The matrix A is defined by $A = \begin{bmatrix} 1 & 2 & 3 \\ 3 & 1 & 2 \\ 2 & 3 & 1 \end{bmatrix}$. Show that $A^3 - 3A^2 - 16A - 16I = 2I - A$ and express $(2I - A)^{-1}$ as a quadratic polynomial in A. **Ans.** $(17I + A - A^2)/52$.

11. If $\mathbf{A} = \begin{bmatrix} 1 & 2 \\ -1 & 3 \end{bmatrix}$. Verify Cayley-Hamilton theorem. Hence express $\mathbf{A}^6 - 4\mathbf{A}^5 + 8\mathbf{A}^4 - 12\mathbf{A}^3 + 14\mathbf{A}^2$, as a linear polynomial in **A**. **Ans.** $-4\mathbf{A} + 5\mathbf{I}$.

12. Using Cayley-Hamilton theorem, find $\mathbf{A}^8$, if $\mathbf{A} = \begin{bmatrix} 1 & 2 \\ 2 & -1 \end{bmatrix}$. *(PTU 2003; Anna 2003;)* **Ans.** $625\mathbf{I}$.

13. (*i*) Find the characteristic equation of the matrix $\mathbf{A} = \begin{bmatrix} 2 & 1 & 1 \\ 0 & 1 & 0 \\ 1 & 1 & 2 \end{bmatrix}$ and hence compute $\mathbf{A}^{-1}$. Also find the matrix represented by $\mathbf{A}^8 - 5\mathbf{A}^7 + 7\mathbf{A}^6 - 3\mathbf{A}^5 + \mathbf{A}^4 - 5\mathbf{A}^3 + 8\mathbf{A}^2 - 2\mathbf{A} + \mathbf{I}$.

(Rajasthan, 2005; AMIETE, June 2004; UPTU 2003, MDU 2002)

Ans. $\lambda^3 - 5\lambda^2 + 7\lambda - 3 = 0,\ \mathbf{A}^{-1} = \frac{1}{3}\begin{bmatrix} 2 & -1 & -1 \\ 0 & 3 & 0 \\ -1 & -1 & 2 \end{bmatrix},\ \mathbf{A}^2 + \mathbf{A} + \mathbf{I} = \begin{bmatrix} 8 & 5 & 5 \\ 0 & 3 & 0 \\ 5 & 5 & 8 \end{bmatrix}$.

(*ii*) Find the characteristic equation of the matrix $\mathbf{A} = \begin{bmatrix} 2 & 1 & 1 \\ 0 & 1 & 0 \\ 1 & 1 & 2 \end{bmatrix}$ and hence find the matrix $\mathbf{A}^7 - 5\mathbf{A}^6 + 9\mathbf{A}^5 - 13\mathbf{A}^4 + 17\mathbf{A}^3 - 21\mathbf{A}^2 - 8\mathbf{I}$. **Ans.** $\lambda^3 - 5\lambda^2 + 7\lambda - 3 = 0$; **I**.

14. If $\mathbf{A} = \begin{bmatrix} 1 & 0 & 2 \\ 0 & -1 & 1 \\ 0 & 1 & 0 \end{bmatrix}$. Verify Cayley-Hamilton theorem. Compute $2\mathbf{A}^8 - 3\mathbf{A}^5 + \mathbf{A}^4 + \mathbf{A}^2 - 4\mathbf{I}$.

[**Hint:** The characteristic equation of **A** is $\lambda^3 - 2\lambda + 1 = 0$. By Cayley-Hamilton theorem $\mathbf{A}^3 - 2\mathbf{A} + \mathbf{I} = \mathbf{0}$.

$$\begin{aligned}
2\mathbf{A}^8 - 3\mathbf{A}^5 + \mathbf{A}^4 + \mathbf{A}^2 - 4\mathbf{I} &= 2\mathbf{A}^5(\mathbf{A}^3 - 2\mathbf{A} + \mathbf{I}) + 4\mathbf{A}^6 - 5\mathbf{A}^5 + \mathbf{A}^4 + \mathbf{A}^2 - 4\mathbf{I} \\
&= 4\mathbf{A}^3(\mathbf{A}^3 - 2\mathbf{A} + \mathbf{I}) - 5\mathbf{A}^5 + 9\mathbf{A}^4 - 4\mathbf{A}^3 + \mathbf{A}^2 - 4\mathbf{I} \\
&= -5\mathbf{A}^2(\mathbf{A}^3 - 2\mathbf{A} + \mathbf{I}) + 9\mathbf{A}^4 - 14\mathbf{A}^3 + 6\mathbf{A}^2 - 4\mathbf{I} \\
&= 9\mathbf{A}(\mathbf{A}^3 - 2\mathbf{A} + \mathbf{I}) - 14\mathbf{A}^3 + 24\mathbf{A}^2 - 9\mathbf{A} - 4\mathbf{I} \\
&= -14(\mathbf{A}^3 - 2\mathbf{A} + \mathbf{I}) + 24\mathbf{A}^2 - 37\mathbf{A} + 10\mathbf{I} = 24\mathbf{A}^2 - 37\mathbf{A} + 10\mathbf{I} \\
&= 24\begin{bmatrix} 1 & 2 & 2 \\ 0 & 2 & -1 \\ 0 & -1 & 1 \end{bmatrix} - 37\begin{bmatrix} 1 & 0 & 2 \\ 0 & -1 & 1 \\ 0 & 1 & 0 \end{bmatrix} + \begin{bmatrix} 10 & 0 & 0 \\ 0 & 10 & 0 \\ 0 & 0 & 10 \end{bmatrix} = \begin{bmatrix} -3 & 48 & -26 \\ 0 & 95 & -61 \\ 0 & -61 & 34 \end{bmatrix}.
\end{aligned}$$
]

15. Show that $\mathbf{A} = \begin{bmatrix} 1 & 2 & 0 \\ 2 & -1 & 0 \\ 0 & 0 & -1 \end{bmatrix}$ satisfies its own characteristic equation. Hence or otherwise evaluate $\mathbf{A}^{-2}$.

Ans. $1/5\begin{bmatrix} 1 & 0 & 0 \\ 0 & 1 & 0 \\ 0 & 0 & 5 \end{bmatrix}$.

16. If $\mathbf{A} = \begin{bmatrix} 1 & 0 & 0 \\ 1 & 0 & 1 \\ 0 & 1 & 0 \end{bmatrix}$, then show that $\mathbf{A}^n = \mathbf{A}^{n-2} + \mathbf{A}^3 - \mathbf{A}$ for $n \geq 4$. Hence, find $\mathbf{A}^{20}$. **Ans.** $\begin{bmatrix} 1 & 0 & 0 \\ 10 & 1 & 0 \\ 10 & 0 & 1 \end{bmatrix}$.

17. Find eigenvalues and corresponding eigenvectors for each of the following matrices.

(*i*) $\begin{bmatrix} 2 & 2 \\ -1 & 5 \end{bmatrix}$, (*ii*) $\begin{bmatrix} \cos\theta & -\sin\theta \\ \sin\theta & \cos\theta \end{bmatrix}$, (*iii*) $\begin{bmatrix} 8 & -6 & 2 \\ -6 & 7 & -4 \\ 2 & -4 & 3 \end{bmatrix}$, (*JNTU*, 2006; *AMIE*, 2005; *AMIETE*, 2005; *VTU* 2004; *MDU* 2003; *PTU* 2003)

(*iv*) $\begin{bmatrix} 1 & 1 & 3 \\ 1 & 5 & 1 \\ 3 & 1 & 1 \end{bmatrix}$, (*AMIETE*; *June* 2011; 2009, *Raipur*, 2005) (*v*) $\begin{bmatrix} 2 & 2 & -7 \\ 2 & 1 & 2 \\ 0 & 1 & -3 \end{bmatrix}$, (*vi*) $\begin{bmatrix} -2 & 2 & -3 \\ 2 & 1 & -6 \\ -1 & -2 & 0 \end{bmatrix}$, (*AMIETE, Dec, 2010, 2009; NIT Kurukshetra, 2008; UPTU 2006; AMIE 2005; MDU 2005*)

(*vii*) $\begin{bmatrix} 3 & 1 & 4 \\ 0 & 2 & 6 \\ 0 & 0 & 5 \end{bmatrix}$, (*AMIETE, June 2010, 2009; UPTU 2005*) (*viii*) $\begin{bmatrix} 2 & 2 & 1 \\ 1 & 3 & 1 \\ 1 & 2 & 2 \end{bmatrix}$,

(*ix*) $\begin{bmatrix} 3 & 10 & 5 \\ -2 & -3 & -4 \\ 3 & 5 & 7 \end{bmatrix}$, (*x*) $\begin{bmatrix} 1 & 2 & 1 \\ 6 & -1 & 0 \\ -1 & -2 & -1 \end{bmatrix}$, (*xi*) $\begin{bmatrix} 2 & 2 & 1 \\ 4 & 8 & 1 \\ -1 & -2 & 0 \end{bmatrix}$,

(*xii*) $\begin{bmatrix} 3 & 2 & 4 \\ 2 & 0 & 2 \\ 4 & 2 & 3 \end{bmatrix}$, (*xiii*) $\begin{bmatrix} 3 & -4 & 0 \\ 4 & 3 & 0 \\ 0 & 0 & 5 \end{bmatrix}$, (*xiv*) $\begin{bmatrix} 1 & 0 & 0 \\ 0 & 2 & 1 \\ 2 & 0 & 3 \end{bmatrix}$, (*AMIE, W-2010, S-2008*)

(*xv*) $\begin{bmatrix} 2 & 0 & 1 \\ 0 & 2 & 0 \\ 1 & 0 & 2 \end{bmatrix}$ (*Kurukshetra, 2005; GGSIPU 2004; JNTU 2005; MDU 2002*)

Ans. (*i*) 3, 4; $[2, 1]^T$, $[1, 1]^T$ (*ii*) $e^{\pm i\theta}$; $[1, -i]^T$, $[1, i]^T$
(*iii*) 0, 3, 15; $[1, 2, 2]^T$, $[2, 1, -2]^T$, $[2, -2, 1]^T$ (*iv*) –2, 3, 6; $[-1, 0, 1]^T$, $[1, -1, 1]^T$, $[1, 2, 1]^T$
(*v*) 1, 3, –4; $[-1, 4, 1]^T$, $[5, 6, 1]^T$, $[3, -2, 2]^T$ (*vi*) 5, –3, –3; $[1, 2, -1]^T$, $[2, -1, 0]^T$, $[3, 0, 1]^T$
(*vii*) 2, 3, 5; $[1, -1, 0]^T$, $[1, 0, 0]^T$, $[3, 2, 1]^T$ (*viii*) 5, 1, 1; $[1, 1, 1]^T$, $[1, 0, -1]^T$, $[2, -1, 0]^T$;
(*ix*) 3, 2, 2; $[1, 1, -2]^T$, $[5, 2, -5]^T$ (*x*) 0, 3, –4; $[1, 6, -13]^T$, $[2, 3, -2]^T$, $[-1, 2, 1]^T$
(*xi*) 1, 3, 6; $[1, 1, -3]^T$, $[1, 1, -1]^T$, $[2, 5, -2]^T$ (*xii*) 8, –1, –1; $[2, 1, 2]^T$, $[1, -2, 0]^T$, $[0, -2, 1]^T$
(*xiii*) 5, 3 ± 4*i*; $[0, 0, 1]^T$, $[1, -i, 0]^T$, $[1, i, 0]^T$. (*xiv*) 1, 2, 3; $[-1, -1, 1]^T$, $[0, 1, 0]^T$, $[0, -1, 1]^T$
(*xv*) 1, 2, 3; $[1, 0, -1]^T$, $[0, 1, 0]^T$, $[1, 0, 1]^T$

18. Examine whether **A** is similar to **B**, where $\mathbf{A} = \begin{bmatrix} 5 & 5 \\ -2 & 0 \end{bmatrix}$ and $\mathbf{B} = \begin{bmatrix} 1 & 2 \\ -3 & 4 \end{bmatrix}$ (*AMIETE, Dec-2006*)

Ans. Matrices **A** and **B** are similar. $\mathbf{P} = \begin{bmatrix} 1 & 1 \\ 1 & 2 \end{bmatrix}$.

19. Show that the following matrices **A** are diagonalizable. Find the matrix **P** such that $\mathbf{P}^{-1}\mathbf{AP}$ is a diagonal matrix.

(*i*) $\begin{bmatrix} 5 & 4 \\ 1 & 2 \end{bmatrix}$, (*AMIETE Dec. 2004*), (*ii*) $\begin{bmatrix} -5 & 9 \\ -6 & 10 \end{bmatrix}$, (*iii*) $\begin{bmatrix} -1 & 1 & 2 \\ 0 & -2 & 1 \\ 0 & 0 & -3 \end{bmatrix}$, *(iv)* $\begin{bmatrix} 5 & 7 & -5 \\ 10 & 4 & -1 \\ 2 & 8 & -3 \end{bmatrix}$, *(v)* $\begin{bmatrix} 0 & 2 & 1 \\ -4 & 6 & 1 \\ -1 & -2 & -2 \end{bmatrix}$, *(vi)* $\begin{bmatrix} 5 & -6 & -6 \\ -1 & 4 & 2 \\ 3 & -6 & -4 \end{bmatrix}$,

(vii) $\begin{bmatrix} 3 & 1 & -1 \\ -2 & 1 & 2 \\ 0 & 1 & 2 \end{bmatrix}$ *(AMIETE, June 2010, Dec. 2007)*, *(viii)* $\begin{bmatrix} 3 & 2 & 1 \\ 0 & 2 & 0 \\ 1 & 2 & 3 \end{bmatrix}$ *(AMIETE June 2006)*,

(ix) $\begin{bmatrix} -1 & 2 & -2 \\ 1 & 2 & 1 \\ -1 & -1 & 0 \end{bmatrix}$ *(UPTU, 2006; Raipur, 2004; AMIETE, June 2001)* *(x)* $\begin{bmatrix} 2 & 1 \\ 1 & 0 \end{bmatrix}$, *(AMIE-W- 2010)*

Ans. *(i)* $\lambda = 6$: $[4, 1]^T$; $\lambda = 1$: $[1, -1]^T$. $\mathbf{P} = \begin{bmatrix} 4 & 1 \\ 1 & -1 \end{bmatrix}$; $\mathbf{P}^{-1} = \dfrac{1}{-5}\begin{bmatrix} -1 & -1 \\ -1 & 4 \end{bmatrix}$

(ii) $\lambda = 1$: $[3, 2]^T$; $\lambda = 4$: $[1, 1]^T$ $\mathbf{P} = \begin{bmatrix} 3 & 1 \\ 2 & 1 \end{bmatrix}$; $\mathbf{P}^{-1} = \begin{bmatrix} 1 & -1 \\ -2 & 3 \end{bmatrix}$

(iii) $\lambda = -1$: $[1, 0, 0,]^T$; $\lambda = -2$: $[1, -1, 0]^T$; $\lambda = -3$: $[1, 2, -2]^T$. $\mathbf{P} = \begin{bmatrix} 1 & 1 & 1 \\ 0 & -1 & 2 \\ 0 & 0 & -2 \end{bmatrix}$; $\mathbf{P}^{-1} = \dfrac{1}{2}\begin{bmatrix} 2 & 2 & 3 \\ 0 & -2 & -2 \\ 0 & 0 & -1 \end{bmatrix}$,

(iv) $\lambda = 1$: $[2, 1, 3]^T$; $\lambda = 2$: $[1, 1, 2]^T$; $\lambda = 3$: $[-1, 1, 1]^T$. $\mathbf{P} = \begin{bmatrix} 2 & 1 & -1 \\ 1 & 1 & 1 \\ 3 & 2 & 1 \end{bmatrix}$; $\mathbf{P}^{-1} = \begin{bmatrix} -1 & -3 & 2 \\ 2 & 5 & -3 \\ -1 & -1 & 1 \end{bmatrix}$.

(v) $\lambda = 4$: $[2, 5, -2]^T$; $\lambda = -1$: $[1, 1, -3]^T$; $\lambda = 1$: $[1, 1, -1]^T$. $\mathbf{P} = \begin{bmatrix} 2 & 1 & 1 \\ -5 & 1 & 1 \\ -2 & -3 & -1 \end{bmatrix}$; $\mathbf{P}^{-1} = 1/6\begin{bmatrix} -2 & 2 & 0 \\ -3 & 0 & -3 \\ 13 & -4 & 3 \end{bmatrix}$.

(vi) $\lambda = 1$: $[3, -1, 3]^T$; $\lambda = 2, 2$: $[2, 1, 0]^T$, $[2, 0, 1]^T$. $\mathbf{P} = \begin{bmatrix} 3 & 2 & 2 \\ -1 & 1 & 0 \\ 3 & 0 & 1 \end{bmatrix}$; $\mathbf{P}^{-1} = \begin{bmatrix} -1 & 2 & 2 \\ -1 & 3 & 2 \\ 3 & -6 & -5 \end{bmatrix}$.

(vii) $\lambda = 1$: $[1, -1, 1]^T$; $\lambda = 2$: $[1, 0, 1]^T$; $\lambda = 3$: $[0, 1, 1]^T$ $\mathbf{P} = \begin{bmatrix} 1 & 1 & 0 \\ -1 & 0 & 1 \\ 1 & 1 & 1 \end{bmatrix}$; $\mathbf{P}^{-1} = \begin{bmatrix} -1 & -1 & 1 \\ 2 & 1 & -1 \\ -1 & 0 & 1 \end{bmatrix}$.

(viii) $\lambda = 2, 2; [1, 0, -1]^T, [-2, 1, 0]^T; \lambda = 4 : [1, 0, 1]^T$. $\mathbf{P} = \begin{bmatrix} 1 & -2 & 1 \\ 0 & 1 & 0 \\ -1 & 0 & 1 \end{bmatrix}$; $\mathbf{P}^{-1} = 1/2\begin{bmatrix} 1 & 2 & -1 \\ 0 & 2 & 0 \\ 1 & 2 & 1 \end{bmatrix}$.

(ix) $\lambda = 1 : [1, 0, -1]^T; \lambda = \sqrt{5} : \left[\sqrt{5} - 1, 1, -1\right]^T; \lambda = -\sqrt{5} : \left[\sqrt{5} + 1, -1, 1\right]^T$,

$$\mathbf{P} = \begin{bmatrix} 1 & \sqrt{5}-1 & \sqrt{5}+1 \\ 0 & 1 & -1 \\ -1 & -1 & 1 \end{bmatrix}; \mathbf{P}^{-1} = \begin{bmatrix} 0 & -1 & -1 \\ 1/2\sqrt{5} & (2+\sqrt{5})/2\sqrt{5} & 1/2\sqrt{5} \\ 1/2\sqrt{5} & (2-\sqrt{5})/2\sqrt{5} & 1/2\sqrt{5} \end{bmatrix}.$$

(*x*) [**Hint:** The characteristic equation of A is given by

$$|\mathbf{A}-\lambda I| = \begin{vmatrix} 2-\lambda & 1 \\ 1 & -\lambda \end{vmatrix} = 0 \quad \text{or} \quad \lambda^2 - 2\lambda - 1 = 0 \quad \text{or} \quad \lambda = 1 \pm \sqrt{2}$$

Since the matrix **A** has two distinct eigen values, it has two linearly independent eigen vectors and hence it is diagonalizable.

Corresponding to the eigen value $\lambda = 1+\sqrt{2}$, we have eigen vector $[1+\sqrt{2}, 1]^T$.

Corresponding to the eigen value $\lambda = 1-\sqrt{2}$, we have eigen vector $[1-\sqrt{2}, 1]^T$.

The model matrix **P** is given be

$$\mathbf{P} = \begin{bmatrix} 1+\sqrt{2} & 1-\sqrt{2} \\ 1 & 1 \end{bmatrix} \text{ and } \mathbf{P}^{-1} = \frac{1}{2\sqrt{2}}\begin{bmatrix} 1 & \sqrt{2}-1 \\ -1 & 1+\sqrt{2} \end{bmatrix}.$$

It can be verified that

$$\mathbf{P}^{-1}\mathbf{AP} = \begin{bmatrix} 1+\sqrt{2} & 0 \\ 0 & 1-\sqrt{2} \end{bmatrix} = \text{diag}\,(1+\sqrt{2}, 1-\sqrt{2})].$$

20. (*i*) Reduce the matrix $\mathbf{A} = \begin{bmatrix} -1 & 2 & -2 \\ 1 & 2 & 1 \\ -1 & -1 & 0 \end{bmatrix}$ to the diagonal form.

(*AMIETE, Dec 2009; UPTU 2006, Raipur 2004*)

(*ii*) Diagonalise the matrix $\mathbf{A} = \begin{bmatrix} 1 & 1 & 2 \\ -1 & 2 & 1 \\ 0 & 1 & 3 \end{bmatrix}$.

(*iii*) Find a matrix P which transforms the matrix $\mathbf{A} = \begin{bmatrix} 3 & -1 & 1 \\ -1 & 5 & -1 \\ 1 & -1 & 3 \end{bmatrix}$ to diagonal form. Also, write the diagonal matrix. (*AMIE, June 2011*)

Ans. (*i*) $\mathbf{D} = \begin{bmatrix} 1 & 0 & 0 \\ 0 & \sqrt{5} & 0 \\ 0 & 0 & -\sqrt{5} \end{bmatrix}$,

(*ii*) The eigen values of **A** are 1, 2 and 3 with corresponding eigen vectors $(-1, -2, 1)^T$, $(1, -1, 1)^T$ and $(1, 0, 1)^T$ respectively.

$$\mathbf{P} = \begin{bmatrix} -1 & 1 & 1 \\ -2 & -1 & 0 \\ 1 & 1 & 1 \end{bmatrix}; \quad \mathbf{P}^{-1}\mathbf{AP} = diag\,(1, 2, 3) = \begin{bmatrix} 1 & 0 & 0 \\ 0 & 2 & 0 \\ 0 & 0 & 3 \end{bmatrix}.$$

(*iii*) Modal matrix $\mathbf{P} = \begin{bmatrix} 1 & 1 & 1 \\ 0 & 1 & -2 \\ -1 & 1 & 1 \end{bmatrix}$, diag (2, 3, 6).

21. Find the matrix **A** whose eigenvalues and the corresponding eigenvectors are as given below:

(*i*) Eigenvalues: 1, 1, 3; Eigenvectors: $[1, 0, -1]^T$, $[0, 1, -1]^T$, $[1, 1, 0]^T$. (*AMIETE, June 2008*)

(*ii*) Eigenvalues: 0, –4, 3; Eigenvectors: $[1, 6, -13]^T$, $[-1, 2, 1]^T$, $[2, 3, -2]$.

(*iii*) Eigenvalues: 1, 2, 3; Eigenvectors: $[-1, -1, 1]^T$, $[0, 1, 0]^T$, $[0, -1, 1]^T$. (*AMIE, S-2009*)

Ans. (*i*) $\mathbf{P}=\begin{bmatrix}1&0&1\\0&1&1\\-1&-1&0\end{bmatrix}$; $\mathbf{P}^{-1}=\frac{1}{2}\begin{bmatrix}1&-1&-1\\-1&1&-1\\1&1&1\end{bmatrix}$; $\mathbf{A}=\mathbf{PDP}^{-1}=\begin{bmatrix}2&1&1\\1&2&1\\0&0&1\end{bmatrix}$

(*ii*) $\mathbf{P}=\begin{bmatrix}1&-1&2\\6&2&3\\-13&1&-2\end{bmatrix}$; $\mathbf{P}^{-1}=\frac{1}{84}\begin{bmatrix}-7&0&-7\\-27&24&9\\32&12&8\end{bmatrix}$; $\mathbf{A}=\mathbf{PDP}^{-1}=\begin{bmatrix}1&2&1\\6&-1&0\\-1&-2&-1\end{bmatrix}$.

(*iii*) $\mathbf{A}=\begin{bmatrix}1&0&0\\0&2&1\\2&0&3\end{bmatrix}$.

22. If λ be an eigen value of a non-singular matrix **A**, show that $|\mathbf{A}|/\lambda$ is an eigen value of the matrix adj.**A**. (*PTU 2003; UPTU 2001*).

[**Hint:** Since λ is an eigenvalue of a non-singular matrix **A**.
Therefore, $\lambda \neq 0$ and hence there exists a non-zero column vector **X** such that $\mathbf{AX} = \lambda\mathbf{X}$.
Pre-multiplying adj. **A** to both sides

$$(\text{adj.}\mathbf{A})\,(\mathbf{AX}) = \lambda\,[(\text{adj. }\mathbf{A})\mathbf{X}]$$

$$\Rightarrow \quad [(\text{adj. }\mathbf{A})\,\mathbf{A}]\mathbf{X} = \lambda\,[(\text{adj. }\mathbf{A})]\mathbf{X}$$

$$\Rightarrow \quad [|\mathbf{A}|\mathbf{I}]\mathbf{X} = \lambda\,[\text{adj. }\mathbf{A}]\mathbf{X}$$

$$[\text{adj. }\mathbf{A}]\mathbf{X} = \left[\frac{|\mathbf{A}|}{\lambda}\right]\mathbf{X}.$$

Thus $|\mathbf{A}|/\lambda$ is an eigen-value of adj. **A**]

23. Verify that $\mathbf{x}_1 = [4, 1, -1]^T$, $\mathbf{x}_2 = [1, 0, 4]^T$ and $\mathbf{x}_3 = [1, -4, 0]^T$ are eigenvectors for the symmetric matrix

$\mathbf{A}=\begin{bmatrix}7&4&-4\\4&-8&-1\\-4&-1&-8\end{bmatrix}$ corresponding to the eigenvalues $\lambda_1 = 9$ and $\lambda_2 = \lambda_3 = -9$, respectively.

24. Find $\mathbf{A}^n$ in the following cases:

(*i*) $\mathbf{A}=\begin{pmatrix}-3&10\\-3&8\end{pmatrix}$; (*ii*) $\mathbf{A}=\begin{pmatrix}-13&6\\-35&16\end{pmatrix}$.

Ans. (*i*) $\begin{bmatrix}3.2^{n+1}-5.3^n & -5.2^{n+1}+10.3^n\\ 3.2^n-3^{n+1} & -5.2^n+2.3^{n+1}\end{bmatrix}$; (*ii*) $\begin{bmatrix}-7.2^{n+1}+15 & 3.2^{n+1}-6\\ -35.2^n+35 & 15.2^n-14\end{bmatrix}$.

25. Show that the transformation $y_1 = x_1\cos\theta + x_2\sin\theta$, $y_2 = -x_1\sin\theta + x_2\cos\theta$ is orthogonal.

[**Hint:** The given transformation can be written as $\begin{pmatrix}y_1\\y_2\end{pmatrix}=\begin{pmatrix}\cos\theta&\sin\theta\\-\sin\theta&\cos\theta\end{pmatrix}\begin{pmatrix}x_1\\x_2\end{pmatrix}$.

Here the matrix of transformation $\mathbf{A}=\begin{pmatrix}\cos\theta&\sin\theta\\-\sin\theta&\cos\theta\end{pmatrix}$.

Since $\mathbf{A}^{-1} = \mathbf{A}^{\mathbf{T}}$, the transformation in orthogonal.]

26. Determine the values of α, β, γ when $\mathbf{A}=\begin{bmatrix}0&2\beta&r\\\alpha&\beta&-r\\\alpha&-\beta&r\end{bmatrix}$ is orthogonal.

Ans. $\alpha=\pm 1/\sqrt{2}, \beta=\pm 1/\sqrt{6}, \gamma=\pm 1/\sqrt{3}$.

27. Prove that the inverse of an orthogonal matrix is orthogonal and its transpose is also orthogonal.
28. Show that the following matrices are orthogonal

(*i*) $\mathbf{A}=\begin{bmatrix}\cos\theta & -\sin\theta & 0\\ \sin\theta & \cos\theta & 0\\ 0 & 0 & 1\end{bmatrix}$, (*ii*) $\mathbf{B}=\frac{1}{9}\begin{bmatrix}-8 & 4 & 1\\ 1 & 4 & -8\\ 4 & 7 & 4\end{bmatrix}$, (*AMIETE, W-2008*)

(*iii*) $\mathbf{A}=\frac{1}{3}\begin{bmatrix}2 & 2 & 1\\ -2 & 1 & 2\\ 1 & -2 & 2\end{bmatrix}$, (*iv*) $\mathbf{B}=\frac{1}{7}\begin{bmatrix}6 & -3 & 2\\ -3 & -2 & 6\\ 2 & 6 & 3\end{bmatrix}$.

29. If $\mathbf{A}=\frac{1}{3}\begin{bmatrix}1 & 2 & a\\ 2 & 1 & b\\ 2 & -2 & c\end{bmatrix}$ is orthagonal, find a, b, c and $\mathbf{A}^{-1}$. (Bombay, 2006)

Ans. $a = 2,\ b = 2,\ c = 1,\ \mathbf{A}^{-1}=\frac{1}{3}\begin{bmatrix}1 & 2 & 2\\ 2 & 1 & -2\\ 2 & 2 & 1\end{bmatrix}$.

30. Let $\mathbf{A}=\begin{bmatrix}1 & 0\\ -1 & 1\end{bmatrix}$. Prove that $\mathbf{A}^2 = 2\mathbf{A} - \mathbf{I}$ and compute $\mathbf{A}^{100}$. **Ans.** $\begin{bmatrix}1 & 0\\ -100 & 1\end{bmatrix}$.

31. If $\mathbf{A}=\begin{bmatrix}1 & 2\\ 5 & 4\end{bmatrix}$, find a non singular matrix **P** such that $\mathbf{P}^{-1}\mathbf{AP}=\begin{bmatrix}6 & 0\\ 0 & -1\end{bmatrix}$. **Ans.** $\mathbf{P}=\begin{bmatrix}2 & 1\\ 5 & -1\end{bmatrix}$.

32. Two eigenvalues of the matrix $\mathbf{A}=\begin{bmatrix}2 & 2 & 1\\ 1 & 3 & 1\\ 1 & 2 & 2\end{bmatrix}$ are equal to 1 each. Find the eigenvalues of $\mathbf{A}^{-1}$.

(*AMIE, S-2009*) **Ans.** 1, 1, 1/5.

6.7 COMPLEX MATRICES

If one or more elements of a matrix are complex, Then it is called a complex matrix.

For example $\begin{bmatrix}3 & 2+4i\\ 6-2i & 3+8i\end{bmatrix}, \begin{bmatrix}2 & i+4\\ 3i-5 & 6\end{bmatrix}, \begin{bmatrix}2i+1 & 5+3i\\ 1-3i & 4+i\end{bmatrix}$ are complex matrices.

6.7.1 Conjugate Matrix

The conjugate of a complex matrix $\mathbf{A} = (a_{ij})$ is the matrix obtained from **A** by taking the conjugate of each entry in **A** and is denoted by $\bar{\mathbf{A}} = (\overline{a_{ij}})$.

For example, if $\mathbf{A}=\begin{bmatrix}2 & 1-3i & 5+8i\\ -6i & 1+4i & 2-3i\end{bmatrix}$. Then $\bar{\mathbf{A}}=\begin{bmatrix}2 & 1+3i & 5-8i\\ 6i & 1-4i & 2+3i\end{bmatrix}$.

6.7.2 Transpose of the Conjugate of a Matrix

Consider a square matrix **A**. Its conjugate is $\bar{\mathbf{A}}$. Then the transpose of $\bar{\mathbf{A}}$ is the matrix $(\bar{\mathbf{A}})^{\mathrm{T}}$. It can be easily seen that $(\bar{\mathbf{A}})^{\mathrm{T}} = (\overline{\mathbf{A}^{\mathrm{T}}})$, that is, the transpose of the conjugate of a square matrix is the same as the conjugate of its transpose. The special notation $\overset{*}{\mathbf{A}}, \overset{\mathrm{H}}{\mathbf{A}}$, or $\overset{\theta}{\mathbf{A}}$ is used for the conjugate-transpose of **A**. That is, $\overset{*}{\mathbf{A}}, \overset{\mathrm{H}}{\mathbf{A}}$, or $\overset{\theta}{\mathbf{A}} = (\bar{\mathbf{A}})^{\mathrm{T}} = (\overline{\mathbf{A}^{\mathrm{T}}})$.

For Example, if $\mathbf{A} = \begin{bmatrix} 2+3i & 4-9i \\ 8 & 3+7i \\ -5-i & 6i \end{bmatrix}$. Then $\overset{\mathrm{H}}{\mathbf{A}} = (\bar{\mathbf{A}})^{\mathrm{T}} = \begin{bmatrix} 2-3i & 8 & -5+i \\ 4+9i & 3-7i & -6i \end{bmatrix}$.

6.7.3 Hermitian and Skew-Hermitian Matrices

A complex matrix **A** is called an **Hermitian matrix** if $\bar{\mathbf{A}} = \mathbf{A}^{\mathrm{T}}$ or $\mathbf{A} = (\bar{\mathbf{A}})^{\mathrm{T}}$, that is, $\overset{\mathrm{H}}{\mathbf{A}} = \mathbf{A}$.

A complex matrix **A** is skew Hermitian matrix if $\bar{\mathbf{A}} = -\overset{\mathrm{T}}{\mathbf{A}}$ or $\mathbf{A} = -(\bar{\mathbf{A}})^{\mathrm{T}}$, that is, $\mathbf{A}^{\mathrm{H}} = \mathbf{A}$. Note that **A** must be square. If **A** is a real matrix, then are Hermitian matrix is same as a symmetric matrix and a skew-Hermitian matrix is same as a skew-symmetric matrix.

Let a matrix $\mathbf{A} = (a_{ij})$ be an Hermitian matrix. Then, by definition, $a_{ii} = \bar{a}_{ii}$. Let $a_{ii} = \alpha + i\beta$. Now $a_{ii} = \bar{a}_{ii}$ gives $\alpha + i\beta = \alpha - i\beta$, or $2i\beta = 0$, or $\beta = 0$.

Hence, $a_{ii} = \alpha$. Therefore, all the diagonal element are real.

Now, let a matrix $\mathbf{A} = (a_{ij})$ be *a* skew-Hermitian matrix. Then by definition, $a_{ii} = \bar{a}_{ii}$.

Again, let $a_{ii} = \alpha + i\beta$. Thus $\alpha + i\beta = -(\alpha - i\beta)$, or $2\alpha = 0$, or $\alpha = 0$.

Hence, $a_{ii} = i\beta$. Therefore, the diagonal elements of a skew Hermitian matrix are either 0 or pure imaginary numbers. For Example,

$$\begin{bmatrix} 5 & 3-2i \\ 3+2i & 6 \end{bmatrix}, \begin{bmatrix} 1 & 2+3i & 3+i \\ 2-3i & 2 & 1-2i \\ 3-i & 1+2i & 5 \end{bmatrix} \text{ are Hermitian matrices and}$$

$$\begin{bmatrix} 3i & 4+i \\ -4+i & -i \end{bmatrix}, \begin{bmatrix} i & 2-3i & 4+5i \\ -2-3i & 0 & 2i \\ -4+5i & 2i & -3i \end{bmatrix} \text{ are skew Hermitian matrices.}$$

We note the following:

1. A complex matrix **A** is said to be *normal* if it commutes with its conjugate, transpose, $\overset{*}{\mathbf{A}}$, that is, if $\mathbf{A}\overset{*}{\mathbf{A}} = \overset{*}{\mathbf{A}}\mathbf{A}$. Note that **A** must be a square matrix.

   Example If $\mathbf{A} = \begin{bmatrix} 2+3i & 1 \\ i & 1+2i \end{bmatrix}$, then $\mathbf{A}\overset{*}{\mathbf{A}} = \begin{bmatrix} 2+3i & 1 \\ i & 1+2i \end{bmatrix}\begin{bmatrix} 2-3i & -i \\ 1 & 1-2i \end{bmatrix} = \begin{bmatrix} 14 & 4-4i \\ 4+4i & 6 \end{bmatrix}$

   and $\overset{*}{\mathbf{A}}\mathbf{A} = \begin{bmatrix} 2-3i & -i \\ 1 & 1-2i \end{bmatrix}\begin{bmatrix} 2+3i & 1 \\ i & 1+2i \end{bmatrix} = \begin{bmatrix} 14 & 4-4i \\ 4+4i & 6 \end{bmatrix}$.

   Since $\mathbf{A}\overset{*}{\mathbf{A}} = \overset{*}{\mathbf{A}}\mathbf{A}$, the complex matrix **A** is normal.
   The definition reduces to that for real matrices when **A** is real.

2. For any complex square matrix **A**, the matrix $\mathbf{A} + (\bar{\mathbf{A}})^{\mathrm{T}}$ is always an Hermitian matrix and the matrix $\mathbf{A} - (\bar{\mathbf{A}})^{\mathrm{T}}$ is always a Skew-Hermitian matrix. Therefore, a complex square matrix **A** can be written as the sum of an Hermitian matrix and a skew-Hermitian matrix, that is

   $$\mathbf{A} = \frac{1}{2}(\mathbf{A} + \bar{\mathbf{A}}^{\mathrm{T}}) + \frac{1}{2}(\mathbf{A} - \bar{\mathbf{A}}^{\mathrm{T}}) = (\mathbf{P} + \mathbf{Q}) \text{ (say).}$$

   Proof: $\bar{\mathbf{P}} = \frac{1}{2}\overline{(\mathbf{A} + \bar{\mathbf{A}}^{\mathrm{T}})} = \frac{1}{2}(\bar{\mathbf{A}} + \mathbf{A}^{\mathrm{T}}), (\bar{\mathbf{P}})^{\mathrm{T}} = \frac{1}{2}(\bar{\mathbf{A}}^{\mathrm{T}} + \mathbf{A}) = \mathbf{P}.$

   Hence **P** is an Hermitian matrix.

$$\bar{\mathbf{Q}} = \frac{1}{2}\overline{(\mathbf{A} - \bar{\mathbf{A}}^{\mathrm{T}})} = \frac{1}{2}(\bar{\mathbf{A}} - \mathbf{A}^{\mathrm{T}}), (\bar{\mathbf{Q}})^{\mathrm{T}} = \frac{1}{2}(\bar{\mathbf{A}}^{\mathrm{T}} - \mathbf{A}) = -\frac{1}{2}(\mathbf{A} - \bar{\mathbf{A}}^{\mathrm{T}}) = -\mathbf{Q}.$$

Hence, **Q** is a skew-Hermitian matrix.

6.7.4 Unitary Matrices

A complex matrix **A** is said to be an unitary matrix if $(\bar{\mathbf{A}})^{\mathrm{T}}\mathbf{A} = \mathbf{A}(\bar{\mathbf{A}})^{\mathrm{T}} = \mathbf{I}$, that is, if $(\bar{\mathbf{A}})^{\mathrm{T}} = \mathbf{A}^{-1}$. Note that **A** must be necessarily be square and invertible. If **A** is a real, then an unitary matrix is same as an orthogonal matrix.

Remarks

(*i*) The inverse and the transpose of an unitary matrix are unitary.

(*ii*) The product of two or more unitary matrices is an unitary matrix.

ILLUSTRATIVE EXAMPLES

EXAMPLE 6.34 Find real numbers α, β, γ such that **X** is Hermitian, where $\mathbf{X} = \begin{bmatrix} 3 & \alpha+2i & \beta i \\ 3-2i & 0 & 1+\gamma i \\ \beta i & 1-\alpha i & -1 \end{bmatrix}$.

Solution We have $\bar{\mathbf{X}} = \begin{bmatrix} 3 & \alpha-2i & -\beta i \\ 3+2i & 0 & 1-\gamma i \\ -\beta i & 1+\alpha i & -1 \end{bmatrix}$.

Therefore $(\bar{\mathbf{X}})^{\mathrm{T}} = \begin{bmatrix} 3 & 3+2i & -\beta i \\ \alpha-2i & 0 & 1+\alpha i \\ -\beta i & 1-\gamma i & -1 \end{bmatrix} = \begin{bmatrix} 3 & \alpha+2i & \beta i \\ 3-2i & 0 & 1+\gamma i \\ \beta i & 1-\alpha i & -1 \end{bmatrix}$.

By definition of equality of matrices, the following entries must be equal. Thus:

$$\alpha + 2i = 3 + 2i, \qquad 1 + \alpha i = 1 + \gamma i, \qquad -\beta i = \beta i$$

The above equations yields $\alpha = 3, \beta = 0, \gamma = 3$.

EXAMPLE 6.35 If $\mathbf{A} = \begin{bmatrix} 3 & 5+2i & -3 \\ 5-2i & 7 & 4i \\ -3 & -4i & 5 \end{bmatrix}$. Show that **A** is a Hermitian matrix. Verify that *i***A** is a skew-Hermitian matrix.

Solution We have $\bar{\mathbf{A}} = \begin{bmatrix} 3 & 5-2i & -3 \\ 5+2i & 7 & -4i \\ -3 & 4i & 5 \end{bmatrix}$. Therefore $\overset{\mathrm{H}}{\mathbf{A}} = (\bar{\mathbf{A}})^{\mathrm{T}} = \begin{bmatrix} 3 & 5+2i & -3 \\ 5-2i & 7 & 4i \\ -3 & -4i & 5 \end{bmatrix} = \mathbf{A}$.

Hence **A** is an Hermitian matrix.

Now $i\mathbf{A} = \begin{bmatrix} 3i & -2+5i & -3i \\ 2+5i & 7i & -4 \\ -3i & 4 & 5i \end{bmatrix}$. Therefore $\overline{i\mathbf{A}} = \begin{bmatrix} -3i & -2-5i & 3i \\ 2-5i & -7i & -4 \\ 3i & 4 & -5i \end{bmatrix}$.

$$(i\mathbf{A})^{\mathrm{H}} = (\overline{i\mathbf{A}})^{\mathrm{T}} = \begin{bmatrix} -3i & 2-5i & 3i \\ -2-5i & -7i & 4 \\ 3i & -4 & -5i \end{bmatrix} = -\begin{bmatrix} 3i & -2+5i & -3i \\ 2+5i & 7i & -4 \\ -3i & 4 & 5i \end{bmatrix} = -(i\mathbf{A}).$$

Thus *i***A** is a skew-Hermitian matrix.

EXAMPLE 6.36 Express the matrix $\mathbf{A}=\begin{bmatrix}1+2i & 2 & 5-5i\\ 2i & 2+i & 4+2i\\ -1+i & -4 & 7\end{bmatrix}$ as the sum of Hermitian and a Skew-Hermitian matrix.

Solution We have $\mathbf{A}=\begin{bmatrix}1+2i & 2 & 5-5i\\ 2i & 2+i & 4+2i\\ -1+i & -4 & 7\end{bmatrix}$.

Therefore, $\bar{\mathbf{A}}=\begin{bmatrix}1-2i & 2 & 5+5i\\ -2i & 2-i & 4-2i\\ -1-i & -4 & 7\end{bmatrix}$, and $\overset{\mathrm{H}}{\mathbf{A}}=(\bar{\mathbf{A}})^{\mathrm{T}}=\begin{bmatrix}1-2i & -2i & -1-i\\ 2 & 2-i & -4\\ 5+5i & 4-2i & 7\end{bmatrix}$.

Now $\frac{1}{2}(\mathbf{A}+\overset{\mathrm{H}}{\mathbf{A}})=\begin{bmatrix}1 & 1-i & 2-3i\\ 1+i & 2 & i\\ 2+3i & -i & 7\end{bmatrix}$, and $\frac{1}{2}(\mathbf{A}-\overset{\mathrm{H}}{\mathbf{A}})=\begin{bmatrix}2i & 1+i & 3-2i\\ -1+i & i & 4+i\\ -(3+2i) & -4+i & 0\end{bmatrix}$.

Thus
$$\mathbf{A}=\frac{1}{2}(\mathbf{A}+\overset{\mathrm{H}}{\mathbf{A}})+\frac{1}{2}(\mathbf{A}-\overset{\mathrm{H}}{\mathbf{A}})=\underbrace{\begin{bmatrix}1 & 1-i & 2-3i\\ 1+i & 2 & i\\ 2+3i & -i & 7\end{bmatrix}}_{\substack{\mathbf{R}\ (say)\\ \text{Hermitian matrix}}}+\underbrace{\begin{bmatrix}2i & 1+i & 3-2i\\ -1+i & i & 4+i\\ -(3+2i) & -4+i & 0\end{bmatrix}}_{\substack{\mathbf{S}\ (say)\\ \text{Skew-Hermitian matrix}}}$$

We observe that $\mathbf{R}^{\mathbf{H}}=\mathbf{R}$, that is, $\mathbf{R}$ is an Hermitian matrix and $\mathbf{S}^{\mathbf{H}}=-\mathbf{S}$, that is, $\mathbf{S}$ is skew-Hermitian matrix.

EXAMPLE 6.37 If $\mathbf{A}=\begin{bmatrix}2+i & 3 & -1+3i\\ -5 & i & 4-2i\end{bmatrix}$, show that $\mathbf{A}\overset{*}{\mathbf{A}}$ is a Hermitian matrix, where $\overset{*}{\mathbf{A}}$ is the conjugate transpose of $\mathbf{A}$.

(AMEITE, June 2010; JNTU, 2005; UPTU 2003)

Solution We have $\bar{\mathbf{A}}=\begin{bmatrix}2-i & 3 & -1-3i\\ -5 & -i & 4+2i\end{bmatrix}$.

$(\bar{\mathbf{A}})^{\mathrm{T}}=\overset{*}{\mathbf{A}}=\begin{bmatrix}2-i & -5\\ 3 & -i\\ -1-3i & 4+2i\end{bmatrix}$. Therefore $\mathbf{A}\overset{*}{\mathbf{A}}=\underset{(2\times 3)}{\begin{bmatrix}2+i & 3 & -1+3i\\ -5 & i & 4-2i\end{bmatrix}}\underset{(3\times 2)}{\begin{bmatrix}2-i & -5\\ 3 & -i\\ -1-3i & 4+2i\end{bmatrix}}$.

$$=\begin{bmatrix}4-i^2+9+1-9i^2 & -10-5i-3i-10+10i\\ -10+5i+3i-10-10i & 25-i^2+16-4i^2\end{bmatrix}=\begin{bmatrix}24 & -20+2i\\ -20-2i & 46\end{bmatrix},$$

which is an Hermitian Matrix.

EXAMPLE 6.38 Prove the following:

(*a*) If **A** is skew-Hermitian matrix, then $\pm\, i\mathbf{A}$ is Hermitian.

(*b*) If **A** and **B** are Hermitian, then **AB** – **BA** is skew-Hermitian.

(*c*) $\overset{\mathrm{H}}{\mathbf{B}}\mathbf{AB}$ is Hermitian or Skew-Hermitian according as **A** is Hermitian or skew-Hermitian.

Solution (*a*) Consider $\mathbf{B}=-i\mathbf{A}$. Since **A** is skew-Hermitian, $(\bar{\mathbf{A}})^{\mathrm{T}}=-\mathbf{A}$.

Then $(\bar{\mathbf{B}})^{\mathrm{T}}=(\overline{-i\mathbf{A}})^{\mathrm{T}}=i(\bar{\mathbf{A}})^{T}=i(-\mathbf{A})=-i\mathbf{A}=\mathbf{B}$ and **B** is Hermitian.

Similarly it can be shown that $\mathbf{B}=i\mathbf{A}$ is Hermitian.

(*b*) Given $\quad \mathbf{A}^H = \mathbf{A}$ and $\mathbf{B}^H = \mathbf{B}$

Now
$$(\mathbf{AB} - \mathbf{BA})^H = (\mathbf{AB})^H - (\mathbf{BA})^H = \mathbf{B}^H\mathbf{A}^H - \mathbf{A}^H\mathbf{B}^H = \mathbf{BA} - \mathbf{AB}$$
$$= -(\mathbf{AB} - \mathbf{BA}).$$ Thus $\mathbf{AB} - \mathbf{BA}$ is skew-Hermitian..

(*c*) (*i*) Let **A** is Hermitian, therefore $\mathbf{A}^H = \mathbf{A}$. We have to prove that $\mathbf{B}^H\mathbf{AB}$ is Hermitian, that is,

$$(\mathbf{B}^H\mathbf{AB})^H = \mathbf{B}^H\mathbf{AB}. \qquad \text{...}(i)$$

$$\text{L.H.S.} = (\mathbf{B}^H\mathbf{AB})^H = \mathbf{B}^H \cdot \mathbf{A}^H \cdot (\mathbf{B}^H)^H = \mathbf{B}^H\mathbf{AB} = \text{R.H.S.}$$

Therefore, $\mathbf{B}^H\mathbf{AB}$ is Hermitian according as **A** is Hermitian.

(*ii*) Let **A** be skew-Hermitian. Therefore $\mathbf{A}^H = -\mathbf{A}$. We have to prove that $(\mathbf{B}^H\mathbf{AB})^H$ is skew-Hermitian,

that is,
$$(\mathbf{B}^H\mathbf{AB})^H = -(\mathbf{B}^H\mathbf{AB}).$$

$$\text{L.H.S.} = (\mathbf{B}^H\mathbf{AB})^H = \mathbf{B}^H \cdot \mathbf{A}^H \cdot (\mathbf{B}^H)^H = \mathbf{B}^H(-\mathbf{A})\mathbf{B} = -(\mathbf{B}^H\mathbf{AB}) = \text{R.H.S.}$$

Hence, $\mathbf{B}^H\mathbf{AB}$ is Skew-Hermitian according as **A** is Skew-Hermitian.

EXAMPLE 6.39 Prove that the matrix $\mathbf{A} = \begin{bmatrix} \frac{1}{2}(1+i) & \frac{1}{2}(-1+i) \\ \frac{1}{2}(1+i) & \frac{1}{2}(1-i) \end{bmatrix}$ is unitary and find $\mathbf{A}^{-1}$. (*Bombay 2006*)

Solution We have
$$\bar{\mathbf{A}} = \begin{bmatrix} \frac{1}{2}(1-i) & \frac{1}{2}(-1-i) \\ \frac{1}{2}(1-i) & \frac{1}{2}(1+i) \end{bmatrix}.$$

$\therefore$
$$(\bar{\mathbf{A}})^T = \begin{bmatrix} \frac{1}{2}(1-i) & \frac{1}{2}(1-i) \\ -\frac{1}{2}(1+i) & \frac{1}{2}(1+i) \end{bmatrix}$$

and
$$(\bar{\mathbf{A}})^T \cdot \mathbf{A} = \mathbf{A}(\bar{\mathbf{A}})^T = \begin{bmatrix} \frac{1}{2}(1-i) & \frac{1}{2}(1-i) \\ -\frac{1}{2}(1+i) & \frac{1}{2}(1+i) \end{bmatrix} \begin{bmatrix} \frac{1}{2}(1+i) & -\frac{1}{2}(1-i) \\ \frac{1}{2}(1+i) & \frac{1}{2}(1-i) \end{bmatrix}$$

$$= \begin{bmatrix} \frac{1}{4}(1-i^2) + \frac{1}{4}(1-i^2) & -\frac{1}{4}(1-i)^2 + \frac{1}{4}(1-i)^2 \\ -\frac{1}{4}(1+i)^2 + \frac{1}{4}(1+i)^2 & \frac{1}{4}(1-i^2) + \frac{1}{4}(1-i^2) \end{bmatrix} = \begin{bmatrix} 1 & 0 \\ 0 & 1 \end{bmatrix} = \mathbf{I}.$$

Therefore, the given matrix is an unitary matrix.

Hence,
$$\mathbf{A}^{-1} = (\bar{\mathbf{A}})^T = \begin{bmatrix} \frac{1}{2}(1-i) & \frac{1}{2}(1-i) \\ -\frac{1}{2}(1+i) & \frac{1}{2}(1+i) \end{bmatrix}.$$

EXAMPLE 6.40 Given that $\mathbf{A}=\begin{bmatrix} 0 & 1+2i \\ -1+2i & 0 \end{bmatrix}$, show that $(\mathbf{I}-\mathbf{A})(\mathbf{I}+\mathbf{A})^{-1}$ is a unitary matrix.

[AMIETE, June 2009; June 2008, KUK, 2008]

Solution We have $\mathbf{I}-\mathbf{A}=\begin{bmatrix} 1 & -1-2i \\ 1-2i & 1 \end{bmatrix}$.

$\therefore$ $$\mathbf{I}+\mathbf{A}=\begin{bmatrix} 1 & 1+2i \\ -1+2i & 1 \end{bmatrix}, \quad |\mathbf{I}+\mathbf{A}| = 1+(1-4i^2) = 6.$$

$$(\mathbf{I}+\mathbf{A})^{-1}=\frac{1}{6}\begin{bmatrix} 1 & -1-2i \\ 1-2i & 1 \end{bmatrix}.$$

$\therefore$ $$(\mathbf{I}-\mathbf{A})(\mathbf{I}+\mathbf{A})^{-1}=\frac{1}{6}\begin{bmatrix} 1 & -1-2i \\ 1-2i & 1 \end{bmatrix}\begin{bmatrix} 1 & -1-2i \\ 1-2i & 1 \end{bmatrix}=\frac{1}{6}\begin{bmatrix} 1-(1-4i^2) & -2-4i \\ 2-4i & -(1-4i^2)+1 \end{bmatrix}$$

$$=\frac{1}{6}\begin{bmatrix} -4 & -2-4i \\ 2-4i & -4 \end{bmatrix}=\mathbf{U}\text{ (say)}.$$

Conjugate transpose of $\mathbf{U}=(\bar{\mathbf{U}})^{\mathrm{T}}=\frac{1}{6}\begin{bmatrix} -4 & 2+4i \\ -2+4i & -4 \end{bmatrix}$.

Therefore, $$\mathbf{U}\cdot(\bar{\mathbf{U}})^{\mathrm{T}}=\frac{1}{6}\begin{bmatrix} -4 & -2-4i \\ 2-4i & -4 \end{bmatrix}\cdot\frac{1}{6}\begin{bmatrix} -4 & 2+4i \\ -2+4i & -4 \end{bmatrix}$$

$$=\frac{1}{36}\begin{bmatrix} 16+4-16i^2 & -8-16i+8+16i \\ -8+16i+8-16i & 4-16i^2+16 \end{bmatrix}=\frac{1}{36}\begin{bmatrix} 36 & 0 \\ 0 & 36 \end{bmatrix}=\begin{bmatrix} 1 & 0 \\ 0 & 1 \end{bmatrix}=\mathbf{I}$$

Similarly, it can be shown that $(\bar{\mathbf{U}})^{\mathrm{T}}\cdot\mathbf{U}=\mathbf{I}$. Hence, $(\mathbf{I}-\mathbf{A})(\mathbf{I}+\mathbf{A})^{-1}$ is a unitary matrix.

Theorem - I

(*a*) The eigenvalues of an Hermitian matrix are real. (Refer Fig. 6.1).
(*b*) The eigenvalues of a Skew-Hermitian matrix are zero or pure imaginary.
(*c*) The eigenvalues of an unitary matrix have absolute value 1.

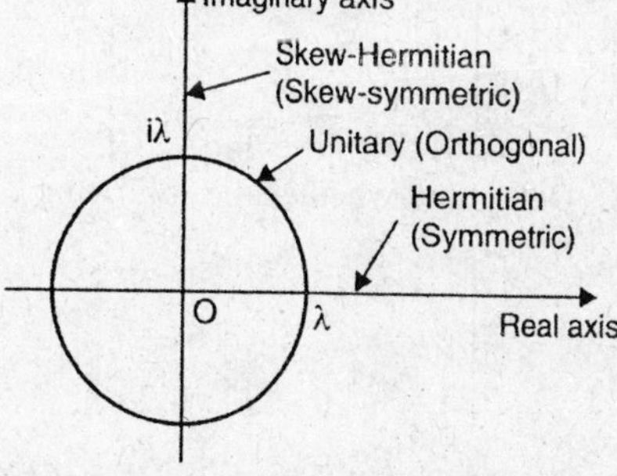

Fig. 6.1: Eigen values of various matrices

Proof

Let λ be an eigenvalue of $\mathbf{A}$ and $\mathbf{x}$ be the corresponding eigenvector. Thus $\mathbf{Ax}=\lambda\mathbf{x}$. ...(6.12)

Premultiplying both sides (6.12) by $\bar{\mathbf{x}}^{\mathrm{T}}$, we get

$$\bar{\mathbf{x}}^{\mathrm{T}}\mathbf{Ax}=\bar{\mathbf{x}}^{\mathrm{T}}\lambda\mathbf{x}=\lambda\bar{\mathbf{x}}^{\mathrm{T}}\mathbf{x} \quad \text{or} \quad \lambda=\frac{\bar{\mathbf{x}}^{\mathrm{T}}\mathbf{Ax}}{\bar{\mathbf{x}}^{\mathrm{T}}\mathbf{x}}. \qquad ...(6.13)$$

Note that $\bar{\mathbf{x}}^{\mathrm{T}}\mathbf{Ax}$ and $\bar{\mathbf{x}}^{\mathrm{T}}\mathbf{x}$ are scalars.

The denominator $\bar{\mathbf{x}}^{\mathrm{T}}\mathbf{x}=\bar{x}_1x_1+\bar{x}_2x_2+........+\bar{x}_nx_n=|x_1|^2+|x_2|^2+......+|x_n|^2$ is real, and is positive. Therefore, λ is real if the numerator is real. The numerator is real by showing that it is equal to its complex conjugate.

(*a*) Let **A** be an Hermitian matrix, that is, $\bar{\mathbf{A}}^{\mathrm{T}} = \mathbf{A}$ or $\bar{\mathbf{A}} = \mathbf{A}^{\mathrm{T}}$. Now

$$\bar{\mathbf{x}}^{\mathrm{T}}\mathbf{A}\mathbf{x} = (\bar{\mathbf{x}}^{\mathrm{T}}\mathbf{A}\mathbf{x})^{\mathrm{T}} = \mathbf{x}^{\mathrm{T}}\mathbf{A}^{\mathrm{T}}\bar{\mathbf{x}} = \mathbf{x}^{\mathrm{T}}\bar{\mathbf{A}}\,\bar{\mathbf{x}} = \overline{(\bar{\mathbf{x}}^{\mathrm{T}}\mathbf{A}\mathbf{x})}. \qquad \text{...(6.14)}$$

Hence, $\bar{\mathbf{x}}^{\mathrm{T}}\mathbf{A}\mathbf{x}$ is real. From this and Eq (6.13), whose denominator is real, we conclude that λ is real.

(*b*) If **A** is skew-Hermitian, then $\bar{\mathbf{A}}^{\mathrm{T}} = -\mathbf{A}$ or $\bar{\mathbf{A}} = -\mathbf{A}^{\mathrm{T}}$, so we get minus sign in Eq. (6.14), that is,

$$\bar{\mathbf{x}}^{\mathrm{T}}\mathbf{A}\mathbf{x} = -\overline{(\bar{\mathbf{x}}^{\mathrm{T}}\mathbf{A}\mathbf{x})}. \qquad \text{...(6.15)}$$

This is a complex number $C = \alpha + i\beta$ that equals minus its conjugate $\bar{C} = \alpha - i\beta$. that is, $\alpha + i\beta = -(\alpha - i\beta)$. Hence, $\alpha = -\alpha = 0$ so that C is pure imaginary or zero. Division by the real $\bar{\mathbf{x}}^{\mathrm{T}}\mathbf{x}$ in (6.13) now gives a pure imaginary λ or λ equal to 0.

(*c*) Let A be an unitary matrix. We take (6.12) and its conjugate transpose,

$$\mathbf{A}\mathbf{x} = \lambda\mathbf{x} \quad \text{and} \quad (\bar{\mathbf{A}}\bar{\mathbf{x}})^{\mathrm{T}} = (\bar{\lambda}\bar{x})^{\mathrm{T}} = \bar{\lambda}\bar{x}^{T}.$$

Multiply the two left sides and the two right sides,

$$(\bar{\mathbf{A}}\bar{\mathbf{x}})^{\mathrm{T}}\bar{\mathbf{A}}\bar{\mathbf{x}} = \bar{\lambda}\,\lambda\,\bar{\mathbf{x}}^{\mathrm{T}}\mathbf{x} = |\lambda|^{2}\,\bar{\mathbf{x}}^{\mathrm{T}}\mathbf{x}. \qquad \text{...(6.16)}$$

But **A** is unitary, $\bar{\mathbf{A}}^{\mathrm{T}} = \bar{\mathbf{A}}^{1}$, so that on the left hand side, we obtain

$$(\bar{\mathbf{A}}\bar{\mathbf{x}})^{\mathrm{T}}\mathbf{A}\mathbf{x} = \bar{\mathbf{x}}^{\mathrm{T}}\bar{\mathbf{A}}^{\mathrm{T}}\mathbf{A}\mathbf{x} = \bar{\mathbf{x}}^{\mathrm{T}}\bar{\mathbf{A}}^{1}\mathbf{A}\mathbf{x} = \bar{\mathbf{x}}^{\mathrm{T}}\mathbf{I}\mathbf{x} = \bar{\mathbf{x}}^{\mathrm{T}}\mathbf{x}.$$

Eq. (5.16) may be written as $\bar{\mathbf{x}}^{\mathrm{T}}\mathbf{x} = |\lambda|^{2}\,\bar{\mathbf{x}}^{\mathrm{T}}\mathbf{x}$.

Since $\mathbf{x} \neq \mathbf{0}$, we have $\bar{\mathbf{x}}^{\mathrm{T}}\mathbf{x} \neq 0$. We now divide by $\bar{\mathbf{x}}^{\mathrm{T}}\mathbf{x}$ to get $|\lambda|^2 = 1$ or $|\lambda| = 1$. Hence the result.

EXERCISE 6.2

1. Show that (*i*) $\begin{bmatrix} 1 & 1+i & 2+3i \\ 1-i & 2 & -i \\ 2-3i & i & 0 \end{bmatrix}$ is Hermitian (*ii*) $\begin{bmatrix} 0 & 1-i & 2-3i \\ 1+i & 1 & -2 \\ 2+3i & -2 & 0 \end{bmatrix}$ is Hermitian.

(*iii*) $\begin{bmatrix} i & 1+i & 2-3i \\ -1+i & 2i & 1 \\ -2-3i & -1 & 0 \end{bmatrix}$ is Skew-Hermitian. (*iv*) $\begin{bmatrix} 2i & 1+i & 2+3i \\ -1+i & 5i & 2 \\ -2-3i & -2 & 0 \end{bmatrix}$ is Skew-Hermitian.

2. Determine whether the following matrices are Hermitian:

(*i*) $\begin{bmatrix} 2 & 2+3i & 4-5i \\ 2-3i & 5 & 6+2i \\ 4+5i & 6-2i & -7 \end{bmatrix}$, (*ii*) $\begin{bmatrix} 3 & 2+i & 4-i \\ 2+i & 7 & -i \\ 4-i & -i & 6 \end{bmatrix}$, (*iii*) $\begin{bmatrix} 5 & 4 & -3 \\ 4 & 1 & 6 \\ -3 & 6 & 2 \end{bmatrix}$.

Ans. (*a*) yes, (*b*) No, (*c*) yes. Infact, a real matrix is Hermitian if and only if it is symmetric.

3. Prove that if **A** is an Hermitian matrix, then (*i***A**) is a skew-Hermitian matrix.

4. If $\mathbf{A} = \begin{bmatrix} -1 & 2+i & 5-3i \\ 2-i & 7 & 5i \\ 5+3i & -5i & 2 \end{bmatrix}$, show that **A** is an Hermitian matrix and *i***A** is a skew-Hermitian matrix.

(Sambalpur, 2002)

5. Prove that every Hermitian matrix can be written as **A** + *i***B**, where **A** is real and symmetric and **B** is real and skew-symmetric.

6. Express $\mathbf{A} = \begin{bmatrix} 2i & 2+i & 1-i \\ -2+i & -i & 3i \\ -1-i & 3i & 0 \end{bmatrix}$, as $\mathbf{P} + i\mathbf{Q}$ where $\mathbf{P}$ is real and skew symmetric and $\mathbf{Q}$ is real and symmetric. *(Bombay, 2006)* **Ans.** $\begin{bmatrix} 0 & 2 & 1 \\ -2 & 0 & 0 \\ -1 & 0 & 0 \end{bmatrix} + i\begin{bmatrix} 2 & 1 & -1 \\ 1 & -1 & 3 \\ -1 & 3 & 0 \end{bmatrix}$.

7. If $\mathbf{S} = \begin{bmatrix} 1 & 1 & 1 \\ 1 & a^2 & a \\ 1 & a & a^2 \end{bmatrix}$, where $a = e^{2i\pi/3}$, prove that $\mathbf{S}^{-1} = \frac{1}{3}\bar{\mathbf{S}}$. *(Kurukshetra, 2006; JNTU, 2001)*

8. (*i*) If $\mathbf{A} = \begin{bmatrix} 2+i & 3 & -1+3i \\ -5 & i & 4-2i \end{bmatrix}$, verify that $\overset{\text{H}}{\mathbf{A}}\mathbf{A}$ is an Hermitian matrix.

[**Hint:** We have $\overset{\text{H}}{\mathbf{A}}\mathbf{A} = \begin{bmatrix} 2-i & -5 \\ 3 & -i \\ -1-3i & 4+2i \end{bmatrix}\begin{bmatrix} 2+i & 3 & -1+3i \\ -5 & i & 4-2i \end{bmatrix} = \begin{bmatrix} 30 & 6-8i & -19+17i \\ 6+8i & 10 & -5+5i \\ -19-17i & -5-5i & 30 \end{bmatrix} = \mathbf{B}\text{(say)}.$

Now $\overset{\text{H}}{\mathbf{B}} = (\bar{\mathbf{B}})^{\text{T}} = \begin{bmatrix} 30 & 6+8i & -19-17i \\ 6-8i & 10 & -5-5i \\ -19+17i & -5+5i & 30 \end{bmatrix}^{\text{T}} = \begin{bmatrix} 30 & 6-8i & -19+17i \\ 6+8i & 10 & -5+5i \\ -19-17i & -5-5i & 30 \end{bmatrix} = \mathbf{B}$].

(*ii*) If $\mathbf{A} = \begin{bmatrix} 2+i & 3 & -1+3i \\ -5 & i & 4-2i \end{bmatrix}$, show that $\mathbf{A}\overset{*}{\mathbf{A}}$ is a Hermitian matrix, where $\overset{*}{\mathbf{A}}$ is the conjugate transpose of $\mathbf{A}$. *(JNTU, 2005; UPTU, 2003)*

9. Express the following matrix as the sum of an Hermitian and a skew Hermitian matrix

$\begin{bmatrix} 3 & 2i & 1 \\ 1+i & 4 & -1 \\ 2 & 3i & 5 \end{bmatrix}$. **Ans.** $\frac{1}{2}\left\{\begin{bmatrix} 6 & 1+i & 3 \\ 1-i & 8 & -1-3i \\ 3 & 3i-1 & 10 \end{bmatrix} + \begin{bmatrix} 0 & 3i-1 & -1 \\ 1+3i & 0 & -1+3i \\ 1 & 3i+1 & 0 \end{bmatrix}\right\}$.

10. Show that the following matrices are unitary:

(*i*) $\begin{bmatrix} i/2 & -\sqrt{3}/2 \\ \sqrt{3}/2 & -i/2 \end{bmatrix}$, (*ii*) $\begin{bmatrix} 1/\sqrt{2} & i/\sqrt{2} \\ -i/\sqrt{2} & -1/\sqrt{2} \end{bmatrix}$, (*iii*) $\begin{bmatrix} \frac{1}{3}-\frac{2}{3}i & \frac{2}{3}i \\ -\frac{2}{3}i & -\frac{1}{3}-\frac{2}{3}i \end{bmatrix}$, (*iv*) $\frac{1}{\sqrt{3}}\begin{bmatrix} 1 & 1+i \\ 1-i & -1 \end{bmatrix}$

(UPTU, 2002)

11. Show that the matrix $\begin{bmatrix} \alpha+i\gamma & -\beta+i\delta \\ \beta+i\delta & \alpha+i\gamma \end{bmatrix}$ is a unitary matrix, if $\alpha^2 + \beta^2 + \gamma^2 + \delta^2 = 1$. *(UPTU, 2006)*

12. Find the values of k for which $\begin{bmatrix} 1/k & i/k \\ -i/k & -1/k \end{bmatrix}$ is unitary. **Ans.** $k = \pm i\sqrt{2}$.

QUADRATIC FORMS

Q be a real polynomial in variables $x_1, x_2, \ldots\ldots, x_n$ such that every term in Q has degree two, that is,

$Q(x_1, x_2, \ldots\ldots, x_n) = \sum_{i=1}^{n}\sum_{j=1}^{n} b_{ij}\, x_i\, x_j$. Take $n = 3$, we have

$$Q = b_{11} x_1^2 + b_{12} x_1 x_2 + b_{13} x_1 x_3 + b_{21} x_2 x_1 + b_{22} x_2^2 + b_{23} x_2 x_3 + b_{31} x_3 x_1 + b_{32} x_3 x_2 + b_{33} x_3^2$$

$$= b_{11} x_1^2 + (b_{12} + b_{21}) x_1 x_2 + (b_{13} + b_{31}) x_1 x_3 + b_{22} x_2^2 + (b_{23} + b_{32}) x_2 x_3 + b_{33} x_3^2$$

$$= \begin{bmatrix} x_1 & x_2 & x_3 \end{bmatrix} \begin{bmatrix} b_{11} & b_{12} & b_{13} \\ b_{21} & b_{22} & b_{23} \\ b_{31} & b_{32} & b_{33} \end{bmatrix} \begin{bmatrix} x_1 \\ x_2 \\ x_3 \end{bmatrix} = \mathbf{x}^T \mathbf{B} \mathbf{x}, \qquad \text{...(6.17)}$$

where $\mathbf{B} = (b_{ij})$, $\mathbf{x} = \begin{bmatrix} x_1 & x_2 & x_3 \end{bmatrix}^T$ is the column vector of a suitable order of the variable.

Then Q is called a quadratic form of **B**.

Now set $a_{ij} = \frac{1}{2}(b_{ij} + b_{ji})$. The matrix $\mathbf{A} = (a_{ij})$ is symmetric that is $\mathbf{A}^T = \mathbf{A}$, since $a_{ji} = \frac{1}{2}(b_{ji} + b_{ij}) = a_{ij}$.

Hence, the quadratic form Q in Eq. (6.17) can be written as,

$$Q = \begin{bmatrix} x_1 & x_2 & x_3 \end{bmatrix} \begin{bmatrix} a_{11} & a_{12} & a_{13} \\ a_{21} = a_{12} & a_{22} & a_{23} \\ a_{31} = a_{13} & a_{32} = a_{23} & a_{33} \end{bmatrix} \begin{bmatrix} x_1 \\ x_2 \\ x_3 \end{bmatrix} = \mathbf{x}^T \mathbf{A} \mathbf{x},$$

where **A** is a symmetric matrix. **A** is called the matrix of the quadratic form.

Now $\quad \mathbf{x}^T\mathbf{A} = [a_{11}x_1 + a_{12}x_2 + a_{13}x_3 \quad a_{12}x_1 + a_{22}x_2 + a_{23}x_3 \quad a_{13}x_1 + a_{23}x_2 + a_{33}x_3]$

$$\therefore \quad \mathbf{x}^T\mathbf{A}\mathbf{x} = [a_{11}x_1 + a_{12}x_2 + a_{13}x_3 \quad a_{12}x_1 + a_{22}x_2 + a_{23}x_3 \quad a_{13}x_1 + a_{23}x_2 + a_{33}x_3] \times \begin{bmatrix} x_1 \\ x_2 \\ x_3 \end{bmatrix}$$

$$= (a_{11}x_1 + a_{12}x_2 + a_{13}x_3)x_1 + (a_{12}x_1 + a_{22}x_2 + a_{23}x_3)x_2 + (a_{13}x_1 + a_{23}x_2 + a_{33}x_3)x_3$$

$$= a_{11} x_1^2 + a_{22} x_2^2 + a_{33} x_3^2 + 2a_{12} x_1 x_2 + 2a_{13} x_1 x_3 + 2a_{23} x_2 x_3 = Q$$

Then $\quad Q = a_{11} x_1^2 + a_{22} x_2^2 + a_{33} x_3^2 + 2a_{12} x_1 x_2 + 2a_{13} x_1 x_3 + 2a_{23} x_2 x_3 \qquad$ ...(6.18)

is the quadratic form of **A**.

The rule to form '**Q**' when **A** is given and vice versa is clear from (6.18).

Thus if $\quad \mathbf{A} = \begin{bmatrix} 2 & 3 & -5 \\ 3 & 4 & 9 \\ -5 & 9 & 6 \end{bmatrix}$, $Q = 2x_1^2 + 4x_2^2 + 6x_3^2 + 2(3) x_1 x_2 + 2(-5) x_1 x_3 + 2(9) x_2 x_3$.

that is, $\quad Q(x_1, x_2, x_3) = 2x_1^2 + 4x_2^2 + 6x_3^2 + 6x_1 x_2 - 10 x_1 x_3 + 18 x_2 x_3$.

Also if $\quad Q(x_1, x_2, x_3) = x_1^2 - 4x_2^2 + 5x_3^2 + 4x_1 x_2 + 6 x_1 x_3 + 12 x_2 x_3$,

then the procedure for obtaining the symmetric matrix **A** for the quadratic form is as under.

We have $\quad b_{11} = 1,\ b_{12} + b_{21} = 4,\ b_{13} + b_{31} = 6,\ b_{23} + b_{32} = 12,\ b_{22} = -4,\ b_{33} = 5.$

Hence, $\quad a_{11} = 1,\ a_{12} = a_{21} = (1/2)(b_{12} + b_{21}) = 2,\ a_{13} = a_{31} = (1/2)(b_{13} + b_{31}) = 3,$

$a_{23} = a_{32} = (1/2)(b_{23} + b_{32}) = 6,\ a_{22} = -4,\ a_{33} = 5.$

EXAMPLE 6.41 Find the symmetric matrix that corresponds to each of the following quadratic forms:

(*i*) $Q(x_1, x_2, x_3) = 2x_1^2 - 8x_1x_2 + x_2^2 - 16x_1 x_3 + 14 x_2 x_3 + 5 x_3^2$,

(*ii*) $Q(x_1, x_2, x_3, x_4) = x_1^2 + 2x_2^2 - 7x_3^2 + x_4^2 - 4x_1 x_2 + 8x_1 x_3 - 6x_3 x_4$.

Solution The symmetric matrix $\mathbf{A} = [a_{ij}]$ that represents $Q(x_1, x_2, \ldots\ldots, x_n)$ has the diagonal entry a_{ii} equal to coefficient of the square terms x_i^2 and the non-diagonal entries a_{ij} and a_{ji} each equal to half of the coefficient of the cross product term $x_i \cdot x_j$. Thus

$$(i)\ \mathbf{A} = \begin{bmatrix} 2 & -4 & -8 \\ -4 & 1 & 7 \\ -8 & 7 & 5 \end{bmatrix}, \qquad (ii)\ \mathbf{B} = \begin{bmatrix} 1 & -2 & 4 & 0 \\ -2 & 2 & 0 & 0 \\ 4 & 0 & -7 & -3 \\ 0 & 0 & -3 & 1 \end{bmatrix}.$$

EXAMPLE 6.42 Find the quadratic form $Q(x_1, \ldots\ldots x_n)$ that corresponds to each of the following symmetric matrices:

$$(i)\ \mathbf{A} = \begin{bmatrix} 3 & -5 \\ -5 & 8 \end{bmatrix}, \qquad (ii)\ \mathbf{B} = \begin{bmatrix} 1 & -3 & 2 \\ -3 & 7 & -5 \\ 2 & -5 & 8 \end{bmatrix}, \qquad (iii)\ \mathbf{C} = \begin{bmatrix} 1 & -1/2 & -1/2 & 1 \\ -1/2 & -6 & 13/2 & -11/2 \\ -1/2 & 13/2 & -6 & 9/2 \\ 1 & -11/2 & 9/2 & -3 \end{bmatrix}.$$

Solution The quadratic form $Q(x_1, \ldots, x_n)$ that corresponds to a symmetric matrix is defined by $Q(x_1, \ldots., x_n) = \mathbf{x}^T\mathbf{A}\mathbf{x}$, where $\mathbf{x} = [x_i]$ is the column vector of unknowns.

$$(i)\ Q(x_1, x_2) = \mathbf{x}^T\mathbf{A}\mathbf{x} = \begin{bmatrix} x_1 & x_2 \end{bmatrix} \begin{bmatrix} 3 & -5 \\ -5 & 8 \end{bmatrix} \begin{bmatrix} x_1 \\ x_2 \end{bmatrix} = \begin{bmatrix} 3x_1 - 5x_2 & -5x_1 + 8x_2 \end{bmatrix} \begin{bmatrix} x_1 \\ x_2 \end{bmatrix}$$

$$= 3x_1^2 - 5x_1 x_2 - 5x_1 x_2 + 8x_2^2 = 3x_1^2 - 10x_1 x_2 + 8x_2^2.$$

The coefficient 3 of the square term x_1^2 and the coefficient 8 of the square term x_2^2 are the diagonal elements of **A**, and the coefficient –10 of the cross product term x_1x_2 is the sum of the nondiagonal elements –5 and –5 of **A**, or twice the nondiagonal element –5, since **A** is symmetric.

(*i*) Since **B** is a 3-square matrix, there are three unknowns, x_1, x_2, x_3. Then

$$Q(x_1, x_2, x_3) = x_1^2 - 6x_1x_2 + 7x_2^2 + 4x_1 x_3 - 10 x_2 x_3 + 8x_3^2.$$

Here we use the fact that the coefficients of the square terms x_1^2, x_2^2, x_3^2 are the respective diagonal elements 1, 7, 8 of **B** and the coefficient of the cross product term $x_i \cdot x_j$ in the sum of the nondiagonal elements b_{ij} and b_{ji} (or twice b_{ij}, since $b_{ij} = b_{ji}$).

(*iii*) Since **C** is a 4-square matrix, there are four unknowns. Hence

$$Q(x_1, x_2, x_3, x_4) = x_1^2 - 6x_2^2 - 6x_3^2 - 3x_4^2 - x_1 x_2 - x_1 x_3 + 2x_1 x_4 + 13x_2 x_3 - 11x_2 x_4 + 9x_3 x_4.$$

EXAMPLE 6.43 Express $Q(x_1, x_2, x_3) = 2x_1^2 + 3x_2^2 + 4x_3^2 - 9x_1x_3 - 8x_1 x_2$ as a product of matrices.

Solution Since there are three unknowns, so **A** is a 3-square matrix.

$$\text{Hence}\quad Q = \mathbf{x}^T\mathbf{A}\mathbf{x} = \begin{bmatrix} x_1 & x_2 & x_3 \end{bmatrix} \begin{bmatrix} 2 & -4 & -9/2 \\ -4 & 3 & 0 \\ -9/2 & 0 & 4 \end{bmatrix} \begin{bmatrix} x_1 \\ x_2 \\ x_3 \end{bmatrix}.$$

6.8.1 Type/Nature of the Quadratic Form

Let **A** be the matrix of the quadratic from $Q = \mathbf{x}^T\mathbf{A}\mathbf{x}$, where **A** is a symmetric matrix. The quadratic form Q is said to be.

(*i*) ***Positive definite***. If all the eigenvalues of **A** are real and positive ($\lambda > 0$).
(*ii*) ***Negative definite***. If all the eigenvalues of **A** are real and negative ($\lambda < 0$).

(*iii*) ***Semi-positive definite.*** If all the eigenvalues of **A** are real and non-negative ($\lambda \geq 0$).
(*iv*) ***Semi-negative definite.*** If all the eigenvalues of **A** are real and non-positive ($\lambda \leq 0$).
(*v*) ***Indefinite.*** If some eigenvalues of **A** are positive and some are negative.

We define the following:

Rank of the quadratic form: The rank (r) of a symmetric matrix **A** is called the rank of the quadratic form, that is, the number of non-zero eigenvalues.

Index of the quadratic form: The number of positive eigenvalues is the index (k) of the quadratic form.

Signature of the quadratic form: We define

$$\text{Signature} = (\text{Number of positive eigenvalues}) - (\text{Number of negative eigenvalues})$$
$$= k - (r - k) = 2k - r, \text{ that is, } 2(\text{index}) - \text{rank}.$$

Signature can be a negative integer.

EXAMPLE 6.44 Identify the nature of the quadratic forms:

(*i*) $3x_1^2 - 2x_2^2 - x_3^2 - 4x_1x_2 + 12x_2x_3 + 8x_1x_3$, (*ii*) $2x_1^2 + 5x_2^2 + 3x_3^2 + 4x_1x_2$.

Find the rank, index and signature of the quadratic form.

Solution (*i*) We have $\mathbf{A} = \begin{bmatrix} 3 & -2 & 4 \\ -2 & -2 & 6 \\ 4 & 6 & -1 \end{bmatrix}$.

The eigenvalues are given by

$$|\mathbf{A} - \lambda\mathbf{I}| = \begin{bmatrix} 3-\lambda & -2 & 4 \\ -2 & -2-\lambda & 6 \\ 4 & 6 & -1-\lambda \end{bmatrix} = 0, \text{ or } \lambda^3 - 63\lambda + 162 = 0.$$

The eigenvalues are $\lambda = 3, 6, -9$. Therefore, the quadratic form is indefinite and rank = 3, index = 2, signature = 2 – (1) = 1.

(*ii*) We have $\mathbf{A} = \begin{bmatrix} 2 & 2 & 0 \\ 2 & 5 & 0 \\ 0 & 0 & 3 \end{bmatrix}$.

The eigenvalues are given by

$$|\mathbf{A} - \lambda\mathbf{I}| = \begin{bmatrix} 2-\lambda & 2 & 0 \\ 2 & 5-\lambda & 0 \\ 0 & 0 & 3-\lambda \end{bmatrix} = 0, \text{ or } (2-\lambda)[(5-\lambda)(3-\lambda)] - 2[2(3-\lambda)] = 0$$

or $\quad (2-\lambda)(\lambda^2 - 8\lambda + 15) - 12 + 4\lambda = 0, \text{ or } \lambda^3 - 10\lambda^2 + 27\lambda - 18 = 0,$

or $\quad (\lambda - 1)(\lambda^2 - 9\lambda + 18) = 0, \text{ or } (\lambda - 1)(\lambda - 3)(\lambda - 6) = 0.$

The eigenvalues are $\lambda = 1, 3, 6$. The quadratic form is positive definite. We have rank = 3, index = 3, signature = 3.

Theorem - II

Under a non-singular linear transformation, a quadratic form $\mathbf{x}^T\mathbf{AX}$, remains a quadratic form.

Proof

Let $\mathbf{x}^T\mathbf{AX}$ be a quadratic form, where **A** is symmetric and $\mathbf{x} = (x_1, x_2, \ldots\ldots, x_n)^T$. Let $\mathbf{x} = \mathbf{Py}$ be a non-singular linear transformation, which transforms the quadratic form from the variables $x_1, x_2, \ldots\ldots, x_n$ to $y_1, y_2, \ldots\ldots y_n$.

Then $\quad \mathbf{x}^T\mathbf{Ax} = (\mathbf{Py})^T\mathbf{A}(\mathbf{Py}) = \mathbf{y}^T(\mathbf{P}^T\mathbf{AP})\mathbf{y} = \mathbf{y}^T\mathbf{By}$, where $\mathbf{B} = \mathbf{P}^T\mathbf{AP}$.

Now $\quad \mathbf{B}^T = (\mathbf{P}^T\mathbf{AP})^T = \mathbf{P}^T\mathbf{A}^T\mathbf{P} = \mathbf{P}^T\mathbf{AP} = \mathbf{B}.$

Hence, **B** is symmetric and $\mathbf{y}^T\mathbf{By}$ is also a quadratic form.

This proves the invariance of a quadratic form under a non-singular linear transformation.

6.9 CANONICAL FORM OF A QUADRATIC FORM

A quadratic form $Q = \mathbf{x}^T\mathbf{A}\mathbf{X}$ is said to be in *canonical form* if all the mixed terms such as x_1x_2, x_1x_3, ... are absent, that is, $a_{ij} = 0$, $i \neq J$. We may also say that a canonical form is a *sum of squares form*. The canonical form is, therefore, given by

$$Q = a_1y_1^2 + a_2y_2^2 + \ldots\ldots + a_ry_r^2$$

If rank $(\mathbf{A}) = r < n$ and

$$Q = a_1y_1^2 + a_2y_2^2 + \ldots\ldots + a_ny_n^2$$

If rank $(\mathbf{A}) = n$, where a_1, a_2, ..., a_n are any real values.

Since the matrix **A** is symmetric, it is diagonalizable. Hence, every quadratic form $Q = \mathbf{x}^T\mathbf{A}\mathbf{X}$ can be reduced to a sum of squares form. The number of square terms is equal to the rank r.

Remarks

(*a*) The rank of a symmetric matrix **A** is called the *rank of Q*, the quadratic form of **A**.

(*b*) A quadratic form Q of *rank r* can be reduced to the form

$$\sum_{i=1}^{r} a_i\, y_i^2 \qquad \ldots(i)$$

(*c*) The *number of positive terms* in (*i*) above is called the *index* of Q.

(*d*) (The number of positive terms) minus (number of a negative terms) in (*i*) is called *signature* of Q.

(*e*) Obviously *signature* of Q = 2 (*index*) – *rank*.

6.9.1 Reduction of a Quadratic form to a Canonical Form

The following methods can be used to reduce a quadratic form to the sum of squares form.

1. Rearrange the terms and produce sum of squares. If the quadratic form contains the variables x_1, x_2, x_3, the first square term (containing x_1, x_2, x_3) shall eliminate the product terms $x_1\,x_2$ and $x_1\,x_3$. The second square term (containing x_2, x_3) shall eliminate the product term $x_2\,x_3$. The third term is a square term in x_3.
2. ***Lagrange reduction*** Let the quadratic form contain the variables x_1, x_2, x_3. Lagrange reduction is same as in (1) except that we write the non-singular transformation as

$$\left.\begin{aligned} y_1 &= x_1 + px_2 + qx_3 \\ y_2 &= \qquad x_2 + rx_3 \\ y_3 &= \qquad\qquad x_3 \end{aligned}\right\}, \quad \text{or} \quad \mathbf{y} = \begin{bmatrix} 1 & p & q \\ 0 & 1 & r \\ 0 & 0 & 1 \end{bmatrix}\mathbf{x}, \quad \text{or} \quad \mathbf{y} = \mathbf{P}\mathbf{x} \qquad \ldots(6.19)$$

The sum of squares form is $ay_1^2 + by_2^2 + cy_3^2$. Substitute (6.19) in $ay_1^2 + by_2^2 + cy_3^2$, simplify and compare with the given quadratic form. We determine the values of a, b, c, p, q and r. Eq. (6.19) gives the required transformation.
3. ***Reduction to diagonal form*** Use elementary row and column transformations to reduce **A** to diagonal form. In this case, it is difficult to write the matrix of transformation.
4. ***Orthogonalisation*** Find the eigenvalues and eigenvectors of **A**. Obtain the normalised modal matrix **P**. Under the transformation $\mathbf{x} = \mathbf{P}\mathbf{y}$, the quadratic form reduces to

$$\lambda_1y_1^2 + \lambda_2y_2^2 + \ldots\ldots + \lambda_ry_r^2, \quad \text{or} \quad \lambda_1y_1^2 + \lambda_2y_2^2 + \ldots\ldots + \lambda_ny_n^2, \text{ depending on the rank of } \mathbf{A}.$$

ILLUSTRATIVE EXAMPLES

EXAMPLE 6.45 Is **A** Hermitian or skew-Hermitian? Find $\overline{\mathbf{x}}^T\mathbf{A}\mathbf{x}$.

(*i*) $\mathbf{A} = \begin{bmatrix} 4 & 3-2i \\ 3+2i & -4 \end{bmatrix}$, $\mathbf{x} = \begin{bmatrix} -2i \\ 1+i \end{bmatrix}$, (*ii*) $\mathbf{A} = \begin{bmatrix} -i & 1 & 2+i \\ -1 & 0 & 3i \\ -2+i & 3i & i \end{bmatrix}$, $\mathbf{x} = \begin{bmatrix} 0 \\ 1 \\ 2 \end{bmatrix}$.

Solution (*i*) we have $\bar{\mathbf{A}} = \begin{bmatrix} 4 & 3+2i \\ 3-2i & -4 \end{bmatrix}$, $\bar{\mathbf{A}}^{\mathrm{T}} = \begin{bmatrix} 4 & 3-2i \\ 3+2i & -4 \end{bmatrix} = \mathbf{A}$. Hence, the matrix **A** is an Hermitian.

$$\bar{\mathbf{x}}^{\mathrm{T}}\mathbf{A}\mathbf{x} = \begin{bmatrix} 2i & 1-i \end{bmatrix} \begin{bmatrix} 4 & 3-2i \\ 3+2i & -4 \end{bmatrix} \begin{bmatrix} -2i \\ 1+i \end{bmatrix} = \begin{bmatrix} 2i & 1-i \end{bmatrix} \begin{bmatrix} 4(-2i)+(3-2i)(1+i) \\ (3+2i)(-2i)+(-4)(1+i) \end{bmatrix}$$

$$= \begin{bmatrix} 2i & 1-i \end{bmatrix} \begin{bmatrix} -7i+5 \\ -10i \end{bmatrix} = 4.$$

(*ii*)

$$\bar{\mathbf{A}} = \begin{bmatrix} i & 1 & 2-i \\ -1 & 0 & -3i \\ -2-i & -3i & -i \end{bmatrix}, \quad \bar{\mathbf{A}}^{\mathrm{T}} = \begin{bmatrix} i & -1 & -2-i \\ 1 & 0 & -3i \\ 2-i & -3i & -i \end{bmatrix} = -\begin{bmatrix} -i & 1 & 2+i \\ -1 & 0 & 3i \\ -2+i & 3i & i \end{bmatrix} = -\mathbf{A}.$$

Hence, the matrix is skew-Hermitian.

$$\bar{\mathbf{x}}^{\mathrm{T}}\mathbf{A}\mathbf{x} = \begin{bmatrix} 0 & 1 & 2 \end{bmatrix} \begin{bmatrix} -i & 1 & 2+i \\ -1 & 0 & 3i \\ -2+i & 3i & i \end{bmatrix} \begin{bmatrix} 0 \\ 1 \\ 2 \end{bmatrix} = \begin{bmatrix} 0 & 1 & 2 \end{bmatrix} \begin{bmatrix} 5+2i \\ 6i \\ 5i \end{bmatrix} = 16i.$$

EXAMPLE 6.46 Transform the quadratic form

$$\mathbf{x}^{\mathrm{T}}\mathbf{A}\mathbf{x} = x_1^2 + 4x_1x_2 - 6x_1x_3 - 18x_2x_3 + 7x_2^2 + 7x_3^2$$

to sum of squares form. Write the transformation. Find the rank, index and signature.

Solution Collecting the terms, we write

$$Q = (x_1 + 2x_2 - 3x_3)^2 - 4x_2^2 - 9x_3^2 + 12x_2x_3 - 18x_2x_3 + 7x_2^2 + 7x_3^2$$

$$= (x_1 + 2x_2 - 3x_3)^2 + 3x_2^2 - 6x_2x_3 - 2x_3^2$$

$$= (x_1 + 2x_2 - 3x_3)^2 + 3(x_2 - x_3)^2 - 3x_3^2 - 2x_3^2$$

$$= (x_1 + 2x_2 - 3x_3)^2 + 3(x_2 - x_3)^2 - 5x_3^2$$

$$= y_1^2 + 3y_2^2 - 5y_3^2, \text{ where } y_1 = x_1 + 2x_2 - 3x_3, \quad y_2 = x_2 - x_3, \quad y_3 = x_3.$$

The non-singular tranformation is $\mathbf{y} = \mathbf{P}\mathbf{x}$, where

$$\mathbf{P} = \begin{bmatrix} 1 & 2 & -3 \\ 0 & 1 & -1 \\ 0 & 0 & 1 \end{bmatrix}.$$

Rank = 3, Index = Number of positive coefficients = 2,

Signature = (Number of +ve coefficients) – (Number of –ve coefficients) = 2 – 1 = 1.

EXAMPLE 6.47 Find the matrix of the quadratic form

$$Q = 6x_1^2 + 3x_2^2 + 3x_3^2 - 4x_1x_2 + 4x_1x_3 - 2x_2x_3$$

and find the linear transformation $\mathbf{y} = \mathbf{P}\mathbf{x}$ which transforms this quadratic form to sum of squares. Find the index and signature of the quadratic form.

Solution Obviously, the matrix of the quadratic form is:

$$\mathbf{A} = \begin{bmatrix} 6 & -2 & 2 \\ -2 & 3 & -1 \\ 2 & -1 & 3 \end{bmatrix}.$$

The given quadratic form Q, can be reduced to sum of squares form by Lagrange's reduction.

We have $$Q = 6x_1^2 + 3x_2^2 + 3x_3^2 - 4x_1x_2 + 4x_1x_3 - 2x_2x_3$$

Let the transformation be

$$\left.\begin{aligned} y_1 &= x_1 + px_2 + qx_3 \\ y_2 &= x_2 + rx_3 \\ y_3 &= x_3 \end{aligned}\right\}, \quad \text{or} \quad \mathbf{y} = \begin{bmatrix} 1 & p & q \\ 0 & 1 & r \\ 0 & 0 & 1 \end{bmatrix}\mathbf{x}, \quad \text{or} \quad \mathbf{y} = \mathbf{Px},$$

and the sum of squares form be $ay_1^2 + by_2^2 + cy_3^2$. Hence, $a(x_1 + px_2 + qx_3)^2 + b(x_2 + rx_3)^2 + cx_3^2$

$$= ax_1^2 + 2apx_1x_2 + 2aqx_1x_3 + (ap^2 + b)x_2^2 + (2apq + 2br)\,x_2x_3 + (aq^2 + br^2 + c)\,x_3^2$$

$$= 6x_1^2 - 4x_1x_2 + 4x_1x_3 + 3x_2^2 - 2x_2x_3 + 3x_3^2.$$

Comparing the coefficients, we obtain

$$a = 6; \quad 2ap = -4, \quad p = -1/3; \quad 2aq = 4; \quad q = 1/3; \quad ap^2 + b = 3; \quad b = 7/3;$$

$$2apq + 2br = -2, \quad 14r/3 = -2/3, \quad r = -1/7; \quad aq^2 + br^2 + c = 3, \quad c = 3 - \frac{2}{3} - \frac{1}{21} = \frac{16}{7}.$$

Hence, the sum of squares form is $6y_1^2 + \frac{7}{3}y_2^2 + \frac{16}{7}y_3^2$.

The transformation is $\mathbf{y} = \mathbf{Px}$, where $\mathbf{P} = \begin{bmatrix} 1 & -1/3 & 1/3 \\ 0 & 1 & -1/7 \\ 0 & 0 & 1 \end{bmatrix}$.

Rank = 3, index = 3, signature = 3.

EXAMPLE 6.48 Find the matrix of the quadratic form

$$Q = 6x_1^2 + 3x_2^2 + 3x_3^2 - 4x_1x_2 + 4x_1x_3 - 2x_2x_3$$

and find the linear transformation $\mathbf{x} = \mathbf{Py}$ which transforms this quadratic form to sum of squares. *(KUK, 2006)*

Solution Obviously, the matrix of the quadratic form is:

$$\mathbf{A} = \begin{bmatrix} 6 & -2 & 2 \\ -2 & 3 & -1 \\ 2 & -1 & 3 \end{bmatrix}.$$

To determine the matrix **P** we reduce **P** to a diagonal form by proceeding as follows:

(*i*) Write the matrix **A** and **I** side by side.

(*ii*) Perform identical row followed by column transformations on **A** but perform only corresponding row transformations **I**.

(*iii*) The form that **I** assumes, when **A** has been reduced to diagonal form, is $\mathbf{P}^T$ (i.e., transpose of **P**). Thus we can get **P** from $\mathbf{P}^T$.

Thus we have $\left[\begin{array}{ccc|ccc} 6 & -2 & 2 & 1 & 0 & 0 \\ -2 & 3 & -1 & 0 & 1 & 0 \\ 2 & -1 & 3 & 0 & 0 & 1 \end{array}\right]$. Operate $R_2 \to R_2 + \frac{1}{3}R_1$, $R_3 \to R_3 - \frac{1}{3}R_1$. (on **A** and **I**)

$\approx \left[\begin{array}{ccc|ccc} 6 & -2 & 2 & 1 & 0 & 0 \\ 0 & 7/3 & -1/3 & 1/3 & 1 & 0 \\ 0 & -1/3 & 7/3 & -1/3 & 0 & 1 \end{array}\right]$ Operate $C_2 \to C_2 + \frac{1}{3}C_1, C_3 \to C_3 - \frac{1}{3}C_1$. (on **A** only)

$\approx \left[\begin{array}{ccc|ccc} 6 & 0 & 0 & 1 & 0 & 0 \\ 0 & 7/3 & -1/3 & 1/3 & 1 & 0 \\ 0 & -1/3 & 7/3 & -1/3 & 0 & 1 \end{array}\right]$ Operate $R_3 \to R_3 + \frac{1}{7}R_2$,

$\approx \left[\begin{array}{ccc|ccc} 6 & 0 & 0 & 1 & 0 & 0 \\ 0 & 7/3 & -1/3 & 1/3 & 1 & 0 \\ 0 & 0 & 16/7 & -2/7 & 1/7 & 1 \end{array}\right]$ Operate $C_3 \to C_3 + \frac{1}{7}C_2$. (On **A** only)

$$\approx \left[\begin{array}{ccc|ccc} 6 & 0 & 0 & 1 & 0 & 0 \\ 0 & 7/3 & 0 & 1/3 & 1 & 0 \\ 0 & 0 & 16/7 & -2/7 & 1/7 & 1 \end{array}\right]$$

Now $\quad \mathbf{P}^T = \begin{bmatrix} 1 & 0 & 0 \\ 1/3 & 1 & 0 \\ -2/7 & 1/7 & 1 \end{bmatrix}. \quad \therefore \mathbf{P} = \begin{bmatrix} 1 & 1/3 & -2/7 \\ 0 & 1 & 1/7 \\ 0 & 0 & 1 \end{bmatrix}.$

Thus the transformation $\mathbf{x} = \mathbf{P}\mathbf{y} = \begin{bmatrix} 1 & 1/3 & -2/7 \\ 0 & 1 & 1/7 \\ 0 & 0 & 1 \end{bmatrix} \begin{bmatrix} y_1 \\ y_2 \\ y_3 \end{bmatrix}$ ...(*i*)

reduces $\mathbf{Q}$ to the required form $6y_1^2 + \frac{7}{3}y_2^2 + \frac{16}{7}y_3^2$.

Remark:

From (i), we find that by putting

$x_1 = y_1 + \frac{1}{3}y_2 - \frac{2}{7}y_3, \quad x_2 = y_2 + \frac{1}{7}y_3, \quad x_3 = y_3$ in Q the sum of squares from is $6y_1^2 + \frac{7}{3}y_2^2 + \frac{16}{7}y_3^2$.

EXAMPLE 6.49 Reduce the quadratic form to canonical form by Lagrange reduction. Write the linear transformation. Find the index and signature of the quadratic form

$$\mathbf{x}^T\mathbf{A}\mathbf{x} = 2x_1^2 + 3x_2^2 + 15x_3^2 + 4x_1x_2 - 8x_1x_3 - 2x_2x_3.$$

Solution Let the transformation be

$$y_1 = x_1 + px_2 + qx_3, \quad y_2 = x_2 + rx_3, \quad y_3 = x_3, \text{ and the}$$

sum of squares form be $ay_1^2 + by_2^2 + cy_3^2$. Hence $a(x_1 + px_2 + qx_3)^2 + b(x_2 + rx_3)^2 + cx_3^2$

$$= ax_1^2 + 2apx_1x_2 + 2aqx_1x_3 + (ap^2 + b)x_2^2 + (2apq + 2br)\,x_2x_3 + (aq^2 + br^2 + c)\,x_3^2$$

$$= 2x_1^2 + 4x_1x_2 - 8x_1x_3 + 3x_2^2 - 2x_2x_3 + 15x_3^2.$$

Comparing the coefficients, we obtain

$a = 2; \quad 2ap = 4, \quad p = 1; \quad 2aq = -8, \quad q = -2; \quad ap^2 + b = 3, \quad b = 1;$

$2(apq + br) = -2, \quad r = -1 + 4$ [illegible], $\quad aq^2 + br^2 + c = 15, \quad c = 15 - 8 - 9 = -2.$

Hence, the sum of squares form is $2y_1^2$ [illegible] $-2y_3^2$.

The transformation is $\mathbf{y} = \mathbf{P}\mathbf{x}$, where $\mathbf{P} = \begin{bmatrix} 1 & p & q \\ 0 & 1 & r \\ 0 & 0 & 1 \end{bmatrix} = \begin{bmatrix} 1 & 1 & -2 \\ 0 & 1 & 3 \\ 0 & 0 & 1 \end{bmatrix}.$

Rank = 3, index = number of positive coefficients = 2,

signature = (number of +ve coefficients) – (number of –ve coefficients) = 2 – 1 = 1.

EXAMPLE 6.50 Reduce the quadratic forms to canonical form through orthogonal transformation.

(*i*) $x_1^2 + 6x_1x_2 - 7x_2^2$, (*ii*) $11x_1^2 - 16x_1x_2 + 8x_1x_3 - x_2^2 - 4x_2x_3 - 4x_3^2$,

(*iii*) $6x_1^2 - 4x_1x_2 + 4x_1x_3 + 3x_2^2 - 2x_2x_3 + 3x_3^2$.

Solution (*i*) The matrix of the quadratic form is $\mathbf{A} = \begin{bmatrix} 1 & 3 \\ 3 & -7 \end{bmatrix}$.

The eigenvalues are given by

$$|\mathbf{A} - \lambda\mathbf{I}| = \begin{bmatrix} 1-\lambda & 3 \\ 3 & -7-\lambda \end{bmatrix} = 0, \text{ or } (1-\lambda)(-7-\lambda) - 9 = 0.$$

We obtain $\lambda = 2, -8$. The eigenvalues are $\lambda = 2, -8$.

We obtain the eigenvectors from the equations

$$(1 - \lambda)\, x_1 + 3x_2 = 0, \quad 3x_1 - (7 + \lambda)x_2 = 0.$$

For $\lambda = 2$, we get the equations $-x_1 + 3x_2 = 0$, $3x_1 - 9x_2 = 0$, whose solution is $x_1 = 3$, $x_2 = 1$. Hence, the eigenvector can be taken as $\mathbf{V}_1 = [3, \; 1]^T$. The normalised eigenvector is

$$\hat{\mathbf{V}}_1 = \left[3/\sqrt{10} \quad 1/\sqrt{10}\right]^T.$$

For $\lambda = -8$, we get the equations $9x_1 + 3x_2 = 0$, $3x_1 + x_2 = 0$, whose solution is $x_1 = -1$, $x_2 = 3$. The eigenvector can be written as $\mathbf{V}_2 = [-1, \; 3]^T$. The normalised eigenvector is

$$\hat{\mathbf{V}}_2 = \left[-1/\sqrt{10} \quad 3/\sqrt{10}\right]^T.$$

The orthogonal transformation is $\mathbf{x} = \mathbf{Py}$ where

$$\mathbf{P} = \begin{bmatrix} 3/\sqrt{10} & -1/\sqrt{10} \\ 1/\sqrt{10} & 3/\sqrt{10} \end{bmatrix}.$$

We have $\quad \mathbf{P}^{-1}\mathbf{AP} = \mathbf{P}^T\mathbf{AP} = \mathbf{D}$, where $\mathbf{D} = \text{diag}\,(2, -8) = \begin{bmatrix} 2 & 0 \\ 0 & -8 \end{bmatrix}.$

Therefore $\quad \mathbf{x}^T\mathbf{Ax} = (\mathbf{Py})^T\mathbf{A}(\mathbf{Py}) = \mathbf{y}^T(\mathbf{P}^T\mathbf{AP})\mathbf{y} = \mathbf{y}^T\mathbf{Dy} = 2y_1^2 - 8y_2^2$, is the sum of squares form.

(*ii*) The matrix of the quadratic form is $\mathbf{A} = \begin{bmatrix} 11 & -8 & 4 \\ -8 & -1 & -2 \\ 4 & -2 & -4 \end{bmatrix}.$

The eigenvalues are given by $|\mathbf{A} - \lambda\mathbf{I}| = \begin{vmatrix} 11-\lambda & -8 & 4 \\ -8 & -1-\lambda & -2 \\ 4 & -2 & -4-\lambda \end{vmatrix} = 0,$

or $\quad (11 - \lambda)\,[(1 + \lambda)\,(4 + \lambda) - 4] + 8\,(32 + 8\lambda + 8) + 4\,(16 + 4 + 4\lambda) = 0,$

or $\quad (11 - \lambda)\,(\lambda^2 + 5\lambda) + 80\lambda + 400 = 0, \quad$ or $\quad \lambda^3 - 6\lambda^2 - 135\lambda - 400 = 0.$

We obtain $\lambda = -5, -5, 16$. The eigenvalues are $\lambda = -5, -5, 16$.

We obtain the eigenvectors from the equations.

$$(11 - \lambda)x_1 - 8x_2 + 4x_3 = 0, \quad -8x_1 - (1 + \lambda)\,x_2 - 2x_3 = 0, \quad 4x_1 - 2x_2 - (4 + \lambda)\,x_3 = 0,$$

For $\lambda = -5$ (the equal roots), we get the equations

$$16x_1 - 8x_2 + 4x_3 = 0, \quad -8x_1 + 4x_2 - 2x_3 = 0, \quad 4x_1 - 2x_2 + x_3 = 0.$$

Clearly only one equation $4x_1 - 2x_2 + x_3 = 0$ is independent, the other two being multiples of that. The solution is for $x_1 = 0$, $x_2 = 1$, we get $x_3 = 2$. The eigenvector can be taken as $\mathbf{V}_1 = [0 \;\; 1 \;\; 2]^T$.

The normalized eigenvector is $\hat{\mathbf{V}}_1 = \left[0, 1/\sqrt{5}, \;\; 2/\sqrt{5}\right]^T$.

We seek a second solution $\mathbf{V}_2 = [a, \; b, \; c]^T$, which is orthogonal to $\mathbf{V}_1 = [0 \;\; 1 \;\; 2]^T$, that is, such that $4a - 2b + c = 0$, and also $b + 2c = 0$. One such solution is $a = -5$, $b = -8$, $c = 4$. The eigenvector can be written as $\mathbf{V}_2 = [-5, \; -8, \; 4]^T$.

The normalised eigenvector is $\hat{\mathbf{V}}_2 = \left[-5/\sqrt{105}, \;\; -8/\sqrt{105} \;\; 4/\sqrt{105}\right]^T$.

For $\lambda = 16$, we get the equations:

$$-5x_1 - 8x_2 + 4x_3 = 0, \quad -8x_1 - 17x_2 - 2x_3 = 0, \quad 4x_1 - 2x_2 - 20x_3 = 0.$$

This system yields a non-zero solution $x_1 = 4$, $x_2 = -2$, $x_3 = 1$.

The eigenvector can be written as $\mathbf{V}_3 = [4, \ -2, \ 1]^T$.

The normalised eigenvector is $\hat{\mathbf{V}}_3 = \left[4/\sqrt{21}, \ \ -2/\sqrt{21}, \ \ 1/\sqrt{21}\right]^T$.

The orthogonal transformation is $\mathbf{x} = \mathbf{P}\mathbf{y}$ where

$$\mathbf{P} = \begin{bmatrix} 0 & -5/\sqrt{105} & 4/\sqrt{21} \\ 1/\sqrt{5} & -8/\sqrt{105} & -2/\sqrt{21} \\ 2/\sqrt{5} & 4/\sqrt{105} & 1/\sqrt{21} \end{bmatrix}.$$

We have $\mathbf{P}^{-1}\mathbf{A}\mathbf{P} = \mathbf{P}^T\mathbf{A}\mathbf{P} = \mathbf{D}$, where $\mathbf{D} = \text{diag}\,(-5, -5, 16) = \begin{bmatrix} -5 & & \\ & -5 & \\ & & 16 \end{bmatrix}$.

Therefore, $\mathbf{x}^T\mathbf{A}\mathbf{x} = (\mathbf{P}\mathbf{y})^T\,\mathbf{A}\,(\mathbf{P}\mathbf{y}) = \mathbf{y}^T\,(\mathbf{P}^T\mathbf{A}\mathbf{P})\mathbf{y} = \mathbf{y}^T\mathbf{D}\mathbf{y}$

$= -5y_1^2 - 5y_2^2 + 16y_3^2$ is the sum of the squares form.

Note: Rank = 3, index = 1, signature = –1.

(*iii*) The matrix of the quadratic form is $\mathbf{A} = \begin{bmatrix} 6 & -2 & 2 \\ -2 & 3 & -1 \\ 2 & -1 & 3 \end{bmatrix}$.

The eigenvalues are given by $|\mathbf{A} - \lambda\mathbf{I}| = \begin{vmatrix} 6-\lambda & -2 & 2 \\ -2 & 3-\lambda & -1 \\ 2 & -1 & 3-\lambda \end{vmatrix} = 0.$

Operating $C_2 \to C_2 + C_1$, we obtain

$$\begin{vmatrix} 6-\lambda & 0 & 2 \\ -2 & 2-\lambda & -1 \\ 2 & 2-\lambda & 3-\lambda \end{vmatrix} = 0 \quad \text{or} \quad (2-\lambda)\begin{vmatrix} 6-\lambda & 0 & 2 \\ -2 & 1 & -1 \\ 2 & 1 & 3-\lambda \end{vmatrix} = 0,$$

or $\qquad (2 - \lambda)\,[(6 - \lambda)\,(3 - \lambda + 1) + 2\,(-2 - 2)] = 0$

or $\qquad (2 - \lambda)\,(\lambda^2 - 10\lambda + 16) = 0 \quad \text{or} \quad (2 - \lambda)\,(\lambda - 2)\,(\lambda - 8) = 0.$

We obtain $\lambda = 2, 2, 8$. The eigenvalues are $\lambda = 2, 2, 8$.

We obtain the eigenvectors from the equations:

$$(6 - \lambda)x_1 - 2x_2 + 2x_3 = 0, \quad -2x_1 + (3 - \lambda)x_2 - x_3 = 0, \quad 2x_1 - x_2 + (3 - \lambda)\,x_3 = 0.$$

For $\lambda = 2$ (the equal roots) we get the equations

$$4x_1 - 2x_2 + 2x_3 = 0, \quad -2x_1 + x_2 - x_3 = 0, \quad 2x_1 - x_2 + x_3 = 0.$$

Clearly only one equation $2x_1 - x_2 + x_3 = 0$ is independent, the other two being multiples of that. The solution is for $x_1 = -1$, $x_2 = 0$, we get $x_3 = 2$. The eigenvector can be taken as $\mathbf{V}_1 = [-1, \ 0, \ 2]^T$.

The normalised eigenvector is $\hat{\mathbf{V}}_1 = \left[-1/\sqrt{5}, \ \ 0 \ \ 2/\sqrt{5}\right]^T$.

We seek a second solution $\mathbf{V}_2 = [a, \ b, \ c]^T$, which is orthogonal to $\mathbf{V}_1 = [-1, \ 0, \ 2]^T$, that is, such that $2a - b + c = 0$, and also $-a + 2c = 0$. One such solution is $a = 2$, $b = 5$, $c = 1$. The eigenvector can be written as $\mathbf{V}_2 = [2, \ 5, \ 1]^T$. The normalised eigenvector is $\hat{\mathbf{V}}_2 = \left[2/\sqrt{30}, \ \ 5/\sqrt{30}, \ \ 1/\sqrt{30}\right]^T$.

For $\lambda = 8$, we get the equations

$$-2x_1 - 2x_2 + 2x_3 = 0, \quad -2x_1 - 5x_2 - x_3 = 0, \quad 2x_1 - x_2 - 5x_3 = 0$$

or $\qquad x_1 + x_2 - x_3 = 0, \quad x_2 + x_3 = 0.$

Choosing $x_3 = 1$, we get $x_1 = 2$, $x_2 = -1$. The eigenvector is $\mathbf{V}_3 = [2, \ -1, \ 1]^T$.

The normalised eigenvector is $\hat{\mathbf{V}}_3 = \left[2/\sqrt{6}, \ -1/\sqrt{6}, \ 1/\sqrt{6}\right]^T$.

The orthogonal transformation is $\mathbf{x} = \mathbf{P}\mathbf{y}$, where

$$\mathbf{P} = \begin{bmatrix} -1/\sqrt{5} & 2/\sqrt{30} & 2/\sqrt{6} \\ 0 & 5/\sqrt{30} & -1/\sqrt{6} \\ 2/\sqrt{5} & 1/\sqrt{30} & 1/\sqrt{6} \end{bmatrix}.$$

We have $\mathbf{P}^{-1}\mathbf{A}\mathbf{P} = \mathbf{P}^T\mathbf{A}\mathbf{P} = \mathbf{D}$, where $\mathbf{D} = \text{diag}(2, 2, 8) = \begin{bmatrix} 2 & & \\ & 2 & \\ & & 8 \end{bmatrix}$.

Therefore, $\mathbf{x}^T\mathbf{A}\mathbf{x} = (\mathbf{P}\mathbf{y})^T \mathbf{A} (\mathbf{P}\mathbf{y}) = \mathbf{y}^T (\mathbf{P}^T\mathbf{A}\mathbf{P})\mathbf{y} = \mathbf{y}^T\mathbf{D}\mathbf{y}$

$= 2y_1^2 + 2y_2^2 + 8y_3^2$ is the sum of the squares form.

Note: Rank = 3, index = 3, signature = 3.

EXERCISE 6.3

1. Find the symmetric matrix that corresponds to each of the following quadratic forms:

(i) $Q(x_1, x_2) = x_1^2 - 4x_1x_2 + 4x_2^2$, (ii) $Q(x_1, x_2, x_3) = 3x_1^2 + 2x_1x_2 - 4x_1x_3 + 8x_2x_3 + x_2^2$,

(iii) $Q(x_1, x_2, x_3) = 4x_1x_3 + 2x_2x_3 + x_3^2$, (iv) $Q(x_1, x_2, x_3) = x_1x_2 + x_2^2 + 4x_1x_3 + x_3^2$,

(v) $Q(x_1, x_2, x_3, x_4) = 2x_1^2 - 7x_2^2 + 3x_3^2 + x_4^2 + 8x_1x_2 - 2x_1x_3 + 10x_1x_4 - 12x_2x_3 + 16x_2x_4 + 18x_3x_4$.

Ans. (i) $\begin{bmatrix} 1 & -2 \\ -2 & 4 \end{bmatrix}$, (ii) $\begin{bmatrix} 3 & 1 & -2 \\ 1 & 1 & 4 \\ -2 & 4 & 0 \end{bmatrix}$, (iii) $\begin{bmatrix} 0 & 0 & 2 \\ 0 & 0 & 1 \\ 2 & 1 & 1 \end{bmatrix}$, (iv) $\begin{bmatrix} 0 & 1/2 & 2 \\ 1/2 & 1 & 0 \\ 2 & 0 & 1 \end{bmatrix}$, (v) $\begin{bmatrix} 2 & 4 & -1 & 5 \\ 4 & -7 & -6 & 8 \\ -1 & -6 & 3 & 9 \\ 5 & 8 & 9 & 1 \end{bmatrix}$

2. Find the quadratic from Q that corresponds to each of the following symmetric matrices:

(i) $\mathbf{A} = \begin{bmatrix} 3 & 1 \\ 1 & -1 \end{bmatrix}$, (ii) $\mathbf{B} = \begin{bmatrix} 4 & -3 & 5 \\ -3 & 2 & 1 \\ 5 & 1 & -6 \end{bmatrix}$, (iii) $\mathbf{C} = \begin{bmatrix} 1 & -1 & 0 & 2 \\ -1 & 2 & 1 & 0 \\ 0 & 1 & 1 & 2 \\ 2 & 0 & 2 & -1 \end{bmatrix}$

Ans. (i) $Q(x_1, x_2) = 3x_1^2 + 2x_1x_2 - x_2^2$,

(ii) $Q(x_1, x_2, x_3) = 4x_1^2 - 6x_1x_2 + 2x_2^2 + 10x_1x_3 + 2x_2x_3 - 6x_3^2$,

(iii) $Q(x_1, x_2, x_3, x_4) = x_1^2 + 2x_2^2 + x_3^2 - x_4^2 - 2x_1x_2 + 4x_1x_4 + 2x_2x_3 + 4x_3x_4$.

3. Express $Q(x_1, x_2, x_3) = 2x_1^2 + 3x_2^2 + 4x_3^2 - 9x_1x_3 - 8x_1x_2$ as a product of matrices.

Ans. $Q = [x_1 \ x_2 \ x_3] \begin{bmatrix} 2 & -4 & -9/2 \\ -4 & 3 & 0 \\ -9/2 & 0 & 4 \end{bmatrix} \begin{bmatrix} x_1 \\ x_2 \\ x_3 \end{bmatrix}$.

4. Let $Q(x_1, x_2) = 2x_1^2 - 6x_1x_2 - 3x_2^2$ and $x_1 = s + 2t$, $x_2 = 3s - t$. Rewrite $Q(x_1, x_2)$ in matrix notation, and find the matrix $\mathbf{A}$ representing the quadratic form. Also find $Q(s, t)$.

Ans. $Q(x_1, x_2) = \mathbf{x}^T\mathbf{A}\mathbf{x} = [x_1 \quad x_2]\begin{bmatrix} 2 & -3 \\ -3 & -3 \end{bmatrix}\begin{bmatrix} x_1 \\ x_2 \end{bmatrix}$; $\mathbf{A} = \begin{bmatrix} 2 & -3 \\ -3 & -3 \end{bmatrix}$; $Q(s, t) = -43s^2 - 4st + 17t^2$.

5. Identify the nature of the quadratic forms:

(*i*) $2x_1x_2 + 2x_1x_3 + 2x_2x_3$, (*ii*) $x_1^2 + 9x_2^2 + x_3^2 - 6x_1x_2 + 2x_1x_3 - 6x_2x_3$.

Find the rank, index and signature of the quadratic form.

Ans. (*i*) Eigenvalues are : –1, –1, 2; Indefinite; 3, 1, –1.

(*ii*) Eigenvalues are : 0, 0, 11; Semi-positive definite; 1, 1, 1.

6. Use Lagrange reduction to reduce the quadratic form $x_1^2 + 3x_2^2 + 3x_3^2 - 2x_2x_3$ to sum of squares form. Find the index and signature. *(Andhra 2000)* **Ans.** $y_1^2 + 2y_2^2 + 4y_3^2; 3, 3.$

7. Find the transformation that will transform $2x_1^2 + 3x_2^2 + 15x_3^2 + 4x_1x_2 - 8x_1x_3 - 2x_2x_3$ into a sum of squares and find the reduced form. Also find the rank, index and signature.

Ans. $2y_1^2 + y_2^2 - 2y_3^2;\quad \mathbf{P} = \begin{bmatrix} 1 & 1 & -2 \\ 0 & 1 & 3 \\ 0 & 0 & 1 \end{bmatrix};\ 3, 2, 1.$

8. Find the transformation which will transform $4x_1^2 + 3x_2^2 + x_3^2 - 8x_1x_2 - 6x_2x_3 + 4x_1x_3$ into sum of squares and find the reduced form. Also find the rank, index and signature.

Ans. $4y_1^2 - y_2^2 + y_3^2;\quad \mathbf{P} = \begin{bmatrix} 1 & 1 & -3/2 \\ 0 & 1 & -1 \\ 0 & 0 & 1 \end{bmatrix};\ 3, 2, 1.$

9. Find the rank, index and signature of the quadratic form $x_1^2 + 2x_2^2 - 3x_3^2$. *(Madras 2006)* **Ans,** 3, 2, 1.

10. Reduce the following quadratic forms into a sum of squares by an orthogonal transformation and give the matrix of transformation. Also state the nature of quadratic form and determine the rank, index and signature of each of these.

(*i*) $2x_1^2 + 4x_1x_2 + 8x_1x_3 + 5x_2^2 + 16x_2x_3 + 17x_3^2$, (*ii*) $x_1^2 + 3x_2^2 + 3x_3^2 - 2x_2x_3$,

(*iii*) $3x_1^2 - 2x_1x_2 + 2x_1x_3 + 5x_2^2 - 2x_2x_3 + 3x_3^2$, (*iv*) $2x_1x_2 + 2x_1x_3 - 2x_2x_3$. *(Madras, 2006)*

Ans. (*i*) $y_1^2 + y_2^2 + 484y_3^2;\quad \mathbf{P} = \begin{bmatrix} 2/\sqrt{6} & 2/\sqrt{14} & 1/\sqrt{21} \\ 1/\sqrt{6} & -3/\sqrt{14} & 2/\sqrt{21} \\ -1/\sqrt{6} & 1/\sqrt{14} & 4/\sqrt{21} \end{bmatrix}$, positive definite, rank = 3, index = 3, signature=3.

(*ii*) $y_1^2 + 2y_2^2 + 4y_3^2;\quad \mathbf{P} = \begin{bmatrix} 1 & 0 & 0 \\ 0 & 1/\sqrt{2} & 1/\sqrt{2} \\ 0 & 1/\sqrt{2} & -1/\sqrt{2} \end{bmatrix}$, positive definite, rank = 3, index = 3, signature = 3.

(*iii*) $2y_1^2 + 3y_2^2 + 6y_3^2;\quad \mathbf{P} = \begin{bmatrix} 1/\sqrt{2} & 1/\sqrt{3} & 1/\sqrt{6} \\ 0 & 1/\sqrt{3} & -2/\sqrt{6} \\ -1/\sqrt{2} & 1/\sqrt{3} & 1/\sqrt{6} \end{bmatrix}$, positive definite, rank = 3, index = 3, signature = 3.

(*iv*) $-2y_1^2 + y_2^2 + y_3^2;\quad \mathbf{P} = \begin{bmatrix} -1/\sqrt{3} & 1/\sqrt{2} & 1/\sqrt{6} \\ 1/\sqrt{3} & 0 & 2/\sqrt{6} \\ 1/\sqrt{3} & 1/\sqrt{2} & -1/\sqrt{6} \end{bmatrix}$, indefinite, rank = 3, index = 2, signature = 1.

CHAPTER

7

Ordinary Differential Equations of First Order

7.1 INTRODUCTION

The study of differential equations is of paramount importance in engineering mathematics. This is primarily due to the reason that many physical laws and relations are expressible mathematically in the form of differential equations. It is from these equations that we obtain relationship among the variables of the problems. The interpretation and analysis of the results provides us with the information concerning the manner in which the physical quantities involved depend upon one another. Before we classify and examine different kinds of differential equations, let us consider the following differential equations:

(*i*) $\frac{dy}{dx}+2y+x=0,$ (*ii*) $(2x+y)\,dx+(\sin y+x)\,dy=0,$

(*iii*) $\frac{d^2y}{dx^2}+c^2y=0,$ (*iv*) $\rho=\left\{1+\left(\frac{dy}{dx}\right)^2\right\}^{3/2}\Big/\frac{d^2y}{dx^2},$

(*v*) $\frac{d^3y}{dx^3}+5\frac{d^2y}{dx^2}+5\frac{dy}{dx}+y=0,$ (*vi*) $\frac{\partial^2u}{\partial t^2}=c^2\frac{\partial^2u}{\partial x^2},$

(*vii*) $x\frac{\partial u}{\partial x}+y\frac{\partial u}{\partial y}=nu,$ (*viii*) $\frac{dx}{dt}+py=a\cos\omega t,\ \frac{dy}{dt}+px=b\sin\omega t.$

From the above equations, it is clear that all of them contain either the differential coefficients or differentials. A *differential equation* can be defined as an equation containing derivatives of various orders and the variables. Symbolically a differential equation may be written as follows:

$$f(x,y,y',y'',\dots,y^{(n)})=0 \text{ or } f\left(x,y,\frac{dy}{dx},\frac{d^2y}{dx^2},\dots,\frac{d^ny}{dx^n}\right)=0.$$

Differential equations which involve one independent variable are called *ordinary differential equations*. Equations (*i*) to (*v*) are all examples of ordinary differential equations.

If the differential equation involves more than one independent variable and partial derivatives of the dependent variable with respect to them, then it is called a *partial differential equation.*

Equations (*vi*) and (*vii*) are partial differential equations.

The differential equations which have two or more dependent variables and a single independent variable are called *simultaneous differential equations*. Equations (*viii*) is an example of simultaneous equations.

Other Definitions

(1) **Order:** The order of a differential equation is the order of the highest order derivative occurring in the equation. Equations (*i*) and (*ii*) are of the first order while equation (*iii*) is of the second order and equation (*v*) is of the three order.

(2) **Degree:** The degree of a differential equation is the degree (or power) of the highest order derivative occurring in the equation after the equations has been made free from fractions and the radicals as far as derivatives are concerned. Equations (*i*) and (*ii*) are ordinary, first order and first degree equations; equation (*iii*) is an ordinary, second order and first degree; equation (*v*) is an ordinary, third order and first degree; equation (*iv*) is an ordinary, second order and second degree as it can be written as:

$$\rho^2 \left(\frac{d^2 y}{dx^2}\right)^2 = \left\{1 + \left(\frac{dy}{dx}\right)^2\right\}^3.$$

Similarly equation (*vi*) is a second order, first degree partial differential equation and so on.

(3) **Linear and Non-linear differential equation**

Linear: A differential equation is linear, when the dependent variable and all its derivatives are of the first degree (that is, the power of each term involving y is 1) and no products of the dependent variable and its derivatives or of various order derivatives occur. The form of a linear ordinary differential equation, in general, is

$$c_0 y^{(n)} + c_1 y^{(n-1)} + c_2 y^{(n-2)} + \ldots + c_{n-1} y' + c_n y = g(x),$$

where $c_i = c_i(x)$ are some functions of x or are constants.

The following equations

$$y' = 4x^2, \quad y'' + 3y' + 4y = 2x, \quad x^2 \frac{d^2 y}{dx^2} + x\frac{dy}{dx} + 3y = \log x$$

and $y''' + 6y'' + 3y' + y = 0$ are some examples of linear differential equations.

Non-linear: A differential equation which is not linear, is called a non-linear differential equation. Some examples of non-linear ordinary differential equations are given below.

$$(y')^2 + 3xy' + y = 0, \quad \left[1 + (y')^2\right]^{1/2} = 8y, \quad y'' + 4yy' + 2y = \cos x, \quad y'' + \sin y = 0.$$

EXAMPLE 7.1 Find the order and the degree of the following differential equations. State also whether they are linear or non-linear.

(*a*) $y'' + 4xy' + 2y = \cos x$, (*b*) $y''' + xy'' + 2y\,(y')^2 + xy = 0$, (*c*) $y'' + y' + \cos y = 0$,

(*d*) $\dfrac{d(xy')}{dx} + x^2 y = 0$, (*e*) $y' = \sqrt{x} + \sqrt{y}$, (*f*) $y'y'' + y' + 5y = \sin x$, (*g*) $y + \dfrac{dy}{dx} = \dfrac{1}{2}\int y\,dx$.

Solution (*a*) Order two; degree one; linear.

(*b*) Order three; degree one; non-linear (because of the presence of first derivative having degree two).

(*c*) Order two; degree one; non-linear (because of the presence of cos y, which is a non-linear function of y).

(*d*) $\dfrac{d(xy')}{dx} + x^2 y = 0 \Rightarrow xy'' + y' + x^2 y = 0$. Order two; degree one; linear.

(*e*) Order one; degree one; non-linear.

(*f*) Order two; degree one; non-linear (because of the occurrence of the product of first and second order derivatives).

(*g*) Differentiating w.r.t. x, we get $dy/dx + d^2y/dx^2 = y/2$. Order two; degree one; linear.

There is no discipline in engineering and other branches of science where differential equations are not finding place. Problems dealing with the vibration of strings, drying of wood and baking of bread by the simultaneous heat and mass transfer, conduction of heat in solids, vibrations of electrical and mechanical systems, bending of beams and velocity of chemical reactions are a few of the many areas where differential equations are extensively used. It is therefore, imperative for an engineering student to know the differential equations in depth and in detail. However, an engineers' approach to the study of differential equations is different from that of a student of mathematics. An engineer is not much concerned with the mathematical rigor of the solution but views the differential equation as a tool by which he derives the solution to authenticate the practical results. The study of differential equations from an engineer's view point can thus be classified into the following stages.

(*a*) Setting up of the differential equation from the given physical problem.

(*b*) To obtain the solution and evaluate the arbitrary constants with the help of the initial and boundary conditions.

(*c*) To analyse and interpret the results from the mathematical solution.

7.2 FORMATION OF DIFFERENTIAL EQUATIONS

An ordinary differential equation is formed by the elimination of arbitrary constants which occur in the relationship between the variables.

Let y and x be the dependent and the independent variables respectively. The equation

$$f(x, y, c) = 0 \qquad ...(7.1)$$

containing one arbitrary constant c, represents a family of curves. The equation

$$\phi(x, y, c, d) = 0 \qquad ...(7.2)$$

containing two arbitrary constants c and d also represents a family of curves. We often say that it represents a two parameter family of curves.

To eliminate the arbitrary constant c in equation (7.1), we need two equations. One equation is given by (7.1) itself and the second equation is obtained by differentiating equation (7.1) with respect to x. On eliminating c from the two equations, we obtain an equation containing x, y *and* y' which is a first order differential equation.

Similarly, to eliminate the arbitrary constants c and d in equation (7.2), we need three equations. One equation is given by equation (7.2), and the remaining two equations are obtained by differentiating equation (7.2) with respect to x two times.

On eliminating c and d from these three equations, we obtain a second order differential equation.

EXAMPLE 7.2 Find the differential equation of the family of curves given by the equation $x^2 - y^2 + 2axy = 1$, where a is a parameter.

Solution Differentiating the equation of the curve w.r.t. x, we obtain

$$2x - 2y\frac{dy}{dx} + 2a\left(y + x\frac{dy}{dx}\right) = 0. \qquad ...(i)$$

Substituting for $2a$ from the original equation in (i), we get

$$2\left(x - y\frac{dy}{dx}\right) + \left(\frac{1 - x^2 + y^2}{xy}\right)\left(y + x\frac{dy}{dx}\right) = 0$$

or

$$(x^2 + y^2)\left(y - x\frac{dy}{dx}\right) + \left(x\frac{dy}{dx} + y\right) = 0. \qquad ...(ii)$$

Relation (ii) is the required differential equation.

EXAMPLE 7.3 Form the differential equation not involving A, B, C from the equation $2x^2 + 3y^2 + Ax + By + C = 0$.

Solution This equation has three arbitrary constants. Differentiating successively three times w.r.t. x, we obtain

$$4x + 6y\frac{dy}{dx} + A + B\frac{dy}{dx} = 0 \quad ...(i), \qquad 4 + 6y\frac{d^2y}{dx^2} + 6\left(\frac{dy}{dx}\right)^2 + B\frac{d^2y}{dx^2} = 0. \qquad ...(ii)$$

$$6y\frac{d^3y}{dx^3} + 6\left(\frac{dy}{dx}\right)\frac{d^2y}{dx^2} + 12\left(\frac{dy}{dx}\right)\left(\frac{d^2y}{dx^2}\right) + B\frac{d^3y}{dx^3} = 0. \qquad ...(iii)$$

Eliminating (B) from (ii) and (iii), the eliminant is given by

$$\frac{d^3y}{dx^3}\left\{2 + 3\left(\frac{dy}{dx}\right)^2\right\} = 9\frac{dy}{dx}\left(\frac{d^2y}{dx^2}\right)^2.$$

EXAMPLE 7.4 Form the differential equation satisfied by all circles with their centres on the line $y = 2x$.

Solution The family of circles is given by $(x - a)^2 + (y - 2a)^2 = r^2$, where a and r are arbitrary constants (parameters). Differentiating this twice, we obtain

$$(x - a) + (y - 2a)\,y' = 0 \quad ...(i), \qquad \text{and} \qquad 1 + (y - 2a)\,y'' + y'^2 = 0. \qquad ...(ii)$$

From the first equation, $a = \dfrac{x + yy'}{1 + 2y'}$.

Substituting this in the second equation, we get the differential equation as

$$1 + \left[y - \frac{2(x + yy')}{1 + 2y'} \right] y'' + y'^2 = 0. \quad \text{or} \quad (y - 2x)y'' + (1 + y'^2)(1 + 2y') = 0.$$

EXAMPLE 7.5 Find the differential equation of all ellipses with centres at origin.

Solution The two-parameter family of such ellipses is given by $\dfrac{x^2}{a^2} + \dfrac{y^2}{b^2} = 1$.

Differentiating twice, $\dfrac{x}{a^2} + \dfrac{yy'}{b^2} = 0; \quad \dfrac{1}{a^2} + \dfrac{yy'' + y'^2}{b^2} = 0.$

From both of these equations, we get $\dfrac{-b^2}{a^2} = \dfrac{yy'}{x} = yy'' + y'^2$.

Thus the differential equation is $xyy'' + xy'^2 - yy' = 0$.

7.3 SOLUTION OF A DIFFERENTIAL EQUATION

By the solution of the differential equation is meant the relation between the dependent and the independent variables not involving derivatives. We have seen in the examples of Art. 7.2 that when we eliminate arbitrary constants from a given relation, the order of the resulting differential equation is the same as the number of arbitrary constants. It is therefore, logical that the *solution(integral)* of the differential equation must contain, in general, the same number of arbitrary constants as the order of the equation. Such a solution is known as the *general solution* or the *complete solution* of differential equation. The general solution is also called the *complete integral* or *complete primitive* of the differential equation. For example $y = A \cos \alpha t + B \sin \alpha t$ is the complete solution of the differential equation

$\dfrac{d^2 y}{dt^2} + \alpha^2 y = 0$ because the number of arbitrary constants A and B corresponds exactly with the order of the equation. It is important to note that the general solution of a differential equation can be expressed in different (but equivalent) forms.

A *particular solution* is one which is obtained from the general solution by assigning particular values (definite values) to the arbitrary constants. Thus $y^2 = 4x$ is a particular solution of $dy/dx = 2/y$ as it can be derived from the general solution $y^2 = 4Ax$ by assuming that the solution curve passes through the point (1, 2).

A differential equation may sometimes have an additional solution that cannot be obtained from the general solution (by assigning specific values to its arbitrary constant) and is then called a **singular solution.** This is not of great engineering interest.

For example the differential equation $y'^2 + xy' - y = 0$ has the general solution $y = cx + c^2$, which may be verified by differentiation and substitution.

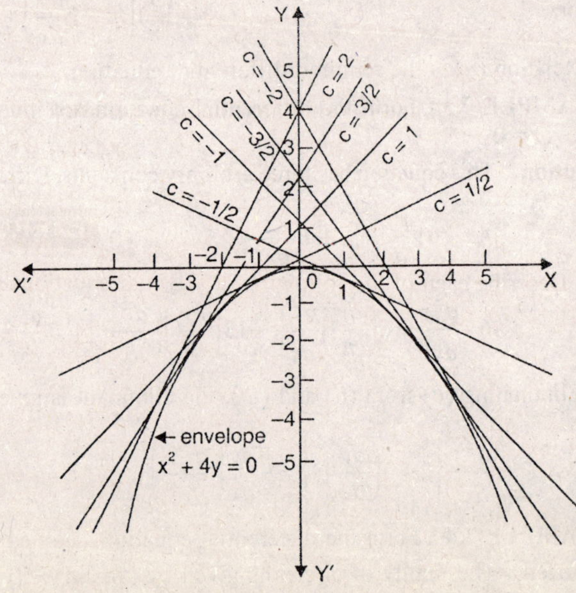

Fig. 7.1: Singular solution of $y'^2 + xy' - y = 0$.

This represents a family of straight lines, one straight line for each value of c. These are the particular solutions shown in Fig. 7.1. However, we find that $x^2 = -4y$, which is the equation of the parabola, is also a solution of the differential equation. This is a singular solution of above differential equation because we cannot find it from $y = cx + c^2$ by choosing a suitable value of c. It may be noted that this singular solution is the envelope of the family of the straight lines represented by the general solution.

A differential equation together with the condition is called an *initial value problem* (I.V.P.) and condition is called the *initial condition*. It may be noted that initial condition for example $y = \pi/4$ when $x = \sqrt{2}$ may be written as $y(\sqrt{2}) = \pi/4$. The given initial condition is used to determine a value of arbitrary constant in the general solution and hence to find a particular solution.

EXAMPLE 7.6 Verify that $y = Ae^{3x} + Bxe^{3x}$ is a general solution of the differential equation $y'' - 6y' + 9y = 0$.

Solution We have $y = Ae^{3x} + Bxe^{3x}$.

Therefore, $y' = 3Ae^{3x} + Be^{3x} + 3Bxe^{3x}$, $y'' = 9Ae^{3x} + 6Be^{3x} + 9Bxe^{3x}$.

Then $y'' - 6y' + 9y = (9Ae^{3x} + 6Be^{3x} + 9Bxe^{3x}) - 6(3Ae^{3x} + Be^{3x} + 3Bxe^{3x}) + 9(Ae^{3x} + Bxe^{3x}) = 0$,

and so that the given relation is a general solution since the order of the differential equation and the number of arbitrary constants in the solution are both equal to 2.

EXAMPLE 7.7 Show that $y = c_1e^x + c_2e^{-3x} + \sin x$ is a general solution of the differential equation $y'' + 2y' - 3y = 2\cos x - 4\sin x$. Also find the particular solution of the differential equation satisfying the conditions

$$y(0) = 2,\ y'(0) = -5.$$

Solution We have $y = c_1e^x + c_2e^{-3x} + \sin x$.

$\therefore$ $y' = c_1e^x + 3c_2e^{-3x} + \cos x$, $y'' = c_1e^x + 9c_2e^{-3x} - \sin x$.

Substitute y, y', y'' in the differential equation, we obtain

$$y'' + 2y' - 3y = (c_1e^x + 9c_2e^{-3x} - \sin x) + 2(c_1e^x - 3c_2e^{-3x} + \cos x) - 3(c_1e^x + c_2e^{-3x} + \sin x) = 2\cos x - 4\sin x$$

and so the given relation is a general solution since the order of the differential equation and the number of arbitrary constants in the solution are both equal to 2.

The general solution is $y = y(x) = c_1e^x + c_2e^{-3x} + \sin x$. ...(*i*)

At $x = 0$, $y(0) = c_1 + c_2 = 2$ that is, $c_1 + c_2 = 2$.

Differentiating (*i*) w.r.t. x, we obtain $y'(x) = c_1e^x + 3c_2e^{-3x} + \cos x$.

At $x = 0$, $y'(0) = c_1 - 3c_2 + 1 = -5$ that is, $c_1 - 3c_2 = -6$. ...(*ii*)

Solving (*i*) and (*ii*) simultaneously, we find $c_1 = 0$, $c_2 = 2$. The required particular solution is

$$y_p(x) = 2e^{3x} + \sin x.$$

EXERCISE 7.1

Describe each of the following differential equations giving its order and degree. State also whether they are linear or non-linear.

1. $y''' + 6y'' + 4y' + y = e^x$.
2. $y^2 + 4yy' + 2y = \cos x$.
3. $(y')^2 + 3xy' + 6y = 0$.
4. $L\dfrac{d^2Q}{dt^2} + R\dfrac{dQ}{dt} + \dfrac{Q}{C} = 0$.
5. $(y'')^2 + (y')^3 + 3y = x^2$.
6. $y''' + 2(y'')^2 + y' = \cos x$.
7. $(dy/dx)^2 + 5y^{1/3} = x$.
8. $(dy/dx)^{1/2} + y = x^{1/3}$.

Ans. 1. Three, one, linear. 2. Two, one, non-linear. 3. One, two; non-linear. 4. Two, one, linear. 5. Two, two, non-linear. 6. Three, one, non-linear. 7. One, two, non-linear. 8. One, one, non-linear.

Form the differential equations from the following equations:

9. $x^2 + y^2 + 2gx + 2fy + c = 0$, not containing f, g, c. **Ans.** $\left\{1+\left(\frac{dy}{dx}\right)^2\right\}\frac{d^3y}{dx^3} - 3\frac{dy}{dx}\left(\frac{d^2y}{dx^2}\right)^2 = 0$.

10. $y^2 = m(a^2 - x^2)$, not containing m and a. **Ans.** $x\{(dy/dx)^2 + y(d^2y/dx^2)\} - y(dy/dx) = 0$.

11. $x^2 + y^2 - 2ky = k^2$, not containing k. **Ans.** $(x^2 - 2y^2)(dy/dx)^2 - 4xy(dy/dx) - x^2 = 0$.

12. $x = A \sin(\omega t + \alpha)$, not containing A and α. **Ans.** $d^2x/dt^2 + \omega^2 x = 0$.

13. $(x - h)^2 + (y - k)^2 = a^2$, not containing h and k.
Alternative form: Obtain the differential equation of all circles of radius 'a' and centre at (h, k).
Ans. $(1 + y'^2)^3 = a^2 y''^2$.

14. $y = ae^{2x} + be^{-3x} + ce^x$, not containing a, b, and c. **Ans.** $(d^3y/dx^3) - 7(dy/dx) + 6y = 0$.

15. $xy = Ae^x + Be^{-x} + x^2$, not containing A and B. (*AMIE, S-2005*) **Ans.** $xy'' + 2y' + x^2 - xy - 2 = 0$.

16. $y = e^x(A \cos x + B \sin x)$, not containing A and B. (*PTU 2003*) **Ans.** $y'' - 2y' + 2y = 0$.

17. Find the differential equation of the family of circles of fixed radius r with centres on the x-axis.
(*AMIETE, June 2007*) **Ans.** $y^2 [1 + (y')^2] = r^2$.

18. Form the differential equation of all parabolas with principal axis along the x-axis. **Ans.** $yy'' + y'^2 = 0$.

19. Form the differential equation satisfied by all parabolas whose axes are parallel to x-axis.
[**Hint:** The three-parameter formily of the given parabolas is given by $(y - b)^2 = 4c(x - a)$, where a, b, c being arbitrary. Differentiating three times.] **Ans.** $y'y''' = 3y''$.

20. Find the differential equation of the family of parabolas with foci at the origin and axes along the x-axis.
Ans. $yy'^2 + 2xy' - y = 0$.

21. Find the differential equation of all straight lines at a unit distance from the origin.
Ans. $(xy' - y)^2 = 1 + (y')^2$.

22. Find the differential equation of the family of cardiods $r = a(1 - \cos\theta)$.
Ans. $(1 - \cos\theta)dr = r \sin\theta\, d\theta$.

23. Verify that $y = ce^{-x} + x^2 - 2x$ is solution of the differential equation $dy/dx + y = x^2 - 2$.

24. Show that $y = e^{-x}(a \cos x + b \sin x)$ is the solution of the differential equation $y'' + 2y' + 2y = 0$.

25. Show that $y = 2(x^2 - 1) + ce^{-x^2}$ is a solution of $y' + 2xy = 4x^3$.

26. Show that $Ax^2 + By^2 = 1$ is a solution of the differential equation $x(yy'' + y'^2) = yy'$.

27. Show that $y = a \cos(\log x) + b \sin(\log x)$ is a solution of the differential equation $x^2y'' + xy' + y = 0$.
Ans. $(1 + y'^2)^3 = a^2 y''^2$.

7.4 ANALYTICAL SOLUTION OF A DIFFERENTIAL EQUATION OF THE FIRST ORDER AND FIRST DEGREE

A differential equation of the first order and fist degree can be written as

$$f\left(x, y, \frac{dy}{dx}\right) = 0 \quad \text{or} \quad \frac{dy}{dx} = \phi(x, y) \quad \text{or} \quad M(x, y)dx + N(x, y)dy = 0 \qquad ...(7.3)$$

It may be remarked that it is not always possible to solve every first order differential equation in a closed form. We will discuss the solutions of a few types of equations which can be classified into any one of the following forms.

(*a*) Equations in the variables separable form.
(*b*) Equations which are homogeneous.
(*c*) Equations which are linear.
(*d*) Equations which are exact.

7.5 VARIABLES SEPARABLE FORM

In this case differential equation can be written in the form $M(x)\, dx + N(y)\, dy = 0$ that is x appears only in the co-efficient of dx and y appears only in the co-efficient of dy.

Integrating, we have the general solution.

$$\int M(x)dx+\int N(y)dy=c, \text{ a constant.}$$

EXAMPLE 7.8 Solve $\dfrac{dy}{dx}+\sqrt{\dfrac{1-y^2}{1-x^2}}=0.$

Solution The equation can be written in the form

$$\frac{dx}{\sqrt{1-x^2}}+\frac{dy}{\sqrt{1-y^2}}=0. \text{ Integrating, we get } \sin^{-1}x+\sin^{-1}y=c_1$$

or $\qquad \sin^{-1}\left\{x\sqrt{1-y^2}+y\sqrt{1-x^2}\right\}=c_1 \qquad \left[\because \sin^{-1}A+\sin^{-1}B=\sin^{-1}\left(A\sqrt{1-B^2}+B\sqrt{1-A^2}\right)\right]$

or $\qquad x\sqrt{1-y^2}+y\sqrt{1-x^2}=c,$ where $c=\sin c_1$ is an arbitrary constant.

$\therefore \qquad x\sqrt{1-y^2}+y\sqrt{1-x^2}=c,$ is the required solution.

EXAMPLE 7.9 Solve the initial value problem $x\dfrac{dy}{dx}+\cot y=0,\ y(\sqrt{2})=\dfrac{\pi}{4}.$

Solution The equation can be written as $\dfrac{\sin y}{\cos y}dy+\dfrac{dx}{x}=0.$

Integrating, we obtain $-\log|(\cos y)|+\log|x|=c_1$ or $\log\left(\dfrac{x}{\cos y}\right)=c_1$ or $\dfrac{x}{\cos y}=c,$ where $c=e^{c_1}$.

$\therefore\ x=c\cos y$ is the general solution of the given differential equation.

Using the given conditions $y=\pi/4$, when $x=\sqrt{2}$, we get c = 2.

The particular solution of initial value problem is $x=2\cos y$.

EXAMPLE 7.10 Solve the differential equation $3e^{2x}\sec^2 3y\,dy+2(e^{2x}-1)\tan 3y\,dx=0.$

Solution The given equation may be written as $\dfrac{3\sec^2 3y}{\tan 3y}dy+\dfrac{2(e^{2x}-1)}{e^{2x}}dx=0.$

Integrating, we obtain $\log|\tan 3y|+2\left(x+\dfrac{e^{-2x}}{2}\right)=c$ or $\log|\tan 3y|+2x+e^{-2x}=c$, the required solution.

7.5.1 Equations Reducible to Separable Form

The following form of the differential equation can be reduced to separable form by substitution.

Equation of the form dy/dx = f(ax + bx + c)

Substituting $ax+by+c=u$, we get $a+b\dfrac{dy}{dx}=\dfrac{du}{dx}$ or $\dfrac{dy}{dx}=\dfrac{1}{b}\left(\dfrac{du}{dx}-a\right).$

The above differential equation simplifies to

$$\frac{1}{b}\left(\frac{du}{dx}-a\right)=f(u) \text{ or } \frac{du}{dx}=a+bf(u) \text{ or } \frac{du}{a+bf(u)}=dx.$$

Integrating, we obtain $\displaystyle\int\frac{du}{a+bf(u)}=x+c.$

EXAMPLE 7.11 Solve the initial-value problem $\dfrac{dy}{dx}=(y-2x)^2-7,\ y(0)=0.$

Solution Let $u = y - 2x$, then $\frac{du}{dx} = \left(\frac{dy}{dx}\right) - 2$ so the given differential equation is transformed into

$$\frac{du}{dx} + 2 = u^2 - 7 \text{ or } \frac{du}{dx} = u^2 - 9 \text{ or } \frac{du}{(u+3)(u-3)} = dx.$$

Using partial fractions, $\frac{1}{6}\left[\frac{1}{u-3} - \frac{1}{u+3}\right]du = dx$. Integrating, we obtain

$$\frac{1}{6}\log\left|\frac{u-3}{u+3}\right| = x + c_1 \text{ or } \frac{u-3}{u+3} = e^{6x+6c_1} = ce^{6x} \text{ or } u = \frac{3(1+ce^{6x})}{1-ce^{6x}}$$

Resubstituting $u = y - 2x$, we get $y = 2x + \frac{3(1+ce^{6x})}{1-ce^{6x}}$.

Now applying the initial condition $y(0) = 0$, we get $c = -1$.

Therefore, the particular solution is $y = 2x + 3\{(1 - e^{6x})/(1 + e^{6x})\}$.

EXAMPLE 7.12 Solve $x^4 \frac{dy}{dx} + x^3 y + \operatorname{cosec}(xy) = 0$. *(NIT Kurukshetra, 2009, KUK, 2006)*

Solution The given differential equation can be written as $x^3\left(x\frac{dy}{dx} + y\right) = -\operatorname{cosec}(xy)$ ...(*i*)

Substituting $xy = u$ and $x(dy/dx) + y = du/dx$ in (*i*), we obtain

$$x^3(du/dx) = -\operatorname{cosec} u \quad \text{or} \quad x^{-3}dx = -\sin u\, du.$$

Integrating, we obtain $-\frac{1}{2x^2} = \cos u + c_1$ or $\cos xy + \frac{1}{2x^2} = c$, the required solution of the given equation.

EXERCISE 7.2

Find the general solution of the following differential equations.

1. $(1 + 2x)y' = 1 - x$. **Ans.** $2x + 4y = 3\log(2x + 1) + c$.
2. $dy/dx = (1 + y^2)/(1 + x^2)$. **Ans.** $y = (x + c)/(1 - cx)$.
3. $(x + 1)y' + 1 = 2e^{-y}$. *(Madras, 2000S)* **Ans.** $(x + 1)(2 - e^y) = c$.
4. $(1 - y)\, y' = 1 + y^2$. **Ans.** $x = \tan^{-1} y - \frac{1}{2}\log(1 + y^2) + c$.
5. $\sec^2 x \tan y\, dx + \sec^2 y \cdot \tan x\, dy = 0$. *(PTU 2003)* **Ans.** $\tan x \tan y = c$.
6. $(x^2 - yx^2)\, dy + (y^2 + xy^2)\, dx = 0$. **Ans.** $\log(x/y) - (y + x)/xy = c$.
7. $xdx + ye^{-x^2} dy = 0$. **Ans.** $e^{x^2} + y^2 = c$.
8. $\frac{y}{x} \cdot y' = \sqrt{1 + x^2 + y^2 + x^2y^2}$. *(VTU, 2006)* **Ans.** $\sqrt{1 + y^2} = \frac{2(1 + x^2)^{3/2}}{3} + c$.
9. $dy/dx = e^{x-y} + x^2 e^{-y}$. *(GGSIPU 2006; AMIE, S-2002)* **Ans.** $e^y = e^x + (x^3/3) + c$.
10. $y \log y\, dx + x \log x\, dy = 0$.. *(UPTU-2004)* **Ans.** $y = e^{c/\log x}$.
11. $(1 + e^x)yy' - e^x(1 + y) = 0$. **Ans.** $(1 + e^x)(1 + y) = ce^y$.
12. $(x + y)^2 y' = a^2$. *(AMIETE June 2010, Dec 2007; AMIE, S-2005)* **Ans.** $x + y = a\tan\left(\frac{y + c}{a}\right)$.
13. $y' = \sin(x + y) + \cos(x + y)$. *(VTU, 2005)* **Ans.** $\log\left|1 + \tan\left(\frac{x + y}{2}\right)\right| = x + c$.

14. $xy^3y' = 1 - x^2 + y^2 - x^2y^2$. **Ans.** $\log x - \frac{x^2}{2} - \frac{y^2}{2} + \frac{1}{2}\log(1+y^2) = c$.

15. $(x + y)^2 (xy' + y) = xy (1 + y')$. **Ans.** $\frac{1}{x+y} + \log xy = c$

16. $\frac{dy}{dx} + \frac{1+y^3}{xy^2(1+x^2)} = 0$. **Ans.** $\frac{x^6(1+y^3)^2}{(1+x^2)^3} = c$

17. $\log\left(\frac{dy}{dx}\right) = ax + by$. *(AMIETE; W-2002)* **Ans.** $\frac{e^{ax}}{a} + \frac{e^{-by}}{b} = c$.

18. $\cos (x + y)\, dy = dx$. *(PTU, 2003)* **Ans.** $y = c + \tan\left(\frac{x+y}{2}\right)$.

19. $x \cos x \cos y + (\sin y)\, dy/dx = 0$. *(PTU, 2002)* **Ans.** $x \sin x + \cos x + \log \cos y + c$.

20. $xy\,dy + \frac{1+y^2}{1+x^2}dx = 0$ **Ans.** $x^2(1 + y^2) = c(1 + x^2)$.

21. $xyy' = 1 + x + y + xy$. *(PTU, 2003)* **Ans.** $x - y + \log x (1 + y) = c$.

22. $(x + y + 1)y' = 1$. **Ans.** $x = ce^y - y - 2$.

23. $y - xy' = a (y^2 + y')$. *(AMIE, S-2007)* **Ans.** $cy = (x + a) (1 - ay)$.

24. $3e^x \tan y\, dx + (1 - e^x) \sec^2 y\, dy = 0$. *(VTU, 2004)* **Ans.** $\tan y = c(1-e^x)^3$

25. $x^4y' + x^3y = - \sec (xy)$. *(AMIETE, Dec. 2003)* **Ans.** $\sin (xy) = (1/2\, x^2) + c$.

26. $xy\frac{dy}{dx} = \frac{1+y^2}{1+x^2}(1+x+x^2)$. *(AMIE, W-2007)* **Ans.** $1 + y^2 = (cx^2)e^{2\tan^{-1}x}$.

Solve the following initial value problems.

27. $(x \log x)\, y' = 2y$, given $y(2) = (\log 2)^2$. **Ans.** $y = (\log x)^2$.

28. $y^2y' = \cos^2 x$, given that $y = 0$, when $x = 0$. **Ans.** $\frac{y^3}{3} = \frac{1}{2}\left(x + \frac{\sin 2x}{2}\right)$

29. $\cos x(e^{2y} - y)\, y' = e^y \sin 2x$, $y(0) = 0$. **Ans.** $e^y + ye^{-y} + e^{-y} = 4 - 2 \cos x$.

30. $\frac{dy}{dx} = \frac{x(2\log x+1)}{\sin y + y\cos y}$, given that $y = 0$, when $x = 1$. *(Raipur 2004; VTU, 2000)*

Ans. $y \sin y = x^2 \log x$.

31. $y - x\frac{dy}{dx} = a\left(y^2 + \frac{dy}{dx}\right)$, $y (0) = a$. *(AMIE, W-2010)* **Ans.** $(1 - a^2)y = (x + a)(1 - ay)$.

32. $\frac{dy}{dx} = (4x+y+1)^2$, $y(0) = 1$. *(AMIE, S-2009; S-2002)* **Ans.** $(4x + y + 1) = 2\tan\left[2x + \left(\frac{\pi}{4}\right)\right]$.

33. $\frac{dy}{dx} = \frac{x}{y} - \frac{x}{1+y}$, $y(0) = 2$. **Ans.** $2y^3 + 3y^2 = 3x^2 + 28$.

34. $3e^x \tan y\, dx + (1 + e^x) \sec^2 y\, dy = 0$, given $y (0) = \pi/4$. **Ans.** $\tan y(1 + e^x)^3 = 8$.

35. $e^x (\cos y\, dx - \sin y\, dy) = 0$, $y(0) = 0$. *(G.G.SIPU, 2006)* **Ans.** $y = \cos^{-1}(e^{-x})$.

36. Find the equation of the curve which satisfies the differential equation $(1 + y^2)\, dx - xy\, dy = 0$ and passes through the point $(1, \sqrt{3})$. Also name the curve. *(AMIE., S-2005)* **Ans.** $4x^2 - y^2 = 1$; Hyperbola.

7.6 HOMOGENEOUS EQUATIONS

If a function $f(x, y)$ possesses the property $f(\lambda x, \lambda y) = \lambda^n f(x,y)$ for some real number n, then $f(x, y)$ is said to be a **homogeneous function** of degree n. For example, $f(x, y) = x^4 + y^4$ is a homogeneous function of degree 4 since

$$f(\lambda x, \lambda y) = (\lambda x)^4 + (\lambda y)^4 = \lambda^4(x^4 + y^4) = \lambda^4 f(x, y),$$

whereas $f(x, y) = x^2 + y^2 + 4$ is seen not to be homogeneous.

A first order differential equation in differential form

$$M(x, y)dx + N(x, y)\,dy = 0 \qquad \text{...(7.4)}$$

is said to be **homogeneous** if both coefficients M and N are homogeneous functions of the *same* degree. In other words, (7.4) is homogeneous if

$$M(\lambda x, \lambda y) = \lambda^n M(x, y) \text{ and } N(\lambda x, \lambda y) = \lambda^n N(x, y).$$

If M and N are homogeneous functions of degree n, we can also write

$$M(x, y) = x^n M(1, v) \text{ and } N(x, y) = x^n N(1,v) \text{ where } v = y/x, \qquad \text{...(7.5)}$$

and $M(x, y) = y^n M(u, 1)$ and $N(x, y) = y^n N(u, 1)$ where $u = x/y$. ...(7.6)

Either of the substitution $y = vx$ or $x = uy$, where v and u are new dependent variables will reduce a homogeneous equation to a *separable* first-order differential equation.

As a consequence of (7.5) a homogeneous equation

$M(x, y)\,dx + N(x, y)\,dy = 0$ can be rewritten as

$$x^n M(1, v)\,dx + x^n N(1, v)\,dy = 0 \text{ or } M(1, v)\,dx + N(1, v)\,dy = 0 \qquad \text{...(7.7)}$$

where $v = y/x$ or $y = vx$. By substituting the differential $dy = v\,dx + x\,dv$ in (7.7) and gathering terms, we obtain a separable differential equation in the variables v and x.

$$M(1, v)\,dx + N(1, v)\,[v\,dx + x\,dv] = 0$$

or
$$[M(1, v) + vN(1, v)]\,dx + xN(1, v)\,dv = 0$$

or
$$\frac{dx}{x} + \frac{N(1,v)\,dv}{M(1,v) + vN(1,v)} = 0.$$

Integrating, we obtain

$$\int \frac{N(1,v)\,dv}{M(1,v) + vN(1,v)} + \log|x| = \log|c|.$$

After integrating, we replace v by y/x to obtain the general solution of the given differential equation. To solve the homogeneous equation

$M(u, 1)\,dx + N(u, 1)\,dy = 0$, we use the substitution

$x = uy$ and $dx = u\,dy + y\,du$.

EXAMPLE 7.13 Find the solution of the differential equation

$$x\frac{dy}{dx} = y(\log y - \log x + 1). \qquad \textit{(AMIETE, June 2001)}$$

Solution The given differential equation $\dfrac{dy}{dx} = \dfrac{y}{x}\left\{\log\left(\dfrac{y}{x}\right) + 1\right\}$ is homogeneous.

Substituting $y = vx$ and $\dfrac{dy}{dx} = v + x\left(\dfrac{dv}{dx}\right)$, we obtain $v + x\left(\dfrac{dv}{dx}\right) = v\log v + v$ or $x\left(\dfrac{dv}{dx}\right) = v\log v$.

Separating the variables, $\dfrac{dv}{v\log v} = \dfrac{dx}{x}$. Integrating, we obtain

$$\log|\log v| = \log|x| + \log|c| \text{ or } \log v = cx.$$

Replacing v by y/x, we obtain $\log(y/x) = cx$ or $y = xe^{cx}$.

EXAMPLE 7.14 Find the solution of differential equation $x(y - x)\, dy = y(x + y)\, dx$.

Solution The given differential equation $\dfrac{dy}{dx} = \dfrac{(xy + y^2)}{(xy - x^2)}$, is homogeneous.

Substituting $y = vx$ and $\dfrac{dy}{dx} = v + x\left(\dfrac{dv}{dx}\right)$, we obtain $v + x\dfrac{dv}{dx} = \dfrac{v + v^2}{v - 1}$ or $x\dfrac{dv}{dx} = \dfrac{2v}{v-1}$.

Separating the variables, we get $\dfrac{v-1}{2v}dv = \dfrac{dx}{x}$ or $\left(\dfrac{1}{2} - \dfrac{1}{2v}\right)dv = \dfrac{dx}{x}$.

Integrating, we obtain $\dfrac{1}{2}v - \dfrac{1}{2}\log|v| = \log|x| + \log|c_1|$,

or $v - \log|v| = 2\log|x| + 2\log(c_1)$ or $v = \log vx^2c$, where $c = c_1^2$.

Replacing v by y/x, we get $y = x(\log xyc)$ or $y/x = \log xyc$.

EXAMPLE 7.15 Solve the initial value problem $(3xy + y^2)\, dx + (x^2 + xy)\, dy = 0$, $y(1) = 1$.

Solution The given differential equation $\dfrac{dy}{dx} = -\dfrac{3xy + y^2}{x^2 + xy}$ is homogeneous.

Substituting $y = vx$ and $\dfrac{dy}{dx} = v + x\dfrac{dv}{dx}$, we obtain

$$v + x\frac{dv}{dx} = -\frac{3v + v^2}{1 + v} \text{ or } x\frac{dv}{dx} = -\frac{3v + v^2}{1 + v} - v = -\frac{2v(v+2)}{v+1}.$$

Separating the variables, we get

$$\frac{(v+1)}{v(v+2)}dv = -2\frac{dx}{x} \text{ or } \frac{1}{2}\left[\frac{1}{v} + \frac{1}{v+2}\right]dv = -\frac{2}{x}dx.$$

Integrating, we obtain

$$\frac{1}{2}[\log|v| + \log|v+2|] = -2\log|x| + \log|c_1|$$

or $$\log|v(v+2)| + 4\log|x| = 2\log|c_1|$$

or $v(v+2)x^4 = c$, where $c = c_1^2$. Replacing, v by y/x, we get $x^2y(2x + y) = c$.

Using the initial condition $y(1) = 1$, we obtain $c = 3$.

The particular solution is $x^2y(2x + y) = 3$.

EXAMPLE 7.16 Solve the differential equation $(1 + 3e^{x/y})\, dx + 3e^{x/y}\{1 - (x/y)\}dy = 0$.

Solution The given equation is homogeneous. The appearance of x/y throughout the equation suggests the use of the transformation $x = uy$, $\dfrac{dx}{dy} = u + y\dfrac{du}{dy}$ to obtain $(1 + 3e^u)\left[u + y\left(\dfrac{du}{dy}\right)\right] + 3e^u(1 - u) = 0$

or $$u + y\frac{du}{dy} = \frac{3e^u(u-1)}{1+3e^u} \text{ or } y\frac{du}{dy} = \frac{-3e^u - u}{1 + 3e^u}.$$

Separating the variables, we get

$\dfrac{1+3e^u}{u+3e^u}du = -\dfrac{dy}{y}$. Integrating, we obtain $\log|u + 3e^u| = -\log|y| + \log|c|$

or $u + 3e^u = (c/y)$. Replacing u by (x/y) we obtain

$$(x/y) + 3e^{x/y} = (c/y) \quad \text{or } x + 3ye^{x/y} = c.$$

EXAMPLE 7.17 Find the solution of differential equation $xy \log(x/y)\,dx + \{y^2 - x^2 \log(x/y)\}\,dy = 0$, given that $y(1) = e$. *(AMIETE, June 2001)*

Solution The given equation can be rewritten as

$$-xy \log(y/x)\,dx + \{y^2 + x^2 \log(y/x)\}\,dy = 0,$$

or $$\frac{dy}{dx} = \frac{xy \log(y/x)}{y^2 + x^2 \log(y/x)}, \text{ which is homogeneous.}$$

Using the transformation $y = vx$ and $\dfrac{dy}{dx} = v + x\dfrac{dv}{dx}$ to obtain

$v + x\dfrac{dv}{dx} = \dfrac{v \log v}{v^2 + \log v}$ or $x\dfrac{dv}{dx} = \dfrac{-v^3}{v^2 + \log v}$. Separating the variables, $\dfrac{dx}{x} + \dfrac{(v^2 + \log v)}{v^3}dv = 0.$

Integrating and replacing v by y/x, we obtain

$$\log|x| + \log|v| - \left(\frac{\log v}{2} + \frac{1}{4}\right)\frac{1}{v^2} = c \quad \text{or} \quad \log y - \left[\frac{1}{2}\log\left(\frac{y}{x}\right) + \frac{1}{4}\right]\left(\frac{x^2}{y^2}\right) = c$$

Using the initial condition $y(1) = e$, we obtain $c = (4e^2 - 3)/4e^2$.

The particular solution is $\log y - \dfrac{x^2}{4y^2}\left\{2\log\left(\dfrac{y}{x}\right) + 1\right\} = \dfrac{4e^2 - 3}{4e^2}.$

EXAMPLE 7.18 Solve the differential equation $\{x \tan(y/x) - y \sec^2(y/x)\}\,dx + \{x \sec^2(y/x)\}\,dy = 0$ *(VTU, 2006; AMIE, W-2005; PTU-2003)*

Solution The given equation is homogeneous and can be rewritten as

$$\{\tan(y/x) - (y/x)\sec^2(y/x)\} + \{\sec^2(y/x)\}\frac{dy}{dx} = 0.$$

Using the transformation $y = vx$ and $\dfrac{dy}{dx} = v + x\left(\dfrac{dv}{dx}\right)$ to obtain

$$\tan v - v\sec^2 v + \sec^2 v\left[v + x\left(\frac{dv}{dx}\right)\right] = 0 \quad \text{or} \quad \tan v + x\sec^2 v\left(\frac{dv}{dx}\right) = 0.$$

Separating the variables, we obtain $\dfrac{dx}{x} + \dfrac{\sec^2 v}{\tan v}dv = 0.$

Integrating, we obtain $\log|x| + \log|\tan v| = \log|c|$ or $x \tan v = c$.

Replacing v by y/x, we get $x \tan(y/x) = c$, the required solution.

EXERCISE 7.3

Find the general solution of the following differential equations.

1. $(x + y)\,dx + (y - x)\,dy = 0$. *(PTU 2004)* **Ans.** $\log(x^2 + y^2) = 2\tan^{-1}(y/x) + c.$
2. $(2x + y)\,dx + (2y + x)\,dy = 0$. *(AMIETE, 2002)* **Ans.** $x^2 + xy + y^2 = c.$
3. $(x - y)\,dy - (2x - y)\,dx = 0$. **Ans.** $2x^2 - 2xy + y^2 = c.$
4. $y^2 - (xy + x^2)\,y' = 0$. **Ans.** $cx - y = x \log y.$
5. $xy' + (y^2/x) = y$. **Ans.** $cx = e^{x/y}.$
6. $xy^2y' - y^3 = x^3$. **Ans.** $\log x = (y^3/3x^3) + c.$
7. $(y^2 + 2xy)dx + (2x^2 + 3xy)\,dy = 0$. **Ans.** $xy^2(x + y) = c.$
8. $2x^2y' = y(x + y)$. **Ans.** $(x - y)^2 = cxy^2.$
9. $x(x - y)y' = y(x + y)$. *(GGSIPU 2006)* **Ans.** $(x/y) + \log(xy) = c.$
10. $y^2 + x^2y' = xyy'$. **Ans.** $y = x \log(cy).$

11. $(1 + e^{x/y})\,dx + e^{x/y}\{1 - (x/y)\}\,dy = 0$. (*AMIETE, June 2009, PTU, 2006; VTU, 2003*) **Ans.** $x + ye^{x/y} = c$.

12. $\{x\cos(y/x) + y\sin(y/x)\}ydx + \{x\cos(y/x) - y\sin(y/x)\}xdy = 0$.

Alternative form: $x\cos(y/x)(ydx + xdy) = y\sin(y/x)(xdy - ydx)$ **Ans.** $xy\cos(y/x) = c$.

[**Hint:** The given equation may be rewritten as $\dfrac{dy}{dx} = \dfrac{y(y\sin(y/x) + x\cos(y/x))}{x\{y\sin(y/x) - x\cos(y/x)\}}$].

13. $x\sin(y/x)\{ydx + xdy\} + \cos(y/x)\{xdy - ydx\} = 0$. **Ans.** $xy\sin(y/x) = c$.

14. $(x^2 - y^2)dx - xy\,dy = 0$. (*AMIETE, S-2010*) **Ans.** $x^2(x^2 - 2y^2) = c$.

15. $x^3y' = y^3 + y^2\sqrt{y^2 - x^2}$. **Ans.** $xy = c\left(y + \sqrt{y^2 - x^2}\right)$.

16. $xdy - ydx = \sqrt{x^2 + y^2}\,dx$. (*PTU 2003*) **Ans.** $y + \sqrt{x^2 + y^2} = cx^2$.

17. $\dfrac{y}{x}\cos\dfrac{y}{x}dx - \left(\dfrac{x}{y}\sin\dfrac{y}{x} + \cos\dfrac{y}{x}\right)dy = 0$. **Ans.** $y\sin\left(\dfrac{y}{x}\right) = c$.

18. $\{x\tan(y/x) - y\sec^2(y/x)\}dx + x\sec^2(y/x)dy = 0$. (*PTU 2003*) **Ans.** $x\tan(y/x) = c$.

19. $2xy\,dy = (x^2 + y^2)\,dx$. (*AMIE, W-2010*) **Ans.** $x^2 - y^2 = cx$.

20. $xy' - y = x\sin(y/x)$. (*VTU, 2000S*) **Ans.** $y = 2x\tan^{-1}(cx)$.

21. $(y^3 - 3xy^2)\,dx + (2x^2y - xy^2)\,dy = 0$. (*AMIETE, 2001*) **Ans.** $x^3e^{y/x} = cy^2$.

22. $(2x - y)e^{y/x}dx + (y + xe^{y/x})dy = 0$. **Ans.** $y^2 + 2x^2e^{y/x} = c$.

23. $(2xy + x^2)y' = 3y^2 + 2xy$. (*AMIETE, Dec. 2006*) **Ans.** $y(x + y) = cx^3$.

Solve the following initial value problems.

24. $(x^2 + 2y^2)\,dx - xy\,dy = 0$, $y(1) = 0$. **Ans.** $y^2 = x^4 - x^2$.

25. $(x^3 - 3y^3)\,dx + 3xy^2\,dy = 0$, $y(1) = 2$. **Ans.** $y^3 = x^3(8 - \log|x|)$.

26. $(x + ye^{y/x})\,dx - xe^{y/x}\,dy = 0$, $y(1) = 0$. **Ans.** $\log|x| = e^{y/x} - 1$.

27. $(3xy + y^2)\,dx + (x^2 + xy)\,dy = 0$, $y(1) = 1$. **Ans.** $x^2y(2x + y) = 3$.

28. $x^2dy + y(x + y)\,dx = 0$ given that $y = 1$ when $x = 1$. (*UPTU 2006*) **Ans.** $x^2y = (2x + y)/3$.

29. $(x^2 - y^2)\,dx + 2xy\,dy = 0$, $y(1) = 1$. **Ans.** $x^2 + y^2 = 2x$.

7.7 EQUATIONS OF THE FORM

$$dy/dx = (a_1x + b_1y + c_1)/(a_2x + b_2y + c_2)$$

The differential equation

$$\frac{dy}{dx} = \frac{a_1x + b_1y + c_1}{a_2x + b_2y + c_2}, \quad \text{...(7.8)}$$ is not homogeneous.

Case I. If $\dfrac{a_1}{a_2} = \dfrac{b_1}{b_2} = t$, then equation (7.8) can be written as

$$\frac{dy}{dx} = \frac{t(a_2x + b_2y) + c_1}{a_2x + b_2y + c_2}. \quad \text{...(7.9)}$$

Substituting $a_2x + b_2y = v$, we obtain $a_2 + b_2y' = v'$. Then Eq. (7.9) becomes

$$\frac{dy}{dx} = \frac{1}{b_2}(v' - a_2) = \frac{tv + c_1}{v + c_2} \quad \text{or} \quad v' = a_2 + \frac{b_2(tv + c_1)}{v + c_2} = \frac{a_2(v + c_2) + b_2(tv + c_1)}{v + c_2}$$

or $$\frac{(v + c_2)\,dv}{a_2(v + c_2) + b_2(tv + c_1)} = dx,$$ which is in separable form.

Integrating and replacing v by $a_2x + b_2y$, we obtain the general solution.

Case II. If $\frac{a_1}{a_2} \neq \frac{b_1}{b_2}$, then we substitute $x = X + h$, $y = Y + k$ in the differential equation to get

$$\frac{dy}{dx} = \frac{dY}{dX} = \frac{a_1(X+h)+b_1(Y+k)+c_1}{a_2(X+h)+b_2(Y+k)+c_2} = \frac{a_1X+b_1Y+(a_1h+b_1k+c_1)}{a_2X+b_2Y+(a_2h+b_2k+c_2)}. \quad ...(7.10)$$

Choose h and k such that $a_1h + b_1k + c_1 = 0$ and $a_2h + b_2k + c_2 = 0$. Then, equation (7.10) simplifies to

$$\frac{dY}{dX} = \frac{a_1X+b_1Y}{a_2X+b_2Y},$$

which is a homogeneous equation in the variables, X, Y. We solve this equation and substitute $X = x - h$ and $Y = y - k$ to obtain the general solution.

EXAMPLE 7.19 Find the solution of the differential equation $(4x + 2y + 5)\, dy - (2x + y - 1)\, dx = 0$.

Solution We have $\frac{dy}{dx} = \frac{2x+y-1}{2(2x+y)+5}$.

Substituting $2x + y = v$, we get $2 + y' = v'$. Hence $y' = v' - 2 = \frac{v-1}{2v+5}$ or $v' = 2 + \frac{v-1}{2v+5} = \frac{5v+9}{2v+5}$.

Separating the variables, we obtain $\frac{(2v+5)}{5v+9}dv = dx$ or $\frac{2}{5}\left[1+\frac{7/10}{v+(9/5)}\right]dv = dx$.

Integrating, we obtain $\frac{2}{5}(v) + \frac{7}{25}\log\left|v+\frac{9}{5}\right| = x + c$.

Replacing v by $(2x + y)$, we obtain $\frac{2}{5}(2x+y) + \frac{7}{25}\log\left|2x+y+\frac{9}{5}\right| - x = c$.

EXAMPLE 7.20 Find the solution of the differential equation

$$(x - 2y - 3)\, dy + (x - y - 2)\, dx = 0.$$

(Rajasthan, 2006)

Solution The given differential equation is $\frac{dy}{dx} = \frac{y-x+2}{x-2y-3}$.

To convert it into a homogeneous equation, make the substitution $x = X + h$, $y = Y + k$. Then

$$\frac{dY}{dX} = \frac{Y-X+(k-h+2)}{X-2Y+(h-2k-3)}.$$

Choose h, k such that $k - h + 2 = 0$, $h - 2k - 3 = 0$. Solving, we obtain $h = 1$ and $k = -1$.

For this choice of h, k, we get $\frac{dY}{dX} = \frac{Y-X}{X-2Y}$

Substituting $Y = vX$, we obtain

$$\frac{dY}{dX} = v + X\frac{dv}{dX} = \frac{vX-X}{X-2vX} = \frac{v-1}{1-2v} \quad \text{or} \quad X\frac{dv}{dX} = \frac{v-1}{1-2v} - v = \frac{2v^2-1}{1-2v}.$$

Separating the variables, we get $\frac{dX}{X} = \frac{2v-1}{1-2v^2}dv = \frac{(-1/2)(-4v)-1}{1-2v^2}dv$.

Integrating, we obtain $\log|X| = -\frac{1}{2}\log|1-2v^2| - \int\frac{dv}{1-2v^2} + \log|c_1|$

$$= -\frac{1}{2}\log|1-2v^2| - \frac{1}{2\sqrt{2}}\log\left|\frac{1+v\sqrt{2}}{1-v\sqrt{2}}\right| + \log|c_1|,$$

or $$X^2(1-2v^2)\left(\frac{1+v\sqrt{2}}{1-v\sqrt{2}}\right)^{1/\sqrt{2}} = c, \text{ where } c = c_1^2.$$

Substituting $v = \frac{Y}{X} = \frac{y-k}{x-h} = \frac{y+1}{x-1}$ and simplifying, we get the solution as

$$(x-1)^2\left[1-2\left(\frac{y+1}{x-1}\right)^2\right]\left[\frac{1+\{(y+1)/(x-1)\}\sqrt{2}}{1-\{(y+1)/(x-1)\}\sqrt{2}}\right]^{1/2} = c$$

or $$(x-1)^2 - 2(y+1)^2 = c\left(\frac{x-y\sqrt{2}-\sqrt{2}-1}{x+y\sqrt{2}+\sqrt{2}-1}\right)^{1/\sqrt{2}}.$$

EXAMPLE 7.21 Solve $\frac{dy}{dx} = \frac{2y-x-4}{y-3x+3}$. *(AMIETE, Dec-2010)*

Solution We have $\frac{dy}{dx} = \frac{2y-x-4}{y-3x+3}$. To convert it into a homogeneous equation, make the substitution $x = x + h$, $y = y + k$. Then

$$\frac{dy}{dx} = \frac{2y-x+(2k-h-4)}{y-3x+(k-3h+3)}.$$

Choose h, k such that $2k - h - 4 = 0$, $k - 3h + 3 = 0$. Solving, we obtain $h = 2$ and $h = 3$.

For this choice of h, k, we get $$\frac{dY}{dX} = \frac{2Y - X}{Y - 3X}.$$

Substituting $Y = vX$, we obtain $$\frac{dY}{dX} = v + X\frac{dv}{dX} = \frac{2vX - X}{vX - 3X} = \frac{2v-1}{v-3}$$

or $$X\frac{dv}{dX} = \frac{2v-1}{v-3} - v = \frac{-1+5v-v^2}{v-3}.$$

Separating the variables, we get $$\frac{dX}{X} + \frac{v-3}{v^2-5v+1}dv = 0.$$

Integrating, we obtain $$\log X + \int\frac{v-3}{v^2-5v+1}dv = c_1. \qquad ...(i)$$

Now $$\int\frac{v-3}{v^2-5v+1}dv = \frac{1}{2}\int\frac{(2v-5)-1}{v^2-5v+1}dv = \frac{1}{2}\log(v^2-5v+1) - \frac{1}{2}\int\frac{dv}{v^2-5v+1}$$

$$= \frac{1}{2}\log(v^2-5v+1) - \frac{1}{2}\int\frac{dv}{\left(v-\frac{5}{2}\right)^2 - \left(\frac{\sqrt{21}}{2}\right)^2}$$

$$= \frac{1}{2}\log(v^2-5v+1) - \frac{1}{2}\cdot\frac{1}{2\cdot(\sqrt{21}/2)}\log\frac{\left(v-\frac{5}{2}\right)-\frac{\sqrt{21}}{2}}{v-\frac{5}{2}+\frac{\sqrt{21}}{2}}.$$

(*i*) become $\log X + \frac{1}{2}\log(v^2 - 5v + 1) - \frac{1}{2\sqrt{21}}\log\frac{2v-(5+\sqrt{21})}{2v-(5-\sqrt{21})} = c_1.$

Substituting $v = Y/X$, we get

$$\log X + \frac{1}{2}\log\frac{Y^2 - 5YX + X^2}{X^2} - \frac{1}{2\sqrt{21}}\log\frac{2Y-(5+\sqrt{21})X}{2Y-(5-\sqrt{21})X}$$

or $$(Y^2 - 5YX + X^2)^{\sqrt{21}}\ \frac{2Y-(5-\sqrt{21})X}{2Y-(5+\sqrt{21})X} = c_1.$$

Now put $X = x - 2$ and $Y = y - 3$, the required solution is

$$\frac{(x^2 - 5xy + y^2 + 11x + 4y - 17)^{\sqrt{21}}\quad \{2(y-3)-(5-\sqrt{21})(x-2)\}}{2(y-3)-(5+\sqrt{21})(x-2)} = c_1.$$

EXERCISE 7.4

Find the general solution of the following differential equations.

1. $(2x + y + 1)dx + (4x + 2y - 1)\,dy = 0.$ **Ans.** $x + 2y + \log(2x + y - 1) = c.$
2. $(2x - y + 2)\,dx + (4x - 2y - 1)\,dy = 0.$ **Ans.** $x + 2y + \log(2x - y) = c.$
3. $(6x + 9y + 6)y' = (2x + 3y - 1).$ **Ans.** $2x + 3y + 1 = c\exp(x - 3y).$
4. $(2x + 4y + 3)y' = x + 2y + 1.$ **Ans.** $\log(4x + 8y + 5) = 4x - 8y + c.$
5. $(6x - 4y + 1)\,dy - (3x - 2y + 1)\,dx.$ (*AMIETE., W-2006*) **Ans.** $x - 2y + \frac{1}{4}\log\left|3x - 2y + \frac{1}{4}\right| = c.$
6. $(2x + 3y + 4)dx - (4x + 6y + 5)\,dy = 0.$ (*Madras 2000S*) **Ans.** $7x - 14y + 3\log(14x + 21y + 22) = c.$
7. $(3x - 6y + 2)dx - (x - 2y + 1)\,dy = 0.$ **Ans.** $\frac{1}{5}(x - 2y) + \frac{2}{25}\log\left|x - 2y + \frac{3}{5}\right| + x = c.$
8. $y' + \frac{2x + 3y}{y + 2} = 0.$ **Ans.** $(2x + y - 4)^2 = c(x + y - 1).$
9. $(3x - y - 2)\,y' = x + 3y - 4.$ **Ans.** $\log[(x-1)^2 + (y-1)^2] = 6\tan^{-1}\left(\frac{y-1}{x-1}\right) + c.$
10. $y' = \frac{7x - 3y - 7}{7y - 3x + 3}.$ (*AMIETE, June 2004*) **Ans.** $(x + y - 1)^5\,(x - y - 1)^2 = c.$
11. $(x + 2y - 3)dx = (2x + y - 3)\,dy.$ (*NIT Kurukshetra, 2007; VTU, 2005*) **Ans.** $(x + y - 2) = c(x - y)^3.$
12. $y'\,(x + y - 2) = x - y + 1.$ **Ans.** $x^2 - 2xy - y^2 + 2x + 4y = c.$
13. $y' = (y + x - 2)/(y - x - 4).$ (*AMIETE, June 2011*) **Ans.** $x^2 + 2xy - y^2 - 4x + 8y = c.$
14. $(x + 2y)(dx - dy) = dx + dy.$ (*VTU, 2001*) **Ans.** $(3x + 6y - 1)^{2/3} = c_1 e^{x - y}.$
15. $(y - x + 1)\,dy - (y + x + 2)dx = 0.$

    (*AMIETE, June 2007; Jammu Univ., 2002*) **Ans.** $2\,(y - x + 1)^2 - (2x + 1)^2 = c_1.$

7.8 LINEAR FIRST ORDER DIFFERENTIAL EQUATIONS

Definition: *When the dependent variable y and its derivative dy/dx occur in the differential equation in the first degree and are not multiplied together, the equation is known as a linear differential equation of the first order.*

A typical linear equation of the first order is

$$\frac{dy}{dx} + yP(x) = Q(x), \qquad ...(7.11)$$

where $P(x)$ and $Q(x)$ are given continuous functions of x (or are constants). Eq. (7.11) is also sometimes called **Leibnitz linear equation.**

If $Q(x) = 0$, then the equation is easily solved, as it is in the separable form. In this case, the equation $y' + yP(x)\,y = 0$ is also called a *homogeneous first order equation.*

To find the solution of (7.11), we multiply both sides by $e^{\int P(x)dx}$ and obtain

$$\frac{dy}{dx}e^{\int P(x)dx} + yP(x)e^{\int P(x)dx} = Q(x)\cdot e^{\int P(x)dx} \quad \text{or} \quad \frac{d}{dx}(ye^{\int P(x)dx}) = Q(x)e^{\int P(x)dx},$$

$e^{\int P(x)dx}$ is an *integrating factor* (I.F.) of (7.11) and its solution is

$$y(x)e^{\int P(x)dx} = \int Q(x)\cdot e^{\int P(x)dx}dx + c.$$

Hence we write the solution of (7.11) as $\mathbf{y(x)\times I.F. = \int [Q(x)\times (I.F.)]dx + c}$ where **c** is a constant.

EXAMPLE 7.22 Solve the initial value problem $y' + y \tan x = \sin 2x$, $y(0) = 1$.

Solution An integrating factor is $\text{I.F.} = e^{\int \tan x dx} = e^{\log|\sec x|} = \sec x.$

The general solution of equation is $y\cdot \sec x = \int (\sin 2x)\sec x\, dx + c$

or $$y(x) = \cos x\left[2\int \sin x\, dx + c\right] = c\cos x - 2\cos^2 x.$$

Applying the initial condition, $y(0) = 1$, we get $1 = c - 2$ or $c = 3$.

Hence, the solution of IVP is $y = 3\cos x - 2\cos^2 x$.

EXAMPLE 7.23 Solve: $(1 + y^2)\, dx = (\tan^{-1} y - x)\, dy$. *(AMIE, W-2010; AMIETE., S-2010; UPTU, 2005)*

Solution This equation is linear if we take y as independent and x as dependent variable. Thus it is linear differential equation when written in the form

$$\frac{dx}{dy} + \frac{x}{1+y^2} = \frac{\tan^{-1} y}{1+y^2}.$$

An integrating factor is I.F. $= e^{\int dy/(1+y^2)} = e^{\tan^{-1} y}$

The general solution of equation is $x\, e^{\tan^{-1} y} = \int \frac{\tan^{-1} y(e^{\tan^{-1} y})}{1+y^2}dy + c.$

Putting $\tan^{-1} y = t$ and $\frac{dy}{1+y^2} = dt$, we get

$$\int\left(\frac{\tan^{-1} y}{1+y^2}\right)\cdot e^{\tan^{-1} y}dy = \int t e^t\, dt = t\, e^t - e^t = e^t(t-1) = (\tan^{-1} y - 1)e^{\tan^{-1} y}.$$

Hence the solution is

$$x\, e^{\tan^{-1} y} = (\tan^{-1} y - 1)e^{\tan^{-1} y} + c \quad \text{or} \quad x = \tan^{-1} y - 1 + c\, e^{-\tan^{-1} y}.$$

EXAMPLE 7.24 Solve $y \log y\, dx + (x - \log y)\, dy = 0.$ *(UPTU, 2000)*

Solution The equation, with x taken as dependent variable, may be put in the form $\frac{dx}{dy} + \frac{1}{y\log y}x = \frac{1}{y}.$

An integrating factor is $\text{I.F.} = e^{\int dy/(y\log y)} = e^{\log(\log y)} = \log y.$

The general solution of equation is

$$x \cdot \log y = \int \frac{1}{y} \cdot \log y \, dy + c_1 = \frac{1}{2} \log^2 y + c_1 \quad \text{or} \quad 2x \log y = (\log y)^2 + c.$$

EXAMPLE 7.25 Solve $2(1+x)y' - (1+2x)y = x^2\sqrt{1+x}$.

Solution The given equation is $\dfrac{dy}{dx} - \dfrac{(1+2x)}{2(1+x)} y = \dfrac{x^2}{2\sqrt{1+x}}$.

An integrating factor is $\text{I.F.} = e^{\int P\,dx} = e^{\int -\frac{(1+2x)}{2(1+x)}dx} = e^{\int \left\{-1+\frac{1}{2(1+x)}\right\}dx} = e^{\left\{-x+\frac{1}{2}\log(1+x)\right\}} = \sqrt{1+x} \cdot e^{-x}$

The general solution of the equation is

$$y \cdot \sqrt{1+x} \cdot e^{-x} = \int \frac{x^2}{2\sqrt{1+x}} \cdot \sqrt{1+x} \cdot e^{-x} dx + c = \frac{1}{2} \int x^2 e^{-x} dx + c$$

$$= \frac{1}{2}\left[x^2(-e^{-x}) - (2x)e^{-x} + 2(-e^{-x})\right] + c \quad \text{or} \quad y\sqrt{1+x} = ce^x - \frac{1}{2}(x^2 + 2x + 2).$$

EXERCISE 7.5

1. Find the general solution of linear differential equation of first order $(dx/dy) + Px = Q$ (where P and Q are constants or functions of y). *(AMIETE, June 2009)* **Ans.** $x \cdot e^{\int P dy} = \int Q \cdot e^{\int P dy} dy + c.$

Find the general solution of the following first order linear differential equations.

2. $xy' - 3y = x - 1$. **Ans.** $y = cx^3 - (x/2) + 1/3$.
3. $(1 + x^2)\, y' + 2xy = x^2$. *(AMIETE, Dec, 2010)* **Ans.** $y(1 + x^2) = (x^3/3) + c$.
4. $(x^2 + 3)\, y' + 4xy = x$. **Ans.** $y(x^2 + 3)^2 = (x^4/4) + (3x^2/2) + c$
5. $(x \log x)\dfrac{dy}{dx} + y = \log x^2$. **Ans.** $y \log x = (\log x)^2 + c$.
6. $(x + 2y^3)\, y' + y$. *(Raipur, 2004)* **Ans.** $x = y^3 + cy$.
   [**Hint:** Take x as the dependent variable. The equation can be written as $(dx/dy) - (x/y) = 2y^2$.].
7. $y' + (2y/x) = \sin x$. **Ans.** $x^2 y = (2 - x^2) \cos x + 2x \sin x + c$.
8. $e^{-y} \sec^2 y\, dy = dx + x dy$. *(AMIE, S-2002; AMIETE, Dec. 2002)* **Ans.** $xe^y = \tan y + c$.
9. $(1 - x^2)\dfrac{dy}{dx} - xy = 1$. **Ans.** $y\sqrt{(1 - x^2)} = \sin^{-1} x + c$.
10. $dr + (2r \cot \theta + \sin 2\theta)\, d\theta = 0$. *(JNTU, 2003)* **Ans.** $2r \sin^2\theta + \sin^4\theta = c$.
11. $x^2 y' = 3x^2 - 2xy + 1$. **Ans.** $y = x + x^{-1} + cx^{-2}$.
12. $\dfrac{dy}{dx} = \dfrac{y}{2y \log y + y - x}$. **Ans.** $x = y \log y + cy^{-1}$.
13. $\sqrt{1 - y^2}\, dx = (\sin^{-1} y - x)dy$. **Ans.** $x = \sin^{-1} y - 1 + ce^{-\sin^{-1} y}$.
14. $\dfrac{dy}{dx} = \dfrac{\log x - 3x^3 y}{x^4}$. **Ans.** $x^3 y = c + \dfrac{1}{2}(\log x)^2$.
15. $ye^y\, dx = (y^3 + 2xe^y)dy$. **Ans.** $xy^{-2} = c - e^{-y}$.
16. $(x + 1)y' - y = e^{3x}(x + 1)^2$. *(AMIETE, June 2011; AMIE, S-2005)* **Ans.** $y = \left(\dfrac{1}{3}e^{3x} + c\right)(x + 1)$.

17. $(1+y^2)dx+(x-e^{\tan^{-1}y})dy=0.$ **Ans.** $xe^{\tan^{-1}y}=\frac{1}{2}e^{2\tan^{-1}y}+c.$

18. $\cot 3x\frac{dy}{dx}-3y=\cos 3x+\sin 3x, 0<x<\pi/2.$ *(AMIE, S-2008; AMIETE; June 2007)*

Ans. $y \cos 3x = [6x - \cos 6x - \sin 6x]/12 + c.$

19. $y' \sec x = y + \sin x.$ *(AMIETE; W-2005)* **Ans.** $y = ce^{\sin x} - \sin x - 1.$

20. $y' + y \cot x = 2 \cos x.$ *(AMIETE; June 2003)* **Ans.** $y\sin x=c-\frac{1}{2}\cos 2x.$

21. $(1+y^2)dx+(x-e^{-\tan^{-1}y})dy=0.$ (*VTU*, 2006) **Ans.** $xe^{\tan^{-1}y}=\tan^{-1}y+c.$

Solve the following initial value problems:

22. $y'+\frac{2xy}{1+x^2}=\frac{1}{(1+x^2)^2},\quad y(1)=0.$ **Ans.** $y(1 + x^2) = \tan^{-1} x - (\pi/4).$

23. $(1-x^2)y'+2xy=x\sqrt{1-x^2}, y(0)=0.$ **Ans.** $y=\sqrt{1-x^2}+x^2-1.$

24. $x^2y' + 2xy - x + 1 = 0, y(1) = 0.$ **Ans.** $yx^2 = (x^2/2) - x + 0.5.$

25. $y' + y \cot x = 5 e^{\cos x}, y(\pi/2) = -4.$ **Ans.** $y \sin x + 5 e^{\cos x} = 1.$

26. $2\cos x\cdot\frac{dy}{dx}+4y\sin x=\sin 2x, y(\pi/3)=0.$ Show that the maximum value of $y = 1/8$. (*VTU*, 2003)

Ans. $y = -2 \cos^2 x + \cos x.$

27. $y' + y \cot x = 4 x \operatorname{cosec} x, y(\pi/2) = 0.$ **Ans.** $y \sin x = 2 x^2 - (\pi^2/2).$

28. $xy' - 3y = x - 1, y(1) = 0.$ **Ans.** $y = (x^3 - 3x + 2)/6.$

29. $dy/dx=y/(x+\sqrt{xy})$ *(NIT, Kurukshetra, 2010)* **Ans.** $\sqrt{x}=\sqrt{y}\,(\log\sqrt{y}+c).$

7.9 EQUATIONS REDUCIBLE TO LINEAR FORM

The equation $\frac{dy}{dx}+yP(x)=y^nQ(x), \quad (n\neq 0,1).$ ...(7.12)

is known as **Bernoulle's equation** after the swiss mathematician JAKOB BERNOULLI (1654-1705). This equation is non-linear, but it can be changed to a linear form by suitable substitution and the solution of the transformed equation is obtained by the above method.

To find the solution of (7.12), consider the following transformation

$$z(x) = [y(x)]^{1-n}. \quad ...(7.13)$$

We obtain $dz/dx=(1-n)y^{-n}(dy/dx)=(1-n)y^{-n}\left[y^nQ(x)-yP(x)\right]$

$=(1-n)\left[Q(x)-y^{1-n}P(x)\right]=(1-n)[Q(x)-zP(x)]$, using equation (7.13).

Hence, equation (7.12) reduces to the linear equation

$$\frac{dz}{dx}+(1-n)zP(x)=(1-n)Q(x). \quad ...(7.14)$$

When the solution of this linear equation is obtained and $z(x)$ is replaced by y^{1-n}, we get the general solution of the Bernoulli equation.

Remark An equation of the form $f'(y)\frac{dy}{dx}+f(y)P(x)=Q(x)$ is a linear equation of the first order. To solve this equation, we put $f(y) = z$ etc.

EXAMPLE 7.26 Solve the differential equation $y' - 4y - xy^{3/2} = 0.$

Solution Write the given equation as $\frac{dy}{dx}-4y=xy^{3/2}$, which is a Bernoulli equation with $n = 3/2$. Consider the transformation $z = y^{1-(3/2)} = y^{-1/2}$, we obtain

$$\frac{dz}{dx}=-\frac{1}{2}y^{-3/2}\frac{dy}{dx}=-\frac{1}{2}y^{-3/2}\left[4y+xy^{3/2}\right]=\frac{-2}{y^{1/2}}-\frac{1}{2}x \quad \text{or} \quad \frac{dz}{dx}+2z=-\frac{x}{2}.$$

An integrating factor of this equation is $\text{I.F.}=e^{\int 2\,dx}=e^{2x}$.

The general solution of the transformed equation is obtained as

$$z\cdot e^{2x}=-\frac{1}{2}\int xe^{2x}dx+c=-\frac{1}{2}\left(x\frac{e^{2x}}{2}-\frac{1}{4}e^{2x}\right)+c=\frac{1}{8}e^{2x}-\frac{1}{4}xe^{2x}+c$$

or $$y^{-1/2}e^{2x}=\frac{1}{8}e^{2x}-\frac{1}{4}xe^{2x}+c_1 \quad \text{or} \quad y=\frac{64e^{4x}}{(e^{2x}-2xe^{2x}+c)^2}.$$

EXAMPLE 7.27 Solve the differential equation $xy(x^2y^2+1)\frac{dy}{dx}=1$. *(AMIE, W-2005)*

Solution Write the given equation as $(dx/dy)-yx=x^3y^3$. ...(*i*)

Here the dependent variable is x and the independent variable is y. The equation (*i*) is Bernoulli equation with $n = 3$. Set $z = x^{1-3} = x^{-2}$, we obtain

$$\frac{dz}{dy}=-2x^{-3}\frac{dx}{dy}=-2x^{-3}(xy+x^3y^3)=-2x^{-2}y-2y^3 \quad \text{or} \quad \frac{dz}{dy}+2y\cdot z=-2y^3.$$

An integrating factor is $\text{I.F.}=e^{\int 2y\cdot dy}=e^{y^2}$.

The general solution of the transformed equation is obtained as

$$z\cdot e^{y^2}=\int(-2y^3)e^{y^2}dy+c=-\int te^t\,dt+c \qquad (\text{put } y^2=t \text{ and } 2ydy=dt)$$

$$=c-e^t(t-1)=c-e^{y^2}(y^2-1) \quad \text{or} \quad x^{-2}e^{y^2}=c-e^{y^2}(y^2-1)$$

or $$x^2=\frac{e^{y^2}}{c-e^{y^2}(y^2-1)}.$$

EXAMPLE 7.28 Solve $\frac{dz}{dx}+\left(\frac{z}{x}\right)\log z=\frac{z}{x}(\log z)^2$. *(AMIETE, June 2003)*

Solution The given differential equation is $\frac{1}{z}\frac{dz}{dx}+\frac{1}{x}\log z=\frac{1}{x}(\log z)^2$ ...(*i*)

Put $\log z = t$ so that $\frac{1}{z}\frac{dz}{dx}=\frac{dt}{dx}$. Therefore (*i*) become,

$$\frac{dt}{dx}+\frac{t}{x}=\frac{t^2}{x} \quad \text{or} \quad \frac{1}{t^2}\frac{dt}{dx}+\frac{1}{xt}=\frac{1}{x} \qquad ...(ii)$$

Now set $\frac{1}{t}=v$ so that $-\frac{1}{t^2}\frac{dt}{dx}=\frac{dv}{dx}$. The equation (*ii*) reduces to

$$-\frac{dv}{dx}+\frac{v}{x}=\frac{1}{x} \quad \text{or} \quad \frac{dv}{dx}-\frac{v}{x}=-\frac{1}{x}.$$ This is Leibnitz's linear equation in v.

An integrating factor of this equation is $\text{I.F.}=e^{\int -\frac{1}{x}dx}=e^{-\log x}=e^{\log x^{-1}}=\frac{1}{x}$.

The solution of the transformed equation is $v\cdot\frac{1}{x}=-\int\frac{1}{x}\cdot\frac{1}{x}dx+c=\frac{1}{x}+c.$

Replacing v by $1/\log z$, we obtain

$(x\log z)^{-1}=x^{-1}+c$ or $(\log z)^{-1}=1+cx$, the required solution.

EXAMPLE 7.29 Solve the differential equation $x\,dy-[y+xy^3(1+\sin x)]\,dx=0$.

Solution Write the given equation as

$\frac{dy}{dx}-\frac{y}{x}=y^3(1+\sin x)$, which is a Bernoulli equation with $n=3$.

Set $z=y^{1-3}=y^{-2}.$

Therefore, $\frac{dz}{dx}=-2y^{-3}\frac{dy}{dx}=-2y^{-3}\left[\frac{y}{x}+y^3(1+\sin x)\right]=\frac{-2y^{-2}}{x}-2(1+\sin x)$

or $\frac{dz}{dx}+\frac{2}{x}z=-2(1+\sin x)$. An integrating factor of this equation is

$$\text{I.F.}=e^{\int\frac{2}{x}dx}=e^{2\log x}=e^{\log x^2}=x^2.$$ The solution is

$$z\,x^2=-2\int(1+\sin x)x^2dx=-2\int(\underset{\text{I}}{x^2}+\underset{\text{II}}{x^2\sin x})\,dx$$

$$=-\frac{2x^3}{3}-2\left[x^2(-\cos x)-\int 2x(-\cos x)\,dx\right]=-\frac{2x^3}{3}+2x^2\cos x-4\int x\cos x\,dx$$

$$=-\frac{2x^3}{3}+2x^2\cos x-4\left[x\sin x-\int\sin x\,dx\right]=-\frac{2x^3}{3}+2x^2\cos x-4x\sin x-4\cos x+c$$

or $$\frac{x^2}{y^2}=-\frac{2}{3}x^3+2x^2\cos x-4x\sin x-4\cos x+c.$$

EXERCISE 7.6

Find the solution of the following equations.

1. $y'+\frac{y}{x}=\frac{y^2}{x^2}.$ **Ans.** $\frac{1}{y}=cx+\frac{1}{2x}.$

2. $y'+\frac{xy}{1-x^2}=xy^{1/2}.$ *(AMIE, S-2010, S-2005)* **Ans.** $y^{1/2}=c(1-x^2)^{1/4}-\{(1-x^2)/3\}$

3. $x(dy/dx)+y=x^3y^6.$ *(AMIETE, S-2010)* **Ans.** $1=(2.5+cx^2)\,x^3y^5.$

   [**Hint:** Dividing throughout by xy^6 and put $y^{-5}=z$]

4. $xy(1+xy^2)y'=1.$ *(AMIETE, W-2010; AMIE, S-2006)* **Ans.** $1/x=(2-y^2)+ce^{-y^2/2}$

   [**Hint:** Rewrite the given equation as $\frac{dx}{dy}-yx=y^3x^2$]

5. $3\frac{dy}{dx}+3\frac{y}{x}=2x^4y^4.$ **Ans.** $y^{-3}=-x^5+cx^3.$

6. $y'+y\tan x=y^3\cos x.$ **Ans.** $\cos^2 x=y^2\left(c-2\sin x+\frac{2}{3}\sin^3 x\right).$

7. $y'-\frac{\tan y}{1+x}=(1+x)e^x\sec y.$ *(NIT, Kurukshetra, 2010; Raipur 2004)* **Ans.** $\sin y=(1+x)(e^x+c).$

8. $r\sin\theta - \dfrac{dr}{d\theta}\cos\theta = r^2$. *(VTU, 2005)* **Ans.** $\dfrac{1}{r} = \sin\theta + c\cos\theta$.

9. $y' + x\sin 2y = x^3\cos^2 y$. *(PTU 2002)* **Ans.** $\tan y = \dfrac{1}{2}(x^2 - 1) + ce^{-x^2}$.

***Alternative form*:** $\sec^2 y\dfrac{dy}{dx} + 2x\tan y = x^3$.

10. $y^2 + x^2\dfrac{dy}{dx} = xy\dfrac{dy}{dx}$. **Ans.** $y/x = \log y + c$.

11. $(1 + x)y' + 1 = 2e^{-y}$. *(PTU 2001)* **Ans.** $(x + 1)e^y = 2x + c$.

12. $2xy' = y^2(y^3 + x^2)$. **Ans.** $x^2 = ce^{y^2/3} - y^3 - 3$.

13. $\tan y\dfrac{dy}{dx} + \tan x = \cos y\cos^2 x$. *(KUK, 2003)* **Ans.** $\sec y\sec x = \sin x + c$

14. $\tan x\cos y\,dy + \sin y\,dx + e^{\sin x}dx = 0$. **Ans.** $\sin x\sin y = c - e^{\sin x}$.

***Alternative form*:** $\dfrac{dy}{\cot x} + \dfrac{dx}{\cot y} + \dfrac{e^{\sin x}}{\cos y}dx = 0$.

15. $3y^2\dfrac{dy}{dx} + 2xy^3 = 4xe^{-x^2}$. **Ans.** $y^3e^{x^2} = 2x^2 + c$.

16. $\dfrac{dy}{dx} + \dfrac{y}{x}\log y = \dfrac{y}{x^2}(\log y)^2$. *(AMIETE, Dec 2003, 2001)* **Ans.** $\dfrac{1}{x\log y} = \dfrac{1}{2x^2} + c$.

[Hint: The given equation may be written as $\dfrac{1}{y(\log y)^2}y' + \dfrac{1}{x(\log y)} = \dfrac{1}{x^2}$. Put $\dfrac{1}{\log y} = z$.]

17. $xy\dfrac{dy}{dx} + (x^2 + x + y^2) = 0$. **Ans.** $6x^2y^2 + 3x^4 + 4x^3 = c$.

[Hint: The given equation may be written as $yy' + (y^2/x) = -(x + 1)$. Put $y^2 = z$]

18. $dx - xy(1 + xy^2)\,dy = 0$. *(AMIE W-2005)* **Ans.** $1 + x(y^2 - 2 + ce^{-y^2/2}) = 0$.

[Hint: The equation may be written as $\dfrac{1}{x^2}\dfrac{dx}{dy} - \dfrac{y}{x} = y^3$]

19. $y' + \sin y + x\cos y + x = 0$. **Ans.** $\tan\dfrac{y}{2} = 1 - x + ce^{-x}$.

[Hint: Write the equation in the form :

$$y' + 2\sin\frac{y}{2}\cos\frac{y}{2} + x\left(2\cos^2\frac{y}{2}\right) = 0 \Rightarrow y'\frac{\sec^2(y/2)}{2} + \tan\frac{y}{2} = -x. \text{ Put } \tan y/2 = t.]$$

20. $e^y(y' + 1) = e^x$. *(PTU 2002; VTU, 2000S)* **Ans.** $2e^y = e^x + ce^{-x}$.

21. $xy' + y = y^2\log x$. **Ans.** $1 = (1 + \log x)y - cxy$.

22. $(x - y^2)\,dx + 2xy\,dy = 0$. **Ans.** $y^2/x = c - \log x$.

23. $(xy^2 - e^{1/x^3})dx - x^2y\,dy = 0$. *(PTU-2001)* **Ans.** $3y^2 = 2x^2e^{1/x^3} + cx^2$.

24. $y'\tan y + \tan x = \cos y\cos^2 x$. **Ans.** $\sec y = (c + \sin x)\cos x$.

25. $ydx - xdy + 3x^2y^2e^{x^3}dx = 0$. **Ans.** $(x/y) + e^{x^3} = c$.

26. $3\dfrac{dy}{dx} + xy = xy^{-2}$. *(AMIETE, June 2009)* **Ans.** $y = \left[1 + ce^{-x^2/2}\right]^{1/3}$.

27. $(1-x^2)\dfrac{dy}{dx}+xy=xy^2.$ *(AMIE, S-2011)* **Ans.** $\sqrt{x^2-1}\left(\dfrac{1}{y}-1\right)=c.$

[**Hint:** Dividing each term by $y^2\,(1-x^2)$, we obtain

$\dfrac{1}{y^2}\dfrac{dy}{dx}+\dfrac{x}{1-x^2}\cdot\dfrac{1}{y}=\dfrac{x}{1-x^2}.$ ...(*i*) Setting $\dfrac{1}{y}=v$ or $-\dfrac{1}{y^2}\dfrac{dy}{dx}=\dfrac{dv}{dx}$ in (*i*),

we get $-\dfrac{dv}{dx}+\dfrac{x}{1-x^2}v=\dfrac{x}{1-x^2}$ or $\dfrac{dv}{dx}+\dfrac{x}{x^2-1}v=\dfrac{x}{x^2-1}.$ I.F $=\sqrt{x^2-1}.$]

7.10 EXACT DIFFERENTIAL EQUATIONS

The differential of a function of one or more variables is called an exact differential. For example $3x^2y\,dx + x^3\,dy$ is an exact differential, being the differential of x^3y.

Definition: *An equation obtained by equating an exact differential to zero, without any operation of reduction or elimination, is called an exact differential equation.*

Thus $$3x^2y\,dx + x^3\,dy = 0 \qquad ...(i)$$

is an exact differential equation because it has been obtained by equating an exact differential to zero while $3y\,dy + x\,dy = 0$ is not because it has been obtained from (*i*) by dividing both sides by x^2.

7.11 CONDITION FOR M (x, y) dx + N (x, y) dy = 0 TO BE EXACT

Necessary Condition: In case $M(x,y)dx + N(x,y)dy = 0$ is an exact differential equation then $M(x,y)dx + N(x,y)dy$ is an exact differential of some function u of x and y that is, $u(x, y)$.

Hence $$du = M(x,y)dx + N(x,y)dy. \qquad ...(7.15a)$$

The total or exact differential of $u(x, y)$ is expressed as

$$du = \frac{\partial u}{\partial x}dx+\frac{\partial u}{\partial y}dy \qquad ...(7.15b)$$

Comparing Eqs. (7.15a) and (7.15b), we get

$$\therefore \quad \frac{\partial u}{\partial x} = M(x,y) \text{ and } \frac{\partial u}{\partial y}=N(x,y) \qquad ...(7.16)$$

Now from Eq. (7.16), we obtain

$$\frac{\partial^2 u}{\partial y\partial x} = \frac{\partial M}{\partial y} \text{ and } \frac{\partial^2 u}{\partial x\partial y}=\frac{\partial N}{\partial x}.$$

Since $$\frac{\partial^2 u}{\partial y\partial x} = \frac{\partial^2 u}{\partial x\partial y}, \text{ we have } \frac{\partial \mathbf{M}}{\partial \mathbf{y}}=\frac{\partial \mathbf{N}}{\partial \mathbf{x}} \qquad ...(7.17)$$

which is, therefore, the *necessary condition* of exactness.

Sufficient Condition.

If $\dfrac{\partial M}{\partial y}=\dfrac{\partial N}{\partial x}$, we shall prove that

$$M(x, y)\,dx + N(x, y)\,dy = 0 \text{ is exact.}$$

Set $\int M(x,y)dx = u$, then $\dfrac{\partial u}{\partial x}=M(x,y)$ and hence $\dfrac{\partial^2 u}{\partial y\partial x}=\dfrac{\partial M}{\partial y}=\dfrac{\partial N}{\partial x}$ by equation (7.17),

that is, $\dfrac{\partial N}{\partial x}=\dfrac{\partial}{\partial x}\left(\dfrac{\partial u}{\partial y}\right).$

Integration, we get $N=\dfrac{\partial u}{\partial y}+\phi(y)$, where $\phi(y)$ is a function of y alone.

Hence, $M(x,y)\,dx + N(x,y)\,dy = \frac{\partial u}{\partial x}dx + \left[\frac{\partial u}{\partial y} + \phi(y)\right]dy$

$$= \left(\frac{\partial u}{\partial x}dx + \frac{\partial u}{\partial y}dy\right) + \phi(y)dy = du + \phi(y)dy = d\left[u + \int \phi(y)\,dy\right]$$

which shows that $M(x, y)dx + N(x, y)dy$ is an exact differential and hence $M(x, y)dx + N(x, y)dy = 0$ is an exact differential equation. By proving the sufficiency part of the condition, we have found out the method of solution of an exact equation.

For when $\frac{\partial M}{\partial y} = \frac{\partial N}{\partial x}$, $M(x,y)\,dx + N(x,y)\,dy = d\left[u + \int \phi(y)dy\right]$.

$\therefore$ $M(x,y)\,dx + N(x,y)\,dy = 0$ gives $d\left[u + \int \phi(y)\,dy\right] = 0$.

Integrating, $u + \int \phi(y)\,dy = c$ or $\int M(x,y)\,dx + \int \phi(y)\,dy = c$, ...(7.18)

which is the required solution.

Here $\int M(x,y)\,dx$ stands for integral of M w.r.t. x regarding y as a constant and $\phi(y)$ stands for those terms of N that donot involve x.

Method to solve an exact equation *Mdx + Ndy* = 0

(*i*) *Integrate M w.r.t. x, regarding y as a constant.*
(*ii*) *Integrate w.r.t. y those terms in N that don't contain x.*
(*iii*) *Equate the sum of integrals thus obtained to a constant which is the required solution.*

EXAMPLE 7.30 Check the equation $(1 + 6y^2 - 3x^2y)\,dy = (3xy^2 - x^3)dx$ for exactness. If it is exact, find the solution.

Solution The given equation may be written as $(3xy^2 - x^3)dx - (1 + 6y^2 - 3x^2y)dy = 0$.

We have $M(x, y) = 3xy^2 - x^3$, $N(x, y) = -(1 + 6y^2 - 3x^2y)$.

$$\frac{\partial M}{\partial y} = 6xy,\ \frac{\partial N}{\partial x} = 6xy.\ \text{Since } \frac{\partial M}{\partial y} = \frac{\partial N}{\partial x}.$$

Therefore the given equation is exact and its solution is given by

$$\int_{(y\text{ const.})} Mdx + \int (\text{terms in } N \text{ not containing } x)\,dy = c.$$

that is, $\int_{y\text{ const.}} (3xy^2 - x^3)\,dx - \int (1 + 6y^2)\,dy = c$ or $(3x^2y^2/2) - (x^4/4) - y - 2y^3 = c$.

EXAMPLE 7.31 Solve the equation $(2y \sin x - \cos y)dx + (x \sin y - 2 \cos x)\,dy = 0$.

Solution We have $M(x, y) = 2y \sin x - \cos y$, $N(x, y) = x \sin y - 2 \cos x$.

$$\partial M/\partial y = 2\sin x + \sin y, \quad \partial N/\partial x = \sin y + 2\sin x.$$

Since $\partial M/\partial y = \partial N/\partial x$, the given equation is exact. Therefore its solution is given by

$$\int_{(y\text{ const.})} (2y\sin x - \cos y)\,dx + \int (\text{terms in } N \text{ not containing } x)\,dy = c$$

that is, $-2y\cos x - x\cos y = \text{const.}$ or $2y\cos x + x\cos y = c$.

EXAMPLE 7.32 Solve the equation $(y^2e^{xy^2} + 4x^3)\,dx + (2xye^{xy^2} - 3y^2)\,dy = 0$. *(KUK, 2004)*

Solution We have $M(x, y) = y^2e^{xy^2} + 4x^3$, $N = 2xye^{xy^2} - 3y^2$.

$$\partial M/\partial y = 2ye^{xy^2} + y^2e^{xy^2}\cdot 2xy, \quad \partial N/\partial x = 2ye^{xy^2} + 2xye^{xy^2}\cdot y^2.$$

Since $\partial M/\partial y = \partial N/\partial x$, the given equation is exact. Therefore its solution is

$$\underset{(y\text{ const.})}{\int M\,dx} + \int(\text{terms in } N \text{ not containing } x)dy = c$$

that is, $\underset{(y\text{ const.})}{\int(y^2e^{xy^2} + 4x^3)dx} + \int(-3y^2)dy = c$ or $e^{xy^2} + x^4 - y^3 = c.$

EXAMPLE 7.33 Solve $\dfrac{dy}{dx} + \dfrac{y\cos + \sin y + y}{\sin x + x\cos y + x} = 0,\ y = 0$ when $x = 0$. *(AMIE, W-2010)*

Solution The given equation may be written as $(y\cos x + \sin y + y)dx + (\sin x + x\cos y + x)dy = 0$.

We have $M(x, y) = y\cos x + \sin y + y,\ N(x, y) = \sin x + x\cos y + x.$

$\partial M/\partial y = \cos x + \cos y + 1, \quad \partial N/\partial x = \cos x + \cos y + 1.$

Since $\partial M/\partial y = \partial N/\partial x$, the given equation is exact.

Therefore, its solution is given by

$$\underset{(y\text{ const.})}{\int(y\cos x + \sin y + y)\,dx} + \int(\text{term in } N \text{ not containing } x)dy = c.$$

That is, $y\sin x + x\sin y + xy = c$. Using the initial condition $y(0) = 0$, we get $c = 0$.

Hence, $y\sin x + x\sin y + xy = 0$ is the solution of initial value problem.

EXAMPLE 7.34 Determine for what values of a and b, the following differential equation is exact and obtain the general solution of the exact equation $(y + x^3)\,dx + (ax + by^3)dy = 0$.

Solution We have $M(x, y) = y + x^3, \quad N(x, y) = ax + by^3$.

$\partial M/\partial y = 1,\ \partial N/\partial x = a$. Hence if $a = 1$, the equation is exact that is, the equation is exact for $a = 1$, irrespective of the value of b. The solution is given by

$$\underset{(y\text{ const.})}{\int M\,dx} + \int(\text{terms in } N \text{ not containing } x)dy = c \quad \text{that is,} \quad \int(y + x^3)dx + b\int y^3dy = c$$

or $xy + (x^4/4) + (by^4/4) = c$ for all b and c is the arbitrary constant.

EXERCISE 7.7

Solve the following differential equations:

1. $(x^2 + y^2 - a^2)\,x\,dx + (x^2 - y^2 - b^2)\,y\,dy = 0.$ *(KUK, 2005; MDU 2003)* **Ans.** $x^4 + 2x^2y^2 - 2a^2x^2 - y^4 - 2b^2y^2 = c$
2. $\dfrac{dy}{dx} = \dfrac{y+1}{(y+2)e^y - x}.$ **Ans.** $(y + 1)(x - e^y) = c$
3. $(2x^2 + 6xy - y^2)dx + (3x^2 - 2xy + y^2)\,dy = 0.$ **Ans.** $2x^3 + 9x^2y - 3xy^2 + y^3 = c.$
4. $(x^2 - ay)dx = (ax - y^2)\,dy.$ *(Kurukshetra, 2004)* **Ans.** $x^3 + y^3 - 3axy = c.$
5. $(1 + e^{x/y})\,dx + e^{x/y}[1 - (x/y)]\,dy = 0.$ *(AMIE S-2005; VTU; 2001)* **Ans.** $x + ye^{x/y} = c.$
6. $\{y(1 + x^{-1}) + \cos y\}\,dx + (x + \log x - x\sin y)\,dy = 0.$ *(VTU 2006)* **Ans.** $y(x + \log x) + x\cos y = c.$
7. $(\sec x\tan x\tan y - e^x)\,dx + \sec x\sec^2y\,dy = 0.$ *(MDU 2003)* **Ans.** $\sec x\tan y - e^x = c.$
8. $y\sin 2x\,dx - (y^2 + \cos^2x)dy = 0.$ **Ans.** $3y\cos 2x + 2y^3 + 3y = c.$
9. $(5x^4 + 3x^2y^2 - 2xy^3)\,dx + (2x^3y - 3x^2y^2 - 5y^4)\,dy = 0.$ **Ans.** $x^5 + x^3y^2 - x^2y^3 - y^5 = c.$
10. $(x^2 - 4xy - 2y^2)\,dx + (y^2 - 4xy - 2x^2)\,dx = 0.$ **Ans.** $x^3 - 6x^2y - 6xy^2 + y^3 = c.$

11. $(2xy + y - \tan y)\, dx + (x^2 - x \tan^2 y + \sec^2 y)\, dy = 0$. (*Osmania 2000S*)
Ans. $x^2y + xy - x \tan y + \tan y = c$.

12. $(e^{2y} - y \cos xy)\, dx + (2xe^{2y} - x \cos xy + 2y)\, dy = 0$. **Ans.** $xe^{2y} - \sin xy + y^2 + c = 0$.

13. $(ye^{xy} - 2y^3)\, dx + (xe^{xy} - 6xy^2 - 2y)\, dy = 0$. **Ans.** $e^{xy} - 2xy^3 - y^2 = c$.

14. $\dfrac{dy}{dx} = \dfrac{xy^2 - \cos x \sin x}{y(1-x^2)}$, $y(0) = 2$. **Ans.** $y^2(1 - x^2) - \cos^2 x = 3$.

15. $(xe^{xy} + 2y)\, y' + ye^{xy} = 0$. **Ans.** $e^{xy} + y^2 = c$.

16. $\left(\dfrac{y}{x}\sec y - \tan y\right)dx + (\sec y \log x - x)\, dy = 0$. (*AMIETE, W-2005*) **Ans.** $y \log x - x \sin y = c$.

[**Hint:** We may rewrite the given equation as $\left(\dfrac{y}{x} - \sin y\right)dx + (\log x - x\cos y)\,dy = 0$.]

17. Find the value of λ for which the differential equation $(xy^2 + \lambda x^2 y)\, dx + (x + y)\, x^2 dy = 0$ is exact.
(*GGSIPU 2006*) **Ans.** $\lambda = 3$.

18. The differential equation $(x + x^2 + ay^2)\, dx + (y^3 - y + bxy)\, dy = 0$ is exact when $b = 2a$. Is this correct?
Ans. Yes.

7.12 INTEGRATING FACTOR FOUND BY INSPECTION

Many times when the equation $M(x, y)\, dx + N(x, y)\, dy = 0$ is not exact, it can be changed into an exact form by multiplying it by a suitable factor. Such a multiplier is called an *integrating factor* (I.F.) of the differential equation. For example equation $ydx - xdy + y^3\, dy = 0$ is not exact. However, if we divide it by y^2, we get

$$\frac{y\,dx - x\,dy}{y^2} + y\,dy = 0 \quad \text{or} \quad \frac{d}{dx}\left(\frac{x}{y}\right) + \frac{d}{dy}\left(\frac{y^2}{2}\right) = 0.$$

This is an exact equation and the solution is given by $\dfrac{x}{y} + \dfrac{y^2}{2} = \text{constant}$. In this case $1/y^2$ is called an integrating factor (I.F.).

Thus it may be possible to guess an integrating factor in the given equation by grouping properly the terms of the equation. A few combinations of most common occurence which will be of immense help in reducing a non-exact equation to an exact form are:

(*i*) $x\, dy + y\, dx = d(xy)$;

(*ii*) $\dfrac{xdy - ydx}{x^2} = d\left(\dfrac{y}{x}\right)$

(*iii*) $\dfrac{xdy - ydx}{y^2} = -d\left(\dfrac{x}{y}\right)$

(*iv*) $\dfrac{xdy - ydx}{x^2 + y^2} = \dfrac{(xdy - ydx)/x^2}{1 + (y/x)^2} = d\left[\tan^{-1}(y/x)\right]$

(*v*) $\dfrac{xdy - ydx}{x^2 + y^2} = \dfrac{(xdy - ydx)/y^2}{1 + (x/y)^2} = -d\left[\tan^{-1}(x/y)\right]$

(*vi*) $\dfrac{xdy - ydx}{xy} = d\left(\log\left|\dfrac{y}{x}\right|\right)$

(*vii*) $\dfrac{ydx - xdy}{xy} = d(\log|x/y|)$

(*viii*) $\dfrac{xdx + ydy}{x^2 + y^2} = \dfrac{1}{2}d\left[\log(x^2 + y^2)\right]$

(*ix*) $\dfrac{2xydx - x^2dy}{y^2} = d\left(\dfrac{x^2}{y}\right)$

(*x*) $\dfrac{2xy\,dy - y^2dx}{x^2} = d\left(\dfrac{y^2}{x}\right)$

(*xi*) $\dfrac{x\,dy + y\,dx}{xy} = d\left[\log(xy)\right]$

(*xii*) $\dfrac{xdy - ydx}{x^2 - y^2} = \dfrac{1}{2}d\left(\log\dfrac{x+y}{x-y}\right)$

(*xiii*) $d\left(\dfrac{e^x}{y}\right) = \dfrac{ye^x dx - e^x dy}{y^2}$

EXAMPLE 7.35 Solve the following differential equations by finding an integrating factor by inspection.

(i) $y\,dx - x\,dy + e^{1/x}dx = 0.$ (ii) $xdx + ydy = \dfrac{a^2\,(xdy - ydx)}{x^2 + y^2}.$

(*UPTech, 2008; PTU 2005; AMIE, S-2001*)

(iii) $(1 + xy)y\,dx + (1 - xy)\,x\,dy = 0.$ (iv) $y(2xy + e^x)dx = e^x dy.$ (*AMIE, W-2008*)

(v) $x\,dy - y\,dx - (1 - x^2)\,dx = 0.$

Solution (i) Multiplying the differential equation throughout by an integrating factor $1/x^2$, we obtain

$$\frac{ydx - xdy}{x^2} + \frac{e^{1/x}}{x^2}dx = 0 \quad \text{or} \quad -d(y/x) - d(e^{1/x}) = 0.$$

Integrating, we obtain $(y/x) + e^{1/x} = c$ or $y + xe^{1/x} = cx.$

(ii) The given differential equation may be rewritten as:

$$xdx + ydy - \left\{a^2\,\frac{(xdy - ydx)}{x^2}\Big/ 1 + \left(\frac{y}{x}\right)^2\right\} = 0$$

or $$x\,dx + y\,dy - a^2 \cdot d(\tan^{-1}y/x) = 0.$$

Integrating, we get $(x^2/2) + (y^2/2) - a^2\tan^{-1}(y/x) = c_1$

or $$x^2 + y^2 - 2a^2\tan^{-1}(y/x) = c, \text{ the required result.}$$

(iii) The given equation may be rewritten as:

$$(y\,dx + x\,dy) + xy(ydx - xdy) = 0. \qquad ...(1)$$

If we multiply equation (1) by $1/x^2y^2$, the equation becomes exact.

$$\frac{(ydx + xdy)}{x^2y^2} + \frac{1}{xy}(ydx - xdy) = 0$$

or $$\frac{d\,(xy)}{x^2y^2} + \frac{ydx - xdy}{xy} = 0 \quad \text{or} \quad d\left(-\frac{1}{xy}\right) + \left(\frac{dx}{x} - \frac{dy}{y}\right) = 0.$$

Integrating, we get the solution as $-\dfrac{1}{xy} + \log x - \log y = c$

or $$-\frac{1}{xy} + \log\left(\frac{x}{y}\right) = c \text{ or } \log\left(\frac{x}{y}\right) = c + \left(\frac{1}{xy}\right).$$

(iv) Dividing each term by y^2, we get $2x\,dx + (e^x/y)\,dx - (e^x/y^2)\,dy = 0$

or $$2x\,dx + \left(\frac{ye^x dx - e^x dy}{y^2}\right) = 0 \quad \text{or} \quad 2x\,dx + d(e^x/y) = 0.$$

Integrating, we get the solution as $x^2 + (e^x/y) = c,$

or $$x^2y + e^x = cy \text{ is the required solution.}$$

(v) Here $1/x^2$ an integrating factor. Upon introducing it, the equation becomes $\dfrac{xdy - ydx}{x^2} - \left(\dfrac{1}{x^2} - 1\right)dx = 0.$

Integrating, we get the solution as $(y/x) + (1/x) + x = c$ or $y + x^2 + 1 = cx.$

7.13 RULES FOR FINDING INTEGRATING FACTORS

Rule 1 If the functions $M(x, y)$ and $N(x, y)$ in the equation $M(x, y)\,dx + N(x, y)\,dy = 0$ are homogeneous functions of degree n and $Mx + Ny \neq 0$, then $\dfrac{1}{Mx + Ny}$ is an integrating factor. If $Mx + Ny = 0$, then $1/(xy)$, or $1/x^2$ or $1/y^2$ is an integrating factor.

EXAMPLE 7.36 Solve $y^2dx + (x^2 - xy - y^2)dy = 0$.

Solution The equation is homogeneous and $Mx + Ny \neq 0$, then $\dfrac{1}{Mx+Ny} = \dfrac{1}{y(x^2-y^2)}$ is an integrating factor.

Using this I.F. the given differential equation becomes $\dfrac{y}{x^2-y^2}dx + \dfrac{x^2-xy-y^2}{y(x^2-y^2)}dy = 0$, which is exact.

Integrating it, the solution is

$$\underset{(y \text{ const.})}{\int \frac{y}{x^2-y^2}dx} + \int \frac{1}{y}dx = c_2 \qquad \left(\because \frac{x^2-xy-y^2}{y(x^2-y^2)} = \frac{1}{y} - \frac{x}{x^2-y^2}\right)$$

or $$\frac{1}{2}\int\left(\frac{1}{x-y} - \frac{1}{x+y}\right)dx + \int \frac{1}{y}dx = c_2 \quad \text{or} \quad \frac{1}{2}\log\frac{x-y}{x+y} + \log y = \log c_1$$

or $$(x-y)y^2 = c(x+y), \text{ the required result.}$$

EXAMPLE 7.37 Solve: $(x^2y - 2xy^2)\, dx - (x^3 - 3x^2y)dy = 0$. *(PTU 2003)*

Solution The equation is homogeneous with $M(x, y) = x^2y - 2xy^2$ and $N(x, y) = -(x^3 - 3x^2y)$.

An I.F. is $\dfrac{1}{Mx+Ny} = \dfrac{1}{x(x^2y-2xy^2) - y(x^3-3x^2y)} = \dfrac{1}{x^2y^2}$.

Multiplying the differential equation throughout by an I.F $= 1/x^2y^2$ the given differential equation becomes

$\left(\dfrac{1}{y} - \dfrac{2}{x}\right)dx - \left(\dfrac{x}{y^2} - \dfrac{3}{y}\right)dy = 0$ which can be seen as exact.

Integrating it, the solution is $\dfrac{x}{y} - 2\log x + 3\log y = \log c$ or $\log\dfrac{y^3}{cx^2} = -\dfrac{x}{y}$ or $y^3 = cx^2e^{-x/y}$

Rule 2 For the non-exact equation $M(x, y)\, dx + N(x, y)\, dy = 0$, if $\left(\dfrac{\partial M}{\partial y} - \dfrac{\partial N}{\partial x}\right)\Big/N$ is a function of x alone, say $\phi(x)$, then an integrating factor is given by I.F. $= e^{\int \phi(x)dx}$.

EXAMPLE 7.38 Solve $(xy^2 - e^{1/x^3})dx - x^2y\,dy = 0$. *(AMIE, W-2005)*

Solution We have $M(x,y) = xy^2 - e^{1/x^3}$, $N(x, y) = -x^2y$.

$\partial M/\partial y = 2xy$, $\partial N/\partial x = -2xy$. The equation is not exact.

Since, $\dfrac{\left(\dfrac{\partial M}{\partial y}\right) - \left(\dfrac{\partial N}{\partial x}\right)}{N} = \dfrac{2xy-(-2xy)}{-x^2y} = \dfrac{-4}{x} = f(x)$ a function of x alone, $e^{\int f(x)dx}$ is an integration factor.

We have I.F. $= e^{\int f(x)dx} = e^{\int \frac{-4}{x}dx} = e^{-4\log x} = x^{-4}$.

Multiplying the differential equation throughout by an I.F. $1/x^4$, we obain $\left(\dfrac{y^2}{x^3} - \dfrac{1}{x^4}e^{1/x^3}\right)dx - \dfrac{y}{x^2}dy = 0$, which is an exact equation.

The solution is $\int\left(\frac{y^2}{x^3}-\frac{1}{x^4}e^{1/x^3}\right)dx+0=c$

or $-\frac{y^2x^{-2}}{2}+\frac{1}{3}\int e^{x^{-3}}(-3x^{-4})\,dx=c$ or $\frac{1}{3}e^{x^{-3}}-\frac{1}{2}\frac{y^2}{x^2}=c$

EXAMPLE 7.39 Solve: $(x^2+y^2+x)\,dx+xy\,dy=0$. *(AMIE, W-2005)*

Solution We have $M(x,y)=x^2+y^2+x$, $N(x,y)=xy$.

$\partial M/\partial y=2y$, $\partial N/\partial x=y$, the equation is not exact.

Since, $\frac{\left(\frac{\partial M}{\partial y}\right)-\left(\frac{\partial N}{\partial x}\right)}{N}=\frac{2y-y}{xy}=\frac{1}{x}=f(x)$, a function of x alone, $e^{\int f(x)dx}$ is an integrating factor.

We have I.F. $=e^{\int f(x)dx}=e^{\int\frac{1}{x}dx}=e^{\log x}=x$.

Multiplying the differential equation throughout by the I.F., we get $x(x^2+y^2+x)\,dx+x^2y\,dy=0$, which being exact. Integrating, we obtain the solution as $\frac{x^4}{4}+\frac{x^2y^2}{2}+\frac{x^3}{3}=c_1$ or $3x^4+4x^3+6x^2y^2=c$.

Rule 3 For the non-exact equation $M(x,y)\,dx+N(x,y)\,dy=0$, if $\frac{\left(\frac{\partial N}{\partial x}\right)-\left(\frac{\partial M}{\partial y}\right)}{M}$ is a function of y alone, say, $g(y)$, then the integrating factor is given by I.F. $=e^{\int g(y)\,dy}$.

EXAMPLE 7.40 Solve the equation: $(3x^2y^4+2xy)\,dx+(2x^3y^3-x^2)\,dy=0$.

Solution We have $M(x,y)=3x^2y^4+2xy$, $\partial M/\partial y=12x^2y^3+2x$.

$N(x,y)=2x^3y^3-x^2$, $\partial N/\partial x=6x^2y^3-2x$

The equation is not exact.

We have $\frac{\left(\frac{\partial N}{\partial x}\right)-\left(\frac{\partial M}{\partial y}\right)}{M}=\frac{6x^2y^3-2x-12x^2y^3-2x}{xy(3xy^3+2)}=-\frac{2}{y}=g(y)$ which is a function of y alone.

The integrating factor is I.F. $=e^{\int g(y)dy}=e^{\int(-2/y)dy}=1/y^2$.

Multiplying the given differential equation throughout by the integrating factor $1/y^2$, we get

$$\left(3x^2y^2+\frac{2x}{y}\right)dx+\left(2x^3y-\frac{x^2}{y^2}\right)dy=0.$$

This equation is exact and its solution is $x^3y^2+(x^2/y)=c$, a constant or $x^3y^3+x^2=cy$.

EXAMPLE 7.41 Solve $(xy^3+y)dx+2(x^2y^2+x+y^4)\,dy=0$. *(AMIE, S-2011, W-2005)*

Solution We have $M(x,y)=xy^3+y$, $N(x,y)=2(x^2y^2+x+y^4)$.

$\partial M/\partial y=3xy^2+1$, $\partial N/\partial x=4xy^2+2$. The equation is not exact.

Since $\frac{1}{M}\left(\frac{\partial N}{\partial x}-\frac{\partial M}{\partial y}\right)=\frac{1}{y(xy^2+1)}(4xy^2+2-3xy^2-1)=\frac{1}{y}=g(y)$ which is a function of y alone.

The integrating factor is I.F. $=e^{\int g(y)dy}=e^{\int\frac{1}{y}dy}=e^{\log y}=y$.

Multiplying the given differential equation throughout by the integrating factor y, we get

$(xy^4 + y^2)\,dx + (2x^2y^3 + 2\,xy + 2y^5)\,dy = 0$, which is an exact equation.

Integrating, we obtain the solution as $3x^2y^4 + 6xy^2 + 2y^6 = c$.

Rule 4 If the non-exact equation $M(x, y)\,dx + N(x, y)\,dy = 0$ is of the form $yf_1(xy)\,dx + xf_2(xy)\,dy = 0$, then $\dfrac{1}{Mx - Ny}$ is an integrating factor, provided $Mx - Ny \neq 0$.

EXAMPLE 7.42 Solve the equation:

$$(x^3y^3 + x^2y^2 + xy + 1)y\,dx + (x^3y^3 - x^2y^2 - xy + 1)x\,dy = 0.$$

Solution The equation is of the form $yf_1(xy)\,dx + xf_2(xy)\,dy = 0$ and $1/(Mx - Ny)$ is an integrating factor.

We have $M(x, y) = y(x^3y^3 + x^2y^2 + xy + 1), \quad N(x, y) = x(x^3y^3 - x^2y^2 - xy + 1).$

An integrating factor is

$$\frac{1}{Mx - Ny} = \frac{1}{xy(x^3y^3 + x^2y^2 + xy + 1) - (x^3y^3 - x^2y^2 - xy + 1)xy} = \frac{1}{2x^2y^2(xy + 1)}.$$

Multiplying the given differential equation throughout by this factor, we obtain

$$\left(\frac{x^2y^2 + 1}{x^2y^2}\right)ydx + \left[\frac{(x^2y^2 - xy + 1) - xy}{x^2y^2}\right]xdy = 0$$

or
$$(ydx + xdy) + \frac{ydx + xdy}{x^2y^2} - \frac{2x^2y}{x^2y^2}dy = 0$$

or
$$d(xy) + \frac{d(xy)}{x^2y^2} - \frac{2}{y}dy = 0 \quad \text{or} \quad dz + \frac{dz}{z^2} - \frac{2}{y}dy = 0, \text{ where } z = xy.$$

Integrating, we obtain $z - (1/z) - 2\log y = c$ or $xy - (1/xy) - 2\log y = c$,

EXAMPLE 7.43 Solve $(1 + xy)y\,dx + (1 - xy)\,x\,dy = 0$.

Solution The given equation is of the form $yf_1(xy)\,dx + xf_2(xy)\,dy = 0$ and $1/(Mx - Ny)$ is an integrating factor.

We have $M(x, y) = (1 + xy)y,\ N(x, y) = (1 - xy)x.$

$$\text{The integrating factor, I.F.} = \frac{1}{Mx - Ny} = \frac{1}{(1 + xy)xy - (1 - xy)xy} = \frac{1}{2x^2y^2}$$

Multiplying the differential equation throughout by the I.F. $1/2\,x^2y^2$,

we obtain
$$\left(\frac{1}{2x^2y} + \frac{1}{2x}\right)dx + \left(\frac{1}{2xy^2} - \frac{1}{2y}\right)dy = 0, \text{ which is an exact equation.}$$

The solution is
$$\frac{1}{2y}\left(-\frac{1}{x}\right) + \frac{1}{2}\log x - \frac{1}{2}\log y = c_1 \quad \text{or} \quad \log\frac{x}{y} - \frac{1}{xy} = c.$$

EXAMPLE 7.44 Solve: $(xy \sin xy + \cos xy)\,y\,dx + (xy \sin xy - \cos xy)\,x\,dy = 0$

Solution The given equation is of the form $f_1(xy)\,y\,dx + f_2(xy)\,x\,dy = 0$ and $1/(Mx - Ny)$ is an integrating factor.

We have $M(x, y) = (xy \sin xy + \cos xy)y, \quad N(x, y) = (xy \sin xy - \cos xy)x.$

$$\text{The integrating factor is I.F.} = \frac{1}{Mx - Ny} = \frac{1}{(xy\sin xy + \cos xy)xy - (xy\sin xy - \cos xy)xy} = \frac{1}{2xy\cos xy}.$$

Multiplying the given differential equation by the integrating factor $1/2\,xy\cos xy$, we get

$$\frac{1}{2}\left(\tan xy + \frac{1}{xy}\right)ydx + \frac{1}{2}\left(\tan xy - \frac{1}{xy}\right)xdy = 0$$

or $$\tan xy(ydx + xdy) + \frac{1}{x}dx - \frac{1}{y}dy = 0 \quad \text{or} \quad (\tan xy)\,d(xy) + \frac{1}{x}dx - \frac{1}{y}dy = 0$$

or $$\tan z\,dz + \frac{1}{x}dx - \frac{1}{y}dy = 0, \text{ where } z = xy.$$

Integrating it, the solution is $\log|\sec z| + \log x - \log y = \log c$

or $$\log\left\{\frac{x\sec z}{y}\right\} = \log c \quad \text{or} \quad \frac{x}{y}\sec z = c \quad \text{or} \quad x\sec(xy) = cy.$$

Rule 5 For the equation of the type $x^a y^b(my\,dx + nx\,dy) + x^{a'} y^{b'}(m'y\,dx + n'x\,dy) = 0$, an integrating factor is $x^h y^k$

where $$\frac{a+h+1}{m} = \frac{b+k+1}{n}, \quad \frac{a'+h+1}{m'} = \frac{b'+k+1}{n'}.$$

EXAMPLE 7.45 Solve $x^3y^3(2y\,dx + x\,dy) - (5\,y\,dx + 7\,x\,dy) = 0$

Solution Comparing the given equation with $x^a y^b(mydx + nxdy) + x^{a'} y^{b'}(m'ydx + n'xdy) = 0$.

we have $a = b = 3$, $m = 2$, $n = 1$; $a' = b' = 0$, $m' = -5$, $n' = -7$.

The integrating factor is I.F $= x^h y^k$

where $$\frac{a+h+1}{m} = \frac{b+k+1}{n}, \quad \frac{a'+h+1}{m'} = \frac{b'+k+1}{n'}$$

that is, $$\frac{3+h+1}{2} = \frac{3+k+1}{1}, \quad \frac{0+h+1}{-5} = \frac{0+k+1}{-7}$$

or $$h - 2k = 4,\ 7h - 5k = -2.$$

Solving these, we get $h = -8/3$ and $k = -10/3$. $\therefore$ I.F. $= 1/x^{8/3}y^{10/3}$.

Multiplying the given differential equation by the integrating factor $1/x^{8/3}\,y^{10/3}$, we get

$$x^{1/3}\,y^{-1/3}(2ydx + xdy) - \frac{1}{x^{8/3}\,y^{10/3}}(5ydx + 7x\,dy) = 0$$

or $$(2x^{1/3}\,y^{2/3} - 5x^{-8/3}y^{-7/3})dx + (x^{4/3}y^{-1/3} - 7x^{-5/3}y^{-10/3})\,dy = 0$$

which is an exact equation. The solution is

$$\frac{3}{2}x^{4/3}y^{2/3} + 3x^{-5/3}y^{-7/3} = c, \quad \text{or} \quad x^3y^3 + 2 = cx^{5/3}y^{7/3}.$$

EXAMPLE 7.46 Solve $(y^2 + 2x^2y)\,dx + (2x^3 - xy)\,dy = 0$. *(Rajasthan, 2005; KUK, 2005; VTU 2001)*

Solution Rewriting the equation as $y(y\,dx - x\,dy) + 2\,x^2\,(y\,dx + x\,dy) = 0$ and comparing with

$$x^a y^b(mydx + nxdy) + x^{a'} y^{b'}(m'ydx + n'xdy) = 0,$$

we have $a = 0$, $b = 1$, $m = 1$, $n = -1$; $a' = 2$, $b' = 0$, $m' = n' = 1$.

The integrating factor is *I.F.* $= x^h y^k$

where $$\frac{a+h+1}{m} = \frac{b+k+1}{n}, \quad \frac{a'+h+1}{m'} = \frac{b'+k+1}{n'}$$

that is, $$\frac{0+h+1}{1} = \frac{1+k+1}{-1}, \quad \frac{2+h+1}{1} = \frac{0+k+1}{1} \quad \text{or} \quad h + k = -3 \text{ and } h - k = -2.$$

Solving these, we get $h = -5/2$; $k = -1/2$. $\therefore$ I.F. $= 1/x^{5/2}y^{1/2}$.

Multiplying the given differential equation by the integrating factor $1/x^{5/2}y^{1/2}$, we get

$(x^{-5/2}y^{3/2} + 2x^{-1/2}y^{1/2})dx + (2x^{1/2}y^{-1/2} - x^{-3/2}y^{1/2})dy = 0$, which is an exact equation.

The solution is $-\frac{2}{3}x^{-3/2}y^{3/2} + 4x^{1/2}y^{1/2} = c$ or $6(xy)^{1/2} - (y/x)^{3/2} = c.$

EXERCISE 7.8

1. Find the value of α so that $e^{\alpha x^2}$ is an integrating factor of following differential equations:

   (a) $x(1 - y)dx - dy = 0$. *(AMIETE, June 2005)* **Ans.** 1/2.

   (b) $(e^{-y^2/2} - xy)dy - dx = 0$. *(AMIETE, Dec. 2005)* **Ans.** 1/2.

2. Given that the integrating factor of the equation $y\sec^2 x dx + \left[3\tan x - \left(\frac{\sec y}{y}\right)^2\right]dy = 0$ is of the form y^n. Find n and hence solve the equation. **Ans.** $n = 2; y^3 \tan x - \tan y = c.$

Solve the following differential equations:

3. $x^2y\,dx - (x^3 + y^3)\,dy = 0.$ **Ans.** $x^3 = 3y^3(\log y - c).$
4. $xy^2\,dy - (x^3 + y^3)\,dx = 0.$ *(KUK, 2009)* **Ans.** $y^3 = 3x^3(\log x - c).$
5. $(3xy^2 - y^3)dx - (2x^2y - xy^2)\,dy = 0.$ **Ans.** $3\log x - 2\log y + (y/x) = c.$
6. $(3xy - 2ay^2)\,dx + (x^2 - 2\,axy)dy = 0.$ *(KUK, 2010; Andhra 2000)* **Ans.** $x^2(ay - x)\,y = c.$
7. $(x^4 + y^4)\,dx - xy^3dy = 0.$ **Ans.** $y^4 = 4x^4\log x + cx^4.$
8. $(x + y^2)\,dx - 2\,xy\,dy = 0.$ **Ans.** $x = ce^{y^2/x}.$
9. $(x^2 + y^2 + 2x)dx + 2y\,dy = 0.$ **Ans.** $(x^2 + y^2)e^x = c.$
10. $(2xy^2 - 3y^3)\,dx + (7 - 3xy^2)\,dy = 0.$ **Ans.** $x^2 - (7/y) - 3xy = c.$
11. $2xy\log y dx + (x^2 + y^2\sqrt{y^2+1})dy = 0.$ **Ans.** $x^2\log y + \{(y^2+1)^{3/2}/3\} = c.$
12. $(xy^2 - x^2)\,dx + (3x^2y^2 + x^2y - 2x^3 + y^2)\,dy = 0.$ **Ans.** $e^{6y}(54x^2y^2 - 36x^3 + 18y^2 - 6y + 1) = c.$
13. $(x^2y^2 + xy + 1)y\,dx + (x^2y^2 - xy + 1)x\,dy = 0.$ *(MDU 2005, 2003)* **Ans.** $xy + \log(x/y) - (1/xy) = c.$
14. $y(xy + 2x^2y^2)dx + x(xy - x^2y^2)dy = 0.$ **Ans.** $x^2 = cye^{1/xy}.$
15. $(y\,dx + 3x\,dy) + 2y\,(3y\,dx + 4x\,dy) = 0.$ **Ans.** $6y + 1 = c\,x^{3/5}\,y^{9/5}.$
16. $(3x + 2y^2)\,y\,dx + 2x\,(2x + 3y^2)\,dy = 0.$ **Ans.** $x^3y^4 + x^2y^6 = c.$
17. $(x^4e^x - 2m\,xy^2)dx + 2\,mx^2y\,dy = 0.$ **Ans.** $e^x + m(y/x)^2 = c.$
18. $(x^4y^4 + x^2y^2 + xy)y\,dx + (x^4y^4 - x^2y^2 + xy)\,x\,dy = 0.$ *(AMIE W-2004)*

    Ans. $\{x^2y^2/2\} + \log(x/y) - (1/xy) = c.$
19. $(3x^2y^3e^y + y^3 + y^2)\,dx + (x^3y^3e^y - xy)\,dy = 0.$ *(AMIETE, June 2008)* **Ans.** $y\,(x^3e^y + x) + x = cy.$
20. $(1 + xy + x^2y^2)y\,dx + (1 - xy + x^2y^2)x\,dy = 0.$ *(MDU, 2005, 2003)* **Ans.** $(-1/xy) + \log x + xy - \log y = c.$
21. $(2x^2y^2 + y)dx - (x^3y - 3x)\,dy = 0.$ *(NIT, Kurukshetra, 2008; MDU 2002)* **Ans.** $4x^2y = 5 + cx^{4/7}y^{12/7}.$
22. $(8y\,dx + 8x\,dy) + x^2y^3\,(4y\,dx + 5\,x\,dy) = 0.$ **Ans.** $4x^2y^2 + x^4y^5 = c.$
23. $\sin y(dy/dx) = \cos x(2\cos y - \sin^2 x).$ **Ans.** $\cos y = \frac{\sin^2 x}{2} - \frac{\sin x}{2} + \frac{1}{4} + \lambda e^{-2\sin x}.$
24. $xdx + ydy + \{(xdy - ydx)/(x^2 + y^2)\} = 0.$ *(AMIE S-2010)* **Ans.** $x^2 - 2\tan^{-1}(x/y) + y^2 = c.$
25. $(y^3 - 2x^2y)dx + (2xy^2 - x^3)dy = 0.$ **Ans.** $x^2y^2(y^2 - x^2) = c.$
26. $(5x^3 + 12x^2 + 6y^2)\,dx + 6xy\,dy = 0.$ *(AMIE., S-2008)* **Ans.** $x^5 + 3x^4 + 3x^2y^2 = c.$

7.14 DIFFERENTIAL EQUATIONS OF THE FIRST ORDER AND HIGHER DEGREE

Introduction: In the preceeding sections we have discussed the solution of the differential equations of the first order and first degree. Here we study the solution of the differential equations of the first order but with a degree higher than the first. For convenience, we will represent dy/dx by p.

A most general form of the differential equation of the first order and nth degree is

$$p^n + P_1 p^{n-1} + P_2\, p^{n-2} + \dots + P_n = 0, \qquad \text{...(7.19)}$$

where $P_1, P_2, \dots\dots$ are functions of x and y.

We shall discuss the solution of Eq. (7.19) under the following cases:

(*a*) Equations solvable for p.
(*b*) Equations solvable for y.
(*c*) Equations solvable for x.

7.15 EQUATIONS SOLVABLE FOR p

Since Eq. (7.19) is a polynomial of degree n in p, we resolve it into n linear factors of the form

$$(p - F_1)(p - F_2) \dots (p - F_n) = 0, \qquad \text{...(7.20)}$$

where $F_1, F_2, \dots F_n$ are functions of x and y.

Equation (7.20) is equivalent to

$$p - F_1 = 0,\ p - F_2 = 0,\ \dots\dots,\ p - F_n = 0$$

or

$$\frac{dy}{dx} = F_1;\ \frac{dy}{dx} = F_2; \dots \frac{dy}{dx} = F_n. \qquad \text{...(7.21)}$$

Each of the equations in (7.21) is a differential equation of the first order and first degree and therefore, can be easily integrated.

Let the solutions of these equations be

$$f_1(x, y, c_1) = 0;\quad f_2(x, y, c_2) = 0 \dots \quad \text{and} \quad f_n(x, y, c_n) = 0 \qquad \text{...(7.22)}$$

where $c_1, c_2, \dots c_n$ are arbitrary constants.

Combining all the solutions found out in Eq. (7.22), the solution of Eq. (7.19) is given by

$$f_1(x, y, c) \cdot f_2(x, y, c) \dots f_n(x, y, c) = 0. \qquad \text{...(7.23)}$$

There is no loss of generality by replacing n arbitrary constants $c_1, c_2, \dots c_n$ by a single constant c as every particular solution can be obtained by giving suitable value to c.

EXAMPLE 7.47 Solve the equation $p^2 - 5p + 6 = 0 \quad (p = dy/dx)$.

Solution The given equation can be written as $(p - 2)(p - 3) = 0$.

Hence $p = 2$ or 3. When $p = 2$ that is, $dy/dx = 2$. Integrating this, we get $y = 2x + c$. Similarly $p = 3$ gives $y = 3x + c$. Hence the complete solution of the equation is given by

$$(y - 2x - c)(y - 3x - c) = 0.$$

EXAMPLE 7.48 Solve equation $x^2(dy/dx)^2 - 2xy(dy/dx) + 2y^2 - x^2 = 0$.

Solution The given equation can be written in terms of p as $x^2p^2 - 2xyp + 2y^2 - x^2 = 0$.

Solving for p, we get $p = \dfrac{y \pm \sqrt{x^2 - y^2}}{x} = \dfrac{dy}{dx}$.

This is a homogeneous equation and therefore putting $y = vx$, we get

$$v + x\left(\frac{dv}{dx}\right) = v \pm \sqrt{1 - v^2} \quad \text{or} \quad x\left(\frac{dv}{dx}\right) = \pm\sqrt{1 - v^2} \quad \text{or} \quad \frac{dv}{\sqrt{1 - v^2}} = \pm\frac{dx}{x}.$$

Integration gives $\sin^{-1} v = \pm \log x + c$ or $\sin^{-1}(y/x) = \pm \log x + c$.

Hence the solution of the given equation is $\left(\sin^{-1}\dfrac{y}{x} - \log x - c\right)\left(\sin^{-1}\dfrac{y}{x} + \log x - c\right) = 0$.

EXERCISE 7.9

Solve the following differential equations:

1. $p^2 - 9p + 18 = 0$. **Ans.** $(y - 3x - c)(y - 6x - c) = 0$.
2. $p^3 + 2xp^2 - y^2p^2 - 2xy^2p = 0$. *(Madras, 2003)* **Ans.** $(y - c)(y + x^2 - c)(xy - cy + 1) = 0$.
3. $p^2 + 2py \cot x = y^2$. *(Bhillai, 2005; Kerala, 2005)* **Ans.** $\left(y - \dfrac{c}{1+\cos x}\right)\left(y - \dfrac{c}{1-\cos x}\right) = 0$.
4. $y = x(p + \sqrt{1+p^2})$. **Ans.** $x^2 + y^2 = cx$.
5. $p^2 + x^3y - x^3p - yp = 0$. **Ans.** $(y - ce^x)(4y - x^4 + c) = 0$.
6. $p - \dfrac{1}{p} = \dfrac{x}{y} - \dfrac{y}{x}$. **Ans.** $(xy - c)(x^2 - y^2 - c) = 0$.
7. $p(p + y) = x(x + y)$. **Ans.** $(2y - x^2 + c)(x + y - 1 + ce^{-x}) = 0$.
8. $xyp^2 + p(3x^2 - 2y^2) - 6xy = 0$. **Ans.** $(y - cx^2)(y^2 + 3x^2 - c) = 0$.
9. $xyp^2 - (x^2 + y^2)p + xy = 0$. **Ans.** $(y - cx)(y^2 - x^2 - c) = 0$.
10. $xy^2(p^2 + 2) = 2py^3 + x^3$. **Ans.** $(x^2 - y^2 + c)(x^2 - y^2 + cx^4) = 0$.

7.16 EQUATIONS SOLVABLE FOR y

Sometimes it is possible to express y explicitly as a function of x and p in the given equation $F(x, y, p) = 0$, so that it can be put in the form

$$y = f(x, p). \qquad \text{...(7.24)}$$

Differentiation with respect to x leads to an equation of the form

$$\frac{dy}{dx} = p = \frac{\partial f}{\partial x} + \frac{\partial f}{\partial p}\frac{\partial p}{\partial x}. \qquad \text{...(7.25)}$$

Equation (7.25) is a first order differential equation in x and p. Let the solution of this equation be

$$\phi(x, p, c) = 0 \qquad \text{...(7.26)}$$

Elimination of p between equation (7.24) and (7.26) yields the solution of the given equation.

In certain problems it may be tedious to eliminate p between (7.24) and (7.26). In such cases, (7.24) and (7.26) together constitute the solution giving x and y in terms of the parameter p.

EXAMPLE 7.49 Solve the differential equation $y = 2px + p^4x^2$.

Solution Differentiating with respect to x and denoting dy/dx by p, we get

$$p = 2p + 2x\left(\frac{dp}{dx}\right) + 2xp^4 + 4x^2p^3\left(\frac{dp}{dx}\right) \quad \text{or} \quad \left[p + 2x\left(\frac{dp}{dx}\right)\right](1 + 2xp^3) = 0.$$

This gives $\quad p + 2x\dfrac{dp}{dx} = 0 \quad$ and $\quad 1 + 2xp^3 = 0$.

Case I: $\quad p + 2x\left(\dfrac{dp}{dx}\right) = 0 \quad$ or $\quad 2\left(\dfrac{dp}{p}\right) + \left(\dfrac{dx}{x}\right) = 0.$

Integration gives $2 \log p + \log x = \log c$ or $p^2 = c/x$.

Eliminating p from this relation and the given equation, we get

$$y = 2x\sqrt{\frac{c}{x}} + c^2 \quad \text{or} \quad (y - c^2)^2 = 4cx.$$

Case II: $\quad 1 + 2xp^3 = 0$. This gives $p = \left(-\dfrac{1}{2x}\right)^{1/3}$.

Eliminating p from this relation and the given equation, we get

$$y = \left(-\frac{1}{2x}\right)^{1/3}\left\{2x - \frac{x}{2}\right\}.$$

This also gives a solution but as it does not contain any arbitrary constant, it is called a singular solution.

EXAMPLE 7.50 Solve the differential equation $y = (1 + p)x + p^2$.

Solution Differentiating with respect to x, we get $\frac{dy}{dx} = p = 1 + p + x\left(\frac{dp}{dx}\right) + 2p\left(\frac{dp}{dx}\right)$.

This gives $(x + 2p)\frac{dp}{dx} = -1$ or $\frac{dx}{dp} + x = -2p$.

This is a linear equation. An integrating factor is $\text{I.F.} = e^{\int dp} = e^p$.

The solution of this equation is

$$xe^p = -2\int pe^p\,dp + c = -2(pe^p - e^p) + c \quad \text{or} \quad x = ce^{-p} + 2(1 - p). \qquad \text{...}(i)$$

Substituting this value of x in the given equation, we get

$$y = (1 + p)\{2(1 - p) + ce^{-p}\} + p^2. \qquad \text{...}(ii)$$

Equations (*i*) and (*ii*) together with p as parameter constitute the solution of the given equation.

NOTE The above equation is a particular case of more general class of equation $y = xf(p) + g(p)$ which is called Lagrange's equation.

EXAMPLE 7.51 Solve $y - 2px = \tan^{-1}(xp^2)$.

Solution We have $y = 2px + \tan^{-1}(xp^2)$. ...(*i*)

Differentiating both sides with respect to x,

$$\frac{dy}{dx} = p = 2\left[p + x\frac{dp}{dx}\right] + \frac{1}{1 + x^2p^4}\cdot\left[p^2 + 2xp\frac{dp}{dx}\right] \quad \text{or} \quad \left(p + 2x\frac{dp}{dx}\right) + \left(p + 2x\frac{dp}{dx}\right)\frac{p}{1 + x^2p^4} = 0$$

or $\left(p + 2x\frac{dp}{dx}\right)\left(1 + \frac{p}{1 + x^2p^4}\right) = 0$. This gives $p + 2x\frac{dp}{dx} = 0$.

Separating the variables and integrating, we have $\int\frac{dx}{x} + 2\int\frac{dp}{p} = a$ constant.

or $\log x + 2\log p = \log c$ or $\log xp^2 = \log c$ or $xp^2 = c$ or $p = \sqrt{c/x}$. ...(*ii*)

Eliminating p from (*i*) and (*ii*), we get $y = 2\sqrt{cx} + \tan^{-1}c$, which is the general solution of (*i*).

NOTE The factor $\left(1 + \frac{p}{1 + x^2p^4}\right) = 0$, has not been considered as it concerns 'singular solution' of (*i*), whereas we are interested only in finding *general solution.*

EXERCISE 7.10

Solve the following differential equations:

1. $y = 3x + \log p$. **Ans.** $y = 3x + \log\frac{3}{1 - ce^{3x}}$.

2. $p^3 + mp^2 = a(y + mx)$.

Ans. The required solution is the p-eliminant of $ax = c + \frac{3p^2}{2} - mp + m^2\log(p + m)$ and $p^3 + mp^2 = a(y + mx)$.

3. $xp + y = x^4p^2$. **Ans.** $xy = c^2x + c$.

4. $y = 2px + f(xp^2)$. **Ans.** $y = 2c\sqrt{x} + f(c^2)$.

5. $y = p \tan p + \log \cos p$. **Ans.** $x = \tan p + c;\ y = p \tan p + \log \cos p$.

6. $y = \dfrac{1}{\sqrt{1+p^2}} + b$. **Ans.** $(x + c)^2 + (y - b)^2 = 1$.

7. $y = x + a \tan^{-1} p$. **Ans.** $x = c + \dfrac{a}{2}\left\{\log \dfrac{p-1}{\sqrt{1+p^2}} - \tan^{-1} p\right\}$ with the given relation.

8. $y = (1 + p)x + e^p$. **Ans.** $x = ce^{-p} - \dfrac{1}{2}e^p;\ y = c(1+p)e^{-p} + \dfrac{1}{2}(1-p)e^p$.

9. $y = p^2x + p$. **Ans.** $x = (\log p - p + c)(p - 1)^{-2}$, with the given relation.

10. $y = p \sin p + \cos p$. **Ans.** $x = \sin p + c$, with the given relation.

7.17 EQUATIONS SOLVABLE FOR x

Let $F(x, y, p) = 0$ be the given equation. Suppose it is possible to solve it for x in terms of y and p so that

$$x = f(y, p). \qquad ...(7.27)$$

Differentiate (7.27) with respect to y and denote $dx/dy = 1/p$, we get

$$\frac{dx}{dy} = \frac{1}{p} = \frac{\partial f}{\partial y} + \frac{\partial f}{\partial p}\frac{dp}{dy}. \qquad ...(7.28)$$

Equation (7.28) is a differential equation of first order and first degree in y and p. Let its solution be

$$\phi\,(y, p, c) = 0 \qquad ...(7.29)$$

Elimination of p from (7.27) and (7.29) gives the solution of the given equation. In case it is not possible to eliminate p then (7.27) and (7.29) together constitute the solution with p as a parameter.

EXAMPLE 7.52 Solve the differential equation $x = y + p^2$.

Solution Differentiating with respect to y and denoting dx/dy by $1/p$, we get

$$\frac{dx}{dy} = \frac{1}{p} = 1 + 2p\left(\frac{dp}{dy}\right) \quad \text{or} \quad \frac{1-p}{p} = 2p\frac{dp}{dy}.$$ Separating the variables, we get $-\dfrac{2p^2}{p-1}dp = dy$.

Integration gives $-2 \log (p - 1) - 2p - p^2 = y + c$. ...(i)

Since it is not possible to eliminate p from (i) and the given equation, we denote the solution by

$$y = c - 2\left[\log(p-1) + p + \frac{p^2}{2}\right] \text{ and } x = y + p^2.$$

EXERCISE 7.11

Solve the following differential equations:

1. $x + \left(\dfrac{p}{\sqrt{1+p^2}}\right) = a$. **Ans.** $(x - a)^2 + (y + c)^2 = 1$

2. $p^3 - 2xyp + 4y^2 = 0$. **Ans.** $16y = c(2x - c)^2$

3. $p = \tan\left(x - \dfrac{p}{1+p^2}\right)$. **Ans.** $y + (1 + p^2)^{-1} = c$, with the given relation

4. $p^2y + 2px - y = 0$. **Ans.** $y^2 = 2cx + c^2$

5. $p^3 - 4xyp + 8y^2 = 0$. **Ans.** $y = c(x - c)^2$

6. $y = 2px + y^2p^3$. **Ans.** $y^2 = 2cx + c$

7.18 CLAIRAUT'S EQUATION

A non-linear differential equation of the form $y = px + f(p)$ is called the *Clairaut's equation.* Differentiating with respect to x, we get

$$\frac{dy}{dx} = p = x\left(\frac{dp}{dx}\right) + p + f'(p)\left(\frac{dp}{dx}\right) \qquad ...(7.30)$$

or

$$[x + f'(p)]\frac{dp}{dx} = 0 \qquad ...(7.31)$$

Now either $\qquad dp/dx = 0$ or $x + f'(p) = 0.$

Let $dp/dx = 0$. Integration gives $p = c$. Elimination of p between this relation and the given equation yields

$$y = cx + f(c) \qquad ...(7.32)$$

The other factor $x + f'(p) = 0$ leads to a singular solution and is therefore not discussed.

Rule In order to solve Clairaut's equation, we replace p by c and this gives the solution.

EXAMPLE 7.53 Solve $y = (x - a)\,p - p^2$.

Solution The given equation is written as $y = px - ap - p^2$. ...(*i*)

Since (*i*) is a Clairaut's equation, its solution is $y = (x - a)\,c - c^2$.

EXAMPLE 7.54 Solve $p = \sin(y - px)$.

Solution The given equation may be written as

$\sin^{-1}p = y - px$ or $y = px + \sin^{-1}p$ which is the Clairaut's equation.

$\therefore$ Its solution is $y = cx + \sin^{-1}c$.

7.19 EQUATIONS REDUCIBLE TO CLAIRAUT'S FORM

There are certain equations of the first order and degree greater than one which can be solved by transforming them into Clairaut's form by appropriate substitutions.

EXAMPLE 7.55 Solve the differential equation: $x^2(y - px) = p^2y$.

Solution Put $x^2 = u$ and $y^2 = v$. This gives $2xdx = du$; $2ydy = dv$

or $\quad \dfrac{y}{x}\cdot\dfrac{dy}{dx} = \dfrac{dv}{du}$ that is, $\dfrac{y}{x}p = \dfrac{dv}{du}$ or $p = \dfrac{x}{y}\cdot\dfrac{dv}{du}$.

Substituting for p in the given equation, we get

$$x^2\left(y - \frac{x^2}{y}\frac{dv}{du}\right) = \frac{x^2}{y^2}\left(\frac{dv}{du}\right)^2 \cdot y$$

or

$$\left(y^2 - x^2\frac{dv}{du}\right) = \left(\frac{dv}{du}\right)^2 \quad \text{or} \quad v = u\frac{dv}{du} + \left(\frac{dv}{du}\right)^2.$$

This is a Clairaut's equation and so its solution is

$$v = uc + c^2 \quad \text{or} \quad y^2 = cx^2 + c^2.$$

EXERCISE 7.12

Find the general solution of the following differential equations:

1. $(px - y)(py + x) = a^2p$; (use $x^2 = u$ and $y^2 = v$). (*JNTU, 2006*) **Ans.** $y^2 = cx^2 - a^2c/(c + 1)$.
2. $a\,x\,y\,p^2 + (x^2 - ay^2 - b)\,p - xy = 0$; use $x^2 = u$ and $y^2 = v$. **Ans.** $y^2 = cx^2 - bc/(ac + 1)$.
3. $xyp^2 - (x^2 + y^2 - 1)\,p + xy = 0$; use $x^2 = u$ and $y^2 = v$. **Ans.** $y^2 = cx^2 + \left(\dfrac{c}{c-1}\right)$.

4. $y = 2px + y^2p^3$; use $y^2 = v$. **Ans.** $y^2 = cx + \left(\frac{c^3}{8}\right)$.

5. $4yp^2 + 2xp - y = 0$; Put $y^2 = v$. **Ans.** $y^2 = cx + c^2$.

6. $(px + y)^2 = py^2$; Put $xy = v$. **Ans.** $xy = cy - c^2$.

7. $e^{3x}(p - 1) + p^3e^{2y} = 0$; Put $e^x = u$ and $e^y = v$. **Ans.** $e^y = ce^x + c^3$.

8. $y = px + \sqrt{1 + p^2}$. **Ans.** $y = cx + \sqrt{1+c^2}$.

9. $y = px + \left(\frac{ap}{\sqrt{1+p^2}}\right)$. **Ans.** $y = cx + \left(\frac{ac}{\sqrt{1+c^2}}\right)$.

10. $p^3x - p^2y - 1 = 0$. **Ans.** $y = cx - \left(\frac{1}{c^2}\right)$.

11. $(x - a)(dy/dx)^2 + (x - y)(dy/dx) - y = 0$. **Ans.** $(y - cx)(1 + c) + ac^2 = 0$.

12. $(y - px)(p - 1) = p$. **Ans.** $(y - cx)(c - 1) = c$.

13. $\sin(px - y) = p$. *(PTU, 2006)* **Ans.** General solution: $y = cx - \sin^{-1}c$.

14. $p = \log(px - y)$ **Ans.** General solution: $y = cx - e^c$.

7.20 ORTHOGONAL TRAJECTORIES

Two families of curves are said to be orthogonal trajectories when every member of one family cuts every member of the other family at right angles. Families of curves related in this manner occur very frequently in many engineering applications. For instance, in two-dimensional problems in heat flow, the lines of heat-flow in a body are everywhere perpendicular to the isothermal curves. In electrostatics, the equipotential lines and the lines of electric force are orthogonal trajectories of each other. Similarly in problems dealing with the flow of electricity, the paths along which the current flows move at right angles to the equipotential curves.

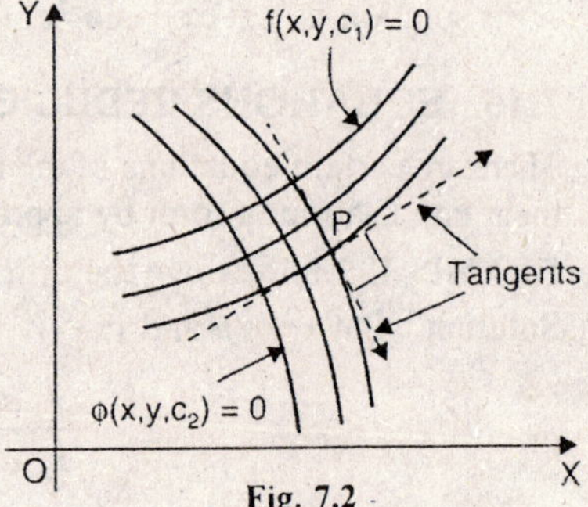

Fig. 7.2

7.20.1 Orthogonal Trajectories of Family of Curves in Cartesian Coordinates

Let $$f(x, y, c) = 0, \quad ...(7.33)$$

represent a one-parameter family of curves and c is called the *parameter* of the family.

Differentiating with respect to x and eliminating c, we can obtain the first order differential equation for which $f(x, y, c) = 0$ are the integral curves.

Let the differential equation be $F(x, y, y') = 0$. ...(7.34)

We know that when two curves intersect at right angles, the product of their slopes at their point of intersection is equal to -1. Using this fact, the slope of the orthogonal trajectory is given by $-1/y'$, since the slope of an integral curve given by, eq. (6.33) is y'.

Replacing y' by $-1/y'$ in $F(x, y, y') = 0$. Hence we get the differential equation

$$F\left(x, y, -\frac{1}{y'}\right) = 0. \quad ...(7.35)$$

governing the family of orthogonal curves.

Solving this equation, we obtain the one parameter family of orthogonal curves.

EXAMPLE 7.56 Find the orthogonal trajectories of the family of parabolas $x^2 = cy$.

Solution Differentiating the given equation w.r.t. x, we get $y' = 2x/c$. Eliminating c, we obtain $y'/y = 2/x$, which is the differential equation governing the given family.

Substituting $-1/y'$ for y', we obtain the differential equation governing the orthogonal family as

$$-\frac{1}{yy'} = \frac{2}{x} \quad \text{or} \quad ydy = -\frac{x}{2}dx.$$

Integration of the equation gives $\dfrac{x^2}{4} + \dfrac{y^2}{2} = c_1^2$.

Hence, the orthogonal trajectories of the given family of parabolas will be represented by a certain family of ellipses with semi-axes $2c_1$ (in the x-direction) and $c_1\sqrt{2}$. Refer Fig. 7.3.

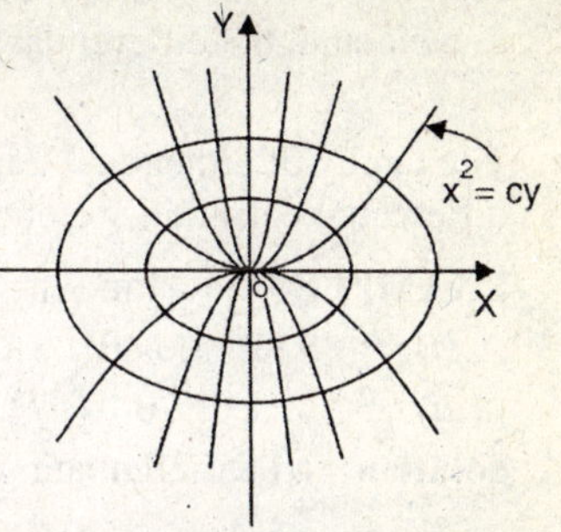

Fig. 7.3

EXAMPLE 7.57 Show that the one parameter family of curves $x^2 = 4a(y + a)$ are self-orthogonal. (*Kerala 2005*).

Solution A family of curves is self-orthogonal if it is its own orthogonal family.

The given equation represents a family of parabolas with y-axis as the axis of the family. Differentiating this equation, we get $x = 2ay'$.

Eliminating a, we obtain $x^2 = \dfrac{2x}{y'}\left(y + \dfrac{x}{2y'}\right)$ or $xy'^2 - 2yy' - x = 0$, ...(*i*)

which is the differential equation governing the given family.

Replacing y' by $-1/y'$, we obtain the differential equation governing the orthogonal family as

$$\frac{x}{y'^2} + \frac{2y}{y'} - x = 0 \quad \text{or} \quad xy'^2 - 2yy' - x = 0 \quad \text{which is same as } (i)$$

Therefore, the orthogonal trajectories are the given curves themselves.

EXAMPLE 7.58 Show that the family of conics

$$\frac{x^2}{a^2 + \lambda} + \frac{y^2}{b^2 + \lambda} = 1 \qquad ...(i)$$

is self-orthogonal, λ being the parameter of the family. (*NIT Kurukshetra, 2010*)

Solution Differentiating the given equation with respect to x, we obtain $\dfrac{x}{a^2 + \lambda} + \dfrac{y}{b^2 + \lambda} \cdot y' = 0$.

Eliminate λ, we obtain $\lambda = -\dfrac{b^2 x + a^2 yy'}{x + yy'} \Rightarrow a^2 + \lambda = \dfrac{(a^2 - b^2)x}{x + yy'},\ b^2 + \lambda = \dfrac{(b^2 - a^2)yy'}{x + yy'}$.

Substituting these values in the given equation, we get

$$(x + yy')\left(x - \frac{y}{y'}\right) = a^2 - b^2, \qquad ...(ii)$$

as the differential equation of the given gamily. Changing y' to $-1/y'$, we obtain the differential equation governing the orthogonal family as $\left(x - \dfrac{y}{y'}\right)(x + yy') = 0$ which is same as (*ii*). Thus we see that the family (*i*) is self orthogonal, that is every member of the family (*i*) cuts every other member of the same family orthogonally.

7.20.2 Orthogonal Trajectories of Family of Curves in Polar Coordinates

Let $$f(r, \theta, c) = 0, \qquad ...(7.36)$$

represents a family of curves in polar coordinates form, where c is a parameter. Differentiating and eliminating c, we obtain the differential equation governing this family as

$$F(r, \theta, r') = 0, \text{ where } r' = dr/d\theta$$

The family of orthogonal trajectories is governed by the differential equation $F(r, \theta, (-r^2/r')] = 0$. Solving this differential equation, we obtain the orthogonal trajectories of the given family of curves.

EXAMPLE 7.59 Find the orthogonal trajectory of the following family of curves:

(*i*) $r = a(1 + \cos\theta)$ (*KUK, 2008; JNTU, 2006*) (*ii*) $r = a(\sec\theta + \tan\theta)$

(*iii*) $r^2 = a^2 \cos 2\theta$ (*VTU, 2003S*) (*iv*) $r^n = a \sin n\theta$ (*VTU, 2006*)

Solution (*i*) Differentiating $r = a(1 + \cos\theta)$ with resect to θ, we get $r' = -a \sin\theta$.

Eliminating a, we obtain $r' = -r\left(\dfrac{\sin\theta}{1+\cos\theta}\right)$, which is the differential equation of the given family.

Replacing r' by $-r^2/r'$, we obtain the differential equation governing the orthogonal trajectories as

$$\frac{-r^2}{r'} = -r\left(\frac{\sin\theta}{1+\cos\theta}\right) \quad \text{or} \quad r' = r\left(\frac{1+\cos\theta}{\sin\theta}\right) = r(\operatorname{cosec}\theta + \cot\theta).$$

Separating the variables, we get $dr/r = (\operatorname{cosec}\theta + \cot\theta)\, d\theta$.

Integrating, we obtain $$\log r = \log\tan\frac{\theta}{2} + \log\sin\theta + \log c = \log\left(c\sin\theta\tan\frac{\theta}{2}\right).$$

$$\Rightarrow \quad r = c\sin\theta\tan\frac{\theta}{2} = 2c\sin\frac{\theta}{2}\cos\frac{\theta}{2}\tan\frac{\theta}{2} = 2c\sin^2\frac{\theta}{2} = c(1-\cos\theta).$$

Hence the orthogonal trajectories of the given family of curves are given by $r = c(1 - \cos\theta)$, c is a parameter.

(*ii*) Differentiating $r^2 = a^2\cos 2\theta$ with respect to θ, we get $rr' = -a^2 \sin 2\theta$. Eliminating the parameter a we obtain the differential equation of the given family of curves $r' = -r\tan 2\theta$. Replacing r' by $-r^2/r'$, we find the differential equation of the family of orthogonal trajectories $-r^2/r' = -r\tan 2\theta$ or $r' = r\cot 2\theta$. Separating the variables, we get $dr/r = \cot 2\theta\, d\theta$.
Integrating, we find

$$\log r = \frac{1}{2}\log\sin 2\theta + \log c \Rightarrow r^2 = c^2\sin 2\theta.$$

Hence, the orthogonal trajectories of the system of lemniscates are lemniscates whose axes of symmetry make with the polar axis an angle of ± 45°.

(*iii*) Differentiating $r = a(\sec\theta + \tan\theta)$ with respect to θ,

we get $$r' = a\sec\theta(\sec\theta + \tan\theta).$$

Eliminating a, we obtain $r' = r\sec\theta$, which is the differential equation governing the given family of curves. Replacing r' by $-r^2/r'$, we obtain the differential equation governing the orthogonal trajectories as $$r' = -r\cos\theta.$$

Separating the variables, we get $dr/r = -\cos\theta\, d\theta$.

Integrating, we obtain $$\log r = -\sin\theta + k,$$

or $$r = be^{-\sin\theta}, \text{ where } b = e^k \text{ is an arbitrary constant.}$$

Hence the orthogonal trajectories of the given family of curves are given by $r = be^{-\sin\theta}$.

(*iv*) The given family of curves is $r^n = a \sin n\theta,$

or $n \log r = \log a + \log \sin n\theta.$ ...(1)

Differentiating (1) with respect to θ, we get

$$\frac{n}{r} r' = n \cot n\,\theta, \quad \text{or} \quad r' = r \cot n\,\theta,$$

which is the differential equation governing the given family of curves.

Replacing r' by $-r^2/r'$, we obtain the differential equation governing the orthogonal trajectories as

$$r' = -r \tan n\,\theta.$$

Separating the variables, we get $\frac{dr}{r} = -\tan n\,\theta\, d\,\theta.$

Integrating, we obtain $\log r = \frac{1}{n} \log \cos n\theta + \log c$

or $r^n = b \cos n\,\theta$, where $b = c^n$ is an arbitrary constant.

Hence the orthogonal trajectories of the given family of curves are given by $r^n = b \cos n\theta$.

EXAMPLE 7.60 Find the equation of the system of orthogonal trajectories of a series of confocal and coaxial parabolas $r = 2a/(1+\cos\theta)$. (*Osmania, 2003*)

Solution Differentiating the given equation with respect to θ, we get

$$r' = \frac{dr}{d\theta} = \frac{2a \sin\theta}{(1+\cos\theta)^2}.$$

Eliminating a, we obtain $r' = \frac{r \sin\theta}{1+\cos\theta}$, which is the differential equation governing the given family of curves. Replacing r' by $-r^2/r'$, we obtain the differential equation governing the orthogonal trajectories is

$$\frac{-r^2}{r'} = \frac{r\sin\theta}{1+\cos\theta}, \quad \text{or} \quad r' = -\frac{r(1+\cos\theta)}{\sin\theta} = -r\cot\left(\frac{\theta}{2}\right).$$

Separating the variables, we get $\frac{dr}{r} = -\cot\left(\frac{\theta}{2}\right)d\theta.$

Integrating, we obtain $\log r = -2\log\sin\left(\frac{\theta}{2}\right) + \log c \Rightarrow r = \frac{1}{\sin^2(\theta/2)}\cdot c.$

$\Rightarrow \quad r = \frac{2c}{2\sin^2(\theta/2)} = \frac{c_1}{1-\cos\theta}$ is the equation of the required trajectories, c_1 is a parameter.

EXERCISE 7.13

1. Find the orthogonal trajectories of the family of:
   (*i*) hyperbolas $x^2 - y^2 = \alpha$, (α is the parameter). **Ans.** $xy = c.$
   (*ii*) semi-cubical parabolas $ay^2 = x^3$, where a is the parameter. (*JNTU, 2005*) **Ans.** $2x^2 + 3y^2 = c.$
2. Show that the differential equation of the family of circles $x^2 + y^2 - 2ay = 0$ is $y' = 2xy/(x^2 - y^2)$. Hence write down the differential equation of the orthogonal family and by solving it, show that the equation of the family is $x^2 + y^2 - 2bx = 0$.
3. Show that the one parameter family of curves $y^2 = 4a\,(a + x)$ is self orthogonal.

4. Find the orthogonal trajectories of the family of confocal conies $\frac{x^2}{a^2} + \frac{y^2}{b^2 + \lambda} = 1$, where λ is the parameter. *(KUK, 2006; VTU, 2000S)* **Ans.** $x^2 + y^2 = 2a^2 \log x + c$.
5. Find the orthogonal trajectories of the cardioids $r = a(1 - \cos\theta)$. *(Kurukshetra, 2005)* **Ans.** $r = c(1 + \cos\theta)$.

Find the orthogonal trajectories of the following family of curves of parameter a.

6. $r = a\theta$ **Ans.** $r = ce^{-\theta^2/2}$.
7. $r^n \cos n\theta = a^n$ **Ans.** $r^n \sin n\theta = c^n$.
8. $r^n = a^n \cos n\theta$ **Ans.** $r^n = c^n \sin n\theta$.
9. $r = a(1 + \sin\theta)$ **Ans.** $r = b(1 - \sin\theta)$.
10. $r = a(1 + \cos n\theta)$ **Ans.** $r^{n^2} = c(1 - \cos^2 n\theta)$.

CHAPTER

8

Applications of Differential Equations of First Order

8.1 INTRODUCTION

In this chapter we shall discuss a few physical applications problems which lead to the differential equations of the first order and first degree.

A brief discussion of the fundamental principles required in the formulation of the differential equations is presented, to enable the reader to have a clear grasp of the problems.

8.2 PHYSICAL APPLICATIONS

8.2.1 Mechanics

Consider a body of mass m which moves with a varying velocity, that is, with an accelerated motion, along a straight line. Let it move through a distance x in time t under the action of a force F. Then

(*i*) its velocity v is given by $v = \frac{dx}{dt}$. ...(8.1)

(*ii*) its acceleration f is given by $f = \frac{d^2x}{dt^2} = \frac{dv}{dt} = v\frac{dv}{dx}$ ...(8.2)

(iii) the net effective force F, from Newton's second law of motion, is given by

$$F = mass \times acceleration = mf = m\frac{d^2x}{dt^2} = m\frac{dv}{dt} = mv\frac{dv}{dx} \quad ...(8.3)$$

If the body moves along a curved path through a distance s in time t, its velocity and acceleration are given respectively by

$$v = \frac{ds}{dt} \text{ and } f = \frac{d^2s}{dt^2}. \quad ...(8.4)$$

Resisted Motion

EXAMPLE 8.1 A body which weights w kg moves in a straight line under a force $\frac{w}{g}n^2x$ kg wt, towards $x = 0$ and an air-resistance $\frac{w}{g}kv^2$ kg wt, where v is the velocity. Find its equation of motion and determine v in terms of x given that $v = u$ at $x = 0$.

Solution From Newton's second law of motion, the net effective force $mv\frac{dv}{dx}$ must balance all the forces, m being the mass of the body.

Therefore, $$mv\frac{dv}{dx} = -\frac{w}{g}n^2x - \frac{w}{g}kv^2.$$

Since the weight $w = mg$, the above equation can be written as

$$v\frac{dv}{dx} + kv^2 = -n^2x \quad ...(i)$$

This is the equation of motion. To solve the above equation, put

$$v^2=\theta \quad\text{and}\quad v\frac{dv}{dx}=\frac{1}{2}\frac{d\theta}{dx} \text{ in } (i), \text{ we get } \frac{d\theta}{dx}+2k\theta=-2n^2x.$$

This is a linear equation. An integrating factor (I.F.) is $e^{\int 2kdx}=e^{2kx}$. Its solution is given by

$$\theta\cdot e^{2kx}=-2n^2\int x\cdot e^{2kx}\,dx+c=-2n^2\left[x\cdot\frac{e^{2kx}}{2k}-\frac{e^{2kx}}{4k^2}\right]+c.$$

At $x=0$, $v=u$, therefore $\theta=u^2=\dfrac{n^2}{2k^2}+c$ or $c=u^2-\dfrac{n^2}{2k^2}$.

Substituting for c and θ, we obtain $v^2=\left(u^2-\dfrac{n^2}{2k^2}\right)e^{-2kx}+\left(\dfrac{n^2}{2k^2}-\dfrac{n^2x}{k}\right)$.

Resisted Vertical Motion

EXAMPLE 8.2 A body of mass m, falling from rest in subject to the force of gravity and an air resistance of n^2/g times the square of velocity. Show that the distance travelled by the body in t seconds in $(g/n^2)\log\cosh nt$.

Solution Let v be the velocity of the body after falling from rest a distance x in time t.

From Newton's second law of motion, the equation of motion in given by

$$mv\frac{dv}{dx}=mg-m\frac{n^2}{g}v^2, \quad\text{that is,}\quad v\frac{dv}{dx}+\frac{n^2}{g}v^2=g \qquad ...(i)$$

This is a differential equation of the first order. To find its solution, put $v^2=\theta$ and $v\dfrac{dv}{dx}=\dfrac{1}{2}\dfrac{d\theta}{dx}$.

Eqn. (i) is transformed as $\dfrac{d\theta}{dx}+\dfrac{2n^2}{g}\theta=2g$ ...(ii)

Equation (ii) is a linear differential equation. The I.F. $=e^{\int 2\frac{n^2}{g}dx}=e^{2n^2x/g}$.

Hence the solution of Eqn. (ii) is $\theta\cdot e^{2n^2x/g}=\int 2g\cdot e^{2n^2x/g}\,dx=\dfrac{g^2}{n^2}e^{2n^2x/g}+c$.

Now at $x=0$, $v=0$, that is, $\theta=0$. Therefore, $c=-g^2/n^2$.

Hence, $\theta=v^2=\dfrac{g^2}{n^2}(-e^{-2n^2x/g}+1)$, that is, $v=\dfrac{g}{n}\sqrt{-e^{-2n^2x/g}+1}=\dfrac{dx}{dt}$ ...(iii)

Separating the variables, Eqn (iii) can be written as $\dfrac{g}{n}dt=\dfrac{e^{n^2x/g}}{\sqrt{-e^{-2n^2x/g}-1}}$.

Integrating both sides, we get $\dfrac{g}{n}t+c=\dfrac{g}{n^2}\cosh^{-1}(e^{n^2x/g})$.

At $t=0$, $x=0$, therefore $c=0$.

Hence, we obtain $e^{n^2x/g}=\cosh nt$ or $x=\dfrac{g}{n^2}\log\cosh nt$.

EXAMPLE 8.3 A body of mass m, falling from rest in subject to the force of gravity and an air resistance proportional to the square of the velocity (v^2). If it falls through a distance x and possesses velocity v at that instant, prove that $\dfrac{2k}{m}x=\log\left(\dfrac{a^2}{a^2-v^2}\right)$, where $mg=ka^2$. *(AMIETE, June 2009)*

Solution The equation of motion is

$$mv\frac{dv}{dx} = mg - kv^2 \quad \text{or} \quad mv\frac{dv}{dx} = k(a^2 - v^2). \qquad [\because mg = ka] \qquad ...(i)$$

Separating the variables and integrating, we get

$$\int \frac{vdv}{a^2 - v^2} = \int \frac{k}{m} dx + c \quad \text{or} \quad -\frac{1}{2}\log(a^2 - v^2) = \frac{kx}{m} + c. \qquad ...(ii)$$

Initially, when $x = 0$, $v = 0$. Therefore $-\frac{1}{2}\log a^2 = c$.

Substituting the value of c in *(ii)*, we obtain

$$-\frac{1}{2}\log(a^2 - v^2) = \frac{kx}{m} - \frac{1}{2}\log a^2 \quad \text{or} \quad \frac{1}{2}\log\left(\frac{a^2}{a^2 - v^2}\right) = \frac{kx}{m} \quad \text{or} \quad \frac{2k}{m}x = \log\left(\frac{a^2}{a^2 - v^2}\right).$$

8.2.2 Law of Natural Growth

(i) Population Growth

Thomas Mathus's Law states that the time rate at which the population of a country grows is proportional to the total population of the country present at that time.

Mathematically if $y(t)$ denotes the total population at time t years and y_0 the population at time $t = 0$ then this assumption can be expressed as

$$\frac{dy}{dt} = ky(t), \qquad y(t_0) = y_0 \qquad ..(8.5a)$$

where k is the constant of proportionality. If $k > 0$, we say that *y grows exponentially* and k is called the *growth constant.* The constant y_0 is called the *initial value*.

If $k < 0$, it is a decay law.

The solution of this equation is $\quad y(t) = y_0 e^{kt} \qquad ...(8.5b)$

EXAMPLE 8.4 f the population of a country doubles in 50 years, in how many years will it triple, assuming that the rate of increase is proportional to the numbers of inhabitants?

Solution Let y denote the population at time t years and y_0 the population at time $t = 0$.

By equation (8.5b), the population $y = y_0 e^{kt}$.

At $t = 50$, $y = y_0$. Therefore $2y_0 = y_0 e^{50k}$ or $e^{50k} = 2$. When $y = 3y_0$, (8.5b) gives $3 = e^{kt}$.

Then $3^{50} = (e^{50k})^t = 2^t$. Taking logarithm, we obtain $50 \log 3 = t \log 2$

or $\qquad t = (50 \log 3) / \log 2 = 50\ (0.4771 / 0.3010) = 79$ years.

EXAMPLE 8.5 If the world population in 1990 was 4.5 billion and it is growing exponentially with growth constant $k = (\log_e 3)/8$, estimate the world population in the year *(a)* 2014; *(b)* 2020.

Solution *(a)* $t = 2014 - 1990 = 24$ yrs.

By equation (8.5b), the population $y = y_0 e^{kt} = 45\,(10)^8 [e^{(\log_e 3)/8}]^{24}$

$$= 45\,(10)^8\,(e^{\log 3})^3 = 45\,(10)^8\,(3)^3 = 1215\,(10^8).$$

Thus, the population will reach 121.5 billion people in 2014.

(b) $\qquad t = 2020 - 1990 = 30$ yrs.

$$y = y_0 e^{kt} = 45\,(10)^8 [e^{(\log_e 3)/8}]^{30}$$

$$= 45\,(10)^8\,(e^{\log 3^{3.75}}) = 45\,(10)^8\,(3^{3.75})$$

$$= 45\,(10)^8\,(61.546) = 2769.57\,(10)^8$$

$$= 277 \text{ billion people.}$$

(ii) Bacterial Growth

EXAMPLE 8.6 In a certain culture of bacteria the rate of increase is proportional to the number present.

(*a*) If it is found that the number doubles in 3 hours, how many may be expected at the end of 9 hours?

(*b*) If there are 10^4 at the end of 4 hours and 2×10^4 at the end of 6 hours, how many were there in the beginning?

Solution (*a*) Let $y(t)$ denote the number of bacteria at time t hours and y_0 the number of bacteria at time $t = 0$.

Then $$\frac{dy}{dt} = ky \quad \text{or} \quad \frac{dy}{y} = k\,dt. \qquad \text{...(i)}$$

Integrating, we obtain $$\log y = kt + \log c \quad \text{or} \quad y = ce^{kt}. \qquad \text{...(ii)}$$

Since $y(t_0) = y_0$ at time $t = 0$, therefore $$c = y_0 \text{ and } y = y_0 e^{kt}. \qquad \text{...(iii)}$$

At time 3 hr, $y = 2y_0$. Then $$2y_0 = y_0 e^{3k} \text{ and } e^{3k} = 2.$$

When $t = 9$, $$y = y_0 e^{9k} = y_0 (e^{3k})^3 = y_0 (2)^3 = 8y_0,$$

that is, there are 8 times the original number.

(*b*) When $t = 4$, $y = 10^4$. Then from (*iii*), $$y_0 = 10^4/e^{4k} \qquad \text{...(iv)}$$

When $t = 6$, $y = 2 \times 10^4$. Hence $$y_0 = 2 \times 10^4/e^{6k} \qquad \text{...(v)}$$

Equating the values of y_0, we obtain $$\frac{10^4}{e^{4k}} = \frac{2 \times 10^4}{e^{6k}} \quad \text{or} \quad e^{2k} = 2.$$

Thus, the original numbers is $$y_0 = \frac{10^4}{(e^{2k})^2} = \frac{10^4}{4} = 2500 \text{ bacteria.}$$

8.2.3 Radioactive Decay

Experiments show that a radioactive substance decomposes at a rate proportional to the amount present.

The *half life* of a radioactive substance is the time required for half of the radio active nuclei present in a sample to decay. If y_0 is the number of radioactive nuclei present at time zero, the number still present at any later time t will be

$$y(t) = y_0 e^{kt}. \qquad \text{...(8.6)}$$

Here $k < 0$, we say that y decays exponentially and k is called decay constant.

We now determine at what time T will only half of the original quantity remain?

Employing (8.6), at time T, we have

$$\frac{1}{2} y_0 = y_0 e^{kt} \quad \text{or} \quad \frac{1}{2} = e^{kt}.$$ Taking logarithm, we obtain $-\log 2 = kT$ or $T = (-\log 2)/k$.

T is called the half-life of the substance and it is related to the decay constant k. The longer the half-life of a substance the more stable it is.

EXAMPLE 8.7 Find the half-life of a radioactive substance if three quarters of it is present after 8 hours.

Solution Employing (8.6), we have $\frac{3}{4} y_0 = y_0 e^{8k}$ or $\frac{3}{4} = e^{8k}$ or $e^k = (3/4)^{1/8}$.

Again $$\frac{1}{2} y_0 = y_0 e^{kT} \quad \text{or} \quad \frac{1}{2} = (e^k)^T = \left(\frac{3}{4}\right)^{T/8}.$$

Taking logarithm, $$-\log 2 = \frac{T}{8}(\log 3 - \log 4) \quad \text{or} \quad T = \frac{8 \log 2}{\log 4 - \log 3}$$

$$= \frac{8\,(0.3010)}{0.6021 - 0.4771} = \frac{8\,(0.3010)}{0.125} = 19.26 \text{ hours.}$$

EXAMPLE 8.8 The half-life of a radium is 1690 years. How much will be left of 32 gm of radium after 6760 years?

Solution Employing $T = -\dfrac{\log 2}{k}$ we get $k = -\dfrac{\log 2}{1690}$.

The quantity of radium is given by $y = y_0 e^{-(\log 2)^{t/1690}}$.

Substituting $y_0 = 32$ and $t = 6760$, we get the quantity

$$y = 32\, e^{-(\log 2)^{6760/1690}} = \frac{32}{e^{(\log 2)^4}} = \frac{32}{2^4} = 2 \text{ gm}.$$

EXAMPLE 8.9 Uranium disintegrates at a rate proportional to the amount present at any instant. If M_1 and M_2 grams of uranium are present at times T_1 and T_2 respectively, show that the half-life of a uranium is

$$\frac{(T_2 - T_1)\log 2}{\log (M_1/M_2)}.$$

Solution Let the mass of uranium at any time t be m grams.

Then the equation of disintegration of uranium is $\dfrac{dm}{dt} = -\,km$, where k is a constant.

Integrating, we obtain $\displaystyle\int \frac{dm}{m} = -k \int dt + c$ or $\log m = c - kt$ ...(*i*)

Initially, when $t = 0$, $m = M$ (say), so that $c = \log M$.

Therefore, equation (*i*) becomes $kt = \log M - \log m$. ...(*ii*)

Also when $t = T_1$, $m = M_1$ and when $t = T_2$, $m = M_2$.

Therefore, from (*ii*), we get $kT_1 = \log M - \log M_1$ ...(*iii*)

and $kT_2 = \log M - \log M_2$ ...(*iv*)

Subtracting (*iii*) from (*iv*), we get

$$k(T_2 - T_1) = \log M_1 - \log M_2 = \log (M_1/M_2)$$

or
$$k = \frac{\log (M_1/M_2)}{T_2 - T_1}. \quad ...(v)$$

Let the mass reduce to half its initial value in time T, that is, when $t = T$, $m = M/2$.

$\therefore$ From (*ii*), we get $kt = \log M - \log (M/2) = \log 2$

or
$$T = \frac{1}{k}\log 2 = \frac{(T_2 - T_1)\log 2}{\log (M_1/M_2)}, \text{ from } (v).$$

8.2.4 Newton's Law of Cooling

Newton's law of cooling states that the rate of change of the temperature of an object is proportional to the difference of the temperature between the object and the surroundings. If $\theta(t)$ is the temperature of the object, at time t, kept in a medium which is maintained at a constant temperature θ_0,

then $\dfrac{d\theta}{dt} \propto (\theta - \theta_0)$. That is, $\dfrac{d\theta}{dt} = k(\theta - \theta_0)$, ...(8.7)

where k is the constant of proportionality.

Solving this equation, $\dfrac{d\theta}{\theta - \theta_0} = k\, dt$ which on integration, gives

$$\log (\theta - \theta_0) = kt + \log c, \text{ that is, } \theta - \theta_0 = ce^{kt}, \Rightarrow \theta = \theta_0 + ce^{kt}.$$

If two conditions are given about the temperature of the object, then we find the arbitrary constant c and the constant of proportionality k. A typical condition is that when $t = 0$, $\theta = \theta_I$, the initial temperature of the object. In such a case $\theta_I = \theta_0 + c$. Thus $c = \theta_I - \theta_0$ and $\theta = \theta_0 + (\theta_I - \theta_0)e^{kt}$.

The other typical condition is that when $t = t_1$, $\theta = \theta_1$, which enables us to calculate k. We are thus in a position to calculate θ for any given time t, or the time taken to reach a given temperature.

EXAMPLE 8.10 The temperature of a cup of coffee is 92°C, when freshly poured, the room temperature being 23°C. In one minute it has cooled to 87°C. How much time must elapse before the temperature of the cup becomes 65°C.

Solution By Newton's law of cooling, $d\theta/dt = (\theta - 23)k$, where $\theta(t)$ is the temperature of the coffee in degrees centigrade at time t minutes.

On integration, we get $\theta = 23 + ce^{kt}$,

when $t = 0$, $\theta = 92°\text{C}$ (given) we get $c = \theta - 23 = 92 - 23 = 69$.

Then $\theta = 23 + 69\, e^{kt}$.

When $t = 1$, $\theta = 87°\text{C}$. Thus, $87° = 23 + 69\, e^{k} \Rightarrow k = \log_e (64/69)$.

When $\theta = 65°\text{C}$, $65 = 23 + 69\, e^{kt}$, which gives $t = \log_e (42/69)/k = \log_e (42/69)/\log_e (64/69)$

$$= \frac{1.62325 - 1.83885}{1.80618 - 1.82885} = \frac{21560}{3267} = 6.6 \text{ minutes}.$$

EXAMPLE 8.11 A body originally at 80°C cools down to 60°C in 20 minutes, the temperature of the air being 40°C. What will be the temperature of the body after 40 minutes from the original?

Solution By Newton's law of cooling, $d\theta/dt = (\theta - 40)k$, where $\theta(t)$ is the temperature of the body in °C at time t minutes.

On integration, we get $\theta = 40 + ce^{kt}$.

When $t = 0$, $\theta = 80°\text{C}$ (given) we get $c = \theta - 40 = 80 - 40 = 40$.

Then $\theta = 40 + 40\, e^{kt}$.

When $t = 20$ min, $\theta = 60°\text{C}$. Thus, $60° = 40 + 40\, e^{20k} \Rightarrow k = (1/20) \log_e (1/2)$.

When $t = 40$ min, $\theta = 40 + 40\, e^{40k}$

$$\Rightarrow \quad \log_e \left(\frac{\theta - 40}{40}\right) = 40k = 40\left(\frac{1}{20}\log_e 2^{-1}\right) = -2\log_e 2$$

$$\text{or} \quad \log_e \left(\frac{40}{\theta - 40}\right) = \log_e 2^2 \quad \text{or} \quad \frac{40}{\theta - 40} = 4 \quad \text{or} \quad \theta = 50°\text{C}.$$

8.2.5 Heat Conduction

When different points of a solid body are at different temperatures, heat flows from regions at higher temperature to those at lower temperatures. The quantity Q of heat that will be conducted in unit time will be given by

$$Q = -k\,A\frac{dT}{dx}, \qquad ...(8.8)$$

where A is the area of the body through which the heat flows, dt/dx in the temperature gradient at any point, k is constant called the coefficient of thermal conductivity of the solid and depend essentially on the material of the body. The quantity $\frac{Q}{A} = -k\,\frac{dT}{dx}$ is called the flux of heat across the surface and negative sign is prefixed because with x increasing, T decreases.

EXAMPLE 8.12 A steam pipe 20 cm in diameter is protected with a covering 6 cm thick for which k = 0.0003 cal/sec cm °C. In the steady state, find

(*a*) the heat loss per hour through a meter length of the pipe, if the surface of the pipe is 200°C and that of the outer surface of the covering is at 30°C.

(*b*) the temperature at a distance $x > 10$ cm from the centre of the pipe.

Solution

(*a*) Here the isothermal surfaces are circular cylinders, the axis of each one of them is the axis of the pipe. Consider one such cylinder whose radius is x cm and whose length is 1 metre. The lateral surface of the cylinder is $200\,\pi\,x$ cm^2. For steady state, the quantity of heat passing through this surface is constant. Let Q be the quantity of heat flowing across this surface, then

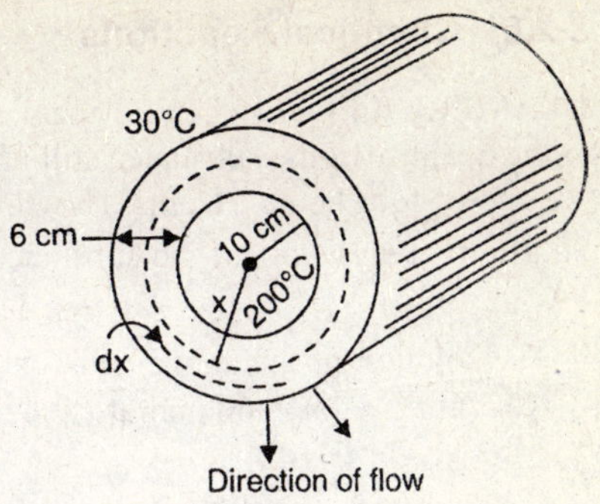

30°C
6 cm
10 cm
200°C
x
dx
Direction of flow

Fig. 8.1

$$Q = -k\,A\frac{dT}{dx} = -\,200\,\pi\,x\,k\frac{dT}{dx}$$

or $$Q\frac{dx}{x} = -\,200\,\pi\,k\,dT \qquad \text{...}(i)$$

Now at $x = 10$ cm, $T = 200$°C and at $x = 16$ cm, $T = 30$°C.
Integrating (i) under the given conditions, we get

$$Q\int_{10}^{16}\frac{dx}{x} = -200\,k\,\pi\int_{200}^{30} dT \quad \text{or} \quad Q(\log_e 16 - \log_e 10) = 200\,k\,\pi\,(200 - 30)$$

or $$Q\log_e 1.6 = 34000\,k\,\pi \quad \text{or} \quad Q = 34000\,k\,\pi\,/\,\log_e 1.6$$

$$= 34000 \times 0.0003 \times 22\,/\,7 \times \log_e 1.6 \text{ cal/sec.}$$

Now the quantity of heat flowing in one hour

$$= \frac{34000 \times 0.0003 \times 22 \times 3600}{7 \times 0.47 \times 1000}\,k \text{ cal} = 245.54\,k \text{ cal.}$$

(*b*) Integrating $200\,\pi\,k\,dT = \dfrac{-34000\,k\,\pi}{\log_e 1.6}\dfrac{dx}{x}$ between the limits $T = 30$, $x = 16$ and $T = T$, $x = x$, we get

$$\int_{30}^{T} dT = \frac{-170}{\log_e 1.6}\int_{16}^{x}\frac{dx}{x} \quad \text{or} \quad T - 30 = -\frac{170}{\log_e 1.6}\log_e\frac{x}{16}$$

or $$T = \left(30 + \frac{170}{\log_e 1.6}\log_e\frac{16}{x}\right)^{\circ} C.$$

Check: When $x = 10$ cm, $T = 30 + 170 = 200$°C. When $x = 16$ cm, $T = 30 + 0 = 30$°C.

EXAMPLE 8.13 Calculate the amount of heat per sq. cm. per hour passing through a refrigerator wall, if the thickness of the wall is 6 cm and the temperature inside the refrigerator is 0°C, while outside it is 20°C. Assume $k = 0.0002$ cal s^{-1} cm^{-1} °C^{-1}.

Solution Let x denote the distance of any point within the wall from the outer surface. Let Q be the quantity of heat flowing at that point. Then

$$Q = -k\,A\frac{dT}{dx}.$$

At $x = 0$, $T = 20$ and when $x = 6$, $T = 0$. Integrating the above equation within the given limits, we get

$$Q\int_0^6 dx = -kA\int_{20}^{0} dT \quad \text{or} \quad Q[x]_0^6 = -kA\,[T]_{20}^{0} \quad \text{or} \quad 6Q = 20kA$$

Therefore, $$Q = \frac{20 \times 0.0002}{6}\,cal/\text{sec}\cdot cm^2 = 0.000667\,cal/\text{sec}\cdot cm^2$$

$$= 3600 \times 0.000667 = 2.4 \text{ cal/cm}^2 \cdot \text{hour}$$

8.2.6 Chemical Reactions

EXAMPLE 8.14 In a certain chemical reaction, the rate of conversion of a substance at time t is proportional to the quantity of the substance still untransformed at the instant. At the end of one hour 60 gms remain and at the end of four hours 21 gms. How many grams of the first substance were there initially?

Solution Let m gms of the substance remain after t hours. Then the differential equation of the reactions is

$$dm/dt = -km \text{ where } k \text{ is a constant.}$$

The solution of the above equation is given by $m = ce^{-kt}$ ...(*i*)

Now at $t = 1$, $m = 60$ and at $t = 4$, $m = 21$. Substituting these values in (*i*), we get

$$60 = ce^{-k}, \text{ and } 21 = ce^{-4k} = c\left(\frac{60}{c}\right)^4.$$

This gives $21\,c^3 = 60 \times 60 \times 60 \times 60$ or $c = 85.14$

Now initially $t = 0$, therefore, $m = c = 85$ gms approximately.

8.2.7 Mixing Problem

EXAMPLE 8.15 A tank contains 5 cubic metres of fresh water. Brine containing 10 kg of salt/cubic metre runs in at the rate of 0.1 cubic meter/min. The mixture kept uniform by stirring, runs out at the rate of 0.05 cubic metres/min. Find the amount of salt present in the tank at time t.

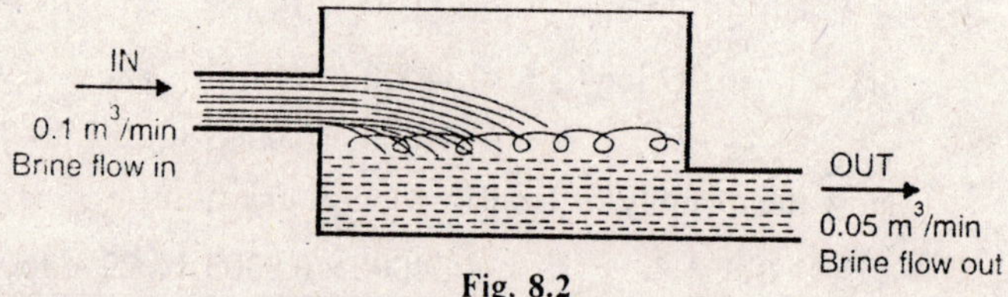

Fig. 8.2

Solution Let A kg be the quantity of salt in the tank at time t minutes.

Since dA/dt is the rate of change of A with respect to t, we have

dA/dt = (rate of substance entering) – (rate of substance leaving). ...(*i*)

Now substance entering per minute = $10 \times 0.1 = 1$ kg. ...(*ii*)

Brine flows in at the rate of 0.1 cubic metre/min and runs out at the rate of 0.05 cubic metres/min and therefore brine in the tank increases at the rate of 0.05 cu metres/min.

Thus brine in the tank at time t minutes is $(5 + 0.05t)m^3$ and concentration of salt at that time is $A/(5 + 0.05t)$.

As 0.05 cubic metres of brine leaves the tank per minute, amount of substance lost per minute $= \left(\frac{0.05A}{5 + 0.05t}\right) kg$ ...(*iii*)

Using (*ii*) and (*iii*), the equation (*i*) can be written as $\frac{dA}{dt} = 1 - \frac{0.05\,A}{5 + 0.05\,t} = 1 - \frac{A}{100 + t}$

that is, $\frac{dA}{dt} + \frac{A}{100 + t} = 1$ ......(*iv*) which is a linear differential equation.

$$\text{I.F.} = e^{\int 1/(100+t)\,dt} = e^{\log(100+t)} = 100 + t.$$

Therefore, the solution in (*iv*) is $(100 + t)\,A = \int 1 \cdot (100 + t)\,dt + c = 100\,t + (t^2/2) + c.$

Now $A = 0$ when $t = 0$. Thus $c = 0$.

$\therefore$ $(100 + t)\,A = 100\,t + (t^2/2)$

or $A = t\,(200 + t)/2\,(100 + t)$, the required result.

8.2.8 Electric Circuits

Electric circuits can be regarded to be made up of

(i) a source of electric energy (electromotive force) such as a generator or a battery,

(ii) three passive elements namely resistance, inductance and capacitance.

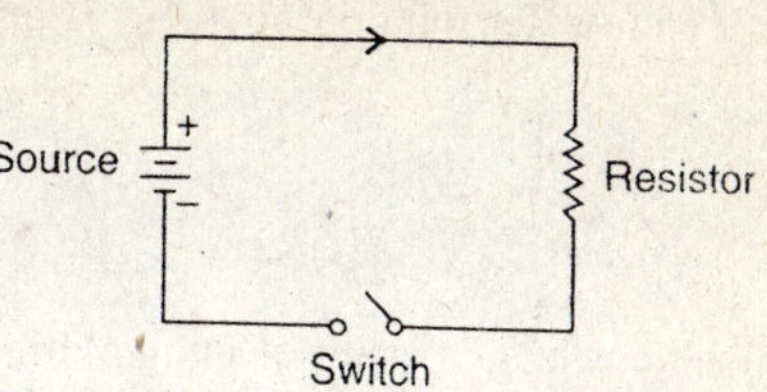

Fig. 8.3: Simple electric circuit

The simplest electric circuit is a series circuit which consists of a resistor and a source of energy as shown in Fig 8.3. If we close the switch, a current i will flow through the resistor and the voltage drop E_R by Ohm's law is given by

$$E_R = Ri \qquad \text{...(8.9)}$$

The more complicated electric circuits deal with inductors which oppose a change in current and capacitors which oppose a change in voltage or which store energy. It has been established experimentally that for an inductor, and for a capacitor, the voltage drop E_L and E_C are given respectively by

$$E_L = L\frac{di}{dt}, \qquad \text{...(8.10)}$$

$$E_C = \frac{1}{C}q = \frac{1}{C}\int_{t_0}^{t} i\,dt \qquad \text{...(8.11)}, \qquad \text{since } i = dq/dt$$

where R is the resistance in ohms, L is the inductance in henries, C is the capacitance in farads, i is the current in amperes, q is the charge in coulombs and E is the electromotive force (e.m.f.) in volts, white the time t is measured in seconds.

The study of the circuit built by above elements is governed by the following two laws of Kirchhoff's.

1. In a closed circuit the algebraic sum of the voltage drops across each element of the circuit is equal to the resultant electromotive force.
2. At any point of a circuit, the sum of the inflowing currents is equal to the sum of outflowing currents.

Remark The readers are advised to remember the following two integrals which are generally used in Electric circuits problems.

(i)
$$\int e^{ax}\sin bx\,dx = \frac{e^{ax}}{a^2+b^2}(a\sin bx - b\cos bx)$$
$$= \frac{e^{ax}}{\sqrt{a^2+b^2}}\sin[bx - \tan^{-1}(b/a)].$$

(ii)
$$\int e^{ax}\cos bx\,dx = \frac{e^{ax}}{a^2+b^2}(a\cos bx + b\sin bx)$$
$$= \frac{e^{ax}}{\sqrt{a^2+b^2}}\cos[bx - \tan^{-1}(b/a)].$$

Differential Equations of Electric Circuits

RL-Circuits : Let us consider an electric circuit containing resistance R and inductance L and let an e.m.f. E be applied to it as shown in Fig. 8.4. Applying Kirchhoff's first law, we have the total voltage drop across R and L equal to E.

That is,
$$L\frac{di}{dt} + Ri = E(t)$$

or
$$\frac{di}{dt} + \frac{R}{L}i = \frac{E}{L} \qquad \text{...(8.12)}$$

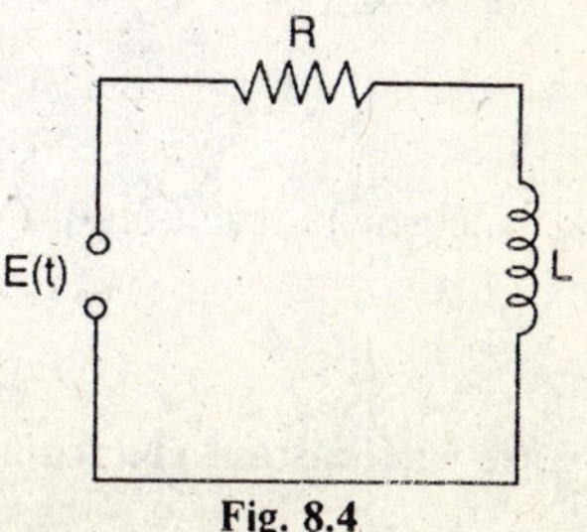

Fig. 8.4

Equation (8.12) is a linear differential equation of the first order and its integrating factor is $e^{\int \frac{R}{L}dt} = e^{\frac{Rt}{L}}$.

Hence the solution of (8.12) is given by

$$i(t) = e^{(-Rt)/L} \int \frac{E}{L} e^{Rt/L}\, dt + \lambda e^{(-Rt)/L}$$

$$= \frac{E}{R} + \lambda e^{(-Rt)/L}, \qquad \text{...(8.13)}$$

where λ is a constant of integration to be determined from the initial conditions.

We shall consider the following cases.

Case A: Constant electromotive force $E(t) = E_0$. Let at time $t = 0$, the circuit be connected to a constant electromotive force $E(t) = E_0$.

Equation (8.13) gives the solution

$$i(t) = \frac{E_0}{R} + \lambda e^{(-Rt)/L} \qquad \text{...(8.14)}$$

From (8.14) we see that $i(0) = E_0/R + \lambda$. Hence for the initial condition $i(0) = 0$ we get $\lambda = -E_0/R$ and from (8.14) the particular solution

$$i(t) = \frac{E_0}{R}(1 - e^{-Rt/L}) \qquad \text{...(8.15)}$$

Case B: Periodic electromotive force $E(t) = E_0 \sin \omega t$.

For this E(t), equation (8.12) is

$$i(t) = e^{-Rt/L}\left[\frac{E_0}{L}\int e^{(Rt)/L} \sin \omega t\, dt + \lambda\right],$$

Integration by parts yields

$$i(t) = \lambda e^{-(R/L)t} + \frac{E_0}{R^2 + \omega^2 L^2}(R \sin \omega t - \omega L \cos \omega t) \qquad \text{...(8.16)}$$

This may also be re-written as

$$i(t) = \lambda e^{-(R/L)t} + \frac{E_0}{\sqrt{R^2 + \omega^2 L^2}} \sin(\omega t - \phi), \qquad \text{...(8.17)}$$

where $\phi = \tan^{-1}(\omega L/R)$.

RC-Circuits Let an electromotive force $E(t)$ be applied to an electric circuit containing resistance R in series with a capacitor having capacitance C. Using Kirchhoff's law, the equation of the circuit is given by

$$Ri + \frac{1}{C} q = E(t) \qquad \text{...(8.18)}$$

Differentiating (8.18) with respect to t, remembering that current i and charge q are related by $i = dq/dt$ so we can write (8.18) in the form

$$R\frac{di}{dt} + \frac{1}{C} i = \frac{dE}{dt} \qquad \text{...(8.19)}$$

Dividing (8.19) by R, we obtain

$$\frac{di}{dt} + \frac{1}{RC} i = \frac{1}{R}\frac{dE}{dt} \qquad \text{...(8.19A)}$$

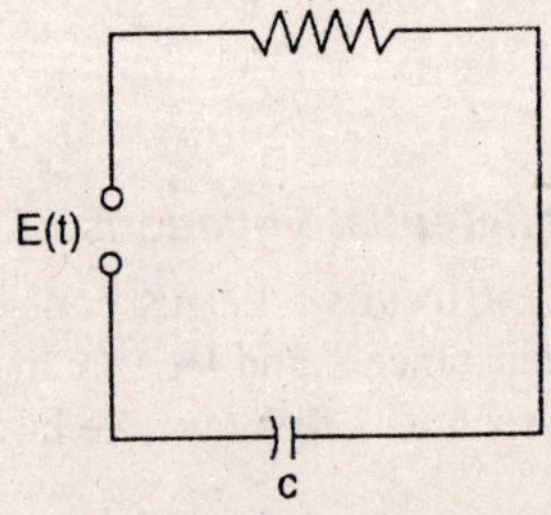

R

E(t)

C

Fig. 8.5: RC-circuit

The general solution of (8.19A) is given by

$$i(t) = e^{-t/(RC)}\left[\frac{1}{R}\int e^{t/(RC)} \frac{dE}{dt}\, dt + \lambda\right] \qquad \text{...(8.20)}$$

Case A: Constant electromotive force. If E = constant, then $dE/dt = 0$ and Eq. (8.20) become

$$i(t) = \lambda e^{-t/(RC)} \qquad \text{...(8.21)}$$

Case B: Periodic electromotive force $E(t) = E_0 \sin \omega t$.

For this $E(t)$, we have $$dE/dt = \omega E_0 \cos \omega t.$$

Substituting this value in (8.20) and integrating by parts, we obtain

$$i(t) = \lambda e^{-t/(RC)} + \frac{\omega E_0 C}{1+(\omega RC)^2}(\cos \omega t + \omega RC \sin \omega t) \quad ...(8.21)$$

$$= \lambda e^{-t/(RC)} + \frac{\omega E_0 C}{\sqrt{1+(\omega RC)^2}} \sin(\omega t - \phi), \quad ...(8.22)$$

where $\tan \phi = -1/(\omega RC)$.

RLC-Circuits: Consider a circuit, Fig 8.6, containing resistor of resistance R (ohms) an inductor of inductance L (henrys) and a capacitor of capacitance C (farads) are connected in series to a source of electromotive force $E(t)$ (volts), where t is time.

The voltage drops across the inductor is $L\frac{di}{dt}$, across the resistor is Ri and aross the capacitor is $\frac{1}{C}\int i(t)\,dt$. The differential equation of an RLC electric circuit is therefore,

$$L\frac{di}{dt} + Ri + \frac{1}{C}q = E(t) \quad ...(8.23)$$

Since $i = dq/dt$, $di/dt = d^2q/dt^2$

equation (8.23) can be written as $$L\frac{d^2q}{dt^2} + R\frac{dq}{dt} + \frac{q}{C} = E(t) \quad ...(8.24)$$

from which $q = q(t)$ may be found. (Refer chapter 9).

By differentiating Eq. (8.23) and using $dq/dt = i$, we have

$$L\frac{d^2i}{dt^2} + R\frac{di}{dt} + \frac{i}{C} = E'(t)$$

from which $i = i(t)$ may be found. (Refer chapter 9).

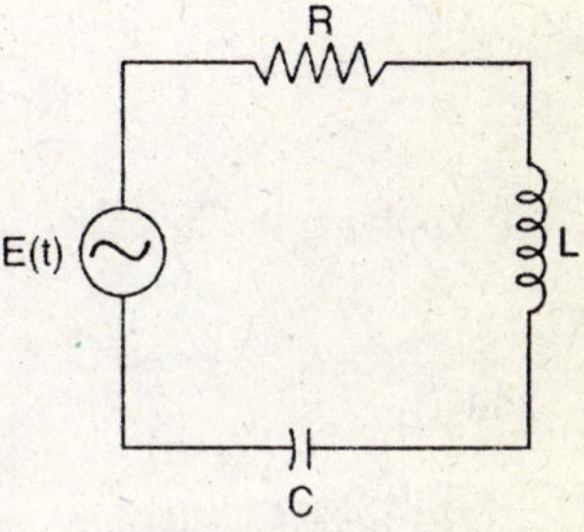

Fig. 8.6: RLC-circuit

EXAMPLE 8.16 The initial value problem governing the current i flowing in a series RL circuit when a voltage $v(t) = t$ is applied is given by $iR + L\frac{di}{dt} = t$, $t \geq 0$, $i(0) = 0$, where R and L are constants. Find the current $i(t)$ at time t. *(AMIETE, June, 2009; Dec 2008)*

Solution The given differential equation may be rewritten as

$$\frac{di}{dt} + \frac{R}{L}i = \frac{1}{L}t. \quad ...(i)$$

An integrating factor is $e^{\int (R/L)dt} = e^{(Rt)/L}$. The solution of equation (i) is given by

$$i(t) = e^{-(Rt)/L}\left[\int \frac{1}{L}\underset{\text{II}}{e^{(Rt)/L}}\;\underset{\text{I}}{t}\,dt + \lambda\right]$$

$$= e^{-(Rt)/L}\frac{1}{L}\left[\frac{e^{(Rt)/L}}{(R/L)}t - \int 1\cdot\frac{e^{(Rt)/L}}{(R/L)}dt\right] + \lambda e^{-(Rt)/L} = \frac{1}{R}\left[t - \frac{L}{R}\right] + \lambda e^{-(Rt)/L}.$$

Applying the initial condition $i(0) = 0$, we obtain

$$0 = -\frac{L}{R^2} + \lambda, \quad \text{or} \quad \lambda = \frac{L}{R^2}.$$

Hence, the current $i(t)$ at time t is given by

$$i(t) = \frac{t}{R} + \frac{L}{R^2}[e^{-(Rt)/L} - 1].$$

EXAMPLE 8.17 One of the basic equation in electric circuits is

$$L\frac{di}{dt} + Ri = E(t), \qquad \text{...}(i)$$

where L (henries) is called the inductance, R (ohm) the resistance, i (amperes) the current, and E (volts) the electromotive force. Solve eq. (i) when $L = 3$ henries, $R = 15$ ohm, $E(t) = E_0 \sin \omega t = 110 \sin 120\,\pi\, t$.

Solution We have

$$3\frac{di}{dt} + 15\,i = 110 \sin 120\,\pi\,t \quad \text{or} \quad \frac{di}{dt} + 5i = \frac{110}{3} \sin 120\,\pi\,t.$$

An integrating factor is $e^{\int 5dt} = e^{5t}$. The solution of equation (i) is given by

$$i(t) = e^{-5t} \int \frac{110}{3} (\sin 120\,\pi t) \cdot e^{5t}\, dt + \lambda e^{-5t}$$

$$= \frac{110}{3} \cdot \frac{5 \sin (120\,\pi t) - 120\,\pi \cos (120\,\pi\,t)}{25 + 14400\,\pi^2} + \lambda e^{-5t}$$

$$= \frac{22}{3} \cdot \frac{\sin (120\,\pi t) - 24\,\pi \cos (120\,\pi\,t)}{1 + 576\,\pi^2} + \lambda e^{-5t}.$$

Applying the initial condition i(0) = 0, we obtain $\lambda = \dfrac{(22)\,(24\pi)}{3\,(1 + 576\,\pi^2)}$,

and

$$i(t) = \frac{22}{3} \cdot \left[\frac{\sin (120\,\pi t) - 24\,\pi \cos (120\,\pi\,t) + 24\,\pi\, e^{-5t}}{(1 + 576\,\pi^2)}\right]$$

The above result may be written in an alternative way:

$$i(t) = \frac{22}{3} \cdot \frac{1}{\sqrt{1 + 576\,\pi^2}} \left[\sin (120\,\pi t) \cdot \frac{1}{\sqrt{1 + 576\,\pi^2}} - \cos (120\,\pi\,t) \cdot \frac{24\,\pi}{\sqrt{1 + 576\,\pi^2}}\right] + \frac{176\,\pi\, e^{-5t}}{1 + 576\,\pi^2}$$

$$= \frac{22}{3\sqrt{1 + 576\,\pi^2}} \left[\sin (120\,\pi t) \cdot \cos \phi - \cos (120\,\pi\,t) \cdot \sin \phi\right] + \frac{176\,\pi\, e^{-5t}}{1 + 576\,\pi^2}$$

$$= \frac{22}{3\sqrt{1 + 576\,\pi^2}} \sin (120\,\pi t - \phi) + \frac{176\,\pi\, e^{-5t}}{1 + 576\,\pi^2},$$

where $\sin \phi = 24\,\pi \big/ \sqrt{1 + 576\,\pi^2}$ and $\cos \phi = 1 \big/ \sqrt{1 + 576\,\pi^2}$.

EXAMPLE 8.18 The equation of an electromotive force in terms of current i for an electrical circuit having resistance R and a condenser of capacity given by C, in series is given by

$$E = Ri + \int \frac{i}{C}\, dt.$$

Find the current i at any time t, when $E = E_0 \sin \omega t$. (*NIT, Kurukshetra, 2007; PTU, 2006*)

Solution Replacing E by $E_0 \sin \omega t$ in the given equation and differentiating with respect to t, we obtain

$$R\frac{di}{dt} + \frac{i}{C} = \omega E_0 \cos \omega t \quad \text{or} \quad \frac{di}{dt} + \frac{i}{RC} = \frac{\omega E_0}{R} \cos \omega t \qquad \text{...}(i)$$

Equation (i) is a linear differential equation of the first order.

An integrating factor is $e^{t/RC}$. Hence its solution is given by

$$i(t)\, e^{t/RC} = \frac{\omega E_0}{R} \int e^{t/RC} \cos \omega t \, dt + \lambda, \quad \lambda \text{ is constant of integration}$$

$$= \frac{\omega E_0 C}{1+\omega^2 R^2 C^2} e^{t/RC} (\cos \omega t + \omega RC \sin \omega t) + \lambda$$

$$\therefore \qquad i(t) = \frac{\omega E_0 C}{1+\omega^2 R^2 C^2} [\cos \omega t + (\omega RC) \sin \omega t] + \lambda e^{-t/RC}$$

or

$$i(t) = \frac{\omega E_0 C}{\sqrt{1+\omega^2 R^2 C^2}} \sin (\omega t + \phi) + \lambda e^{-t/RC}, \text{ where } \phi = \tan^{-1} [1/(\omega RC)].$$

EXERCISE 8.1

1. The velocity v of a parachute falling vertically satisfies the equation $v\frac{dv}{dx} = g\left(1-\frac{v^2}{k^2}\right)$, where g and k are constants. If v and x be both initially zero, find the expression for v^2 in terms of x.
Ans. $v^2 = k^2 (1-e^{-2gx/k^2})$.
2. The acceleration and velocity of a body falling in the air approximately satisfy the equation $f = g - kv^2$, where g and k are constants. Find the distance traversed as a function of the time, if the body falls from rest. **Ans.** $s = (1/k) \log \cosh (t\sqrt{gk})$.
3. A particle of mass m in projected vertically upwards with initial velocity v_0. The resisting force at any time is k times the velocity. Formulate the differential equation of motion and show that the distance s covered by the particle at any time t is given by
$$s = \left(\frac{g}{k^2}+\frac{v_0}{k}\right)(1-e^{-kt}) - \frac{g}{k}t.$$
4. A particle falls under gravity in a resisting medium whose resistance varies with velocity. Find the relation between distance and velocity if initially the particle starts from rest. *(UPTU, 2003)*
Ans. $s = \frac{g}{\lambda^2}\log\left(\frac{g}{g-\lambda v}\right) - \frac{v}{\lambda}$, where λ is constant of proportionality.
5. A body falls vertically under the action of gravity and suffers a deceleration proportional to its velocity. Find its velocity as a function of time. **Ans.** $v = (g/k)(1 - e^{-kt})$
6. The population of a village increases continuously at the rate proportional to the number of its inhabitants present at any time. If the population of the village was 20,000 in 2005 and 25,000 in the year 2010, what will be the population of the village in 2015? **Ans.** 31250.
7. If the population of a country is 100 million people and the population is increasing exponentially with a growth constant $k = \log 2$, calculate precisely the population after 5 years.
Ans. 3.2 billion people.
8. The number N of bacteria in a culture grew at a rate proportional to N. The value of N was initially 100 and increased to 332 in one hour. What would be the value of N after 1.5 hours? *(JNTU, 2003)*
Ans. 605.
9. The rate at which bacteria multiply is proportional to the instantaneous number present. If the original number doubles in 2 hours, in how many hours will it triple? *(Andhra, 2000)*
Ans. 3.17 hour, $(= 2 \log_e 3/\log_e 2)$.

10. If in a culture of yeast the rate of growth $y'(t)$ is proportional to the amount $y(t)$ present at time t, and if $y(t)$ doubles in one day, how much can be expected (*i*) after 3 days at the same rate of growth (*ii*) after one week? **Ans.** 8; 128.

11. Neutrons in an atomic pile increases at a rate proportional to the number of neutrons present at any instant. If N_0 neutrons are present initially and N_1 and N_2 neutrons are present at time t_1 and t_2 respectively show that

$$(N_2/N_0)^{t_1} = (N_1/N_0)^{t_2}.$$

[**Hint:** The law of growth gives us $dN/dt = kN$, solution of which is $N = ce^{kt}$.
At time $t = 0$, $N = N_0$, then $c = N_0$. Then at any time t, $N = N_0\,e^{kt}$.
By the given conditions $N_1 = N_0 e^{kt_1}$ and $N_2 = N_0 e^{kt_2}$.
These two imply, $(N_1/N_0)^{t_2} = e^{kt_1t_2} = (N_2/N_0)^{t_1}$. Hence the result.]

12. A radioactive substance disintegrates at a rate proportional to its mass. When the mass is 10 mgm, the rate of disintegration is 0.051 mgm/day. How long will it take for the mass to be reduced from 10 mgm to 5 mgm. **Ans.** 136 days.

13. After 2 days, 10 gms of radioactive chemical is present. Three days later 5 grams is present. How much of the chemical was present initially assuming the rate of disintegration is proportional to the instantaneous amount which is present. **Ans.** 21.54 gm.

14. How long does it take for 90% of a given quantity of the radioactive element to decay, given that its half-life in 5.3 years. **Ans.** About 17.6 years.

15. In a certain chemical reaction, the rate of conversion of a substance at time t is proportional to the quantity of the substance still untransformed at that instant. At the end of one hour 60 gm remain and at the end of four hours 21 gm. How many grams of the first substance were there initially. **Ans.** 85 gm (approx).

16. According to Newton's law of cooling, the rate at which a substance cools in moving air is proportional to the difference between the temperature of the substance and that of the air. If the temperature of the air is 27°C and the substance cools from 97°C to 67°C in 15 minutes, find when the temperature will be 37°C. **Ans.** 52 minutes.

17. If the air is maintained at 17°C and the temperature of a body cools from 97°C to 57°C in 10 minutes, find the temperature of the body after 40 minutes. **Ans.** 22°C.

18. A body cools in 7 minutes from 60°C to 40°C. What will be its temperature after the next 7 minutes? The temperature of surroundings is 10°C. **Ans.** 28°C

19. A cup of coffee at temperature 100°C is placed in a room whose temperature is 15°C and it cools to 60°C in 5 minutes. Find its temperature after a further interval of 5 minutes. **Ans.** 38.9°C.

20. An object whose temperature 75°C cools in an atmosphere of constant temperature 25°C at the rate of kθ, being the excess temperature over that of the atmosphere. If after 10 minutes, the temperature of the object falls to 65°C, find its temperature after 20 minutes. Also find the time required to cool down to 55°C. **Ans.** 57°C; 22.9 minutes.

21. A steam pipe 20 cm in diameter contains steam at 150°C and is covered by a layer of insulation 5 cm thick. The outside temperature is kept at 60°C. By how much should the thickness of covering be increased in order that the rate of heat loss should be decreased by 25%? **Ans.** 2.16 cm.

22. A pipe 20 cm in diameter contain steam at 150°C and is protected with a covering 5 cm, thick for which $k = 0.0025$ cal · cm^{-1} °C^{-1} sec^{-1}. If the temperature of the outer surface of the covering is 40°C, find the temperature half way through the covering under steady state conditions. **Ans.** $T = 89.5$°C.

23. A pipe 10 cm in diameter contain steam at 100°C. It is to be covered with two coats of insulating material, each being 2.5 cm thick, one of asbestos ($k = 0.0006$ cal · cm^{-1} °C^{-1} sec^{-1}) and the other of magnesia ($k = 0.00017$ cal · cm^{-1} °C^{-1} sec^{-1}). If the outside surface temperature is 30°C, find the heat loss per hour from a metre length of pipe when the asbestos is inside and magnesia outside, and vice versa and compare the two results for efficiency of insulation. **Ans.** 67,000 cal/hour, 55,000 cal/hour.

24. A tank contains 0.5 cubic metres of brine in which 30 kgm of salt is dissolved. Fresh water runs into the tank at the rate of 15×10^{-5} cubic metres/sec and the mixture kept uniform by stirring runs out at the same rate. How much salt is in the tank after 1 hour? **Ans.** $30e^{-1.08}$ kgm.

[**Hint:** If m kgm is the amount of salt after t secs.

$$\frac{dm}{dt} = \frac{0 - 15(10^{-5})m}{0.5} = -3\,m\,(10^{-4}).$$

On solving the differential equation, we obtain

$$\log m = -3\,(10^{-4})\,t + \log c \quad \text{or} \quad m = ce^{-3(10^{-4})t}$$

At $t = 0$, $m = 30$ kgm and hence $c = 30$.

Therefore, $m = 30\,e^{-3(10^{-4})t}$. When $t = 3600$, $m = 30e^{-1.08}$ kg].

25. A tank contains 5 cu meters of brine made by dissolving 30 kg of salt in water. Salt water containing 10 kg of salt/cubic metre runs in at the rate of 0.1 cubic metre/min and the mixture, kept uniform by stirring, runs out at the rate of 0.15 cubic metres/min. Find the amount of salt in the tank after 1 hour.

[**Hint:** $$\frac{dm}{dt} = 1 - \frac{m}{5 - 0.05\,t} \times 0.15 = 1 - \frac{3m}{100 - t}$$

or $$\frac{dm}{dt} + \frac{3m}{100 - t} = 1 \text{ which gives}$$

$$\frac{m}{(100 - t)^3} = \int 1 \cdot \frac{dt}{(100 - t)^3} + c = \frac{1}{2\,(100 - t)^2} + c$$

When $t = 0$, $m = 30$ then $c = \dfrac{-20}{(100)^3}$ etc.]

26. Other conditions remaining the same in the above problem, suppose that the mixture is running out at the same rate at which it is running in, namely 0.1 cubic metres/min. Find the amount of salt at the end of 1 hour. **Ans.** 43.976 kg.

27. In a chemical reaction, a given substance is being converted into another at a rate proportional to the amount of the unconverted. If one-fifth of the original amount has been transformed in four minutes, how much time will be required to transform one-half. **Ans.** 12.4 minutes.

28. A tank initially contain 100 gallons of brine holding 200 pound of salt in solution. Fresh brine containing 1 pound of salt per gallon flows into the tank at the rate of 3 gallon per min, and the mixture; kept uniform by stirring, flows out at the same rate. Find the amount of salt in the tank at the end of 1 hour and 30 minutes. **Ans.** 106.72 lb (approx.).

29. Under certain conditions, cane-sugar in water is converted into dextrose at a rate proportion to the amount that is unconverted at any time. If, of 75 grams at time $t = 0$, 8 grams are converted during the first 30 minutes, find the amount converted in one hour and 30 minutes. **Ans.** 21.6 grams (approx.).

30. A resistor of 10 ohms, an inductor of 2 henries and a battery of E volts are connected in series with a switch S. At $t = 0$ the switch is closed and the current $i = 0$. Find i for $t > 0$ if (*a*) $E(t) = 20e^{-3t}$, (*b*) $E(t) = 50 \sin 5t$. (See Fig. 8.7)

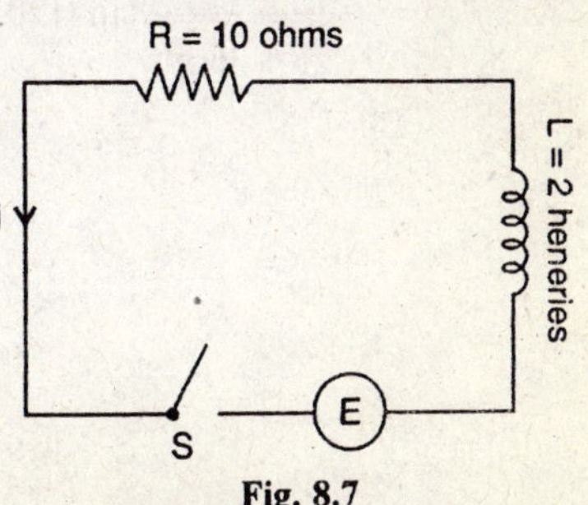

Fig. 8.7

Ans. (*a*) $i(t) = 5(e^{-3t} - e^{-5t})$, (*b*) $i(t) = \dfrac{5}{2}(\sin 5t - \cos 5t) + \dfrac{5}{2}e^{-5t}$.

31. A constant electromotive force $E(t)$ volts in applied to a circuit containing a constant resistance R ohms in series and a constant inductance L henries. If the initial current in zero, show that the current builds up to half its theoretical maximum in $(L \log 2) / R$

seconds.

[**Hint:** We have $\dfrac{di}{dt}+\dfrac{R}{L}i=\dfrac{E}{L}$, $\quad i(t)=\dfrac{E}{R}(1-e^{(-Rt)/L})$, $\quad i(0)=0$.

$i(t)$ attain the maximum value E/R.

Let the current in the circuit be half its theoretical maximum after a time T seconds.

$$\frac{1}{2}\frac{E}{R}=\frac{E}{R}(1-e^{-RT/L}) \quad \text{or} \quad e^{-RT/L}=1/2 \quad \text{or} \quad \frac{-RT}{L}=-\log 2 \quad \text{or} \quad T=(L\log 2)/R.]$$

32. The initial value problem governing the current i, flowing in a series RL circuit when a sinusoidal voltage $v(t)$ = 10 sin t is applied, is given by

$$L\frac{di}{dt}+iR=10\sin t, \qquad t\geq 0, \qquad i(0)=0,$$

where R and L are constants. Find the current $i(t)$ at time t.

Ans. $i(t)=\dfrac{10}{R^2+L^2}(R\sin t-L\cos t+Le^{-Rt/L})$.

33. The current, i in a circuit containing an inductance L, resistance R and voltage $v(t)=E\sin\omega t$ is given by differential equation

$$L\frac{di}{dt}+Ri = E\sin\omega t.$$

If $i=0$ at $t=0$, show that at any time t

$$i(t)=\frac{E}{\sqrt{R^2+L^2\omega^2}}\{\sin(\omega t-\phi)+e^{-Rt/L}\cdot\sin\phi\},$$

where $\tan\phi = L\omega/R$. *(Kurukshetra, 2005)*

34. A condenser of capacitance C is charged through a resistance R by a steady voltage V. The charge q on the plate is given by

$$R\frac{dq}{dt}+\frac{q}{c}=V.$$

If initially the plate is chargeless, find the charge and the current at time t.

Ans. $q=VC(1-e^{-t/RC})$, $\quad i=\dfrac{V}{R}e^{-t/RC}$

35. In an electric circuit containing a resistance R and a condenser of capacitance C in series and an e.m.f. E, the charge q on the condenser is given by

$$R\frac{dq}{dt}+\frac{q}{C}=E.$$

If R = 10 ohms, $C=10^{-3}$ farad and $E(t)$ = 100 sin 120 πt volts; find q, given that $q=0$ when $t=0$.

Ans. $q=\dfrac{1}{2\sqrt{25+36\pi^2}}\sin(120\pi t-\phi)+\dfrac{3\pi e^{-100t}}{25+36\pi^2}$, $\qquad$ where $\theta=\tan^{-1}\left(\dfrac{6\pi}{5}\right)$.

CHAPTER

9

Linear Differential Equations of Second and Higher Order

9.1 DEFINITIONS

A differential equation in which the dependent variable and its derivatives occur in the first degree and are also not multiplied together is known as a linear differential equation. The most general form of a linear differential equation of nth order is given by

$$P_0 \frac{d^n y}{dx^n} + P_1 \frac{d^{n-1} y}{dx^{n-1}} + P_2 \frac{d^{n-2} y}{dx^{n-2}} + \ldots + P_{n-1} \frac{dy}{dx} + P_n y = X, \; P_0 \neq 0 \qquad \ldots(9.1)$$

where the function X and the coefficients $P_0, P_1, P_2, \ldots, P_{n-1}, P_n$ are all functions of x. If $X = 0$, the resulting equation is called a *homogeneous linear equation,* otherwise *non-homogeneous linear differential equation.*

If the coefficients $P_0, P_1, P_2, \ldots, P_{n-1}, P_n$ are all constants, eqn. (9.1) can be written in the form

$$a_0 \frac{d^n y}{dx^n} + a_1 \frac{d^{n-1} y}{dx^{n-1}} + a_2 \frac{d^{n-2} y}{dx^{n-2}} + \ldots + a_{n-1} \frac{dy}{dx} + a_n y = X, \; a_0 \neq 0. \qquad \ldots(9.2)$$

Eq. (9.2) is called a non-homogeneous linear differential equation of nth order with constant coefficients. Such equations are of common occurrence in many engineering disciplines like electrical circuits, mechanical vibrations etc.

9.2 DIFFERENTIAL OPERATOR D

The symbol D is called a *differential operator* because it transforms a differentiable function into another function. We have

$$\frac{d}{dx} \equiv D, \; \frac{d^2}{dx^2} \equiv D^2, \ldots \frac{d^n}{dx^n} \equiv D^n,$$

so that $\qquad \dfrac{dy}{dx} = Dy, \; \dfrac{d^2 y}{dx^2} = D^2 y, \ldots$ and in general $\dfrac{d^n y}{dx^n} = D^n y.$

Thus in terms of the operator notation D, the differential equation (9.2) can be written as

$$a_0 D^n y + a_1 D^{n-1} y + a_2 D^{n-2} y + \ldots + a_{n-1} Dy + a_n y = X$$

or

$$(a_0 D^n + a_1 D^{n-1} + a_2 D^{n-2} + \ldots + a_{n-1} D + a_n) y = X$$

or

$$f(D) y = X \qquad \ldots(9.3)$$

where $\qquad f(D) = a_0 D^n + a_1 D^{n-1} + a_2 D^{n-2} + \ldots + a_{n-1} D + a_n.$

It is obvious that $f(D)$ is a polynomial of degree n in D. In many ways the operator D can be treated as an algebraical symbol and it obeys almost all laws of algebra. *i.e.,*

(*i*) $D(u + v) = Du + Dv$; (Distributive law)

(*ii*) $D^p D^q u = D^q D^p u$; (Commutative law)

(*iii*) $D(\lambda u) = \lambda Du$; (Commutative law for constant λ)

(*iv*) $D^p D^q u = D^{p+q} u$; (the Index law).

It may be mentioned that the commutative law does not hold good in respect of variables *i.e.,* $Du \neq uD$ and $D(uv) \neq uDv$.

Keeping in view the above properties of symbol D, the function $f(D)$ in Eq. (9.3) can be factorised into n linear factors which can be written in any order. For example

$$\frac{d^2y}{dx^2} - 5\frac{dy}{dx} + 6y = (D^2 - 5D + 6)\,y = (D-3)\,(D-2)y = (D-2)\,(D-3)y.$$

Eq. (9.3) with $X = 0$ can be written as

$$(a_0 D^n + a_1 D^{n-1} + a_2 D^{n-2} + \ldots + a_{n-1} D + a_n)y = 0 \text{ or } f(D)y = 0. \qquad \ldots(9.4)$$

9.3 SOLUTIONS OF LINEAR DIFFERENTIAL EQUATIONS

If $y_1(x)$ and $y_2(x)$ are two independent solutions of Linear homogeneous equation (9.4) on I, then a linear combination of the solutions $c_1y_1(x) + c_2y_2(x)$ is also a solution of Eq. (9.4), where c_1 and c_2 are arbitrary constants.

Proof. Since $y_1(x)$ and $y_2(x)$ are two independent solutions of Eq. (9.4), they must satisfy this equation. This yields

$$f(D)y_1 = 0 \qquad \ldots(9.5)$$

and

$$f(D)y_2 = 0 \qquad \ldots(9.6)$$

Multiplying Eq. (9.5) by c_1 and Eq. (9.6) by c_2 and adding, we obtain.

$$c_1 f(D)y_1 + c_2 f(D)y_2 = 0$$

or

$$f(D)[c_1y_1(x) + c_2y_2(x)] = 0 \qquad \ldots(9.7)$$

Eq. (9.7) shows that $y = c_1y_1(x) + c_2y_2(x)$ satisfies Eq. (9.4) and hence this is also a solution of Eq. (9.4).

The result of the above theorem can now be extended to any finite number of independent solutions of Eq. (9.4). Thus if $y = y_1(x)$, $y = y_2(x)$, ...and $y = y_k(x)$ are k independent solutions of Eq. (9.4) on I, then a linear combination of the solutions

$$c_1y_1(x) + c_2y_2(x) + \ldots + c_ky_k(x),$$

where $c_1, c_2, \ldots c_k$ are arbitrary constants, is also a solution of Eq. (9.4) on I.

The complete solution $y = y_c$ is given by $y_c = c_1y_1(x) + c_2y_2(x) + \ldots + c_ky_k(x)$, ...(9.8)

and

$$f(D)y_c = 0 \qquad \ldots(9.9)$$

Again let $y = y_p$ be any function of x, not containing any arbitrary constant, which satisfies Eq. (9.3).

Then

$$f(D)y_p = X \qquad \ldots(9.10)$$

Adding Eqs. (9.9) and (9.10), we have $f(D)[y_c + y_p] = X$. ...(9.11)

We observe from Eq. (9.11) that $y(x) = y_c(x) + y_p(x)$ satisfies Eq. (9.3). Since $y_c(x)$ contains n arbitrary constants, the solution $y(x) = y_c(x) + y_p(x)$ also contains n arbitrary constants. Hence $y(x) = y_c(x) + y_p(x)$ is the general or complete solution of Eq. (9.3).

The part $y_c(x)$ is called the *complementary function* (*C.F.*) of Eq. (9.3) and is obtained from the solution of this equation after putting $X = 0$. The part $y_p(x)$ is called the *particular integral* (*P.I.*) of Eq. (9.3). It is some function of x, not involving any constant, which satisfies Eq. (9.3). Hence the general or complete solution (C.S.) of Eq. (9.3) is given by

$$y(x) = \text{C.F.} + \text{P.I.}$$

EXAMPLE 9.1 Show the e^{3x}, e^{-2x} and their linear combination $c_1e^{3x} + c_2e^{-2x}$ are solutions of the homogeneous equation $y'' - y' - 6y = 0$.

Solution For $y_1 = e^{3x}$, we have

$$y_1' = 3e^{3x},\ y_1'' = 9e^{3x},$$

$$y_1'' - y_1' - 6y_1 = 9e^{3x} - 3e^{3x} - 6e^{3x} = 0.$$

For $y_2 = e^{-2x}$, we have

$$y_2' = -2e^{-2x},\ y_2'' = 4e^{-2x},$$

$$y_2'' - y_2' - 6y_2 = 4e^{-2x} + 2e^{-2x} - 6e^{-2x} = 0.$$

Hence, e^{3x}, e^{-2x} are solution of $y'' - y' - 6y = 0$.

Substituting $y = c_1e^{3x} + c_2e^{-2x} = c_1y_1 + c_2y_2$, we obtain

$$\begin{aligned} y'' - y' - 6y &= (c_1y_1 + c_2y_2)'' - (c_1y_1 + c_2y_2)' - 6(c_1y_1 + c_2y_2) \\ &= c_1(y_1'' - y_1' - 6y_1) + c_2(y_2'' - y_2' - 6y_2) \\ &= c_1(0) + c_2(0) = 0. \end{aligned}$$

9.3.1 Linear Dependence and Linear Independence

A set of functions $f_1(x), f_2(x), \ldots\ldots, f_n(x)$ is said to be *linearly dependent* on some interval I (where they are defined), if and only if there exists to constants $c_1, c_2, \ldots\ldots, c_n$ at least one of which is not zero, such that

$$c_1f_1(x) + c_2f_2(x) + \ldots\ldots + c_nf_n(x) = 0$$

holds for every x in the interval.

In this case, one or more functions can be expressed as a linear combination of the remaining functions. For example if $c_1 \neq 0$, then

$$f_1(x) = -\frac{1}{c_1}\left[c_2f_2(x) + \ldots\ldots + c_nf_n(x)\right].$$

If the set of functions is not linearly dependent on I, it is said to be *linearly independent.*

In other words, a set of functions is linear independent on an interval I if the equation.

$$c_1f_1(x) + c_2f_2(x) + \ldots\ldots + c_nf_n(x) = 0$$

implies $c_1 = 0 = c_2 = \ldots\ldots = c_n$.

Illustration

The functions $f_1(x) = x, f_2(x) = x^2, f_3(x) = 4x - 3x^2$ are linearly dependent on the interval $(-\infty, \infty)$.

Since $f_3(x)$ can be written as a linear combination of f_1 and f_2. Observe that $f_3(x) = 4.f_1(x) - 3.f_2(x)$ for every x in the interval $(-\infty, \infty)$.

EXAMPLE 9.2 Show that the functions $x^2 - x, 3x^2 + x + 1, 9x^2 - x + 2$ are linearly dependent for $x \in (0, \infty)$.

Solution The given functions are linearly dependent if the equation

$$c_1(x^2 - x) + c_2(3x^2 + x + 1) + c_3(9x^2 - x + 2) = 0 \qquad \ldots(i)$$

hold for c_1, c_2, c_3 not all zero.

Eqn(i) can be written as

$$(c_1 + 3c_2 + 9c_3)x^2 + (-c_1 + c_2 - c_3)x + (c_2 + 2c_3) - 0, \text{ for all } x.$$

We have $c_2 - c_1 - c_3 = 0$, and $c_2 + 2c_3 = 0$. The solutions of these equations is $c_1 = -3c_3, c_2 = -2c_3$, where c_3 is arbitrary. For example, if $c_3 = 1$, then

$$c_1 = -3, c_2 = -2 \text{ and } f_3(x) = 3(x^2 - x) + 2(3x^2 + x + 1).$$

The given functions are linearly dependent.

A very elegant procedure to test the linear dependence or independence of a given set of functions is the application of *wronskians*. Let $f_1(x), f_2(x), \ldots\ldots, f_n(x)$ be n functions. Suppose each of the functions $f_1(x), f_2(x), \ldots\ldots, f_n(x)$ possess at least $n - 1$ derivatives on the interval I. The determinent

$$W(f_1, f_2, \ldots\ldots, f_n) = \begin{vmatrix} f_1 & f_2 & \cdots & f_n \\ f_1' & f_2' & \cdots & f_n' \\ \vdots & \vdots & \cdots & \vdots \\ f_1^{(n-1)} & f_2^{(n-1)} & \cdots & f_n^{(n-1)} \end{vmatrix} = W(x),$$

is called the *wronskian* of n functions.

We have the following result for testing the linear dependence or independence of the solutions of the linear homegeneous differential equation

$$a_0y^{(n)} + a_1y^{(n-1)} + \ldots\ldots + a_{n-1}y' + a_ny = 0, \quad a_0 \neq 0 \qquad \ldots(A)$$

Let $y_1, y_2, ..., y_n$ be n solutions of the homogeneous linear nth-order differential equation (A) on an interval I. Then the set of solutions is *linearly independent* on I if and only if $W(y_1, y_2, ..., y_n) \neq 0$ for every x in the interval.

The n linearly independent solutions $y_1, y_2, ..., y_n$ of the homogeneous linear nth-order differential equation (A) on an interval I is said to be a *fundamental set of solutions* on the interval, or a basis for all solutions, of the equation on that interval.

EXAMPLE 9.3 Show that the functions 1, sin x, cos x are linearly independent.

(*AMIETE, Dec 2009, June 2007*)

Solution The wronskian of the functions is

$$W(x) = \begin{vmatrix} 1 & \sin x & \cos x \\ 0 & \cos x & -\sin x \\ 0 & -\sin x & -\cos x \end{vmatrix} = -1 \neq 0.$$

Hence, the given functions are linearly independent on any interval I.

EXAMPLE 9.4 Show that the functions x, x^2, x^3 defined on any interval I, not containing zero are always linearly independent. (*AMIETE, Dec 2006*)

Solution The wronskian of the functions is

$$W(x) = \begin{vmatrix} x & x^2 & x^3 \\ 1 & 2x & 3x^2 \\ 0 & 2 & 6x \end{vmatrix}$$

$$= x\,(12x^2 - 6x^2) - 1\,(6x^3 - 2x^3) = 2x^3.$$

Therefore, $w(x) \neq 0$ on any interval not containing zero. Hence, the functions are linearly independent.

EXAMPLE 9.5 Evaluate the wronskian determinant for the functions:

$$y_1(x) = \sin x, \quad y_2(x) = \sin\left(x + \frac{\pi}{8}\right), \quad y_3(x) = \sin\left(x - \frac{\pi}{8}\right).$$

Solution We have

$$W(y_1, y_2, y_3) = \begin{vmatrix} \sin x & \sin\left(x + \frac{\pi}{8}\right) & \sin\left(x - \frac{\pi}{8}\right) \\ \cos x & \cos\left(x + \frac{\pi}{8}\right) & \cos\left(x - \frac{\pi}{8}\right) \\ -\sin x & -\sin\left(x + \frac{\pi}{8}\right) & -\sin\left(x - \frac{\pi}{8}\right) \end{vmatrix} = 0.$$

Since the first and the last row of the determinant are proportional.

EXAMPLE 9.6 Show that the set of functions {sin $2x$, cos $2x$, e^{2x}} forms a basis of the third order equation

$$y''' - 2y'' + 4y' - 8y = 0.$$

Solution We have

$$y_1(x) = \sin 2x, \quad y_1' = 2\cos 2x, \quad y_1'' = -4\sin 2x, \quad y_1''' = -8\cos 2x,$$

and $$y_1''' - 2y_1'' + 4y_1' - 8y_1 = -8\cos 2x + 8\sin 2x + 8\cos 2x - 8\sin 2x = 0$$

$$y_2(x) = \cos 2x, \quad y_2' = -2\sin 2x, \quad y_2'' = -4\cos 2x, \quad y_2''' = 8\sin 2x,$$

and $$y_2''' - 2y_2'' + 4y_2' - 8y_2 = 8\sin 2x + 8\cos 2x - 8\sin 2x - 8\cos 2x = 0.$$

$$y_3(x) = e^{2x},\ y_3' = 2e^{2x},\ y_3'' = 4e^{2x},\ y_3''' = 8e^{2x}$$

and $$y_3''' - 2y_3'' + 4y_3' - 8y_3 = 8e^{2x} - 8e^{2x} + 8e^{2x} - 8e^{2x}\ 0.$$

Hence, $y_1(x)$, $y_2(x)$ and $y_3(x)$ are solution of the given equation.

The wronskian is given by

$$W(y_1, y_2, y_3) = \begin{vmatrix} \sin 2x & \cos 2x & e^{2x} \\ 2\cos 2x & -2\sin 2x & 2e^{2x} \\ -4\sin 2x & -4\cos 2x & 4e^{2x} \end{vmatrix}$$

$$= 8[\sin 2x(-\sin 2x + \cos 2x)e^{2x} - \cos 2x(\cos 2x + \sin 2x)e^{2x} + e^{2x}(-\cos^2 2x - \sin^2 2x)]$$

$$= 8[-\sin^2 2x + \sin 2x \cos 2x - \cos^2 2x - \cos 2x \sin 2x - \cos^2 2x - \sin^2 2x]e^{2x}$$

$$= 8[-2]e^{2x} = -16e^{2x} \neq 0 \text{ for any value of } x.$$

The functions y_1, y_2, y_3 cannot be linearly dependent on the whole of the x-axis.

Therefore, the set $\{y_1(x), y_2(x), y_3(x)\}$ forms a basis of the equation.

EXAMPLE 9.7 Show that the set of functions $\{x, 1/x\}$ forms a basis of the equation $x^2y'' + xy' - y = 0$. Obtain a particular solution when $y(1) = 1$, $y'(1) = 2$. *(AMIETE, Dec 2006)*

Solution We have

$$y_1(x) = x,\ y'_1 = 1,\ y''_1 = 0 \quad \text{and} \quad x^2y''_1 + xy'_1 - y_1 = x - x = 0.$$

$$y_2(x) = 1/x,\ y'_2 = -1/x^2,\ y''_2 = 2/x^3$$

and $$x^2y''_2 + xy'_2 - y_2 = x^2\left(\frac{2}{x^3}\right) + x\left(-\frac{1}{x^2}\right) - \frac{1}{x} = 0.$$

Hence $y_1(x)$ and $y_2(x)$ are solutions of the given equation. The wronskian is given by

$$W(y_1, y_2) = \begin{vmatrix} x & 1/x \\ 1 & -1/x^2 \end{vmatrix} = -\frac{2}{x} \neq 0, \quad \text{for } x \geq 1.$$

Therefore the set $\{y_1(x), y_2(x)\}$ forms a basis of the equation. The general solution is

$$y(x) = c_1y_1(x) + c_2y_2(x) = c_1x + \frac{c_2}{x} \qquad \text{...}(i)$$

Substituting in (*i*) the given conditions, we obtain

$$y(1) = 1 = c_1 + c_2, \quad y'(1) = 2 = c_1 - c_2.$$

Solving these equations, we obtain $c_1 = 3/2$, $c_2 = -1/2$.

The particular solution is $y(x) = \frac{1}{2}\left(3x - \frac{1}{x}\right)$.

9.4 THE COMPLEMENTARY FUNCTION (C.F.)

To find the complementary function, we solve the homogeneous equation

$$a_0\frac{d^ny}{dx^n} + a_1\frac{d^{n-1}y}{dx^{n-1}} + \ldots + a_{n-1}\frac{dy}{dx} + a_ny = 0,\ a_0 \neq 0. \qquad \text{...(9.12)}$$

In operator notation, Eq. (9.12) can be written as

$$(a_0D^n + a_1D^{n-1} + a_2D^{n-2} + \ldots + a_{n-1}D + a_n)y = 0$$

or $$f(D)y = 0. \qquad \text{...(9.13)}$$

Since $f(D)$ is a polynomial of degree n in D and D obeys the algebraical laws, we can in general factorise $f(D)$ into n linear factors and Eq. (9.13) can be written as

$$(D - m_1)(D - m_2)(D - m_3)...(D - m_n)y = 0 \qquad ...(9.14)$$

where $m_1, m_2, ...m_n$ are real or complex numbers.

The linear factors in Eq. (9.14) can be written in any order and the solution corresponding to each one of them can be obtained from the linear equations

$$(D - m_1)y = 0; (D - m_2)y = 0; ...(D - m_n)y = 0.$$

The differential equation $(D - m)y = 0$ for any m, has $y = e^{mx}$ as its solution. Substituting e^{mx} for y and its derivatives in Eq. (9.13), we get

$$(a_0m^n + a_1m^{n-1} + a_2m^{n-2} + ...+ a_n)e^{mx} = 0. \qquad ...(9.15)$$

Since $e^{mx} \neq 0$. Eq. (9.15) gives

$$a_0m^n + a_1m^{n-1} + a_2m^{n-2} + ...+ a_n = 0 \qquad ...(9.16)$$

Eq. (9.16) is called the *characteristic equation or the auxiliary equation* of Eq. (9.13) and is obtained by putting $D = m$ in $f(D) = 0$. Being an algebraic equation of degree n in m, Eq. (9.16) determines n roots $m_1, m_2 ...m_n$ as needed in Eq. (9.14).

These roots may be (*i*) real and different (*ii*) equal or (*iii*) imaginary.

Case I. *Characteristic equation having real and different roots.*

Let $m_1, m_2, ...,m_n$ be n real and different roots of Eq. (9.16). The solution corresponding to each one of them is given by $e^{m_1x}, e^{m_2x}, ..., e^{m_nx}$. Associating n arbitrary constants, the general solution of Eq. (9.13) is given by

$$y = c_1e^{m_1x} + c_2e^{m_2x} + ... + c_ne^{m_nx} \qquad ...(9.17)$$

Case II. *Characteristic equation having equal roots*

Let two of the n roots m_1 and m_2 of Eq. (9.13) be equal. Its solution as given in (9.17) above can be written as

$$y = (c_1 + c_2)e^{m_1x} + c_3e^{m_3x} + ... + c_ne^{m_nx}$$

or $$y = ce^{m_1x} + c_3e^{m_3x} + ...c_ne^{m_nx} \qquad ...(9.18) \qquad \text{where} \qquad c = c_1 + c_2.$$

This solution contains only $(n - 1)$ arbitrary constants and is therefore, not a complete solution of the given equation. To obtain complete solution in this case, the given equation can be written as

$$(D - m_1)(D - m_2)(D - m_3) ... (D - m_n)y = 0. \qquad ...(9.19)$$

For the repeated factor, the corresponding part of the equation is given by

$$(D - m_1)^2y = 0 \quad \text{or} \quad (D - m_1)(D - m_1)y = 0. \qquad ...(9.20)$$

Let $$(D - m_1)y = v. \qquad ...(9.21)$$

Eq. (9.20) can be written as $(D - m_1)v = 0$ or $dv/dx - m_1v = 0$. ...(9.22)

Eq. (9.22) is a linear first order differential equation and its solution is given by $v = Ae^{m_1x}$. Substituting for v in Eq. (9.21), we get

$$(D - m_1)y = Ae^{m_1x} \quad \text{or} \quad \frac{dy}{dx} - m_1y = Ae^{m_1x}. \qquad ...(9.23)$$

Eq. (9.23) is a linear differential equation of the first order. The integrating factor is e^{-m_1x}. Hence the solution of (9.23) is given by

or $$ye^{-m_1x} = \int A\,dx + B = Ax + B \quad \text{or} \quad y = (Ax + B)e^{m_1x}. \qquad ...(9.24)$$

Thus the complete solution of Eq. (9.13) is given by

$$y(x) = (Ax + B)e^{m_1x} + c_3e^{m_3x} + ... + c_ne^{m_nx}. \qquad ...(9.25)$$

It can similarly be shown that if the given equation $f(D)y = 0$ has three equal roots, each equal to m_1, then the solution corresponding to the repeated factor is given by the equation

$$(D - m_1)^3y = 0. \qquad ...(9.26)$$

Solving Eq. (9.26) on similar lines as for double roots above, its solution is given by

$$y(x) = (Ax^2 + Bx + C)\,e^{m_1x}. \qquad ...(9.27)$$

The complete solution of the given equation is given by

$$y(x) = (Ax^2 + Bx + C)e^{m_1x} + c_4e^{m_4x} + ... + c_ne^{m_nx}. \qquad ...(9.28)$$

Case III. *Characteristic equation having imaginary roots*

We know that imaginary roots of an algegraic equation with real coefficients always occur in conjugate pairs. Let the two roots m_1 and m_2 of the characteristic equation (9.16) be denoted by $m_1 = \alpha + i\beta$ and $m_2 = \alpha - i\beta$. Since the roots are different, using case I, we can write the solution

$$y(x) = c_1e^{(\alpha+i\beta)x} + c_2e^{(\alpha-i\beta)x} + c_3e^{m_3x} + ... + c_ne^{m_nx}$$

$$= e^{\alpha x}\left[c_1e^{i\beta x} + c_2e^{-i\beta x}\right] + c_3e^{m_3x} + ... + c_ne^{m_nx}$$

$$= e^{\alpha x}\left[(c_1 + c_2)\cos\beta x + i(c_1 - c_2)\sin\beta x\right] + c_3e^{m_3x} + ... + ...c_ne^{m_nx}$$

[Using Euler's theorem $e^{i\theta} = \cos\theta + i\sin\theta$]

$$= e^{\alpha x}(A_1\cos\beta x + A_2\sin\beta x) + c_3e^{m_3x} + ...+ c_ne^{m_nx} \qquad ...(9.29)$$

where $A_1 = c_1 + c_2$; $\quad A_2 = i(c_1 - c_2)$.

Note: The form $e^{\alpha x}(A_1\cos\beta x + A_2\sin\beta x)$ can also be represented by

$$A_1e^{\alpha x}\cos(\beta x + A_2) \text{ or by } A_1e^{\alpha x}\sin(\beta x + A_2).$$

Case IV. *Characteristic equation having repeated imaginary roots*

Let $m_1 = \alpha + i\beta$ be repeated twice and $m_2 = \alpha - i\beta$ be also repeated twice. Since this is case of two repeated roots. Using case II we can write the solution as

$$y(x) = (A_1x + A_2)e^{(\alpha+i\beta)x} + (A_3x + A_4)e^{(\alpha-i\beta)x} + c_5e^{m_5x} + ... + c_ne^{m_nx}$$

$$= e^{\alpha x}\{(A_1'x + A_2')\cos\beta x + (A_3'x + A_4')\sin\beta x\} + c_5e^{m_5x} + ... + c_ne^{m_nx}$$

where $A_1' = A_1 + A_3$, $A_2' = A_2 + A_4$, $A_3' = i(A_1 - A_3)$, $A_4' = i(A_2 - A_4)$.

Working Procedure to find the C.F. of

$$\frac{d^ny}{dx^n} + a_1\frac{d^{n-1}y}{dx^{n-1}} + ... + a_ny = X$$

Step 1. Write the characteristic equation of the corresponding homogeneous equation.

$$m^n + a_1m^{n-1} + a_2m^{n-2} + ... + a_n = 0 \text{ and solve it for } m.$$

Step 2. Then write the complementary function according to the following scheme.

Table 9.1

Nature of the roots of the characteristic equation	Complementary Function (C.F)
1. All roots real and distinct $m_1, m_2, m_3, ...m_n$	$c_1e^{m_1x} + c_2e^{m_2x} + c_3e^{m_3x} + \cdots\cdots$
2. Two roots real and equal and remaining distinct $m_1, m_1, m_3, m_4, ...m_n$	$(c_1 + c_2x)e^{m_1x} + c_3e^{m_3x} + ...$
3. Three roots real and equal and remaining distinct $m_1, m_1, m_1, m_4, ...m_n$	$(c_1 + c_2x + c_3x^2)e^{m_1x} + c_4e^{m_4x} + ...$
4. A pair of roots complex and remaining distinct $\alpha \pm i\beta, m_3, m_4, \cdots\cdots m_n$	$e^{\alpha x}(c_1\cos\beta x + c_2\sin\beta x) + c_3e^{m_3x} + ...$
5. Two pairs of complex roots equal and remaining distinct. $\alpha \pm i\beta, \alpha \pm i\beta, m_5, m_6 ......m_n$	$e^{\alpha x}\{(c_1 + c_2x)\cos\beta x + (c_3 + c_4x)\sin\beta x\} + c_5e^{m_5x} + ...$

ILLUSTRATIVE EXAMPLES

EXAMPLE 9.8 Solve the differential equation $\frac{d^2y}{dx^2}+3a\frac{dy}{dx}-4a^2y=0$.

Solution The given homogeneous equation in operator notation is

$$(D^2 + 3aD - 4a^2)y = 0.$$

The characteristic equation of the differential equation is $m^2 + 3am - 4a^2 = 0$ or $(m - a)(m + 4a) = 0$. Its roots are $m = a, -4a$.

Hence a complete solution is $y(x) = c_1e^{ax} + c_2e^{-4ax}$.

EXAMPLE 9.9 Solve the differential equation $y''' - 3y'' + 4y = 0$.

Solution The given homogeneous equation in operator notation is

$$(D^3 - 3D^2 + 4)\,y = 0.$$

The characteristic equation of the differential equation is

$$m^3 - 3m^2 + 4 = 0 \quad \text{or} \quad (m + 1)(m - 2)^2 = 0.$$

Its roots are $m = -1, 2, 2$. Hence a complete solution of the equation is $y(x) = c_1e^{-x} + (c_2 + c_3x)e^{2x}$.

EXAMPLE 9.10 Solve the differential equation $y^{iv}+32y''+256y=0$. *(AMIETE, Dec. 2004)*

Solution The given homogeneous equation in operator notation is

$$(D^4 + 32D^2 + 256)\,y = 0.$$

The characteristic equation of the differential equation is

$$m^4 + 32\,m^2 + 256 = 0 \quad \text{or} \quad (m^2 + 16)^2 = 0.$$

The roots of this equation are $m = \pm 4i, \pm 4i$.

Hence a complete solution is $y(x) = (c_1 + c_2\,x)\cos 4x + (c_3 + c_4x)\sin 4x$.

EXAMPLE 9.11 Solve the differential equation $y^{iv}+a^4y=0$. *(NIT Kurukshetra, 2007)*

Solution The given homogeneous equation in operator notation is $(D^4 + a^4)y = 0$.

The characteristic equation of the differential equation is $m^4 + a^4 = 0$

or $\quad m^4+2m^2a^2+a^4-2m^2a^2=0 \quad$ or $\quad (m^2+a^2)^2-(\sqrt{2}am)^2=0$

or $\quad (m^2+a^2-\sqrt{2}\,am)(m^2+a^2+\sqrt{2}\,am)=0.$

Its roots are $m=\frac{a(1\pm i)}{\sqrt{2}}, \frac{a(-1\pm i)}{\sqrt{2}}$.

Hence a complete solution of the equation is

$$y(x)=e^{ax/\sqrt{2}}\left[c_1\cos(ax/\sqrt{2})+c_2\sin(ax/\sqrt{2})\right]+e^{-ax/\sqrt{2}}\left[c_3\cos(ax/\sqrt{2})+c_4\sin(ax/\sqrt{2})\right].$$

EXAMPLE 9.12 Solve the differential equation $\frac{d^3s}{dt^3}+9\frac{ds}{dt}=0$.

Solution The given homogeneous equation in operator notation is $(D^3 + 9D)\,s = 0$.

The characteristic equation of the differential equation is $m^3 + 9m = 0$ or $m\,(m^2 + 9) = 0$. Its roots are $m = 0, \pm 3i$. Hence a complete solution of the equation is $s(t) = c_1 + c_2\cos 3t + c_3\sin 3t$.

EXAMPLE 9.13 Solve the following initial value problems.

(*i*) $4y'' - 8y' + 3y = 0;\ y(0) = 1,\ y'(0) = 3.$

(*ii*) $y'' + 4y' + 13y = 0;\ y(0) = 0,\ y'(0) = 1.$

(*iii*) $y''' + 3y'' - 4y = 0;\ y(0) = 1,\ y'(0) = 0,\ y''(0) = 1/2.$

Solution (*i*) The characteristic equation of the given differential equation is $4m^2 - 8m + 3 = (2m - 3)(2m - 1) = 0$. Its roots are $m = 3/2, 1/2$. Hence a complete solution is $y(x) = c_1 e^{(3x)/2} + c_2 e^{x/2}$.

By differentiating this, we find $y'(x) = \frac{3}{2} c_1 e^{(3x)/2} + \frac{1}{2} c_2 e^{x/2}$.

Substituting the given data into the equations for $y(x)$ and $y'(x)$ respectively, we have

$$y(0) = 1 = c_1 + c_2 \text{ and } y'(0) = 3 = \frac{3}{2} c_1 + \frac{c_2}{2}.$$

The solution of this system is $c_1 = 5/2$, $c_2 = -3/2$. Hence, the particular solution is

$$y(x) = [5e^{(3x)/2} - 3e^{x/2}]/2.$$

(*ii*) The characteristic equation of the given differential equation is given by

$$m^2 + 4m + 13 = 0, \text{ or } m = \frac{-4 \pm \sqrt{16 - 52}}{2} = -2 \pm 3i = \alpha \pm i\beta.$$

The complete solution is given by

$$y(x) = (c_1 \cos\beta x + c_2 \sin\beta x)e^{\alpha x} = (c_1 \cos 3x + c_2 \sin 3x)e^{-2x}.$$

The first initial condition gives $y(0) = 0 = c_1$, and so

$$y'(x) = c_2 e^{-2x}(3 \cos 3x - 2 \sin 3x).$$

The second initial condition gives $y'(0) = 1 = 3c_2$ or $c_2 = 1/3$.

Hence, the particular solution is $y(x) = (e^{-2x} \sin 3x)/3$.

(*iii*) The characteristic equation of the given differential equation is $m^3 + 3m^2 - 4 = 0$ or $(m - 1)(m^2 + 4m + 4) = 0$ or $(m - 1)(m + 2)^2 = 0$. The roots of this equation are $m = 1, -2, -2$. The complete solution is

$$y(x) = c_1 e^x + (c_2 + c_3 x)e^{-2x}. \quad \ldots(i)$$

By successive differentiating (*i*), we find

$$y'(x) = c_1 e^x + c_3 e^{-2x} - 2(c_2 + c_3 x)e^{-2x},$$

and

$$y''(x) = c_1 e^x - 4c_3 e^{-2x} + 4(c_3 x + c_2)e^{-2x}.$$

Substituting the given data into the equations for $y(x)$, $y'(x)$ and $y''(x)$ respectively, we have

$$y(0) = 1 = c_1 + c_2,\ y'(0) = 0 = c_1 - 2c_2 + c_3,$$

and

$$y''(0) = 1/2 = c_1 + 4c_2 - 4c_3.$$

The solution of this system is $c_1 = c_2 = c_3 = 1/2$.

Hence, the particular solution is $y(x) = [e^x + (x + 1)e^{-2x}]/2$.

EXAMPLE 9.14 Find a complete solution of each of the following equations.

(*i*) $\frac{d^6 y}{dx^6} - 64y = 0$, (*ii*) $(D^8 + 6D^6 - 32D^2)y = 0$. (*AMIETE, June 2005*)

Solution (*i*) The characteristic equaion of the given differential equation is

$$m^6 - 64 = 0 \quad \text{or} \quad (m^2 - 4)(m^4 + 4m^2 + 16) = 0.$$

Solving $m^4 + 4m^2 + 16 = 0$ for m^2, we get

$$m^2 = \frac{-4 \pm \sqrt{16 - 64}}{2} = -2 \pm 2\sqrt{3}\, i.$$

Let $m = \alpha + i\beta \Rightarrow m^2 = (\alpha + i\beta)^2 = \alpha^2 - \beta^2 + 2i\alpha\beta = -2 + 2\sqrt{3}\, i$.

By comparing the real and imaginiary parts, we have $\alpha^2 - \beta^2 = -2$ and $\alpha\beta = \sqrt{3}$.

Now $(\alpha^2 + \beta^2)^2 = (\alpha^2 - \beta^2)^2 + 4\alpha^2\beta^2 \Rightarrow \alpha^2 + \beta^2 = 4$.

Solving the equations $\alpha^2 + \beta^2 = 4$ and $\alpha^2 - \beta^2 = -2$, we get

$\alpha = \pm 1$ and $\beta = \pm\sqrt{3}$. Therefore, the roots of the auxiliary equation are $m = \pm 2, 1 \pm i\sqrt{3}, -1 \pm i\sqrt{3}$.

A complete solution of the given equation is.

$$y(x) = c_1 e^{2x} + c_2 e^{-2x} + (c_3 \cos\sqrt{3}x + c_4 \sin\sqrt{3}x)e^{x} + (c_5 \cos\sqrt{3}x + c_6 \sin\sqrt{3}x)e^{-x}.$$

(*ii*) The characteristic equation is

$$m^8 + 6m^6 - 32m^2 = m^2(m^6 + 6m^4 - 32) = 0.$$

when $m^2 = 2$, the polynomial inside the bracket vanishes, it has $m^2 - 2$ as a factor. Dividing $m^6 + 6m^4 - 32$ by $m^2 - 2$ to obtain $m^4 + 8m^2 + 16$. The characteristic equation can be re-written as

$$m^2(m^2 - 2)(m^4 + 8m^2 + 16) = m^2(m + \sqrt{2})(m - \sqrt{2})(m^2 + 4)^2 = 0,$$

which has unrepeated roots $m_1 = -\sqrt{2}$ and $m_2 = \sqrt{2}$ and the repeated roots $m_3 = 0$, $m_4 = 0 \pm 2i$.

A complete solution of the given equation is

$$y(x) = c_1 e^{-\sqrt{2}x} + c_2 e^{\sqrt{2}x} + c_3 + c_4 x + (c_5 + c_6 x)\cos 2x + (c_7 + c_8 x)\sin 2x.$$

EXERCISE 9.1

1. Verify that the functions $y_1(x) = x_2$ and $y_2(x) = x^2 \log x$ are both solutions of the homogeneous linear equation $x^3y''' - 2xy' + 4y = 0$ on the interval $(0, \infty)$. Verify also that a linear combination of these functions is also a solution.

2. Evaluate the wronskian of e^x, e^{-x}, and e^{2x} for a general value of x. **Ans.** $-6e^{2x}$.

Determine whether the given set of functions is linearly dependent or linearly independent for $x \in (0, \infty)$.

3. $f_1(x) = \sqrt{x} + 5, \quad f_2(x) = \sqrt{x} + 5x, \quad f_3(x) = x - 1, \quad f_4(x) = x^2.$ **Ans.** Linearly dependent

[**Hint:** Observe that $f_2(x) = 1 \cdot f_1(x) + 5 \cdot f_3(x) + 0 \cdot f_4(x)$ for every x in the interval $(0, \infty)$.]

4. $f_1(x) = \sin x, f_2(x) = \sin 2x, f_3(x) = \sin 3x.$ **Ans.** Linearly independent

Verify that the given functions form a fundamental set of solutions of the differential equation on the indicated interval. Form the general solution.

5. $e^x, e^{2x}, e^{3x}, (-\infty, \infty);\quad y''' - 6y'' + 11y' - 6y = 0.$

Ans. The function satisfy the third-order differential equation and are linearly independent on the interval since $W(e^x, e^{2x}, e^{3x}) = 2e^{6x} \neq 0;\quad y(x) = c_1e^x + c_2e^{2x} + c_3e^{3x}.$

6. $x, x^{-2}, x^{-2}\log x, (0, \infty);\ x^3y''' + 6x^2y'' + 4xy' - 4y = 0.$

Ans. The functions satisfy the third order differential equations and are linearly independent on the interval since $W(x, x^{-2}, x^{-2}\log x) = 9x^{-6} \neq 0$; $y(x) = c_1x + c_2x^{-2} + c_3x^{-2}\log x$.

7. Show that $y_1(x) = \sin x$ and $y_2(x) = 4\sin x - 2\cos x$ are linearly independent solutions of second order D.E. of $y'' + y = 0$. Write the solution $y_3(x) = \cos x$ as a linear combination of y_1 and y_2.
Ans. $W(y_1, y_2) = 2 \neq 0$; $y_3 = 2y_1 - (y_2/2)$.

Find a complete solution of each of the following differential equations.

8. $2\dfrac{d^2y}{dx^2} + 5\dfrac{dy}{dx} + 2y = 0.$ **Ans.** $y(x) = c_1e^{-x/2} + c_2e^{-2x}$.
9. $4y'' + 12y' + 9y = 0.$ **Ans.** $y(x) = (c_1 + c_2x)e^{-3x/2}$.
10. $y''' - 2y'' - 5y' + 6y = 0.$ **Ans.** $y(x) = c_1e^x + c_2e^{-2x} + c_3e^{3x}$.
11. $y''' - 3y'' + 3y' - y = 0.$ **Ans.** $y(x) = (c_1 + c_2x + c_3x^2)e^x$.
12. $y''' + y'' + 4y' + 4 = 0.$ **Ans.** $y(x) = c_1e^{-x} + c_2\cos 2x + c_3\sin 2x$.
13. $4y''' + 4y'' + y' = 0.$ **Ans.** $y(x) = c_1 + (c_2 + c_3x)e^{-x/2}$.
14. $\dfrac{d^3y}{dx^3} + 2\dfrac{d^2y}{dx^2} + \dfrac{dy}{dx} = 0.$ (*AMIETE; June 2009*) **Ans.** $y(x) = c_1 + (c_2 + c_3x)e^{-x}$.

15. $d^3y/dx^3 + y = 0.$ **Ans.** $y(x) = c_1e^{-x} + e^{x/2}\left(c_2 \cos\frac{\sqrt{3}}{2}x + c_3 \sin\frac{\sqrt{3}}{2}x\right).$

16. $(D^3 - 8)\,y = 0.$ (*AMIETE., Dec 2008*) **Ans.** $y = e^{-x}(c_1 \cos\sqrt{3}x + c_2 \sin\sqrt{3}x) + c_3e^{2x}.$

17. $4y^{iv} - 12y''' - y'' + 27y' - 18y = 0.$ **Ans.** $y(x) = c_1e^x + c_2e^{2x} + c_3e^{-3x/2} + c_4e^{3x/2}.$

18. $y^{iv} + 8y'' + 16y = 0.$ (*JNTU 2005*) **Ans.** $y(x) = (Ax + B)\cos 2x + (Cx + D)\sin 2x.$

19. $(D^4 + 2D^3 + 3D^2 + 2D + 1)y = 0.$ **Ans.** $y(x) = e^{-x/2}\left[(c_1 + c_2x)\cos\frac{\sqrt{3}}{2}x + (c_3 + c_4x)\sin\frac{\sqrt{3}}{2}x\right].$

20. $(D^4 - 6D^3 + 12D^2 - 8D)y = 0.$ (*AMIETE., Dec 2008*) **Ans.** $y(x) = A + (B + Cx + Dx^2)e^{2x}.$

Solve the following initial value problems.

21. $y'' - 4y' + 4y = 0;\ y(0) = 3,\ y'(0) = 4$ **Ans.** $y(x) = (3 - 2x)e^{2x}.$

22. $y'' - 10y' + (25 + \pi^2)y = 0;\ y(0) = 0,\ y'(0) = \pi e.$ **Ans.** $y(x) = e^{5x+1}\sin \pi x.$

23. $(4D^2 + 16D + 17)y = 0;\ y = 1$ when $t = 0$ and $y = 0$ when $t = \pi.$

Ans. $y(t) = e^{-2t}\left[c_1 \cos(t/2) + c_2 \sin(t/2)\right].$

24. $y''' - 6y'' + 11y' - 6y = 0;\ y(0) = 0,\ y'(0) = -4,\ y''(0) = -18.$

[**Hint:** The auxiliary equation is $m^3 - 6m^2 + 11m - 6 = 0$, or $(m - 1)(m - 2)(m - 3) = 0$. The general solution is $y(x) = c_1e^x + c_2e^{2x} + c_3e^{3x}$. Find $y'(x)$, $y''(x)$ and substituting the initial conditions and then solving the equations obtained to get c_1, c_2 etc.] **Ans.** $y(x) = e^x + 2e^{2x} - 3e^{3x}$.

9.5 A PARTICULAR INTEGRAL (P.I.)

We have seen that particular integral y_p is that value of y which does not involve any arbitrary constant and satisfies the entire equation $f(D)y = X$. To make this concept further clear we consider the differential equation $\frac{d^2y}{dx^2} + 2\frac{dy}{dx} + 5y = 2 + 5x$. It is obvious that $y = x$ is a solution of this equation. This is called the particular integral. Similarly the equation

$$\frac{d^3y}{dx^3} - 7\frac{d^2y}{dx^2} + 5\frac{dy}{dx} + 6y = 5e^x$$

is satisfied by $y = e^x$ which is the particular integral of the equation.

9.6 THE INVERSE OPERATOR $\frac{1}{f(D)}$

Definition: $\frac{1}{f(D)}X$ defines that function of x which when operated on by $f(D)$ gives X that is,

$$f(D)\left[\frac{1}{f(D)}X\right] = X \qquad ...(9.30)$$

In other words $\frac{1}{f(D)}X$ is the solution of $f(D)\,y = X$ and is therefore, called the particular integral of this equation.

Thus the particular integral y_p of the equation $f(D)y = X$ is symbolically given by $y_p = \frac{1}{f(D)}X$.

It must be noted that $\frac{1}{f(D)}$ is an operator and must always precede the function operated on. It is incorrect to write y_p as $X \cdot \frac{1}{f(D)}$ or $\frac{X}{f(D)}$. $f(D)$ and $1/f(D)$ are called inverse operators.

As particular cases, we will show that

$$(i)\ \frac{1}{D}X = \int X\,dx\ ; \quad (ii)\ \frac{1}{D-\alpha}X = e^{\alpha x}\int e^{-\alpha x}X\,dx \qquad ...(9.31)$$

Case: (i) $\frac{1}{D}X$. Let $\frac{1}{D}X = y$.

Operating on both sides by D and remembering D and $1/D$ are inverse operators, we get

$$D\left[\frac{1}{D}X\right] = Dy \quad \text{or} \quad X = \frac{dy}{dx}. \quad \text{Therefore, } y = \int X\,dx. \quad \text{Thus } \frac{1}{D}X = \int X\,dx.$$

Case: (ii) $\frac{1}{D-\alpha}X$. Let $\frac{1}{D-\alpha}X = y$. Operating on both sides by $D - \alpha$, we get

$$(D-\alpha)\left[\frac{1}{D-\alpha}X\right] = (D-\alpha)y \quad \text{or} \quad X = (D-\alpha)y = \frac{dy}{dx} - \alpha y.$$

This is a linear first order differential equation. Its I.F. is $e^{-\alpha x}$. Hence its solution is

$$ye^{-\alpha x} = \int e^{-\alpha x}X\,dx \text{ or } y = e^{\alpha x}\int e^{-\alpha x}X\,dx \qquad \text{(no constant to be added).}$$

9.7 METHODS OF FINDING THE PARTICULAR INTEGRALS (P.I.)

Let us consider the differential equation

$$(a_0D^n + a_1D^{n-1} + a_2D^{n-2} + ... + a_{n-1}D + a_n)y = X \text{ or } f(D)y = X.$$

Its particular integral is given by $y_p(x) = \frac{1}{f(D)}X$.

We now discuss the methods for obtaining $y_p(x)$.

Case I: General Methods.

(i) *Method of Factors.* Since $f(D)$ is a polynomial of degree n in D, it can be factorised into n linear factors that is, $f(D) = (D - m_1)(D - m_2) \ldots (D - m_n)$.

$$\therefore \quad y_p(x) = \frac{1}{(D-m_1)(D-m_2)...(D-m_n)}X.$$

Now $$\frac{1}{D-m_n}X = e^{m_n x}\int e^{-m_n x}X\,dx. \qquad \text{(by Eq. (9.31))}$$

The above result is next operated by $\frac{1}{D-m_{n-1}}$. Performing integration successively by taking the symbolic factors in the denominator from right to left, we get $y_p(x)$.

(ii) *Method of Partial Fractions.* It is sometimes possible to express $\frac{1}{f(D)}$ into partial fractions so that

$$y_p(x) = \frac{1}{f(D)}X = \left(\frac{A_1}{x-m_1} + \frac{A_2}{x-m_2} + ... + \frac{A_n}{x-m_n}\right)X.$$

Using result (ii) of Eq. (9.31), we get

$$y_p(x) = A_1e^{m_1x}\int e^{-m_1x}X\,dx + ... + A_ne^{m_nx}\int e^{-m_nx}X\,dx.$$

It may be mentioned that the use of second method is advisible when the factors are all distinct.

EXAMPLE 9.15 Solve the differential equation $y'' - 3y' + 2y = e^{3x}$. *(AMIETE, June 2010)*

Solution The given equation in operator notation is $(D^2 - 3D + 2)y = e^{3x}$.

The characteristic equation of the corresponding homogeneous equation is $m^2 - 3m + 2 = 0$ or $(m - 1)(m - 2) = 0$. Its roots are $m = 1, 2$. The complementary function is $y_c(x) = c_1e^x + c_2e^{2x}$.

The particular integral is $y_p(x) = \dfrac{1}{(D-1)(D-2)} \cdot e^{3x} = \dfrac{1}{(D-1)} \cdot \dfrac{1}{(D-2)} e^{3x}$.

Using the method of factors

Here we first evaluate $\dfrac{1}{D-2} e^{3x} = e^{2x} \int e^{3x} e^{-2x}\, dx = e^{2x} \int e^x dx = e^{3x}$.

$$\therefore \quad \frac{1}{D-1}\left\{\frac{1}{D-2} e^{3x}\right\} = \frac{1}{D-1}\{e^{3x}\} = e^x \int e^{3x} e^{-x} dx = \frac{e^{3x}}{2}.$$

Thus
$$\frac{1}{(D-1)} \cdot \frac{1}{(D-2)} e^{3x} = \frac{e^{3x}}{2}.$$

Using the method of partial fractions

$$y_p(x) = \frac{1}{(D-1)(D-2)} e^{3x} = \left(\frac{1}{D-2} - \frac{1}{D-1}\right) e^{3x} = \frac{1}{D-2} e^{3x} - \frac{1}{D-1} e^{3x}.$$

$$= e^{2x} \int e^{-2x} e^{3x}\, dx - e^x \int e^{-x} e^{3x} dx = e^{3x} - \frac{e^{3x}}{2} = \frac{e^{3x}}{2}, \text{ which is same as above.}$$

Hence the general solution of equation is $y(x) = y_c(x) + y_p(x) = c_1e^x + c_2e^{2x} + (e^{3x}/2)$.

Case II. Special Methods.

Depending upon the particular form of the function X, we shall now develop the methods of obtaining the particular integral, $y_p(x)$, of the differential equation $f(D)y = X$.

(*i*) X = $e^{\alpha x}$.

Since $D\, e^{\alpha x} = \alpha e^{\alpha x}$; $D^2\, e^{\alpha x} = \alpha^2\, e^{\alpha x}$,...., $D^{n-1} e^{\alpha x} = \alpha^{n-1} e^{\alpha x}$, $D^n e^{\alpha x} = \alpha^n\, e^{\alpha x}$.

Therefore $f(D)e^{\alpha x} = (a_0\alpha^n + a_1\alpha^{n-1} + a_2\alpha^{n-2} + ... + a_{n-1}\alpha + a_n)e^{\alpha x}$.

Operating on both sides by $\dfrac{1}{f(D)}$, we obtain

$$\frac{1}{f(D)}\left[f(D)e^{\alpha x}\right] = \frac{1}{f(D)}\left[f(\alpha)e^{\alpha x}\right] \text{ or } e^{\alpha x} = f(\alpha)\frac{1}{f(D)} e^{\alpha x}.$$

Thus
$$\mathbf{\frac{1}{f(D)} e^{\alpha x} = \frac{1}{f(\alpha)} e^{\alpha x}, \text{provided } f(\alpha) \neq 0.} \quad ...(9.32)$$

Case of Failure. If $f(\alpha) = 0$, then $f(D)$ has $(D - \alpha)$ as a factor, and the above method fails. To calculate $y_p(x)$ in this case, we write

$$f(D) = (D - \alpha)\, \phi(D) \text{ where } \phi(D) \neq 0 \text{ at } D = \alpha.$$

$$\therefore \quad y_p(x) = \frac{1}{f(D)} e^{\alpha x} = \frac{1}{\phi(D)(D-\alpha)} e^{\alpha x} = \frac{1}{\phi(\alpha)(D-\alpha)} e^{\alpha x} = \frac{1}{\phi(\alpha)} e^{\alpha x} \int e^{-\alpha x} \cdot e^{\alpha x} dx = \frac{1}{\phi(\alpha)} e^{\alpha x}.$$

[Since $f(D) = (D - \alpha)\, \phi(D)$, differentiating both sides w.r.t. D, we get

$$f'(D) = (D - \alpha)\phi'(D) + \phi(D)$$

$$\therefore \quad f'(D)|_{D=\alpha} = \phi(\alpha), \quad \therefore f'(\alpha) = \phi(\alpha)].$$

$$\therefore \quad \mathbf{\frac{1}{f(D)} e^{\alpha x} = x\frac{1}{\phi(\alpha)} e^{\alpha x} = x\frac{1}{f'(\alpha)} e^{\alpha x}, \text{provided } f'(\alpha) \neq 0.} \quad ...(9.33)$$

If $f'(\alpha) = 0$, then $(D - \alpha)^2$ is a factor of $f(D)$ and the above method fails. Then we write

$$f(D) = (D - \alpha)^2 \phi(D).$$

Hence the particular integral $y_p(x)$ in this case is given by

$$y_p(x) = \frac{1}{\phi(D)(D-\alpha)^2} e^{\alpha x} = \frac{1}{\phi(\alpha)(D-\alpha)^2} e^{\alpha x}$$

$$= x^2 \frac{1}{2!\phi(\alpha)} e^{\alpha x} = x^2 \frac{1}{f''(\alpha)} e^{\alpha x}, \text{provided } f''(\alpha) \neq 0.$$

In general if $(D - \alpha)^r$ occur as a factor in $f(D)$, then

$f(D) = (D - \alpha)^r \phi(D)$ and $y_p(x)$ is given by

$$y_p(x) = \frac{1}{f(D)} e^{\alpha x} = \frac{1}{(D-\alpha)^r \phi(D)} e^{\alpha x} = \frac{x^r e^{\alpha x}}{r!\,\phi(\alpha)} = x^r \frac{1}{f^r(\alpha)} e^{\alpha x}. \qquad \text{...(9.34)}$$

EXAMPLE 9.16 Find the particular integral of the following differential equations

(*i*) $(D^3 - 2D^2 - 5D + 6)y = e^{4x}$, (*ii*) $(D^3 - 3D^2 + 4)y = e^{2x}$.

Solution (*i*) The particular integral is $y_p(x) = \dfrac{1}{D^3 - 2D^2 - 5D + 6} \cdot e^{4x}$

$$= \frac{1}{(D-1)(D-3)(D+2)} e^{4x} = \frac{1}{(4-1)(4-3)(4+2)} e^{4x} = \frac{1}{18} e^{4x}.$$

(*ii*) The particular integral is $y_p(x) = \dfrac{1}{D^3 - 3D^2 + 4} e^{2x}$

$f(D) = D^3 - 3D^2 + 4$ vanishes when D is replaced by 2. It is a case of failure. We multiply the numerator by x and differentiate the denominator w.r.t. D.

$$\therefore\ y_p(x) = x \cdot \frac{1}{3D^2 - 6D} e^{2x}.$$

It is again a case of failure. We multiply the numerator by x and differentiate the denominator w.r.t. D.

$$y_p(x) = x^2 \frac{1}{6D-6} e^{2x} = \frac{x^2}{6(2)-6} e^{2x} = \frac{x^2}{6} e^{2x}.$$

(*ii*) When X = sin(ax + b) or cos (ax + b)

We have
$$D \sin(ax + b) = a \cos(ax + b),$$
$$D^2 \sin(ax + b) = -a^2 \sin(ax + b),$$
$$D^4 \sin(ax + b) = (-a^2)^2 \sin(ax + b).$$

In general
$$(D^2)^k \sin(ax + b) = (-a^2)^k \sin(ax + b).$$

Since $f(D^2)$ is a polynomial expression in D^2, we have

$$f(D^2) \sin(ax + b) = f(-a^2) \sin(ax + b).$$

Operating on both sides of the above relation by $\dfrac{1}{f(D^2)}$, we obtain

$$\frac{1}{f(D^2)}\left[f(D^2)\sin(ax+b)\right] = \frac{1}{f(D^2)}\left[f(-a^2)\sin(ax+b)\right]$$

or
$$\sin(ax+b) = f(-a^2)\frac{1}{f(D^2)}\sin(ax+b).$$

On rewriting the terms, we get

$$\frac{1}{f(D^2)}\sin(ax+b) = \frac{1}{f(-a^2)}\sin(ax+b), \text{ provided } f(-a^2) \neq 0 \qquad \text{...(9.35)}$$

Case of Failure. If $f(-a^2) = 0$, then $D^2 + a^2$ is a factor of $f(D)$ and the above method fails and we proceed as follows.

Since $e^{i(ax+b)} = \cos(ax+b) + i\sin(ax+b)$.

$$\therefore \quad \frac{1}{f(D^2)}\sin(ax+b) = \frac{1}{f(D^2)}\text{imaginary part of } e^{i(ax+b)} = x\cdot\frac{1}{f'(D^2)}, \text{ imaginary part of } e^{i(ax+b)}$$

where $D^2 = -a^2$ and $\quad f'(D^2) = \dfrac{d}{dD}f(D^2) \quad [\because f(-a^2) = 0]$

$$\mathbf{\frac{1}{f(D^2)}\sin(ax+b) = x\frac{1}{f'(-a^2)}\sin(ax+b), \quad provided\ f'(-a^2) \neq 0}$$

If $f'(-a^2) = 0$, then the particular integral is given by

$$\mathbf{\frac{1}{f(D)^2}\sin(ax+b) = x^2\frac{1}{f''(-a^2)}\sin(ax+b), \ provided\ f''(-a^2) \neq 0.}$$

Similarly $$\frac{1}{f(D^2)}\cos(ax+b) = \frac{1}{f(-a^2)}\cos(ax+b), \qquad ...(9.36)$$

provided $f(-a^2) \neq 0$.

If $f(-a^2) = 0$, then $$\frac{1}{f(D)^2}\cos(ax+b) = x\frac{1}{f'(-a^2)}\cos(ax+b), \qquad ...(9.37)$$

provided $f'(-a^2) \neq 0$ and so on.

Illustration: The particular integral of $(D^2 + a^2)y = \sin ax$ is $\dfrac{-x}{2a}\cos ax$. *(AMIETE., June 2009)*

EXAMPLE 9.17 Find the particular integral of the following differential equations

(*i*) $(D^3 + 1)y = \cos(2x + 3)$. (*ii*) $(D^3 + 4D)y = \sin 2x$. (*iii*) $\dfrac{d^2y}{dx^2} + 4y = \cos 2x$. *(AMIE, W-2005)*

Solution (*i*) The particular integral is

$$y_p(x) = \frac{1}{D^3+1}\cos(2x+3) = \frac{1}{D\cdot D^2+1}\cos(2x+3)$$

$$= \frac{1}{D(-2^2)+1}\cos(2x+3) = \frac{1}{1-4D}\cos(2x+3) = \frac{1+4D}{1-16D^2}\cos(2x+3)$$

$$= \frac{1+4D}{1-16(-2^2)}\cos(2x+3) = \frac{1}{65}[\cos(2x+3) - 8\sin(2x+3)].$$

(*ii*) The particular integral is $y_p(x) = \dfrac{1}{D(D^2+4)}\sin 2x$

$$= x\frac{1}{3D^2+4}\sin 2x = x\frac{1}{3(-2^2)+4}\sin 2x = -\frac{x}{8}\sin 2x.$$

(*iii*) The particular integral is $y_p(x) = \dfrac{1}{D^2+4}\cdot\cos 2x$

$$= x\cdot\frac{1}{2D}\cos 2x = \frac{x}{2}\int\cos 2x\,dx = \frac{x\sin 2x}{4}.$$

(*iii*) when X = x^m, m being a positive integer.

The particular integral $y_p(x) = \dfrac{1}{f(D)} x^m = [f(D)]^{-1} x^m$.

We expand $[f(D)]^{-1}$ by Binomial theorem in ascending powers of D upto the mth power and evaluate the result by direct differentiation of x^m. The powers of D beyond m need not be considered as their derivatives will all be zero. This is illustrated in the following problem.

EXAMPLE 9.18 Find the particular integral of $(D^2 + 5D + 4)y = x^2 + 7x + 9$.

Solution The particular integral is $y_p(x) = \dfrac{1}{D^2 + 5D + 4}(x^2 + 7x + 9)$

$$= \frac{1}{4\left(1 + \frac{5D}{4} + \frac{D^2}{4}\right)}(x^2 + 7x + 9) = \frac{1}{4}\left[1 + \left(\frac{5D}{4} + \frac{D^2}{4}\right)\right]^{-1} (x^2 + 7x + 9)$$

$$= \frac{1}{4}\left[1 - \left(\frac{5D}{4} + \frac{D^2}{4}\right) + \left(\frac{5D}{4} + \frac{D^2}{4}\right)^2 - \ldots\right](x^2 + 7x + 9)$$

$$= \frac{1}{4}\left[1 - \frac{5D}{4} - \frac{D^2}{4} + \frac{25D^2}{16} + \ldots\ldots\right](x^2 + 7x + 9)$$

$$= \frac{1}{4}\left[1 - \frac{5D}{4} + \frac{21D^2}{16} + \ldots\ldots\right](x^2 + 7x + 9)$$

$$= \frac{1}{4}\left[(x^2 + 7x + 9) - \frac{5}{4}D(x^2 + 7x + 9) + \frac{21}{16}D^2(x^2 + 7x + 9)\right]$$

$$= \frac{1}{4}\left[(x^2 + 7x + 9) - \frac{5}{4}(2x + 7) + \frac{21}{16}(2)\right] = \frac{1}{4}\left(x^2 + \frac{9}{2}x + \frac{23}{8}\right).$$

(*iv*) X = $e^{\alpha x}$ V, where V is a function of x.

Since when $\quad D(e^{\alpha x}V) = \alpha e^{\alpha x}V + e^{\alpha x}DV = e^{\alpha x}(D + \alpha)V.$

$$D^2(e^{\alpha x}V) = \alpha^2 e^{\alpha x}V + \alpha e^{\alpha x}DV + \alpha e^{\alpha x}DV + e^{\alpha x}D^2V$$

$$= e^{\alpha x}(D^2 + 2D\alpha + \alpha^2)V = e^{\alpha x}(D + \alpha)^2 V.$$

Similarly $\quad D^3(e^{\alpha x}V) = e^{\alpha x}(D + \alpha)^3 V$

In general, $\quad D^n(e^{\alpha x}V) = e^{\alpha x}(D + \alpha)^n V.$

Hence $\quad f(D)(e^{\alpha x}V) = e^{\alpha x}f(D + \alpha)V.$ ...(9.38)

Now $\quad$ let $f(D + \alpha)V = V_1$.

Operating on both sides by $\dfrac{1}{f(D+\alpha)}$, we get

$$\frac{1}{f(D+\alpha)}[f(D+\alpha)V] = \frac{1}{f(D+\alpha)}V_1 \quad \text{or} \quad V = \frac{1}{f(D+\alpha)}V_1. \qquad \ldots(9.39)$$

Substituting for V from Eq. (9.39) in Eq. (9.38), we get

$$f(D)\left(e^{\alpha x}\frac{1}{f(D+\alpha)}V_1\right) = e^{\alpha x}V_1.$$

Operating on both sides by $\dfrac{1}{f(D)}$, we have

$$\frac{1}{f(D)}\left[f(D)\left(e^{\alpha x}\frac{1}{f(D+\alpha)}V_1\right)\right]=\frac{1}{f(D)}e^{\alpha x}V_1$$

Rewriting the terms, we obtain

$$\frac{1}{f(D)}e^{\alpha x}V_1=e^{\alpha x}\frac{1}{f(D+\alpha)}V_1$$

Now since V_1 is any function of x, we have the rule

$$\mathbf{\frac{1}{f(D)}e^{\alpha x}V=e^{\alpha x}\frac{1}{f(D+\alpha)}V.} \qquad \text{...(9.40)}$$

EXAMPLE 9.19 Find the particular integral of $(D^2 - 2D + 5)y = e^{2x}\sin x$.

Solution The particular integral is

$$y_p(x)=\frac{1}{D^2-2D+5}e^{2x}\sin x=e^{2x}\frac{1}{(D+2)^2-2(D+2)+5}\sin x$$

$$=e^{2x}\frac{1}{D^2+2D+5}\sin x=e^{2x}\frac{1}{-1+2D+5}\sin x$$

$$=\frac{e^{2x}}{2}\frac{1}{D+2}\sin x=\frac{e^{2x}}{2}\frac{(D-2)}{D^2-4}\sin x=\frac{e^{2x}}{2}\frac{(D-2)}{-5}\sin x=-\frac{e^{2x}}{10}(\cos x-2\sin x)$$

(v) When $X = xV$, where V is a function of x.

We have by successive differentiation, $D(xV) = xDV + V$,

$$D^2(xV) = xD^2V + 2DV,$$

$$D^3(xV) = xD^3V + 3D^2V,$$

....

....

In general, $D^n(xV) = xD^nV + nD^{n-1}V = xD^nV + \frac{d}{dD}(D^n)V.$

Hence $f(D)\,[xV] = xf(D)\,V + f'(D)V$ ...(9.41), where $f'(D)=\frac{d}{dD}f(D)$.

Now let $V_1 = f(D)V$ so that $V=\frac{1}{f(D)}V_1$. Substituting in (9.41), we get

$$f(D)\left[x\cdot\frac{1}{f(D)}V_1\right]=xf(D)\frac{1}{f(D)}V_1+f'(D)\frac{1}{f(D)}V_1,$$

or

$$xV_1=f(D)\left[x\cdot\frac{1}{f(D)}V_1\right]-f'(D)\frac{1}{f(D)}V_1$$

and

$$\frac{1}{f(D)}xV_1=x\frac{1}{f(D)}V_1-\frac{1}{f(D)}f'(D)\frac{1}{f(D)}V_1=x\frac{1}{f(D)}V_1-\frac{f'(D)}{[f(D)]^2}V_1$$

and since V_1 is any function of x, we have

$$\mathbf{\frac{1}{f(D)}[xV]=x\frac{1}{f(D)}V-\frac{f'(D)}{[f(D)]^2}V.} \qquad \text{...(9.42)}$$

EXAMPLE 9.20 Find the particular integral of $(D^2 - 2D + 1)y = x \sin x$.

Solution The particular integral is

$$y_p(x) = \frac{1}{D^2 - 2D + 1} \cdot x \sin x = x \frac{1}{D^2 - 2D + 1} \sin x - \frac{(2D-2)}{(D^2 - 2D + 1)^2} \sin x$$

$$= x \frac{1}{-1 - 2D + 1} \sin x - \frac{(2D-2)}{(-1 - 2D + 1)^2} \sin x$$

$$= -\frac{1}{2} x \int \sin x \, dx - \frac{1}{2D^2}(D-1) \sin x = \frac{1}{2} x \cos x - \frac{1}{2} \cdot \frac{1}{D} \int (\cos x - \sin x)\, dx$$

$$= \frac{1}{2} x \cos x - \frac{1}{2} \cdot \frac{1}{D}(\sin x + \cos x) = \frac{1}{2} x \cos x - \frac{1}{2} \int (\sin x + \cos x)\, dx = \frac{1}{2} x \cos x - \frac{1}{2}(-\cos x + \sin x)$$

$$= \frac{1}{2}(x \cos x + \cos x - \sin x).$$

Working Procedure to find the P.I.

The particular integral, $y_p(x)$, of the differential equation $f(D)\, y = X$ is symbolically written as

$$y_p(x) = \frac{1}{f(D)} X.$$

Table 9.2	
Form of X	***Particular integral***
(1) e^{ax}	
$y_{p(x)} = \frac{1}{f(D)} e^{ax} = \frac{1}{f(a)} e^{ax}$,	provided $f(a) \neq 0$.
$y_p(x) = x \cdot \frac{1}{f'(a)} e^{ax}$,	when $f(a) = 0$, provided $f'(a) \neq 0$, where $f'(a) = \frac{d}{dD} f(D)\big\vert_{D=a}$.
$y_p(x) = x^2 \cdot \frac{1}{f''(a)} e^{ax}$,	when $f(a) = f'(a) = 0, f''(a) \neq 0$.
(2) $\sin(ax + b)$ **or** $\cos(ax + b)$	
$y_p(x) = \frac{1}{f(D^2)} \begin{matrix}\sin\\ \cos\end{matrix} (ax+b) = \frac{1}{f(-a^2)} \begin{matrix}\sin\\ \cos\end{matrix} (ax+b)$, provided $f(-a^2) \neq 0$	
$= x \frac{1}{f'(-a^2)} \begin{matrix}\sin\\ \cos\end{matrix} (ax+b)$, $f(-a^2) = 0$, $f'(-a^2) \neq 0$	
$= x^2 \frac{1}{f''(-a^2)} \begin{matrix}\sin\\ \cos\end{matrix} (ax+b)$, $f(-a^2) = f'(-a^2) = 0, f''(-a^2) \neq 0$.	
(3) x^m **(m = positive integer)** $y_p(x) = [f(D)]^{-1} x^m$. Expand $[f(D)]^{-1}$ in ascending powers of D upto D^m and differentiate term by term.	
(4) $e^{ax} V$	$y_p(x) = \frac{1}{f(D)} e^{ax} V = e^{ax} \cdot \frac{1}{f(D+a)} V$
(5) xV	$y_p(x) = \frac{1}{f(D)}[xV] = x \frac{1}{f(D)} V - \frac{f'(D)}{[f(D)]^2} V$

ILLUSTRATIVE EXAMPLES

EXAMPLE 9.21 Find the general solution of the following differential equations.

(*i*) $y''' - 2y'' - 5y' + 6y = e^{4x} + 2$, (*ii*) $y''' - y' = e^x + e^{-x}$,

(*iii*) $\dfrac{d^3y}{dx^3} - 5\dfrac{d^2y}{dx^2} + 8\dfrac{dy}{dx} - 4y = e^{2x} + 3e^{-x}$, (*iv*) $(2D + 1)^2 y = 4e^{-x/2}$,

(*v*) $\dfrac{d^3y}{dx^3} + \dfrac{d^2y}{dx^2} - \dfrac{dy}{dx} - y = \cosh x$, (*vi*) $\dfrac{d^2y}{dx^2} - 4y = \cosh(2x - 1) + 3^x$.

(*Kurukshetra, 2004; VTU 2000*)

Solution (*i*) The given equation in operator notation is $(D^3 - 2D^2 - 5D + 6)y = e^{4x} + 2$.

The characteristic equation of the correponding homogeneous equation is $m^3 - 2m^2 - 5m + 6 = (m - 1)(m - 3)(m + 2) = 0$. Its roots are $m = 1, 3, -2$.

The complementary function is given by $y_c(x) = c_1e^x + c_2e^{3x} + c_3e^{-2x}$.

The particular integral is

$$y_p(x) = \frac{1}{(D-1)(D-3)(D+2)}e^{4x} + 2 = \frac{1}{(D-1)(D-3)(D+2)}e^{4x} + 2\cdot\frac{1}{(D-1)(D-3)(D+2)}e^{0x}$$

$$= \frac{1}{(4-1)(4-3)(4+2)}e^{4x} + 2\cdot\frac{1}{(-1)(-3)(2)}e^{0x} = \frac{e^{4x}}{18} + \frac{1}{3}.$$

The general solution is $y(x) = y_c(x) + y_p(x) = c_1e^x + c_2e^{3x} + c_3e^{-2x} + \dfrac{e^{4x}}{18} + \dfrac{1}{3}$.

(*ii*) The characteristic equation of the corresponding homogeneous equation $y''' - y' = 0$ is $m^3 - m = 0$ or $m(m + 1)(m - 1) = 0$

The roots of this equation are $m = 0, -1, 1$.

The complementary function is $y_c(x) = c_1 + c_2e^{-x} + c_3e^x$.

The particular integral is $y_p(x) = \dfrac{1}{D(D+1)(D-1)}(e^x + e^{-x})$

$$= \frac{1}{D(D+1)(D-1)}e^x + \frac{1}{D(D+1)(D-1)}e^{-x}$$

$$= \frac{1}{2(D-1)}e^x + \frac{1}{2(D+1)}e^{-x} = \frac{xe^x}{2} + \frac{xe^{-x}}{2}.$$

The general solution of the equation is

$$y(x) = y_c(x) + y_p(x) = c_1 + c_2e^{-x} + c_3e^x + \frac{xe^x}{2} + \frac{xe^{-x}}{2}.$$

(*iii*) The given equaiton in operator notation is

$$(D^3 - 5D^2 + 8D - 4)\, y = e^{2x} + 3e^{-x}.$$

The characteristic equation of the corresponding homogeneous equation $y''' - 5y'' + 8y' - 4y = 0$ is $m^3 - 5m^2 + 8m - 4 = 0$ or $(m - 1)(m - 2)^2 = 0$. Its roots are $m = 1, 2, 2$.

The complementary function is $y_c(x) = c_1 e^x + (c_2 + c_3x)e^{2x}$.

The particular integral is $y_p(x) = \dfrac{1}{(D-1)(D-2)^2}(e^{2x} + 3e^{-x})$

$$= \frac{1}{(D-1)(D-2)^2} e^{2x} + 3\frac{1}{(D-1)(D-2)^2} e^{-x}$$

$$= \frac{1}{(2-1)(D-2)^2} e^{2x} + 3\frac{1}{(-1-1)(-1-2)^2} e^{-x} = \frac{1}{(D-2)^2} e^{2x} - \frac{e^{-x}}{6}$$

$$= \frac{x^2 e^{2x}}{2} - \frac{e^{-x}}{6}.$$ The general solution of the equation is

$$y(x) = y_c(x) + y_p(x) = c_1 e^x + (c_2 + c_3 x)e^{2x} + \frac{(x^2 e^{2x})}{2} - \frac{(e^{-x})}{6}.$$

(*iv*) The characteristic equation of the corresponding homogeneous equation is $(2m + 1)^2 = 0$. The roots of this equation are $m = -1/2, -1/2$.
The complementary function is $y_c(x) = (c_1 + c_2 x)e^{-x/2}$.

The particular integral is $y_p(x) = \frac{1}{(2D+1)^2} 4e^{-x/2} = 4\frac{1}{(2D+1)^2} e^{-x/2}$

$$= 4 \cdot \frac{1}{4} \frac{1}{(D+(1/2))^2} e^{-x/2} = \frac{(x^2 e^{-x/2})}{2}.$$

The general solution of the equation is $y(x) = y_c(x) + y_p(x) = (c_1 + c_2 x)e^{-x/2} + \frac{(x^2 e^{-x/2})}{2}$.

(*v*) The given equation in operator notation is $(D^3 + D^2 - D - 1)y = \cosh x = (e^x + e^{-x})/2$.
The characteristic equation of the corresponding homogeneous equation is $m^3 + m^2 - m - 1 = 0$ or $(m - 1)(m + 1)^2 = 0$. Its roots are $m = 1, -1, -1$.
The complementary function is $y_c(x) = c_1 e^x + (c_2 + c_3 x)e^{-x}$.

The particular integral is $y_p(x) = \frac{1}{(D-1)(D+1)^2} \cosh x$

$$= \frac{1}{(D-1)(D+1)^2} \cdot \frac{e^x + e^{-x}}{2} = \frac{1}{2}\left[\frac{1}{(D-1)(D+1)^2} e^x + \frac{1}{(D+1)^2(D-1)} e^{-x}\right]$$

$$= \frac{1}{2}\left[\frac{1}{(D-1)(1+1)^2} e^x + \frac{1}{(D+1)^2(-1-1)} e^{-x}\right]$$

$$= \frac{1}{2}\left[\frac{1}{4} \cdot \frac{1}{D-1} e^x - \frac{1}{2} \cdot \frac{1}{(D+1)^2} e^{-x}\right] = \frac{1}{8} x e^x - \frac{1}{4}\frac{x^2}{2} e^{-x} = \frac{1}{8} x e^x - \frac{1}{8} x^2 e^{-x}.$$

The general solution of the equation is $y(x) = y_c(x) + y_p(x) = c_1 e^x + (c_2 + c_3 x)e^{-x} + \frac{xe^x}{8} - \frac{x^2 e^{-x}}{8}$.

(*vi*) The given equation in operator notation is $(D^2 - 4)y = \cosh(2x - 1) + 3^x$.
The characteristic equation of the corresponding homogeneous equation is $m^2 - 4 = 0$. Its roots are $m = \pm 2$.
The complementary function is $y_c(x) = c_1 e^{2x} + c_2 e^{-2x}$.

The particular integral is $y_p(x) = \frac{1}{D^2 - 4} \cosh(2x - 1) + 3^x$.

Now $\dfrac{1}{D^2-4}\cosh(2x-1) = \dfrac{1}{D^2-4}\left[\dfrac{e^{(2x-1)}+e^{-(2x-1)}}{2}\right]$

$$= \frac{e^{-1}}{2}\frac{1}{D^2-4}e^{2x} + \frac{e}{2}\frac{1}{D^2-4}e^{-2x} = \frac{1}{2e}x\frac{1}{2D}e^{2x} + \frac{e}{2}x\frac{1}{2D}e^{-2x}$$

$$= \frac{x}{2e}\frac{e^{2x}}{4} - \frac{xe}{2}\frac{e^{-2x}}{4} = \frac{x}{4}\left[\frac{e^{2x-1}-e^{-(2x-1)}}{2}\right] = \frac{x}{4}\sinh(2x-1),$$

and $\dfrac{1}{D^2-4}3^x = \dfrac{1}{D^2-4}e^{(\log 3)x}$ $\quad$ ($\because$ 3^x may be written as $e^{\log 3^x} \Rightarrow e^{x\log 3}$)

$$= \frac{e^{(\log 3)x}}{(\log 3)^2-4} = \frac{3^x}{(\log 3)^2-4},\quad \text{(Rewrite } e^{(\log 3)x}=3^x\text{)}$$

The general solution of the equation is $y(x) = y_c(x) + y_p(x)$

$$= c_1e^{2x} + c_2e^{-2x} + \frac{x}{4}\sinh(2x-1) + \frac{3^x}{(\log 3)^2-4}.$$

EXAMPLE 9.22 Find the general solution of the following differential equations.

(*i*) $y'' - 5y' + 4y = 65\sin 2x$, (*AMIETE, June 2005*) (*ii*) $y'' + 3y' + 2y = \cos 2x$,

(*iii*) $y''' + 6y'' + 11y' + 6y = 2\sin x$, (*iv*) $(D-1)^2(D^2+1)y = e^x + \sin^2\dfrac{x}{2}$,

(*v*) $(D^4 - 3D^2 - 4)y = 24\sin 2x - 40e^{-2x}$, (vi) $(D^3+1)y = \cos(2x-1)$.

Solution (*i*) The given equation in operator notation is $(D^2 - 5D + 4)y = 65\sin 2x$. The characteristic equation of the corresponding homogeneous equation is $m^2 - 5m + 4 = 0$ or $(m-1)(m-4) = 0$. Its roots are $m = 1, 4$.

The complementary function is $y_c(x) = c_1e^x + c_2e^{4x}$.

The particular integral is $y_p(x) = \dfrac{1}{D^2-5D+4}\cdot 65\sin 2x$

$$= 65\cdot\frac{1}{-2^2-5D+4}\sin 2x = -13\frac{\sin 2x}{D} = \frac{13}{2}\cos 2x.$$

The general solution of the equation is $y(x) = y_c(x) + y_p(x)$.

$$y(x) = c_1e^x + c_2e^{4x} + \frac{(13\cos 2x)}{2}.$$

(*ii*) The given equation in operator notation is $(D^2 + 3D + 2)y = \cos 2x$. The characteristic equation of the corresponding homogeneous equation is $m^2 + 3m + 2 = (m+1)(m+2) = 0$. Its roots are $m = -1, -2$.

The complementary function is $y_c(x) = c_1e^{-x} + c_2e^{-2x}$.

The particular integral is $y_p(x) = \dfrac{1}{D^2+3D+2}\cos 2x.$

$$= \frac{1}{(-2^2)+3D+2}\cos 2x = \frac{\cos 2x}{3D-2} = \frac{(3D+2)}{9D^2-4}\cos 2x = \frac{(3D+2)\cos 2x}{9(-4)-4}$$

$$= -\frac{1}{40}(-6\sin 2x + 2\cos 2x) = \frac{1}{20}(3\sin 2x - \cos 2x).$$

The general solution of the equation is

$$y(x) = y_c(x) + y_p(x) = c_1e^{-x} + c_2e^{-2x} + \frac{(3\sin 2x - \cos 2x)}{20}.$$

(*iii*) The given equation in operator notation is $(D^3 + 6D^2 + 11D + 6)y = 2\sin x$. The characteristic equation of the corresponding homogeneous equation is

$$m^3 + 6m^2 + 11m + 6 = 0 \quad \text{or} \quad (m + 1)(m + 2)(m + 3) = 0.$$

Its root are $m = -1, -2, -3$.

The complementary function is $y_c(x) = c_1e^{-x} + c_2e^{-2x} + c_3e^{-3x}$.

The particular integral is $y_p(x) = \dfrac{1}{D^3 + 6D^2 + 11D + 6} 2\sin x$

$$= \frac{1}{D(D^2 + 11) + 6(D^2 + 1)} 2\sin x = \frac{1}{D(-1^2 + 11) + 6(-1^2 + 1)} 2\sin x = \frac{2\sin x}{10D} = -\frac{1}{5}\cos x.$$

The general solution of the equation is

$$y(x) = y_c(x) + y_p(x) = c_1e^{-x} + c_2e^{-2x} + c_3e^{-3x} - \frac{(\cos x)}{5}.$$

(*iv*) The characteristic equation of the correstponding homogeneous equation is $(m - 1)^2 (m^2 + 1) = 0$. Its roots are $m = 1, 1, \pm i$.

The complementary function is $y_c(x) = (c_1 + c_2x)e^x + c_3 \cos x + c_4 \sin x$.

The particular integral is $y_p(x) = \dfrac{1}{(D-1)^2(D^2+1)}\left(e^x + \sin^2 \dfrac{x}{2}\right)$.

Now $\dfrac{1}{(D-1)^2(D^2+1)} e^x = \dfrac{1}{2} \cdot \dfrac{1}{(D-1)^2} e^x = \dfrac{1}{2} \cdot \dfrac{x^2}{2} e^x = \dfrac{x^2}{4} e^x$,

and $\dfrac{1}{(D-1)^2(D^2+1)} \sin^2 \dfrac{x}{2} = \dfrac{1}{2} \dfrac{1}{(D^2+1)(D-1)^2} (1 - \cos x)$

$$= \frac{1}{2}\left[\frac{1}{(D^2+1)(D-1)^2} e^{ox} - \frac{1}{(D^2+1)(D-1)^2} \cos x\right]$$

$$= \frac{1}{2}\left[1 - \frac{1}{(D^2+1)} \cdot \frac{1}{-2D} \cos x\right] = \frac{1}{2}\left[1 + \frac{1}{2} \frac{1}{(D^2+1)} \sin x\right]$$

$$= \frac{1}{2} + \frac{1}{4} \cdot \frac{1}{D^2+1} \sin x = \frac{1}{2} + \frac{1}{4} \cdot x \frac{\sin x}{2D} = \frac{1}{2} - \frac{1}{8} x\cos x$$

The general solution of the equation is $y(x) = y_c(x) + y_p(x)$

$$= (c_1 + c_2x)e^x + (c_3 \cos x + c_4 \sin x) + (x^2e^x/4) + (1/2) - (x\cos x)/8.$$

(*v*) The characteristic equation of the corresponding homogenous equation is

$m^4 - 3m^2 - 4 = 0$ or $(m^2 - 4)(m^2 + 1) = 0$. Its roots are $m = \pm 2, \pm i$.

The complementary function is $y_c(x) = c_1e^{2x} + c_2e^{-2x} + c_3 \cos x + c_4 \sin x$.

The particular integral is $y_p(x) = \dfrac{1}{D^4 - 3D^2 - 4} 24\sin 2x - 40e^{-2x}$.

Now $\dfrac{1}{D^4 - 3D^2 - 4} \sin 2x = \dfrac{1}{(-2^2)^2 - (3)(-2^2) - 4} \sin 2x = \dfrac{\sin 2x}{24}$,

and $\dfrac{1}{D^4 - 3D^2 - 4} e^{-2x} = x \cdot \dfrac{1}{4D^3 - 6D} e^{-2x} = x \dfrac{1}{4(-2)^3 - 6(-2)} e^{-2x} = \dfrac{xe^{-2x}}{-20}$.

$\therefore \qquad y_p(x) = \sin 2x + 2xe^{-2x}$.

The general solution of the equation is $y(x) = y_c(x) + y_p(x)$

$= c_1e^{2x} + c_2e^{-2x} + c_3 \cos x + c_4 \sin x + \sin 2x + 2xe^{-2x}$.

(*vi*) The characteristic equation of the corresponding homogenous equation is

$m^3 + 1 = 0 \Rightarrow (m + 1)(m^2 - m + 1) = 0$. Its roots are $-1,\ (1 \pm \sqrt{3}\,i)/2$.

The complementary function is $y_c(x) = c_1e^{-x} + e^{x/2}\left(c_2 \cos \frac{\sqrt{3}}{2}x + c_3 \sin \frac{\sqrt{3}}{2}x\right)$.

The particular integral is $y_p(x) = \frac{1}{D^3+1}\cos(2x-1) = \frac{1}{1-4D}\cos(2x-1)$

$$= \frac{1+4D}{1-16D^2}\cos(2x-1) = \frac{1}{65}[\cos(2x-1) - 8\sin(2x-1)].$$

The general solution of the equation is $y(x) = y_c(x) + y_p(x)$

$$= c_1e^{-x} + e^{x/2}\left(c_2 \cos \frac{\sqrt{3}}{2}x + c_3 \sin \frac{\sqrt{3}}{2}x\right) + \frac{1}{65}[\cos(2x-1) - 8\sin(2x-1)].$$

EXAMPLE 9.23 An inductor of 2 henries, resistor of 16 ohms and capacitor of 0.02 farads are connected in series with a battery of e.m.f, $E(t) = 100 \sin 3t$. At $t = 0$ the charge on the capacitor and current in the circuit are zero. Find the (*i*) charge and (*ii*) current at $t > 0$.

Solution By kirchhoff's laws, we have

$$L\frac{di}{dt} + Ri + \frac{1}{C}q = E(t) \qquad ...(1)$$

Substituting $L = 2$, $R = 16$, $C = 0.02$ and $E(t) = 100 \sin 3t$ in (1), we find

$$2\frac{di}{dt} + 16\,i + \frac{q}{0.02} = 100 \sin 3\,t.$$

Since $i = dq/dt$, therefore $\frac{d^2q}{dt^2} + 8\frac{dq}{dt} + 25\,q = 50 \sin 3\,t \qquad ...(2)$

Equation (2) in D operator form is

$$(D^2 + 8D + 25)q = 50 \sin 3t, \text{ where } D = d/dt.$$

The characteristic equation of the corresponding homogeneous equation is $m^2 + 8m + 25 = 0$

or $$m = \frac{-8 \pm \sqrt{64-100}}{2} = -4 \pm 3\,i.$$

The complementary function is $q_c(t) = e^{-4t}(c_1 \cos 3t + c_2 \sin 3t)$.

The particular integral is $q_p(t) = \frac{1}{D^2+8D+25}50 \sin 3\,t$.

$$= \frac{50}{8}\frac{1}{D+2}\cdot\frac{D-2}{D-2}\cdot\sin 3\,t.$$

$$= -\frac{50}{8}\cdot\frac{(D-2)\sin 3\,t}{13}$$

$$= -\frac{25}{52}(3\cos 3t - 2\sin 3t).$$

The general solution is $q(t) = e^{-4t}(c_1 \cos 3t + c_2 \sin 3t) - \frac{25}{52}(3\cos 3t - 2\sin 3t)$.

Substituting in, the initial conditions, $q = 0$, $dq/dt = 0$ at $t = 0$, we have

$$q(t) = \frac{25}{52}(2\sin 3t - 3\cos 3t) + \frac{25}{52}e^{-4t}(3\cos 3t + 2\sin 3t).$$

Therefore, $$i(t) = \frac{dq}{dt} = \frac{75}{52}(2\cos 3t + 3\sin 3t) - \frac{25}{52}e^{-4t}(17\sin 3t + 6\cos 3t).$$

EXAMPLE 9.24 Find the general solution of the following differential equations.

(*i*) $(D^3 - 2D + 4)y = 3x^2 - 5x + 2$. (*ii*) $(D^3 + 4D)y = 24x^2 + 12 + 8\sin 2x$.

(*iii*) $(D^2 - 4D + 3)y = x^3$. (*iv*) $(D-1)^2(D+1)^2 y = \sin^2\frac{x}{2} + e^x + x$. (*MDU 2005*)

(*v*) $(D^4 + 16D^2)y = x^2 + 5$. (*NIT, Kurukshetra, 2006*) (*vi*) $(D^2 + 3D + 2)y = e^{e^x}$.

Solution (*i*) The characteristic equation of the corresponding homogeneous equation is

$$m^3 - 2m + 4 = 0 \text{ or } (m+2)(m^2 - 2m + 2) = 0. \text{ It roots are } m = -2,\ 1 \pm i.$$

The complementary function is $y_c(x) = c_1 e^{-2x} + e^x(c_2 \cos x + c_3 \sin x)$.

The particular integral is $y_p(x) = \frac{1}{D^3 - 2D + 4}(3x^2 - 5x + 2)$

$$= \frac{1}{4}\left\{1 - \left(\frac{D}{2} - \frac{D^3}{4}\right)\right\}^{-1}(3x^2 - 5x + 2) = \frac{1}{4}\left(1 + \frac{D}{2} + \frac{D^2}{4} + \ldots\right)(3x^2 - 5x + 2)$$

$$= \frac{1}{4}\left[(3x^2 - 5x + 2) + \left(3x - \frac{5}{2}\right) + \frac{3}{2}\right] = \frac{1}{4}(3x^2 - 2x + 1).$$

The general solution of the equation is $y(x) = y_c(x) + y_p(x)$

$$= c_1 e^{-2x} + e^x(c_2 \cos x + c_3 \sin x) + \frac{(3x^2 - 2x + 1)}{4}.$$

(*ii*) The characteristic equation of the corresponding homogeneous equation is

$$m^3 + 4m = 0 \text{ or } m(m^2 + 4) = 0. \text{ Its roots are } m = 0, \pm 2i.$$

The complementary function is $y_c(x) = c_1 + c_2 \cos 2x + c_3 \sin 2x$.

The particular integral is $y_p(x) = \frac{1}{D^3 + 4D}(24x^2 + 12 + 8\sin 2x)$

$$= \frac{1}{D^3 + 4D}(24x^2 + 12) + \frac{1}{D^3 + 4D}8\sin 2x$$

$$= \frac{12}{4D}\left(1 + \frac{D^2}{4}\right)^{-1}(2x^2 + 1) + 8x\frac{1}{3D^2 + 4}\sin 2x$$

$$= \frac{3}{D}\left(1 - \frac{D^2}{4} + \ldots\ldots\right)(2x^2 + 1) + 8x \cdot \frac{1}{3(-2^2) + 4}\sin 2x$$

$$= \frac{3}{D}(2x^2 + 1 - 1) - x\sin 2x = 2x^3 - x\sin 2x.$$

The general solution of the equation is $y(x) = y_c(x) + y_p(x)$

$$= c_1 + c_2 \cos 2x + c_3 \sin 2x + 2x^3 - x\sin 2x.$$

(*iii*) The characteristic equation of the corresponding homogeneous equation is

$$m^2 - 4m + 3 = 0 \text{ or } (m-1)(m-3) = 0. \text{ Its roots are } m = 1, 3.$$

The complementary function is $y_c(x) = c_1e^x + c_2e^{3x}$.

The particular integral is $y_p(x)=\dfrac{1}{D^2-4D+3}x^3=\dfrac{1}{3\left(1-\dfrac{4D-D^2}{3}\right)}x^3$

$$=\frac{1}{3}\left(1-\frac{4D-D^2}{3}\right)^{-1}x^3=\frac{1}{3}\left[1+\frac{4D-D^2}{3}+\frac{16D^2-8D^3}{9}+\frac{64D^3}{27}+\ldots\right]x^3$$

$$=\frac{1}{3}\left[x^3+\frac{12x^2-6x}{3}+\frac{96x-48}{9}+\frac{128}{9}\right]=\frac{1}{3}\left(x^3+4x^2+\frac{26x}{3}+\frac{80}{9}\right)=(9x^3+36x^2+78x+80)/27.$$

The general solution of the equation is $y(x) = y_c(x) + y_p(x)$

$$=c_1e^x+c_2e^{3x}+\frac{(9x^3+36x^2+78x+80)}{27}.$$

(*iv*) The characteristic equation of the corresponding homogeneous equation is $(m-1)^2(m+1)^2=0$. Its roots are $m = 1, 1, -1, -1$.

The complementary function is $y_c(x)=(c_1+c_2x)e^x+(c_3+c_4x)e^{-x}$.

The particular integral is $y_p(x)=\dfrac{1}{(D-1)^2(D+1)^2}\sin^2\dfrac{x}{2}+e^x+x$

$$=\frac{1}{(D-1)^2(D+1)^2}\sin^2\frac{x}{2}+\frac{1}{(D-1)^2(D+1)^2}e^x+\frac{1}{(D-1)^2(D+1)^2}x$$

Now $\dfrac{1}{(D-1)^2(D+1)^2}\sin^2\dfrac{x}{2}=\dfrac{1}{(D-1)^2(D+1)^2}\dfrac{(1-\cos x)}{2}$

$$=\frac{1}{2}\frac{1}{(D-1)^2(D+1)^2}e^{ox}-\frac{1}{2}\frac{1}{(D^2-2D+1)(D^2+2D+1)}\cos x$$

$$=\frac{1}{2}-\frac{1}{2}\cdot\frac{1}{4}\cos x=\frac{1}{2}-\frac{1}{8}\cos x,$$

$$\frac{1}{(D-1)^2(D+1)^2}e^x=\frac{1}{4}\cdot\frac{1}{(D-1)^2}e^x=\frac{1}{4}x\cdot\frac{1}{2(D-1)}e^x$$

$$=\frac{1}{8}x\frac{1}{(D-1)}e^x=\frac{1}{8}x^2e^x,\ \text{and}\ \frac{1}{(D-1)^2(D+1)^2}x=\frac{1}{D^4-2D^2+1}x$$

$$=\left[1-(2D^2-D^4)\right]^{-1}x=(1+2D^2-D^4+\ldots.)x=x.$$

$$\therefore\quad y_p(x)=\frac{1}{2}-\frac{1}{8}\cos x+\frac{x^2e^x}{8}+x.$$

The general solution of the equation is $y(x) = y_c(x) + y_p(x)$

$$=(c_1+c_2x)e^x+(c_3+c_4x)e^{-x}+\frac{1}{2}-\frac{1}{8}\cos x+\frac{x^2e^x}{8}+x.$$

(*v*) The characteristic equation of the corresponding homogeneous equation is $m^4 + 16m^2 = 0$ or $m^2(m^2 + 16)$. Its roots are $m = 0, 0, \pm 4i$.

The complementary function is $y_c(x)=c_1+c_2x+(c_3\cos 4x+c_4\sin 4x)$.

The particular integral is $y_p(x)=\dfrac{1}{D^4+16D^2}(x^2+5)$

$$= \frac{1}{16D^2}\left(1+\frac{1}{16}D^2\right)^{-1}(x^2+5) = \frac{1}{16D^2}\left(1-\frac{1}{16}D^2+\ldots\ldots\right)(x^2+5)$$

$$= \frac{1}{16D^2}\left(x^2+5-\frac{1}{16}\cdot 2\right) = \frac{1}{16}\frac{1}{D}\int\left(x^2+\frac{39}{8}\right)dx = \frac{1}{16}\frac{1}{D}\left(\frac{x^3}{3}+\frac{39}{8}x\right)$$

$$= \frac{1}{16}\int\left(\frac{x^3}{3}+\frac{39}{8}x\right)dx = \frac{1}{16}\left(\frac{x^4}{12}+\frac{39x^2}{16}\right) = \frac{x^2}{64}\left(\frac{x^2}{3}+\frac{39}{4}\right).$$

The general solution of the equation is $y(x) = y_c(x) + y_p(x)$

$$= c_1 + c_2x + (c_3 \cos 4x + c_4 \sin 4x) + \frac{x^2}{64}\left(\frac{x^2}{3}+\frac{39}{4}\right).$$

(*vi*) The characteristic equation of the corresponding homogeneous equation is
$m^2 + 3m + 2 = 0$. Its roots are $m = -1, -2$.

The complementary function is $y_c(x) = c_1e^{-x} + c_2e^{-2x}$.

The particular integral is $y_p(x) = \frac{1}{D^2+3D+2}e^{e^x} = \frac{1}{(D+1)(D+2)}e^{e^x} = \left(\frac{1}{D+1}-\frac{1}{D+2}\right)e^{e^x}$

$$= \left[e^{-x}\int e^x e^{e^x}\,dx - e^{-2x}\int e^{2x}e^{e^x}\,dx\right] \qquad \text{(Putting } e^x = t,\ e^x dx = dt \Rightarrow dx = dt/t)$$

$$= e^{-x}\int\frac{te^t}{t}dt - e^{-2x}\int t^2\frac{e^t}{t}dx = e^{-x}e^t - e^{-2x}\int te^t\,dt$$

$$= e^{-x}e^{e^x} - e^{-2x}(t-1)e^t = e^{-x}e^{e^x} - e^{-2x}(e^x-1)e^{e^x} = e^{-2x}e^{e^x}.$$

The general solution of the equation is $y(x) = y_c(x) + y_p(x) = c_1e^{-x} + c_2e^{-2x} + e^{-2x}e^{e^x}$.

EXAMPLE 9.25 Find the general solution of the following differential equations.

(*i*) $\frac{d^2y}{dx^2} - 4\frac{dy}{dx} + 4y = 8x^2e^{2x}\sin 2x$, (*UPTU 2007; NIT Kurukshetra, 2007; JNTU, 2006; DU 2002*)

(*ii*) $(D^2 - 4D + 3)y = 2xe^{3x} + 3e^x \cos 2x$, (*SVTU 2007; JNTU 2006*)

(*iii*) $(D^3 - 3D^2 - 6D + 8)y = xe^{-3x}$,

(*iv*) $(D^2 - 2D + 1)y = xe^x \sin x$. (*AMIE, W-2010; PTU, 2009; UPTU 2005*)

(*v*) $(D^3 + 2D^2 + D)y = x^2e^{2x} + \sin^2 x$. (*NIT Kurukshetra, 2008; PTU, 2003*)

Solution (*i*) The given equation in operator notation is $(D^2 - 4D + 4)y = 8x^2 e^{2x} \sin 2x$.

The characteristic equation of the corresponding homogeneous equation is
$m^2 - 4m + 4 = 0$ or $(m - 2)^2 = 0$. Its roots are $m = 2, 2$.

The complementary function is $y_c(x) = (c_1 + c_2 x)e^{2x}$.

The particular integral is $y_p(x) = \frac{1}{(D-2)^2}8x^2e^{2x}\sin 2x$

$$= 8e^{2x}\frac{1}{(D+2-2)^2}x^2\sin 2x = 8e^{2x}\frac{1}{D^2}x^2\sin 2x.$$

Now $\frac{1}{D^2}x^2\sin 2x = \frac{1}{D}\int \underset{\text{I}}{x^2}\,\underset{\text{II}}{\sin 2x}\,dx = \frac{1}{D}\left[\left(\frac{1}{4}-\frac{x^2}{2}\right)\cos 2x + \frac{x}{2}\sin 2x\right]$

$$= \int \left[\left(\frac{1}{4} - \frac{x^2}{2} \right) \cos 2x + \frac{x}{2} \sin 2x \right] dx = \frac{3 - 2x^2}{8} \sin 2x - \frac{x \cos 2x}{2}.$$

$\therefore \quad y_p(x) = e^{2x} [(3 - 2x^2) \sin 2x - 4x \cos 2x]$.

The general solution of the equation is $y(x) = y_c(x) + y_p(x)$

$$= (c_1 + c_2 x) e^{2x} + [(3 - 2x^2) \sin 2x - 4x \cos 2x] e^{2x}.$$

(ii) The characteristic equation of the corresponding homogeneous equation is

$m^2 - 4m + 3 = 0$. or $(m - 1)(m - 3) = 0$. Its roots are $m = 1, 3$.

The complementary function is $y_c(x) = c_1 e^x + c_2 e^{3x}$.

The particular integral is $y_p(x) = \dfrac{1}{D^2 - 4D + 3} (2xe^{3x} + 3e^x \cos 2x)$

$$= 2 \frac{1}{D^2 - 4D + 3} xe^{3x} + 3 \frac{1}{D^2 - 4D + 3} e^x \cos 2x$$

$$= 2e^{3x} \frac{1}{(D+3)^2 - 4(D+3) + 3} x + 3e^x \frac{1}{(D+1)^2 - 4(D+1) + 3} \cos 2x$$

$$= 2e^{3x} \frac{1}{D^2 + 2D} x + 3e^x \frac{1}{D^2 - 2D} \cos 2x = \left(2e^{3x} \cdot \frac{1}{D} \cdot \frac{1}{D+2} x \right) + \left(3e^x \frac{1}{-4 - 2D} \cos 2x \right)$$

$$= 2e^{3x} \frac{1}{D} \left(\frac{1}{2} - \frac{1}{4} D \right) x - \frac{3}{2} e^x \frac{D-2}{D^2 - 4} \cos 2x$$

$$= \frac{1}{2} e^{3x} \frac{1}{D} (2x - 1) + \frac{3}{16} e^x (D - 2) \cos 2x = \frac{1}{2} e^{3x} (x^2 - x) - \frac{3}{8} e^x (\cos 2x + \sin 2x).$$

The general solution of the equation is $y(x) = y_c(x) + y_p(x)$

$$= c_1 e^x + c_2 e^{3x} + e^{3x} (x^2 - x)/2 - 3e^x (\cos 2x + \sin 2x)/8.$$

(iii) The characteristic equation of the corresponding homogeneous equation is

$m^3 - 3m^2 - 6m + 8 = 0$, or $(m - 1)(m + 2)(m - 4) = 0$. Its roots are $m = 1, -2, 4$.

The complementary function is $y_c(x) = c_1 e^x + c_2 e^{-2x} + c_3 e^{4x}$.

The particular integral is $y_p(x) = \dfrac{1}{D^3 - 3D^2 - 6D + 8} \cdot xe^{-3x}$

$$= e^{-3x} \frac{1}{(D-3)^3 - 3(D-3)^2 - 6(D-3) + 8} x$$

$$= e^{-3x} \frac{1}{D^3 - 12D^2 + 39D - 28} x = e^{-3x} \left(-\frac{1}{28} - \frac{39}{784} D \right) x = e^{-3x} \left(-\frac{1}{28} x - \frac{39}{784} \right).$$

The general solution of the equation is $y(x) = y_c(x) + y_p(x)$

$$= c_1 e^x + c_2 e^{-2x} + c_3 e^{4x} - e^{-3x} (28x + 39)/784.$$

(iv) The characteristic equation of the corresponding homogeneous equation is

$m^2 - 2m + 1 = 0$ or $(m - 1)^2 = 0$. Its roots are 1, 1.

The complementary function is $y_c(x) = (c_1 + c_2 x) e^x$.

The particular integral is $y_p(x) = \frac{1}{(D-1)^2} e^x \cdot x\sin x = e^x \cdot \frac{1}{(D+1-1)^2} x\sin x$

$$= e^x \frac{1}{D^2} x\sin x = e^x \cdot \frac{1}{D} \int \underset{\text{I}}{x} \underset{\text{II}}{\sin x}\, dx$$

$$= e^x \cdot \frac{1}{D}\left[x(-\cos x) - \int 1(-\cos x)dx\right] = e^x \frac{1}{D}(-x\cos x + \sin x)$$

$$= e^x \int (-x\cos x + \sin x)dx = e^x\left[-(x\sin x - \int \sin x dx) - \cos x\right]$$

$$= e^x(-x\sin x - \cos x - \cos x) = -e^x(x\sin x + 2\cos x).$$

The general solution of the equation is $y(x) = y_c(x) + y_p(x) = (c_1 + c_2x)e^x - e^x(x\sin x + 2\cos x)$.

(*v*) The characteristic equation of the corresponding homogeneous equation is

$m^3 + 2m^2 + m = 0$ or $m(m^2 + 2m + 1) = 0$ or $m(m + 1) = 0$. Its roots are $m = 0, -1, -1$.

The complementary function is $y_c(x) = c_1 + (c_2 + c_3x)e^{-x}$.

The particular integral is $y_p(x) = \frac{1}{D^3 + 2D^2 + D}(x^2e^{2x} + \sin^2 x)$

Now $$\frac{1}{D^3 + 2D^2 + D} x^2e^{2x} = e^{2x} \frac{1}{(D+2)^3 + 2(D+2)^2 + (D+2)} x^2$$

$$= e^{2x} \frac{1}{18\left[1 + \left(\frac{7}{6}D + \frac{4}{9}D^2 + ...\right)\right]} x^2 = \frac{e^{2x}}{18}\left[1 + \left(\frac{7}{6}D + \frac{4}{9}D^2 + ...\right)\right]^{-1} x^2$$

$$= \frac{e^{2x}}{18}\left[1 - \left(\frac{7}{6}D + \frac{4}{9}D^2 + ...\right) + \left(\frac{7}{6}D\right)^2 + ...\right]x^2 = \frac{e^{2x}}{18}\left[1 - \frac{7}{6}D + \frac{11}{12}D^2 + ...\right]x^2$$

$$= \frac{e^{2x}}{18}\left(x^2 - \frac{7}{3}x + \frac{11}{6}\right).$$

and $$\frac{1}{D^3 + 2D^2 + D}\sin^2 x = \frac{1}{2}\frac{1}{D^3 + 2D^2 + D}(1 - \cos 2x)$$

$$= \frac{1}{2D}\frac{1}{D^2 + 2D + 1} e^{ox} - \frac{1}{2}\frac{1}{DD^2 + 2D^2 + D}\cos 2x$$

$$= \frac{1}{2}x + \frac{1}{2(3D + 8)}\cos 2x = \frac{1}{2}x + \frac{1}{2}\frac{3D - 8}{9D^2 - 64}\cos 2x$$

$$= \frac{1}{2}x + \frac{1}{200}(6\sin 2x + 8\cos 2x) = \frac{x}{2} + \frac{1}{100}(3\sin 2x + 4\cos 2x).$$

The general solution of the equation is $y(x) = y_c(x) + y_p(x)$

$$= c_1 + (c_2 + c_3x)e^{-x} + \frac{x}{2} + \frac{1}{100}(3\sin 2x + 4\cos 2x) + \frac{e^{2x}}{18}\left(x^3 - \frac{7}{3}x + \frac{11}{6}\right).$$

EXAMPLE 9.26 Find the general solution of the following differential equations.

(*i*) $y'' + 3y' + 2y = x \sin 2x$,

(*ii*) $y'' - y = x^2 \sin 3x$,

(*iii*) $y'' + 4y = x \sin x + 2^x$, (*AMIE; W-2005*)

(*iv*) $(D^4 + 2D^2 + 1)y = x^2 \cos x$, (*AMIE; W-2008*)

(*v*) $(D^2 - 1)y = x \sin x + x^2e^x$.

Solution (*i*) The given equation in operator notation is $(D^2 + 3D + 2)\, y = x\sin 2x$.

The characteristic equation of the corresponding homogeneous equation is

$m^2 + 3m + 2 = 0$ or $(m + 1)(m + 2) = 0$. Its roots are $m = -1, -2$.

The complementary function is $y_c(x) = c_1e^{-x} + c_2e^{-2x}$.

The particular integral is $y_p(x) = \dfrac{1}{D^2+3D+2} x\sin 2x$

$$= x\frac{1}{D^2+3D+2}\sin 2x - \frac{2D+3}{(D^2+3D+2)^2}\sin 2x$$

$$= x\frac{1}{3D-2}\sin 2x - \frac{2D+3}{D^4+6D^3+13D^2+12D+4}\sin 2x$$

$$= x\frac{1}{3D-2}\sin 2x - \frac{2D+3}{(-4)^2+6(-4)D+13(-4)+12D+4}\sin 2x$$

$$= x\frac{3D+2}{9D^2-4}\sin 2x + \frac{1}{4}\frac{(2D+3)(3D-8)}{9D^2-64}\sin 2x = \frac{-x(3\cos 2x+\sin 2x)}{20} + \frac{24\sin 2x + 7\cos 2x}{200}.$$

The general solution of the equation is $y(x) = y_c(x) + y_p(x)$

$$= c_1e^{-x} + c_2e^{-2x} - \left(\frac{30x-7}{200}\right)\cos 2x - \left(\frac{5x-12}{100}\right)\sin 2x.$$

(*ii*) The given equation in operator notation is $(D^2 - 1)y = x^2 \sin 3x$.

The characteristic equation of the corresponding homogeneous equation is $m^2 - 1 = 0$.

Its roots are $m = 1, -1$. The complementary function is $y_c(x) = c_1e^{x} + c_2e^{-x}$.

The particular integral is $y_p(x) = \dfrac{1}{D^2-1}x^2\sin 3x$

$$= x\frac{1}{D^2-1}x\sin 3x - \frac{2D}{(D^2-1)^2}x\sin 3x = \left\{x^2\frac{1}{D^2-1}\sin 3x - x\frac{2D}{(D^2-1)^2}\sin 3x\right\}$$

$$-2D\left\{x\frac{1}{D^4-2D^2+1}\sin 3x - \frac{4D^3-4D}{(D^4-2D^2+1)^2}\sin 3x\right\}$$

$$= x^2\frac{1}{D^2-1}\sin 3x - x\frac{2D}{(D^2-1)^2}\sin 3x - 2D\left\{x\frac{1}{(D^2-1)^2}\sin 3x\right\} + \frac{8D^2}{(D^2-1)^3}\sin 3x$$

$$= -\frac{1}{10}x^2\sin 3x - \frac{3}{50}x\cos 3x - \frac{1}{50}D(x\sin 3x) + \frac{9}{125}\sin 3x$$

$$= -\frac{1}{10}x^2\sin 3x - \frac{3}{25}x\cos 3x + \frac{13}{250}\sin 3x.$$

The general solution of the equation is $y(x) = y_c(x) + y_p(x)$

$$= c_1e^{x} + c_2e^{-x} - \frac{25x^2-13}{250}\sin 3x - \frac{3}{25}x\cos 3x.$$

(*iii*) The given equation in operator notation is $(D^2 + 4)y = x \sin x + 2^x$.

The characteristic equation of the corresponding homogeneous equation is $m^2 + 4 = 0$.

Its roots are $m = 2i, -2i$.

The complementary function is $y_c(x) = c_1 \cos 2x + c_2 \sin 2x$.

The particular integral is $y_p(x)=\dfrac{1}{D^2+4}(x\sin x+2^x)$

$$=\frac{1}{D^2+4}x\sin x+\frac{1}{D^2+4}2^x=x\frac{1}{D^2+4}\sin x-\frac{2D}{(D^2+4)^2}\sin x+\frac{1}{D^2+4}e^{\log 2^x}$$

$$=\frac{x}{3}\sin x-\frac{2}{9}\cos x+\frac{1}{D^2+4}e^{(\log 2)x}=\frac{x}{3}\sin x-\frac{2}{9}\cos x+\frac{1}{(\log 2)^2+4}2^x.$$

The general solution of the equation is $y(x) = y_c(x) + y_p(x)$

$$=c_1\cos 2x+c_2\sin 2x+\frac{x}{3}\sin x-\frac{2}{9}\cos x+\frac{1}{(\log 2)^2+4}\cdot 2^x$$

(*iv*) The characteristic equation of the corresponding homogeneous equation is

$m^4 + 2m^2 + 1 = 0$ or $(m^2 + 1)^2 = 0$. which has roots $\pm i, \pm i$.

The complementary function is $y_c(x)=(c_1+c_2x)\cos x+(c_3+c_4x)\sin x.$

The particular integral is $y_p(x)=\dfrac{1}{D^4+2D^2+1}x^2\cos x$

$$=\frac{1}{(D^2+1)^2}x^2\left(\frac{e^{ix}+e^{-ix}}{2}\right)=\frac{1}{2}\left[\frac{1}{(D^2+1)^2}x^2e^{ix}+\frac{1}{(D^2+1)^2}x^2e^{-ix}\right].$$

Now $\dfrac{1}{(D^2+1)^2}x^2e^{ix}=e^{ix}\dfrac{1}{\{(D+i)^2+1\}^2}x^2=e^{ix}\dfrac{1}{(D^2+2Di)^2}x^2=e^{ix}\dfrac{1}{4i^2D^2}\dfrac{1}{\left(1+\dfrac{D}{2i}\right)^2}x^2$

$$=-\frac{1}{4}e^{ix}\cdot\frac{1}{D^2}\left(1+\frac{D}{2i}\right)^{-2}x^2=-\frac{1}{4}e^{ix}\cdot\frac{1}{D^2}\left(1-\frac{D}{i}-\frac{3}{4}D^2+\ldots\right)x^2$$

$$=-\frac{1}{4}e^{ix}\cdot\frac{1}{D^2}\left(x^2-\frac{2x}{i}-\frac{3}{2}\right)=-\frac{1}{4}e^{ix}\cdot\frac{1}{D}\left(\frac{x^3}{3}-\frac{x^2}{i}-\frac{3x}{2}\right)=-\frac{1}{4}e^{ix}\left(\frac{x^4}{12}-\frac{x^3}{3i}-\frac{3x^2}{4}\right).$$

Similarly, or on changing i to $-i$ in the above, we obtain

$$\frac{1}{(D^2+1)^2}x^2e^{-ix}=-\frac{1}{4}e^{-ix}\left(\frac{x^4}{12}+\frac{x^3}{3i}-\frac{3x^2}{4}\right).$$

Therefore, the particular integral is

$$y_p(x)=-\frac{1}{8}\left[\frac{x^4}{6}\cdot\frac{e^{ix}+e^{-ix}}{2}-\frac{2x^3}{3}\frac{e^{ix}-e^{-ix}}{2i}-\frac{3x^2}{2}\cdot\frac{e^{ix}+e^{-ix}}{2}\right]=-\frac{1}{8}\left[\frac{x^4}{6}\cos x-\frac{2x^3}{3}\sin x-\frac{3x^2}{2}\cos x\right].$$

The general solution of the equation is $y(x) = y_c(x) + y_p(x)$

$$=(c_1+c_2x)\cos x+(c_3+c_4x)\sin x+\frac{1}{12}x^3\sin x+\frac{1}{48}(9x^2-x^4)\cos x.$$

(*v*) The characteristic equation of the corresponding homogeneous equation is

$m^2 - 1 = 0$. Its roots are $m = \pm 1$.

The complementary function is $y_c(x)=c_1e^x+c_2e^{-x}$.

The particular integral is $y_p(x)=\dfrac{1}{D^2-1}(x\sin x+x^2e^x)$

$$=\text{Imaginary part of }\frac{1}{D^2-1}xe^{ix}+e^x\frac{1}{(D+1)^2-1}x^2$$

$$= \text{I.P. of } e^{ix}\frac{1}{(D+i)^2-1}x + e^x\frac{1}{D^2+2D}x^2$$

$$= \text{I.P. of } e^{ix}\frac{1}{D^2+2iD-2}x + e^x\frac{1}{D(D+2)}x^2$$

$$= \text{I.P. of } \left\{-\frac{e^{ix}}{2}\left(1-\frac{(2iD+D^2)}{2}\right)^{-1}x\right\} + \frac{e^x}{2}\left(\frac{1}{D}-\frac{1}{D+2}\right)x^2$$

$$= \text{I.P. of } \left\{-\frac{e^{ix}}{2}(1+iD+\ldots\ldots)x\right\} + \frac{e^x}{2}\left[\frac{x^3}{3}-\frac{1}{2}\left(1+\frac{D}{2}\right)^{-1}x^2\right]$$

$$= \text{I.P. of } \left(-\frac{1}{2}\right)(\cos x + i\sin x)(x+i) + \frac{e^x}{2}\left[\frac{x^3}{3}-\frac{1}{2}\left(1-\frac{D}{2}+\frac{D^2}{4}+\ldots\ldots\right)x^2\right]$$

$$= -\frac{1}{2}(\cos x + x\sin x) + \frac{e^x}{2}\left(\frac{x^3}{3}-\frac{x^2}{2}+\frac{x}{2}-\frac{1}{4}\right).$$

The general solution of the equation is $y(x) = y_c(x) + y_p(x)$

$$= c_1e^x + c_2e^{-x} + \frac{e^x}{2}\left(\frac{x^3}{3}-\frac{x^2}{2}+\frac{x}{2}-\frac{1}{4}\right) - \frac{1}{2}(\cos x + x\sin x)$$

$$= c_1e^x + c_2e^{-x} + \frac{e^x}{2}\left(\frac{x^3}{3}-\frac{x^2}{2}+\frac{x}{2}\right) - \frac{1}{2}(\cos x + x\sin x).$$

($\because$ $-e^x/8$ can be combined with the first term of complementary function)

EXAMPLE 9.27 Find the general solution of the following differential equations

(*i*) $y'' + y = \text{cosec } x$. (*AMIE, W-2007; MDU-2002*)

(*ii*) $y'' + a^2y = \sec ax$. (*AMIE-2005; VTU, 2001*) (*iii*) $(D^2 + 4)y = \tan 2x$. (*MDU 2002*)

Solution (*i*) The given equation in operator notation is $(D^2 + 1)y = \text{cosec } x$.

The characteristic equation of the corresponding homogeneous equation is $(m^2 + 1) = 0$.

Its roots are $m = i, -i$.

The complementary function is $y_c(x) = c_1 \cos x + c_2 \sin x$.

The particular integral is $y_p(x) = \dfrac{1}{D^2+1}\text{cosec}\,x = \dfrac{1}{(D+i)(D-i)}\text{cosec}\,x$

$$= \frac{1}{2i}\left(\frac{1}{D-i}-\frac{1}{D+i}\right)\text{cosec}\,x = \frac{1}{2i}\left(\frac{1}{D-i}\text{cosec}\,x - \frac{1}{D+i}\text{cosec}\,x\right).$$

Now $$\frac{1}{D-i}\text{cosec}\,x = \frac{1}{D-i}e^{ix}\frac{e^{-ix}}{\sin x} = e^{ix}\frac{1}{D+i-i}\frac{e^{-ix}}{\sin x} = e^{ix}\frac{1}{D}\frac{\cos x - i\sin x}{\sin x}$$

$$= e^{ix}\int(\cot x - i)\,dx = e^{ix}[\log|\sin x| - ix].$$

Changing i to $-i$, we have $\dfrac{1}{D+i}\text{cosec}\,x = e^{-ix}[\log|\sin x| + ix]$.

$\therefore$ $$y_p(x) = \frac{1}{2i}[e^{ix}(\log|\sin x| - ix) - e^{-ix}(\log|\sin x| + ix)]$$

$$= \log|\sin x|\left(\frac{e^{ix}-e^{-ix}}{2i}\right) - \frac{ix}{i}\left(\frac{e^{ix}+e^{-ix}}{2}\right) = \sin x\log|\sin x| - x\cos x.$$

The general solution of the equation is $y(x) = y_c(x) + y_p(x)$

$= c_1 \cos x + c_2 \sin x - x \cos x + (\sin x) \log |\sin x|$.

(*ii*) The given equation in operator notation is $(D^2 + a^2)y = \sec ax$.

The characteristic equation of the corresponding homogeneous equation is $m^2 + a^2 = 0$.

Its roots are $m = \pm ia$.

The complementary function is $y_c(x) = c_1 \cos ax + c_2 \sin ax$.

The particular integral is $y_p(x) = \dfrac{1}{D^2 + a^2} \sec ax$

$$= \frac{1}{2ia}\left[\frac{1}{D - ia} - \frac{1}{D + ia}\right] \sec ax = \frac{1}{2ia}\left[\frac{1}{D - ia}\sec ax - \frac{1}{D + ia}\sec ax\right].$$

Now $\dfrac{1}{D - ia}\sec ax = e^{iax}\displaystyle\int \sec ax \cdot e^{-iax} dx$

$$= e^{iax}\int \frac{\cos ax - i\sin ax}{\cos ax} dx = e^{iax}\int (1 - i\tan ax)\, dx = e^{iax}\left(x + \frac{i}{a}\log\cos ax\right).$$

Changing i to $-i$, we have $\dfrac{1}{D + ia}\sec ax = e^{-iax}\left(x - \dfrac{i}{a}\log\cos ax\right)$.

$\therefore$ The particular integral is $y_p(x) = \dfrac{1}{2ia}\left[e^{iax}\left(x + \dfrac{i}{a}\log\cos ax\right) - e^{-iax}\left(x - \dfrac{i}{a}\log\cos ax\right)\right]$

$$= \frac{x}{a}\frac{e^{iax} - e^{-iax}}{2i} + \frac{1}{a^2}\log\cos ax \cdot \frac{e^{iax} + e^{-iax}}{2} = \frac{x}{a}\sin ax + \frac{1}{a^2}(\cos ax \cdot \log\cos ax).$$

The general solution of the equation is

$$y(x) = c_1 \cos ax + c_2 \sin ax + (1/a)\, x \sin ax + (1/a^2) \cos ax \log (\cos ax).$$

(*iii*) The characteristic equation is $m^2 + 4 = 0$. Its roots are $m = 2i, -2i$.

The complementary function is $y_c(x) = c_1 \cos 2x + c_2 \sin 2x$.

The particular integral is $y_p(x) = \dfrac{1}{D^2 + 4} \cdot \tan 2x = \dfrac{1}{(D + 2i)(D - 2i)}\tan 2x$

$$= \frac{1}{4i}\left[\frac{1}{D - 2i} - \frac{1}{D + 2i}\right]\tan 2x = \frac{1}{4i}\left[\frac{1}{D - 2i}\tan 2x - \frac{1}{D + 2i}\tan 2x\right].$$

Now $\dfrac{1}{D - 2i}\tan 2x = e^{2ix}\displaystyle\int \tan 2x \cdot e^{-2ix} dx$

$$= e^{2ix}\int \tan 2x(\cos 2x - i\sin 2x)\, dx = e^{2ix}\int [\sin 2x - i(\sin^2 2x/\cos 2x)]dx$$

$$= e^{2ix}\int [\sin 2x - i(1 - \cos^2 2x/\cos 2x)]dx = e^{2ix}\int [\sin 2x - i(\sec 2x - \cos 2x)]dx$$

$$= \frac{e^{2ix}}{2}[-\cos 2x - i\log|\sec 2x + \tan 2x| + i\sin 2x]$$

$$= -\frac{e^{2ix}}{2}[(\cos 2x - i\sin 2x) + i\log|\sec 2x + \tan 2x|]$$

$$= -\frac{e^{2ix}}{2}[e^{-2ix} + i\log|\sec 2x + \tan 2x|] = -\frac{1}{2}[1 + ie^{2ix}\log|\sec 2x + \tan 2x|].$$

Changing i to $-i$, we have

$$\frac{1}{D + 2i}\tan 2x = -\frac{1}{2}[1 - ie^{-2ix}\log|\sec 2x + \tan 2x|].$$

Thus $y_p(x) = \frac{1}{4i}\left[\frac{-1}{2}\{1 + ie^{2ix}\log|\sec 2x + \tan 2x|\} + \frac{1}{2}\{1 - ie^{-2ix}\log|\sec 2x + \tan 2x|\}\right]$

$$= -\frac{1}{4}\left(\frac{e^{2ix} + e^{-2ix}}{2}\right)\log|\sec 2x + \tan 2x| = -\frac{1}{4}\cos 2x \log|\sec 2x + \tan 2x|.$$

The general solution of the equation is $y(x) = y_c(x) + y_p(x)$

$$= c_1 \cos 2x + c_2 \sin 2x - [\cos 2x \log|\sec 2x + \tan 2x|]/4.$$

EXERCISE 9.2

Find the general solution of the following differential equations.

1. $y'' - 3y' + 2y = e^{3x}$. *(AMIETE, W-2007)* **Ans.** $y(x) = c_1e^x + c_2e^{2x} + (e^{3x}/2)$.
2. $y''' - 6y'' + 11y' - 6y = e^{-x}$. **Ans.** $y(x) = c_1e^x + c_2e^{2x} + c_3e^{3x} - (e^{-x}/24)$.
3. $y'' + 4y' + 4y = 12e^{-2x}$. **Ans.** $y(x) = (c_1 + c_2 x)e^{-2x} + 6x^2e^{-2x}$.
4. $y'' - 2y' - 3y = 3e^{2x}$. **Ans.** $y(x) = c_1e^{-x} + c_2e^{3x} - e^{2x}$.
5. $y''' - 2y'' - 5y' + 6y = 4e^{-x} - e^{2x}$. **Ans.** $y(x) = c_1e^x + c_2e^{-2x} + c_3e^{3x} + (e^{-x}/2) + (e^{2x}/4)$
6. $y'' - 9y = 54e^{3x}$. **Ans.** $y(x) = c_1e^{3x} + c_2e^{-3x} + 9xe^{3x}$.
7. $[D^2 + 5D + 6]\,y = e^x$ *(AMIETE, Dec, 2010)* **Ans.** $y(x) = c_1e^{-2x} + c_2e^{-3x} + e^x/12$.
8. $y'' + 5y' + 6y = 3e^{-2x} + e^{3x}$. **Ans.** $y(x) = c_1\, e^{-2x} + c_2e^{-3x} + 3xe^{-2x} + (e^{3x}/30)$.
9. $\frac{d^2y}{dx^2} - \frac{dy}{dx} - 6y = e^x \cosh 2x$. **Ans.** $y(x) = c_1e^{3x} + c_2e^{-2x} + \frac{xe^{3x}}{10} - \frac{e^{-x}}{8}$.

   [Hint: $e^x \cosh 2x = e^x\left(\frac{e^{2x} + e^{-2x}}{2}\right) = \frac{e^{3x} + e^{-x}}{2}$]
10. $4y'' - 4y' + y = e^{x/2}$. **Ans.** $y(x) = (c_1 + c_2x)e^{x/2} + (x^2e^{x/2})/8$.
11. $(D^3 - 5D^2 + 8D - 4)y = e^{2x} + 2e^x + 3e^{-x} + 2$.

Ans. $y(x) = c_1e^x + (c_2 + c_3x)e^{2x} + (x^2e^{2x}/2) + 2xe^x - (e^{-x}/6) - (1/2)$..

[Hint: P.I. of $\frac{1}{D^3 - 5D^2 + 8D - 4}2 = \frac{1}{(D-1)(D-2)^2}2 = 2\frac{1}{(D-1)(D-2)^2}\cdot e^{0x} = 2\frac{1}{(-1)(-2)^2} = -\frac{1}{2}$.]

12. $(D + 2)(D - 1)^2y = e^{-2x} + 2 \sinh x$. **Ans.** $y(x) = c_1e^{-2x} + (c_2 + c_3x)e^x + \frac{(xe^{-2x})}{9} + \left(\frac{x^2e^x}{6}\right) + \frac{e^{-x}}{4}$.
13. $\frac{d^2y}{dx^2} + 6\frac{dy}{dx} + 9y = 50e^{2x}$. **Ans.** $y(x) = (c_1 + c_2x)e^{-3x} + 2e^{2x}$.
14. $\frac{d^3y}{dx^3} - 5\frac{d^2y}{dx^2} + 8\frac{dy}{dx} - 4y = e^{2x} + 3e^{-x}$. **Ans.** $y(x) = c_1e^x + (c_2 + c_3x)e^{2x} + \left(\frac{x^2e^{2x}}{2}\right) - \frac{e^{-x}}{6}$
15. $y'' + 4y' + 8y = (1 + e^x)^2$. **Ans.** $y(x) = (c_1 \cos 2x + c_2 \sin 2x)e^{-2x} + \left(\frac{2e^x}{13}\right) + \frac{e^{2x}}{20} + \frac{1}{8}$.
16. $(D^2 + 4)y = \cos 2x$. *(GGSIPU 2006)* **Ans.** $y(x) = c_1 \cos 2x + c_2 \sin 2x + (x \sin 2x)/4$.
17. $(D^2 + 9)y = \cos 3x$. *(AMIETE, Dec 2010)* **Ans.** $y(x) = c_1 \cos 3x + c_2 \sin 3x + (x \sin 3x)/6$.
18. $y'' + 3y' + 2y = \sin 2x$. *(AMIETE, June 2002)* **Ans.** $y(x) = c_1e^{-x} + c_2e^{-2x} - (6\cos 2x + 2\sin 2x)/40$.

19. $(D^3 + D^2 + D + 1)y = \sin 2x$. **Ans.** $y(x) = c_1 e^{-x} + (c_2 \cos x + c_3 \sin x) + (2\cos 2x - \sin 2x)/15$.

20. $\dfrac{d^3y}{dx^3} + \dfrac{d^2y}{dx^2} - \dfrac{dy}{dx} - y = \sin(2x-3)$. *(AMIE, S-2009)*

Ans. $y(x) = c_1 e^x + (c_2 + c_3 x)e^{-x} + \dfrac{1}{25}[2\cos(2x-3) - \sin(2x-3)]$.

21. $\dfrac{d^2y}{dx^2} + 5\dfrac{dy}{dx} + 6y = e^{-2x} + \sin x$. **Ans.** $y(x) = c_1 e^{-2x} + c_2 e^{-3x} + xe^{-2x} + (\sin x - \cos x)/10$.

22. $(D^3 - 3D^2 + 4D - 2)y = e^x + \cos x$. *(UPTU 2006)*

Ans. $y(x) = c_1 e^x + e^x(c_2 \cos x + c_3 \sin x) + xe^x + (3\sin x + \cos x)/10$.

23. $\dfrac{d^3y}{dx^3} + 2\dfrac{d^2y}{dx^2} + \dfrac{dy}{dx} = e^{-x} + \sin 2x$. *(VTU 2004)*

Ans. $y(x) = c_1 + (c_2 + c_3 x)e^{-x} - (x^2 e^{-x}/2) + (3\cos 2x - 4\sin 2x)/50$.

24. $(D^2 + 4)y = e^x + \sin 3x$. *(MDU 2003)* **Ans.** $y(x) = c_1 \cos 2x + c_2 \sin 2x + (e^x - \sin 3x)/5$.

25. $y''' + y = \sin 3x - \cos^2(x/2)$. *(MDU 2003; Madras 2000)*

Ans. $y(x) = c_1 e^{-x} + e^{x/2}\left(c_2 \cos\dfrac{\sqrt{3}}{2}x + c_3 \sin\dfrac{\sqrt{3}}{2}x\right) + \dfrac{1}{730}(\sin 3x + 27\cos 3x) - \dfrac{1}{2} - \dfrac{1}{4}(\cos x - \sin x)$.

26. $\dfrac{d^2y}{dx^2} + 2\dfrac{dy}{dx} + y = e^{2x} - \cos^2 x$. *(Delhi, 2002)* [**Hint :** $\cos^2 x = (1 + \cos 2x)/2$]

Ans. $y(x) = (c_1 + c_2 x)e^{-x} + \dfrac{1}{9}e^{2x} + \dfrac{1}{50}(3\cos 2x - 4\sin 2x) - \dfrac{1}{2}$.

27. $(D^2 + 9)y = \cos^3 x$. [**Hint :** $y_p = \dfrac{1}{D^2+9}\dfrac{1}{4}(\cos 3x + 3\cos x)$]

Ans. $y(x) = c_1 \cos 3x + c_2 \sin 3x + \dfrac{x}{24}\sin 3x + \dfrac{3}{32}\cos x$.

28. $y''' + 2y'' + y' = e^{2x} + \sin 2x$. **Ans.** $y(x) = c_1 + (c_2 + c_3 x)e^{-x} + (e^{2x}/18) + (6\cos 2x - 8\sin 2x)/100$.

29. $(D^2 - 4D + 4)y = e^{2x} - \sin 2x$. *(AMIETE, Dec 2010)* **Ans.** $y(x) = (c_1 + c_2 x)e^{2x} + (x^2 e^{2x}/2) - (\cos 2x)/8$.

30. $(D^2 - 4D + 3)\, y = \sin 3x \cos 2x$. *(MDU 2003; Madras, 2000)*

[**Hint:** We may write $\sin 3x \cos 2x = (\sin 5x + \sin x)/2$].

Ans. $y(x) = c_1 e^x + c_2 e^{3x} + (10\cos 5x - 11\sin 5x)/884 + (\sin x + 2\cos x)/20$.

31. $y'' + a^2 y = \cos ax$. **Ans.** $y(x) = c_1 \cos ax + c_2 \sin ax + (x \sin ax)/2a.$.

32. $y'' + 2y' + 401y = \sin 20x + 40\cos 20x$. **Ans.** $y(x) = e^{-x}(c_1 \cos 20x + c_2 \sin 20x) + \sin 20x$.

33. $(2D^2 + 2D + 3)\, y = x^2 + 2x - 1$. **Ans.** $y(x) = e^{-x/2}\left(c_1 \cos\dfrac{1}{2}\sqrt{5}x + c_2 \sin\dfrac{1}{2}\sqrt{5}x\right) + \dfrac{1}{3}x^2 + \dfrac{2}{9}x - \dfrac{25}{27}$.

34. $(D^4 - 2D^3 + D^2)y = x^3$. *(AMIE, W-2008)*

Ans. $y(x) = c_1 + c_2 x + (c_3 + c_4 x)e^x + (x^5 + 10x^4 + 60x^3 + 240x^2)/20$.

35. $(D^2 - 4D + 4)y = 8(e^{2x} + \sin 2x + x^2)$. *(GGSIPU-2006; AMIETE-June 2004)*

Ans. $y(x) = (c_1 + c_2 x)e^{2x} + 4x^2 e^{2x} + \cos 2x + 2x^2 + 4x + 3$.

36. $y'' + 4y = e^x + \sin 2x + x^2$. **Ans.** $y(x) = c_1 \cos 2x + c_2 \sin 2x + \dfrac{1}{5}e^x - \dfrac{x}{4}\cos 2x + \dfrac{1}{4}\left(x^2 - \dfrac{1}{2}\right)$.

37. $\dfrac{d^2y}{dx^2}+2\dfrac{dy}{dx}+4y=2x^2+3e^{-x}$. (*AMIETE, June 2011*)

Ans. $y(x)=e^{-x}(c_1\cos\sqrt{3}\,x+c_2\sin\sqrt{3}\,x)+\frac{1}{2}x(x-1)+e^{-x}$.

38. $(D^2+3D+2)y=e^{2x}+x^2+\sin x$. (*DCE, 2004*)

Ans. $y(x)=c_1e^{-x}+c_2e^{-2x}+(e^{2x}/12)+(2x^2-6x+7)/4-(3\cos x-\sin x)/10$.

39. $y'''+2y''+y'=e^{2x}+x^2+x$. **Ans.** $y(x)=c_1+(c_2+c_3x)e^{-x}+(e^{2x}+6x^3-27x^2+72x)/18$.

40. $y'''+y=5e^{2x}+x^3+3$. **Ans.** $y(x)=c_1e^{-x}+e^{x/2}\left(c_2\cos\frac{\sqrt{3}}{2}x+c_3\sin\frac{\sqrt{3}}{2}x\right)+\frac{5}{9}e^{2x}+x^3-3$.

41. $y'''+2y''+y'=x^2+x$. **Ans.** $y(x)=c_1+(c_2+c_3x)e^{-x}+(x^3/3)-(3x^2/2)+4x$.

42. $y''-3y'+2y=x^2+e^x$. **Ans.** $y(x)=c_1e^x+c_2e^{2x}+\frac{1}{2}(x^2+3x+3.5-2xe^x)$.

43. $(D^4+2D^3-3D^2)y=x^2$. **Ans.** $y(x)=c_1+c_2x+c_3e^x+c_4e^{-3x}-x^2(3x^2+8x+28)/108$.

44. $y^{iv}+4y''=96x^2$. **Ans.** $y(x)=c_1+c_2x+c_3\cos 2x+c_4\sin 2x+2x^4-6x^2$.

45. $y^{iv}+y'''+y''=5x^2+\cos x$.

Ans. $y(x)=c_1+c_2x+e^{-x/2}\left(c_3\cos\frac{\sqrt{3}}{2}x+c_4\sin\frac{\sqrt{3}}{2}x\right)+\frac{5}{12}x^4-\frac{5}{3}x^3+10x-\sin x-10$.

46. $y''-2y'+5y=e^{2x}\sin x$. **Ans.** $y(x)=e^x(c_1\cos 2x+c_2\sin 2x)-e^{2x}(\cos x-2\sin x)/10$.

47. $(D^2-4D+3)y=e^x\cos 2x$. **Ans.** $y(x)=c_1e^x+c_2e^{3x}-[e^x(\cos 2x+\sin 2x)]/8$.

48. $y''-2y'=e^x\sin x$. (*VTU 2000*) **Ans.** $y(x)=c_1+c_2e^{2x}-(e^x\sin x)/2$.

49. $y'''+2y''+y'=x^2e^{2x}+\sin^2x$.

Ans. $y(x)=c_1+(c_2+c_3x)e^{-x}+\dfrac{e^{2x}}{18}\left(x^2-\dfrac{7x}{3}+\dfrac{11}{6}\right)+\dfrac{x}{2}+\dfrac{(3\sin 2x+4\cos 2x)}{100}$.

50. $y''-2y'+2y=x+e^x\cos x$. (*UPTU 2002*)

Ans. $y(x)=e^x(c_1\cos x+c_2\sin x)+(x+1)/2+(xe^x\sin x)/2$.

51. $y''+y=e^{-x}+x^3+e^x\sin x+7$.

Ans. $y(x)=c_1\cos x+c_2\sin x+e^{-x}/2+(x^3-6x)-e^x(2\cos x-\sin x)/5+(7)$.

52. $16y''+8y'+y=48xe^{-x/4}$. (*AMIETE, June 2006*) **Ans.** $y(x)=(c_1+c_2x)e^{-x/4}+(x^3e^{-x/4})/2$.

53. $y''-4y=x^2e^{3x}$. **Ans.** $y(x)=c_1e^{2x}+c_2e^{-2x}+e^{3x}(25x^2-60x+62)/125$.

54. $\dfrac{d^3y}{dx^3}-3\dfrac{d^2y}{dx^2}+3\dfrac{dy}{dx}-y=(x+1)e^x$. **Ans.** $y(x)=(c_1+c_2x+c_3x^2)e^x+\dfrac{x^3e^x}{6}\left(1+\dfrac{x}{4}\right)$.

55. $(D^2-4D+13)y=18e^{2x}\sin 3x$. (*AMIETE, June 2010, Dec. 2008*)

Ans. $y(x)=e^{2x}(c_1\cos 3x+c_2\sin 3x)-3x\,e^{2x}\cos 3x$.

56. $y'''+2y''+y'=x^2e^{2x}+\sin^2x$. (*PTU 2003*)

Ans. $y(x)=c_1+(c_2+c_3x)e^{-x}+e^{2x}(6x^2-14x+11)/108+(x/2)+3\sin 2x+4\cos 2x/100$.

57. $\dfrac{d^2y}{dx^2}-y=(1+x^2)e^x$. (*AMIE, S-2011*) **Ans.** $y(x)=c_1e^x+c_2e^{-x}+\dfrac{xe^x}{12}(2x^2-3x+9)$.

58. $(D^2 + 4D + 3)y = e^{-x} \sin x + xe^{3x}$. (*Anna 2002S*)

Ans. $y(x) = c_1e^{-x} + c_2e^{-3x} - [e^{-x}(2\cos x + \sin x)]/5 - e^{-3x}(x^2 + x)/4.$

59. $y'' + 3y' + 2y = \cosh 2x \sin x$.

Ans. $y(x) = c_1e^{-x} + c_2e^{-2x} - e^{2x}(7\cos x - 11\sin x)/340 + e^{-2x}(\cos x - \sin x)/4.$

60. $(D^2 - 4D + 1)y = e^{2x} \sin 2x$. (*AMIETE, Dec. 2005*) **Ans.** $y(x) = c_1e^{(2+\sqrt{3})x} + c_2e^{(2-\sqrt{3})x} - (e^{2x} \sin 2x)/7.$

61. $(D^2 + 4)y = x \sin x$. (*AMIE, S-2010; Raipur, 2004*)

Ans. $y(x) = c_1 \cos 2x + c_2 \sin 2x + (3x\sin x - 2\cos x)/9.$

62. $y'' + 4y' + 3y = x\sin 2x$. (*AMIETE., June 2009, 2008*)

Ans. $y(x) = Ae^{-x} + Be^{-3x} - \dfrac{1}{4225}[65x(8\cos 2x + \sin 2x) - 188\cos 2x - 316\sin 2x]$

63. $y'' + 4y = x\sin^2 x$. **[Hint :** $x\sin^2 x = x(1 - \cos 2x)/2 = (x/2) - (x\cos 2x)/2$].

Ans. $y(x) = c_1 \cos 2x + c_2 \sin 2x + (x/8) - (x\cos 2x)/128 - (x^2 \sin 2x)/16.$

64. $(D^4 - 1)y = x \sin x$. (*Kurukshetra 2006*)

Ans. $y(x) = c_1e^x + c_2e^{-x} + c_3 \cos x + c_4 \sin x + (x^2 \cos x - 3x \sin x)/8.$

65. $y^{iv} + 2y'' + y = x^2 \cos x$. (*AMIE., W-2008; Osmania, 2003S*)

Ans. $y(x) = (c_1 + c_2x) \cos x + (c_3 + c_4 x) \sin x + [4x^3 \sin x - x^2(x^2 - 9) \cos x]/48.$

66. $y'' + 4y = x \cos x$. (*DCE, 2005*) **Ans.** $y(x) = c_1\cos 2x + c_2 \sin 2x + (3x \cos x + 2 \sin x)/9.$

67. $(D^2 + 5D + 6)y = e^{-2x}(\sec^2 x)(1 + 2\tan x)$. (*AMIETE., June 2003*)

Ans. $y(x) = c_1e^{-3x} + c_2e^{-2x} + e^{-2x} \tan x.$

68. $y'' + 3y' + 2y = xe^x \sin x$. (*Osmania, 2003; VTU, 2001*)

Ans. $y(x) = c_1e^{-x} + c_2e^{-2x} - e^x[5x(\sin x + \cos x) + 2\sin x - 5\cos x]/50.$

69. $(D^2 - 1)y = x \sin x + (1 + x^2)e^x$. (*Kurukshetra, 2010; MDU, 2002*)

Ans. $y(x) = c_1e^x + c_2e^{-x} - (x \sin x + \cos x)/2 + xe^x(2x^2 - 3x + 9)/12.$

Find the solution of the following differential equations satisfying the given conditions.

70. $y''' - 6y'' + 11y' - 6y = 0,\ y(0) = 0,\ y'(0) = -4,\ y''(0) = -18.$ **Ans.** $y(x) = e^x + 2e^{2x} - 3e^{3x}$

71. $y'' - 7y' + 6y = e^{2x},\ y(0) = 0.$ (*GGSIPU 2006*) **Ans.** $y(x) = c(e^x - e^{6x}) + \{e^{2x}(e^{4x} - 1)\}/4.$

72. $y'' + 2y' + 10y + 37 \sin 3x = 0,\ y(0) = 3;\ y'(0) = 0.$ Also find $y(\pi/2)$.

Ans. $y(x) = -3e^{-x}\cos 3x + 6\cos 3x - \sin 3x;\ 1.$

73. $y'' + 4y' + 5y = -2\cosh x,\ y(0) = 0,\ y'(0) = 1.$ **Ans.** $y(x) = 3e^{-2x}(\cos x + 3\sin x)/5 - (e^x/10) - (e^{-x}/2).$

74. $y^{iv} + 13y'' + 36y = 0,\ y(0) = 0,\ y''(0) = 0,\ y(\pi/2) = -1,\ y'(\pi/2) = -4.$ **Ans.** $y(x) = 2 \sin 2x + \sin 3x.$

75. $(D^2 + 6D + 10)y = 50x;\ y(0) = 0,\ y'(0) = 1.$ **Ans.** $y(x) = 5x - 3 + (3 \cos x + 5 \sin x)\, e^{-3x}.$

76. $(D^2 + 1)^2y = 24x \cos x,\ y(0) = 0,\ y'(0) = y''(0) = 0$ and $y'''(0) = 12.$ **Ans.** $y(x) = 3x^2 \sin x - x^3 \cos x.$

77. If the roots of the auxiliary equation corresponding to the differential equation are 2, 2, 3, $\pm 2i$ then form the differential equation and find its solution.

Ans. $(D^4 - 10D^3 + 41D^2 - 76D + 52)y = 0;\ y = (c_1 + c_2x)\, e^{2x} + (c_3 \cos 2x + c_4 \sin 2x)\, e^{3x}.$

78. A circuit consists of an inducatance of 0.05 henry, a resistance of 20 ohms, a condenser of capacitance 100 microfarads, and an e.m.f of E = 100 volts. Find $i(t)$ and $q(t)$, given the initial conditions $q = 0$, $i = 0$ when $t = 0$.

Ans. $q(t) = e^{-200t}(-0.01 \cos 400\,t - 0.005 \sin 400\,t) + 0.01$, and $i(t) = 5e^{-200\,t} \sin 400\,t.$

9.8 TWO IMPORTANT METHODS FOR DETERMINING A PARTICULAR INTEGRAL OF $F(D)Y = X$ ARE AVAILABLE

9.8.1 Method of Variation of Parameters

This is a very elegant method for finding the particular integral (P.I.) of a linear differential equation whose complementary function (C.F.) is known. Though the method is general, we will illustrate it by applying it to a second order linear differential equation with constant coefficients. Consider the equation.

$$\frac{d^2y}{dx^2} + p\frac{dy}{dx} + qy = X \qquad ...(9.43)$$

where p, q, are all constants and X is functions of x. It gives

$$\text{P.I.} = -y_1\int \frac{y_2 X}{W}dx + y_2\int \frac{y_1 X}{W}dx, \qquad ...(9.44)$$

where y_1 and y_2 are the solutions of differential equation

$$\frac{d^2y}{dx^2} + p\frac{dy}{dx} + qy = 0 \qquad ...(9.45)$$

and $W = \begin{vmatrix} y_1 & y_2 \\ y_1' & y_2' \end{vmatrix}$ is called the *Wronskian* of (y_1, y_2).

Proof: Let the complementary function of Eq. (9.43) be $y_c = c_1 y_1 + c_2 y_2$ ...(9.46)

Replacing c_1, c_2 (regarded as parameters) by unknown functions $u(x)$ and $v(x)$, let the P.I. be

$$y = uy_1 + vy_2 \qquad ...(9.47)$$

Differentiating Eq. (9.47) w.r.t. x we get

$$y' = uy'_1 + vy'_2 + u'y_1 + v'y_2$$
$$= uy'_1 + vy'_2. \qquad ...(9.48)$$

on assuming that u, v satisfy the equation

$$u'y_1 + v'y_2 = 0. \qquad ...(9.49)$$

Differentiating Eq. (9.46) w.r.t., x, we obtain

$$y'' = uy''_1 + u'y'_1 + vy''_2 + v'y'_2.$$

Substituting the values of y, y' and y'' in Eq. (9.42), we get

$$(uy''_1 + u'y'_1 + vy''_2 + v'y'_2) + p(uy'_1 + vy'_2) + q(uy_1 + vy_2) = X$$

or $$u(y''_1 + py'_1 + qy_1) + v(y''_2 + py'_2 + qy_2) + u'y'_1 + v'y'_2 = X \text{ or } u'y'_1 + v'y'_2 = X \qquad ...(9.50)$$

Solving Eq. (9.48) and Eq. (9.49) we get

$$u' = \begin{vmatrix} 0 & y_2 \\ X & y_2' \end{vmatrix} \div \begin{vmatrix} y_1 & y_2 \\ y_1' & y_2' \end{vmatrix} = -\frac{y_2 X}{W} \quad \text{and} \quad v' = \begin{vmatrix} y_1 & 0 \\ y_1' & X \end{vmatrix} \div \begin{vmatrix} y_1 & y_2 \\ y_1' & y_2' \end{vmatrix} = \frac{y_1 X}{W}$$

where $W = \begin{vmatrix} y_1 & y_2 \\ y_1' & y_2' \end{vmatrix}$ is called the *Wronskian* of (y_1, y_2).

Integrating, $u = -\int \frac{y_2 X}{W}dx, \quad v = \int \frac{y_1 X}{W}dx.$

Substituting these in Eq. (9.47), the P.I. is known.

Remark As the solution is obtained by varying the arbitrary constants c_1, c_2 of the complementary function, the method is known as *variation of parameters*.

EXAMPLE 9.28 Solve the following differential equations by the method of variation of parameters

(*i*) $(D^2 - 1)y = 2/(1 + e^x)$. (*VTU, 2005; AMIETE, June 2001*) (*ii*) $(D^2 - 1)y = e^{-x}\sin(e^{-x}) + \cos(e^{-x})$. (*AMIETE, Dec. 2002*)

(*iii*) $y'' - 2y' + y = e^x \log x$. (*Kurukshetra, 2005; Madras 2003*)

(*iv*) $y'' + 4y = \tan 2x$. (*AMIETE, Dec-2010*)

(*v*) $y'' - 2y' + y = x^{3/2} e^x$. (*SVTU, 2006S; Kurukshetra, 2005*)

Solution (*i*) The characteristic equation of the corresponding homogeneus equation is

$$m^2 - 1 = (m-1)(m+1) = 0. \text{ Its roots are } m_1 = 1 \text{ and } m_2 = -1.$$

The complementary function is $y_c(x) = c_1 e^x + c_2 e^{-x}$.

To determine $y_p(x)$, proceed as under:

Here $y_1 = e^x$, $y_2 = e^{-x}$ and $X = 2/(1 + e^x)$.

The Wronskian $W(y_1, y_2)$ is given by $W(e^x, e^{-x}) = \begin{vmatrix} e^x & e^{-x} \\ e^x & -e^{-x} \end{vmatrix} = -2$.

Thus $y_p(x) = -y_1\int \frac{y_2 X}{W}dx + y_2\int \frac{y_1 X}{W}dx = e^x\int \frac{e^{-x}}{1+e^x}dx - e^{-x}\int \frac{e^x}{1+e^x}dx.$

To evaluate $\int \frac{dx}{e^x(1+e^x)}$, put $e^x = t$, $dx = \frac{dt}{t}$.

$$\therefore \quad \int \frac{dx}{e^x(1+e^x)} = \int \frac{dt}{t^2(1+t)} = \int\left(\frac{1}{t^2} - \frac{1}{t} + \frac{1}{1+t}\right)dt$$

$$= -\frac{1}{t} - \log t + \log(1+t) = -\frac{1}{e^x} + \log\left\{\frac{1+e^x}{e^x}\right\} + c_3,$$

and $$\int \frac{e^x}{1+e^x}dx = \log(1+e^x) + c_4.$$

$$\therefore \quad y_p(x) = e^x\left[-\frac{1}{e^x} + \log(1+e^x) - x + c_3\right] - e^{-x}[\log(1+e^x) + c_4].$$

The general solution of the equation is

$$y(x) = c_1 e^x + c_2 e^{-x} + e^x\left[-\frac{1}{e^x} + \log(1+e^x) - x + c_3\right] - e^{-x}[\log(1+e^x) + c_4]$$

$$= e^x[-e^{-x} + \log(1+e^x) - x + c_1] + e^{-x}[c_2 - \log(1+e^x)].$$

($c_3 e^x$ and $c_4 e^{-x}$ can be combined with the first and second term respectively of complements fucntion.)

(*ii*) The characteristic equation of the corresponding homogeneous equation is

$$(m^2 - 1) = (m+1)(m-1) = 0. \text{ Its roots are } m_1 = 1 \text{ and } m_2 = -1.$$

The complementary function is $y_c(x) = c_1 e^x + c_2 e^{-x}$.

To determine $y_p(x)$, proceed as under:

Here $y_1 = e^x$, $y_2 = e^{-x}$ and $X = e^{-x}\sin(e^{-x}) + \cos(e^{-x})$

The Wronskian $W(y_1, y_2)$ is given by $W(e^x, e^{-x}) = \begin{vmatrix} e^x & e^{-x} \\ e^x & -e^{-x} \end{vmatrix} = -2.$

Thus $$y_p(x) = -y_1\int \frac{y_2 X}{W}dx + y_2\int \frac{y_1 X}{W}dx$$

$$= \frac{e^x}{2}\int e^{-x}[e^{-x}\sin(e^{-x}) + \cos(e^{-x})]\,dx - \frac{e^{-x}}{2}\int e^x[e^{-x}\sin(e^{-x}) + \cos(e^{-x})]\,dx$$

$$= \frac{e^x}{2}\int [e^{-2x}\sin(e^{-x}) + e^{-x}\cos(e^{-x})]\,dx - \frac{e^{-x}}{2}\int [\sin(e^{-x}) + e^x\cos(e^{-x})]\,dx.$$

Now $\int [e^{-2x}\sin(e^{-x})+e^{-x}\cos(e^{-x})]\,dx = e^{-x}\cos(e^{-x})-\sin(e^{-x})-\sin(e^{-x})$

$= e^{-x}\cos(e^{-x})-2\sin(e^{-x})$ and $\int [\sin e^{-x}+e^{x}\cos(e^{-x})]\,dx = \dfrac{\cos e^{-x}}{e^{-x}}$.

$$\therefore \quad y_p(x) = \frac{e^x}{2}[e^{-x}\cos(e^{-x})-2\sin(e^{-x})]-\frac{e^{-x}}{2}\left[\frac{\cos(e^{-x})}{e^{-x}}\right]$$

$$= \frac{\cos(e^{-x})}{2}-e^x\sin(e^{-x})-\frac{\cos(e^{-x})}{2} = -e^x\sin(e^{-x}).$$

The general solution of the equation is $y(x) = c_1e^x + c_2e^{-x} - e^x\sin(e^{-x})$.

(*iii*) The characteristic equation of the corresponding homogeneous equation is $m^2 - 2m + 1 = (m-1)^2 = 0$, and its roots are $m = 1, 1$, which is a repeated root.
The complementary function is $y_c = (c_1 + c_2x)e^x$.
To find $y_p(x)$, proceed as under:
Here $y_1 = e^x$, $y_2 = xe^x$, $X = e^x \log x$.

The wronskian $W(y_1, y_2)$ is given by $W(e^x, x\,e^x) = \begin{vmatrix} e^x & x\,e^x \\ e^x & (1+x)e^x \end{vmatrix} = e^{2x}$.

Thus $y_p(x) = -y_1\displaystyle\int \frac{y_2X}{W}dx + y_2\int \frac{y_1X}{W}dx$

$$= -e^x\int \frac{xe^x\cdot e^x\log x}{e^{2x}}dx + xe^x\int \frac{e^x\cdot e^x\log x}{e^{2x}}dx = -e^x\int \underset{\text{II}}{x}\,\underset{\text{I}}{\log x}\,dx + xe^x\int \log x\,dx$$

$$= -e^x\left(\frac{x^2}{2}\log x - \int \frac{1}{x}\cdot\frac{x^2}{2}dx\right) + x\cdot e^x\left(x\log x - \int \frac{1}{x}\cdot x\,dx\right)$$

$$= -e^x\left(\frac{x^2}{2}\log x - \frac{x^2}{4}\right) + xe^x(x\log x - x) = \frac{1}{4}x^2e^x(2\log x - 3).$$

Hence the general solution of the equation is $y(x) = (c_1 + c_2x)\,e^x + \frac{1}{4}x^2e^x(2\log x - 3)$.

(*iv*) The characteristic equation of the corresponding homogeneous equation is $m^2 + 4 = 0$. Its roots are $m = \pm 2i$. The complementary function is $y_c(x) = c_1\cos 2x + c_2\sin 2x$.
To find $y_p(x)$, proceed as under:
Here $y_1 = \cos 2x$, $y_2 = \sin 2x$ and $X = \tan 2x$.

The Wronskian $W(y_1, y_2)$ is given by $W(\cos 2x, \sin 2x) = \begin{vmatrix} \cos 2x & \sin 2x \\ -2\sin 2x & 2\cos 2x \end{vmatrix} = 2$.

Thus $y_p(x) = -y_1\displaystyle\int \frac{y_2X}{W}dx + y_2\int \frac{y_1X}{W}dx = -\cos 2x\int \frac{\sin 2x\tan 2x}{2}dx + \sin 2x\int \frac{\cos 2x\tan 2x}{2}dx$

$$= -\frac{1}{2}\cos 2x\int (\sec 2x - \cos 2x)\,dx + \frac{1}{2}\sin 2x\int \sin 2x\,dx$$

$$= -\frac{1}{4}\cos 2x[\log|\sec 2x + \tan 2x| - \sin 2x] - \frac{1}{4}\sin 2x\cos 2x$$

$$= -\frac{1}{4}\cos 2x\log|\sec 2x + \tan 2x|.$$

The general solution of the equation is

$$y(x) = c_1\cos 2x + c_2\sin 2x - \frac{1}{4}\cos 2x\log|\sec 2x + \tan 2x|.$$

(*v*) The given equation in operator noitation is $(D^2 - 2D + 1)y = x^{3/2}e^x$.
The characteristic equation of the corresponding homogeneous equation is $m^2 - 2m + 1 = 0$. Its roots are $m = 1, 1$. The complementary function is $y_c(x) = c_1e^x + c_2e^x$.

$$\text{The Wronskian } W(x) = \begin{vmatrix} e^x & x\,e^x \\ e^x & (x+1)\,e^x \end{vmatrix} = e^{2x}.$$

$$\text{Thus } y_p(x) = -e^x \int \frac{xe^x \cdot x^{3/2}\, e^x}{e^{2x}}\, dx + xe^x \int \frac{e^x \cdot x^{3/2}\, e^x}{e^{2x}}\, dx$$

$$= -\frac{2}{7} x^{7/2}\, e^x + \frac{2}{5} x^{5/2}\, xe^x = \frac{4}{35} x^{7/2}\, e^x.$$

The general solution of the equation is $y(x) = c_1e^x + c_2xe^x + \dfrac{4}{35}x^{7/2}\, e^x$.

9.8.2 Method of Undetermined Coefficients

In this method we assume a *trial solution* (assumed particular solution) containing unknown constants (indicated by A, B, C, E...) which are to be determined by substitution in the given equation. The trial solution to be assumed in each case depends on the special form of X containing constants, polynomials, exponentials, sines and cosines functions, sums or products of these functions. These set of functions has the remarkable property that derivatives of their sums and products are again sums and products of constants polynomials, exponentials, sines and cosines. The method of underdetermined coefficients is not applicable to the sets of functions that contains $\log x$, x^{-1}, $\tan x$, $\sec x$, $\sin^{-1}x$ etc. since the number of terms obtained by differenting each of these function is infinite. The special form of X and assumed trial solution is shown in the following table.

Table 9.3: Trial Particular Solutions

X	Form of $y_p(x)$
1. 4 (any constant)	A
2. $6x + 9$	$Ax + B$
3. $2x^2 - 3$	$Ax^2 + Bx + C$
4. $x^3 - x + 1$	$Ax^3 + Bx^2 + Cx + E$
5. $\sin 3x$	$A \cos 3x + B \sin 3x$
6. $\cos 3x$	$A \cos 3x + B \sin 3x$
7. e^{4x}	$A\, e^{4x}$
8. $(9x - 3)e^{4x}$	$(Ax + B)e^{4x}$
9. x^2e^{4x}	$(Ax^2 + Bx + C)e^{4x}$
10. $e^{3x} \sin 4x$	$e^{3x}(A \cos 4x + B \sin 4x)$
11. $6x^2 \sin 3x$	$(Ax^2 + Bx + C) \cos 3x + (Ex^2 + Fx + G) \sin 3x$
12. $xe^{2x} \cos 3x$	$(Ax + B)\, e^{2x} \cos 3x + (Cx + E)\, e^{2x} \sin 3x$

The above method holds in case no terms in the assumed trial solution appears in the complementary solution. If any term of the assumed trial solution does appear in the complementary solution, we must multiply this trial solution by the smallest positive integral power of x which is large enough so that none of the terms which are then present appear in the complementary solution.

EXAMPLE 9.29 Find the general solution of the following differential equations by the method of undetermined coefficients.

(*i*) $y'' + 2y' + 4y = 2x^2 + 3e^{-x}$ (*ii*) $y'' - 2y' + 3y = x^3 + \sin x$.

(*iii*) $y'' - 2y' - 3y = 4x - 5 + 6xe^{2x}$. (*VTU-2000*)

(*iv*) $y'' - 2y' = e^x \sin x$. (*NIT, Kurukshetra, 2009; VTU, 2006*)

Solution (*i*) The characteristic equation of the homogeneous equation is $m^2 + 2m + 4 = 0$. The roots of this equation are $m = -2 \pm \sqrt{4-16}/2 = -1 \pm i\sqrt{3}$.

The complementary function is $y_c(x) = e^{-x}(c_1 \cos\sqrt{3}x + c_2 \sin\sqrt{3}x)$.

Since $X = 2x^2 + 3e^{-x}$, we choose the particular integral as $y_p(x) = (Ax^2 + Bx + C) + Ee^{-x}$.

Substituting y_p and the derivatives

$y'_p(x) = 2Ax + B - Ee^{-x}$ and $y''_p(x) = 2A + Ee^{-x}$ into the given differential equation, we obtain

$y''_p + 2y'_p + 4y_p = 4Ax^2 + (4A + 4B)x + (2A + 2B + 4C) + 3Ee^{-x} = 2x^2 + 3e^{-x}$.

Equating coefficients of like terms, we obtain $A = 1/2$, $B = -1/2$, $C = 0$ and $E = 1$.

Hence, the particular integral is $y_p(x) = \frac{1}{2}x^2 - \frac{1}{2}x + e^{-x}$.

The general solution of the given equation is $y(x) = e^{-x}(c_1 \cos\sqrt{3}x + c_2 \sin\sqrt{3}x) + \frac{1}{2}x^2 - \frac{1}{2}x + e^{-x}$.

(*ii*) The characteristic equation of the homogeneous equation is $m^2 - 2m + 3 = 0$.

The roots of this equation are $m = \{2 \pm \sqrt{4-12}\}/2 = 1 \pm i\sqrt{2}$.

The complementary function is $y_c(x) = e^x(c_1 \cos\sqrt{2}x + c_2 \sin\sqrt{2}x)$.

We choose the particular integral as $y_p(x) = (Ax^3 + Bx^2 + Cx + E) + (F \sin x + G \cos x)$.

Then $y'_p(x) = 3Ax^2 + 2Bx + C + F \cos x - G \sin x$, $y''_p(x) = 6Ax + 2B - F \sin x - G \cos x$.

Substituting in the given equation, we obtain $y''_p - 2y'_p + 3y_p$

$= 3Ax^3 + 3(B - 2A)x^2 + (3C - 4B + 6A)x + (3E - 2C + 2B) + 2(F + G)\sin x + 2(G - F)\cos x$

$= x^3 + \sin x$.

Equating coefficients of like terms, we obtain $3A = 1$ and $A = 1/3$; $B - 2A = 0$ and $B = 2/3$; $3C - 4B + 6A = 0$ and $C = 2/9$; $3E - 2C + 2B = 0$ and $E = -8/27$; $2(F + G) = 1$, $G - F = 0$ and $F = G = 1/4$.

Thus, the particular integral is $y_p = \frac{1}{3}x^3 + \frac{2}{3}x^2 + \frac{2}{9}x - \frac{8}{27} + \frac{1}{4}(\sin x + \cos x)$.

The general solution of the given equation is

$$y(x) = e^x(c_1 \cos\sqrt{2}x + c_2 \sin\sqrt{2}x) + \frac{1}{27}(9x^3 + 18x^2 + 6x - 8) + \frac{1}{4}(\sin x + \cos x).$$

(*iii*) The characteristic equation of the homogeneous equation is

$m^2 - 2m - 3 = 0$ or $(m + 1)(m - 3) = 0$. Its roots are $m_1 = 3$ and $m_2 = -1$.

The complementary function is $y_c(x) = c_1e^{3x} + c_2e^{-x}$.

We choose the particular integral as $y_p(x) = (Ax + B) + (Cx + E)e^{2x}$.

Then $y'_p(x) = A + Ce^{2x} + 2(Cx + E)e^{2x}$,

$y''_p(x) = 2Ce^{2x} + 2Ce^{2x} + 4(Cx + E)e^{2x}$,

and $y''_p - 2y'_p - 3y_p = -3Ax - 2A - 3B - 3Cxe^{2x} + (2C - 3E)e^{2x} = 4x - 5 + 6xe^{2x}$.

Equating coefficients of like terms, we get $-3A = 4$ and $A = -4/3$; $2A + 3B = 5$ and $B = 23/9$; $-3C = 6$ and $C = -2$; $2C - 3E = 0$ and $E = -4/3$.

Hence a particular integral is $y_p(x) = -\frac{4}{3}x + \frac{23}{9} - 2xe^{2x} - \frac{4}{3}e^{2x}$.

The general solution of the given equation is $y(x) = c_1e^{3x} + c_2e^{-x} - \frac{4}{3}x + \frac{23}{9} - \left(2x + \frac{4}{3}\right)e^{2x}$.

(*iv*) The characteristic equation of the homogeneous equation is $m^2 - 2m = 0$. Its roots are $m_1 = 0$ and $m_2 = 2$. The complementary function is $y_c(x) = c_1 + c_2e^{2x}$. We choose the particular integral as

$y_p(x) = e^x(A \sin x + B \cos x)$.

Then $y'_p = (A - B)e^x \sin x + (A + B)e^x \cos x$, $y''_p = -2B e^x \sin x + 2A e^x \cos x$,

and $y''_p - 2y'_p = -2Ae^x \sin x - 2Be^x \cos x = e^x \sin x$.

Equating coefficients of like terms, we obtain $-2A = 1$ and $A = -1/2$; $-2B = 0$ and $B = 0$.

Hence, a particular integral of the differential equation is $y_p(x) = -(e^x \sin x)/2$.

The general solution of the given equation is $y(x) = c_1 + c_2e^{2x} - (e^x \sin x)/2$.

EXAMPLE 9.30 Using the method of undetermined coefficients, solve the given initial value problems

(*i*) $5y'' + y' = -6x,\quad y(0) = 0,\quad y'(0) = -10,$ (*ii*) $y'' + 2y' + y = e^{-x},\quad y(0) = -1,\quad y'(0) = 1,$

(*iii*) $y'' - y' - 2y = 10\sin x,\quad y(0) = 1,\quad y'(0) = 3.$

Solution (*i*) The characteristic equation of the homogeneous equation is $5m^2 + m = 0$ or $m(5m + 1) = 0$ and its roots are $0, -1/5$.

The complementary function is $y_c(x) = c_1 + c_2e^{-x/5}$.

We choose the particular integral as $y_p(x) = x(Ax + B) = Ax^2 + Bx$. We have $y'_p(x) = 2Ax + B,\ y''_p = 2A$.

Substituting in the given equation, we obtain $5y''_p + y'_p = 5(2A) + (2Ax + B) = 2Ax + (10A + B) = -6x$.

Equating coefficients of like terms, we get $2A = -6$ and $A = -3$; $10A + B = 0$ and $B = 30$.

Thus, the particular integral is $y_p(x) = -3x^2 + 30x$.

The general solution is $y(x) = c_1 + c_2e^{-x/5} - 3x^2 + 30x$.

Thus $y(0) = c_1 + c_2 = 0$, from the first initial condition.

By differentiating and from the second initial condition

$y'(x) = (-1/5)c_2e^{-x/5} - 6x + 30,\quad y'(0) = (-c_2/5) + 30 = -10 \Rightarrow c_2 = 200.$

$\therefore$ $c_1 = -200$. The solution of the given initial value problem is $y(x) = -200 + 200e^{-x/5} - 3x^2 + 30x$.

(*ii*) The characteristic equation of the homogeneous equation is $m^2 + 2m + 1 = 0$ or $(m+1)^2 = 0$ and its roots are $-1, -1$:

The complementary function is $y_c(x) = (c_1 + c_2x)e^{-x}$.

We choose the particular integral as $y_p(x) = Ax^2e^{-x}$.

We have $y'_p(x) = A(2x - x^2)e^{-x}$, $y''_p(x) = A(2 - 4x + x^2)e^{-x}$.

Substituting in the given equation, we get

$y''_p + 2y'_p + y_p = A(2 - 4x + x^2)e^{-x} + 2A(2x - x^2)e^{-x} + Ax^2e^{-x} = 2Ae^{-x} = e^{-x}$. Hence $A = 1/2$.

This gives the general solution of the given equation, $y(x) = (c_1 + c_2x)e^{-x} + (1/2)x^2e^{-x}$.

Thus $y(0) = c_1 = -1$, from the first initial condition. By differentiation and from the second initial condition $y'(x) = (c_2 - c_1 - c_2x)e^{-x} + \left(x - \dfrac{x^2}{2}\right)e^{-x}$, $y'(0) = c_2 - c_1 = 1 \Rightarrow c_2 = 0$.

This gives the answer $y(x) = \left(\dfrac{x^2}{2} - 1\right)e^{-x}$.

(*iii*) The characteristic equation of the homogeneous equation is $m^2 - m - 2 = 0$ or $(m + 1)(m - 2) = 0$ and its roots are $-1, 2$. The complementary function is $y_c(x) = c_1e^{-x} + c_2e^{2x}$.

We choose the partcular integral as $y_p(x) = A \cos x + B \sin x$.

We have $y'_p(x) = -A \sin x + B \cos x$, $y''_p(x) = -A \cos x - B \sin x$.

Substituting in the given equation, we get

$$y''_p - y'_p - 2y_p = -(A\cos x + B\sin x) + (A\sin x - B\cos x) - (2A\cos x + 2B\sin x)$$

$$= (A-3B)\sin x - (3A+B)\cos x = 10\sin x.$$

Equating coefficients of like terms, we get $A - 3B = 10; 3A + B = 0$.

On solving, we obtain $A = 1$ and $B = -3$. Thus, the particular integral is $y_p(x) = \cos x - 3\sin x$.

The general solution is $y(x) = c_1 e^{-x} + c_2 e^{2x} + \cos x - 3\sin x$. (*i*)

Substituting the initial conditions in (*i*) we get

$y(0) = 1 = c_1 + c_2 + 1, y'(x) = -c_1 e^{-x} + 2c_2 e^{2x} - \sin x - 3\cos x$, $y'(0) = 3 = -c_1 + 2c_2 - 3$.

The solution is $c_1 = -2$, $c_2 = 2$.

The solution of the given initial value problem is $y(x) = -2e^{-x} + 2e^{2x} + \cos x - 3\sin x$.

EXAMPLE 9.31 Find the general solution of the following differential equations by the method of undetermined coefficients

(*i*) $y'' - 6y' + 9y = 6x^2 + 2 - 12e^{3x}$, (*ii*) $y''' - 6y'' + 12y' - 8y = 12e^{2x} + 27e^{-x}$,

(*iii*) $y'' + 9y = \cos 3x$, (*AMIETE., Dec 2008*) (*iv*) $y'' - 4y' + 13y = 12e^{2x} \sin 3x$. (*AMIETE, June 2007*)

Solution (*i*) The characteristic equation of the homogeneous equation is $m^2 - 6m + 9 = 0$, or $(m-3)^2 = 0$ and its roots are $m_1 = m_2 = 3$.

The complementary function is $y_c(x) = (c_1 + c_2 x)e^{3x}$.

We note that e^{3x} and xe^{3x} are present in the complementary function (due to the double root $m = 3$) and e^{3x} is also a term on the right hand side X.

Therefore, we choose the particular integral as $y_p(x) = (Ax^2 + Bx + C) + Ex^2e^{3x}$.

We have $y'_p(x) = 2Ax + B + 2Exe^{3x} + 3Ex^2e^{3x}$,

$y''_p(x) = 2A + 2Ee^{3x} + 12Exe^{3x} + 9Ex^2e^{3x}$.

Substituting in the given equation, we get

$$y''_p - 6y'_p + 9y_p = 9Ax^2 + (-12A + 9B)x + 2A - 6B + 9C + 2Ee^{3x} = 6x^2 + 2 - 12e^{3x}.$$

Equating coefficients of like terms, we get $9A = 6$ and $A = 2/3$; $-12A + 9B = 0$ and $B = 8/9$;
$2A - 6B + 9C = 2$ and $C = 2/3$, $2E = 12$ and $E = -6$.

Thus, the particular integral is $y_p(x) = \frac{2}{3}x^2 + \frac{8}{9}x + \frac{2}{3} - 6x^2e^{3x}$.

Hence the general solution is $y(x) = y_c(x) + y_p(x) = (c_1 + c_2 x)e^{3x} + \frac{2}{3}x^2 + \frac{8}{9}x + \frac{2}{3} - 6x^2e^{3x}$.

(*ii*) The characteristic equation of the homogeneous equation is $m^3 - 6m^2 + 12m - 8 = (m-2)^3 = 0$.

Its roots are $m_1 = m_2 = m_3 = 2$.

The complementary function is $y_c(x) = (c_1 + c_2x + c_3x^2)e^{2x}$.

We note that e^{2x}, xe^{2x} and x^2e^{2x} are present in the complementary function (due to the triple roots $m = 2$) and e^{2x} is also a term on the right hand side X.

Therefore, we choose the particular integral as $y_p(x) = Ax^3e^{2x} + Be^{-x}$.

We have $y'_p(x) = A(3x^2 + 2x^3)e^{2x} - Be^{-x}$,

$y''_p(x) = A(6x + 12x^2 + 4x^3)e^{2x} + Be^{-x}$,

$y'''_p(x) = A(6 + 36x + 36x^2 + 8x^3)e^{2x} - Be^{-x}$.

Substituting in the given equation, we get

$$y'''_p - 6y''_p + 12y'_p - 8y_p = Ae^{2x}[(6 + 36x + 36x^2 + 8x^3) - 6(6x + 12x^2 + 4x^3)$$
$$+ 12(3x^2 + 2x^3) - 8x^3] + Be^{-x}[-1 - 6 - 12 - 8]$$
$$= 6Ae^{2x} - 27Be^{-x} = 12e^{2x} + 27e^{-x}.$$

Equating coefficients of like terms, we get $6A = 12$ and $A = 2$; $-27B = 27$ and $B = -1$.
Therefore, the particular integral is $y_p(x) = 2x^3e^{2x} - e^{-x}$.
The general solution is $y(x) = (c_1 + c_2 x + c_3x^2)\, e^{2x} + 2x^3\, e^{2x} - e^{-x}$.

(*iii*) The characteristic equation of the homogeneous equation is $m^2 + 9 = 0$ and its roots are $m = \pm 3i$.
The complementary function is $y_c(x) = c_1 \cos 3x + c_2 \sin 3x$.
We note that $\cos 3x$ appears as a term both in complementary function and the right hand side X.
Therefore, we choose the particular integral as $y_p(x) = x(A \cos 3x + B \sin 3x)$.

We have $\quad y'_p(x) = A \cos 3x + B \sin 3x + 3x(-A \sin 3x + B \cos 3x)$,

$$y''_p(x) = 6(-A \sin 3x + B \cos 3x) + 9x(-A \cos 3x - B \sin 3x).$$

Substituting in the given equation, we get

$$y''_p + 9y_p = \sin 3x[-6A - 9xB + 9xB] + \cos 3x[6B - 9xA + 9xA] = \cos 3x$$

or $\quad -6A \sin 3x + 6B \cos 3x = \cos 3x$.

Equating coefficients of like terms, we get A = 0; 6B = 1 and B = 1/6.
The particular integral is $y_p(x) = (x \sin 3x)/6$.
The general solution is $y(x) = c_1 \cos 3x + c_2 \sin 3x + (x \sin 3x)/6$.

(*iv*) The characteristic equation of the homogeneous equation is $m^2 - 4m + 13 = 0$.
The roots of this equation are $m = 4 \pm \sqrt{16 - 52}/2 = 2 \pm 3i$.
The complementary function is $y_c(x) = e^{2x}(c_1 \cos 3x + c_2 \sin 3x)$.
We note that $e^{2x} \sin 3x$ appears both in the complementary function and the right hand side X.
Therefore, we choose $y_p(x) = x\, e^{2x}(A \cos 3x + B \sin 3x)$.

We have $\quad y'_p(x) = (1 + 2x)e^{2x}\, (A \cos 3x + B \sin 3x) + 3xe^{2x}(-A \sin 3x + B \cos 3x)$,

$$y''_p(x) = (4 + 4x)e^{2x}\, (A \cos 3x + B \sin 3x) + 6(1 + 2x)e^{2x}\, (-A \sin 3x + B \cos 3x) + 9xe^{2x}(-A \cos 3x - B \sin 3x).$$

Substituting in the given equation, we get

$$y''_p - 4y'_p + 13y_p = e^{2x} \cos 3x[A(4 + 4x) + 6B(1 + 2x) - 9Ax - 4A(1 + 2x) - 12xB + 13Ax] + e^{2x} \sin 3x[B(4 + 4x) - 6A(1 + 2x) - 9Bx - 4B(1 + 2x) + 12Ax + 13xB] = 12e^{2x} \sin 3x$$

or $\quad 6Be^{2x} \cos 3x - 6Ae^{2x} \sin 3x = 12e^{2x} \sin 3x$.

Equating coefficients of like terms, we get $-6A = 12$ and $A = -2$; $6B = 0$ and $B = 0$.
Therefore, the particular integral is $y_p(x) = -2xe^{2x} \cos 3x$.
The general solution is $y(x) = e^{2x}(c_1 \cos 3x + c_2 \sin 3x - 2x \cos 3x)$.

EXERCISE 9.3

Find the general solution of the following differential equations, using the method of variation of parameters.

1. $y'' + a^2y = \sec ax$. (*VTU, 2001*)
   Ans. $y(x) = c_1 \cos ax + c_2 \sin ax + (1/a^2) \cos ax \log |\cos ax|) + (1/a)\, x \sin ax$.
2. $(D^2 + 4)y = \cos 2x$. (*Osmania, 2000S*) **Ans.** $y(x) = c_1 \cos 2x + c_2 \sin 2x + (x/4) \sin 2x$.
3. $y'' + y = \tan x$. (*UPTU, 2009; NIT Kurukshetra, 2008*)
   Ans. $y(x) = c_1 \cos x + c_2 \sin x - \cos x \log |\sec x + \tan x|$.
4. $y'' + y = x \sin x$. (*SVTU 2007; JNTU, 2005*) **Ans.** $y(x) = c_1 \cos x + c_2 \sin x + \dfrac{x \sin x}{4} - \dfrac{x^2 \cos x}{4}$.
5. $y'' + 4y = 4\sec^2 2x$. (*AMIE, S-2011*) **Ans.** $y(x) = c_1 \cos 2x + c_2 \sin 2x - 1 + \sin 2x \log |\sec 2x + \tan 2x|$.

6. $y'' - 6y' + 9y = (e^{3x}/x^2)$. *(AMIETE, S-2010; Raipur, 2005; AMIE, W-2005)*

[Hint: $W(x) = \begin{bmatrix} e^{3x} & xe^{3x} \\ 3e^{3x} & e^{3x} + 3xe^{3x} \end{bmatrix} = e^{6x}$]. **Ans.** $y(x) = (c_1 + c_2 x)e^{3x} - e^{3x} \log x$.

7. $(D^2 - 1)y = (1 + e^{-x})^{-2}$. **Ans.** $y(x) = c_1 e^x + c_2 e^{-x} - 1 + e^{-x} \log |1 + e^x|$.

8. $y'' - 2y' + 2y = e^x \tan x + 3x$.

Ans. $y(x) = \{c_1 \sin x + [c_2 - \log|\sec x + \tan x|]\cos x\}e^x + (3x/2) + (3/2)$.

9. $y'' - 4y' + 4y = (x + 1)\, e^{2x}$. **Ans.** $y(x) = (c_1 + c_2 x)e^{2x} + (1/6)x^3 e^{2x} + (1/2)x^2 e^{2x}$

10. $y'' + 2y' + y = e^{-x} \cos x$. *(AMIE, S-2006)* **Ans.** $y(x) = (c_1 + c_2 x - \cos x)\, e^{-x}$

11. $y'' + 16y = 32 \sec 2x$. *(AMIETE, June 2010, Dec. 2007)*

Ans. $y(x) = c_1 \cos 4x + c_2 \sin 4x + 8\cos 2x - 4\sin 4x \log|\sec 2x + \tan 2x|$.

12. $y'' + 2y' + y = e^{-x} \log x$. **Ans.** $y(x) = c_1 e^{-x} + c_2 x e^{-x} + \left(\frac{x^2}{2}\log x - \frac{3}{4}x^2\right)e^{-x}$.

13. $y'' + 3y' + 2y = 2e^x$. *(AMIETE, June 2011, 09; AMIE, Dec 2010)* **Ans.** $y(x) = c_1 e^{-x} + c_2 e^{-2x} + (e^x/3)$.

14. $y'' - 2y' + y = e^x / x^3$. **Ans.** $y(x) = [c_1 + c_2 x + (1/2x)]e^x$.

15. $y'' + y = 3x - 8\cot x$. **Ans.** $y(x) = c_1 \sin x + c_2 \cos x + 8\sin x \log|\operatorname{cosec} x + \cot x| + 3x$.

16. $(D^2 - 2D + 1)y = e^x$. *(NIT Kurukshetra, 2006)* **Ans.** $y(x) = (c_1 + xc_2)\, e^x + \frac{1}{2}x^2 e^x$.

17. $y'' - 4y' + 4y = e^{2x} \sec^2 x$. **Ans.** $y(x) = (c_1 + c_2 x - \log\cos x)e^{2x}$.

18. $y'' - 8y' + 16y = 12e^{4x}/x^4$. **Ans.** $y(x) = (c_1 + c_2 x + 2x^{-2})e^{4x}$.

19. $y'' + y = \cos ax$. *(Osmania, 2000S)* **Ans.** $y(x) = c_1 \cos x + c_2 \sin x - x\cos x + \sin x \log|\sin x|$.

Find the general solution of the following differential equations by the method of undetermined coefficients.

20. $y'' + 4y = 8x^2$; **Ans.** $y(x) = c_1 \cos 2x + c_2 \sin 2x + 2x^2 - 1$.

21. $y'' + y = x^2 + x$. **Ans.** $y(x) - c_1 \cos x + c_2 \sin x + x^2 + x - 2$.

22. $y'' + 4y' + 3y = 5e^{2x}$. **Ans.** $y(x) = c_1 e^{-3x} + c_2 e^{-x} + (e^{2x}/3)$.

23. $y'' - 3y' + 2y = e^x$. **Ans.** $y(x) = c_1 e^x + c_2 e^{2x} - xe^x$.

24. $y'' + 5y' + 6y = 3e^{-2x} + e^{3x}$. **Ans.** $y(x) = c_1 e^{-2x} + c_2 e^{-3x} + 3xe^{-2x} + (e^{3x}/30)$.

25. $y'' + 2y' + y = e^{-x}$. **[Hint:** Choose $y_p = Ax^2 e^{-x}$] **Ans.** $y(x) = (c_1 + c_2 x)e^{-x} + (x^2 e^{-x}/2)$.

26. $y'' - 2y' - 3y = 6e^{-x} - 8e^x$. **Ans.** $y(x) = c_1 e^{-x} + c_2 e^{3x} - (3xe^{-x}/2) + 2e^x$.

27. $y'' + 4y' + 4y = 12e^{-2x}$. **Ans.** $y(x) = (c_1 + c_2 x)e^{-2x} + 6x^2 e^{-2x}$.

28. $y'' - y' - 2y = 4 \sin 3x$. **Ans.** $y(x) = c_1 e^{2x} + c_2 e^{-x} + (6\cos 3x - 22 \sin 3x)/65$.

29. $y'' + y = 3 \sin x$. **Ans.** $y(x) = c_1 \cos x + c_2 \sin x - (3x\cos x)/2$.

30. $y'' + y = 2\cos x$. *(VTU, 2000S)* **Ans.** $y(x) = c_1 \cos x + c_2 \sin x - x \sin x$.

31. $y'' + 2y' + 4y = 2x^2 + 3e^{-x}$. **Ans.** $y(x) = e^{-x}(c_1 \cos\sqrt{3}x + c_2 \sin\sqrt{3}x) + \frac{1}{2}x^2 - \frac{1}{2}x + e^{-x}$.

32. $y'' + 3y' + 2y = 10\, e^{3x} + 4x^2$. **Ans.** $y(x) = c_1 e^{-2x} + c_2 e^{-x} + (e^{3x})/2 + (2x^2 - 6x + 7)$.

33. $y'' + 2y' + 5y = \frac{5}{4}e^{x/2} + 40\cos 4x - 55 \sin 4x$. **[Hint:** Choose $y_p = Ae^{x/2} + B\cos 4x + C\sin 4x$].

Ans. $y(x) = e^{-x}(c_1 \cos 2x + c_2 \sin 2x) + (e^{x/2}/5) + 5\sin 4x$.

34. $y'' - 2y' + 2y = 2e^x \cos x.$ **Ans.** $y(x) = (c_1 \cos x + c_2 \sin x)\, e^x + xe^x \sin x.$

35. $y'' - 2y' = e^x \sin x.$ *(VTU, 2000)* **Ans.** $y(x) = c_1 + c_2 e^{2x} - (e^x \sin x)/2.$

36. $y'' - y = e^{3x} \cos 2x - e^{2x} \sin 3x.$ *(AMIETE, Dec 2009)*

Ans. $y(x) = c_1 e^x + c_2 e^{-x} + \frac{1}{30} e^{2x} (2 \cos 3x + \sin 3x) + \frac{1}{40} e^{3x} (\cos 2x + 3 \sin 2x).$

9.9 LINEAR EQUATIONS WITH VARIABLE COEFFICIENTS—THE EULER-CAUCHY AND LEGENDRE LINEAR EQUATIONS

We will now study two types of differential equations with variable coefficients which can be transformed into differential equations with constant coefficients by appropriate substitutions.

9.9.1 The Euler-Cauchy Linear Equation

An equation of the form

$$a_0 x^n \frac{d^n y}{dx^n} + a_1 x^{n-1} \frac{d^{n-1} y}{dx^{n-1}} + a_2 x^{n-2} \frac{d^{n-2} y}{dx^{n-2}} + \ldots\ldots + a_{n-1} x \frac{dy}{dx} + a_n y = X \qquad \ldots(9.51)$$

where $a_0, a_1, a_2, \ldots\ldots, a_n$ are constants and X is a function of x is called *Euler-Cauchy's equation.* Eq. (9.51) can be solved by changing it into a linear equation with constant coefficients by the following transformation.

Under the transformation $x = e^z$ or $z = \log x$, we have

$$y' = \frac{dy}{dx} = \frac{dy}{dz} \cdot \frac{dz}{dx} = \frac{1}{x} \frac{dy}{dz}. \qquad \ldots(9.52)$$

$$y'' = \frac{d(y')}{dx} = \frac{d}{dx}\left(\frac{dy}{dx}\right) = \frac{d}{dx}\left(\frac{1}{x} \frac{dy}{dz}\right) = -\frac{1}{x^2} \frac{dy}{dz} + \frac{1}{x} \frac{d}{dx}\left(\frac{dy}{dz}\right)$$

$$= -\frac{1}{x^2} \frac{dy}{dz} + \frac{1}{x} \frac{d}{dz}\left(\frac{dy}{dz}\right)\frac{dz}{dx} = -\frac{1}{x^2} \frac{dy}{dz} + \frac{1}{x^2} \frac{d^2 y}{dz^2} = \frac{1}{x^2}\left(\frac{d^2 y}{dz^2} - \frac{dy}{dz}\right) \qquad \ldots(9.53)$$

Similarly $$y''' = -\frac{2}{x^3}\left(\frac{d^2 y}{dz^2} - \frac{dy}{dz}\right) + \frac{1}{x^3}\left(\frac{d^3 y}{dz^3} - \frac{d^2 y}{dz^2}\right) = \frac{1}{x^3}\left(\frac{d^3 y}{dz^3} - 3\frac{d^2 y}{dz^2} + 2\frac{dy}{dz}\right) \qquad \ldots(9.54)$$

and so on.

In operator notation, set $D = d/dx, D^2 = d^2/dx^2, D^3 = d^3/dx^3$, $\mathcal{D} = d/dz, \mathcal{D}^2 = d^2/dz^2, \mathcal{D}^3 = d^3/dz^3$. etc. Then, Eqs. (9.52), (9.53) and (9.54) can be written as

$$y' = \frac{dy}{dx} = Dy = \frac{1}{x}\mathcal{D}y, \text{ or } xDy = \mathcal{D}y,$$

$$y'' = \frac{d^2 y}{dx^2} = D^2 y = \frac{1}{x^2}(\mathcal{D}^2 y - \mathcal{D}y), \text{ or } x^2 D^2 y = \mathcal{D}(\mathcal{D} - 1)y,$$

and $$y''' = \frac{d^3 y}{dx^3} = D^3 y = \frac{1}{x^3}(\mathcal{D}^3 y - 3\mathcal{D}^2 y + 2\mathcal{D}y), \text{ or } x^3 D^3 y = \mathcal{D}(\mathcal{D} - 1)(\mathcal{D} - 2)y.$$

These suggest the generalization $x^n D^n y = \mathcal{D}(\mathcal{D} - 1)\ldots\ldots[\mathcal{D} - (n-1)]y.$

Substituting for $x\frac{dy}{dx}, x^2 \frac{d^2 y}{dx^2}, \ldots\ldots,$ Eq. (9.51) is reduced to linear equation with constant coefficients and thus can be solved by the methods already discussed.

9.9.2 The Legendre's Linear Equation

This equation is of the form

$$a_0(ax+b)^n \frac{d^n y}{dx^n} + a_1(ax+b)^{n-1} \frac{d^{n-1} y}{dx^{n-1}} + \ldots + a_{n-1}(ax+b)\frac{dy}{dx} + a_n y = X \qquad \ldots(9.55)$$

where $a_0, a_1, a_2, \ldots\ldots, a_n$ are constants and X is a function of x.

Such equation can straight away be reduced to linear equation with constant coefficients by putting $ax + b = e^z$ or $z = \log(ax + b)$.

$$y' = \frac{dy}{dx} = Dy = \frac{dy}{dz}\cdot\frac{dz}{dx} = \frac{a}{ax+b}\frac{dy}{dz}, \text{ or } (ax+b)\,Dy = a\frac{dy}{dz} = a\,\mathcal{D}\,y,$$

$$y'' = \frac{d^2 y}{dx^2} = D^2 y = \frac{a^2}{(ax+b)^2}\left(\frac{d^2 y}{dz^2} - \frac{dy}{dz}\right) \text{ or } (ax+b)^2 D^2 y = a^2\,\mathcal{D}(\mathcal{D}-1)\,y, \text{ and so on.}$$

Substituting for $x\dfrac{dy}{dx}, x^2\dfrac{d^2y}{dx^2}$......, Eq. (9.55) is reduced to linear equation with constant coefficients.

ILLUSTRATIVE EXAMPLES

EXAMPLE 9.32 Find the general solution of the following homogeneous differential equations

(*i*) $2x^2y'' + xy' - 6y = 0$. (*ii*) $(x^3D^3 + 3x^2D^2 - 2xD + 2)y = 0$.

(*iii*) $x^2y'' + xy' + 9y = 0$.

Solution (*i*) Using the transformation $x = e^z$, so that $z = \log x$. This reduces the given equation (in operator notation) to $\{2\mathcal{D}(\mathcal{D}-1) + \mathcal{D} - 6\}\,y = 0$ or $(2\mathcal{D}^2 - \mathcal{D} - 6)y = 0$ or $(\mathcal{D}-2)(2\mathcal{D}+3)y = 0$, where $\mathcal{D} = d/dz$.

The solution is $y(z) = c_1e^{2z} + c_2e^{-3z/2}$.

Substituting $e^z = x$; we get the general solution as $y(x) = c_1x^2 + (c_2/x\sqrt{x})$.

(*ii*) The transformation $x = e^z$ reduces the given equation to

$$\{\mathcal{D}(\mathcal{D}-1)(\mathcal{D}-2) + 3\mathcal{D}(\mathcal{D}-1) - 2\mathcal{D} + 2\}\,y = (\mathcal{D}^3 - 3\mathcal{D} + 2)y$$

$$= (\mathcal{D}-1)(\mathcal{D}^2 + \mathcal{D} - 2)y = (\mathcal{D}-1)^2(\mathcal{D}+2)y = 0, \text{ (where } \mathcal{D} = d/dz\text{)},$$

whose solution is $y(z) = (c_1 + c_2z)e^z + c_3e^{-2z}$.

Substituting $e^z = x$, we obtain the general solution as $y(x) = c_1x + c_2x\log x + (c_3/x^2)$.

(*iii*) The transformation $x = e^z$ reduces the given equation to

$$\{\mathcal{D}(\mathcal{D}-1) + \mathcal{D} + 9\}\,y = (\mathcal{D}^2 + 9)\,y = 0,$$

whose solution is $y(z) = c_1\cos 3z + c_2\sin 3z$.

Replacing z by $\log x$, the general solution of the given equation is

$$y(x) = c_1\cos(3\log x) + c_2\sin(3\log x).$$

EXAMPLE 9.33 Find the general solution of the following differential equations

(*i*) $x^2y'' - 2xy' - 4y = x^2 + 2\log x$, *(Bhopal 2003S)*

(*ii*) $x^2y'' - 2xy' + 2y = x^3 + \sin(5\log x)$, (*iii*) $x^2y'' + 3xy' + y = (1-x)^{-2}$. *(PTU-2003)*

Solution (*i*) Using the transformation $x = e^z$ so that $z = \log x$. We obtain the equation (in operator notation)

as $\{\mathcal{D}(\mathcal{D}-1) - 2\mathcal{D} - 4\}y = e^{2z} + 2z$

or $(\mathcal{D}^2 - 3\mathcal{D} - 4)\,y = e^{2z} + 2z$, where $\mathcal{D} = d/dz$.

The characteristic equation of the corresponding homogeneous equation is

$$m^2 - 3m - 4 = 0, \text{ or } (m+1)(m-4) = 0, \text{ or } m = -1, 4.$$

The complementary function is given by $y_c(z) = c_1e^{-z} + c_2e^{4z}$.

The particular integral is

$$y_p(z) = \frac{1}{\mathcal{D}^2 - 3\mathcal{D} - 4}(e^{2z} + 2z) = \frac{1}{\mathcal{D}^2 - 3\mathcal{D} - 4}e^{2z} + 2\frac{1}{\mathcal{D}^2 - 3\mathcal{D} - 4}z$$

$$= \frac{1}{2^2 - 3(2) - 4}e^{2z} - \frac{1}{2}\left(1 + \frac{3\mathcal{D}}{4} - \frac{\mathcal{D}^2}{4}\right)^{-1} z = -\frac{1}{6}e^{2z} - \frac{1}{2}\left(1 - \frac{3\mathcal{D}}{4} + \ldots\right)z = -\frac{1}{6}e^{2z} - \frac{1}{2}\left(z - \frac{3}{4}\right).$$

The general solution is $y(z) = y_c + y_p = c_1 e^{-z} + c_2 e^{4z} - \frac{e^{2z}}{6} - \frac{z}{2} + \frac{3}{8}$.

Substituting $e^z = x$, we obtain $y(x) = c_1 x^{-1} + c_2 x^4 - \frac{x^2}{6} - \frac{1}{2}\log x + \frac{3}{8}$.

(*ii*) The transformation $x = e^z$ reduces the equation (in operator notation) to

$\{\mathcal{D}(\mathcal{D} - 1) - 2\mathcal{D} + 2\}\, y = e^{3z} + \sin 5z$

or $\quad (\mathcal{D}^2 - 3\mathcal{D} + 2)\, y = (\mathcal{D} - 1)(\mathcal{D} - 2)\, y = e^{3z} + \sin 5z$, where $\mathcal{D} = d/dz$.

The characteristic equation of the corresponding homogeneous equation is $(m - 1)(m - 2) = 0$ or $m = 1, 2$.

The complementary function is $y_c(z) = c_1 e^z + c_2 e^{2z}$. The particular integral is

$$y_p(z) = \frac{1}{\mathcal{D}^2 - 3\mathcal{D} + 2}(e^{3z} + \sin 5z) = \frac{1}{\mathcal{D}^2 - 3\mathcal{D} + 2}e^{3z} + \frac{1}{\mathcal{D}^2 - 3\mathcal{D} + 2}\sin 5z$$

$$= \frac{1}{3^2 - 3.3 + 2}e^{3z} + \frac{1}{-25 - 3\mathcal{D} + 2}\sin 5z = \frac{1}{2}e^{3z} - \frac{1}{23 + 3\mathcal{D}} \cdot \frac{(23 - 3\mathcal{D})}{(23 - 3\mathcal{D})}\sin 5z$$

$$= \frac{1}{2}e^{3z} - \frac{(23 - 3\mathcal{D})\sin 5z}{529 - 9\mathcal{D}^2} = \frac{1}{2}e^{3z} - \frac{(23\sin 5z - 15\cos 5z)}{529 - 9(-25)} = \frac{1}{2}e^{3z} - \frac{(23\sin 5z - 15\cos 5z)}{754}.$$

The general solution is $y(z) = y_c(z) + y_p(z) = c_1 e^z + c_2 e^{2z} + \frac{1}{2}e^{3z} - \frac{1}{754}(23\sin 5z - 15\cos 5z)$.

Substituting $e^z = x$, we obtain

$$y(x) = c_1 x + c_2 x^2 + \frac{1}{2}x^3 - \frac{1}{754}[23\sin(5\log x) - 15\cos(5\log x)].$$

(*iii*) The transformation $x = e^z$ reduces the equation (in operator notation) to

$\{\mathcal{D}(\mathcal{D} - 1) + 3\mathcal{D} + 1\}\, y = (1 - e^z)^{-2}$

or $\quad \{\mathcal{D}^2 + 2\mathcal{D} + 1\}\, y = (\mathcal{D} + 1)^2 y = \frac{1}{(1 - e^z)^2}$, where $\mathcal{D} = d/dz$.

The characteristic equation of the corresponding homogeneous equation is $(m + 1)^2 = 0$ or $m = -1, -1$.

The complementary function is $y_c(z) = (c_1 + c_2 z)e^{-z}$.

The particular integral is $y_p(z) = \frac{1}{(\mathcal{D} + 1)^2} \cdot \frac{1}{(1 - e^z)^2} = \frac{1}{\mathcal{D} + 1} \cdot \frac{1}{\mathcal{D} + 1} \cdot \frac{1}{(1 - e^z)^2}$

$$= \frac{1}{\mathcal{D} + 1} \cdot e^{-z} \int \frac{e^z}{(1 - e^z)^2}dz = \frac{1}{\mathcal{D} + 1}e^{-z}\left(\frac{1}{1 - e^z}\right)$$

$$= e^{-z} \cdot \int e^{-z}\frac{e^z}{1 - e^z}dz = e^{-z}\int \frac{dt}{t(1 - t)}, \text{ where } t = e^z,$$

$$= e^{-z}\int\left(\frac{1}{t}+\frac{1}{1-t}\right)dt = e^{-z}[\log t - \log(1-t)] = e^{-z}[z - \log(1-e^z)].$$

The term ze^{-z} occurring in the particular integral is absorbable in the complementary function. The general solution is $y(z) = y_c(z) + y_p(z)$

$$= (c_1 + c_2 z)e^{-z} - e^{-z}\log(1-e^z) = [c_1 + c_2 z - \log(1-e^z)]e^{-z}$$

Substituting $e^z = x$, we obtain $y(x) = [c_1 + c_2 \log x - \log(1-x)]x^{-1}$.

EXAMPLE 9.34 Find the general solution of the following differential equations

(*i*) $x^2y'' - xy' + 4y = \cos(\log x) + x\sin(\log x)$, (*ii*) $x^3y''' + 2xy' - 2y = x^2\log x + 3x$,

(*iii*) $x^2y'' - 3xy' + y = \log x\dfrac{[\sin(\log x)+1]}{x}$, (*AMIETE, June 2007*)

(*iv*) $x^3y''' + 5x^2y'' + 5xy' + y = x^2 + \log x,\ x > 0.$ (*AMIETE, June, 2010; Dec. 2007*)

(*v*) $x^2\dfrac{d^2y}{dx^2} + 4x\dfrac{dy}{dx} + 2y = x + \sin x.$ (*AMIETE, Dec 2010*)

Solution (*i*) Using the transformation $x = e^z$, we get (in operator notation)

$$\{\mathcal{D}(\mathcal{D}-1) - \mathcal{D} + 4\}y = \cos z + e^z\sin z$$

or $(\mathcal{D}^2 - 2\mathcal{D} + 4)y = \cos z + e^z \sin z$, where $\mathcal{D} = d/dz$.

The characteristic equation of the corresponding homogeneous equation is

$$m^2 - 2m + 4 = 0 \quad \text{or} \quad m = 1 \pm i\sqrt{3}.$$

The complementary function is $y_c(z) = e^z(c_1\cos\sqrt{3}z + c_2\sin\sqrt{3}z)$.

The particular integral is

$$y_p(z) = \frac{1}{\mathcal{D}^2 - 2\mathcal{D} + 4}\cos z + \frac{1}{\mathcal{D}^2 - 2\mathcal{D} + 4}\cdot e^z\sin z$$

$$= \frac{1}{3-2\mathcal{D}}\cos z + e^z\frac{1}{(\mathcal{D}+1)^2 - 2(\mathcal{D}+1) + 4}\sin z$$

$$= \frac{1}{3-2\mathcal{D}}\cos z + e^z\frac{1}{\mathcal{D}^2+3}\sin z = \frac{3+2\mathcal{D}}{9-4\mathcal{D}^2}\cos z + e^z\cdot\frac{1}{-1^2+3}\sin z$$

$$= \frac{(3+2\mathcal{D})}{9-4(-1)}\cos z + \frac{1}{2}e^z\sin z = \frac{1}{13}(3\cos z - 2\sin z) + \frac{1}{2}e^z\sin z.$$

The general solution is $y(z) = e^z(c_1\cos\sqrt{3}z + c_2\sin\sqrt{3}z) + \dfrac{1}{13}(3\cos z - 2\sin z) + \dfrac{1}{2}e^z\sin z$

Substituting $e^z = x$, we obtain

$$y(x) = x[c_1\cos\sqrt{3}(\log x) + c_2\sin\sqrt{3}(\log x)] + \frac{1}{13}[3\cos(\log x) - 2\sin(\log x)] + \frac{1}{2}x\sin(\log x).$$

(*ii*) Using the transformation $x = e^z$, we get (in operator notation)

$$\{\mathcal{D}(\mathcal{D}-1)(\mathcal{D}-2) + 2\mathcal{D} - 2\}y = ze^{2z} + 3e^z$$

or $(\mathcal{D}-1)(\mathcal{D}^2 - 2\mathcal{D} + 2)\mathcal{D} = ze^{2z} + 3e^z$, where $\mathcal{D} = d/dz$.

The characteristic equation of the corresponding homogeneous equation is

$$(m-1)(m^2 - 2m + 2) = 0 \quad \text{or} \quad m = 1, 1 \pm i.$$

The complementary function is $y_c(z) = c_1e^z + e^z(c_2\cos z + c_3\sin z)$.

The particular integral is $y_p(z) = \dfrac{1}{\mathcal{D}^3 - 3\mathcal{D}^2 + 4\mathcal{D} - 2} ze^{2z} + 3e^z$

$$= e^{2z} \frac{1}{(\mathcal{D}+2)^3 - 3(\mathcal{D}+2)^2 + 4(\mathcal{D}+2) - 2} z + 3 \cdot \frac{1}{\mathcal{D}^3 - 3\mathcal{D}^2 + 4\mathcal{D} - 2} e^z$$

$$= e^{2z} \frac{1}{\mathcal{D}^3 + 3\mathcal{D}^2 + 4\mathcal{D} + 2} z + 3z \frac{1}{3\mathcal{D}^2 - 6\mathcal{D} + 4} e^z$$

$$= \frac{e^{2z}}{2}(1 + 2\mathcal{D} + \cdots)^{-1} z + 3ze^z = \frac{e^{2z}}{2}(1 - 2\mathcal{D} - \cdots) z + 3ze^z = \frac{e^{2z}}{2}(z-2) + 3ze^z$$

The general solution is $y(z) = y_c(z) + y_p(z)$

$$= c_1 e^z + e^z(c_2 \cos z + c_3 \sin z) + \frac{1}{2} e^{2z}(z-2) + 3ze^z.$$

Substituting $e^z = x$, we obtain

$$y(x) = c_1 x + x[c_2 \cos(\log x) + c_3 \sin(\log x)] + \frac{1}{2} x^2 (\log x - 2) + 3x \log x.$$

(*iii*) Using the transformation $x = e^z$, we get (in operator notation)

$$\{\mathcal{D}(\mathcal{D}-1) - 3\mathcal{D} + 1\} y = z \frac{(1 + \sin z)}{e^z}$$

or $\quad (\mathcal{D}^2 - 4\mathcal{D} + 1)y = e^{-z} z(1 + \sin z)$, where $\mathcal{D} = d/dz$.

The characteristic equation of the corresponding homogeneous equation is

$$m^2 - 4m + 1 = 0 \text{ or } m = 2 \pm \sqrt{3}.$$

The complementary function is

$$y_c(z) = c_1 e^{(2+\sqrt{3})z} + c_2 e^{(2-\sqrt{3})z} = e^{2z}(c_1 e^{\sqrt{3}z} + c_2 e^{-\sqrt{3}z}).$$

The particular integral is $y_p(z) = \dfrac{1}{\mathcal{D}^2 - 4\mathcal{D} + 1} e^{-z} z(1 + \sin z)$

$$= e^{-z} \frac{1}{(\mathcal{D}-1)^2 - 4(\mathcal{D}-1) + 1} z(1 + \sin z) = e^{-z}\left\{\frac{1}{\mathcal{D}^2 - 6\mathcal{D} + 6} z + \frac{1}{\mathcal{D}^2 - 6\mathcal{D} + 6} z \sin z\right\}.$$

Now $\quad \dfrac{1}{\mathcal{D}^2 - 6\mathcal{D} + 6} z = \dfrac{1}{6}\left(1 - \dfrac{6\mathcal{D} - \mathcal{D}^2}{6}\right)^{-1} z = \dfrac{1}{6}(1 + \mathcal{D} + \ldots\ldots) z = \dfrac{1}{6}(z+1), \qquad \ldots(i)$

and $\quad \dfrac{1}{\mathcal{D}^2 - 6\mathcal{D} + 6} z \sin z = z \dfrac{1}{\mathcal{D}^2 - 6\mathcal{D} + 6} \sin z - \dfrac{2\mathcal{D} - 6}{(\mathcal{D}^2 - 6\mathcal{D} + 6)^2} \sin z$

$$= z \cdot \frac{1}{5 - 6\mathcal{D}} \cdot \frac{5 + 6\mathcal{D}}{5 + 6\mathcal{D}} \sin z - \frac{2\mathcal{D} - 6}{\mathcal{D}^4 - 12\mathcal{D}^3 + 48\mathcal{D}^2 - 72\mathcal{D} + 36} \sin z$$

$$= z \cdot \frac{(5 + 6\mathcal{D}) \sin z}{25 - 36\mathcal{D}^2} - \frac{(2\mathcal{D} - 6)}{(\mathcal{D}^2)^2 - 12\mathcal{D}^2(\mathcal{D}) + 48\mathcal{D}^2 - 72\mathcal{D} + 36} \sin z$$

$$= \frac{z(5 \sin z + 6 \cos z)}{61} + \frac{(2 \cos z - 6 \sin z)}{60\mathcal{D} + 11}$$

$$= \frac{1}{61} z(5 \sin z + 6 \cos z) + \frac{2}{60\mathcal{D} + 11} \cdot \frac{60\mathcal{D} - 11}{60\mathcal{D} - 11} \cos z - \frac{6}{60\mathcal{D} + 11} \cdot \frac{60\mathcal{D} - 11}{60\mathcal{D} - 11} \sin z$$

$$=\frac{1}{61}z(5\sin z+6\cos z)+\frac{120\sin z+22\cos z}{3721}+\frac{360\cos z-66\sin z}{3721}$$

$$=\frac{1}{61}z(5\sin z+6\cos z)+\frac{2}{3721}(27\sin z+191\cos z). \quad ...(ii)$$

Adding (*i*) and (*ii*), and multiply the result by e^{-z}, we get

$$y_p(z)=e^{-z}\left[\frac{1}{6}(z+1)+\frac{z}{61}(5\sin z+6\cos z)+\frac{2}{3721}(27\sin z+191\cos z)\right].$$

The general solution of the given equation $y(z)=y_c(z)+y_p(z)$

$$=e^{2z}(c_1e^{\sqrt{3}z}+c_2e^{-\sqrt{3}z})+e^{-z}\left[\frac{1}{6}(z+1)+\frac{z}{61}(5\sin z+6\cos z)+\frac{2}{3721}(27\sin z+191\cos z)\right].$$

Substituting $e^z=x$, we get the general solution as

or $$y(x)=x^2(c_1x^{\sqrt{3}}+c_2x^{-\sqrt{3}})+\frac{1}{x}\left[\frac{1}{6}(1+\log x)+\frac{\log x}{61}\{5\sin(\log x)+6\cos(\log x)\}\right.$$

$$\left.+\frac{2}{3721}\{27\sin(\log x)+191\cos(\log x)\}\right].$$

(*iv*) Using the transformation $x=e^z$, we get (in operator notation)

$$\{\mathcal{D}(\mathcal{D}-1)(\mathcal{D}-2)+5\mathcal{D}(\mathcal{D}-1)+5\mathcal{D}+1\}y=e^{2z}+z$$

or $$\{\mathcal{D}^3+2\mathcal{D}^2+2\mathcal{D}+1\}y=e^{2z}+z,\ \text{where } \mathcal{D}=d/dz.$$

The characteristic equation of the corresponding homogeneous equation is

$$m^3+2m^2+2m+1=0,\quad \text{or}\quad (m+1)(m^2+m+1)=0.$$

Its roots are $m=-1,\ (-1\pm i\sqrt{3})/2$. The complementary function is

$$y_c(z)=c_1e^{-z}+e^{-z/2}\{c_2\cos(\sqrt{3}z/2)+c_3\sin(\sqrt{3}z/2)\}$$

The particular integral is $y_p(z)=\dfrac{1}{\mathcal{D}^3+2\mathcal{D}^2+2\mathcal{D}+1}e^{2z}+\dfrac{1}{\mathcal{D}^3+2\mathcal{D}^2+2\mathcal{D}+1}z$

$$=\frac{1}{8+8+4+1}e^{2z}+(1+2\mathcal{D}+\cdots)^{-1}z=\frac{1}{21}e^{2z}+(1-2\mathcal{D}+\cdots)z=\frac{1}{21}e^{2z}+z-2.$$

The general solution is

$$y(z)=c_1e^{-z}+e^{-z/2}\{c_2\cos(\sqrt{3}z/2)+c_3\sin(\sqrt{3}z/2)\}+\frac{1}{21}e^{2z}+z-2.$$

Substituting $e^z=x$, we get

$$y(x)=\frac{c_1}{x}+\frac{1}{\sqrt{x}}\left[c_2\cos(\sqrt{3}\log x/2)+c_3\sin(\sqrt{3}\log x/2)\right]+\frac{x^2}{21}+\log x-2.$$

(*v*) Using the transformation $x=e^z$, we get (in operator notation)

$$\{\mathcal{D}(\mathcal{D}-1)+4\mathcal{D}+2\}y=e^z+\sin e^z$$

or $$(\mathcal{D}^2+3\mathcal{D}+2)y=e^z+\sin e^z,\ \text{where } \mathcal{D}=d/dz.$$

The characteristic equation of the corresponding homogeneous equation is

$$m^2+3m+2=0,\quad \text{or}\quad m=-1,-2.$$

The complementary function is $y_c(z)=c_1e^{-z}+c_2e^{-2z}$

The particular integral is $y_p(z)=\dfrac{1}{(\mathcal{D}+1)(\mathcal{D}+2)}\sin e^z = \left(\dfrac{1}{\mathcal{D}+1}-\dfrac{1}{\mathcal{D}+2}\right)\sin e^z.$

Now use $\dfrac{1}{D+a}X = e^{-ax}\int e^{ax}X\,dx.$ We have

$$\frac{1}{\mathcal{D}+1}\sin e^z = e^{-z}\int e^z \sin e^z\,dz = -e^{-z}\cos e^z,$$

and $$\frac{1}{\mathcal{D}+2}\sin e^z = e^{-2z}\int e^{2z}(\sin e^z)\,dz = -e^{-z}\cos e^z + e^{-2z}\sin e^z.$$

Therefore, $y_p(z) = -e^{-z}\cos e^z + e^{-z}\cos e^z - e^{-2z}\sin e^z = -e^{-2z}\sin e^z.$

The general solution is $y(z) = y_c(z)+y_p(z) = c_1e^{-z}+c_2e^{-2z}-e^{-2z}\sin e^z.$

Substituting $e^z = x$, we get $y(x) = c_1x^{-1}+c_2x^{-2}-x^{-2}\sin x = \dfrac{c_1}{x}+\dfrac{c_2}{x}-\dfrac{\sin x}{x^2}.$

EXAMPLE 9.35 Solve the following differential equations, by the method of variation of parameters.

(*i*) $x^2y''-2x(1+x)\dfrac{dy}{dx}+2(x+1)y=x^3$, given that x and xe^{2x} are solutions of the corresponding homogeneous equation.

(*ii*) $x^2y''-3xy'+3y=2x^4e^x$.

Solution (i) The given equation may be written as under

$$y''-\frac{2(1+x)}{x}y'+\frac{2(1+x)}{x^2}y=x.$$

The complementary function is $y_c(x) = c_1x + c_2x\,e^{2x}$.

$y_1 = x$ and $y_2 = xe^{2x}$ are linearly independent over any interval. Here $X = x$,

$$W(x)=\begin{vmatrix} x & xe^{2x} \\ 1 & e^{2x}(2x+1)\end{vmatrix} = 2x^2e^{2x}\neq 0,$$ over any interval not including 0. In any such interval,

Thus $$y_p(x) = -y_1\int\frac{y_2X}{W}dx + y_2\int\frac{y_1X}{W}dx$$

$$= -x\int\frac{(xe^{2x})x}{2x^2e^{2x}}dx + xe^{2x}\int\frac{x\,x}{2x^2e^{2x}}dx = -\frac{x^2}{2}-\frac{x}{4}.$$

The general solution of the equation is $y(x) = c_1x+c_2x\,e^{2x}-\dfrac{x^2}{2}-\dfrac{x}{4}.$

(*ii*) Using the transformation $x = e^z$, we get (in operator notation)

$$\{\mathcal{D}(\mathcal{D}-1)-3\mathcal{D}+3\}y = 2e^{4z}\cdot e^{e^z} = 2e^{4z+e^z}$$

or $$(\mathcal{D}^2-4\mathcal{D}+3)y = 2e^{4z+e^z},\ \text{where } \mathcal{D} = d/dz.$$

The characteristic equation of the corresponding homogeneous equation is

$m^2 - 4m + 3 = 0$ or $(m-1)(m-3) = 0$. Its roots are $m = 1, 3$.

The complementary function is $y_c(z) = c_1e^z + c_2e^{3z}$.

To determine $y_p(z)$, proceed as under:

Here $y_1 = e^z, y_2 = e^{3z}, X = 2e^{4z+e^z}$.

The Wronskian $W(y_1, y_2)$ is given by $W(e^z, e^{3z}) = \begin{vmatrix} e^z & e^{3z} \\ e^z & 3e^{3z} \end{vmatrix} = 3e^{4z} - e^{4z} = 2e^{4z}$.

Employing the formula:

$$y_p(x) = -y_1 \int \frac{y_2 X}{W} dx + y_2 \int \frac{y_1 X}{W} dx, \text{ we obtain}$$

$$y_p(z) = -e^z \int \frac{e^{3z} \cdot 2e^{4z+e^z}}{2e^{4z}} dz + e^{3z} \int \frac{e^z \cdot 2e^{4z+e^z}}{2e^{4z}} dz = -e^z \int e^{3z} \cdot e^{e^z} dz + e^{3z} \int e^z \cdot e^{e^z} dz$$

$$= -e^z (e^{2z} \cdot e^{e^z} - 2e^z e^{e^z} + 2e^{e^z}) + e^{3z} \cdot e^{e^z} = 2e^{e^z}(e^{2z} - e^z).$$

The general solution is

$$y(z) = y_c(z) + y_p(z) = c_1 e^z + c_2 e^{3z} + 2e^{e^z}(e^{2z} - e^z).$$

Substituting $e^z = x$, we obtain $y(x) = c_1 x + c_2 x^3 + 2x^2 e^x - 2xe^x$.

EXAMPLE 9.36 Solve the following differential equations:

(*i*) $(3x+2)^2 \dfrac{d^2y}{dx^2} + 3(3x+2)\dfrac{dy}{dx} - 36y = 3x^2 + 4x + 1,$ (*Sambalpur 2002*)

(*ii*) $(1+x)^2 \dfrac{d^2y}{dx^2} + (1+x)\dfrac{dy}{dx} + y = 4\cos[\log(1+x)],$ (*Kurukshetra, 2010; MDU 2005*)

(*iii*) $(2x+3)^2 \dfrac{d^2y}{dx^2} - (2x+3)\dfrac{dy}{dx} - 12y = 6x.$ (*VTU, 2007; Kerala, 2005; Anna, 2002S*)

Solution (*i*) Put $3x + 2 = e^z$ so that $z = \log(3x + 2)$ and $dz/dx = 3/(3x + 2)$.

Then $\dfrac{dy}{dx} = \dfrac{dy}{dz} \cdot \dfrac{dz}{dx} = \dfrac{3}{3x+2} \cdot \dfrac{dy}{dz} \Rightarrow (3x+2)\dfrac{dy}{dx} = 3\mathcal{D}y$, where $\mathcal{D} = \dfrac{d}{dz}$.

Now $\dfrac{d^2y}{dx^2} = \dfrac{d}{dx}\left(\dfrac{3}{3x+2} \cdot \dfrac{dy}{dz}\right) = -\dfrac{3^2}{(3x+2)^2}\dfrac{dy}{dz} + \dfrac{3}{3x+2} \cdot \dfrac{d}{dx}\left(\dfrac{dy}{dz}\right)$

$$= \frac{-3^2}{(3x+2)^2}\frac{dy}{dz} + \frac{3}{3x+2} \cdot \frac{d}{dz}\left(\frac{dy}{dz}\right)\frac{dz}{dx} = -\frac{3^2}{(3x+2)^2}\frac{dy}{dz} + \frac{3^2}{(3x+2)^2}\frac{d^2y}{dz^2}$$

or $\quad (3x+2)^2 \dfrac{d^2y}{dx^2} = 3^2[\mathcal{D}(\mathcal{D}-1)y]$.

The differential equation in operator notation is

$$[3^2\mathcal{D}(\mathcal{D}-1) + 3^2\mathcal{D} - 36]y = \frac{(e^z-2)^2}{3} + \frac{4}{3}(e^z - 2) + 1 = \frac{e^{2z}-1}{3}$$

or $\quad 9(\mathcal{D}^2 - 4)y = (e^{2z} - 1)/3$.

The characteristic equation of the corresponding homogeneous equation is $9(m^2 - 4) = 0$. Its roots are $m = \pm 2$. The complementary function is $y_c(z) = c_1 e^{2z} + c_2 e^{-2z}$.

The particular integral is $y_p(z) = \dfrac{1}{27}\left(\dfrac{1}{\mathcal{D}^2-4}e^{2z} - \dfrac{1}{\mathcal{D}^2-4}e^{0z}\right) = \dfrac{1}{27} \cdot z\dfrac{e^{2z}}{4} + \dfrac{1}{27} \cdot \dfrac{1}{4} = \dfrac{1}{108}(ze^{2z} + 1)$.

The general solution is $y(z) = y_c(z) + y_p(z) = c_1e^{2z} + c_2e^{-2z} + \frac{1}{108}(ze^{2z} + 1)$.

Substituting $e^z = 3x + 2$, we obtain

$$y(x) = c_1(3x+2)^2 + c_2(3x+2)^{-2} + \frac{1}{108}[(3x+2)^2 \log(3x+2) + 1].$$

(*ii*) Put $1 + x = e^z$ or $z = \log(1 + x)$, so that

$$(1+x)\frac{dy}{dx} = \mathcal{D}y \text{ and } (1+x)^2\frac{d^2y}{dx^2} = \mathcal{D}(\mathcal{D}-1)\,y, \text{ where } \mathcal{D} = \frac{d}{dz}.$$

The given equation in operator notation is

$$\mathcal{D}(\mathcal{D} - 1)y + \mathcal{D}y + y = 4\cos z \quad \text{or} \quad (\mathcal{D}^2 + 1)y = 4\cos z.$$

The characteristic equation of the corresponding homogeneous equation is $(m^2 + 1) = 0$.

Its roots are $m = \pm i$.

The complementary function is $y_c(z) = c_1 \cos z + c_2 \sin z$.

The particular integral is

$$y_p(z) = 4\cdot\frac{1}{\mathcal{D}^2+1}\cos z = 4z\cdot\frac{1}{2\mathcal{D}}\cos z = 2z \sin z.$$

The general solution is

$$y(z) = y_c(z) + y_p(z) = c_1 \cos z + c_2 \sin z + 2z \sin z.$$

Substituting $e^z = 1 + x$, we obtain

$$y(x) = c_1 \cos[\log(1 + x)] + c_2 \sin[\log(1 + x)] + 2\log(1 + x)\sin[\log(1 + x)].$$

(*iii*) Put $2x + 3 = e^z$ so that $z = \log(2x + 3)$ and $dz/dx = 2/(2x + 3)$.

Then $\frac{dy}{dx} = \frac{2}{2x+3}\cdot\frac{dy}{dz} \Rightarrow (2x+3)\frac{dy}{dx} = 2\mathcal{D}y$, where $\mathcal{D} = \frac{d}{dz}$.

$$(2x+3)^2\frac{d^2y}{dx^2} = 2^2[\mathcal{D}(\mathcal{D}-1)\,y].$$

The differential equation in operator notation is $[2^2\mathcal{D}(\mathcal{D}-1) - 2\mathcal{D} - 12]y = 3(e^z - 3)$.

That is, $[4\mathcal{D}^2 - 6\mathcal{D} - 12]y = 3(e^z - 3)$.

The characteristic equation of the corresponding homogeneous equation is $4m^2 - 6m - 12 = 0$.

Its roots are $m = \frac{3 \pm \sqrt{57}}{4}$.

The complementary function is $y_c(z) = c_1\, e^{\frac{(3+\sqrt{57})}{4}z} + c_2\, e^{\frac{(3-\sqrt{57})}{4}z}$.

The particular integral is

$$y_p(z) = \frac{1}{2(2\mathcal{D}^2 - 3\mathcal{D} - 6)}3(e^z - 3)$$

$$= \frac{3}{2(-7)}e^z - \frac{9}{2}\frac{1}{2\mathcal{D}^2 - 3\mathcal{D} - 6}e^{oz} = -\frac{3}{14}e^z + \frac{3}{4}.$$

The general solution is $y(z) = y_c(z) + y_p(z)$

$$= c_1\, e^{\frac{(3+\sqrt{57})}{4}z} + c_2\, e^{\frac{(3-\sqrt{57})}{4}z} - \frac{3}{14}e^z + \frac{3}{4}.$$

Substituting $e^z = (2x + 3)$, we obtain

$$y(x) = c_1\,(2x+3)^{(3+\sqrt{57})/4} + c_2\,(2x+3)^{(3-\sqrt{57})/4} - \frac{3}{14}(2x+3) + \frac{3}{4}.$$

EXERCISE 9.4

Find the general solution of the following homogeneous differential equations.

1. $x^2y'' + xy' - y = 0.$ **Ans.** $y = c_1x + c_2x^{-1}$

2. $x^2y'' - 2xy' - 4y = 0.$ **Ans.** $y = c_1x^4 + c_2x^{-1}$

3. $4x^2y'' + 8xy' + y = 0.$ **Ans.** $y = c_1x^{-1/2} + c_2x^{-1/2} \log x$

4. $x^2y'' + 7xy' + 13y = 0.$ **Ans.** $y = x^{-3}[c_1 \cos(2 \log x) + c_2 \sin (2 \log x)]$

5. $x^3y''' + 5x^2y'' + 7xy' + 8y = 0$ **Ans.** $y = c_1x^{-2} + c_2 \cos (2 \log x) + c_3 \sin(2 \log x).$

6. $y''' = 6y/x^3.$ **[Hint:** The given equation is $x^3y''' - 6y = 0$]. **Ans.** $y = c_1x^3 + c_2 \cos(\sqrt{2} \log x + c_3).$

Find the general solution of the following differential equations.

7. $2x^2y'' + 7xy' - 3y = x^2 + x^{-2}.$ **Ans.** $y = c_1\sqrt{x} + \dfrac{c_2}{x^3} + \dfrac{x^2}{15} - \dfrac{1}{5x^2}$

8. $x^3y''' - x^2y'' + 2xy' - 2y = x^3 + 3x.$ *(AMIETE, June 2006)*

Ans. $y = (c_1 + c_2 \log x)x + c_3x^2 + \dfrac{x^3}{4} - \dfrac{3}{2}x(\log x)^2.$

9. $x^3 y''' + 2x^2y'' + 2y = 10(x + x^{-1}).$ *(GGSIPU-2006; DCE, 2005; PTU 03)*

Ans. $y = c_1x^{-1} + x[c_2 \cos (\log x) + c_3 \sin (\log x)] + 5x + 2x^{-1} \log x.$

10. $2x^2y'' + 3xy' - 3y = x^3.$ *(AMIETE, Dec. 2005)* **Ans.** $y = c_1x + \dfrac{c_2}{x\sqrt{x}} + \dfrac{x^3}{18}.$

11. $x^2D^2y - xDy + y = \log x$, where $D = d/dx.$ *(AMIETE, June, 2010)*

Ans. $y = (c_1 + c_2 \log x)\, x + \log x + 2.$

12. $x^3y''' + 3x^2 y'' + xy' + y = x + \log x.$ *(UPTU 2001)*

Ans. $y = c_1x^{-1} + \sqrt{x}\left[c_2 \cos\left(\dfrac{\sqrt{3}}{2}\log x\right) + c_3 \sin\left(\dfrac{\sqrt{3}}{2}\log x\right)\right] + \dfrac{1}{2}x + \log x.$

13. $x^2y'' - 2xy' - 4y = x^2 + 2 \log x;\ x > 0.$ *(KUK, 2010)* **Ans.** $y = c_1x^{-1} + c_2x^4 - \dfrac{x^2}{6} - \dfrac{1}{2}\log x + \dfrac{3}{8}.$

14. $x^2y'' - xy' - 3y = x^2 \log x$ *(AMIE, S-2010)* **Ans.** $y = c_1x^3 + c_2x^{-1} - \dfrac{x^2}{9}(3\log x + 2).$

15. $x^2y'' + 2xy' - 12y = x^3 \log x.$ *(UPTU-2004)* **Ans.** $y = c_1x^3 + c_2x^{-4} + \{x^3 \log x(7\log x - 2)\}/98.$

16. $x^2y'' + xy' + y = (\log x)\cdot \sin(\log x),\ x > 0.$ *(KUK 2006; Madras 2006; Kerala 2005; PTU 2003)*

Ans. $y = c_1 \cos(\log x) + c_2 \sin(\log x) + \dfrac{\log x}{4}\sin(\log x) - \dfrac{(\log x)^2}{4}\cos(\log x).$

17. $y'' + \dfrac{1}{x}y' = \dfrac{12\log x}{x^2}.$ **Ans.** $y = c_1 \log x + c_2 + 2(\log x)^3.$

18. $x^2y'' - 2y = x^2 + x^{-1}.$ **Ans.** $y = c_1x^2 + c_2x^{-1} + \{\log x\,(x^2 - x^{-1})\}/3.$

19. $x^2y'' + 2xy' - 20y = (x + 1)^2$ **Ans.** $y = c_1x^{-5} + c_2x^4 - \left(\dfrac{x^2}{14} + \dfrac{x}{9} + \dfrac{1}{20}\right).$

20. $x^2y'' - 3xy' + y = \dfrac{\sin(\log x)}{x}$. **Ans.** $y = x^2(c_1x^{\sqrt{3}} + c_2x^{-\sqrt{3}}) + \dfrac{1}{61x}(5\sin\log x + 6\cos\log x)$

21. $x^2y'' - 3xy' + 3y = x\log x$. **Ans.** $y = c_1x + c_2x^3 - \dfrac{x}{4}[(\log x)^2 + \log x]$.

22. $x^2y'' - xy' + 2y = x\log x$. **Ans.** $y = x(c_1\cos\log x + c_2\sin\log x) + x\log x$.

23. $x^2y'' - 2xy' + 2y = x + x^2\log x + x^3$. **Ans.** $y = c_1x + c_2x^2 - x\log x + \dfrac{x^3}{2} + x^2\left[\dfrac{1}{2}(\log x)^2 - \log x\right]$.

24. $x^2y'' - 3xy' + 5y = x^2\sin(\log x)$. *(AMIE, S-2011; MDU 2002)*

Ans. $y = x^2[c_1\cos(\log x) + c_2\sin(\log x)] - \{x^2\log x\cdot\cos(\log x)\}/2$.

25. $(x^2D^2 + 4xD + 2)y = e^x$. *(KUK, 2005; UPTU, 2005)* **Ans.** $y = c_1x^{-1} + x^{-2}(c_2 + e^x)$.

Using the method of variation of parameters, find the general solution of the following differential equations:

26. $x^2D^2y + xDy - y = x^2e^x$. *(AMIETE, Dec. 2005)* **Ans.** $y = c_1x + c_2x^{-1} + e^x - x^{-1}e^x$.

27. $x^2y'' - 2xy' + 2y = x^4$. **Ans.** $y = c_1x + c_2x^2 + x^4/6$.

28. $x^2y'' + 3xy' + y = (1 - x^2)$. **Ans.** $y = (c_1 + c_2\log x)x^{-1} + 1 - \dfrac{1}{2}x + \dfrac{1}{9}x^2$.

29. $x^2y'' + xy' - y = x^3/(1 + x^2)$. *(AMIETE., Dec. 2006)*

Ans. $y = c_1x + c_2x^{-1} + \dfrac{x}{4}\log(1 + x^2) + \dfrac{x}{4} - \dfrac{1}{4x}\log(1 + x^2)$.

Find the solutions of the following differential equations, which satisfy the given conditions.

30. $x^2y'' - 3xy' + 4y = 0$, $y(1) = 0$ and $y'(1) = 3$. **Ans.** $y(x) = 3x^2\log x$.

31. $x^3y''' - 3x^2y'' + 6xy' - 6y = 0$, $y(1) = 2$, $y'(1) = 1$, $y''(1) = -4$. **Ans.** $y = 2x + x^2 - x^3$.

32. $x^3y'' - x^2y' + xy = 1$; $y(1) = 1/4$, $y(e) = e + (1/4e)$. *(AMIETE, Dec. 2004)*

Ans. $y = (1/4x) + x\log x$, $x > 0$.

33. $(x^3D^2 + x^2D - x)y = 3x^3$; $y(1) = 1$, $y'(1) = 2$. **Ans.** $y = x^2$, $x > 0$.

34. $(x^2D^2 \quad 3xD + 3)y = \log x$; $y(1) = 1$, $y'(1) = 2$. **Ans.** $y = \dfrac{1}{9}(5x^3 + 4 + \log x^3)$, $x > 0$.

Find the general solution of the following differential equations.

35. $(x + 1)^2y'' - 3(x + 1)y' + 4y = x^2$.

Ans. $y = [c_1 + c_2\log(x + 1)](x + 1)^2 + \dfrac{1}{2}(x + 1)^2\{\log(x + 1)\}^2 - 2(x + 1) + \dfrac{1}{4}$.

36. $(x + 2)^2y'' - (x + 2)y' + y = 3x + 4$. **Ans.** $y = (x + 2)\left[c_1 + c_2\log|x + 2| + \dfrac{3}{2}\log^2|x + 2|\right] - 2$

37. $(2x - 1)^2y'' - 4(2x - 1)y' + 8y = 8x$.

Ans. $y = c_1(2x - 1) + c_2(2x - 1)^2 - (2x - 1)\log(2x - 1) + 1/2$.

38. $(x + 1)^2y'' + (x + 1)y' = (2x + 3)(2x + 4)$. **Ans.** $y = c_1 + c_2\log(x + 1) + [\log(x + 1)]^2 + x^2 + 8x$.

39. $(2x + 3)^2y'' - 2(2x + 3)y' - 12y = 6x$. *(VTU 2003S)*

Ans. $y = c_1(2x + 3)^{-1} + c_2(2x + 3)^3 - (3x/2) + (3/4)$.

40. $(1 + x)^2\dfrac{d^2y}{dx^2} + (1 + x)\dfrac{dy}{dx} + y = \sin[2\log(1 + x)]$. *(PTU, 2006; VTU, 2004)*

Ans. $y = c_1\cos[\log(1 + x)] + c_2\sin[\log(1 + x)] - \dfrac{1}{3}\sin\{2\log(1 + x)\}$.

41. $(1+x)^2 y'' + (1+x)y' + y = 2\sin[\log(1+x)]$. (*JNTU, 2005; Kerala, 2005*)

Ans. $y = c_1 \cos[\log(1+x)] + c_2 \sin[\log(1+x)] - \log(1+x)\cos[\log(1+x)]$.

42. $(1+2x)^2 \dfrac{d^2y}{dx^2} - 6(1+2x)\dfrac{dy}{dx} + 16y = 8(1+2x)^2$.

Ans. $y = (1+2x)^2[\{\log(1+2x)\}^2 + c_1 \log(1+2x) + c_2]$.

9.10 SIMULTANEOUS LINEAR EQUATIONS

Many problems in engineering give rise to differential equations in which there are two or more dependent variables and a single independent variable usually the time t. Such equations are termed as *simultaneous linear equations*. We consider here a system of linear differential equations with constant coefficients only. We will discuss the solution of a system of two linear equations in two linear dependent variables x and y and one independent variable t. The solution of such system can be obtained by eliminating one of the variables and solving the resulting linear equation for the second variable. We illustrate the method of obtaining the solution through the following solved examples.

EXAMPLE 9.37 Solve the system of equations $\dfrac{dx}{dt} + 4x + 3y = t, \dfrac{dy}{dt} + 2x + 5y = e^t$. (*UPTU, 2006*)

Solution The given equations in the operator notation are given by

$$(D+4)\,x + 3y = t, \quad \text{...(i)}$$

$$2x + (D+5)\,y = e^t. \quad \text{...(ii)}$$

Operate on (*i*) by $(D+5)$ and multiply (*ii*) by 3 and subtract, we obtain

$$(D^2 + 9D + 14)x = 1 + 5t - 3e^t. \quad \text{...(iii)}$$

Eq. (*iii*) is a linear differential equation of 2nd order. Its characteristic equation is given by

$m^2 + 9m + 14 = 0$, gives $m_1 = -2$ and $m_2 = -7$.

The complementary function is given by $x_c(t) = c_1e^{-2t} + c_2e^{-7t}$.

The particular integral is

$$x_p(t) = \frac{1}{D^2 + 9D + 14}(e^{ot} + 5t - 3e^t) = \frac{1}{14} + \frac{5}{14}\left(t - \frac{9}{14}\right) - \frac{3}{24}e^t = -\frac{31}{196} + \frac{5t}{14} - \frac{1}{8}e^t$$

Hence the general solution of the equation is

$$x(t) = c_1e^{-2t} + c_2e^{-7t} + \frac{5t}{14} - \frac{1}{8}e^t - \frac{31}{196}. \quad \text{...(iii)}$$

Differentiating (*iii*), $dx/dt = -2c_1e^{-2t} - 7c_2e^{-7t} + (5/14) - (e^t/8)$.

Substituting for dx/dt and x in the first equation, we get

$$-2c_1e^{-2t} - 7c_2e^{-7t} + (5/14) - (e^t/8) + 4c_1e^{-2t} + 4c_2e^{-7t} - (31/49) + (10t/7) - (e^t/2) - t = -3y.$$

$$\therefore \quad y(t) = -\frac{2}{3}c_1e^{-2t} + c_2e^{-7t} - \frac{1}{7}t + \frac{5}{24}e^t + \frac{9}{98}. \quad \text{...(iv)}$$

The required solution is given by equations (*iii*) and (*iv*).

EXAMPLE 9.38 Solve the simultaneous equations $\dfrac{dx}{dt} + 5x - 2y = t, \; \dfrac{dy}{dt} + 2x + y = 0$ subject to $x(0) = 0$, $y(0) = 0$. (*AMIETE., June 2009; UPTU, 2008; Kurukshetra, 2005*)

Solution The given equations in the operator notation are given by

$$(D + 5)\,x - 2y = t, \qquad ...(i)$$

$$2x + (D + 1)\,y = 0. \qquad ...(ii)$$

Multiplying (*i*) by 2 and operate on (*ii*) by $(D + 5)$ and then subtracting, we obtain

$$(D^2 + 6D + 9)\,y = -2t.$$

Its characteristic equation is $m^2 + 6m + 9 = 0 \Rightarrow (m + 3)^2 = 0$, giving $m_1 = m_2 = -3$.

The complementary function is given by $y_c(t) = (c_1 + c_2 t)e^{-3t}$.

The particular integral is $y_p(t) = \dfrac{1}{(D+3)^2}(-2t) = -\dfrac{2}{9}\left(1+\dfrac{D}{3}\right)^{-2} t = -\dfrac{2}{9}\left(1-\dfrac{2D}{3}+...\right)t = -\dfrac{2}{9}t+\dfrac{4}{27}$.

Hence $y(t) = (c_1 + c_2 t)e^{-3t} - \dfrac{2}{9}t + \dfrac{4}{27}$. ...(*iii*)

Now to find *x*, substitute the value of *y* in (*ii*). From (*iii*)

$$D_y(t) = c_2 e^{-3t} + (c_1 + c_2 t)(-3)\,e^{-3t} - \frac{2}{9}.$$

Substituting for *y* and *Dy* in (*ii*), we obtain

$$x(t) = -\frac{1}{2}[Dy + y] = \left\{\left(c_1 - \frac{1}{2}c_2\right) + c_2 t\right\}e^{-3t} + \frac{t}{9} + \frac{1}{27} \qquad ...(iv)$$

The stipulated initial conditions then apply, the equation (*iii*) and (*iv*) gives

$$0 = c_1 + \frac{4}{27} \text{ and } c_1 - \frac{1}{2}c_2 + \frac{1}{27} = 0 \text{ whence } c_1 = -4/27 \text{ and } c_2 = -2/9.$$

Thus the solution of the initial-value problem is

$$x(t) = -\frac{1}{27}(1+6t)e^{-3t} + \frac{1}{27}(1+3t),\quad y(t) = -\frac{2}{27}(2+3t)e^{-3t} + \frac{2}{27}(2-3t).$$

EXAMPLE 9.39 Find the solution of the system of equations

$$\frac{d^2x}{dt^2} - 3x - 4y = 0,\ \frac{d^2y}{dt^2} + x + y = 0. \qquad (AMIE,\ W\text{-}2008)$$

Solution The given equations in the operator notation are written as:

$$(D^2 - 3)x - 4y = 0,\ x + (D^2 + 1)\,y = 0.$$

Operate on the first by $D^2 + 1$ and multiply second by 4 and adding, we get

$$(D^4 - 2D^2 + 1)x = 0 \text{ or } (D^2 - 1)^2 = 0.$$

The characteristic equation is $(m^2 - 1)^2 = 0$, giving $m_1, m_2 = 1$, and $m_3, m_4 = -1$.

Hence the solution is $x = (c_1 + c_2 t)e^t + (c_3 + c_4 t)e^{-t}$.

As done in the above example, we calculate d^2x/dt^2 and substitute for *x*, d^2x/dt^2 and get *y*. But this is laborious and we calculate *y* directly.

Operate on the second equation by $(D^2 - 3)$ and subtract first from the result of the second, we get

$$(D^4 - 2D^2 + 1)\,y = 0.$$

The characteristic equation is $m^4 - 2m^2 + 1 = 0$ giving $m_1, m_2 = 1$, and $m_3, m_4 = -1$.

$$\therefore \qquad y = (k_1 + k_2 t)e^t + (k_3 + k_4 t)e^{-t}.$$

Substituting for *y* and *x* in the first equation, we get

$$k_1 = \frac{c_2 - c_1}{2},\ k_2 = -\frac{c_2}{2},\ k_3 = -\frac{c_4 + c_3}{2},\ k_4 = -\frac{c_4}{2}.$$

Hence $y = \left(\frac{c_2 - c_1}{2} - \frac{c_2 t}{2}\right)e^t - \left(\frac{c_3 + c_4}{2} + \frac{c_4 t}{2}\right)e^{-t}.$

It may be noted that by solving two equations we got eight constants. These constants are not all independent as is obvious from above. The number of independent arbitrary constants is equal to the exponent of the highest powers of D obtained in the expansion of the determinant formed by the operator coefficients in the dependent variable. In this case, the determinant is

$$\begin{vmatrix} D^2 - 3 & -4 \\ 1 & D^2 + 1 \end{vmatrix} = D^4 - 2D^2 + 1$$

and the exponent of the highest power of D is 4. So there are 4 independent constants.

EXAMPLE 9.40 The coordinates x and y of the position of a particle at time t satisfy the differential equations $\frac{dy}{dt} + 2x = \sin 2t$ and $\frac{dx}{dt} - 2y = \cos 2t$. If $x(0) = 1$ and $y(0) = 0$, show that the path of the particle is given by $4x^2 + 4xy + 5y^2 = 4$.

Solution Elimineting x from the given the equations,

$$\frac{d^2y}{dt^2} + 2\frac{dx}{dt} = 2 \cos 2t \qquad ...(i)$$

$$-\frac{2\,dx}{dt} + 4y = -2 \cos 2t \qquad ...(ii)$$

Adding (*i*) and (*ii*), we obtain $\frac{d^2y}{dt^2} + 4y = 0$. This equation in D-operator form is $(D^2 + 4)\,y = 0$, where $D \equiv d/dt$. The characteristic equation of the corresponding homogenous equation is $(m^2 + 4) = 0$. Its roots are $m = \pm\, 2i$. The solution is

$$y(t) = c_1\cos 2t + c_2\sin 2t.$$

$$x(t) - \frac{1}{2}\sin 2t - \frac{1}{2}\frac{d}{dt}(c_1 \cos 2t + c_2 \sin 2t)$$

$$= \frac{1}{2}\sin 2t - \frac{1}{2}\,(-2\,c_1 \sin 2t + 2\,c_2 \cos 2t)$$

$$= \frac{1}{2}\sin 2t + c_1 \sin 2t - c_2 \cos 2t$$

Applying the initial conditions $x(0) = 1$ and $y(0) = 0$.

When $y(0) = 0 \ \Rightarrow\ c_1 = 0,\ x(0) = 1 \ \Rightarrow\ c_2 = -1.$

Thus the coordinates of the particle at time t are $x = \frac{1}{2}\sin 2t + \cos 2t$ and $y = -\sin 2t$.

Eliminating t from these two relations, we obtain

$$y + 2x = 2 \cos 2t \Rightarrow (y + 2x)^2 = 4\cos^2 2t = 4\,(1 - \sin^2 2t) = 4\,(1 - y^2)$$

$\Rightarrow$ $4x^2 + 4xy + 5y^2 = 4$, the desired result.

EXERCISE 9.5

Solve the following simultaneous equations

1. $\dfrac{dx}{dt} - 7x + y = 0, \quad \dfrac{dy}{dt} - 2x - 5y = 0.$

Ans. $x = e^{6t}(c_1 \cos t + c_2 \sin t),\ y = e^{6t}[(c_1 - c_2)\cos t + (c_1 + c_2)\sin t].$

2. $(D+5)x - 2y = t,\ 2x + (D+1)y = 0$; given that $x = 0 = y$ when $t = 0$.

Ans. $x = -\dfrac{1}{27}(1+6t)e^{-3t} + \dfrac{1}{27}(1+3t),\ y = -\dfrac{2}{27}(2+3t)e^{-3t} + \dfrac{2}{27}(2-3t).$

3. $\dfrac{dx}{dt} + \dfrac{dy}{dt} + 2x + y = 0, \quad \dfrac{dy}{dt} + 5x + 3y = 0.$ *(GGSIPU-2006)*

Ans. $x = c_1 \cos t + c_2 \sin t,\ y = \dfrac{1}{2}[(c_2 - 3c_1)\cos t - (c_1 + 3c_2)\sin t].$

4. $(D+5)x + y = e^t, \quad (D+3)y - x = e^{2t}.$

Ans. $x = e^{-4t}(c_1 + c_2 t) + \dfrac{4e^t}{25} - \dfrac{1}{36}e^{2t},\ y = -c_2 e^{-4t} - e^{-4t}(c_1 + c_2 t) + \dfrac{1}{25}e^t + \dfrac{7}{36}e^{2t}.$

5. $4\dfrac{dx}{dt} + 9\dfrac{dy}{dt} + 2x + 31y = e^t, \quad 3\dfrac{dx}{dt} + 7\dfrac{dy}{dt} + x + 24y = 3.$

Ans. $x = -\dfrac{e^{-4t}}{2}(c_1 \cos t + c_2 \sin t) - \dfrac{e^{-4t}}{2}(c_1 \sin t - c_2 \cos t) + \dfrac{31}{26}e^t - \dfrac{93}{17},$

$y = e^{-4t}(c_1 \cos t + c_2 \sin t) + \dfrac{6}{17} - \dfrac{2e^t}{13}.$

6. $4\dfrac{dx}{dt} + 9\dfrac{dy}{dt} + 44x + 49y = t, \quad 3\dfrac{dx}{dt} + 7\dfrac{dy}{dt} + 34x + 38y = e^t.$

Ans. $x = c_1 e^{-t} + c_2 e^{-6t} + \dfrac{19}{3}t - \dfrac{56}{9} - \dfrac{29}{7}e^t,\ y = -c_1 e^{-t} + 4c_2 e^{-6t} - \dfrac{17}{3}t + \dfrac{55}{9} + \dfrac{24}{7}e^t$

7. $\dfrac{dx}{dt} + \dfrac{dy}{dt} - 2y = 2\cos t - 7\sin t, \quad \dfrac{dx}{dt} - \dfrac{dy}{dt} + 2x = 4\cos t - 3\sin t.$ *(UPTU, 2001)*

Ans. $x = c_1 e^{\sqrt{2}t} + c_2 e^{-\sqrt{2}t} + 3\cos t,\ y = (\sqrt{2}+1)c_1 e^{\sqrt{2}t} + (1-\sqrt{2})c_2 e^{-\sqrt{2}t} + 2\sin t.$

8. $\dfrac{dx}{dt} = y, \quad \dfrac{dy}{dt} = -x,\ x(0) = 0,\ y(0) = 0.$ *(UPTU 2002)* **Ans.** $x(t) = 0 = y(t)$.

9. $\dfrac{dx}{dt} - \dfrac{dy}{dt} + 2y = \cos 2t, \quad \dfrac{dx}{dt} + \dfrac{dy}{dt} - 2x = \sin 2t.$ *(DCE, 2004)*

Ans. $x = e^t(c_1 \cos t + c_2 \sin t) - \dfrac{1}{2}\cos 2t,\ y = e^t(c_1 \sin t - c_2 \cos t) - \dfrac{1}{2}\sin 2t.$

10. $Dx + 2y = e^t,\ Dy - 2x = e^{-t}.$

Ans. $x = \dfrac{1}{5}e^t + \dfrac{2}{5}e^{-t} - c_1 \sin 2t + c_2 \cos 2t,\ y = \dfrac{2}{5}e^t + \dfrac{1}{5}e^{-t} + c_1 \cos 2t + c_2 \sin 2t.$

11. $(3D+1)x+3Dy=3t+1, \quad (D-3)x+Dy=2t.$ **Ans.** $x=\frac{(1-3t)}{10},\ y=\frac{11}{20}t^2+\frac{6}{10}t+c.$

12. $\frac{d^2x}{dt^2}+y=\sin t, \quad \frac{d^2y}{dt^2}+x=\cos t.$ (*UPTU 2004*)

Ans. $x=c_1e^t+c_2e^{-t}+c_3\cos t+c_4\sin t+\frac{t}{4}(\sin t-\cos t),$

$y=-c_1e^t-c_2e^{-t}+c_3\cos t+c_4\sin t+\frac{1}{4}(2+t)(\sin t-\cos t).$

13. $\frac{dx}{dt}+2x+3y=0, \quad 3x+\frac{dy}{dt}+2y=2e^{2t}.$ (*DU 2002*)

Ans. $x=c_1e^t+c_2e^{-5t}-\frac{6}{7}e^{2t},\ y=c_2e^{-5t}-c_1e^t+\frac{8}{7}e^{2t}.$

14. A mechanical system with two degrees of freedom satisfies the equations.

$$2\frac{d^2x}{dt^2}+3\frac{dy}{dt}=4, \quad 2\frac{d^2y}{dt^2}-3\frac{dx}{dt}=0.$$

Obtain expressions for x and y in terms of t, given x, y, dx/dt, dy/dt all vanish at $t=0$.

Ans. $x=\frac{8}{9}\left(1-\cos\frac{3t}{2}\right),\ y=\frac{4}{3}t-\frac{8}{9}\sin\frac{3}{2}t.$

15. $\frac{d^2x}{dt^2}+4x+5y=t^2, \quad \frac{d^2y}{dt^2}+5x+4y=t+1.$ (*DCE 2004*)

Ans. $x(t)=c_1e^t+c_2e^{-t}+c_3\cos 3t+c_4\sin 3t-\frac{1}{9}\left(4t^2-5t+\frac{37}{9}\right),$

$y(t)=-c_1e^t-c_2e^{-t}+c_3\cos 3t+c_4\sin 3t+\frac{1}{9}\left(5t^2-4t+\frac{44}{9}\right).$

16. $\frac{dx}{dt}+\frac{dy}{dt}+3x=\sin t, \quad \frac{dx}{dt}+y-x=\cos t.$ (*UPTU, 2003*)

Ans. $x=c_1e^{3t}+c_2e^{-t}+\frac{1}{5}(2\sin t-\cos t),\ y=\frac{1}{5}(2\cos t+\sin t)-2c_1e^{3t}+2c_2e^{-t}.$

17. $\frac{dx}{dt}+2x-3y=t, \frac{dy}{dt}-3x+2y=e^{2t}.$

Ans. $x=c_1e^{-5t}+c_2e^t+\frac{3}{7}e^{2t}-\frac{2t}{5}-\frac{13}{25},\ y=-c_1e^{-5t}+c_2e^t+\frac{4}{7}e^{2t}-\frac{3t}{5}-\frac{12}{25}.$

CHAPTER

10

Series Solution of Differential Equations

10.1 INTRODUCTION

In chapter 9 we discussed the solution of linear differential equations with constant coefficients. These equations have been solved by algebraic methods and the solutions have been obtained in terms of known functions which include exponential functions, trigonometric and hyperbolic functions, etc. However, in the case of differential equations with variable coefficients the problems become intricate as it is generally not possible to discuss their solutions in closed form. One has mostly resort to numerical methods for the solutions of such problems. There are, however, situations where we can discuss the solutions of such equations in the form of infinite power series when these series satisfy certain conditions. Bessel's equation, Legendre's equation and Hypergeometric equation, etc., are the class of equations where solutions have been discussed in the form of infinite power series. The series solutions methods can be classified into two categories

(*a*) power series method and

(*b*) general series solution method (*Frobenius method*).

10.2 POWER SERIES METHOD

The basic concept of power series method is simple and we will apply this technique to the solution of the first and the second order differential equations namely

$$P_0(x)y' + P_1(x)y = Q(x) \qquad ...(10.1)$$

and $$P_0(x)y'' + P_1(x)y' + P_2(x)y = 0, \qquad P_0(x) \neq 0. \qquad ...(10.2)$$

Dividing the equation (10.2) by $P_0(x)$, we can write it in the standard form as

$$y'' + p(x)y' + q(x)y = 0, \qquad ...(10.3)$$

where $$p(x) = \frac{P_1(x)}{P_0(x)} \quad \text{and} \quad q(x) = \frac{P_2(x)}{P_0(x)}.$$

To find the solution of (10.1) or (10.2), we assume a series for $y(x)$ of the form

$$y(x) = a_0 + a_1x + a_2x^2 + \ldots\ldots \qquad ...(10.4)$$

Substituting for $y(x)$, $y'(x)$ and $y''(x)$ from (10.4) in, say, (10.3) and rearranging the terms of different powers of x, we will obtain an algebraic equation of the type.

$$\lambda_0 + \lambda_1x + \lambda_2x^2 + \ldots\ldots = 0 \qquad ...(10.5)$$

Since (10.5) holds for all values of x identically, we obtain

$$\lambda_0 = 0, \ \lambda_1 = 0, \ \lambda_2 = 0 \ \ldots\ldots$$

From these equations we can determine the coefficients a_0, a_1, a_2, ..., etc.

EXAMPLE 10.1 Find the power series solution about $x = 0$, of the differential equation $y'(x) - y(x) = 2x - x^2$.

Solution Let the power series solution about the point $x = 0$ be

$$y(x) = a_0 + a_1x + a_2x^2 + a_3x^3 + \ldots\ldots + a_nx^n + \ldots\ldots$$

Differentiating, we obtain

$$y'(x) = a_1 + 2a_2x + 3a_3x^2 + \ldots\ldots + na_nx^{n-1} + \ldots\ldots$$

Substituting for $y(x)$ and $y'(x)$ in the given differential equation and rearranging the terms, we get

$$(a_1 - a_0) + (2a_2 - a_1 - 2)x + (3a_3 - a_2 + 1)x^2 + (4a_4 - a_3)x^3 + \ldots\ldots + \{(n+1)a_{n+1} - a_n\}\, x^n + \ldots\ldots = 0.$$

Equating the coefficients of various powers of x to zero, we obtain

$$a_1 = a_0, \quad 2a_2 = a_1 + 2 = a_0 + 2 \quad \text{or} \quad a_2 = (a_0/2) + 1,$$

$$a_3 = \frac{1}{3}(a_2 - 1) = \frac{1}{3}\left\{\frac{a_0}{2} + 1 - 1\right\} = \frac{a_0}{6} = \frac{a_0}{3!}, \text{ and so on .}$$

Substituting for a's in the expression for $y(x)$, we obtain

$$y(x) = a_0 + a_0 x + \frac{1}{2}(a_0 + 2)x^2 + \frac{a_0}{3!}x^3 + \ldots\ldots = a_0\left(1 + x + \frac{x^2}{2} + \frac{x^3}{3!} + \ldots\ldots\right) + x^3.$$

The closed solution of the above equation is $y = ce^x + x^2$.

We find that the two solutions are exactly equal.

EXAMPLE 10.2 Find the power series solution about $x = 0$, of the differential equation

$$(1 - x^2)\, y'' - 2x\, y' + 2y = 0.$$

Solution Assume the power series solution be

$$y(x) = a_0 + a_1 x + a_2 x^2 + a_3 x^3 + \ldots\ldots \tag{A}$$

We have

$$y'(x) = a_1 + 2a_2 x + 3a_3 x^2 + 4a_4 x^3 + \ldots\ldots,$$

$$y''(x) = 2a_2 + 6a_3 x + 12a_4 x^2 + \ldots\ldots,$$

Substituting for $y(x)$, $y'(x)$ and $y''(x)$ in the given differential equation and simplifying, we get

$$2(a_0 + a_2) + 6a_3 x + 4(3a_4 - a_2)x^2 + 10(2a_5 - a_3)x^3 + \ldots\ldots = 0$$

Equating the coefficients of various powers of x to zero, we obtain

$$a_2 = -a_0; \quad a_3 = 0; \quad a_4 = a_2/3 = -a_0/3; \quad a_5 = 0 \ldots\ldots$$

Substituting for a's in (A), the power series solution is

$$y(x) = a_0\left(1 - x^2 - \frac{x^4}{3} - \ldots\ldots\right) + a_1 x, \text{ the required solution.}$$

EXERCISE 10.1

Find the series solutions of the following differential equations.

1. $y' = 2xy.$ **Ans.** $y = a_0\left(1 + x^2 + \dfrac{x^4}{2!} + \dfrac{x^6}{3!} + \ldots\ldots\right) = a_0 e^{x^2}.$

2. $(1 + x)\, y' - (2x + 3)\, y = 0.$ **Ans.** $y = a_0\left(1 + 3x + 4x^2 + \dfrac{10}{3}x^3 + 2x^4 + \ldots\ldots\right).$

3. $(1 + x^2)y'' + xy' - y = 0.$ (*Delhi, 2002*) **Ans.** $y = a_0\left(1 + \dfrac{x^2}{2} - \dfrac{x^4}{8} + \ldots\ldots\right) + a_1 x.$

4. $y'' + x^2 y = 0.$

Ans. $y = a_0\left(1 - \dfrac{x^4}{3.4} - \dfrac{x^8}{3.4.7.8} - \dfrac{x^{12}}{3.4.7.8.11.12} + \ldots\ldots\right) + a_1\left(x - \dfrac{x^5}{4.5} + \dfrac{x^9}{4.5.8.9} - \dfrac{x^{13}}{4.5.8.9.12.13} + \ldots\ldots\right).$

5. $y'' + xy = 0.$ (*AMIETE, June, 2010, 2009*)

Ans. $y = a_0\left(1 - \frac{1^2}{3!}x^3 + \frac{1.4}{6!}x^6 - \frac{1.4.7}{9!}x^9 + \ldots\ldots\right) + a_1\left(x - \frac{2}{4!}x^4 + \frac{2.5}{7!}x^7 - \frac{2.5.8}{10!}x^{10} + \ldots\ldots\right).$

6. $y'' - xy' + y = 0.$ **Ans.** $y = a_0\left(1 - \frac{x^2}{2!} - \frac{1}{4!}x^4 - \frac{1.3}{6!}x^6 \ldots\ldots\right) + a_1 x.$

7. Find the power series solution about the point $x_0 = 1$ of the differential equation $xy' = 1 - x + 2y.$

Ans. $y(x) = a_0\,[1 + 2(x-1) + (x-1)^2] + \frac{1}{2} + (x-1).$

10.3 VALIDITY OF THE POWER SERIES METHODS

Let us consider a differential equation

$$x^2\frac{d^2y}{dx^2} + (x^2 - x)\frac{dy}{dx} + 2y = 0 \qquad \ldots(10.6)$$

If we assume a solution of the form

$$y = a_0 + a_1x + a_2x^2 + a_3x^3 + \ldots\ldots$$

and solve the equation by the method discussed above, we find that

$$a_0 = 0,\ a_1 = 0,\ a_2 = 0\ \ldots\ldots$$

and hence there is no series satisfying the given equation. The question that now arises on the basis of above conclusions is whether we straightway reject that the above equation has no series solution and if it is not so then what should be the conditions under which the above equation admits of the series solution.

The following definitions and theorems help us in establishing the validity of the series methods.

Definitions

Regular functions *A function f (x) is said to be regular at x = 0, if it can be developed as a power series of the form*

$$f(x) = \sum_{k=0}^{\infty} a_k x^k$$

That is, power of x are positive integral power.

Ordinary point *A point x = 0 is said to be an ordinary point of the differential equation (10.3) if both p(x) and q(x) are regular functions at x = 0, that is, both can be expressed as power series.*

For the differential equation $(1 - x^2)y'' - 2xy' + 2y = 0$ ...... (A)

$$p(x) = -\frac{2x}{1 - x^2} = -2x\,(1 + x^2 + x^4 + \ldots\ldots)$$

$$q(x) = \frac{2}{1 - x^2} = 2\,(1 + x^2 + x^4 + \ldots\ldots)$$

Thus $x = 0$ is an ordinary point of the differential equation (A).

Singular point *When p(x) and q(x) defined (10.3) are not regular functions at x = 0 or they are both not finite at x = 0, then x = 0 is known as the singular point of the equation.* In equation (10.6)

$$p(x) = -\frac{x^2 - x}{x^2} \to \infty \quad \text{as} \quad x \to 0$$

$$q(x) = \frac{2}{x^2} \to \infty \quad \text{as} \quad x \to 0.$$

Both $p(x)$ and $q(x)$ are infinite at $x = 0$ and so $x = 0$ is a singular point of the equation.

Regular singular point *$x = 0$ is called a regular singular point of the equation if $xp(x)$ and $x^2q(x)$ tend to finite values at $x = 0$.* Equation (10.6) satisfies these conditions although $x = 0$ is a singular point but this point is a regular singularity.

Irregular singular point *A point which is not a regular singular point of the equation is known as an irregular singular point.*

On the basis of above definitions, the following theorems will be helpful in establishing usefulness of series solutions.

Theorem 1

If $x = 0$ is an ordinary point of the differential equation (10.2), that is, $P_0(x) \neq 0$ at $x = 0$, then the solution of (10.2) can be expressed as a series of the form

$$y = a_0 + a_1x + a_2x^2 + \ldots\ldots = \sum_{k=0}^{\infty} a_k x^k \qquad \ldots(10.7)$$

Theorem 2

If $x = 0$ is a regular singular point of (10.2), that is, $P_0(x) = 0$ and $xp(x)$ and $x^2q(x)$ are finite at $x = 0$ then at least one solution can be expressed in a series of the form

$$y = x^m (a_0 + a_1x + a_2x^2 + \ldots\ldots) = \sum_{k=0}^{\infty} a_k x^{m+k} \qquad \ldots(10.8)$$

m may be a positive or a negative integer or a fraction and is called the index of the series solution. This method of solution was suggested by George Frobenius (1849-1917) and is called the **Frobenius Method.**

It is also called extended power series method.

Theorem 3

The series given in (10.7) and (10.8) above are convergent at every point within the circle drawn in the complex plane with centre at $x = 0$ and radius $|R|$ where R is the singular point nearest the point $x = 0$.

Theorem 4

If $x = 0$ is an irregular singular point of the equation then the solution of the equation cannot be expressed in the form of series. For the equation

$$x^4 \frac{d^2y}{dx^2} + 2x\frac{dy}{dx} + y = 0,$$

$x = 0$ is an irregular singular point. It is not possible to discuss the solution of this equation in series of the form (10.6) although it has a solution

$$y = c_1 \cos\frac{1}{x} + c_3 \sin\frac{1}{x}.$$

The proofs of the above theorems are beyond the scope of present discussions.

ILLUSTRATIVE EXAMPLES

Case 1 *When x = 0 is an ordinary point*

EXAMPLE 10.3 Find the power series solution about $x = 2$, of the initial value problem

$$4y'' - 4y' + y = 0, \qquad y(2) = 0, \qquad y'(2) = 1/e.$$

Express the solution in the closed form. *(AMIETE, Dec, 2006)*

Solution Employing Taylor' series to determine the solution as

$$y(x) = y(2) + (x-2)\, y'(2) + \frac{(x-2)^2}{2!}\, y''(2) + \frac{(x-3)^3}{3!}\, y'''(2) + \ldots\ldots$$

From the given equation,

we have $\quad y'' = \frac{1}{4}(4y' - y).$ ...(*i*)

Differentiating (*i*) $(n-2)$ times, we obtain $y^{(n)} = \frac{1}{4}[4y^{(n-1)} - y^{(n-2)}]$

Setting $x = 2$, we obtain $y^{(n)}(2) = \frac{1}{4}[4y^{(n-1)}(2) - y^{(n-2)}(2)],\ n = 2, 3, \ldots\ldots$

Using the values $y(2) = 0$ and $y'(2) = 1/e$, we obtain

$$y''(2) = \frac{1}{4}[4y'(2) - y(2)] = \frac{1}{4}\left[\frac{4}{e} - 0\right] = \frac{1}{e},$$

$$y'''(2) = \frac{1}{4}[4y''(2) - y'(2)] = \frac{1}{4}\left[\frac{4}{e} - \frac{1}{e}\right] = \frac{3}{4e},$$

$$y^{iv}(2) = \frac{1}{4}[4y'''(2) - y''(2)] = \frac{1}{4}\left[\frac{3}{e} - \frac{1}{e}\right] = \frac{1}{2e},$$

$$y^{v}(2) = \frac{1}{4}[4y^{iv}(2) - y'''(2)] = \frac{1}{4}\left[\frac{2}{e} - \frac{3}{4e}\right] = \frac{5}{16e},$$

The solution is given by

$$y(x) = 0 + (x-2)\left(\frac{1}{e}\right) + \frac{(x-2)^2}{2!}\left(\frac{1}{e}\right) + \frac{(x-2)^3}{3!}\left(\frac{3}{4e}\right) + \frac{(x-2)^4}{4!}\left(\frac{1}{2e}\right) + \frac{(x-2)^5}{5!}\left(\frac{5}{16e}\right) + \ldots\ldots$$

$$= \frac{(x-2)}{e}\left[1 + \frac{\{(x-2)/2\}}{1!} + \frac{\{(x-2)/2\}^2}{2!} + \frac{\{(x-2)/2\}^3}{3!} + \frac{\{(x-2)/2\}^4}{4!} + \ldots\ldots\right]$$

$$= \frac{1}{e}(x-2)\, e^{(x-2)/2}.$$

This solution can be verified. The given equation in D operator form is $(4D^2 - 4D + 1)y = 0$. The characteristic equation of the homogeneous equation is

$4m^2 - 4m + 1 = 0$, or $(2m-1)^2 = 0$. Its roots are $m = ½, ½$.

The solution of the differential equation is

$y(x) = (c_1 + c_2 x)e^{x/2}$. Applying the initial conditions, we get

$$y(2) = 0 = (c_1 + 2c_2)\, e, \qquad \text{or} \qquad c_1 + 2c_2 = 0$$

$$y'(2) = \frac{1}{e} = (c_2)\, e + (c_1 + 2c_2)\frac{1}{2} e = \left(\frac{c_1}{2} + 2c_2\right) e \qquad \text{or} \qquad \frac{c_1}{2} + 2c_2 = \frac{1}{e^2}.$$

On solving, we get $c_1 = -2/e^2$ and $c_2 = 1/e^2$. Hence, we have

$$y(x) = \left(\frac{1}{e^2}x - \frac{2}{e^2}\right) e^{x/2} = \frac{1}{e^2}(x-2)\, e^{x/2} = \frac{1}{e}(x-2)\, e^{(x-2)/2}.$$

EXAMPLE 10.4 Find the first five non-vanishing terms in the power series solution of the initial value problem $(1 - x^2)y'' + 2xy' + y = 0$, $y(0) = 1$, $y'(0) = 1$. (*AMIETE, June 2010; Dec 2008, 2007*)

Solution Assume the power series solution about the point $x = 0$ be

$$y(x) = a_0 + a_1x + a_2x^2 + a_3x^3 + a_4x^4 + a_5x^5 + a_6x^6 + \ldots\ldots:$$

We have
$$y'(x) = a_1 + 2a_2x + 3a_3x^2 + 4a_4x^3 + 5a_5x^4 + 6a_6x^5 + \ldots\ldots$$

$$y''(x) = 2a_2 + 6a_3x + 12a_4x^2 + 20a_5x^3 + 30a_6x^4 + \ldots\ldots$$

Substituting for $y(x)$, $y'(x)$ and $y''(x)$ in the given differential equation, we obtain

$$(1 - x^2)[2a_2 + 6a_3x + 12a_4x^2 + 20a_5x^3] + 2x[a_1 + 2a_2x + 3a_3x^2 + 4a_4x^3 + \ldots\ldots]$$
$$+ [a_0 + a_1x + a_2x^2 + a_3x^3 + \ldots\ldots]$$

or
$$(a_0 + 2a_2) + 3(2a_3 + a_1)x + 3(4a_4 + a_2)x^2 + (20a_5)x^3 + a_3 = 0.$$

Equating the coefficients of various powers of x to zero, we obtain

$$a_2 = \frac{-1}{2}a_0, \quad a_3 = \frac{-a_1}{2}, \quad a_4 = \frac{-a_2}{4} = \frac{a_0}{8}, \quad a_5 = \frac{-a_3}{20} = \frac{a_1}{40},$$ where a_0 and a_1 are arbitrary constant.

The power series solution is

$$y(x) = a_0 + a_1x - \frac{1}{2}a_0x^2 - \frac{a_1}{2}x^3 + \frac{a_0}{8}x^4 + \frac{a_1}{40}x^5 + \ldots\ldots$$

$$= a_0\left(1 - \frac{1}{2}x^2 + \frac{1}{8}x^4 + \ldots\ldots\right) + a_1\left(x - \frac{x^3}{2} + \frac{x^5}{40} + \ldots\ldots\right)$$

The initial condition $y(0) = 1$ gives $a_0 = 1$ and $y'(0) = 1$ gives $a_1 = 1$.

The first five non-vanishing terms in the power series solution are

$$y(x) = 1 + x - \frac{1}{2}x^2 - \frac{1}{2}x^3 + \frac{1}{8}x^4.$$

EXAMPLE 10.5 Find the power series solution about $x = 0$, of the differential equation

$$(2 - x^2)y'' + 2xy' - 2y = 0.$$

Solution Since $x = 0$ is an ordinary point, we assume the solution in the form

$$y(x) = \sum_{k=0}^{\infty} a_k x^k.$$

Substituting for $y(x)$, $y'(x)$ and $y''(x)$ in the given equation, we get

$$(2 - x^2)\Sigma a_k k(k-1)x^{k-2} + 2x\Sigma a_k kx^{k-1} - 2\Sigma a_k x^k = 0.$$

Regrouping the terms, we obtain

$$2\sum_{k=2}^{\infty} a_k k(k-1)x^{k-2} = \sum_{k=0}^{\infty} a_k [k(k-1) - 2k + 2]x^k = 0$$

or
$$\sum_{k=0}^{\infty} [2a_{k+2}(k+2)(k+1) - a_k(k^2 - 3k + 2)]x^k = 0.$$

Equating to zero, the coefficient of x^k, the recurrence relation is given by

$$a_{k+2} = \frac{(k-1)(k-2)}{2(k+1)(k+2)}a_k.$$

This gives
$$a_2 = \frac{2}{2.1.2}a_0 = \frac{a_0}{2}, \quad a_3 = 0; \quad a_4 = 0; \quad a_5 = 0; \quad a_6 = 0\ldots\ldots$$

This shows that all the coefficients beyond a_2 are zero. Hence the solution is given by

$$y = a_0 + a_1 x + a_2 x^2 = a_0\left(1+\frac{x^2}{2}\right) + a_1 x.$$

Since this involves two arbitrary constants a_0 and a_1, this is the general solution.

EXAMPLE 10.6 Find the power series solution about $x = 0$, of the differential equation $y'' - xy' + x^2y = 0$.

Solution As $x = 0$ is an ordinary point of the equation, we assume a solution of the form

$$y(x) = \sum_{k=0}^{\infty} a_k x^k.$$

Substituting for $y(x)$, $y'(x)$ and $y''(x)$ in the given equation, we get

$$\sum_{k=2}^{\infty} a_k\, k\,(k-1)\, x^{k-2} - \sum_{k=0}^{\infty} a_k\, kx^k + \sum_{k=0}^{\infty} a_k\, x^{k+2} = 0$$

or

$$\sum_{k=0}^{\infty} \{a_{k+2}\,(k+2)\,(k+1) - ka_k\}x^k + \sum_{k=0}^{\infty} a_k\, x^{k+2} = 0.$$

Equating to zero, the coefficient of x^{k+2}, the recurrence relation is given by

$$a_{k+4}\,(k+4)\,(k+3) - a_{k+2}\,(k+2) + a_k = 0$$

which gives

$$12a_4 - 2a_2 + a_0 = 0,$$
$$20a_5 - 3a_3 + a_1 = 0,$$
$$30a_6 - 4a_4 + a_2 = 0.$$

From the summation $\sum_{k=0}^{\infty} [a_{k+2}\,(k+2)\,(k+1) - k\, a_k]\,x^k$, we get $2a_2 = 0$, when $k = 0$.

This gives $\quad a_2 = 0;\quad 6a_3 - a_1 = 0 \quad$ or $\quad a_3 = a_1/6$, (when $k = 1$.)

$\therefore \quad 12a_4 = -a_0$ or $a_4 = -\dfrac{a_0}{12}$ when $k = 2$.

$$20a_5 = 3a_3 - a_1 = \frac{a_1}{2} - a_1 = -\frac{a_1}{2} \quad \text{or} \quad a_5 = -\frac{a_1}{40}.$$

$$30a_6 = 4a_4 \quad \text{or} \quad a_6 = \frac{4a_4}{30} = -\frac{4}{30}\cdot\frac{a_0}{12} = -\frac{a_0}{90}.$$

Hence the solution of the equation is given by

$$y = a_0 + a_1 x + a_2 x^2 + a_3 x^3 + a_4 x^4 + a_5 x^5 + a_6 x^6 + \ldots\ldots$$

$$= a_0\left(1 - \frac{x^4}{12} - \frac{x^6}{90} - \ldots\right) + a_1\left(x + \frac{x^3}{6} - \frac{x^5}{40} + \ldots\right)$$

Case 2 *When x = 0 is a regular singular point. (Frobenius Method)*

Since $x = 0$ is a regular singular point of the equation we assume a solution in series of the form

$$y(x) = \sum_{k=0}^{\infty} a_k x^{m+k}.$$

Working rule for the solution of equation (10.2)

1. Substitute for $y(x)$, $y'(x)$ and $y''(x)$ in the given differential equation.

2. Rearrange the terms in powers of x and equate to zero the coefficient of lowest power of x. This gives us a quadratic equation in m which is called the ***indicial equation***. The two roots of the indicial equation determine two solutions.
3. Equate to zero the coefficients of various powers of x to determine a_0, a_1, a_2, ... The coefficient of x^{m+k} equated to zero is called the **recurrence relation**.
4. Substitute for the values of a_0, a_1, ... in the assumed solution of y to get the solution of the problem.

EXAMPLE 10.7 Find the power series solution about the point $x_0 = 2$ of the equation

$$y'' + (x-1)y' + y = 0. \qquad \textit{(AMIETE, Dec 2009, June 2007)}$$

Solution Setting $x = v + 2$ in the given equation, we obtain

$$\frac{d^2y}{dv^2} + (v+1)\frac{dy}{dv} + y = 0.$$

Assume the series to be of the form

$$y = a_0 + a_1 v + a_2 v^2 + a_3 v^3 + a_4 v^4 + \ldots\ldots + a_n v^n + \ldots\ldots .$$

Then

$$dy/dv = a_1 + 2a_2 v + 3a_3 v^2 + 4a_4 v^3 + \ldots\ldots + na_n v^{n-1} + \ldots\ldots,$$

$$d^2y/dv^2 = 2a_2 + 6a_3 v + 12a_4 v^2 + \ldots\ldots + n(n-1)a_n v^{n-2} + \ldots\ldots,$$

and

$$\frac{d^2y}{dv^2} + (v+1)\frac{dy}{dv} + y = (2a_2 + a_1 + a_0) + (6a_3 + 2a_1 + 2a_2)v + (12a_4 + 3a_2 + 3a_3)v^2$$
$$+ \ldots\ldots + [(n+2)(n+1)a_{n+2} + (n+1)a_n + (n+1)a_{n+1}]v^n + \ldots\ldots = 0.$$

Equating the coefficient of powers of v to zero, we obtain

$$a_2 = -(a_0 + a_1)/2, \quad a_3 = -\frac{1}{3}(a_1 + a_2) = \frac{1}{6}(a_0 - a_1), \quad a_4 = -\frac{1}{4}(a_2 + a_3) = \frac{1}{12}(a_0 + 2a_1), \ldots\ldots$$

$$(n+2)(n+1)a_{n+2} + (n+1)a_n + (n+1)a_{n+1} = 0 \text{ and } a_{n+2} = -\frac{1}{n+2}(a_n + a_{n+1}).$$

Thus, noting that $v = x - 2$, the power series solution is

$$y(x) = a_0 + a_1(x-2) + a_2(x-2)^2 + a_3(x-2)^3 + a_4(x-2)^4 + \ldots\ldots$$

$$= a_0 + a_1(x-2) - \frac{1}{2}(a_0 + a_1)(x-2)^2 + \frac{1}{6}(a_0 - a_1)(x-2)^3 + \frac{1}{12}(a_0 + 2a_1)(x-2)^4 + \ldots\ldots$$

$$= a_0\left[1 - \frac{1}{2}(x-2)^2 + \frac{1}{6}(x-2)^3 + \frac{1}{12}(x-2)^4 - \ldots\ldots\right]$$

$$+ a_1\left[(x-2) - \frac{1}{2}(x-2)^2 - \frac{1}{6}(x-2)^3 + \frac{1}{6}(x-2)^4 - \ldots\ldots\right].$$

EXAMPLE 10.8 Find the series solution about $x = 0$, of the differential equation

$$2x^2 y'' - xy' + (x^2 + 1)y = 0.$$

Solution Since $x = 0$ is a regular singular point, we assume the solution in the form

$$y(x) = \sum_{k=0}^{\infty} a_k x^{m+k}.$$

Substituting for $y(x)$, $y'(x)$ and $y''(x)$ in the given equation, and rearranging the terms in powers of x, we get

$$\sum a_k [2(m+k)(m+k-1) - (m+k) + 1]x^{m+k} + \sum a_k x^{m+k+2} = 0$$

Equating the coefficient of x^m to zero, the indicial equation is written as

$$a_0[2m(m-1) - (m-1)] = 0.$$

Since a_0 is arbitrary and, therefore, $\neq 0$, this gives us $m = 1, 1/2$.

Equating the coefficient of x^{m+1} to zero, we get $a_1[2m(m+1) - m] = 0$.

Since the quantity within the brackets is not zero for any of the above values of m, this gives $a_1 = 0$.

Equating the coefficient of x^{m+k} to zero, the recurrence relation is given by

$$a_k = \frac{1}{(m+k-1)(2m+2k-1)} a_{k-2}, \quad k = 2, 3, \ldots\ldots$$

Since $a_1 = 0$, the above relation gives $a_3 = a_5 = \ldots\ldots = 0$

Also
$$a_2 = -\frac{1}{(m+1)(2m+3)} a_0$$

$$a_4 = -\frac{1}{(m+3)(2m+7)} a_2 = \frac{a_0}{(m+1)(m+3)(2m+3)(2m+7)}.$$

For $m = 1$,
$$a_2 = -\frac{1}{2.5} a_0, \qquad a_4 = \frac{1}{2.5.4.9} a_0$$

For $m = ½$,
$$a_2 = -\frac{1}{(3/2).4} a_0 = -\frac{a_0}{6}, \qquad a_4 = \frac{1}{(3/2).4.(7/2)} a_0 = \frac{a_0}{168}.$$

Hence, the solution of the equation is

$$y = Ax\left(1 - \frac{x^2}{10} + \frac{x^4}{360} + \ldots\ldots\right) + Bx^{1/2}\left(1 - \frac{x^2}{6} + \frac{x^4}{168} + \ldots\ldots\right),$$

where A and B are arbitrary constants.

EXAMPLE 10.9 Find the power solution about the origin of the differential equation

$$2x(1-x)y'' + (5-7x)y' - 3y = 0.$$

Solution We find that $x = 0$ is a regular singular point of the equation and we assume the solution in the from

$$y(x) = \sum_{k=0}^{\infty} a_k x^{m+k}.$$

Substituting for $y(x)$, $y'(x)$, $y''(x)$ and rearranging the terms, we get

$$\sum a_k[(m+k)(2m+2k+3)]x^{m+k-1} - \sum a_k(m+k+1)(2m+2k+3)x^{m+k} = 0$$

Equating to zero the coefficient of x^{m-1}, the indicial equation is written as

$$a_0 m(2m+3) = 0.$$

Since $a_0 \neq 0$,
$$m = 0 \quad \text{or} \quad -3/2.$$

Equating to zero the coefficient of x^{m+k}, the recurrence relation is given by

$$a_{k+1} = \frac{(m+k+1)(2m+2k+3)}{(m+k+1)(2m+2k+5)} a_k \quad \text{or} \quad a_{k+1} = \frac{2m+2k+3}{2m+2k+5} a_k,$$

This gives
$$a_1 = \frac{2m+3}{2m+5} a_0, \qquad a_2 = \frac{2m+5}{2m+7} a_1 = \frac{2m+3}{2m+7} a_0, \qquad a_3 = \frac{2m+7}{2m+9} a_2 = \frac{2m+3}{2m+9} a_0.$$

For $m = 0$,
$$a_1 = \frac{3}{5} a_0; \quad a_2 = \frac{3}{7} a_0; \quad a_3 = \frac{3}{9} a_0 \ldots\ldots$$

Hence for $m = 0$,
$$y_1 = a_0\left(1 + \frac{3}{5}x + \frac{3}{7}x^2 + \frac{3}{9}x^3 + \ldots\ldots\right)$$

For $m = -3/2$, $\quad a_1 = 0$, and therefore, $a_2 = 0 = a_3 = a_4 \ldots\ldots$

Hence, for $m = -3/2$, the second solution is $y_2 = a_0 x^{-3/2}$.

Thus the complete solution is $y = Ay_1 + By_2$.

Thus if the roots of indicial equation differ by a quantity not an integer, we get two independent solution by substituting these values of m in the series for y.

EXAMPLE 10.10 Find the series solution about $x = 0$, of the differential equation $(x - x^3)\, y'' + (1 - 5x)\, y' - 4y = 0$, by the Frobenius method.

Solution We find that $x = 0$ is a regular singular point of the equation. We assume the solution in the form

$$y(x) = \sum_{k=0}^{\infty} a_k x^{m+k}.$$

Substituting for $y(x)$, $y'(x)$ and $y''(x)$ in the given equation and rearranging the terms, we get

$$\sum a_k\,(m+k)^2\, x^{m+k-1} - \sum a_k\,(m+k+2)^2\, x^{m+k} = 0$$

Setting the coefficient of the lowest power to x to zero, the indicial equation is given by

$$a_0 m^2 = 0, \text{ which gives } m = 0, 0.$$

Since both the values of m are equal, this gives us only one independent solution.

Equating to zero the coefficient of x^{m+k}, we get

$$a_{k+1} = \frac{(m+k+2)^2}{(m+k+1)} a_k$$

This gives

$$a_1(m+1)^2 = (m+2)^2 a_0,$$
$$a_2(m+2)^2 = (m+3)^2 a_1,$$
$$a_3(m+3)^2 = (m+4)^2 a_2,$$
$$\cdots \qquad \cdots$$
$$\cdots \qquad \cdots$$

Hence the solution is given by

$$y_m = a_0 x^m \left[1 + \frac{(m+2)^2}{(m+1)^2} x + \frac{(m+3)^2}{(m+1)^2} x^2 + \frac{(m+4)^2}{(m+1)^2} x^3 + \ldots\ldots\right].$$

For $m = 0$, this gives us only one solution. To get the second solution, we proceed as follow.

The left hand side of the given differential equation after substituting the value of $y(x)$ gives

$$(x - x^2)\, y'' + (1 - 5x)\, y' - 4y = \sum a_k\,(m+k)^2\, x^{m+k-1} - \sum a_k\,(m+k+2)^2\, x^{m+k}$$

$$= a_0 m^2 x^{m-1} + a_1(m+1)^2 x^m - a_0(m+2)^2 x^m + a_2(m+2)^2 x^{m+1} - a_1(m+3)^2 x^{m+1}$$
$$+ a_3(m+3)^2 x^{m+2} - a_2(m+4)^2 x^{m+2} + \ldots\ldots$$

$$= a_0 m^2\, x^{m-1}.$$ [all other terms are zero due to recurrence relation.]

Now differentiate both sides of the equation partially w.r.t. m, we get

$$(x - x^2)\frac{d^2}{dx^2}\left(\frac{\partial y}{\partial m}\right) + (1 - 5x)\frac{d}{dx}\left(\frac{\partial y}{\partial m}\right) - 4\frac{\partial y}{\partial m} = a_0[2m\, x^{m-1} + m^2\, x^{m-1} \log x].$$

If $m = 0$, we see that the right side is zero.

Hence $\left.\dfrac{\partial y}{\partial m}\right|_{m=0}$ is also a solution of the equation.

Thus to get the second solution, we differentiate y_m w.r.t. m and set $m = 0$ in the differentiated result.

Now
$$\frac{\partial y_m}{\partial m} = a_0 x^m \log x \left(1 + \frac{(m+2)^2}{(m+1)^2} x + \frac{(m+3)^2}{(m+1)^2} x^2 + \ldots\ldots\right)$$
$$+ a_0 x^m \left[\frac{(m+2)^2}{(m+1)^2}\left(\frac{2}{m+2} - \frac{2}{m+1}\right) x + \frac{(m+3)^2}{(m+1)^2}\left(\frac{2}{m+3} - \frac{2}{m+1}\right) x^2 + \ldots\ldots\right]$$

Therefore, $y_m\big|_{m=0} = a_0\,(1+2^2x+3^2x^2+4^2x^3+5^2x^4+\ldots\ldots)$

$$\left.\frac{\partial y_m}{\partial m}\right|_{m=0} = a_0 \log x\,(1+2^2x+3^2x^2+4^2x^3+\ldots\ldots)+a_0\left[2^2\,(1-2)x+3^2\left(\frac{2}{3}-\frac{2}{1}\right)x^2+\ldots\ldots\right]$$

Hence the complete solution is

$$y = A\,y_m\Big|_{m=0} + B\left.\frac{\partial y_m}{\partial m}\right|_{m=0}.$$

Thus we find that when the roots of the indicial equation are equal, each equal to say α, the solution of the differential equation can be obtained by the following rule.

Let y_m be the series which satisfies all the recurrence relation except the indicial equational; the two solutions are given by

$$y_1 = y_m\Big|_{m=\alpha} \quad \text{and} \quad y_2 = \left.\frac{\partial y_m}{\partial m}\right|_{m=\alpha}.$$

The complete solution is $y = Ay_1 + By_2$.

EXAMPLE 10.11 Solve in series: $x(1-x)\,y''-3x\,y'-y=0$.

Solution Since $x = 0$ is a regular singular point, we assume the solution

$$y(x) = \sum_{k=0}^{\infty} a_k x^{m+k}.$$

Substituting for $y(x)$, $y'(x)$ and $y''(x)$ and rearranging the terms in power of x, we obtain

$$x(1-x)\sum a_k\,(m+k)\,(m+k-1)\,x^{m+k-2}-3\sum a_k\,(m+k)\,x^{m+k}-\sum a_k\,x^{m+k}=0$$

or

$$\sum a_k\,(m+k)\,(m+k-1)\,x^{m+k-1}-\sum a_k\,(m+k+1)^2\,x^{m+k}=0.$$

Equating to zero the coefficient of lowest power of x, the indicial equation is written as

$$a_0 m\,(m-1) = 0.$$

Since a_0 is arbitrary, this gives $m = 0$ or 1.

Equating the coefficient of x^{m+k} to zero, we get

$$a_{k+1} = \frac{(m+k+1)^2}{(m+k)\,(m+k+1)}\,a_k = \frac{m+k+1}{m+k}\,a_k.$$

This gives

$$a_1 = \frac{m+1}{m}\,a_0, \quad a_2 = \frac{m+2}{m+1}\,a_1 = \frac{m+2}{m}\,a_0$$

$$a_3 = \frac{m+3}{m+2}\,a_2 = \frac{m+3}{m}\,a_0, \text{ and so on.}$$

For a value of $m = 1$, the solution of the equation is

$$y_1 = a_0\,x(1+2x+3x^2+\ldots\ldots)$$

For $m = 0$, we find a_1, a_2, ... are all infinite and, therefore, the general solution is

$$y_m = a_0x^m\left(1+\frac{m+1}{m}\,x+\frac{m+2}{m}\,x^2+\ldots\ldots\right).$$

To get the second solution, we multiply both sides of y by $(m-0)$ and differentiate the result w.r.t. m.

This gives $\dfrac{\partial}{\partial m}(my_m) = a_0x^m \log x[m+(m+1)x+(m+2)x^2+\ldots\ldots]+a_0x^m\,(1+x+x^2+\ldots)$

Putting $m = 0$, we get the second solution as,

$$y_2 = a_0 \log x\,(x + 2x^2 + 3x^3 + ...) + a_0\,(1 + x + x^2 + ...).$$

Hence the general solution of the equation is given by

$$y = Ay_1 + By_2.$$

Thus in general, if the roots α and β of the indicial equation differ by an integer (α > β), the two solutions of the differential equation are given by

$$y_1 = y_m\big|_{m=\alpha} \text{ and } y_2 = \left[\frac{\partial}{\partial m}\{(m-\beta)\,y_m\}\right]_{m=\beta},$$

where y_m is the series solution satisfying all the recurrence relations except the indicial equation.

The complete solution is $y = Ay_1 + By_2$.

Remarks

1. We should note that when roots of indicial equation are equal then one solution is of the form $\Sigma\, a^k x_{m+k}$ which second solution must contain a logarithmic term.
2. When the roots differ by an integer one solution is of the form $\Sigma\, a^k x^{m+k}$ while second may or may not contain a logarithmic term as is clear from the following examples.

EXAMPLE 10.12 Solve by extended series method:

$$x(1 - x)\,y'' - 3\,y' + 2y = 0.$$

Solution Since $x = 0$ is a regular singular point, we assume the solution in the form

$$y(x) = \sum_{k=0}^{\infty} a_k x^{m+k}.$$

Substituting for $y(x)$, $y'(x)$ and $y''(x)$ in the given equation and rearranging the terms, we have

$$\sum (m+k)(m+k-4)\,a_k x^{m+k-1} - \sum (m+k-2)(m+k+1)\,a_k\, x^{m+k} = 0.$$

Equating to zero the coefficient of lowest degree term (that is, x^{m-1}), we get the indicial equation

$m\,(m - 4)\,a_0 = 0$, which gives $m = 0, 4$ ($\because a_0$ is arbitrary).

Equating to zero the coefficient of x^{m+k}, we have $a_{k+1} = \dfrac{m+k-2}{m+k-3}\,a_k$.

Therefore, $a_1 = \dfrac{m-2}{m-3}a_0$, $a_2 = \dfrac{m-1}{m-2}a_1$, $a_3 = \dfrac{m}{m-3}a_0$,

$a_4 = \dfrac{m+1}{m}a_3$, $a_5 = \dfrac{m+2}{m+1}a_4$, $a_6 = \dfrac{m+3}{m+2}a_5$ etc

For $m = 0$, $a_1 = \dfrac{2}{3}a_0$, $a_2 = \dfrac{1}{2}a_1 = \dfrac{1}{3}a_0$, $a_3 = 0$, a_4 is indeterminate = c(say).

$\therefore$ $a_5 = 2c$, $a_6 = 3c$ etc.

Hence $y = a_0\left(1 + \dfrac{2}{3}x + \dfrac{1}{3}x^2\right) + cx^4\,(1 + 2x + 3x^2 + 4x^3 + ...)$.

EXAMPLE 10.13 Find the power series solution about the origin, of the differential equation

$$x^2y'' + 6\,xy' + (6 + x^2)y = 0.$$ *(AMIETE, June 2008, Dec 2005)*

Solution The point $x = 0$ is a regular singular point of the equation. Let the solution be written as

$$y(x) = \sum_{k=0}^{\infty} a_k x^{m+k}.$$

Differentiating, we obtain

$$y'(x) = \sum_{k=0}^{\infty} (m+k) a_k x^{m+k-1}, \quad y''(x) = \sum_{k=0}^{\infty} (m+k)(m+k-1) a_k x^{m+k-2}.$$

Substituting in the differential equation, we get

$$\sum_{k=0}^{\infty} (m+k)(m+k-1) a_k x^{m+k} + 6\sum_{k=0}^{\infty} (m+k)\, a_k x^{m+k} + 6\sum_{k=0}^{\infty} a_k x^{m+k} + \sum_{k=0}^{\infty} a_k x^{m+k+2} = 0.$$

or $$\sum_{k=0}^{\infty} [(m+k)(m+k+5)+6]\, a_k x^{m+k} + \sum_{k=0}^{\infty} a_k x^{m+k+2} = 0.$$

The lowest degree term is the term containing x^m. Equating the coefficient of x^m to zero, the indicial equation is written as

$[m\,(m+5)+6]\, a_0 = 0$ or $(m+2)\,(m+3) = 0$, since a_0 is arbitrary and therefore not equal to zero. This gives us indicial roots $m = -2, -3$.

Equating the coefficient of x^{m+1} to zero we obtain

$$[(m+1)\,(m+6)+6]\; a_1 = 0.$$

For $m = -2$, $a_1 = 0$ and for $m = -3$, a_1 is arbitrary. Therefore, the indicial root $m = -3$ gives the complete solution.

Equating the coefficient of x^{m+k} to zero, the recurrence relation is given by

$$a_k = -\frac{1}{[(m+k)\,(m+k+5)+6]}\, a_{k-2}, \; k = 2, 3, \ldots\ldots$$

We have $$a_2 = -\frac{1}{(m+2)\,(m+7)+6}\, a_0, \quad a_3 = -\frac{a}{(m+3)\,(m+8)+6},$$

$$a_4 = -\frac{a_2}{(m+4)\,(m+9)+6}, \quad a_5 = -\frac{a_3}{(m+5)\,(m+10)+6}, \ldots\ldots$$

For $m = -3$, we get

$$a_2 = -a_0/2, \; a_3 = -a_1/6, \; a_4 = -\frac{a_2}{12} = \frac{a_0}{24}, \; a_5 = -\frac{a_3}{20} = \frac{a_1}{120}, \ldots\ldots$$

The solution is given by

$$y(x) = x^{-3}\left[a_0\left(1 - \frac{x^2}{2!} + \frac{x^4}{4!} - \ldots\ldots\right) + a_1\left(x - \frac{x^3}{3!} + \frac{x^5}{5!} - \ldots\ldots\right)\right] = a_0 y_1(x) + a_1 y_2(x).$$

For $m = -2$, we get

$$a_1 = 0, \; a_2 = -a_0/6, \; a_3 = -\frac{a_1}{12} = 0, \; a_4 = -\frac{a_2}{20} = \frac{a_0}{120}, \ldots\ldots$$

Therefore, the solution is

$$y(x) = a_0 x^{-2}\left[1 - \frac{x^2}{3!} + \frac{x^4}{5!} - \ldots\ldots\right] = a_0 x^{-3}\left[x - \frac{x^3}{3!} + \frac{x^5}{5!} - \ldots\ldots\right] = a_0 y_2(x).$$

We find that the indicial root $m = -2$ produced a linearly dependent solution.

EXAMPLE 10.14 Find the power series solution about the origin, of the equation

$$xy'' + y' + xy = 0 \quad \ldots(i)$$

(AMIETE, Dec 2009, June 2002, SVTU., 2007)

Solution Taking $y = a_0 x^m + a_1 x^{m+1} + a_2 x^{m+2} + \ldots\ldots$

We get on substitution

$$x\{m(m-1)a_0x^{m-2}+(m+1)m\,a_1x^{m-1}+(m+2)(m+1)a_2x^m+\ldots\ldots\}$$

$$+\{m\,a_0x^{m-1}+(m+1)a_1x^m+(m+2)a_2x^{m+1}+\ldots\ldots\}+x\{a_0x^m+a_1x^{m+1}+\ldots\ldots\}=0.$$

or $$m^2a_0x^{m-1}+(m+1)^2a_1x^m+\{(m+2)^2a_2+a_0\}x^{m+1}+\ldots\ldots=0.$$

The lowest power of x is x^{m-1}. Its coefficient equated to zero gives $m^2a_0=0$ as $a_0\neq 0$ the indicial equation gives two equal roots $m=0$.

The coefficients of x^m, x^{m+1}, ... equated to zero give

$$(m+1)^2a_1=0,\quad (m+2)^2a_2+a_0=0,\quad (m+3)^2a_3+a_1=0, \text{ etc.}$$

The coefficients a_1, a_3, a_5, ... are all zero.

Also, $$a_2=-\frac{a_0}{(m+2)^2},\qquad a_4=-\frac{a_2}{(m+4)^2}=\frac{a_0}{(m+2)^2(m+4)^2}\text{ etc.}$$

Therefore, $$y_m=a_0x^m\left[1-\frac{x^2}{(m+2)^2}+\frac{x^4}{(m+2)^2(m+4)^2}-\ldots\ldots\right].\qquad \ldots(ii)$$

On putting $m=0$, this gives only one solution of equation (*i*), that is

$$y_1=a_0\left[1-\frac{x^2}{2^2}+\frac{x^4}{2^2.4^2}-\frac{x^6}{2^2.4^2.6^2}+\ldots\ldots\right].\qquad \ldots(iii)$$

To obtain a second solution, we partially differentiate (*ii*) with respect to m and then put $m=0$.

$$\frac{\partial y_m}{\partial m}=a_0x^m\log x\left[1-\frac{x^2}{(m+2)^2}+\frac{x^4}{(m+2)^2(m+4)^2}-\ldots\ldots\right]$$

$$+a_0x^m\frac{\partial}{\partial m}\left[1-\frac{x^2}{(m+2)^2}+\frac{x^4}{(m+2)^2(m+4)^2}-\ldots\ldots\right].\qquad \ldots(iv)$$

Let $t=\dfrac{1}{(m+2)^2}$. Therefore $\log t=-2\log(m+2)$.

Differentiating, $$\frac{1}{t}\frac{\partial t}{\partial m}=\frac{-2}{m+2}.$$

Therefore, $$\frac{\partial t}{\partial m}=\frac{-2t}{m+2}=\frac{-2}{m+2}\cdot\frac{1}{(m+2)^2}.\text{ That is, }\frac{\partial}{\partial m}\left\{\frac{1}{(m+2)^2}\right\}=\frac{1}{(m+2)^2}\cdot\frac{-2}{m+2}.$$

Similarly, $$\frac{\partial}{\partial m}\left\{\frac{1}{(m+2)^2(m+4)^2}\right\}=\frac{1}{(m+2)^2(m+4)^2}\left\{\frac{-2}{m+2}-\frac{2}{m+4}\right\}\text{ and so on.}$$

Hence (*iv*) becomes $$\frac{\partial y_m}{\partial m}=a_0x^m\log x\left[1-\frac{x^2}{(m+2)^2}+\frac{x^4}{(m+2)^2(m+4)^2}-\ldots\ldots\right]$$

$$+a_0x^m\left[0+\frac{x^2}{(m+2)^2}\cdot\frac{2}{m+2}-\frac{2x^4}{(m+2)^2(m+4)^2}\left(\frac{1}{m+2}+\frac{1}{m+4}\right)+\ldots\ldots\right]\qquad \ldots(v)$$

Putting $m=0$, the second solution is $y_2=\left(\dfrac{\partial y_m}{\partial m}\right)_{m=0}$

$$= a_0 \log x \left[1 - \frac{x^2}{2^2} + \frac{x^4}{2^2.4^2} - \ldots\ldots\right] + a_0 \left[\frac{x^2}{2^2} - \frac{x^4}{2^2.4^2}\left(1+\frac{1}{2}\right) + \frac{x^6}{2^2.4^2.6^2}\left(1+\frac{1}{2}+\frac{1}{3}\right) - \ldots\ldots\right] \quad \ldots(vi)$$

Combining (*iii*) and (*vi*), the complete solution can be given a

$$y = c_1 y_1 + c_2 y_2$$

$$= c_1 a_0 \left[1 - \frac{x^2}{2^2} + \frac{x^4}{2^2.4^2} - \frac{x^6}{2^2.4^2.6^2} + \ldots\ldots\right] + c_2 a_0 \log x \left[1 - \frac{x^2}{2^2} + \frac{x^4}{2^2.4^2} - \ldots\ldots\right]$$

$$+ c_2 a_0 \left[\frac{x^2}{2^2} - \frac{x^4}{2^2.4^2}\left(1+\frac{1}{2}\right) + \frac{x^6}{2^2.4^2.6^2}\left(1+\frac{1}{2}+\frac{1}{3}\right) - \ldots\ldots\right]$$

$$= (A + B\log x)\left[1 - \frac{x^2}{2^2} + \frac{x^4}{2^2.4^2} - \ldots\ldots\right] + B\left[\frac{x^2}{2^2} - \frac{x^4}{2^2.4^2}\left(1+\frac{1}{2}\right) + \frac{x^6}{2^2.4^2.6^2}\left(1+\frac{1}{2}+\frac{1}{3}\right) - \ldots\ldots\right],$$

where A and B are arbitrary constants.

Remark The differential equation (*i*) of problem 10.14 is called *Bessel's equation of zeroth order. The first solution y(x) is called Bessel function of the first kind of order zero and is denoted by $J_0(x)$. The second solution $y_2(x)$ is called the Bessel function of the second kind of order zero or the Neumann function of the second kind and is denoted by $y_0(x)$.* Thus the complete solution of the *Bessel's equation of order zero* is

$$\boldsymbol{y(x) = AJ_0(x) + BY_0(x).}$$

[Refer chapter 12]

EXERCISE 10.2

Find the series solutions about $x = 0$, of the following differential equations.

1. $(1 - x^2)\, y'' - xy' + 4y = 0.$ (*Rajasthan, 2006; AMIETE, Dec 2002*)

Ans. $y(x) = A(1 - 2x^2) + Bx\left(1 - \frac{1}{2}x^2 - \frac{1}{2.4}x^4 - \frac{1.3}{2.4.6}x^6 - \frac{1.3.5}{2.4.6.8}x^8 - \ldots\ldots\right).$

2. $(1 - x^2)\, y'' - 2xy' + 2y = 0.$ **Ans.** $y(x) = A\left(1 - x^2 - \frac{x^4}{3} - \frac{x^6}{5} - \ldots\ldots\right) + Bx.$

3. $2x(1 - x)\, y'' + (1 - x)\, y' + 3y = 0.$ (*UPTU, 2004*)

Ans. $y(x) = A\sqrt{x}\,(1 - x) + B\left(1 - 3x + \frac{3x^2}{1.3} + \frac{3x^3}{3.5} + \frac{3x^4}{5.7} + \ldots\ldots\right) + Bx.$

4. $3x\, y'' + 2\, y' + y = 0.$ **Ans.** $y(x) = A\left(1 - \frac{x}{2} + \frac{x^2}{20} - \frac{x^3}{480} + \ldots\ldots\right) + Bx^{1/3}\left(1 - \frac{x}{4} + \frac{x^2}{56} - \ldots\ldots\right).$

5. $9x(1 - x)\, y'' - 12\, y' + 4y = 0.$ (*AMIETE, June 2009; Madras 2006; Roorkee, 2000*)

Ans. $y(x) = A\left(1 + \frac{1}{3}x + \frac{1.4}{3.6}x^2 + \frac{1.4.7}{3.6.9}x^3 + \ldots\ldots\right) + Bx^{7/3}\left(1 + \frac{8}{10}x + \frac{8.11}{10.13}x^2 + \frac{8.11.14}{10.13.16}x^3 + \ldots\ldots\right).$

6. $4x\, y'' + 2\, y' + y = 0.$ (*PTU, 2005*) **Ans.** $y(x) = A\cos\sqrt{x} + B\sin\sqrt{x}.$

7. $2x^2\, y'' - xy' + (1 - x^2)\, y = 0.$

Ans. $y(x) = Ax\left(1 + \frac{x^2}{2.5} + \frac{x^4}{2.4.5.9} + \ldots\ldots\right) + B\sqrt{x}\left(x + \frac{x^2}{2.3} + \frac{x^4}{2.4.3.7} + \ldots\ldots\right).$

8. $4x\,y'' + 2(1-x)\,y' - y = 0.$

$$\textbf{Ans. } y(x) = A\left(1 + \frac{x}{2.1!} + \frac{x^2}{2^2.2!} + \frac{x^3}{2^3.3!} + \ldots\ldots\right) + B\sqrt{x}\left(1 + \frac{x}{1.3} + \frac{x^2}{1.3.5} + \frac{x^3}{1.3.5.7} + \ldots\ldots\right).$$

9. $2x\,y'' + y' + y = 0.$

$$\textbf{Ans. } y(x) = A\left(1 - x + \frac{x^2}{2!(1.3)} - \frac{x^3}{3!(1.3.5)} + \ldots\ldots\right) + B\sqrt{x}\left(1 - \frac{1}{3}x + \frac{x^2}{2!(3.5)} - \frac{x^3}{3!(3.5)} + \ldots\ldots\right).$$

10. $x\dfrac{d^2y}{dx^2} + \dfrac{dy}{dx} - y = 0.$

$$\textbf{Ans. } y(x) = (A + B\log x)\left(1 + x + \frac{x}{(2!)^2} + \frac{x^3}{(3!)^2} + \ldots\ldots\right)$$

$$-\,2B\left[x + \frac{1}{(2!)^2}\left(1 + \frac{1}{2}\right)x^2 + \frac{1}{(3!)^2}\left(1 + \frac{1}{2} + \frac{1}{3}\right)x^3 + \ldots\ldots\right]$$

11. $x^2y'' + xy' + (x^2 - 1)\,y = 0.$

$$\textbf{Ans. } y(x) = (A + B\log x)\left(-\frac{1}{2}x + \frac{1}{2^2 4}x^3 + \ldots\ldots\right) + B\left(x^{-1} + \frac{1}{2^2}x + \ldots\ldots\right).$$

12. $x(1-x)y'' - (1+3x)y' - y = 0.$ (*AMIETE, June 2004*)

$$\textbf{Ans. } y(x) = (A + B\log x)\,(1.2x^2 + 2.3x^3 + 3.4\,x^4 + \ldots\ldots) + B(-1 + x + 5x^2 + 11x^3 + \ldots\ldots).$$

13. $x(1-x)y'' + 4y' + 2y = 0$ (*AMIETE, June 2001*)

$$\textbf{Ans. } y(x) = A\left(1 - \frac{1}{2}x + \frac{1}{10}x^2\right) + Bx^{-3}\,(1 - 5x + 10x^2).$$

CHAPTER 11

Gamma, Beta and Other Special Functions

11.1 BETA AND GAMMA FUNCTIONS: INTRODUCTION

Beta and Gamma functions are improper integrals which are commonly encountered in many science and engineering applications. These functions are used in evaluating many definite integrals.

11.2 GAMMA FUNCTION

The *Gamma function* denoted by $\Gamma(n)$ is defined by the improper integral

$$\Gamma(n) = \int_0^\infty e^{-x} x^{n-1} dx \qquad \text{...(11.1)}$$

This integral converges for all positive values of n that is, $n > 0$ and is also known as *Euler's second integral* or *Euler's integral of the second kind.*

Integrating (11.1) by parts, we have

$$\Gamma(n) = \{-e^{-x}x^{n-1}\}_0^\infty + \int_0^\infty e^{-x}(n-1)x^{n-2}dx = 0 + (n-1)\Gamma(n-1)$$

Thus we obtain

$$\Gamma(n = (n-1)\Gamma(n-1) \qquad \text{...(11.2)}$$

The function relation (11.2) is known as recurrence relation for the Gamma function. Using successive reduction in (11.2) above, we get

$\Gamma n = (n-1)(n-2)\Gamma(n-2) = (n-1)(n-2)(n-3)\Gamma(n-3)$ and so on.

In particular if n is a positive integer, then

$$\Gamma n = (n-1)(n-2)\ldots 3.2.1\,\Gamma 1. \text{ Now } \Gamma 1 = \int_0^\infty e^{-x}dx = [-e^{-x}]_0^\infty = 1.$$

Hence for positive integral values of n, $\Gamma n = (n-1)!$ ...(11.3)

It is for this reason that Γn is also sometimes called the ***factorial function***.

It may be mentioned that the values of Γn are known from the tables of Gamma functions for the interval $1 < n \le 2$. The functional relation 11.2 helps us to compute the values of Γn for the unit interval $2 < n \le 3$, and these values further enable us to determine the values of Γn in the next unit interval $3 < n \le 4$. The process is continued for the successive unit intervals and in this way we can obtain the values of Γn for non-integral values of $n > 1$.

The recurrence relation (11.2) can be rewritten as

$$\Gamma n = \frac{\Gamma n+1}{n} = \frac{\Gamma n+2}{n(n+1)} = \ldots = \frac{\Gamma n+k+1}{n(n+1)\ldots(n+k)} \qquad (11.4)$$

This relation helps us to define Γn for $0 < n \le 1$ since the values of Γn for the unit interval $1 < n \le 2$ are known from the table. After having known the values of Γn for $0 < n \le 1$, the values of Γn can be computed for the range $-1 < n \le 0$ and this procedure is used repetitively to compute Γn for non-integral negative values of n, choosing for k the smallest integer such that $n + k + 1 > 0$.

It is thus found that the formulae (11.1) and (11.4) together give a complete definition of Γn for all n not equal to zero or a negative integer. The graph of Gamma function is shown in Figure 11.1 along with the table of values of n for $1 \le n \le 2$.

Table of values and graph of the Gamma Function

n	Γn
1.0	1.0000
1.1	0.9514
1.2	0.9182
1.3	0.8975
1.4	0.8873
1.5	0.8862
1.6	0.8935
1.7	0.9086
1.8	0.9314
1.9	0.9618
2.0	1.0000

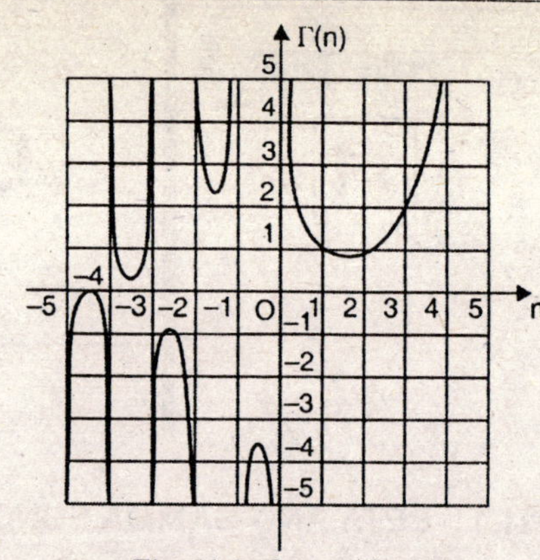

Fig. 11.1: Gamma function

It can also be seen from relation (11.4) that when $n = 0$ or any negative integral value, the Gamma of that value is infinite, that is, $\Gamma 0 = \infty$, $\Gamma(\text{negative integers}) = \infty$.

11.3 EVALUATION OF $\Gamma(1/2)$

(VTU 2006)

Putting $n = 1/2$ in the definition of Γn, we obtain

$$\Gamma 1/2 = \int_0^{\infty} e^{-x}\, x^{-1/2}\, dx$$

Put $x = y^2$, so that $dx = 2y\, dy$. This gives

$$\Gamma 1/2 = 2\int_0^{\infty} e^{-y^2}\, dy \qquad ...(i)$$

Rewriting (*i*), with x in place of y, we have

$$\Gamma 1/2 = 2\int_0^{\infty} e^{-x^2}\, dx \qquad ...(ii)$$

Multiplying (*i*) and (*ii*), we get

$$(\Gamma 1/2)^2 = 4\left(\int_0^{\infty} e^{-y^2}\, dy\right)\left(\int_0^{\infty} e^{-x^2}\, dx\right) = 4\int_0^{\infty}\int_0^{\infty} e^{-(x^2+y^2)}\, dx\, dy$$

Changing to polar co-ordinates, we get

$$(\Gamma 1/2)^2 = 4\int_0^{\pi/2}\int_0^{\infty} e^{-r^2}\, r\, dr\, d\theta = 4\int_0^{\pi/2}\left[-\frac{e^{-r^2}}{2}\right]_0^{\infty} d\theta = 2\int_0^{\pi/2} d\theta = \pi$$

Hence $\qquad \Gamma(1/2) = \sqrt{\pi}.$

From the foregoing we have the following important results

$$\mathbf{\Gamma(n) = (n-1)\,\Gamma n - 1 \text{ for all } n \text{ (except 0 and negative integers)}}$$

$$\mathbf{\Gamma(n) = (n-1)! \text{ when } n \text{ is a positive integer}}$$

$$\mathbf{\Gamma 1 = 1, \quad \Gamma 0 = \infty, \quad \Gamma(-\text{ve integer}) = \infty}$$

$$\mathbf{\Gamma(1/2) = \sqrt{\pi} = 1.7728 \qquad \Gamma(-1/2) = -2\sqrt{\pi}.}$$

EXAMPLE 11.1 Evaluate the following:

(*i*) $\dfrac{\Gamma(7)}{2\Gamma(4)\Gamma(3)}$, (*ii*) $\dfrac{\Gamma(3)\Gamma(3/2)}{\Gamma(9/2)}$, (*iii*) $\Gamma 3.5$, (*AMIETE, June 2010*)

(*iv*) $\Gamma 1/2\,\Gamma(3/2)\,\Gamma(5/2)$, (*v*) $\Gamma(-2.5)$, (*vi*) $\Gamma(-1/3)$, (*vii*) $\Gamma(-7/2)$.

Solution

(*i*) $\dfrac{\Gamma(7)}{2\Gamma(4)\Gamma(3)} = \dfrac{6!}{2.3!2!} = \dfrac{6.5.4.3!}{2.3!2} = 30$. (*ii*) $\dfrac{\Gamma(3)\Gamma(3/2)}{\Gamma(9/2)} = \dfrac{2!(1/2)\Gamma(1/2)}{7/2\cdot 5/2\cdot 3/2\cdot 1/2\Gamma(1/2)} = \dfrac{16}{105}$.

(*iii*) $\Gamma 7/2 = 5/2\cdot 3/2\cdot 1/2\sqrt{\pi} = \dfrac{15}{8}\sqrt{\pi}$. (*iv*) $\Gamma(1/2)\Gamma(3/2)\Gamma(5/2) = \sqrt{\pi}\cdot 1/2\sqrt{\pi}\cdot 3/2\cdot 1/2\cdot\sqrt{\pi} = \dfrac{3}{8}\pi^{3/2}$.

(*v*) When n is negative, $\Gamma(n)$ is given by $\Gamma(n) = \dfrac{\Gamma(n+1)}{n}$. ...(*a*)

Putting $n = -2.5, -1.5, -.5$ in (*a*), we obtain

$$\Gamma(-2.5) = \frac{\Gamma(-1.5)}{-2.5} = \frac{\Gamma(-.5)}{(-2.5)(-1.5)} = \frac{\Gamma .5}{(-2.5)(-1.5)(-.5)} = \frac{-8\sqrt{\pi}}{15}.$$

(*vi*) $\Gamma(-1/3) = \dfrac{\Gamma((-1/3)+1)}{-1/3} = -3\Gamma(2/3)$.

(*vii*) $\Gamma(-7/2) = \dfrac{\Gamma(-5/2)}{(-7/2)} = \dfrac{\Gamma(-3/2)}{(-7/2)(-5/2)} = \dfrac{\Gamma(-1/2)}{(-7/2)(-5/2)(-3/2)} = \dfrac{\Gamma(1/2)}{(-7/2)(-5/2)(-3/2)(-1/2)} = \dfrac{16\sqrt{\pi}}{105}$.

EXAMPLE 11.2 Evaluate the following:

(*i*) $\dfrac{\Gamma 3\,\Gamma 2.5}{3\Gamma 5.5}$, (*ii*) $\dfrac{7\Gamma 11/3}{\Gamma 3/2\,\Gamma 2/3}$, (*iii*) $\Gamma -5.5$, (*iv*) $\dfrac{\Gamma -1/2}{2\Gamma 2.5}$,

Solution (*i*) $\dfrac{\Gamma 3\,\Gamma 2.5}{3\Gamma 5.5} = \dfrac{2!(1.5)(0.5)\Gamma .5}{3(4.5)(3.5)(2.5)(1.5)(.5)\Gamma .5} = \dfrac{16}{945}$.

(*ii*) $\dfrac{7\Gamma 11/3}{\Gamma 3/2\,\Gamma 2/3} = \dfrac{7\cdot 8/3\cdot 5/3\cdot 2/3\Gamma 2/3}{1/2\,\Gamma 1/2\,\Gamma 2/3} = \dfrac{1120}{27\sqrt{\pi}}$.

(*iii*) $\Gamma -5.5 = \dfrac{\Gamma -5.5+1}{-5.5} = \dfrac{\Gamma(-5.5+5+1)}{(-5.5)(-4.5)(-3.5)(-2.5)(-1.5)(-.5)} = \dfrac{\Gamma 1/2\times 2^6}{11\cdot 9\cdot 7\cdot 5\cdot 3\cdot 1} = \dfrac{64\sqrt{\pi}}{10395}$.

(*iv*) $\dfrac{\Gamma -1/2}{2\Gamma 2.5} = \dfrac{\Gamma(-1/2)+1}{2\times(-1/2)\times 1.5\times .5\Gamma .5} = -\dfrac{4}{3}$.

11.4 TRANSFORMATION OF GAMMA FUNCTION

From the definition of Gamma function, we have $\Gamma n = \int_0^\infty e^{-x}x^{n-1}dx$

(*i*) Put $x = y^2$, we have $\Gamma n = \int_0^\infty e^{-x}x^{n-1}dx = \int_0^\infty e^{-y^2}(y^2)^{n-1}2y\,dy = 2\int_0^\infty e^{-y^2}y^{2n-1}dy$.

(*ii*) Put $x = cy$, $dx = c\,dy$, we get

$$\Gamma n = \int_0^\infty e^{-cy}c^n y^{n-1}dy = c^n\int_0^\infty e^{-cy}y^{n-1}dy.$$

Therefore,
$$\int_0^\infty e^{-cy}y^{n-1}dy = \frac{\Gamma(n)}{c^n}.$$

(*iii*) Put $x = \log\dfrac{1}{y}$ or $y = e^{-x}$, $dy = e^{-x}dx$, we get

$$\Gamma n = -\int_1^0\left(\log\frac{1}{y}\right)^{n-1}dy = \int_0^1\left(\log\frac{1}{y}\right)^{n-1}dy. \quad \text{(JNTU 2003; Madras 2003)}$$

(*iv*) Put $x^n = y$, $nx^{n-1}dx = dy$. This gives

$$\Gamma n = \frac{1}{n}\int_0^\infty e^{-(y)^{1/n}}\,dy$$

or
$$\int_0^\infty e^{-(y)^{1/n}}\,dy = n\Gamma n = \Gamma(n+1).$$

Putting $n = 1/2$, we get $\int_0^\infty e^{-y^2}\,dy = \Gamma(3/2) = 1/2\,\Gamma(1/2) = (1/2)\sqrt{\pi}.$

EXAMPLE 11.3 Evaluate each of the following integrals.

(*i*) $\int_0^\infty \sqrt{x}e^{-x^3}\,dx$, (*JNTU, 2003*), (*ii*) $\int_0^\infty 4x^4 e^{-x^4}\,dx$,

(*iii*) $\int_0^\infty \frac{x^c}{c^x}\,dx$, (*UPTU, 2006*), (*iv*) $\int_0^1 x^4[\log(1/x)]^3\,dx$, (*Madras, 2000*).

(*v*) $\int_0^1 x^m(\log x)^n\,dx$ where n is a positive integer and $m > -1$. (*SVTU, 2006; Bhopal, 2002 S*).

Solution (*i*) $I = \int_0^\infty \sqrt{x}\,e^{-x^3}\,dx$. Putting $x^3 = y$, then $3x^2dx = dy$, we get

$$I = \int_0^\infty e^{-y}\,y^{1/6}\,\frac{1}{3}\,y^{-2/3}\,dy = \frac{1}{3}\int_0^\infty y^{-1/2}\,e^{-y}\,dy = \frac{1}{3}\Gamma(1/2) = \frac{\sqrt{\pi}}{3}.$$

(*ii*) $I = \int_0^\infty 4x^4\,e^{-x^4}\,dx$. Putting $x^4 = y$, $4x^3dx = dy$, we get $I = \int_0^\infty y^{\frac{1}{4}}\,e^{-y}dy = \int_0^\infty y^{5/4-1}\,e^{-y}dy = \Gamma(5/4)$

(*iii*) $\int_0^\infty \frac{x^c}{c^x}\,dx = \int_0^\infty \frac{x^c}{e^{x\log c}}\,dx = \int_0^\infty (e^{-x\log c})\,x^c\,dx$ [Put $x \log c = t$, $dx = dt/\log c$]

$$= \int_0^\infty e^{-t}\left(\frac{t}{\log c}\right)^c \frac{dt}{\log c} = \frac{1}{(\log c)^{c+1}}\int_0^\infty e^{-t}t^c\,dt = \frac{1}{(\log c)^{c+1}}\int_0^\infty e^{-t}t^{(c+1)-1}dt = \frac{\Gamma(c+1)}{(\log c)^{c+1}}, \qquad c > 1.$$

(*iv*) $\int_0^1 x^4\left[\log(1/x)\right]^3 dx = -\int_\infty^0 e^{-4t/5}\cdot\frac{t^3}{125}\cdot\frac{1}{5}e^{-t/5}dt$ [Put $x = e^{-t/5}$, $dx = (-1/5)\,e^{-t/5}\,dt$, $\log(1/x) = t/5$]

$$= \frac{1}{625}\int_0^\infty e^{-t}t^3dt = \frac{\Gamma 4}{625} = \frac{6}{625}.$$

(*v*) $I = \int_0^1 x^m(\log x)^n\,dx$. Letting $x = e^{-y}$, $dx = -e^{-y}\,dy$, we get

$$I = \int_\infty^0 -e^{-my}(-1)^n\,y^n\,e^{-y}dy = \int_0^\infty (-1)^n y^n\,e^{-(m+1)y}dy$$

If $(m+1)y = t$, $(m+1)dy = dt$. This gives

$$(-1)^n\int_0^\infty \frac{t^n}{(m+1)^n}e^{-t}\,\frac{1}{m+1}dt = \frac{(-1)^n}{(m+1)^{n+1}}\Gamma(n+1) = \frac{(-1)^n n!}{(m+1)^{n+1}}.$$

11.5 THE BETA FUNCTION

The *Beta function* denoted by $\beta(m, n)$ is defined by the integral

$$\beta(m, n) = \int_0^1 x^{m-1}(1-x)^{n-1}\,dx.$$

This integral converges for positive values of m and n and is also known as *Euler's first integral* or *Euler's integral of the first kind.* It may be remarked that Beta functions are symmetrical with respect to m and n i.e., $B(m, n) = \beta(n, m)$. This can be easily proved by using the property of definite integral $\int_0^a f(x)\,dx = \int_0^a f(a-x)dx$.

11.6 TRANSFORMATION OF BETA FUNCTION

The *Beta function* has been defined by the integral

$$\beta(m, n) = \int_0^1 x^{m-1}(1-x)^{n-1}dx. \tag{11.5}$$

(*i*) Use the transformation $x = 1 - y$, $dx = -dy$; when $x = 0$, $y = 1$ and when $x = 1$, $y = 0$, the integral (11.5), becomes

$$\beta(m, n) = -\int_1^0 (1-y)^{m-1} y^{n-1} dy = \int_0^1 y^{n-1}(1-y)^{m-1}dy = \beta(n, m).$$

This property shows that Beta functions are symmetrical with respect to m and n. [This result can also be proved by applying the property of definite integral $\int_0^a f(x)dx = \int_0^a f(a-x)\,dx$.]

(*ii*) Put $x = \frac{1}{1+y}$, $dx = -\frac{1}{(1+y)^2}dy$; when $x = 0$, $y = \infty$; when $x = 1$, $y = 0$. Substituting for x in (11.5), we get

$$\beta(m, n) = -\int_\infty^0 \left(\frac{1}{1+y}\right)^{m-1} \frac{y^{n-1}}{(1+y)^{n-1}} \frac{1}{(1+y)^2} dy = \int_0^\infty \frac{y^{n-1}}{(1+y)^{m+n}} dy$$

$$= \int_0^\infty \frac{y^{m-1}}{(1+y)^{m+n}} dy \quad \text{(by symmetrical property)} \qquad (VTU,\ 2003)$$

(*iii*) Put $x = \sin^2\theta$, $dx = 2\sin\theta\cos\theta\, d\theta$; when $x = 0$, $\theta = 0$; when $x = 1$, $\theta = \pi/2$. This substitution gives

$$\beta(m, n) = \int_0^{\pi/2} (\sin^2\theta)^{m-1}(\cos^2\theta)^{n-1} 2\sin\theta\cos\theta\, d\theta$$

$$= 2\int_0^{\pi/2} \sin^{2m-1}\theta \cos^{2n-1}\theta\, d\theta.$$

(*iv*) Put $x = \frac{y-b}{a-b}$, $dx = \frac{dy}{a-b}$; when $x = 0$, $y = b$; when $x = 1$, $y = a$. This transformation gives

$$\beta(m, n) = \int_b^a \left(\frac{y-b}{a-b}\right)^{m-1} \left(\frac{a-y}{a-b}\right)^{n-1} \frac{dy}{a-b} = \int_b^a \frac{1}{(a-b)^{m+n-1}} (y-b)^{m-1}(a-y)^{n-1}dy$$

$$= \frac{1}{(a-b)^{m+n-1}} \int_b^a (y-b)^{m-1}(a-y)^{n-1}dy.$$

11.7 RELATION BETWEEN BETA AND GAMMA FUNCTIONS

To prove that $\beta(m, n) = \frac{\Gamma m\, \Gamma n}{\Gamma(m+n)}$ $\quad m, n > 0$.

We have $\Gamma m = \int_0^\infty e^{-t} \cdot t^{m-1}dt$. Letting $t = x^2$, we get $\Gamma m = \int_0^\infty 2e^{-x^2} x^{2m-1}dx$ ...(11.6)

Similarly $\Gamma n = \int_0^\infty 2e^{-y^2} y^{2n-1}dy$. Then $\Gamma m\, \Gamma n = 4\left(\int_0^\infty e^{-x^2} x^{2m-1}dx\right)\left(\int_0^\infty e^{-y^2} y^{2n-1}dy\right)$

$$= 4\int_0^\infty \int_0^\infty e^{-(x^2+y^2)} x^{2m-1}\, y^{2n-1} dx\, dy.$$

Transforming to polar co-ordinates $x = r\cos\theta$, $y = r\sin\theta$ and $dx\ dy = r\,d\theta\,dr$, we obtain

$$\overline{|m}\,\overline{|n} = 4\int_{r=0}^{\infty}\int_{\theta=0}^{\pi/2} e^{-r^2} r^{2m+2n-1}\cos^{2m-1}\theta\sin^{2n-1}\theta\,dr\,d\theta$$

$$= 4\int_0^{\infty} e^{-r^2} r^{2m+2n-1}dr\int_0^{\pi/2}\cos^{2m-1}\theta\sin^{2n-1}\theta\,d\theta$$

$$= \left(2\int_0^{\infty} e^{-r^2} r^{2m+2n-1}dr\right)\left(2\int_0^{\pi/2}\cos^{2m-1}\theta\sin^{2n-1}\theta\,d\theta\right)$$

$$= \overline{|m+n}\,\beta(m,n) \quad \text{(by Art (11.6) (iii) and Eq. (11.6) above)}$$

Hence we get

$$\beta(m,n) = \frac{\overline{|m}\,\overline{|n}}{\overline{|m+n}} \qquad ...(11.7)\quad (JNTU,\ 2006)$$

This result finds useful applications in the evaluation of many definite integrals.

Cor. 1: If $m + n = 1$, so that $n = 1 - m$, we get

$$\overline{|m}\,\overline{|1-m} = \beta(m, 1-m) = \int_0^{\infty}\frac{x^{m-1}}{1+x}\,dx.$$

Cor. 2: $\beta(m,n) = 2\int_0^{\pi/2}\sin^{2m-1}\theta\cos^{2n-1}\theta\,d\theta = \dfrac{\overline{|m}\,\overline{|n}}{\overline{|m+n}}$.

Putting $2m - 1 = p$, $2n - 1 = q$, we get

$$\int_0^{\pi/2}\sin^p\theta\cos^q\theta\,d\theta = \frac{\overline{\left|\frac{1+p}{2}\right.}\,\overline{\left|\frac{1+q}{2}\right.}}{2\,\overline{\left|\frac{p+q+2}{2}\right.}} = \frac{1}{2}\beta\left(\frac{1+p}{2},\frac{1+q}{2}\right).$$

As particular cases, we have

$$\int_0^{\pi/2}\sin^p\theta\,d\theta = \frac{\overline{\left|\frac{1+p}{2}\right.}\,\overline{\left|\frac{1}{2}\right.}}{2\,\overline{\left|\frac{p+2}{2}\right.}} = \frac{\overline{\left|\frac{1+p}{2}\right.}}{2\,\overline{\left|\frac{p+2}{2}\right.}}\sqrt{\pi},$$

and

$$\int_0^{\pi/2}\cos^q\theta\,d\theta = \frac{\overline{\left|\frac{1+q}{2}\right.}\,\overline{|1/2}}{2\,\overline{\left|\frac{q+2}{2}\right.}} = \frac{\overline{\left|\frac{1+q}{2}\right.}}{2\,\overline{\left|\frac{q+2}{2}\right.}}\sqrt{\pi}.$$

Since Beta functions are symmetrical w.r.t p and q, we have

$$\int_0^{\pi/2}\sin^p\theta\,d\theta = \int_0^{\pi/2}\cos^p\theta\,d\theta.$$

EXAMPLE 11.4 Evaluate the following

(*i*) $\beta(3.5, 1.5)$, (*ii*) $\int_0^1\sqrt{(1-x^4)}dx$, (*iii*) $\int_0^{\infty} x^{2n-1}e^{-ax^2}dx$.

Solution (*i*) $\beta(3.5, 1.5) = \dfrac{\overline{|(3.5)}\,\overline{|(1.5)}}{\overline{|(3.5+1.5)}} = \dfrac{\overline{|(3.5)}\,\overline{|(1.5)}}{\overline{|5}} = \dfrac{(2.5)(1.5)(.5)\sqrt{\pi}(0.5)\sqrt{\pi}}{4!} = \dfrac{5\pi}{128}$.

(*ii*) Put $x = \sqrt{\sin\theta}$, $dx = \frac{1}{2}(\sin\theta)^{-1/2}\cos\theta\,d\theta$; then

$$\int_0^1 \sqrt{1-x^4}\,dx = \frac{1}{2}\int_0^{\pi/2} (\sin\theta)^{-1/2}\cos^2 d\theta = \frac{1}{2}\,\frac{\Gamma(1/4)\,\Gamma(3/2)}{2\,\Gamma(7/4)} = \frac{1}{2}\,\frac{\Gamma(1/4)\cdot\frac{1}{2}\sqrt{\pi}}{2\cdot\frac{3}{4}\Gamma(3/4)} = \frac{\sqrt{\pi}\,\Gamma(1/4)}{6\,\Gamma(3/4)}.$$

(*iii*) Let $ax^2 = t \Rightarrow x^2 = t/a$, $2x\,dx = dt/a$. The given integral become

$$\int_0^\infty x^{2n-1} e^{-ax^2} dx = \frac{1}{2}\int_0^\infty x^{2n-2} e^{-ax^2}\cdot 2x\,dx = \frac{1}{2}\int_0^\infty \left(\frac{t}{a}\right)^{n-1} e^{-t}\frac{dt}{a} = \frac{1}{2a^n}\int_0^\infty e^{-t}t^{n-1}dt = \frac{1}{2a^n}\Gamma(n).$$

EXAMPLE 11.5 Evaluate the following

(*i*) $\int_0^{\pi/2} \sin^6\theta\,d\theta$, (*ii*) $\int_0^{\pi/2} \sin^2\theta\cos^5\theta\,d\theta$, (*iii*) $\int_0^{\pi/2} \sin^6\theta\cos^4\theta\,d\theta$.

Solution (*i*) $\int_0^{\pi/2} \sin^6\theta\,d\theta = \int_0^{\pi/2} \sin^6\theta\cos^0\theta\,d\theta = \dfrac{\Gamma(1/2)\,\Gamma(7/2)}{2\Gamma 4} = \dfrac{\sqrt{\pi}\cdot\frac{5}{2}\cdot\frac{3}{2}\cdot\frac{1}{2}\sqrt{\pi}}{2.3!} = \dfrac{5\pi}{32}.$

(*ii*) $\int_0^{\pi/2} \sin^2\theta\cos^5\theta\,d\theta = \dfrac{\Gamma(3)\cdot\Gamma(3/2)}{2\Gamma(9/2)} = \dfrac{2 1(.5)\sqrt{\pi}}{2(3.5)(2.5)(1.5)(.5)\sqrt{\pi}} = \dfrac{8}{105}.$

(*iii*) $\int_0^{\pi/2} \sin^6\theta\cos^4\theta\,d\theta = \dfrac{\Gamma(7/2)\cdot\Gamma(5/2)}{2\Gamma 6} = \dfrac{\frac{5}{2}\cdot\frac{3}{2}\cdot\frac{1}{2}\sqrt{\pi}\cdot\frac{3}{2}\cdot\frac{1}{2}\cdot\sqrt{\pi}}{(2)\,5!} = \dfrac{3\pi}{512}.$

EXAMPLE 11.6 Show that

(*i*) $\int_0^1 y^{q-1}\left(\log\frac{1}{y}\right)^{p-1} dy = \dfrac{\Gamma(p)}{q^p}$, where $p > 0$, $q > 0$. *(Rohtak, 2006 S)*

(*ii*) $\beta(p,q) = \int_0^1 \dfrac{x^{p-1}+x^{q-1}}{(1+x)^{p+q}}dx$, *(Osmania, 2003)*

(*iii*) $\beta(m, n) = \beta(m+1, n) + \beta(m, n+1)$, *(JNTU, 2006; Madras, 2003)*

(*iv*) $\int_0^\infty \dfrac{x^{m-1}}{(a+bx)^{m+n}}dx = \dfrac{\beta(m,n)}{a^n b^m}$, where m, n, a, b are positive

(*v*) $\int_0^1 \dfrac{dx}{\sqrt{1-x^n}} = \dfrac{\Gamma(1/n)\sqrt{\pi}}{n\Gamma\left(\frac{1}{n}+\frac{1}{2}\right)}$, *(Anna, 2005)* (*vi*) $\int_0^\infty 3^{-4x^2}dx$.

Solution (*i*) Setting $\log(1/y) = x$ so that $1/y = e^x$ or $y = e^{-x}$ and $dy = -e^{-x}dx$.

Therefore,

$$\int_0^1 y^{q-1}\log\left(\frac{1}{y}\right)^{p-1} dy = \int_\infty^0 e^{-(q-1)x}\cdot x^{p-1}(-e^{-x})\,dx = \int_0^\infty e^{-qx}x^{p-1}dx$$

$$= \int_0^\infty e^{-t}\cdot\left(\frac{t}{q}\right)^{p-1}\frac{dt}{q} \qquad \text{(where } qx = t\text{)}$$

$$= \frac{1}{q^p}\int_0^\infty e^{-t}t^{p-1}dt = \frac{\Gamma(p)}{q^p}.$$

(*ii*) We know $\beta(p,q) = \int_0^1 x^{p-1}(1-x)^{q-1}dx$.

Setting $x = 1/(1+y)$, $dx = \{-1/(1+y)^2\}\, dy$.

Now, $$\beta(p, q) = -\int_{\infty}^{0}\left(\frac{1}{1+y}\right)^{p-1}\left(\frac{y}{1+y}\right)^{q-1}\cdot\frac{1}{(1+y)^2}dy = \int_0^{\infty}\frac{y^{q-1}}{(1+y)^{p+q}}dy$$

$$= \int_0^1\frac{y^{q-1}}{(1+y)^{p+q}}dy + \int_1^{\infty}\frac{y^{q-1}}{(1+y)^{p+q}}dy. \qquad ...(i)$$

Now, putting $y = 1/z$ in the second integral, we have

$$\int_1^{\infty}\frac{y^{q-1}}{(1+y)^{p+q}}dy = -\int_1^{0}\frac{(1/z)^{q-1}}{\{1+(1/z)\}^{p+q}}\left(\frac{1}{z^2}\right)dz = \int_0^1\frac{z^{p-1}}{(1+z)^{p+q}}dz.$$

Therefore, from (*i*) we have

$$\beta(p, q) = \int_0^1\frac{y^{q-1}}{(1+y)^{p+q}}dy + \int_0^1\frac{z^{p-1}}{(1+z)^{p+q}}dz = \int_0^1\frac{x^{p-1}+x^{q-1}}{(1+x)^{p+q}}dx.$$

(*iii*) $$\beta(m+1, n) + \beta(m, n+1) = \int_0^1 x^m(1-x)^{n-1}dx + \int_0^1 x^{m-1}(1-x)^n dx$$

$$= \int_0^1\{x^m(1-x)^{n-1} + x^{m-1}(1-x)^n\}dx$$

$$= \int_0^1 x^{m-1}(1-x)^{n-1}\{x+(1-x)\}dx = \int_0^1 x^{m-1}(1-x)^{n-1}dx = \beta(m, n).$$

(*iv*) Put $bx = at$, $dx = (a/b)dt$

$$\therefore \int_0^{\infty}\frac{x^{m-1}}{(a+bx)^{m+n}}dx = \int_0^{\infty}\frac{(at/b)^{m-1}}{(a+at)^{m+n}}\cdot\frac{a}{b}dt = \frac{1}{a^n b^m}\int_0^{\infty}\frac{t^{m-1}}{(1+t)^{m+n}}dt = \frac{1}{a^n\, b^m}\beta\,(m, n).$$

(v) Setting $x^n = t \Rightarrow x = (t)^{1/n}$, $dx = \frac{1}{n}(t)^{\frac{1}{n}-1}\, dt$.

Therefore, $$\int_0^1\frac{dx}{\sqrt{1-x^n}} = \int_0^1\frac{\frac{1}{n}(t)^{\frac{1}{n}-1}}{\sqrt{1-t}}dt = \frac{1}{n}\int_0^1(t)^{\frac{1}{n}-1}\,(1-t)^{-1/2}\, dt$$

$$= \frac{1}{n}\int_0^1(t)^{\frac{1}{n}-1}\,(1-t)^{\frac{1}{2}-1}\, dt = \frac{1}{n}\beta\left(\frac{1}{n}, \frac{1}{2}\right) = \frac{1}{n}\cdot\frac{\overline{|(1/n)}\,\overline{|(1/2)}}{\overline{\left|\left(\frac{1}{n}+\frac{1}{2}\right)\right.}} = \frac{\sqrt{\pi}\,\overline{|1/n}}{n\overline{\left|\left(\frac{1}{n}+\frac{1}{2}\right)\right.}}.$$

(*vi*) $$\int_0^{\infty}3^{-4x^2}\, dx = \int_0^{\infty}(e^{\log 3})^{(-4x^2)}\, dx = \int_0^{\infty}e^{-(4\log 3)(x^2)}\, dx.$$

Let $(4\log 3)\, x^2 = t$. Then the given integral becomes

$$\int_0^{\infty}e^{-t}\, d\left(\frac{t^{1/2}}{\sqrt{4\log 3}}\right) = \frac{1}{2\sqrt{4\log 3}}\int_0^{\infty}e^{-t}\, t^{-1/2}\, dt = \frac{1}{4\sqrt{\log 3}}\int_0^{\infty}e^{-t}\, t^{\frac{1}{2}-1}\, dt$$

$$= \frac{1}{4\sqrt{\log 3}}\cdot\overline{|1/2} = \frac{\sqrt{\pi}}{4\sqrt{\log 3}}.$$

EXAMPLE 11.7 Express the following in terms of Beta functions:

(*i*) $\int_{-1}^{1}(1-x^2)^n\,dx$ (*ii*) $\int_0^1 x^m(1-x^p)^n\,dx$, and hence

evaluate the integral $\int_0^1 x^{3/2}(1-\sqrt{x})^{1/2}dx$.

Solution (*i*) We have $I=\int_{-1}^{1}(1-x^2)^n\,dx=\int_{-1}^{1}(1+x)^n(1-x)^n\,dx$.

Let $1+x=2t$. Then $1-x=1-(2t-1)=2(1-t)$ and $dx=2dt$.

At $x=-1$, $t=0$; and when $x=1$, $t=1$.

$$\therefore\quad I=\int_{-1}^{1}(1+x)^n(1-x)^n\,dx=\int_0^1(2t)^n\,\{2(1-t)\}^n\cdot 2dt=2^{2n+1}\int_0^1 t^n(1-t)^n\,dt$$

$$=2^{2n+1}\int_0^1 t^{\overline{n+1}-1}(1-t)^{\overline{n+1}-1}dt=2^{2n+1}\beta(n+1,n+1).$$

(*ii*) Let $x^p=y\Rightarrow x=(y)^{1/p}$. Then $dx=(1/p)\,(y)^{(1/p)-1}\,dy$.

$$\therefore\quad I=\int_0^1 x^m(1-x^p)^n\,dx=\int_0^1 y^{m/p}(1-y)^n\cdot\frac{1}{p}y^{(1/p)-1}dy=\frac{1}{p}\int_0^1 y^{\frac{m+1}{p}-1}(1-y)^n\,dy$$

$$=\frac{1}{p}\int_0^1 y^{\frac{m+1}{p}-1}(1-y)^{\overline{n+1}-1}dy=\frac{1}{p}\beta\left(\frac{m+1}{p},n+1\right).$$

Now, comparing the integral $\int_0^1 x^{3/2}(1-\sqrt{x})^{1/2}dx$ with the given integral, we find that $m=3/2$, $p=1/2$ and $n=1/2$. Therefore,

$$\int_0^1 x^{3/2}(1-\sqrt{x})^{1/2}dx=2\beta(5,3/2)=\frac{2\,\Gamma(5)\,\Gamma(3/2)}{\Gamma(13/2)}=\frac{2.4!\,\Gamma(3/2)}{\frac{11}{2}\cdot\frac{9}{2}\cdot\frac{7}{2}\cdot\frac{5}{2}\cdot\frac{3}{2}\,\Gamma\left(\frac{3}{2}\right)}=\frac{512}{3465}$$

EXAMPLE 11.8 Evaluate $\int_0^\infty e^{-ax}\,x^{m-1}\sin bx\,dx$ in terms of Gamma function.

(*AMIETE, W-2010; UPTU, 2003*)

Solution We know $\Gamma m=\int_0^\infty e^{-x}\,x^{m-1}dx$ (let $x=ay$, $dx=a\,dy$)

$$=\int_0^\infty e^{-ay}\,a^m y^{m-1}dy\quad\text{or}\quad\int_0^\infty e^{-ay}\,y^{m-1}dy=\frac{\Gamma m}{a^m}.\qquad\ldots(i)$$

Now $\int_0^\infty e^{-ax}x^{m-1}\sin bx\,dx=\int_0^\infty e^{-ax}x^{m-1}$ (imaginary part of e^{ibx}) dx ($\because e^{ibx}=\cos bx+i\sin bx$)

$$=\text{I.P. of}\int_0^\infty e^{-(a-ib)x}x^{m-1}dx=\text{I.P. of}\left\{\frac{\Gamma m}{(a-ib)^m}\right\}\qquad\ldots\text{ by }(i)$$

$$=\text{I.P. of}\left\{\frac{\Gamma m}{[r^m(\cos\theta-i\sin\theta)^m]}\right\}\qquad\text{where } a=r\cos\theta,\ b=r\sin\theta$$

$$=\text{I.P. of}\left\{\frac{\Gamma m}{[r^m(\cos m\theta-i\sin m\theta)]}\right\},\qquad\text{(using Demoivre's theorem)}$$

$$= \text{I.P. of } \left\{ \frac{\overline{|m}\,(\cos m\theta + i\sin m\theta)}{r^m(\cos m\theta + i\sin m\theta)(\cos m\theta - i\sin m\theta)} \right\}$$

$$= \text{I.P. of } \left\{ \frac{\overline{|m}\,(\cos m\theta + i\sin m\theta)}{r^m \cdot 1} \right\} = \frac{\overline{|m}}{r^m}\sin m\theta, \qquad \text{where } r = \sqrt{a^2+b^2}, \theta = \tan^{-1}(b/a).$$

EXAMPLE 11.9 Establish the following relations

(*i*) $\int_0^1 \frac{dx}{\sqrt{(1-x^4)}} = \frac{\sqrt{\pi}}{4}\frac{\overline{|1/4}}{\overline{|3/4}}$, *(AMIETE, JUNE 2010)*

(*ii*) $\int_0^{\pi/2} \sqrt{(\cot\theta)}\,d\theta = \frac{1}{2}\overline{|1/4}\;\overline{|3/4} = \frac{\pi}{\sqrt{2}}$, *(Osmania, 2003S; VTU, 2001)*

(*iii*) $\int_0^{\pi/2} \sqrt{(\sin\theta)}\,d\theta \times \int_0^{\pi/2} \frac{d\theta}{\sqrt{(\sin\theta)}} = \pi$. *(AMIETE, June 2009; VTU 2007)*

Solution

(*i*) Let $I = \int_0^1 \frac{dx}{\sqrt{(1-x^4)}}$. Put $x = \sin^{1/2}\theta$, so that $dx = (1/2)\sin^{-1/2}\theta\cos\theta\,d\theta$. The given integral after substitution becomes $I = \frac{1}{2}\int_0^{\pi/2} \sin^{-1/2}\theta\,d\theta = \frac{1}{2}\frac{\overline{|1/4}\,\overline{|1/2}}{2\overline{|3/4}} = \frac{1}{4}\frac{\overline{|1/4}}{\overline{|3/4}}\sqrt{\pi}$.

(*ii*) $\int_0^{\pi/2} \cos^{1/2}\theta\sin^{-1/2}\theta\,d\theta = \frac{1}{2}B\left(\frac{3}{4},\frac{1}{4}\right) = \frac{\overline{|3/4}\,\overline{|1/4}}{2\,\overline{|1}} = \frac{1}{2}\overline{|3/4}\,\overline{|1/4} = \frac{1}{2}\frac{\pi}{\sin(\pi/4)} = \frac{\pi\sqrt{2}}{2}$.

(*iii*) $\int_0^{\pi/2} \sqrt{(\sin\theta)}\,d\theta \times \int_0^{\pi/2} \sin^{-1/2}\theta\,d\theta = \frac{\overline{|3/4}\,\overline{|1/2}}{2\overline{|5/4}} \times \frac{\overline{|1/4}\,\overline{|1/2}}{2\overline{|3/4}} = \frac{\sqrt{\pi}}{2\cdot\frac{1}{4}\overline{|1/4}} \times \frac{\overline{|1/4}\sqrt{\pi}}{2} = \pi$.

EXAMPLE 11.10 Using Beta and Gamma functions show that for any positive integer '*m*'

(*i*) $\int_0^{\pi/2} \sin^{2m-1}(\theta)\,d\theta = \frac{(2m-2)(2m-4)\ldots\ldots 2}{(2m-1)(2m-3)\ldots\ldots 3}$, *(AMIETE, June 2011)*

(*ii*) $\int_0^{\pi/2} \sin^{2m}(\theta)\,d\theta = \frac{(2m-1)(2m-3)\ldots\ldots 1}{(2m)(2m-2)\ldots\ldots\ldots 2}\,\frac{\pi}{2}$. *(AMIETE, June 2011)*

Solution Refer Article 11.6, we have proved

$$\beta(m,n) = 2\int_0^{\pi/2} \sin^{2m-1}\theta\cos^{2n-1}\theta\,d\theta.$$

Therefore, $\int_0^{\pi/2} \sin^{2m-1}(\theta)\,d\theta = \frac{1}{2}\beta(m,\frac{1}{2})$ and $\int_0^{\pi/2} \sin^{2m}(\theta)\,d\theta = \frac{1}{2}\beta\left(m+\frac{1}{2},\frac{1}{2}\right)$.

(*i*) Let $I_1 = \int_0^{\pi/2} \sin^{2m-1}(\theta)\,d\theta = \frac{1}{2}\beta\left(m,\frac{1}{2}\right) = \frac{\overline{|(m)}\,\overline{|(1/2)}}{2\overline{|(m+1/2)}}$.

We have $\overline{|m} = (m-1)!$ and $\overline{\left|\left(m+\frac{1}{2}\right)\right.} = \left(m-\frac{1}{2}\right)\left(m-\frac{3}{2}\right)\ldots\left(\frac{1}{2}\right)\Gamma\left(\frac{1}{2}\right)$

$$= \frac{1}{2^m}[(2m-1)(2m-3)\ldots 3.1]\,\Gamma\left(\frac{1}{2}\right).$$

Therefore,
$$I = \frac{(m-1)!\; 2^m\; \Gamma(1/2)}{2\,(2m-1)\,(2m-3)\ldots\ldots 3.1.\,\Gamma(1/2)} = \frac{2^{m-1}\,[(m-1)\,(m-2)\ldots\ldots 2.1]}{(2m-1)\,(2m-3)\ldots\ldots 3.1}$$
$$= \frac{(2m-2)\,(2m-4)\ldots\ldots 4.2}{(2m-1)\,(2m-3)\ldots\ldots 3.1}.$$

(*ii*) Let
$$I_2 = \int_0^{\pi/2} \sin^{2m}(\theta)\,d\theta = \frac{1}{2}\beta\left(m+\frac{1}{2},\frac{1}{2}\right) = \frac{\Gamma(m+1/2)\;\Gamma(1/2)}{2\,\Gamma(m+1)}$$
$$= \frac{1}{2\,(m!)}\left[\frac{(2m-1)\,(2m-3)\ldots\ldots 3.1}{2^m}\right](\sqrt{\pi})^2$$
$$= \frac{(2m-1)\,(2m-3)\ldots\ldots 3.1}{2^{m+1}\,[m\,(m-1)\ldots\ldots 2.1]}(\pi) = \frac{(2m-1)\,(2m-3)\ldots\ldots 3.1}{(2m)\,(2m-2)\ldots\ldots 4.2}\,\frac{\pi}{2}.$$

EXAMPLE 11.11 Prove that (*i*) $\overline{|m}\;\overline{|(m+1/2)} = \dfrac{\sqrt{\pi}}{2^{2m-1}}\,\overline{|2m}$ and (*ii*) $\overline{|(1/4)}\;\overline{|(3/4)} = \pi\,\sqrt{2}.$

(AMIETE, June 2011, 2010)

Solution (*i*) We know that $\beta(m, m) = \int_0^1 x^{m-1}(1-x)^{m-1}dx.$ ...(*a*)

Setting $x = \sin^2\theta$, $dx = 2\sin\theta\cos\theta\,d\theta$

Therefore, $$\beta(m, m) = \int_0^{\pi/2} \sin^{2m-2}\theta\,\cos^{2m-2}\theta\cdot 2\sin\theta\cos\theta\,d\theta$$
$$= 2\int_0^{\pi/2} \sin^{2m-1}\theta\,\cos^{2m-1}\theta\,d\theta = 2\int_0^{\pi/2}\frac{(2\sin\theta\cos\theta)^{2m-1}}{2^{2m-1}}d\theta = 2\int_0^{\pi/2}\frac{(\sin 2\theta)^{2m-1}}{2^{2m-1}}d\theta.$$

Replacing 2θ by ϕ, the above integrals become

$$\beta(m, m) = \frac{1}{2^{2m-1}}\int_0^{\pi}\sin^{2m-1}\phi\,d\phi = \frac{2}{2^{2m-1}}\int_0^{\pi/2}\sin^{2m-1}\phi\,d\phi = \frac{1}{2^{2m-1}}\,\frac{\overline{|m}\;\overline{|(1/2)}}{\overline{\left|\left(m+\frac{1}{2}\right)\right.}}. \quad \text{...}(b)$$

But $$\beta(m, m) = \frac{\overline{|m}\;\overline{|m}}{\overline{|2m}}. \quad \text{...}(c)$$

Relations (*b*) and (*c*) give $\dfrac{\overline{|m}}{\overline{|2m}} = \dfrac{\sqrt{\pi}}{2^{2m-1}\,\overline{\left|m+\frac{1}{2}\right.}}$ or $\overline{|m}\;\overline{|(m+1/2)} = \dfrac{\sqrt{\pi}}{2^{2m-1}}\,\overline{|2m}.$ ...(*d*)

(*b*) Setting $m = 1/4$ in (*d*), we obtain

$$\overline{|1/4}\;\overline{|3/4} = \frac{\sqrt{\pi}\;\overline{|1/2}}{2^{(1/2)-1}} = \sqrt{\pi}\cdot\sqrt{\pi}\cdot\sqrt{2} = \pi\sqrt{2}\,.$$

EXAMPLE 11.12 Show that $\int_{-1}^{1}(1+x)^{p-1}\,(1-x)^{q-1}dx = 2^{p+q-1}\dfrac{\overline{|p}\;\overline{|q}}{\overline{|p+q}}.$

Also establish the duplication formula $\overline{\left|n+\frac{1}{2}\right.} = \dfrac{\overline{|(2n+1)}}{2^{2n}\,\overline{|(n+1)}}\sqrt{\pi}\,.$

Solution From the definition of Beta function, we have

$$\beta(m, n) = \int_0^1 x^{m-1}(1-x)^{n-1}\, dx$$

Put $x = (1 + y)/2$, $dx = (1/2)\, dy$; when $x = 0$, $y = -1$ and when $x = 1$, $y = 1$ and replacing m by p and n by q, we have

$$\beta(p, q) = \int_{-1}^1 \left(\frac{1+y}{2}\right)^{p-1}\left(\frac{1-y}{2}\right)^{q-1}\frac{1}{2}dy = \frac{1}{2^{p+q-1}}\int_{-1}^1 (1+y)^{p-1}(1-y)^{q-1}dy$$

or
$$\int_{-1}^1 (1+y)^{p-1}(1-y)^{q-1}dy = 2^{p+q-1}\beta(p, q) = 2^{p+q-1}\frac{\Gamma p\,\Gamma q}{\Gamma(p+q)}.$$

Again
$$\beta(m, n) = 2\int_0^{\pi/2}\cos^{2m-1}\theta\sin^{2n-1}\theta\, d\theta$$

or
$$\int_0^{\pi/2}\cos^{2m-1}\theta\sin^{2n-1}\theta\, d\theta = \frac{1}{2}\beta(m, n) = \frac{\Gamma m\,\Gamma n}{2\Gamma(m+n)}. \qquad ...(i)$$

Putting $m = 1/2$ in the above integral, we get

$$\int_0^{\pi/2}\sin^{2n-1}\theta\, d\theta = \frac{\Gamma n\,\Gamma(1/2)}{2\Gamma\left(n+\frac{1}{2}\right)}. \qquad ...(ii)$$

Again putting $m = n + 1$, the integral (i) reduces to

$$\int_0^{\pi/2}\sin^{2n-1}\theta\,\cos^{2n+1}\theta\, d\theta = \frac{\Gamma n\,\Gamma(n+1)}{2\Gamma(2n+1)}. \qquad ...(iii)$$

The integral (*iii*) can be written as

$$\int_0^{\pi/2}\sin^{2n-1}\theta\cos^{2n-1}\theta\cos^2\theta\, d\theta = \frac{1}{2^{2n-1}}\int_0^{\pi/2}(2\sin\theta\cos\theta)^{2n-1}\cos^2\theta\, d\theta$$

$$= \frac{1}{2^{2n}}\int_0^{\pi/2}(\sin 2\theta)^{2n-1}(1+\cos 2\theta)d\theta$$

$$= \frac{1}{2^{2n}}\int_0^{\pi/2}(\sin 2\theta)^{2n-1}d\theta + \frac{1}{2^{2n}}\int_0^{\pi/2}(\sin 2\theta)^{2n-1}\cos 2\theta\, d\theta.$$

Replacing 2θ by ϕ, the above integrals become

$$\frac{1}{2^{2n}}\int_0^{\pi}(\sin\phi)^{2n-1}\frac{1}{2}d\phi + \frac{1}{2^{2n}}\int_0^{\pi}(\sin\phi)^{2n-1}(\cos\phi)\frac{1}{2}d\phi$$

$$= \frac{1}{2^{2n}}\int_0^{\pi/2}(\sin\phi)^{2n-1}d\phi + \frac{1}{2^{2n+1}}\left[\frac{(\sin\phi)^{2n}}{2n}\right]_0^{\pi} = \frac{1}{2^{2n}}\int_0^{\pi/2}\sin^{2n-1}\theta\, d\theta.$$

This gives

$$\int_0^{\pi/2}\sin^{2n-1}\theta\cos^{2n+1}\theta\, d\theta = \frac{1}{2^{2n}}\int_0^{\pi/2}\sin^{2n-1}\theta\, d\theta. \qquad ...(iv)$$

Relations (*ii*), (*iii*) and (*iv*) give

$$\frac{\Gamma n\,\Gamma(1/2)}{2\Gamma(n+1/2)} = \frac{2^{2n}\,\Gamma n\,\Gamma(n+1)}{2\Gamma(2n+1)}$$

or
$$\Gamma(n+1/2) = \frac{\Gamma(2n+1)}{2^{2n}\Gamma(n+1)}\sqrt{\pi}.$$

11.8 DIRICHLET'S INTEGRAL

If V represents the closed region in the first octant bounded by the surface $\left(\frac{x}{a}\right)^p+\left(\frac{y}{b}\right)^q+\left(\frac{z}{c}\right)^r=1$ and the coordinate planes, then if all constants are positive,

$$\iiint_V x^{\alpha-1}\,y^{\beta-1}\,z^{\gamma-1}\,dx\,dy\,dz=\frac{a^\alpha b^\beta c^\gamma}{pqr}\,\frac{\Gamma(\alpha/p)\,\Gamma(\beta/q)\,\Gamma(\gamma/r)}{\Gamma\left(1+\frac{\alpha}{p}+\frac{\beta}{q}+\frac{\gamma}{r}\right)}.$$

This integral is known as Dirichlets integral and is often useful in evaluating many multiple integrals.

Let $\left(\frac{x}{a}\right)^p=u,\ \left(\frac{y}{b}\right)^q=v,\ \left(\frac{z}{c}\right)^r=w,$ then

$x=au^{1/p},\ y=bv^{1/q},\ z=cw^{1/r}$ and $dx=\frac{a}{p}u^{\frac{1}{p}-1}\,du,\ dy=\frac{b}{q}v^{\frac{1}{q}-1}\,dv,\ dz=\frac{c}{r}w^{\frac{1}{r}-1}\,dw.$

Substituting for x, y, z in the given integral, we obtain

$$I=\iiint_R a^{\alpha-1}u^{\frac{\alpha-1}{p}}\,b^{\beta-1}v^{\frac{\beta-1}{q}}\,c^{r-1}w^{\frac{\gamma-1}{r}}\,\frac{a}{p}\cdot\frac{b}{q}\cdot\frac{c}{r}\,u^{\frac{1}{p}-1}\,v^{\frac{1}{q}-1}\,w^{\frac{1}{r}-1}\,du\,dv\,dw$$

$$=\frac{a^\alpha b^\beta c^\gamma}{pqr}\iiint_R u^{\frac{\alpha}{p}-1}\,v^{\frac{\beta}{q}-1}\,w^{\frac{\gamma}{r}-1}\,du\,dv\,dw,$$

where R is the region in the uvw space bounded by the plane $u+v+w=1$ and the coordinate planes of this space. Thus

$$I=\frac{a^\alpha b^\beta c^\gamma}{pqr}\int_{u=0}^{1}\int_{v=0}^{1-u}\int_{w=0}^{1-u-v}u^{\frac{\alpha}{p}-1}\,v^{\frac{\beta}{q}-1}\,w^{\frac{\gamma}{r}-1}\,du\,dv\,dw.$$

Integrating the above integral and substituting the limits, we get

$$I=\frac{a^\alpha b^\beta c^\gamma}{pqr}\int_0^1\int_0^{1-u}\frac{r}{\gamma}[1-u-v]^{\frac{\gamma}{r}}\,u^{\frac{\alpha}{p}-1}\,v^{\frac{\beta}{q}-1}\,du\,dv$$

$$=\frac{a^\alpha b^\beta c^\gamma}{pqr}\,\frac{r}{\gamma}\int_0^1 u^{\frac{\alpha}{p}-1}\left[\int_0^{1-u}v^{\frac{\beta}{q}-1}(1-u-v)^{\frac{\gamma}{r}}\,dv\right]du.$$

Putting $v=(1-u)t$, so that $dv=(1-u)dt$, we have

$$\int_{v=0}^{1-u}v^{\frac{\beta}{q}-1}(1-u-v)^{\frac{\gamma}{r}}\,dv=(1-u)^{\frac{\beta}{q}+\frac{\gamma}{r}}\int_{t=0}^{1}t^{\frac{\beta}{q}-1}(1-t)^{\frac{\gamma}{r}}\,dt$$

$$=(1-u)^{\frac{\beta}{q}+\frac{\gamma}{r}}\,\frac{\Gamma\left(\frac{\beta}{q}\right)\Gamma\left(\frac{\gamma}{r}+1\right)}{\Gamma\left(\frac{\beta}{q}+\frac{\gamma}{r}+1\right)}=(1-u)^{\frac{\beta}{q}+\frac{\gamma}{r}}\,\frac{\Gamma\left(\frac{\beta}{q}\right)\times\frac{\gamma}{r}\Gamma\left(\frac{\gamma}{r}\right)}{\Gamma\left(\frac{\beta}{q}+\frac{\gamma}{r}+1\right)}$$

so that I can be written as

$$I = \frac{a^\alpha b^\beta c^\gamma}{pqr}\frac{r}{\gamma}\int_0^1 u^{\frac{\alpha}{p}-1}(1-u)^{\frac{\beta}{q}+\frac{\gamma}{r}}\frac{\Gamma\left(\frac{\beta}{q}\right)\times\Gamma\left(\frac{\gamma}{r}\right)\Gamma\left(\frac{\gamma}{r}\right)}{\Gamma\left(\frac{\beta}{q}+\frac{\gamma}{r}+1\right)}du$$

$$= \frac{a^\alpha b^\beta c^\gamma}{pqr}\frac{\Gamma\left(\frac{\beta}{q}\right)\Gamma\left(\frac{\gamma}{r}\right)}{\Gamma\left(1+\frac{\beta}{q}+\frac{\gamma}{r}\right)}\int_0^1 u^{\frac{\alpha}{p}-1}(1-u)^{\frac{\beta}{q}+\frac{\gamma}{r}}du$$

$$= \frac{a^\alpha b^\beta c^\gamma}{pqr}\frac{\Gamma\left(\frac{\beta}{q}\right)\Gamma\left(\frac{\gamma}{r}\right)\Gamma\left(\frac{\alpha}{p}\right)\Gamma\left(\frac{\beta}{q}+\frac{\gamma}{r}+1\right)}{\Gamma\left(\frac{\beta}{q}+\frac{\gamma}{r}+1\right)\Gamma\left(1+\frac{\alpha}{p}+\frac{\beta}{q}+\frac{\gamma}{r}\right)} = \frac{a^\alpha b^\beta c^\gamma}{pqr}\frac{\Gamma\left(\frac{\alpha}{p}\right)\Gamma\left(\frac{\beta}{q}\right)\Gamma\left(\frac{\gamma}{r}\right)}{\Gamma\left(1+\frac{\alpha}{p}+\frac{\beta}{q}+\frac{\gamma}{r}\right)}. \quad ...(11.8)$$

EXAMPLE 11.13 Evaluate the Dirichlet integral $I = \iiint_V x^{1/2}y^{1/2}z^{1/2}\,dx\,dy\,dz$, where V is the region in the first octant bounded by the plane $x + y + z = 1$ and the coordinate planes.

Solution Comparing the given integral with Dirichlet integral equation, we obtain
$\alpha = \beta = \gamma = 3/2$, $p = q = r = 1$, $a = b = c = 1$.

$$\therefore \qquad I = \frac{\left[\Gamma(3/2)\right]^3}{\Gamma(11/2)} = \frac{\left[(1/2)\Gamma(1/2)\right]^3}{(9/2)(7/2)(5/2)(3/2)(1/2)\Gamma(1/2)} = \frac{4\pi}{945}.$$

EXAMPLE 11.14 Find the volume of the region bounded by $x^m + y^m + z^m = a^m$, where $m > 0$.

Solution The volume bounded by the region is 8 times the volume bounded by the first octant and is given by

$$I = 8\iiint dx\,dy\,dz.$$

The given equation can be written as

$$\left(\frac{x}{a}\right)^m + \left(\frac{y}{a}\right)^m + \left(\frac{z}{a}\right)^m = 1.$$

Let $\left(\frac{x}{a}\right)^m = u, \left(\frac{y}{a}\right)^m = v, \left(\frac{z}{a}\right)^m = w$ so that $u + v + w = 1$.

Now the volume in uvw space bounded by the plane $u + v + w + 1$ and the coordinate planes of this space is given by

$$I = \frac{8a^3}{m^3}\int_{u=0}^{1}\int_{v=0}^{1-u}\int_{w=0}^{1-u-v} u^{\frac{1}{m}-1} v^{\frac{1}{m}-1} w^{\frac{1}{m}-1}\,du\,dv\,dw$$

$$= \frac{8a^3}{m^3}\int_0^1\int_0^{1-u} u^{\frac{1}{m}-1} v^{\frac{1}{m}-1}(1-u-v)^{\frac{1}{m}}\,m\,du\,dv.$$

Putting $v = (1-u)t$, $dv = (1-u)\,dt$, we get

$$I = \frac{8a^3}{m^2}\int_0^1 u^{\frac{1}{m}-1}(1-u)^{\frac{2}{m}}du\int_0^1 t^{\frac{1}{m}-1}(1-t)^{\frac{1}{m}}dt$$

$$= \frac{8a^3}{m^2} \frac{\Gamma\left(\frac{1}{m}\right)\Gamma\left(\frac{2}{m}+1\right)}{\Gamma\left(\frac{3}{m}+1\right)} \times \frac{\Gamma\left(\frac{1}{m}\right)\Gamma\left(\frac{1}{m}+1\right)}{\Gamma\left(\frac{2}{m}+1\right)} = \frac{8a^3}{m^2} \frac{\Gamma\left(\frac{1}{m}\right)\Gamma\left(\frac{1}{m}\right) \times \frac{1}{m}\Gamma\left(\frac{1}{m}\right)}{\frac{3}{m}\Gamma\left(\frac{3}{m}\right)} = \frac{8a^3}{m^2} \frac{\left(\Gamma(1/m)\right)^3}{3\Gamma(3/m)}.$$

EXERCISE 11.1

1. Evaluate the following:

(*i*) $\frac{\Gamma(4)\Gamma(2.5)}{\Gamma(5.5)}$, (*ii*) $\Gamma(-2.5)$, (*iii*) $\frac{\Gamma 3\,\Gamma(3/2)}{\Gamma(9/2)}$, (*iv*) $\Gamma(-0.5)$,

(*v*) $\beta(3, 5/2)$, (*vi*) $\beta(3/2, 2)$, (*vii*) $\beta(9/2, 7/2)$, (*Andhra, 2000*)

(*viii*) $\Gamma\left(\frac{1}{5}\right)\Gamma\left(\frac{2}{5}\right)$, (*ix*) $\beta(1/3, 2/3)$.

Ans. (*i*) $16/105$, (*ii*) $-8\sqrt{\pi}/15$, (*iii*) $16/105$, (*iv*) $-2\sqrt{\pi}$, (*v*) $16/315$, (*vi*) $4/15$, (*vii*) $5\pi/2048$, (*viii*) $4\pi^2/\sqrt{5}$, (*ix*) $2\pi/\sqrt{3}$.

2. Evaluate the following integrals:

(*i*) $\int_{-\infty}^{\infty} e^{-x^2}\,dx$, (*AMIETE, June 2009*) (*ii*) $\int_0^{\infty} e^{-x^4}\,dx$, (*iii*) $\int_0^{\infty} e^{-x^2/2}\,dx$,

(*iv*) $\int_0^{\infty} e^{-x^3}\,dx$, (*v*) $\int_0^{\infty} e^{-kx}\,x^{p-1}dx, k > 0$, (*VTU, 2002*)

(*vi*) $\int_0^{\infty} e^{-4x} x^{3/2}\,dx$, (*vii*) $\int_0^{\infty} \frac{e^{-\sqrt{x}}}{x^{7/4}}\,dx$, (*viii*) $\int_0^{\infty} y^3 e^{-2y}\,dy$,

(*ix*) $\int_0^{\infty} x^2 e^{-2x^2}\,dx$, (*x*) $\int_0^1 \{\log(1/x)\}^{-1/2}\,dx$, (*xi*) $\int_0^{\infty} \sqrt{x}\, e^{-x^2}\,dx$,

(*xii*) $\int_0^{\infty} x^{1/3}\, e^{-x^2}\,dx$, (*xiii*) $\int_0^1 \frac{dx}{\sqrt{-\log x}}$, (*xiv*) $\int_0^{\infty} \sqrt{x}\, e^{-\sqrt[3]{x}}\,dx$.

Ans. (*i*) $\sqrt{\pi}$, (*ii*) $\Gamma(5/4)$, (*iii*) $\sqrt{\pi}/2$, (*iv*) $\frac{1}{3}\Gamma(1/3)$, (*v*) $\frac{\Gamma(p)}{k^p}$, (*vi*) $\frac{3}{128}\sqrt{\pi}$, (*vii*) $8\pi/3$, (*viii*) $3/8$, (*ix*) $\sqrt{2\pi}/16$, (*x*) $\sqrt{\pi}$, (*xi*) $\frac{1}{2}\Gamma(3/4)$, (*xii*) $\frac{1}{2}\Gamma(2/3)$, (*xiii*) $\sqrt{\pi}$, (*xiv*) $\frac{315}{16}\sqrt{\pi}$.

[**Hint:** (*xiv*) Put $x^{1/3} = t$, $dx = 3t^2dt$. Now the given integral

$$= \int_0^{\infty} e^{-t} \cdot t^{3/2}\,(3t^2)\,dt = 3\int_0^{\infty} e^{-t} \cdot t^{\frac{7}{2}}\,dt = 3\int_0^{\infty} e^{-t} t^{\left(\frac{9}{2}-1\right)} dt$$

$$= 3\Gamma 9/2 = 3 \cdot \frac{7}{2} \cdot \frac{5}{2} \cdot \frac{3}{2} \cdot \frac{1}{2}\sqrt{\pi} = \frac{315}{16}\sqrt{\pi}.]$$

3. Evaluate each of the following integrals using the Beta and Gamma functions.

(*i*) $\int_0^1 t^{-1/3}(1-t)^{2/3}\,dt$, (*ii*) $\int_0^4 \frac{dx}{\sqrt{4t - t^2}}$, (*iii*) $\int_0^{\pi/2} \sin^4\theta \cos^5\theta\, d\theta$,

(*iv*) $\int_0^2 \frac{x^2}{\sqrt{2-x}}\,dx$, (*v*) $\int_0^2 \sqrt{x(2-x)}\,dx$, (*vi*) $\int_0^{\pi/2} \sqrt{\tan\theta}\, d\theta$, (*Madras, 2006*) (*vii*) $\int_0^{\pi/2} \sin^3 x \cos^{5/2} x\, dx$, (*viii*) $\int_0^{\infty} x^n e^{-a^2x^2} dx$, (*ix*) $\int_0^{\infty} x^6 e^{-2x} dx$,

(*x*) $\int_0^{\pi/2} \tan^p \theta\, d\theta$, (*xi*) $\int_0^{\infty} \frac{dx}{1+x^4}$.

Ans. (*i*) $\frac{4\pi}{3\sqrt{3}}$, (*ii*) π, (*iii*) $\frac{8}{315}$, (*iv*) $\frac{64\sqrt{2}}{15}$, (*v*) $\frac{\pi}{2}$,

[**Hint:** (*v*) Put $x = 2t$, the integral becomes

$$\int_0^1 \sqrt{4t(1-t)}\, 2dt = 4\int_0^1 t^{1/2}(1-t)^{1/2} dt = 4\beta(3/2, 3/2)$$

$$= \frac{4\overline{|(3/2)}\,\overline{|(3/2)}}{\overline{|3}} = \frac{4\cdot\left(\frac{1}{2}\sqrt{\pi}\right)\left(\frac{1}{2}\sqrt{\pi}\right)}{2} = \frac{\pi}{2}.]$$

(*vi*) $\frac{1}{2}\overline{|(1/4)}\,\overline{|(3/4)} = \frac{\pi\sqrt{2}}{2}$, (*vii*) $\frac{8}{77}$, (*viii*) $\frac{1}{2a^{n+1}}\overline{\left|\frac{(n+1)}{2}\right.}, n > -1$, (*ix*) $\frac{45}{8}$,

(*x*) $\frac{\pi}{2}\sec\frac{p\pi}{2}$, (*xi*) $\frac{1}{4}\beta\left(\frac{1}{4}, \frac{3}{4}\right) = \frac{1}{4}\overline{|(1/4)}\,\overline{|(3/4)} = \frac{\pi}{2\sqrt{2}}$.

4. Prove that

(*i*) $\int_0^1 \frac{x dx}{\sqrt{1-x^5}} = \frac{1}{5}\beta\left(\frac{2}{5}, \frac{1}{2}\right)$, (*Raipur, 2006*) (*ii*) $\int_0^1 \frac{dx}{\sqrt{1+x^4}} = \frac{1}{4\sqrt{2}}\beta\left(\frac{1}{4}, \frac{1}{2}\right)$, (*VTU, 2003*)

(*iii*) $\int_0^2 (8-x^3)^{-\frac{1}{3}} dx = \frac{1}{3}\beta\left(\frac{1}{3}, \frac{2}{3}\right)$, (*iv*) $\int_0^1 x^3(1-\sqrt{x})^5 dx = 2\beta(8, 6)$, (*JNTU, 2006*)

(*v*) $\int_0^2 \sqrt{x}(4-x^2)^{-1/4} dx = \beta\left(\frac{3}{4}, \frac{3}{4}\right)$, (*vi*) $\int_0^1 (1-x^3)^{-1/3} dx = \frac{1}{3}\beta\left(\frac{1}{3}, \frac{2}{3}\right)$.

5. Establishing the following:

(*i*) $\beta(n, n) = 2\int_0^{1/2} (t-t^2)^{n-1} dt = \frac{\sqrt{\pi}\,\overline{|(n)}}{2^{2n-1}\overline{|(n+1/2)}}$,

(*ii*) $1\cdot 3\cdot 5\cdot \ldots (2n-1)\sqrt{\pi} = 2^n\overline{|(n+1/2)}$, (*iii*) $\beta\left(m, \frac{1}{2}\right) = 2^{2m-1}\beta(m, m)$,

(*AMIETE, June 2011; VTU, 2004*)

(*iv*) $\frac{\beta(m+1, n)}{m} = \frac{\beta(m, n+1)}{n} = \frac{\beta(m, n)}{m+n}$, (*v*) $\int_0^{\infty} \frac{x^{m-1}+x^{n-1}}{(1+x)^{m+n}} dx = 2\beta(m, n)$.

[**Hint:** (*iii*) We know that $\beta(m, n) = 2\int_0^{\pi/2} \sin^{2m-1}\theta \cos^{2n-1}\theta\, d\theta$ ...(1)

Setting $n = 1/2$, we have $\beta(m, 1/2) = 2\int_0^{\pi/2} \sin^{2m-1}\theta\, d\theta$

Again putting $n = m$ in (1), we obtain

$$\beta(m, m) = 2\int_0^{\pi/2} (\sin\theta\cos\theta)^{2m-1}\, d\theta = \frac{1}{2^{2m-2}}\int_0^{\pi/2} \sin^{2m-1} 2\theta\, d\theta$$

$$= \frac{1}{2^{2m-1}}\int_0^{\pi} \sin^{2m-1}\phi\, d\phi, \qquad \text{putting } 2\theta = \phi$$

$$= \frac{1}{2^{2m-1}}\cdot 2\int_0^{\pi/2} \sin^{2m-1}\phi\, d\phi$$

or $2^{2m-1}\beta(m, m) = 2\int_0^{\pi/2} \sin^{2m-1}\theta\, d\theta = \beta\left(m, \frac{1}{2}\right)$]

6. Prove the following:

(i) $\displaystyle\int_0^{\pi/2} \frac{\sin^{2m-1}\theta\cos^{2n-1}}{(a\sin^2\theta + b\cos^2\theta)^{m+n}} = \frac{\Gamma(m)\Gamma(n)}{2a^m b^n \Gamma(m+n)}$,

(ii) $\displaystyle\int_0^{\infty} \frac{x^{m-1} - x^{n-1}}{(1+x)^{m+n}}\, dx = 0,\; m > 0,\, n > 0,$

(iii) $\displaystyle\int_0^{\pi/2} (\sin x)^{8/3}(\sec x)^{1/2}\, dx = \frac{60\Gamma(5/6)\Gamma(1/4)}{13\Gamma(1/12)}$,

(iv) $\displaystyle\int_{-\infty}^{\infty} \frac{e^{2x}}{ae^{3x} + b}\, dx = \frac{2\pi}{3\sqrt{3}\, a^{2/3} b^{1/3}}$, where $a, b > 0$.

Hence or otherwise show that $\displaystyle\int_{-\infty}^{\infty} \frac{e^{2x}}{(e^{3x}+1)^2}\, dx = \frac{2\pi}{9\sqrt{3}}$.

7. Show that

(i) $\displaystyle\int_0^{\infty} \frac{x^4(1+x^5)}{(1+x)^{15}}\, dx = \frac{1}{5005}$, (ii) $\displaystyle\int_0^{\infty} \frac{x^8(1-x^6)}{(1+x)^{24}} = 0$,

(iii) $\displaystyle\int_0^{4} y^{3/2}(16-y^2)^{1/2}\, dy = \frac{64}{21}\left\{\Gamma(1/4)\right\}^2$,

[**Hint:** (*iii*) Put $y^2 = 16t$, that is, $y = 4\sqrt{t}$, $dy = 2t^{-1/2}\, dt$. Then the integral becomes

$$\int_0^1 \{8t^{3/4}\}\{4(1-t)^{1/2}\}\{2t^{-1/2}\, dt\} = 64\int_0^1 t^{1/4}(1-t)^{1/2}\, dt$$

$$= 64\,\beta\left(\frac{5}{4}, \frac{3}{2}\right) = 64\,\frac{\Gamma(5/4)\;\Gamma(3/2)}{\Gamma(11/4)} = \frac{64\,(1/4)\,\Gamma(1/4)\,(1/2)\,\Gamma(1/2)}{\frac{7}{4}\cdot\frac{3}{4}\,\Gamma(3/4)}$$

$$= \frac{128\sqrt{\pi}\,\Gamma(1/4)}{21\,\Gamma(3/4)} = \frac{128\sqrt{\pi}}{21}\,\frac{\{\Gamma(1/4)\}^2}{\Gamma(1/4)\,\Gamma(3/4)} = \frac{64}{21}\left[\frac{2}{\pi}\{\Gamma(1/4)\}^2.\right]$$

(iv) $\displaystyle\int_0^1 \frac{x^2 dx}{\sqrt{(1-x^4)}} \times \int_0^1 \frac{dx}{\sqrt{(1+x^4)}} = \frac{\pi}{4\sqrt{2}}$, *(AMIETE, Dec. 2009)*

[Hint: (*iv*) Put $x^2 = \sin\theta$ in the first integral, we have

$$\int_0^1 \frac{x^2 dx}{\sqrt{1-x^4}} = \frac{1}{2}\int_0^{\pi/2} \sin^{1/2}\theta\, d\theta = \frac{1}{4}\beta\left(\frac{3}{4},\frac{1}{2}\right) = \frac{1}{4}\frac{\Gamma(3/4)\,\Gamma(1/2)}{\Gamma(5/4)} = \frac{\Gamma(3/4)\,\Gamma(1/2)}{\Gamma(1/4)} \quad ...(i)$$

Now, put $x^2 = \tan\theta$, in the second integral, we have

$$\int_0^1 \frac{dx}{\sqrt{(1+x^4)}} = \frac{1}{\sqrt{2}}\int_0^{\pi/4} \frac{d\theta}{\sqrt{\sin 2\theta}} = \frac{1}{2\sqrt{2}}\int_0^{\pi/2} \sin^{-1/2}\phi\, d\phi$$

$$= \frac{1}{4\sqrt{2}}\beta\left(\frac{1}{4},\frac{1}{2}\right) = \frac{1}{4\sqrt{2}}\frac{\Gamma(1/4)\,\Gamma(1/2)}{\Gamma(3/4)} \quad ...(ii)$$

Multiplying (*i*) and (*ii*), we obtain the desired result.]

(*v*) $\int_{-1}^1 (1-x^2)^n dx$, (where n is a positive integer) $= \dfrac{2^{2n+1}(n!)^2}{(2n+1)!}$,

(*vi*) $\int_0^1 \dfrac{dx}{\sqrt{1-x^4}} = \dfrac{1}{8}\sqrt{\dfrac{2}{\pi}}\left[\Gamma(1/4)\right]^2$, (*vii*) $\int_0^1 \dfrac{dx}{(1-x^n)^{1/n}} = \dfrac{\pi}{n\sin(\pi/n)}$,

(*viii*) $\int_0^\infty \dfrac{e^{-\sqrt{x}}}{x^{7/4}}\, dx = \dfrac{8}{3}\sqrt{\pi}$.

8. (*i*) Given $\int_0^\infty \dfrac{x^{n-1}}{1+x}\, dx = \dfrac{\pi}{\sin n\pi}$, show that $\Gamma(n)\,\Gamma(1-n) = \dfrac{\pi}{\sin n\pi}$, where $0 < n < 1$. (*Delhi, 2006*)

Hence, evaluate $\int_0^\infty \dfrac{dx}{1+x^4}$. (*VTU, 2006; JNTU, 2005*) **Ans.** $\pi/4\sqrt{2}$.

[Hint: Let $\dfrac{x}{1+x} = y$. Then $x = \dfrac{y}{1-y}$ and $dx = \dfrac{1}{(1-y)^2}dy$.

Therefore, $\int_0^\infty \dfrac{x^{n-1}}{1+x}dx = \int_0^1 y^{n-1}(1-y)^{-n}\,dy = \int_0^1 y^{n-1}(1-y)^{(1-n)-1}\,dy = \beta(n, 1-n)$

$$= \frac{\Gamma(n)\,\Gamma(1-n)}{\Gamma 1} = \Gamma(n)\,\Gamma(1-n) = \frac{\pi}{\sin n\pi}.]$$

(*ii*) Find $\beta\left(\dfrac{2}{3},\dfrac{1}{3}\right)$.

[Hint: $\beta\left(\dfrac{2}{3},\dfrac{1}{3}\right) = \dfrac{\Gamma 2/3\ \Gamma 1/3}{\Gamma(2/3+1/3)} = \dfrac{\Gamma 1/3\ \Gamma\left(1-\dfrac{1}{3}\right)}{\Gamma 1} = \dfrac{\pi}{\sin(\pi/3)} = \dfrac{2\pi}{\sqrt{3}}$.]

9. Show that if $n > -1$, $\int_0^\infty x^n e^{-k^2x^2} dx = \dfrac{1}{2k^{n\times 1}}\Gamma\left(\dfrac{n+1}{2}\right)$. Hence or otherwise evaluate $\int_{-\infty}^\infty e^{-k^2x^2} dx$.

Ans. $\sqrt{\pi}/k$.

10. Express $\int_0^1 x^m (1-x^n)^p\, dx$ in terms of Beta function and hence evaluate the integral $\int_0^1 x^5 (1-x^3)^{10}\, dx$.

Ans. $\dfrac{1}{n}\beta\left(\dfrac{m+1}{n}, p+1\right)$, $\dfrac{1}{396}$.

11. Prove that $\int_0^1 \frac{x^{m-1}(1-x)^{n-1}}{(a+bx)^{m+n}}\,dx = \frac{1}{a^n(a+b)^m}\beta(m, n)$. **[Hint:** Let $x/(a+bx) = z/(a+b)$.]

12. Show that $\iint x^{l-1} y^{m-1}\,dy\,dx\ (l, m > 0)$, taken over the area $x \ge 0$, $y \ge 0$, $x + y \le 1$ equals $\Gamma(l)\,\Gamma(m)/\Gamma(l+m+1)$.

[Hint: Given integral $= \int_0^1\int_0^{1-x} x^{l-1} y^{m-1}\,dy\,dx = \frac{1}{m}\int_0^1 x^{l-1}[y^m]_0^{1-x}\,dx = \frac{1}{m}\int_0^1 x^{l-1}(1-x)^m\,dx$

$$= \frac{1}{m}\beta(l, m+1) = \frac{1}{m}\frac{\Gamma(l)\,\Gamma(m+1)}{\Gamma(l+m+1)} = \frac{\Gamma(l)\,\Gamma(m)}{\Gamma(l+m+1)}.]$$

13. Find the mass of the region in the xy-plane bounded by $x + y = 1$, $x = 0$, $y = 0$ if the density is $\sigma = \sqrt{xy}$. **Ans.** $\pi/24$.

14. Show that $\iint x^{m-1} y^{n-1}\,dx\,dy$ over the positive quadrant of the ellipse $x^2/a^2 + y^2/b^2 = 1$ is $\frac{a^m b^n}{2n}\beta\left(\frac{m}{2}, \frac{m}{2}+1\right)$.

15. Find the mass of the region bounded by $x^2 + y^2 + z^2 = a^2$ if the density σ of the region is given by $\sigma = x^2y^2z^2$. **Ans.** $4\pi a^9/945$.

16. Find the mass of an octant of the ellipsoid $(x/a)^2 + (y/b)^2 + (z/c)^2 = 1$, the density at any point being $p = k\,xyz$. *(UPTU, 2002)* **Ans.** $ka^2b^2c^2/48$.

17. Evaluate $\iiint_V x^{l-1} y^{m-1} z^{n-1}\,dV$, where V is the volume cut from first octant by the plane $x + y + z = 1$.

Ans. $\Gamma(l)\,\Gamma(m)\,\Gamma(n)\,/\,\Gamma(l+m+n+1)$.

18. Evaluate $I = \iiint_V x^{l-1} y^{m-1} z^{n-1}\,dx\,dy\,dz$, where V is the region in the first octant bounded by the sphere $x^2 + y^2 + z^2 = 1$ and the coordinate planes. **Ans.** $\Gamma(l/2)\,\Gamma(m/2)\,\Gamma(n/2)\Big/ 8\,\Gamma\frac{(l+m+n+1)}{2}$.

[Hint: Put $x^2 = u$, $y^2 = v$, $z^2 = w$ so that $x = u^{1/2}$ etc.

Then the given integral $= \frac{1}{8}\iiint_{V_1} u^{\frac{l}{2}-1} v^{\frac{m}{2}-1} w^{\frac{n}{2}-1}\,du\,dv\,dw$, where V_1 is the region in the first octant bounded by the plane $u + v + w = 1$ and the coordinate planes $u = 0$, $v = 0$, $w = 0$.]

19. Find the volume of the solid bounded by the coordinate planes and the surface $(x/a)^{1/2} + (y/b)^{1/2} + (z/c)^{1/2} = 1$. **Ans.** $abc/90$.

20. The plane $x/a + y/b + z/c = 1$ meets the axes in A, B, C. Apply Dirichlet's integral to find the volume of the tetrahedron $OABC$. Also find its mass if the density at any point is $kxyz$. *(UPTU, 2004)* **Ans.** $abc/6$; $ka^2b^2c^2/720$.

[Hint: Let $x/a = u$, $y/b = v$, $z/c = w$.

Volume of the tetrahedron $= \iiint_D dx\,dy\,dz = \iiint_{D'} abc\,du\,dv\,dw = abc\int u^{1-1} v^{1-1} w^{1-1}\,du\,dv\,dw$

$$= \frac{abc\,\Gamma 1\,\Gamma 1\,\Gamma 1}{\Gamma(1+1+1+1)} = \frac{abc}{6}.$$

$$\text{Mass} = \iiint k\,xyz\,dx\,dy\,dz = k\iiint (au)(bv)(cw)\,abc\,du\,dv\,dw$$

$$= k\,a^2b^2c^2 \iiint u^{2-1}\,v^{2-1}\,w^{2-1}\,du\,dv\,dw = k\,a^2b^2c^2\,\frac{\Gamma 2\,\Gamma 2\,\Gamma 2}{\Gamma(2+2+2+1)}$$

$$= \frac{k\,a^2b^2c^2}{6!} = \frac{k}{720}\,a^2b^2c^2.]$$

21. Find the volume of the region bounded by $x^{2/3} + y^{2/3} + z^{2/3} = 1$. **Ans.** $4\pi/35$.

11.9 ELLIPTIC INTEGRALS

The integral

$$u = F(k, \phi) = \int_0^{\phi} \frac{dx}{\sqrt{(1-k^2\sin^2 x)}}, \qquad 0 < k < 1. \qquad \text{...(11.9)}$$

is defined as an *incomplete elliptic integral of the first kind* where ϕ is called the amplitude of $F(k, \phi)$ or u and k is its modulus. This integral is also known as *Legendre's form for the elliptic integral of the first kind.*

If $\phi = \pi/2$. the integral is defined as the complete integral of the first kind and as denoted by $K(k)$.

The integral

$$E(k, \phi) = \int_0^{\phi} \sqrt{(1-k^2\sin^2 x)}\;dx, \qquad 0 < k < 1. \qquad \text{...(11.10)}$$

is defined as *Legendre's form for the elliptic integral of the second kind.* If $\phi = \pi/2$, the integral (11.10) is called the *complete elliptic integral of the second kind* and is denoted by $E(k)$. Thus, we have

$$K(k) = \int_0^{\pi/2} \frac{dx}{\sqrt{(1-k^2\sin^2 x)}}, \qquad k^2 < 1 \qquad \text{...(11.11)}$$

$$\text{and} \quad E(k) = \int_0^{\pi/2} \sqrt{(1-k^2\sin^2 x)}\;dx, \qquad k^2 < 1 \qquad \text{...(11.12)}$$

It is appropriate to mention that the integral (11.10) occurs in the determination of the length of the arc of an ellipse and this suggests the reason for giving this integral the name elliptic integral. For instance, consider the ellipse

$$x = a\sin\phi, \qquad y = b\cos\phi, \qquad (a > b > 0)$$

The length of the arc AP of the ellipse is given by

Fig. 11.2

$$s = \int_0^{\phi} \sqrt{\left(\frac{dx}{d\phi}\right)^2 + \left(\frac{dy}{d\phi}\right)^2}\;d\phi$$

$$= \int_0^{\phi} \sqrt{\{(a\cos\phi)^2 + (-b\sin\phi)^2\}}\;d\phi$$

$$= \int_0^{\phi} \sqrt{\{a^2 - (a^2-b^2)\sin^2\phi\}}\;d\phi$$

$$= a\int_0^{\phi} \sqrt{\{1 - e^2\sin^2\phi\}}\;d\phi = a\,E(e, \phi), \qquad \text{....(11.13)}$$

where $e^2 = \dfrac{a^2-b^2}{a^2}$ is the square of the eccentricity of the ellipse.

It may be remarked that the integral (11.13) cannot be evaluated in terms of known elementary functions and hence a need for the definition of a new function.

If $\phi = \pi/2$, the total perimeter of the ellipse, is given by

$$s = 4a \int_0^{\pi/2} \sqrt{\{1 - e^2 \sin^2 \phi\}}\, d\phi = 4a\, E(e, \pi/2) = 4a\, E(e).$$

Consider now the complete integral of the first kind

$$K(k) = \int_0^{\pi/2} \frac{dx}{\sqrt{\{1 - k^2 \sin^2 x\}}}.$$

Expanding the integrand, we obtain

$$K(k) = \int_0^{\pi/2} \left(1 + \frac{1}{2} k^2 \sin^2 x + \frac{1.3}{2.4} k^4 \sin^4 x + \ldots\ldots\right) dx.$$

The infinite series in this integral is known to be a uniformly convergent series and therefore, can be integrated term by term. Thus, we get on integration

$$K(k) = \frac{\pi}{2} \left(1 + \left(\frac{1}{2}\right)^2 k^2 + \left(\frac{1.3}{2.4}\right)^2 k^4 + \left(\frac{1.3.5}{2.4.6}\right)^2 k^6 + \ldots\ldots\right).$$

If we are given the values of k, $K(k)$ can be computed from the above result.

Tables for elliptic integrals are available for $F(k, \phi)$ and $E(k, \phi)$ for values of k and ϕ, i.e., $0 \le \phi \le \pi/2$ and $0 \le k \le 1$. (Ref. *Tables of Functions* by Jahnake and Emde).

EXAMPLE 11.15 Show that the time period of oscillations of a simple pendulum of length l is given by $4\sqrt{(l/g)}\, K(\sin \alpha/2)$, where α is the angle through which the bob swings on either side of the mean position A.

Solution Let a pendulum of length l be suspended from 0 and let A be the initial position of the mass. Let the bob at any time t be at P where $\angle AOP = \theta$ and the arc $AP = s = l\theta$. The equation of motion of the pendulum is given by

$$m \frac{d^2 (l\theta)}{dt^2} = -\, \text{mg} \sin \theta$$

or

$$\frac{d^2\theta}{dt^2} = -\frac{g}{l} \sin \theta. \qquad \ldots(i)$$

Multiplying both sides of (*i*) by $2(d\theta/dt)$ and integrating, we get

$$\left(\frac{d\theta}{dt}\right)^2 = \frac{2g}{l} \cos \theta + \lambda, \quad \text{where } \lambda \text{ is constant.} \qquad \ldots(ii)$$

Since at $t = 0$, the angular displacement $\theta = \alpha$ and $d\theta/dt = 0$, we obtain $\lambda = -\,(2g/l) \cos \alpha$. Substituting for λ in the above equation, we get

$$\left(\frac{d\theta}{dt}\right)^2 = \frac{2g}{l} (\cos \theta - \cos \alpha)$$

or

$$\frac{d\theta}{dt} = \sqrt{\frac{2g}{l}} (\cos \theta - \cos \alpha)^{1/2} = \sqrt{\frac{2g}{l}} \left(2 \sin^2 \frac{\alpha}{2} - 2 \sin^2 \frac{\theta}{2}\right)^{1/2}.$$

Integrating this equation after separating the variables, we obtain

$$t = \frac{1}{2} \sqrt{\frac{l}{g}} \int \frac{d\theta}{\sqrt{\left(\sin^2 \frac{\alpha}{2} - \sin^2 \frac{\theta}{2}\right)}}.$$

Fig. 11.3

The period of oscillations from $\theta = 0$ to $\theta = \alpha$ which correspond to one quarter of a period T, is given by

$$\frac{T}{4} = \frac{1}{2}\sqrt{\frac{l}{g}}\int_0^{\alpha}\frac{d\theta}{\sqrt{\left(\sin^2\frac{\alpha}{2} - \sin^2\frac{\theta}{2}\right)}}. \qquad ...(iii)$$

Put $\sin\frac{\theta}{2} = \sin\frac{\alpha}{2}\sin\phi$, we get $\frac{1}{2}\cos\frac{\theta}{2}\,d\theta = \sin\frac{\alpha}{2}\cos\phi\,d\phi$

Substituting in equation (*iii*) and simplifying, we obtain

$$\frac{T}{4} = \sqrt{\frac{l}{g}}\int_0^{\pi/2}\frac{d\phi}{\sqrt{\left(1-\sin^2\frac{\alpha}{2}\sin^2\phi\right)}} \quad \text{or} \quad T = 4\sqrt{\frac{l}{g}}\,K\left(\sin\frac{\alpha}{2}\right).$$

EXAMPLE 11.16 Express $\int_0^{\pi/2}\frac{dx}{\sqrt{\sin x}}$ in terms of elliptic integrals.

Solution Letting $x = (\pi/2) - y$, the given integral can be written as

$$\int_0^{\pi/2}\frac{dx}{\sqrt{\sin x}} = \int_0^{\pi/2}\frac{dy}{\sqrt{\cos y}} = \int_0^{\pi/2}\frac{dy}{\sqrt{(1-2\sin^2(y/2)}} \qquad ...(i)$$

Put $\sqrt{2}\sin\frac{y}{2} = \sin\phi$. Then $\sqrt{2}\cos\left(\frac{y}{2}\right)\cdot\frac{1}{2}dy = \cos\phi\,d\phi$.

The given integral (*i*) after effecting the above substitution can be written as

$$\int_0^{\pi/2}\frac{\sqrt{2}\cos\phi\,d\phi}{\sqrt{\left(1-\frac{1}{2}\sin^2\phi\right)}\cos\phi} = \sqrt{2}\,K\left(\sqrt{\frac{1}{2}}\right).$$

EXAMPLE 11.17 Evaluate $\int_1^{\infty}\frac{du}{(3u^2+1)\sqrt{(u^2-1)(u^2-3)}}$.

Solution We have $I = \int_1^{\infty}\frac{du}{(3u^2+1)\sqrt{(u^2-1)(u^2+3)}} = \int_1^{\infty}\frac{(1/u^4)\,du}{\left(3+\frac{1}{u^2}\right)\sqrt{\left(1-\frac{1}{u^2}\right)\left(1+\frac{3}{u^2}\right)}}$.

Let $u = \frac{1}{\cos\theta}$ so that $\frac{1}{u^2}du = \sin\phi\,d\phi$. Then the integral become

$$I = \int_0^{\pi/2}\frac{\cos^2\phi\sin\phi\,d\phi}{(3+\cos^2\phi)\sqrt{(1-\cos^2\phi)(1+3\cos^2\phi)}} = \int_0^{\pi/2}\frac{(3+\cos^2\phi)-3}{(3+\cos^2\phi)\sqrt{1+3\cos^2\phi}}\,d\phi$$

$$= \int_0^{\pi/2}\frac{d\phi}{\sqrt{4-3\sin^2\phi}} - 3\int_0^{\pi/2}\frac{d\phi}{(4-\sin^2\phi)\sqrt{4-3\sin^2\phi}} = \frac{1}{2}K\left(\frac{1}{2}\sqrt{3}\right) - \frac{3}{8}\Pi\left(-\frac{1}{4},\frac{1}{2}\sqrt{3},\frac{1}{2}\pi\right).$$

11.10 JACOBI'S FORM FOR THE ELLIPTIC INTEGRALS

Effecting the transformation $v = \sin x$ and denoting $z = \sin\phi$, the Legendre's form for the elliptic integrals (11.9) and (11.10) can be respectively expressed as

$$F_1(k, z) = \int_0^z \frac{dv}{\sqrt{\{(1-v^2)(1-k^2v^2)\}}} \quad \text{...(11.14)}$$

$$E_1(k, z) = \int_0^z \sqrt{\left(\frac{1-k^2v^2}{1-v^2}\right)}\, dv \quad \text{...(11.15)}$$

Integrals (11.14) and (11.15) are defined *Jacobi's form for elliptic integrals of first and second kind* respectively.

These integrals are helpful in the evaluation of integrals of the type $\int \frac{dx}{\sqrt{P_4(x)}}$, where $P_4(x)$ is a polynomial of degree 4 in x with real zeroes. For example, the integral

$$\int \frac{dx}{\sqrt{\{(x-\alpha)(x-\beta)(x-\gamma)(x-\delta)\}}} \quad \text{...(11.16)}$$

can be changed to k_1 (k, ϕ) by using the *fractional linear transformation*

$$x = \frac{at+b}{ct+d}$$

and evaluating a, b, c, d under the conditions that the points α, β, γ, δ correspond respectively to the points

$$t = -1/k, \ -1, \ 1, \ 1/k.$$

EXAMPLE 11.18 Express the integral $\int \frac{dx}{\sqrt{(x-1)(x-2)(x-3)(x-4)}}$ in terms of elliptic integrals. Hence or otherwise evaluate $\int_1^\infty \frac{du}{\sqrt{(u^2-1)(u^2+3)}}$.

Solution Using the transformation $x = \frac{at+b}{ct+d}$ and choosing a, b, c, d so that x = 1, 2, 3 correspond respectively to t = 0, 1, ∞, we obtain $1 = \frac{b}{d}$, $2 = \frac{a+b}{c+d}$, $3 = \frac{a}{c}$. This leads to $x = \frac{3t+1}{t+1}$. Substituting for x, the given integral can be written as $\int \frac{dt}{\sqrt{t(t-1)(t+3)}}$. If $t = u^2$, we get $2\int \frac{du}{\sqrt{(u^2-1)(u^2+3)}}$. Let u = sec θ, the integral is transformed to $2\int \frac{\sec\theta \tan\theta\, d\theta}{\sqrt{(\sec^2\theta - 1)(\sec^2\theta+3)}} = 2\int \frac{d\theta}{\sqrt{(1+3\cos^2\theta)}} = 2\int \frac{d\theta}{\sqrt{(4-3\sin^2\theta)}}$

$= \int \frac{d\theta}{\sqrt{\left(1-\frac{3}{4}\sin^2\theta\right)}}$ which can be expressed in terms of elliptic integrals.

Since $\int_1^\infty \frac{du}{\sqrt{(u^2-1)(u^2+3)}} = \frac{1}{2}\int_0^{\pi/2} \frac{d\theta}{\sqrt{\left(1-\frac{3}{4}\sin^2\theta\right)}} = \frac{1}{2}K\left(\frac{\sqrt{3}}{2}, \frac{\pi}{2}\right)$.

11.11 ELLIPTIC FUNCTIONS

It is apparent from above that the upper limit z in Jacobi's integral of the first kind is related to the upper limit ϕ in Legendre's form by $z = \sin\phi$. Since ϕ = ampl (u), this gives z = sin (ampl (u)). Thus we are led to define the *elliptic functions*.

$$z = \sin(\text{ampl}(u)) = \text{sn}\, u = \sin\phi \quad ...(11.17)$$

$$\sqrt{1-z^2} = \cos(\text{ampl}(u)) = \text{cn}\, u = \cos\phi \quad ...(11.18)$$

$$\sqrt{1-k^2z^2} = \sqrt{1-k^2 sn^2 u} = \text{dn}\, u = \sqrt{1-k^2\sin^2\phi} \quad ...(11.19)$$

$$\frac{z}{\sqrt{1-z^2}} = \frac{sn\, u}{cn\, u} = \text{tn}\, u = \tan\phi \quad ...(11.20)$$

It is obvious that the above functions have properties analogous to trigonometric functions. Following are some of the various properties of these functions.

1. $sn(0) = 0,\quad cn(0) = 1,\; dn(0) = 1,\quad sn(-u) = -sn\, u,\quad cn(-u) = cn\, u,\quad dn(-u) = dn\, u.$
2. $\frac{d}{du} sn\, u = cn\; dn\, u,\quad \frac{d}{du} cn\, u = -sn\, u\, dn\, u,\quad \frac{d}{du} dn\, u = -k^2\, sn\, u\, cn\, u.$

It is also possible to define *inverse elliptic functions*. For instance, if $z = sn\, u$, then $u = sn^{-1} z = sn^{-1}(z, k)$ because u depends on both z and k.

EXAMPLE 11.19 Prove the following relations

(*i*) $\frac{d}{du} sn\, u = cn\, u\, dn\, u.$ (*ii*) $\frac{d}{du} dn\, u = -k^2\, sn\, u\, cn\, u.$

(*iii*) $\frac{d}{dz} sn^{-1}(z, k) = \frac{1}{\sqrt{(1-z^2)(1-k^2z^2)}}.$

Solution (*i*) By definition $u = \int_0^{\phi} \frac{dx}{\sqrt{(1-k^2\sin^2 x)}}$ and $sn\, u = \sin\phi$.

Thus $\frac{d}{du} sn\, u = \frac{d}{du}\sin\phi = \cos\phi \frac{d\phi}{du} = \text{cn}\, u \frac{d\phi}{du}.$

Now $\frac{du}{d\phi} = \frac{1}{\sqrt{(1-k^2\sin^2\theta)}}$ and thus $\frac{d\phi}{du} = \sqrt{(1-k^2\sin^2\theta)} = dn\, u.$

Hence, $\frac{d}{du} sn\, u = cn\, u \frac{d\phi}{du} = cn\, u\sqrt{(1-k^2\sin^2\theta)} = cn\, u\, dn\, u.$

(*ii*) $\frac{d}{du} dn\, u = -k^2\, sn\, u\, cn\, u$

Now $\frac{d}{du} dn\, u = \frac{d}{du}\sqrt{(1-k^2\sin^2\phi)} = \frac{d}{du}\sqrt{(1-k^2sn^2u)}$

$$= \frac{1}{2}(1-k^2sn^2u)^{-1/2}\frac{d}{du}(-k^2sn^2u) = \frac{-k^2}{2\, dn\, u}\cdot 2\, sn\, u \frac{d}{du}(sn\, u)$$

$$= -\frac{k^2}{dn\, u} sn\, u\, cn\, u\, dn\, u = -k^2\, sn\, u\, cn\, u.$$

(*iii*) $\frac{d}{dz} sn^{-1}(z, k) = \frac{d}{dz} sn^{-1} z$, the dependence on the modulus k being understood.

Since $z = sn\, u$, $dz/du = cn\, u\; dn\, u = \sqrt{(1-z^2)}\sqrt{(1-k^2z^2} = \sqrt{(1-z^2)(1-k^2z^2)}$.

Hence, $\frac{du}{dz} = \frac{d}{dz}(sn^{-1}z) = \frac{1}{\sqrt{(1-z^2)(1-k^2z^2)}}.$

EXAMPLE 11.20 Evaluate the integral $\int_1^3 \frac{dx}{\sqrt{(x^4+4x^2+3)}}$.

Solution The given integral can be written as $\int_1^3 \frac{dx}{\sqrt{(x^2+3)(x^2+1)}}$.

Let $x = \sqrt{3}\tan\theta$, $dx = \sqrt{3}\sec^2\theta\, d\theta$, the integral becomes

$$\int_{\pi/6}^{\pi/3} \frac{\sqrt{3}\sec^2\theta\, d\theta}{\sqrt{3}\sec\theta\sqrt{(3\tan^2\theta+1)}} = \int_{\pi/6}^{\pi/3} \frac{d\theta}{\sqrt{(3-2\cos^2\theta)}} = \frac{1}{\sqrt{3}}\int_{\pi/6}^{\pi/3} \frac{d\theta}{\sqrt{\left(1-\frac{2}{3}\cos^2\theta\right)}}.$$

Now put $\theta = (\pi/2) - \phi$, $d\theta = -d\phi$, when $\theta = \pi/6$, $\phi = \pi/3$ and $\theta = \pi/3$, $\phi = \pi/6$.
Hence the given integral becomes

$$\frac{1}{\sqrt{3}}\int_{\phi=\pi/6}^{\pi/3} \frac{d\phi}{\sqrt{\left(1-\frac{2}{3}\sin^2\phi\right)}} = \frac{1}{\sqrt{3}}\left[\int_0^{\pi/3} \frac{d\phi}{\sqrt{\left(1-\frac{2}{3}\sin^2\phi\right)}} - \int_0^{\pi/6} \frac{d\phi}{\sqrt{\left(1-\frac{2}{3}\sin^2\phi\right)}}\right]$$

$$= \frac{1}{\sqrt{3}}\left[F\left(\sqrt{\frac{2}{3}}, \frac{\pi}{3}\right) - F\left(\sqrt{\frac{2}{3}}, \frac{\pi}{6}\right)\right].$$

EXERCISE 11.2

Express the following integrals in terms of elliptic integrals.

1. $\int_0^{\pi/2} \frac{dx}{\sqrt{(\sin^2 x + 2\cos^2 x)}}$ **2.** $\int_0^{\pi/4} \frac{\sec x\, dx}{\sqrt{(1-\tan^2 x)}}$ **3.** $\int_0^{\pi/2} \frac{dx}{\sqrt{(1+2\sin x)}}$

4. $\int_0^{\pi/2} \frac{dx}{\sqrt{(1+3\sin^2 x)}}$ *(Kerala, 2005)* **Ans.** $\frac{1}{2}K(\sqrt{3}/2)$.

5. $\int_0^{\pi/2} \frac{dx}{\sqrt{(2-\cos x)}}$ **Ans.** $\frac{2}{\sqrt{3}}\left\{K\left(\sqrt{\frac{2}{3}}\right) - F\left(\sqrt{\frac{2}{3}}, \frac{1}{4}\pi\right)\right\}$

6. $\int_0^{\pi/2} \frac{dx}{\sqrt{7-\sin x}}$ **Ans.** $\frac{1}{\sqrt{2}}\left\{K\left(\frac{1}{2}\right) - F\left(\frac{1}{2}, \frac{1}{4}\pi\right)\right\}$

7. $\int_0^2 \frac{dx}{\sqrt{(4-x^2)(9-x^2)}}$ **8.** $\int_0^1 \frac{dx}{\sqrt{x(1-x)(1+x)}}$ **9.** $\int_4^6 \frac{dx}{\sqrt{(x-1)(x-2)(x-3)}}$

10. $\int_4^\infty \frac{dx}{\sqrt{(x-1)(x-2)(x-3)(x-4)}}$ **Ans.** $\frac{2}{3}K\left(\frac{1}{3}\right)$.

11. $\int_0^1 \frac{dx}{\sqrt{1-x^4}}$ **Ans.** $(1/\sqrt{2})\,K\,(1/\sqrt{2})$ **12.** $\int_1^\infty \frac{dt}{\sqrt{t^4-1}} = \frac{1}{2}K\left(\frac{1}{\sqrt{2}}\right)$ **13.** $\int_0^\infty \frac{dx}{\sqrt{(x^4+x^2+1)}}$

14. $\int_0^{\pi/2} \sqrt{\cos x}\, dx$ **Ans.** $2\sqrt{2}\,E\,(1/\sqrt{2}) - \sqrt{2}\,K\,(1/\sqrt{2})$. **15.** $\int_0^{\pi/2} \sqrt{1+4\sin^2 x}\, dx$

16. Represent $\int_0^x \sqrt{(1-4\sin^2\theta)}\, d\theta$ in terms of incomplete elliptic integrals where $0 \le x \le \pi/6$.

17. Show that $\int_0^x \frac{dv}{\sqrt{(1-v^2)(1-k^2v^2)}} = \text{sn}^{-1}(x, k) = F(k, \sin^{-1} x)$.

18. By means of the substitution $x = \frac{a}{2}(1-\sin\theta)$, show that $\int_0^{a/2} \frac{dx}{\sqrt{\{(2ax-x^2)(a^2-x^2)\}}} = \frac{2}{3a} K\left(\frac{1}{3}\right)$.

19. Show that the length of the arc of the sine curve $y = \sin x$ where $0 \le x \le \pi$ is given by $2\sqrt{2}\, E\left(\frac{1}{\sqrt{2}}\right)$.

20. Show that the integral $\int_0^{\pi/4} \frac{dx}{\sqrt{(5+3\cos x)}}$ is elliptic and determine its value.

Ans. $\frac{1}{\sqrt{2}} F\left(\frac{1}{2}\sqrt{3}, \frac{1}{8}\pi\right)$.

21. A simple pendulum 30 cm long is displaced through a right angle from the vertical and then released. Show that an error of about 15.33% would be made in its period of oscillations under the assumption of small oscillation.

22. If $k > 1$, show by substituting $\sin x = k \sin\phi$ that

$$\int_0^{\phi} \sqrt{(1-k^2\sin^2\phi)}\, d\phi = \left(\frac{1}{k}-k\right) F\left(\frac{1}{k}, x\right) + kE\left(\frac{1}{k}, x\right)$$

provided $k\sin\phi \le 1$.

11.12 ERROR FUNCTION OR PROBABILITY INTEGRAL

The *error function* is defined by the integral

$$erf\,(x) = \frac{2}{\sqrt{\pi}} \int_0^x e^{-t^2}\, dt \qquad ...(11.21)$$

Expanding the integrand and integrating term by term, we obtain

$$erf\,(x) = \frac{2}{\sqrt{\pi}} \left(x - \frac{x^3}{3(1!)} + \frac{x^5}{5(2!)} - \frac{x^7}{7(3!)} + ...\right) \qquad ...(11.22)$$

Another form

Let $t^2 = u$. Then $dt = \frac{1}{2t} du = \frac{1}{2\sqrt{u}} du$, and $erf(x) = \frac{1}{\sqrt{\pi}} \int_0^{x^2} u^{-1/2} e^{-u}\, du$. ...(11.23)

This is another form of the *error function*. Using the definition, we obtain

$$erf(\infty) = \frac{1}{\sqrt{\pi}} \int_0^{\infty} u^{-1/2} e^{-u}\, du = \frac{1}{\sqrt{\pi}} \Gamma(1/2) = \frac{1}{\sqrt{\pi}} (\sqrt{\pi}) = 1. \qquad ...(11.24)$$

11.12.1 Complementary Error Function erfc (x)

Using the definition of the error function given in Eqs. (11.23) and (11.24), we write

$$erf(x) = \frac{1}{\sqrt{\pi}} \int_0^{x^2} u^{-1/2} e^{-u}\, du = \boxed{\frac{1}{\sqrt{\pi}} \int_0^{\infty} u^{-1/2} e^{-u}\, du} - \frac{1}{\sqrt{\pi}} \int_{x^2}^{\infty} u^{-1/2} e^{-u}\, du$$

$$= 1 - \frac{1}{\sqrt{\pi}} \int_{x^2}^{\infty} u^{-1/2} e^{-u}\, du = 1 - erfc(x), \quad \text{...(11.25)}$$

where $$erfc(x) = \frac{1}{\sqrt{\pi}} \int_{x^2}^{\infty} u^{-1/2} e^{-u}\, du. \quad \text{...(11.26)}$$

and this function erfc(x) is called the *complementary error function.*

Another form

$$erfc(x) = 1 - erf(x) = 1 - \frac{2}{\sqrt{\pi}} \int_0^x e^{-t^2}\, dt. \quad \text{[Using Eq. (11.21)]}$$

$$= \frac{2}{\sqrt{\pi}} \int_0^{\infty} e^{-t^2}\, dt - \frac{2}{\sqrt{\pi}} \int_0^x e^{-t^2}\, dt = \frac{2}{\sqrt{\pi}} \int_x^{\infty} e^{-t^2}\, dt. \quad \text{...(11.27)}$$

Equations (11.21) and (11.27) are the commonly used definitions of *error function* and *complementary error function* respectively. The error function finds lot of application in the theory of probability and also in the solution of partial differential equations occurring in diffusion problems. The graphs of $erf(x)$ and $erfc(x)$ for $x \geq 0$ are given in Fig. 11.4. From the figure it is obvious that $erf(0) = 0$, $erfc(0) = 1$.

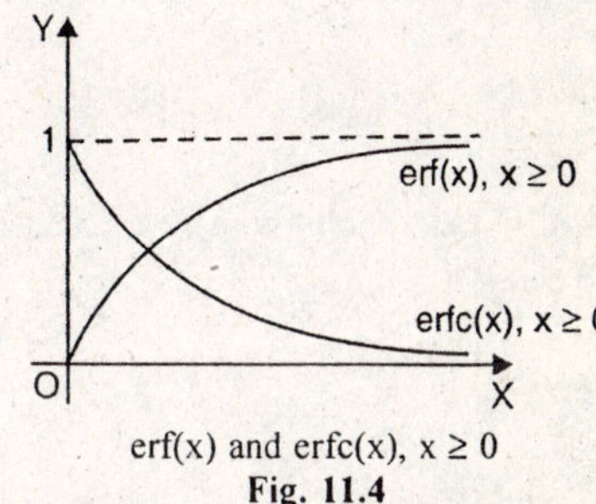

erf(x) and erfc(x), x ≥ 0
Fig. 11.4

11.12.2 Some Properties of Error Functions

1. erf(–x) = –erf (x).

Using the definition given in Eq. (11.21), we get

$$erf(-x) = \frac{2}{\sqrt{\pi}} \int_0^{-x} e^{-t^2}\, dt. \quad (\text{Let } t = -u,\ dt = -du)$$

$$= \frac{2}{\sqrt{\pi}} \int_0^{x} e^{-u^2} (-du) = -\frac{2}{\sqrt{\pi}} \int_0^{x} e^{-u^2}\, du = -erf(x).$$

2. erfc(–x) = 1 + erf (x) = 2 – erfc(x).

Using equation (11.25), we get

$$erfc(-x) = 1 - erf(-x) = 1 + erf(x)$$
$$= 1 + [1 - erfc(x)] = 2 - erfc(x).$$

3. Derivative of error function.

(*i*) $$\frac{d}{dx}[erf(\alpha x)] = \frac{2\alpha}{\sqrt{\pi}} e^{-\alpha^2 x^2}. \quad \text{(Osmania, 2003)}$$

(*ii*) $$\frac{d}{dx}[erfc(\alpha x)] = -\frac{2\alpha}{\sqrt{\pi}} e^{-\alpha^2 x^2}.$$

Proof (*ii*) $$\frac{d}{dx}[erfc(\alpha x)] = \frac{d}{dx}\left[\frac{2}{\sqrt{\pi}} \int_{\alpha x}^{\infty} e^{-t^2}\, dt\right]$$

On applying the rule of differentiation under integral sign, we obtain

$$\frac{d}{dx}[erfc(\alpha x)] = \frac{2}{\sqrt{\pi}}\left[\int_{\alpha x}^{\infty}\left(\frac{\partial}{\partial x}e^{-t^2}\right)dt + \frac{d}{dx}(\infty)\,e^{-\infty} - \frac{d}{dx}(\alpha x)\,e^{-\alpha^2x^2}\right]$$

$$= \frac{2}{\sqrt{\pi}}\left[0+0-\alpha e^{-\alpha^2x^2}\right] = \frac{-2\alpha}{\sqrt{\pi}}e^{-\alpha^2x^2}.$$

4. Integral of error function.

(*i*) $$\int_0^t erf\,(\alpha x)\,dx = t\,erf\,(\alpha t) + \frac{1}{\alpha\sqrt{\pi}}[e^{-\alpha^2t^2} - 1].$$ (*Osmania, 2003*)

(*ii*) $$\int_0^t erfc\,(\alpha x)\,dx = t\,erfc\,(\alpha t) - \frac{1}{\alpha\sqrt{\pi}}[e^{-\alpha^2t^2} - 1].$$

Proof (*i*) Integrating the left hand side by parts, we obtain

$$\int_0^t \underset{\text{II}}{1}\cdot \underset{\text{I}}{erf\,(\alpha x)}\,dx = [x\,erf\,(\alpha x)]_0^t - \int_0^t x\frac{d}{dx}[erf\,(\alpha x)\,dx]$$

$$= t\,erf\,(\alpha t) - \frac{2\alpha}{\sqrt{\pi}}\int_0^t x\,e^{-\alpha^2x^2}\,dx.$$

Let $\alpha^2 x^2 = u$. Then, $2\alpha^2 x\,dx = du$ or $x\,dx = du/2\alpha^2$.

Hence,
$$\int_0^t erf\,(\alpha x)\,dx = t\,erf\,(\alpha t) - \left(\frac{2\alpha}{\sqrt{\pi}}\right)\left(\frac{1}{2\alpha^2}\right)\int_0^{\alpha^2t^2} e^{-u}\,du$$

$$= t\,erf\,(\alpha t) + \frac{1}{\alpha\sqrt{\pi}}(e^{-\alpha^2t^2} - 1).$$

EXAMPLE 11.21 Show that $\displaystyle\int_0^\infty e^{-t^2-2\alpha t}dt = \frac{\sqrt{\pi}}{2}e^{\alpha^2}[1 - erf\,(\alpha)].$

Solution We have $\displaystyle\int_0^\infty e^{-t^2-2\alpha t}dt = \int_0^\infty e^{-t^2-2\alpha t-\alpha^2+\alpha^2}dt = \int_0^\infty e^{-(t+\alpha)^2}\cdot e^{\alpha^2}\,dt.$

$$= e^{\alpha^2}\left[\int_\alpha^\infty e^{-u^2}du\right] \quad \text{put } t+\alpha = u,\ dt = du, \text{ when } \begin{cases} t=0, & u=\alpha \\ t=\infty, & u=\infty \end{cases}$$

$$= e^{\alpha^2}\left[\int_0^\infty e^{-u^2}du - \int_0^\alpha e^{-u^2}du\right] = e^{\alpha^2}\left[\frac{\sqrt{\pi}}{2}\{erf\,(\infty) - erf\,(\alpha)\}\right]$$

$$= e^{\alpha^2}\left[\frac{\sqrt{\pi}}{2}(1 - erf\,(\alpha))\right] = \frac{\sqrt{\pi}}{2}e^{\alpha^2}[1 - erf\,(\alpha)].$$

CHAPTER

12

Bessel's Differential Equation, Bessel's Functions — Legendre Differential Equation and Legendre Polynomials

12.1 BESSEL'S DIFFERENTIAL EQUATION

The differential equation

$$x^2 \frac{d^2 y}{dx^2} + x \frac{dy}{dx} + (x^2 - n^2)\, y = 0 \qquad ...(12.1)$$

where n is a given number, real or complex, is known as Bessel's differential equation of *order n*, named after the German mathematician *Friedrich Wilhelm Bessel* (1784–1846). This equation arises in many practical problems in electrical engineering, acoustics, hydrodynamics and the theory of elasticity. Problem involving vibration of membranes as in loud speakers; problems dealing with wind tunnel interference; loading electrical transmission lines and the boundary value problems in heat flow in regions concerning cylindrical configurations make extensive use of Bessel functions. The point x = 0 is a regular singularity of the equation. We seek its solution of the form

$$y(x) = \sum_{k=0}^{\infty} a_k\, x^{m+k}, \qquad (a_0 \neq 0) \quad ...(12.2)$$

Substituting for y, dy/dx and d^2y/dx^2 in equation (12.1) from equation (12.2) and rearranging the terms in powers of x, we obtain

$$x^2 \sum_{k=0}^{\infty} a_k\,(m+k)(m+k-1)\, x^{m+k-2} + x \sum_{k=0}^{\infty} a_k\,(m+k)\, x^{m+k-1} + (x^2 - n^2) \sum_{k=0}^{\infty} a_k\, x^{m+k} = 0$$

or

$$\sum a_k \{(m+k)^2 - n^2\}\, x^{m+k} + \sum_{k=0}^{\infty} a_k\, x^{m+k+2} = 0. \qquad ...(12.3)$$

Equating to zero the coefficient of lowest power of x by taking $k = 0$, the indicial equation is written as

$$a_0\,(m^2 - n^2) = 0. \qquad ...(12.4)$$

Since a_0 is the coefficient of the starting term of the series solution and is arbitrary, it cannot be equal to zero. Equation (12.4), therefore, gives $m^2 - n^2 = 0$ or $m = +n$ or $-n$.

Equation to zero the coefficient of next term that is, x^{m+1} we get

$$a_1\, \{(m+1)^2 - n^2\} = 0. \qquad ...(12.5)$$

Since the expression within the bracket in (12.5) does not vanish for values of m determined above, this gives us $a_1 = 0$.

Setting the coefficient of general term that is, x^{m+k+2} equal to zero, the recurrence relation is given by

$$a_{k+2}\, \{(m+k+2)^2 - n^2\} + a_k = 0, \quad k = 0, 1, 2 \ldots$$

or

$$a_{k+2} = \frac{1}{(m-n+k+2)(m+n+k+2)}\, a_k. \qquad ...(12.6)$$

Putting $k = 1, 3, 5, \ldots$ Since $a_1 = 0$, Eq. (12.6) gives us $a_3 = a_5 = a_7 = \ldots = 0$.

Now putting $k = 0, 2, 4, \ldots$, we get

$$a_2 = -\frac{1}{(m-n+2)(m+n+2)} a_0$$

$$a_4 = -\frac{1}{(m-n+4)(m+n+4)} a_2$$

$$= \frac{1}{(m-n+4)(m+n+4)(m-n+2)(m+n+2)} a_0$$

and so on.

Therefore, $$y(x) = a_0 x^m \left[1 - \frac{x^2}{(m+2)^2 - n^2} + \frac{x^4}{[(m+2)^2 - n^2][(m+4)^2 - n^2]} - \ldots\right].$$

Here for the value of $m = n$, the first solution of the given differential equation is given by

$$y_1(x) = a_0 x^n \left[1 - \frac{x^2}{2(2n+2)} + \frac{x^4}{[2.4(2n+2)(2n+4)]} - \ldots\right]. \quad \text{...(12.7)}$$

If n is not an integer, the second solution is obtained corresponding to $m = -n$ by replacing n by $-n$ in (12.7). If however, n is zero or an integer the second solution will have to be obtained. See solved Example 10.14, when n is zero.

12.2 BESSEL FUNCTIONS, $J_n(x)$

Equation (12.7) in Art. (12.1) gives us one solution of the Bessel's equation. As a_0 is arbitrary, we choose it in a manner that the equation is further simplified. The particular solution of Bessel's equation by setting $a_0 = \dfrac{1}{2^n \Gamma(n+1)}$ is known as Bessel function of the first kind and of order n. It is denoted by $J_n(x)$. Thus

$$J_n(x) = \frac{x^n}{2^n \Gamma(n+1)} \left[1 - \frac{x^2}{2(2n+2)} + \frac{x^4}{2 \cdot 4(2n+2)(2n+4)} - \ldots\right]$$

or $$J_n(x) = \sum_{k=0}^{\infty} (-1)^k \frac{(x/2)^{n+2k}}{k!\,\Gamma(n+k+1)}. \quad (12.8)$$

Depending upon the values of n, we get different types of solutions.

Case 1: When n is neither zero nor an integer.

If n is not an integer, the second solution is obtained by replacing n by $-n$ in (12.8) and is given by

$$\boldsymbol{J_{-n}(x) = \sum_{k=0}^{\infty} (-1)^k \frac{(x/2)^{-n+2k}}{k!\,\Gamma(-n+k+1)}}, \quad \text{...(12.9)}$$

which is called Bessel's function of the first kind of order $-n$.

It is clear from (12.8) and (12.9) that the two solutions $J_n(x)$ and $J_{-n}(x)$ are linearly independent so long as n is a non-integer. It can also be established by ratio test that the two series in (12.8) and (12.9) are uniformly convergent. Hence for n, other than zero or an integer, the complete solution of the differential equation (12.1) is given by

$$y(x) = A\,J_n(x) + B\,J_{-n}(x) \quad \text{...(12.10)}$$

where A and B are arbitrary constants.

12.2.1 Bessel's Function of Order Zero

Setting $n = 0$ in Eq. (12.8), we obtain

$$J_0(x) = \sum_{k=0}^{\infty} \frac{(-1)^k}{(k!)^2}\left(\frac{x}{2}\right)^{2k} = 1 - \frac{1}{(1!)^2}\left(\frac{x}{2}\right)^2 + \frac{1}{(2!)^2}\left(\frac{x}{2}\right)^4 - \frac{1}{(3!)^2}\left(\frac{x}{2}\right)^6 + \cdots \qquad \text{...(12.11)}$$

We find that $J_0(0) = 1$ so that the graph of $y = J_0(x)$ cuts the y-axis at (0, 1). Further, $|J_0(x)| \le 1$.

The presence of positive and negative terms shows that the graph of $y = J_0(x)$ oscillates and decays fast as $x \to \infty$. The graph of $J_0(x)$ is given in Fig. 12.1.

12.2.2 Bessel's Function of Order 1

Setting $n = 1$ in Eq. (12.8), we obtain

$$J_1(x) = \sum_{k=0}^{\infty} \frac{(-1)^k}{k!(k+1)!}\left(\frac{x}{2}\right)^{2k+1} = \frac{x}{2} - \frac{x^3}{(2^3)(1!)(2!)} + \frac{x^5}{(2^5)(2!)(3!)} - \cdots \qquad \text{...(12.12)}$$

$$= \frac{x}{2}\left[1 - \frac{1}{1!2!}\left(\frac{x}{2}\right)^2 + \frac{1}{2!3!}\left(\frac{x}{2}\right)^4 - \frac{1}{3!4!}\left(\frac{x}{2}\right)^6 + \cdots\right]. \qquad \textit{(BPTU, 2005)}$$

We find that $J_1(0) = 0$ so that the graph of $y = J_1(x)$ passes through the origin. The curve $y = J_1(x)$ oscillates and decays rapidly as $x \to \infty$. The graph of $J_1(x)$ is also given in Fig. 12.1 .

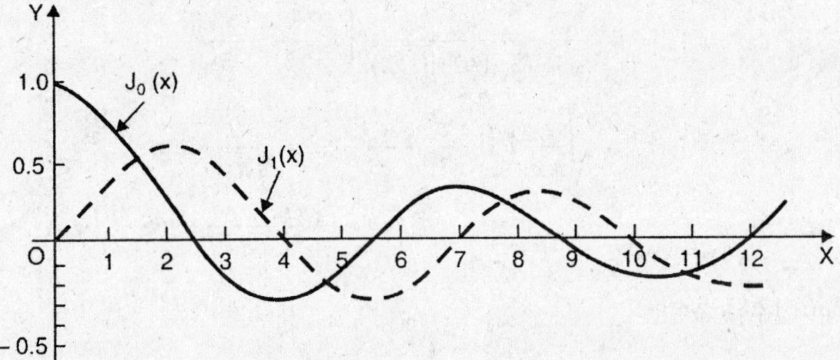

Fig. 12.1: Graphs of $J_0(x)$ and $J_1(x)$

Case 2: When n is an integer.

If n is an integer, the two functions $J_{-n}(x)$ and $J_n(x)$ are not linearly independent but are connected by the relation

$$J_{-n}(x) = (-1)^n J_n(x). \qquad (n = 1, 2, \ldots) \quad \text{...(12.13)}$$

Proof:

$$J_{-n}(x) = \sum_{k=0}^{\infty} \frac{(-1)^k (x/2)^{-n+2k}}{k!\,\Gamma(-n+k+1)}$$

$$= \sum_{k=0}^{n-1} (-1)^k \frac{(x/2)^{-n+2k}}{k!\,\Gamma(-n+k+1)} + \sum_{k=n}^{\infty} \frac{(-1)^k (x/2)^{-n+2k}}{k!\,\Gamma(-n+k+1)}.$$

It is obvious from above that so long as $k \le (n-1)$, all terms in the first summation are zero because the gamma of zero or a negative integer is infinite. Hence,

$$J_{-n}(x) = \sum_{k=n}^{\infty} \frac{(-1)^k (x/2)^{-n+2k}}{k!\,\Gamma(-n+k+1)}.$$

Setting $k = n + p$, we see that when $k = n$, $p = 0$ and when $k \to \infty$, $p \to \infty$. Therefore,

$$J_{-n}(x) = \sum_{p=0}^{\infty} \frac{(-1)^{n+p} (x/2)^{-n+2n+2p}}{(n+p)!\Gamma(p+1)}.$$

Since p and n are integers $\Gamma(p+1) = p!$ and $(n+p)! = \Gamma(n+p+1)$

Therefore, $$J_{-n}(x) = (-1)^n \sum_{p=0}^{\infty} \frac{(-1)^p (x/2)^{n+2p}}{p!\Gamma(n+p+1)} = (-1)^n J_n(x), \quad (n = 1, 2, \ldots) \quad \ldots(12.14)$$

(*VTU 2006; Bhopal 2002S*)

12.2.3 Bessel's Function of Order Half

Case 1: When n = 1/2.

Putting $n = 1/2$ in (12.8), we obtain

$$J_{1/2}(x) = \frac{x^{1/2}}{2^{1/2}\Gamma(3/2)}\left(1 - \frac{x^2}{2.3} + \frac{x^4}{2.3.4.5} - \ldots\right)$$

$$= \frac{x}{\sqrt{x}\sqrt{2}\cdot 1/2\cdot\sqrt{(1/2)}}\left(1 - \frac{x^2}{3!} + \frac{x^4}{5!} - \ldots\right)$$

$$= \sqrt{\frac{2}{\pi x}}\left(x - \frac{x^3}{3!} + \frac{x^5}{5!} - \ldots\right) = \sqrt{\frac{2}{\pi x}}\sin x. \quad \ldots(12.15)$$

(*AMIETE, June 2011; VTU 2006, JUTU 2003*)

Case 2: When n = – 1/2.

Putting $n = -1/2$ in (12.8), we get

$$J_{-1/2}(x) = \frac{x^{-1/2}}{2^{-1/2}\Gamma(1/2)}\left(1 - \frac{x^2}{2(-1+2)} + \frac{x^4}{2.4(-1+2)(-1+4)} - \ldots\right)$$

$$= \sqrt{\frac{2}{\pi x}}\left(1 - \frac{x^2}{2!} + \frac{x^4}{4!} - \ldots\right) = \sqrt{\frac{2}{\pi x}}\cos x. \quad \ldots(12.16)$$

(*AMIETE, June 2009; Anna 2005; WBTU 2005; VTU 2003*)

12.3 DERIVATIVES AND INTEGRALS OF BESSEL FUNCTIONS

We prove the following properties of Bessel functions. These properties greatly help us in the solutions of problems involving Bessel functions.

I. (*a*) $\frac{d}{dx}[x^n J_n(x)]$ **or** $[x^n J_n(x)]' = x^n J_{n-1}(x)$. (*AMIETE, June 2010, 2002*)

(*b*) $\int x^n J_{n-1}(x)\,dx = x^n J_n(x) + c.$

Proof:

(*a*) We know that

$$J_n(x) = \sum_{k=0}^{\infty} (-1)^k \frac{(x/2)^{n+2k}}{k!\Gamma(n+k+1)}.$$

Multiplying the Bessel's function $J_n(x)$ by x^n, we obtain

$$x^n J_n(x) = \sum_{k=0}^{\infty} \frac{(-1)^k\, x^{2n+2k}}{2^{n+2k}\, k!\,\Gamma(n+k+1)}.$$

Therefore,
$$\frac{d}{dx}\left[x^n J_n(x)\right] = \frac{d}{dx}\sum_{k=0}^{\infty} \frac{(-1)^k\, x^{2n+2k}}{2^{n+2k}\, k!\,\Gamma(n+k+1)}$$

$$= \sum_{k=0}^{\infty} \frac{(-1)^k\,(2n+2k)\, x^{2n+2k-1}}{2^{n+2k}\, k!\,\Gamma(n+k+1)}$$

$$= \sum_{k=0}^{\infty} \frac{(-1)^k\, x^{2n+2k-1}}{2^{n+2k-1}\, k!\,\Gamma(n-1+k+1)}$$

$$= x^n \sum_{k=0}^{\infty} \frac{(-1)^k\,(x/2)^{n-1+2k}}{k!\,\Gamma(n-1+k+1)}$$

$$= x^n J_{n-1}(x).$$ (Madras, 2006; VTU, 2005; UPTU 2005)

Setting $n = 0$, we get $J_0'(x) = J_{-1}(x) = -J_1(x)$, since $J_{-n}(x) = (-1)^n J_n(x)$

Integrating right hand side of formula 1(*a*), we obtain

$$\int x^n J_{n-1}(x)\,dx = \int [x^n J_n(x)]'\,dx + c = x^n J_n(x) + c.$$

II. (*a*) $\frac{d}{dx}[x^{-n} J_n(x)] = -x^{-n} J_{n+1}(x).$ $(n = 0, 1, 2, \ldots)$

(*b*) $\int x^{-n} J_{n+1}(x)\,dx = -x^{-n} J_n(x) + c.$

Proof:

(*a*) Multiplying the Bessel's function $J_n(x)$ by x^{-n} and differentiating, we get

$$\frac{d}{dx}[x^{-n} J_n(x)] = \frac{d}{dx}\sum_{k=0}^{\infty} \frac{(-1)^k\, x^{2k}}{2^{n+2k}\, k!\,\Gamma(n+k+1)}$$

$$= \sum_{k=0}^{\infty} \frac{(-1)^k\, 2k\, x^{2k-1}}{2^{n+2k}\, k!\,\Gamma(n+k+1)}$$

$$= \sum_{k=1}^{\infty} \frac{(-1)^k\, x^{2k-1}}{2^{n+2k-1}\,(k-1)!\,\Gamma(n+k+1)}$$

$$= \sum_{p=0}^{\infty} \frac{(-1)^{1+p}\, x^{2p+1}}{2^{n+2p+1}\, p!\,\Gamma(n+1+p+1)} \qquad (k = 1 + p)$$

$$= -x^{-n} \sum_{p=0}^{\infty} \frac{(-1)^p\, x^{n+2p+1}}{2^{n+2p+1}\, p!\,\Gamma(n+1+p+1)}$$

$$= -x^{-n} J_{n+1}(x).$$

Hence,
$$\frac{d}{dx}[x^{-n} J_n(x)] = -x^{-n} J_{n+1}(x).$$ (PTU 2006; BPTU 2005)

Integrating, we get

$$\int x^{-n} J_{n+1}(x)\,dx = -\int [x^{-n} J_n(x)]'\,dx + c = -x^{n} J_n(x) + c.$$

12.4 BESSEL'S FUNCTION OF THE FIRST KIND

Bessel's functions of the first kind is one of the most important special functions and satisfies a large number of recurrence relations. We shall prove some of these relations.

12.4.1 Recurrence Formulae of Bessel Functions, $J_n(x)$

(a) $J_n'(x) = J_{n-1}(x) - \frac{n}{x} J_n(x)$, (b) $J_n'(x) = \frac{n}{x} J_n(x) - J_{n+1}(x)$,

(c) $J_n'(x) = \frac{1}{2}[J_{n-1}(x) - J_{n+1}(x)]$, (d) $J_n(x) = \frac{x}{2n}[J_{n-1}(x) + J_{n+1}(x)]$.

(*AMIETE, June, 2010*)

Proof:

From formula I(a) and II(a) Article 12.3, we have

$$\frac{d}{dx}[x^{n} J_n(x)] = x^{n} J_{n-1}(x), \qquad ...(12.17)$$

and

$$\frac{d}{dx}[x^{-n} J_n(x)] = -x^{-n} J_{n+1}(x). \qquad ...(12.18)$$

Differentiating the left hand sides of (12.17) and (12.18), we get

$$x^{n}J'_n(x) + nx^{n-1} J_n(x) = x^{n} J_{n-1}(x), \qquad ...(i)$$

and

$$x^{-n}J'_n(x) - nx^{-n-1} J_n(x) = -x^{-n} J_{n+1}(x). \qquad ...(ii)$$

Equations (*i*) and (*ii*) may be rewritten as

or

$$J'_n(x) = J_{n-1}(x) - \frac{n}{x} J_n(x), \qquad ...(12.19)$$

and

$$J'_n(x) = \frac{n}{x} J_n(x) - J_{n+1}(x). \qquad ...(12.20)$$

(*JNTU, 2006; Anna, 2005*)

Adding (12.19) and (12.20), we get

$$J'_n(x) = \frac{1}{2}[J_{n-1}(x) - J_{n+1}(x)]. \qquad ...(12.21)$$

(*SVTU, 2007; PTU, 2005*)

Subtracting (12.20) from (12.19), we get

$$J_n(x) = \frac{x}{2n}[J_{n-1}(x) + J_{n+1}(x)]. \qquad ...(12.22)$$

(*Anna, 2005S; VTU, 2003S*)

Relation (12.22) is very important as it connects Bessel functions of three successive orders.

ILLUSTRATIVE EXAMPLES

EXAMPLE 12.1 Find the solution of the differential equation $x^2y'' + xy' + (x^2 - 1/4)y = 0$.

Solution The given equation is the Bessel's differential equation with $n = 1/2$. Therefore, the complete solution is

$$y(x) = A\,J_n(x) + B\,J_{-n}(x) = A\,J_{1/2}(x) + B\,J_{-1/2}(x)$$

$$= A\sqrt{\frac{2}{\pi x}}\sin x + B\sqrt{\frac{2}{\pi x}}\cos x = \sqrt{\frac{2}{\pi x}}(A\sin x + B\cos x).$$

EXAMPLE 12.2 Transform the equation $9x^2y'' + 9xy' + (81x^2 - 1)y = 0$ using the substitution $3x = z$ and hence find the general solution of the equation.

Solution When $3x = z$, we have

$$\frac{dy}{dx} = \frac{dy}{dz}\cdot\frac{dz}{dx} = 3\frac{dy}{dz}, \text{ and } \frac{d^2y}{dx^2} = 3\frac{d^2y}{dz^2}\cdot\frac{dz}{dx} = 9\frac{d^2y}{dz^2}.$$

Now, the given equation is transformed to

$$9\left(\frac{z^2}{9}\right)(9)\frac{d^2y}{dz^2} + 9\left(\frac{z}{3}\right)(3)\frac{dy}{dz} + (9z^2 - 1)\,y = 0$$

or

$$z^2\frac{d^2y}{dz^2} + z\frac{dy}{dz} + \left(z^2 - \frac{1}{9}\right)y = 0.$$

The equation is the Bessel's differential equation with $n = 1/3$.

Hence, the general solution is

$$y(z) = AJ_{1/3}(z) + BJ_{-1/3}(z) \quad \text{or} \quad y(x) = AJ_{1/3}(3x) + BJ_{-1/3}(3x).$$

EXAMPLE 12.3 Express $J_5(x)$ in terms of $J_0(x)$ and $J_1(x)$. *(AMIETE, June 2007, 2004; V.T.U, 2001)*

Solution We know that

$$J_n(x) = \frac{x}{2n}[J_{n-1}(x) + J_{n+1}(x)] \text{ or } J_{n+1}(x) = \frac{2x}{x}J_n(x) - J_{n-1}(x).$$

Putting $n = 1, 2, 3, 4$ successively, we obtain

$$J_2(x) = \frac{2}{x}J_1(x) - J_0(x), \quad ...(i) \qquad J_3(x) = \frac{4}{x}J_2(x) - J_1(x), \quad ...(ii)$$

$$J_4(x) = \frac{6}{x}J_3(x) - J_2(x), \quad ...(iii) \qquad J_5(x) = \frac{8}{x}J_4(x) - J_3(x). \quad ...(iv)$$

Substituting the value of $J_2(x)$ in (ii), we obtain

$$J_3(x) = \frac{4}{x}\left[\frac{2}{x}J_1(x) - J_0(x)\right] - J_1(x) = \left(\frac{8}{x^2} - 1\right)J_1(x) - \frac{4}{x}J_0(x). \quad ...(v)$$

Now substituting the value of $J_3(x)$ from (v) and $J_2(x)$ from (*i*) in (*iii*), we obtain

$$J_4(x) = \left(\frac{48}{x^3} - \frac{8}{x}\right)J_1(x) + \left(1 - \frac{24}{x^2}\right)J_0(x). \quad ...(vi)$$

Finally putting the values of $J_4(x)$ from (*vi*) and $J_3(x)$ from (*v*) in (*iv*), we get

$$J_5(x) = \left(\frac{384}{x^4} - \frac{72}{x^2} + 1\right)J_1(x) + \left(\frac{12}{x} - \frac{192}{x^3}\right)J_0(x).$$

EXAMPLE 12.4 Express $J_{3/2}(x)$ and $J_{5/2}(x)$ in terms of sine and cosine functions.

Solution Put $n = 1/2$ in relation (12.22), we obtain

$$\frac{1}{x}J_{1/2}(x) = J_{-1/2}(x) + J_{3/2}(x).$$

Therefore,

$$J_{3/2}(x) = \frac{1}{x}J_{1/2}(x) - J_{-1/2}(x).$$

Substituting of $J_{1/2}(x)$ and $J_{-1/2}(x)$ from (12.15) and (12.16), we get

$$J_{3/2}(x) = \sqrt{\frac{2}{\pi x}}\left(\frac{\sin x}{x} - \cos x\right). \qquad \textit{(AMIETE June, 2006; VTU, 2006)}$$

Again put $n = 3/2$ in (12.22), we obtain

$$\frac{3}{x} J_{3/2}(x) = J_{1/2}(x) + J_{5/2}(x).$$

Therefore,
$$J_{5/2}(x) = \frac{3}{x} J_{3/2}(x) - J_{1/2}(x)$$

$$= \frac{3}{x}\sqrt{\frac{2}{\pi x}}\left(\frac{\sin x}{x} - \cos x\right) - \sqrt{\frac{2}{\pi x}} \sin x.$$

$$= \sqrt{\frac{2}{\pi x}}\left(\frac{3 - x^2}{x^2} \sin x - \frac{3}{x} \cos x\right).$$

(AMIETE, June 2010, Dec 08; JNTU 2006)

EXAMPLE 12.5 Prove that $4J_0'''(x) + 3J_0'(x) + J_3(x) = 0$. *(Osmania, 2003)*

Solution We have $J_0'(x) = -J_1(x)$.

Differentiating w.r.t. x we get

$$J_0''(x) = -J_1'(x) = -\frac{1}{2}[J_0(x) - J_2(x)]. \quad \text{[from (12.21)]}$$

Differentiating once again w.r.t. x, we get

$$J_0'''(x) = -\frac{1}{2}[J_0'(x) - J_2'(x)]$$

$$= -\frac{1}{2} J_0'(x) + \frac{1}{4}[J_1(x) - J_3(x)] = -\frac{1}{2} J_0'(x) - \frac{1}{4} J_0'(x) - \frac{1}{4} J_3(x).$$

Transferring all the terms to the left hand side, we get

$$4J_0'''(x) + 3J_0'(x) + J_3(x) = 0.$$

EXAMPLE 12.6 Show that $\frac{d}{dx}\{J_n^2(x)\} = \frac{x}{2n}[J_{n-1}^2(x) - J_{n+1}^2(x)]$. *(Kerala 2005; Osmania, 2003S)*

Solution
$$\frac{d}{dx}\{J_n^2(x)\} = 2J_n(x)\, J_n'(x)$$

$$= \frac{x}{n}[J_{n-1}(x) + J_{n+1}(x)]\ \frac{1}{2}[J_{n-1}(x) - J_{n+1}(x)]$$

$$= \frac{x}{2n}[J_{n-1}^2(x) - J_{n+1}^2(x)].$$

Alternatively:
$$\frac{d}{dx}\{J_n^2(x)\} = \frac{d}{dx}\{x^{-n} J_n(x)\, x^n J_n(x)\}$$

$$= -x^{-n} J_{n+1}(x) \cdot x^n J_n(x) + x^{-n} J_n(x)\, x^n J_{n-1}(x)$$

$$= J_n(x)[J_{n-1}(x) - J_{n+1}(x)]$$

$$= \frac{x}{2n}[J_{n-1}(x) + J_{n+1}(x)]\ [J_{n-1}(x) - J_{n+1}(x)]$$

$$= \frac{x}{2n}[J_{n-1}^2(x) - J_{n+1}^2(x)].$$

EXAMPLE 12.7 Evaluate the following integral in term of the Bessel's function

(*i*) $\int \frac{1}{x} J_2(x)\, dx$, (*AMIETE, June 2007, June 06*) (*ii*) $\int x^2 J_1(x)\, dx$, (*AMIETE, June 2010, Dec 2009*)

(*iii*) $\int x^3 J_0(x)\, dx$, (*iv*) $\int x^2 J_0(x)$, (*v*) $\int x J_0^2(x)\, dx$. (*AMIETE, Dec 2007, Dec 2003*)

Solution (*i*) Using equation $\int x^{-n} J_{n+1}(x)\, dx = -x^{-n} J_n(x) + c$, we obtain

$$\int x^{-1} J_2(x)\, dx = -x^{-1} J_1(x) + c.$$

(*ii*) Using equation $\int x^n J_{n-1}(x)\, dx = x^n J_n(x) + c$, we obtain

$$\int x^2 J_1(x)\, dx = x^2 J_2(x) + c.$$

(*iii*) We write the given integral as $\int \underset{\text{I}}{x^2} \underset{\text{II}}{(x J_0)}\, dx$. Integrating by parts, we obtain

$$\int x^2 (x J_0)\, dx = x^2 (xJ_1) - \int (x J_1)(2x)\, dx$$

$$= x^3 J_1 - 2\int x^2 J_1\, dx = x^3 J_1 - 2x^2 J_2 + c.$$

(*iv*)

$$\int x^2 J_0(x)\, dx = \int x (x J_0(x))\, dx = x (x J_1(x)) - \int 1.(x J_1(x))\, dx$$

$$= x^2 J_1(x) - \int x J_1(x)\, dx = x^2 J_1(x) + \int x J_0'(x)\, dx$$

$$= x^2 J_1(x) + x J_0(x) - \int 1 \cdot J_0(x)\, dx$$

$$= x^2 J_1(x) + x J_0(x) - \int J_0(x)\, dx + c.$$

(*v*) Integrating by parts, we obtain

$$\int x J_0^2(x)\, dx = \frac{x^2}{2} J_0^2(x) - \int \frac{x^2}{2} 2 J_0(x) J_0'(x) \cdot dx$$

$$= \frac{x^2}{2} J_0^2(x) + \int x^2 J_0(x) J_1(x)\, dx \qquad [\because J_0'(x) = -J_1(x)]$$

$$= \frac{x^2}{2} J_0^2(x) + \int (x J_1(x)) \cdot \frac{d}{dx}[x J_1(x)]\, dx \qquad [\because \{x J_1(x)\}' = x J_0(x)]$$

$$= \frac{x^2}{2} J_0^2(x) + \frac{1}{2}[x J_1(x)]^2 = \frac{1}{2} x^2 [J_0^2(x) + J_1^2(x)] + c.$$

EXAMPLE 12.8 Integrate $I = \int_1^2 x^{-3} J_4(x)\, dx$. Given $J_0(1) = 0.7652$, $J_0(2) = 0.2239$, $J_1(1) = 0.4401$, $J_1(2) = 0.5767$.

Solution We have $\int_1^2 x^{-3} J_4(x)\, dx = -\left[x^{-3} J_3(x)\right]_1^2 = \left[\frac{1}{x^3} J_3(x)\right]_2^1$

$$= J_3(1) - \frac{1}{8} J_3(2). \qquad ...(i)$$

We know $$J_3(x) = \left(\frac{8}{x^2} - 1\right) J_1(x) - \frac{4}{x} J_0(x).$$

Therefore, $$J_3(2) = J_1(2) - 2J_0(2) = 0.5767 - 0.4478 = 0.1289.$$

and $$J_3(1) = 7\,J_1(1) - 4J_0(1) = 7(0.4401) - 4(0.7652)$$
$$= 3.0807 - 3.0608 = 0.0199.$$

Therefore, $$\int_1^2 x^{-3} J_4(x)\,dx = 0.0199 - \frac{1}{8}(0.1289) = 0.0038, \quad \text{[from } (i)\text{]}.$$

EXAMPLE 12.9 Prove that $\int J_3(x)\,dx + J_2(x) + \frac{2}{x} J_1(x) = 0.$ (*AMIETE; JUNE 2011, 2009*)

Solution We have $$\int J_3(x)\,dx = \int [\underset{\text{II}}{x^{-2} J_3(x)}]\, \underset{\text{I}}{x^2}\, dx.$$

Now, $$\frac{d}{dx}[x^{-n} J_n(x)] = -x^{-n} J_{n+1}(x)$$

or $$\int x^{-n} J_{n+1}(x)\,dx = -x^{-n} J_n(x) + c.$$

Therefore, $$\int J_3(x)\,dx = -x^{-2} J_2(x)\cdot x^2 + \int x^{-2} J_2(x)\cdot 2x\,dx \qquad \text{(Integration by parts)}$$
$$= -J_2(x) + \int 2x^{-1} J_2(x)\,dx$$

or $$\int J_3(x)\,dx + J_2(x) + \frac{2}{x} J_1(x) = 0.$$

[***NOTE:*** We can show that $\int J_3(x)\,dx = -2J_2(x) - J_0(x)$.

From (12.21), $$2J_2'(x) = J_1(x) - J_3(x).$$

Therefore, $$\int J_3(x)\,dx = -2\,J_2(x) + \int J_1(x)\,dx = -2\,J_2(x) - J_0(x) + c.$$

Similarly, $$\int J_5(x)\,dx = -2\,J_4(x) + \int J_3(x)\,dx = -2\,J_4(x) - 2\,J_2(x) - J_0(x) + c.$$

Similarly, $\int J_7(x)\,dx$ and $\int J_9(x)\,dx$ etc. can be found.]

EXAMPLE 12.10 Show that $\int x\,J_0^2(x)\,dx = \frac{1}{2} x^2 \{J_0^2(x) + J_1^2(x)\} + c$ where c is a constant.

(*AMIETE, Dec 2007, 2003; UPTU 2004; Osmania 2002*)

Solution Integrating by parts, we obtain

$$\int \underset{\text{II}}{x}\, \underset{\text{I}}{J_0^2(x)}\,dx = \frac{1}{2}x^2 J_0^2(x) - \int \frac{1}{2}x^2 \cdot (2J_0(x)\,J_0'(x)\,dx$$
$$= \frac{1}{2}x^2 J_0^2(x) + \int x^2 J_0(x)\,J_1(x)\,dx \qquad [\text{Since } J_0'(x) = -J_1(x)]$$
$$= \frac{1}{2}x^2 J_0^2(x) + \int x\,J_0(x)\cdot x\,J_1(x)\,dx$$

$$= \frac{1}{2} x^2 J_0^2(x) + \int x J_1(x) \cdot \frac{d}{dx}[x J_1(x)]\, dx$$

$$= \frac{1}{2} x^2 J_0^2(x) + \frac{1}{2} x^2 J_1^2(x) = \frac{1}{2} x^2 [J_0^2(x) + J_1^2(x)] + c.$$

EXAMPLE 12.11 Show that $J_0^2 + 2(J_1^2 + J_2^2 + J_3^2 + \cdots) = 1$, where $J_n(x)$ is the Bessel function of the nth order. *(AMIETE, June 2008, Dec 2006)*

Solution Consider first the following derivative

$$\frac{d}{dx}[J_n^2 + J_{n+1}^2] = 2 J_n J_n' + 2 J_{n+1} J_{n+1}'. \qquad ...(i)$$

From the relation $x J_n' = n J_n - x J_{n+1}$, we obtain $J_n' = \frac{n}{x} J_n - J_{n+1}$. ...(*ii*)

Further, from the relation $x J_n' = x J_{n-1} - n J_n$, we obtain $J_n' = J_{n-1} - \frac{n}{x} J_n$.

Setting $n = n + 1$, we get $J_{n+1}' = J_n - \frac{(n+1)}{x} J_{n+1}$. ...(*iii*)

Substituting (*ii*) and (*iii*) in (*i*), we get

$$\frac{d}{dx}[J_n^2 + J_{n+1}^2] = 2 J_n \left[\frac{n}{x} J_n - J_{n+1}\right] + 2 J_{n+1}\left[J_n - \frac{(n+1)}{x} J_{n+1}\right]$$

$$= 2\left[\frac{n}{x} J_n^2 - \frac{(n+1)}{x} J_{n+1}^2\right].$$

Setting $n = 0, 1, 2, \ldots$, we get

$$[J_0^2 + J_1^2]' = 2\left[0 - \frac{1}{x} J_1^2\right]$$

$$[J_1^2 + J_2^2]' = 2\left[\frac{1}{x} J_1^2 - \frac{2}{x} J_2^2\right]$$

$$[J_2^2 + J_3^2]' = 2\left[\frac{2}{x} J_2^2 - \frac{3}{x} J_3^2\right]$$

....

$$[J_n^2 + J_{n+1}^2]' = 2\left[\frac{n}{x} J_n^2 - \frac{(n+1)}{x} J_{n+1}^2\right] \qquad ...(iv)$$

Adding, we obtain

$\left[J_0^2 + 2(J_1^2 + J_2^2 + J_3^2 + \cdots)\right]' = 0$, since $J_n \to 0$ as $n \to \infty$. Integrating, we obtain $J_0^2 + 2(J_1^2 + J_2^2 + J_3^2 + \cdots) = c$. Let $x = 0$, since $J_0(0) = 1$ and $J_n(0) = 0$, $n > 0$, we obtain $c = 1$. Therefore, we have $J_0^2 + 2(J_1^2 + J_2^2 + J_3^2 + \cdots) = 1$.

EXERCISE 12.1

Find the solution of the following differential equations in terms of Bessel's functions.

1. $16x^2y'' + 16xy' + (16x^2 - 1)y = 0$. **Ans.** $y(x) = AJ_{1/4}(x) + BJ_{-1/4}(x)$.

2. $x^2y'' + xy' + \left(x^2 - \frac{1}{9}\right)y = 0$. **Ans.** $y(x) = AJ_{1/3}(x) + BJ_{-1/3}(x)$.

3. Using the indicated substitution, transform the given equation and hence find the general solution in terms of Bessel's functions.

$9x^2y'' + 9xy' + (81x^2 - 1)\,y = 0,\ z = 3x.$ **Ans.** $y(x) = AJ_{1/3}(3x) + BJ_{-1/3}(3x).$

4. Show that:

(*i*) $J_2(x) = \dfrac{2}{x}J_1(x) - J_0(x).$ (*AMIETE Dec 2005*) (*ii*) $J_1(x) + J_3(x) = \dfrac{4}{x}J_2(x).$ (*PTU 2003*)

(*iii*) $J_3(x) = \left(\dfrac{8}{x^2} - 1\right)J_1(x) - \dfrac{4}{x}J_0(x).$ (*AMIETE, W2009; WBTU, 2005; Madras, 2003*)

(*iv*) $J_4(x) = \left(\dfrac{48}{x^3} - \dfrac{8}{x}\right)J_1(x) + \left(1 - \dfrac{24}{x^2}\right)J_0(x).$ (*AMIETE, Dec 2010, June 03, June 02, VTU, 2003S*)

(*v*) $J_{3/2}(x)\sin x - J_{-3/2}(x)\cos x = \sqrt{2}\,\pi/x^3.$

5. Show that:

(*i*) $J_{-3/2}(x) = -\sqrt{\dfrac{2}{\pi x}}\left(\dfrac{\cos x}{x} + \sin x\right).$ (*ii*) $J_{-5/2}(x) = \sqrt{\dfrac{2}{\pi x}}\left(\dfrac{3}{x}\sin x + \dfrac{3 - x^2}{x^2}\cos x\right).$ (*VTU 2000*)

6. Prove the following:

(*i*) $J_1'(x) = J_0(x) - \dfrac{1}{x}J_1(x).$

(*ii*) $J_2'(x) = \left(1 - \dfrac{4}{x^2}\right)J_1(x) + \dfrac{2}{x}J_0(x).$ (*AMIETE, Dec 2010, Dec 03*)

[**Hint:** Employing recurrence formulas

$$J_n'(x) = J_{n-1}(x) - \frac{n}{x}J_n(x) \qquad ...(i)$$

$$J_n'(x) = \frac{n}{x}J_n(x) - J_{n+1}(x) \qquad ...(ii)$$

On Putting $n = 2$ in (*i*) we obtain

$$J_2'(x) = J_1(x) - \frac{2}{x}J_2(x) \qquad ...(iii)$$

On equating (*i*) and (*ii*) and then putting $n = 1$, we obtain

$$J_0(x) - \frac{1}{x}J_1(x) = \frac{1}{x}J_1(x) - J_2(x)$$

$$\Rightarrow \quad J_2(x) = \frac{2}{x}J_1(x) - J_0(x). \qquad ...(iv)$$

Now, Eq. (*iii*) becomes

$$J_2'(x) = J_1(x) - \frac{2}{x}\left[\frac{2}{x}J_1(x) - J_0(x)\right] \qquad \text{[from } (iv)\text{]}$$

$$= \left(1 - \frac{4}{x^2}\right)J_1(x) + \frac{2}{x}J_0(x).]$$

(*iii*) $J_3'(x) = \left(\dfrac{12}{x^2} - 1\right)J_0(x) - \left(\dfrac{24}{x^3} - \dfrac{5}{x}\right)J_1(x).$

(*iv*) $2J_0''(x) = [J_2(x) - J_0(x)].$ (*v*) $J_1''(x) = \dfrac{1}{x}J_2(x) - J_1(x).$ (*AMIETE, Dec 2006*)

7. Show that:

(*i*) $\frac{d}{dx}[x\,J_n(x)\,J_{n+1}(x)] = x[J_n^2(x) - J_{n+1}^2(x)]$. (*VTU 2006*)

(*ii*) $\frac{d}{dx}[J_n^2(x)\,J_{n+1}^2(x)] = 2\left[\frac{n}{x}J_n^2(x) - \frac{n+1}{x}J_{n+1}^2(x)\right]$. (*VTU, 2000S; UPTU 2005*)

(*iii*) $J_{1/2}^2(x) + J_{-1/2}^2(x) = 2/(\pi x)$.

(*iv*) $J'_{1/2}(x)\,J_{-1/2}(x) - J'_{-1/2}(x)\,J_{1/2}(x) = 2/(\pi x)$. (*Delhi 2002*)

8. Prove the following:

(*i*) $\int J_2(x)\,dx = -2J_1(x) + \int J_0(x)\,dx + c$. (*ii*) $\int J_5(x)\,dx = c - [J_0(x) + 2J_2(x) + 2J_4(x)]$.

(*iii*) $\int x^{-3} J_4(x)\,dx = -x^{-3} J_3(x) + c$.

(*iv*) $\int x^3 J_0(x)\,dx = x(x^2 - 4)\,J_1(x) + 2x^2 J_0(x) + c$.

(*v*) $\int J_5(x)\,dx = -J_4(x) - \frac{4}{x}J_3(x) - \frac{8}{x^2}J_2(x) + c$.

9. Prove the following:

(*i*) $\int x^{-1} J_4(x)\,dx = -\frac{1}{x}J_3(x) - \frac{2}{x^2}J_2(x) + c$. (*AMIETE, Dec 2005*)

(*ii*) $\int_0^{\alpha} x\,J_0(\lambda x)\,dx = \frac{\alpha}{\lambda}J_1(\alpha\lambda)$, α, λ are constants.

(*iii*) $\int J_0(x)\,J_1(x)\,dx = -\frac{1}{2}J_0^2(x) + c$.

(*iv*) $\int_0^x t\,J_n^2(t)\,dt = \frac{1}{2}x^2\,\{J_n^2(x) - J_{n-1}(x)\,J_{n+1}(x)\}$.

(*v*) $\sqrt{\left(\frac{\pi x}{2}\right)}\,J_{3/2}(x) = \frac{1}{x}\sin x - \cos x$.

10. Prove the following relation:

(*i*) $J''_n(x) = \frac{1}{4}[J_{n-2}(x) - 2\,J_n(x) + J_{n+2}(x)]$. (*AMIETE, Dec 2009, JNTU, 2006*)

(*ii*) $x^2\,J''_n(x) = (n^2 - n - x^2)\,J_n(x) + xJ_{n+1}(x)$, where $J_n(x)$ is the Bessel's function of first kind and of order n and dash denotes differentiation w.r.t. x. (*AMIETE-June 2001*)

11. Prove that $\int_0^x x^{-n}\,J_{n+1}(x)\,dx = \frac{1}{2^n\,\Gamma(n+1)} - x^{-n}\,J_n(x), n > -1$. (*AMIETE, Dec 2002*)

12.5 GENERATING FUNCTION OF BESSEL FUNCTION $J_n(x)$

We shall now show that Bessel functions of various orders can be derived as coefficients of various powers of t in the expansion of the function $e^{\frac{x}{2}\left(t-\frac{1}{t}\right)}$.

That is

$$e^{\frac{x}{2}\left(t-\frac{1}{t}\right)} = \sum_{n=-\infty}^{\infty} t^n\,J_n(x).$$

The function on the left hand side is called the *generated function* of the Bessel's functions $J_n(x)$.
We know that

$$e^{\frac{xt}{2}} = 1+\frac{xt}{2}+\frac{1}{2!}\cdot\left(\frac{xt}{2}\right)^2+\frac{1}{3!}\cdot\left(\frac{xt}{2}\right)^3+\ldots\ldots+\frac{1}{n!}\cdot\left(\frac{xt}{2}\right)^n+\frac{1}{(n+1)!}\cdot\left(\frac{xt}{2}\right)^{n+1}+\ldots\ldots$$

and
$$e^{\frac{-x}{2t}} = 1-\frac{x}{2t}+\frac{1}{2!}\cdot\left(\frac{x}{2t}\right)^2-\frac{1}{3!}\cdot\left(\frac{x}{2t}\right)^3+\ldots\ldots+(-1)^n\frac{1}{n!}\cdot\left(\frac{x}{2t}\right)^n+\ldots\ldots$$

Multiplying these two series and collecting the coefficient of t^n, we obtain

$$\frac{1}{n!}\cdot\left(\frac{x}{2}\right)^n-\frac{1}{(n+1)!}\cdot\left(\frac{x}{2}\right)^{n+2}+\frac{1}{2!}\cdot\left(\frac{x}{2}\right)^{n+4}-\ldots\ldots$$

$$=\frac{x^n}{2^n}\frac{1}{n!}\left\{1-\frac{x^2}{2(2n+2)}+\frac{x^4}{2.4\,(2n+2)\,(2n+4)}-\ldots\right\}=J_n(x).$$

Again if we collect the coefficient of t^{-n} from the product of the two series, we obtain the same result except that this will be multiplied by the factor $(-1)^n$.

Hence,
$$e^{\frac{x}{2}\left(t-\frac{1}{t}\right)} = J_0(x)+J_1(x)\,t+J_2(x)\,t^2+\ldots+J_n(x)\,t^n+\ldots$$
$$+J_{-1}(x)\,t^{-1}+J_{-2}(x)\,t^{-2}+\ldots+J_{-n}(x)\,t^{-n}+\ldots$$
$$=\sum_{n=-\infty}^{\infty}t^n J_n(x). \qquad [\because J_{-n}=(-1)^nJ_n] \qquad \ldots(12.23)$$

12.5.1 Integral Representation for Bessel Function $J_n(x)$

Regrouping the terms in the expansion (12.23) we obtain

$$e^{\frac{x}{2}\left(t-\frac{1}{t}\right)} = J_0(x)+(t-t^{-1})\,J_1(x)+(t^2+t^{-2})\,J_2(x)+\ldots\ldots \qquad \ldots(12.24)$$

Substitute $t=\cos\theta+i\sin\theta$ and $(1/t)=\cos\theta-i\sin\theta$ in (12.23) we get

$$e^{ix\sin\theta} = J_0(x)+2i\sin\theta\,J_1(x)+2\cos 2\theta\,J_2(x)+\ldots\ldots$$

We know that $e^{i(x\sin\theta)}=\cos(x\sin\theta)+i\sin(x\sin\theta)$.

Separating real and imaginary parts in the above relation, we get

$$\cos(x\sin\theta) = J_0(x)+2\,[J_2(x)\cos 2\theta+J_4(x)\cos 4\theta+\ldots\ldots] \qquad \ldots(12.25)$$

$$\sin(x\sin\theta) = 2\,[J_1(x)\sin\theta+J_3(x)\sin 3\theta+\ldots\ldots] \qquad \ldots(12.26)$$

The series in (12.25) and (12.26) are known as *Jacobi Series*.

Multiplying both sides of (12.25) by $\cos n\theta$ and (12.26) by $\sin n\theta$ and integrating w.r.t. θ from 0 to π, we get

$$\frac{1}{\pi}\int_0^{\pi}\cos(x\sin\theta)\cos n\theta\,d\theta = \begin{cases}0, & \text{when } n \text{ is odd}\\ J_n(x) & \text{when } n \text{ is even}\end{cases} \qquad \ldots(12.27)$$

and
$$\frac{1}{\pi}\int_0^{\pi}\sin(x\sin\theta)\sin n\theta\,d\theta = \begin{cases}J_n(x), & \text{when } n \text{ is odd}\\ 0 & \text{when } n \text{ is even}\end{cases} \qquad \ldots(12.28)$$

Adding (12.27) and (12.28), we get

$$\frac{1}{\pi}\int_0^{\pi}\{\cos(x\sin\theta)\cos n\theta+\sin(x\sin\theta)\sin n\theta\}\,d\theta=J_n(x),$$

where n is an integer odd or even. Hence for any integral n

$$J_n(x) = \frac{1}{\pi}\int_0^{\pi} \cos(n\theta - x\sin\theta)\, d\theta, \ n \text{ being an integer.} \qquad (VTU,\ 2006) \qquad ...(12.29)$$

This equation provides the integral representation the Bessel function of the first kind $J_n(x)$ and is called the *Bessel integral.* For $n = 0$ the above equation gives the integral representation of $J_0(x)$

$$J_0(x) = \frac{1}{\pi}\int_0^{\pi} \cos(x\sin\theta)\, d\theta .$$

EXAMPLE 12.12 Prove that $J_0^2(x) + 2[J_1^2(x) + J_2^2(x) + J_3^2(x) + \cdots] = 1.$

Solution We know that

$$e^{\frac{x}{2}\left(t-\frac{1}{t}\right)} = J_0(x) + \left(t - \frac{1}{t}\right)J_1(x) + \left(t^2 + \frac{1}{t^2}\right)J_2(x) + \left(t^3 - \frac{1}{t^3}\right)J_3(x) + \cdots \qquad ...(i)$$

Changing x into $-x$ in Eq. (i), we have

$$e^{-\frac{x}{2}\left(t-\frac{1}{t}\right)} = J_0(-x) + \left(t - \frac{1}{t}\right)J_1(-x) + \left(t^2 + \frac{1}{t^2}\right)J_2(-x) + \left(t^3 - \frac{1}{t^3}\right)J_3(-x) + \cdots$$

$$= J_0(x) - \left(t - \frac{1}{t}\right)\cdot J_1(x) + \left(t^2 + \frac{1}{t^2}\right)J_2(x) - \left(t^3 - \frac{1}{t^3}\right)J_3(x) + \cdots \qquad ...(ii)$$

$[\because J_n(-x) = (-1)^n J_n(x)]$

Multiplying (i) and (ii) together, we obtain

$$e^{\frac{x}{2}\left(t-\frac{1}{t}\right)} \cdot e^{-\frac{x}{2}\left(t-\frac{1}{t}\right)} = \left[J_0 + \left(t - \frac{1}{t}\right)J_1 + \left(t^2 + \frac{1}{t^2}\right)J_2 + \cdots\right] \times \left[J_0 - \left(t - \frac{1}{t}\right)J_1 + \left(t^2 + \frac{1}{t^2}\right)J_2 - \cdots\right]$$

or

$$1 = J_0^2 - \left(t - \frac{1}{t}\right)^2 J_1^2 + \left(t^2 + \frac{1}{t^2}\right)^2 J_2^2 - \ldots\ldots +$$

terms involving the products of the Bessel functions ...(iii)

In the right hand side of (iii), the constant term occurs only in the portion containing squares of the Bessel functions.

$$\text{The constant term} = J_0^2 - (-2)J_1^2 + 2J_2^2 - (-2)J_3^2 + \ldots\ldots$$

$$= J_0^2 + 2(J_1^2 + J_2^2 + J_3^2 + \cdots)$$

The constant term in the left hand side of (iii) is 1.

Therefore, we obtain $J_0^2 + 2(J_1^2 + J_2^2 + J_3^2 + \cdots) = 1.$

EXAMPLE 12.13 Show that $\frac{1}{\pi}\int_0^{\pi} \cos(x\cos\phi)\, d\phi = J_0(x).$ (*Madras, 2006; VTU, 2000S*)

Solution Setting $\theta = (\pi/2) + \phi$ in (12.25), we get

$$\cos(x\cos\phi) = J_0(x) - 2J_2(x)\cos 2\phi + 2J_4(x)\cos 4\phi - \ldots.$$

Integrating w.r.t ϕ on both sides within the limits 0 to π, we get

$$\int_0^{\pi} \cos(x\cos\phi)\, d\phi = \pi J_0(x), \text{ since the other terms vanish}$$

or

$$J_0(x) = \frac{1}{\pi}\int_0^{\pi} \cos(x\cos\phi)\, d\phi.$$

EXERCISE 12.2

1. Show that:

(*i*) $\cos x = J_0(x) - 2J_2(x) + 2J_4(x) - 2J_6(x) + \ldots$ (*Kerala, 2005; AMIETE, Dec 2004*)

(*ii*) $\sin x = 2[J_1(x) - J_3(x) + J_5(x) - J_7(x) + \ldots]$ (*Kerala, 2005; Anna, 2005S; AMIETE, Dec 2004*)
[Put $\theta = \pi/2$ in Jacobi series].

(*iii*) $1 = J_0 + 2[J_2 + J_4 + J_6 + \ldots]$.

2. Show that:

(*i*) $\cos(x\cos\theta) = J_0(x) - 2J_2(x)\cos 2\theta + 2J_4(x)\cos 4\theta - \ldots$

(*ii*) $\sin(x\cos\theta) = 2[J_1(x)\cos\theta - J_3(x)\cos 3\theta + J_5(x)\cos 5\theta - \ldots]$ (*Madras 2003S*)
[**Hint:** Replace θ by $(\pi/2 - \theta)$ in Jacabi series]

3. Show that:

(*i*) $J_0(x) = \frac{1}{\pi}\int_0^{\pi} \cos(x\sin\theta)\,d\theta.$ (*ii*) $\int_0^{\pi} \sin(x\sin\theta)\,d\theta = 4\sum_{n=0}^{\infty} \frac{J_{2n+1}(x)}{2n+1}.$

(*iii*) $J_0(x) = \frac{1}{2\pi}\int_0^{2\pi} \exp(ix\cos\phi)\,d\phi = \frac{1}{2\pi}\int_0^{2\pi} \cos(x\cos\theta)\,d\theta.$

4. Using the result $J_0(x) = \frac{1}{\pi}\int_0^{\pi} \cos(x\cos\theta)\,d\theta$, prove that $\int_0^{\infty} e^{-bx} J_0(ax)\,dx = (a^2 + b^2)^{-1/2}$ and hence deduce that $\int_0^{\infty} J_0(ax)\,dx = 1/a$. (*AMIETE, Dec 2003*)

[**Hint:** $J_0(x) = \frac{1}{\pi}\int_0^{\pi} \cos(x\cos\theta)\,d\theta = \frac{2}{\pi}\int_0^{\pi/2} \cos(x\cos\theta)\,d\theta$.

Hence $\int_0^{\infty} e^{-bx} J_0(ax)\,dx = \frac{2}{\pi}\int_0^{\infty} dx \int_0^{\pi/2} e^{-bx}\cos(ax\cos\theta)\,d\theta$

$$= \frac{2}{\pi}\int_0^{\pi/2} d\theta \int_0^{\infty} e^{-bx}\cos(ax\cos\theta)\,d\theta. \quad \ldots(i)$$

Using the result $\int_0^{\infty} e^{-bx}\cos kx\,dx = \frac{b}{b^2 + k^2}$.

We have $\int_0^{\infty} e^{-bx}\cos(ax\cos\theta)\,d\theta = \frac{b}{b^2 + a^2\cos^2\theta}$.

From (i), we obtain $\int_0^{\infty} e^{-bx} J_0(ax)\,dx = \frac{2}{\pi}\int_0^{\pi/2} \frac{b}{b^2 + a^2\cos^2\theta}\,d\theta = \frac{2}{\pi}\int_0^{\pi/2} \frac{b\sec^2\theta}{(a^2 + b^2) + b^2\tan^2\theta}\,d\theta$

$$= \frac{1}{\sqrt{a^2 + b^2}} \quad \text{(Setting } b\tan\theta = t\text{)}$$

If $b = 0$, then $\int_0^{\infty} J_0(ax)\,dx = \frac{1}{a}$.]

12.6 EQUATIONS REDUCIBLE TO BESSEL'S EQUATION

A number of second order differential equations with variable coefficients can be solved by changing them into Bessel form by suitable transformations. In the discussion below, we derive very general form of Bessel equation and will use it for writing the solution of the given equation by suitably adjusting the parameters.

Let us consider the differential equation

$$t^2 \frac{d^2u}{dt^2} + t\frac{du}{dt} + (t^2 - n^2)u = 0. \qquad ...(12.30)$$

We change the independent variable t to z by the relation $t = \lambda z$, and equation (12.30) can be transformed to

$$z^2 \frac{d^2u}{dz^2} + z\frac{du}{dz} + (\lambda^2 z^2 - n^2)u = 0. \qquad ...(12.31)$$

A solution of (12.31) is given by $\qquad u = J_n(\lambda z). \qquad ...(12.32)$

We again change the independent variable from z to x by the transformation $z = x^\beta$ so that

$$\frac{dz}{z} = \beta \frac{dx}{x}.$$

Equation (12.31) can be written as

$$z\frac{d}{dz}\left(z\frac{du}{dz}\right) + (\lambda^2 z^2 - n^2)\,u = 0$$

and is transformed as

$$\frac{x}{\beta}\frac{d}{dx}\left(\frac{x}{\beta}\frac{du}{dx}\right) + (\lambda^2 x^{2\beta} - n^2)\,u = 0.$$

Now changing the dependent variable from u to y by the relation $u = x^{-\alpha}y$, the above equation can be written as

$$x\frac{d}{dx}\left[x\left(x^{-\alpha}\frac{dy}{dx} - \alpha x^{-\alpha-1}\,y\right)\right] + \beta^2(\lambda^2 x^{2\beta} - n^2)\,x^{-\alpha}y = 0.$$

Differentiating the first bracket and dividing by $x^{-\alpha}$, the transformed equation is written as

$$\mathbf{x^2 \frac{d^2y}{dx^2} + (1-2\alpha)\,x\frac{dy}{dx} + \left[\lambda^2\beta^2 x^{2\beta} + (\alpha^2 - n^2\beta^2)\right]y = 0.} \qquad ...(12.33)$$

We see that a solution of (12.33) is given by

$$y = x^\alpha J_n(\lambda x^\beta).$$

Equation (11.33) is fairly a very general form of Bessel's equation and the solution of a given equation can be obtained by comparison by evaluating the parameters α, β, and λ.

EXAMPLE 12.14 Solve the equation $4\dfrac{d^2y}{dx^2} + 9xy = 0.$

Solution The given equation can be written as $x^2\dfrac{d^2y}{dx^2} + \dfrac{9}{4}x^3y = 0.$

Comparing this equation with the standard transformed equation(12.33), we obtain

$1 - 2\alpha = 0, \quad 2\beta = 3, \quad \lambda^2\beta^2 = 9/4, \quad \alpha^2 - n^2\beta^2 = 0,$

giving $\alpha = 1/2, \quad \beta = 3/2, \quad \lambda = 1, \quad n = \pm 1/3.$

Hence the solution is given by

$$y = Ax^{1/2} J_{1/3}(x^{3/2}) + Bx^{1/2} J_{-1/3}(x^{3/2})$$

or

$$y = \sqrt{x}\,[AJ_{1/3}(x^{3/2}) + BJ_{-1/3}(x^{3/2})].$$

EXAMPLE 12.15 Solve the equation $x\frac{d^2y}{dx^2} - 3\left(\frac{dy}{dx}\right) + xy = 0.$

Solution The given equation can be written as $x^2\left(\frac{d^2y}{dx^2}\right) - 3x\left(\frac{dy}{dx}\right) + x^2y = 0.$

Comparing with the general equation, we get

$$1 - 2\alpha = -3, \quad 2\beta = 2, \quad \lambda^2\beta^2 = 1, \quad \alpha^2 - n^2\beta^2 = 0,$$

$$\Rightarrow \quad \alpha = 2, \quad \beta = 1, \quad \lambda = 1, \quad n = \pm 2.$$

Hence a solution of the given equation is $y = Ax^2J_2(x)$.

12.7 BESSEL'S FUNCTION OF THE SECOND KIND $Y_0(x)$

We have seen that if n is neither zero nor an integer, the complete solution of the Bessel differential equation of order n is given by $y = AJ_n(x) + BJ_{-n}(x)$, where A and B are arbitrary constants.

When n is an integer, we have shown that $J_n(x)$ and $J_{-n}(x)$ are linearly dependent since $J_{-n}(x) = (-1)^n J_n(x)$. Therefore we have only one independent solution. We have to obtain the second independent solution to find the general solution. Also, when $n = 0$, we find that $J_{-n}(x) = J_n(x)$. Hence, we have to find the second solution of Bessel's equation of order n, when n is zero or an integer.

Bessel's equation of zeroth order is

$$xy'' + y' + xy = 0. \qquad ...(12.34)$$

In this case the indicial equation has the double root $m = 0$. The first solution is $y_1(x) = J_0(x)$ while the second solution has the form

$$y_2(x) = y_1(x)\log x + \sum_{k=1}^{\infty} a_k x^k \qquad ...(12.35)$$

$$= J_0(x)\log x + [a_1x + a_2x^2 + a_3x^3 + \cdots], \text{ since } y_1(x) = J_0(x).$$

We substitute y_2 and its derivatives

$$y_2'(x) = J_0' \log x + \frac{1}{x}J_0 + [a_1 + 2a_2x + 3a_3x^2 + \cdots]$$

$$y_2''(x) = J_0'' \log x + \frac{2}{x}J_0' - \frac{1}{x^2}J_0 + [2a_2 + 6a_3x + \cdots]$$

into Eq. (12.34). We obtain

$$[xJ_0'' + J_0' + xJ_0]\log x + 2J_0' + [a_1 + 4a_2x + (a_1 + 9a_3)x^2 + \cdots + (a_{k-1} + (k+1)^2 a_{m+1})x^k + \cdots] = 0$$

The first term vanishes as $J_0(x)$ is a solution of Eq. (12.34). Substituting for J_0'

$$J_0' = \sum_{k=1}^{\infty} \frac{(-1)^k\,(2k)x^{2k-1}}{(k!)^2\,2^{2k}} = \sum_{k=1}^{\infty} \frac{(-1)^k\,x^{2k-1}}{k!(k-1)!\,2^{2k-1}}$$

$$= -\frac{x}{2} + \frac{x^3}{(2!)\,(2)^3} - \frac{x^5}{(3!)\,(2!)\,(2^5)} + \cdots,$$

we obtain

$$\left[-x + \frac{x^3}{(2!)\,(2^2)} - \frac{x^5}{(3!)\,(2!)\,(2^4)} + \cdots\right] + [a_1 + 4a_2x + (a_1 + 9a_3)x^2 + (a_2 + 16a_4)x^3 + \cdots] = 0.$$

Equating the coefficients of even powers of x to zero, we obtain

$a_1 = 0$, $a_1 + 9a_3 = 0$, or $a_3 = 0$, $a_{2k-1} + (2k+1)^2 a_{2k+1} = 0$, $k = 1, 2, \ldots\ldots$.

Therefore, $a_1 = 0 = a_3 = a_5 = \dots$.

Equating the coefficient of x to zero, we obtain $-1 + 4a_2 = 0$, or $a_2 = 1/4 = 1/2^2$.

Equating the coefficient of x^{2k+1} to zero, we obtain

$$\frac{(-1)^{k+1}}{k!(k+1)!2^{2k}} + a_{2k} + (2k+2)^2 a_{2k+2} = 0, \qquad k = 1, 2, \dots$$

For $k = 1$, we get

$$\frac{1}{(2!)(2^2)} + a_2 + 16a_4 = 0 \text{ or } a_4 = -\frac{1}{16}\left[\frac{1}{(2!)(2^2)} + \frac{1}{2^2}\right] = -\frac{1}{2^2 \cdot 4^2}\left[1 + \frac{1}{2}\right].$$

For $k = 2$, we get

$$-\frac{1}{(3!)(2!)(2^4)} + a_4 + 36a_6 = 0$$

or
$$a_6 = \frac{1}{6^2}\left[\frac{1}{(3!)(2!)(2^4)} + \frac{1}{2^2 \cdot 4^2}\left(1 + \frac{1}{2}\right)\right] = \frac{1}{2^2 \cdot 4^2 \cdot 6^2}\left[1 + \frac{1}{2} + \frac{1}{3}\right], \dots\dots$$

Therefore,

$$y_2(x) = J_0(x) \log x + \left[\frac{1}{2^2}x^2 - \frac{1}{2^2 \cdot 4^2}\left(1 + \frac{1}{2}\right)x^4 + \frac{1}{2^2 \cdot 4^2 \cdot 6^2}\left(1 + \frac{1}{2} + \frac{1}{3}\right)x^6 - \cdots\right]$$

is the second linearly independent solution of Eq. (12.34). However, any other linear combination of $y_1(x)$ and $y_2(x)$ can also be taken as the second linearly independent solution. Another independent solution is $a(y_2 + bJ_0)$, where $a(\neq 0)$ and b are constants. It is customary to choose $a = 2/\pi$ and

$$b = \lim_{n \to \infty}\left[1 + \frac{1}{2} + \frac{1}{3} + \cdots + \frac{1}{n} - \log n\right] - \log 2$$

$$= \gamma - \log 2, \quad (\gamma = 0.5772 \dots\dots \text{ is called Euler's constant}).$$

The solution $a(y_2 + bJ_0)$ this obtained is called **Bessel function of the second kind** of *order zero* (Fig. 12.2) or **Neumann's function** of *order zero* after the German mathematician Carl Neumann (1832–1925) and is denoted by $Y_0(x)$.

$$\text{Thus } Y_0(x) = \frac{2}{\pi}\left[J_0(x)\left(\log\frac{x}{2} + \gamma\right) + \left\{\frac{1}{2^2}x^2 - \frac{1}{2^2 \cdot 4^2}\left(1 + \frac{1}{2}\right)x^4 + \frac{1}{2^2 \cdot 4^2 \cdot 6^2}\left(1 + \frac{1}{2} + \frac{1}{3}\right)x^6 - \cdots\right\}\right].$$

For small $x > 0$ the function $Y_0(x)$ behaves about like $\log x$ and $Y_0(x) \to -\infty$ as $x \to 0$. The graphs of the functions $Y_0(x)$ and $Y_1(x)$ are shown in Fig. 12.2.

Hence the complete solution of Eq. (12.34) is of the form

$$y = c_1 J_0(x) + c_2 Y_0(x). \qquad \dots(12.36)$$

In a similar way, the general solution of Bessel's equation of the nth order that is, the equation

$$x^2\frac{d^2y}{dx^2} + x\frac{dy}{dx} + (x^2 - n^2)y = 0, \qquad n \geq 0$$

can be taken in the form $y = c_1 J_n(x) + c_2 Y_n(x)$ ...(12.37)

where c_1 and c_2 are arbitrary constants.

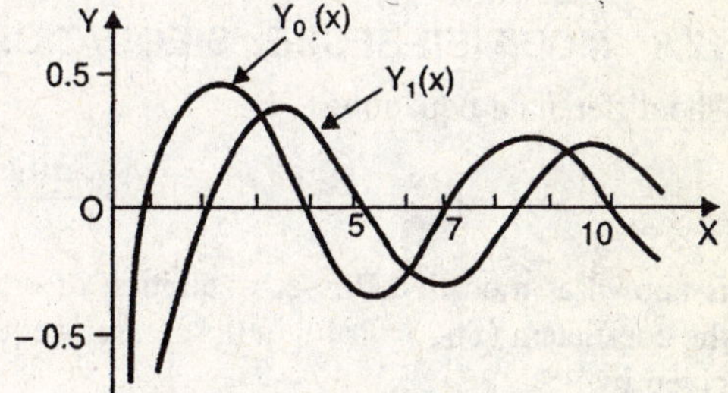

Fig. 12.2: Bessel functions of the second kind

A standard practice is to defined the second solution of Bessel's equation by the function

$$Y_n(x) = \frac{J_n(x)\cos n\pi - J_{-n}(x)}{\sin n\pi}, \text{ when } n \neq \text{integer, and } Y_n(x) = \lim_{\nu \to n}\frac{J_\nu(x)\cos \nu\pi - J_{-\nu}(x)}{\sin \nu\pi}, \text{ when } n \text{ is an integer.}$$

This function is known as **Bessel function of the second kind** of *order n* or **Neumann's function** of *order n*. Being a linear combination of two solutions of Bessel's equation, $Y_n(x)$ is also a solution. When n is an integer, we see on applying L'Hospital rule that

$$Y_n(x) = \lim_{\nu \to n} \frac{-\pi J_\nu \sin \nu\pi + \dfrac{\partial J_\nu}{\partial \nu} \cos \nu\pi - \dfrac{\partial J_{-\nu}}{\partial \nu}}{\pi \cos \nu\pi} = \lim_{\nu \to n} \frac{1}{\pi}\left[\frac{\partial J_\nu}{\partial \nu} - (-1)^n \frac{\partial J_{-\nu}}{\partial \nu}\right].$$

The Neumann's function $Y_n(x)$ satisfies recurrence formulae of the same type as those for $J_n(x)$.

EXAMPLE 12.16 Find the solution of the differential equation $\dfrac{d^2y}{dx^2} + \left(9x - \dfrac{20}{x^2}\right)y = 0.$

Solution The given equation may be written as

$$x^2y'' + (9x^3 - 20)y = 0.$$

Comparing this with the standard transformed Eq. (12.33), we get

$$1 - 2\alpha = 0,\ \lambda^2\beta^2 = 9,\ 2\beta = 3,\ \alpha^2 - n^2\beta^2 = -20$$

This gives $\alpha = 1/2,\ \beta = 3/2,\ \lambda = 2,\ n = 3.$

Hence the solution is

$$y = \sqrt{x}\,[AJ_3(2x^{3/2}) + BY_3(2x)^{3/2}].$$

EXAMPLE 12.17 Find the solution of the differential equation $x^2y'' + xy' + \frac{1}{4}(x-1)y = 0$ using the substitution $z = \sqrt{x}$.

Solution When $z = \sqrt{x}$, we have $\dfrac{dy}{dx} = \dfrac{dy}{dz}\cdot\dfrac{dz}{dx} = \dfrac{1}{2\sqrt{x}}\dfrac{dy}{dz}$ and $\dfrac{d^2y}{dx^2} = \dfrac{1}{4x}\dfrac{d^2y}{dz^2} - \dfrac{1}{4x\sqrt{x}}\dfrac{dy}{dz}.$

Substituting in the given differential equation and using $\sqrt{x} = z$, we get

$$x^2\left(\frac{1}{4x}\frac{d^2y}{dz^2} - \frac{1}{4x\sqrt{x}}\frac{dy}{dz}\right) + \frac{x}{2\sqrt{x}}\frac{dy}{dz} + \frac{1}{4}(x-1)y = 0 \text{ or } z^2\frac{d^2y}{dz^2} + z\frac{dy}{dz} + (z^2-1)y = 0.$$

This equation is the Bessel equation of order one. Therefore, the solution is

$$y(z) = AJ_1(z) + BY_1(z), \text{ or } y(x) = AJ_1\left(\sqrt{x}\right) + BY_1\left(\sqrt{x}\right).$$

12.8 MODIFIED BESSEL'S EQUATION

The differential equation

$$x^2\frac{d^2y}{dx^2} + x\frac{dy}{dx} - (x^2 + n^2)\,y = 0 \qquad \text{...(12.38)}$$

is known as modified Bessel's equation of order n. This equation is similar to Bessel's equation except that the coefficient $(x^2 - n^2)$ of y in the Bessel's equation is replaced by $(i^2x^2 - n^2)$. The solution of this equation is given by

$$J_n(ix) = \sum_{k=0}^{\infty} \frac{(-1)^k\,(ix/2)^{n+2k}}{k!\,\Gamma(n+k+1)} = i^n \sum_{k=0}^{\infty} \frac{(-1)^k\,(i)^{2k}\,(x/2)^{n+2k}}{k!\,\Gamma(n+k+1)}$$

or

$$i^{-n} J_n(ix) = \sum_{k=0}^{\infty} \frac{(x/2)^{n+2k}}{k!\,\Gamma(n+k+1)}. \qquad \left(i = \sqrt{-1}\right) \quad \text{...(12.39)}$$

The series $\sum_{k=0}^{\infty} \frac{(x/2)^{n+2k}}{k!\,\Gamma(n+k+1)}$ which is a positive term series is denoted by $I_n(x)$. It is a real function and is called the *modified Bessel function of first kind and of order n*. Thus

$$I_n(x) = i^{-n} J_n(ix) \qquad ...(12.40)$$

which is also solution of the equation (12.38) because i^{-n} is a constant.

Another fundamental solution of the above equation is called the modified Bessel function of the second kind and of order n. It is defined by

$$K_n(x) = \frac{\pi/2}{\sin n\pi}[1_{-n}(x) - I_n(x)],\ n \neq \text{integral}.$$

Hence the complete solution of Equation (12.38) may be written in the form

$$y = A\, I_n(x) + B\, K_n(x),$$

where A and B are arbitrary constants.

In contrast to Bessel functions, the functions $I_n(x)$ and $k_n(x)$ are non-oscillatory, but their behaviour is similar to exponential functions. $I_0(x)$ is approximately 1 when x is small but for large values of x,

$$I_0(x) \approx \exp.(x)/\sqrt{2\pi x}.$$

12.9 Ber AND Bei FUNCTIONS

We now discuss another differential equation which finds application in certain problems in electrical engineering and also in problems involving eddy-currents. The differential equation is defined by

$$x\frac{d^2y}{dx^2} + \frac{dy}{dx} - ixy = 0. \qquad ...(12.41)$$

Equation (12.41) is a special case of equation (12.33) Art. 12.6 with $\alpha = 0$, $n = 0$, $\beta = 1$ and $\lambda^2 = -i = i^3$, so that $\lambda = i^{3/2}$.

Hence a solution of equation (12.41) is given by $J_0(i^{3/2}x)$.

The function $J_0(i^{3/2}x)$ is a complex function and can be decomposed into real and imaginary parts on expansion. Art 12.2 Eq. (12.11). This gives

$$J_0(i^{3/2}x) = 1 - \frac{i^3x^2}{2^2} + \frac{i^6x^4}{2^2\cdot 4^2} - \frac{i^9x^6}{2^2\cdot 4^2\cdot 6^2} + \frac{i^{12}x^8}{2^2\cdot 4^2\cdot 6^2\cdot 8^2} + \ldots$$

$$= \left[1 - \frac{x^4}{2^2\cdot 4^2} + \frac{x^8}{2^2\cdot 4^2\cdot 6^2\cdot 8^2} - \ldots\right] + i\left[\frac{x^2}{2^2} - \frac{x^6}{2^2\cdot 4^2\cdot 6^2} + \frac{x^{10}}{2^2\cdot 4^2\cdot 6^2\cdot 8^2\cdot 10^2} - \ldots\right]$$

$$= 1 + \sum_{k=1}^{\infty}(-1)^k \frac{x^{4k}}{2^2\cdot 4^2\cdot 6^2\ldots(4k)^2} + i\left[-\sum_{k=1}^{\infty}\frac{(-1)^k x^{4k-2}}{2^2\cdot 4^2\cdot 6^2\ldots(4k-2)^2}\right]. \qquad ...(12.42)$$

We define the ber (Bessel-real) and bei (Bessel-imaginary) functions by the following relations:

$$\text{ber } x = 1 + \sum_{k=1}^{\infty}(-1)^k \frac{x^{4k}}{2^2\cdot 4^2\cdot 6^2\ldots(4k)^2} \text{ or } \sum_{k=0}^{\infty}\frac{(-1)^k x^{4k}}{2^{4k}[(2k)!]^2}, \qquad ...(12.43)$$

and

$$\text{bei } x = -\sum_{k=1}^{\infty}(-1)^k \frac{x^{4k-2}}{2^2\cdot 4^2\cdot 6^2\ldots(4k-2)^2} \text{ or } \sum_{k=0}^{\infty}\frac{(-1)^k x^{4k+2}}{2^{4k+2}[(2k+1)!]^2}. \qquad ...(12.44)$$

Hence a solution of (12.41) is

$$y = J_0(i^{3/2}x) = \text{ber } x + i \text{ bei } x. \qquad ...(12.45)$$

EXAMPLE 12.18 Prove that:

(*i*) $\dfrac{d}{dx}(x \text{ ber}' x) = -x \text{ bei } x,$ (*ii*) $\dfrac{d}{dx}(x \text{ bei}' x) = x \text{ ber } x.$

Solution (*i*) We know that

$$\text{ber } x = 1 - \frac{x^4}{2^2 \cdot 4^2} + \frac{x^8}{2^2 \cdot 4^2 \cdot 6^2 \cdot 8^2} - \ldots\ldots \infty$$

and

$$\text{bei } x = \frac{x^2}{2^2} - \frac{x^6}{2^2 \cdot 4^2 \cdot 6^2} + \frac{x^{10}}{2^2 \cdot 4^2 \cdot 6^2 \cdot 8^2 \cdot 10^2} - \ldots\ldots \infty$$

Now

$$\text{ber}' x = \frac{-x^3}{2^2 \cdot 4} + \frac{x^7}{2^2 \cdot 4^2 \cdot 6^2 \cdot 8} - \ldots\ldots \infty$$

Therefore,

$$x \text{ ber}' x = -\frac{x^4}{2^2 \cdot 4} + \frac{x^8}{2^2 \cdot 4^2 \cdot 6^2 \cdot 8} - \ldots\ldots \infty$$

$$\frac{d}{dx}(x \text{ ber}' x) = \frac{-x^3}{2^2} + \frac{x^7}{2^2 \cdot 4^2 \cdot 6^2} - \ldots\ldots \infty$$

$$= -x\left(\frac{x^2}{2^2} - \frac{x^6}{2^2 \cdot 4^2 \cdot 6^2} + \ldots\ldots \infty\right) = -x \text{ bei } x.$$

(*ii*) Also,

$$\text{bei}'(x) = \frac{x}{2} - \frac{x^5}{2^2 \cdot 4^2 \cdot 6} + \frac{x^9}{2^2 \cdot 4^2 \cdot 6^2 \cdot 8^2 \cdot 10} - \ldots\ldots$$

Therefore,

$$x \text{ bei}' x = \frac{x^2}{2} - \frac{x^6}{2^2 \cdot 4^2 \cdot 6} + \frac{x^{10}}{2^2 \cdot 4^2 \cdot 6^2 \cdot 8^2 \cdot 10} - \ldots\ldots$$

$$\frac{d}{dx}(x \text{ bei}' x) = x - \frac{x^5}{2^2 \cdot 4^2} + \frac{x^9}{2^2 \cdot 4^2 \cdot 6^2 \cdot 8^2} - \ldots\ldots$$

$$= x\left[1 - \frac{x^4}{2^2 \cdot 4^2} + \frac{x^8}{2^2 \cdot 4^2 \cdot 6^2 \cdot 8^2} - \ldots\ldots\right] = x \text{ ber } x.$$

EXAMPLE 12.19 Show that $\displaystyle\int_0^p x(\text{ber}^2 x + \text{bei}^2 x)\,dx = p(\text{ber } p \text{ bei}' p - \text{bei } p \text{ ber}' p)$.

Solution We have proved in the above Example that

$$\frac{d}{dx}(x \text{ bei}' x) = x \text{ ber } x$$

and

$$\frac{d}{dx}(x \text{ ber}' x) = -x \text{ bei } x.$$

Therefore,

$$\int_0^p x(\text{ber}^2 x + \text{bei}^2 x)\,dx = \int_0^p x(\text{ber } x) \text{ ber } x\,dx + \int_0^p x(\text{bei } x) \text{ bei } x\,dx$$

$$= [\text{ber } x \cdot x \text{ bei}' x]_0^p - \int_0^p \text{ber}' x \cdot x \text{ bei}' x\,dx + [\text{bei } x(-x \text{ ber}' x)]_0^p - \int_0^p \text{bei}' x(-x \text{ ber}' x)\,dx$$

$= p(\text{ber } p \text{ bei}' p - \text{bei } p \text{ ber}' p)$, other terms cancel each other.

12.10 ORTHOGONALITY OF BESSEL FUNCTIONS

If α and β are two distinct roots of $J_n(x) = 0$, that is $J_n(\alpha) = J_n(\beta) = 0$. ...(12.46)

then
$$\int_0^1 xJ_n(\alpha x)\,J_n(\beta x)\,dx = \begin{cases} 0, & \alpha \neq \beta \\ \frac{1}{2}[J_{n+1}(\alpha)]^2, & \alpha = \beta. \end{cases} \quad ...(12.47)$$

(AMIETE, Dec 2010, June 2009)

This relation defines the orthogonality property of the functions $J_n(\alpha x)$ on the interval (0, 1) with respect to the weight function x. To prove this, we see that $J_n(\alpha x)$ and $J_n(\beta x)$ are the solutions of the Bessel's equations.

$$x^2\frac{d^2u}{dx^2} + x\frac{du}{dx} + (\alpha^2x^2 - n^2)u = 0 \quad ...(12.48)$$

and
$$x^2\frac{d^2v}{dx^2} + x\frac{dv}{dx} + (\beta^2x^2 - n^2)v = 0 \quad ...(12.49)$$

that is, $u = J_n(\alpha x)$ and $v = J_n(\beta x)$.

Multiplying (12.48) by v/x and (12.49) by u/x and subtracting, we get

$$x(u''v - uv'') + (u'v - uv') + (\alpha^2 - \beta^2)\,xuv = 0$$

or
$$\frac{d}{dx}[x(u'v - uv')] = (\beta^2 - \alpha^2)\,xuv. \quad ...(12.50)$$

Integrating both sides w.r.t. x between the limits 0 to 1, we get

$$(\beta^2 - \alpha^2)\int_0^1 x\,uv\,dx = [x(u'v - uv')]_0^1. \quad ...(12.51)$$

Now since $u = J_n(\alpha x)$, therefore

$$\frac{du}{dx} = \frac{d}{dx}[J_n(\alpha x)] = \frac{d}{d(\alpha x)}[J_n(\alpha x)]\cdot\frac{d(\alpha x)}{dx} = \alpha\,J_n'(\alpha x),$$

where the prime denoted derivative w.t.t. αx.

Similarly, $dv/dx = \beta J_n'(\beta x)$.

Substituting for u, v, du/dx and dv/dx in (12.51), we obtain

$$(\beta^2 - \alpha^2)\int_0^1 x\,J_n(\alpha x)\,J_n(\beta x)\,dx = [\alpha\,J_n(\beta x)\,J_n'(\alpha x) - \beta J_n(\alpha x)\,J_n'(\beta x)]_{x=1}$$

or
$$\int_0^1 x\,J_n(\alpha x)\,J_n(\beta x)\,dx = \frac{1}{\beta^2 - \alpha^2}[\alpha\,J_n(\beta x)\,J_n'(\alpha x) - \beta J_n(\alpha x)\,J_n'(\beta x)]_{x=1}. \quad ...(12.52)$$

If α and β are the distinct roots of the equation

$J_n(x) = 0$, then $J_n(\alpha) = 0 = J_n(\beta)$, (12.52) can be written as,

$$\int_0^1 x\,J_n(\alpha x)\,J_n(\beta x)\,dx = 0. \quad ...(12.53)$$

If however, $\alpha = \beta$ the value of the integral is 0/0, which is indeterminate. To find the value of this expression we assume that α is the root of $J_n(x) = 0$ and β is a value in the neighbourhood of α. The value of (12.52) can be given by

$$\lim_{\beta\to\alpha}\int_0^1 x\,J_n(\alpha x)\,J_n(\beta x)\,dx = \lim_{\beta\to\alpha}\frac{\alpha\,J_n(\beta)\,J_n'(\alpha)}{\beta^2-\alpha^2} \qquad \left(\frac{0}{0}\right)$$

$$= \lim_{\beta\to\alpha}\frac{\alpha\,J_n'(\beta)\,J_n'(\alpha)}{2\beta} = \frac{1}{2}[J_n'(\alpha)]^2.$$

Therefore,
$$\int_0^1 x\,J_n^2(\alpha x)\,dx = \frac{1}{2}[J_n'(\alpha)]^2. \qquad ...(12.54)$$

Since $\qquad x\,J_n'(x) = n\,J_n(x) - xJ_{n+1}(x),\quad$ from (12.20)

therefore, $\qquad \alpha\,J_n'(\alpha) = n\,J_n(\alpha) - \alpha J_{n+1}(\alpha) = -\,\alpha J_{n+1}(\alpha)$, since $J_n(\alpha) = 0$.

Thus, $\qquad J_n'(\alpha) = -\,J_{n+1}(\alpha)$.

Eq. (12.54) can be written as

$$\int_0^1 x\,J_n^2(\alpha x)\,dx = \frac{1}{2}\,J^2{}_{n+1}(\alpha). \qquad ...(12.55)$$

If however the interval under reference is from 0 to a, it can be shown that

$$\int_0^a x\,J_n^2(\alpha x)\,dx = \frac{a^2}{2}\,J^2{}_{n+1}(a\alpha), \qquad ...(12.56)$$

where α's are the roots of $J_n(a\alpha) = 0$.

The integral relations (12.53) and (12.55) are very important in the sense that they enable us to determine the coefficients in the expansion of a function of x in terms of Bessel functions.

12.11 FOURIER-BESSEL SERIES

If a function $f(x)$ is continuous and has a finite number of oscillations in the interval $0 \le x \le a$, then the function can be developed as an infinite series in the form

$$f(x) = c_1J_n(\alpha_1 x) + c_2J_n(\alpha_2 x) + c_3J_n(\alpha_3 x) + \cdots + c_nJ_n(\alpha_n x) + \cdots \qquad (12.57)$$

where $\alpha_1, \alpha_2, \ldots$ are the positive roots of $J_n(x) = 0$. To determine the coefficients c_n, we multiply both sides of (12.57) by $xJ_n(\alpha_n x)$ and integrate with respect to x within the limits 0 to a, assuming that termwise integration is valid. From the orthogonal property of Bessel functions, it will be seen that all integrals on the right handside of (12.57) will vanish due to (12.53) except the one containing c_n and we have

$$\int_0^a x\,f(x)\,J_n(\alpha_n x)dx = c_n\int_0^a x\,J_n^2(\alpha_n x)\,dx = c_n\frac{a^2}{2}J_{n+1}^2(a\alpha_n), \qquad \text{from (12.56)}.$$

Therefore,
$$c_n = \frac{2}{a^2J_{n+1}^2(a\alpha_n)}\int_0^a x\,f(x)\,J_n(\alpha_n x)\,dx. \qquad ...(12.58)$$

Giving n the values 0, 1, 2, ... we can find out c's and hence the function $f(x)$ in (12.57).

EXAMPLE 12.20 Show that $1 = \sum_{i=1}^{\infty}\frac{2}{\alpha_i}\cdot\frac{J_0(\alpha_i x)}{J_1(\alpha_i)}$.

Solution Let $f(x) = \sum_{i=1}^{\infty} c_iJ_n(\alpha_i x) \qquad ...(i)$

then $c_i = \dfrac{2}{a^2 J_{n+1}^2(a\alpha_i)} \int_0^a x\, f(x)\, J_n(\alpha_i \cdot x)\, dx.$

Here $f(x) = 1$, $a = 1$ and $n = 0$. Therefore, we have to develope 1 in series of Bessel functions of order zero in the interval $0 \le x \le 1$. From (12.58)

$$c_i = \frac{2}{J_1^2(\alpha_i)} \int_0^1 x J_0(\alpha_i x) dx = \frac{2}{J_1^2(\alpha_i)} \left[\frac{x J_1(\alpha_i x)}{\alpha_i}\right]_0^1 = \frac{2}{J_1^2(\alpha_i)} \cdot \left[\frac{J_1(\alpha_i)}{\alpha_i}\right] = \frac{2}{\alpha_i J_1(\alpha_i)}.$$

Now putting the value of c_i in $f(x) = \sum_{i=1}^{\infty} c_i J_0(\alpha_i x)$, we have $1 = \sum_{i=1}^{\infty} \dfrac{2J_0(\alpha_i x)}{\alpha_i J_1(\alpha_i)}$

or $$\frac{J_0(\alpha_1 x)}{\alpha_1 J_1(\alpha_1)} + \frac{J_0(\alpha_2 x)}{\alpha_2 J_1(\alpha_2)} + \ldots\ldots = \frac{1}{2}.$$

EXAMPLE 12.21 Show that the Fourier-Bessel series in $J_2(\alpha_n x)$ for $f(x) = x^2$ $(0 < x < 2)$, where α_n are determined by $J_2(2\alpha_n) = 0$ is

$$x^2 = 4 \sum_{n=1}^{\infty} \frac{J_2(\alpha_n x)}{\alpha_n J_3(2\alpha_n)}.$$

Solution Let the Fourier-Bessel series representing $f(x) = x^2$ be given by

$$x^2 = \sum_{n=1}^{\infty} c_n J_2(\alpha_n x).$$

Multiplying both sides by $x J_2(\alpha_n x)$ and integrating w.r.t. x between 0 and 2, we obtain

$$\int_0^2 x^3 J_2(\alpha_n x)\, dx = c_n \int_0^2 x J_2^2(\alpha_n x)\, dx$$

or $$\left[\frac{x^3 J_3(\alpha_n x)}{\alpha_n}\right]_0^2 = 2c_n J_3^2(2\alpha_n)$$

or $$\frac{8J_3(2\alpha_n)}{\alpha_n} = 2c_n J_3^2(2\alpha_n). \quad \text{Therefore,} \quad c_n = \frac{4}{\alpha_n J_3(2\alpha_n)}.$$

Hence, $$x^2 = 4 \sum_{n=1}^{\infty} \frac{1}{\alpha_n} \cdot \frac{J_2(\alpha_n x)}{J_3(2\alpha_n)}.$$

EXAMPLE 12.22 Expand $f(x) = x^3$ in the interval $0 < x < 3$ in terms of Bessel functions $J_1(\alpha_n x)$, where α_n are determined by $J_1(3\alpha_n) = 0$.

Solution Let the Fourier-Bessel series for $f(x) = x^3$ be given by

$$x^3 = \sum_{n=1}^{\infty} c_n J_1(\alpha_n x).$$

Multiplying both sides of the above relation by $x\, J_1(\alpha_n x)$ and integrating w.r.t. x within the limits 0 to 3, we get

$$\int_0^3 x^4 J_1(\alpha_n x)\,dx = c_n \int_0^3 x J_1^2(\alpha_n x)\,dx, \text{ all other integral vanish due to Eq. (12.53).}$$

Now
$$\int_0^3 x J_1^2(\alpha_n x)\,dx = \frac{9}{2} J_2^2(3\alpha_n), \quad \text{from Eq. (12.56)}$$

$\therefore$
$$c_n = \frac{2}{9J_2^2(3\alpha_n)} \int_0^3 x^4 J_1(\alpha_n x)\,dx$$

$$= \frac{2}{9J_2^2(3\alpha_n)} \left\{ \left[x^2 \cdot \frac{x^2 J_2(\alpha_n x)}{\alpha_n} \right]_0^3 - \int_0^3 2x \cdot \frac{x^2 J_2(\alpha_n x)\,dx}{\alpha_n} \right\}$$

$$= \frac{2}{9J_2^2(3\alpha_n)} \left\{ \frac{81}{\alpha_n} J_2(3\alpha_n) - \frac{2}{\alpha_n} \left[\frac{x^3 J_3(\alpha_n x)}{\alpha_n} \right]_0^3 \right\}$$

$$= \frac{2}{9J_2^2(3\alpha_n)} \left[\frac{81}{\alpha_n} J_2(3\alpha_n) - \frac{2}{\alpha_n^2} \cdot 27 J_3(3\alpha_n) \right]$$

$$= \frac{6.1}{\alpha_n^2 J_2^2(3\alpha_n)} [3\alpha_n J_2(3\alpha_n) - 2J_3(3\alpha_n)].$$

Hence,
$$x^3 = \sum_{n=1}^{\infty} \frac{6}{\alpha_n^2} \frac{1}{J_2^2(3\alpha_n)} [3\alpha_n J_2(3\alpha_n) - 2J_3(3\alpha_n)] J_1(\alpha_n x).$$

EXERCISE 12.3

1. Show that under the transformation $y = u/\sqrt{x}$, Bessel's equation becomes

$$\frac{d^2u}{dx^2} + \left(1 + \frac{1-4n^2}{4x^2}\right) u = 0.$$

Hence find the solution of this equation. (*AMIETE, June 2005*)

Ans. $u(x) = c_1 \sqrt{x}\, J_n(x) + c_2 \sqrt{x}\, J_{-n}(x)$.

Find the transformation which will reduce the following equations into Bessel's equation and solve the transformed equation. Primes denote differentiation w.r.t. x.

2. $x^2 y'' + xy' + 4(x^4 - n^2)\, y = 0.$ **Ans.** $y = AJ_n(x^2) + BJ_{-n}(x^2)$.

3. $xy'' + (9/4)\, x^2 y = 0.$ **Ans.** $y = \sqrt{x}\,[AJ_{1/3}(x^{3/2}) + BJ_{-1/3}(x^{3/2})]$.

4. $xy'' + y' + \left(3x - \frac{1}{4x}\right) y = 0.$ **Ans.** $y = AJ_{1/2}(\sqrt{3}x) + BJ_{-1/2}(\sqrt{3}x)$.

5. $\frac{d^2y}{dx^2} + xy = 0.$ **Ans.** $y = Ax^{1/2} J_{1/3}\left(\frac{2}{3}x^{3/2}\right) + Bx^{1/2} J_{-1/3}\left(\frac{2}{3}x^{3/2}\right)$.

6. $x^2y'' - xy' + 4x^2y = 0.$ **Ans.** $y = AxJ_1(2x) + BxY_1(2x).$

7. Find the solution of the differential equation $xy'' + y' + y = 0$ using the substitution $z = 2\sqrt{x}$.

Ans. $y(x) = AJ_0(2\sqrt{x}) + BY_0(2\sqrt{x})$.

8. $y'' + \frac{2}{x}y' + k^2y = \frac{p^2}{x^2}y.$ **Ans.** $y = x^{-1/2}[AJ_n(kx) + BJ_{-n}(kx)], n^2 = p^2 + \frac{1}{4}.$

9. Reduce the differential equation $x\frac{d^2y}{dx^2} + a\frac{dy}{dx}\ k^2xy = 0$ to Bessel's equation. (*Madras, 2006*)

Ans. $y = x^n[c_1J_n(kx) + c_2J_{-n}(kx)]$, n is non-integral.

or $y = x^n[c_1J_n(kx) + c_2Y_n(kx)]$, n is integral, where $n = (1 - a)/2$.

10. $xy'' + y' + \frac{1}{4}y = 0.$ (*Anna, 2005*) **Ans.** $y = c_1J_0(\sqrt{x}) + c_2Y_0(\sqrt{x})$.

11. Show that $\int_0^1 x(1 - x^2)J_0(\alpha x)\,dx = \frac{4}{\alpha^3}J_1(\alpha) - \frac{2}{\alpha^2}J_0(\alpha).$

12. Prove that $\int x^{2n+1}J_n^2(x)\,dx = x\frac{x^{2n+2}}{4n+2}\left[J_n^2(x) + J_{n+1}^2(x)\right].$

13. Prove that $\int x\log x\,J_0(x)\,dx = J_0(x) + x\log x\,J_i(x).$

If α_n is the positive root of $J_1(x) = 0$, establish the following relations.

14. If $\alpha_1, \alpha_2, \ldots \alpha_n$ are the successive positive roots of $J_0(x) = 0$, prove that

$$x^2 = 2\sum_{n=1}^{\infty}\frac{\alpha_n^2 - 4}{\alpha_n^3 J_1(\alpha_n)}J_0(\alpha_n x).$$

15. Show that $J_0(kx) = 2J_0(k)\sum_{n=1}^{\infty}\frac{\alpha_n J_0(\alpha_n x)}{(\alpha_n^2 - k^2)J_1(\alpha_n)}$, where k is not a root of $J_0(x) = 0$.

16. Show that $\log x = -2\sum_{n=1}^{\infty}\frac{J_0(\alpha_n x)}{\alpha_n^2 J_1^2(\alpha_n)}.$

17. Expand $f(x) = 4x - x^3$ over the interval (0, 2) in terms of Bessel functions of first kind of order one which satisfy the condition $[J_1(\alpha x)]_{x=2} = 0$.

Ans. $4x - x^3 = 8\sum_{n=1}^{\infty}\frac{J_3(2\alpha_n)}{\alpha_n^2 J_2^2(2\alpha_n)}\cdot J_1(\alpha_n x).$

If α_n is the positive root of $J_1(x) = 0$, establish the following relations.

18. $x^2 = \frac{1}{2} + 4\sum_{n=1}^{\infty}\frac{J_0(\alpha_n x)}{\alpha_n^2 J_0(\alpha_n)}.$

19. $(1 - x^2) = \frac{1}{3} - 64\sum_{n=1}^{\infty}\frac{J_0(\alpha_n x)}{\alpha_n^4 J_0(\alpha_n)}.$

[**Hint:** For problems (18) and (19), we use the following result.

If $0, \alpha_1, \alpha_2, \alpha_3, \ldots$ are roots of $J_1(x) = 0$, then we can express $f(x)$ in the form of Bessel-Fourier series, in the interval (0, 1), of the form

$$f(x) = A_0 + A_1J_0(\alpha_1 x) + A_2J_0(\alpha_2 x) + A_3J_0(\alpha_3 x) + \cdots$$

where $$A_0 = 2\int_0^1 xf(x)\,dx,$$

and $$A_n = \frac{2}{J_0^2(\alpha_n)}\int_0^1 x\,f(x)\,J_0(\alpha_n x)\,dx.$$

Proof: Multiplying both sides by $x\,dx$ and integrating between 0 and 1,

$$\int_0^1 x\,f(x)\,dx = A_0\int_0^1 x\,dx + A_1\int_0^1 x\,J_0(\alpha_1 x)dx + \cdots$$

$$= \left[\frac{x^2}{2}\right]_0^1 A_0 + A_1\left[\frac{xJ_1(\alpha_1 x)}{\alpha_1}\right]_0^1 + \cdots = \frac{1}{2}A_0.$$

Therefore, $$A_0 = 2\int_0^1 xf(x)\,dx.$$

Multiplying both sides by $x\,J_0(\alpha_n x)$ and integrating between 0 and 1, we get

$$\int_0^1 xJ_0(\alpha_n x)\,f(x)\,dx = A_0\int_0^1 xJ_0(\alpha_n x)\,dx + A_1\int_0^1 xJ_0(\alpha_n x)\,J_0(\alpha_1 x)dx + \cdots + A_n\int_0^1 xJ_0^2(\alpha_n x)\,dx + \cdots$$

$$= A_0\left[\frac{xJ_1(\alpha_n x)}{\alpha_n}\right]_0^1 + A_1\left[J_0(\alpha_1 x)\cdot\frac{xJ_1(\alpha_n x)}{\alpha_n}\right]_0^1 - \int_0^1 \frac{\alpha_1}{\alpha_n}J_0'(\alpha_1 x)xJ_1(\alpha_n x)\,dx$$

$$+\left\{\left[J_0(\alpha_n x)\cdot\frac{xJ_1(\alpha_n x)}{\alpha_n}\right]_0^1 - \int_0^1 \alpha_n J_0'(\alpha_n x)\,xJ_1(\alpha_n x)\cdot\frac{1}{\alpha_n}\,dx\right\} + \cdots$$

$$= 0 + \frac{\alpha_1}{\alpha_n}A_1\int_0^1 xJ_1(\alpha_1 x)\,J_1(\alpha_n x)\,dx + \cdots + A_n\int_0^1 xJ_1^2(\alpha_n x)\,dx + \cdots$$

$$= 0 + 0 + \cdots + \frac{A_n}{2}J_2^2(\alpha_n).$$

Therefore, $$A_n = \frac{2}{J_2^2(\alpha_n)}\int_0^1 xJ_0(\alpha_n x)\,f(x)\,dx.$$

Now $\dfrac{2n}{x}J_n(n) = J_{n-1}(x) + J_{n+1}(x)$.

Putting $n = 1$ and $x = \alpha_n$ we have

$$\frac{2}{\alpha_n}J_1(\alpha_n) = J_0(\alpha_n) + J_2(\alpha_n) \Rightarrow J_2(\alpha_n) = -J_0(\alpha_n).$$

Therefore, $$A_n = \frac{2}{J_0^2(\alpha_n)}\int_0^1 xJ_0\,(\alpha_n x)\,f(x)\,dx\,].$$

12.12 LEGENDRE'S DIFFERENTIAL EQUATION

The differential equation of the form

$$(1-x^2)\frac{d^2y}{dx^2}-2x\frac{dy}{dx}+n(n+1)\,y=0, \qquad ...(12.59)$$

where n is any real number is known as Legendre's equation. This equation is of considerable importance in applied mathematics and finds extensive use in boundary value problems involving spherical configurations. In most physical applications only integral values of n are needed.

Equation (12.59) can also be written as

$$\frac{d}{dx}\left\{(1-x^2)\frac{dy}{dx}\right\}+n(n+1)\,y=0. \qquad ...(12.60)$$

Since $x=0$ is an ordinary point of the differential equation, we can assume the solution of (12.59) in a series of the form

$$y=\sum_{k=0}^{\infty} a_k\, x^k. \qquad ...(12.61)$$

Substituting for y, $\frac{dy}{dx}$ and $\frac{d^2y}{dx^2}$ from (12.61) in (12.59) and rearranging the terms in powers of x, we obtain

$$\sum a_k k\,(k-1)x^{k-2}-\sum a_k\,\{k\,(k+1)-n\,(n+1)\}\,x^k=0. \qquad ...(12.62)$$

Equating to zero the coefficient of x^k, the ***recurrence relation*** or ***recursion formula*** is given by

$$a_{k+2}=\frac{(k-n)\,(k+n+1)}{(k+2)\,(k+1)}\,a_k,\ k=0,1,2. \qquad ...(12.63)$$

The relation (12.63) gives even and odd coefficients in terms of the one immediately preceding it, except for a_0 and a_1 which are arbitrary. From (12.63) therefore, we get

$$a_2=-\frac{n(n+1)}{2!}a_0,$$

$$a_4=\frac{(n-2)\,n(n+1)\,(n+3)}{4!}a_0,$$

$$a_6=-\frac{(n-4)\,(n-2)n\,(n+1)\,(n+3)\,(n+5)}{6!}a_0,$$

$$a_3=-\frac{(n-1)\,(n+2)}{3!}a_1,$$

$$a_5=-\frac{(n-3)\,(n+4)}{5.4}a_3=\frac{(n-3)\,(n-1)\,(n+2)\,(n+4)}{5!}a_1,$$

and so on.

Inserting these coefficients in (12.61) the solution of equation (12.59) can be written as

$$y=a_0\left[1-\frac{n(n+1)}{2!}x^2+\frac{(n-2)n\,(n+1)\,(n+3)}{4!}x^4-\cdots\right]$$

$$+a_1\left[x-\frac{(n-1)\,(n+2)}{3!}x^3+\frac{(n-3)\,(n-1)\,(n+2)\,(n+4)}{5!}x^5-\cdots\right] \qquad ...(12.64)$$

$$=a_0\,y_1(x)+a_1\,y_2(x).$$

It is clear from above that the ratio of $y_1(x)$ and $y(x)$ is not constant and therefore $y_1(x)$ and $y_2(x)$ are two independent solutions and (12.64), therefore, gives the complete solution of the problem.

From the recurrence relation (12.63), it is obvious that when $k = n$, $a_{k+2} = 0 = a_{k+4} = a_{k+6} \dots$. Thus, when n is even, solution $y_1(x)$ turns out to be a polynomial and when n is odd, same holds true for y_2. These polynomials with either a_0 or a_1 adjusted suitably* are known as Legendre polynomials, and denoted by $P_n(x)$. The other non-terminating solution with a_1 suitably adjusted is known as the Legendre function of second kind and is denoted by $Q_n(x)$.

From (12.63) above we get

$$a_k = -\frac{(k+2)(k+1)}{(n-k)(n+k+1)} a_{k+2},\ k \le (n-2) \qquad \text{...(12.65)}$$

We will now express all the non-vanishing coefficients in terms of a_n which is the coefficient of the highest power of x in the polynomial solution. The coefficient a_n will be arbitrary.

Let us choose $a_n = 1$ when $n = 0$ and $a_n = \dfrac{(2n)!}{2^n (n!)^2}$ for $n = 1, 2, \dots$

The choices of assuming the above values of a_n is necessitated from the consideration that the value of the polynomial is 1 when $x = 1$.

From (12.65) we get

$$a_{n-2} = -\frac{n(n-1)}{2(2n-1)} a_n = -\frac{n(n-1)}{2(2n-1)} \cdot \frac{(2n)!}{2^n (n!)^2}$$

$$= -\frac{n(n-1)\,2n(n-1)}{2(2n-1)2^n\, n(n-1)} \cdot \frac{(2n-2)!}{(n-2)!n!} = -\frac{(2n-2)!}{2^n (n-1)!(n-2)!}.$$

Similarly

$$a_{n-4} = -\frac{n(n-2)(n-3)}{4(2n-3)} a_{n-2}$$

$$= \frac{(n-2)(n-3)(2n-2)!}{4.2^n (n-1)!(n-2)!(2n-3)}$$

$$= \frac{(n-2)(n-3)(2n-2)(2n-3)\cdot(2n-4)!}{4.2^n (n-1)(n-2)(n-3)\cdot(n-2)!\cdot(n-4)!\cdot(2n-3)}$$

$$= \frac{(2n-4)!}{2!2^n (n-2)!\ (n-4)!} \quad \text{and so on}$$

and in general when $n - 2k \ge 0$,

$$a_{n-2k} = \frac{(-1)^k\,(2n-2k)!}{2^n k!(n-k)!(n-2k)!}.$$

Thus, we get

$$P_n(x) = a_n x^n + a_{n-2} x^{n-2} + a_{n-4} x^{n-4} + \dots\dots$$

$$= \frac{(2n)!}{2^n (n!)^2} x^n - \frac{(2n-2)!}{2^n (n-1)!(n-2)!} x^{n-2} + \frac{(2n-4)!\,x^{n-4}}{2^n 2!(n-2)!(n-4!)} + \dots\dots$$

that is,

$$P_n(x) = \sum_{k=0}^{N} \frac{(-1)^k\,(2n-2k)!}{2^n k!(n-k)!(n-2k)!} x^{n-2k} \qquad \text{...(12.66)}$$

where $N = \dfrac{n}{2}$ when n is even and $(n-1)/2$ when n is odd.

---

*a_0 or a_1 are so choosen that $P_n(x) = 1$ for $x = 1$.

12.13 SOLUTION OF LEGENDRE'S EQUATION — SERIES DESCENDING POWERS OF x

In the foregoing discussion, we have studied the solution assuming a series in ascending powers of x and represented various coefficients in the polynomial solution in terms of a_n, the coefficient of the highest power of x. The final form of the polynomial for $P_n(x)$ as given in (12.66) is a series in descending power of x. Guided by this result we discuss below an alternative method of solving (12.59) by assuming a series for y in descending powers of x.

Let us assume that

$$y = \sum_{k=0}^{\infty} a_k x^{m-k}. \qquad ...(12.67)$$

Substituting for y, $\frac{dy}{dx}$ and $\frac{d^2y}{dx^2}$ from (12.67) in (12.59) and rearranging the terms in powers of x, we obtain

$$\sum_{k=0}^{\infty} a_k (m-k)(m-k-1)x^{m-k-2} - \sum_{k=0}^{\infty} a_k[(m-k)(m-k+1) - n(n+1)]x^{m-k} = 0$$

or

$$\sum_{k=0}^{\infty} a_k (m-k)(m-k-1)x^{m-k-2} - \sum_{k=0}^{\infty} a_k (m-k+n+1)(m-k-n)x^{m-k} = 0. \qquad ...(12.68)$$

Equating to zero the coefficient of highest power of x, that is, x^m the indicial equation is written as

$$a_0(m+n+1)(m-n) = 0. \qquad ...(12.69)$$

This gives $m = n$ or $m = -(n+1)$, since a_0 is arbitrary. Equating to zero the coefficient of x^{m-1}, we get

$$a_1(m+n)(m-n-1) = 0.$$

This gives $a_1 = 0$, as $(m+n)$ and $(m-n-1)$ are not zero by virtue of the values of m obtained above.

To find a relation between the successive coefficients we equate to zero the coefficient of x^{m-k-2}, the recurrence relation is given by

$$a_{k+2} = \frac{(m-k)(m-k-1)}{(m-k+n-1)(m-k-n-2)} a_k. \qquad ...(12.70)$$

Since $a_1 = 0$, therefore from (12.70), we get

$$a_3 = a_5 = ... = 0.$$

The two values of m, that is, $m = n$ and $m = -(n+1)$ determine two solutions of (12.59).

***Case 1:* *When* $m = n$**

For $m = n$, equation (12.70) gives

$$a_{k+2} = -\frac{(n-k)(n-k-1)}{(2n-k-1)(k+2)} a_k.$$

Setting $k = 0, 2, 4, ...$ we get

$$a_2 = -\frac{n(n-1)}{2(2n-1)} a_0,$$

$$a_4 = -\frac{(n-2)(n-3)}{4(2n-3)}, \; a_2 = \frac{n(n-1)(n-2)(n-3)}{2.4(2n-1)(2n-3)} a_0 \text{ etc.}$$

Inserting the values of $a_2, a_4, ...$ in (12.67), the solution y_1 corresponding to $m = n$ is given by

$$y_1(x) = a_0\left[x^n - \frac{n(n-1)}{2(2n-1)}x^{n-2} + \frac{n(n-1)(n-2)(n-3)}{2.4(2n-1)(2n-3)}x^{n-4} + \cdots\right] \qquad ...(12.71)$$

Choosing $a_0 = \dfrac{(2n)!}{2^n \cdot (n!)^2}$, the solution $y_1(x)$ is called the ***Legendre polynomial*** of *degree n* and is denoted by $P_n(x)$.

Thus, $$P_n(x) = \frac{(2n)!}{2^n \cdot (n!)^2}\left[x^n - \frac{n(n-1)}{2(2n-1)}x^{n-2} + \frac{n(n-1)(n-2)(n-3)}{2.4(2n-1)(2n-3)}x^{n-4} - \ldots\ldots\right] \qquad \ldots(12.72)$$

Case 2: *When* $m = -(n+1)$.

For $m = -(n+1)$, equation (12.70) gives

$$a_{k+2} = \frac{(n+k+1)(n+k+2)}{(k+2)(2n+k+3)}a_k.$$

Setting $k = 0, 2, 4, \ldots$, we get

$$a_2 = \frac{(n+1)(n+2)}{2(2n+3)}a_0,$$

$$a_4 = \frac{(n+3)(n+4)}{4(2n+5)}a_2 = \frac{(n+1)(n+2)(n+3)(n+4)}{2.4(2n+3)(2n+5)}a_0.$$

Inserting the values of a_2, a_4, ... in (12.67), the solution $y_2(x)$ corresponding to $m = -(n+1)$ is given by

$$y_2(x) = a_0\left[x^{-n-1} + \frac{(n+1)(n+2)}{2(2n+3)}x^{-n-3} + \frac{(n+1)(n+2)(n+3)(n+4)}{2.4(2n+3)(2n+5)}x^{-n-5} + \cdots\right] \qquad \ldots(12.73)$$

Choosing $a_0 = \dfrac{n!}{1.3.5\ldots(2n+1)}$, the solution $y_2(x)$ in (12.73) is called the Legendre function of the second kind and is denoted by $Q_n(x)$. Therefore,

$$Q_n(x) = \frac{n!}{1.3.5\ldots(2n+1)}\left[x^{-n-1} + \frac{(n+1)(n+2)}{2(2n+3)}x^{-n-3} + \frac{(n+1)(n+2)(n+3)(n+4)}{2.4(2n+3)(2n+5)}x^{-n-5} + \cdots\right]. \qquad \ldots(12.74)$$

Since n is a positive integer, the series (12.72) for $Q_n(x)$ is a non-terminating series. The complete solution of (12.59) is therefore, given by

$$y = AP_n(x) + BQ_n(x), \qquad \ldots(12.75)$$

where A and B are arbitrary constants.

12.14 LEGENDRE POLYNOMIALS $P_n(x)$

Giving n the values 0, 1, 2, 3, 4, ... in (12.72), we have

$$P_0(x) = 1,$$
$$P_1(x) = x,$$
$$P_2(x) = \frac{1}{2}(3x^2 - 1),$$
$$P_3(x) = \frac{1}{2}(5x^3 - 3x),$$
$$P_4(x) = \frac{1}{8}(35x^4 - 30x^2 + 3),$$
$$P_5(x) = \frac{1}{8}(63x^5 - 70x^3 + 15x),$$
$$P_6(x) = \frac{1}{16}(231x^6 - 315x^4 + 105x^2 - 5) \text{ etc.} \qquad \ldots(12.76)$$

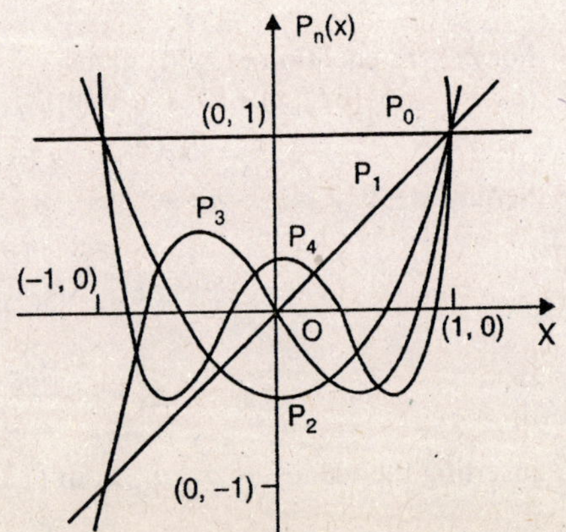

Fig. 12.3: Legendre polynomials — for $n = 0, 1, 2, 3, 4$.

These are known as Legendre polynomials and each satisfies the differential equation (12.59). It can be easily verified that the value of each polynomial is 1 for $x = 1$.

EXAMPLE 12.23 Express the sum of Legendre polynomials in terms of powers of x.

$$P(x) = 5P_4(x) + 10P_3(x) + 2P_2(x) + P_1(x),$$

where $P_m(x)$ is the Legendre polynomial of order m.

Solution Substituting the expressions for P_1, P_2, P_3 and P_4, we obtain

$$P(x) = 5\cdot\frac{1}{8}(35x^4 - 30x^2 + 3) + 10\cdot\frac{1}{2}(5x^3 - 3x) + 2\cdot\frac{1}{2}(3x^2 - 1) + x$$

$$= \frac{1}{8}(175x^4 - 150x^2 + 15 + 200x^3 - 120\,x + 24x^2 - 8 + 8x)$$

$$= \frac{1}{8}(175x^4 + 200x^3 - 126\,x^2 - 112x + 7).$$

EXAMPLE 12.24 (*i*) Express $f(x) = x^4 + 2x^3 - 6x^2 + 5x - 3$ in terms of Legendre polynomials.

(*AMIETE Dec 2009; June 2007, Dec 2005*)

(*ii*) Expand $1 - 2x + x^2 + 5x^3$ in a series of Legendre polynomials.

Solution (*i*) Writing various powers of x in terms of Legendre polynomials we get

$$1 = P_0(x),\quad x = P_1(x),\quad x^2 = \frac{1}{3}\{2P_2(x) + 1)\} = \frac{1}{3}[2P_2(x) + P_0(x)]$$

$$x^3 = \frac{1}{5}[2P_3(x) + 3x] = \frac{1}{5}[2P_3(x) + 3P_1(x)]$$

$$x^4 = \frac{1}{35}(8P_4(x) + 30x^2 - 3) = \frac{1}{35}[8P_4(x) + 10\{2P_2(x) + P_0(x)\} - 3P_0(x)]$$

$$= \frac{1}{35}[8P_4(x) + 20P_2(x) + 7P_0(x)].$$

Therefore,

$$f(x) = \frac{1}{35}[8P_4(x) + 20P_2(x) + 7P_0(x)] + \frac{2}{5}[(2P_3(x) + 3P_1(x)]$$

$$- 2[(2P_2(x) + P_0(x)] + 5P_1(x) - 3P_0(x)$$

$$= \frac{1}{35}[8P_4(x) + 28P_3(x) - 120P_2(x) + 217P_1(x) - 168P_0(x)].$$

(*ii*) $f(x) = P_0(x) - 2P_1(x) + \frac{1}{3}[2P_2(x) + P_0(x)] + [2P_3(x) + 3P_1(x)] = \frac{4}{3}P_0(x) + P_1(x) + \frac{2}{3}P_2(x) + 2P_3(x)$

$$= \frac{1}{3}[6P_3(x) + 2P_2(x) + 3P_1(x) + 4P_0(x)].$$

12.15 RODRIGUES'S FORMULA

The relation $\mathbf{P_n(x) = \frac{1}{2^n n!}\frac{d^n}{dx^n}[(x^2 - 1)^n]}$ is known as *Rodrigues's formula*. (*AMIETE, June 2011, 2009*)

To prove this result, let

$$u = (x^2 - 1)^n, \text{ then } \frac{du}{dx} = 2nx(x^2 - 1)^{n-1}.$$

Multiplying both sides by $(x^2 - 1)$ and transferring all the terms on one side, we get

$(x^2 - 1)\dfrac{du}{dx} - 2nxu = 0$. Therefore, $(1 - x^2)\dfrac{du}{dx} + 2nxu = 0$.

Differentiating this $(n + 1)$ times by Leibnitz's theorem, we obtain

$$(1 - x^2)\frac{d^{n+2}u}{dx^{n+2}} - 2x\frac{d^{n+1}u}{dx^{n+1}} + n(n+1)\frac{d^n u}{dx^n} = 0.$$

Denoting $\dfrac{d^n u}{dx^n}$ by u_n, the above equation can be written as

$(1 - x^2)\dfrac{d^2 u_n}{dx^2} - 2x\dfrac{du_n}{dx} + n(n+1)u_n = 0$, which is the Legendre differential equation for $y = u_n$. Hence u_n is a solution of this equation, or $u_n = cP_n(x)$.

But $u_n = \dfrac{d^n}{dx^n}(x^2 - 1)^n$. Therefore, $cP_n(x) = \dfrac{d^n}{dx^n}(x^2 - 1)^n$, ...(12.77)

where c is an arbitrary constant.

To evaluate the constant c, we will use the property $P_n(1) = 1$.

Now $\quad cP_n(x) = \dfrac{d^n}{dx^n}[(x-1)^n (x+1)^n]$

$$= (x-1)^n \frac{d^n}{dx^n}(x+1)^n + {}^n c_1\, n(x-1)^{n-1}\frac{d^{n-1}}{dx^{n-1}}(x+1)^n + \frac{n^2(n-1)^2}{2!}(x-1)^{n-2}$$

$$\frac{d^{n-2}}{dx^{n-2}}(x+1)^n + \cdots + (x+1)^n\frac{d^n}{dx^n}(x-1)^n.$$

Setting $x = 1$ on both the sides, we get

$cP_n(1) = \left[(x+1)^n \cdot \dfrac{d^n}{dx^n}(x-1)^n\right]_{x=1}$, all other terms being zero because of the factor $(x - 1)$.

Therefore, $c = 2^n n!$ [as $P_n(1) = 1$].

Hence from (12.77), we obtain

$$P_n(x) = \frac{1}{2^n n!}\frac{d^n}{dx^n}\left[(x^2 - 1)^n\right].$$

We can obtain the various Legendre's polynomials with the help of Rodrigue's formula.

For example, on taking $n = 0$, we get $P_0(x) = 1$. Taking successively, $n = 1, 2, 3, \ldots$ etc, we get

$$P_1(x) = \frac{1}{2}\frac{d}{dx}(x^2 - 1) = x,$$

$$P_2(x) = \frac{1}{2^2(2!)}\frac{d^2}{dx^2}(x^2 - 1)^2 = \frac{1}{8}\frac{d^2}{dx^2}(x^4 - 2x^2 + 1) = \frac{1}{2}(3x^2 - 1),$$

$$P_3(x) = \frac{1}{2^3(3!)}\frac{d^3}{dx^3}(x^2 - 1)^3 = \frac{1}{48}\frac{d^3}{dx^3}(x^6 - 3x^4 + 3x^2 - 1) = \frac{1}{2}(5x^3 - 3x),$$

$$P_4(x) = \frac{1}{2^4(4!)}\frac{d^4}{dx^4}\left[(x^2 - 1)^4\right] = \frac{1}{(16)(24)}\frac{d^4}{dx^4}(x^8 - 4x^6 + 6x^4 - 4x^2 + 1)$$

$$= \frac{1}{(16)(24)}(1680x^4 - 1440x^2 + 144) = \frac{1}{8}(35x^4 - 30x^2 + 3)$$

Similarly, we get

$$P_5 = \frac{1}{8}(63x^5 - 70x^3 + 15x),$$

$$P_6 = \frac{1}{16}(231x^6 - 315x^4 + 105x^2 - 5) \text{ etc}$$

The graphs of $P_0(x)$, $P_1(x)$, $P_2(x)$, $P_3(x)$ and $P_4(x)$ are given in Fig. 12.3 we note that $|P_n(x)| \le 1, x \in (-1, 1)$.

EXAMPLE 12.25 Using Rodrigues's formula, prove that $P'_{n+1}(x) - P'_{n-1}(x) = (2n+1) P_n(x)$. Hence show that $\int_x^1 P_n(x)\, dx = \frac{P_{n-1}(x) - P_{n+1}(x)}{2n+1}$.

Solution $$P_n(x) = \frac{1}{2^n n!} \frac{d^n}{dx^n}(x^2 - 1)^n.$$

Therefore, $$P_{n+1}(x) = \frac{1}{2^{n+1}(n+1)!} \frac{d^{n+1}}{dx^{n+1}}(x^2 - 1)^{n+1}.$$

$$P'_{n+1}(x) = \frac{1}{2^{n+1}(n+1)!} \frac{d^{n+2}}{dx^{n+2}}(x^2 - 1)^{n+1}$$

$$= \frac{1}{2^{n+1}(n+1)!} \frac{d^n}{dx^n}[D^2(x^2 - 1)^{n+1}]$$

$$= \frac{1}{2^{n+1}(n+1)!} D^n[D(n+1)\, 2x(x^2 - 1)^n]$$

$$= \frac{1}{2^n n!} D^n[(x^2 - 1)^n + 2nx^2(x^2 - 1)^{n-1}]$$

$$= \frac{1}{2^n \cdot n!} D^n[(x^2 - 1)^n + \frac{2nD^n}{2^n n!}[(x^2 - 1 + 1)(x^2 - 1)^{n-1}]$$

$$= \frac{1}{2^n \cdot n!} D^n(x^2 - 1)^n + 2n \cdot \frac{1}{2^n n!} D^n(x^2 - 1)^n + \frac{1}{2^{n-1}(n-1)!} D^n(x^2 - 1)^{n-1}$$

$$= P_n + 2nP_n + P'_{n-1} = (2n+1)P_n + P'_{n-1} \quad \text{or} \quad P_n = \frac{P'_{n+1} - P'_{n-1}}{2n+1}.$$

Therefore, $$\int_x^1 P_n(x)\, dx = \frac{1}{2n+1}\left[P_{n+1}(x) - P_{n-1}(x)\right]_x^1 = \frac{P_{n-1}(x) - P_{n+1}(x)}{2n+1}.$$

12.16 THE GENERATING FUNCTION FOR LEGENDRE POLYNOMIALS

We shall prove that

$$G(z, x) = (1 - 2xz + z^2)^{-1/2} = \sum_{n=0}^{\infty} P_n(x) z^n, \quad z \ne 1. \qquad ...(12.78)$$

The function on the left hand side is called the *generating function* of the Legendre's polynomials. Further Legendre's polynomials can also be obtained by using a generating function.

Expanding the function on the left side by binomial theorem for negative fractional index, we obtain

$$\{1 - z(2x - z)\}^{-1/2} = 1 + \frac{1}{2} z\,(2x - z) + \frac{1.3}{2.4} z^2\,(2x - z)^2 + \frac{1.3.5}{2.4.6} z^3\,(2x - z)^3 + \ldots\ldots$$

$$\ldots\ldots + \frac{1.3.5\ldots(2n-3)}{2.4.6\ldots(2n-2)} z^{n-1}(2x - z)^{n-1} + \frac{1.3.5\ldots(2n-1)}{2.4.6\ldots 2n} z^n(2x - z)^n + \cdots$$

The coefficient of z^n in $\dfrac{1.3.5...(2n-1)}{2.4.6...2n} z^n(2x-z)^n$ is $\dfrac{1.3.5...(2n-1)}{2.4.6...2n}(2x)^n$.

The coefficient of z^n in $\dfrac{1.3.5...(2n-3)}{2.4.6...(2n-2)} z^{n-1}(2x-z)^{n-1}$ is $\dfrac{1.3.5...(2n-3)}{2.4.6...(2n-2)}[-{}^{n-1}C_1(2x)^{n-2}]$

$$= -\frac{1.3.5...(2n-3)\,(n-1)2^{n-2}x^{n-2}}{2.4.6...(2n-2)}$$

$$= -\frac{1.3.5...(2n-3)\,(2n-1)\,2n(n-1)}{2^n n!\,(2n-1)}\,2^{n-2}\,x^{n-2}$$

$$= -\frac{1.3.5...(2n-1)}{n!} \bullet \frac{n(n-1)}{2(2n-1)}x^{n-2}$$

The coefficient z^n in $\dfrac{1.3.5...(2n-5)}{2.4.6...(2n-4)} z^{n-2}(2x-z)^{n-2}$ is $\dfrac{1.3.5...(2n-1)}{n!}\,\dfrac{n(n-1)\,(n-2)\,(n-3)}{2.4(2n-1)\,(2n-3)}x^{n-4}$.

Proceeding like this, the coefficient of z^n in the expansion of $(1-2xz+z^2)^{-1/2}$ is given by

$$\frac{1.3.5...(2n-1)}{n!}\left[x^n - \frac{n(n-1)}{2(2n-1)}x^{n-2} + \frac{n(n-1)\,(n-2)\,(n-3)}{2.4(2n-1)\,(2n-3)}x^{n-4}\cdots\right] \text{ that is } P_n(x).$$

Thus, we see that in the expansion of $(1-2xz+z^2)^{-1/2}$, $P_1(x)$, $P_2(x)$, $P_3(x)$, ..., $P_n(x)$ etc. are the coefficients of z^1, z^2, z^3, ... etc. respectively. Hence we have

$$(1-2xz+z^2)^{-1/2} = 1 + zP_1(x) + z^2P_2(x) + \cdots + z^nP_n(x) + \cdots$$

or
$$\mathbf{(1-2xz+z^2)^{-1/2} = \sum_{n=0}^{\infty} P_n(x)z^n, \quad z \neq 1.} \qquad ...(12.79) \quad (UPTU,\ 2005)$$

This proves that $P_n(x)$ can be obtained by the generating function.

EXAMPLE 12.26 Show that

(*i*) $P_n(1) = 1$, (*VTU, 2003S; AMIETE; June, 2010, 2005*)

(*ii*) $P_n(-x) = (-1)^n P_n(x)$. Hence find $P_n(-1)$.

Solution (*i*) Substituting $x = 1$ in the generating function, we get

$$(1-2z+z^2)^{-½} = \sum_{n=0}^{\infty} P_n(1)z^n$$

or
$$\sum_{n=0}^{\infty} P_n(1)z^n = (1-z)^{-1} = 1 + z + z^2 + \cdots + z^n + \ldots\ldots$$

Comparing the coefficient of z^n on both sides, we get $P_n(1) = 1$, $n = 0, 1, \ldots\ldots$.

(*ii*) Replacing x and z by $-x$ and $-z$ respectively in the generating function given in Eq. (12.78), we get

$$(1-2xz+z^2)^{-½} = \sum_{n=0}^{\infty} P_n(-x)\,(-z)^n = \sum_{n=0}^{\infty} (-1)^n\, P_n(-x)\, z^n.$$

Since the left hand side is $\sum_{n=0}^{\infty} P_n(x)\, z^n$, we obtain

$$\sum_{n=0}^{\infty} P_n(x)z^n = \sum_{n=0}^{\infty} (-1)^n\, P_n(-x)\, z^n.$$

Comparing the coefficients of z^n on both sides, we get

$$P_n(x) = (-1)^n P_n(-x) \text{ or } P_n(-x) = (-1)^n P_n(x).$$

Setting $x = 1$, we get $P_n(-1) = (-1)^n P_n(1) = (-1)^n$. *(BPTU, 2005S; VTU, 2003)*

EXAMPLE 12.27 Show that

(i) $P_{2m}(0) = (-1)^m \dfrac{(2m)!}{2^{2m}(m!)^2}$ (ii) $P_{2m+1}(0) = 0$ *(AMIETE, June 2003)*

Solution (i) Putting $x = 0$ in (12.79), we get

$$\sum P_n(0)z^n = (1+z^2)^{-1/2} = 1 - \frac{1}{2}z^2 + \frac{1.3}{2.4}z^4 + \cdots + (-1)^r \frac{1.3.5\ldots(2r-1)}{2.4.6\ldots 2r}z^{2r} + \cdots$$

Comparing the coefficient of z^{2m} on both sides, we get

$$P_{2m}(0) = (-1)^m \frac{1.3.5\ldots(2m-1)}{2.4.6\ldots 2m}$$

$$= (-1)^m \frac{1.2.3.4\ldots(2m-1)\cdot 2m}{2^m \cdot m!\,2^m \cdot m!} = (-1)^m \frac{(2m)!}{(2^m\, m!)^2}.$$

(ii) $P_n(x) = \displaystyle\sum_{k=0}^{N} \frac{(-1)^k (2n-2k)!\,x^{n-2k}}{2^n\, k!\,(n-k)!\,(n-2k)!}$,

where $N = n/2$, when n is even and $N = (n-1)/2$ when n is odd.

Put $n = 2m + 1$, the last term $N = m$.

Therefore, the last term in $P_{2m+1}(x)$ will be

$$\frac{(-1)^m (4m+2-2m)!\,x^{2m+1-2m}}{2^{2m+1} m!\,(m+1)!\,(1)!}.$$

It is clear that the last term and all the terms preceding it contain x. Hence, when $x = 0$, $P_{2m+1}(0) = 0$.

12.17 RECURRENCE RELATIONS FOR LEGENDRE POLYNOMIALS

In many applications use of recurrence relations between Legendre's polynomials simplifies the solution of a given problem. The following recurrence relations have been derived from the generating function for $P_n(x)$. We prove the following recurrence relations.

(*a*) $(n+1)P_{n+1}(x) = (2n+1)x\,P_n(x) - nP_{n-1}(x)$, *(SPTU, 2007; VTU, 2003; JNTU, 2000S)*

(*b*) $nP_n(x) = xP_n'(x) - P_{n-1}'(x)$, *(JNTU, 2006; VTU, 2006)*

(*c*) $(2n+1)P_n(x) = P_{n+1}'(x) - P_{n-1}'(x)$, *(Madras, 2006)*

(*d*) $P_{n+1}'(x) = xP_n'(x) + (n+1)P_n(x)$,

(*e*) $(1-x^2)P_n'(x) = n[P_{n-1}(x) - xP_n(x)]$.

(*f*) $(1-x^2)P_n'(x) = (n+1)[xP_n(x) - P_{n+1}(x)]$, where the primes denote differentiation with respect to x.

Proofs

(*a*) The generating function is

$$(1 - 2xz + z^2)^{-1/2} = \sum_{n=0}^{\infty} P_n(x)\,z^n. \quad \ldots(12.80)$$

Differentiating both sides partially w.r.t. z, we get

$$-\frac{1}{2}(1-2xz+z^2)^{-3/2}(-2x+2z) = \sum_{n=0}^{\infty} nP_n(x)\,z^{n-1} \quad \ldots(12.81)$$

or $$(x-z)(1-2xz+z^2)^{-1/2} = (1-2xz+z^2)\sum_{n=0}^{\infty} nP_n(x)\, z^{n-1}$$

or $$(x-z)\sum_{n=0}^{\infty} P_n(x)\, z^n = (1-2xz+z^2)\sum_{n=0}^{\infty} nP_n(x)\, z^{n-1}.$$

Equating the coefficient of z^n on both sides, we obtain

$$xP_n(x) - P_{n-1}(x) = (n+1)P_{n+1}(x) - 2xnP_n(x) + (n-1)\, P_{n-1}(x)$$

or $$\mathbf{(n+1)\, P_{n+1}(x) = (2n+1)\, xP_n(x) - nP_{n-1}(x).} \qquad ...(12.82)$$

NOTE: Equating the coefficients of z^{n-1} from both sides, we get another relation

$$\mathbf{nP_n(x) = (2n-1)xP_{n-1}(x) - (n-1)\, P_{n-2}(x).}$$

(*b*) Differentiating Eq. (12.80) partially with respect to x, we get

$$-\frac{1}{2}(1-2xz+z^2)^{-3/2}(-2z) = \sum_{n=0}^{\infty} P_n'(x)\, z^n \quad or \quad z\,(1-2xz+z^2)^{-3/2} = \sum_{n=0}^{\infty} P_n'(x)\, z^n. \qquad ...(12.83)$$

Again differentiating Eq. (12.80) with respect to z, we get

$$(x-z)(1-2xz+z^2)^{-3/2} = \sum_{n=0}^{\infty} nP_n(x)\, z^{n-1}. \qquad ...(12.84)$$

Divide (12.84) by (12.83), we get

$$\frac{x-z}{z} = \frac{\sum_{n=0}^{\infty} nP_n(x)\, z^{n-1}}{\sum_{n=0}^{\infty} P_n'(x)\, z^n} \quad \text{or} \quad (x-z)\sum_{n=0}^{\infty} P_n'(x)\, z^n = \sum_{n=0}^{\infty} nP_n(x)\, z^n.$$

Equating the coefficients of z^n on both sides, we get

$$\mathbf{xP_n'(x) - P_{n-1}'(x) = nP_n(x).}$$

(*c*) Differentiating Eq. (12.82) w.r.t. x, we get

$$(n+1)\, P_{n+1}'(x) = (2n+1)\, P_n(x) + (2n+1)x\, P_n'(x) - nP_{n-1}'(x). \qquad ...(12.85)$$

Substituting for $xP_n'(x)$ from formula (*b*) in the above result, we get

$$(n+1)\, P_{n+1}'(x) = (2n+1)\, P_n(x) + (2n+1)[P_{n-1}'(x) + nP_n(x)] - nP_{n-1}'(x)$$

or $$\mathbf{P_{n+1}'(x) - P_{n-1}'(x) = (2n+1)\, P_n(x).}$$

(*d*) The Eq. (12.85) may be written as

$$(n+1)\, P_{n+1}'(x) = (2n+1)\, P_n(x) + (n+1)\, xP_n'(x) + n[xP_n'(x) - P_{n-1}'(x)].$$

Using the result of formula (*b*), we get

$$(n+1)\, P_{n+1}'(x) = (2n+1)\, P_n(x) + (n+1)\, xP_n'(x) + n\cdot nP_n(x)$$

$$= (n+1)\, xP_n'(x) + (n^2+2n+1)\, P_n(x)$$

or $$\mathbf{P_{n+1}'(x) = xP_n'(x) + (n+1)\, P_n(x).}$$

(*e*) Replacing n by $(n-1)$ in formula (*d*) above, we get

$$P_n'(x) - xP_{n-1}'(x) = nP_{n-1}(x). \qquad ...(12.86)$$

From formula (b), we have

$$xP_n'(x) - P_{n-1}'(x) = nP_n(x). \qquad ...(12.87)$$

Multiplying (12.87) by x and subtracting from (12.86), we get

$$\mathbf{(1-x^2)\, P_n'(x) = n[P_{n-1}(x) - xP_n(x)].}$$

(*f*) From formula (*a*)

$$(n+1)\,P_{n+1}(x) = (2n+1)\,xP_n(x) - nP_{n-1}(x), \text{ we have}$$

$$(n+1)\,xP_n(x) - (n+1)\,P_{n+1}(x) = nP_{n-1}(x) - nxP_n(x)$$

or $$(n+1)\,[xP_n(x) - P_{n+1}(x)] = n[P_{n-1}(x) - xP_n(x)]$$

or $$\mathbf{(n+1)\,[xP_n(x) - P_{n+1}(x)] = (1-x^2)\,P_n'(x)}, \text{ from formula } (e).$$

Remarks

1. Eq. (12.83) may be written as

$$z(1-2xz+z^2)^{-1/2} = (1-2xz+z^2)\sum_{n=0}^{\infty} P_n'(x)\,z^n$$

or $$z\sum_{n=0}^{\infty} P_n(x)\,z^n = (1-2xz+z^2)\sum_{n=0}^{\infty} P_n'(x)\,z^n.$$

Equating coefficients of z^n on both sides, we get

$$\mathbf{P_{n-1}(x) = P_n'(x) - 2xP_{n-1}'(x) + P_{n-2}'(x).} \qquad \text{...(12.88)}$$

(12.88) may also be written as

$$\mathbf{P_n(x) = P_{n+1}'(x) - 2xP_n'(x) + P_{n-1}'(x)} \quad \text{(Replacing } n \text{ by } n+1\text{)} \qquad \text{...(12.88p)}$$

2. Adding Eq. (12.88p) and Eq. $nP_n(x) = xP_n'(x) - P_{n-1}'(x)$, [formula (*b*)]

we obtain $$(n+1)\,P_n(x) = P_{n+1}'(x) - xP_n'(x) \qquad \text{...(12.89)}$$

or $$\mathbf{nP_{n-1}(x) = P_n'(x) - xP_{n-1}'(x)} \quad \text{(on replacing } n \text{ by } n-1\text{)} \qquad \text{...(12.89q)}$$

3. The following results are fundamental. The other relations can be obtained from them.

$$\left\{\begin{aligned} \mathbf{(n+1)P_{n+1}(x)} &= \mathbf{(2n+1)xP_n(x) - nP_{n-1}(x).} \\ \mathbf{P_{n-1}(x)} &= \mathbf{P_n'(x) - 2xP_{n-1}'(x) + P_{n-2}'(x).} \\ \mathbf{nP_n(x)} &= \mathbf{xP_n'(x) - P_{n-1}'(x).} \end{aligned}\right\}$$

EXAMPLE 12.28 Using the recurrence relation

$$(n+1)P_{n+1}(x) = (2n+1)\,xP_n(x) - nP_{n-1}(x),$$

generate the Legendre's polynomials P_2, P_3, P_4, P_5 given that $P_0(x) = 1$ and $P_1(x) = x$.

Solution For $n = 1$, we get

$$2P_2(x) = 3xP_1(x) - P_0(x) = 3x(x) - 1, \text{ since, we are given that } P_0(x) = 1,\ P_1(x) = x.$$

or $$P_2(x) = \frac{1}{2}(3x^2 - 1).$$

For $n = 2$, we get $$3P_3(x) = 5xP_2(x) - 2P_1(x)$$

$$= \frac{5}{2}x(3x^2 - 1) - 2x = \frac{1}{2}(15x^3 - 9x),$$

or $$P_3(x) = \frac{1}{2}(5x^3 - 3x).$$

For $n = 3$, we get $$4P_4(x) = 7xP_3(x) - 3P_2(x) = \frac{7}{2}x(5x^3 - 3x) - \frac{3}{2}(3x^2 - 1)$$

$$= \frac{1}{2}(35x^4 - 30x^2 + 3),$$

or $$P_4(x) = \frac{1}{8}(35x^4 - 30x^2 + 3).$$

For $n = 4$, we get
$$5P_5(x) = 9xP_4(x) - 4P_3(x)$$
$$= \frac{9}{8}x(35x^4 - 30x^2 + 3) - 2(5x^3 - 3x)$$
$$= \frac{1}{8}(315x^5 - 350\,x^3 + 75x),$$
or
$$P_5(x) = \frac{1}{8}(63x^5 - 70x^3 + 15x).$$

EXAMPLE 12.29 Using the recurrence relation for Legendre's polynomial $(n+1)\,P_{n+1}(x) = (2n+1)\,xP_n(x) - nP_{n-1}(x)$, evaluate $P_2(1.5)$ and $P_3(2.1)$. *(AMIETE, June 2008; Dec, 2005)*

Solution Setting $n = 1$, in the given relation, we get

$2P_2(x) = 3xP_1(x) - P_0(x)$. Since, $P_0(x) = 1$ for all x and $P_1(x) = x$, we get $P_1(1.5) = 1.5$.

Therefore,
$$P_2(1.5) = \frac{1}{2}[3(1.5)\,(1.5) - 1] = \frac{23}{8} = 2.875.$$

Now, setting $n = 2$, we get
$$3P_3(x) = 5xP_2(x) - 2P_1(x) \text{ and } 2P_2(x) = 3xP_1(x) - P_0(x).$$

Now $\quad P_0(2.1) = 1, \qquad P_1(2.1) = 2.1.$

Therefore, $\quad P_2(2.1) = \frac{1}{2}[3(2.1)\,(2.1) - 1] = 6.115,$

and $\quad P_3(2.1) = \frac{1}{3}[5(2.1)\,(6.115) - 2(2.1)] = 20.0025.$

12.18 ORTHOGONAL PROPERTY OF LEGENDRE'S POLYNOMIALS

Like trigonometric functions cos mx and sin mx, the Legendre polynomials are orthogonal functions. The Legendre polynomials $P_m(x)$, $P_n(x)$ satisfy the following orthogonal property

$$\int_{-1}^{1} P_m(x)\,P_n(x)\,dx = \begin{cases} 0, & \text{if } m \neq n \quad \text{...(12.90)} \\ \dfrac{2}{2n+1}, & \text{if } m = n \quad \text{...(12.91)} \end{cases}$$

[AMIETE, June, 2010, Dec., 2008, 07, June 04; VTU, 2007, JNTU, 2006)

Case 1: When m ≠ n.

To derive Eq. (12.90), we use the fact that Legendre polynomials $P_m(x)$ and $P_n(x)$ satisfy the differential equation
$$(1 - x^2)\,y''(x) - 2xy'(x) + n(n+1)\,y(x) = 0$$
that is,
$$(1 - x^2)\,P_m'' - 2xP_m' + m(m+1)\,P_m = 0 \quad \text{...}(i)$$
and
$$(1 - x^2)\,P_n'' - 2xP_n' + n(n+1)\,P_n = 0. \quad \text{...}(ii)$$

Multiplying the first equation by P_n and the second equation by $-P_m$ and adding, we get
$$(1 - x^2)[P_m''\,P_n - P_n''P_m] - 2x(P_m'\,P_n - P_n'\,P_m) + [m(m+1 - n(n+1)]P_m\,P_n = 0$$
which can be written as
$$\frac{d}{dx}[(1 - x^2)\{P_m'\,P_n - P_n'\,P_m\}] = (n - m)\,(m + n + 1)P_mP_n.$$

Integrating both sides between the limits –1 to 1, we get
$$(n - m)\,(m + n + 1)\int_{-1}^{1} P_m(x)\,P_n(x)\,dx = (1 - x^2)\,[P_nP_m' - P_mP_n']_{-1}^{1}.$$

The right side vanishes at $x = \pm 1$. Since $m \neq n$, we get

$$\int_{-1}^{1} \mathbf{P_m(x)\, P_n(x)\, dx = 0, m \neq n.}$$

This establishes that the Legendre polynomials are *orthogonal functions* over the interval –1 to 1.

Case 2: *When m = n.*

From the generating function

$$(1 - 2xz + z^2)^{-1/2} = \sum_{n=0}^{\infty} P_n(x)\, z^n,$$

we have on squaring both sides

$$(1 - 2xz + z^2)^{-1} = \left[\sum_{n=0}^{\infty} P_n(x)\, z^n\right]^2.$$

Then by integrating from –1 to 1, we have

$$\int_{-1}^{1} \frac{dx}{1 - 2xz + z^2} = \int_{-1}^{1} \left[\sum_{n=0}^{\infty} P_n(x)\, z^n\right]^2 dx.$$

From the left hand side, we obtain

$$\int_{-1}^{1} \frac{dx}{1 - 2xz + z^2} = -\frac{1}{2z}[\log(1 - 2xz + z^2)]_{-1}^{1} = -\frac{1}{2z}[\log(1 - 2z + z^2) - \log(1 + 2z + z^2)]$$

$$= -\frac{1}{2z}[\log(1 - z)^2 - \log(1 + z)^2] = \frac{1}{z}[\log(1 + z) - \log(1 - z)]$$

$$= \frac{1}{z}\left[\left(z - \frac{z^2}{2} + \frac{z^3}{3} - \frac{z^4}{4} + \cdots\right) - \left(-z - \frac{z^2}{2} - \frac{z^3}{3} - \cdots\right)\right]$$

$$= 2\left[1 + \frac{z^2}{3} + \frac{z^4}{5} + \cdots + \frac{z^{2n}}{2n+1} + \cdots\right]. \quad \text{...(A)}$$

Using the orthogonal property given in Eq. (12.90), we obtain from the right hand side

$$\int_{-1}^{1} \left[\sum_{n=0}^{\infty} P_n(x)\, z^n\right]^2 dx = \sum_{n=0}^{\infty} \int_{-1}^{1} P_n^2(x)\, z^{2n} dx = \sum_{n=0}^{\infty} z^{2n} \int_{-1}^{1} P_n^2(x)\, dx. \quad \text{...(B)}$$

Comparing the coefficients of z^{2n}, in equations (A) and (B), we obtain

$$\int_{-1}^{1} \boldsymbol{P_n^2(x)\, dx = \frac{2}{2n+1}}.$$

EXAMPLE 12.30 Show that $\displaystyle\int_{-1}^{1} x^2 P_{n-1}(x)\, P_{n+1}(x)\, dx = \frac{2^n (n+1)}{(2n-1)(2n+1)(2n+3)}$.

(AMIETE, Dec. 2008; JNTU, 2006)

Solution We have from the recurrence relations. Article (12.17) formula (a), refer note thereon.

$$xP_{n-1} = \frac{1}{2n-1}[nP_n + (n-1)P_{n-2}],$$

and $\quad xP_{n+1} = \dfrac{1}{2n+3}[(n+2)\,P_{n+2} + (n+1)P_n]$.

Therefore, $x^2 P_{n-1}\,P_{n+1} = \dfrac{1}{(2n-1)\,(2n+3)}[n(n+2)\,P_n P_{n+2} + n(n+1)P_n^2 + (n-1)\,(n+2)P_{n-2}\,P_{n+2}$

$$+ (n^2-1)P_n P_{n-2}].$$

Integrating both sides from –1 to 1 and using orthogonality of Legendre's polynomials, we get

$$\int_{-1}^{1} x^2 P_{n-1}\,P_{n+1}\,dx = \frac{n(n+1)}{(2n-1)\,(2n+3)}\int_{-1}^{1} P_n^2\,dx = \frac{2n(n+1)}{(2n-1)\,(2n+1)\,(2n+3)}.$$

EXAMPLE 12.31 Find the value of the following integrals

(*i*) $\displaystyle\int_{-1}^{1} x^3 P_4(x)\,dx$, $\qquad$ (*ii*) $\displaystyle\int_{-1}^{1} x^5 P_4(x)\,dx$, $\qquad$ (*iii*) $\displaystyle\int_{-1}^{1} x^6 P_4(x)\,dx$.

Solution Employing the formula:

$$\int_{-1}^{1} f(x)\,P_n(x)\,dx = \frac{1}{2^n n!}\int_{-1}^{1} f^{(n)}(x)\,(1-x^2)^n\,dx,$$ where f is any function integrable on interval [–1, 1].

(*i*) $$\int_{-1}^{1} x^3 P_4(x)\,dx = \frac{1}{2^4 4!}\int_{-1}^{1}[D^4(x^3)]\,(1-x^2)^4\,dx = 0. \qquad (\text{as } D^4x^3 = 0).$$

(*ii*) $$\int_{-1}^{1} x^5 P_4(x)\,dx = \frac{1}{2^4 4!}\int_{-1}^{1}[D^4(x^5)]\,(1-x^2)^4\,dx$$

$$= \frac{1}{2^4 4!}\int_{-1}^{1} 120x(1-x^2)^4\,dx = 0. \qquad (\because \text{ integrand is odd function of } x).$$

(*iii*) $$\int_{-1}^{1} x^6 P_4(x)\,dx = \frac{1}{2^4 4!}\int_{-1}^{1}[D^4(x^6)]\,(1-x^2)^4\,dx$$

$$= \frac{360}{2^4 4!}\int_{-1}^{1} x^2(1-x^2)^4\,dx \qquad (\text{Put } x = \sin\theta,\ dx = \cos\theta\,d\theta)$$

$$= \frac{360}{2^4 4!}\int_{-\pi/2}^{\pi/2} \sin^2\theta\cos^9\theta\,d\theta = \frac{15}{8}\int_{0}^{\pi/2} \sin^2\theta\cos^9\theta\,d\theta$$

$$= \frac{15}{8}\,\frac{8.6.4.2}{11.9.7.5.3.1} = \frac{16}{231}.$$

12.19 FOURIER-LEGENDRE SERIES

Let a function $f(x)$ is sectionally continuous in the interval (–1, 1) and if its derivative $f'(x)$ is sectionally continuous in every interval interior to (–1, 1), then it admits of a series expansion of the form

$$f(x) = \sum_{n=0}^{\infty} a_n P_n(x) = a_0 P_0(x) + a_1 P_1(x) + a_2 P_2(x) + \cdots + a_n P_n(x) + \cdots \qquad ...(12.92)$$

To obtain the value of the coefficient a_n, we multiply both sides of (12.92) by $P_n(x)$ and integrate with respect to x over the interval (–1, 1). We obtain on using the orthogonal property of $P_n(x)$

$$\int_{-1}^{1} f(x)\, P_n(x)\, dx = \sum_{n=0}^{\infty} a_n \int_{-1}^{1} [P_n(x)]^2\, dx = a_n\left(\frac{2}{2n+1}\right).$$

The general coefficient a_n is given by

$$a_n = \frac{2n+1}{2}\int_{-1}^{1} f(x)\, P_n(x)\, dx, \quad n = 0, 1, 2, \ldots\ldots \qquad \ldots(12.93)$$

If $f(x)$ is a polynomial, then it is easy to evaluate a_n using the Rodrigue's formula (Article 12.16). In this case, we have

$$a_n = \frac{2n+1}{2}\int_{-1}^{1} \frac{1}{2^n\, n!}\frac{d^n}{dx^n}[(x^2-1)^n]\, f(x)\, dx$$

$$= \frac{(2n+1)}{2^{n+1} n!}\int_{-1}^{1} \frac{d^n}{dx^n}[(x^2-1)^n]\, f(x)\, dx.$$

Integrating by parts, we obtain

$$a_n = \frac{(2n+1)(-1)^n}{2^{n+1} n!}\int_{-1}^{1} (x^2-1)^n\, f^{(n)}(x)\, dx.$$

If $f^n(x)$ is an odd function, then $a_n = 0$.

If $f^n(x)$ is an even function, then $a_n = \dfrac{(-1)^n (2n+1)}{2^n n!}\displaystyle\int_{0}^{1} (x^2-1)^n\, f^{(n)}(x)\, dx.$

If $f(x)$ is a polynomial, then at some stage $f^n(x)$ = constant, and $c_i = 0$, $i > n$.

EXAMPLE 12.32 Prove that for all values of x

$$x^4 = \frac{8}{35}P_4(x) + \frac{4}{7}P_2(x) + \frac{1}{5}P_0(x).$$

Solution From Art 12.19, let $\quad x^4 = \displaystyle\sum_{n=0}^{\infty} a_n P_n(x)$

$$= a_0 P_0(x) + a_1 P_1(x) + a_2 P_2(x) + \cdots + a_n P_n(x) + \cdots$$

Multiplying both sides by $P_n(x)$ and integrating w.r.t. x between –1 and 1, we get

$$\int_{-1}^{1} x^4 P_n(x)\, dx = a_n \int_{-1}^{1} P_n^2(x)\, dx,$$

all other integrals are zero due to orthogonality.

Therefore, $\quad \displaystyle\int_{-1}^{1} x^4 P_n(x)\, dx = \frac{2}{2n+1} a_n, \qquad \ldots(A)$

or $\quad a_n = \dfrac{2n+1}{2}\displaystyle\int_{-1}^{1} x^4 P_n(x)\, dx.$

Now $\displaystyle\int_{-1}^{1} x^4 P_n(x)\, dx = 0$ if $n > 4$ (Rodrigue's formula)

Therefore, $\quad a_5 = a_6 = \ldots\ldots = 0.$

Again $\int_{-1}^{1} x^4 P_3(x)\,dx$ and $\int_{-1}^{1} x^4 P_1(x)\,dx$ are zero because integrands are odd functions of x.

Therefore, $\quad a_1 = a_3 = 0.$

Finally $\quad a_0 = \frac{1}{2}\int_{-1}^{1} x^4 \cdot P_0(x)\,dx = \frac{1}{5}$ from (A) above.

$$a_2 = \frac{5}{2}\int_{-1}^{1} x^4 P_2(x)\,dx = \frac{5}{2}\int_{-1}^{1} x^4\left(\frac{3x^2-1}{2}\right)dx$$

$$= \frac{5}{4}\left(\frac{3x^7}{7} - \frac{x^5}{5}\right)_{-1}^{1} = \frac{2\times 5}{4}\left(\frac{3}{7} - \frac{1}{5}\right) = \frac{5}{2}\cdot\frac{8}{35} = \frac{4}{7}.$$

$$a_4 = \frac{9}{2}\int_{-1}^{1} x^4 P_4(x)\,dx = \frac{9}{2}\int_{-1}^{1} x^4\left\{\frac{35x^4 - 30x^2 + 3}{8}\right\}dx$$

$$= \frac{9}{16}\left[\frac{35x^9}{9} - \frac{30x^7}{7} + \frac{3x^5}{5}\right]_{-1}^{1} = \frac{2\times 9}{16}\left(\frac{35}{9} - \frac{30}{7} + \frac{3}{5}\right) = \frac{8}{35}.$$

Therefore, $\quad x^4 = a_0P_0(x) + a_2P_2(x) + a_4P_4(x)$

$$= \frac{1}{5}P_0(x) + \frac{4}{7}P_2(x) + \frac{8}{35}P_4(x).$$

EXAMPLE 12.33 If $f(x) = \begin{cases} 0, & -1 < x \le 0 \\ x, & 0 < x < 1 \end{cases}$ show that $f(x) = \frac{1}{4}P_0(x) + \frac{1}{2}P_1(x) + \frac{5}{16}P_2(x) - \frac{3}{32}P_4(x) + \cdots.$

(*AMIETE, June 2011, 2007, 2004; UPTU, 2003*)

Solution Let $f(x)$ be expanded in the form $f(x) = \sum_{n=0}^{\infty} a_n P_n(x)$.

Then the coefficient a_n is given by

$$a_n = \frac{2n+1}{2}\int_{-1}^{1} f(x)\,P_n(x)\,dx.$$

Therefore,

$$a_0 = \frac{1}{2}\left[\int_{-1}^{0} 0\times 1\,dx + \int_{0}^{1} x\cdot 1\,dx\right] = \frac{1}{4},$$

$$a_1 = \frac{3}{2}\int_{0}^{1} x\,P_1(x)\,dx = \frac{3}{2}\int_{0}^{1} x^2\,dx = \frac{1}{2},$$

$$a_2 = \frac{5}{2}\int_{0}^{1} x\,P_2(x)\,dx = \frac{5}{2}\int_{0}^{1} x\,\frac{3x^2-1}{2}\,dx$$

$$= \frac{5}{4}\left[\frac{3x^4}{4} - \frac{x^2}{2}\right]_{0}^{1} = \frac{5}{4}\cdot\frac{1}{4} = \frac{5}{16},$$

$$a_3 = \frac{7}{2}\int_0^1 xP_3 dx = \frac{7}{2}\int_0^1 x \cdot \frac{5x^3 - 3x}{2} dx = \frac{7}{4}\left[\frac{5x^5}{5} - \frac{3x^3}{3}\right]_0^1 = 0,$$

$$a_4 = \frac{9}{2}\int_0^1 xP_4(x)\, dx = \frac{9}{2}\int_0^1 x \cdot \frac{35x^4 - 30x^2 + 3}{8} dx = \frac{9}{16}\left[\frac{35x^6}{6} - \frac{30x^4}{4} + \frac{3x^2}{2}\right]_0^1 = -\frac{3}{32}, \text{ and so on.}$$

Hence, $f(x) = \frac{1}{4}P_0(x) + \frac{1}{2}P_1(x) + \frac{5}{16}P_2(x) - \frac{3}{32}P_4(x) + \ldots\ldots$

EXAMPLE 12.34 Find the Fourier-Legendre series of the function $f(x) = 3x^3 + 2x^2 + x + 3$.

Solution We have $f(x) = a_0P_0(x) + a_1P_1(x) + a_2P_2(x) + \ldots\ldots$, where

$$a_n = \frac{2n+1}{2}\int_{-1}^{1} f(x)\, P_n(x)\, dx = \frac{(-1)^n\,(2n+1)}{2^{n+1}\cdot n!}\int_{-1}^{1}(x^2-1)^n\, f^{(n)}(x)\, dx.$$

We have for $f(x) = 3x^3 + 2x^2 + x + 3$, $\quad f^{(4)}(x) = 0 = f^{(5)}(x) = \ldots$, we get

for $n = 0$
$$a_0 = \frac{1}{2}\int_{-1}^{1} f(x)\, dx = \int_0^1 (2x^2 + 3)\, dx = \frac{11}{3},$$

for $n = 1$
$$a_1 = -\frac{3}{4}\int_{-1}^{1}(x^2 - 1)\, f'(x)\, dx = -\frac{3}{4}\int_{-1}^{1}(x^2 - 1)(9x^2 + 4x + 1)\, dx$$

$$= -\frac{3}{2}\int_0^1 (9x^4 - 8x^2 - 1)\, dx = -\frac{3}{2}\left(\frac{9}{5} - \frac{8}{3} - 1\right) = \frac{14}{5},$$

for $n = 2$
$$a_2 = \frac{5}{16}\int_{-1}^{1}(x^2 - 1)^2 f''(x)\, dx = \frac{5}{16}\int_{-1}^{1}(x^2 - 1)^2\,(18x + 4)dx$$

$$= \frac{5}{8}\int_0^1 (4x^4 - 8x^2 + 4)dx = \frac{5}{8}\left(\frac{4}{5} - \frac{8}{3} + 4\right) = \frac{4}{3},$$

for $n = 3$
$$a_3 = -\frac{7}{96}\int_{-1}^{1}(x^2 - 1)^3\, f'''(x)\, dx = -\frac{7}{96}\int_{-1}^{1}(x^2 - 1)^3\,(18)\, dx$$

$$= -\frac{21}{8}\int_0^1 (x^6 - 3x^4 + 3x^2 - 1)\, dx = -\frac{21}{8}\left(\frac{1}{7} - \frac{3}{5}\right) = \frac{6}{5}.$$

Hence,
$$f(x) = 3x^3 + 2x^2 + x + 3 = \left[\frac{11}{3}P_0(x) + \frac{14}{5}P_1(x) + \frac{4}{3}P_2(x) + \frac{6}{5}P_3(x)\right].$$

EXERCISE 12.4

1. Express the following sums of Legendre polynomials in terms of powers of x

(*i*) $3P_3(x) + 2P_2(x) + 4P_1(x) + 5P_0(x)$, (*ii*) $8P_4(x) + 2P_2(x) + P_0(x)$,

(*iii*) $2P_2(x) + 4P_1(x) + 5P_0(x)$. *(AMIETE, Dec, 2009)*

Ans. (*i*) $\frac{1}{2}(15x^3 + 6x^2 - x + 8)$, (*ii*) $35x^4 - 27x^2 + 3$, (*iii*) $3x^2 + 4x + 4$.

2. Express the following polynomials in terms of Legendre polynomials.

(*i*) $f(x) = x^4 + 3x^3 - x^2 + 5x - 2$

(*AMIETE, June, 2010; SVTU., 2007; VTU, 2003 S; JNTU, 2000*)

(*ii*) $f(x) = 7x^4 + 6x^3 + 3x^2 + x - 6$ (*AMIETE, June 2006*)

(*iii*) $f(x) = 1 + x - x^2$

Ans. (*i*) $\frac{8}{35}P_4(x)+\frac{6}{5}P_3(x)-\frac{2}{2!}P_2(x)+\frac{34}{5}P_1(x)-\frac{32}{15}P_0(x)$,

(*ii*) $\frac{1}{5}[8P_4(x)+12P_3(x)+30P_2(x)+23P_1(x)-18P_0(x)]$,

(*iii*) $\frac{1}{3}[2P_0(x)+3P_1(x)-2P_2(x)]$.

3. Prove that

(*i*) $P_n'(1) = n(n+1)/2$, (*UPTU, 2003*) (*ii*) $P_n'(-1) = (-1)^{n-1}n(n+1)/2$. (*AMIETE, Dec2002*)

4. Show that $P_n(x) = P_{n+1}'(x) - 2xP_n'(x) + P_{n-1}'(x)$.

5. Show that

(*i*) $P_{n+1}(x) = \frac{2n+1}{n+1}xP_n(x) - \frac{n}{n+1}P_{n-1}(x)$, (*ii*) $P_n(x) = P_{n+1}'(x) - 2xP_n'(x) + P_{n-1}'(x)$.

6. Prove that

(*i*) $P_{2n}'(0) = 0$, (*ii*) $P_{2n+1}'(0) = \frac{(-1)^n\,(2n+1)!}{2^{2n}(n!)^2}$.

7. Prove that $xP_9' - P_8' = 9P_9$.

8. Prove that

(*i*) $P_{2n}'(0) = 0$, (*ii*) $P_{2n+1}'(0) = \frac{(-1)^n\,(2n+1)!}{2^{2n}(n!)^2}$.

9. Prove that $\frac{1-z^2}{(1-2xz+z^2)^{3/2}} = \sum_{n=0}^{\infty}(2n+1)\,P_n(x)\,z^n$. (*AMIETE, Dec, 2002*)

10. Show that $\frac{1+z}{z(1-2xz+z^2)^{1/2}} - \frac{1}{z} = \sum_{n=0}^{\infty}\{P_n(x)+P_{n+1}(x)\}z^n$, (*AMIETE, June 2001*)

where $P_n(x)$ is the Legendre polynomial of degree n.

[**Hint:** L.H.S. $= \frac{1+z}{z(1-2xz+z^2)^{1/2}} - \frac{1}{z} = \left(1+\frac{1}{z}\right)\sum_{n=0}^{\infty}P_n(x)\,z^n - \frac{1}{z}$

$$= \sum_{n=0}^{\infty}P_n(x)z^n + \sum_{n=0}^{\infty}P_n(x)z^{n-1} - \frac{1}{z}$$

$$= \sum_{n=0}^{\infty}P_n(x)z^n + P_0(x)z^{-1} + \sum_{n=1}^{\infty}P_n(x)z^{n-1} - \frac{1}{z}$$

$$= \sum_{n=0}^{\infty}P_n(x)z^n + \frac{1}{z} + \sum_{m=0}^{\infty}P_{m+1}(x)\,z^m - \frac{1}{z}, \quad \text{where } n = m+1$$

$$= \sum_{n=0}^{\infty}\{P_n(x)+P_{n+1}(x)\}z^n = \text{R.H.S.} \quad \text{Hence proved]}$$

11. Show that

(i) $\int_{-1}^{1} x^3 P_3(x)\, dx = \frac{4}{35}$, (AMIETE, Dec., 2003; June 2001)

(ii) $\int_{-1}^{1} P_3(x)\, P_4(x)\, dx = 0$, AMIETE, June 2007) (iii) $\int_{-1}^{1} x^6 P_4(x)\, dx = \frac{16}{231}$.

12. Show that $\int P_n(x)\, dx = \frac{1}{2n+1}[P_{n+1}(x) - P_{n-1}(x)]$. (AMIETE, Dec, 2008)

[**Hint:** Using the recurrence relations:

$$\left.\begin{aligned} nP_n(x) &= xP_n'(x) - P_{n-1}'(x) \\ P_{n+1}'(x) &= xP_n'(x) + (n+1)\, P_n(x) \end{aligned}\right\}, \text{ we obtain}$$

$$P_{n+1}'(x) - P_{n-1}'(x) = (2n+1)\, P_n(x).$$

Integrating, we get the required result.]

13. Show that:

(i) $\int_{-1}^{1} P_n(x)\, dx = 0$, $(n \neq 0)$

(ii) $\int_{-1}^{1} x^m P_n(x)\, dx = 0$, ($m$ being an integer $< n$) (AMIETE, June 2008)

(iii) $\int_{-1}^{1} P_0(x)\, dx = 2$, (iv) $\int_{-1}^{1} x^n P_n(x)\, dx = \frac{2^{n+1}(n!)^2}{(2n+1)!}$.

14. Prove that

(i) $\int_{-1}^{1} x\, P_n(x)\, P_n'(x)\, dx = \frac{2n}{2n+1}$,

(ii) $\int_{-1}^{1} xP_{n-1}(x)\, P_n(x)\, dx = \frac{2n}{4n^2-1}$, $n = 1, 2, \ldots$ (AMIETE, Dec. 2004)

(iii) $\int_{-1}^{1} x^2 P_{n-1}(x)\, P_{n+1}(x)\, dx = \frac{2n(n+1)}{(2n-1)(2n+1)(2n+3)}$. (Bhopal 2000)

15. If m and n are integers, show that $\int_{-1}^{1} xP_n(x)\, P_m'(x)\, dx$ is either zero, or 2, or $2n/(2n+1)$.

16. Prove that $\int_{-1}^{1} (1-x^2)\, P_m'(x)\, P_n'(x)\, dx = \begin{cases} 0, & \text{if } m \neq n \\ \frac{2n(n+1)}{2n+1}, & \text{if } m = n. \end{cases}$ (AMIETE, Dec. 2009, 2006)

17. Show that $\int_{-1}^{1} P_n(x)\,(1 - 2xz + z^2)^{-1/2}\, dx = 2z^n/(2n+1)$.

18. Using Rodrigues's formula for Legendre polynomials $P_n(x)$, show that

$$\int_{-1}^{1} f(x)\, P_n(x)\, dx = \frac{(-1)^n}{2^n n!} \int_{-1}^{1} (x^2-1)^n f^n(x)\, dx,$$ where f is any function integrable on interval $[-1, 1]$.

Hence show that $\int_{-1}^{1} P_m(x)\, P_n(x)\, dx = 0, \quad m \neq n.$ *(AMIETE, June, 2005)*

19. Show that

(*i*) $x^2 = \frac{1}{3}[P_0(x) + 2P_2(x)]$, (*ii*) $x^3 = \frac{1}{5}[3P_1(x) + 2P_3(x)]$,

(*iii*) $x^5 = \frac{8}{63}\left[P_5(x) + \frac{7}{2}P_3(x) + \frac{27}{8}P_1(x)\right]$.

20. Obtain the Fourier-Legendre series of the following functions defined by

(*i*) $f(x) = \begin{cases} 0, & -1 < x < 0 \\ 1, & 0 < x < 1 \end{cases}$. (*ii*) $f(x) = \begin{cases} -1, & -1 < x < 0 \\ 1, & 0 < x < 1. \end{cases}$

(*iii*) $f(x) = \begin{cases} 0, & \text{if } -1 < x < 0 \\ x, & \text{if } 0 < x < 1 \end{cases}$. *(AMIETE, June, 2011)*

Ans. (*i*) $f(x) = \frac{1}{2}P_0(x) + \frac{3}{4}P_1(x) - \frac{7}{16}P_3(x) + \frac{11}{32}P_5(x) + ...$

(*ii*) $f(x) = \frac{3}{2}P_1(x) - \frac{7}{8}P_3(x) + \frac{11}{16}P_5(x) + ...$

(*iii*) $f(x) = \frac{1}{4}P_0(x) + \frac{1}{2}P_1(x) + \frac{5}{16}P_2(x) - \frac{3}{32}P_4(x) + \frac{13}{256}P_6(x) + ...$

21. Expand $f(x) = x^3 - x$, $-1 \le x \le 1$, as a Fourier–Legendre series.

Ans. $f(x) = x^3 - x = [-2P_1(x) + 2P_3(x)]/5$.

22. Expand $f(x) = x^3 + x$, $-1 \le x \le 1$, as a Fourier–Legendre series.

Ans. $f(x) = x^3 + x = [8P_1(x) + 2P_3(x)]/5$.

23. Develop Fourier–Legendre series of the function $f(x) = 15 - 42x^2 + 35x^4$.

Ans. $f(x) = 8[P_0(x) - P_2(x) + P_4(x)]$.

CHAPTER

13

Numerical Methods

13.1 INTRODUCTION

In many engineering problems we are required to find a root or all the roots of a given equation $f(x) = 0$. When the equation is simple, for example a quadratic, we can easily solve it. But it may not be so easy to solve when the equation is algebraic and of a higher degree or when it is transcendental equation

We consider in this chapter

(i) numerical methods for solving algebraic and transcendental equations,

(ii) some of the most important numerical methods for solving linear systems of equations,

(iii) determination of eigen values by iteration (Power method) and

(iv) numerical solution of ordinary differential equations.

With the advent of computers these methods have assumed great importance and hence become indispensable tools in the hands of engineers and scientists to tackle a large number of problems they are confronted with.

13.2 SOLUTION OF ALGEBRAIC AND TRANSCENDENTAL EQUATIONS

In many engineering applications, we need to find a root or all the roots of a polynomial equation or a root of a transcendental equation, which is a combination of polynomials, exponential functions, trigonometric functions etc.

Zero or root: The value of x which satisfy the equation $f(x) = 0$ is called a *root* or a *zero of* $f(x) = 0$. Geometrically, the root of an equation $f(x) = 0$ is the value of x at which the graph of the function $y = f(x)$ intersects the x-axis (see Fig. 13.1). We assume that $f(x)$ is a continuous function in some interval which contains the root.

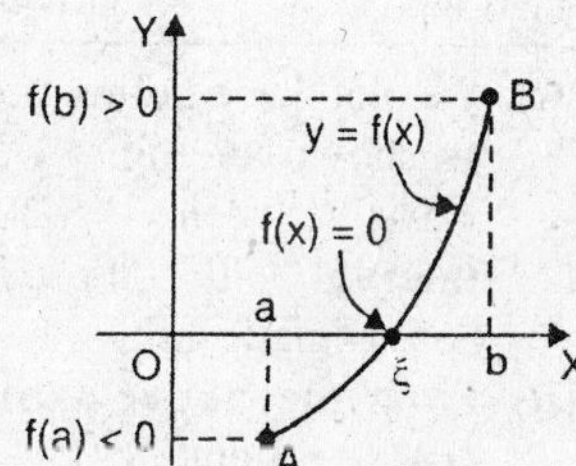

Fig. 13.1: Root of $f(x) = 0$.

The methods for finding a root are classified as

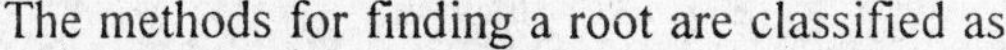

(i) direct methods, and

(ii) iterative methods

Direct methods These methods give the solution in a finite number of steps. However, there are no direct methods available for finding a root of a transcendental equation.

Iterative methods These methods are based on successive approximations. We start with an approximation to the true solution and by applying the method repeatedly we get a better and better approximation. Since, we cannot perform infinite number of iterations, we stop the iteration when $|x_{k+1} - x_k| \le \varepsilon$, where ε is the prescribed error and x_k, x_{k+1} are two consecutive iterates. Hence, to use an iterative method, we need one or more initial approximations and the iteration function $\phi(x)$.

13.2.1 Initial Approximations

We, generally use the following methods to obtain one or more initial approximations to the root.

Graphic method: We sketch a rough graph of the function $y = f(x)$. This is done by preparing a table of values of x and y and plotting the different points. From this graph we find its point of intersection with the x-axis. Any point in its neighbourhood may be taken as an initial approximation. If we write the equation $f(x) = 0$ in an equivalent form, $\phi_1(x) = \phi_2(x)$, where $\phi_1(x)$ and $\phi_2(x)$ are simpler functions compared to $f(x)$, then the abscissa of the point of intersection of the graphs of the functions $y = \phi_1(x)$ and $y = \phi_2(x)$ gives the root of $f(x) = 0$. Any point in the neighbourhood may be taken as an initial approximation to the root. For example, the equation $f(x) = e^x \sin x - 1 = 0$ may be written in an equivalent form as $\sin x = e^{-x}$. The abscissa of the point of intersection of the graphs of the functions $y = \sin x$ and $y = e^{-x}$ gives the root of $f(x) = e^x \sin x - 1 = 0$. However, the method is not a practical one.

The most suitable method to find an approximation to the root is the intermediate value theorem, which we states below:

Intermediate value theorem: *If $f(x)$ is continuous in some interval $[a, b]$ and $f(a)$ and $f(b)$ have opposite signs, (that is, $f(a)\,f(b) < 0$) then the equation $f(x) = 0$ has atleast one root or an odd number of real roots in the open interval (a, b)* [see Fig. 13.2].

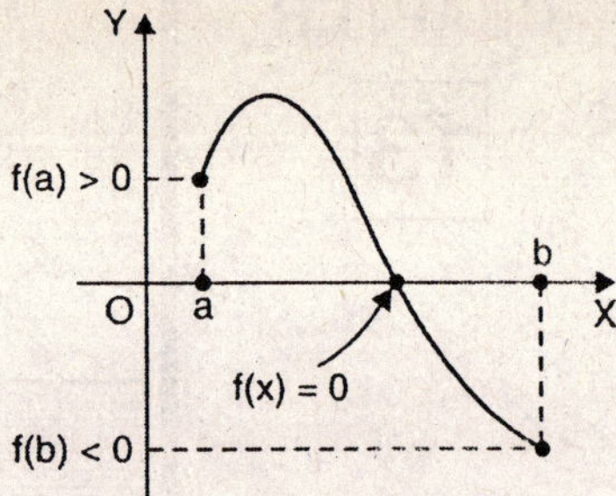

Fig. 13.2: $f(x) = 0$ has one root between $x = a$ and $x = b$.

We prepare a table of values of the function $y = f(x)$ for various values of x. From the changes of sign of $f(x)$ in this table of values, we obtain an interval (a, b) which contains the root. We may take either of the end points or a suitable point in the interval (a, b) as an initial approximation. However, the values of $f(x)$ at the end point a, b usually suggest a suitable choice of the approximation. For example, if $f(a) = -1$ and $f(b) = 3$, the root is closer to a then to b.

EXAMPLE 13.1 Find an initial approximation to a root of equation $f(x) = 2x^3 + 3x^2 + 2x + 5 = 0$. Using (*i*) graphic method, (*ii*) intermediate value theorem.

Solution We prepare a table of values of the function

$$y = f(x) = 2x^3 + 3x^2 + 2x + 5$$

x	–3	–2	–1	0	1	2
$f(x)$	–28	–3	4	5	12	37

(*i*) We plot these points and obtain the graph of $y = f(x)$ as shown in Fig. 13.3. The graph intersects the x-axis between $x = -2$ and $x = -1$. Any point in the interval $(-2, -1)$ can be taken as an initial approximation.

(*ii*) From the above table of values, we find that $f(-2) = -3$ and $f(-1) = 4$. Since $f(-2)\,f(-1) < 0$, we conclude that a root exists in the interval $(-2, -1)$.

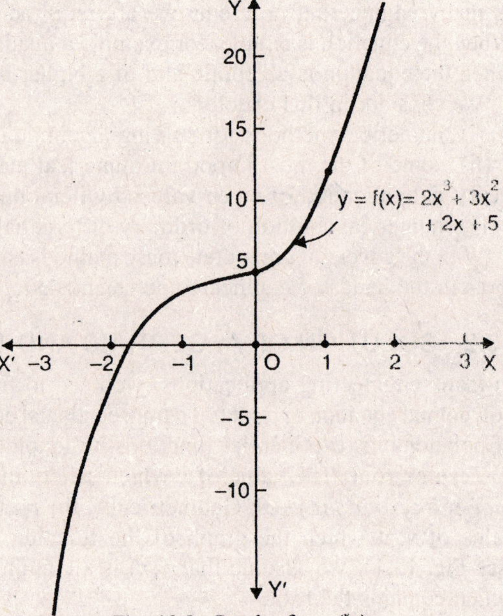

Fig. 13.3: Graph of $y = f(x)$.

EXAMPLE 13.2 Find the approximate value of the root of equation $3x - \cos x - 1 = 0$ and find an interval which contains the root of the equation.

Solution We plot the graph of the functions $y = 3x - 1$ and $y = \cos x$ as given in Fig. 13.4. We find that the two graphs intersect at a point P. The abscissa of the point of intersection of the graphs of these equations is seen to be about 0.6. Hence, a root of the given equation lies in the interval (0, 1).

13.2.2 Bisection Method

This method is based on the repeated application of the intermediate value theorem to obtain an approximation to the root. Let the root of $f(x) = 0$ lies in the interval (a_0, b_0), that is $f(a_0)\,f(b_0) < 0$. We bisect this interval and obtain $c_1 = (a_0 + b_0)/2$. Then the root lies in the interval (a_0, c_1) provided $f(a_0)\,f(c_1) < 0$. Otherwise it lies in the interval (c_1, b_0). Repeat the process until the root is known to the desired accuracy.

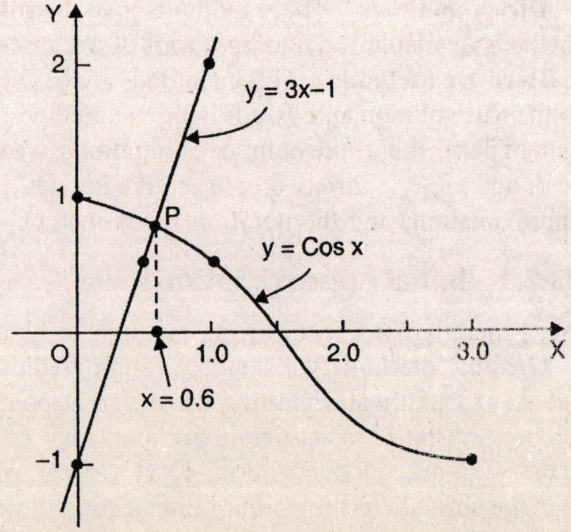

Fig. 13.4: Graphs of $y = 3x - 1$ and $y = \cos x$.

The number of iterations required to achieve the desired accuracy is calculated in the following way.

Let n be the required number of iterations. After the nth iteration, we have $\frac{b_0 - a_0}{2^n} \leq \varepsilon$.

Taking logarithms of both sides, we obtain

$$n \geq \log\left(\frac{b_0 - a_0}{\varepsilon}\right)\Big/2. \qquad (13.1)$$

For example, if $b_0 = 0.8$, $a_0 = 0.5$ then $b_0 - a_0 = 0.3$ and for an accuracy of 10^{-3} (that is, $\varepsilon = 0.001$), by the bisection method we have to carry on n bisections. Thus

$$n \geq \frac{\log 300}{\log 2} \simeq \frac{2.47712}{0.30103} = 8.2$$

That is, 9 bisections are needed.

EXAMPLE 13.3 Perform three iterations of the bisection method to obtain the root of the equation $x^3 - 5x + 1 = 0$ which lies between 2 and 3.

Solution Let $f(x) = x^3 - 5x + 1$. Since, $f(2) = -1 < 0$ and $f(3) = 13 > 0$, the root of $f(x) = 0$ lies in the interval (2, 3).

First iteration Let $a_0 = 2$ and $b_0 = 3$. Then, $c_1 = \frac{1}{2}(a_0 + b_0) = \frac{1}{2}(2+3) = 2.5$.

Since, $f(c_1) = (2.5)^3 - 5(2.5) + 1 = 15.625 - 12.5 + 1 = 4.125 > 0$

and $f(a_0)\, f(c_1) = (-1)\,(4.125) < 0$, the root lies in the interval (2, 2.5).

Second iteration Let $a_1 = 2$ and $b_1 = 2.5$. Then, $c_2 = \frac{1}{2}(a_1 + b_1) = 2.25$.

Since, $f(c_2) = (2.25)^3 - 5(2.25) + 1 = 11.39 - 11.25 + 1 = 1.14 > 0$

and $f(a_1)\, f(c_2) = (-1)\,(1.14) < 0$, the root lies in the interval (2, 2.25).

Third iteration Let $a_2 = 2$ and $b_2 = 2.25$ then $c_3 = 1/2(a_2 + b_2) = 2.125$.

Since, $f(c_3) = (2.125)^3 - 5(2.125) + 1 = 9.596 - 10.625 + 1 = -0.029 < 0$

and $f(c_3)\, f(b_2) = (-0.029)\,(1.14) < 0$, the root lies in the interval (2.125, 2.25).

The approximation to the root is taken as $x \approx \frac{1}{2}(2.125 + 2.25) = 2.1875$.

EXAMPLE 13.4 Perform three iterations of the bisection method to obtain the smallest root of the equation $x + e^x = 0$.

Solution Let $f(x) = x + e^x$. Since $f(0) = 1 > 0$ and $f(-1) = -1 + \frac{1}{e} = -1 + \frac{1}{2.71828} = -0.63212 < 0$, the root lies in (–1, 0).

First iteration: Let $a_0 = -1$ and $b_0 = 0$. Then $c_1 = \frac{1}{2}(a_0 + b_0) = \frac{1}{2}(-1+0) = -0.5$.

Since, $f(c_1) = -0.5 + e^{-0.5} = -0.5 + 0.60653 = 0.10653 > 0$ and $f(a_0)\, f(c_1) < 0$, the root lies in the interval (–1, – 0.5).

Second iteration: Let $a_1 = -1$ and $b_1 = -0.5$. Then $c_2 = \frac{1}{2}(a_1 + b_1) = -0.75$.

Since $f(c_2) = -0.75 + e^{-0.75} < 0$, and $f(b_1)\, f(c_2) < 0$, the root lies in the interval (– 0.75, – 0.50).

Third iteration: Let $a_2 = -0.75$ and $b_2 = -0.50$. Then $c_3 = \frac{1}{2}(a_2 + b_2) = \frac{1}{2}(-0.75 - 0.50) = -0.625$.

Since, $f(c_3) = f(-0.625) = -0.625 + e^{-0.625} < 0$ and $f(c_3)\, f(b_2) < 0$, the root lies in the interval (–0.625, 0.50).

The approximation to the root is taken as $x = \frac{1}{2}(-0.625 - 0.50) = -0.5625$.

13.2.3 Regula–Falsi Method

In this method, we require two initial approximations to the root. The initial approximations x_0 and x_1 to the root are so chosen that $f(x_0)f(x_1) < 0$ and for each k, $f(x_{k-1})f(x_k) < 0$. That is, at every stage of iteration, the root lies in the interval (x_{k-1}, x_k).

Refer Fig. 13.5(a). The equation of straight line passing through two points $P(x_0, f(x_0))$ and $Q(x_1, f(x_1))$ is

$$y - f(x_0) = \frac{f(x_1) - f(x_0)}{x_1 - x_0}(x - x_0).$$

This line meet the x-axis, where $y = 0$ that is,

$$x_2 = \frac{x_0 f(x_1) - x_1 f(x_0)}{f(x_1) - f(x_0)}.$$

Successive application of this process gives $x_3, x_4, \ldots$ leading to an improved value of x close to an exact value of x (= ξ) as shown in the figure.

If the root lies initially in (x_0, x_1) then one of the end points is fixed for all iteration. Refer Fig. 13.5(a), the end point x_0 is fixed, while in Fig. 13.5(b) the end point x_1 is fixed.

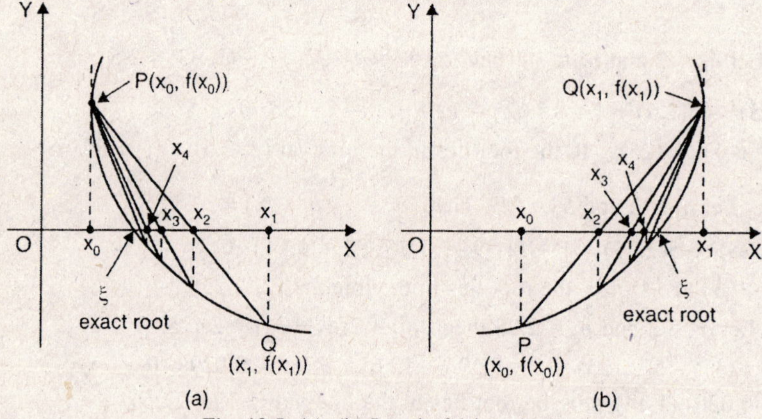

Fig. 13.5: (a), (b) Regula-falsi method.

In the light of above If x_0 is fixed, then the method is of the form

$$x_{k+1} = \frac{x_0 f_k - x_k f_{k-1}}{f_k - f_{k-1}}, \quad k = 1, 2, 3, \ldots \tag{13.2}$$

The method (13.2) with this condition is called the *regula-falsi method* or *the method of false position*. [See Figure (13.5), (*a*, *b*)].

EXAMPLE 13.5 The equation $f(x) = x^3 - 3x + 4 = 0$ has a root lies in the interval (–3, –2). Obtain the root correct to three decimal places using regula falsi method.

Solution Since the accuracy of three decimal places is required, we iterate until $(x_{k+1} - x_k) \leq 0.0005$. We have to retain atleast 5 decimal digits in our computation.

The root lies in (–3, –2). We have $x_0 = -3$, $x_1 = -2$, $f_0 = f(-3) = (-3)^3 - 3(-3) + 4 = -14$, $f_1 = f(-2) = 2$.

Employing the formula (13.2), we get the following results.

$k = 1$: $$x_2 = \frac{x_0 f_1 - x_1 f_0}{f_1 - f_0} = \frac{(-3)(2) - (-2)(-14)}{2 - (-14)} = -2.125,$$

$$f_2 = f(-2.125) = 0.77930$$

Since $f_0 f_2 < 0$, the root lies in the interval (x_0, x_2).

$k = 2$: $$x_3 = \frac{x_0 f_2 - x_2 f_0}{f_2 - f_0} = \frac{(-3)(0.77930) - (-2.125)(-14)}{0.77930 - (-14)} = \frac{-32.0879}{14.77930} = -2.17114.$$

$|x_3 - x_2| = 0.04614, f_3 = f(-2.17114) = 0.27944.$

Since $f_0 f_3 < 0$, the root lies in the interval (x_0, x_3).

$k : 3$: $$x_4 = \frac{x_0 f_3 - x_3 f_0}{f_3 - f_0} = \frac{(-3)(0.27944) - (-2.17114)(-14)}{0.27944 - (-14)} = -2.18732.$$

$|x_4 - x_3| = 0.01622, f_4 = f(-2.18732) = 0.09702.$

Since $f_0 f_4 < 0$, the root lies in the interval (x_0, x_4).

$k = 4$: $$x_5 = \frac{x_0 f_4 - x_4 f_0}{f_4 - f_0} = \frac{(-3)(0.09702) - (-2.18732)(-14)}{0.09702 - (-14)} = -2.19291.$$

$|x_5 - x_4| = 0.00559, f_5 = f(-2.19291) = 0.03335.$

Since $f_0 f_5 < 0$, the root lies in the interval (x_0, x_5).

$k = 5$: $$x_6 = \frac{x_0 f_5 - x_5 f_0}{f_5 - f_0} = \frac{(-3)(0.03335) - (-2.19291)(-14)}{0.03335 - (-14)} = -2.19484.$$

$|x_6 - x_5| = 0.00193, f_6 = f(-2.19484) = 0.01127.$

Since $f_0 f_6 < 0$, the root lies in the interval (x_0, x_6).

$k = 6$: $$x_7 = \frac{x_0 f_6 - x_6 f_0}{f_6 - f_0} = \frac{(-3)(0.01127) - (-2.19484)(-14)}{0.01127 - (-14)} = -2.19548.$$

$|x_7 - x_6| = 0.00065 \cdot f_7 = f(-2.19548) = 0.00394.$

Since $f_0 f_7 < 0$, the root lies in the interval (x_0, x_7).

$k = 7$: $$x_8 = \frac{x_0 f_7 - x_7 f_0}{f_7 - f_0} = \frac{(-3)(0.00394) - (-2.19548)(-14)}{0.00394 - (-14)} = -2.19571.$$

$|x_8 - x_7| = 0.00023 < 0.0005.$

Hence, the root correct to three decimal places is –2.195. Note that left end point x_0 of the initial interval is fixed in all iterations.

EXAMPLE 13.6 A root of the equation $x \log_{10} x - 1.2 = 0$ lies in the interval (2, 3). Determine this root correct to three decimal places using regula-falsi method.

Solution Since the accuracy of three decimal places is required, we iterate until $|x_{k+1} - x_k| \le 0.0005$.

The root lies in (2, 3). We have $x_0 = 2, x_1 = 3$.

$f_0 = f(2) = 2 \log_{10} 2 - 1.2 = 2(0.30103) - 1.2 = -0.59794,$

$f_1 = f(3) = 3 \log_{10} 3 - 1.2 = 3(0.47712) - 1.2 = 0.23136.$

Employing the formula (13.2), we get the following results

$k = 1$: $$x_2 = \frac{x_0 f_1 - x_1 f_0}{f_1 - f_0} = \frac{2(0.23136) - 3(-0.59794)}{0.23136 - (-0.59794)} = \frac{2.25654}{0.8293} = 2.721, \ |x_2 - x_1| = 0.279.$$

$$f_2 = f(2.721) = 2.721 \log_{10}(2.721) - 1.2 = 2.721(0.43473) - 1.2 = -0.0171.$$

Since $f_1 f_2 < 0$, the root lies in the interval (x_2, x_1).

$k = 2$: $$x_3 = \frac{x_1 f_2 - x_2 f_1}{f_2 - f_1} = \frac{3(-0.0171) - 2.721(0.23136)}{-0.0171 - 0.23136} = \frac{0.68083}{0.24846} = 2.7402,$$

$$|x_3 - x_2| = 0.0192, f_3 = f(2.7402) = 2.7402 \log_{10}(2.7402) - 1.2$$
$$= 2.7402(0.4378) - 1.2 = -0.00034.$$

Since $f_1 f_3 < 0$, the root lies in the interval (x_3, x_1).

$k : 3$: $$x_4 = \frac{x_1 f_3 - x_3 f_1}{f_3 - f_1} = \frac{3(-0.00034) - 2.7402(0.23136)}{-0.00034 - 0.23136} = 2.7405,$$

$$|x_4 - x_3| = 0.0003 < 0.0005.$$

Hence the root correct to three decimal places is 2.740.

Note that the right end point x_1, of the initial interval, is fixed in all iterations.

EXAMPLE 13.7 Find by the method of regula-falsi a root of the equation $x^3 + x^2 - 3x - 3 = 0$ lying between 1 and 2. *(AMIETE, June 2011)*

Solution The root lies in (1, 2). We have $x_0 = 1$, $x_1 = 2$, $f_0 = f(1) = 1^3 + 1^2 - 3(1) - 3 = -4 < 0$, $f_1 = f(2) = 2^3 + 2^2 - 3(2) - 3 = 3 > 0$.

$k = 1$: $$x_2 = \frac{x_0 f_1 - x_1 f_0}{f_1 - f_0} = \frac{1(3) - 2(-4)}{3-(-4)} = \frac{11}{7} = 1.57143,$$

$$f_2 = f(1.57143) = -1.36465.$$

Since $f_1 f_2 < 0$, the root lies in the interval (x_2, x_1).

$k = 2$: $$x_3 = \frac{x_1 f_2 - x_2 f_1}{f_2 - f_1} = \frac{2(-1.36465) - 1.57143(3)}{-1.36465 - 3} = 1.70541,$$

$$f_3 = f(1.70541) = -0.24776.$$

Since $f_1 f_3 < 0$, the root lies in the interval (x_3, x_1).

$k = 3$: $$x_4 = \frac{x_1 f_3 - x_3 f_1}{f_3 - f_1} = \frac{2(-0.24776) - (1.70541)(3)}{-0.24776 - 3} = 1.72788,$$

$$f_4 = f(1.72788) = -0.03936.$$

Since $f_1 f_4 < 0$, the root lies in the interval (x_4, x_1).

$k = 4$: $$x_5 = \frac{x_1 f_4 - x_4 f_1}{f_4 - f_1} = \frac{2(-0.03936) - 1.72788(3)}{-0.03936 - 3} = 1.73140,$$

$$f_5 = f(1.73140) = -0.0062.$$

Since $f_1 f_5 < 0$, the root lies in the interval (x_5, x_1).

$k = 5$: $$x_6 = \frac{x_1 f_5 - x_5 f_1}{f_5 - f_1} = \frac{2(-0.0062) - 1.7314(3)}{-0.0062 - 3} = 1.7319,$$

$$|x_6 - x_5| = 0.0005.$$

Hence, the root correct to three decimal places is 1.731.

13.2.4 Newton-Raphson Method

Consider the graph of $y = f(x)$ shown in Fig. 13.6. The exact root ξ occurs where the graph crosses the x-axis. Let x_0 be an initial approximation near the exact root ξ of the equation $f(x) = 0$.

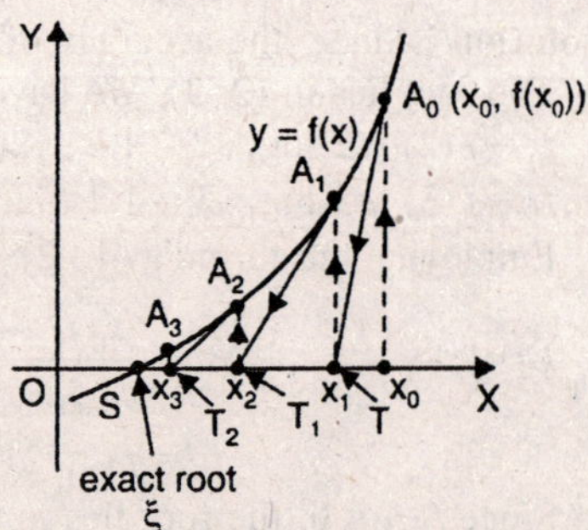

Fig. 13.6: Newton-Raphson method.

$A_0(x_0, f(x_0))$ is a point on the curve $y = f(x)$. Draw tangent to the curve at A_0. The underlying idea is that we approximate the curve $y = f(x)$ by the tangent to the curve at the point A_0 and take the point of intersection of this tangent with the x-axis as the next approximation to the root. The equation of the tangent to the curve $y = fx)$ at the point $(x_0, f(x_0))$ is given by

$$y - f(x_0) = f'(x_0)(x - x_0).$$

Setting $y = 0$ and solving for x, we obtain

$$x = x_0 - \frac{f(x_0)}{f'(x_0)}, \quad f'(x_0) \neq 0.$$

Hence, we obtain the next approximation to the root as

$$x_1 = x_0 - \frac{f(x_0)}{f'(x_0)}.$$

The iteration method is written as

$$x_{k+1} = x_k - \frac{f_k}{f'_k}, \quad k = 0, 1, 2, \ldots\ldots \tag{13.3}$$

Refer Fig. 13.6. OS represents the real root ξ of the equation $y = f(x)$ whose graph is drawn. Hence the quantities OT, OT_1, OT_2, ... are successive approximation to the desired root.

This method is called the *Newton-Raphson method* or the *tangent method.*

Remarks **(1)** The method requires one initial approximation.

(2) The method may fail if the initial approximation x_0 is far away from the root.

(3) The method has second order or quadratic convergence.

EXAMPLE 13.8 Perform three iterations of the Newton-Raphson method to find a root of the equation $x^4 - 12x + 7 = 0$, which is close to 2.

Solution We have $f(x) = x^4 - 12x + 7, \qquad f'(x) = 4x^3 - 12.$

Newton-Raphson method gives the iteration

$$x_{k+1} = x_k - \frac{f_k}{f'_k} = x_k - \frac{(x_k^4 - 12x_k + 7)}{(4x_k^3 - 12)}, \quad k = 0, 1, 2, \ldots$$

We obtain the following results

$k = 0$: $\quad x_0 = 2, f_0 = 2^4 - 12(2) + 7 = -1, \qquad f'_0 = 4(2)^3 - 12 = 20,$

$$x_1 = 2 - \frac{(-1)}{20} = 2.05.$$

$k = 1$: $\quad x_1 = 2.05, f_1 = (2.05)^4 - 12(2.05) + 7 = 0.06101, \qquad f'_1 = 4(2.05)^3 - 12 = 22.4605,$

$$x_2 = 2.05 - \frac{(0.06101)}{22.4605} = 2.04728.$$

$k = 2$: $\quad x_2 = 2.04728, f_2 = (2.04728)^4 - 12(2.04728) + 7 = 0.0001,$

$$f'_2 = 4(2.04728)^3 - 12 = 22.32351,$$

$$x_3 = 2.04728 - \frac{(0.001)}{22.32351} = 2.04724. \quad |x_3 - x_2| = 0.00004 < 0.0005$$

Hence, the root correct to three decimal places is 2.047.

EXAMPLE 13.9 Find the root of the equation $f(x) = x \sin x + \cos x = 0$ which is near $x = \pi$ correct to three decimal places using Newton-Raphson method. *(AMIETE, June 2009; VTU 2006)*

Solution We have $f(x) = x \sin x + \cos x, f'(x) = \sin x + x \cos x - \sin x = x \cos x.$

Newton-Raphson method gives the iteration

$$x_{k+1} = x_k - \frac{f_k}{f'_k} = x_k - \frac{(x_k \sin x_k + \cos x_k)}{x_k \cos x_k}, \quad k = 0, 1, 2, \ldots$$

We obtain the following results.

$k = 0$: $\quad x_0 = \pi = 3.1416, f_0 = \pi \sin \pi + \cos \pi = -1.0,$

$$f'_0 = \pi \cos \pi = -\pi = -3.1416,$$

$$x_1 = 3.1416 - \frac{(-1.0)}{(-3.1416)} = 2.8233.$$

$k = 1$: $\quad x_1 = 2.8233, \; f_1 = 2.8233 \sin 161.69^0 + \cos 161 \cdot 69^0$

$$= 2.8233(0.31399) - 0.94943 = -0.0629,$$

$$f'_1 = 3.1416 \cos 161.69^\circ = -3.1416(0.94943) = -2.98273,$$

$$x_2 = 2.8233 - \frac{(-0.0629)}{(-2.98273)} = 2.8022.$$

$k = 2$: $x_2 = 2.8022, f_2 = 2.8022 \sin 160.49° + \cos 160.49°$
$= 2.8022(0.33408) - 0.94254 = -0.00638,$

$f_2' = 2.8022 \cos 160.49° = -2.6412,$

$$x_3 = 2.8022 - \frac{(-0.00638)}{(-2.6412)} = 2.7997, \text{ etc.}$$

EXAMPLE 13.10 The equation $f(x) = \log x - x + 3 = 0$ has a root in the interval (4, 5). Obtain the root correct to three decimal places using Newton-Raphson method with $x_0 = 4.5$.

Solution We have $f(x) = \log_e x - x + 3,\ f'(x) = \frac{1}{x} - 1$

Newton-Raphson method gives the iteration

$$x_{k+1} = x_k - \frac{f_k}{f_k'} = x_k - \frac{(\log_e x_k - x_k + 3)}{\frac{1}{x_k} - 1}, \quad k = 0, 1, 2, \ldots$$

We obtain the following results

$k = 0$: $x_0 = 4.5,\ f_0 = \log_e 4.5 - 4.5 + 3 = 2.30258(0.65321) - 1.5 = 0.00407,$

$$f_0' = \frac{1}{4.5} - 1 = -0.77778, \quad x_1 = 4.5 - \frac{0.00407}{-0.77778} = 4.5052.$$

$k = 1$: $x_1 = 4.5052, f_1 = \log_e 4.5052 - 4.5052 + 3 = 2.30258(0.65371) - 1.5052 = -0.00002.$

$$f_1' = \frac{1}{4.5052} - 1 = -0.77803,$$

$$x_2 = 4.5052 - \left(\frac{-0.00002}{-0.77803}\right) = 4.5052.$$

Hence, the root correct to three decimal places is 4.505.

EXAMPLE 13.11 Using Newton-Raphson method, derive formulas to find (*i*) $1/N$, (*ii*) $N^{1/q}$, (*iii*) $(1/N)^{1/q}$, $N > 0$, q integer. Hence, find $1/17$, $(12)^{1/3}$, $(1/2)^{1/4}$ to four decimals. Use suitable initial approximations.

Solution (*i*) Let $x = \frac{1}{N}$, or $\frac{1}{x} - N = 0$. Taking $f(x) = \frac{1}{x} - N = 0$.

We have $f'(x) = -\frac{1}{x^2}$. Newton-Raphson method gives the iteration

$$x_{k+1} = x_k - \frac{f(x_k)}{f'(x_k)} = x_k - \frac{[(1/x_k) - N]}{(-1/x_k^2)} = 2x_k - Nx_k^2, \quad (k = 0, 1, 2, \ldots\ldots)$$

We have $N = 17$, $x = 1/17$. Let $x_0 = 0.05$. The method gives

$$x_{k+1} = 2x_k - 17x_k^2, \qquad k = 0, 1, 2, \ldots\ldots$$

We obtain

$$x_1 = 2x_0 - 17x_0^2 = 2(0.05) - 17(0.05)^2 = 0.0575,$$

$$x_2 = 2x_1 - 17x_1^2 = 0.115 - 0.05621 = 0.0588,$$

$$x_3 = 2x_2 - 17x_2^2 = 0.1176 - 0.05878 = 0.05882.$$

Since $|x_3 - x_2| = 0.00002$, we get $(1/17) \approx 0.0588$

(ii) Let $x = N^{1/q}$, or $x^q - N = 0, f(x) = x^q - N = 0.$

We have $f'(x) = qx^{q-1}$.

Newton-Raphson method gives the iteration

$$x_{k+1} = x_k - \frac{x_k^q - N}{qx_k^{q-1}} = \frac{(q-1)x_k^q + N}{qx_k^{q-1}}. \quad \text{...}(i)$$

Since $(8)^{1/3} = 2$ and $(27)^{1/3} = 3$, we have $N = 12$, $q = 3$.
Let $x_0 = 2.5$. Equation (i) become

$$x_{k+1} = \frac{2x_k^3 + 12}{3x_k^2}, \quad k = 0, 1, 2, \ldots$$

We obtain,

$$x_1 = \frac{2x_0^3 + 12}{3x_0^2} = \frac{2(2.5)^3 + 12}{3(2.5)^2} = \frac{43.25}{18.75} = 2.3066,$$

$$x_2 = \frac{2x_1^3 + 12}{3x_1^2} = \frac{2(2.3066)^3 + 12}{3(2.3066)^2} = \frac{36.544}{15.961} = 2.2895,$$

$$x_3 = \frac{2x_2^3 + 12}{3x_2^2} = \frac{36.0022}{15.7254} = 2.2894.$$

Since $|x_3 - x_2| = 0.0001$, we get the approximation as $\sqrt[3]{12} \approx 2.2894$.

(iii) Let $x = \left(\frac{1}{N}\right)^{\frac{1}{q}}$, $f(x) = N - x^{-q}$. We have $f'(x) = qx^{-q-1}$.

Newton-Raphson method gives

$$x_{k+1} = x_k - \frac{f(x_k)}{f'(x_k)} = x_k - \frac{N - x_k^{-q}}{qx_k^{-q-1}} = \frac{x_k(q + 1 - Nx_k^q)}{q}, \quad k = 0, 1, 2, \ldots$$

which is the iteration formula for finding the value of $(1/N)^{1/q}$.
We have $N = 2$, $q = 4$, take $x_0 = 0.8$.

We obtain

$$x_1 = \frac{x_0[5 - 2(x_0)^4]}{4} = \frac{0.8}{4}[5 - 2(0.8)^4] = 0.836,$$

$$x_2 = \frac{x_1}{4}[5 - 2(x_1)^4] = \frac{0.836}{4}[5 - 2(0.836)^4] = 0.84,$$

$$x_3 = \frac{x_2}{4}[5 - 2(x_2)^4] = \frac{0.84}{4}[5 - 2(0.84)^4] = 0.84.$$

We get the approximation of $(1/2)^{1/4}$ as 0.84.

13.2.5 Secant Method

Newton's method is based on approximating the graph of $y = f(x)$ with a tangent line and on then using the root of this straight line as an approximation to the root ξ of $f(x)$. But the evaluation of derivative involved may sometimes be difficult or time consuming. From Newton's method the *secant method* follows by replacing the derivative $f'(x_k)$ by the difference quotient

$$f'(x_k) \approx \frac{f(x_k) - f(x_{k-1})}{x_k - x_{k-1}}.$$

Doing so, we have the secant method

$$x_{k+1} = x_k - f(x_k)\frac{x_k - x_{k-1}}{f(x_k) - f(x_{k-1})}, \quad k \geq 1 \quad \text{...}(13.4)$$

This method is a two-point iteration method since two approximate values are needed to obtain an improved value and is called the *secant method* or the *chord method*.

Geometrically, we intersect the x-axis at x_{k+1} with the secant of $f(x)$ passing through P_{k-1} and P_k in Fig. 13.7. We need two starting values x_0 and x_1. Evaluation of derivatives is now avoided.

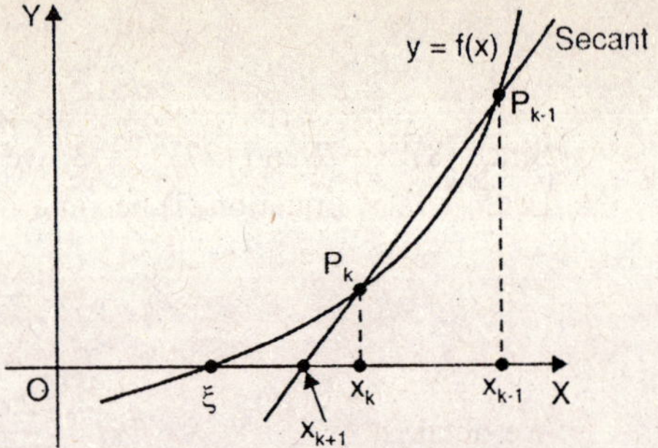

Fig. 13.7: Secant method.

Remark The method require two initial approximations to the root. The bisection method is also a two-point method, but the secant method will almost always converge faster than bisection.

EXAMPLE 13.12 Find the positive solution of $f(x) = \cos x - xe^x = 0$ by the secant method, starting from $x_0 = 0$, $x_1 = 1$.

Solution We have $f(0) = 1$, $f(1) = \cos 1 - e = \cos 57°18' - 2.71828 = -2.17798$.

Using secant method, we get the following results

$k = 1$:
$$x_2 = x_1 - f(x_1)\,\frac{x_1 - x_0}{f(x_1) - f(x_0)}$$
$$= 1 - (-2.17798)\frac{1}{-2.17798 - 1} = 0.31467,$$
$$|x_2 - x_1| = 0.68533,\; f(x_2) = f(0.31467) = 0.51987.$$

$k = 2$:
$$x_3 = x_2 - f(x_2)\,\frac{x_2 - x_1}{f(x_2) - f(x_1)}$$
$$= 0.31467 - (0.51987)\,\frac{-0.68533}{0.51987 + 2.17798} = 0.44673,$$
$$|x_3 - x_2| = 0.13206,\; f(x_3) = f(0.44673) = 0.20354.$$

$k = 3$:
$$x_4 = x_3 - f(x_3)\,\frac{x_3 - x_2}{f(x_3) - f(x_2)}$$
$$= 0.44673 - (0.20354)\,\frac{0.13206}{0.20354 - 0.51987} = 0.53171.$$

Repeating this process, the successive iteration are $x_5 = 0.51690$, $x_6 = 0.51775$, $x_7 = 0.51776$.
Hence the root is 0.5177.

13.2.6 General Iteration Method

The method is also called *successive approximation method* or *fixed point iteration method.* We write the equation $f(x) = 0$ in an equivalent form as

$$x = \phi(x) \qquad \text{...(13.5)}$$

and write an iteration method (recurrence relation) as

$$x_{k+1} = \phi(x_k), \quad k = 0, 1, 2, \ldots \qquad \text{...(13.6)}$$

The function $\phi(x)$ is called the *iteration function.*

Note that we can write the equation $f(x) = 0$ in the form $x = \phi(x)$ in several ways and can have several iteration methods of the form (13.6).

For example, consider the equation $x^3 - 5x - 11 = 0$, which has a root in the interval (2, 3). We write the given equation in the form $x = \phi(x)$ and the corresponding iteration method in the following ways:

(*i*) $x = \frac{1}{5}(x^3 - 11)$, and $x_{k+1} = \frac{1}{5}(x_k^3 - 11)$, $\quad k = 0, 1, 2, \ldots\ldots$ ...(A)

(*ii*) $x = (5x + 11)^{1/3}$, and $x_{k+1} = (5x_k + 11)^{1/3}$, $\quad k = 0, 1, 2, \ldots\ldots$ ...(B)

(*iii*) $x = x^3 - 4x - 11$, and $x_{k+1} = x_k^3 - 4x_k - 11$, $\quad k = 0, 1, 2, \ldots\ldots$ ...(C)

If we take $x_0 = 2{,}5$, we obtain from

Method (A): $x_1 = 0.925$, $x_2 = -2.04$, $x_3 = -3.898$, ... which is not converging to the root in (2, 3).

Method (B): $x_1 = 2.8642$, $x_2 = 2.9364$, $x_3 = 2.9503$, $x_4 = 2.9530$, ... which is converging to the root in (2, 3).

Method (C): $x_1 = -5.375$, $x_2 = -144.78$, ..., which is not converging to any root.

Hence, the convergence of the method of the form (13.6) depends on the suitable choice of iteration function $\phi(x)$ and the initial approximation x_0.

Condition for convergence

Method (A): $$\phi(x) = \frac{1}{5}(x^3 - 11),\ \phi'(x) = \frac{1}{5}\cdot(3x^2),$$
$$|\phi'(x)| > 1, \quad 2 < x < 3.$$

The iteration do not converge to the root in (2, 3).

Method (B): $$\phi(x) = (5x + 11)^{1/3},\ \phi'(x) = \frac{5}{3}(5x + 11)^{-2/3},$$
$$|\phi'(x)| < 1, \quad 2 < x < 3.$$

Therefore, the method converges to the root in (2, 3).

Method (C): $$\phi(x) = x^3 - 4x - 11,\ \phi'(x) = 3x^2 - 4,$$
$$|\phi'(x)| > 1, \quad 2 < x < 3.$$

The iteration do not converge to the root in (2, 3).

Thus the root of $x^3 - 5x - 11 = 0$ after four iterations work out to 2.953. (Refer *method* (B)).

Remark A simple method to put $f(x) = 0$ in the form $x = \phi(x)$ which has a root in the interval (α, β) and satisfying the condition of convergence, that is $|\phi'(x)| < 1$, $\alpha < x < \beta$ is as follows:

Put the equation $f(x) = 0$ in any one way in the form
$$x = g(x). \qquad ...(i)$$

If $|g'(x_0)| < 1$ for x_0 near the root, then multiply both sides of (i) by k. This gives
$$kx = kg(x). \qquad ...(ii)$$

Subtracting (*ii*) from (*i*), we obtain
$$x - kx = (1 - k)\, g(x) \text{ or } x = kx + (1 - k)\, g(x) = \phi(x), \text{ say} \qquad ...(iii)$$

Now we choose k such that $|\phi'(x_0)| < 1$. That is, $|k + (1 - k)\, g'(x_0)| < 1$

or in particular $k + (1 - k)\, g'(x_0) = 0$ which gives $k = g'(x_0)/[g'(x_0) - 1]$.

Substituting this value of k in right hand side of Eq. (iii), we get the required $\phi(x)$.

EXAMPLE 13.13 The equation $x^3 - 2x^2 - 5 = 0$ has a root in the interval (2, 3). Write this equation in an equivalent form $x = \phi(x)$ so that the general iteration method $x_{k+1} = \phi(x_k)$ is convergent. Hence, perform six iteration of this method starting with $x_0 = 3$, first verifying that the root lies in (2, 3).

Solution Let $f(x) = x^3 - 2x^2 - 5$. Since $f(2) = 8 - 8 - 5 = -5 < 0$ and $f(3) = 27 - 18 - 5 = 4 > 0$.

Therefore, the root of $f(x) = 0$ lies in the interval (2, 3).

We write the given equation in the form $x = 2 + (5/x^2)$. Here $\phi(x) = 2 + (5/x^2)$.

Therefore, $\phi'(x) = -10/x^3$. $\quad |\phi'(x_0)| = |-10/x_0^3| < 1$, taking $x_0 = 3$.

Therefore, the method converges to the root in (2, 3).

$$x_1 = 2 + \frac{5}{(3)^2} = 2.56; \quad x_2 = 2 + \frac{5}{(2.56)^2} = 2.763;$$
$$x_3 = 2 + \frac{5}{(2.763)^2} = 2.655; \quad x_4 = 2 + \frac{5}{(2.655)^2} = 2.709;$$
$$x_5 = 2 + \frac{5}{(2.709)^2} = 2.681; \quad x_6 = 2 + \frac{5}{(2.681)^2} = 2.695.$$

Hence root after six iterations = 2.695.

EXAMPLE 13.14 Find by the method of general iteration, the root near 3.8 of the equation $2x - \log_{10}x = 7$, correct to four decimal places.

Solution The given equation can be written in the form

$$x = \frac{1}{2}(\log_{10} x + 7), \text{ and } x_{k+1} = \frac{1}{2}(\log_{10} x_k + 7), \quad k = 0, 1, 2, \ldots$$

The iteration function, $\phi(x) = \frac{1}{2}(\log_{10} x + 7) = \frac{1}{2}(0.4343 \log_e x + 7)$,

$$\phi'(x) = \frac{1}{2}\left(\frac{0.4343}{x}\right) = \frac{0.2171}{x},$$

$$|\phi'(x_0)| = \left|\frac{0.2171}{x_0}\right| < 1, \quad x_0 = 3.8.$$

Therefore, the method converges to the root near 3.8.

We obtain from $x_{k+1} = \phi(x_k)$, $k = 0, 1, 2, \ldots$

$$x_1 = \frac{1}{2}[\log_{10} 3.8 + 7] = 3.79, \quad x_2 = \frac{1}{2}[\log_{10} 3.79 + 7] = 3.7893,$$

$$x_3 = \frac{1}{2}[\log_{10} 3.7893 + 7] = 3.7892, \ldots$$

We take $x = 3.7892$ as the correct root up to four decimal places.

EXAMPLE 13.15 Use the general iteration method to find the smallest positive root of the equation $2x - \cos x - 3 = 0$, which has a root in the interval $(0, \pi/2)$ correct to three decimal places. *(AMIE, S 2004)*

Solution We rewrite the equation in the form $x = \frac{1}{2}(\cos x + 3)$ so that $\phi(x) = \frac{1}{2}(\cos x + 3)$.

$$\phi'(x) = \frac{-\sin x}{2}, \quad |\phi'(x)| = \left|\frac{\sin x}{2}\right| < 1, \quad \left(0 < x < \frac{\pi}{2}\right).$$

Hence, the iteration method can be applied and we take $x_0 = \pi/2$. The successive iterates are as follows:

$$x_1 = \frac{1}{2}\left(\cos\frac{\pi}{2} + 3\right) = \frac{1}{2}(3) = 1.5, \quad x_2 = \frac{1}{2}(\cos 85°57' + 3) = \frac{1}{2}(0.07074 + 3) = 1.535 \qquad (\because 1.5r = 85°57')$$

$x_3 = 1.518$ $\qquad x_4 = 1.526$ $\qquad x_5 = 1.522$

$x_6 = 1.524$ $\qquad x_7 = 1.523$ $\qquad x_8 = 1.524$

..., which is a convergent sequence.

Hence, we take the root as 1.524, correct to three decimal places.

EXERCISE 13.1

1. Perform three iterations of the bisection method to obtain the root of the following equations.
   (*i*) $x^3 - 2x - 5 = 0$ *(PTU, 2005)*
   (*ii*) $x^3 - 5x + 1 = 0$ lies in the interval (0, 1)
   (*iii*) $x^3 - 9x + 1 = 0$ lies in the interval (2, 4)

   Ans. (*i*) 2.687, (*ii*) 0.1875, (*iii*) 2.937

2. Using regula-falsi method, find the real root of the following equations correct to three decimal places.
   (*i*) $x^3 - 4x - 9 = 0$ *(VTU, 2007)*
   (*ii*) $xe^x - 2 = 0$ *(SVTU 2007; NIT, Kurukshetra 2007)*
   (*iii*) $x^3 - 2x - 5 = 0$ *(Manipal, 2005)*
   (*iv*) $2x - \log_{10}x = 7$, root lies between 3.5 and 4.0
   (*v*) $xe^x = \cos x$ in (0, 1) *(NIT Kurukshetra 2008)*

   Ans. (*i*) 2.706, (*ii*) 0.853, (*iii*) 2.094, (*iv*) 3.789, (*v*) 0.518

3. Find a real root of the following equations correct to three decimal places by Newton-Raphson method.
   (*i*) $x^3 - 3x + 4 = 0$,

(*ii*) $xe^x = \cos x$ (*VTU 2003; Madras 2000*)
(*iii*) $e^x - 3x = 0$ lying between 0.4 and 0.9
(*iv*) $x \log_{10} x = 1.2$ (*VTU 2005; Burdwan 2003*)
(*v*) $3x = \cos x + 1$ (*SVTU 2007; VTU 2003 S; Bhopal 2002 S*)
(*vi*) $e^{-x} - \sin x = 0$,
(*vii*) $x^3 - 7x + 10 = 0$, root lies between $x = 0$ and $x = 1$.

Ans. (*i*) –2.196, (*ii*) 0.518, (*iii*) 0.619, (*iv*) 2.741, (*v*) 0.607, (*vi*) 0.594, (*vii*) 0.143.

4. Using Newton-Raphson method, establish the formula $x_{k+1} = \frac{1}{2}\left(x_k + \frac{N}{x_k}\right)$ to calculate the square root of N.
5. Evaluate the following correct to three decimal places by using the Newton-Raphson method
(*i*) 1/14 (*ii*) 1/18 (*AMIETE, Dec, 2010*)
(*iii*) $\sqrt{29}$ (*iv*) $(10)^{1/3}$
(*v*) $(18)^{1/3}$ (*AMIETE, Dec, 2010*) (*vi*) $(1/3)^{1/4}$ (*vii*) $(32)^{1/4}$
Ans. (*i*) 0.0714, (*ii*) 0.056, (*iii*) 5.385, (*iv*) 2.154, (*v*) 2.621, (*vi*) 0.759 (*vii*) 2.378.
6. Find the positive root correct to three decimal places of $f(x) = x - 2 \sin x = 0$ by the secant method starting from $x_0 = 2$, $x_1 = 1.9$. **Ans.** 1.895
7. A root of the equation $xe^x - 1 = 0$ lies in the interval (0.5, 1.0). Determine this root correct to three decimal places using secant method. **Ans.** 0.567
8. Find a real root of the following equation correct to three decimal places using general iteration method.
(*i*) $f(x) = x^3 + x^2 - 1 = 0$, (*ii*) $x^3 - 3x + 4 = 0$
(*iii*) $x^3 - 5x - 1 = 0$, root lies in the interval (–1, 0)
(*iv*) $x^3 - 7x + 1 = 0$, root lies in the interval (2, 3) (*v*) $\cos x = 3x$.
Ans. (*i*) 0.607, (*ii*) –2.196, (*iii*) –0.202, (*iv*) 2.570, (*v*) 0.317.
9. Show that there is a root of the equation $1 + \log_e r = 0.5\, r$ between $r = 5$ and $r = 6$ and evaluate this root correct to two decimal places. **Ans.** $r = 5.36$
10. By using the first two terms of the series for $\cos x$, find an approximate value of the positive root of the equation $\cos x = 3x$. Find this root correct to four decimal places. **Ans.** 0.3167

[**Hint:** Using the series for $\cos x$, that is, $1 - \frac{x^2}{2} + \cdots$ and set this equal to $3x$, we get $x^2 + 6x - 2 = 0$.

Therefore, $x = -3 \pm \sqrt{11} = -3 \pm 3.317$. Positive root is approximately 0.317.

$$x_2 = x_1 - \frac{f(x_1)}{f'(x_1)} = 0.317 - \frac{\cos(0.317) - 3\,(0.317)}{-\sin(0.317) - 3}$$

$$= 0.317 - \frac{0.9501 - 0.9510}{-3.3118} \qquad (\because 0.317^r = 18°10')$$

$$= 0.3167].$$

11. Find an approximate value of the root of $f(x) = xe^{-x} - \cos x = 0$ in the neighbourhood of $x = 1$. **Ans.** 1.20047

13.3 SOLUTION OF LINEAR SYSTEM OF EQUATIONS

The numerical methods for solution of linear system of equations can be classified as (*i*) *direct methods* and (*ii*) *indirect or iterative methods*.

Direct methods These methods give exact solutions after a finite number of arithmetical operations.

Indirect or iterative methods These methods are based on successive approximations. We start with an approximation to the true solution and by applying the method repeatedly we get a better and better approximation till required accuracy has been achieved.

Direct methods

Diagonal system We illustrate the methods for a 3 × 3 system of equations. First, consider the system of equations.

$$\begin{aligned} a_{11}x_1 \qquad\qquad &= b_1 \\ a_{22}x_2 \qquad &= b_2 \\ a_{33}x_3 &= b_3 \end{aligned} \qquad ...(i)$$

or **Dx** = **b**, where **D** is a diagonal matrix. The solution of the system is $x_i = b_1/a_{ii}$, $i = 1, 2, 3$.

Lower triangular system Consider the system of equations

$$\begin{aligned} a_{11}x_1 \qquad\qquad\qquad\qquad &= b_1 \\ a_{21}x_1 + a_{22}x_2 \qquad\qquad &= b_2 \\ a_{31}x_1 + a_{32}x_2 + a_{33}x_3 &= b_3 \end{aligned} \qquad ...(ii)$$

or **Lx** = **b**, where **L** is a lower triangular matrix. The solution of the system is obtained by *forward substitution*, as.

$$x_1 = \frac{b_1}{a_{11}}, \quad x_2 = \frac{1}{a_{22}}(b_2 - a_{21}x_1), \quad x_3 = \frac{1}{a_{33}}(b_3 - a_{31}x_1 - a_{32}x_2).$$

Upper triangular system

Consider the system of equations

$$\begin{aligned} a_{11}x_1 + a_{12}x_2 + a_{13}x_3 &= b_1 \\ a_{22}x_2 + a_{23}x_3 &= b_2 \\ a_{33}x_3 &= b_3 \end{aligned} \qquad ...(iii)$$

or **Ux** = **b**, where **U** is an upper triangular matrix. The solution of the system is obtained by *backward substitution* as

$$x_3 = \frac{b_3}{a_{33}}, \quad x_2 = \frac{1}{a_{22}}(b_2 - a_{23}x_3), \quad x_1 = \frac{1}{a_{11}}(b_1 - a_{12}x_2 - a_{13}x_3).$$

Hence, if the given system is in either of the three form (i) to (iii), then the solution is directly obtained. All the direct methods reduce the given system of equations to one of the above three forms and then the solution is obtained.

We shall discuss the following **direct methods** for solution of linear simultaneous equations.

(*a*) Gauss elimination method

(*b*) Gauss-Jordan method

(*c*) LU-factorization or decomposition method

Further, Doolittle's, Crout's and cholesky's methods are variants of the Gauss elimination. They use the idea of **A** = **LU**-factorization or decomposition.

13.3.1 Gauss Elimination Method

In this method, we write the augmented matrix [**A**|**b**] and perform the elementary row transformation to reduce the system to the form [**U**|**c**], where **U** is an upper triangular matrix and **c** is the new column vector. Using backward substitution method, we obtain the solution values $x_n, x_{n-1}, ..., x_1$. We illustrate the method for the following 3 × 3 system of equations.

$$\begin{aligned} a_{11}x_1 + a_{12}x_2 + a_{13}x_3 &= b_1, \\ a_{21}x_1 + a_{22}x_2 + a_{23}x_3 &= b_2, \\ a_{31}x_1 + a_{32}x_2 + a_{33}x_3 &= b_3. \end{aligned}$$

We write the augmented matrix as

$$[\mathbf{A}|\mathbf{b}] = \left[\begin{array}{ccc|c} a_{11} & a_{12} & a_{13} & b_1 \\ a_{21} & a_{22} & a_{23} & b_2 \\ a_{31} & a_{32} & a_{33} & b_3 \end{array}\right].$$

First step: The element $a_{11} \neq 0$ in the 1×1 position is called the *first pivot*. We make all elements in the first column, below a_{11} as zeros, using elementary row operations. We obtain the equivalent system as

$$[\mathbf{A}|\mathbf{b}] \approx \left[\begin{array}{ccc|c} a_{11} & a_{12} & a_{13} & b_1 \\ 0 & a_{22}^{(1)} & a_{23}^{(1)} & b_2^{(1)} \\ 0 & a_{32}^{(1)} & a_{33}^{(1)} & b_3^{(1)} \end{array}\right].$$

Second step: The element $a_{22}^{(1)} \neq 0$ in the 2×2 position is called the *second pivot*. We make all elements in the second column below $a_{22}^{(1)}$ as zeros, using elementary row operations. We obtain the equivalent system as

$$[\mathbf{A}|\mathbf{b}] = \left[\begin{array}{ccc|c} a_{11} & a_{12} & a_{13} & b_1 \\ 0 & a_{22}^{(1)} & a_{23}^{(1)} & b_2^{(1)} \\ 0 & 0 & a_{33}^{(2)} & b_3^{(2)} \end{array}\right] \approx [\mathbf{U}\,|\,\mathbf{c}]$$

The element $a_{33}^{(2)} \neq 0$ is called the *third pivot*.

We solve the system by backward substitution method.

From the third row, we obtain $a_{33}^{(2)} x_3 = b_3^{(2)}$, $x_3 = b_3^{(2)} / a_{33}^{(2)}$.

From the second row, we obtain $a_{22}^{(1)} x_2 + a_{23}^{(1)} x_3 = b_2^{(1)}$, $\quad x_2 = \dfrac{1}{a_{22}^{(1)}}[b_2^{(1)} - a_{23}^{(1)} x_3]$.

From the first row, we obtain $a_{11}x_1 + a_{12}x_2 + a_{13}x_3 = b_1$, $\quad x_1 = \dfrac{1}{a_{11}}[b_1 - a_{12}x_2 - a_{13}x_3]$.

If at any stage of elimination, the pivot element become zero, then we interchange this row with any other row below it such that we obtain a non-zero pivot element.

EXAMPLE 13.16 Solve the following system of equations using Gauss elimination method.

(*i*) $2x_1 + x_2 + x_3 = 10,$
$3x_1 + 2x_2 + 3x_3 = 18,$
$x_1 + 4x_2 + 9x_3 = 16.$

(*ii*) $2x_1 + 2x_2 + x_3 = 12,$
$3x_1 + 2x_2 + 2x_3 = 8,$
$5x_1 + 10x_2 - 8x_3 = 10.$

Solution (*i*) We write the augmented matrix $[\mathbf{A}|\mathbf{b}]$ and perform the elementary row transformations to reduce the system to the form $[\mathbf{U}|\mathbf{c}]$.

$$[\mathbf{A}|\mathbf{b}] = \left[\begin{array}{ccc|c} 2 & 1 & 1 & 10 \\ 3 & 2 & 3 & 18 \\ 1 & 4 & 9 & 16 \end{array}\right], \; R_1 \leftrightarrow R_3 \approx \left[\begin{array}{ccc|c} 1 & 4 & 9 & 16 \\ 2 & 1 & 1 & 10 \\ 3 & 2 & 3 & 18 \end{array}\right], \; R_2 - 2R_1,\; R_3 - 3R_1$$

$$\approx \left[\begin{array}{ccc|c} 1 & 4 & 9 & 16 \\ 0 & -7 & -17 & -22 \\ 0 & -10 & -24 & -30 \end{array}\right], \; R_3 - \frac{10}{7}R_2 \approx \left[\begin{array}{ccc|c} 1 & 4 & 9 & 16 \\ 0 & -7 & -17 & -22 \\ 0 & 0 & 2/7 & 10/7 \end{array}\right].$$

Using back substitution method, we obtain the solution as

$$x_3 = 5, \quad x_2 = \frac{1}{7}[22 - 17(5)] = -9, \quad x_1 = 16 - 4(-9) - 9(5) = 7.$$

The solution vector is given by $\mathbf{x} = [7, -9, 5]^T$.

(ii) We perform elementary row transformations on the augmented matrix $[\mathbf{A}/\mathbf{b}]$ to reduce the system to the form $[\mathbf{U}|\mathbf{c}]$

$$[\mathbf{A}|\mathbf{b}] = \left[\begin{array}{ccc|c} 2 & 2 & 1 & 12 \\ 3 & 2 & 2 & 8 \\ 5 & 10 & -8 & 10 \end{array}\right], \; R_2 - \frac{3}{2}R_1, R_3 - \frac{5}{2}R_1 \approx \left[\begin{array}{ccc|c} 2 & 2 & 1 & 12 \\ 0 & -1 & 1/2 & -10 \\ 0 & 5 & \dfrac{-21}{2} & -20 \end{array}\right] \; R_3 + 5R_2$$

$$\approx \left[\begin{array}{ccc|c} 2 & 2 & 1 & 12 \\ 0 & -1 & 1/2 & -10 \\ 0 & 0 & -8 & -70 \end{array}\right].$$

Using back substitution method, we obtain the solution as

$$x_3 = \frac{35}{4}, \qquad x_2 = 10 + \frac{1}{2}\left(\frac{35}{4}\right) = \frac{115}{8}, \qquad x_1 = \frac{1}{2}\left[12 - 2\left(\frac{115}{8}\right) - \frac{35}{4}\right] = -\frac{51}{4}.$$

The solution vector is given by $\mathbf{x} = \left[\dfrac{-51}{4}, \dfrac{115}{8}, \dfrac{35}{4}\right]^T$.

EXAMPLE 13.17 Solve the system of equations $\begin{bmatrix} 2 & 1 & 2 & 1 \\ 1 & -1 & 1 & 2 \\ 4 & 3 & 3 & -3 \\ 2 & 2 & -1 & 1 \end{bmatrix} \begin{bmatrix} x_1 \\ x_2 \\ x_3 \\ x_4 \end{bmatrix} = \begin{bmatrix} 6 \\ 6 \\ -1 \\ 10 \end{bmatrix}$,

using Gauss elimination method.

Solution We write the augmented matrix **[A|b]** and perform the elementary row transformations to reduce the system to the form **[U|c]**.

$$[\mathbf{A}|\mathbf{b}] = \left[\begin{array}{cccc|c} 2 & 1 & 2 & 1 & 6 \\ 1 & -1 & 1 & 2 & 6 \\ 4 & 3 & 3 & -3 & -1 \\ 2 & 2 & -1 & 1 & 10 \end{array}\right], R_1 \leftrightarrow R_2 \approx \left[\begin{array}{cccc|c} 1 & -1 & 1 & 2 & 6 \\ 2 & 1 & 2 & 1 & 6 \\ 4 & 3 & 3 & -3 & -1 \\ 2 & 2 & -1 & 1 & 10 \end{array}\right], \; R_2 - 2R_1, \; R_3 - 4R_1, \; R_4 - 2R_1$$

$$\approx \left[\begin{array}{cccc|c} 1 & -1 & 1 & 2 & 6 \\ 0 & 3 & 0 & -3 & -6 \\ 0 & 7 & -1 & -11 & -25 \\ 0 & 4 & -3 & -3 & -2 \end{array}\right], \; R_3 - \frac{7}{3}R_2, R_4 - \frac{4}{3}R_2 \approx \left[\begin{array}{cccc|c} 1 & -1 & 1 & 2 & 6 \\ 0 & 3 & 0 & -3 & -6 \\ 0 & 0 & -1 & -4 & -11 \\ 0 & 0 & -3 & 1 & 6 \end{array}\right], \; R_4 - 3R_3$$

$$\approx \left[\begin{array}{cccc|c} 1 & -1 & 1 & 2 & 6 \\ 0 & 3 & 0 & -3 & -6 \\ 0 & 0 & -1 & -4 & -11 \\ 0 & 0 & 0 & 13 & 39 \end{array}\right].$$

Using back substitution method, we obtain

$x_4 = 3$, $x_3 = 11 - 4(3) = -1$, $x_2 = \frac{1}{3}[-6 + 3(3)] = 1$, $x_1 = 6 + 1 - (-1) - 2(3) = 2$.

The solution vector is given by $\mathbf{x} = [2, 1, -1, 3]^T$

EXAMPLE 13.18 Using the Gauss elimination method, show that the following system of equations is inconsistent

$$\begin{bmatrix} 2 & -3 & 7 \\ 3 & 1 & -3 \\ 2 & 19 & -47 \end{bmatrix} \begin{bmatrix} x_1 \\ x_2 \\ x_3 \end{bmatrix} = \begin{bmatrix} 5 \\ 13 \\ 32 \end{bmatrix}.$$

Solution We write the augmented matrix [**A**|**b**] and perform the elementary row transformations to reduce the system to the form [**U**|**c**].

$$[\mathbf{A}|\mathbf{b}] = \left[\begin{array}{ccc|c} 2 & -3 & 7 & 5 \\ 3 & 1 & -3 & 13 \\ 2 & 19 & -47 & 32 \end{array}\right],\ R_2 - \frac{3}{2}R_1,\ R_3 - R_1 \approx \left[\begin{array}{ccc|c} 2 & -3 & 7 & 5 \\ 0 & 11/2 & -27/2 & 11/2 \\ 0 & 22 & -54 & 27 \end{array}\right],\ R_3 - 4R_2$$

$$\approx \left[\begin{array}{ccc|c} 2 & -3 & 7 & 5 \\ 0 & 11/2 & -27/2 & 11/2 \\ 0 & 0 & 0 & 5 \end{array}\right].$$

The third row of the last augmented matrix means $0x_1 + 0x_2 + 0x_3 = 5$. Since this implies $0 = 5$, no numbers x_1, x_2, x_3 can satisfy this equation. We conclude that the system has no solution inconsistent system.

13.3.2 Gauss-Jordan Method

In this method, we reduce the augmented matrix [**A**|**b**] of the given system of equations to the form [**I**|**c**] by using elementary row transformations where **I** is the identity matrix and **c** is the solution vector. The reduction is equivalent to finding the solution as $\mathbf{x} = \mathbf{A}^{-1}\mathbf{b}$.

The first step is the same as in the Gauss elimination method. From second step onwards, we make elements below and above the pivot as zeros, using elementary row transformation. Finally, we divide each row by its pivot to obtain the form [**I**|**c**]. Then, **c** is the solution vector.

It has been seen that Gauss-Jordan elimination method require larger operations than Gaussian elimination. Hence, we do not normally use the Gauss-Jordan method for finding the solution of a system. However, this method is very useful for finding the inverse ($\mathbf{A}^{-1}$) of a matrix **A**. We consider the augmented matrix [**A**|**I**] and reduce it to the form [$\mathbf{I}|\mathbf{A}^{-1}$] using elementary row transformations. The aim is to reduce **A** to **I**. As soon as this is achieved, other matrix gives $\mathbf{A}^{-1}$. The solution vector is determined by $\mathbf{x} = \mathbf{A}^{-1}\mathbf{b}$.

EXAMPLE 13.19 Solve the following systems of equations (if possible) using Gauss-Jordon method.

$$(i)\ \begin{bmatrix} 1 & -1 & -1 \\ 2 & 3 & 5 \\ 1 & -2 & 3 \end{bmatrix}\begin{bmatrix} x_1 \\ x_2 \\ x_3 \end{bmatrix} = \begin{bmatrix} -3 \\ 7 \\ -11 \end{bmatrix},\quad (ii)\ \begin{bmatrix} 1 & -1 & -1 \\ 1 & -1 & 1 \\ -1 & 1 & 1 \end{bmatrix}\begin{bmatrix} x_1 \\ x_2 \\ x_3 \end{bmatrix} = \begin{bmatrix} 8 \\ 3 \\ 4 \end{bmatrix},\quad (iii)\ \begin{bmatrix} 1 & 3 & -2 \\ 4 & 1 & 3 \\ 2 & -5 & 7 \end{bmatrix}\begin{bmatrix} x_1 \\ x_2 \\ x_3 \end{bmatrix} = \begin{bmatrix} -7 \\ 5 \\ 19 \end{bmatrix}.$$

Solution (*i*) We write the augmented matrix [**A**|**b**] and perform the elementary row transformations to reduce the system to the form [**I**|**c**].

$$[\mathbf{A}|\mathbf{b}] = \left[\begin{array}{ccc|c} 1 & -1 & -1 & -3 \\ 2 & 3 & 5 & 7 \\ 1 & -2 & 3 & -11 \end{array}\right],\ R_2 - 2R_1,\ R_3 - R_1$$

$$\approx \left[\begin{array}{ccc|c} 1 & -1 & -1 & -3 \\ 0 & 5 & 7 & 13 \\ 0 & -1 & 4 & -8 \end{array}\right],\ R_2/5 \approx \left[\begin{array}{ccc|c} 1 & -1 & -1 & -3 \\ 0 & 1 & 7/5 & 13/5 \\ 0 & -1 & 4 & -8 \end{array}\right],\ R_1 + R_2,\ R_3 + R_2$$

$$\approx \left[\begin{array}{ccc|c} 1 & 0 & 2/5 & -2/5 \\ 0 & 1 & 7/5 & 13/5 \\ 0 & 0 & 27/5 & -27/5 \end{array}\right],\ \frac{5}{27}R_3 \approx \left[\begin{array}{ccc|c} 1 & 0 & 2/5 & -2/5 \\ 0 & 1 & 7/5 & 13/5 \\ 0 & 0 & 1 & -1 \end{array}\right],\ R_1 - \frac{2}{5}R_3,\ R_2 - \frac{7}{5}R_3$$

$$\approx \left[\begin{array}{ccc|c} 1 & 0 & 0 & 0 \\ 0 & 1 & 0 & 4 \\ 0 & 0 & 1 & -1 \end{array}\right].$$ The solution vector is given by $\mathbf{x} = [0, 4, -1]T$.

$$(ii)\ [\mathbf{A}|\mathbf{b}] = \left[\begin{array}{ccc|c} 1 & -1 & -1 & 8 \\ 1 & -1 & 1 & 3 \\ -1 & 1 & 1 & 4 \end{array}\right],\ R_2 - R_1, R_3 + R_1 \approx \left[\begin{array}{ccc|c} 1 & -1 & -1 & 8 \\ 0 & 0 & 2 & -5 \\ 0 & 0 & 0 & 12 \end{array}\right].$$

Gauss Jordon gives the last row of the system as 0, 0, 0, 12. Since the implies 0 = 12, the system has no solution-inconsistent.

$$(iii)\ [\mathbf{A}|\mathbf{b}] = \left[\begin{array}{ccc|c} 1 & 3 & -2 & -7 \\ 4 & 1 & 3 & 5 \\ 2 & -5 & 7 & 19 \end{array}\right],\ R_2 - 4R_1, R_3 - 2R_1 \approx \left[\begin{array}{ccc|c} 1 & 3 & -2 & -7 \\ 0 & -11 & 11 & 33 \\ 0 & -11 & 11 & 33 \end{array}\right],\ -R_2/11, -R_3/11$$

$$= \left[\begin{array}{ccc|c} 1 & 3 & -2 & -7 \\ 0 & 1 & -1 & -3 \\ 0 & 1 & -1 & -3 \end{array}\right],\ R_1 - 3R_2, R_3 - R_2 \approx \left[\begin{array}{ccc|c} 1 & 0 & 1 & 2 \\ 0 & 1 & -1 & -3 \\ 0 & 0 & 0 & 0 \end{array}\right].$$

The system is consistent and has infinite number of solutions. We find that the last equation is satisfied for all values of x_1, x_2, x_3. From the second equation, we get $x_2 - x_3 = -3$ or $x_2 = x_3 - 3$. From the first equation, we get $x_1 + x_3 = 2$, or $x_1 = 2 - x_3$. Therefore, we obtain the solution $x_1 = 2 - x_3, x_2 = x_3 - 3$ and x_3 is arbitrary.

EXAMPLE 13.20 Find the inverse of the following matrices using Gauss-Jordan method.

$$(i)\ \mathbf{A} = \begin{bmatrix} 4 & 2 & -1 \\ 3 & -2 & 1 \\ 2 & 4 & -3 \end{bmatrix}, \qquad (ii)\ \mathbf{A} = \begin{bmatrix} 2 & 3 & 4 \\ 4 & 3 & 1 \\ 1 & 2 & 4 \end{bmatrix}.$$

Solution (*i*) We perform elementary row transformations on the augmented matrix **[A|I]** to reduce it to the form **[I|B]**. Then $\mathbf{B} = \mathbf{A}^{-1}$.

$$[\mathbf{A}|\mathbf{I}] = \left[\begin{array}{ccc|ccc} 4 & 2 & -1 & 1 & 0 & 0 \\ 3 & -2 & 1 & 0 & 1 & 0 \\ 2 & 4 & -3 & 0 & 0 & 1 \end{array}\right],\ R_1/4 \approx \left[\begin{array}{ccc|ccc} 1 & 1/2 & -1/4 & 1/4 & 0 & 0 \\ 3 & -2 & 1 & 0 & 1 & 0 \\ 2 & 4 & -3 & 0 & 0 & 1 \end{array}\right],\ R_2 - 3R_1\ R_3 - 2R_1$$

$$\approx \left[\begin{array}{ccc|ccc} 1 & 1/2 & -1/4 & 1/4 & 0 & 0 \\ 0 & -7/2 & 7/4 & -3/4 & 1 & 0 \\ 0 & 3 & -5/2 & -1/2 & 0 & 1 \end{array}\right],\ R_2 \leftrightarrow R_3 \approx \left[\begin{array}{ccc|ccc} 1 & 1/2 & -1/4 & 1/4 & 0 & 0 \\ 0 & 3 & -5/2 & -1/2 & 0 & 1 \\ 0 & -7/2 & 7/4 & -3/4 & 1 & 0 \end{array}\right],\ R_2/3$$

$$\approx \left[\begin{array}{ccc|ccc} 1 & 1/2 & -1/4 & 1/4 & 0 & 0 \\ 0 & 1 & -5/6 & -1/6 & 0 & 1/3 \\ 0 & -7/2 & 7/4 & -3/4 & 1 & 0 \end{array}\right],\ R_1 - \frac{1}{2}R_2,\ R_3 + \frac{7}{2}R_2$$

$$\approx \left[\begin{array}{ccc|ccc} 1 & 0 & 1/6 & 1/3 & 0 & -1/6 \\ 0 & 1 & -5/6 & -1/6 & 0 & 1/3 \\ 0 & 0 & -7/6 & -4/3 & 1 & 7/6 \end{array}\right],\ -\frac{6}{7}R_3 \approx \left[\begin{array}{ccc|ccc} 1 & 0 & 1/6 & 1/3 & 0 & -1/6 \\ 0 & 1 & -5/6 & -1/6 & 0 & 1/3 \\ 0 & 0 & 1 & 8/7 & -6/7 & -1 \end{array}\right],\ R_1 - \frac{1}{6}R_3, R_2 + \frac{5}{6}R_3$$

$$\approx \left[\begin{array}{ccc|ccc} 1 & 0 & 0 & 1/7 & 1/7 & 0 \\ 0 & 1 & 0 & 11/14 & -5/7 & -1/2 \\ 0 & 0 & 1 & 8/7 & -6/7 & -1 \end{array}\right] = [\mathbf{I}\,|\,\mathbf{B}].$$

$$\text{Hence, } \mathbf{A}^{-1} = \mathbf{B} = \begin{bmatrix} 1/7 & 1/7 & 0 \\ 11/14 & -5/7 & -1/2 \\ 8/7 & -6/7 & -1 \end{bmatrix} = \frac{1}{14}\begin{bmatrix} 2 & 2 & 0 \\ 11 & -10 & -7 \\ 16 & -12 & -14 \end{bmatrix}.$$

(*ii*) We have $[\mathbf{A}|\mathbf{I}] = \left[\begin{array}{ccc|ccc} 2 & 3 & 4 & 1 & 0 & 0 \\ 4 & 3 & 1 & 0 & 1 & 0 \\ 1 & 2 & 4 & 0 & 0 & 1 \end{array}\right], R_1 \leftrightarrow R_3$

$$\approx \left[\begin{array}{ccc|ccc} 1 & 2 & 4 & 0 & 0 & 1 \\ 4 & 3 & 1 & 0 & 1 & 0 \\ 2 & 3 & 4 & 1 & 0 & 0 \end{array}\right], R_2 - 4R_1,\ R_3 - 2R_1$$

$$\approx \left[\begin{array}{ccc|ccc} 1 & 2 & 4 & 0 & 0 & 1 \\ 0 & -5 & -15 & 0 & 1 & -4 \\ 0 & -1 & -4 & 1 & 0 & -2 \end{array}\right], -R_2/5, -R_3$$

$$\approx \left[\begin{array}{ccc|ccc} 1 & 2 & 4 & 0 & 0 & 1 \\ 0 & 1 & 3 & 0 & -1/5 & 4/5 \\ 0 & 1 & 4 & -1 & 0 & 2 \end{array}\right], R_1 - 2R_2, R_3 - R_2$$

$$\approx \left[\begin{array}{ccc|ccc} 1 & 0 & -2 & 0 & 2/5 & -3/5 \\ 0 & 1 & 3 & 0 & -1/5 & 4/5 \\ 0 & 0 & 1 & -1 & 1/5 & 6/5 \end{array}\right], R_1 + 2R_3, R_2 - 3R_3$$

$$\approx \left[\begin{array}{ccc|ccc} 1 & 0 & 0 & -2 & 4/5 & 9/5 \\ 0 & 1 & 0 & 3 & -4/5 & -14/5 \\ 0 & 0 & 1 & -1 & 1/5 & 6/5 \end{array}\right] = [\mathbf{I}|\mathbf{B}]$$

Hence, $\mathbf{A}^{-1} = \mathbf{B} = \begin{bmatrix} -2 & 4/5 & 9/5 \\ 3 & -4/5 & -14/5 \\ -1 & 1/5 & 6/5 \end{bmatrix} = \dfrac{1}{5}\begin{bmatrix} -10 & 4 & 9 \\ 15 & -4 & -14 \\ -5 & 1 & 6 \end{bmatrix}$

EXAMPLE 13.21 Find the inverse of the matrix using Gauss-Jordan method, $\mathbf{A} = \begin{bmatrix} 4 & 1 & 1 \\ 1 & 3 & 1 \\ 2 & 1 & 5 \end{bmatrix}$.

Hence, solve the system of equations $\mathbf{Ax} = \mathbf{b}$, where $\mathbf{b} = [7, 7, 1]^T$.

Solution We perform elementary row transformations on the augmented matrix [A|I] to reduce it to the form [I|B]. Then $\mathbf{B} = \mathbf{A}^{-1}$.

$$[\mathbf{A}|\mathbf{I}] = \left[\begin{array}{ccc|ccc} 4 & 1 & 1 & 1 & 0 & 0 \\ 1 & 3 & 1 & 0 & 1 & 0 \\ 2 & 1 & 5 & 0 & 0 & 1 \end{array}\right], R_2 \leftrightarrow R_1$$

$$\approx \left[\begin{array}{ccc|ccc} 1 & 3 & 1 & 0 & 1 & 0 \\ 4 & 1 & 1 & 1 & 0 & 0 \\ 2 & 1 & 5 & 0 & 0 & 1 \end{array}\right], R_2 - 4R_1, R_3 - 2R_1$$

$$\approx \left[\begin{array}{ccc|ccc} 1 & 3 & 1 & 0 & 1 & 0 \\ 0 & -11 & -3 & 1 & -4 & 0 \\ 0 & -5 & 3 & 0 & -2 & 1 \end{array}\right], -R_2/11 \approx \left[\begin{array}{ccc|ccc} 1 & 3 & 1 & 0 & 1 & 0 \\ 0 & 1 & 3/11 & -1/11 & 4/11 & 0 \\ 0 & -5 & 3 & 0 & -2 & 1 \end{array}\right], R_1 - 3R_2, R_3 + 5R_2$$

$$\approx \left[\begin{array}{ccc|ccc} 1 & 0 & 2/11 & 3/11 & -1/11 & 0 \\ 0 & 1 & 3/11 & -1/11 & 4/11 & 0 \\ 0 & 0 & 48/11 & -5/11 & -2/11 & 1 \end{array}\right], \frac{11}{48}R_3 \approx \left[\begin{array}{ccc|ccc} 1 & 0 & 2/11 & 3/11 & -1/11 & 0 \\ 0 & 1 & 3/11 & -1/11 & 4/11 & 0 \\ 0 & 0 & 1 & -5/48 & -1/24 & 11/48 \end{array}\right], R_2 - \frac{3}{11}R_3, R_1 - \frac{2}{11}R_3$$

$$\approx \left[\begin{array}{ccc|ccc} 1 & 0 & 0 & 7/24 & -1/12 & -1/24 \\ 0 & 1 & 0 & -1/16 & 3/8 & -1/16 \\ 0 & 0 & 1 & -5/48 & -1/24 & 11/48 \end{array}\right] = [\mathbf{I} \mid \mathbf{B}].$$

Hence,
$$\mathbf{A}^{-1} = \mathbf{B} = \begin{bmatrix} 7/24 & -1/12 & -1/24 \\ -1/16 & 3/8 & -1/16 \\ -5/48 & -1/24 & 11/48 \end{bmatrix} = \frac{1}{48}\begin{bmatrix} 14 & -4 & -2 \\ -3 & 18 & -3 \\ -5 & -2 & 11 \end{bmatrix}.$$

The solution vector is given by $\mathbf{x} = \mathbf{A}^{-1}\mathbf{b}$.

$$\mathbf{x} = \frac{1}{48}\begin{bmatrix} 14 & -4 & -2 \\ -3 & 18 & -3 \\ -5 & -2 & 11 \end{bmatrix}\begin{bmatrix} 7 \\ 7 \\ 1 \end{bmatrix} = \frac{1}{48}\begin{bmatrix} 98 - 28 - 2 \\ -21 + 126 - 3 \\ -35 - 14 + 11 \end{bmatrix} = \begin{bmatrix} 17/12 \\ 17/8 \\ -19/24 \end{bmatrix}.$$

13.3.3 LU Decomposition Method

This method is also called the *factorization method* or *triangularization method*. This method may be used to solve a system of equations or to find the inverse of a matrix. In this method, the matrix **A** is decomposed or factorized as the product of a lower triangular matrix **L** and an upper triangular matrix **U**.

We illustrative the method for the following 3 × 3 system of equations.

$$a_{11}x_1 + a_{12}x_2 + a_{13}x_3 = b_1,$$
$$a_{21}x_1 + a_{22}x_2 + a_{23}x_3 = b_2,$$
$$a_{31}x_1 + a_{32}x_2 + a_{33}x_3 = b_3.$$

The system may be written as $\mathbf{Ax} = \mathbf{b}$. We write the matrix **A** as

$$\mathbf{A} = \mathbf{LU} \qquad \text{...(13.7)}$$

where
$$\mathbf{L} = \begin{bmatrix} 1 & 0 & 0 \\ l_{21} & 1 & 0 \\ l_{31} & l_{32} & 1 \end{bmatrix} \text{ and } \mathbf{U} = \begin{bmatrix} u_{11} & u_{12} & u_{13} \\ 0 & u_{22} & u_{23} \\ 0 & 0 & u_{33} \end{bmatrix}.$$

From the Eq. 13.7, it is obvious that main diagonal entries of matrix **L**, that is, $l_{ii} = 1$, $i = 1, 2, 3$ (the method is called the *Doolittle* method).

In Eq. (13.7) we may also take $\mathbf{L} = \begin{bmatrix} l_{11} & 0 & 0 \\ l_{21} & l_{22} & 0 \\ l_{31} & l_{33} & l_{33} \end{bmatrix}$ and $\mathbf{U} = \begin{bmatrix} 1 & u_{12} & u_{13} \\ 0 & 1 & u_{23} \\ 0 & 0 & 1 \end{bmatrix}$.

The main diagonal entries of above matrix **U**, that is, $u_{ii} = 1$, $i = 1, 2, 3$. (the method is called the *Crout's* method.)

EXAMPLE 13.22 Solve the system of equations $\mathbf{Ax} = \mathbf{b}$, where

$$\mathbf{A} = \begin{bmatrix} 1 & -4 & 2 \\ -4 & 25 & 4 \\ 2 & 4 & 24 \end{bmatrix}, \quad \mathbf{b} = \begin{bmatrix} 81 \\ -153 \\ 324 \end{bmatrix}$$

by factorization method. Take diagonal entries. $u_{ii} = 1$, $i = 1, 2, 3$. Also find $\mathbf{A}^{-1}$.

Solution We write

$$\begin{bmatrix} 1 & -4 & 2 \\ -4 & 25 & 4 \\ 2 & 4 & 24 \end{bmatrix} = \begin{bmatrix} l_{11} & 0 & 0 \\ l_{21} & l_{22} & 0 \\ l_{31} & l_{32} & l_{33} \end{bmatrix} \begin{bmatrix} 1 & u_{12} & u_{13} \\ 0 & 1 & u_{23} \\ 0 & 0 & 1 \end{bmatrix}$$

$$= \begin{bmatrix} l_{11} & l_{11}u_{12} & l_{11}u_{13} \\ l_{21} & l_{21}u_{12} + l_{22} & l_{21}u_{13} + l_{22}u_{23} \\ l_{31} & l_{31}u_{12} + l_{32} & l_{31}u_{13} + l_{32}u_{23} + l_{33} \end{bmatrix}.$$

Comparing the corresponding elements on both sides, we obtain

first column of **L**: $l_{11} = 1;\ l_{21} = -4;\ l_{31} = 2;$

first row of **U**: $l_{11}u_{12} = -4,\ u_{12} = -4/1 = -4;\ l_{11}u_{13} = 2,\ u_{13} = 2/1 = 2;$

second column of **L**: $l_{21}u_{12} + l_{22} = 25,\ l_{22} = 25 - (-4)(-4) = 9;\ l_{31}u_{12} + l_{32} = 4,\ l_{32} = 4 - (2)(-4) = 12;$

second row of **U**: $l_{21}u_{13} + l_{22}u_{23} = 4;\ u_{23} = [4 - (-4)(2)]/9 = 4/3;$

third column of **L**: $l_{31}u_{13} + l_{32}u_{23} + l_{33} = 24,\ l_{33} = 24 - (2)(2) - (12)(4/3) = 4.$

Thus the factorization of **A** = **LU** is, that is

$$\underset{\mathbf{A}}{\begin{bmatrix} 1 & -4 & 2 \\ -4 & 25 & 4 \\ 2 & 4 & 24 \end{bmatrix}} = \underset{\mathbf{L}}{\begin{bmatrix} 1 & 0 & 0 \\ -4 & 9 & 0 \\ 2 & 12 & 4 \end{bmatrix}} \underset{\mathbf{U}}{\begin{bmatrix} 1 & -4 & 2 \\ 0 & 1 & 4/3 \\ 0 & 0 & 1 \end{bmatrix}}.$$

We first solve **Lz** = **b**, that is

$$\begin{bmatrix} 1 & 0 & 0 \\ -4 & 9 & 0 \\ 2 & 12 & 4 \end{bmatrix} \begin{bmatrix} z_1 \\ z_2 \\ z_3 \end{bmatrix} = \begin{bmatrix} 81 \\ -153 \\ 324 \end{bmatrix},$$

we obtain by forward substitution

$z_1 = 81,\ -4z_1 + 9z_2 = -153,\ z_2 = (-153 + 324)/9 = 19,\ 2z_1 + 12z_2 + 4z_3 = 324,$
$z_3 - [162\ \ 81 - 6(19)]/2 = -33/2.$ Solution $\mathbf{z} = [81\ \ 19\ \ -33/2]^T.$

Then we solve **Ux** = **z**, that is

$$\begin{bmatrix} 1 & -4 & 2 \\ 0 & 1 & 4/3 \\ 0 & 0 & 1 \end{bmatrix} \begin{bmatrix} x_1 \\ x_2 \\ x_3 \end{bmatrix} = \begin{bmatrix} 81 \\ 19 \\ -33/2 \end{bmatrix},$$

we obtain by backward substitution

$x_3 = -33/2,\ x_2 + (4/3)x_3 = 19,\ x_2 = 19 + 22 = 41,$
$x_1 - 4x_2 + 2x_3 = 81,\ x_1 = 81 + 4(41) + 2(33/2) = 278.$ Solution $\mathbf{x} = [278\ \ 41\ \ -33/2]^T.$

The inverse of matrix **A** is obtained as $\mathbf{A}^{-1} = \mathbf{U}^{-1}\mathbf{L}^{-1}$.

Now, from **LL′** = **I**

$$\begin{bmatrix} 1 & 0 & 0 \\ -4 & 9 & 0 \\ 2 & 12 & 4 \end{bmatrix} \begin{bmatrix} l'_{11} & 0 & 0 \\ l'_{21} & l'_{22} & 0 \\ l'_{31} & l'_{32} & l'_{33} \end{bmatrix} = \begin{bmatrix} 1 & 0 & 0 \\ 0 & 1 & 0 \\ 0 & 0 & 1 \end{bmatrix}$$

we get by forward substituting

$$\mathbf{L}^{-1} = \begin{bmatrix} 1 & 0 & 0 \\ 4/9 & 1/9 & 0 \\ -11/6 & -1/3 & 1/4 \end{bmatrix} = \frac{1}{36} \begin{bmatrix} 36 & 0 & 0 \\ 16 & 4 & 0 \\ -66 & -12 & 9 \end{bmatrix}.$$

From $\mathbf{UU}^{-1} = \mathbf{I}$,

$$\begin{bmatrix} 1 & -4 & 2 \\ 0 & 1 & 4/3 \\ 0 & 0 & 1 \end{bmatrix} \begin{bmatrix} u'_{11} & u'_{12} & u'_{13} \\ 0 & 1 & u'_{23} \\ 0 & 0 & u'_{33} \end{bmatrix} = \begin{bmatrix} 1 & 0 & 0 \\ 0 & 1 & 0 \\ 0 & 0 & 1 \end{bmatrix}$$

we get by backward substitution

$$\mathbf{U}^{-1} = \begin{bmatrix} 1 & 4 & -22/3 \\ 0 & 1 & -4/3 \\ 0 & 0 & 1 \end{bmatrix} = \frac{1}{3}\begin{bmatrix} 3 & 12 & -22 \\ 0 & 3 & -4 \\ 0 & 0 & 3 \end{bmatrix}.$$

Now, $$\mathbf{A}^{-1} = \frac{1}{108}\begin{bmatrix} 3 & 12 & -22 \\ 0 & 3 & -4 \\ 0 & 0 & 3 \end{bmatrix}\begin{bmatrix} 36 & 0 & 0 \\ 16 & 4 & 0 \\ -66 & -12 & 9 \end{bmatrix} = \frac{1}{36}\begin{bmatrix} 584 & 104 & -66 \\ 104 & 20 & -12 \\ -66 & -12 & 9 \end{bmatrix}.$$

Check $\mathbf{AA}^{-1} = \mathbf{I}$. We have

$$\mathbf{AA}^{-1} = \frac{1}{36}\begin{bmatrix} 1 & -4 & 2 \\ -4 & 25 & 4 \\ 2 & 4 & 24 \end{bmatrix}\begin{bmatrix} 584 & 104 & -66 \\ 104 & 20 & -12 \\ -66 & -12 & 9 \end{bmatrix} = \begin{bmatrix} 1 & 0 & 0 \\ 0 & 1 & 0 \\ 0 & 0 & 1 \end{bmatrix} = \mathbf{I}.$$

EXAMPLE 13.23 Solve the system of equations $\mathbf{Ax} = \mathbf{b}$, where $\mathbf{A} = \begin{bmatrix} 5 & 9 & 2 \\ 9 & 4 & 1 \\ 2 & 1 & 1 \end{bmatrix}$, $\mathbf{b} = \begin{bmatrix} 24 \\ 25 \\ 11 \end{bmatrix}$ by factorization method. Take diagonal entries $l_{ii} = 1$, $i = 1, 2, 3$. Also find $\mathbf{A}^{-1}$.

Solution We write

$$\begin{bmatrix} 5 & 9 & 2 \\ 9 & 4 & 1 \\ 2 & 1 & 1 \end{bmatrix} = \begin{bmatrix} 1 & 0 & 0 \\ l_{21} & 1 & 0 \\ l_{31} & l_{32} & 1 \end{bmatrix}\begin{bmatrix} u_{11} & u_{12} & u_{13} \\ 0 & u_{22} & u_{23} \\ 0 & 0 & u_{33} \end{bmatrix} = \begin{bmatrix} u_{11} & u_{12} & u_{13} \\ l_{21}u_{11} & l_{21}u_{12} + u_{22} & l_{21}u_{13} + u_{23} \\ l_{31}u_{11} & l_{31}u_{12} + l_{32}u_{22} & l_{31}u_{13} + l_{32}u_{23} + u_{33} \end{bmatrix}.$$

Comparing the corresponding elements on both sides, we get

first row of **U**: $u_{11} = 5$; $u_{12} = 9$; $u_{13} = 2$;

first column of **L**: $l_{21}u_{11} = 9$, $l_{21} = 9/5$; $l_{31}u_{11} = 2$, $l_{31} = 2/5$;

second row of **U**: $l_{21}u_{12} + u_{22} = 4$, $u_{22} = 4 - (81/5) = -(61/5)$; $l_{21}u_{13} + u_{23} = 1$, $u_{23} = 1 - (18/5) = -13/5$;

second column of **L**: $l_{31}u_{12} + l_{32}u_{22} = 1$, $l_{32} = [1 - (18/5)]/u_{22} = -(13/5)/(-61/5) = 13/61$;

third row of **U**: $l_{31}u_{13} + l_{32}u_{23} + u_{33} = 1$, $u_{33} = 1 - (4/5) + (169/305) = 46/61$.

Thus the factorization of $\mathbf{A} = \mathbf{LU}$ is

$$\underset{\mathbf{A}}{\begin{bmatrix} 5 & 9 & 2 \\ 9 & 4 & 1 \\ 2 & 1 & 1 \end{bmatrix}} = \underset{\mathbf{L}}{\begin{bmatrix} 1 & 0 & 0 \\ 9/5 & 1 & 0 \\ 2/5 & 13/61 & 1 \end{bmatrix}}\underset{\mathbf{U}}{\begin{bmatrix} 5 & 9 & 2 \\ 0 & -61/5 & -13/5 \\ 0 & 0 & 46/61 \end{bmatrix}}.$$

We first solve $\mathbf{Lz} = \mathbf{b}$, that is

$$\begin{bmatrix} 1 & 0 & 0 \\ 9/5 & 1 & 0 \\ 2/5 & 13/61 & 1 \end{bmatrix}\begin{bmatrix} z_1 \\ z_2 \\ z_3 \end{bmatrix} = \begin{bmatrix} 24 \\ 25 \\ 11 \end{bmatrix},$$

we obtain by forward substitution

$$z_1 = 24,\ z_2 = 25 - \frac{9}{5}(24) = -\frac{91}{5},\ z_3 = 11 - \frac{2}{5}z_1 - \frac{13}{61}z_2 = 11 - \frac{48}{5} + \left(\frac{13}{61}\right)\left(\frac{91}{5}\right) = \frac{322}{61}.$$

Solution $\mathbf{z} = [24 \quad -91/5 \quad 322/61]^{\mathbf{T}}$.

Then we solve $\mathbf{Ux} = \mathbf{z}$, that is

$$\begin{bmatrix} 5 & 9 & 2 \\ 0 & -61/5 & -13/5 \\ 0 & 0 & 46/61 \end{bmatrix}\begin{bmatrix} x_1 \\ x_2 \\ x_3 \end{bmatrix} = \begin{bmatrix} 24 \\ -91/5 \\ 322/61 \end{bmatrix},$$

we obtain by backward substitution

$$x_3 = \left(\frac{322}{61}\right)\left(\frac{61}{46}\right) = 7,\ -\frac{61}{5}x_2 - \frac{13}{5}x_3 = -\frac{91}{5}, \quad x_2 = \left(-\frac{91}{5} + \frac{91}{5}\right)\frac{5}{61} = 0,$$

$5x_1 + 9x_2 + 2x_3 = 24$, $x_1 = (24 - 14)/5 = 2$. Solution $\mathbf{x} = [24 \quad 0 \quad 7]^T$.

The inverse of the matrix **A** is obtained as

$$\mathbf{A}^{-1} = (\mathbf{LU})^{-1} = \mathbf{U}^{-1}\mathbf{L}^{-1}.$$

We have $|\mathbf{L}| = 1(1 - 0) = 1$. $\quad \text{adj } \mathbf{L} = \begin{bmatrix} 1 & -9/5 & -1/61 \\ 0 & 1 & -13/61 \\ 0 & 0 & 1 \end{bmatrix}^T.$

Therefore, $\quad \mathbf{L}^{-1} = \frac{1}{|\mathbf{L}|}\text{adj}(\mathbf{L}) = \begin{bmatrix} 1 & 0 & 0 \\ -9/5 & 1 & 0 \\ -1/61 & -13/61 & 1 \end{bmatrix}.$

We have $|\mathbf{U}| = 5 - \left(\frac{46}{5}\right) = -46$, $\text{adj } \mathbf{U} = \begin{bmatrix} -46/5 & 0 & 0 \\ -414/61 & 230/61 & 0 \\ 1 & 13 & -61 \end{bmatrix}^T$, $\mathbf{U}^{-1} = \frac{1}{46}\begin{bmatrix} 46/5 & 414/61 & -1 \\ 0 & -230/61 & -13 \\ 0 & 0 & 61 \end{bmatrix}.$

Therefore, $\mathbf{A}^{-1} = \frac{1}{46}\begin{bmatrix} 46/5 & 414/61 & -1 \\ 0 & -230/61 & -13 \\ 0 & 0 & 61 \end{bmatrix}\begin{bmatrix} 1 & 0 & 0 \\ -9/5 & 1 & 0 \\ -1/61 & -13/61 & 1 \end{bmatrix} = \frac{1}{46}\begin{bmatrix} -3 & 7 & -1 \\ 7 & -1 & -13 \\ -1 & -13 & 61 \end{bmatrix}.$

Check: $\quad \mathbf{AA}^{-1} = \frac{1}{46}\begin{bmatrix} 5 & 9 & 2 \\ 9 & 4 & 1 \\ 2 & 1 & 1 \end{bmatrix}\begin{bmatrix} -3 & 7 & -1 \\ 7 & -1 & -13 \\ -1 & -13 & 61 \end{bmatrix} = \begin{bmatrix} 1 & 0 & 0 \\ 0 & 1 & 0 \\ 0 & 0 & 1 \end{bmatrix} = \mathbf{I}.$

13.3.4 Choleski Decomposition Method

When the coefficient matrix **A** of the system $\mathbf{Ax} = \mathbf{b}$ is symmetric, then the decomposition can be written as

$$\mathbf{A} = \mathbf{LL}^{\mathrm{T}} \qquad ...(13.8)$$

where **L** is the lower triangular matrix.

This method is also called the *square root* method. Let us illustrate the decomposition for a 3 × 3 system. We write

$$\overset{\mathbf{A}}{\begin{bmatrix} a_{11} & a_{12} & a_{13} \\ a_{12} & a_{22} & a_{23} \\ a_{13} & a_{23} & a_{33} \end{bmatrix}} = \overset{\mathbf{L}}{\begin{bmatrix} l_{11} & 0 & 0 \\ l_{21} & l_{22} & 0 \\ l_{31} & l_{32} & l_{33} \end{bmatrix}}\overset{\mathbf{L}^T}{\begin{bmatrix} l_{11} & l_{21} & l_{31} \\ 0 & l_{22} & l_{32} \\ 0 & 0 & l_{33} \end{bmatrix}} = \begin{bmatrix} l_{11}^2 & l_{11}l_{21} & l_{11}l_{31} \\ l_{21}l_{11} & l_{21}^2 + l_{22}^2 & l_{21}l_{31} + l_{22}l_{32} \\ l_{31}l_{11} & l_{31}l_{21} + l_{32}l_{22} & l_{31}^2 + l_{32}^2 + l_{33}^2 \end{bmatrix}.$$

Comparing the elements, we obtain

$$l_{11}^2 = a_{11},\ l_{11} = \sqrt{a_{11}};\ l_{11}l_{21} = a_{12},\ l_{21} = a_{12}/l_{11};$$

$$l_{11}l_{31} = a_{13},\ l_{31} = a_{13}/l_{11};\ l_{21}^2 + l_{22}^2 = a_{22}, l_{22} = \sqrt{a_{22} - \frac{a_{12}^2}{a_{11}}} = \sqrt{\frac{a_{11}a_{22} - a_{12}^2}{a_{11}}}$$

$$l_{21}l_{31} + l_{22}l_{32} = a_{23},\ l_{32} = \frac{1}{l_{22}}[a_{23} - l_{21}l_{31}];\quad l_{31}^2 + l_{32}^2 + l_{33}^2 = a_{33},\ l_{33} = \sqrt{a_{33} - l_{31}^2 - l_{32}^2}.$$

The remaining part of the solution procedure is same as in **LU** factorization method.

We have $\mathbf{LL}^T\mathbf{x} = \mathbf{b}$. Set $\mathbf{L}^T\mathbf{x} = \mathbf{z}$. Then, we have $\mathbf{Lz} = \mathbf{b}$.

We first solve $\mathbf{Lz} = \mathbf{b}$. Then, solve $\mathbf{L}^T\mathbf{x} = \mathbf{z}$. Again, one forward substitution and one backward substitution is required to obtain $\mathbf{x}$.

To determine $\mathbf{A}^{-1}$, we proceed as under:

We have $\mathbf{A}^{-1} = (\mathbf{LL}^T)^{-1} = (\mathbf{L}^T)^{-1}\,\mathbf{L}^{-1} = (\mathbf{L}^{-1})^T\,\mathbf{L}^{-1}$.

Thus, the inverse of the lower triangular matrix $\mathbf{L}$ is required to find $\mathbf{A}^{-1}$.

Remark A sufficient condition which guarantees the $\mathbf{LL}^T$ factorization of the matrix **A** is that the matrix is positive definite, that is, all the leading minors should be positive.

EXAMPLE 13.24 Solve the system of equations

$$\begin{bmatrix} 1 & 2 & 3 \\ 2 & 8 & 22 \\ 3 & 22 & 82 \end{bmatrix}\begin{bmatrix} x_1 \\ x_2 \\ x_3 \end{bmatrix} = \begin{bmatrix} 5 \\ 28 \\ 96 \end{bmatrix}$$

(AMIETE, June 2006)

by the Choleski decomposition method.

Solution We decompose the coefficient matrix as $\mathbf{A} = \mathbf{LL}^T$. That is,

$$\begin{bmatrix} 1 & 2 & 3 \\ 2 & 8 & 22 \\ 3 & 22 & 82 \end{bmatrix} = \begin{bmatrix} l_{11} & 0 & 0 \\ l_{21} & l_{22} & 0 \\ l_{31} & l_{32} & l_{33} \end{bmatrix}\begin{bmatrix} l_{11} & l_{21} & l_{31} \\ 0 & l_{22} & l_{32} \\ 0 & 0 & l_{33} \end{bmatrix} = \begin{bmatrix} l_{11}^2 & l_{11}l_{21} & l_{11}l_{31} \\ l_{21}l_{11} & l_{21}^2 + l_{22}^2 & l_{21}l_{31} + l_{22}l_{32} \\ l_{31}l_{11} & l_{31}l_{21} + l_{32}l_{22} & l_{31}^2 + l_{32}^2 + l_{33}^2 \end{bmatrix}.$$

Comparing the coefficients, we obtain

$$l_{11}^2 = 1,\quad l_{11} = 1;\quad l_{11}l_{21} = 2,\quad l_{21} = 2;\quad l_{11}l_{31} = 3,\quad l_{31} = 3;$$

$$l_{21}^2 + l_{22}^2 = 8,\quad l_{22}^2 = 8 - 4 = 4,\quad l_{22} = 2;\quad l_{21}l_{31} + l_{22}l_{32} = 22,\quad l_{32} = [22 - (2)(3)]/2 = 8;$$

$$l_{31}^2 + l_{32}^2 + l_{33}^2 = 82,\quad l_{33}^2 = 82 - (3)^2 - (8)^2 = 9,\quad l_{33} = 3.$$

Hence,
$$\mathbf{L} = \begin{bmatrix} 1 & 0 & 0 \\ 2 & 2 & 0 \\ 3 & 8 & 3 \end{bmatrix}.$$

We now solve $\mathbf{Lz} = \mathbf{b}$, that is,

$$\begin{bmatrix} 1 & 0 & 0 \\ 2 & 2 & 0 \\ 3 & 8 & 3 \end{bmatrix}\begin{bmatrix} z_1 \\ z_2 \\ z_3 \end{bmatrix} = \begin{bmatrix} 5 \\ 28 \\ 96 \end{bmatrix},$$

we obtain $z_1 = 5$; $2z_1 + 2z_2 = 28$, $z_2 = (28 - 10)/2 = 9$; $3z_1 + 8z_2 + 3z_3 = 96$, $z_3 = (96 - 15 - 72)/3 = 3$.

Solution $\mathbf{z} = [5\quad 9\quad 3)^T$. Solving $\mathbf{L}^T\mathbf{x} = \mathbf{z}$, that is

$$\begin{bmatrix} 1 & 2 & 3 \\ 0 & 2 & 8 \\ 0 & 0 & 3 \end{bmatrix}\begin{bmatrix} x_1 \\ x_2 \\ x_3 \end{bmatrix} = \begin{bmatrix} 5 \\ 9 \\ 3 \end{bmatrix}$$

we get $x_3 = 1$; $2x_2 + 8x_3 = 9$, $x_2 = (9 - 8)/2 = 1/2$; $x_1 + 2x_2 + 3x_3 = 5$, $x_1 = 5 - 1 - 3 = 1$.

Solution $\mathbf{x} = [1\quad 1/2\quad 1]^T$.

EXAMPLE 13.25 Obtain the Choleski factorisation of the form $\mathbf{A} = \mathbf{LL}^T$, where $\mathbf{L}$ is a lower triangular matrix and $\mathbf{A} = \begin{bmatrix} 1 & 2 & -1 \\ 2 & 13 & -5 \\ -1 & -5 & 6 \end{bmatrix}$. Hence, determine the matrix $\mathbf{A}^{-1}$. (*AMIETE, June 2005*)

Solution We decompose the coefficient matrix as

$$\begin{bmatrix} 1 & 2 & -1 \\ 2 & 13 & -5 \\ -1 & -5 & 6 \end{bmatrix} = \begin{bmatrix} l_{11} & 0 & 0 \\ l_{21} & l_{22} & 0 \\ l_{31} & l_{32} & l_{33} \end{bmatrix}\begin{bmatrix} l_{11} & l_{21} & l_{31} \\ 0 & l_{22} & l_{32} \\ 0 & 0 & l_{33} \end{bmatrix} = \begin{bmatrix} l_{11}^2 & l_{11}l_{21} & l_{11}l_{31} \\ l_{21}l_{11} & l_{21}^2 + l_{22}^2 & l_{21}l_{31} + l_{22}l_{32} \\ l_{31}l_{11} & l_{31}l_{21} + l_{32}l_{22} & l_{31}^2 + l_{32}^2 + l_{33}^2 \end{bmatrix}.$$

Comparing the elements, we obtain

$l_{11}^2 = 1,\ l_{11} = 1;\quad l_{11}l_{21} = 2,\ l_{21} = 2;\quad l_{11}l_{31} = -1,\ l_{31} = -1;$

$l_{21}^2 + l_{22}^2 = 13,\quad l_{22}^2 = 13 - 4 = 9,\quad l_{22} = 3;\quad l_{21}l_{31} + l_{22}l_{32} = -5,\ 2(-1) + 3(l_{32}) = -5,$

$l_{32} = (-5+2)/3 = -1;\quad l_{31}^2 + l_{32}^2 + l_{33}^2 = 6,\quad l_{33}^2 = 6 - 1 - 1 = 4,\quad l_{33} = 2.$

Hence, $\mathbf{L} = \begin{bmatrix} 1 & 0 & 0 \\ 2 & 3 & 0 \\ -1 & -1 & 2 \end{bmatrix},\ \mathbf{L}^T = \begin{bmatrix} 1 & 2 & -1 \\ 0 & 3 & -1 \\ 0 & 0 & 2 \end{bmatrix}.$

Thus, the decomposition can be written as

$$\begin{bmatrix} 1 & 2 & -1 \\ 2 & 13 & -5 \\ -1 & -5 & 6 \end{bmatrix} = \begin{bmatrix} 1 & 0 & 0 \\ 2 & 3 & 0 \\ -1 & -1 & 2 \end{bmatrix}\begin{bmatrix} 1 & 2 & -1 \\ 0 & 3 & -1 \\ 0 & 0 & 2 \end{bmatrix}.$$

To determine $\mathbf{A}^{-1}$, we proceed as under.

We have $\mathbf{A}^{-1} = (\mathbf{LL}^T)^{-1} = (\mathbf{L}^T)^{-1}\,\mathbf{L}^{-1} = (\mathbf{L}^{-1})^T\mathbf{L}^{-1}$.

We require the inverse of the lower triangular matrix $\mathbf{L}$ to find $\mathbf{A}^{-1}$.

We have $$\mathbf{L} = \begin{bmatrix} 1 & 0 & 0 \\ 2 & 3 & 0 \\ -1 & -1 & 2 \end{bmatrix}.\quad |\mathbf{L}| = 1\begin{vmatrix} 3 & 0 \\ -1 & 2 \end{vmatrix} = 6$$

$$\text{adj } \mathbf{L} = \begin{bmatrix} 6 & -4 & 1 \\ 0 & 2 & 1 \\ 0 & 0 & 3 \end{bmatrix}^T = \begin{bmatrix} 6 & 0 & 0 \\ -4 & 2 & 0 \\ 1 & 1 & 3 \end{bmatrix}.$$

Therefore, $$\mathbf{L}^{-1} = \frac{1}{|\mathbf{L}|}\text{adj}(\mathbf{L}) = \frac{1}{6}\begin{bmatrix} 6 & 0 & 0 \\ -4 & 2 & 0 \\ 1 & 1 & 3 \end{bmatrix},\quad (\mathbf{L}^{-1})^T = \frac{1}{6}\begin{bmatrix} 6 & -4 & 1 \\ 0 & 2 & 1 \\ 0 & 0 & 3 \end{bmatrix}.$$

Hence, $$\mathbf{A}^{-1} = \frac{1}{36}\begin{bmatrix} 6 & -4 & 1 \\ 0 & 2 & 1 \\ 0 & 0 & 3 \end{bmatrix}\begin{bmatrix} 6 & 0 & 0 \\ -4 & 2 & 0 \\ 1 & 1 & 3 \end{bmatrix} = \frac{1}{36}\begin{bmatrix} 53 & -7 & 3 \\ -7 & 5 & 3 \\ 3 & 3 & 9 \end{bmatrix}.$$

Indirect or iterative method

We have so far discussed some direct methods for the solution of the system of equation $\mathbf{Ax} = \mathbf{b}$ and we have seen that these methods yield the solution after an amount of computation that is known in advance. We shall now describe the following indirect or the iterative methods.

1. Gauss-Jacobi Iteration Method
2. Gauss-Seidel Iteration Method

The second of these two is an improvement of the first.

In iterative methods, an initial approximation to the solution vector $\mathbf{x}^{(0)}$ is assumed, and we generate the sequence of iterates $x^{(1)}, x^{(2)}, ..., x^{(k)}, x^{(k+1)}$ .... which is the limit converges to the exact solution.

Both the methods converge when the coefficient matrix **A** is *diagonally dominant*. That is,

$$\sum_{\substack{j=1 \\ j \neq i}}^{n} |a_{ij}| \leq |a_{ii}|, \quad i = 1, 2, ..., n$$

and strict inequality prevails at least for one value of i. (This implies that diagonal elements a_{ii} are non-zero).

In view of the above we should make sure, before starting the iterations, that the coefficient matrix is *diagonally dominant*.

For example, we can see by mere inspection that the coefficient matrix of the system given by

$$\left.\begin{aligned} 2x - 10y + 2z + 154 &= 0 \\ 10x - 2y - 3z - 205 &= 0 \\ 2x + y - 10z + 120 &= 0 \end{aligned}\right\}$$

is not diagonally dominant. But if we interchange the first two equations and write the new system in the form:

$$\left.\begin{aligned} 10x - 2y - 3z - 205 &= 0 \\ 2x - 10y + 2z + 154 &= 0 \\ 2x + y - 10z + 120 &= 0 \end{aligned}\right\}$$

the condition of diagonal dominance is satisfied.

13.3.5 Jacobi-Iteration Method

We assume that the diagonal elements a_{ii} of the matrix **A**, are non-zero. We write the system of equations as

$$\begin{aligned} a_{11}x_1 &= b_1 - (a_{12}x_2 + a_{13}x_3 + \cdots + a_{1n}x_n) \\ a_{22}x_2 &= b_2 - (a_{21}x_1 + a_{23}x_3 + \cdots + a_{2n}x_n) \\ &\cdots\cdots\cdots\cdots\cdots\cdots\cdots\cdots \\ &\cdots\cdots\cdots\cdots\cdots\cdots\cdots\cdots \\ a_{nn}x_n &= b_n - (a_{n1}x_1 + a_{n2}x_2 + \cdots + a_{n,\,n-1}\,x_{n-1}) \end{aligned}$$

The *Gauss-Jacobi iteration method* or simply the *Jacobi-iteration method* is defined as

$$\begin{aligned} x_1^{(k+1)} &= \frac{1}{a_{11}}[b_1 - (a_{12}x_2^{(k)} + a_{13}x_3^{(k)} + \cdots + a_{1n}x_n^{(k)})] \\ x_2^{(k+1)} &= \frac{1}{a_{22}}[b_2 - (a_{21}x_1^{(k)} + a_{23}x_3^{(k)} + \cdots + a_{2n}x_n^{(k)})] \\ &\vdots \\ x_n^{(k+1)} &= \frac{1}{a_{nn}}[b_n - (a_{n1}x_1^{(k)} + a_{n2}x_2^{(k)} + \cdots + a_{n,\,n-1}x_{n-1}^{(k)})] \end{aligned} \qquad \text{...(13.9)}$$

$k = 0, 1, 2, ...$ We start with an initial approximation $\mathbf{x}^{(0)} = [x_1^{(0)}, x_2^{(0)}, ..., x_n^{(0)}]^T$, and generate the sequence of iterates $\mathbf{x}^{(1)}, \mathbf{x}^{(2)}, ...$ which in the limit converges to the exact solution **x**. It may be noted that convergence of the sequence does not depend upon the starting values. If the initial approximation is close to the true solution, the number of iteration required for a specified accuracy is relatively less. Since we replace all the elements simultaneously, this method is also called the *method of simultaneous displacements*.

EXAMPLE 13.26 Solve the system of linear equations up to two places of decimal in the computation by Jacobi iteration method,

$$\begin{bmatrix} 27 & 6 & -1 \\ 6 & 15 & 2 \\ 1 & 1 & 54 \end{bmatrix} \begin{bmatrix} x_1 \\ x_2 \\ x_3 \end{bmatrix} = \begin{bmatrix} 85 \\ 72 \\ 110 \end{bmatrix}.$$

Take the initial approximation as $\mathbf{x}(0) = [x_1^{(0)}, x_2^{(0)}, ..., x_3^{(0)}]^T = [0, 0, 0]^T$.

Solution We write the Jacobi iteration method as

$$x_1^{(k+1)} = \frac{1}{27}[85 - 6x_2^{(k)} + x_3^{(k)}], \quad x_2^{(k+1)} = \frac{1}{15}[72 - 6x_1^{(k)} - 2x_3^{(k)}], \quad x_3^{(k+1)} = \frac{1}{54}[110 - x_1^{(k)} - x_2^{(k)}],$$

$k = 0, 1, 2, 3, \ldots\ldots$. Starting with $x_1^{(0)} = 0$, $x_2^{(0)} = 0$, $x_3^{(0)} = 0$, we get the following results.

$k = 0$; $\quad x_1^{(1)} = \frac{1}{27}[85 - 0 + 0] = 3.15, \quad x_2^{(1)} = \frac{1}{15}[72 - 0 + 0] = 4.8, \quad x_3^{(1)} = \frac{1}{54}[110 - 0 - 0] = 2.04.$

$k = 1$; $\quad x_1^{(2)} = \frac{1}{27}[85 - 6x_2^{(1)} + x_3^{(1)}] = \frac{1}{27}[85 - 6(4.8) + 2.04] = 2.16,$

$$x_2^{(2)} = \frac{1}{15}[72 - 6x_1^{(1)} - 2x_3^{(1)}] = \frac{1}{15}[72 - 6(3.15) - 2(2.04)] = 3.27,$$

$$x_3^{(2)} = \frac{1}{54}[110 - x_1^{(1)} - x_2^{(1)}] = \frac{1}{54}[110 - 3.15 - 4.8] = 1.89.$$

Thus $\quad [x_1^{(2)}, x_2^{(2)}, x_3^{(2)}] = [2.16, 3.27, 1.89].$

$k = 2$: $\quad x_1^{(3)} = \frac{1}{27}[85 - 6(3.27) + 1.89] = 2.49, \quad x_2^{(3)} = \frac{1}{15}[72 - 6(2.16) - 2(1.89)] = 3.68,$

$$x_3^{(3)} = \frac{1}{54}[110 - 2.16 - 3.27] = 1.94.$$

$k = 3$: $\quad x_1^{(4)} = \frac{1}{27}[85 - 6(3.68) + 1.94] = 2.40, \quad x_2^{(4)} = \frac{1}{15}[72 - 6(2.49) - 2(1.94)] = 3.55,$

$$x_3^{(4)} = \frac{1}{54}[110 - 2.49 - 3.68] = 1.92.$$

$k = 4$: $\quad x_1^{(5)} = \frac{1}{27}[85 - 6(3.55) + 1.92] = 2.43, \quad x_2^{(5)} = \frac{1}{15}[72 - 6(2.40) - 2(1.92)] = 3.58,$

$$x_3^{(5)} - \frac{1}{54}[110 - 2.40 - 3.55] = 1.93.$$

$k = 5$: $\quad x_1^{(6)} = \frac{1}{27}[85 - 6(3.58) + 1.93] = 2.42, \quad x_2^{(6)} = \frac{1}{15}[72 - 6(2.43) - 2(1.93)] = 3.57,$

$$x_3^{(6)} = \frac{1}{54}[110 - 2.43 - 3.58] = 1.93.$$

$k = 6$: $\quad x_1^{(7)} = \frac{1}{27}[85 - 6[3.57) + 1.93] = 2.43, \quad x_2^{(7)} = \frac{1}{15}[72 - 6(2.42) - 2(1.93)] = 3.57,$

$$x_3^{(7)} = \frac{1}{54}[110 - 2.42 - 3.57] = 1.93.$$

$k = 7$: $\quad x_1^{(8)} = \frac{1}{27}[85 - 6(3.57) + 1.93] = 2.43, \quad x_2^{(8)} = \frac{1}{15}[72 - 6(2.43) - 2(1.93)] = 3.57,$

$$x_3^{(8)} = \frac{1}{54}[110 - 2.43 - 3.57] = 1.93.$$

The last vector can be taken as the solution vector with two places of accuracy. This took eight iterations.

EXAMPLE 13.27 Perform four iterations of the Gauss-Jacobi iteration method for solving the system of equations

$$4x_1 - x_2 + x_3 = 7,$$
$$4x_1 - 8x_2 + x_3 = -21,$$
$$-2x_1 + x_2 + 5x_3 = 15.$$

Take the initial approximation as $\mathbf{x}^{(0)} = [1.5,\ 3.5,\ 2.5]^T$.

Solution We write the Jacobi iteration method as

$$x_1^{(k+1)} = \frac{1}{4}[7 + x_2^{(k)} - x_3^{(k)}],\quad x_2^{(k+1)} = \frac{1}{8}[21 + 4x_1^{(k)} + x_3^{(k)}],\ x_3^{(k+1)} = \frac{1}{5}[15 + 2x_1^{(k)} - x_2^{(k)}],$$

$k = 0, 1, 2, 3, ...,$

Starting with $x_1^{(0)} = 1.5$, $x_2^{(0)} = 3.5$, $x_3^{(0)} = 2.5$, we get the following results.

$k = 0$:
$$x_1^{(1)} = \frac{1}{4}[7 + x_2^{(0)} - x_3^{(0)}] = \frac{1}{4}[7 + 3.5 - 2.5] = 2.0,$$
$$x_2^{(1)} = \frac{1}{8}[21 + 4x_1^{(0)} + x_3^{(0)}] = \frac{1}{8}[21 + 4(1.5) + 2.5] = 3.69,$$
$$x_3^{(1)} = \frac{1}{5}[15 + 2x_1^{(0)} - x_2^{(0)}] = \frac{1}{5}[15 + 2(1.5) - 3.5] = 2.90.$$

$k = 1$:
$$x_1^{(2)} = \frac{1}{4}[7 + x_2^{(1)} - x_3^{(1)}] = \frac{1}{4}[7 + 3.69 - 2.90] = 1.95,$$
$$x_2^{(2)} = \frac{1}{8}[21 + 4x_1^{(1)} + x_3^{(1)}] = \frac{1}{8}[21 + 4(2) + 2.90] = 3.99,$$
$$x_3^{(2)} = \frac{1}{5}[15 + 2x_1^{(1)} - x_2^{(1)}] = \frac{1}{5}[15 + 2(2) - 3.69] = 3.06.$$

$k = 2$:
$$x_1^{(3)} = \frac{1}{4}[7 + x_2^{(2)} - x_3^{(2)}] = \frac{1}{4}[7 + 3.99 - 3.06] = 1.98,$$
$$x_2^{(3)} = \frac{1}{8}[21 + 4x_1^{(2)} + x_3^{(2)}] = \frac{1}{8}[21 + 4(1.95) + 3.06] = 3.98,$$
$$x_3^{(3)} = \frac{1}{5}[15 + 2x_1^{(2)} - x_2^{(2)}] = \frac{1}{5}[15 + 2(1.95) - 3.99] = 2.98.$$

$k = 3$:
$$x_1^{(4)} = \frac{1}{4}[7 + x_2^{(3)} - x_3^{(3)}] = \frac{1}{4}[7 + 3.98 - 2.98] = 2.00,$$
$$x_2^{(4)} = \frac{1}{8}[21 + 4x_1^{(3)} + x_3^{(3)}] = \frac{1}{8}[21 + 4(1.98) + 2.98] = 3.98,$$
$$x_3^{(4)} = \frac{1}{5}[15 + 2x_1^{(3)} - x_2^{(3)}] = \frac{1}{5}[15 + 2(1.98) - 3.98] = 3.00.$$

Thus
$$\mathbf{x}^{(4)} = [x_1^{(4)}, x_2^{(4)}, x_3^{(4)}]^T = [2.0, 3.98, 3.0]^T$$

13.3.6 Gauss-Seidel Iteration Method

In the Jacobi iteration method, when we evaluate $x_2^{(k+1)}$, both the values, $x_1^{(k)}$ and $x_1^{(k+1)}$ are available. But, we use only the previous iteration value $x_1^{(k)}$. Similarly, when we evaluate $x_3^{(k+1)}$, though the values $x_1^{(k+1)}$, $x_2^{(k+1)}$ are available, we still use the previous iteration values $x_1^{(k)}$ and $x_2^{(k)}$. This is the disadvantage of the Jacobi method.

In Gauss-Seidel method, within an iteration, we use successively, the latest available values of each variable in (13.9). Thus, this method is known as the *method of successive displacements*, since the elements are replaced successively.

This method is known to converge faster than the Jacobi method. We write the method as

$$x_1^{(k+1)} = \frac{1}{a_{11}}[b_1 - (a_{12}x_2^{(k)} + a_{13}x_3^{(k)} + \cdots + a_{1n}x_n^{(k)})]$$

$$x_2^{(k+1)} = \frac{1}{a_{22}}[b_2 - (a_{21}x_1^{(k+1)} + a_{23}x_3^{(k)} + \cdots + a_{2n}x_n^{(k)})]$$

$$\vdots \qquad \qquad \vdots$$

$$x_n^{(k+1)} = \frac{1}{a_{nn}}[b_n - (a_{n1}x_1^{(k+1)} + a_{n2}x_2^{(k+1)} + \cdots + a_{n,n-1}x_{n-1}^{(k+1)})] \qquad \text{...(13.10)}$$

EXAMPLE 13.28 Perform four iteration of the Gauss-Seidel method for finding the solution of the linear system of equations

$$2x_1 + x_2 + 4x_3 = 3,$$
$$x_1 + 5x_2 + x_3 = 5,$$
$$4x_1 + x_2 + 2x_3 = -1.$$

Take the initial approximation $\mathbf{x}^{(0)} = \mathbf{0}$. Compare with the exact solution, $\mathbf{x} = [-1, 1, 1]^T$.

Solution After exchanging the first and third equations, the new system of equations becomes

$$\begin{bmatrix} 4 & 1 & 2 \\ 1 & 5 & 1 \\ 2 & 1 & 4 \end{bmatrix} \mathbf{x} = \begin{bmatrix} -1 \\ 5 \\ 3 \end{bmatrix}.$$

We write the Gauss-Seidel method as

$$x_1^{(k+1)} = \frac{1}{4}[-1 - x_2^{(k)} - 2x_3^{(k)}], \quad x_2^{(k+1)} = \frac{1}{5}[5 - x_1^{(k+1)} - x_3^{(k)}],$$

$$x_3^{(k+1)} = \frac{1}{4}[3 - 2x_1^{(k+1)} - x_2^{(k+1)}], \quad k = 0, 1, 2, 3, \ldots\ldots$$

The initial approximation is $x_1^{(0)} = x_2^{(0)} = x_3^{(0)} = 0$. For $k = 0, 1, 2, 3$, we have the following results.

$k = 0$: $\quad x_1^{(1)} = \frac{1}{4}[-1 - 0 - 0] = -0.25, \qquad x_2^{(1)} = \frac{1}{5}[5 + 0.25 - 0] = 1.05,$

$$x_3^{(1)} = \frac{1}{4}[3 + 2(0.25) - 1.05] = 0.6125.$$

$k = 1$: $\quad x_1^{(2)} = \frac{1}{4}[-1 - 1.05 - 2(0.6125)] = -0.8188, \qquad x_2^{(2)} = \frac{1}{5}[5 + 0.8188 - 0.6125] = 1.0413,$

$$x_3^{(2)} = \frac{1}{4}[3 + 2(0.8188) - 1.0413] = 0.8991.$$

$k = 2$: $\quad x_1^{(3)} = \frac{1}{4}[-1 - 1.0413 - 2(0.8991)] = -0.9599, \qquad x_2^{(3)} = \frac{1}{5}[5 + 0.9599 - 0.8991] = 1.0122,$

$$x_3^{(3)} = \frac{1}{4}[3 + 2(0.9599) - 1.0122] = 0.9769.$$

$k = 3$: $\quad x_1^{(4)} = \frac{1}{4}[-1 - 1.0122 - 2(0.9769)] = -0.9915, \qquad x_2^{(4)} = \frac{1}{5}[5 + 0.9915 - 0.9769] = 1.0029,$

$$x_3^{(4)} = \frac{1}{4}[3 + 2(0.9915) - 1.0029] = 0.9950.$$

The errors in magnitude of the components of the solution vector are 0.0085, 0.0029, 0.0050. The maximum absolute error is 0.0085.

EXAMPLE 13.29 Solve the system of linear equations

$$2x_1 + 15x_2 + 6x_3 = 72$$
$$54x_1 + x_2 + x_3 = 110$$
$$-x_1 + 6x_2 + 27x_3 = 85$$

by Gauss-Seidel method up to two places of decimal. Starting with $(x_1^{(0)}, x_2^{(0)}, x_3^{(0)}) = (0, 0, 0)$.

Solution After exchanging the first and second equations, the new system of equations becomes

$$\begin{bmatrix} 54 & 1 & 1 \\ 2 & 15 & 6 \\ -1 & 6 & 27 \end{bmatrix} \begin{bmatrix} x_1 \\ x_2 \\ x_3 \end{bmatrix} = \begin{bmatrix} 110 \\ 72 \\ 85 \end{bmatrix}.$$

We write the Gauss-Seidel method as

$$x_1^{(k+1)} = \frac{1}{54}[110 - x_2^{(k)} - x_3^{(k)}], \quad x_2^{(k+1)} = \frac{1}{15}[72 - 2x_1^{(k+1)} - 6x_3^{(k)}],$$

$$x_3^{(k+1)} = \frac{1}{27}[85 + x_1^{(k+1)} - 6x_2^{(k+1)}], \qquad k = 0, 1, 2, \ldots\ldots .$$

The initial approximation is $x_1^{(0)} = x_2^{(0)} = x_3^{(0)} = 0$. For $k = 0, 1, 2$, we have the following results.

$k = 0$: $\quad x_1^{(1)} = \frac{1}{54}[110 - 0 - 0] = 2.04, \quad x_2^{(1)} = \frac{1}{15}[72 - 2(2.04) - 0] = 4.53,$

$$x_3^{(1)} = \frac{1}{27}[85 + 2.04 - 6(4.53)] = 2.22.$$

$k = 1$: $\quad x_1^{(2)} = \frac{1}{54}[110 - 4.53 - 2.22] = 1.91, \quad x_2^{(2)} = \frac{1}{15}[72 - 2(1.91) - 6.(2.22)] = 3.66,$

$$x_3^{(2)} = \frac{1}{27}[85 + 1.91 - 6(3.66)] = 2.41.$$

$k = 2$: $\quad x_1^{(3)} = \frac{1}{54}[110 - 3.66 - 2.41] = 1.92, \quad x_2^{(3)} = \frac{1}{15}[72 - 2(1.92) - 6(2.41)] = 3.58,$

$$x_3^{(3)} = \frac{1}{27}[85 + 1.92 - 6(3.58)] = 2.42.$$

$k = 3$: $\quad x_1^{(4)} = \frac{1}{54}[110 - 3.58 - 2.42] = 1.93, \quad x_2^{(4)} = \frac{1}{15}[72 - 2(1.93) - 6(2.42)] = 3.57,$

$$x_3^{(4)} = \frac{1}{27}[85 + 1.93 - 6(3.57)] = 2.43.$$

$k = 4$: $\quad x_1^{(5)} = \frac{1}{54}[110 - 3.57 - 2.43] = 1.93, \quad x_2^{(5)} = \frac{1}{15}[72 - 2(1.93) - 6(2.43)] = 3.57,$

$$x_3^{(5)} = \frac{1}{27}[85 + 1.93 - 6(3.57)] = 2.43.$$

Thus $[x_1^{(5)}, x_2^{(5)}, x_3^{(5)}]^T = (1.93, 3.57, 2.43)^T$.

The last vector can be taken as the solution vector with two places of accuracy. This took 5 iterations.

EXAMPLE 13.30 Perform five iteration of the Gauss-Seidel method for solving the system of equations

$$\begin{bmatrix} 5 & 2 & 1 \\ 1 & 4 & 2 \\ 1 & 2 & 5 \end{bmatrix} \mathbf{x} = \begin{bmatrix} 12 \\ 15 \\ 20 \end{bmatrix}.$$

Take the components of the approximate initial vector as $x_i^{(0)} = b_i / a_{ii}$, $i = 1, 2, 3$.

Solution We write the Gauss–Seidel method as

$$x_1^{(k+1)} = \frac{1}{5}[12 - 2x_2^{(k)} - x_3^{(k)}], \quad x_2^{(k+1)} = \frac{1}{4}[15 - x_1^{(k+1)} - 2x_3^{(k)}],$$

$$x_3^{(k+1)} = \frac{1}{5}[20 - x_1^{(k+1)} - x_2^{(k+1)}], \qquad k = 0, 1, 2, 3, 4.$$

The initial approximation is $x_1^{(0)} = 12/5 = 2.4$, $x_2^{(0)} = 15/4 = 3.75$, $x_3^{(0)} = 20.5 = 4$.

For $k = 0, 1, 2, 3, 4$ we have the following results.

$k = 0$: $\quad x_1^{(1)} = \frac{1}{5}[12 - 2(3.75) - 4] = 0.1, \qquad x_2^{(1)} = \frac{1}{4}[15 - 0.1 - 2(4)] = 1.725,$

$$x_3^{(1)} = \frac{1}{5}[20 - 0.1 - 1.725] = 3.635.$$

$k = 1$: $\quad x_1^{(2)} = \frac{1}{5}[12 - 2(1.725) - 3.635] = 0.983, \qquad x_2^{(2)} = \frac{1}{4}[15 - 0.983 - 2(3.635)] = 1.687,$

$$x_3^{(2)} = \frac{1}{5}[20 - 0.983 - 1.687] = 3.466.$$

$k = 2$: $\quad x_1^{(3)} = \frac{1}{5}[12 - 2(1.687) - 3.466] = 1.032, \qquad x_2^{(3)} = \frac{1}{4}[15 - 1.032 - 2(3.466)] = 1.759,$

$$x_3^{(3)} = \frac{1}{5}[20 - 1.032 - 1.759] = 3.442.$$

$k = 3$: $\quad x_1^{(4)} = \frac{1}{5}[12 - 2(1.759) - 3.442] = 1.008, \qquad x_2^{(4)} = \frac{1}{4}[15 - 1.008 - 2(3.442)] = 1.777,$

$$x_3^{(4)} = \frac{1}{5}[20 - 1.008 - 1.777] - 3.443.$$

$k = 4$: $\quad x_1^{(5)} = \frac{1}{5}[12 - 2(1.777) - 3.443] = 1.001, \qquad x_2^{(5)} = \frac{1}{4}[15 - 1.001 - 2(3.443)] = 1.778,$

$$x_3^{(5)} = \frac{1}{5}[20 - 1.001 - 1.778] = 3.444.$$

$$[x_1^{(5)}, x_2^{(5)}, x_3^{(5)}]^T = (1.001, 1.778, 3.444)^T.$$

The exact solution $\mathbf{x} = (1, 2, 3)^T$.

EXAMPLE 13.31 Perform three iterations of the Gauss-Seidel method for solving the system of equations

$$\begin{bmatrix} 4 & 0 & 2 \\ 0 & 5 & 2 \\ 5 & 4 & 10 \end{bmatrix} \begin{bmatrix} x_1 \\ x_2 \\ x_3 \end{bmatrix} = \begin{bmatrix} 6 \\ -3 \\ 11 \end{bmatrix}.$$

Take the components of the approximate initial vector as $x_i^{(0)} = b_i / a_{ii}$, $i = 1, 2, 3$. Compare with the exact solution $\mathbf{x} = [1, -1, 1]^T$. *(AMIETE, June, 2011)*

Solution We write the Gauss-Seidel method as

$$x_1^{(k+1)} = \frac{1}{4}[6 - 2x_3^{k}], \qquad x_2^{(k+1)} = \frac{1}{5}[-3 - 2x_3^{(k)}],$$

$$x_3^{(k+1)} = \frac{1}{10}[11 - 5x_1^{(k+1)} - 4x_2^{(k+1)}], \qquad k = 0, 1, 2.$$

The initial approximation is $x_1^{(0)} = 6/4 = 1.5$, $x_2^{(0)} = -3/5 = -0.6$, $x_3^{(0)} = 11/10 = 1.1$.
For $k = 0, 1, 2$, we have the following results.

$k = 0$: $x_1^{(1)} = \frac{1}{4}[6 - 2(1.1)] = 0.95$, $x_2^{(1)} = \frac{1}{5}[-3 - 2(1.1)] = -1.04$,

$x_3^{(1)} = \frac{1}{10}[11 - 5(0.95) - 4(-1.04)] = 1.041.$

$k = 1$: $x_1^{(2)} = \frac{1}{4}[6 - 2(1.041)] = 0.9795$, $x_2^{(2)} = \frac{1}{5}[-3 - 2(1.041)] = -1.0164$,

$x_3^{(2)} = \frac{1}{10}[11 - 5(0.9795) - 4(-1.0164)] = 1.0168.$

$k = 2$: $x_1^{(3)} = \frac{1}{4}[6 - 2(1.0168)] = 0.9916$, $x_2^{(3)} = \frac{1}{5}[-3 - 2(1.0168)] = -1.0067$,

$x_3^{(3)} = \frac{1}{10}[11 - 5(0.9916) - 4(-1.0067)] = 1.0069.$

The errors in magnitude of the components of the solution vector are 0.0084, 0.0067, 0.0069. The maximum absolute error is 0.0084.

13.4 DETERMINATION OF EIGEN VALUES BY ITERATION—THE POWER METHOD

We now describe a numerical method that calculates the eigenvalue of **A** that is largest in size by the power method. In this method an initial approximation to the solution vector $\mathbf{x}^{(0)}$ ($\neq 0$) is assumed and it is improved by iteration. Starting with the initial approximation $\mathbf{x}^{(0)}$ with n components and compute successively $\mathbf{x}^{(1)} = \mathbf{A}\mathbf{x}^{(0)}$, $\mathbf{x}^{(2)} = \mathbf{A}\mathbf{x}^{(1)}$, ..., $\mathbf{x}^{(r)} = \mathbf{A}\mathbf{x}^{(r-1)}$. First of all $\mathbf{A}\mathbf{x}^{(0)}$ is obtained by matrix multiplication and written as $\lambda^{(1)}\,\mathbf{x}^{(1)}$ after normalisation. This gives the first approximation $\lambda^{(1)}$ and $\mathbf{x}^{(1)}$ to the eigenvalue and eigen vector. Starting from $\mathbf{x}^{(1)}$ and calculating $\mathbf{A}\mathbf{x}^{(1)} = \lambda^{(2)}\mathbf{x}^{(2)}$, we get the second approximation. This process is repeated till the difference $\mathbf{x}^{(r)} - \mathbf{x}^{(r-1)}$ becomes negligible. Then $\lambda^{(r)}$ will be the largest eigen values of $\mathbf{A}\mathbf{x} = \lambda\mathbf{x}$ and $\mathbf{x}^{(r)}$, the corresponding eigen vector. This method is applicable to any $n \times n$ matrix **A** that has a ***dominant*** eigenvalue (*a*λ such that $|\lambda|$ is greater than the absolute values of the other eigenvalues). To obtain the smallest eigenvalue, we could invert **A** and solve the equation

$$\mathbf{A}^{-1}\mathbf{x} = (1/\lambda)\mathbf{x}. \qquad ...(13.11)$$

EXAMPLE 13.32 Determine the largest eigen value and the corresponding eigen vector of the following matrices **A**, using power method.

(*i*) $\mathbf{A} = \begin{bmatrix} 9 & 4 \\ 4 & 3 \end{bmatrix}$ (*ii*) $\mathbf{A} = \begin{bmatrix} 2 & -1 & 0 \\ -1 & 2 & -1 \\ 0 & -1 & 2 \end{bmatrix}$ (*AMIETE, June 2010; VTU, 2007*)

Solution (*i*) Let the initial approximation to the eigen vector corresponding to the largest eigen value of **A** be of $\mathbf{x}^{(0)} = [1 \quad 0]^T$.

Then, $$\mathbf{A}\mathbf{x}^{(0)} = \begin{bmatrix} 9 & 4 \\ 4 & 3 \end{bmatrix}\begin{bmatrix} 1 \\ 0 \end{bmatrix} = \begin{bmatrix} 9 \\ 4 \end{bmatrix} = 9\begin{bmatrix} 1 \\ 0.444 \end{bmatrix} = \lambda^{(1)}\,\mathbf{x}^{(1)}.$$

The first approximation to the eigen value is $\lambda^{(1)} = 9$ and the corresponding eigen vector is $\mathbf{x}^{(1)} = [1 \quad 0.444]^T$.

Now $$\mathbf{A}\mathbf{x}^{(1)} = \begin{bmatrix} 9 & 4 \\ 4 & 3 \end{bmatrix}\begin{bmatrix} 1 \\ 0.444 \end{bmatrix} = \begin{bmatrix} 10.776 \\ 5.332 \end{bmatrix} = 10.776\begin{bmatrix} 1 \\ 0.494 \end{bmatrix} = \lambda^{(2)}\,\mathbf{x}^{(2)}.$$

Thus the second approximation to the eigen-value is $\lambda^{(2)} = 10.776$ and the corresponding eigen-vector is $\mathbf{x}^{(2)} = [1 \quad 0.494]^T$. Repeating the above process, we obtain

$$\mathbf{A}\mathbf{x}^{(2)} = \begin{bmatrix} 9 & 4 \\ 4 & 3 \end{bmatrix}\begin{bmatrix} 1 \\ 0.494 \end{bmatrix} = 10.976\begin{bmatrix} 1 \\ 0.499 \end{bmatrix} = \lambda^{(3)}\,\mathbf{x}^{(3)}.$$

$$\mathbf{Ax}^{(3)} = \begin{bmatrix} 9 & 4 \\ 4 & 3 \end{bmatrix}\begin{bmatrix} 1 \\ 0.499 \end{bmatrix} = 10.996\begin{bmatrix} 1 \\ 0.4999 \end{bmatrix} = \lambda^{(4)}\,\mathbf{x}^{(4)}.$$

$$\mathbf{Ax}^{(4)} = \begin{bmatrix} 9 & 4 \\ 4 & 3 \end{bmatrix}\begin{bmatrix} 1 \\ 0.4999 \end{bmatrix} = 10.9996\begin{bmatrix} 1 \\ 0.4999 \end{bmatrix} = \lambda^{(5)}\,\mathbf{x}^{(5)}.$$

Hence the largest eigen value is 11 and the corresponding eigen vector is $[1 \quad 0.5]^T$.

(*ii*) Let the initial eigen vector $\mathbf{x}^{(0)} = [1 \quad 0 \quad 0]^T$.

Then, we have
$$\mathbf{Ax}^{(0)} = \begin{bmatrix} 2 & -1 & 0 \\ -1 & 2 & -1 \\ 0 & -1 & 2 \end{bmatrix}\begin{bmatrix} 1 \\ 0 \\ 0 \end{bmatrix} = \begin{bmatrix} 2 \\ -1 \\ 0 \end{bmatrix} = 2\begin{bmatrix} 1 \\ -0.5 \\ 0 \end{bmatrix} = \lambda^{(1)}\mathbf{x}^{(1)}.$$

The first approximation to the eigen value $\lambda^{(1)} = 2$ and the corresponding eigen vector $\mathbf{x}^{(1)} = [1 \quad -0.5 \quad 0]^T$. Hence, we have

$$\mathbf{Ax}^{(1)} = \begin{bmatrix} 2 & -1 & 0 \\ -1 & 2 & -1 \\ 0 & -1 & 2 \end{bmatrix}\begin{bmatrix} 1 \\ -0.5 \\ 0 \end{bmatrix} = \begin{bmatrix} 2.5 \\ -2 \\ 0.5 \end{bmatrix} = 2.5\begin{bmatrix} 1 \\ -0.8 \\ 0.2 \end{bmatrix} = \lambda^{(2)}\mathbf{x}^{(2)}.$$

Repeating the above procedure, we obtain successively

$$\mathbf{Ax}^{(2)} = 2.8\begin{bmatrix} 1 \\ -1 \\ 0.43 \end{bmatrix} = \lambda^{(3)}\mathbf{x}^{(3)}, \quad \mathbf{Ax}^{(3)} = 3.43\begin{bmatrix} 0.87 \\ -1 \\ 0.54 \end{bmatrix} = \lambda^{(4)}\mathbf{x}^{(4)}, \quad \mathbf{Ax}^{(4)} = 3.41\begin{bmatrix} 0.80 \\ -1 \\ 0.61 \end{bmatrix} = \lambda^{(5)}\mathbf{x}^{(5)},$$

$$\mathbf{Ax}^{(5)} = 3.41\begin{bmatrix} 0.76 \\ -1 \\ 0.65 \end{bmatrix} = \lambda^{(6)}\mathbf{x}^{(6)}, \quad \mathbf{Ax}^{(6)} = 3.41\begin{bmatrix} 0.74 \\ -1 \\ 0.67 \end{bmatrix} = \lambda^{(7)}\mathbf{x}^{(7)}.$$

We note $\lambda^{(6)} = \lambda^{(7)}$ and $\mathbf{x}^{(6)} \approx \mathbf{x}^{(7)}$. Hence the largest eigenvalue is 3.41 and the corresponding eigenvector is $[0.74 \quad -1 \quad 0.67]^T$.

EXERCISE 13.2

1. Solve the following system of equations using Gauss elimination method.

(*i*) $4x_1 - x_2 + 2x_3 = 15,$
$-x_1 + 2x_2 + 3x_3 = 5,$
$5x_1 - 7x_2 + 9x_3 = 8.$

(*ii*) $2x_1 + 6x_2 - x_3 = 23,$
$4x_1 - x_2 + 3x_3 = 9,$
$3x_1 + x_2 + 2x_3 = 13.$

(*iii*) $4x_1 + x_2 + x_3 = 5,$
$2x_1 + 5x_2 - 2x_3 = 9,$
$2x_1 + 3x_2 + 6x_3 = 13.$

(*iv*) $x_1 + x_2 + x_3 = -1,$
$3x_1 + 3x_2 + 5x_3 = -2,$
$2x_1 + 5x_2 - 4x_3 = -11.$

Ans. (*i*) 4, 3, 1; (*ii*) 1, 4, 3; (*iii*) 1/2, 2, 1; (*iv*) 1/2, –2, 1/2.

2. Using the Gauss elimination method, solve the system of equations

$$\begin{bmatrix} 1 & 1 & 6 & 1 \\ 1 & 1 & 1 & 4 \\ 5 & 1 & 1 & 1 \\ 1 & 7 & 1 & 1 \end{bmatrix}\begin{bmatrix} x_1 \\ x_2 \\ x_3 \\ x_4 \end{bmatrix} = \begin{bmatrix} -5 \\ -6 \\ 4 \\ 12 \end{bmatrix}.$$

Ans. 1, 2, –1, –2

3. Using the Gauss elimination method, show that the following system of equations is inconsistent

$$\begin{bmatrix} 0 & 4 & 3 \\ 2 & 0 & -1 \\ 3 & 2 & 0 \end{bmatrix} \begin{bmatrix} x_1 \\ x_2 \\ x_3 \end{bmatrix} = \begin{bmatrix} 8 \\ 2 \\ 5 \end{bmatrix}.$$

Ans. Gauss elimination gives the last row of the augmented matrix 0, 0, 0, 4. Since this implies 0 = 4, the system is inconsistent.

4. Using Gauss-Jordan method, solve the system of equations

(*i*) $x_1 - x_2 + x_3 = 0$, $2x_1 + x_2 - 3x_3 = 4$, $x_1 + x_2 + x_3 = 1$.

(*ii*) $x_1 - x_2 - x_3 = 8$, $x_1 - x_2 + x_3 = 3$, $-x_1 + x_2 + x_3 = 4$.

(*iii*) $x_1 + x_2 + x_3 = 9$, $2x_1 - 3x_2 + 4x_3 = 13$, $3x_1 + 4x_2 + 5x_3 = 40$.

Ans. (*i*) 1, 1/2, –1/2; (*ii*) The system has no solution-inconsistent; (*iii*) 1, 3, 5.

5. Find the inverse of the following matrices using Gauss-Jordan method.

(*i*) $\begin{bmatrix} 1 & -1 & 1 \\ 3 & -1 & 2 \\ 2 & -2 & 3 \end{bmatrix}$, (*ii*) $\begin{bmatrix} 8 & 4 & -3 \\ 2 & 1 & 1 \\ 1 & 2 & 1 \end{bmatrix}$,

(*iii*) $\begin{bmatrix} 1 & 2 & 3 \\ 2 & 4 & 5 \\ 3 & 5 & 6 \end{bmatrix}$, (*AMIETE, June 2009*) (*iv*) $\begin{bmatrix} 0 & 1 & 2 \\ 1 & 2 & 3 \\ 3 & 1 & 1 \end{bmatrix}$.

Ans. (*i*) $\frac{1}{2}\begin{bmatrix} 1 & 1 & -1 \\ -5 & 1 & 1 \\ -4 & 0 & 2 \end{bmatrix}$, (*ii*) $\frac{1}{21}\begin{bmatrix} 1 & 10 & -7 \\ 1 & -11 & 14 \\ -3 & 12 & 0 \end{bmatrix}$, (*iii*) $\begin{bmatrix} 1 & -3 & 2 \\ -3 & 3 & -1 \\ 2 & -1 & 0 \end{bmatrix}$, (*iv*) $\frac{1}{2}\begin{bmatrix} 1 & -1 & 1 \\ -8 & 6 & -2 \\ 5 & -3 & 1 \end{bmatrix}$.

6. Using the LU decomposition method, solve the system of equations

$$3x_1 + 2x_2 + 7x_3 = 4,$$
$$2x_1 + 3x_2 + x_3 = 5,$$
$$3x_1 + 4x_2 + x_3 = 7.$$

Choose $l_{11} = l_{22} = l_{33} = 1$. Also find the inverse of the matrix.

Ans. $\mathbf{x} = [7/8, 9/8, -1/8]^T$; $\mathbf{A}^{-1} = \frac{1}{8}\begin{bmatrix} 1 & -26 & 19 \\ -1 & 18 & -11 \\ 1 & 6 & -5 \end{bmatrix}$.

7. Solve the system of equations **Ax** = **b**, where

$$\mathbf{A} = \begin{bmatrix} 2 & -4 & 3 \\ 4 & 1 & -6 \\ 5 & 8 & -4 \end{bmatrix}, \mathbf{b} = \begin{bmatrix} 1 \\ -1 \\ 9 \end{bmatrix}$$

by factorization method. Choose $u_{ii} = 1$, $i = 1, 2, 3$. Also find $\mathbf{A}^{-1}$.

[**Hint:** We write

$$\begin{bmatrix} 2 & -4 & 3 \\ 4 & 1 & -6 \\ 5 & 8 & -4 \end{bmatrix} = \begin{bmatrix} l_{11} & 0 & 0 \\ l_{21} & l_{22} & 0 \\ l_{31} & l_{32} & l_{33} \end{bmatrix} \begin{bmatrix} 1 & u_{12} & u_{13} \\ 0 & 1 & u_{23} \\ 0 & 0 & 1 \end{bmatrix} = \begin{bmatrix} l_{11} & l_{11}u_{12} & l_{11}u_{13} \\ l_{21} & l_{21}u_{12} + l_{22} & l_{21}u_{13} + l_{22}u_{23} \\ l_{31} & l_{31}u_{12} + l_{32} & l_{31}u_{13} + l_{32}u_{23} + l_{33} \end{bmatrix}.$$

Comparing the corresponding elements on both sides, we obtain

first column of **L**: $l_{11} = 2$; $l_{21} = 4$; $l_{31} = 5$;

first row of **U**: $l_{11}u_{12} = -4$, $u_{12} = -4/2 = -2$; $l_{11}u_{13} = 3$, $u_{13} = 3/2$;
second column of **L**: $l_{21}u_{12} + l_{22} = 1$, $l_{22} = 1 - (4)(-2) = 9$; $l_{31}u_{12} + l_{32} = 8$, $l_{32} = 8 - (5)(-2) = 18$;
second row of **U**: $l_{21}u_{13} + l_{22}u_{23} = -6$, $u_{23} = [-6 - (4)(3/2)]/9 = -4/3$;
third column of **L**: $l_{31}u_{13} + l_{32}u_{23} + l_{33} = -4$, $l_{33} = -4 - (5)(3/2) - (18)(-4/3) = 25/2$.
Therefore, we obtain

$$\mathbf{L} = \begin{bmatrix} 2 & 0 & 0 \\ 4 & 9 & 0 \\ 5 & 18 & 25/2 \end{bmatrix}, \quad \mathbf{U} = \begin{bmatrix} 1 & -2 & 3/2 \\ 0 & 1 & -4/3 \\ 0 & 0 & 1 \end{bmatrix}.$$

From the system **Lz** = **b**

$$\begin{bmatrix} 2 & 0 & 0 \\ 4 & 9 & 0 \\ 5 & 18 & 25/2 \end{bmatrix} \begin{bmatrix} z_1 \\ z_2 \\ z_3 \end{bmatrix} = \begin{bmatrix} 1 \\ -1 \\ 9 \end{bmatrix},$$

we obtain by forward substitution

$$z_1 = \frac{1}{2}, z_2 = \frac{1}{9}[-1 - 4(1/2)] = -\frac{1}{3}, z_3 = \frac{2}{25}\left[9 - 5\left(\frac{1}{2}\right) - 18\left(-\frac{1}{3}\right)\right] = 1.$$

From the system **Ux** = **z**

$$\begin{bmatrix} 1 & -2 & 3/2 \\ 0 & 1 & -4/3 \\ 0 & 0 & 1 \end{bmatrix} \begin{bmatrix} x_1 \\ x_2 \\ x_3 \end{bmatrix} = \begin{bmatrix} 1/2 \\ -1/3 \\ 1 \end{bmatrix},$$

We obtain by backward substitution
$x_3 = 1$, $x_2 = (-1/3) + (4/3) = 1$, $x_1 = (1/2) + 2(1) - (3/2)(1) = 1$. Solution $\mathbf{x} = [1 \quad 1 \quad 1]^T$.
Now, from $\mathbf{LL}^{-1} = \mathbf{I}$

$$\begin{bmatrix} 2 & 0 & 0 \\ 4 & 9 & 0 \\ 5 & 18 & 25/2 \end{bmatrix} \begin{bmatrix} l'_{11} & 0 & 0 \\ l'_{21} & l'_{22} & 0 \\ l'_{31} & l'_{32} & l'_{33} \end{bmatrix} = \begin{bmatrix} 1 & 0 & 0 \\ 0 & 1 & 0 \\ 0 & 0 & 1 \end{bmatrix},$$

we get by forward substitution $\mathbf{L}^{-1} = \dfrac{1}{450}\begin{bmatrix} 225 & 0 & 0 \\ -100 & 50 & 0 \\ 54 & -72 & 36 \end{bmatrix}$.

From $\mathbf{UU}' = \mathbf{I}$,

$$\begin{bmatrix} 1 & -2 & 3/2 \\ 0 & 1 & -4/3 \\ 0 & 0 & 1 \end{bmatrix} \begin{bmatrix} u'_{11} & u'_{12} & u'_{13} \\ 0 & u'_{22} & u'_{23} \\ 0 & 0 & u'_{33} \end{bmatrix} = \begin{bmatrix} 1 & 0 & 0 \\ 0 & 1 & 0 \\ 0 & 0 & 1 \end{bmatrix}.$$

we get by backward substitution

$$\mathbf{U}^{-1} = \frac{1}{6}\begin{bmatrix} 6 & 12 & 7 \\ 0 & 6 & 8 \\ 0 & 0 & 6 \end{bmatrix}.$$

Therefore, $\mathbf{A}^{-1} = \mathbf{U}^{-1}\mathbf{L}^{-1} = \dfrac{1}{2700}\begin{bmatrix} 6 & 12 & 7 \\ 0 & 6 & 8 \\ 0 & 0 & 6 \end{bmatrix}\begin{bmatrix} 225 & 0 & 0 \\ -100 & 50 & 0 \\ 54 & -72 & 36 \end{bmatrix} = \dfrac{1}{225}\begin{bmatrix} 44 & 8 & 21 \\ -14 & -23 & 24 \\ 27 & -36 & 18 \end{bmatrix}$.

Check $\quad \mathbf{AA}^{-1} = \dfrac{1}{225}\begin{bmatrix} 2 & -4 & 3 \\ 4 & 1 & -6 \\ 5 & 8 & -4 \end{bmatrix}\begin{bmatrix} 44 & 8 & 21 \\ -14 & -23 & 24 \\ 27 & -36 & 18 \end{bmatrix} = \begin{bmatrix} 1 & 0 & 0 \\ 0 & 1 & 0 \\ 0 & 0 & 1 \end{bmatrix} = \mathbf{I}.]$

8. The matrix **A** is decomposed as **A** = **LU**, where

$$\begin{bmatrix} 4 & 1 & 1 \\ 1 & 4 & -2 \\ 3 & 2 & -4 \end{bmatrix} = \begin{bmatrix} 4 & 0 & 0 \\ 1 & 15/4 & 0 \\ 3 & a & -4 \end{bmatrix} \begin{bmatrix} 1 & 1/4 & 1/4 \\ 0 & 1 & b \\ 0 & 0 & 1 \end{bmatrix}$$

Find the values of (a, b). **Ans.** (5/4, –3/5)

9. Using the Choleski decomposition method, solve the following system of equations

(*i*) $4x_1 + 2x_2 + 14x_3 = 14,$
$2x_1 + 17x_2 - 5x_3 = -101,$
$14x_1 - 5x_2 + 83x_3 = 155.$

(*ii*) $\begin{bmatrix} 4 & 0 & 1 \\ 0 & 4 & 3 \\ 1 & 3 & 37/2 \end{bmatrix} \begin{bmatrix} x_1 \\ x_2 \\ x_3 \end{bmatrix} = \begin{bmatrix} 5 \\ 13 \\ 59 \end{bmatrix},$

(*iii*) $x + 2y + z = 0,$
$2x + 5y = -3,$
$x + 13z = 14.$

Ans. (*i*) 3, –6, 1; (*ii*) 0.5, 1, 3; (*iii*) 1, –1, 1.

10. The matrix $\mathbf{A} = \begin{bmatrix} 4 & 1 & 1 \\ 1 & 2 & 1 \\ 1 & 1 & 3 \end{bmatrix}$ is given. Find the inverse, $\mathbf{A}^{-1}$, using the Choleski method.

Ans. $\dfrac{1}{17}\begin{bmatrix} 5 & -2 & -1 \\ -2 & 11 & -3 \\ -1 & -3 & 7 \end{bmatrix}.$

11. Find the inverse of the matrix $\begin{bmatrix} 1 & 2 & 3 \\ 2 & 8 & 22 \\ 3 & 22 & 82 \end{bmatrix}$, using Choleski method. (*AMIETE, Dec. 2006*)

Ans. $\dfrac{1}{36}\begin{bmatrix} 172 & -98 & 20 \\ -98 & 73 & -16 \\ 20 & -16 & 4 \end{bmatrix}.$

12. Solve by Gauss–Jacobi iteration method the following system of equations

$$27x + 6y - z = 85,$$
$$6x + 15y + 2z = 72,$$
$$x + y + 54z = 110.$$

Take the initial approximation as $\mathbf{x}^{(0)} = \mathbf{0}$. **Ans.** 2.4255, 2.5730, 1.9260

13. Perform three iterations of the Gauss-Jacobi iteration method for solving the system of equations

$$6x_1 + x_2 + 2x_3 = 6,$$
$$x_1 + 4x_2 + 3x_3 = -4,$$
$$2x_1 + x_2 + 8x_3 = 8.$$

Take the initial approximation as $\mathbf{x}^{(0)} = [1.3, -1.9, 0.8]^T$. Compare with the exact solution $x_1 = 1$, $x_2 = -2$, $x_3 = 1$; and find the maximum absolute error.

Ans. $(x_1^{(3)}, x_2^{(3)}, x_3^{(3)})^T = (1.9984, -1.9878, 0.9892)^T$. The errors in magnitude are 0.0016, 0.0122 and 0.0108. The maximum absolute error is 0.0122.

14. Solve the system of linear equations

$$28x + 4y - z = 32,$$
$$x + 3y + 10z = 24,$$
$$2x + 17y + 4z = 35,$$

by Gauss-Seidel method, up to three places of decimal in the computation. Take the initial approximation $\mathbf{x}^{(0)} = \mathbf{0}$. **Ans.** 0.998, 1.723, 2.024

15. Perform five iterations of the Gauss-Seidel method for solving the system of equations

$$(i)\ \begin{bmatrix} 27 & 6 & -1 \\ 6 & 15 & 2 \\ 1 & 1 & 54 \end{bmatrix}\begin{bmatrix} x_1 \\ x_2 \\ x_3 \end{bmatrix} = \begin{bmatrix} 85 \\ 72 \\ 110 \end{bmatrix}, \qquad (ii)\ \begin{bmatrix} 5 & 2 & 1 \\ 1 & 4 & 2 \\ 1 & 2 & 5 \end{bmatrix}\begin{bmatrix} x_1 \\ x_2 \\ x_3 \end{bmatrix} = \begin{bmatrix} 12 \\ 15 \\ 20 \end{bmatrix}.$$

Take the initial approximation $\mathbf{x}^{(0)} = \mathbf{0}$ in each case. **Ans.** (*i*) 2.425, 3.573, 1.927; (*ii*) 0.996, 2, 3

16. The following system of equations are given

$$4x + 2z = 6,$$
$$5y + 2z = -3,$$
$$5x + 4y + 10z = 11.$$

Iterate three times using the Gauss-Seidel method, starting with initial approximations $x^{(0)} = 1.5$, $y^{(0)} = -0.6$ and $z^{(0)} = 1.1$. **Ans.** 0.9916, –1.0067, 1.0069.

17. Find the solution of the system of equations correct to two places of decimal

$$4x + y + z + w = 1,$$
$$x + 6y - 3z + 2w = 4,$$
$$3x + 4y - 7z + 3w = 13,$$
$$x + y + 2z + 5w = 2,$$

using the Gauss-Seidel method. Assume the initial approximation as $[0.4, -0.55, -1.4, 1.1]^T$.

18. Determine the largest eigenvalue and the corresponding eigenvector of the matrices, using the power method.

(*i*) $\mathbf{A} = \begin{bmatrix} 5 & 4 \\ 1 & 2 \end{bmatrix}$ with initial approximate to the eigen vector as $[1 \quad 0]^T$

(*ii*) $\mathbf{A} = \begin{bmatrix} 1 & 6 & 1 \\ 1 & 2 & 0 \\ 0 & 0 & 3 \end{bmatrix}$ with initial approximate to the eigen vector as $[1 \quad 0 \quad 0]^T$

(*iii*) $\mathbf{A} = \begin{bmatrix} 1 & 3 & -1 \\ 3 & 2 & 4 \\ -1 & 4 & 10 \end{bmatrix}$ with initial approximate to the eigen vector as $[1 \quad 1 \quad 0]^T$

(*iv*) $\mathbf{A} = \begin{bmatrix} 2 & -1 & 0 \\ -1 & 2 & -1 \\ 0 & -1 & 2 \end{bmatrix}$ with initial approximate to the eigen vector as $[1 \quad 1 \quad 1]^T$.

Ans. (*i*) 6, $[1 \quad 0.25]^T$; (*ii*) 4, $[2 \quad 1 \quad 0]^T$; (*iii*) 11.66, $[0.025 \quad 0.422 \quad 1.0]^T$; (*iv*) 3.42, $[0.71 \quad -1 \quad 0.71]^T$.

13.5 NUMERICAL SOLUTION OF ORDINARY DIFFERENTIAL EQUATIONS

An equation involving x, y, dy/dx, d^2y/dx^2 etc., is known as an ordinary differential equation. A majority of such equations defining physical systems do not, in general possess closed form solutions and it, therefore, becomes imperative to discuss their solution by numerical methods.

To discuss the application of various numerical methods we consider the solution of the general, first order, first degree differential equation, namely

$$dy/dx = f(x, y) \qquad ...(13.12)$$

with the initial condition $y = y_0$ when $x = x_0$, that is $y(x_0) = y_0$.

The solutions of Eq. (13.12) discussed below are of the following two forms.

(*i*) The dependent variable y is expressed as a power series in independent variable x. This method is sometimes known as *semi-analytical method.*

(*ii*) The solution is obtained purely by numerical methods by using the set of tabulated values of x and y. The methods used under this class are known as *step-by-step method* or *marching methods* because the values of the dependent variable are calculated by short steps for equal interval values of the independent variable.

The **Picard's method** and the **Taylor's series method** fall under category, (*i*) while **Euler's method, Runge–Kutta method, Milne's method** etc. belong to class (*ii*).

13.5.1 Picard's Iterative Method

Consider the first order *initial value problem* (IVP)

$$y' = f(x, y),\ y(x_0) = y_0. \qquad ...(13.13)$$

On integration from x_0 to x, we obtain

$$\int_{x_0}^{x} y'dx = \int_{x_0}^{x} f(x, y)dx \text{ or } y(x) = y(x_0) + \int_{x_0}^{x} f(x, y)\,dx \qquad ...(13.14)$$

Equation (13.14) is known as an *integral equation* because the unknown y in the function $f(x, y)$ on the right hand side occurs under the sign of integration. The solution of Eq. (13.14) is obtained by an iterative procedure. We assume the first approximation to $y(x)$ as $y_0(x) = y(x_0) = y_0$. The first approximation to $y(x)$ is obtained by putting y_0 for y on the right side of (13.15). This iterate is called $y_1(x)$ and is given by

$$y_1(x) = y_0 + \int_{x_0}^{x} f[x, y_0(x)]dx.$$

The integral on the right hand side is now evaluated to obtain an expression for $y_1(x)$.

The second approximation $y_2(x)$ is given by

$$y_2(x) = y_0 + \int_{x_0}^{x} f[x, y_1(x)]dx.$$

Proceeding in the manner stated above, we obtain higher order approximations $y_3, y_4, ..., y_n, y_{n+1}$, etc. where

$$y_n(x) = y_0 + \int_{x_0}^{x} f[x, y_{n-1}(x)]dx \qquad ...(13.15)$$

where $y_0(x) = y_0$, $y_n(x)$ is called the nth approximation to the solution of (13.13).

EXAMPLE 13.33 Solve initial value problem $y' = x - y$, $y(0) = 1$.

Use Picard's method to obtain y for $x = 0.2$ correct to five places of decimal. Check your answer by comparing the result with the exact solution.

Solution We have $f(x, y) = x - y$, $x_0 = 0$, $y_0 = 1$.

$$y(x) = y(x_0) + \int_{x_0}^{x} f(x, y)\,dx = 1 + \int_{0}^{x} (x - y)\,dx.$$

The iteration is defined by

$$y_{n+1}(x) = 1 + \int_{0}^{x} [x - y_n(x)]dx, \quad y_0 = 1.$$

Setting $n = 0, 1, 2, 3, 4$, we get

$$y_1(x) = 1 + \int_{0}^{x} (x - 1)dx = 1 - x + \frac{x^2}{2}.$$

$$y_2(x) = 1+\int_0^x\left(x-1+x-\frac{x^2}{2}\right)dx = 1-x+x^2-\frac{x^3}{6}.$$

$$y_3(x) = 1+\int_0^x\left(x-1+x-x^2+\frac{x^3}{6}\right)dx = 1-x+x^2-\frac{x^3}{3}+\frac{x^4}{24}.$$

$$y_4(x) = 1+\int_0^x\left(x-1+x-x^2+\frac{x^3}{3}-\frac{x^4}{24}\right)dx$$

$$= 1-x+x^2-\frac{x^3}{3}+\frac{x^4}{12}-\frac{x^5}{120}.$$

$$y_5(x) = 1-x+x^2-\frac{x^3}{3}+\frac{x^4}{12}-\frac{x^5}{60}+\frac{x^6}{720}.$$

Now when $x = 0.2$, the various approximations of y are given by $y_1(0.2) \approx 0.82$, $y_2(0.2) \approx 0.83867$, $y_3(0.2) \approx 0.83740$, $y_4(0.2) \approx 0.83746$, $y_5(0.2) \approx 0.83746$.

Hence to an accuracy of five places of decimal, the value of $y(0.2)$ is equal to 0.83746.

The exact solution of the problem is $y(x) = 2e^{-x} + x - 1$ which on substitution for $x = 0.2$ gives upto five places of decimal the value $y(0.2) = 0.83746$. This value is exactly the same as the value obtained by Picard's method.

It may be remarked that Picard's method can be used with ease only when it does not pose any difficulty in performing the necessary integration in the course of obtaining various approximations.

EXAMPLE 13.34 Find the value of y for $x = 0.1$ and 0.2 by Picard's method, given that

$$y' = (y-x)/(y+x),\ y(0) = 1. \qquad \textit{(PTU 2002)}$$

Solution We have $y(x) = y(x_0) + \int_{x_0}^{x} f(x, y)dx = 1+\int_0^x \frac{y-x}{y+x}dx$.

The iteration is defined by

$$y_{n+1}(x) = 1+\int_0^x\left[\frac{y_n(x)-x}{y_n(x)+x}\right]dx, \quad y_0(x) = 1.$$ Setting $n = 0, 1, 2$, we obtained

$$y_1(x) = 1+\int_0^x\frac{1-x}{1+x}dx = 1+\int_0^x\left[-1+\frac{2}{1+x}\right]dx$$

$$= 1+[-x+2\log(1+x)]_0^x = 1-x+2\log(1+x). \qquad ...(i)$$

$$y_2(x) = 1+\int_0^x\frac{\{1-x+2\log(1+x)\}-x}{\{1-x+2\log(1+x)\}+x}dx = 1+\int_0^x\left[1-\frac{2x}{1+2\log(1+x)}\right]dx$$

which is difficult to integrate.

Hence, we use the first approximation and taking $x = 0.1$ and 0.2 in (*i*).

Therefore, $y(0.1) \approx 1 - 0.1 + 2\log(1.1) = 1.0906$,

$y(0.2) \approx 1 - 0.2 + 2\log(1.2) = 1.16464$.

13.5.2 Taylor's Series Method

We consider the differential equation $y'(= dy/dx) = f(x, y)$, ...(13.12)

with the initial condition $y(x_0) = y_0$.

If $y(x)$ is the exact solution of (13.12), then the Taylor's series expansion for $y(x)$ around $x = x_0$ is given by

$$y(x) = y_0 + \frac{(x-x_0)}{1!}y'(0) + \frac{(x-x_0)^2}{2!}y''(0) + \frac{(x-x_0)^3}{3!}y'''(0) + \cdots \quad ...(13.16)$$

where dashes denote differentiation w.r.t. x.

Differentiating Eq. (13.12) repeatedly w.r.t. x, we get

$$y'' = \frac{d^2y}{dx^2} = \frac{\partial f}{\partial x} + \frac{\partial f}{\partial y}\frac{dy}{dx} = \frac{\partial f}{\partial x} + f\frac{\partial f}{\partial y} = \left(\frac{\partial}{\partial x} + f\frac{\partial}{\partial y}\right)f$$

or $$y'' = f' = f_x + ff_y, \quad ...(13.17)$$

$$y''' = \frac{d^3y}{dx^3} = \frac{d}{dx}\left(\frac{d^2y}{dx^2}\right) = \left(\frac{\partial}{\partial x} + f\frac{\partial}{\partial y}\right)\left(\frac{\partial f}{\partial x} + f\frac{\partial f}{\partial y}\right)$$

$$= \frac{\partial^2 f}{\partial x^2} + \frac{\partial f}{\partial x}\frac{\partial f}{\partial y} + 2f\frac{\partial^2 f}{\partial x \partial y} + f\left(\frac{\partial f}{\partial y}\right)^2 + f^2\frac{\partial^2 f}{\partial y^2}$$

or $$f'' = y''' = f_{xx} + 2ff_{xy} + f^2f_{yy} + f_xf_y + ff_y^2 \quad ....(13.18)$$

and so on.

Putting $x = x_0$ and $y = y_0$ in the expressions for various derivatives and substituting them in Eq. (13.16), we can obtain a power series for $y(x)$ in powers of $x - x_0$.

The method is best illustrated by the following example.

EXAMPLE 13.35 Obtain the approximate value of $y(0.1)$ for the initial value problem $y' = 1 + xy$, $y(0) = 1$. Using Taylor's series method.

Solution Differentiating the given equation repeatedly with respect to x, we obtain

$y'' = xy' + y$, $y''' = xy'' + 2y'$, $y^{iv} = xy''' + 3y''$, $y^{v} = xy^{iv} + 4y'''$ and so on.

Solving the above equations in succession with $x_0 = 0$ and $y_0 = 1$, we get

$y'(0) = 1$, $y''(0) = 1$, $y'''(0) = 2$, $y^{iv}(0) = 3$, $y^{v}(0) = 8$.

The Taylor's series expansion for $y(x)$ near $x = 0$ is given by

$$y(x) = 1 + x + \frac{x^2}{2} + \frac{x^3}{6}(2) + \frac{x^4}{24}(3) + \frac{x^5}{120}(8) + \ldots\ldots\ldots\ldots$$

$$= 1 + x + \frac{x^2}{2} + \frac{x^3}{3} + \frac{x^4}{8} + \frac{x^5}{15} + \ldots\ldots\ldots\ldots$$

To obtain $y(0.1)$, we put $x = 0.1$ in the above series and we have $y(0.1) \approx y_1 = 1.105346$.

It is not out of place to mention that the Taylor's series method is applicable only when the various derivatives of $f(x, y)$ exist. Secondly the value of $(x - x_0)$ in the Taylor's expansion of $y(x)$ near x_0 must be small so that the magnitudes of the successive terms diminish rapidly resulting in the faster convergence of the series.

EXAMPLE 13.36 Find the approximate value of $y(0.2)$ for the initial value problem

$y' = 2y + 3e^x$, $y(0) = 0$. Using Taylor's series method.

Solution We have $y' = 2y + 3e^x$, $y'(0) = 2y(0) + 3e^o = 3$.

Differentiating the given equation successively with respect to x and substituting $x = 0$, $y = 0$, we obtain

$y'' = 2y' + 3e^x$, $y''(0) = 2y'(0) + 3 = 9$.

$y''' = 2y'' + 3e^x$, $y'''(0) = 2y''(0) + 3 = 21$.

$y^{iv} = 2y''' + 3e^x$, $y^{iv}(0) = 2y'''(0) + 3 = 45$.

and so on.

Substituting these values in the Taylor's series.

$$y(x) = y(0) + y'(0)x + \frac{y''(0)}{2!}x^2 + \frac{y'''(0)}{3!}x^3 + \frac{y^{(iv)}(0)}{4!}x^4 + \ldots\ldots\ldots$$

$$= 3x + \frac{9}{2}x^2 + \frac{7}{2}x^3 + \frac{15}{8}x^4 + \ldots\ldots\ldots$$

When $x = 0.2$.

$$y(0.2) \approx y_1 = 3(0.2) + 4.5(0.04) + 3.5(0.008) + 1.875(0.0016) + \ldots\ldots\ldots$$

$$= 0.6 + 0.18 + 0.028 + 0.003 = 0.811.$$

The exact solution of the problem is $y(x) = 3(e^{2x} - e^x)$, which on substitution for $x = 0.2$ gives $y(0.2) = 0.8112$.

13.5.3 Euler's Method

We shall now discuss some purely numerical methods for the solution of the first order differential equations. We begin with Euler's method which is basically an elementary method and which will demonstrate the procedure underlying these methods. More accurate methods often used in practical problems will be discussed later.

Let $y = \phi(x)$ be the solution curve of the given differential equation $y' = f(x, y)$. Let $(x_0, y_0), (x_1, y_1) \ldots (x_n, y_n), (x_{n+1}, y_{n+1}) \ldots$ be the points on the solution curve. We suppose that the values of x are equi-spaced at equal spacing h so that $x_{n+1} = x_0 + (n + 1)h = x_n + h$. Since the point (x_{n+1}, y_{n+1}) lies on the curve $y = \phi(x)$, it must satisfy this equation. This gives

$$y_{n+1} = \phi(x_{n+1}) = \phi(x_n + h) \qquad \ldots(13.19)$$

Expanding $\phi(x_n + h)$ by Taylor's theorem, we get

$$y_{n+1} = \phi(x_n) + h\phi'(x_n) + \frac{h^2}{2!}\phi''(x_n) + \frac{h^3}{3!}\phi'''(x_n) + \cdots$$

$$= y_n + hf(x_n, y_n) + \frac{h^2}{2!}\phi''(x_n) + \cdots \qquad \ldots(13.20)$$

If h is now assumed small so that h^2 and its higher powers are ignored in comparison to h we obtain from Eq. (13.20).

$$y_{n+1} = y_n + hf(x_n, y_n) \qquad \ldots(13.21)$$

Setting $n = 0, 1, 2, \ldots$, we can obtain $y_1, y_2, \ldots$ etc. The values $y_1, y_2, y_3, \ldots$ approximate the values of a solution $y(x)$ of the IVP at $x_1, x_2, x_3, \ldots$. If the initial value (x_0, y_0) is known, the various ordinates can be immediately obtained.

Viewed from geometrical considerations, Eq. (13.21) gives the ordinate at $(n + 1)$th point in terms of the values at the nth point. Since $y' = f(x, y)$ defines the slope of the curve, $f(x_n, y_n)$ gives the slope at the nth point. In other words $f(x_n, y_n)$ will be $\tan \alpha$ where α is the angle which the tangent RT at the point (x_n, y_n) makes with the x-axis. See figure 13.8.

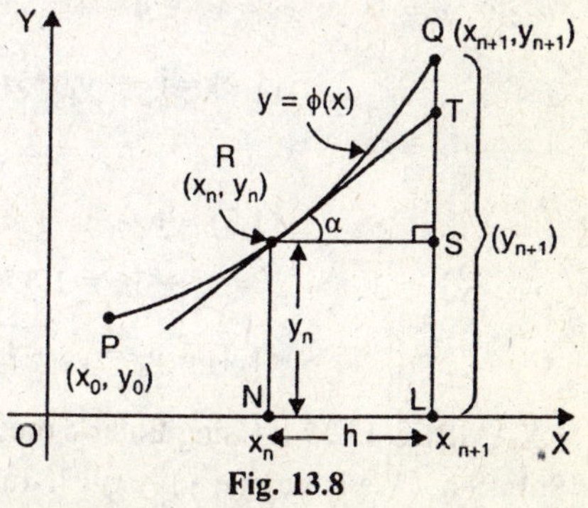

Fig. 13.8

From ΔRST, $TS = h \tan \alpha = hf(x_n, y_n)$.

$\therefore \quad LT = y_n + hf(x_n, y_n) \approx y_{n+1} = LQ$

The assumption that $LT = LQ$ reveals that the slope of the chord RQ has been assumed to be equal to the slope of the tangent RT. Thus the error introduced by the method is TQ. This can be minimised by taking h smaller.

As already remarked this method is of trivial importance and its use is discouraged in practical problems. This is primarily due to the fact that when h is large the omission of h^2 and higher powers of h is not justified and the results obtained will be in serious error.

EXAMPLE 13.37 Obtain the approximate value of $y(0.5)$ for the initial value problem $y' = 2x + y$, $y(0) = 1$, using Euler method with step size $h = 0.1$. Compare with the exact solution.

Solution We have $f(x, y) = 2x + y$, and $h = 0.1$.

Euler method gives

$$y_{n+1} = y_n + hf(x_n, y_n) = y_n + h(2x_n + y_n),\ n = 0, 1, 2, \ldots .$$

The starting values are $x_0 = 0$, $y_0 = 1$. We have the following results.

$n = 0$: $y(0.1) \approx y_1 = y_0 + h(2x_0 + y_0) = 1 + 0.1\,(1) = 1.1.$

$n = 1$: $x_1 = x_0 + h = 0.1,\ y_1 = 1.1.$

$y(0.2) \approx y_2 = y_1 + h(2x_1 + y_1) = 1.1 + 0.1\,(0.2 + 1.1) = 1.23.$

$n = 2$: $x_2 = x_1 + h = x_0 + 2h = 0.2,\ y_2 = 1.23.$

$y(0.3) \approx y_3 = y_2 + h(2x_2 + y_2) = 1.23 + 0.1\,(0.4 + 1.23) = 1.393.$

$n = 3$: $x_3 = x_0 + 3h = 0.3,\ y_3 = 1.393.$

$y(0.4) \approx y_4 = y_3 + h(2x_3 + y_3) = 1.393 + 0.1\,(0.6 + 1.393) = 1.5923.$

$n = 4$: $x_4 = x_0 + 4h = 0.4,\ y_4 = 1.5923.$

$y(0.5) \approx y_5 = y_4 + h(2x_4 + y_4) = 1.5923 + 0.1\,(0.8 + 1.5923) = 1.8315.$

The equation is linear and has integrating factor e^{-x}.

Solving, we find $y = 3e^x - 2x - 2.$

The exact solution is $y(0.5) = 3e^{0.5} - 3 = 3(1.6487) - 3 = 1.9461.$

Remark Better accuracy can be obtained by using smaller values of h or proceeding as per the modification of the method.

EXAMPLE 13.38 Obtain the approximate value of $y(1.4)$ for the initial value problem

$$y' = \sqrt{x^2 + 2y},\ y(1) = 0.2, \text{ using Euler method with } n = 4.$$

Solution We have $f(x, y) = \sqrt{x^2 + 2y}$, and $h = \dfrac{1.4 - 1.0}{4} = 0.1.$

Euler method gives

$$y_{n+1} = y_n + hf(x_n, y_n) = y_n + h\sqrt{x_n^2 + 2y_n},\ n = 0, 1, 2, \ldots .$$

The starting values are $x_0 = 1$, $y_0 = 0.2$. We have the following results.

$n = 0$: $y(1.1) \approx y_1 = y_0 + h\sqrt{(x_0^2 + 2y_0)} = 0.2 + 0.1\sqrt{1 + 0.4} = 0.3183.$

$n = 1$: $x_1 = x_0 + h = 1.1,\ y_1 = 0.3183.$

$y(1.2) \approx y_2 = y_1 + h\sqrt{(x_1^2 + 2y_1)} = 0.3183 + 0.1\sqrt{1.1^2 + 2(0.3183)} = 0.4542.$

$n = 2$: $x_2 = x_0 + 2h = 1.2,\ y_2 = 0.4542.$

$y(1.3) \approx y_3 = y_2 + h\sqrt{(x_2^2 + 2y_2)} = 0.4542 + 0.1\sqrt{1.2^2 + 2(0.4542)} = 0.6074.$

$n = 3$: $x_3 = x_0 + 3h = 1.3,\ y_3 = 0.6074.$

$y(1.4) \approx y_4 = y_3 + h\sqrt{(x_3^2 + 2y_3)} = 0.6074 + 0.1\sqrt{1.3^2 + 2(0.6074)} = 0.7778.$

EXAMPLE 13.39 Using Euler's method, find an approximate value of $y(1.04)$, given that $y' = xy^{1/3}$, $y(1) = 1$.

Solution We have $f(x, y) = xy^{1/3}$. Take $h = 0.01$.

Euler method gives

$$y_{n+1} = y_n + hf(x_n, y_n) = y_n + h[x_n\, y_n^{\,1/3}].$$

The starting values are $x_0 = y_0 = 1$.

The computation work is exhibited in the following table:

n (1)	x_n (2)	y_n (3)	$f(x_n, y_n) = x_n y_n^{1/3}$ (4)	$y_{n+1} = y_n + 0.01\,(x_n y_n^{1/3})$. (5)
0	1.0	1.0	1.0	1.0100
1	1.01	1.0100	1.0133	1.0201
2	1.02	1.0201	1.0268	1.0304
3	1.03	1.0304	1.0403	1.0408
4	1.04	1.0408	etc.	

Procedure: At each step we compute $x_n y_n^{1/3}$ and record it in column 4. This value multiplied by 0.01 is added to y_n to obtain y_{n+1} which is entered in column 5. This value of y_{n+1} is transferred to column 3 for the next stage of computation. Hence $y(1.04) \approx y_4 = 1.0408$.

13.5.4 Euler's Modified Method

The Euler's formula was derived from Taylor's expansion of y_{n+1} by considering only the first two terms of the series. The error arising out of the omission of the terms involving h^2 and beyond is known as truncation error. It is our endeavour here to derive a formula with a smaller error. Differentiating Eq. (13.20) w.r.t. x, we obtain

$$\left(\frac{dy}{dx}\right)_{n+1} = \left(\frac{dy}{dx}\right)_n + h\phi''(x_n) + \frac{h^2}{2!}\phi'''(x_n) + \ldots$$

or
$$f(x_{n+1}, y_{n+1}) = f(x_n, y_n) + h\phi''(x_n) + \frac{h^2}{2!}\phi'''(x_n) + \ldots \qquad \ldots(13.22)$$

Since $\quad dy/dx = f(x, y) = \phi'(x)$.

Multiplying Eq. (13.22) by $h/2$ and subtracting from Eq. (13.20) we get

$$y_{n+1} - \frac{h}{2} f(x_{n+1}, y_{n+1}) = y_n + \frac{h}{2} f(x_n, y_n) - \frac{h^3}{12}\phi'''(x_n) + \ldots$$

Neglecting terms containing h^3 and higher powers, we get

$$y_{n+1} = y_n + h\left(\frac{f(x_n, y_n) + f(x_{n+1}, y_{n+1})}{2}\right) \qquad \ldots(13.23)$$

Eq. (13.23) is known as **Euler's modified formula**. It is named so because it can be obtained from Euler's formula by adding to the latter a length equal to $\frac{TM}{2}$, where RM is parallel to the tangent at Q. See Fig. 13.9.

Again Eq. (13.23) approximates the slope of the chord RQ by the mean slope of the tangents at the points R and Q.

Unfortunately the slope $f(x_{n+1}, y_{n+1})$ of the tangent at the end Q is not known and therefore the value of $f(x_{n+1}, y_{n+1})$ on the right side of Eq. (13.23) can not be found. To circumvent this difficulty we first calculate an approximate y_{n+1}^* from Euler's formula and use it to evaluate $f(x_{n+1}, y_{n+1})$. Thus to trigger each forward step of computation, we use the following two formulae to calculate y_{n+1}.

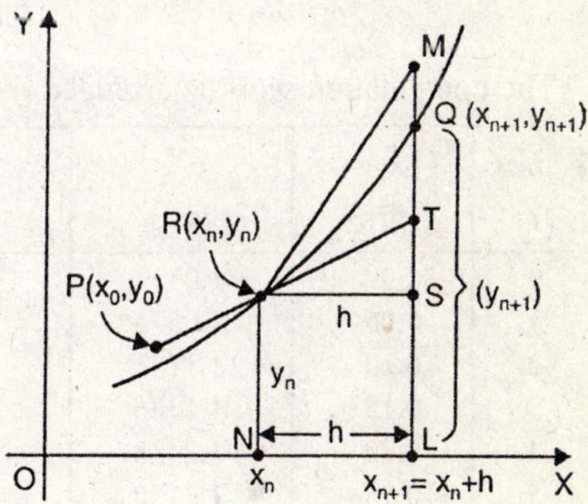

Fig. 13.9

$$y_{n+1} = y_n + \frac{h}{2}[f(x_n, y_n) + f(x_{n+1}, y_{n+1}^*)], \qquad \ldots(13.24)$$

where
$$y_{n+1}^* = y_n + h f(x_n, y_n). \qquad \ldots(13.25)$$

Remark In general the Euler's modified method is an example of a ***predictor-corrector method***. The value of y_{n+1}^* given by (13.25) *predicts* a value of $y(x_n)$, whereas the value of y_{n+1} defined by formula (13.24) *corrects* this estimate.

EXAMPLE 13.40 Use the modified Euler's method to obtain the approximate value of $y(0.3)$ for the solution of the initial-value problem

$y' = x + y$, $y(0) = 1$. Take $h = 0.05$. (*Rohtak, 2005; Bhopal 2002S; Delhi 2002*)

Solution With $x_0 = 0$, $y_0 = 1$, $f(x, y) = x + y$, $f(x_n, y_n) = x_n + y_n$, $h = 0.05$.

We first compute $\overset{*}{y}_{n+1} = y_n + hf(x_n, y_n)$, $n = 0, 1, 2, 3, \ldots$

$n = 0$: $\overset{*}{y}_1 = y_0 + h(x_0 + y_0) = 1 + 0.05\,(0 + 1) = 1.05.$

We use this last value in formula (A)

$$y_{n+1} = y_n + h\left[\frac{f(x_n, y_n) + f(x_{n+1}, \overset{*}{y}_{n+1})}{2}\right] \qquad \ldots(A)$$

along with $x_1 = x_0 + h = 0 + 0.05 = 0.05.$

For $n = 0$: $$y_1 = y_0 + h\left[\frac{x_0 + y_0 + x_1 + \overset{*}{y}_1}{2}\right] = 1 + 0.05\left[\frac{0 + 1 + 0.05 + 1.05}{2}\right] = 1.0525.$$

$n = 1$: $\overset{*}{y}_2 = y_1 + h(x_1 + y_1) = 1.0525 + 0.05\,(0.05 + 1.0525) = 1.1076,$

$x_2 = x_1 + h = 0.05 + 0.05 = 0.10.$

$$y_2 = y_1 + h\left[\frac{x_1 + y_1 + x_2 + \overset{*}{y}_2}{2}\right]$$

$$= 1.0525 + 0.05\left[\frac{0.05 + 1.0525 + 0.10 + 1.1076}{2}\right] = 1.1103.$$

$n = 2$: $\overset{*}{y}_3 = y_2 + h(x_2 + y_2) = 1.1103 + 0.05\,(0.10 + 1.1103) = 1.1708,$

$x_3 = x_2 + h = 0.15.$

$$y_3 = y_2 + h\left[\frac{x_2 + y_2 + x_3 + \overset{*}{y}_3}{2}\right]$$

$$= 1.1103 + 0.05\left[\frac{0.10 + 1.1103 + 0.15 + 1.1708}{2}\right] = 1.1736, \text{ and so on.}$$

The computation work is arranged in the following table.

n	x_n	y_n	$x_n + y_n$	$\overset{*}{y}_{n+1}$	x_{n+1}	$x_{n+1} + \overset{*}{y}_{n+1}$	y_{n+1}
(1)	(2)	(3)	(4)	(5)	(6)	(7)	(8)
0	0.0	1.0	1.0	1.05	0.05	1.10	1.0525
1	0.05	1.0525	1.1025	1.1076	0.10	1.2076	1.1103
2	0.10	1.1103	1.2103	1.1708	0.15	1.3208	1.1736
3	0.15	1.1736	1.3236	1.2398	0.20	1.4398	1.2428
4	0.20	1.2428	1.4428	1.3148	0.25	1.5648	1.3180
5	0.25	1.3180	1.5680	1.3963	0.30	1.6963	1.3997
6	0.30	1.3997	etc				

Procedure At each step we compute $x_n + y_n$ and record it in column 4. This value multiplied by 0.05 and the result added to y_n gives $\overset{*}{y}_{n+1}$. We next calculate $x_{n+1} + \overset{*}{y}_{n+1}$ and record it in column 7. The sum of the 4th and the 7th columns is multiplied by 0.025 and the result is added to y_n to get y_{n+1}.

Hence $y(0.3) \approx y_6 = 1.3997.$

EXAMPLE 13.41 Solve the following by Euler's modified method:

$y' = \log_e (x + y)$, $y(0) = 2$, at $x = 1.2$ and 1.4 with $h = 0.2$. (*SVTU, 2007; VTU, 2006*)

Solution With $x_0 = 0$, $y_0 = 2$, $f(x_n, y_n) = \log(x_n + y_n)$, $h = 0.2$,

we first compute $$y^*_{n+1} = y_n + hf(x_n, y_n),\ n = 0, 1, 2, 3, 4, 5, 6, \ldots$$

$n = 0$: $$y_1^* = y_0 + h \log(x_0 + y_0) = 2 + 0.2 \log(2) = 2.1386.$$

We use this last value in formula (A)

$$y_{n+1} = y_n + h\left[\frac{f(x_n, y_n) + f(x_{n+1}, y_{n+1}^*)}{2}\right] \quad \ldots(A)$$

along with $$x_1 = x_0 + h = 0 + 0.2 = 0.2.$$

For $n = 0$: $$y_1 = y_0 + \frac{h}{2}[f(x_0, y_0) + f(x_1, y_1^*)]$$

$$= 2 + \frac{0.2}{2}[\log 2 + \log(0.2 + 2.1386)] = 2 + 0.1\,(0.6931 + 0.8496) = 2.1543.$$

For $n = 1$: $$y_2^* = 2.1543 + 0.2\,[\log(0.2 + 2.1543)] = 2.1543 + 0.1712 = 2.3255.$$

$$y_2 = y_1 + \frac{h}{2}[\log(x_1 + y_1) + \log(x_2 + y_2^*)]$$

$$= 2.1543 + \frac{0.2}{2}[\log(0.2 + 2.1543) + \log(0.4 + 2.3255)]$$

$$= 2.1543 + 0.1\,(0.8562 + 1.0026) = 2.1543 + 0.1859 = 2.3402.$$

The computation work is arranged in the following table.

n (1)	x_n (2)	y_n (3)	$\log(x_n + y_n)$ (4)	y^*_{n+1} (5)	x_{n+1} (6)	$\log(x_{n+1} + y^*_{n+1})$ (7)	y_{n+1} (8)
0	0.0	2	0.6931	2.1386	0.2	0.8496	2.1543
1	0.2	2.1543	0.8562	2.3255	0.4	1.0028	2.3402
2	0.4	2.3402	1.0081	2.5418	0.6	1.1448	2.5555
3	0.6	2.5555	1.1492	2.7853	0.8	1.2769	2.7981
4	0.8	2.7981	1.2805	3.0542	1.0	1.3998	3.0661
5	1.0	3.0661	1.4207	3.3502	1.2	1.5152	3.3597
6	1.2	3.3597	1.5172	3.6631	1.4	1.6120	3.6726
7	1.4	3.6726	etc				

Procedure: At each step we compute $\log_e(x_n + y_n)$ and record it in column 4. This value multiplied by 0.2 and the result is added to y_n gives y^*_{n+1}. We next calculate $\log(x_{n+1} + y^*_{n+1})$ and record it in column 7. The sum of the 4th and the 7th columns is multiplied by 0.1 and the result is added to y_n to get y_{n+1}.

Hence the required values of $y(1.2) \approx y_6 = 3.3597$ and $y(1.4) \approx y_7 = 3.6726$.

13.5.5 Runge-Kutta Methods

Runge-Kutta methods are designed for greater accuracy and are probably the ones which are frequently used in numerical computation. These methods have the following characteristic, features.

1. They are the one step methods that is, to find y_{m+1} we need only the value at the preceding point (x_m, y_m).
2. They agree with the Taylor's series through the term h^m where m is called the order of the method and is different for different methods.
3. They require the evaluation of the function $f(x, y)$ and not its derivatives as in the case of Taylor's series.

Euler's method and Euler's modified method are Runge-Kutta methods of order one and order two respectively.

Second order method — (Euler's modified method)

Putting n = 0 in Eq. (13.23), we obtain

$$y_1 = y_0 + \frac{h}{2}\{f_0 + f(x_1, y_1)\}, \quad \text{where } f_0 = f(x_0, y_0) \qquad ...(13.26)$$

If we replace y_1 on the right side of Eq. (13.26) by Euler's formula $y_1 = y_0 + hf_0$, we get

$$y_1 = y_0 + \frac{h}{2}\{f_0 + f(x_0 + h, y_0 + hf_0)\} \qquad ...(13.27)$$

Setting $k_1 = hf_0$, $k_2 = hf(x_0 + h, y_0 + k_1)$, Eq. (13.27) can be written as

$$y_1 = y_0 + \frac{1}{2}(k_1 + k_2). \qquad ...(13.28)$$

Eq. (13.28) is called Runge-Kutta formula of order 2 with an error of order h^3.

Third order method — (Runge's method)

Substituting for $dy/dx, d^2y/dx^2, d^3y/dx^3$ from equations (13.13), (13.17) and (13.18) in Eq., (13.16) and combining the resulting equation with Eq. (13.14), we get on replacing $(x - x_0)$ by h,

$$y_1 - y_0 = hf_0 + \frac{h^2}{2}\left\{\frac{\partial f_0}{\partial x} + f_0\frac{\partial f_0}{\partial y}\right\} + \frac{h^3}{6}\left\{\frac{\partial^2 f_0}{\partial x^2} + \frac{\partial f_0}{\partial x}\frac{\partial f_0}{\partial y} + 2f_0 + \frac{\partial^2 f_0}{\partial x \partial y} + f_0\left(\frac{\partial f_0}{\partial y}\right)^2 + f_0^2\frac{\partial^2 f_0}{\partial y^2}\right\} + ...$$

$$= \int_{x_0}^{x_0 + h} f(x, y)\, dx. \qquad ...(13.29)$$

Let us suppose that the solution curve $y = \phi(x)$ of the equation $\frac{dy}{dx} = f(x, y)$ passes through the points (x_0, y_0), $\left(x_0 + \frac{h}{2}, y_1\right)$ and $(x_0 + h, y_2)$. The value of the integral on the right side of Eq. (13.29) by the application of Simpson's rule is given by

$$\int_{x_0}^{x_0 + h} f(x, y)\, dx = \frac{(h/2)}{3}\left[f_0 + 4f\left(x_0 + \frac{h}{2}, y_1\right) + f(x_0 + h, y_2)\right] \qquad ...(13.30)$$

Substituting for the integral from Eq. (13.30) in Eq. (13.29) we obtain

$$k = hf_0 + \frac{h^2}{2}\left(\frac{\partial f_0}{\partial x} + f_0\frac{\partial f_0}{\partial y}\right) + \frac{h^3}{6}\left(\frac{\partial^2 f_0}{\partial x^2} + \frac{\partial f_0}{\partial x}\frac{\partial f_0}{\partial y} + 2f_0\frac{\partial^2 f_0}{\partial x \partial y} + f_0\left(\frac{\partial f_0}{\partial y}\right)^2 + f_0^2\frac{\partial^2 f_0}{\partial y^2}\right) + ...$$

$$= \frac{h}{2}\left\{f_0 + 4f\left[x_0 + \frac{h}{2}f\left(x_0 + \frac{h}{2}, y_1\right) + f(x_0 + h, y_2)\right]\right\} \qquad ...(13.31)$$

where $k = y_1 - y_0$.

In order to expand the functions on the right side of Eq. (13.31) as a two dimensional form of Taylor's series, we must know y_1 and y_2 in terms of y_0. Now the basis of Runge's methods that coefficients of all the terms upto the third power of h on both sides of Eq. (13.31) must be identical. Guided by this consideration and ignoring tedious algebra, the values of y_1 and y_2 are given by

$$y_1 = y_0 + \frac{h}{2}f_0; \quad y_2 = y_0 + f(x_0 + h, y_0 + hf_0) \qquad ...(13.32)$$

The calculations are best performed by using the following terms:

$$k_1 = hf_0; \; k_2 = hf(x_0 + h, y_0 + k_1)$$

$$k_3 = hf(x_0 + h, y_0 + k_2); \quad k_4 = hf\left(x_0 + \frac{h}{2}, y_0 + \frac{k_1}{2}\right)$$

and finally $k = \frac{1}{6}(k_1 + 4k_4 + k_3)$.

This gives the value of y_1 as $$y_1 = y_0 + \frac{1}{6}(k_1 + 4k_4 + k_3). \quad ...(13.33)$$

with an error of the order of h^4 where k is the weighted average of k_1, k_4 and k_3.

Geometrically Runge's method may be described as given in Fig. 13.10.

Figure 13.10 defines the xy-plane and $f(x, y)$ is the slope at any point of this plane.

1. Draw a line PA through (x_0, y_0) with slope $f(x_0, y_0)$. This gives the value of $k_1 = hf_0$.
2. Take a slope corresponding to M, the mid-point of PA and draw another line PB with that slope.

   This gives $k_4 = hf\left(x_0 + \frac{h}{2}, y_0 + \frac{k_1}{2}\right)$.
3. Draw a third line PC with slope corresponding to the point A. This gives the value of

   $k_2 = hf(x_0 + h, y_0 + k_1)$.
4. Finally, draw a line PD with slope corresponding to the point C. This gives the increment k_3.

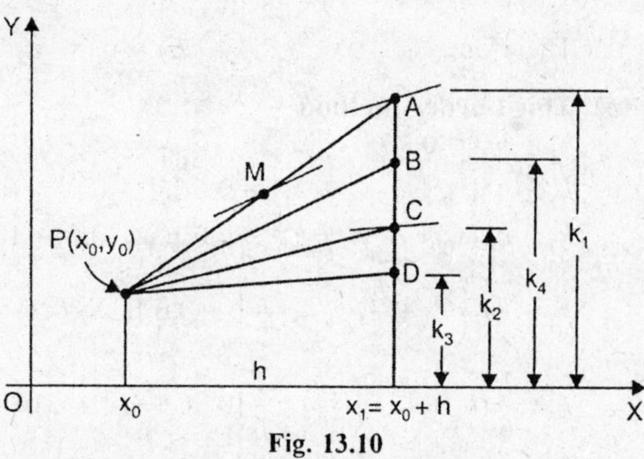

Fig. 13.10

The final value of k is the weighted average of k_1, k_4 and k_3 with an error of the order of h^4.

Fourth order method — (Runge-Kutta method)

Let $dy/dx = f(x, y)$ represents any first order equation, and let h denote the interval between equidistant values of x.

A fourth-order Runge-Kutta method is derived much in the same manner as the third-order method. Without going into the cumber some and complicated mathematical derivation, we define here under a few expressions for the increments needed in the solution of the problem. If the initial values are x_0, y_0, the first increment in y is computed from the formulas

$$\left.\begin{aligned} k_1 &= hf_0 = hf(x_0, y_0), & k_2 &= hf\left(x_0 + \frac{h}{2}, y_0 + \frac{k_1}{2}\right), \\ k_3 &= hf\left(x_0 + \frac{h}{2}, y_0 + \frac{k_2}{2}\right), & k_4 &= hf(x_0 + h, y_0 + k_3) \end{aligned}\right\} \quad ...(13.34)$$

and finally the adopted increment k is given by

$$\boldsymbol{k = \frac{1}{6}(k_1 + 2k_2 + 2k_3 + k_4).} \quad ...(13.35)$$

Eq. (13.35) is the weighted average of k_1, k_2, k_3 and k_4.

The value of y at $x_1 = x_0 + h$ is given by $y_1 = y_0 + k$ with an error of order h^5.

The increment in y for the second interval is computed in a similar manner by means of the formulas:

$$k_1 = hf(x_1, y_1), \qquad k_2 = hf\left(x_1 + \frac{h}{2}, y_1 + \frac{k_1}{2}\right),$$

$$k_3 = hf\left(x_1 + \frac{h}{2}, y_1 + \frac{k_2}{2}\right), \quad k_4 = hf(x_1 + h, y_1 + k_3).$$

The adopted increment k is given by

$$k = \frac{1}{6}(k_1 + 2k_2 + 2k_3 + k_4),$$

and so on for the succeeding intervals.

EXAMPLE 13.42 Apply Runge-Kutta method to approximate y when $x = 1.6$, given that

$$dy/dx = x - y, y(1) = 0.4.$$

Use (*a*) Second-order method, (*b*) Third-order method, (*c*) Fourth-order method.

Solution (*a*) **Second-order method**

Here $(x_0, y_0) = (1, 0.4)$, $h = 0.6$, $f_0 = x_0 - y_0 = 1 - 0.4 = 0.6$.

Then, $k_1 = hf_0 = 0.6\ (0.6) = 0.36,$

$k_2 = hf(x_0 + h, y_0 + k) = 0.6\ [(1 + 0.6) - (0.4 + 0.36)] = 0.6\ (0.84) = 0.504.$

Therefore, $$y(1.6) \approx y_1 = y_0 + \frac{1}{2}(k_1 + k_2) = 0.4 + \frac{1}{2}(0.36 + 0.504) = 0.832.$$

(*b*) **Third-order method**

$k_1 = hf_0 = 0.36,$

$k_2 = hf(x_0 + h, y_0 + k_1) = 0.6\ [(1 + 0.6) - (0.4 + 0.36)] = 0.504,$

$k_3 = hf(x_0 + h, y_0 + k_2) = 0.6\ [(1 + 0.6) - (0.4 + 0.504)] = 0.4176,$

$$k_4 = hf\left(x_0 + \frac{h}{2}, y_0 + \frac{k_1}{2}\right) = 0.6\ [(1 + 0.3) - (0.4 + 0.18)] = 0.432,$$

$$k = \frac{1}{6}(k_1 + 4k_4 + k_3) = \frac{1}{6}[0.36 + 4(0.432) + 0.4176] = 0.4176.$$

Therefore, $y(1.6) \approx y_1 = y_0 + k = 0.8176.$

(*c*) **Fourth-order method**

$k_1 = hf_0 = 0.36,$

$$k_2 = hf\left(x_0 + \frac{h}{2}, y_0 + \frac{k_1}{2}\right) = 0.6\ [(1 + 0.3) - (0.4 + 0.18)] = 0.432,$$

$$k_3 = hf\left(x_0 + \frac{h}{2}, y_0 + \frac{k_2}{2}\right) = 0.6\ [(1 + 0.3) - (0.4 + 0.216)] = 0.4104,$$

$k_4 = hf(x_0 + h, y_0 + k_3) = 0.6\ [(1 + 0.6) - (0.4 + 0.4104)] = 0.4738,$

$$k = \frac{1}{6}(k_1 + 2k_2 + 2k_3 + k_4) = \frac{1}{6}[0.36 + 0.864 + 0.8208 + 0.4738] = \frac{1}{6}(2.5186) = 0.4198 \text{ and}$$

$$y(1.6) \approx y_1 = y_0 + k = 0.4 + 0.4198 = 0.8198.$$

EXAMPLE 13.43 If $\frac{dy}{dx} = 3x + \frac{y}{2}$ with $y(0) = 1$. Find the value of k_1, k_2, k_3, k_4 by taking $h = 0.1$. Using Fourth-order Runge-Kutta method. [*AMIETE, June 2010*]

Solution We have $f(x, y) = 3x + \frac{y}{2}$, $h = 0.1$, $x_0 = 0$, $y_0 = 1$, $f_0 = 3x_0 + \frac{y_0}{2} = 0.5$. Then

$k_1 = hf_0 = 0.1\ (0.5) = 0.05,$

$$k_2 = hf\left[3\left(x_0 + \frac{h}{2}\right), \frac{1}{2}\left(y_0 + \frac{k_1}{2}\right)\right] = 0.1\left[3\ (0.05) + \frac{1}{2}(1 + 0.025)\right] = 0.066,$$

$$k_3 = hf\left[3\left(x_0 + \frac{h}{2}\right), \frac{1}{2}\left(y_0 + \frac{k_2}{2}\right)\right] = 0.1\left[3\ (0.05) + \frac{1}{2}(1 + 0.033)\right] = 0.067,$$

$$k_4 = hf\left[3\ (x_0 + h), \frac{1}{2}(y_0 + k_3)\right] = 0.1\left[3\ (0.1) + \frac{1}{2}(1 + 0.067)\right] = 0.0833.$$

EXAMPLE 13.44 Apply Runge-Kutta method to find an approximate value of y when $x = 0.2$ in steps of 0.1, given that $dy/dx = x + y^2$, $y(0) = 1$. (*VTU, 2006; Osmania 2002; Madras, 2000*)

Solution We have $f(x, y) = x + y^2$, $h = 0.1$, $x_0 = 0$, $y_0 = 1$, $f_0 = x_0 + y_0^2 = 1$. Then

$k_1 = hf_0 = 0.1\ (0 + 1) = 0.1,$

$$k_2 = hf\left(x_0 + \frac{h}{2}, y_0 + \frac{k_1}{2}\right) = 0.1\ [(0 + 0.05) + (1 + 0.05)^2] = 0.11525;$$

$$k_3 = hf\left(x_0 + \frac{h}{2}, y_0 + \frac{k_2}{2}\right) = 0.1\ [(0 + 0.05) + (1 + 0.05763)^2] = 0.11685,$$

$$k_4 = hf(x_0 + h, y_0 + k_3) = 0.1\ [(0 + 0.1) + (1 + 0.11685)^2] = 0.13474.$$

$$k = \frac{1}{6}(k_1 + 2k_2 + 2k_3 + k_4) = \frac{1}{6}[0.1 + 0.2305 + 0.2337 + 0.13474] = 0.1165$$

and therefore, $y(0.1) \approx y_1 = y_0 + k = 1 + 0.1165 = 1.1165.$

For the second step, we calculate new values of k_1, k_2, k_3, k_4 replacing (x_0, y_0) in the first set by (x_1, y_1) respectively. We have $x_1 = 0.1$, $y_1 = 1.1165$.

Therefore, $k_1 = hf(x_1, y_1) = 0.1\ [0.1 + 1.1165^2] = 0.1347,$

$$k_2 = hf\left(x_1 + \frac{h}{2}, y_1 + \frac{k_1}{2}\right) = 0.1\ [(0.1 + 0.05) + (1.1165 + 0.0673)^2] = 0.1551,$$

$$k_3 = hf\left(x_1 + \frac{h}{2}, y_1 + \frac{k_2}{2}\right) = 0.1\ [(0.1 + 0.05) + (1.1165 + 0.0776)^2] = 0.1576,$$

$$k_4 = hf(x_1 + h, y_1 + k_3) = 0.1\ [(0.1 + 0.1) + (1.1165 + 0.1576)^2] = 0.1823.$$

$$k = \frac{1}{6}(k_1 + 2k_2 + 2k_3 + k_4) = \frac{1}{6}[0.1347 + 0.3102 + 0.3152 + 0.1823] = 0.1571$$

and $y(0.2) \approx y_2 = 1.1165 + 0.1571 = 1.2736.$

Thus, $y(0.1) \approx y_1 = 1.1165$ and $y(0.2) \approx y_2 = 1.2736.$

EXAMPLE 13.45 Using Runge-Kutta method of fourth order, solve $\dfrac{dy}{dx} = \dfrac{y^2 - x^2}{y^2 + x^2}$ with $y(0) = 1$ at $x = 0.2, 0.4$.

[*Rohtak, 2005; Madras, 2001S*]

Solution We have $f(x, y) = \dfrac{y^2 - x^2}{y^2 + x^2} = 1 - \dfrac{2x^2}{y^2 + x^2}.$

To find y(0.2), we proceed as under:

We have $x_0 = 0$, $y_0 = 1$, $h = 0.2$, $f_0 = 1 - \dfrac{2x_0^2}{y_0^2 + x_0^2}$. Then

$$k_1 = hf_0 = 0.2\left(1 - \frac{2\,(0)}{1}\right) = 0.2,$$

$$k_2 = hf\left(x_0 + \frac{h}{2}, y_0 + \frac{k_1}{2}\right) = 0.2\left[1 - \frac{2\,(0 + 0.1)^2}{(1 + 0.1)^2 + (0.1)^2}\right] = 0.19672,$$

$$k_3 = hf\left(x_0 + \frac{h}{2}, y_0 + \frac{k_2}{2}\right) = 0.2\left[1 - \frac{2\,(0.1)^2}{(1 + 0.0984)^2 + (0.1)^2}\right] = 0.1967,$$

$$k_4 = hf\ (x_0 + h, y_0 + k_3) = 0.2\left[1 - \frac{2\,(0.2)^2}{(1 + 0.1967)^2 + (0.2)^2}\right] = 0.1891.$$

$$k = \frac{1}{6}(k_1 + 2k_2 + 2k_3 + k_4) = \frac{1}{6}[0.2 + 0.39344 + 0.3934 + 0.1891] = 0.19599$$

and $y(0.2) \approx y_1 = y_0 + k = 1 + 0.19599 = 1.196.$

Now we find y(0.4):

We have $x_1 = 0.2$, $y_1 = 1.196$, $h = 0.2$. We calculate new values of k_1, k_2, k_3, k_4 replacing (x_0, y_0) in the first set by (x_1, y_1) respectively.

$$k_1 = hf(x_1, y_1) = 0.2\left(1 - \frac{2\,(0.2)^2}{(1.196)^2 + (0.2)^2}\right) = 0.1891,$$

$$k_2 = hf\left(x_1 + \frac{h}{2}, y_1 + \frac{k_1}{2}\right) = 0.2\left[1 - \frac{2\,(0.2 + 0.1)^2}{(1.196 + 0.0945)^2 + (0.2 + 0.1)^2}\right] = 0.1795,$$

$$k_3 = hf\left(x_1 + \frac{h}{2}, y_1 + \frac{k_2}{2}\right) = 0.2\left[1 - \frac{2\,(0.2 + 0.1)^2}{(1.196 + 0.0898)^2 + (0.2 + 0.1)^2}\right] = 0.1793$$

$$k_4 = hf\,(x_1 + h, y_1 + k_3) = 0.2\left[1 - \frac{2\,(0.2 + 0.2)^2}{(1.196 + 0.1793)^2 + (0.2 + 0.2)^2}\right] = 0.1688.$$

$$k = \frac{1}{6}(k_1 + 2k_2 + 2k_3 + k_4) = \frac{1}{6}[0.1891 + 0.3590 + 0.3586 + 0.1688] = 0.1792.$$

Hence, $$y(0.4) \approx y_2 = y_0 + k = 1.196 + 0.1792 = 1.3752.$$

EXERCISE 13.3

In the following initial value problems, find the approximate values of $y(x)$ at the given points, using the Picard method.

1. $y' = x + x^4 y$, $y(0) = 3$, $x \in (0.1, 0.2)$. **Ans.** 3.005, 3.02.
2. $y' = x^2 + y^2$, $y(0) = 0$, at $x = 0.4$. **Ans.** 0.0214.
3. $y' = 3x + y^2$, $y(0) = 1$, at $x = 0.1$. **Ans.** 1.12721.
4. $y' = x + y^2$, $y(0) = 1$, at $x = 0.2$. **Ans.** 1.2727.
5. $y' = y + x$, $y(0) = 1$, at $x = 1$. **Ans.** $1 + x + x^2 + \frac{x^3}{3} + \frac{x^4}{12} + \frac{x^5}{60} + \frac{x^6}{720}$, 3.434.
6. $y' = 2x + y$, $y(0) = 1$, at $x = 0.5$. **Ans.** 1.9461.

In the following initial value problems, find the approximate values of $y(x)$ at the given points, using the Taylor series method.

7. $y' = x - y^2$, $y(0) = 1$, $x \in (0.1, 0.2)$. **Ans.** $1 - x + \frac{3}{2}x^2 - \frac{4}{3}x^3 - \frac{17}{12}x^4 - \frac{31}{20}x^5 + \ldots$, 0.9138, 0.8465.
8. $y' = x + y^2$, $y(0) = 1$, $x \in (0.1, 0.2)$. *(Bhopal, 2002)* **Ans.** 1.1164, 1.2735.
9. $y' = (x^2 + y)^{-1}$, $y(4) = 4$, $x \in (4.1, 4.2)$. **Ans.** 4.005, 4.0098.
10. $y' = 3x + y^2$, $y(1) = 1.2$, at $x = 1.1$. **Ans.** 1.7271.
11. $y' = x^2 y - 1$, $y(0) = 1$, $x \in (0.1, 0.2)$. *(Rohtak, 2005)* **Ans.** 0.90033, 0.80226.

In the following initial value problems, find the approximate values of $y(x)$ at the given points, using the Euler method.

12. $y' = 3x - y$, $y(1) = 0$, at $x = 0.5$, using $h = 0.1$. **Ans.** –1.5.
13. $y' = y^2 - \frac{y}{x}$, $y(1) = 1$, at $x = 1.6$, using $h = 0.1$. **Ans.** 1.1351.

14. $y' = 3x + y^2$, $y(1) = 1.2$, at $x = 1.1$, with $n = 4$. **Ans.** 1.7022.
15. $y' = x + y$, $y(0) = 1$, $h = 0.1$ at $x = 1$. **Ans.** 3.18.

In the following initial value problems, find the approximate values of $y(x)$ at the given points, using modified Euler's method.

16. $y' = x^2 + y$, $y(0) = 1$, $h = 0.02$, $x \in [0.04, 0.06]$. **Ans.** 1.0408, 1.0619.
17. $y' = 2 + \sqrt{xy}$, $y(1) = 1$, $h = 0.1$, at $x = 2$. **Ans.** 5.3524.
18. $y' = x + y^2$, $y(0) = 1$, $h = 0.1$, at $x = 0.5$. **Ans.** 2.2069.
19. $y' = \log_e (x + y)$, $y(1) = 2$, $h = 0.2$, $x \in [1.2, 1.4]$. **Ans.** 2.2332, 2.492.
20. Using Runge-Kutta method of (*a*) Second order, (*b*) Third order, (*c*) Fourth order.
    Solve $y' = x - y$, $y(1) = 1$ at $x = 1.1$. **Ans.** (*a*) 1.005, (*b*) 1.00483, (*c*) 1.004837.
21. Use Runge-Kutta method to find y at $x = 0.1, 0.2, 0.3, 0.4$ and 0.5 given that $y' = \dfrac{y^2 - 2x}{y^2 + x}$, $y(0) = 1$.
    (*Delhi, 2002*) **Ans.** 1.0874, 1.557, 1.2104, 1.2544, 1.2892.

    [Hint: $y' = \dfrac{y^2 - 2x}{y^2 + x} = 1 - \dfrac{3x}{y^2 + x} = f(x, y), \text{(say)}$ **]**
22. Use Runge-Kutta method to approximate $y(1.1)$, given that $y' = 3x + y^2$, $y(1) = 1.2$. **Ans.** 1.7278.
23. Apply Runge-Kutta fourth order method to find an approximate value of $y(0.2)$, given that $y' = x + y$, $y(0) = 1$. (*SVTU, 2007; PTU, 2005; Bhopal, 2002*) **Ans.** 1.1103, 1.2428.
24. Use fourth order Runge-Kutta method to approximate y when $x = 0.8$ given that $y' = \sqrt{x + y}$, $y(0.4) = 0.41$. (*SVTU, 2007S*) **Ans.** 0.8489.
25. Using Runge-Kutta method of fourth order, solve for $y(0.1)$, $y(0.2)$ given that $\dfrac{dy}{dx} = xy + y^2$, $y(0) = 1$.
    (*AMIETE, Dec 2009; SVTU, 2007; Madras 2006; PTU, 2002*) **Ans.** 1.1169, 1.2773.
26. Using Runge-Kutta method of fourth order, compute $y(0.2)$ and $y(0.4)$ for $10\dfrac{dy}{dx} = x^2 + y^2$, $y(0) = 1$, taking $h = 0.1$. (*Rohtak, 2003; Bhopal 2002*) **Ans.** 1.0207, 1.0438.
27. Using Runge-Kutta fourth order method with step size $h = 0.2$, obtain an approximate value of $y(2.4)$ for the initial value problem $y' = x(y - x)$, $y(2) = 3$. **Ans.** $y(2.2) = 3.4725$, $y(2.4) = 4.1608$.

Appendices

APPENDIX-1 SOME CONSTANTS, SOME IMPORTANT CONVERSION FACTORS, GREEK ALPHABET

SOME CONSTANTS

$e = 2.71828$

$\pi = 3.14159$

$\sqrt{2} = 1.41421$

$\sqrt{3} = 1.73205$

$\sqrt{10} = 3.16228$

$1° = 60' = 3600'' = 0.01745$ rad

1 radian $= 57°\ 17'\ 44.81''$

$= 57.29578°$

$\log_e 2 = 0.69315$

$\log_e 3 = 1.09861$

$\log_e 10 = 2.30259$

$\log_{10} e = 0.43429$

$\log_e \pi = 1.14473$

$\log_{10} \pi = 0.49715$

π radians = 180 degrees

SOME IMPORTANT CONVERSION FACTORS

1 inch = 2.54000 cm

100 sq. metres – 1 arc

100 hectares = 1 sq. kilometre

1 Quintal = 100 kgm

1 H.P = 178.298 cal/sec = 0.74570 kW

°F = °C (1.8) + 32

1 acre = 4840 yd^2 = 4046.8564 m^2

100 ares = 1 hectare

1 lb – 453.59 gm

1 Kilogm.weight = 2.20462 lb.

1 gallon of water = 4.5460 litres

1 metric Ton (Tonne) = 1000 kgm

1 kW = 1000 watts = 238.662 cal/sec.

GREEK ALPHABET

α	Alpha	ι	Iota	ρ	Rho
β	Beta	κ	Kappa	σ, Σ	Sigma
γ, Γ	Gamma	λ, Λ	Lambda	τ	Tau
δ, Δ	Delta	μ	Mu	υ, Y	Upsilon
$\varepsilon, \in$	Epsilon	ν	Nu	$\varphi, \phi\ \Phi$	Phi
ζ	Zeta	ξ	Xi	χ	Chi
η	Eta	o	Omicron	ψ, Ψ	Psi
$\theta, \vartheta, \Theta$	Theta	π	Pi	ω, Ω	Omega

APPENDIX-2 MATHEMATICAL FORMULAS

ALGEBRA

1. Laws of Exponents

$$a^m . a^n = a^{m+n},\quad (ab)^m = a^m b^m,\quad (a^m)^n = a^{mn},\quad a^{m/n} = \sqrt[n]{a^m}$$

If $a \neq 0,\quad a^m / a^n = a^{m-n},\quad a^0 = 1,\quad a^{-m} = 1/a^m.$

2. Cross-multiplication

If $a_1x + b_1y + c_1 = 0$ and $a_2x + b_2y + c_2 = 0$,

then $$\frac{x}{b_1c_2 - b_2c_1} = \frac{y}{c_1a_2 - c_2a_1} = \frac{1}{a_1b_2 - a_2b_1}$$

$$\therefore \quad x = \frac{b_1c_2 - b_2c_1}{a_1b_2 - a_2b_1},\ y = \frac{c_1a_2 - c_2a_1}{a_1b_2 - a_2b_1}.$$

3. Quadratic Equation

The solutions of the quadratic equation $ax^2 + bx + c = 0$, $a \neq 0$ are given by $x = \dfrac{-b \pm \sqrt{b^2 - 4ac}}{2a}$.

If α and β are the roots of this quardratic equation with $\alpha > \beta$, then $\alpha + \beta = -b/a$, $\alpha\beta = c/a$. The roots are (*i*) real and different, (*ii*) equal (*iii*) imaginary according as the discriminant $b^2 - 4ac$ is (*i*) positive, (*ii*) zero (*iii*) negative.

4. The Cubic equation

If α, β, γ are the roots of cubic equation $ax^3 + bx^2 + cx + d = 0$, then
$\alpha + \beta + \gamma = -b/a$, $\alpha\beta + \beta\gamma + \gamma\alpha = c/a$, $\alpha\beta\gamma = -d/a$.

5. The Factor therorem

If for $x = a$, $f(x) = 0$, then $(x - a)$ is a factor of $f(x)$ or $x = a$ is a root of $f(x) = 0$.

6. Series

If the series in arithmetical progression (A.P.) is $a + (a + d) + (a + 2d) + \ldots$, that is, first term a and the common difference d. Then nth term of A.P. is $a + (n - 1)d$.

S_n = Sum of n terms of the A.P. $= \dfrac{n}{2}(a + l)$, where l is the last term $= \dfrac{n}{2}[2a + (n-1)d]$.

If the series in geometrical progression (G.P.) is $a + ar + ar^2 + \ldots$, a is called the first term and the constant number $r \neq 0$ is called the common ratio of the G.P. Then nth term of G.P. $= ar^{n-1}$.

S_n = Sum of n terms of G.P. $= \dfrac{a(1 - r^n)}{1 - r}$, if $|r| < 1 = \dfrac{a(r^n - 1)}{r - 1}$, if $|r| > 1$.

If $|r| < 1$, $r^n \to 0$ as $n \to \infty$. Therefore S_n or S, sum to infinity $= \dfrac{a}{1-r}$.

Sum of first n natural numbers, that is, $1 + 2 + 3 + \ldots + n$ is $S_1 = \dfrac{n(n+1)}{2}$.

Sum of squares of the first n natural numbers, that is, $1^2 + 2^2 + 3^2 + \ldots + n^2$ is $S_2 = \dfrac{n(n+1)(2n+1)}{6}$.

Sum of cubes of the first n natural numbers, that is, $1^3 + 2^3 + 3^3 + ... + n^3$ is $S_3 = \left[\frac{n(n+1)}{2}\right]^2$.

7. Permutations and Combinations

$$n! = n(n-1)(n-2)...4 \cdot 3 \cdot 2 \cdot 1. \quad n \geq 1, \quad 0! = 1, \quad {}^nP_r = \frac{n!}{(n-r)!}$$

$${}^nC_r = \frac{n!}{r!(n-r)!}, 0 \leq r \leq n \text{ and } n \text{ is a non-negative integer } = \frac{{}^nP_r}{r!}.$$

$${}^nC_0 = 1 = {}^nC_n \, {}^nC_{n-r} = {}^nC_r, \, {}^nC_r + {}^nC_{r-1} = {}^{n+1}C_r.$$

8. Binomial Theorem

(*i*) For any positive integer n,

$$(a+b)^n = a^n + na^{n-1}b + \frac{n(n-1)}{1.2}a^{n-2}b^2 + \frac{n(n-1)(n-2)}{1.2.3}a^{n-3}b^3 + \cdots\cdots + nab^{n-1} + b^n.$$

(*ii*) Certain particular cases of Binomial Formula, where we assume $|x| < 1$

$$(1+x)^{-1} = 1 - x + x^2 - x^3 + \cdots\cdots + (-1)^r x^r + \cdots\cdots$$

$$(1+x)^{-2} = 1 - 2x + 3x^2 - \cdots\cdots + (-1)^r (r+1)x^r + \cdots\cdots$$

$$(1+x)^{-3} = 1 - 3x + 6x^2 - \cdots\cdots + (-1)^r \frac{(r+1)(r+2)}{2}x^r + \cdots\cdots$$

$$(1-x)^{-1} = 1 + x + x^2 + x^3 + \cdots\cdots + x^r + \cdots\cdots$$

$$(1-x)^{-2} = 1 + 2x + 3x^2 + 4x^3 + \cdots\cdots + (r+1)x^r + \cdots\cdots$$

$$(1-x)^{-3} = 1 + 3x + 6x^2 + \cdots\cdots + \frac{(r+1)(r+2)}{2}x^r + \cdots\cdots$$

9. Determinants

(*i*) Det. of order 2: $\begin{vmatrix} a_1 & b_1 \\ a_2 & b_2 \end{vmatrix} = a_1b_2 - a_2b_1.$

(*ii*) Det. of order 3: $\begin{vmatrix} a_1 & b_1 & c_1 \\ a_2 & b_2 & c_2 \\ a_3 & b_3 & c_3 \end{vmatrix} = a_1\begin{vmatrix} b_2 & c_2 \\ b_3 & c_3 \end{vmatrix} - b_1\begin{vmatrix} a_2 & c_2 \\ a_3 & c_3 \end{vmatrix} + c_1\begin{vmatrix} a_2 & b_2 \\ a_3 & b_3 \end{vmatrix}.$

LOGARITHMS

Laws of logarithms

(*i*) $\log_a(mn) = \log_a m + \log_a n$

(*ii*) $\log_a\left(\frac{m}{n}\right) = \log_a m - \log_a n$

(*iii*) $\log_a m^n = n\log_a m$

(*iv*) $\log_b r = \frac{\log_a r}{\log_a b}$ (Base-changing formula)

(Note that we can use any base in place of (a))

Remark $\log_a 1 = 0$, $\log_a a = 1$, $\log_a a^x = x$, $e^{\log x} = x$. $a^x = e^{x \log a}$, for $a > 0$.

MENSURATION

(A) Formulas from Plane Geometry:

Notations

a, b and c are sides of a triangle; $s = (a+b+c)/2$ = semi-perimeter; h = altitude; p = perimeter; $\angle A$, $\angle B$ and $\angle C$ are angles of a triangle, r = radius of a circle, C = circumference, R = radius of a sphere.

Triangle:

1. Area of a triangle $= \dfrac{1}{2} bh$.

$= \sqrt{s(s-a)(s-b)(s-c)}$ (Hero's formula)

$= \dfrac{1}{2} ab \sin C = \dfrac{1}{2} ca \sin B = \dfrac{1}{2} bc \sin A$

$= a^2 \cdot \dfrac{\sqrt{3}}{4}$ (for an equilateral triangle)

$= \dfrac{a^2 \sin B \sin C}{2 \sin A} = \dfrac{b^2 \sin C \sin A}{2 \sin B} = \dfrac{c^2 \sin A \sin B}{2 \sin C}$

$= \dfrac{1}{2} ab$ (right-angled triangle, hypotenuse c)

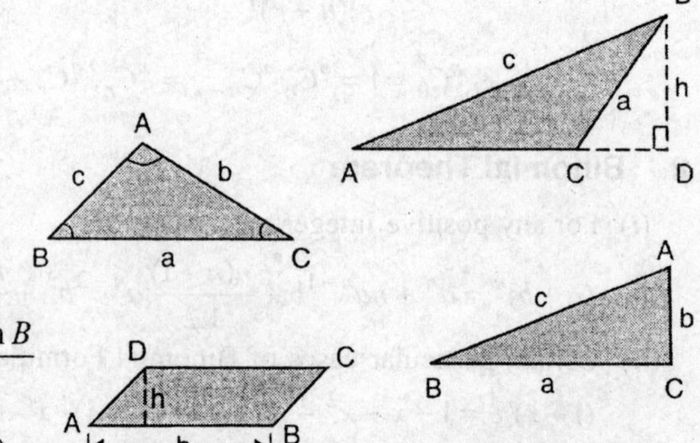

2. Area of a parallelogram = bh

3. Area of a trapezoid $= \dfrac{1}{2}(b_1 + b_2) h$

4. Area of a circle $= \pi r^2, C = 2\pi r$

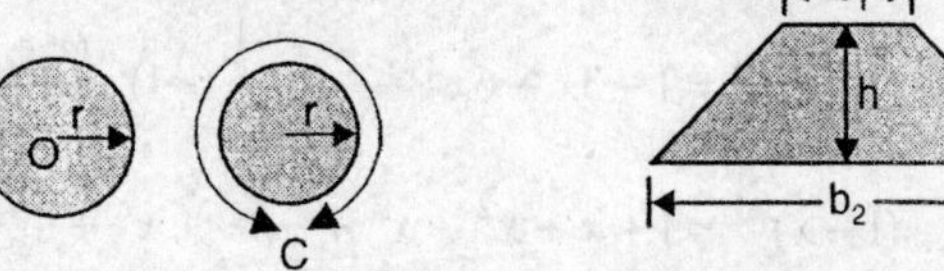

5. Area of a sector of a circle with central angle θ radians $= \dfrac{1}{2} r^2 \theta$, Arc length = $r\theta$, (θ in radians)

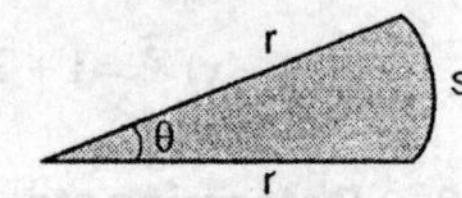

6. Area of a cyclic quadrilateral $= \sqrt{(s-a)(s-b)(s-c)(s-d)}$, where a, b, c, d are sides, and $s = (a+b+c+d)/2$.

7. Area of an ellipse = πab; (semi-major axis a and semi-minor axis b)

(B) Formulas from Solid Geometry:

1. *Right Circular Cylinder:*
Volume = $\pi r^2 h$; Curved surface, S = $2\pi rh$.

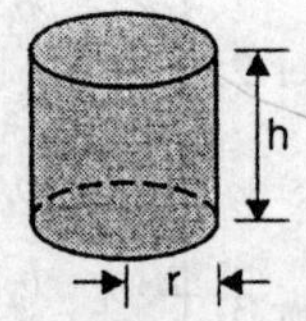

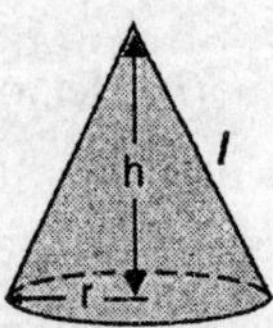

2. *Right Circular Cone:*
Volume $= \pi r^2 h/3$; Curved surface, $S = \pi rl = \pi r\sqrt{r^2 + h^2}$.

3. *Any Cylinder or Prism with Parallel Bases*

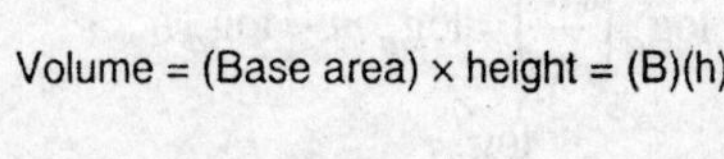

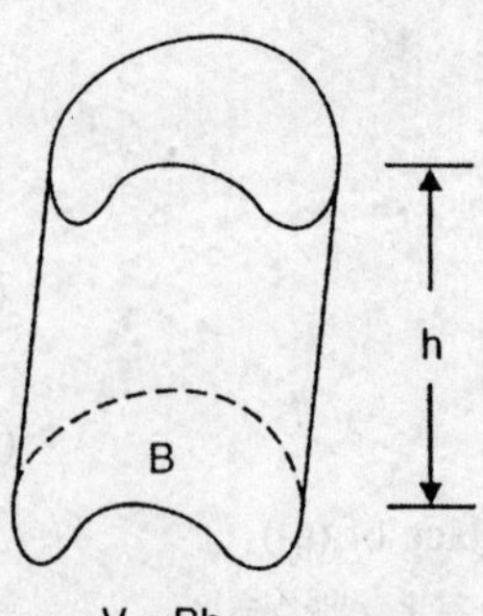

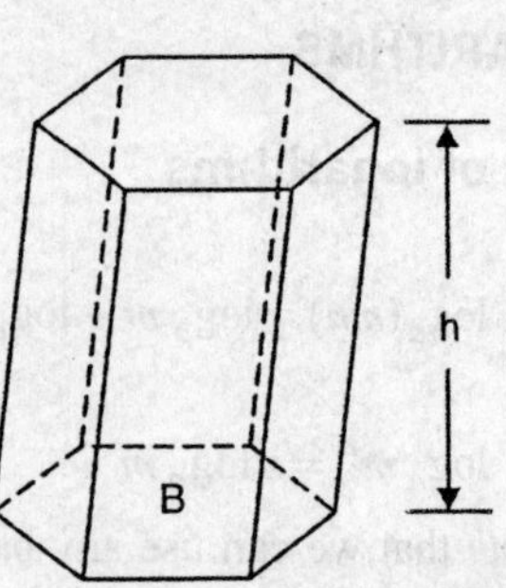

V = Bh

4. Frustum of Pyramids and Cones:

$$\text{Volume} = \frac{h}{3}\left[A_1 + A_2 + \sqrt{A_1 A_2}\right]; \text{ where } h = \text{height} \quad A_1 \text{ and } A_2 \text{ are the areas of ends.}$$

5. Frustum of Right Circular Cone

$$V = \frac{\pi}{3}(r_1^2 + r_1 r_2 + r_2^2)\,h.$$

6. Sphere:

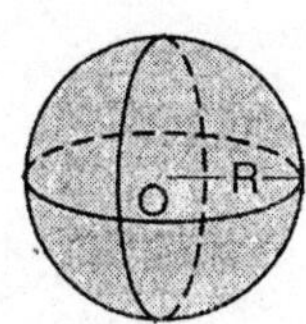

$$\text{Volume} = 4\pi R^3/3, \text{ Surface, } S = 4\,\pi R^2.$$

TRIGONOMETRY

1. Radian: An angle subtended at the centre of a circle by an arc equal in length to the radius of a circle is said to have a measure of 1 radian.
$\theta = l/r$, where l is the length of arc which subtends angle θ (in radians) at the centre of the circle with radius r.

π radians = 180°.

O r r r 1 Radian

2. Pythagorean Identities

$\sin^2\theta + \cos^2\theta = 1,\qquad \sec^2\theta = 1 + \tan^2\theta,\qquad \text{cosec}^2\theta = 1 + \cot^2\theta.$

3. Sign Identities

$\sin(-\theta) = -\sin\theta,\qquad \cos(-\theta) = \cos\theta,\qquad \tan(-\theta) = -\tan\theta.$

sin all
tan cos

4. Complement Identities

$$\sin\left(\frac{\pi}{2} - \theta\right) = \cos\theta,\qquad \cos\left(\frac{\pi}{2} - \theta\right) = \sin\theta,\qquad \tan\left(\frac{\pi}{2} - \theta\right) = \cot\theta.$$

$$\sin\left(\theta - \frac{\pi}{2}\right) = -\cos\theta,\qquad \cos\left(\theta - \frac{\pi}{2}\right) = \sin\theta,\qquad \sin\left(\frac{\pi}{2} + \theta\right) = \cos\theta,\qquad \cos\left(\frac{\pi}{2} + \theta\right) = -\sin\theta.$$

5. Supplement Identities

$$\sin(\pi - \theta) = \sin\theta,\quad \cos(\pi - \theta) = -\cos\theta,\quad \tan(\pi - \theta) = -\tan\theta.$$

$$\sin(\pi + \theta) = -\sin\theta,\quad \cos(\pi + \theta) = -\cos\theta,\quad \tan(\pi + \theta) = \tan\theta.$$

6. Values of trigonometric functions

Angle	0°	π/6	π/4	π/3	π/2	π	3π/2	2π
sin	0	1/2	$1/\sqrt{2}$	$\sqrt{3}/2$	1	0	−1	0
cos	1	$\sqrt{3}/2$	$1/\sqrt{2}$	1/2	0	−1	0	1
tan	0	$1/\sqrt{3}$	1	$\sqrt{3}$	not defined	0	not defined	0

7. Trigonometric Functions of Sum and Difference

$$\sin(A \pm B) = \sin A \cos B \pm \cos A \sin B.$$

$$\cos(A \pm B) = \cos A \cos B \mp \sin A \sin B.$$

$$\tan(A \pm B) = \frac{\tan A \pm \tan B}{1 \mp \tan A \tan B}.$$

8. Trigonometric Functions of Multiples and sub-multiples Angles

$$\sin 2A = 2\sin A \cos A = \frac{2\tan A}{1 + \tan^2 A}.$$

$$\cos 2A = \cos^2 A - \sin^2 A, = 2\cos^2 A - 1, = 1 - 2\sin^2 A = \frac{1-\tan^2 A}{1+\tan^2 A}.$$

$$\sin^2 \frac{A}{2} = \frac{1-\cos A}{2}, \cos^2 \frac{A}{2} = \frac{1+\cos A}{2}.$$

$$\tan 2A = \frac{2\tan A}{1-\tan^2 A}, \sin 3A = 3\sin A - 4\sin^3 A, \cos 3A = 4\cos^3 A - 3\cos A,$$

$$\tan 3A = \frac{3\tan A - \tan^3 A}{1 - 3\tan^2 A}.$$

9. Transformation of a Product into a Sum and vice versa

$$\sin A\cos B = \frac{1}{2}\sin(A+B) + \frac{1}{2}\sin(A-B),$$

$$\cos A\sin B = \frac{1}{2}\sin(A+B) - \frac{1}{2}\sin(A-B),$$

$$\cos A\cos B = \frac{1}{2}\cos(A+B) + \frac{1}{2}\cos(A-B),$$

$$\sin A\sin B = \frac{1}{2}\cos(A-B) - \frac{1}{2}\cos(A+B).$$

$$\sin C + \sin D = 2\sin\frac{C+D}{2}\cos\frac{C-D}{2}, \quad \sin C - \sin D = 2\cos\frac{C+D}{2}\sin\frac{C-D}{2},$$

$$\cos C + \cos D = 2\cos\frac{C+D}{2}\cos\frac{C-D}{2}, \quad \cos C - \cos D = -2\sin\frac{C+D}{2}\sin\frac{C-D}{2},$$

10. Law of Sines or the Sine Formula

If A, B and C are the angles of a triangle and if a, b and c are the lengths of the sides opposite to A, B and C, respectively, then

$$\frac{\sin A}{a} = \frac{\sin B}{b} = \frac{\sin C}{c}.$$

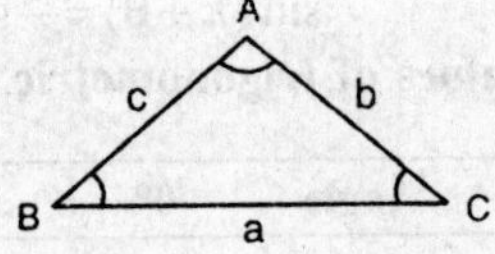

11. Law of Cosines or the Cosine Formulae

$$a^2 = b^2 + c^2 - 2bc\cos A,$$

$$b^2 = c^2 + a^2 - 2ca\cos B,$$

$$c^2 = a^2 + b^2 - 2ab\cos C.$$

12. Projection Formulae

$$a = b\cos C + c\cos B, \quad b = c\cos A + a\cos C, \quad c = a\cos B + b\cos A.$$

13. Circumradius of the triangle, when sides are known:

$$R = \frac{a}{2\sin A} = \frac{b}{2\sin B} = \frac{c}{2\sin C} = \frac{abc}{4\Delta}.$$

14. Inradius of the triangle

$$r = \frac{\Delta}{s} = \frac{\text{area of the triangle}}{\text{semi-perimeter}}.$$

15. Inverse Circular Functions

(a) (*i*) $\sin^{-1}\dfrac{1}{x} = \text{cosec}^{-1}x, x \geq 1,$ (*ii*) $\cos^{-1}\dfrac{1}{x} = \sec^{-1}x, x \geq 1,$ (*iii*) $\tan^{-1}\dfrac{1}{x} = \cot^{-1}x, x > 0.$

(b) (*i*) $\sin^{-1}(-x) = -\sin^{-1}x; x \in [-1, 1],$ (*ii*) $\tan^{-1}(-x) = -\tan^{-1}x, x \in R,$

(*iii*) $\text{cosec}^{-1}(-x) = -\text{cosec}^{-1}x, |x| \geq 1.$

(c) (*i*) $\cos^{-1}(-x) = \pi - \cos^{-1}x, x \in [-1, 1],$ (*ii*) $\sec^{-1}(-x) = \pi - \sec^{-1}x, |x| \geq 1,$

(*iii*) $\cot^{-1}(-x) = \pi - \cot^{-1}x, x \in R.$

(d) (*i*) $\sin^{-1}x + \cos^{-1}x = \dfrac{\pi}{2}, x \in [-1, 1],$ (*ii*) $\tan^{-1}x + \cot^{-1}x = \dfrac{\pi}{2}, x \in R,$

(*iii*) $\text{cosec}^{-1}x + \sec^{-1}x = \dfrac{\pi}{2}, |x| \geq 1,$ (*iv*) $\tan^{-1}x = \dfrac{\pi}{2} - \tan^{-1}\left(\dfrac{1}{x}\right), \quad x > 0.$

(e) $\sin^{-1}x + \sin^{-1}y = \sin^{-1}(x\sqrt{1-y^2} + y\sqrt{1-x^2}).$

(f) $\cos^{-1}x + \cos^{-1}y = \cos^{-1}(xy - \sqrt{1-x^2}\,\sqrt{1-y^2}).$

(g) (*i*) $\tan^{-1}x + \tan^{-1}y = \tan^{-1}\left(\dfrac{x+y}{1-xy}\right), \text{ if } xy < 1,$

(*ii*) $\tan^{-1}x - \tan^{-1}y = \tan^{-1}\left(\dfrac{x-y}{1+xy}\right), \text{ if } xy > -1.$

(h) (*i*) $2\tan^{-1}x = \sin^{-1}\left(\dfrac{2x}{1+x^2}\right), |x| \leq 1,$ (*ii*) $2\tan^{-1}x = \cos^{-1}\left(\dfrac{1-x^2}{1+x^2}\right), x \geq 0.$

(*iii*) $2\tan^{-1}x = \tan^{-1}\left(\dfrac{2x}{1-x^2}\right), -1 < x < 1.$

COMPLEX NUMBERS AND FUNCTIONS

1. Every complex number $x + iy$ can always be expressed in the form $r(\cos\theta + i\sin\theta)$, where $r = \sqrt{x^2 + y^2}$ and $\theta = \tan^{-1}(y/x)$.

2. **De'Moivre's Theorem**
Let $z = r(\cos\theta + i\sin\theta)$. Then $z^n = r^n[\cos n\theta + i\sin n\theta]$,
where n is any positive or negative integer or a real rational number.

$$z^{-n} = r^{-n}[\cos n\theta - i\sin n\theta] = \frac{1}{r^n[\cos n\theta + i\sin n\theta]} = \frac{1}{z^n}.$$

3. **Exponential, circular functions of a complex variable, z**

(*i*) e^z or $\exp(z) = e^{x+iy} = e^x \cdot e^{iy} = e^x(\cos y + i\sin y),$

(*ii*) Exponential form of $z\,(= x + iy) = re^{i\theta}$ (*iii*) $\sin z = \dfrac{e^{iz} - e^{-iz}}{2i},$

(*iv*) $\cos z = \dfrac{e^{iz} + e^{-iz}}{2},$ (*v*) $\tan z = \dfrac{\sin z}{\cos z} = \dfrac{e^{iz} - e^{-iz}}{i(e^{iz} + e^{-iz})}.$

4. **Euler's Theorem:**

$$e^{i\theta} = \cos\theta + i\sin\theta, \text{ where } \theta \text{ is real or complex.}$$

5. **Hyperbolic Functions:** If x be real or complex,

(*a*) $\sinh x = \dfrac{e^x - e^{-x}}{2}$, $\cosh x = \dfrac{e^x + e^{-x}}{2}$, $\tanh x = \dfrac{\sinh x}{\cosh x} = \dfrac{e^x - e^{-x}}{e^x + e^{-x}}$,

$\coth x = 1/\tanh x = \cosh x/\sinh x$, $\operatorname{sech} x = 1/\cosh x$, $\operatorname{cosech} x = 1/\sinh x$.

(*b*) *Fundamental formulae*

$$\cosh^2 x - \sinh^2 x = 1,\ \operatorname{sech}^2 x + \tanh^2 x = 1,\ \coth^2 x - \operatorname{cosech}^2 x = 1.$$

6. Relations between hyperbolic and circular functions

$\sin ix = i\sinh x$; $\cos ix = \cosh x$; $\tan ix = i\tanh x$.

7. Inverse Hyperbolic function

$$\sinh^{-1} x = \log_e(x + \sqrt{x^2+1});\quad \cosh^{-1} x = \log_e(x + \sqrt{x^2-1});\quad \tanh^{-1} x = \frac{1}{2}\log\frac{1+x}{1-x}.$$

EXPONENTIAL AND LOGARITHMIC SERIES

1. Exponential Series

$$e^x = 1 + \frac{x}{1!} + \frac{x^2}{2!} + \frac{x^3}{3!} + \cdots + \frac{x^n}{n!} + \cdots \infty,\quad e = 1 + \frac{1}{1!} + \frac{1}{2!} + \frac{1}{3!} + \cdots \infty$$

2. Logarithmic Series: If $|x| < 1$, then $\log(1+x) = x - \dfrac{x^2}{2} + \dfrac{x^3}{3} - \dfrac{x^4}{4} + \cdots$

Some Important Series

$$\sin x = x - \frac{x^3}{3!} + \frac{x^5}{5!} - \frac{x^7}{7!} + \cdots$$

$$\log(1-x) = -\left(x + \frac{x^2}{2} + \frac{x^3}{3} + \cdots \infty\right)$$

$$\cos x = 1 - \frac{x^2}{2!} + \frac{x^4}{4!} - \frac{x^6}{6!} + \cdots$$

$$\sinh x = x + \frac{x^3}{3!} + \frac{x^5}{5!} + \frac{x^7}{7!} + \cdots \infty$$

$$a^x = e^{x\log a} = 1 + \frac{\log a}{1!}x + \frac{(\log a)^2}{2!}x^2 + \frac{(\log a)^3}{3!}x^3 + \cdots \infty$$

Polar Coordinates

$$x = r\cos\theta,\ y = r\sin\theta,\qquad r = \sqrt{x^2+y^2},\ \theta = \tan^{-1}(y/x),\qquad dxdy = x\,dr\,d\theta.$$

APPENDIX-3 SOME DERIVATIVE AND INTEGRAL FORMULAS

Derivative Formulas

u and v are functions of x, c_1, and c_2 are constants.

1. $\dfrac{d}{dx}(c_1u + c_2v) = c_1u' + c_1v'$ (Sum Rule)

2. $\dfrac{d}{dx}(uv) = uv' + vu'$ (Product Rule)

3. $\dfrac{d}{dx}\left(\dfrac{u}{v}\right) = \dfrac{vu' - uv'}{v^2}$ (Quotient Rule)

4. $\dfrac{dy}{dx} = \dfrac{dy}{du}\cdot\dfrac{du}{dx}$ (Chain Rule)

5. $\dfrac{d}{dx}(x^n) = nx^{n-1}$, $n \neq 0$

6. $\dfrac{d}{dx}(\log_e x) = \dfrac{1}{x}$

7. $\frac{d}{dx}(e^{ax}) = ae^{ax}$

8. $\frac{d}{dx}(a^x) = a^x \log_e a$

9. $\frac{d}{dx}(\log_a x) = \frac{\log_a e}{x} = \frac{1}{x \log_e a}$

10. $\frac{d}{dx}(\sin x) = \cos x$

11. $\frac{d}{dx}(\cos x) = -\sin x$

12. $\frac{d}{dx}(\tan x) = \sec^2 x$

13. $\frac{d}{dx}(\cot x) = -\operatorname{cosec}^2 x$

14. $\frac{d}{dx}(\sec x) = \sec x \tan x$

15. $\frac{d}{dx}(\operatorname{cosec} x) = -\operatorname{cosec} x \cot x$

16. $\frac{d}{dx}(\sinh x) = \cosh x$

17. $\frac{d}{dx}(\cosh x) = \sinh x$

18. $\frac{d}{dx}(\tanh x) = \sec h^2 x$

19. $\frac{d}{dx}(\coth x) = -\operatorname{cosech}^2 x$

20. $\frac{d}{dx}(\operatorname{sech} x) = -\operatorname{sech} x \tanh x$

21. $\frac{d}{dx} \operatorname{cosech} x = -\operatorname{cosech} x \coth x$

22. $\frac{d}{dx}(\sin^{-1} x) = \frac{1}{\sqrt{1-x^2}}$

23. $\frac{d}{dx}(\cos^{-1} x) = \frac{-1}{\sqrt{1-x^2}}$

24. $\frac{d}{dx}(\tan^{-1} x) = \frac{1}{1+x^2}$

25. $\frac{d}{dx} \cot^{-1} x = \frac{-1}{1+x^2}$

26. $\frac{d}{dx} \sec^{-1} x = \frac{1}{x\sqrt{x^2-1}}, x > 1$

27. $\frac{d}{dx} \operatorname{cos ec}^{-1} x = \frac{-1}{x\sqrt{x^2-1}}, x > 1$

28. $\frac{d}{dx} \sinh^{-1} x = \frac{1}{\sqrt{x^2+1}}$

29. $\frac{d}{dx} \cosh^{-1} x = \frac{\pm 1}{\sqrt{x^2-1}}, \ (x^2 > 1)$

30. $\frac{d}{dx} \tanh^{-1} x = \frac{1}{1-x^2}, \ (x^2 < 1)$

31. $\frac{d}{dx} \coth^{-1} x = \frac{-1}{x^2-1}, \ (x^2 > 1)$

32. $\frac{d}{dx} \operatorname{sech}^{-1} x = \frac{\pm 1}{x\sqrt{1-x^2}}, \ (x^2 < 1)$

33. $\frac{d}{dx} \operatorname{cosech}^{-1} x = \frac{\pm 1}{x\sqrt{x^2+1}}.$

Integral Formulas

***u* and *v* are functions of *x*, c_1, and c_2 are constants.**

1. $\int (c_1 u + c_2 v)\, dx = c_1 \int u dx + c_2 \int v dx + c$

2. $\int uv'\, dx = uv - \int u' v dx + c, \begin{pmatrix}\text{integration} \\ \text{by parts}\end{pmatrix}$

3. $\int x^n dx = \frac{x^{n+1}}{n+1} + c \quad (n \neq -1)$

4. $\int \frac{1}{x}\, dx = \log_e |x| + c$

5. $\int e^{ax} dx = \frac{1}{a} e^{ax} + c$

6. $\int a^x dx = \frac{a^x}{\log_e a} + c$

7. $\int \sin ax\, dx = -\frac{1}{a}\cos ax + c$

8. $\int \cos ax\, dx = \frac{1}{a}\sin ax + c$

9. $\int \sec^2 x\, dx = \tan x + c$

10. $\int \operatorname{cosec}^2 x\, dx = -\cot x + c$

11. $\int \sec x \tan x\, dx = \sec x + c$

12. $\int \operatorname{cosec} x \cot x\, dx = -\operatorname{cosec} x + c$

13. $\int \tan ax\, dx = -\frac{1}{a}\log|\cos ax| + c = \frac{1}{a}\log|\sec ax| + c$

14. $\int \cot ax\, dx = \frac{1}{a}\log|\sin ax| + c$

15. $\int \sec x\, dx = \begin{cases} \log|\sec x + \tan x| + c \\ \log\left|\tan\left(\frac{\pi}{4} + \frac{x}{2}\right)\right| + c \end{cases}$

16. $\int \operatorname{cosec} x\, dx = \begin{cases} \log|\operatorname{cosec} x - \cot x| + c \\ \log\left|\tan\frac{x}{2}\right| + c \end{cases}$

17. $\int \frac{dx}{\sqrt{a^2 - x^2}} = \sin^{-1}\left(\frac{x}{a}\right) + c$

18. $\int \frac{dx}{x^2 + a^2} = \frac{1}{a}\tan^{-1}\left(\frac{x}{a}\right) + c$

19. $\int \sinh x\, dx = \cosh x + c$

20. $\int \cosh x\, dx = \sinh x + c$

21. $\int \frac{dx}{\sqrt{x^2 + a^2}} = \begin{cases} \sinh^{-1}\left(\frac{x}{a}\right) + c \\ \log_e\left(x + \sqrt{x^2 + a^2}\right) + c \end{cases}$

22. $\int \frac{dx}{\sqrt{x^2 - a^2}} = \begin{cases} \cosh^{-1}\left(\frac{x}{a}\right) + c \\ \log_e\left|x + \sqrt{x^2 - a^2}\right| + c \end{cases}$

23. $\int \frac{1}{a^2 - x^2}\, dx = \frac{1}{2a}\log\left|\frac{x+a}{x-a}\right| + c$

24. $\int \frac{dx}{x^2 - a^2} = \frac{1}{2a}\log\left|\frac{x-a}{x+a}\right| + c$

25. $\int \sqrt{a^2 - x^2}\, dx = \frac{x}{2}\sqrt{a^2 - x^2} + \frac{a^2}{2}\sin^{-1}\left(\frac{x}{a}\right) + c$

26. $\int \sqrt{x^2 + a^2}\, dx = \frac{x}{2}\sqrt{x^2 + a^2} + \frac{a^2}{2}\log\left|x + \sqrt{x^2 + a^2}\right| + c$

27. $\int \sqrt{x^2 - a^2}\, dx = \frac{x}{2}\sqrt{x^2 - a^2} - \frac{a^2}{2}\log\left|x + \sqrt{x^2 - a^2}\right| + c$

28. $\int \log x\, dx = x\log x - x + c$

29. $\int e^{ax}\sin bx\, dx = \frac{e^{ax}}{a^2 + b^2}(a\sin bx - b\cos bx) + c$

30. $\int e^{ax}\cos bx\, dx = \frac{e^{ax}}{a^2 + b^2}(a\cos bx + b\sin bx) + c.$

APPENDIX-4 SOLVED EXAMINATION PAPER (AMIETE (NEW SCHEME) - JUNE 2011)

AMIETE – ET/CS/IT (NEW SCHEME) – CODE: AE51/AC51/AT51

Subject: ENGINEERING MATHEMATICS – I

Time: 3 Hours **JUNE 2011** **Max. Marks: 100**

Note: There are 9 questions in all.

- **Question 1 is compulsory and carries 20 marks. Answer to Q. 1 must be written in the space provided for it in the answer book supplied and nowhere else.**
- **The answer sheet for the Q. 1 will be collected by the invigilator after 45 minutes of the commencement of the examination.**
- **Out of the remaining EIGHT Questions, answer any FIVE Questions. Each question carries 16 marks.**
- **Any required data not explicity given, may be suitably assumed and stated.**

Q.1. Choose the correct or the best alternative in the following: **(2 × 10)**

a. If in a determinant the corresponding elements of two rows (or columns) are proportional to each other, then the value of the determinant is

(A) unity ✔(B) zero
(C) infinity (D) None of the above

b. In case of matrix multiplication of two matrices **A** and **B**, if **AB = 0** (where **'0'** stands for null matrix), it means that

(A) either A = 0 or B = 0 (B) both of them '0'
✔(C) it is not necessary that either A = 0 or B = 0 (D) None of the above

C. If u = F(x – y, y – z, z – x), then

(A) $\frac{\partial u}{\partial x} - \frac{\partial u}{\partial y} + \frac{\partial u}{\partial z} = 0$ ✔(B) $\frac{\partial u}{\partial x} + \frac{\partial u}{\partial y} + \frac{\partial u}{\partial z} = 0$

(C) $\frac{\partial u}{\partial x} + \frac{\partial u}{\partial y} - \frac{\partial u}{\partial z} = 0$ (D) $\frac{\partial u}{\partial x} - \frac{\partial u}{\partial y} - \frac{\partial u}{\partial z} = 0$

d. The Newton-Raphson method to find a root of the equation f(x) fails when

(A) For a particular value of $x = x_0$ (say), $f(x_0)$ becomes zero
(B) For a particular value of $x = x_0$ (say), $f(x_0)$ becomes unity
✔(C) For a particular value of $x = x_0$ (say), $f'(x_0)$ becomes zero (where f′(x) is the first derivative of f w.r.t. x)
(D) For a particular value of $x = x_0$ (say), $f(x_0)$ becomes equal to $f'(x_0)$

e. If $\left(\frac{\partial M}{\partial y} - \frac{\partial N}{\partial x}\right)\Big/N$ is a function of x, alone say f(x), then integrating factor is

(A) $e^{\int x dx}$ (B) $e^{\int y dy}$ ✔(C) $e^{\int f(x) dx}$ (D) $e^{\int f(y) dy}$

f. The maximum value of $(3x^4 - 2x^3 - 6x^2 + 6x + 1)$ in the interval (0, 2) is

(A) 1 (B) 21 (C) $\frac{1}{2}$ ✔(D) None of the above, exact value = 39/16

g. Value of $\int_0^1 dx \int_0^x e^{y/x} dy$ is

✔(A) $\frac{1}{2}(e-1)$ (B) $\frac{1}{2}(1-e)$ (C) 1 (D) None of the above

h. The value of $J_{1/2}(x)$ is

(A) $\sqrt{\left(\frac{2}{\pi x}\right)\sin x}$ (B) $\sqrt{\left(\frac{2}{\pi x}\right)}\cos x$

✔(C) $\sqrt{\left(\frac{2}{\pi x}\right)}\sin x$ (D) $\sqrt{\left(\frac{2}{\pi x}\right)\cos x}$

i. Value of $\Gamma\left(-\frac{1}{2}\right)$ is

(A) $\sqrt{\pi}$ (B) $-\sqrt{\pi}$ (C) $2\sqrt{\pi}$ ✔(D) $-2\sqrt{\pi}$

j. A matrix 'A' is said to be idempotent matrix if

(A) $A^T A = I$ ✔(B) $A^2 = A$

(C) $A^K = A$, K is any positive integer value (D) $A = A^T$

Answer any FIVE Questions out of EIGHT Questions.
Each question carries 16 marks

Q.2. a. If $u = \sin^{-1}\left[\frac{x+y}{\sqrt{x}+\sqrt{y}}\right]$. Prove that

$$x^2 \frac{\partial^2 u}{dx^2} + 2xy \frac{\partial^2 u}{\partial x \partial y} + y^2 \frac{\partial^2 u}{dy^2} = -\frac{\sin u \cos 2u}{4\cos^3 u}$$ (8)

[**Hint:** Here sin u is a homogeneous function of degree 1/2. We obtain

$$x\frac{\partial}{\partial x}(\sin u) + y\frac{\partial}{\partial u}(\sin u) = \frac{1}{2}\sin u \quad \text{or } x\cos u\frac{\partial u}{\partial x} + y\cos u\frac{\partial u}{\partial y} = \frac{1}{2}\sin u$$

or $$x\frac{\partial u}{\partial x} + y\frac{\partial u}{\partial y} = \frac{1}{2}\tan u.$$

Therefore, $$x^2\frac{\partial^2 u}{\partial x^2} + 2xy\frac{\partial^2 u}{\partial x \partial y} + y^2\frac{\partial^2 u}{\partial y^2} = \left(\frac{1}{2}\tan u\right)\left(\frac{1}{2}\sec^2 u - 1\right)$$

$$= \frac{\sin u}{2\cos u}\left(\frac{1}{2\cos^2 u} - 1\right) = -\frac{\sin u}{4\cos^3 u}(2\cos^2 u - 1)$$

$$= -\frac{\sin u \cos 2u}{4\cos^3 u},$$ which is same as desired.]

b. Find the shortest distance between the line y = 10 – 2x, and the ellipse $\frac{x^2}{4}+\frac{y^2}{9}=1.$ (8)

[See Example 2.54 on Page 2.39.]

Q.3. a. Using the transformation x + y = u, y = uv, show that $\iint [xy(1-x-y)]^{1/2} dxdy = \frac{2\pi}{105}$, integration being taken over the area of the triangle bounded by the lines x = 0, y = 0, x + y = 1. (8)

[**Hint:** we have x = u (1 – v), y = uv, therefore $dxdy = \begin{vmatrix} 1-v & -u \\ v & u \end{vmatrix} dudv = u\,dudv$,

and $[xy(1-x-y)]^{1/2} = [u^2v(1-v)(1-u)]^{1/2} = uv^{1/2}(1-v)^{1/2}(1-u)^{1/2}$.

Therefore, the limits are u = 0, u = 1; v = 0, v = 1. Since u = x + y and v = y/(x + y).

Hence, transformed integral $= \int_0^1\int_0^1 uv^{1/2}(1-v)^{1/2}(1-u)^{1/2}\, u\,dudv$

$= \int_0^1\int_0^1 u^2v^{1/2}(1-v)^{1/2}(1-u)^{1/2}\,dudv$

(Put $u = \sin^2\theta$ and $v = \sin^2\phi$)

Integral $= 4\int_0^{\pi/2}\int_0^{\pi/2} \sin^5\theta\cos^2\theta\sin^2\phi\cos^2\phi\, d\theta d\phi = \frac{4\overline{|3}\,\overline{|3/2}}{2\overline{|9/2}}\cdot\frac{\overline{|3/2}\,\overline{|3/2}}{2\overline{|3}} = \frac{2\pi}{105}$]

b. A rectangular box is open at the top to have a volume of 32 c.c. Find the dimension of the box requiring least material for its construction. [Refer Ex. 2.44 on page 2.34.] (8)

Q.4. a. Find the eigen values and eigen vectors of the matrix

$$\begin{bmatrix} 1 & 1 & 3 \\ 1 & 5 & 1 \\ 3 & 1 & 1 \end{bmatrix}$$

Ans. –2, 3, 6; $[-1, 0, 1]^T$, $[1, -1, 1]^T$, $[1, 2, 1]^T$ (8)

b. Test the consistency of the following system of equations and solve them if possible:

3x + 3y + 2z = 1

x + 2y = 4

10y + 3z = –2

Ans. x = 2, y = 1, z = –4. (8)

Q.5. a. Find by the method of Regula-Falsi a root of the equation
$x^3 + x^2 - 3x - 3 = 0$ lying between 1 and 2. [See Example 13.7 on Page 13.6.] (8)

b. Perform three iterations of the Gauss-Seidel method for solving the system of equations

$$\begin{bmatrix} 4 & 0 & 2 \\ 0 & 5 & 2 \\ 5 & 4 & 10 \end{bmatrix}\begin{bmatrix} x_1 \\ x_2 \\ x_3 \end{bmatrix} = \begin{bmatrix} 6 \\ -3 \\ 11 \end{bmatrix}$$

Take the components of the approximate initial vector as $x_i^{(0)} = b_i/a_{ii}, i = 1, 2, 3.$

Compare with the exact solution $x = [1, -1, 1]^T$. [See Example 13.31 on Page 13.31.] (8)

Q.6. a. Solve $(x+1)\frac{dy}{dx} - y = e^{3x}(x+1)^2$ **Ans.** $y = \left(\frac{1}{3}e^{3x} + c\right)(x+1)$. (8)

b. Solve $\frac{dy}{dx} = \frac{y+x-2}{y-x-4}$ **Ans.** $x^2 + 2xy - y^2 - 4x + 8y = c$ (8)

Q.7. a. Solve $\frac{d^2y}{dx^2} + 2\frac{dy}{dx} + 4y = 2x^2 + 3e^{-x}$ (8)

Ans. $y(x) = e^{-x}(c_1 \cos\sqrt{3}x + c_2 \sin\sqrt{3}x) + \frac{1}{2}x(x-1) + e^{-x}$.

b. Find the general solution of the equation $y'' + 3y' + 2y = 2e^x$, using method of variation of parameters. **Ans.** $y(x) = c_1e^{-x} + c_2e^{-2x} + (e^x/3)$. (8)

Q.8. a. (i) If $f(x) = 0 \quad -1 < x \le 0$

$= x \quad 0 < x < 1$

Show that $f(x) = \frac{1}{4}P_0(x) + \frac{1}{2}P_1(x) + \frac{5}{16}P_2(x) - \frac{3}{32}P_4(x) + ...$ (4)

[See Example 12.33 on Page 12.44.]

(ii) Prove that $\int J_3(x)dx + J_2(x) + \frac{2}{x}J_1(x) = 0$ *[See Example 12.9 on Page 12.10]* (8)

b. Prove that $P_n(x) = \frac{1}{2^n n!}\frac{d^n}{dx^n}[(x^2-1)^n]$ *[See Article 12.15 on Page 12.33]* (8)

Q.9. a. Using Beta and Gamma functions show that for any positive integer 'm'

(i) $\int_0^{\pi/2} \sin^{2m-1}(\theta)\, d\theta = \frac{(2m-2)(2m-4)....2}{(2m-1)(2m-3)....3}$ *[See Example 11.10(i) on Page 11.10.]* (4)

(ii) $\int_0^{\pi/2} \sin^{2m}(\theta)\, d\theta = \frac{(2m-1)(2m-3)....1}{(2m)(2m-2)....2}\cdot\frac{\pi}{2}$ *[See Example 11.10(ii) on Page 11.11.]* (4)

b. Prove that

(i) $\beta\left(m, \frac{1}{2}\right) = 2^{2m-1}\beta(m, m)$ *[See Question 5(iii) on Page 11.16.]* (4)

(ii) $\Gamma(m)\,\Gamma\left(m+\frac{1}{2}\right) = \frac{\sqrt{\pi}}{2^{2m-1}}\Gamma(2m)$ *[See Example 11.11 on Page 11.11.]* (4)